Industrial Maintenance

Second Edition

Industrial Maintenance

Second Edition

By

Michael E. Brumbach

and

Jeffrey A. Clade

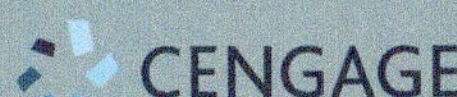

Australia • Brazil • Canada • Mexico • Singapore • United Kingdom • United States

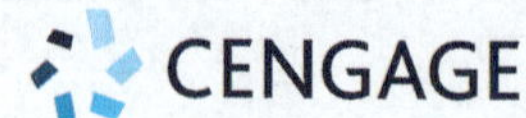

***Industrial Maintenance*, 2E**

Michael E. Brumbach, Jeffrey A. Clade

Vice President, Editorial: Dave Garza

Director of Learning Solutions: Sandy Clark

Acquisitions Editor: Jim DeVoe

Managing Editor: Larry Main

Senior Product Manager: John Fisher

Editorial Assistant: Aviva Ariel

Director, Market Development Management: Debbie Yarnell

Marketing Development Manager: Erin Brennan

Director, Brand Management: Jason Sakos

Marketing Brand Manager: Erin McNary

Senior Production Director: Wendy Troeger

Production Manager: Mark Bernard

Content Project Manager: Barbara LeFleur

Senior Art Director: David Arsenault

Technology Project Manager: Joe Pliss

Library of Congress Control Number: 2012941115

ISBN-13: 978-0-357-67064-4

ISBN-10: 0-357-67064-7

Cengage
200 Pier 4 Boulevard
Boston, MA 02210
USA

Cengage is a leading provider of customized learning solutions with employees residing in nearly 40 different countries and sales in more than 125 countries around the world. Find your local representative at: **www.cengage.com.**

To learn more about Cengage platforms and services, register or access your online learning solution, or purchase materials for your course, visit **www.cengage.com.**

Notice to the Reader

Publisher does not warrant or guarantee any of the products described herein or perform any independent analysis in connection with any of the product information contained herein. Publisher does not assume, and expressly disclaims, any obligation to obtain and include information other than that provided to it by the manufacturer. The reader is expressly warned to consider and adopt all safety precautions that might be indicated by the activities described herein and to avoid all potential hazards. By following the instructions contained herein, the reader willingly assumes all risks in connection with such instructions. The publisher makes no representations or warranties of any kind, including but not limited to, the warranties of fitness for particular purpose or merchantability, nor are any such representations implied with respect to the material set forth herein, and the publisher takes no responsibility with respect to such material. The publisher shall not be liable for any special, consequential, or exemplary damages resulting, in whole or part, from the readers' use of, or reliance upon, this material.

Printed at CLDPC, USA, 02-21

We dedicate this book to
Mr. Curt Shoaf

A colleague, mentor, and friend.
Look up the word multicraft.
The definition will read,
"Curt Shoaf—Mechanix"

Strange how much you've got to know
Before you know how little you know
—Anon

CONTENTS

SECTION 2

MECHANICAL KNOWLEDGE 129

SECTION 3

ELECTRICAL KNOWLEDGE 269

The following appendices F through K can be found on the student companion website at www.cengage.com.

Follow the instructions in the Supplements section of the Preface for access information.

PREFACE

For several years, we have been hearing how there was a need for multicrafted maintenance technicians in the industry. Instead of an electrician or mechanic or welder, there was a need for a multicraft technician who could perform these duties and more. To answer this need, we developed an Industrial Electricity program, an Industrial Mechanics program, and a Welding program. We taught our students the traditional way. We presented them with hand tools, piping systems, safety, print reading, fluid power, power transmission, electrical fundamentals, motor controls, electronics, PLCs, drives, gas welding, arc welding, and more. Each of these subjects was taught as an individual course. Each of these courses required separate texts.

Then one day, an idea formed. What if we could combine these programs into one? We then created an Associate Degree in Industrial Maintenance. The student could now combine courses from all three programs and leave with a multicraft degree. However, this created a challenge. We did not want to continue using separate texts for this program, and the economic burden on our students became great. So, we decided to produce a text that would satisfy the requirements for a vast majority of the courses taught. *Industrial Maintenance* was created to meet this need. For our program, Industrial Maintenance replaces approximately 20 texts that were previously required to complete the courses in the areas of Industrial Electricity, Industrial Mechanics, and Welding.

INTENDED AUDIENCE

Industrial Maintenance, 2E, serves as a vital resource for all individuals involved in the maintenance field. This text will best serve those individuals enrolled in high school vocational/trade programs, two-year technical college/community college programs, adult education programs, and corporate training/certification programs. However, *Industrial Maintenance* should also prove to be a valuable resource for those individuals already employed in the maintenance field—where any maintenance technician can keep a copy of the book in his tool box, locker, or other handy location.

Industrial Maintenance, 2E, addresses the needs of the multicrafted maintenance technician and presents an all-encompassing view of the field of industrial maintenance, which covers a variety of technical skill areas. These include, but are not limited to, mechanics (mechanical installation, fluid power, piping systems, power transmission, print reading, and safety, to name a few), electrical (electrical theory, test equipment, electronics, the *National Electrical Code*®, control circuits, rotating machines, PLCs, and drives, for example), and welding (gas welding and arc welding). A multicrafted maintenance technician must have knowledge and skills in all these areas. *Industrial Maintenance*, 2E, addresses these areas in a format designed with the technician in mind.

NEW TO THIS EDITION

Chapter 1

- All photos are now in color or have been reshot in color.
- All graphics are now in color.
- Additional information concerning lockout/tagout.
- New information concerning NFPA 70E.
- New symbol information for the Class D Fire classification.
- The addition of a section on ladders and scaffolds.

Chapter 2

- All photos are now in color or have been reshot in color.
- Additional information concerning the bimetal hacksaw blade.

Chapter 3

- All photos are now in color or have been reshot in color.
- Additional explanation concerning torqueing of a fastener.

Chapter 4

- A restructuring of Chapter 4 has occurred to include piping, hydraulics, and pneumatics.
    - 4-1 Mechanical Drawings
    - 4-2 Piping Symbols and Drawing
    - 4-3 Hydraulic/Pneumatic Symbols and Drawings
    - 4-4 Electrical Symbols and Drawings
    - 4-5 Welding Symbols and Drawings
- Some graphics have been changed and others are new.
- There is new content concerning hydraulic and pneumatic symbols and drawings.
- Errors concerning weld symbols have been corrected.

Chapter 5

- New chapter concerning rigging and rigging equipment has been added.
- New color graphics and color photos are included.
- Chapter 5 includes:
    - 5-1 Formulae and Weight Estimations
    - 5-2 Load Balancing
    - 5-3 Synthetic Slings
    - 5-4 Fiber Rope and Securing
    - 5-5 Wire Rope and Wire Rope Slings
    - 5-6 Chain and Chain Slings
    - 5-7 Prelift Planning

Chapter 6

- All photos are now in color.
- Some graphics have been redone in color to make the content that is stressed more easily recognizable.
- Additional information concerning belt and pulley wear guides has been added.
- Errors have been corrected.

Chapter 7

- All photos are now in color.

Chapter 8

- All photos are now in color.
- Some graphics have been redone in color to make the content that is stressed more easily recognizable.
- Content concerning bar sag has been added.
- The dial caliper graphic has been revised to include more detail on how to read it.
- New photo showing a laser alignment kit in use has been included.

Chapter 9

- There is added content and graphics concerning automatic oilers.

Chapter 10

- All photos are now in color.
- Some graphics have been redone in color to make the content that is stressed more easily recognizable.

Chapter 11

- All photos are now in color.

Chapter 12

- There is a new photo showing a bladder-type accumulator.

Chapter 13

- All photos are now in color.
- Information on piping symbols and sketches has been moved to Chapter 4.

Chapter 14

- Added color.
- Fixed errors.
- Added explanation of centripetal and centrifugal force.

Chapter 15

- Added color.
- Fixed errors.

Chapter 16

- Added color.
- Fixed errors.

Chapter 17

- Added color.
- Fixed errors.

Chapter 18

- Added color.
- Fixed errors.
- Updated content to the NEC 2011 standard.

Chapter 19

- Added color.
- Fixed errors.
- Added content on Current Transformers (CTs).

Chapter 20

- Added color.
- Fixed errors.
- Added content on.
  - Permanent Magnet motors
  - Brushless DC motors
  - Stepping motors
  - Motor Maintenance

Chapter 21

- Added color.
- Fixed errors.

Chapter 22

- Added color.
- Fixed errors.

Chapter 23

- Added color.
- Fixed errors.

Chapter 24

- Added color.
- Completely re-wrote chapter.

Chapter 25

- Added color.
- Completely re-wrote chapter.

Chapter 26

- Added color.
- Fixed errors.

Chapter 27

- Added color photos to supplement content.
- Some graphics are now in color.

Chapter 28

- Color photos have been added to supplement new content or to support existing content.
- Some graphics have been modified to supplement new content.
- New content on auto-darkening welding hoods is included.
- New content on arc-welding methods has been added.

Chapter 31

- Added color.
- Fixed errors.

TEXT LAYOUT

In order for the student to be able to navigate with ease through the various technical skills required to train multiskilled technicians, *Industrial Maintenance* is divided into five sections.

Section 1: General Knowledge

This section presents the following topics: Safety, Tools, Fasteners, and Industrial Print Reading, and Safety and Installation.

Section 2: Mechanical Knowledge

This section introduces Mechanical Power Transmission, Bearings, Coupled Shaft Alignment, Lubrication, Seals and Packing, Pumps and Compressors, Fluid Power, and Piping Systems.

Section 3: Electrical Knowledge

The third section provides information on Electrical Fundamentals, Test Equipment, Basic Resistive Electrical Circuits, Reactive Circuits and Power Factor Correction, Wiring Methods, Transformers and Power Distribution, Electrical Machinery, Control and Controlled Devices, Motor Control Circuits, Basic Industrial Electronics, Electronic Variable-Speed Drives, Programmable Logic Controllers, and Lighting.

Section 4: Welding Knowledge

This section introduces the essentials of welding and specifically covers Gas Welding and Arc Welding.

Section 5: Preventive Maintenance

This section discusses Developing and Implementing Preventive Maintenance, Mechanical PMs, and Electrical PMs.

Note: Need help organizing an *Industrial Maintenance* course? Please see our course syllabus, appearing after the Preface on page xvii and also available in electronic format on the Instructor Companion Web site that accompanies the text.

SECTION AND CHAPTER FEATURES

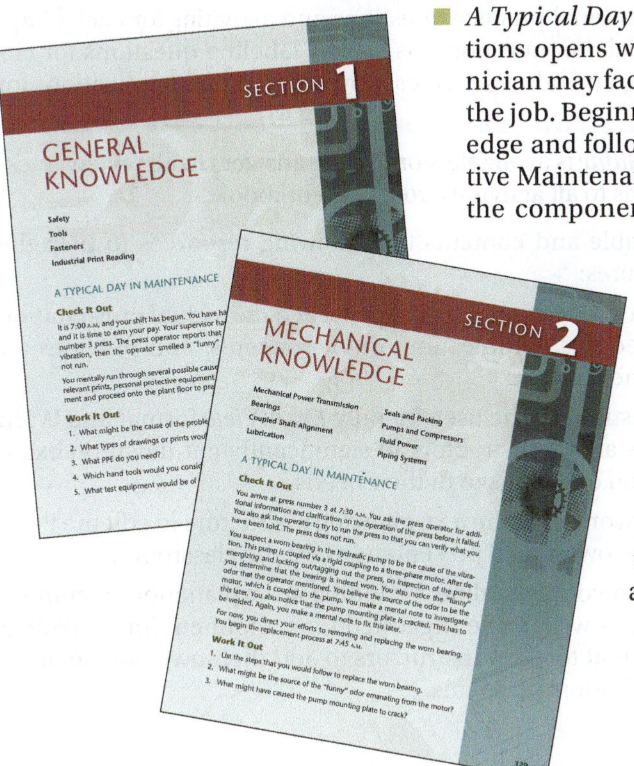

- *A Typical Day in Maintenance*—Each of the five sections opens with an on-going situation that a technician may face during the course of a typical day on the job. Beginning with Section 1 on General Knowledge and following through to Section 5 on Preventive Maintenance, the author walks readers through the components of a faulty machine, and they are encouraged to *Check It Out* and discover the situation at hand. The explanation of the scenario is then followed up by a series of questions, where readers must *Work It Out* and attempt to diagnose and troubleshoot the problem. The faulty machine presents problems along the way that relate to each section that follows, and the goal is to encourage critical thinking skills among aspiring technicians.

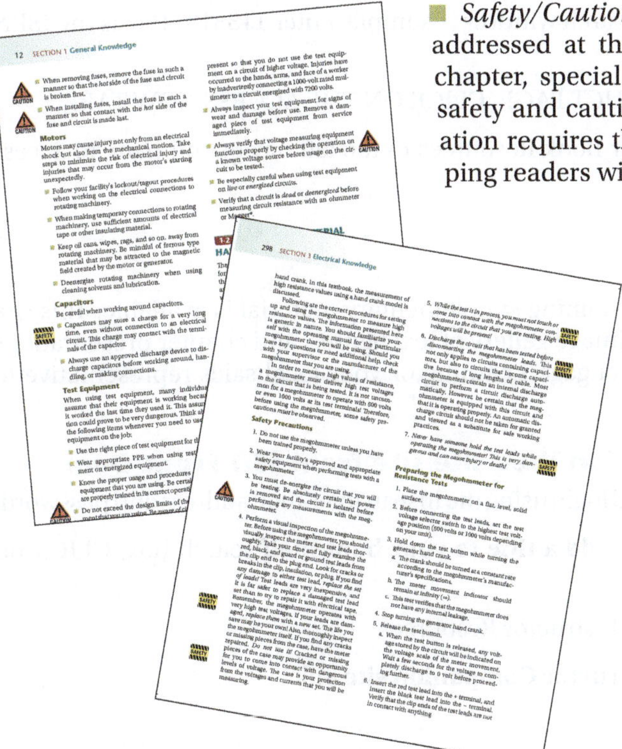

- *Safety/Caution Notes*—Although safety issues are addressed at the beginning of the text in a separate chapter, special attention has been given to integrate safety and caution notes into the text whenever a situation requires the mention of a safety concern. Equipping readers with this proper knowledge helps to warn technicians of dangerous situations on the job that could lead to injury or death.

- *Modular Chapters*—Each of the chapters is written in a modular format, allowing readers who are already familiar with the content to study particular areas of interest or weakness, while also serving as a handy on-the-job reference manual.

TEXT SUPPLEMENT PACKAGE

- An extensive *Workbook* provides additional practice questions and activities for each chapter in the text. Matching and identification questions, as well as labeling questions for circuits, diagrams, tools and schematics, and other activities provide hands-on applications for important concepts.
- To aid the instructor, an *Instructor's Guide* is available, containing answers to all review questions from the text, as well as solutions to all activities from the workbook.
- An Instructor Web site is also available and contains the following resources to help the instructor prepare for classroom lectures:
  - *PowerPoint* presentations—broken down chapter by chapter, each individual presentation highlights important learning objectives, outlines important concepts, and provides art from the text to visually reinforce the content.
  - *Test Bank*—providing over 500 questions in the user-friendly *ExamView* format and Word documents, this bank of questions allows instructors to significantly cut down on exam prep time and lets them edit and add questions to fit their needs.
  - *Image Library*—contains *all* the artwork from the text, allowing instructors to edit the PowerPoint presentations or create their own visual presentations for the classroom.
  - *Sample Course Syllabus*—provides a course outline, including an explanation of competencies for each section of the text, as well as the suggested time allotment for individual chapters. It is available in Word format to allow instructors to add their own course notes to the syllabus and to tailor it to individual programs.

Student Companion Website

This free website contains PDF files of text Appendices F through K.
Accessing a Student Companion Website site from Cengage:

1. GO TO: http://www.cengage.com
2. ENTER author, title or ISBN in the search window (Example: enter 1133131190 as the ISBN for this text.)
3. WHEN YOU ARRIVE AT THE PRODUCT PAGE, CLICK ON THE ACCESS NOW TAB.
4. CLICK on the resources listed under Book Resources in the left navigation pane to access the project files.

INSTRUCTOR SITE

An Instruction Companion Web site containing supplementary material is available. This site contains an Instructor Guide, testbank, image gallery of text figures, and chapter presentations done in PowerPoint. Contact Delmar Cengage Learning or your local sales representative to obtain an instructor account.

Accessing an Instructor Companion Web site site from SSO Front Door

1. GO TO: http://login.cengage.com and login using the Instructor e-mail address and password.
2. ENTER author, title, or ISBN in **the Add a title to your bookshelf** search box, CLICK on **Search** button
3. CLICK **Add to My Bookshelf** to add Instructor Resources
4. At the Product page click on the **Instructor Companion site** link

New Users

If you're new to Cengage.com and do not have a password, contact your sales representative.

INDUSTRIAL MAINTENANCE SYLLABUS

Industrial Maintenance, 2E, was written to assist in the education and training of the multicrafted maintenance technician. As a result, the text is very broad in scope. When developing a course or courses to teach industrial maintenance, the textbook, workbook, PowerPoint, and test bank can be invaluable. This document serves as a guide to using *Industrial Maintenance*, 2E, and the ancillary materials.

Note: Would you like to add your own notes to the syllabus printed here? Please see the Instructor Resource Web site that accompanies this textbook for the electronic version.

COMPETENCIES

Industrial maintenance technicians must be multiskilled (or multicrafted) in order to perform their required job duties. As a result, these individuals must possess knowledge in three key areas: mechanical, electrical, and welding. To facilitate learning in these areas, *Industrial Maintenance*, 2E, is divided into five sections: General Knowledge, Mechanical Knowledge, Electrical Knowledge, Welding Knowledge, and Preventive Maintenance. Mastery of the required competencies in these areas is demonstrated by the completion of assignments in the accompanying workbook and by passing related tests created from the accompanying test bank.

Section 1: General Knowledge

All aspects of industrial maintenance share some common areas of knowledge, such as safety, tools, fasteners, and print reading. Whether working on mechanical, electrical, or welding projects, the maintenance technician must have knowledge of all of these areas. *Industrial Maintenance*, 2E, addresses these areas in Section 1: General Knowledge. By using the textbook and associated PowerPoint presentation, the instructor provides the student with the material necessary to master these topics. Because maintenance requires a hands-on approach to problem solving, using the accompanying workbook allows the student to demonstrate mastery of the competencies of these topics to the instructor. Because it may not be possible to use plant or laboratory facilities to demonstrate mastery of these competencies, the workbook has been designed to provide exercises that simulate the actual equipment and procedures used. The accompanying test bank is used to create tests for assessment and verification of knowledge gained.

Suggested time:	Chapter 1	Safety	two classes
	Chapter 2	Tools	two classes
	Chapter 3	Fasteners	two classes

	Chapter 4	Industrial Print Reading	two classes
	Chapter 5	Rigging and Installation	six classes

Section 2: Mechanical Knowledge

A large part of what a multicrafted industrial maintenance technician does revolves around mechanical skills. *Industrial Maintenance*, 2E, addresses mechanical knowledge in Section 2. By using the textbook and associated PowerPoint presentation, the instructor provides the student with the material necessary to master mechanical power transmission, bearings, coupled shaft alignment, lubrication, seals and packing, pumps and compressors, fluid power, and piping systems. Because maintenance requires a hands-on approach to problem solving, using the accompanying workbook allows the student to demonstrate mastery of the competencies of these topics to the instructor. Because it may not be possible to use plant or laboratory facilities to demonstrate mastery of these competencies, the workbook has been designed to provide exercises that simulate the actual equipment and procedures used. The accompanying test bank is used to create tests for assessment and verification of knowledge gained.

Suggested time:	Chapter 6	Mechanical Power Transmission	two classes
	Chapter 7	Bearings	two classes
	Chapter 8	Coupled Shaft Alignment	two classes
	Chapter 9	Lubrication	one class
	Chapter 10	Seals and Packing	one class
	Chapter 11	Pumps and Compressors	two classes
	Chapter 12	Fluid Power	two classes
	Chapter 13	Piping Systems	two classes

Section 3: Electrical Knowledge

An equally large part of what a multicrafted industrial maintenance technician does includes electrical skills. *Industrial Maintenance*, 2E, addresses electrical knowledge in Section 3. By using the textbook and associated PowerPoint presentation, the instructor provides the student with the material necessary to master electrical fundamentals, test equipment, resistive circuits, reactive circuits, wiring methods, transformers and power distribution, electrical machinery, control and controlled devices, motor control circuits, basic industrial electronics, electronic variable speed drives, programmable logic controllers, and lighting. Because maintenance requires a hands-on approach to problem solving, using the accompanying workbook allows the student to demonstrate mastery of the competencies of these topics to the instructor. Because it may not be possible to use plant or laboratory facilities to demonstrate mastery of these competencies, the workbook has been designed to provide exercises that simulate the actual equipment and procedures used. The accompanying test bank is used to create tests for assessment and verification of knowledge gained.

Suggested time:	Chapter 14	Electrical Fundamentals	two classes
	Chapter 15	Test Equipment	two classes
	Chapter 16	Basic Resistive Electrical Circuits	four classes
	Chapter 17	Reactive Circuits and Power Factor	four classes
	Chapter 18	Wiring Methods	two classes
	Chapter 19	Transformers	four classes
	Chapter 20	Electrical Machinery	two classes
	Chapter 21	Control and Controlled Devices	two classes
	Chapter 22	Motor Control Circuits	two classes

	Chapter 23	Basic Industrial Electronics	two classes
	Chapter 24	Electronic Variable-Speed Drives	four classes
	Chapter 25	Programmable Logic Controllers	two classes
	Chapter 26	Lighting	two classes

Section 4: Welding Knowledge

An additional skill that is important for a multicrafted maintenance technician to possess is the ability to weld. *Industrial Maintenance*, 2E, addresses welding knowledge in Section 4. By using the textbook and associated PowerPoint presentation, the instructor provides the student with the material necessary to learn the fundamentals of gas welding and arc welding. Because welding is an art that requires a hands-on approach, using the accompanying workbook allows the student to demonstrate mastery of the competencies of these topics to the instructor. The accompanying test bank is used to create tests for assessment and verification of knowledge gained. Unlike other topics covered by this textbook, there is no substitute for performing an actual weld. Therefore, it is strongly suggested that arrangements be made to allow the student to perform a weld by using both gas and arc welding techniques.

Suggested time:	Chapter 27	Gas Welding	two classes
	Chapter 28	Arc Welding	two classes

Section 5: Preventive Maintenance

With today's demands for maximum profitability, it has become increasingly important for plant machinery to operate at maximum efficiency and with minimum downtime. To achieve this goal, many facilities implement a preventive maintenance program. *Industrial Maintenance*, 2E, introduces the preventive maintenance program in Section 5. The textbook and associated PowerPoint presentation can be used to help the student understand the importance of the program as well as ways to plan and implement a preventive maintenance program. The accompanying test bank is used to create tests for assessment and verification of knowledge gained.

Suggested time:	Chapter 29	Preventive Maintenance—Developing and Implementing	one class
	Chapter 30	Mechanical PM	one class
	Chapter 31	Electrical PM	one class

ABOUT THE AUTHORS

With a combined total of over 25 years experience in the field and more than 40 years of teaching experience in the subject area, the authors share their knowledge and firsthand experiences with the learner.

MICHAEL E. BRUMBACH

Mike is the Industrial Maintenance Department Chair as well as an instructor in the Industrial Maintenance program at York Technical College in Rock Hill, SC. He has been employed by York Technical College since 1986, having spent the previous 11 years working in industry. Mike possesses an AS degree.

You may e-mail Mike at mbrumbach@comporium.net

JEFFREY (JC) CLADE

JC is an instructor in the Industrial Maintenance program. He has been employed by York Technical College since 1997, having spent the previous 15 years serving with the U.S. Air Force and working in industry. JC possesses an AS degree and is a member of the International Association of Electrical Inspectors (IAEI) and the National Fire Protection Association (NFPA).

You may e-mail JC at jclade@comporium.net

ACKNOWLEDGMENTS

FROM MIKE BRUMBACH

I would like to thank my wife, Kathy; my mother; and my daughter, Jennifer, and her husband, Jason. Working on a project such as this is so much easier when you have the support and love of a great family.

And the other factor that makes working on a project like this easier and fun is having a fantastic person such as Jeffrey (JC) Clade with whom to work. The mechanical and welding portion of this text was written by JC. He impresses me greatly with his electrical, mechanical, and welding knowledge. He has a very bright and promising career ahead of him. I envy him his youth and energy. I also hope his wife, Ruth, will forgive me for planting the seed about writing textbooks in JC's mind. JC has the talent, personality, and people skills that make him an excellent instructor, a mentor, and a fine human being. I am proud and honored to call him my best friend and "little brother." A finer person you will never meet.

—Mike

FROM JC CLADE

I would like to extend my deepest appreciation and love to my wife, son, and mother for their patience and support during this revision. Ruth, my wife, has been extremely supportive throughout this project, and she has made many sacrifices. For this, I would like to thank her. Dru, my son, has been a huge help in this revision. He was my model for a few photographs, at a moment's notice. Thank-you, son, for all of your support!

To my mother, Andi: Thank-you for always being you. You are very special to me! I can only hope that I have made you as proud of me as I am of you. As a published author yourself, you know firsthand the difficulties of writing and how much time is invested in this endeavor, but as a mother, you have had to sacrifice even more as I haven't been there for you as much as I had wished. Thanks for your understanding!

A special thanks goes to my Aunt Mary, who has done so much for Mom in my absence. To Mary: I know that I have relied on you heavily, maybe more than I should have. I can only hope that you realize just how much I appreciate you and love you.

I would like to thank Rick Mack, Rick Childers, and Hezekiah Barnette for their expertise and willingness to help. You guys are aces in my book and a huge help, and I appreciate you all greatly!

Finally, I would like to thank the one individual who has probably helped me the most throughout my career, Mike Brumbach. Mike has been a true friend for going on two decades now. I have always thought of Mike as my "older brother." I have always admired his humility and his ability to reason and to solve problems. He has also always been someone that I know I can count on when I need to talk. He continues to cultivate the best in me. I think he sees my potential, and he tries to encourage me to grow in such a way that I don't recognize the growth until it has occurred. To Mike: Thanks a million for being such a great friend! I'm glad to have worked with you throughout these many years!

—*JC*

We would also like to thank our friends and colleagues who have been so supportive, helpful, and encouraging during this time. In addition, we wish to thank our students, who inspire us and remind us why we teach. Unless you've been there, you cannot appreciate the feeling when past students pay you a visit to tell you about the terrific job they got, the good money they are making, or the promotion they received, and they want to thank-you for all that you did for them, when, as we all know, they are the ones who did it! We also wish to thank Stacy Masucci, Andrea Timpano, and John Fisher of Delmar Cengage Learning. They have been extremely helpful and supportive in allowing us the freedom to create this text. Finally, we wish to thank our reviewers:

William Blackledge
Augusta Technical College
Augusta, GA

Robert Brown
Central Carolina Community College
Sandford, NC

Kevin Donohoo
Gateway Community and Technical College
Florence, KY

Chris Halm
Corning Community College
Elmira, NY

Tim Hisaw
Tennessee Technology Center
Crump, TN

Don Husky
Bainbridge College
Wigham, GA

Ron Meyer
Central Community College
Doniphan, NE

Edward Olson
Laramie County Community College
Cheyenne, WY

Ralph Potter
Bowling Green Technical College
Bowling Green, KY

Stephen Roggy
Greenville Technical College
Greenville, NC

Bill Welborn
Alamance Community College
Graham, NC

SECTION 1

GENERAL KNOWLEDGE

Safety

Tools

Fasteners

Industrial Print Reading

A TYPICAL DAY IN MAINTENANCE

Check It Out

It is 7:00 A.M, and your shift has begun. You have had your breakfast and coffee, and it is time to earn your pay. Your supervisor hands you a work order for the number 3 press. The press operator reports that the press was making a loud vibration, then the operator smelled a "funny" odor, and now the press will not run.

You mentally run through several possible causes of the problem. You gather the relevant prints, personal protective equipment (PPE), hand tools, and test equipment and proceed onto the plant floor to press number 3. It is now 7:20 A.M.

Work It Out

1. What might be the cause of the problem with the press?
2. What types of drawings or prints would be helpful?
3. What PPE do you need?
4. Which hand tools would you consider taking along?
5. What test equipment would be of benefit?

CHAPTER 1

Safety

This text will delve into the field of industrial maintenance, but before we can do that, we must begin with safety. Without safety, maintenance would be a deadly occupation because equipment and machinery are dangerous by nature. A wise man once said, "An accident is nothing more than an unsafe condition and an unsafe act that have been combined. If the unsafe condition had been removed, there would have been no accident. If the unsafe act had been removed, once again there would have been no accident." The maintenance mechanic needs to be safety oriented to avoid unnecessary accidents and bodily harm when working around equipment and machinery. It takes only one accident, and a very small moment in time, to lose a life.

OBJECTIVES

After studying this chapter, the student should be able to

- Understand the importance of personal protective equipment.
- Demonstrate the proper procedure for lifting.
- Use lockout and/or tagout when needed.
- List several general electrical safety practices.
- Discuss the proper use of safety belts, scaffolds, and ladders.
- Identify all the organizations that govern the safety of hazardous material.
- List the classifications of fire and the proper method of extinguishing a fire.

1-1 WORKPLACE SAFETY

Being aware of the surroundings and potential hazards is the first step in safety. Survey the surrounding area before starting any work. If an unsafe condition or potential hazard is noticed during the survey of the surrounding area, try to correct it, thereby eliminating the unsafe condition. If the unsafe condition cannot be removed, then do not perform any unsafe acts. Horseplay is the most common unsafe act. Horseplay does not belong in the workplace. All horseplay should be eliminated while any maintenance is being performed!

Accidents could kill or **maim** a person. Maiming occurs when some part of the body becomes mutilated, severed, disfigured, or seriously wounded. Injuries can range from a simple cut to becoming crippled. The loss of life and bodily injury must be avoided at all costs.

Work Environment

The working environment needs to be kept clean and safe. The workplace is dangerous enough without adding any unnecessary hazards. Keeping the shop area clean helps create a safe working atmosphere. Creating a cleaning roster for the shop spreads the work out among all shop members so that one person is not doing all of the cleaning. Once a shop is cleaned, try to keep it clean. Keep the floor free of debris and slippery fluids. Place oily, dirty rags in their proper storage containers, and send them out for cleaning on a regular basis. If oily rags are left piled up without ventilation for an extended period of time, combustion may occur. Store hazardous materials in their proper places. For instance, spray cans and other combustible products must be kept in a cool, ventilated place away from a source of ignition. Lighting is important. The working environment should be well lit to provide the best visibility possible. Try to keep all of the areas that you are working in clean and clear of debris. Also keep the machinery and equipment clean. A clean machine is easier to work on than a dirty one. Sometimes it is impossible to keep a machine or a piece of equipment clean due to its process, but it should be kept as clean as possible.

Personal Protective Equipment

Protective apparel may be necessary in some case, depending on the environment that you may be working in. Most facilities mandate the use of personal protective apparel and equipment, but some facilities allow the worker to decide whether to use **personal protective equipment (PPE)**. Either way, it is important to use personal protective equipment in industrial maintenance.

Foot Protection

Steel-toed boots should be used if there is a possibility that something might accidentally fall on the toes, causing them to be injured or even severed. These boots are heavy and sometimes very uncomfortable, but they are worth wearing if they can save your toes from being injured or severed. Most often, safety shoes are necessary, but electricians should not use shoes that use nails to fasten the soles to the shoe because they can provide a path to conduct electrical current. Some safety-toed shoes use a nonconductive material such as fiberglass for the hardened toe instead of steel. These shoes are lighter and more comfortable than steel-toed work shoes and offer the additional protection of reducing the risk of electrical shock.

Head Protection

Hardhats are also made for personal protection. Workers who were working on the Hoover Dam developed the hardhat. They dipped their cloth baseball hats in tar and let them dry. They did this to protect them from debris that fell from above while they were working on the walls of the dam. Hardhats are important if something could fall on your head. This may not be an issue at some facilities. Hardhats not only protect your head from falling objects but also protect your head if it accidentally rams into something above. This can happen when someone is climbing or is being lifted into an area that has pipes and ductwork overhead. Sometimes it becomes difficult to see what is above or behind your head that might cause injury. The hardhat protects against these accidental instances. Wearing the hardhat improperly is as bad as not wearing one at all. A hardhat must be worn according to each manufacturer's specification. How do you wear a hardhat improperly? Wearing a hardhat on the back of your head with the bill facing toward the back of your feet, as shown in **Figure 1-1A**, offers very little or no protection to the forehead. Wear the hardhat properly to offer the most protection possible, as shown in **Figure 1-1B**.

Protective Outer Wear

An apron, jumpsuit, or Tyvek suit may have to be worn to protect the clothing from chemicals or grease. Sometimes protective suits are worn to cover everything except the hands and the head. Finally, an entire ensemble may have to be worn, protecting everything, including the hands and head. For information on the type of apparel that must be worn in a particular situation, see your safety officer.

Hand Protection

Wearing gloves is a great way to protect your hands from being burned or cut. Gloves give an added layer of protection that is needed when a rope slips

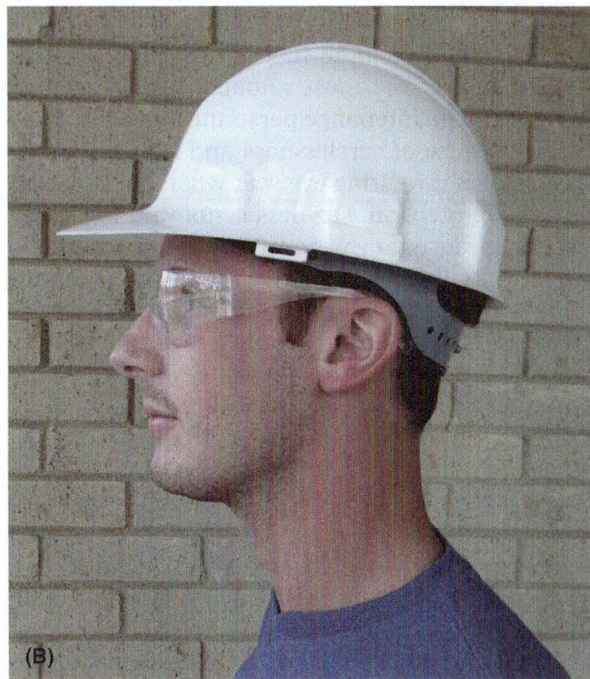

FIGURE 1-1 (A) Improper way to wear a hardhat. (B) Proper way to wear a hardhat.

through your hands, when something hot has to be moved, or when something that has very sharp edges has to be held. Gloves should be worn whenever there is a threat of injury to the hands. Gloves should fit comfortably. Do not wear gloves that are too large or too tight. There are times when gloves can become a safety hazard themselves. It is a good practice to step back and analyze each situation before you put the gloves on. Gloves should not be worn if the glove, along with your hand, could be pulled into a machine or piece of equipment that is moving. Do not get in close proximity to any moving machinery or equipment if at all possible. By getting close to moving machinery, you have committed an unsafe act.

Hearing Protection

Noise levels are measured in decibels (dB). Exposure to noise levels equal to or greater than 85 dB may cause hearing loss. That is why the National Institute for Occupational Safety and Health (NIOSH) recommends that hearing protection devices (HPDs) be used whenever the noise level equals or exceeds 85 dB, regardless of the exposure time. To illustrate what 85 dB may sound like, here are some examples of common sounds that a person may be exposed to. A common hand drill runs at approximately 95 dB; an impact wrench is somewhere around 103 dB; a hammer drill operates at approximately 114 dB; and a 12-gauge shotgun fires at 165 dB. As you can see, it is not difficult to be exposed to dangerous noise levels in our normal lives. In industry, there are many noisy machines and moving pieces of equipment. It is easy to see how people can lose their hearing if certain safety measures are not taken. See **Figure 1-2**. Hearing loss can be prevented with the proper training and the use of hearing protection devices. Dispensers may be mounted in strategic locations throughout the plant where hearing protection is required. These dispensers, as seen in **Figure 1-2**, contain inexpensive, disposable foam earplugs that can be squeezed and inserted into the ear canals, where they then expand, sealing out dangerous noise levels. Hearing is something that many people take for granted. It is too late to protect hearing once it is gone! Most facilities have mandated the use of hearing protection devices. If your facility does not mandate the use of hearing protection devices, take it upon yourself to preserve your hearing. Become a leader, and set an example for others to follow.

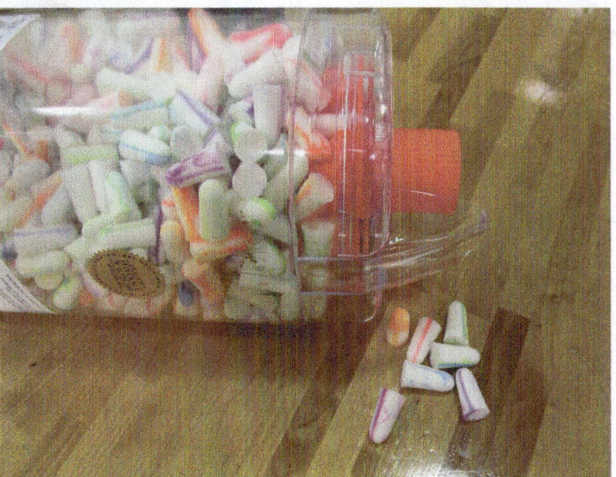

FIGURE 1-2 Earplugs.

Eye Protection

Vision is another thing that is taken for granted. Sight is one of the most valuable senses that we have. Many maintenance personnel have lost their vision because of carelessness and laziness. Carelessness is not wearing eyewear when the situation warrants protection. Laziness is not wearing safety glasses because they were left in the shop and it is too far to walk back. The best thing to do is to find some way to keep your eye protection with you at all times. Eye protection should be worn whenever there is a threat of vision loss due to injury of the eye. Injury can occur by a flying projectile, falling debris, a chemical that has been splashed, or a thousand other ways. An eye injury caused by any of those hazards takes only a fraction of a second to occur, and then it is too late. Protect your eyes *before* you start working by wearing the proper eye protection. Make sure that your eyewear has side shields to protect the eye from foreign objects that may enter from the side. If you are already wearing prescription glasses, you can put the side shields on for added protection. See **Figure 1-3**.

Cold Weather PPE

People working in extreme cold weather may be exposed to conditions that can cause frostbite and hypothermia. Persons working in these conditions can protect themselves from the elements by wearing protective cold weather PPE. These garments may include, but are not limited to, hooded outerwear, jackets, and sweatshirts. In extremely cold climates, it may be necessary to wear a parka. A parka is a very thick coat that is lined with fur to insulate and keep the body heat within the coat. The parka also has an outside lining that is resistant to rain and snow and does not allow moisture to enter the coat, thus keeping the body dry. It also includes a hood with a drawstring that, when pulled tight, protects the face from the elements. It is also essential to wear thick wool socks and gloves to protect the digits from frostbite.

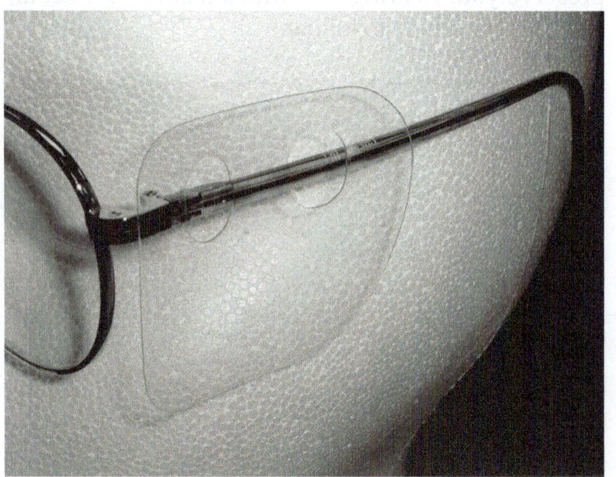

FIGURE 1-3 Disposable side shields for prescription glasses.

Appearance

A long-tailed shirt should be tucked into the pants, and the pants should not be too large for the person who is wearing them. Long-sleeved shirts should be worn with the sleeves rolled down and buttoned.

If you have long hair and are working around machinery, be aware of where your hair is. Some people who have long hair have been pulled into machinery or equipment by their hair. There is absolutely nothing wrong with having long hair and working around machinery—just be smart about it. Tuck long hair under a hat or inside your shirt, or wear a hair net. This prevents an accident from occurring. Some facilities mandate that long hair be in a hair net.

Jewelry is another safety issue. Rings can get caught on a moving piece of equipment and cause the loss of a finger. Many people, every day, lose fingers when rings get caught on something. Make sure to remove rings before starting any work. Long necklaces are also dangerous. As you lean over a moving part of a machine, it may get caught in the machine. Be careful with bracelets as well. Be aware of these dangers if jewelry is worn. A safety-conscious person would not wear jewelry in this environment.

Pinch Points

It is very important that you know about and understand **pinch points**. Pinch points are points on a machine where two or more separate components meet or come together. There are two types of pinch points: rotating and pressing. A **rotating pinch point** is where two rollers, gears, or anything that is rotating meet. A **pressing pinch point** is where two separate parts of a machine that are pressed together meet. Whatever goes into a pinch point becomes compressed. If fingers or hands get pulled into a pinch point, the fingers or hand will certainly be mutilated, and the digits or the entire hand itself could be severed.

Lifting

Many people each year injure themselves while performing the simple act of lifting. The injury comes from lifting too much weight or lifting improperly. The first thing to do is to know your facility's policy on lifting. Most facilities require the use of a lifting girdle or belt. This is to provide support to the back while lifting. Next, determine whether the item is

FIGURE 1-4 Proper footing for lifting.

too heavy to lift. This factor varies from person to person. If it is decided that the load is not too heavy, then proceed with the lifting action. If the load is too heavy, do not be ashamed to ask for help. Here are some very simple steps to ensure safety while lifting. First, place both feet around and beside what is to be lifted, if possible. Make sure that a good footing has been established. Bend down at the knees into a squatted position. This is illustrated in **Figure 1-4**. Grasp the object that is to be lifted, and get a firm grip on it. Straighten the back so it is as close as possible to a 90-degree angle from the floor. Keep the back straight through the entire lift. Do not, at any time, arch the back. Once a firm grip has been established and the back is straight, proceed with the lift by straightening the legs. Be sure to keep the load close to the body while lifting it. Lift slowly. Do not twist or turn the body during the lift. Lift the load straight up with the legs. Once the load has been completely lifted, then you can turn or rotate at the waist if necessary. Reverse the process when setting the load down. Once again, remember to keep the back as straight as possible through the entire process of setting the load down.

Securing the Area

When maintenance is being performed on a machine, the area in which you are working may have to be secured. For instance, a portion of a catwalk or subfloor may have to be removed to gain access to the damaged part of the machine. In this case, a hazardous condition has been created. The area needs to be roped off so that people are aware of the hazard and its location. Awareness prevents falls and injury. If you are working overhead, and there is a chance that something will be dropped or fall, once again, secure the area to prevent people from entering the hazardous area. Securing an area can also prevent curious people from getting too close to you. Your undivided attention must be given when you are working on a machine or a piece of equipment. Any distractions could prove to be dangerous. Securing the area keeps these distractions at a safe distance, allowing you to focus on the task at hand.

Electrical Safety

Physical contact with electricity can be fatal. If not fatal, it can be painful and cause permanent injury, so it must be avoided.

Every year, tens of thousands of people are injured by accidental contact with an energized electrical circuit. Over one thousand of these individuals will lose their lives because of this accidental contact. A common misconception is that it takes a lot of current or voltage to cause injury or death. This is not true! An electrical current of only 0.0025 ampere can be sensed. A current of only 0.025 ampere causes you to be held by the circuit. Death can occur with a current as low as 0.15 ampere. These current amounts, and their effects, depend on many variables. How moist is your skin? How well connected to the circuit are you? What path must the current take through your body? How much voltage is present?

It is vital that you become aware of electrical safety practices and observe caution when working around electrical circuits. By learning about some general electrical safety concerns, and by using some forethought, you will be able to safely perform work on electrical circuits not only for you but for your coworkers as well.

Electrical safety must be everyone's concern. Whether the maintenance technician, operator, or supervisor, all must take responsibility for electrical safety. In this chapter, we discuss some general guidelines for electrical safety. Most facilities have extensive and specific rules regarding electrical safety. You should follow the rules and procedures at your facility to the letter. If you are unclear or uncertain as to the nature or application of any of your facility's procedures, you should not hesitate to discuss the matter with your supervisor immediately.

General Electrical Safety Precautions

The following is a list of general safety precautions that should be observed during work on or around electricity. Many of these things are often overlooked.

- All electrical equipment (motors, generators, pumps, control equipment, conductors, etc.) must be installed so that all energized parts are properly insulated or adequately guarded.
- An adequate clear workspace must be maintained around all energized equipment. This workspace is necessary for the inspection, repair, or replacement of the electrical equipment.
- Ensure that all conductors are adequately protected by cable trays, conduit, or metal-sheathed cable as per the policies and procedures at your facility.
- Identify all circuit conductors, disconnects, switchgear, and so on, with clear identification tags.
- Ensure that all electrical drawings are kept up to date and are accurate.
- Replace all permanent equipment guards that have been removed or damaged.
- Maintain good housekeeping practices before, during, and after any work performed.
- Always follow your facility's policies and procedures and/or the *National Electrical Code.*

Lockout/Tagout

You may think of an electrician when you hear the term *lockout,* but this safety rule should also apply to all maintenance personnel. Every facility should issue all maintenance personnel a lockout with an identification tag on it. However, there are some facilities that do not do this. The purpose of the lockout procedure is to prevent injury or death by removing all hazardous energies that are present on a machine before any repairs are performed. An **electrical lockout** is accomplished by turning off the main disconnect that supplies power to the machine and locking the handle of the disconnect in the off position so that it cannot be turned back on until all lockouts are removed. This action removes two safety hazards. First, if the machine is locked out, there should be no electricity on the machine, thus preventing injury by electric shock. Second, removing the control voltage prevents the machine from being accidentally started while you or your hands are inside the machine. Disconnects are not all that can be locked out. Circuit breakers and valves can also be locked out once they are placed in a safe position. All maintenance personnel who are working on a machine need to place their lockout, with an identification tag, on the disconnecting means

FIGURE 1-5 Multilock hasp.

of the hazardous energy. Using multilock hasps allows room for everyone to attach their personal lock on the disconnecting means. See **Figure 1-5.** Before any work is to be started on a machine that has been locked out, make sure the machine is clear of personnel, and verify that all energies have been isolated. Bleed off any stored energies or pressures (they could be hydraulic, pneumatic, or capacitive), and press all start buttons.

A voltmeter must be used to verify the absence of electricity. If there is verification that all energies have been isolated and removed, then all repairs may begin. Make sure that you are the only person who removes your lockout. The only other person who has the authority to remove a lockout, besides you, is the plant engineer. However, he or she can remove your lockout only in the case of an emergency or if you are not on company grounds, and your immediate supervisor is present at the lockout when it is removed. The purpose of your supervisor's being present is to verify your absence. Most lockout locks today only come with one key, and most plants are extremely strict concerning the losing of a lockout key. It is not uncommon to have a zero-tolerance policy concerning the violation of their lockout/tagout policies. If a key is lost, it typically means that you have lost your job. At no time should a lockout be cut off unless the key is lost and both you and your supervisor are present. Lockout procedures are very strict for your safety. There are strict Occupational Safety and Health Administration (OSHA) regulations that must be followed when using the lockout procedure. For this information, see the safety officer in your facility. **Tagout** is simply the placement of a tag on an energy isolation device. This tag is a prominent warning to others of your presence on the machine. See **Figure 1-6.** *The tagout does not lock out the energy isolation device,*

FIGURE 1-6 Tag used for the tagout procedure.

it just warns of your presence on the machine. Anyone seeing a tagout should investigate to find the location of the person who tagged out the machine before starting the machine. **Figure 1-7** shows a typical lockout cabinet. The devices in the cabinet can be used to lock out all the various types of energy that may injure personnel.

The Buddy System

Whenever possible, the buddy system should be used. This simply means that maintenance personnel need to work in pairs. If something were to go wrong and you were injured, your partner, or buddy, could go get help. Normally, this is not a problem on the first shift because there are usually more maintenance personnel on the first shift, allowing the formation of teams or crews. However, some facilities use what is called a skeleton crew on the off shifts. This simply means that the maintenance department uses as few personnel as possible for maintenance on the second and third shifts. In this case, it is possible for only one or two maintenance people to be on duty during these off shifts. If you are one of these individuals, upon receiving a work request, it is important to notify somebody of your destination and the approximate time of your return. Once notified, they would know where to look for you if an extended period of time were to elapse with no contact from you.

All personnel who are left to work alone should have a two-way radio in their possession in the case of an emergency. The other radio should be left with someone who can help if the maintenance personnel should call because of an emergency.

All facilities should have no fewer than two maintenance personnel on every shift. If an accident were to happen, there would be an immediate response. This is important because when someone is injured, seconds may make the difference between life and death!

Safety Concerns before Starting the Work

The following precautions should be taken before starting the work. They may seem obvious, but it is

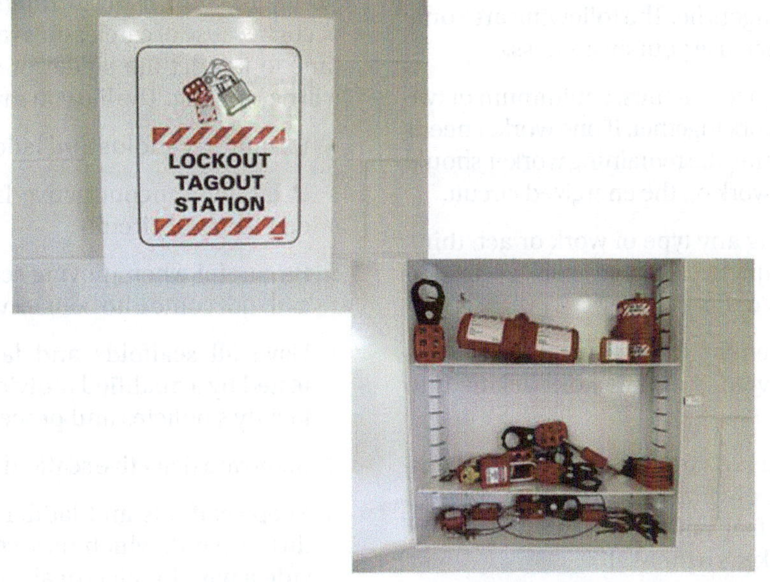

FIGURE 1-7 Lockout cabinet.

amazing how often we overlook these precautions and begin work in an unsafe environment.

- Be certain that all workers involved in the job have the necessary skills and qualifications to perform the work.
- Familiarize yourself with the job at hand.
- Identify all of the obvious safety concerns, and then think about the not so obvious ones. Ask yourself, "What would happen if ... ?"
- Verify that all required safeguards or safety devices are present and in working order.
- Know the location of the nearest fire extinguisher, first-aid kit, and emergency phone.
- Notify all appropriate personnel of the nature of the work to be performed.
- Verify that all personnel involved with the job are familiar with the safety and emergency policies and procedures as related to the work that is to be performed.
- Plan safety into the job.

- Maintain a neat and orderly work site.
- Do not attempt the work when tired or taking medications that cause drowsiness, or if lighting is insufficient.
- Check for atmospheric hazards such as dust, vapors, or gases that may ignite from an electrical arc or spark.

Worker Safety

You need to think about the situation that you are working in and evaluate it to determine how safe it is and where the dangers lie. The following are some items to check to increase your awareness.

- For work on energized circuits, a minimum of two workers should work together. If one worker needs to leave the worksite, the remaining worker should not perform any work on the energized circuit.
- Before performing any type of work or act, think about the consequences of that work or act. Do not take chances or cut corners!
- Always work carefully. Place yourself in the safest position to avoid slips, falls, or backing into energized parts.
- Make certain you are satisfied that you are working in the safest environment possible. Do not continue if you feel uncomfortable because of yours or a coworker's actions.

- Always exercise caution. Treat all circuits as energized, even when you know they are not!
- Remove all jewelry (watches, rings, bracelets, etc.). Gold, for example, is an excellent conductor of electricity, and any jewelry can become caught in moving machinery!
- Do not hesitate to obtain medical treatment for all injuries, no matter how minor they may seem.

Personal Protective Equipment

Review the following list to ensure you are wearing the proper PPE. Wearing safety equipment may be inconvenient or uncomfortable, but your personal safety is more important than comfort.

- Wear a hardhat.
- Wear safety glasses (with side shields).
- Use a safety belt when working in an elevated location.
- Use a lifting belt when lifting a heavy load.
- Have your safety belt inspected and tested by a qualified individual according to your facility's policies and procedures.
- Use approved rubber gloves, leather gloves, rubber blankets, and rubber mats when working on energized circuits or equipment.
- Carefully inspect all PPE each time you use it. Your life depends on it!
- Have all rubber items electrically tested according to your facility's policies and procedures.
- Discard all worn, defective, or otherwise unsafe items.

Scaffolds and Ladders

Accidents from using scaffolds and ladders usually occur because of carelessness and rushing. Take some time to inspect the ladder or scaffold on which you will be working. The life you save may be your own.

- Examine scaffolds and ladders before each use.
- Use only nonconductive ladders when working on electrical circuits.
- Be careful when moving scaffolds and ladders so as to not come into contact with energized circuits.
- Have all scaffolds and ladders inspected and tested by a qualified individual according to your facility's policies and procedures.
- Remove a defective scaffold or ladder from service.
- Keep scaffolds and ladders clean and free from dirt or paint, which may conceal defects or provide a path for electrical current to flow. See section 1-3 for more information on scaffolds and ladders.

Portable Hand and Power Tools

Here are some items that should be considered when working with power tools.

- Use care with all hand and power tools to prevent their coming into contact with energized equipment or circuits.
- Be especially careful when cutting or drilling "blind." Know what is inside the wall or on the other side of the wall to eliminate the risk of cutting or drilling into energized circuits.

CAUTION

- Inspect all hand and power tools for defects, and remove them from service if unsafe.
- Inspect the power cords of all power tools for cracks, breaks, and proper grounding (if not of the double-insulated type).

Working on and around Electrical Circuits

Keep your mind on the task at hand. All it takes is a lapse in concentration, or working without thinking, and a shock or electrocution occurs. Try to remember the following:

- Treat all circuits as *live* or *energized*, even when you know for certain that they are not. Often injury occurs from the surprise reaction of coming into contact with an assumed *dead* circuit. Quite often, the electrical shock does not cause the injury, but the surprised reaction to the electrical current causes you to lose your balance and fall from a ladder or jerk your hand from a conductor only to smash it into a cabinet!
- Familiarize yourself with NFPA 70E. This is the *Standard for Electrical Safety in the Workplace* as published by the NFPA. NFPA 70E not only addresses the hazards of electrical shock, but also the hazards from arc flash. This has resulted in standards for electrical safety including personal protective equipment or PPE that was not considered in years past. The scope and nature of the requirements as set forth by NFPA 70E identify different hazards and risks, the use of rubber gloves and insulated tools, and protective clothing and PPE.
- Deenergize, lockout, and tagout all electrical circuits upon which work is to be performed. If it is not practical to deenergize a circuit, follow your facility's policies and procedures for working on energized circuits.

- Upon completing repairs, make a thorough examination of the circuit before reapplying power.
- Verify that all personnel are clear and it is safe to reenergize the circuit.

Grounding

Proper grounding of equipment is often overlooked. This is an accident waiting to happen. Also, unintentional grounding of personnel causes injury and death.

- Verify that all electrical equipment such as motors, generators, conduits, switchgear, transformers, and so on, are properly and adequately grounded. Any defective or missing grounds should be repaired immediately in accordance with your facility's policies and procedures.
- When working on electrical circuits or equipment, take all necessary precautions to prevent yourself from becoming the path for the electrical current to ground. Use proper PPE in accordance with your facility's policies and procedures. Be especially mindful in damp or wet areas.

Switches

Never assume you know when a switch is off. Personally verify the state of all switches and circuits. Do not get offended when coworkers check behind you. They are just being safe, and you should do the same for them.

- To ensure the protection of yourself and coworkers, always verify the condition of the circuit before the opening or closing of any switches that affect that circuit. Verification minimizes the risk of injury in the event a faulty circuit is encountered or should a worker become exposed to a live circuit.
- Always open or close switches in a firm, positive manner. Use sufficient force to cause the switch contacts to make or break fully and quickly.
- Switches should be opened or closed fully and completely. Partially opened or closed switches may produce arcing or flashover, which may damage the switch and burn or injure the operator.

Fuses and Disconnects

Again, the following precautions are often overlooked. Think about these items every time you work around fuses and disconnects. If you keep safety at the forefront of your thoughts, you minimize the chances of an accident's occurring.

- Never remove a fuse from a circuit under load. Always open the associated disconnect switch before removing the fuse.
- Use the appropriate and approved fuse puller as specified by your facility.

- When removing fuses, remove the fuse in such a manner so that the *hot* side of the fuse and circuit is broken first.

- When installing fuses, install the fuse in such a manner so that contact with the *hot* side of the fuse and circuit is made last.

Motors

Motors may cause injury not only from an electrical shock but also from the mechanical motion. Take steps to minimize the risk of electrical injury and injuries that may occur from the motor's starting unexpectedly.

- Follow your facility's lockout/tagout procedures when working on the electrical connections to rotating machinery.
- When making temporary connections to rotating machinery, use sufficient amounts of electrical tape or other insulating material.
- Keep oil cans, wipes, rags, and so on, away from rotating machinery. Be mindful of ferrous type material that may be attracted to the magnetic field created by the motor or generator.
- Deenergize rotating machinery when using cleaning solvents and lubrication.

Capacitors

Be careful when working around capacitors.

- Capacitors may store a charge for a very long time, even without connection to an electrical circuit. This charge may contact with the terminals of the capacitor.
- Always use an approved discharge device to discharge capacitors before working around, handling, or making connections.

Test Equipment

When using test equipment, many individuals assume that their equipment is working because it worked the last time they used it. This assumption could prove to be very dangerous. Think about the following items whenever you need to use test equipment on the job:

- Use the right piece of test equipment for the job.
- Wear appropriate PPE when using test equipment on energized equipment.
- Know the proper usage and procedures of the test equipment that you are using. Be certain that you are properly trained in its correct operation and use.

- Do not exceed the design limits of the test equipment that you are using. Be aware of circuit voltages present so that you do not use the test equipment on a circuit of higher voltage. Injuries have occurred to the hands, arms, and face of a worker by inadvertently connecting a 1000-volt rated multimeter to a circuit energized with 7200 volts.
- Always inspect your test equipment for signs of wear and damage before use. Remove a damaged piece of test equipment from service immediately.

- Always verify that voltage measuring equipment functions properly by checking the operation on a known voltage source before usage on the circuit to be tested.
- Be especially careful when using test equipment on *live* or *energized* circuits.
- Verify that a circuit is *dead* or *deenergized* before measuring circuit resistance with an ohmmeter or Megger®.

1-2 HAZARDOUS MATERIAL HANDLING

There are many codes and charts that are required for hazardous chemicals and the areas in which they are stored. There are many laws, regulations, and codes that must be adhered to when using or working around hazardous materials. To better understand these laws, regulations and codes, one must know about the organizations that develop these codes and standards to ensure a safe working environment. These organizations are represented by abbreviations such as OSHA, NIOSH, UL®, NFPA, ANSI, and NEMA.

The Occupational Safety and Health Administration (OSHA) is an organization that all people should be aware of. OSHA, on occasion, visits a facility to give periodic inspections and safety audits. OSHA is a safety enforcement organization. As the name implies, OSHA is concerned about the occupational safety and health of all people. For this reason, they have instituted many laws and regulations that must be followed to ensure a safe working environment. OSHA has many training programs available. Most businesses recognize the need for a safety and health program. A good facility that cares about the safety of its employees will institute these safety programs into a regular agenda, thus heightening awareness throughout the facility of many safety issues that otherwise might go unnoticed. If an accident occurs at your facility, the first thing OSHA will ask for is a copy of the safety and health program. If these documents cannot be provided upon request, a hefty fine may result.

The National Institute for Occupational Safety and Health (NIOSH) is an organization that works with OSHA to develop and revise the recommended exposure limits for hazardous materials and conditions that may be in the working environment. NIOSH also concentrates on researching and making recommendations for preventive measures that must be taken to counteract the adverse health affects of hazardous materials and the locations in which they are stored before a serious accident occurs. NIOSH, however, is primarily concerned with research, whereas OSHA is the organization that is responsible for the enforcement of these laws and standards.

Underwriters Laboratories Inc. (UL) is an organization that tests equipment and products. These equipment and products must meet or exceed the minimum standards that are specified in national codes and standards. Underwriters Laboratories Inc. is an independent company that works under contract with many manufacturers to ensure the conformity of their product to the national standards. Any equipment that has been tested by Underwriters Laboratories Inc. and passed the required standards has the UL label on it somewhere.

The National Fire Protection Association (NFPA) is most commonly known for its many publications that are used and adopted as law to be enforced by local authorities having jurisdiction. Some of these publications are the NFPA 70, NFPA 5000, NFPA 72, and NFPA 704. The NFPA 70 is better known as the *National Electrical Code®* (*NEC®*). The *NEC* is a publication that is revised every three years and is usually adopted by local or state governments into law for local authorities to enforce. It gives strict guidelines for electrical installations. The sole purpose for this publication, as stated in *Article 90.1*, is the practical safeguarding of persons and property from the hazards arising from the use of electricity. The NFPA 5000 is the NFPA Building Code. The goal of the NFPA in writing this publication is to have the best scientifically based codes and standards possible, developed in a fully open consensus process. The NFPA has been advised by adopting authorities that they want to adopt a full and coordinated set of codes for building safe and structurally sound buildings. The NFPA 72 is the National Fire Alarm Code. The NFPA 72 sets minimum requirements for fire alarm systems, household fire warning equipment, protected premises fire alarm systems, systems with a supervising station (a facility that receives signals and is always staffed to respond), initiating devices (such as heat detectors, smoke detectors, radiant energy detectors, and fire-gas detectors), and audible and visible notification devices. The National Fire Alarm Code also covers inspection, testing, and maintenance of fire alarm systems.

The NFPA 704 Standard

The NFPA 704 is the standard system for the identification of the hazards of materials for emergency response. The NFPA 704 standard provides a readily recognized and easily understood system for identifying specific hazards and their severity, using spatial, visual, and numerical methods to describe in simple terms the relative hazards of a material. This standards system can be seen in **Figure 1-8**. It addresses the health, flammability, instability, and related hazards that may be presented as short-term, acute exposures that are most likely to occur as a result of fire, spill, or similar emergency. There are three main objectives for this system: (1) to provide an appropriate signal or alert for the protection of both public and private emergency response personnel; (2) to assist in planning for effective fire and emergency control operations, including cleanup; and (3) to assist all personnel in evaluating the hazards of a known hazardous material.

The symbol in **Figure 1-8** is of a diamond shape. It, in turn, has four smaller diamond shapes within it. Each of these smaller diamonds is a different color. To the lower right of the red diamond, there is a yellow diamond. At the bottom, there is a white diamond, and a blue diamond resides to the far left, just below the red diamond. Each colored diamond identifies the different hazards of a hazardous material. These are the flammability hazards, instability or reactivity hazards, the degree of health impact, and any special hazards that may arise during the use of the material. The red diamond represents the flammability; the yellow diamond represents the reactivity or susceptibility to the release of energy

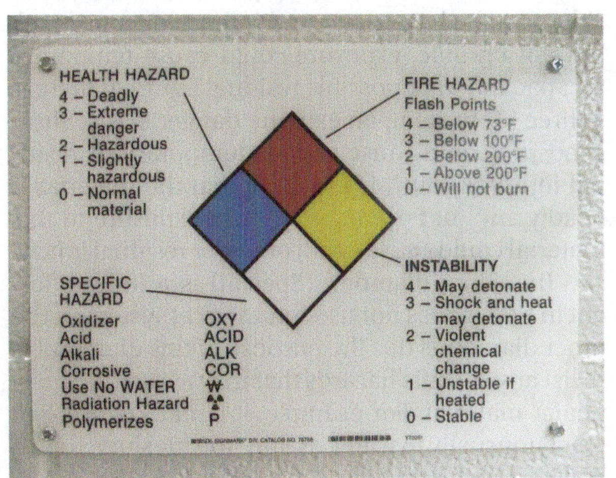

FIGURE 1-8 NFPA Standards 704.

(instability); the white diamond represents the specific hazard and the blue diamond represents the severity of health hazards or possible injury that may arise from handling the hazardous material. With the red, yellow, and blue diamonds, a hazard severity is identified by a numerical rating system that ranges from zero (0) to four (4), four being the most severe.

In the red diamond (Flammability), a zero (0) severity indicates that the material does not burn. If there is a one (1) in the red diamond, then the material has a flash point above 200°F (93°C). If there is a two (2), the material has a flash point below 200°F (93°C) but above 100°F (38°C). If there is a three (3) present, the flash point is below 100°F (38°C) yet above 73°F (23°C). A four (4) indicates that the material has a flash point below 73°F (23°C). A material with a four (4) for flammability has to be kept in a cool environment.

In the yellow diamond (Reactivity or Instability), a zero (0) severity indicates that the material is stable. If there is a one (1) in the yellow diamond, then the material is found to be unstable if heated. If there is a two (2) present, the material may have a violent chemical change at elevated temperatures and pressures or may react violently with water or may form explosive mixtures with water. If there is a three (3) present, shock or heat may cause the material to detonate. A four (4) indicates that the material is very unstable and is capable of detonating itself or has an explosive decomposition or reaction at normal temperatures.

In the blue diamond (Health), a zero (0) severity indicates that the material, on exposure to fire conditions, would offer no hazard beyond that of ordinary combustible material. If there is a one (1) in the blue diamond, then the material is considered slightly hazardous, meaning that if exposed to this, material irritation or minor residual injury may occur. If there is a two (2), the material is considered to be hazardous, and intense or continued, but not chronic, exposure could cause temporary incapacitation or possible residual injury. If there is a three (3) present, an extreme danger exists. Short exposure could cause serious temporary or residual injury. A four (4) indicates that the material is deadly and that even a very short exposure to this material could cause death or major residual injury.

The white diamond (Special) is a little different in that it does not use a numerical system as the other diamonds do. The purpose of this diamond is to list any specific hazards that may exist with a particular material. For example, if a material were to react unusually to water, as does magnesium, then a W with a slash across the center, as seen in **Figure 1-9**, would be seen in the white diamond, indicating

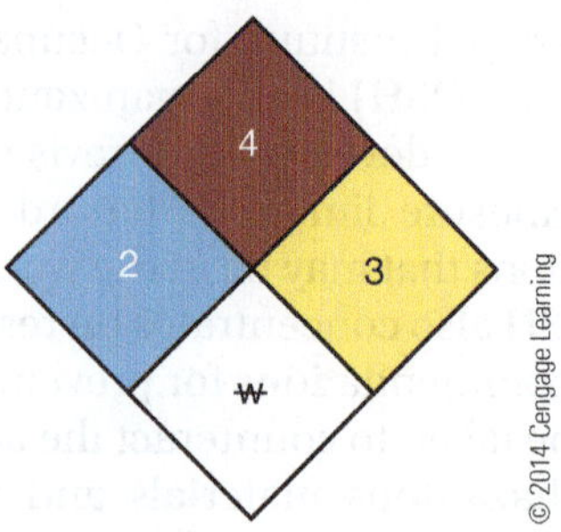

FIGURE 1-9 This symbol shows that a material is reactive to water.

that it would not be desirable to put water on this material. This information is helpful in the case of a fire. The addition of water to this type of hazardous material, in the event of a fire, only worsens the fire and could cause an explosion. **Figure 1-10** shows an example of the NFPA Standards Hazard Identification System on a fence surrounding a gas processing plant. Notice that the fire hazard is rated at four (4), which means that the flash point is below 73°F (23°C); the Health Hazard is a three (3), which indicates Extreme Danger; the Instability is rated at three (3), which conveys that shock and heat may cause detonation, and that the Specific Hazard is an Oxidizer.

Another specific hazard and its code that might be seen in the white diamond is OX, identifying an oxidizer. If a material were acidic, the white diamond would contain the word ACID. An ALK identifies an alkaline material. Corrosive materials are identified by COR, and finally, a material that has a radiation hazard is identified by the recognizable trefoil symbol that is shown in **Figure 1-11**.

It is important to note that the name of the chemical or material will be either above or below the NFPA 704 standard symbol.

FIGURE 1-10 This is an example of the NFPA standards hazard identification system in use.

FIGURE 1-11 This symbol shows that a material is a radiation hazard.

The Hazardous Material Information Guide

The Hazardous Material Information Guide (HMIG), which can be seen in **Figure 1-12**, has the same information as the NFPA 704 standard with one exception; the white section contains the necessary PPE that should be worn when handling the hazardous material instead of the special hazards that were listed for the NFPA 704 standard. A list of PPE ranges from A to K and includes a letter X. This index can be seen in **Figure 1-13**.

The Material Safety Data Sheet

OSHA Regulations Standard 29 CRF 1910.1200 is an OSHA regulation that states that any chemical mixture that is made by a manufacturer must have a Material Safety Data Sheet (MSDS) written for that particular chemical. An example of an MSDS is shown in **Figure 1-14** (see in page 17).

Standard 29 CRF 1910.1200 has a very specific purpose. The purpose of this section is to ensure that the hazards of all chemicals produced or imported are evaluated and that information concerning their hazards is transmitted to employers and employees. This transmittal of information is to be accomplished by means of comprehensive hazard communication programs, which are to include container labeling and other forms of warning, MSDS, and employee training. This is stated in section (a)(1) of the standard.

Section (b)(1) requires chemical manufacturers or importers to assess the hazards of chemicals that they produce or import and requires all employers to provide information to their employees about the hazardous chemicals to which they are exposed, by means of a hazard communication program, labels and other forms of warning, MSDS, and information and training. In addition, this section requires distributors to transmit the required information to employers.

The name of the chemical must be listed at the top of the MSDS. Each MSDS has eight sections following the name of the chemical, all listed in roman numerals, thus providing all of the necessary information that is required in the OSHA Regulations Standard 29 CRF 1910.1200. Each section is listed below, along with the information that should be found in that section. **Figure 1-15** (see in page 18) shows a compliance center that contains all of the MSDS Sheets for all hazardous products that are used in that department.

- Section I gives the name of the manufacturer, the address, and a telephone number for information. This section also contains the date that the MSDS was prepared.
- Section II contains the Hazard Ingredients/Identity Information. This is where all hazardous components and specific chemical identities are listed.
- Section III contains the Physical/Chemical Characteristics. This includes such information as the boiling point, melting point, specific gravity, appearance, color, and solubility in water.
- Section IV gives the Fire and Explosion Hazard Data. This is where the flash point can be found as well as the extinguishing media. Also, any special firefighting procedures are listed. Any unusual fire and explosion hazards that may exist are also listed here.
- Section V has the Reactivity Data. Stability, instability, certain conditions to avoid, material incompatibility, or any hazardous decomposition or components and specific chemical identities. Also, any byproducts that may be produced will be listed in this section.
- Section VI contains the Health Hazard Data. Exposure limits, carcinogenity, signs and symptoms of exposure, and emergency and first-aid procedures are listed here.
- Section VII contains the Precautions for Safe Handling and Use. This lists the steps to be taken in the event of a spill or release, the proper method of disposal, all precautions that need to be considered in handling and storage, or any other precautions that may exist.

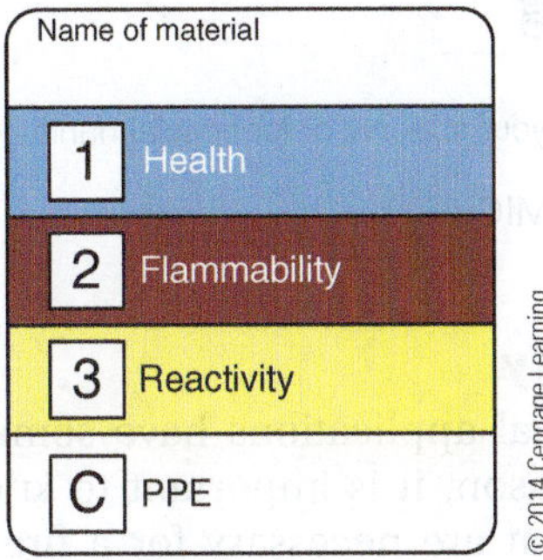

FIGURE 1-12 The Hazardous Material Information Guide (HMIG).

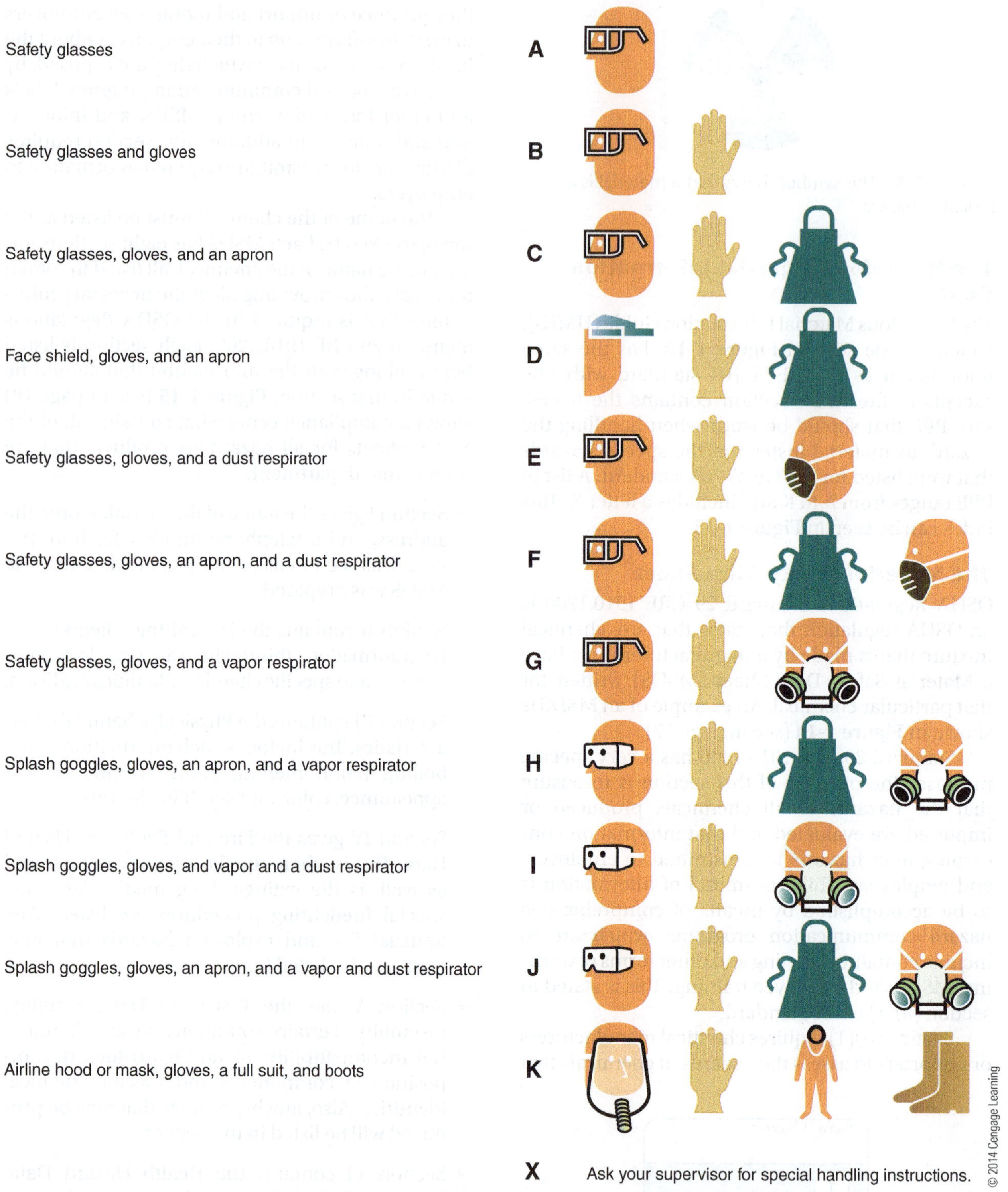

Description	Code
Safety glasses	A
Safety glasses and gloves	B
Safety glasses, gloves, and an apron	C
Face shield, gloves, and an apron	D
Safety glasses, gloves, and a dust respirator	E
Safety glasses, gloves, an apron, and a dust respirator	F
Safety glasses, gloves, and a vapor respirator	G
Splash goggles, gloves, an apron, and a vapor respirator	H
Splash goggles, gloves, and vapor and a dust respirator	I
Splash goggles, gloves, an apron, and a vapor and dust respirator	J
Airline hood or mask, gloves, a full suit, and boots	K

X Ask your supervisor for special handling instructions.

FIGURE 1-13 Recommended personal protective equipment (PPE) in the HMIG.

- Section VIII lists the Control Measures, which includes respiratory equipment, ventilation, and required PPE.
- Section IX is used for any Additional Comments that a manufacturer may have about its product.

Fire Safety

All industrial applications have some risk of fire. For this reason, it is important to know the components that are necessary for a fire to exist, the different classes of fire, and the proper procedure for quenching each. It is important to know this

MATERIAL SAFETY DATA SHEET

Date of Preparation – January 2000

SECTION I—PRODUCT IDENTITY

IDENTITY— ACME A-1 Citrus Cleaner
PART #— 8832456
Manufacturer's Name— ACME Quality Product, 1234 Main Street, Hometown, USA 00000-0000
Emergency Phone (800)-555-1234
Information Phone (800)-555-4321

HMIS CODES—
HEALTH – 2
FLAMMABILITY – 4
REACTIVITY – 0
SPECIAL – 0

SECTION II—HAZARDOUS INGREDIENTS

INGREDIENT	CAS#	OSHA PEL	ACGIH TLV	STEL	%
Propane	74-98-6	1000PPM	2500PPM	760PPM	<5
Isobutane	78-28-5	N/A	N/A		<5
Ethyl Acetate	141–78-6	400PPM	400PPM		<5
Aliphathic vs Hydrocarbons	8052–41-3	100PPM	100PPM		<70

SECTION III — PHYSICAL DATA

PRODUCT WEIGHT—	6.39 lb/gal	765 g/l
SPECIFIC GRAVITY—	0.77	
BOILING POINT—	<0–412 F	<-18–211 C
MELTING POINT—	N/A	
VOLATILE VOLUME—	97%	
EVAPORATE RATE—	Slower than Ether	
VAPOR DENSITY—	Heavier than air	
VAPOR PRESSURE PSIG—	@70 F: 80–90	
SOLUBILITY IN WATER—	Slight	

SECTION IV—FIRE AND EXPLOSION HAZARD DATA

FLASH POINT— –156 F
EXTINGUISHING MEDIA— Carbon dioxide, Dry chemical, Foam, Water
UNUSUAL FIRE AND EXPLOSION HAZARDS
Store away from excessive heat. Do not use in the vicinity of an open flame or sparks. Do not use on surfaces that are hot. Over exposure may cause health hazard. Obtain medical attention if overexposure occurs.
SPECIAL FIRE FIGHTING PROCEDURES
Fire fighters should wear NIOSH approved positive pressure self-contained breathing apparatus. Aerosol containers may burst when heated.

SECTION V—REACTIVITY DATA

STABILITY—Stable
CONDITIONS TO AVOID—All sources of ignition.
INCOMPATIBILITY—None known.
HAZARDOUS DECOMPOSITION BYPRODUCTS—Carbon Dioxide, Carbon Monoxide
HAZARDOUS POLYMERIZATION—Will not occur.

MATERIAL SAFETY DATA SHEET

(continued)

SECTION VI—HEALTH HAZARD DATA

ROUTES OF ENTRY—Inhalation, skin absorption, indigestion
HEALTH HAZARDS (Acute and Chronic)—(ACUTE) Irritation of the skin, eyes, or respiratory system. May cause nervous system depression. Extreme overexposure may result in unconsciousness and possibly death.
(CHRONIC) This product contains no known carcinogen. Overexposure can cause permanent brain and nervous system damage.
SIGNS OF OVEREXPOSURE—
SKIN: Redness
INHALATION: Difficulty in breathing
INDIGESTION: Vomiting
Overexposure may also cause dizziness and loss of concentration.
EMERGENCY FIRST AID PROCEDURES:
SKIN—Wash affected area thoroughly with soap and water. Remove contaminated clothing and launder before reuse.
INHALATION—Remove from exposure into fresh air environment. Restore breathing.
INGESTION—Do NOT induce vomiting. Get medical attention immediately.
EYES—Flush eyes thoroughly with large amounts of water for 15 minutes. Get medical attention immediately.

SECTION VII—PRECAUTIONS FOR SAFE HANDLING AND USE

STEPS TO BE TAKEN IN CASE MATERIAL IS RELEASED OR SPILLED—Remove all sources of ignition. Ventilate and remove with inert absorbent.
WASTE DISPOSAL METHOD—Waste may be hazardous and should be tested for ignitability prior to discarding. Do not incinerate. Always depressurize container. Dispose of in accordance with Federal, State, and Local regulations.
PRECAUTIONS TO BE TAKEN WHEN HANDLING OR STORING—Keep away from any source of ignition. Gloves and safety spectacles with side shields should be worn when handling this product. Provide proper ventilation in the place of storage.

SECTION VIII—CONTROL MEASURES

RESPIRATORY PROTECTION—Self-Contained Breathing Apparatus if the above TLV limits are exceeded.
VENTILATION—Local exhaust.
MECHANICAL—None
OTHER PRECAUTIONS—Intentional misuse by deliberately concentrating and inhaling the contents can be harmful or fatal.

PROTECTIVE CLOTHING—Gloves and safety spectacles with side shields. No other protective clothing is required.

WORK/HYGIENIC PRACTICES—Do not use while smoking. Wash hands after use. Avoid breathing of vapors. Keep out of reach of children. Consult your physician if you have a medical condition prior to use.

FIGURE 1-14 An example of a Material Safety Data Sheet (MSDS).

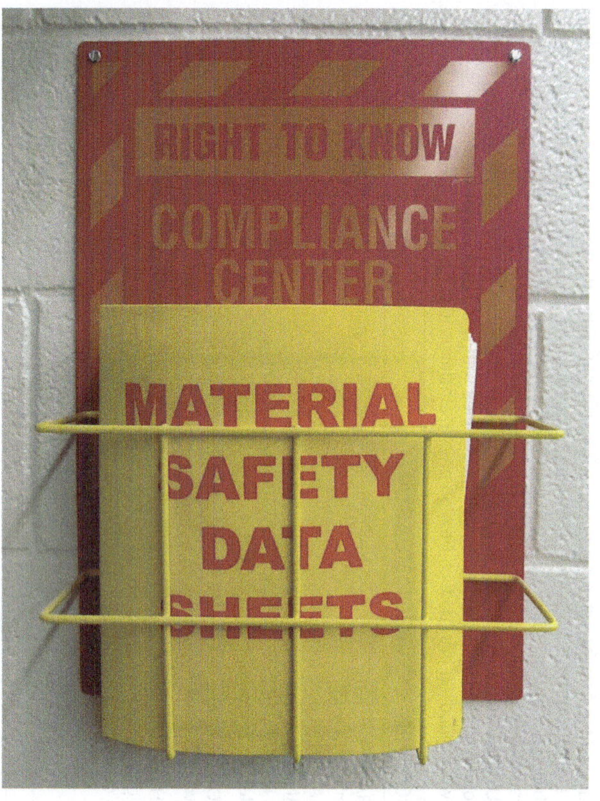

FIGURE 1-15 MSDS Data Sheet Compliance Center.

information before an accident. Trying to learn this information after a fire has started could cost someone his or her life. It is important to always stay prepared and not to lose your head in the case of an emergency. Staying calm is difficult to do if you are not trained and prepared.

For a fire to exist, three components must be present: fuel, heat, and oxygen. If only two of these components exist, there can be no fire. To extinguish a fire that has already started, just remove one of the three elements, and the fire will cease to exist.

Many industrial facilities maintain an emergency response or first responder team that is made up of employees who are working in the maintenance department. As a maintenance employee, you will most likely be asked to put out a fire as the need may arise. Keep in mind, however, that, in the event of a fire, the local fire department should be notified as soon as possible. The facility's emergency response team should be trained to put out fires, but they usually are not going to be trained to the level of a professional firefighter. The object of the emergency response or the first responder team is to extinguish the flame, if possible, or to contain the flame and prevent it from spreading in the event of a larger fire. It should also be the responsibility of the emergency response or first responder team to ensure the safety and proper evacuation of all nonessential personnel from the area in which the fire is being fought. Every employee that is working in an industrial environment must know the location of the nearest fire alarm and fire extinguisher and the class of each fire extinguisher. Often, a fire can be extinguished quickly if someone can get an extinguisher to the site quickly. When an extinguisher cannot be located, a fire can rapidly get out of control.

Fire Classifications

The combustible material classifies the fire; it is also the fuel of the fire. There are four classifications: Class A, Class B, Class C, and Class D. Class A fires are fires that burn ordinary combustibles such as wood, paper, cloth, upholstery, trash, some plastics, or any other carbon-based solid materials that are not metals. There is a pictograph that represents a Class A fire. A Class A classification may also be represented by a green triangle with a capital letter A in it. Both the pictograph and the triangle can be seen in **Figure 1-16**.

Class B fires are fires that burn flammable liquids or gases such as gasoline, oil, grease paints and thinners, propane, acetone, grease, or any other similar gas or liquid that is in a nonmetal state. The pictograph that represents a Class B fire can be seen in **Figure 1-16**. A Class B classification may also be represented by a red square with a capital letter B in it.

A Class C fire is a fire that burns on any electrical or electrically energized equipment. An example would be an electric motor, a drive cabinet, a junction box, or even an appliance. If the electrical equipment that is burning is not electrically energized, it would be considered a Class A or Class B fire; however, it is important to consider all electrical equipment to be electrically energized and

FIGURE 1-16 Fire symbols.

use a Class C fire extinguisher. Most industrial machines have a main disconnect. This is the point at which the machine can be totally removed from its electrical energy source. It is important to remember, if a Class C fire does exist, to pull the main disconnect handle to the off position. Keep in mind, however, that there may be some remaining stored electrical energy in some electrical equipment as you fight the fire. For this reason, it is always important to consider all electrical equipment energized. As with the others, there is a pictograph that represents a Class C fire, which can be seen in **Figure 1-16**. A blue circle with a capital letter C in it may also represent a Class C classification.

A Class D fire is a fire that burns combustible metals, such as magnesium, potassium, powdered aluminum, zinc, sodium, or titanium. Because most combustible metals are highly volatile to water in a high-heat environment, a special dry-powder extinguishing agent or a foam extinguisher is used on this class of fire.

Extinguisher Use

Here is a proposed list of the procedures that should be followed in the event of a fire.

- Locate and activate the nearest fire alarm. If no fire alarm is in the immediate area, then yell the word *fire* as loudly as possible to get the attention of a coworker or a supervisor.
- Make certain that a call to 911 is placed. It is most important to notify the local fire department of the fire.
- Make sure that the facility's emergency response or first responder team is notified.
- Get all nonessential personnel out of harm's way.
- Deenergize all electrical sources that are in the immediate vicinity of the fire, if possible.
- Quickly identify the class of the fire and retrieve the appropriate extinguisher.
- Always make sure that you have an escape route as you are trying to extinguish the fire.
- Use the extinguisher properly to achieve the maximum firefighting capability of the extinguisher. Try to use the P.A.S.S. method.
    - Pull the pin.
    - Aim at the base of the fire.
    - Squeeze the handle or trigger. (Be prepared; some extinguishers are extremely loud when the extinguishing agent is released.)
    - Sweep the extinguisher from side to side as it is used.
- Continue to fight the fire as long as possible, preventing any spreading until the local firefighters arrive.
- If you extinguish the fire, keep a watchful eye until the fire department arrives to ensure that it does not flare up again.
- If a fire becomes too large and gets to a point where it is out of control before the fire department's arrival, evacuate the facility and stand as far away from the fire as possible. Remember, a loss of life due to a fire is not acceptable.

Remember, the safety of all personnel is what is important. These OSHA standards and NFPA standards are for our safety. If a safety violation is noticed, report it as soon as possible to avoid any unnecessary injury. Safety is everyone's job! Take it seriously.

1-3 LADDERS AND SCAFFOLDS

Ladders

OSHA Standard 1926.1051 (General Requirements) states that a stairway or ladder shall be provided at all personnel points of access where there is a break in elevation of 19 in. (48 cm) or more, and no ramp, runway, sloped embankment, or personnel hoist is provided. OSHA Standard 1926.1053 lists the requirements for the manufacturing or construction of a ladder. OSHA Standard 1926.25 covers portable wooden ladders; 1926.26 covers portable metal ladders; 1926.27 covers fixed ladders; and 1926.1053 covers ladders that are used in the construction industry. Regarding ANSI ratings, different safety codes apply, depending on the type of material and type of ladder. The ANSI Codes are as follows:

Wood Ladders	ANSI A14.1
Metal ladders	ANSI A14.2
Fixed ladders	ANSI A14.3
Fiberglass ladders	ANSI A14.5
Steel ladders	ANSI A14.7
Rolling scaffold	ANSI A10.8, 2011 edition, (covered later in this section)

Duty ratings have also been established by ANSI. These ratings identify the ladder's load capacity and its intended use. All ladders must have a rating label.

Rating	Load Capacity	Rated Use
Type IAA	375 lbs (170 kg)	Special duty
Type IA	300 lbs (136 kg)	Extra-heavy duty (industrial)
Type I	250 lbs (113 kg)	Heavy duty (industrial)
Type II	225 lbs (102 kg)	Medium duty (commercial)
Type III	200 lbs (91 kg)	Light duty (household)

What is a ladder? A ladder is a structure consisting of two side rails joined at intervals by steps (which are referred to as rungs) for climbing up and down. Ladders are manufactured in lengths of 3 feet to 50 feet (0.9–15.2 m).

Wood ladders have some disadvantages. They deteriorate with age and are extremely susceptible to weather. Shrinkage can occur, causing loose rungs. Shrinkage typically occurs when the wood becomes dried out. The integrity of a wood ladder must be maintained by regularly coating it with clear shellac or linseed oil throughout its working life. Wood ladders should never be painted because it covers the defects that otherwise would be visible and can be caught during inspection of the ladder. Wood ladders tend to splinter if mistreated. Finally, wood ladders tend be very heavy. The wood ladder does have some advantages. For example, they are nonconductive and do not tend to transfer heat of cold to the worker because of wood's insulating properties. They typically cost less than other ladders and are very durable. Wood ladders must be stored on their edge, away from excessive dampness, dryness, and heat, to reduce the possibility of warping. If possible, the best method of storing a wood ladder is by hanging it horizontally on hooks that are spaced no more than 4 to 6 feet (1.2–1.8 m) apart.

Metal ladders are normally constructed from aluminum. Their disadvantage is the tendency to become very cold and very hot. This is transferred to the worker standing on the ladder. Metal ladders cannot be used within 4 feet (1.2 m) of energized electrical circuits or equipment, as they are made from an excellent conductive material. The advantages are that they are lightweight, extremely tough and durable, and will not splinter or crack when subjected to impact. Also, they are resistive to deterioration with age and require much less maintenance than wood.

Fiberglass ladders are the most popular ladder. However, even these have some disadvantages. They are typically heavier than both wooden and metal ladders. They have been known to crack and fail when overloaded, and they may crack or chip when severely impacted. They are also extremely susceptible to sun rot. The UV rays from the sun tend to break down the resin compounds that form the shape of the ladder. When this occurs, very small fiberglass splinters are exposed and can become lodged into your skin when lifting or carrying the ladder. The advantages are that they do not conduct electricity when dry; they can withstand considerable abuse while maintaining their integrity; they do not require surface finishing and are typically more comfortable to stand on for long periods of time; and finally, they do not transfer heat or cold to the worker.

The fixed ladder is a ladder that is permanently attached to a structure. It is commonly constructed of steel or aluminum. Fabrication of a fixed ladder, including the design, materials and welding, must be done under the supervision of a qualified, licensed structural engineer. The width between side rails is normally 16 inches (40.6 cm) and the spacing between rungs is 12 inches (30.4 cm). Any fixed ladder that has a length of 24 to 50 feet (7.3–15.2 m) must have a cage, well, or ladder safety system. If a cage is installed, it must start no less than 2 feet (0.6 m) and no more than 8 feet (2.4 m) from the ground or platform. Fixed ladders must also have a load capacity rating of 250 lb (113 kg), and they must be attached every 10 feet, 7 inches (3.2 m) off the wall or structure.

Over 11% of all injuries that require time away from work are related to falling from ladders. Safe climbing employs the three-point contact method. This means that contact must be maintained at all times when you are on the ladder. It is also a good practice to use the rungs instead of the rails while climbing or descending on a ladder. Another cause for ladder injury is that the worker will stand on the top rung or even the table (very top) of the ladder so he or she can "reach just a little farther." While reaching, the center of gravity is shifted, causing the ladder to tip, resulting in a fall with injury or possibly even death. Every ladder that is manufactured will have safety labels placed on them stating "DO NOT STAND ON OR ABOVE THIS RUNG."

Extension ladders are ladders that have an inner (fly) section that slides up and down in the outer (base or bed) section. It is raised by a rope that and is kept on track by an interlocking side rail system. See **Figure 1-17** for all of the parts on an extension ladder.

When leaning an extension ladder against a building or structure where the fly section does not extend above the structure, it is a common practice to install end covers to protect the end cap on the ladder as well as the structure that it

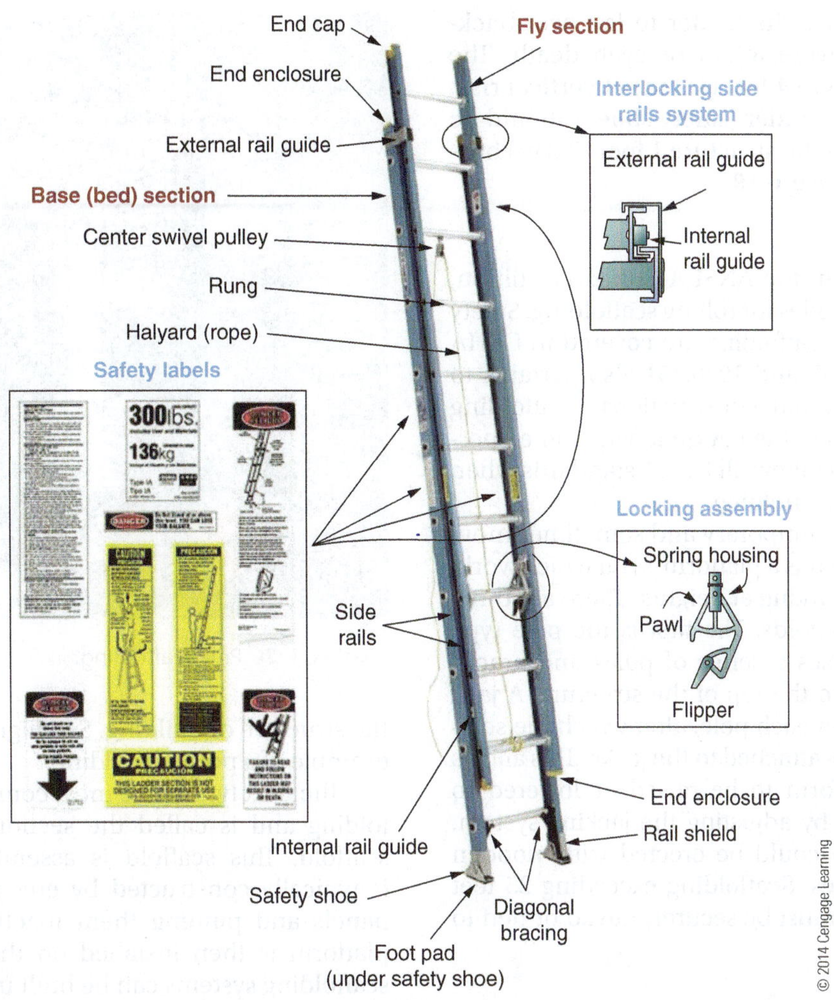

FIGURE 1-17 Extension ladder components.

is leaning upon. A set of end cap covers can be seen in **Figure 1-18**.

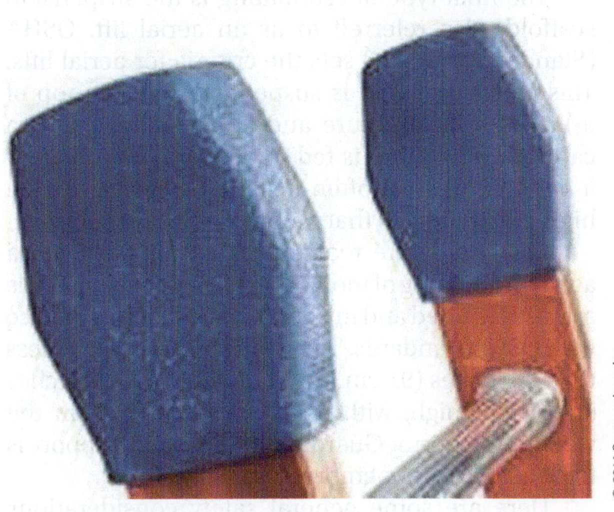

FIGURE 1-18 Extension ladder end caps.

When leaning an extension ladder upon a structure where the fly section rises above the top edge of the structure, the ladder must extend above the structure not less than 3 feet (0.9 m), measured from the highest point of contact. Having this additional length provides a means of support as you mount and dismount the ladder. Never stand on the top three rungs when mounting or dismounting the ladder. To prevent the ladder from kicking out from under you or sliding sideways, the ladder should be secure to the structure.

When extending a ladder, it is important that enough overlap remains to prevent the ladder from folding in the middle. The distance of overlap is determined by the ladder length. For ladders that are 8 to 36 ft (2.4 to 11 m), there must be a minimum of a 3 foot overlap. For ladders that are 36 to 48 ft (11 to 14.6 m), there must be a 4 ft (1.2 m) overlap. And for ladders that are 48 to 60 ft (14.6 to 18.2 m), there must be a minimum of a 5 ft (1.5 m) overlap.

Another thing to be concerned about is angle positioning. If this is not considered, as one climbs higher on the ladder, the shifting center of gravity could cause the ladder to fall over backward, causing severe injury or even death. The rule is that for every 4 feet (1.2 m) of vertical rise, the bottom of the ladder (safety shoes) should be moved away from the structure 1 foot (0.3 m) horizontally. See **Figure 1-19**.

Scaffolds

As mention earlier, the ANSI A10.8, 2011 edition, covers the safety codes for rolling scaffolding. Safety requirements for scaffolding are covered in OSHA (Standard) 1910.28, and 1926.451 also pertains to scaffolds and scaffold constructions. Scaffolding should only be assembled by qualified, trained personnel. Be sure to follow all OSHA standards when constructing any scaffolding.

A scaffold is a temporary and sometimes movable structure that has a platform(s) on which workers stand when working at heights. There are three basic types of scaffolds. The first is the pole type scaffold, which uses a series of poles and braces that are secured to the top of the structure. A jack is then installed on each pole, after which the scaffolding platform is attached to the jacks. This allows the scaffold platform to be raised or lowered to any desired level by adjusting the jacking system. The pole scaffold could be erected with wooden or aluminum poles. Scaffolding exceeding 25 feet (7.6 m) in height must be securely guyed or tied to the structure or building. See **Figure 1-20** to see an example of a pole scaffolding.

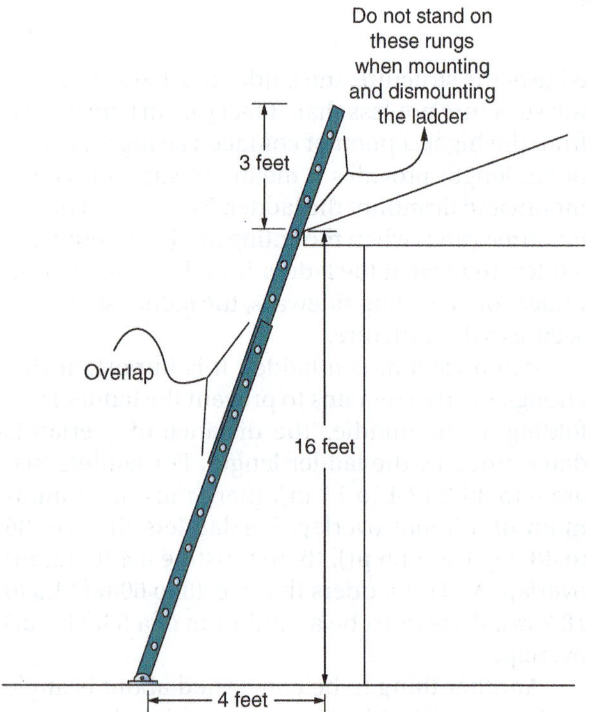

FIGURE 1-19 Proper ladder setup.

FIGURE 1-20 Pole scaffolding.

The next type is the most common type of scaffolding and is called the sectional metal-framed scaffold. This scaffold is assembled on-site and is typically constructed by erecting prefabricated panels and pinning them together. The working platform is then installed on the framing. Many scaffolding systems can be built upon level by level until the desired height has been achieved. These scaffolds must be level when being assembled. Also, each panel should be inspected for damage, defects, or cracks before installing it into the scaffolding system. A damaged scaffold frame or platform could cause the entire structure to collapse if it is not detected.

The final type of scaffolding is the suspension scaffold, also referred to as an aerial lift. OSHA (Standard) 1926.453 sets the criteria for aerial lifts. This type of scaffold is suspended from the top of a building or structure and is lowered through a cabling system that is fed out by either a motor or a hand crank. It is often installed permanently on high-rise buildings that require window washing.

Toe boards are required as a barrier to guard against the falling of tools or other objects. Guardrails are also required and must be installed in accordance with OSHA standards. They must be installed no less than 36 inches (91 cm) and no more than 42 inches (106.6 cm) high, with midrail, measured from the working platform. Guardrail and midrail support is to be at intervals not more than 10 feet (3 m).

Here are some general safety considerations when constructing a scaffold.

- Guardrails, midrails, and toe-boards must be installed on all open sides and ends of platforms more than 10 (3 m) above the ground.
- Platform planks are to be laid with no openings more than 1 inch (2.5 cm) between adjacent planks.
- Overhead protection must be provided for persons on a scaffold who are exposed to overhead hazards.
- Fall protection should be used when working at heights of 10 (3 m) or more (harness, shock absorbing fall lanyard, etc.).
- Safety nets for workers should be used at any level over 25 (7.6 m) when the workers are not otherwise protected by personal fall protective equipment or properly guarded scaffolding work surfaces. Safety nets restricting falling objects must be used when persons are permitted to be underneath a work area.
- Work must not be done on scaffolding during high winds or storms.
- Work must not be done on ice-covered or slippery scaffolds.
- Scaffolds with a height-to-base ratio of more than 4:1 must be restrained by the use of guy lines.
- Mobile scaffolds must be locked in position when in use.
- All tools and materials must be secured to or removed from the platform before the scaffold is moved.

SUMMARY

- Without safety, maintenance would be a deadly occupation. The maintenance mechanic must be very safety oriented to avoid unnecessary accidents and bodily harm when working around equipment and machinery. Remember, an accident is nothing more than an unsafe condition and an unsafe act that have been combined. Horseplay does not belong in the workplace. All horseplay should be eliminated while any maintenance is being performed.
- The working environment has to be kept clean and safe.
- There should be adequate lighting and proper storage of hazardous materials.
- It is important to use personal protective equipment (PPE) in industrial maintenance.
- Exposure to noise levels equal to or greater than 85 dB may cause hearing loss.
- Eye protection should be worn whenever there is a threat of vision loss due to injury of the eye. Protect your eyes *before* you start working. Make sure that your eyewear has side shields to protect the eye from foreign objects that may enter from the side.
- Clothes should be comfortable but, at the same time, they need to be form fitting. Long-tailed shirts should be tucked into the pants, and the pants should not be too large and baggy.
- Tuck long hair under a hat or inside your shirt, or wear a hair net.
- Jewelry should not be worn while working around equipment and machinery.
- Pinch points are points on a machine where two or more separate components come together.
- There are some very simple steps to ensure safety while you are lifting something. Most important, lift with your legs, and remember to keep your back as straight as possible through the entire process.
- All maintenance personnel who are working on a machine need to place their lockout, with an identification tag, on the disconnecting means of the hazardous energy.
- Maintenance personnel should work in pairs.
- The NFPA 704 is used to quickly identify the hazards of a material.
- There are many OSHA regulations and guidelines for the use and handling of hazardous material.
- It is important to know the basics of fire safety such as the classification of the fire, the proper methods to extinguish a particular class of fire, and the proper order of tasks that need to be performed in the event of a fire.

REVIEW QUESTIONS

1. What two things cause accidents?
2. What is the most common unsafe act?
3. What is the difference between lockout and tagout?
4. How much current is considered a lethal amount?
5. What does the OSHA Regulations Standard 29 CRF 1910.1200 regulate?

CHAPTER 2

Tools

This chapter covers many of the common hand and power tools that are used in the field today. It also covers measuring devices. Many of these tools have been around for a very long time. It is important for the correct tool to be selected for the task at hand and that it be used properly. Many tools are used incorrectly, thus causing damage to the tool or injury to the user. Many of these concerns are discussed throughout this chapter.

In this chapter, only the most common hand tools and power tools are reviewed. There are too many hand tools and power tools in existence to be able to go over each one; however, it is important to cover the most common types of tools that are encountered by maintenance personnel. Hand tools and measuring devices are discussed first, and then power tools are discussed.

OBJECTIVES

After studying this chapter, the student should be able to

- Discuss the importance of inspecting a hand tool.
- Demonstrate the proper use of various types of hand tools.
- Discuss the most common types of power tools.
- Discuss the many safety issues when working with electricity and power tools in general.
- List the proper and improper use of each power tool that is mentioned.
- Discuss the importance of inspecting all power tools before and after their use.

2-1 HAND TOOLS

Many hand tools are considered to be common tools because mechanics, electricians, and welders use them, as do many other people. A tool such as the screwdriver is so common that it can be found in just about any home or toolbox. Its use is not restricted to mechanics, electricians, welders, or any other tradesperson. The screwdriver is not the only common hand tool; there are many more. In this chapter, wrenches, screwdrivers, pliers, hammers, chisels, punches, and hacksaws are discussed. Because people are sometimes injured using these hand tools improperly, the safe and proper use of these tools are also discussed. First, it is important to discuss some general care that should be given to all hand tools. Regular maintenance and inspection are vital for the safe use of all hand tools. If a tool is not taken care of, it may start to rust or corrode. Rust and corrosion weaken the integrity of the tool. Keep all tools clean. Avoid putting tools in direct contact with any corrosive chemicals or materials, if possible. If they do come in contact with corrosive chemicals or materials, clean them thoroughly before putting them away. Also avoid placing any tool in or around excessive heat. Excessive heat also weakens the integrity of the tool. If damage is noticed while inspecting the tool, replace the tool. Do not use a damaged tool, because injury may result. Using a damaged tool is an unsafe act and should be avoided.

Wrenches

The mechanic probably uses the wrench more than any other tool. This tool is used to tighten and loosen fasteners and nuts and for many other applications as well. There are many types of wrenches. Open-end, box-end, combination, adjustable, flare nut, socket, and torque wrenches are some of the types that we discuss.

Open End Wrenches

The open-end wrench, as shown in **Figure 2-1**, is a wrench that was designed to approach the fastener from its side. Notice, in Figure 2-1, that the end of the wrench is opened.

The opened end provides easy access to a fastener or nut from its side. The opening of most open-end wrenches is tilted 15° from the centerline of the handle. A right-angle, open-end wrench has the opening at 90° from the centerline of the handle. Tilting the openings allows the fastener to be turned to a greater degree by simply flipping the

FIGURE 2-1 Open-end wrench.

wrench over. For example, a wrench with a 15° tilt can turn a fastener 30° in a confined space. There are times when the open-end wrench is the only wrench that can be used; for example, suppose a nut is in the center of a 24 in. length of all-thread rod and that the all-thread rod is connected to something on both ends. Having the open-end wrench makes it possible to loosen or tighten the nut without removing the all-thread rod from its location at each end. Do not use the open-end wrench for loosening fasteners or nuts that are seized. A seized fastener is usually referred to as a frozen fastener. This type of wrench tends to slip off the fastener or nut if too much pressure is applied while trying to break the fastener or nut free. Applying too much pressure is dangerous because, as the wrench slips off the fastener or nut, the knuckles of the hand tend to strike other objects that are in the area, causing personal injury. If a fastener or nut is seized, then use a box-end wrench to break it free. If it still does not release, then get a breaker-bar to do the job. Both of these types of wrench are discussed later in this chapter. Many people use the open-end wrench improperly by placing a pipe over the handle of the wrench to provide more leverage. Their philosophy is that the extra leverage breaks the fastener or nut free. This is not the manufacturer's intended use of this tool. The wrench was not designed to work under these pressures, and therefore the wrench may crack or break if used in this manner. Injury can also occur if the wrench breaks. The open-end wrench is commonly available in different sizes for standard or metric fasteners or nuts. The size of this wrench is determined by the size of the fastener or nut that it

is used on; for example, if a ½ in. nut needs to be removed from a fastener, a ½ in. open-end wrench is used to remove the nut.

Box-End Wrenches

The box-end wrench is very similar to the open-end wrench, with a couple of exceptions. One difference is that the end of the wrench totally surrounds the fastener or nut. A box-end wrench is shown in **Figure 2-2**.

The name was derived from the fact that the wrench, because of its construction, forms a box around all six sides of a fastener or nut. Because there is no opening at the end of the wrench, it has to approach the fastener or nut from above or below. Its construction does not allow access to the fastener or nut from the side, as does the open-end wrench. Having contact on all six sides of the fastener or nut allows more torque to be applied to the fastener or nut. The open-end wrench makes contact with only four sides of the fastener or nut. Some brands of open-end wrenches may contact only two sides of a fastener or nut. The box-end wrench can come in a 6-point or 12-point configuration, as shown in **Figure 2-3**.

The 12-point configuration is used when there is very little working space for the wrench to travel. The 12-point wrench is capable of only half the travel distance of a 6-point wrench when it is going back for another "bite." *Bite* refers to when the wrench has already turned the nut and the wrench needs to be removed, then placed back in the position that it was in before it started the turn. As with the open-end wrench, the size of the box-end wrench is determined by the size of the fastener that it is used on.

Combination Wrenches

A combination wrench is simply a wrench that has an open-end wrench on one end and a box-end wrench on the other end. Usually, they are the same size. These wrenches are handy because if you are using an open-end wrench and you find that a box-end wrench of the same size is needed, a trip back to the toolbox is not necessary; just flip the wrench around to the other end. For most box-end and combination wrenches, the box-end portion of the wrench is designed with a 15° offset in the handle, allowing clearance so the handle does not strike any other fasteners or objects that may be nearby. This can be seen in **Figure 2-4**.

FIGURE 2-2 Box-end wrench, 12 point.

FIGURE 2-3 Box-end wrench, 6 point and 12 point.

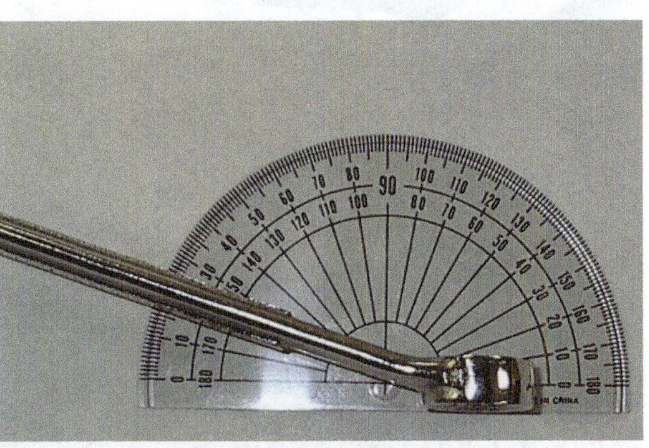

FIGURE 2-4 Combination wrench offset 15°.

Adjustable Wrenches

Adjustable wrenches come in a variety of sizes, from 4 to 24 in. The most common sizes are 6, 8, 10, and 12 in. The length of the wrench determines the size of an adjustable wrench, unlike the open-end or box-end wrenches. There are two types of jaws on the adjustable wrench, an adjustable jaw and a fixed jaw. The adjustable jaw allows this wrench to be used on many different sizes of fasteners and nuts. An adjustable wrench can be seen in **Figure 2-5**.

Notice in the figure a screwlike device is located just below the movable jaw of the adjustable wrench. This device is called a thumbwheel. Spinning this thumbwheel enables the movable jaw to move in or out according to which direction the thumbwheel is turned. Moving this jaw causes the distance between the two jaws to increase or decrease. In order for the wrench to be tightened properly, both jaws must fit squarely on the flats of the fastener or nut. This fit is demonstrated in **Figure 2-6**.

A common problem can occur with this type of wrench. Over time, the thumbwheel can wear out. When it does, it is very difficult for the wrench to maintain a tight fit on the fastener or nut. The movable jaw tends to loosen as pressure is applied to the wrench. As this happens, the jaws are no longer in contact with the entire flat of the fastener or nut. This problem is shown in **Figure 2-7**.

This improper fit could cause injury to the knuckles of the hand if the wrench were to slip off of the fastener or nut. Damage to the fastener or nut could also occur as pressure is applied to the wrench. The corners of the fastener or nut are rounded off.

FIGURE 2-5 Adjustable wrench.

FIGURE 2-6 The proper adjustment of an adjustable wrench against the flats of a bolt.

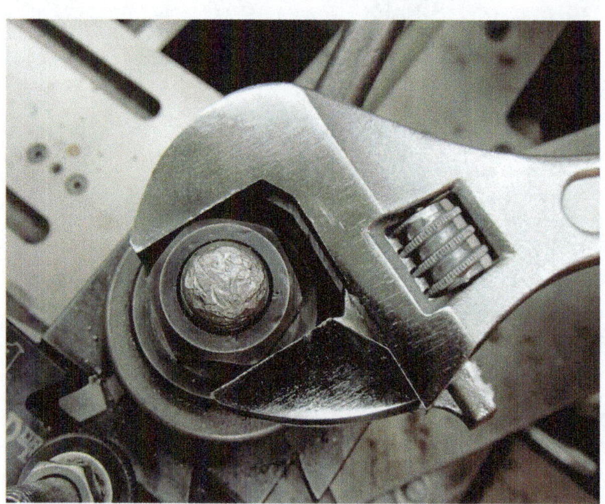

FIGURE 2-7 The improper adjustment of an adjustable wrench against the flats of a bolt.

Once its corners have been rounded, the fastener or nut needs to be replaced. If the thumbwheel starts to wear out, repair it if possible. If it cannot be repaired, replace the wrench with a new one.

When using an adjustable wrench, it is important to use the wrench correctly. Any time the adjustable wrench is used, it should have a tight and secure fit on the fastener or nut, and the adjustable jaw must be placed on the inside of the turn as seen in **Figure 2-8**. This placement allows the pressure to be placed on the fixed jaw, thus minimizing the risk of slipping. Always pull the handle of the adjustable wrench toward the movable jaw when in use.

SAFETY

Never beat on the handle of any wrench, especially an adjustable wrench, with a hammer or any other tool. Beating on the handle of an adjustable wrench can damage the thumbwheel. Some people do this to try to loosen a nut or fastener when it is seized in place. This is not the manufacturer's intended use and can cause damage to the wrench or to the tool that is being used to strike the wrench. It could also lead to personal injury. If a seized fastener or nut is encountered; use a box-end wrench to break it free. If it still does not release, get a breaker-bar to do the job.

Flare Nut Wrenches

A flare nut wrench, sometimes referred to as a split box wrench, looks like a box-end wrench with one exception; there is a slot cut in the very top of the wrench. This is shown in **Figure 2-9**.

This wrench is most commonly used for flared tube fittings. The slot is there to allow the wrench to get to the flared fitting that is on the tubing. Whereas the box-end wrench makes contact with all six corners of a fastener or nut, the flare nut wrench makes contact with only five corners of a nut or fitting, because the slot is cut right where the sixth corner would be. It is important to know the reason for using the flare nut wrench instead of an open-end wrench. Most flared tubing fittings are made of brass, which is a soft metal. Because these fittings are made of a soft metal, the corners of each flat on the fitting are apt to round off when an open-end wrench is used. The flare nut wrench applies pressure to more surface area of the fitting than does the open-end wrench and is less likely to damage the fitting.

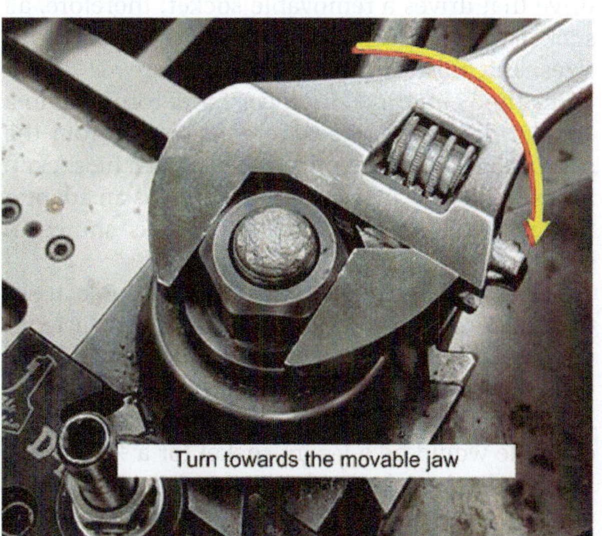

FIGURE 2-8 Rounded corners due to the improper use of an adjustable wrench.

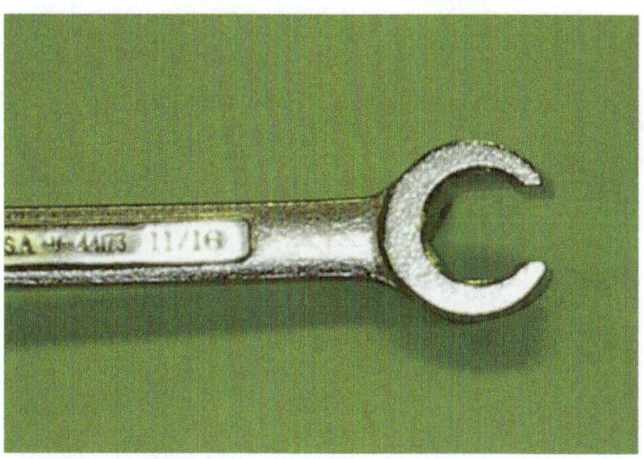

FIGURE 2-9 Flare nut wrench.

Socket Wrenches

A socket wrench is probably the most common wrench that is used by a mechanic. A socket wrench is a very convenient tool to have. These wrenches use an interchangeable socket that is attached to it to tighten or loosen a fastener or nut. We discuss the socket shortly, but first we discuss some of the many types of socket wrenches. Some examples of socket wrenches are reversible ratchet, flex head reversible ratchet, flex head wrench (sometimes referred to as a breaker-bar), the torque wrench, and many more. Those wrenches are the most common socket wrenches used. The reversible ratchet uses a ratcheting mechanism that applies torque when the wrench is turned in one direction and allows slippage when turned in the opposite direction. This wrench is shown in **Figure 2-10**.

There is a small selector on the back of the ratchet head that reverses the direction of the ratcheting mechanism. This mechanism allows the wrench to apply torque in a selected direction.

The flex head reversible ratchet is much like the reversible ratchet. The difference is that the head can pivot to allow more flexibility in hard-to-reach places. This wrench can be seen in **Figure 2-11**.

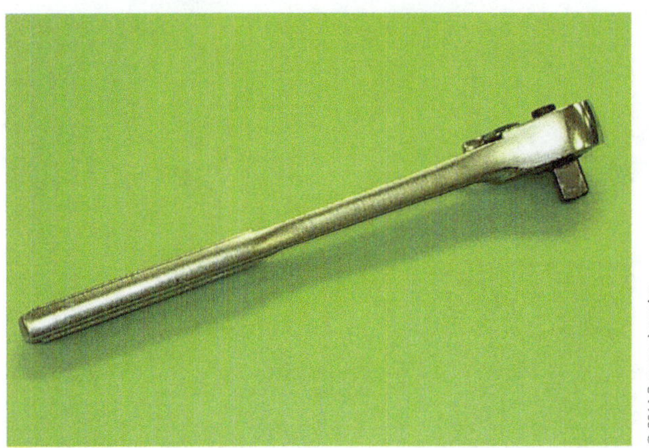

FIGURE 2-10 Ratchet.

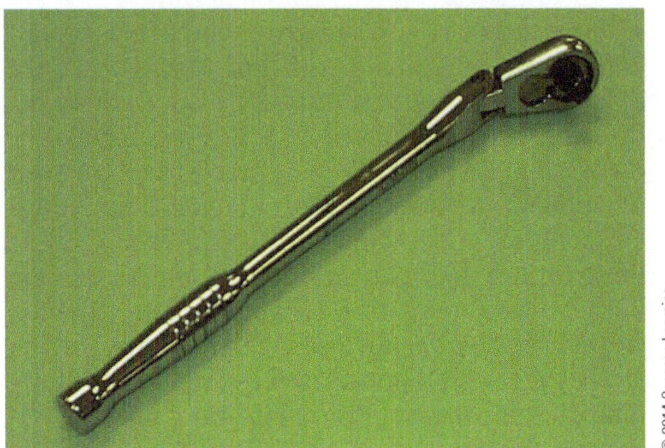

FIGURE 2-11 Flexible head ratchet

Both the reversible ratchet and flex head reversible ratchet usually has a handle that is less than 1 ft long. This size limits the torque that can be placed on the fastener.

A flex head wrench (breaker-bar) does not have a ratcheting mechanism in it; however, it does have a pivoting head. See **Figure 2-12**.

The removal of the ratcheting mechanism allows more torque to be applied to the fastener or nut. If too much pressure is applied to a socket wrench with a ratcheting mechanism, the ratcheting mechanism may break or even shatter. If there were a need for a lot of torque, for example, on a seized fastener, the breaker-bar would be the desired tool. Breaker-bars are usually, but not always, 1 to 2 ft long. This is one of the factors that gives the breaker-bar the extra torque that is needed to break a frozen fastener free. The longer handle means more leverage.

Torque Wrenches

A torque wrench is a type of ratcheting socket wrench. The torque wrench, however, can tell you precisely how much torque is being applied to a

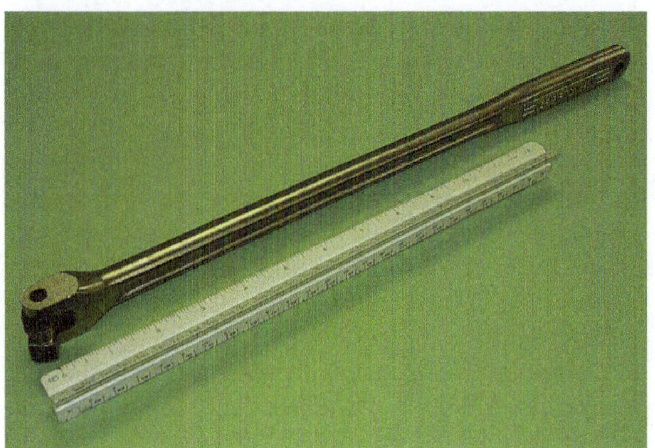

FIGURE 2-12 Flexible head breaker-bar. Notice the 12 in. ruler beside the breaker-bar.

fastener or nut as it is being tightened. All fasteners have torque specifications on them, which, if exceeded, may cause the head to shear off of the fastener. An adjustable micrometer torque wrench, as seen in **Figure 2-13**, has an adjustable setting on it that signals the user when a certain amount of torque has been reached.

If an adjustable micrometer torque wrench is used, a clicking or popping is heard when the desired torque has been reached. This popping of the torque wrench is normal and can also be felt by the hands as it pops, just in case the pop cannot be heard because of a noisy work environment. There are torque wrenches that do not have this popping feature. A deflecting beam or a dial indicating torque wrench simply indicates the amount of torque visually on a scale or dial that is mounted on the handle of the wrench.

All socket wrenches are considered to be the drive that drives a removable socket; therefore, all socket wrenches are one of many different drive sizes. The most common drive sizes are ¼, ⅜, and ½ in. The ½ in. drive produces more torque than a smaller drive. This measurement comes from the size of the square adapter that receives the socket; for example, a ¼ in. ratcheting drive has an adapter that is ¼ in. on all four sides. This is shown in **Figure 2-14**.

All socket sizes refer to the size of the fastener that they are used on; for example, if a ½ in. fastener needed to be tightened, a ½ in. socket would be used. There are sockets for metric fasteners as well. A socket is also sized for a particular drive size. An example would be a $\frac{7}{16}$ in. socket for a ⅜ in. drive receptacle. A socket with a ⅜ in. drive receptacle can be used only on a ⅜ in. drive with-out the use of a drive adapter. An adapter is not recommended because a socket that is made for a ⅜ in. drive is constructed to handle more torque than a socket

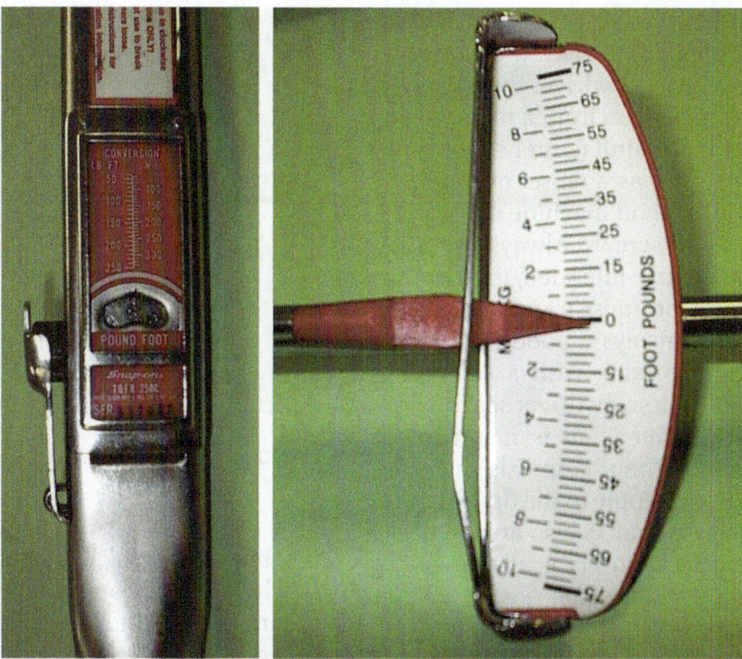

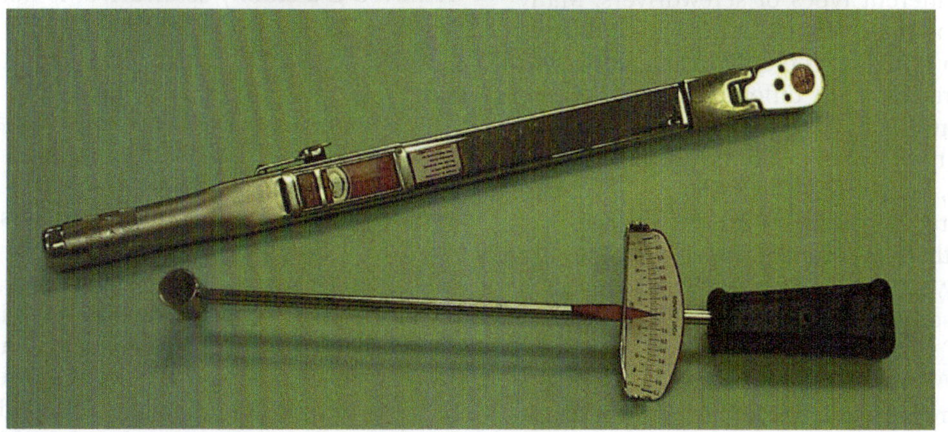

FIGURE 2-13 Micrometer torque wrench (left) and deflecting beam torque wrench (right).

FIGURE 2-14 ¼ in. ratchet.

that is made for a ¼ in. drive even though they both may be used on the same size fastener.

Sockets, as with box-end wrenches, can come in a 6-point or 12-point configuration. Remember, not all socket wrenches have a ratcheting capability.

There are also many different attachments that can be used with the sockets and socket wrenches, such as extension bars, adapters, universal joints, and many more. All of these devices help to reach a fastener that cannot be reached with the wrench alone.

Screwdrivers

There are many types of screwdriver and screwdriver tip styles. Two of the most common screwdriver tip styles are the straight tip (sometimes referred to as a flat tip) and the Phillips tip screwdrivers, both of which are discussed in greater detail later in this chapter. First, let us discuss the proper use of a

screwdriver. Too often, the screwdriver is misused as a chisel or a pry bar. This use is often the cause of mistakes and mishaps. The screwdriver is used to drive screws into or remove screws from a material or mating nuts. It is not the manufacturer's intent for the screwdriver to be used as a chisel or a pry bar. Screwdrivers may be used on machine screws, sheet metal screws, or wood screws. Always use a screwdriver as the manufacturer intended, to avoid damage to the screwdriver or, most important, to avoid personal injury. It is always a good idea to wear safety glasses while working with a screwdriver. The standard length of a screwdriver shank is 4, 6, 8, 10, or 12 in. A screwdriver shank is the metal shaft that extends below the handle and is measured from the bottom of the handle to the tip of the shank. All round screwdriver shanks also are identified by their diameter. These sizes typically range from $\frac{3}{16}$ in. to around $\frac{3}{8}$ in. in diameter.

Before we discuss the different tip styles, let's discuss the different types of screwdrivers. Many people do not realize that there is a difference between a tip style and the type of screwdriver to be used. There are many different types of screwdrivers that are in use today. Some examples of the different types of screwdriver are the stubby, square-shank, round-shank, cabinet, holding, magnetic, offset, ratcheting, electric, and yankee. The screwdrivers that the mechanic is most likely to use are the stubby, square shank, round shank, holding, magnetic, offset, and ratcheting.

Stubby Screwdrivers

The stubby screwdriver is used when a screw has to be inserted in or removed from a location that does not have the room for a standard-size screwdriver. A stubby screwdriver is shown in **Figure 2-15**.

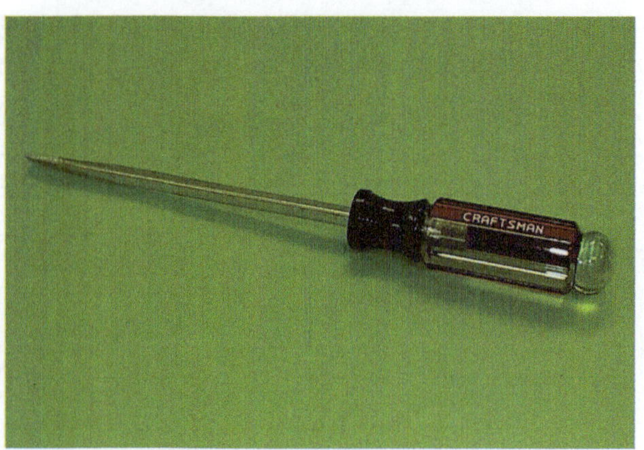

FIGURE 2-16 Square-shank screwdriver.

Most stubby screwdrivers are no more than 3 in. in total length. This total length includes the handle. There are times that a screw cannot be inserted or removed if a stubby screwdriver is not used to do the job.

Square-Shank Screwdrivers

The square-shank screwdriver is usually no less than 6 in. long. The typical length of this type of screwdriver shank is usually 6, 8, 10, or 12 in.; however, some square-shank screwdrivers are longer than 12 in. As the name implies, this screwdriver has a square shank, as shown in **Figure 2-16**.

The square shank allows the mechanic to apply more torque to the screw by placing a wrench on the shank and using the handle of the wrench to turn the screwdriver. The wrench handle gives the mechanic more leverage to break stubborn screws free. The square-shank screwdriver is typically available with a straight tip or a Phillips tip.

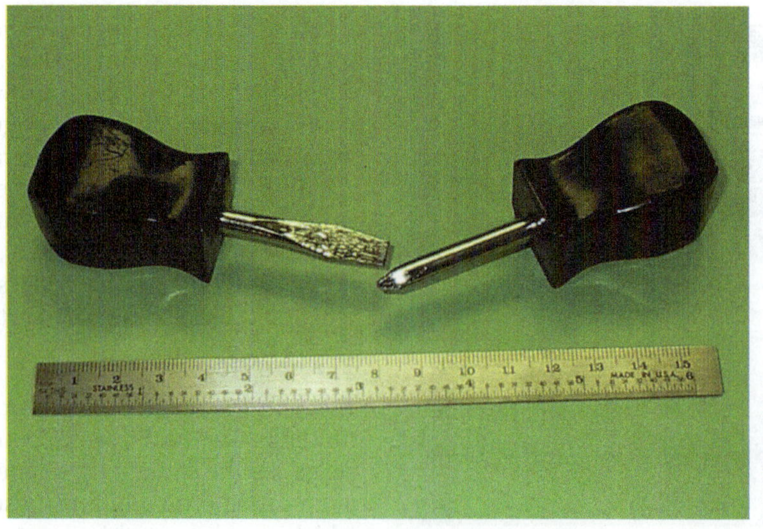

FIGURE 2-15 Stubby screwdrivers. Notice that they are each about 3 in. long.

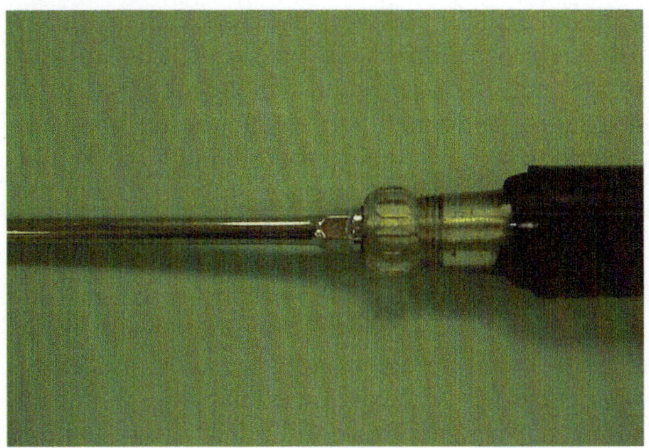

FIGURE 2-17 Round-shank screwdriver with flats at top of shank.

Round-Shank Screwdrivers

The round-shank screwdriver is probably the most common screwdriver in use today. The typical length of this type of screwdriver is anywhere from 4 to 12 in.; however, some round-shank screwdrivers are longer than 12 in. As the name implies, this screwdriver has a round shank. It is not the manufacturer's intent for any other tool to be placed on the shank of this screwdriver for extra leverage. If the manufacturer intends this type of use, there is an area with flats somewhere on the shank, usually just below the handle, as seen in **Figure 2-17**. If you do not see any flats, do not place any other tool on the shank of this screwdriver for extra leverage.

CAUTION

Holding Screwdrivers

The holding screwdriver is another valuable tool. As the name implies, this type of screwdriver uses clips or a sliding collar to hold the screw to the end of the shank. This screwdriver is used when the screw must be placed in a location where there is not enough room for the fingers to hold the screw in place as it is being started. Once the screw has been started, then a standard screwdriver can finish the job. The clip-type holding screwdriver has a clip that looks like a claw, which slides down over the end of the shank. This clip is spring-loaded so as to pull the screw toward the tip of the screwdriver once the screw is placed inside the clip. The sliding-collar holding screwdriver uses a sliding collar that slides down the shank of the screwdriver. It is important to know, however, that the shank on this type of screwdriver is split. It is split in such a way that, as the collar slides down the shank, the tip of the screwdriver becomes bound into the slot of the screw head, thereby holding the screw in place. Once the screw has been started, simply slide the collar up. The shank then unbinds in the screw head and releases the screw.

Magnetic Screwdrivers

A magnetic screwdriver accomplishes the same task as the holding screwdriver without the use of clips or any sliding mechanism. It is simple magnetism that holds the screw in place. This type of screwdriver usually has a magnet in the shaft and uses different style tips that are interchangeable. As the tips are placed in the shaft, the magnet magnetizes the tip. The magnetized tip can then hold a screw. This works great until a nonferrous screw is encountered. For nonferrous screws, use one of the holding screwdrivers. Magnetic screwdrivers also come in handy when a screw has been dropped somewhere that the fingers cannot get to. Simply place the tip of the screwdriver close to the dropped screw, and the tip attracts the screw to it.

Offset Screwdrivers

The offset screwdriver is designed to insert and remove screws that are in a place where it is impossible to use a straight-shank screwdriver (including the stubby). An offset screwdriver can be seen in **Figure 2-18**.

This type of screwdriver is constructed in a manner such that the tips are at right angles to the shaft. This configuration makes the offset screwdriver extremely handy when turning space is limited. The tips are not interchangeable on this type of screwdriver; however, this screwdriver can be bought with many different tip configurations. An example would be a straight tip on one side and a Phillips tip on the other.

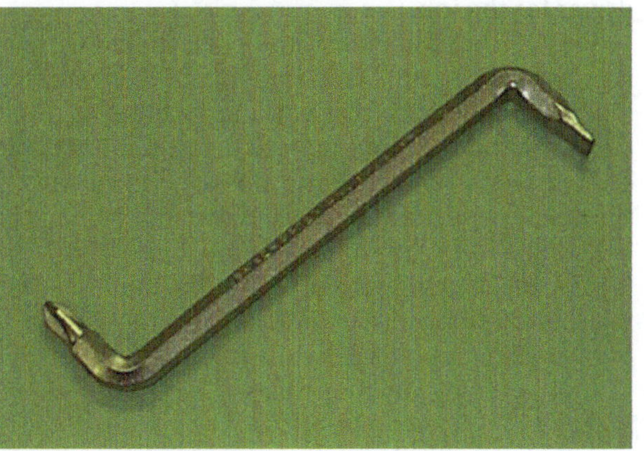

FIGURE 2-18 Offset screwdriver.

Ratcheting Screwdrivers

The ratcheting screwdriver uses a ratcheting mechanism that applies torque when the screwdriver is turned in one direction and allows slippage when turned in the opposite direction. There is a small selector on the handle of the screwdriver that reverses the direction of the ratcheting mechanism. This allows the screwdriver to apply torque in both directions. The tips are interchangeable on this type of screwdriver and function much like those for the magnetic screwdriver. If too much pressure is applied to a screwdriver with a ratcheting mechanism, the ratcheting mechanism may break or even shatter. These screwdrivers are for general use.

There are many different types of screwdriver tips that are available for the ratcheting driver. Some examples of the different types of tips are straight, Phillips, Reed & Prince, Torx, hex socket, and many more. The tips that are listed are the most common types encountered. Most people are familiar with the straight tip and the Phillips tip screwdriver; however, what most people do not know is that the Reed & Prince tip is very often confused with the Phillips tip screwdriver. These two screwdrivers look very similar in their design. The angle of the flukes help determine which type of screwdriver it is. A fluke is one of the four flats on the tip of the screwdriver. If the screwdriver is placed on a table so to stand on the point, the angle of the fluke can be determined. If the fluke is cut at a 30° angle, then it is a Phillips tip screwdriver. If the angle of the fluke is cut at approximately 45°, then it is most likely a Reed & Prince screwdriver. Another quick method of determining the type of screwdriver is to look at the tip of the screwdriver. If it has a thick, blunt point, it is most likely a Phillips tip screwdriver. The Reed & Prince has a thin, sharp point. It is important to keep in mind that these screwdrivers are not to be used interchangeably. Use the correct screwdriver for the job.

It is important to use the correct screwdriver for the task at hand. Make sure that the screwdriver fits in the slot of the screw properly. It needs to be snug, not too tight and not too loose. If the correct screwdriver is used, the tip fills the entire slot of the screw head, but the sides of the screwdriver tip does not extend beyond the screw head. If the tip is too wide, it may damage or scar the material that the screw is being driven into as the screw is driven home. Using a screwdriver tip that is too narrow for the job, may bend or break the tip as pressure is applied. Do not use a screwdriver tip that is rounded, damaged, or worn. A screwdriver that is worn may slip out of the slot as pressure is being applied. If it is possible to repair the tip before it is used, then do so. This repair can be done by redressing the tip with a file, while making sure that the edges are straight as you file the tip back into shape. Be sure not to remove too much metal from the tip as you are filing on the tip of the screwdriver. If the screwdriver is damaged beyond repair, it is best to replace the tool to avoid injury. It is also not a safe practice to hold the work in one hand while using the screwdriver with the other hand. Holding the work is dangerous because as pressure is applied to the screwdriver, it may slip out of the screw slot and puncture the hand that is holding the work. This is a very common mistake. Another thing to remember is to never expose any screwdriver to excessive heat. The excessive heat may reduce the hardness of the screwdriver tip. Always keep all screwdrivers clean and free of grease, especially the handle, because a greasy handle is very hard to grip. Replace any screwdriver that has a damaged handle or grip. A screwdriver with a damaged handle or grip may injure the hand during use. Finally, as mentioned before, never use a screwdriver as a chisel or pry bar.

Pliers

Before we discuss the different types of pliers and their intended use, we discuss some safety issues. Each plier has a specific use and should be used only for the purpose for which it was manufactured. Safety glasses should also be worn any time pliers are in use. Pliers need to be maintained and kept clean. Pliers should never be exposed to excessive heat because heat may change the temper of the pliers. Always make sure, while using a pair of pliers, not to use the pliers with the handles extended to the point where a good grip can be maintained. Always have a firm grip on a pair of pliers while using them. Do not use a set of pliers on nuts and bolts. Wrenches are intended for nuts and bolts. There are times, however, when someone has rounded off the edges of a nut or bolt with a wrench; in this case, it might be necessary to use a set of pliers to remove the fasteners so it can be replaced. This should be the only exception though. Most pliers damage the surface of most fasteners. Never use a set of pliers as a hammer. The linemen's pliers are the type most often misused as a hammer because they are so heavy. Using a set of pliers in this manner is not the manufacturer's intent and can crack or break the tool. When using pliers that have cutting capabilities, never cut hardened material with them unless they were specifically manufactured for this purpose. Finally, it is important to discuss the covering of the handles. Some pliers come with a coating of soft rubber on the handles to provide comfort while using the tool; however, some people may think that this covering protects them from an electrical shock. Unless the tool is specified as having insulated handles, they are not intended to protect you from electrical shock and should not be used on any electrical circuits that may be live. However, there are many types of handle coatings and coverings that may be purchased that provides the proper insulation to protect against electrical shock. These coverings can be found at a local tool supply store.

Each of the many different types of pliers available has a specific use or purpose. Some examples are linemen's pliers, long-nose pliers, diagonal cutting pliers, slip joint pliers, tongue and groove pliers, and locking plier-wrenches. These are the most common types of pliers that a mechanic uses in his or her career.

Linemen's Pliers

The linemen's pliers are usually used by linemen electricians; however, this is such a great tool that most mechanics have one in their toolbox as well. Linemen's pliers are also referred to as side cutters. Linemen's pliers are often confused with ironworker's pliers. The difference can be seen in **Figure 2-19**. Ironworker's pliers can be seen on the top and the electrician's linemen's pliers can be seen on the bottom. Notice that the latter has a rounded nose.

Both of these pliers are popular because they are several tools in one. They have a set of pliers on the end of the tool and are capable of cutting many things with the side cutters, which are located right next to the hinge. Some of the electrician's linemen's pliers are even equipped with crimpers that allow the mechanic to crimp a crimp sleeve if necessary. Because these tools are very heavy and durable, these tools are sometimes misused as a hammer. Because of this use, the linemen's pliers are sometimes referred to as the electrician's hammer. Do not use either of these pliers in this manner! It is not the intended use of this tool by the manufacturer and can cause personal injury.

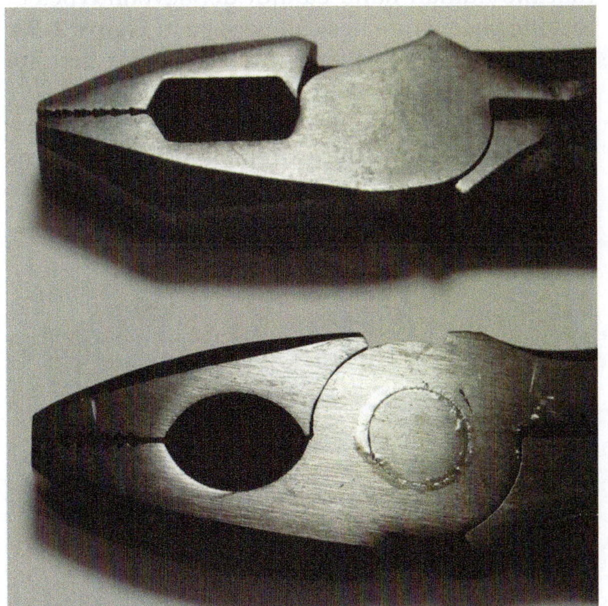

FIGURE 2-19 Electrician's lineman's pliers (bottom) and Ironworker's pliers (top).

Long-Nose Pliers

Long-nose pliers are also referred to as needlenose pliers. There are many different types of needlenose pliers, two of which are shown in **Figure 2-20**.

Needlenose pliers are very helpful when you need to work in awkward places or when working on something that is small or very intricate. These tools usually have side cutters on them as well. The most common mistake that people make with this tool is to try to bend wire or metal that is too stiff. This use can cause the small tips at the end of the pliers to bend.

Diagonal Cutting Pliers

Diagonal cutting pliers are usually referred to as *diagonals* or *dikes*, and are used mostly in cutting wire. Some thin-gauge pins may also be cut with a heavy-duty set of diagonals. It is important to use these pliers as the manufacturer intended to avoid damage to the tool or yourself. Always wear safety glasses when using any tool that has cutting capabilities. It is common for something being cut with a pair of side cutters to fly through the air. This flying projectile can easily find its way into

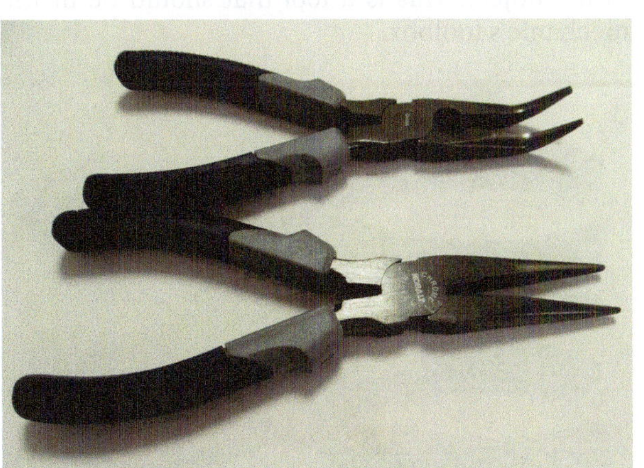

FIGURE 2-20 Straight-nose and curved-nose needle nose pliers.

the eyes and cause injury or blindness. It is also a good habit to point the cutters toward the ground to avoid injury to someone else who may be in the area. Pointing them downward causes the flying projectile to fall to the ground, where it is less likely to cause injury.

Discard any plier that is cracked or broken. If the cutting edges are nicked because something too hard was cut, replace the tool with a new one. Dull cutters can be slightly touched up by using a small honing stone.

Slip Joint Pliers

The slip joint plier is the tool that everybody calls the standard plier. A pair of slip joint pliers can be seen in **Figure 2-21**. These pliers have a slot cut at the hinge to allow them to expand to accommodate a larger job. These tools are used for gripping, turning, and bending.

Tongue and Groove Pliers

The tongue and groove pliers are most commonly referred to as a Channel Lock® plier. Channel Lock® is a registered trademark of a well-known supplier of tongue and groove pliers. There are many other well-known suppliers of tongue and groove pliers who also supply a good-quality tool. The tongue and groove plier has a tongue and groove adjustment design. This can be seen in **Figure 2-22**. This tongue and groove adjustment allows this type of plier to have a different jaw capacity at each groove.

These pliers are usually available in sizes ranging from 4½ in. up to 16 in. in length. These tools are widely used by plumbers, electricians, and mechanics. Because of their construction, these tools are very versatile; they can grip around just about any object whether it is square, flat, or round and can hold the object without causing damage to the object. This is a tool that should be in the mechanic's toolbox.

FIGURE 2-21 Slip-joint pliers.

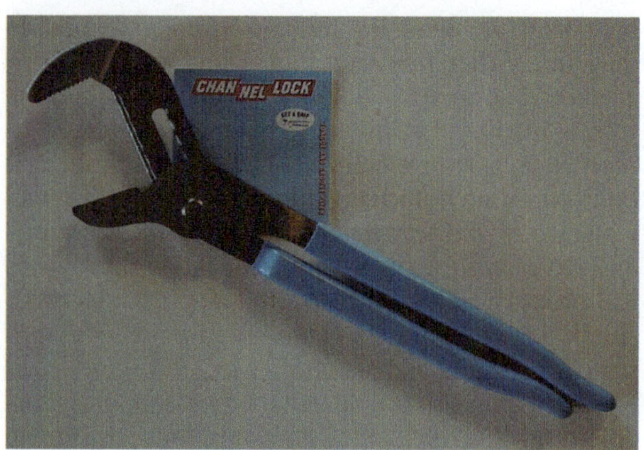

FIGURE 2-22 Channel Lock®.

Locking Plier-Wrenches

The final plier that are discussed is the locking plier-wrench. The mechanic usually refers to this plier as a Vise-Grip®. As is the Channel Lock® plier, the name Vise-Grip® is a registered trademark of a well-known supplier of locking plier-wrenches. This wrench works with a compound leverage system that locks the plier onto the work. There is also an adjustable thumbwheel at the end of the handle, which changes the jaw opening. Most people use this plier to clamp things together while they are working on them. To remove this tool from the work, simply press the release lever in the handle. Releasing it causes the tool to pop open and release the work. These tools can maintain an amazingly large amount of grip and pressure for an infinite amount of time, which is something that the human hand cannot do. Several types of locking plier-wrenches can be seen in **Figure 2-23**. Take notice of the different types of jaws that are available.

These pliers should not be used on fasteners or fittings because the extremely deep serrations

FIGURE 2-23 Various types of locking pliers.

in the jaws causes severe damage. If the serrations on the pliers show signs of wear, the tool should be replaced.

Hammers

The hammer is a striking tool and is probably the most used tool anywhere. It is also one of the most misused tools there is. Most people do not use the hammer correctly, or they use the wrong type of hammer for the task at hand. A hammer is used when something needs to be struck. For this reason, the hammer is called a striking tool. When a hammer is used, it needs to be the correct hammer for the job.

Although there are many types of hammers, when the word hammer is mentioned, most people think of a claw hammer. The claw hammer has a two-pronged claw opposite the striking surface. This claw is used to pull nails out of wood. This is probably not the hammer that is used by the electrician, mechanic, or welder. The most common hammer that is used by these tradespeople is the ball-peen hammer, which is shown in **Figure 2-24**.

It is important to use a hammer as the manufacturer intended. Most people, when using a hammer, hold the hammer improperly by placing the hand high on the handle as the hammer is used to strike an object. This is the incorrect way to hold a hammer while it is being used. The hammer should be held down toward the bottom of the handle. When the handle is held in this manner, more inertia is developed as the hammer is swung toward the object that is to be struck. This is the best way to hold it because most of the weight on a hammer is at the top of the hammer, which is at the opposite end of where the hammer is being held. When the hammer is being used in this manner, it is more efficient because inertia adds to the force that is developed from the arm as the hammer is swung toward the work. Therefore, the hammer produces more work with fewer swings.

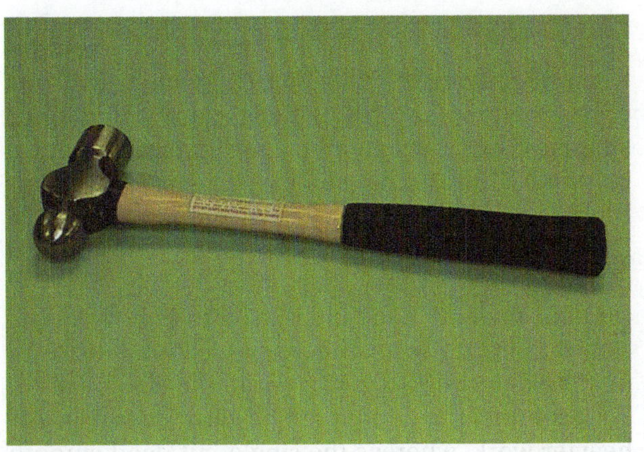

FIGURE 2-24 Ball-peen hammer.

It is important to use the full face of the hammerhead when striking an object. Do not strike the head of the hammer at an angle. Doing so could cause the head of the hammer or the struck tool to chip or break. Hard-faced hammers should not be used on struck tools that are case hardened. They should not, in fact, be used on any hardened metals, because either the hammer or the hardened metal that is being struck will break or chip. Safety glasses should be worn at all times when using a hammer.

Make sure that the proper hammer is used for the task at hand. There are times when something must be struck but cannot be marred or mutilated. In this case, a soft hammer is used. There are many types of soft hammers. There are plastic hammers, wooden hammers, rawhide hammers, dead-blow hammers, rubber hammers, and copper or brass hammers. These hammers are also used in hazardous locations where fumes may ignite due to a spark. If a spark were produced in these areas, an explosion could result, causing injury or even death. There are several different types of metal hammers that are hardened. Some of these are the claw hammer, the ball-peen hammer, the tack hammer, and the welder's chipping hammer. Each hammer is designed for different tasks even though they are all simply striking an object. Select the type of hammer by determining the type of material that is to be struck. As was mentioned earlier, always wear safety glasses when using a hammer of any type.

Chisels, Punches, and Files

Chisels and punches are considered to be "struck tools." The term *struck tool* refers to a tool that is to be struck with a hammer. There are many types of struck tools, but we discuss only chisels and punches because they are the most commonly used struck tools. Files are tools that are used to remove unwanted material. There are various types of files that are in use today, many of which are discussed.

Chisels

There are two types of chisels: wood chisels and hardened metal chisels. Wood chisels are used to chisel wood, as the name implies. Hardened chisels, sometimes referred to as cold chisels, are used to chisel something that is made of metal. Cold chisels are hardened on the cutting edge. As a cold chisel is used over time, it may need to be redressed, or sharpened. Sharpening can be accomplished by using a file or a whetstone to restore the tip to its original shape.

Punches

There are many types of punches as well. The most common punches are the center punch, the prick punch, and the drift punch. Each has its own purpose. Let us first discuss the prick punch, which is used for layout work. It is most commonly used to make a precise indentation on the crosshairs that have previously been laid out. The prick punch has a much sharper point, therefore causing a very precise placement of the tip and a deeper indentation as the tool is struck. The center punch is usually used after the prick punch to make a small indent in a material that will be drilled. When a drill bit is placed on a material such as metal, without the indentation from a center punch, the drill bit tends to move away from the desired position as it starts to rotate. Using a center punch to indent the desired point to be drilled ensures that the drill bit does not move as it starts to rotate, because the tip of the bit sits in the indent. A drift punch is similar to a center punch, with a couple of major differences. The drift punch has a flat tip, and whereas the center punch has a tapered shaft, the drift punch does not. See **Figure 2-25**. Notice how the shaft maintains the same diameter as it rises from the tip. This continues for approximately 2 to 5 in., depending on the size of the drift punch. The drift punch is used to remove dowel pins, shear pins, and taper pins from a retainer.

Always wear safety glasses when using a striking tool and when using all struck tools as well. If the struck tool is showing signs of flaring, most commonly referred to as a mushroomed head, repair it by filing the flared head back to its original condition if possible. Be careful not to overheat the striking surface of the tool. Overheating could change the hardened properties of the tool, causing it to chip or break the next time it is struck. If the tool cannot be effectively repaired using a file or grinding wheel, replace the tool with a new one.

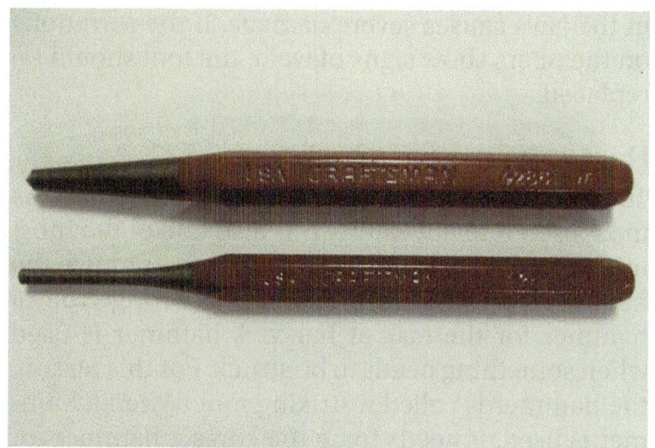

FIGURE 2-25 Center punch (top) and drift punch (bottom).

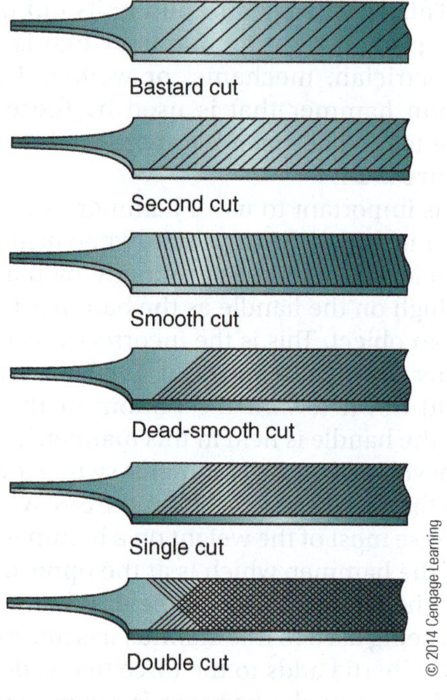

FIGURE 2-26 File cuts.

Files

Files are used to remove unwanted material. Each of the many types is not always named according to the type of **cut** it has. A cut is the grooves that are cut into the hardened metal, which produce a series of very sharp edges across the surface of the file. These sharp edges are what remove the material as the file is moved across it. The four most common files used today are the bastard cut, the second cut, the smooth cut, and the dead smooth cut. All of these are shown in **Figure 2-26**. These files are available in single cut or double cut, which are also shown in Figure 2-26. Single cut means that a single set of teeth are cut into the file. Each cut is parallel to another. Double cut is produced when two cuts cross each other. This pattern causes diamond-shaped teeth, which allow for quicker removal of metal. The double-cut file is usually used for rougher work. Another type of file is the rasp. A rasp is a very rough file that is flat on one side and rounded on the other. This file is usually reserved for use on wood. Each file is used to remove material by filing; however, the type of material that is to be filed determines which type of file is to be used. For example, the double-cut bastard file is used for heavier work, whereas the single-cut dead smooth file is used for very fine finishing work.

When working on cast iron, start with a bastard-cut file and finish with a second-cut file. When filing on soft metal, start with a second-cut file and finish with a smooth-cut file. When filing hard steel, start with a smooth-cut file and finish with a dead-smooth cut file. When filing brass or bronze, start with a bastard-cut file and finish with a second or a smooth-cut file. When filing aluminum, lead, or babbitt metal, use a standard-cut file. When very little material is to be removed, use a small file. For medium-size jobs, use an 8 in. file. For heavy jobs, use a large file.

It is important to know the different parts of a file. These are shown in **Figure 2-27**. The parts include (1) the point, (2) the edge, (3) the face or the cutting teeth, (4) the face, (5) the tang, and (6) the handle.

Files come in different shapes as well. Some are shown in **Figure 2-28**. The different shapes include, but are not limited to, the mill file, the pillar file, the round file, the square file, the taper file, the half-round file, and the triangular file.

A flat file can be used for most work; however, some conditions may require the use of a different shaped file. Use a half-round file when filing on curved surfaces. Use a small triangular file when redressing a damaged thread on a fastener. Use a mill file on precision lathe work. Use a square file or a pillar file to work on slots and keyways. Use a small, round file (sometimes referred to as a rat-tail file) to file within a small-diameter hole.

When using a file on hard metals, apply pressure on the forward stroke only. This method prevents the teeth from dulling quickly. When filing on softer metals, apply pressure on both strokes. This helps clean the file. Use a rocking motion when rounding corners or edges. Applying too much pressure while filing can cause the teeth to break off. Never force a file. Never use a file that has a tang without a handle. Using a file without the handle can cause injury to the palm of the hand. When filing, be careful not to file too quickly. Filing too quickly can cause the file to rock as it is being used. Rocking can cause the material that is being filed to be rounded instead of flat.

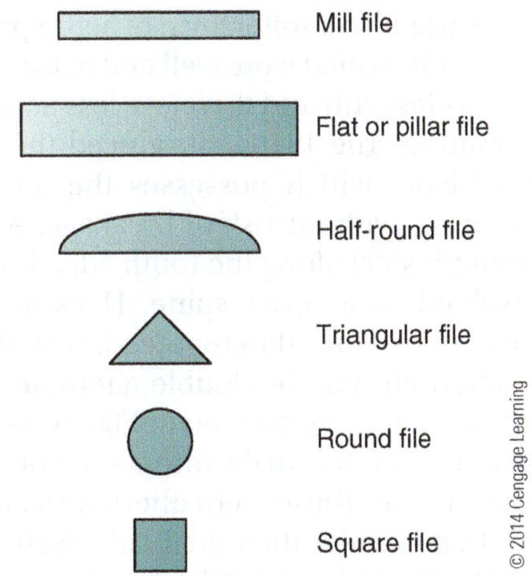

FIGURE 2-28 File shapes. These shapes represent what the file looks like as you are looking at the point of the file.

It is important to keep your files clean and in good working condition. Cleaning can be accomplished by using a card file or a wire brush to clean the file. Never bang the file against a surface to clean it. Banging can cause the teeth to shatter off the file. Always wear safety glasses when working with files.

The Hacksaw

The hacksaw has two major components: the frame and the blade. Most hacksaws are adjustable to accept blades from 8 to 16 in. in length. Hacksaw blades are hardened and tempered. This makes the cutting teeth of the hacksaw blade harder than the material that is to be cut. Hacksaw blades were first available in the "all-hard" form. They cut accurately, but they were extremely brittle. This limited their practicality because they could only be used on something that was firmly clamped in a vice. At that

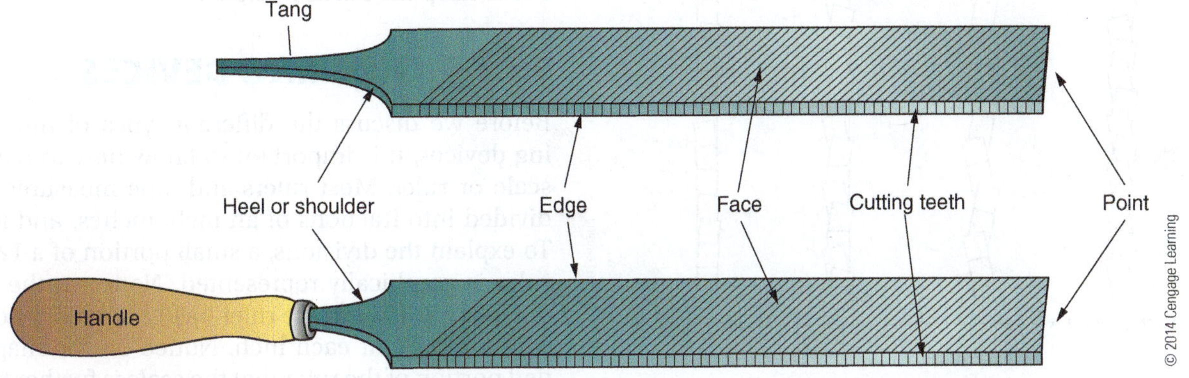

FIGURE 2-27 File parts.

time, there was also a softer form of high-speed steel blade available, which wore well and resisted breakage, but was less stiff and therefore less accurate for precise cutting. The 1980s introduced the bimetal hacksaw blade, which possesses the advantages of both forms, without risk of breakage. A strip of high-strength steel along the tooth edge is electron beam-welded to a softer spine. Hacksaw blades come with one of four different teeth sets: the wave set, the alternate set, the double alternate set, and the raker set. Each can be seen in **Figure 2-29**.

The term **set** refers to the orientation of the teeth on the saw blade. These teeth alternate so as to cut a groove that is wider than the blade itself, thereby keeping the blade from binding as the hacksaw is being used. The teeth also remove the debris from the work as it is being cut. The hacksaw blade has between 14 and 32 teeth per in. Be sure to use the proper hacksaw blade for the task at hand. Use a blade with 14 teeth per in. when cutting on a large section of mild material. Use 18 teeth per in. when cutting on a large section of hard material. Use 24 teeth per in. when cutting on brass, thick wall pipe, angle iron, and copper. Use 32 teeth per in. when cutting on thin wall tubing. One thing to remember is to always make sure that there are two or more cutting teeth on the work at any given time. Failure to do this may result in broken teeth. Now that the hacksaw's construction has been described in detail, it is important to know how to use the hacksaw properly.

One of the most common mistakes is that the hacksaw blade is sometimes placed in the hacksaw backward. The teeth should point away from the handle. Most hacksaw blade manufacturers have an arrow printed on the new blade that directs the user for the proper installation. Placing the blade in the saw correctly ensures that the material is cut away from the user on the downward stroke. Also, a large amount of pressure can be applied while pushing on the hacksaw in a downward motion. If the hacksaw blade were in backward, the teeth would point toward the handle. This means that any cutting would then have to be done on the upward stroke. This would not be an efficient way to cut the material because it is very difficult to achieve the maximum cutting force that is needed as the hacksaw is being pulled up toward the user. It is also important to know that you should not put any pressure on the hacksaw during the return stroke, because this dulls the teeth on the blade more quickly. Another common mistake is to use the hacksaw too rapidly. That is, there are too many strokes per second. There should be no more than one downward stroke per second. Following the one stroke per second rule accomplishes two things. The hacksaw blade does not heat up as much, and the user does not tire as quickly. When a hacksaw blade gets too hot, it tends to crack and may even break. The teeth also tend to break off if they get too hot. Some other safety tips are to always wear gloves and safety glasses while using a hacksaw. Use a holding device, such as a bench vise or a clamp, to hold the material as it is being cut. When a hacksaw blade gets dull, take the time to change it. It probably takes less time to change the blade and finish cutting the material than it would take for the user to continue cutting the material with the dull blade. Always inspect the hacksaw before and after use. If any damage is noticed on the blade, replace the blade at that time. Do not place the hacksaw back in the toolbox with a damaged blade because the next time it is needed, it might be forgotten that the blade was bad and injury may result. If damage is noticed on the hacksaw frame or handle, do not try to repair it. Replace the tool with a new one. This is the best way to remain safe. Also, always remember to keep the hacksaw clean.

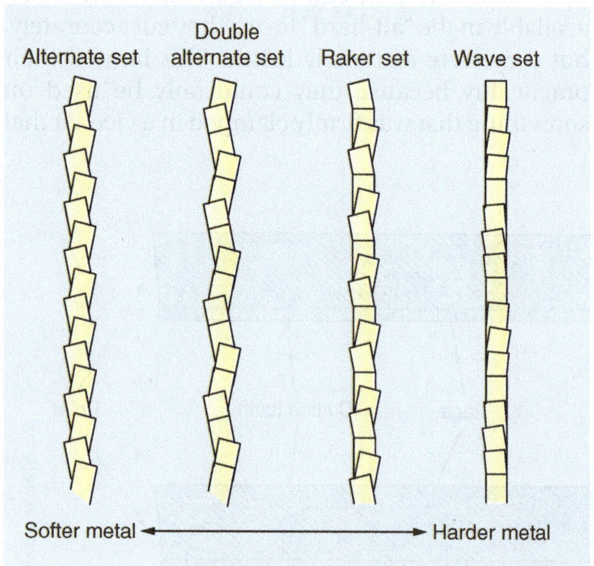

FIGURE 2-29 Hacksaw blade teeth sets.

2-2 MEASURING DEVICES

Before we discuss the different types of measuring devices, it is important to know how to read a scale or ruler. Most rulers and tape measures are divided into fractions of an inch, inches, and feet. To explain the divisions, a small portion of a 12 in. ruler is graphically represented. Notice at the top of **Figure 2-30**, that the ruler has 12 large divisions, which represent each inch. Notice on the magnified portion of the ruler that the scale is further broken down into increments. The markings that are

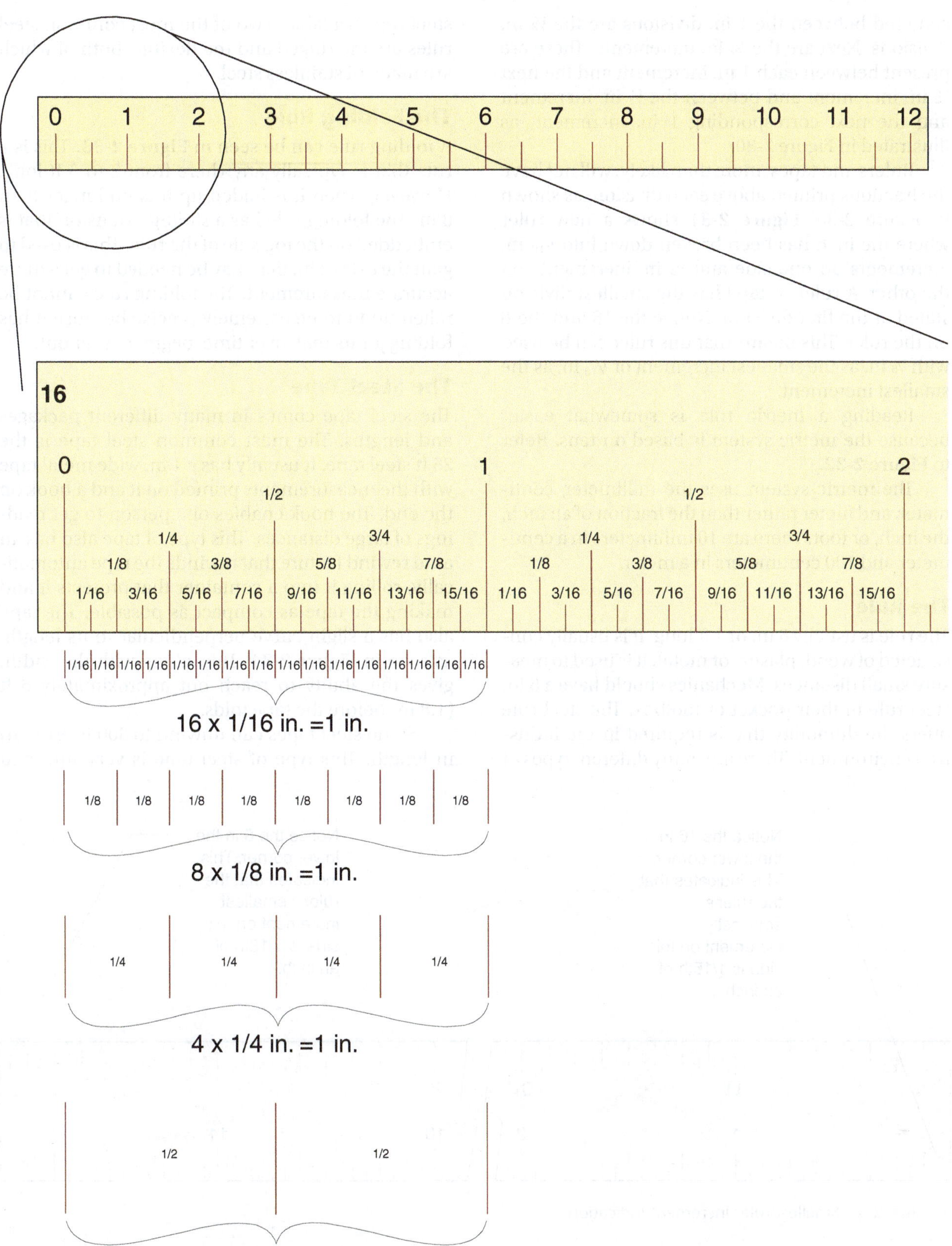

FIGURE 2-30 How to read a ruler.

centered between the 1 in. divisions are the ½ in. divisions. Next are the ¼ in. increments. These are present between each 1 in. increment and the next ½ in. increment and between the ½ in. increment and the next corresponding 1 in. increment, as illustrated in Figure 2-30.

Rulers and tapes, more than likely, will not have the fractions printed above each division, as shown in Figure 2-30. **Figure 2-31** shows a new ruler where the inch has been broken down into ¹⁄₁₆ in. increments on one side and ⅛ in. increments on the other. A ruler or tape has the smallest division stated at the first division. Notice the 16 and the 8 on the ruler. This means that this ruler can be used with ⅛ in. as the smallest increment or ¹⁄₁₆ in. as the smallest increment.

Reading a metric rule is somewhat easier because the metric system is based on tens. Refer to **Figure 2-32**.

The metric system uses the millimeter, centimeter, and meter rather than the fraction of an inch, the inch, or foot. There are 10 millimeters in a centimeter and 100 centimeters in a meter.

The Rule

The rule is usually 6 in. or 1 ft long. It is usually constructed of wood, plastic, or metal. It is used to measure small distances. Mechanics should have a 6 in. steel rule in their pocket or toolbox. The steel rule offers the durability that is required in the industrial environment. There are many different types of steel rule available. Two of the most common steel rules are the ridged and the flexible, both of which are made of stainless steel.

The Folding Rule

A folding rule can be seen in **Figure 2-33**. This is a rule that is typically anywhere from 2 to 6 ft long. However, when it is folded up it is no longer than 8 in. The folding rule has a sliding extension that is embedded on the top side of the rule. This is used to gain the extra 8 in. that may be needed to get a more accurate measurement. The folding rule cannot be relied upon to be extremely precise because it has folding joints that, over time, begin to wear out.

The Steel Tape

The steel tape comes in many different packages and lengths. The most common steel tape is the 25 ft steel tape. It usually has a 1 in. wide metal tape with the measurements printed on it and a hook on the end. The hook enables one person to get readings of large distances. This type of tape also has an auto rewind feature that rewinds the tape automatically, rolling it into a container that protects it and making the tape as compact as possible. The tape also has a slight curve perpendicular to its length, as seen in **Figure 2-34**. This, along with the width, gives the ability to reach out approximately 6 ft (1.8 m) before the tape folds.

Some steel tapes can unwind to 300 ft (91.4 m) in length. This type of steel tape is very thin and

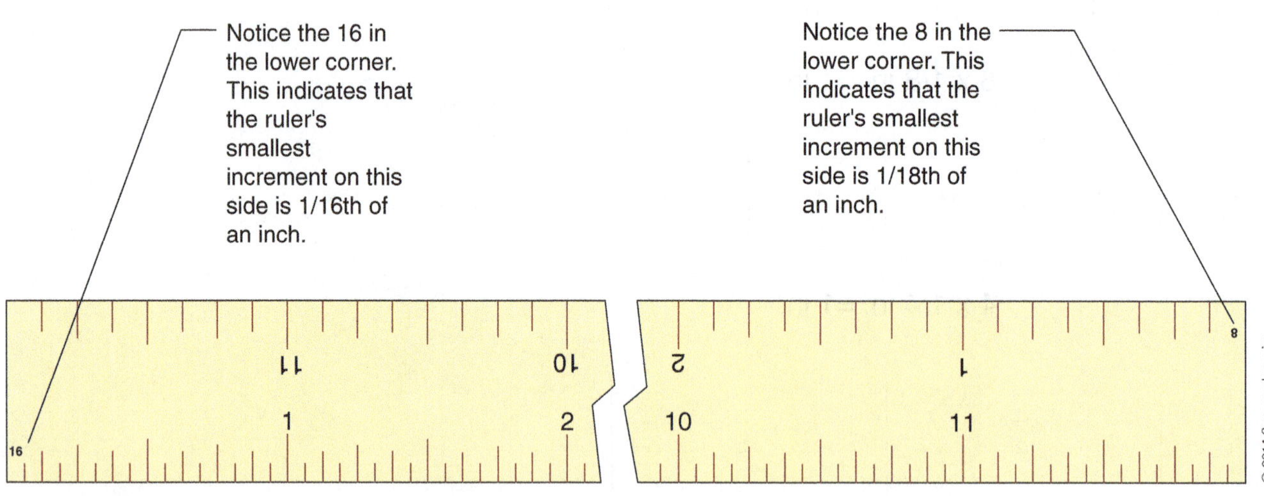

FIGURE 2-31 Smallest ruler increment indication.

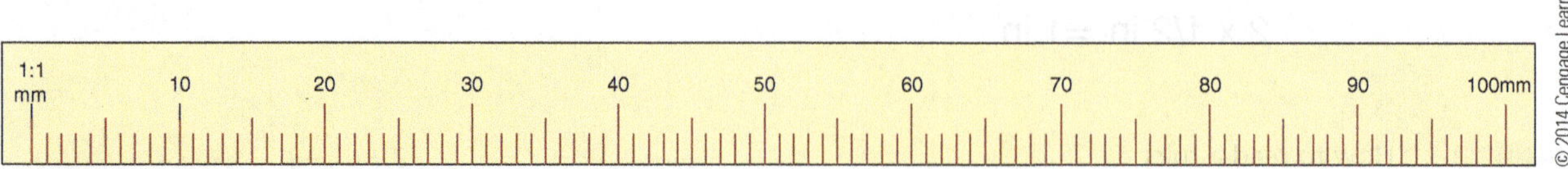

FIGURE 2-32 Metric ruler.

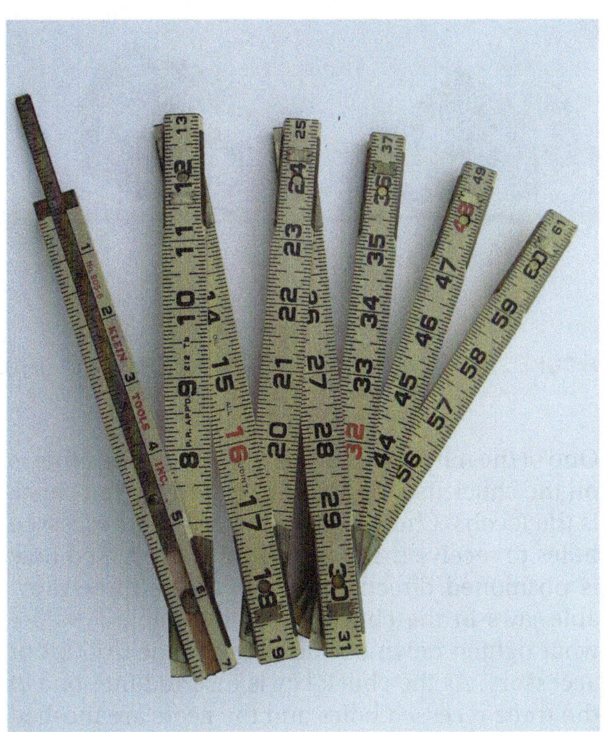

FIGURE 2-33 Folding ruler.

FIGURE 2-34 Steel tape.

narrow and usually does not have an auto rewind feature, so it must be rewound manually. Tapes of this length are usually used for layout work. Some of the longer tapes are not made of steel. They are made of a fiber weave or a thin plastic, making them weigh less.

2-3 POWER TOOLS

Some power tools are so common that they can be found in homes as well as industry. Most people think of a drill when the term *power tool* is heard. This is because it is probably the most common power tool used today. The drill is just one of the power tools that is discussed in this chapter. Others are the portable band saw, the reciprocating saw, the portable side grinder, and the portable circular saw. Before any of these are discussed, however, some very important safety factors must be mentioned.

Always wear safety glasses while operating any power tool. Never operate any power tool without first reading its user's manual and becoming familiar with the tool and its operation. All power tools should be inspected before and after use to ensure that the tool is in good working order and in proper condition. All power tools need to be cleaned before and after use. Always make sure that all vents on a power tool are not restricted by dust, debris, dirt, or grease. Do not cover these vents with the hand while the tools are in use. These vents are needed to keep the power tool cool during operation. If these vents are covered or clogged in any way, the power tool is not able to remove the heat that is produced within the motor. If the heat is not removed, then the life of the power tool will decrease as the insulation on the windings starts to break down.

An ac power tool should have its power cord inspected before and after every use. If a cord is cut or if it is found with a nick in the insulation upon inspection, *do not use it.* Repair or replace the cord according to the user's manual that came with the power tool. Never use an ac power tool while standing in water or any other wet location. If a drill must be used in this environment, a dc cordless drill should be used. This type of drill is referred to as a portable drill, which is covered next. If an ac power tool is used, make sure that the grounding pin has not been removed. Never use a 3-to-2-prong adapter either. The presence of the grounding pin minimizes the possibility that the user will be shocked should a short occur within the tool. Removing or bypassing the ground prong removes that protection. Not all power tools have a grounding pin, however. A double-insulated power tool does not require the grounding pin because all of its metal parts are isolated electrically from the drive motor, thus minimizing the risk of shock. Never plug a cord into an outlet with wet hands. Never remove a plug from an outlet by pulling on the cord. Doing so may cause the conductors of the cord to be pulled out of the plug. Always pull on the plug itself.

When operating a power tool of any kind, keep focused and stay attentive. Always keep in mind that one careless moment while using a power tool can cause serious injury or even death.

Portable Drill

Many manufacturers make portable drills; therefore, portable drills come in many different shapes, sizes, and colors and have different capabilities. It is important to discuss only the basic essentials of a portable drill because some manufacturers' drills have more capabilities than others. There are basically two types of portable drills: ac and cordless (dc). An ac drill is a drill that requires a 120-volt outlet to plug into. The cordless drill is, as the name implies, cordless, meaning that it does not require an ac source to run. A cordless drill requires a charged battery to be operational. The cordless drill is becoming more and more popular for several reasons. One is its portability. An ac portable drill is also portable; however, because an ac outlet is required for operation, its portability is limited to the distance of a drop cord. Consequently, most people prefer the freedom that the cordless drill offers. Another reason most people prefer a cordless drill is because it is usually, with some exceptions, less expensive than an ac portable drill, making it easier to replace if it becomes damaged.

Most portable drills have a variable speed. This feature gives the user more control and makes the tool more flexible. There are times when a job may require a fast rotary motion, and there are times when a job requires a slower rotary motion. The speed is determined by how far the trigger is pulled. If the trigger is pulled only halfway the drill does not run at maximum speed. However, if the trigger is pulled to its maximum distance, the drill runs at maximum speed. It is important to know this because a portable drill can be used for so much more than just drilling. Various accessories and adapters enable the drill to be adapted to do different jobs such as polishing, buffing, sanding, driving screws and fasteners, and mixing.

Most drills also are reversible. This means that the drill can run forward and backward. This feature comes in handy when a drill gets hung or stuck while running in the forward direction. Running the drill in the reverse direction dislodges the drill from the work and it can then be removed.

The maximum size shank diameter that the chuck can hold determines the size of a portable drill. A shank is the portion of the drill bit or accessory that is inserted into the chuck. The shank is measured by its diameter. After the shank has been inserted into the chuck, the chuck then is tightened to keep the drill bit or accessory from slipping or coming out of the drill while it is in use. There are two types of chucks: a keyed chuck (sometimes referred to as a geared chuck) and a keyless chuck. Both can be seen in **Figure 2-35**. A keyed chuck uses miter gears to move three movable jaws in or out.

FIGURE 2-35 Keyed chuck (right) and keyless chuck (left).

One of the miter gears is on the key, and the other is on the chuck itself. The most common keyed chuck is the Jacobs® Chuck. This chuck has three recessed holes to receive the chuck key. Each recessed hole is positioned directly over one of the three movable jaws in the chuck. These moveable jaws are what tighten down on the shank of the drill bit or accessory. As the chuck key is inserted into one of the three recessed holes and the gears are meshed together on the chuck key and chuck, the key can then be turned. This is what causes the three jaws to move in or out, depending on which way the chuck key is turned. An outward motion of the jaws causes the chuck to tighten against the shank of a drill bit or accessory, and an inward motion of the jaws causes the chuck to release the shank of the drill bit or accessory. It is important that the key be inserted into all three holes and tightened before any drilling is done. As the name implies, a keyless chuck requires no key to tighten or release the drill bit. It is designed so that the chuck can be held with the hand as the drill is run in the forward or reverse direction. This chuck has a plastic or rubber coating that allows maximum grip on the chuck. This grip is usually enough to tighten the jaws against the shank of the drill bit or accessory. Most people prefer the keyless chuck to the keyed chuck because the chuck key tends to get lost or misplaced, and using the keyless chuck is usually quicker than using a keyed chuck. However, with a keyless chuck some slippage may occur on the shank as the drill is in use, because the keyless chuck does not tighten as well as the keyed chuck. To prevent this slipping, it might be necessary to use the drill with a keyed chuck if there is some heavy work that needs to be done, such as drilling into stainless steel or a hard material. Now that the shank and chuck have been explained, it is understood what a ⅜ in. drill means. A ⅜ in. drill can receive a drill bit that is ⅜ of an in. or smaller in diameter. A ½ in. drill can receive a drill bit shank that is a ½ in. or smaller in diameter.

Some drills also come with hammer drill capabilities. This means that the drill can be used to drill holes into masonry or concrete. This type of drill has a selector switch that engages the hammer action. The hammer action feels like a vibration in the drill as it rotates; however, simply pulling the trigger does not produce the vibration. Pressure must be applied to the drill before this hammer action starts. It is important to know that if a hammer drill is used, a masonry drill bit must be used. Do not try to drill a hole into concrete or brick with a standard metal bit or a wood bit. Also, as mentioned before, always wear safety glasses while using a drill of any kind.

The Portable Bandsaw

A band saw is a large machine, as seen in **Figure 2-36**, which is usually located in a shop. Because it is difficult to move around, anything that has to be cut on a band saw has to be brought to the shop where the band saw is located. The portable band saw eliminates the trip back to the shop and thus eliminates inefficiency. See **Figure 2-37**.

A band saw makes a very smooth cut when compared to the hacksaw. The blade of a band saw is similar to a hacksaw blade except that a hacksaw blade is a straight blade in which there is a beginning and an end. This construction requires the user to push and pull the hacksaw back and forth across the material that needs to be cut. This back-and-forth motion often causes a cut to be crooked and rough. The band saw blade does not have an end because the ends of the blade are welded together to make a band. Placing this continuous blade on a rotating pair of pulleys enables the metal to be cut by simply pulling a trigger on the band saw. Because the band saw blade continues to rotate in one direction, a very smooth cut is achieved. Over time, however, a band saw blade stretches. There is an adjustment on most portable band saws that takes up the slack caused by the stretching. Some portable band saws have an automatic adjustment on them. Another thing to watch for is the dullness of the blade. Because the band saw is so easy to use, most people do not pay attention to the sharpness of the blade. If a blade is sharp, it cuts through a hard material fairly quickly and with ease. As the blade gets duller, it does not cut as efficiently, and therefore takes longer to cut through the material. Again, do not forget to wear safety glasses when working with the portable band saw.

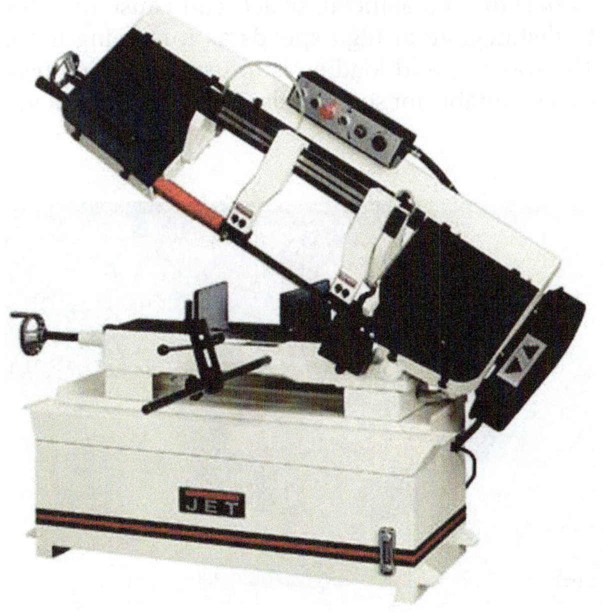

FIGURE 2-36 Stationary band saw.

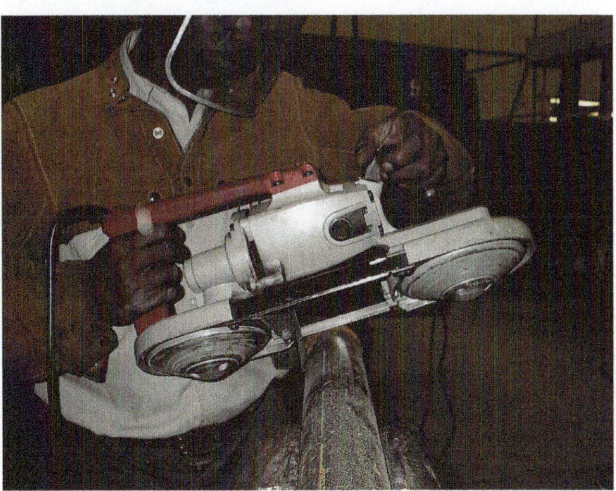

FIGURE 2-37 Portable band saw ("Port-a-band").

The Reciprocating Saw

The reciprocating saw mimics the back-and-forth motion that is made with a hacksaw. The gearing inside the reciprocating saw causes the saw blade to move back and forth across the material that needs to be cut. This saw is usually used to cut wood, plaster, plastic, or some other soft material. An example of a good use for the reciprocating saw is cutting a 4 in. square hole into a piece of wood. The square must first be drawn on the wood with a pencil or scribe. Select a wood bit (a drill bit used to drill into wood) that has a diameter that is large enough to drill a hole that accommodates the reciprocating saw blade. Drill a hole into every corner of the square that is drawn on the wood, being careful not to drill outside the drawn square. After all four holes have been drilled, insert the reciprocating saw blade into one of the holes that has been drilled out and cut along each line on each side of the drawn square. As the last cut is completed, the piece of wood that is left within the drawn square should simply fall out of the middle. Do not forget to true the corners up with the reciprocating saw. As can be seen with this example, there is a purpose for a reciprocating saw. As with the band saw, keep a watchful eye on how sharp the blade is. This is also a tool in which safety glasses need to be worn as it is being used.

The Right-Angle Grinder

The right-angle grinder is a power tool that is used to grind away steel. It is sometimes referred to as a side grinder. There are many side grinder safety issues that must be mentioned, but first it is important to understand the construction of the right-angle grinder. The right-angle grinder is usually available in two sizes, 4½ in. and 7 in. The size of a right-angle grinder is determined by the diameter of the grinding disk it uses. Keep in mind that this measurement is taken from a new grinding disk, not a used one. A used grinding disk is smaller in diameter because as a grinding disk is used, the material that it is made of is removed as the disk comes into contact with the metal being ground. It is important to inspect the grinding disk before and after each use. If it is noticed that a large chunk is missing out of the disk, replace the disk. This missing piece causes the disk to be unbalanced, and the disk could explode at high speeds as it is being used. Think about an unbalanced tire on an automobile. As the speed increases, the tire begins to shake and bounce because the weight is not distributed evenly around the tire. More centrifugal force is applied to one area of the tire instead of the centrifugal force being equal at all points around the tire. The grinding disk can develop these same high centrifugal forces as well. A grinding disk should be as close to round as possible All right-angle grinders should have a guard on them, as can be seen in **Figure 2-38A**. If you spot a right-angle grinder that looks like the one shown in **Figure 2-38B**, do not use it.

These guards should be in place at all times to protect the hands as the grinder is being used. Some people remove these guards to allow them to reach into tighter places; however, this is not the manufacturer's intended operation and should not be done. Most right-angle grinders have a slide switch that is capable of being locked into the run position. If the right angle grinder is going to be used for an extended period of time, this feature should be used. It prevents the hands from getting fatigued and allows the user to keep a better grip on the grinder as it is used.

As a right-angle grinder is used, a stream of sparks flies off the grinding surface. This stream of sparks is a mixture of the material from the disk and molten metal. It is important to secure the area in which these sparks are going to fly. Make sure that there are no people or combustible materials in that area. Always have a fire extinguisher close by. Try to minimize the distance in which the sparks fly by directing the sparks toward the floor. Because of these sparks, it is recommended that a face shield, as well as a pair of safety glasses, be worn while this power tool is in use. A face shield protects the entire head and neck area from flying debris and hot metal, which could burn the skin. It is also a good idea to wear gloves while operating the right-angle grinder. One other safety tip: Aluminum and brass causes the grinding disk to become loaded. This means that as the aluminum or brass is being ground, it tends to get embedded into the grinding disk. As the disc becomes loaded, it produces more heat and less grinding. Loading also causes the disk to become unbalanced, which can cause the disk to disintegrate at high speeds as it is being used. The way to avoid loading is to use a grinding disk that is suitable for soft metals. These disks are made

FIGURE 2-38 (A) Guard on grinder. (B) Guard missing.

with a softer bonding compound than those made for grinding on steel. As the disk is used on the soft metal, because it is made of a softer compound, it wears faster, thus preventing the disc from becoming loaded. A grinding disk has a code letter on it, designating the strength of the bonding material in the compound.

The Circular Saw

The circular saw is a very common power tool that is used often for cutting sheets of plywood, paneling, plastics, and even some metal. It is very easy to change the circular cutting blade on the circular saw. Each blade has its own purpose. Be sure to use the correct blade for the task at hand. Because the circular saw can be used to cut many different types of material (when the blade is changed), it is a desirable tool for working in an environment where many different types of material need to be cut. It is important to know that a different blade is used for each of the materials that was mentioned above. Do not cut plastic or metal with a blade that would be used to cut plywood.

A circular saw does what a table saw does. A table saw is a large, stationary tool that is usually located in a shop. When a table saw is being used, the material has to be moved around on the table portion of the saw as it is being cut. Conversely, the circular saw is guided around on the material as it is being cut. Some materials are quite heavy and have to be supported by the operator when using the table saw. The circular saw allows the operator to place the material on a bench or a set of saw horses while it is being cut. The circular saw also has some adjustments on it that allow for cuts to be made at different angles. This adjustment is usually located in the front of the saw on most models. This adjustment allows for a cut to be made at 45° to 90°. The depth of a cut can also be changed by an adjustment. This adjustment is usually located just below the trigger handle on most models of circular saws. It is usually a good idea to visually inspect any power tool before using it to see whether there is any damage to the saw and that all of the parts are working properly. Do not forget to check the movable guard. Make sure that it is working properly. This guard protects the operator from the rotating blade. If this guard does not work properly or is damaged, do not use the circular saw. Repair or replace the guard before use. Always make sure that the power cord is out of the cutting path before use. It is a common occurrence for the power cord to be cut because the user did not pay attention either when using the power saw or when setting the power saw down. Often the blade of the circular saw is still rotating when the circular saw is set down. Setting the saw down on its power cord while the blade is still rotating usually results in the power cord's being cut. If the power cord has been cut or nicked, it is important to replace the cord. Do not try to repair the cord. A cord that has been repaired was most likely wrapped with an excessive amount of electrical tape to protect the splice or nick. This area of the cord can get caught on the material as the circular saw is being used. This could cause the circular saw to either get in a bind or kick back and cause bodily injury. Injuries that arise from using a circular saw improperly are usually very serious. Try to avoid injuries at all costs. Always wear the proper safety equipment while using the circular saw. Never push a circular saw through the material too fast. Always allow time for the circular saw to cut the material. Pushing a saw through the material too fast can also cause the saw to bind or kick back. When cutting material that is thick and heavy, such as ¾ in. treated plywood, it is a good idea to always have a coworker there to assist in the support of the material as it is being cut. As the saw nears the end of a cut, the material tends to droop, causing the saw to bind. When a coworker is there to assist, this does not happen. Always clean and inspect the circular saw after every use. Cleaning ensures proper operation the next time it is used.

SUMMARY

- There are many different types of hand tools that have been created to make difficult tasks easier to accomplish.
- Wrenches, screwdrivers, pliers, hammers, chisels, punches, and hacksaws are discussed along with the specific and proper use of each of these tools.
- Taking shortcuts, as discussed, or using a hand tool improperly often results in damage to the tool or injury to the user.
- Inspecting and maintaining all hand tools before and after each use ensures that each tool continues to work properly and does not cause injury.
- Safety glasses should be worn when using hand tools and power tools.
- The power cord of an ac power tool should be inspected before and after every use.
- If a cord is cut or if it is found to have a nick in the insulation, replace the cord according to the user's manual that came with the power tool.
- Follow basic electrical safety, such as never using an ac power tool while standing in water or any other wet location.

- A double-insulated power tool does not require the grounding pin because all of its metal parts are isolated electrically from the drive motor, thus minimizing the risk of shock.
- Never remove a plug from an outlet by pulling on the cord. Always pull on the plug itself.
- When operating a power tool of any kind, it is important to keep focused and stay attentive.
- Never use a tool improperly, such as using a drill while covering the ventilation slots. This causes the drill to overheat, which lessens the life of the tool.
- Removing the guard that is on a right-angle grinder increases the risk of injury.
- All power tools must be used as the manufacturer intended, to avoid risk or injury.

REVIEW QUESTIONS

1. Name two ways that a screwdriver is most commonly misused.
2. Do not use _____ on nuts and bolts unless someone has rounded off the edges of the nut or bolt.
3. In reference to a hacksaw blade, what does the term *set* mean?
4. Name two things that must be inspected on a power tool before its use.
5. Describe in detail a problem that arises when using a right-angle side grinder to grind aluminum or brass.

CHAPTER

3

Fasteners

This chapter covers the screw threads that are most common in the field of maintenance. Only fastener threads are discussed in this chapter, not screw threads for transmitting mechanical power.

OBJECTIVES

After studying this chapter, the student should be able to

- Define the terms that are associated with threads.
- Discuss the class of a thread.
- Discuss and demonstrate the proper tapping procedures and the proper method of using a cutting die.
- List several types of fastener.
- Identify the grade of a fastener.
- Use tables to determine the tightening torque of a bolt.

3-1 THREADS

When talking about screw threads, we first must be aware of the Unified Standard, sometimes called the Unified System. The Unified Standard of threads makes it possible to interchange fasteners from one country to another; for example, a damaged fastener on a machine that is installed in the United States but was designed and built in Canada, with Canadian fasteners, can be replaced. The Unified Standard is not the only standard, but it is the most commonly used. These standards cover the cross-sectional shape of a thread, available diameters, and threads per inch for each diameter, all of which is discussed later in this chapter. First, however, it is important to examine the components of the thread.

Thread Construction

There are two types of thread: external and internal. The term *external thread* normally refers to a screw or a bolt. The term *internal thread* usually refers to a nut or a place on a machine that receives the screw or bolt.

External threads consist of grooves that are cut into a shank or a shaft with a cutting die, which is discussed later in this chapter. These grooves have many specifications (**Figure 3-1**). One specification is the **root**, the lowest point of the thread. The root may be flat or rounded. Another specification is the **crest**, which is the highest point of the thread. The distance between the root and the crest is called the **thread depth**. The thread length is measured along the face of the thread. The **face** of the thread is the flat side of the thread that rises at an angle from the root to the crest. The **pitch** of a thread is the distance from any point on a thread to the next corresponding point measured parallel along the axis (on single-thread fasteners only). Threads per inch (TPI) is shown in **Figure 3-2**. If the pitch is known, the TPI can be found by dividing 1 by the pitch:

$$TPI = \frac{1}{Pitch}$$

If the TPI is known, the pitch can be found by dividing 1 by the TPI:

$$Pitch = \frac{1}{TPI}$$

Table 3-1 lists threads per inch for common diameters.

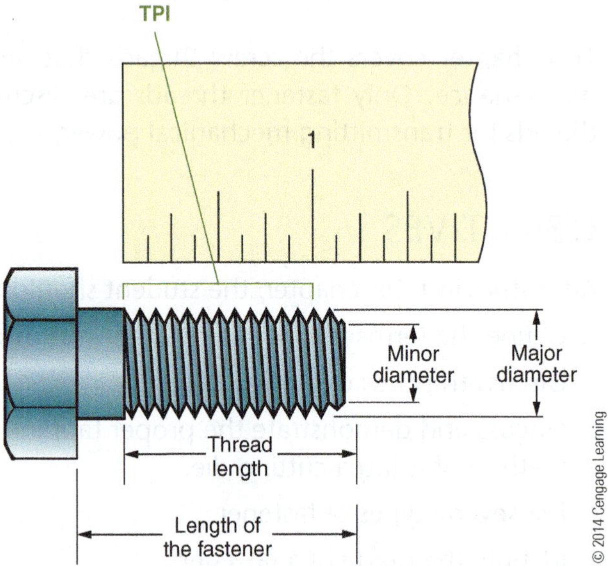

FIGURE 3-2 This fastener has 12 threads per in.

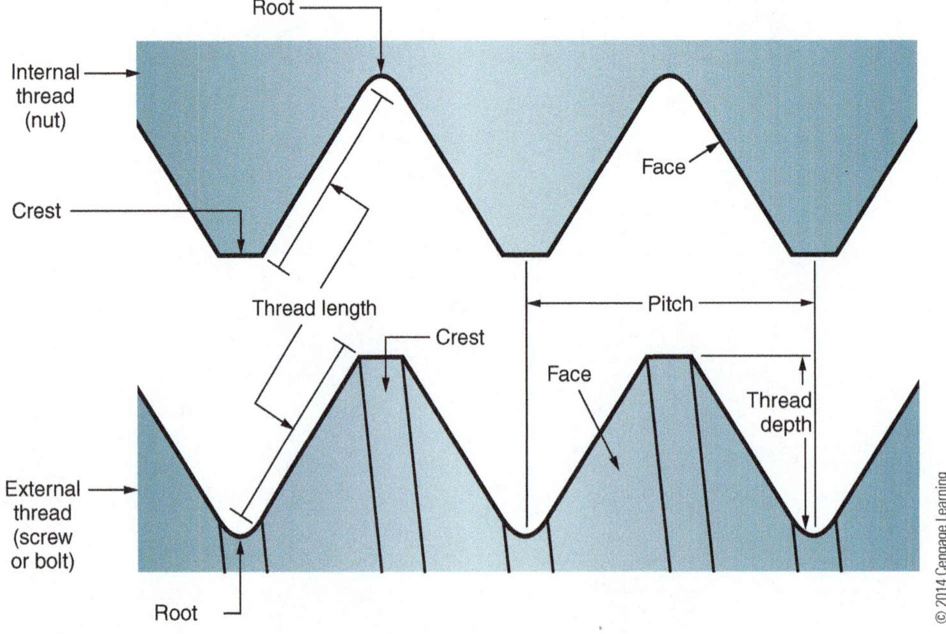

FIGURE 3-1 Thread specifications.

Table 3-1

SCREW THREADS PER INCH FOR COMMON DIAMETERS

Nominal Size (in.)	UNC Threads per Inch	UNF Threads per Inch	UNEF Threads per Inch
4	40	48	—
5	40	44	—
6	32	40	—
8	32	36	—
10	24	32	—
12	24	28	32
1/4	20	28	32
5/16	18	24	32
3/8	16	24	32
7/16	14	20	28
1/2	13	20	28
9/16	12	18	24
5/8	11	18	24
11/16	—	—	24
3/4	10	16	20
13/16	—	—	20
7/8	9	14	20
15/16	—	—	20
1	8	12	20
1 1/16	—	—	20
1 1/8	7	12	18
1 3/16	—	—	18
1 1/4	7	12	18
1 5/16	—	—	18
1 3/8	6	12	18
1 7/16	—	—	18
1 1/2	6	12	18
1 9/16	—	—	18
1 5/8	—	—	18
1 11/16	—	—	18
1 3/4	5	—	—
2	4 1/2	—	—
2 1/4	4 1/2	—	—
2 1/2	4	—	—
2 3/4	4	—	—
3	4	—	—

Thread Series

Fastener threads are categorized as being in one of four series: coarse thread, fine thread, extra-fine thread, or constant pitch.

Screws in the coarse-thread series are identified with the symbol UNC (Unified National Coarse series). These have the largest thread area of the four series. A coarse thread fastener is desirable when it will be removed and installed frequently. The threads are more durable than smaller threads and can still function properly even when slightly damaged. Coarse thread fasteners are used in nonvibrating applications, where temperature does not fluctuate often and where corrosion might occur. Larger threads prolong the life of the screw, bolt, or nut in corrosive conditions.

The fine thread fastener is identified by the symbol UNF (Unified National Fine series). The tensile stress applied to the external threads in this series is greater than that applied to threads in the coarse series because the thread area of the individual threads is not as large. Therefore, fasteners in this series have to be stronger to prevent the threads from being pulled off. Fine-thread screws and bolts are usually short. There are times, however, when longer screws and bolts are encountered when the application requires it. They are used for short-distance applications, where a small lead angle is desired, or where the wall thickness demands a fine pitch. Lead angle refers to the distance necessary to engage the first thread. They can also be used in an environment that has slight vibrations and thermal (temperature) fluctuations.

Fasteners in the extra-fine thread series have the symbol UNEF (Unified National Extra-Fine series). Much like the fine thread series, this thread series is used for short-distance applications and whenever a finer thread is required. This thread series is used when the internal threads are being placed in thin-walled tubes, thin nuts, ferrules, or couplings that possess thin walls.

Constant-pitch fasteners are designated by the symbol UN (United National series). Whereas fasteners in the other three series have only one pitch for each diameter, the thread pitch of these fasteners remains constant. That is, two bolts can have different diameters but still have the same thread pitch. Eight-, 12-, and 16-thread fasteners are the most commonly used.

Thread Diameter

Two more things to be considered are the **major diameter** and the **minor diameter**. These terms apply to both the internal thread and the external thread. With regard to an external thread, the major diameter is the outside diameter of the screw or bolt at the crests of the thread. For an internal thread, it is the maximum diameter that has been cut into the center of the nut, specifically, the root of the thread. This is the place in the nut where the crest of the screw or bolt is. The crest of the screw or bolt does not rest against the root of the thread in the nut; instead, there is some clearance between them. Contact between the screw and the nut should be only on the face of the threads. Refer to **Figure 3-3**.

The minor diameter of an external thread is the smallest diameter of the screw or bolt at the root of the thread. The minor diameter of an internal thread is the diameter of the hole in the nut at the crests of the thread. This is the place in the nut where the root of the screw or bolt is. Again, the root of the screw or bolt does not rest against the crest of the thread in the nut; instead there is some clearance between them, and the only contact between the screw and the nut should be on the face of the threads. The major and minor diameters are illustrated in **Figure 3-4**.

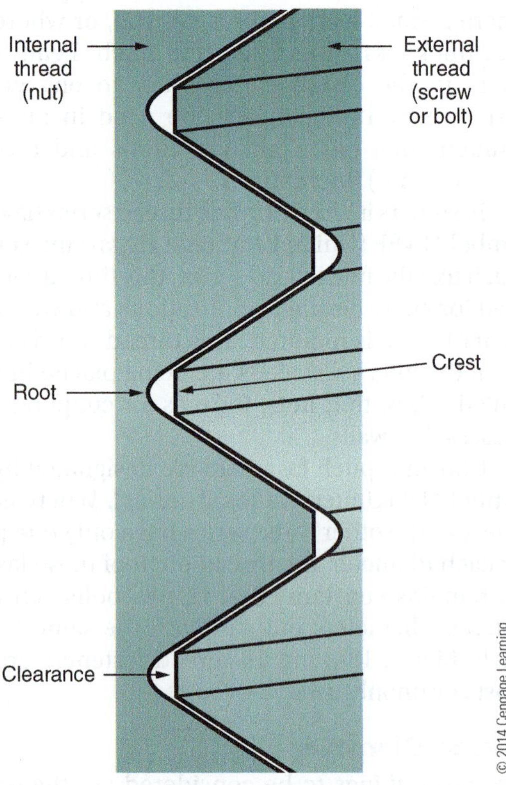

FIGURE 3-3 Contact on the face of a thread.

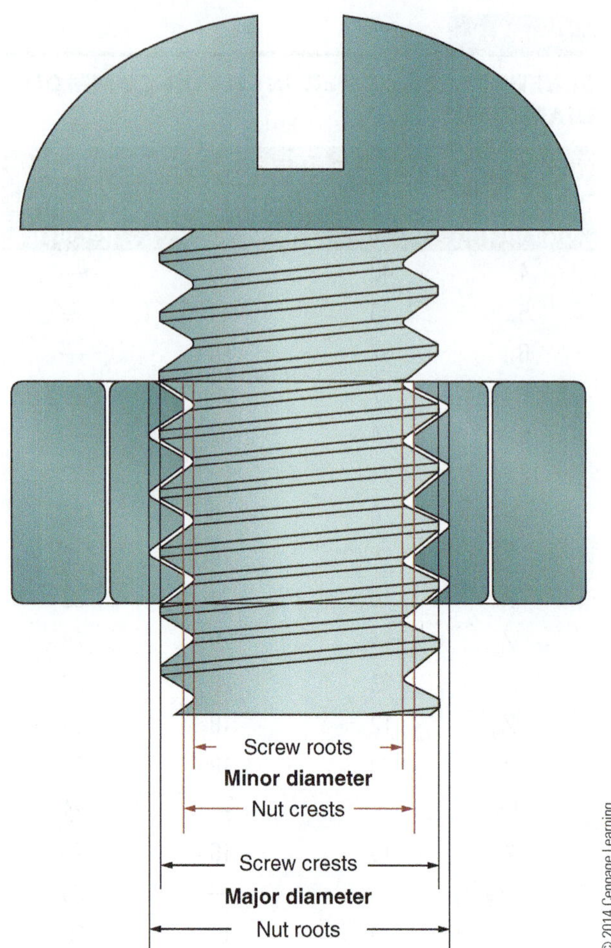

FIGURE 3-4 Major and minor diameters of a bolt and a nut.

Thread Class

Three thread classes have been established to refer to tolerance levels:

- Class 1: Loose
- Class 2: Regular
- Class 3: Interference

External threads have a class designation in one of the following three classes: 1A, 2A, or 3A internal thread tolerance (threaded hole or nut) is designated 1B, 2B, or 3B. A class 1A bolt and a class 1B nut have very relaxed tolerances, whereas a class 3A bolt and a class 3B nut have very close tolerances. Classes 2A and 2B are the most commonly used.

Fastener Specifications

Fasteners are specified by the diameter, the pitch or the number of threads per inch, the series of the thread, and the thread class in a standard sequence of numbers and letters. For example, a ¼ in. screw with 20 coarse threads per inch and extremely close tolerance would be specified as

¼-20UNC-3A

There are special occasions when a left-hand thread is desired. In this case, an LH designates this in the sequence:

¼-20UNC-3A-LH

If the LH (left hand) is not present, then it is a right-hand thread.

Metric Threads

The United States is one of the few countries that uses the English system of measurement (based on the inch); most other countries use the metric system (based on the meter). Once global trading started to be a major part of the U.S. economy, this situation caused problems because the imported machines had metric fasteners. In an effort to establish standards, the American National Standards Institute (ANSI) adopted a screw thread with an International Standardization Organization (ISO) profile (ISO 68 metric thread). The adopted profile is a 60° angle at the roots and crests of the screw thread and is designated as the M-profile (Metric Profile). This profile, unlike the Unified Standard, uses no reference to pitch classification (i.e., UNC, UNF, UNEF, and UN). Whereas the Unified Standard designates the thread pitch in terms of the number of threads in one inch of thread length, the M-profile standard states the specific pitch measurement, in millimeters, from the centerline of one thread to the centerline of the next thread.

The M-profile designates thread tolerance with numbers and letters, just as the Unified Standard does, but the M-profile uses two sets of numbers and letters, whereas the Unified Standard uses only one. The first set represents the **pitch diameter**. The pitch diameter is the diameter of an imaginary cylinder, the surface of which would pass through the threads at such points as to make equal the width of the threads and the width of the spaces cut by the surface of the cylinder. See **Figure 3-5**. The second set represents the tolerance of the crest diameter. The M-profile standard uses a lowercase *g* to represent external threads and a capital *H* to represent internal threads. The *g* tolerance range is 4, 5, 6, 7, 8, and 9; numbers 4, 5, 6, 7, and 8 are used for the *H* tolerance. For example, a metric bolt with an 8 mm diameter, a 1.25 mm pitch, and average tolerance would be specified as

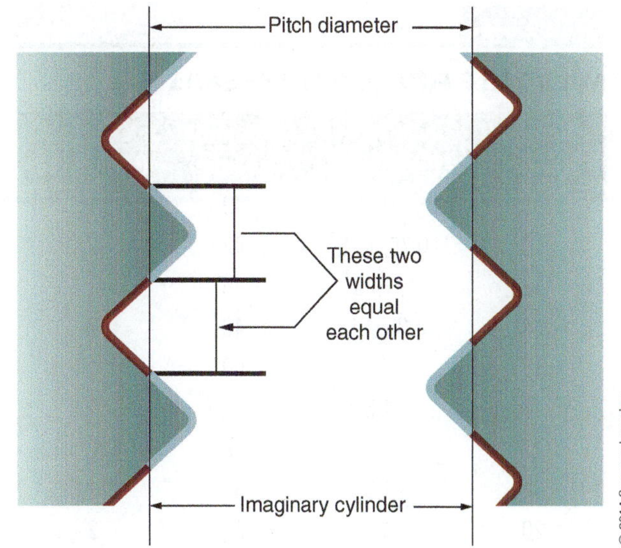

FIGURE 3-5 Pitch diameter.

M8 × 1.25 − 6 g

The M designates metric threads; the 8 represents the nominal diameter in millimeters; the × is used for separation; the pitch is 1.25 mm; the dash is used for separation; and this bolt is of average tolerance. Much like the Unified Standard, the M-profile standard has coarse-pitch (**Table 3-2**) and fine-pitch (**Table 3-3**) categories.

Table 3-2

M-PROFILE COARSE 1-1 PITCH THREADS

Nominal Size (mm)	Pitch (mm)	Nominal Size (mm)	Pitch (mm)
1.6	0.35	20	2.5
2	0.4	24	3
2.5	0.45	30	3.5
3	0.5	36	4
3.5	0.6	42	4.5
4	0.7	48	5
5	0.8	56	5.5
6	1	64	6
8	1.25	72	6
10	1.5	80	6
12	1.75	90	6
14	2	100	6
16	2		

Table 3-3

M-PROFILE FINE-PITCH THREADS

Nominal Size (mm)	Pitch (mm)	Nominal Size (mm)	Pitch (mm)
8	1	40	1.5
10	0.75, 1.25	42	2
12	1, 1.25	45	1.5
14	1.25, 1.5	48	2
15	1	50	1.5
16	1.5	55	1.5
17	1	56	2
18	1.5	60	1.5
20	1	64	2
22	1.5	65	1.5
24	2	70	1.5
25	1.5	72	2
27	2	75	1.5
30	1.5, 2	80	1.5, 2
33	2	85	2
35	1.5	90	2
36	2	95	2
39	2	100	2

3-2 TAPS AND DIES

Taps are used to cut internal threads. Dies are used to cut external threads.

Taps

Screw thread tapping often ends in broken taps and huge frustrations, but these setbacks can be avoided by learning the proper method of tapping. It takes a few times for the necessary skills to be developed, so don't be disappointed if it doesn't work perfectly the first time or two. With practice, you will develop that "feel" that all of the more experienced maintenance personnel tell you about.

Like any other job, before you get started, you must fully understand not only the task at hand but also the tools that you will use. Let's discuss the tap. Because of special applications, there are many types of taps, but here we discuss only the most commonly used. Once the proper tapping procedures have been learned, they can be used with any tap.

Tap Construction

The **tap** has a very simple construction. It is a hardened shank that commonly has four flat sides at the top and a threaded body. The **flats** are used to hold and drive the tap with a tap handle, as shown in **Figure 3-6**. If a tap handle is not available, an adjustable wrench can be used to hold and drive the tap. The threaded body cuts new threads into whatever is being threaded. It consists of **lands**, which do the actual cutting and **flutes** along the axis. Refer to **Figure 3-7**. Flutes allow for the removal of metal chips and permit the cutting fluid or coolant to reach the cutting edges. The most commonly used taps have four flutes.

Tap Selection

The first thing that must be done is to determine what size tap is needed. For the ¼-20UNC-3A fastener used in our earlier discussion, a ¼-20UNC-3A tap would be needed.

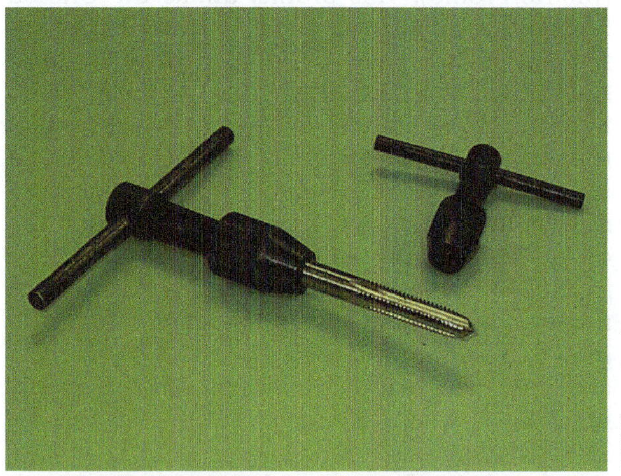

FIGURE 3-6 T-handle tap wrenches. Tap handles tighten against the flats on the tap.

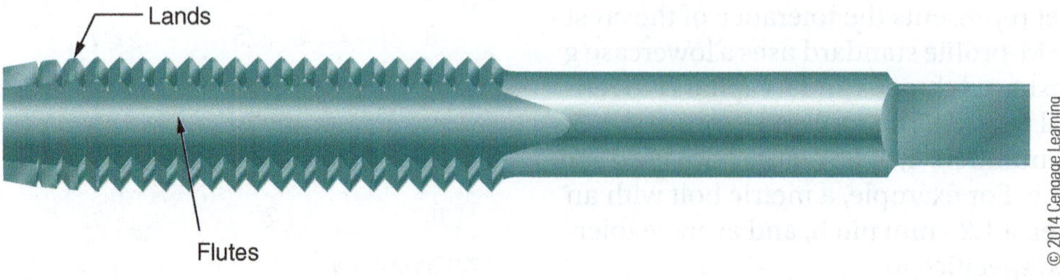

FIGURE 3-7 Lands and flutes on a tap.

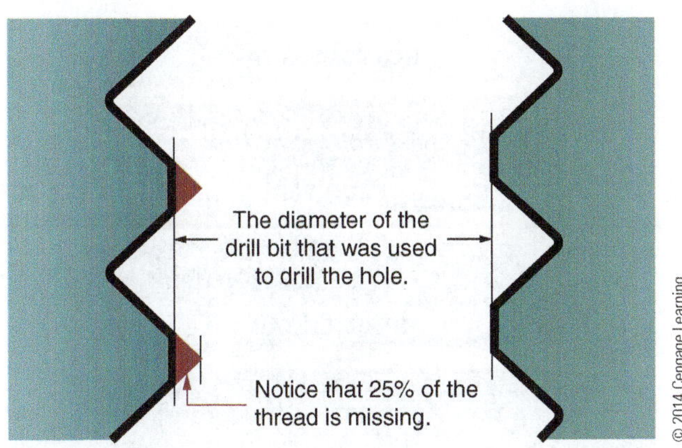
The diameter of the drill bit that was used to drill the hole.
Notice that 25% of the thread is missing.

FIGURE 3-8 Relation of drill bit and thread.

Before the tap can be used, a hole must be drilled, so the next thing to be determined is the size of the drill bit. The drill bit diameter has to be larger than the minor diameter of the thread to be tapped. It is generally accepted that the hole should allow approximately 75% of the full thread to be used. In other words, the bit will be oversized enough so that 25% of the crest of the internal thread will be missing, as shown in **Figure 3-8**. Most maintenance shops have a *Drill and Tap Chart*, which will already have the 75% of the full thread taken into consideration. The chart shown in **Table 3-4** indicates that a No. 7 drill bit is needed for a ¼ in., 20-thread fastener. If the thread is metric, a metric bit must be used to drill the proper size hole. A metric *Drill and Tap Chart*, like the one shown in **Table 3-5**, should be available.

Once the hole has been drilled, the type of tap must be selected. There are three types of tap, as shown in **Figure 3-9**. The first is the taper tap, sometimes referred to as the starter tap. This tap has a taper or angular reduction in diameter, at the leading edge to its point. This taper allows the tap to start cutting threads at a gradual rate over approximately 8 to 10 threads instead of trying to achieve the maximum cut immediately, which might cause the tap to break. This tap is used in through holes and blind holes. Through holes are holes that will penetrate all the way through the material that is being tapped. Blind holes are holes that do not pass all the way through the material but instead stop at a desired depth.

The next tap is called the plug tap. This tap has a tap over three or four threads. Although in some instances threads can be started with this tap, it is usually used after the starter tap. The plug tap can be used on through holes or blind holes.

The final type of tap is the bottom tap, or bottoming tap. This tap is used to install threads all the way to the bottom of a blind hole or to polish a through hole. The bottom tap has no taper or chamfer at the point. For this reason, it would be very difficult to start a thread with this tap.

Table 3-4

DRILL AND TAP CHART FOR COARSE THREAD AND FINE THREAD FASTENERS (STANDARD)

Thread	Drill	Thread	Drill
0-80	3/64	7/16-14	U
1-64	No.53	7/16-20	25/64
1-72	No.53	½-12	27/64
2-56	No.50	½-13	27/64
2-64	No.50	½-20	29/64
3-48	No.47	9/16-12	31/64
3-56	No.45	9/16-18	33/64
4-40	No.43	5/8-11	17/32
4-48	No.42	5/8-18	37/64
5-40	No.38	¾-10	21/32
5-44	No.37	¾-16	11/16
6-32	No.36	7/8-9	49/64
6-40	No.33	7/8-14	13/64
8-32	No.29	1-8	7/8
8-36	No.29	1-12	59/64
10-24	No.25	1-14	59/64
10-32	No.21	1 1/8-7	63/64
12-24	No.16	1 1/8-12	1 3/64
12-28	No.14	1¼-7	1 7/64
¼-20	No.7	1¼-12	1 11/64
¼-28	No.3	1 3/8-6	1 7/32
5/16-18	F	1 3/8-12	1 19/64
5/16-24	I	1½-6	1 11/32
3/8-16	5/16	1½-12	1 27/64
3/8-24	Q	1¾-5	1 9/64

Tapping Procedures

Before you begin drilling, keep safety in mind by wearing the proper eye protection. When drilling the hole, pay particular attention to the bit as it enters the material. The bit must stay at a 90° angle from the material being drilled.

SAFETY

To thread the through hole drilled with the No. 7 drill bit for a ¼ in., 20-thread fastener, first

Table 3-5

DRILL AND TAP CHART (METRIC)

Nominal Size (mm)	Pitch (mm)	Tap Drill Size (mm)	Inch Decimal*
1.6	0.35	1.25	0.05
2	0.40	1.60	0.063
2.5	0.45	2.05	0.081
3	0.50	2.50	0.099
4	0.70	3.30	0.131
5	0.80	4. 20	0.166
6	1.00	5.00	0.198
8	1.00	7.00	0.277
8	1.25	6.80	0.267
10	0.75	9.30	0.365
10	1.25	8.80	0.346
10	1.50	8.50	0.336
12	1.00	11.00	0.434
12	1.25	10.80	0.425
12	1.75	10.30	0.405
14	1.25	12.80	0.503
14	1.50	12.50	0.494
14	2.00	12.00	0.474
16	1.50	14.50	0.572
16	2.00	14.00	0.553
18	1.50	16.50	0.651
20	1.00	19.00	0.749
20	1.50	18.50	0.73
20	2.50	17.50	0.692
22	1.50	20.50	0.809
24	2.00	22.00	0.868
24	3.00	21.00	0.83
25	1.50	23.50	0.927

* Millimeters × 0.03937 = inch decimals.

CAUTION

insert the starter tap into the hole. Then carefully and slowly turn the tapping tool clockwise, providing cutting oil as needed. The cutting oil does two things for the tap; it cools it and lubricates the cutting action. The tap will be hard to control until a few threads have been cut into the material, but close attention must be paid to ensure that the tap is entering the hole at a 90° angle to the material that is being tapped. After the cutting has started, usually at one-half to three-quarters of a turn, slowly turn the tap counterclockwise to remove any burrs and chips from the material. The chips should fall out through the flutes on the tap. (For a blind hole, it may be necessary to completely remove the tap and manually clean the hole.) Resume turning the tap clockwise. You should feel where the tap quit cutting. When you reach that point, continue in the clockwise direction, as you add cutting oil, for approximately one revolution. Be careful as you do this. If the tap starts to feel springy or spongy, then back the tap out, cleaning the burrs and the chips in the process. If you fail to notice the springy feeling, you may break the tap. Continue the cutting, and clean the hole out about once every revolution.

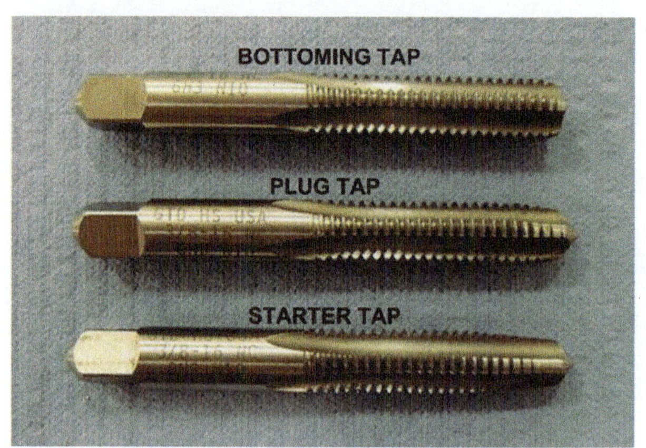

FIGURE 3-9 Three types of tap.

Once the hole has been threaded with the starter tap, polish it with the plug tap and then the bottom tap. Insert the fastener. In the event that a hole is stripped of its threads, redrill and retap the hole for the next fastener size up, and then insert a helicoil (**Figure 3-10**).

FIGURE 3-10 Helicoil inserts.

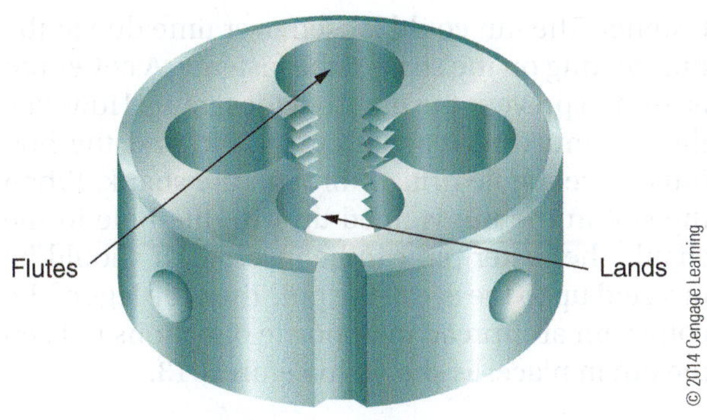

FIGURE 3-11 Lands and flutes on a die.

Dies

External threading is accomplished using threading dies. A **threading die** has the same construction as a tap, including lands and flutes. See **Figure 3-11**. Instead of penetrating the material to be threaded, however, the die surrounds it. External threading with hand dies is similar to tapping in that care must be taken to start the die square. The same "feel" is involved, and as far as the procedure, it is identical. Be careful if you are using a power threader, because you lose the ability to feel the cutting action. If a problem is not noticed or the proper amount of lubrication is not applied, the cutting edges of the die can be damaged, making them useless and dangerous.

3-3 FASTENER TYPES

Because specialty fasteners are being created daily to meet the specific needs of industry, there are many different types of fasteners. However, the bolt still remains the most common fastener. In this section, we discuss some of the different types of fasteners that are commonly used, ranging from the common bolt to specialty fasteners such as pin fasteners and retaining ring fasteners.

Threaded Fasteners

There are many types of threaded fasteners, including the following:

- Bolts
- Machine screws
- Threaded studs
- Double-ended studs

Each type has threads, but each has a different use.

Bolts and Screws

The bolt is usually used in conjunction with a retaining or mating nut. A bolt is tightened and released by the turning of this mating nut. Machine screws are usually threaded into a threaded hole. A screw is tightened and released by the turning of the fastener itself. This is accomplished by turning the head of the fastener. However, because bolts are sometimes used like screws and screws are sometimes used in conjunction with a nut, bolts and screws are not usually looked upon as being different fasteners.

Threaded Studs

Threaded studs are easy to distinguish from bolts and screws. The threaded stud does not have a head. For this reason, the threaded stud relies upon the use of an internally threaded nut, referred to as a mating nut earlier, to produce the tightening action. On a bolt or screw, the head of the fastener provides a surface to bind the load against. Because the stud has no head, it will rely upon the surface of the nut to provide its tightening torque. Threaded studs are manufactured in lengths from 2 to 6 in., but they can be cut to any desired length from a continuously threaded rod that is often referred to as "all-thread." All-thread is usually available in lengths from 24 in. to 96 in. A hacksaw is used to cut a stud from this continuously threaded rod. It is important to redress the end of the stud as it is cut, to allow the retaining nut to advance onto the stud with ease. Failure to redress the end of the stud could result in cross-threading the nut onto the stud, causing damage to the threads.

Double-Ended Studs

A double-ended stud, which can be seen in **Figure 3-12**, is similar to the continuously threaded stud but has a shoulder between two threaded sections. When the threads of a continuously threaded stud are cut, the cutting die starts on one end and runs along the entire length of the fastener. On a double-ended stud, the cutting die is started on one end and cuts only about a third to a half the length of the stud. It is then removed from that end and started again on the opposite end. The threads are again cut to approximately a third to a half the length of the stud. Cutting threads in this manner creates the shoulder.

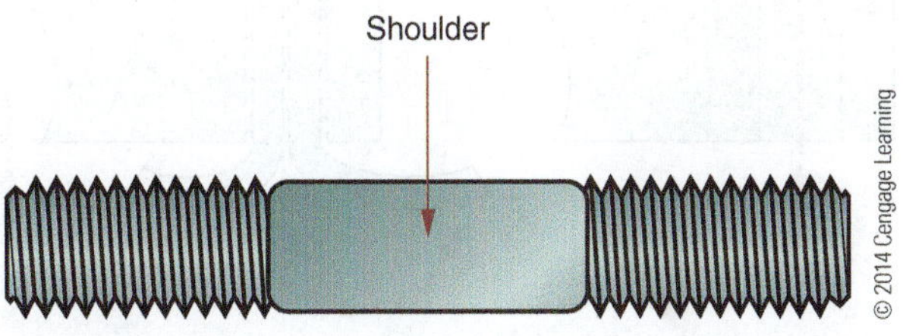

FIGURE 3-12 Double-ended stud.

Pin Fasteners

The most common pin fasteners are listed here:

- Dowel pins
- Taper pins
- Split pins
- Clevis pins
- Cotter pins

The dowell pin is usually used to guide something or to line it up. It can be made of wood or metal. The taper pin is usually used to hold something in place by wedging into the parts that are to be held. As the name implies, the taper pin is tapered. Split pins are hollow pins that have a split along their entire length. The outer diameter is usually only a few thousandths of an inch larger than the hole in which the pin will be used. As the split pin is driven into a hole, it squeezes together at the split. The resulting tension is what keeps the split pin in place. These are also referred to as spring lock pins.

The clevis pin is usually used in a hinge, for example, to attach a small tow-behind trailer to a riding lawn mower. The clevis pin is held in place by a slip pin. Split pins usually have a loop to allow them to be removed with a finger.

Cotter pins (**Figure 3-13**) are most commonly used to hold a nut in place in relation to the fastener. Imagine a mating nut threaded onto the shank of a fastener. The nut could loosen over time due to the nut rotating on the shaft of the fastener. A cotter pin is used to prevent the nut from loosening. However, the nut must have a slot cut into it, and the bolt must have a hole drilled through the shank. When the slot in the nut is lined up with the hole in the shank, the cotter pin can be inserted. It should be inserted up to the eye of the pin. Then the legs of the cotter pin are spread in opposite directions to keep the pin in place, as shown in Figure 3-13.

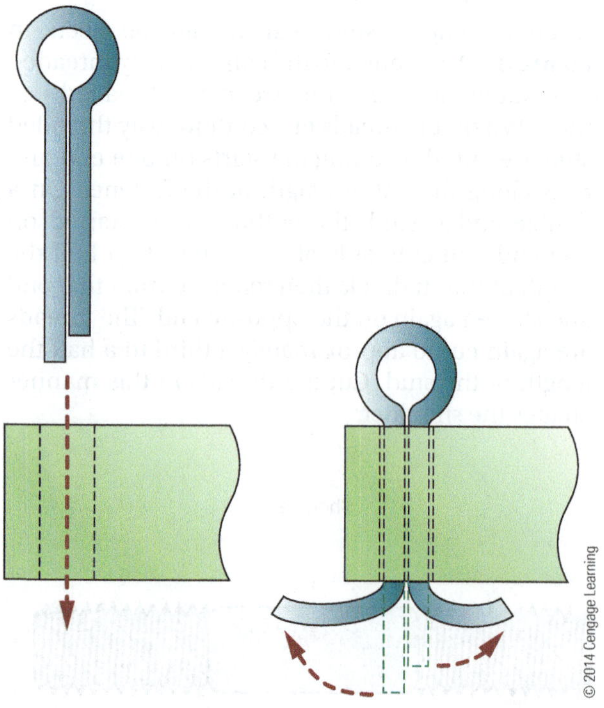

FIGURE 3-13 Cotter pin.

3-4 FASTENER GRADES

It is important to know the strength of the fastener that is being used. A fastener's strength is identified by **grade markings** on the head of the fastener. The American Society for Testing and Materials (ASTM) and the Society of Automotive Engineers (SAE) have developed these head markings for easy identifying the strength of a fastener. The material from which a fastener is manufactured is one of the factors that determines the strength of the fastener. Fasteners are manufactured from different types of steel or from alloys. It is important to use the appropriate grade fastener. Using a fastener with a low grade in a high-torque condition causes the fastener to break or shear as the metal becomes fatigued. The standard ASTM and SAE grade markings are shown in **Figure 3-14**.

Notice that the highest SAE grade in the figure is Grade 8. There are fasteners that are over Grade 8, but they are primarily used on aircraft. If a fastener fails on an aircraft, the plane could crash, so fasteners with a very high tensile strength must be used on all aircraft. These types of fasteners are usually classified as National Aircraft Standard (NAS) or Military Standard (MS) fasteners. The A 354 Grade BD and A 490 fasteners in Figure 3-14 qualify as aircraft quality fasteners. As can be seen in **Figure 3-15**, these two fasteners are made of steel alloy instead of carbon steel. Steel alloy is stronger than carbon steel. The lower grades of fasteners are made of carbon steel.

Also notice in Figure 3-15 that some of the fasteners are heat tempered and quenched. Heating the metal to a high temperature and then quickly cooling the metal with water or oil is called quenching. Quenching the heated metal causes the metal to harden. This action of hardening the metal is called tempering. The tempered hardness is dependent upon the temperature of the metal when it was cooled and how slowly or quickly it was cooled. To slowly cool heated metal is to anneal the metal. This would be in direct contrast to quenching the metal in which the metal would be cooled very quickly. The material and the tempering process used are

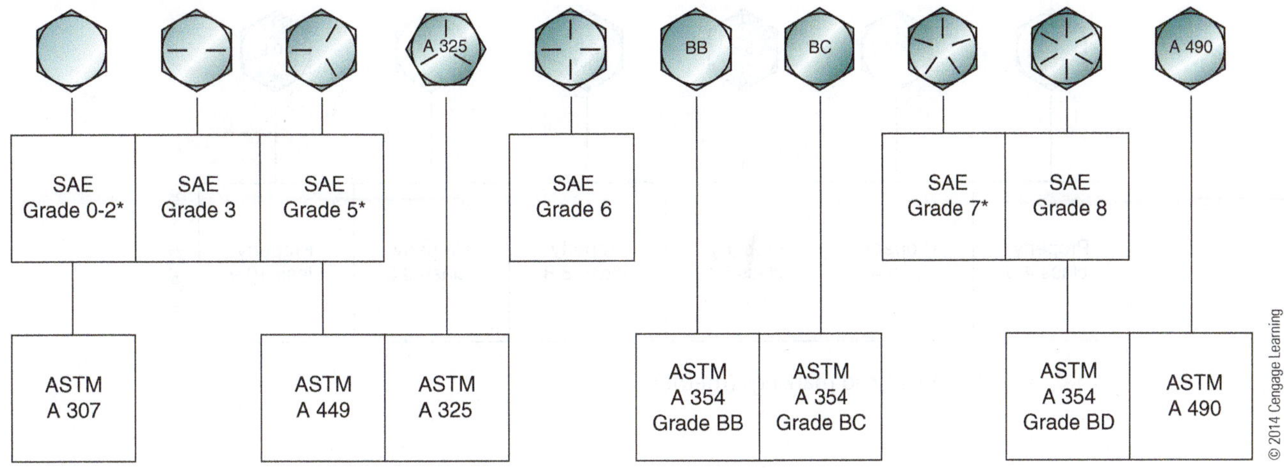

FIGURE 3-14 Fastener grade head markings (standard).

Head marking	Grade	Material
(no marks)	SAE Grade 0	Mild steel
	SAE Grade 1	Low-carbon steel
	ASTM A 307	Low-carbon steel
	SAE Grade 2	Low-carbon steel (better grade steel)
(2 marks)		Medium-carbon steel, cold worked
(3 marks)		Medium-carbon steel, tempered by quenching
A 325		Medium-carbon steel (better grade steel), tempered by quenching
(4 marks)		Medium-carbon steel, tempered by quenching (tempered to a different strength)
BB		Low-alloy steel, tempered by quenching
BC		Low-alloy steel, tempered by quenching (tempered to a different strength)
(5 marks)		Medium-carbon steel, tempered by quenching (threads are cut after tempering)
(6 marks)	SAE Grade 8	Medium-carbon steel alloy
	ASTM A 354 Grade BD	Steel alloy, tempered by quenching
A 490		Steel alloy, tempered by quenching (tempered to a different strength)

FIGURE 3-15 Fastener grade materials (standard).

the two factors that determine the tensile strength of a fastener. The higher the tensile strength, the larger the amount of torque or stress that can be exerted on the fastener.

Figure 3-16 shows the head markings for metric fasteners.

As with the ASTM and SAE standard bolts, metric bolts are also made with different materials and are tempered differently. These are listed in **Figure 3-17**.

3-5 TORQUE SPECIFICATIONS

It is imperative that a nut be of the same grade as the bolt on which it is used. If they are not equal in grade, the tension produced as the fastener tightens may cause the weaker component to fail. It is likened to a chain with a weak link. Each fastener has a torque specification for safe operation. Exceeding this limit causes the fastener to fail.

It is important to tighten a fastener to its specification for several reasons:

- If there is not enough torque, the fastener may loosen owing to the lack of initial tension.
- If too much torque is applied, the fastener shank may fracture or shear.
- If too much torque is applied, the threads of the fastener or nut may become stripped.

Each fastener, due to its construction and method of tempering, has a certain torque specification. A fastener must have a higher initial tension than that of the greatest external load applied to it. If it does not, the fastener becomes fatigued and eventually fails. **Initial tension** is force that is developed by the tightening of the fastener. This is separate from the force that is exerted on the fastener

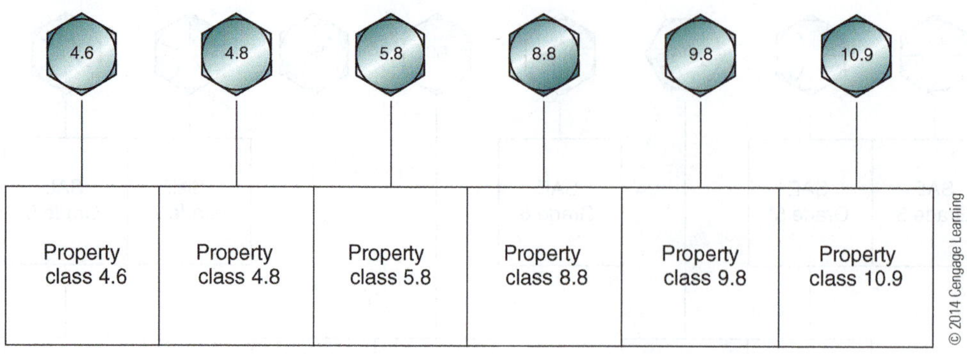

FIGURE 3-16 Fastener head markings (metric).

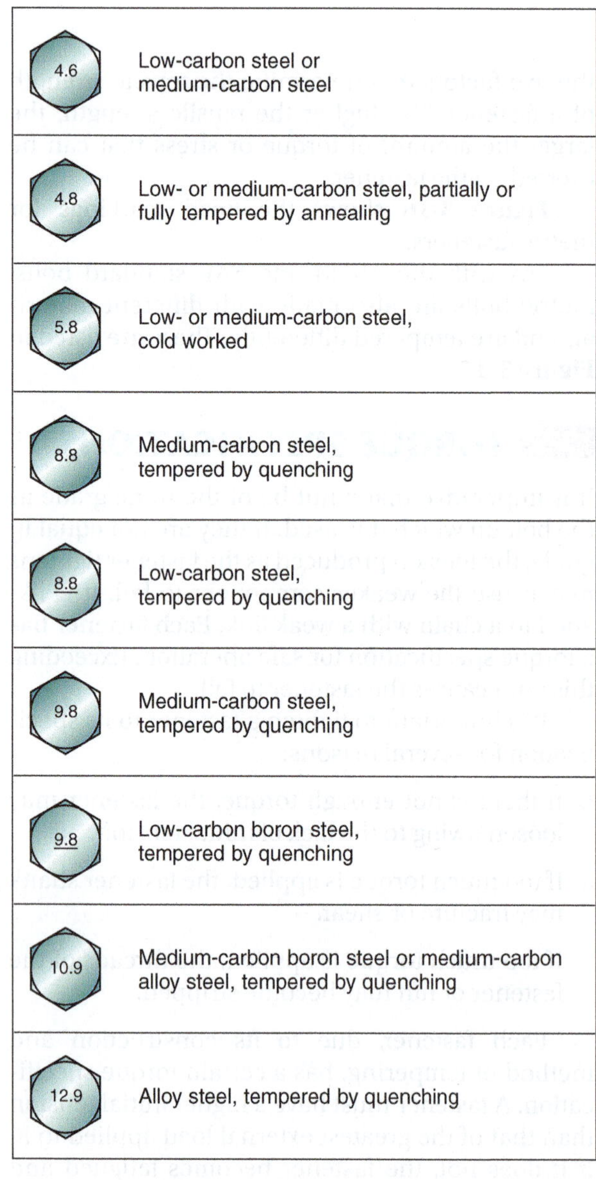

4.6
Low-carbon steel or medium-carbon steel
4.8
Low- or medium-carbon steel, partially or fully tempered by annealing
5.8
Low- or medium-carbon steel, cold worked
8.8
Medium-carbon steel, tempered by quenching
8.8
Low-carbon steel, tempered by quenching
9.8
Medium-carbon steel, tempered by quenching
9.8
Low-carbon boron steel, tempered by quenching
10.9
Medium-carbon boron steel or medium-carbon alloy steel, tempered by quenching
12.9
Alloy steel, tempered by quenching
© 2014 Cengage Learning

FIGURE 3-17 Fastener grades (metric).

from the load. It is important to exert a good clamping force, known as **preload**, on the parts that are to be fastened together. Preloading prevents the parts from shifting or moving. Shifting of the parts causes the fastener to wear out or fail. It is easy to recognize a fastener that was not tightened to its proper preload; the threads are worn away. Preload on bolted fasteners is a function of bolt "stretch." The most accurate preload is obtained by measuring the amount of bolt stretch. Torquing a bolt gives an approximation of the bolt stretch and may be the only method available where both ends of the bolt are not accessible. Bolted casings of large pumps and turbines with bolts larger than about three inches use bolt heaters to lengthen the bolt prior to installation. Once the bolt is installed, a nut is then turned down to a specified number of turns, and the bolt is then allowed to cool, stretching the bolt to the required amount.

Standard Torque Specifications

A torque wrench should be used to tighten a fastener. The amount of torque that is being applied should be known at all times during tightening. Each bolt manufacturer has a recommended torque specification for each fastener it makes. This information should be used whenever possible. If the manufacturer's specification is not known, then it is usually acceptable to use a general table that gives approximate required torque values for given types of fastener. See **Table 3-6**. The values in this table do not take the place of any manufacturer's specifications and should not be used for critical applications. It is also important to keep in mind that these values are for dry fasteners only. Oil on the fastener will change the required torque.

Notice that in Table 3-6 all of the torque values for fine thread fasteners are higher than those for

Table 3-6

TORQUE SPECIFICATIONS (STANDARD)

Nominal Size (in.) or Basic Major Diameter of Thread	Coarse or Fine Thread	SAE Grade of Fastener		
		0-2	5	8
		Torque (in.-lb)		
No. 4-40	C	5	8	12
No. 4-48	F	6	9	13
No. 6-32	C	10	16	22
No. 8-32	C	19	30	41
No. 10-24	C	27	43	60
No. 10-32	F	31	49	68
		Torque (ft-lb)		
¼-20	C	5.5 (66 in.-lbs)	8	12
¼-28	F	6.3 (76 in.-lbs)	10	14
5/16-18	C	11	17	25
5/16-24	F	12	19	27
3/8-18	C	20	30	44
3/8-24	F	23	35	49
7/16-14	C	31	50	70
7/16-20	F	35	55	78
½-13	C	50	75	106
½-20	F	55	85	120
9/16-12	C	70	109	150
9/16-18	F	80	120	171
5/8-11	C	97	150	212
5/8-18	F	110	170	240
¾-10	C	173	266	376
¾-16	F	192	297	420
7/8-9	C	206	430	600
7/8-14	F	225	470	666
1-8	C	250	644	909
1-12	F	274	705	995
1 1/8-7	C	354	794	1288
1 1/8-12	F	397	890	1444
1 ¼-7	C	500	1120	1817
1 ¼-12	F	553	1241	2012
1 3/8-6	C	655	1470	2382
1 3/8-12	F	746	1672	2712
1 ½-6	C	869	1949	3161
1 ½-12	F	978	2193	3557
1 ¾-5	C	1371	3074	4986
2-4.5	C	2061	4621	7497
2 ¼-4.5	C	3014	6759	10,960
2 ½-4	C	4213	9247	14,995
2 ¾-4	C	5597	12,550	20,352
3-4	C	7384	16,559	26,853

Table 3-7

TORQUE DEDUCTIONS

Condition of Fastener	Percentage of Torque to Derate*	
	Coarse Thread	Fine Thread
No lubrication	0% of recommended torque	0% of recommended torque
Plated and cleaned	34% of recommended torque	26% of recommended torque
SAE 20 oil	38% of recommended torque	28% of recommended torque
SAE 40 oil	41% of recommended torque	31% of recommended torque
Plated and SAE 30 oil	45% of recommended torque	35% of recommended torque
White grease	45% of recommended torque	35% of recommended torque
Graphite and oil	55% of recommended torque	49% of recommended torque

*Subtract the applicable percentage from the recommended torque values found in Tables 3-6 and 3-8.

coarse thread fasteners. Because fine thread fasteners have more threads per inch, they have more surface area contact on the threads, therefore requiring more torque to be applied to produce the same fastener tension.

Many factors determine the torque value for a given fastener: the type of material that is used to make the fastener, the tempering of the fastener, the plating of a fastener, and whether the fastener is dry or lubricated. For this reason, certain percentages must be deducted from the dry torque specification. **Table 3-7** shows the deductions that should be taken for some of these factors.

Metric Torque Specifications

The same principles of torque apply to metric fasteners, but a different chart is used for metric fasteners, because torque is measured in newton-meters (N•m). See **Table 3-8**.

If the only torque wrench available reads in foot-pounds, convert from foot-pounds to N•m. One foot-pound equals 1.3558179 N•m, and one N•m equals 0.73756 foot-pound.

3-6 RETAINING RING FASTENERS

A wide variety of retaining ring fasteners are available. Only the most commonly used are discussed in this section. These are the internal and external snap rings, the crescent clip, and the E-clip.

An internal snap ring is shown on the left in **Figure 3-18**. An external snap ring is shown on the right. An external snap ring keeps something from sliding off a shaft, and an internal snap ring keeps something secure within a cylinder. Both of these fasteners come in heavy-duty sizes. A pair of snap

FIGURE 3-18 Internal snap ring (left) and external snap ring (right).

ring pliers is needed to install or remove a snap ring. Most snap ring pliers can be converted from an external set to an internal set. Snap ring pliers usually come with extra tips, which may be needed for larger snap rings.

Crescent clips have a crescent shape, hence the name. The E-clip resembles the capital letter E. An E-clip is shown in **Figure 3-19**. Both types are used in much the same way as the external snap ring, but they can be pressed onto a shaft from the side, whereas the snap ring must be installed by spreading the snap ring with the pliers and sliding it down the shaft to its locking groove. The crescent clip and the E-clip can be removed by simply pressing on both of the ends of the clip just enough to get the inside of the clip off the shaft. Once the clip is off the shaft, gently insert the tip of a standard screwdriver between the shaft and the clip, and twist the screwdriver. It is best to place your hand over the clip so that you will not lose it if it flies off the shaft. **Figure 3-20** shows a variety of E-clips.

Table 3-8

TORQUE SPECIFICATIONS (METRIC)

Thread Diameter	Coarse Thread									
	4.6		5.6		8.8		10.9		12.9	
	N•m	Ft-lb	N•m	Ft-lb	N•m	Ft-lb	N•m	Ft-lb	N•m	Ft-lb
M4	1.02	0.752	1.37	1.01	3	2.21	4.4	3.25	5	3.69
M5	2	1.47	2.70	1.99	5.9	4.4	8.7	6.42	10	7.38
M6	3.5	2.58	4.60	3.39	10	7.38	15	11	18	13
M8	8.4	6.2	11	8.11	25	18	36	27	43	32
M10	17	13	22	16	49	36	72	53	84	62
M12	29	21	39	29	85	63	125	92	145	107
M14	46	34	62	46	135	100	200	148	235	173
M16	71	52	95	70	210	155	310	229	365	269
M18	97	72	130	96	300	221	430	317	500	369
M20	138	102	184	136	425	314	610	450	710	524
M22	186	137	250	184	580	428	820	605	960	708
M24	235	173	315	232	730	538	1050	774	1220	890
M27	350	258	470	347	1100	811	1550	1143	1880	1387
M30	475	350	635	468	1450	1070	2100	1549	2450	1807
M33	645	476	865	638	1970	1453	2770	2043	3330	2456
M36	830	612	1111	819	2530	1866	3560	2626	4280	3157

Thread Diameter	Fine Thread					
	8.8		10.9		12.9	
	N•m	Ft-lb	N•m	Ft-lb	N•m	Ft-lb
M8 × 1.00	22	16	30	22	36	27
M10 × 1.25	42	31	59	44	71	52
M12 × 1.25	76	56	105	77	130	96
M14 × 1.50	120	89	165	122	200	148
M16 × 1.50	180	133	250	184	300	221
M18 × 1.50	260	192	365	269	435	321
M20 × 1.50	360	266	510	376	610	450
M22 × 1.50	480	354	680	502	810	597
M24 × 2.00	610	450	860	634	1050	774

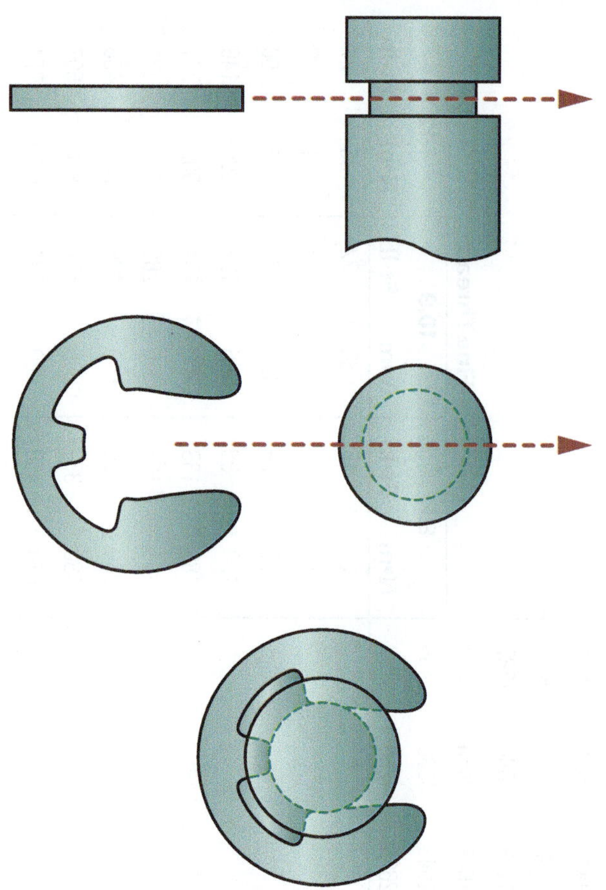

FIGURE 3-19 E-clip.

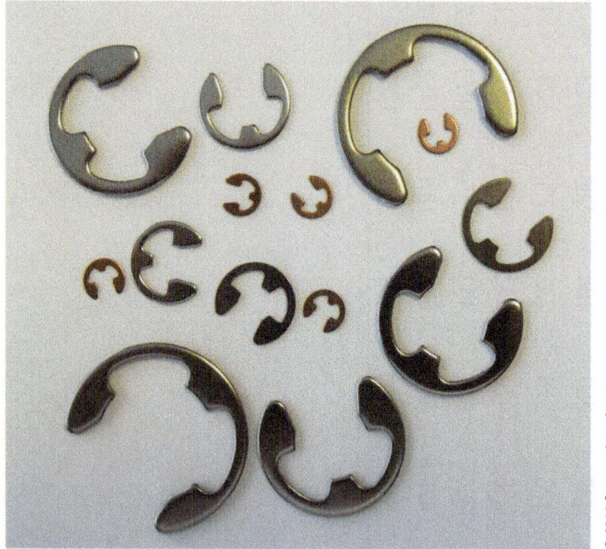

FIGURE 3-20 Various e-clips.

SUMMARY

- The Unified Standard of threads covers the cross-sectional shape of a thread, available diameters, and the threads per inch for each diameter.
- There are two types of threads, external and internal. A screw or a bolt has external threads. A nut or a part of a machine that receives a screw or a bolt has internal threads.
- The root of a thread is the lowest point of the thread; the crest is the highest point. The distance between the root and the crest is called the thread depth. The face of the thread is the flat side of the thread that rises at an angle from the root to the crest. The pitch is the distance from any point on a thread to the next corresponding point of the next thread, measured parallel along the axis. This may also be referred to as threads per inch.
- Fastener threads are categorized in four series: coarse thread (UNC), fine thread (UNF), extra fine thread (UNEF), and constant pitch thread (UN).
- The major diameter of an external thread is the outside diameter at the crests of the thread. The major diameter of an internal thread is the maximum diameter that has been cut into the center of the nut, specifically, the root of the thread. The minor diameter of an external thread is the smallest diameter at the root of the thread. The minor diameter of an internal thread is the diameter of the hole in the nut at the crests of the thread.
- Thread class has been established to refer to different tolerance levels. External threads are classified as 1A, 2A, or 3A. Internal threads are classified as 1B, 2B, or 3B. Designations of 1 represent a relaxed tolerance; those of 3 represent very close tolerance. 2A and 2B are the most commonly used tolerances.
- The designated M-profile (metric profile) has a 60° angle at the roots and crests of the screw thread. This profile uses no reference to pitch classification. Instead, it states the specific pitch measurement, in millimeters, from the centerline of one thread to the centerline of the next thread.
- The tap has a hardened shank with flat sides at the top and a threaded body, which cuts new threads into whatever is being threaded. The threads have lands, which do the actual cutting, and flutes along the axis, which allow for the removal of the metal chips and permit the cutting fluid or coolant to reach the cutting edges. The most commonly used taps have four flutes.
- A threading die has the same construction as the tap, including lands and flutes. The difference is that the cutting die cuts external threads; therefore, the die surrounds what is to be threaded.

- Keep safety in mind by wearing the proper eye protection when using a tap or die.
- A bolt is tightened and released by the turning of a mating nut. Machine screws are usually threaded into a threaded hole. A screw is tightened and released by turning the head.
- A fastener's strength is identified by grade markings on the head of the fastener. The grade markings were developed by the American Society for Testing and Materials (ASTM) and the Society of Automotive Engineers (SAE). The material from which a fastener is manufactured is one of the factors that determines the strength of the fastener.
- The hardening of metal by heating and then cooling is called tempering. The tempered hardness depends on the temperature of the metal when it was cooled and on how slowly or quickly it was cooled.
- Heating the metal to a high temperature and then cooling it quickly with water or oil is called quenching.
- Cooling a metal slowly is called annealing.
- A fastener must have a higher initial tension than that of the greatest external load that will be applied to it.
- Initial tension is force that is developed from the tightening of the fastener. This is separate from the force that is exerted on the fastener from the load.
- The amount of torque that is being placed upon the fastener should be known at all times during tightening. Each bolt manufacturer has a recommended torque specification for each fastener it makes.
- The torque values for fine thread fasteners are higher than those for coarse thread fasteners. Because there are more threads per inch on a fine thread fastener, there is more surface area on the threads, allowing more torque to be applied.

REVIEW QUESTIONS

1. What is the purpose of the Unified Standard?
2. Name the five terms used to describe a thread.
3. What are the four series of threads and their symbols?
4. Of what material is a Grade 2 SAE bolt made?
5. What is the head marking of a metric fastener that is made of low-carbon boron steel?

CHAPTER 4

Industrial Print Reading

Industrial print reading is important in the installation, troubleshooting, and repairing of different types of systems. This chapter introduces the symbols and types of drawings that are used in industry. Some of these are used in later chapters to explain the processes and systems.

OBJECTIVES

After studying this chapter, the student should be able to

- Understand and interpret dimensional drawings.
- Understand and interpret schematics.
- Use the border system to locate key parts or components in the drawing.
- Understand and interpret an exploded view drawing.
- Discuss each of the four basic drawings used to convey information in the electrical field: the single-line drawing, the pictorial diagram, the schematic diagram, and the ladder diagram.
- Understand and interpret some key elements in a welding schematic.

4-1 MECHANICAL DRAWINGS

Maintenance mechanics use many types of drawings in the troubleshooting, repairing, and installing of equipment. This chapter discusses some of these drawings and explains how to effectively use them.

Dimensional Drawings

Figure 4-1 is a simple drawing showing the dimensions of a bracket. This drawing is a common type of drawing that shows the dimensions of an object. It is referred to as a **two-dimensional drawing** and shows the bracket in two planes. To understand what this bracket should look like, refer to **Figure 4-2**, which is a **three-dimensional drawing** of the bracket in Figure 4-1. A three-dimensional drawing shows the component in three planes.

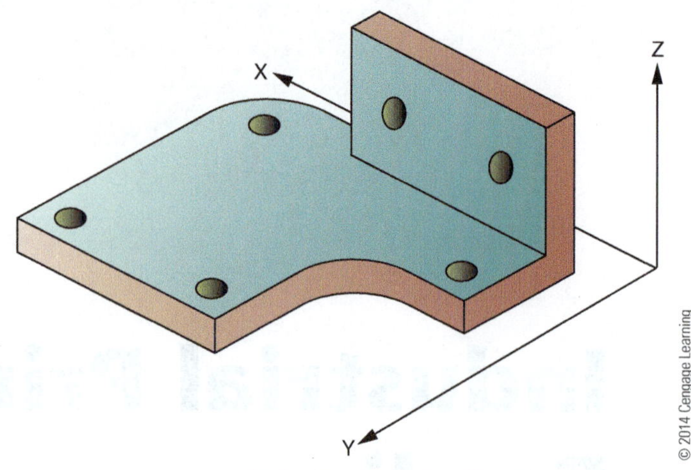

FIGURE 4-2 Three-dimensional drawing of the bracket in Figure 4-1.

Schematics

Another type of drawing is the schematic. A **schematic** shows the connection of elements in a system. **Figure 4-3** is a simplified version of what would be seen on a hydraulic schematic. Schematics tell the maintenance technician how a particular system is supposed to be connected. They also help in troubleshooting a system when a problem occurs.

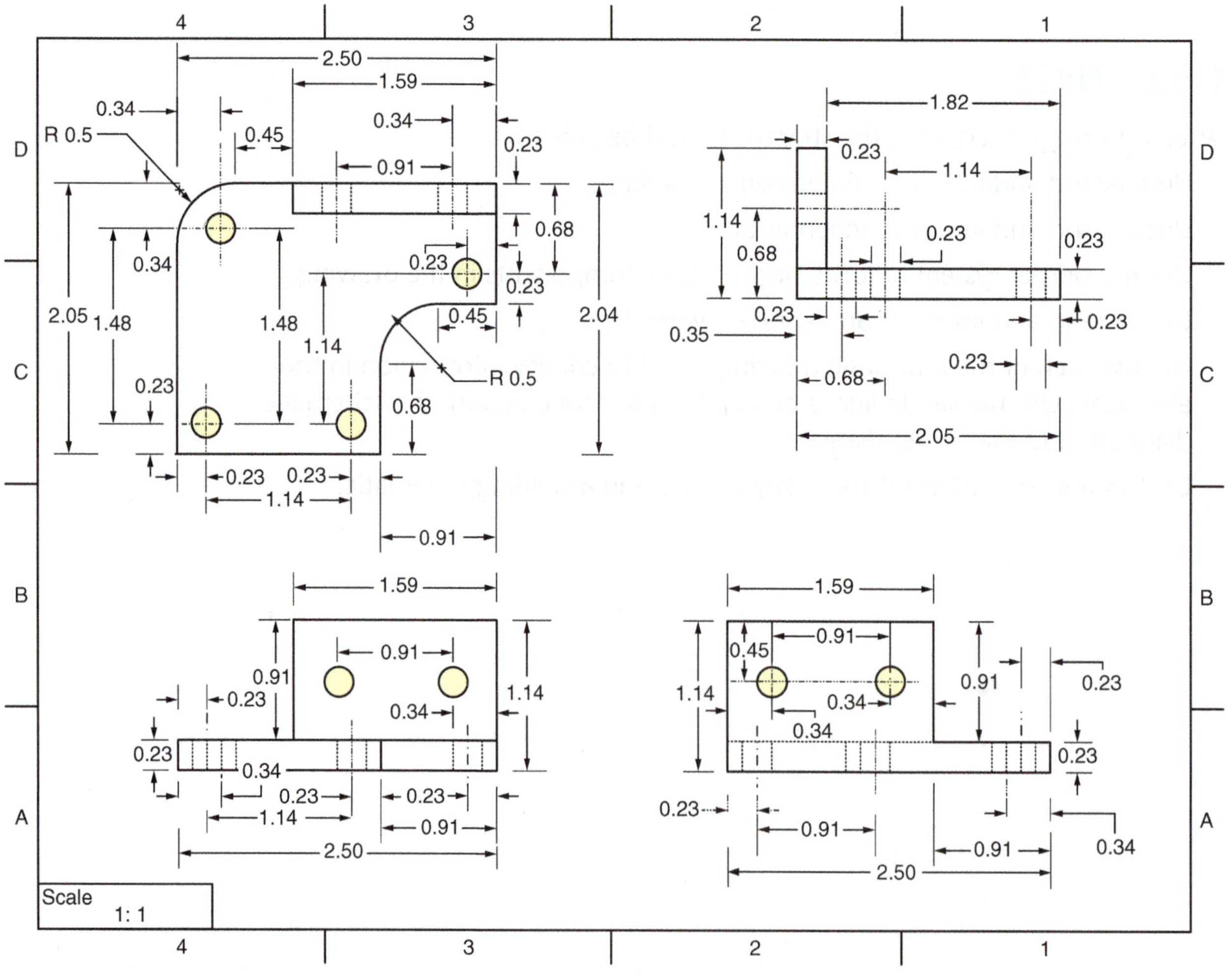

FIGURE 4-1 Dimensional drawing of a bracket.

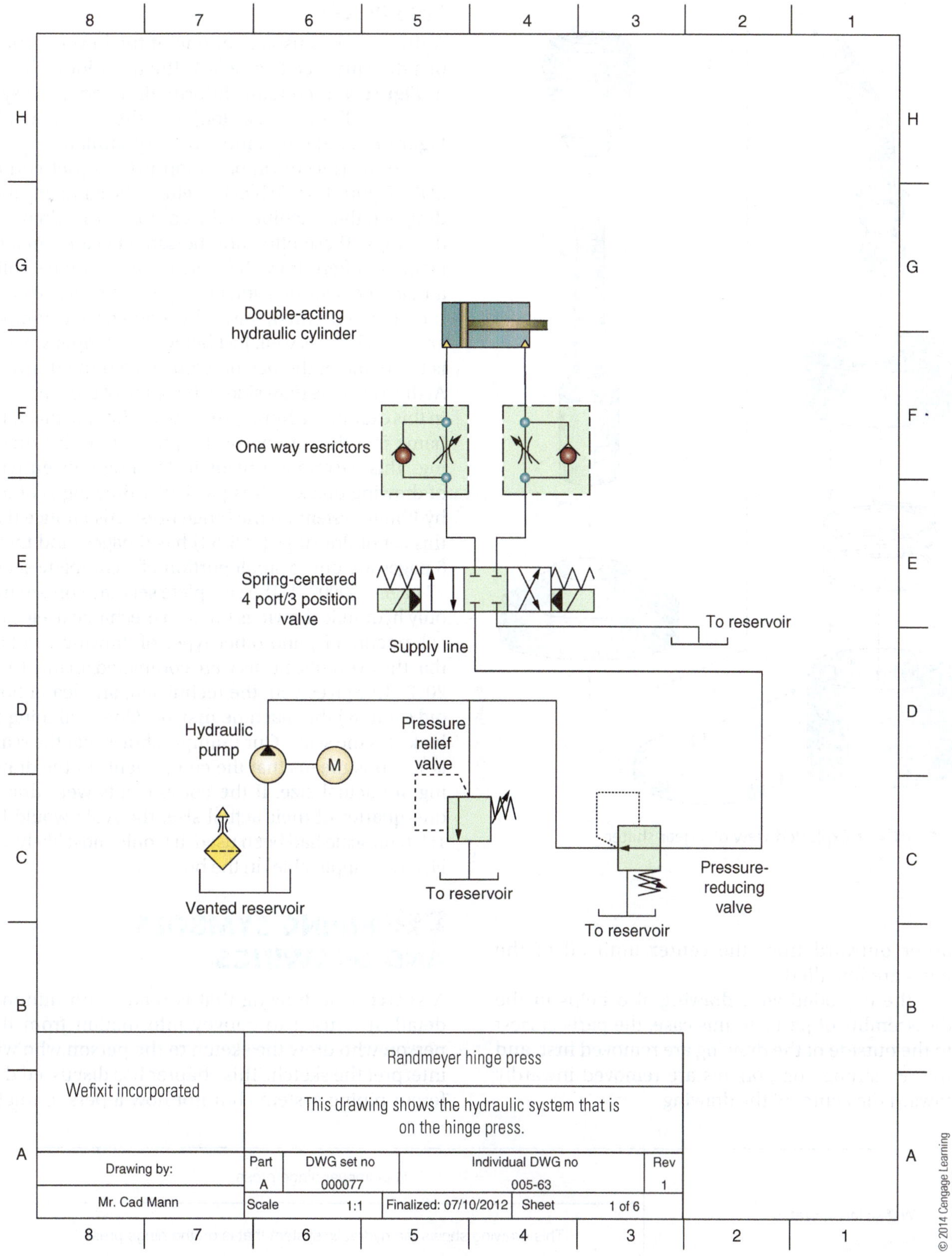

FIGURE 4-3 Hydraulic schematic.

Exploded View Drawings

Another type of drawing is the **exploded view drawing**, which shows the intended assembly of the mechanical parts. Refer to **Figure 4-4**. This type of drawing shows all the components of the assembly and how they fit together. Usually, the components closest to the center are assembled first; components are assembled in

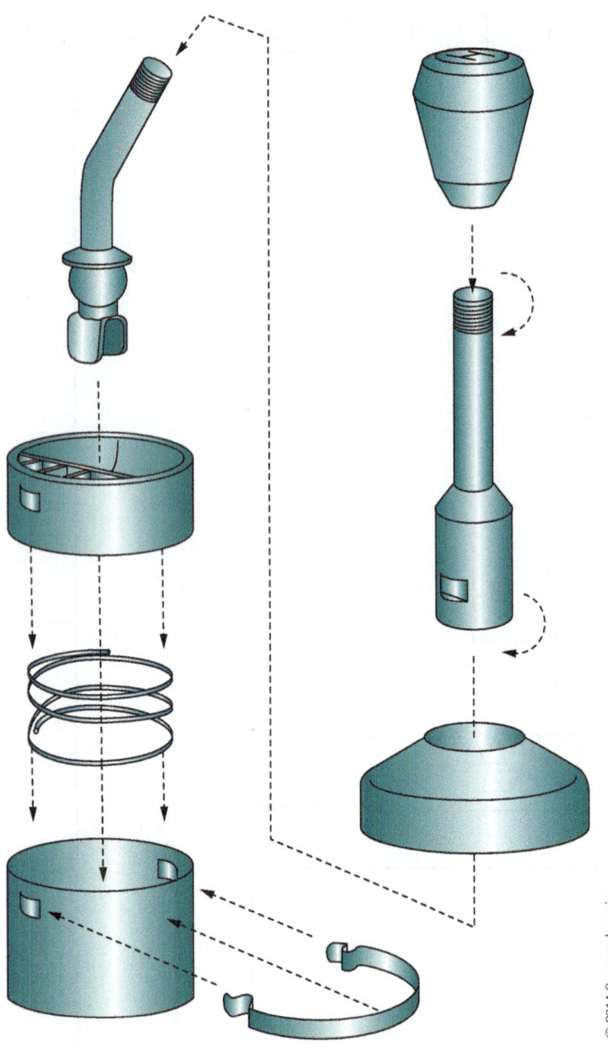

FIGURE 4-4 Exploded view of a gear shifter.

order outward from the center until all of the parts are installed.

The exploded view drawing also helps in the disassembly of parts. In this case, the parts closest to the outside of the drawing are removed first, and the remaining components are removed inwardly toward the center of the drawing.

Title Blocks

A **title block** is usually in one of the lower corners of a drawing. See Figure 4-3. The title block, shown in **Figure 4-5**, contains information about the system. Not all title blocks look like the one shown in Figure 4-5, but they should be very similar.

The first piece of information, in the top left corner of the Figure 4-5 title block, is either the company that designed the machine or the company that drew the drawings. These often are the same because manufacturers often draw their own prints as they build a machine. Our ficticious company is called Wefixit Incorporated. Often, the address and phone number are given here as well. Just below the company name is the name of the person who created the drawing. At the top of the right side is the name of the machine, in this example, a hinge press. Just below the machine name is a description of the purpose of this drawing. This description helps in locating a given type of drawing quickly. This particular drawing is of the hydraulic system on the hinge press. Also notice that this set of drawings (005-63) has 6 pages, and these 6 pages are only a small portion of a complete set of 77 pages (000077). The complete set may contain not only hydraulic drawings but also mechanical, electrical, pneumatic, and other types of drawings. Notice that the hydraulic prints were completed on July 10th, 2012. This gives you, the technician, an idea of how old (or new) the machine may be. One final thing to look at is the scale. Our example shows that the scale is 1:1. This means that the components in the drawing are actual size. If the components were drawn one-quarter of their actual size, the scale would be 1:4. If no scale had been used, it would most likely say N/A (not applicable) in the box.

4-2 PIPING SYMBOLS AND DRAWINGS

A **sketch** is a drawing that is made with minimal detail. It is used to convey information from the person who drew the sketch to the person who will interpret the sketch. This chapter has discussed different piping systems but not how a person might

<table>
<tr><td rowspan="2">Wefixit incorporated</td><td colspan="6">Randmeyer hinge press</td></tr>
<tr><td colspan="6">This drawing shows the hydraulic system that is on the hinge press.</td></tr>
<tr><td rowspan="2">Drawing by:</td><td>Part</td><td>DWG set no</td><td colspan="3">Individual DWG no</td><td>Rev</td></tr>
<tr><td>A</td><td>000077</td><td colspan="3">005-63</td><td>1</td></tr>
<tr><td>Mr. Cad Mann</td><td>Scale</td><td>1:1</td><td>Finalized: 07/10/2012</td><td>Sheet</td><td colspan="2">1 of 6</td></tr>
</table>

FIGURE 4-5 Title block.

know what is required in a piping system or where the piping system might be installed. This is the type of information that is conveyed through a sketch or drawing. Two types of single-line drawings are used in piping sketches: orthographic and isometric.

The **orthographic drawing** is made by using two separate views. One view is from the top of the piping system (looking down) showing all of the horizontal runs that are made. This is referred to as a **plan view**. The other orthographic view is from the side to show all of the vertical runs. This is referred to as an **elevated view**. In order to get a complete picture of the piping system using an orthographic drawing, two separate drawings must be completed. An **isometric drawing** is considered a three-dimensional type of drawing. It has both the horizontal and the vertical runs of the piping system included on one drawing. All vertical runs are illustrated in the drawing as vertical lines and symbols, whereas a line that has a 30° slope represents all horizontal runs. A comparison of an isometric sketch and an orthographic sketch is shown in **Figure 4-6**.

The only major problem that results from using the orthographic sketch is in determining whether a fixture, fitting, or pipe, is turning toward the interpreter or away from the interpreter. This is accomplished through the use of common symbols, as shown in **Figure 4-7**.

Notice on a plan view that when a pipe makes a 90° vertical turn upward, it is depicted in the drawing as a full circle. An incomplete circle obscured by the centerline of the pipe is depicted when a pipe makes a 90° vertical turn downward. On an elevated view, when a pipe makes a 90° horizontal turn toward the interpreter, it is depicted in the drawing as a full circle. An incomplete circle obscured by the centerline of the pipe is depicted when a pipe makes a 90° horizontal turn away from the interpreter.

It is important to know that the line in a sketch represents the centerline of the piping system. Fittings

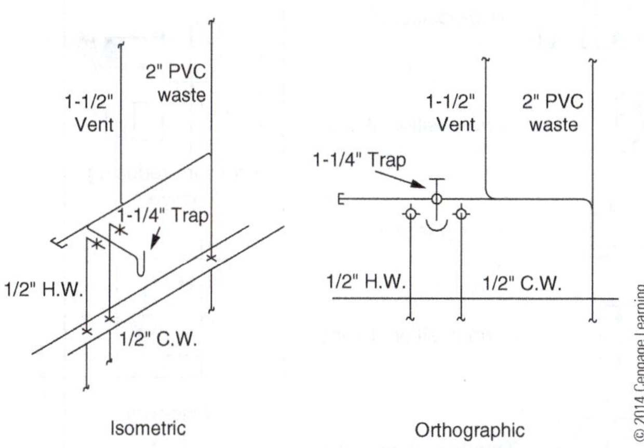

FIGURE 4-6 A comparison of an isometric sketch and an orthographic sketch.

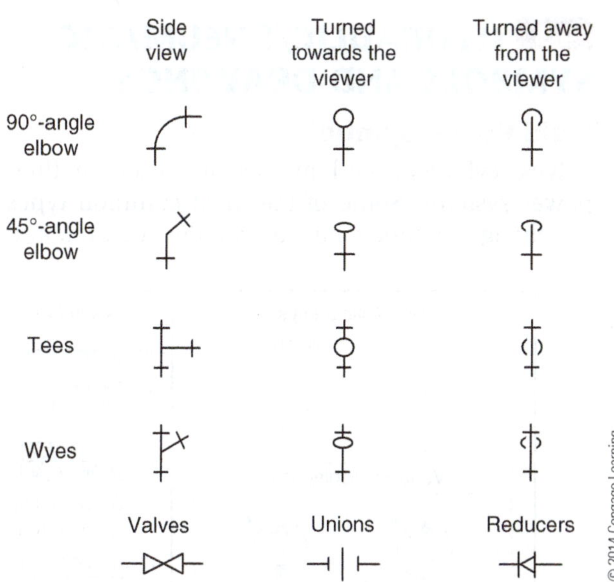

FIGURE 4-7 Common piping symbols.

are indicated by short, intersecting lines that cross a pipeline. When referring to an isometric sketch, it is important to convey such information as where a pipe may penetrate a wall or floor. Placing a small cross on the pipe centerline where it penetrates the wall shows this. This is also illustrated in Figure 4-6. Notice in the figure that the hot water (H.W.) and cold water (C.W.) lines originate from below the floor. Two crosses indicate the penetration of the sole plate. Find the hot and cold pipes in the figure as they extend through the wall to supply the fixture (sink).

Pipe dimensions also have to be given on a sketch to tell the interpreter how long or short a pipe or nipple must be and where to place the necessary fittings. This is illustrated in **Figure 4-8**.

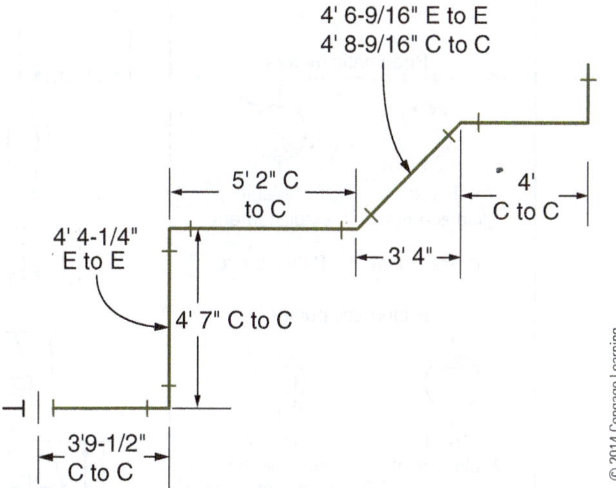

FIGURE 4-8 Piping sketch dimensions.

4-3 HYDRAULIC/PNEUMATIC SYMBOLS AND DRAWINGS

Fluid Power Symbol

Valves, cylinders, and motors, are used in fluid power systems. Some of the most common types of valving are directional control, flow control, and pressure control. A valve is identified by its internal components, the number of ports, and the method that is used to actuate the valve. Another parameter for valve description may be normally closed or normally open. This information is transmitted to the reader with the use of symbols. **Figure 4-9** shows the many different types of American National Standards

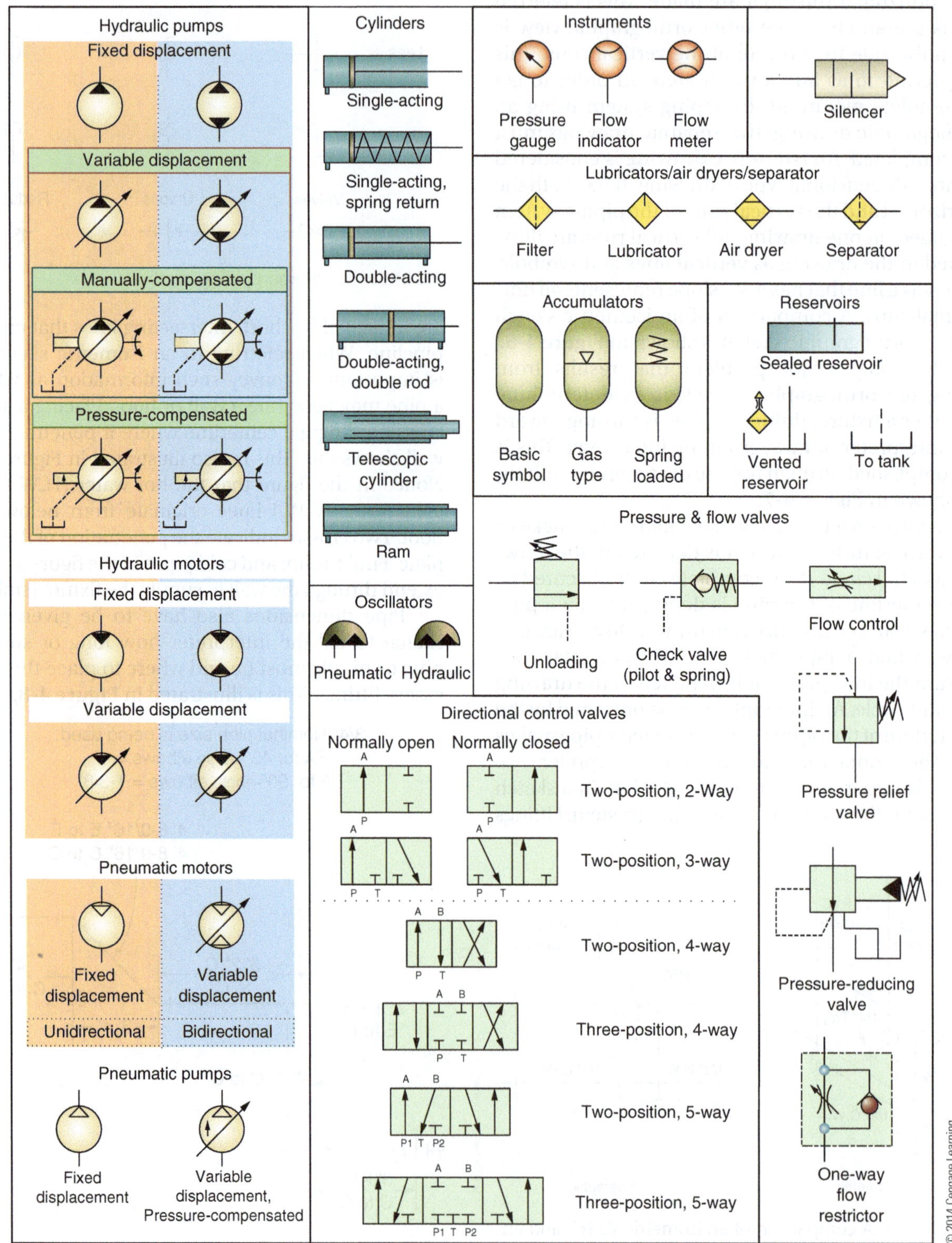

FIGURE 4-9 ANSI symbols for fluid power.

Institute (ANSI) symbols that are used in fluid power systems. As mentioned earlier, it is a common practice to use a symbol to identify how many ports that a valve may have. For example, a valve that has two ports is referred to as a two-way valve. A three-way valve will be a valve that has three ports. There are also four-way and five-way valves, all of which can be seen in Figure 4-9. Valves will have a minimum of two positions and not more than three. Symbols help determine how many spool positions the valve has. Valves can be actuated by manual lever, pushbutton, foot pedal, cam or mechanical operation, solenoid, or pilot operated. For a deeper understanding of the operation of each valve, cylinder, or motor shown in this section, read Chapter 12, where an explanation of operation is given in more detail. Symbols are also used to represent hydraulic and pneumatic cylinders and motors, which can also be seen in Figure 4-9. The ANSI symbols for most of the actuating methods are shown in **Figure 4-10**.

Figure 4-11 shows a hydraulic pump that pumps fluid out of a vented reservoir pressurizing the system. The maximum system pressure is set with the pressure relief valve. If the pressure increased to extreme levels without a pressure relief valve, the system would burst at the weakest point. The pressurized fluid would then pass through a pressure-reducing valve, which sets the working pressure of the system. The working pressure is always lower than the pressure relief setting. Working pressure is discussed in greater detail in Chapter 12. After the fluid flows through the pressure-reducing valve, it goes to the three-position, four-way valve that is in the center of the schematic. If the valve is shifted to the left, the fluid passes through the valve, through the flow control valve (one-way restrictor), and into the cap end (rear side) (left) of the hydraulic cylinder. Any fluid in the rod end (front side) of the cylinder will be pushed out through the check valve in the right-hand flow control (restrictor) and through the valve (which is still shifted to the left). It is then drained back to the reservoir to be cycled through the system again.

As you can see, a drawing explains the operation of a system and provides information necessary for troubleshooting. Drawings are usually surrounded by borders that contain letters and numbers. These are used to locate a key part or component in the drawing. For example, to locate the pressure-reducing valve in **Figure 4-12** (see in page 75), locate the letter C on

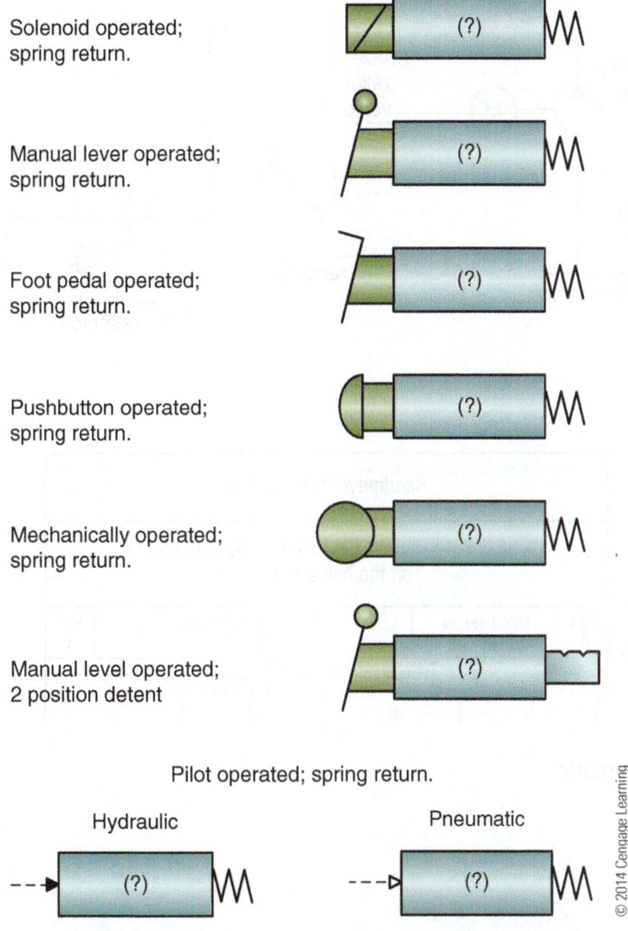

FIGURE 4-10 ANSI symbols showing the different actuating methods.

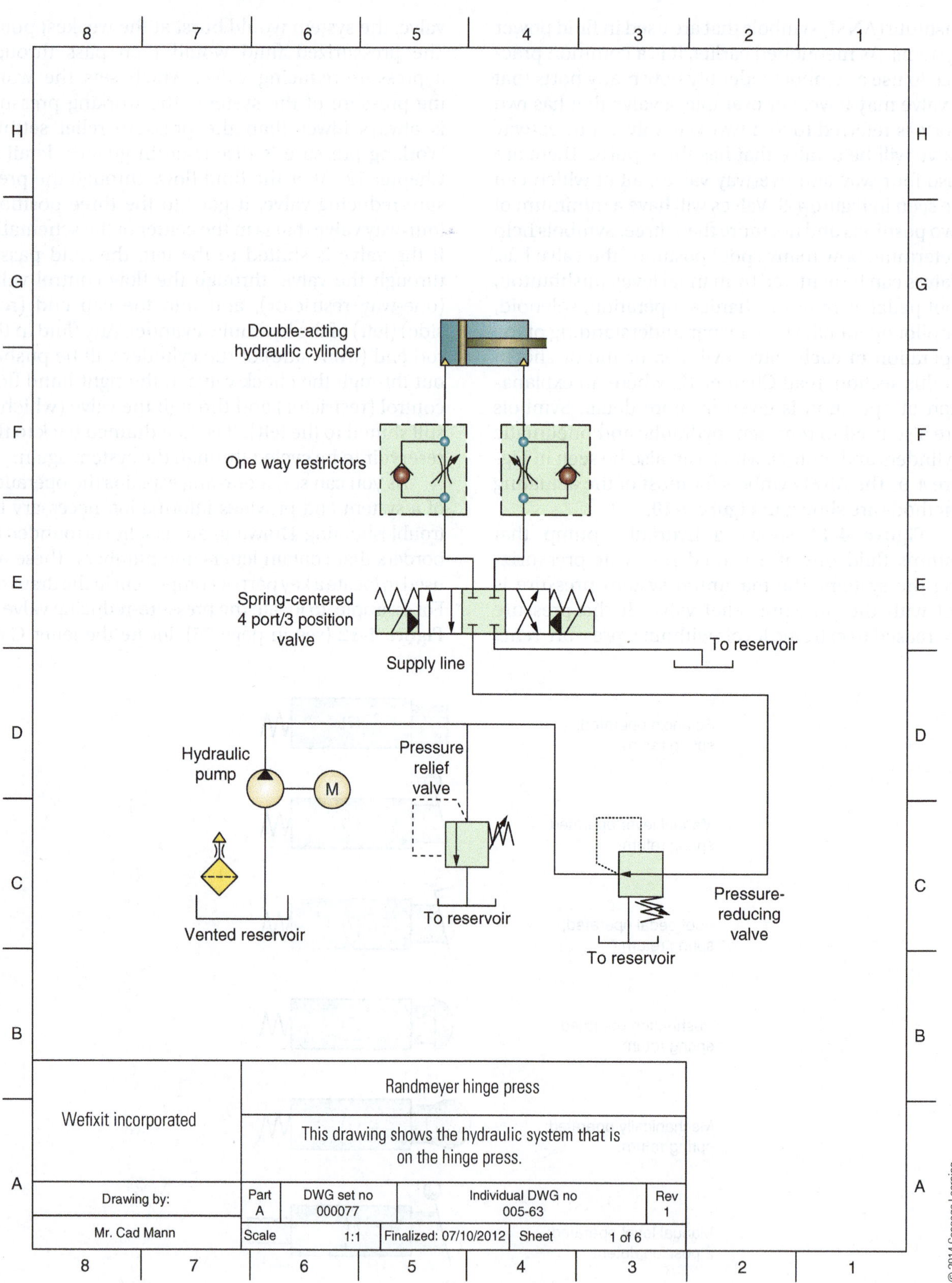
8
7
6
5
4
3
2
1
H
G
F
E
D
C
B
A
Double-acting
hydraulic cylinder
One way restrictors
Spring-centered
4 port/3 position
valve
Supply line
To reservoir
Hydraulic
pump
M
Pressure
relief
valve
Vented reservoir
To reservoir
Pressure-
reducing
valve
To reservoir
Wefixit incorporated
Randmeyer hinge press
This drawing shows the hydraulic system that is
on the hinge press.
Drawing by:
Mr. Cad Mann
Part
A
DWG set no
000077
Individual DWG no
005-63
Rev
1
Scale
1:1
Finalized: 07/10/2012
Sheet
1 of 6

FIGURE 4-11 Hydraulic schematic.

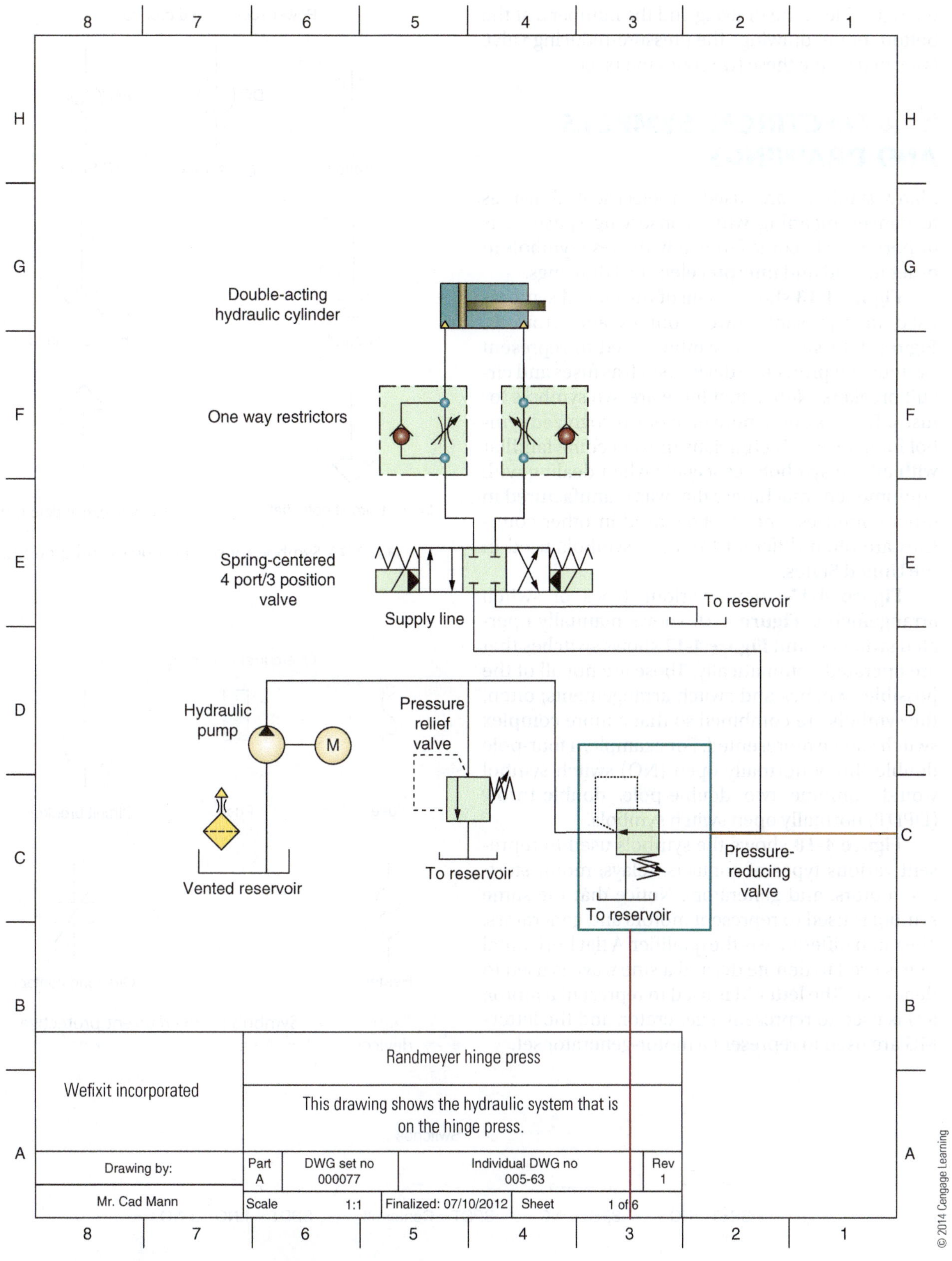

FIGURE 4-12 Locating a component in a schematic.

the right side of the drawing and the number 3 at the bottom of the drawing. The pressure-reducing valve is located where these two zones intersect.

4-4 ELECTRICAL SYMBOLS AND DRAWINGS

Many symbols are used in electrical drawings to convey meaning while conserving space. It is important to become familiar with these symbols in order to read and interpret electrical drawings.

Figure 4-13 shows some of the typical symbols used to represent power sources and grounds. **Figure 4-14** shows the symbols used to represent overcurrent protective devices such as fuses and circuit breakers. Notice that there are two symbols for fuses. There is often more than one recognized symbol for a device. Technicians must become familiar with all the symbols, especially when dealing with equipment or machinery that was manufactured in other countries. The symbols used in other countries are often different from the symbols used in the United States.

CAUTION

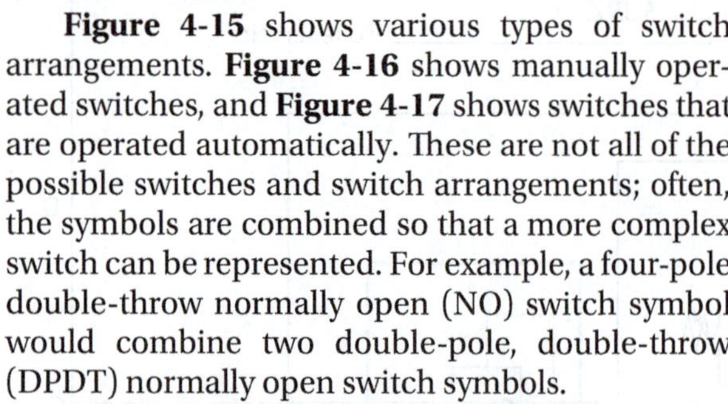

Figure 4-15 shows various types of switch arrangements. **Figure 4-16** shows manually operated switches, and **Figure 4-17** shows switches that are operated automatically. These are not all of the possible switches and switch arrangements; often, the symbols are combined so that a more complex switch can be represented. For example, a four-pole double-throw normally open (NO) switch symbol would combine two double-pole, double-throw (DPDT) normally open switch symbols.

Figure 4-18 shows the symbols used to represent various types of contacts, relays, motor starters, motors, and generators. Notice that the same symbol is used to represent motors and generators. The main difference is the qualifier. A flat horizontal line is used to denote dc, and a sine wave is used to denote ac. The letter M is used to represent a motor, a G is used to represent a generator, and the letters MG are used to represent a motor-generator set.

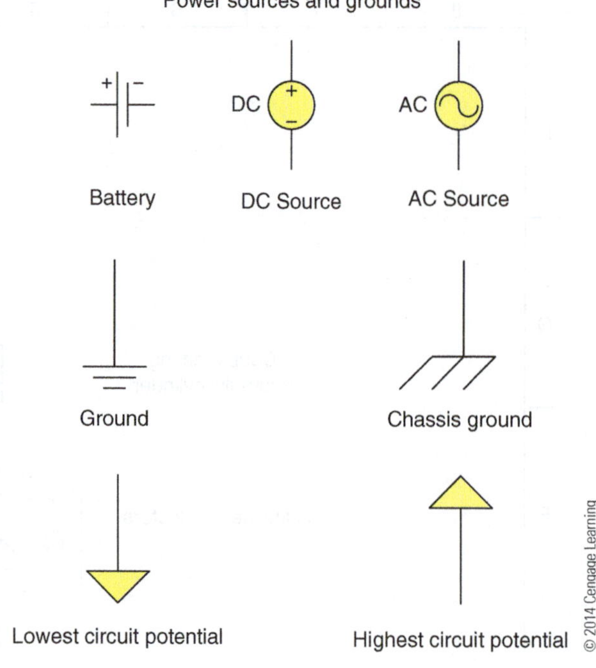

FIGURE 4-13 Symbols for power sources and grounds.

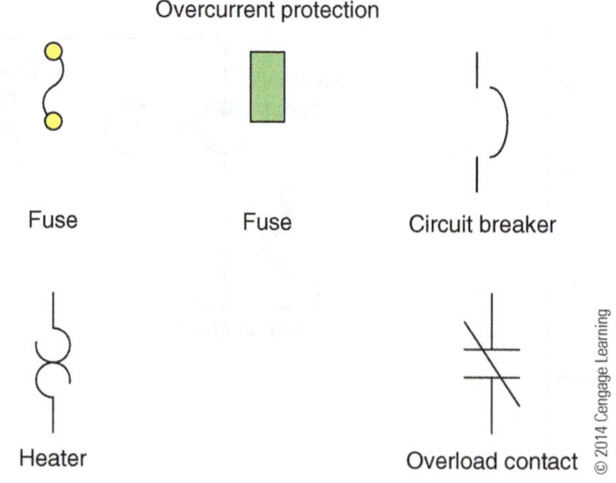

FIGURE 4-14 Symbols for overcurrent protective devices.

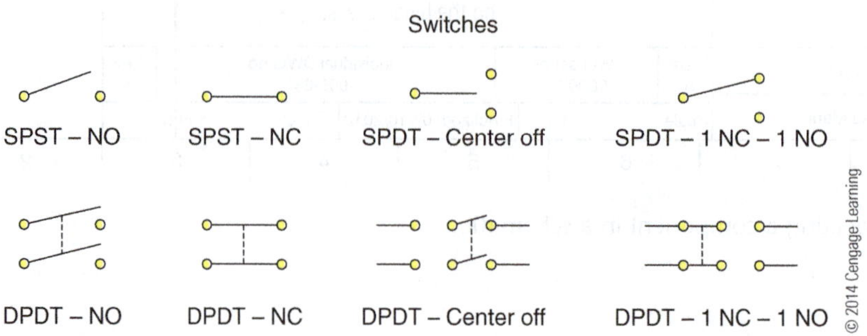

FIGURE 4-15 Symbols of various switching arrangements.

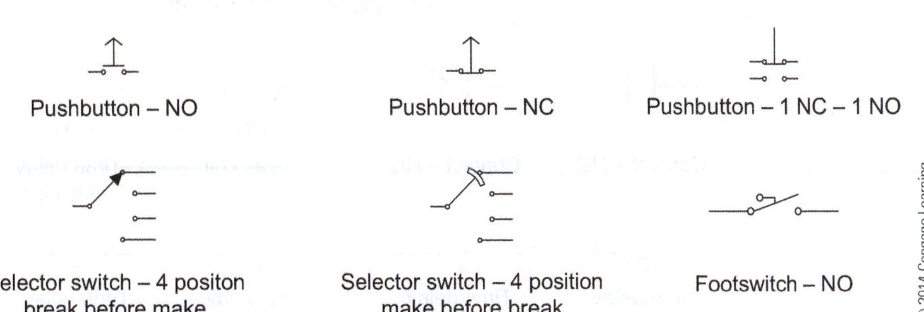

FIGURE 4-16 Symbols for manual switches.

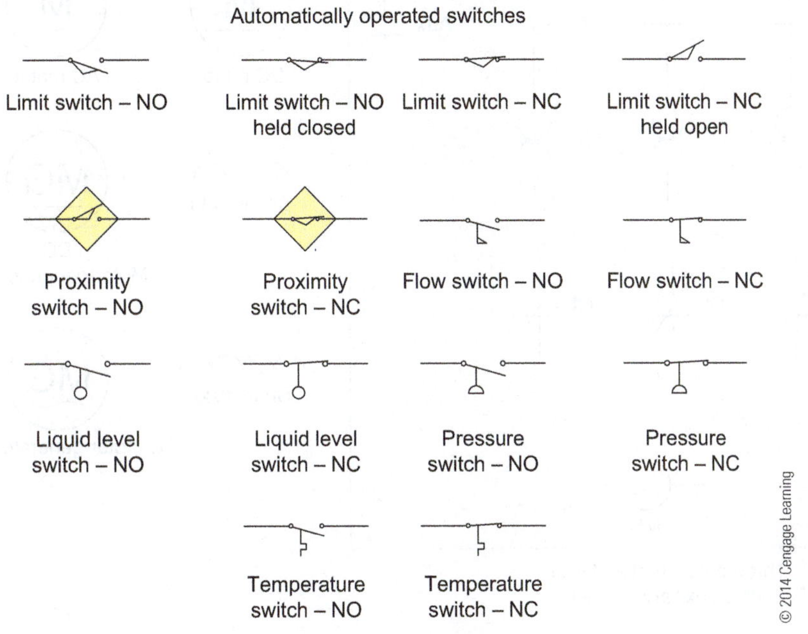

FIGURE 4-17 Symbols for automatic switches.

Resistors and capacitors are shown in **Figure 4-19**. Notice the alternative symbols for these devices. Symbols for inductors and transformers are shown in **Figure 4-20**. Again, notice the alternative symbols. **Figures 4-21A** and **4-21B** (see in page 80)show the different types of symbols used to represent various solid-state devices. Finally, integrated circuits and digital logic gates are shown in **Figure 4-22** on p. 81.

Remember, these figures show the essential symbols used in the electrical field. Many more symbols are in use. You need to become familiar with the symbols used on the machinery and equipment in your facility. If you have any questions about a particular symbol, ask your facility's supervisor or the equipment manufacturer's representative.

Electrical symbols, by themselves, convey little information. Appropriate symbols are combined to form a drawing. An electrical drawing can provide the following information:

- Circuit operation
- Component location
- Electrical connections
- Component function or purpose
- Manufacturer's information
- Wire gauge
- Wire length
- Component specifications
- Circuit specifications
- Motor specifications
- Power specifications

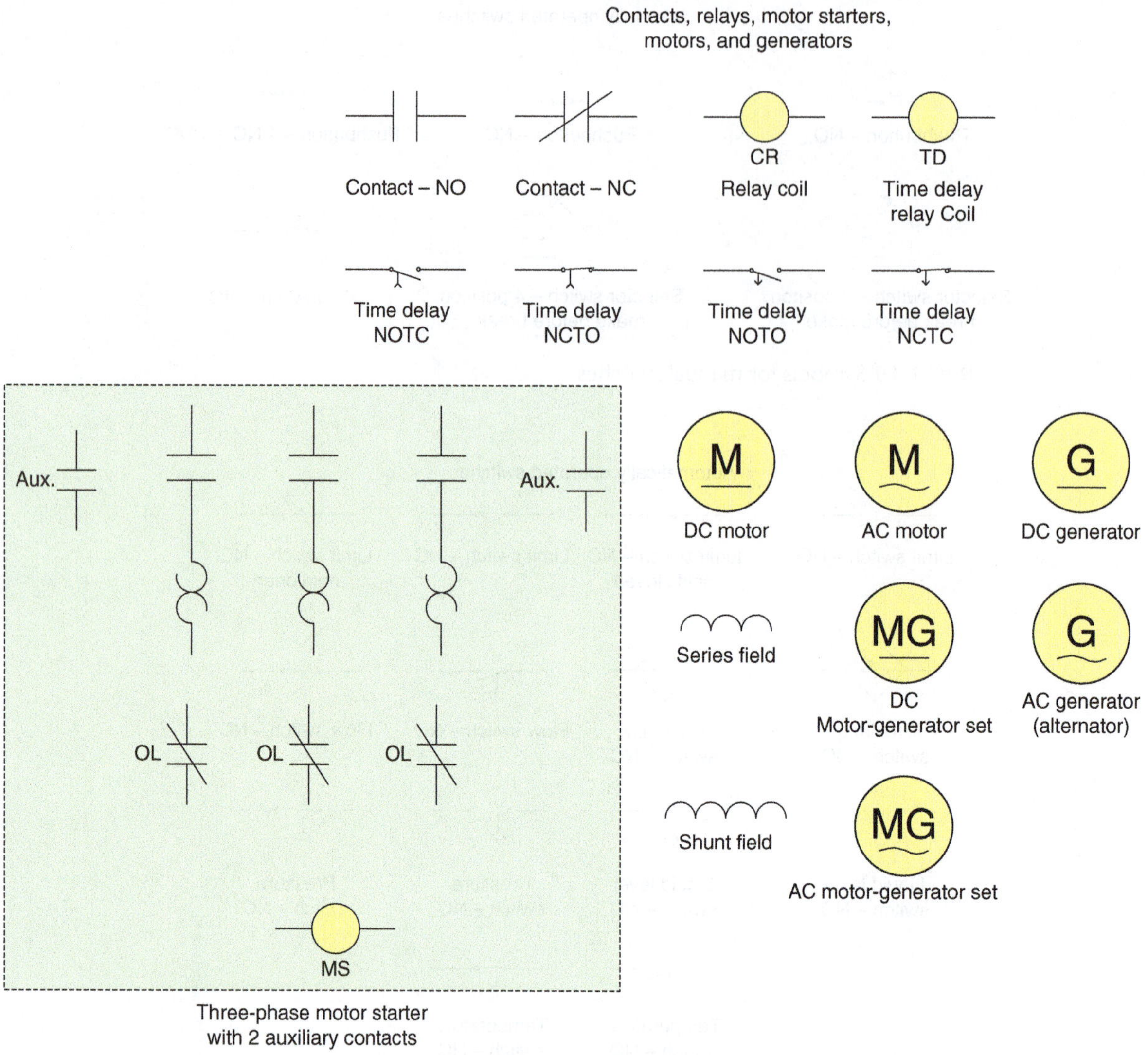

Contacts, relays, motor starters, motors, and generators

Contact – NO
Contact – NC
CR Relay coil
TD Time delay relay Coil

Time delay NOTC
Time delay NCTO
Time delay NOTO
Time delay NCTC

Aux.
Aux.
OL
OL
OL
MS

Three-phase motor starter with 2 auxiliary contacts

M DC motor
M AC motor
G DC generator
Series field
MG DC Motor-generator set
G AC generator (alternator)
Shunt field
MG AC motor-generator set

FIGURE 4-18 Symbols for contacts, relays, motor starters, motors, and generators.

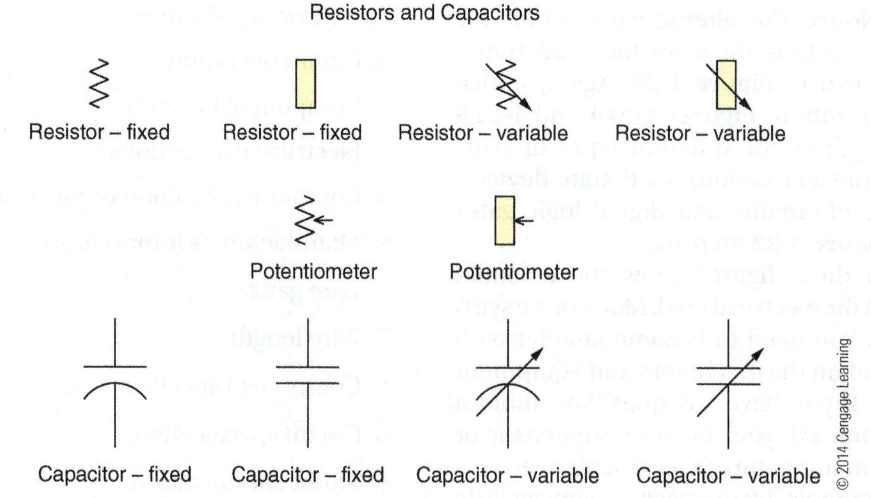

FIGURE 4-19 Symbols for resistors and capacitors.

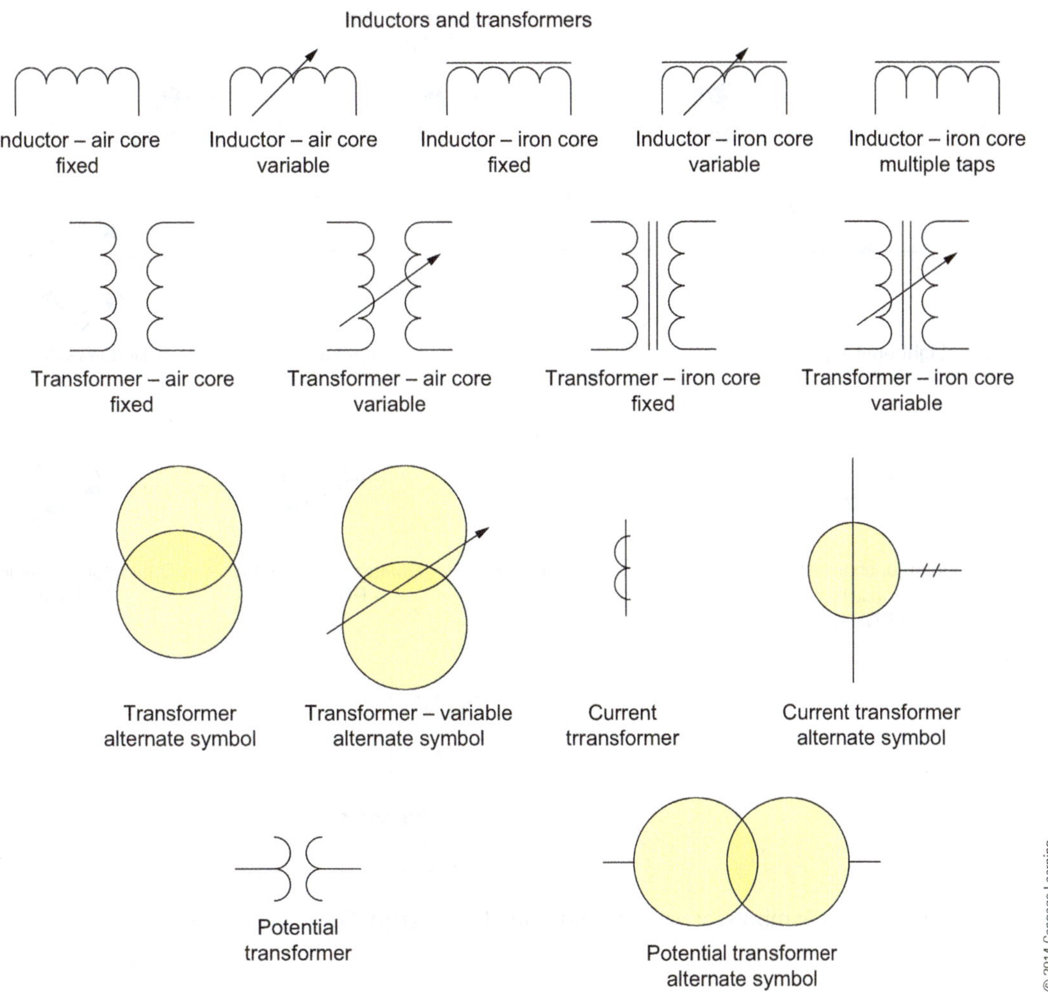

FIGURE 4-20 Symbols for inductors and transformers.

Electrical drawings are typically drawn with all components de-energized. There are four basic types of drawings used to convey information in the electrical field: the single-line drawing, the pictorial diagram, the schematic diagram, and the ladder diagram.

Single-Line Drawings

The **single-line drawing**, shown in **Figure 4-23**, is generally used to provide an overview but not a lot of detail. Single-line drawings do not show the actual electrical connections or the actual physical location of the devices, but they do show that some type of connection exists among components.

Probably the most common use for single-line drawings is to show the power distribution within a facility. This information is helpful when, for example, a particular section of the plant must be isolated for maintenance or repair. Using a single-line drawing, the technician can determine where to electrically isolate one particular section for maintenance without interrupting other portions of the facility.

The single-line drawing in Figure 4-23 shows the power distribution grid of a manufacturing facility. Notice that all of the wires are shown as a single line, hence the name single-line drawing. Notice also that there is not a lot of detail. The components have been arranged so that the highest voltage rating is located at either the top or on the left side of the drawing. Lower voltage ratings are located below or to the right. Consequently, the devices with the highest voltage rating are at the top or left, and the ones with the lowest voltage rating are at the bottom or right.

The simple nature of a single-line drawing allows you to easily trace the flow of power through

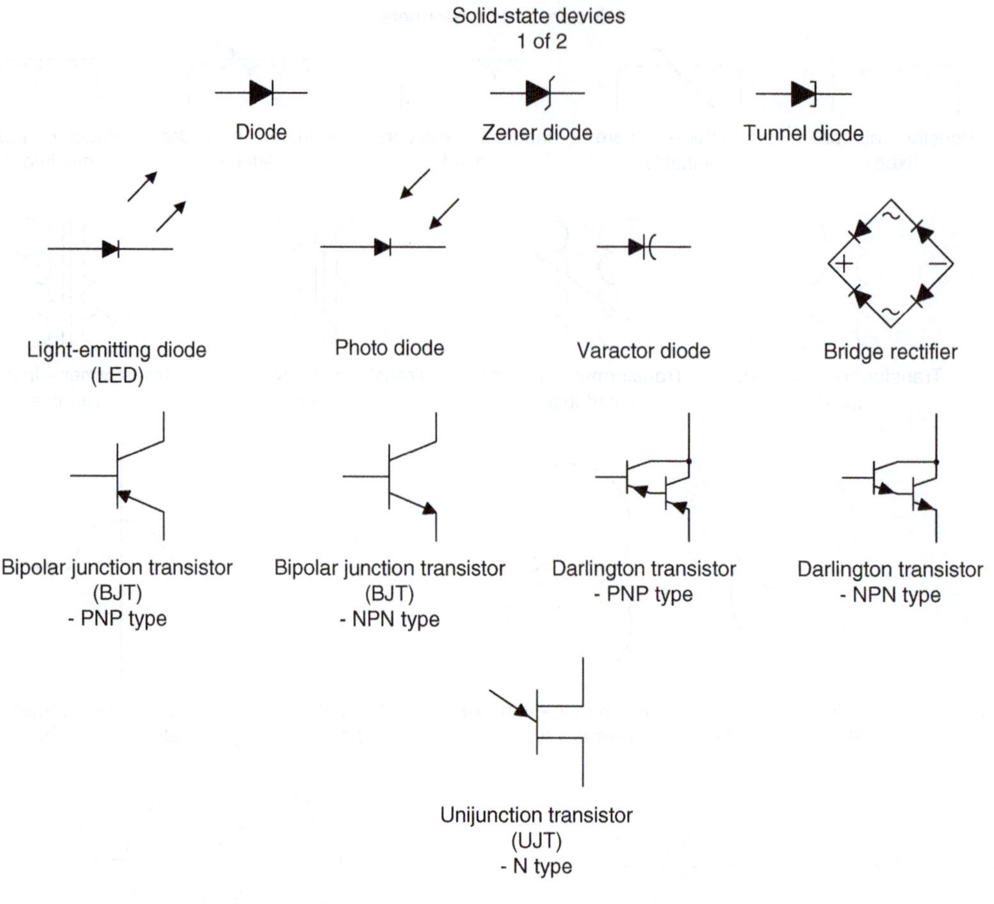

FIGURE 4-21A Symbols for various solid-state devices (Part 1).

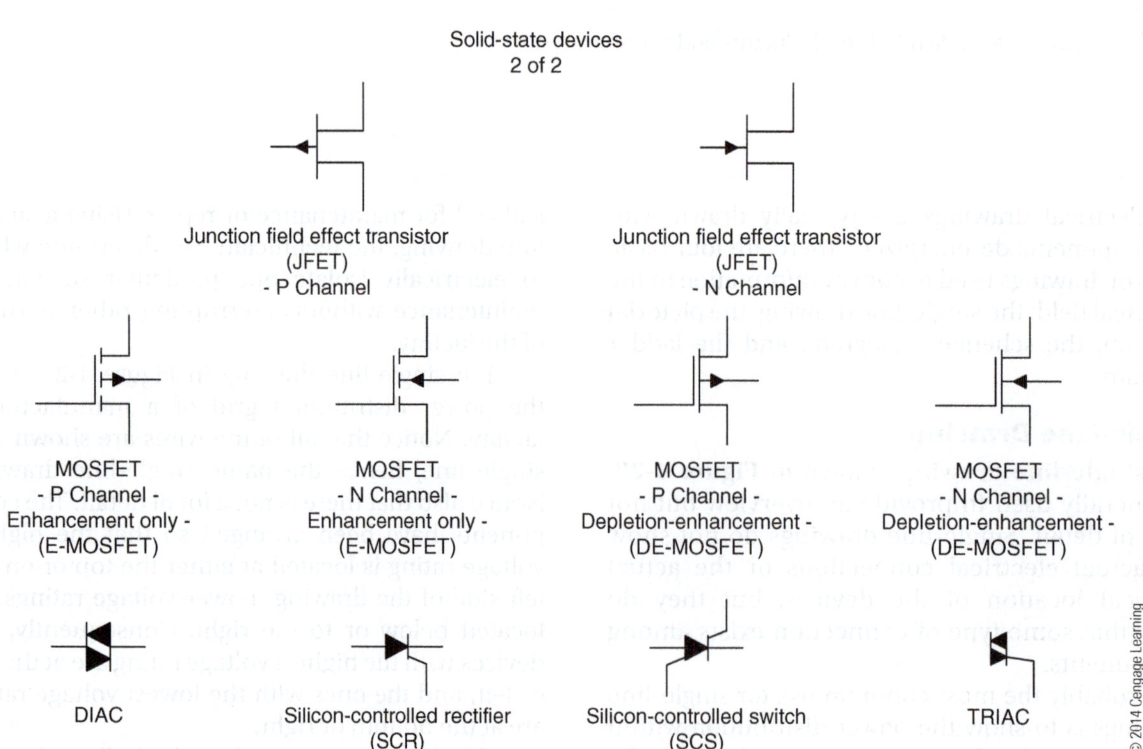

FIGURE 4-21B Symbols for various solid-state devices (Part 2).

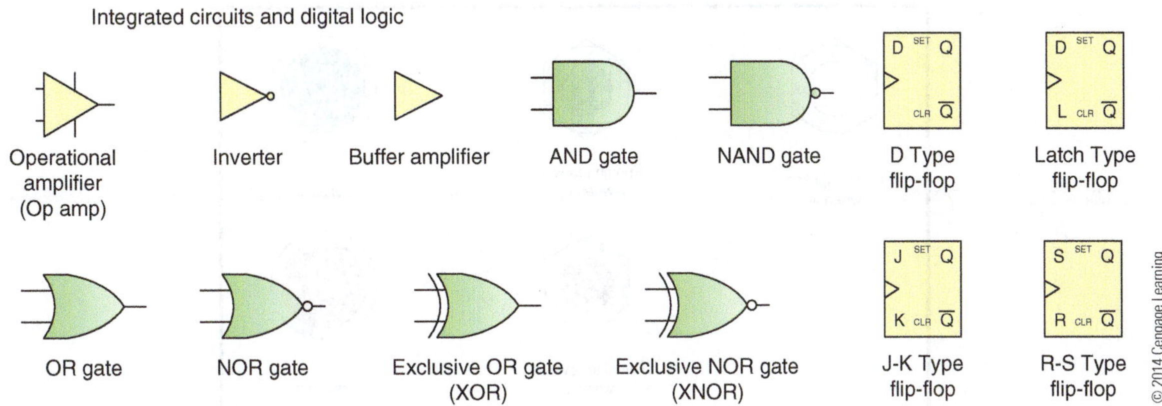

FIGURE 4-22 Symbols for integrated circuits and digital logic gates.

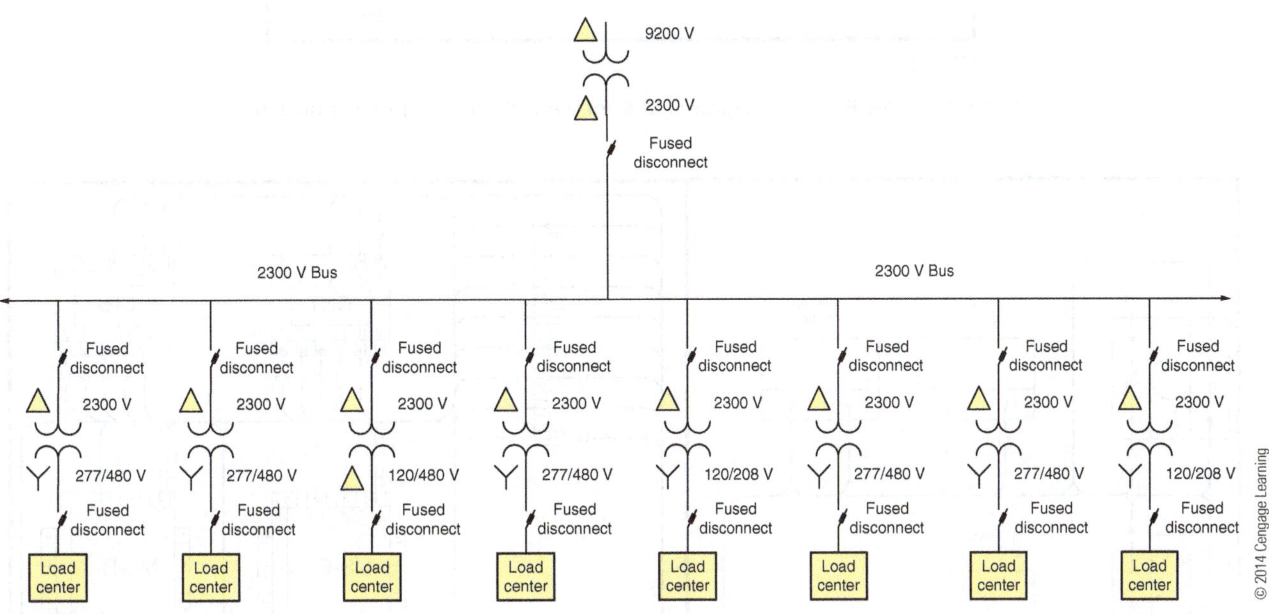

FIGURE 4-23 Single-line drawing.

the drawing. However, you can also use a single-line drawing in reverse. For example, to determine the source of power for a specific component, begin at that component and trace upward (or to the left) until you arrive at the source of power. In this way, the single-line drawing helps identify the power source and any disconnects that can be used to isolate the component from the remainder of the circuit.

Pictorial Diagrams

The **pictorial diagram** is also called a wiring diagram. A pictorial diagram shows the relative physical location of the components, wires, and termination points. A pictorial diagram does not tell you how the circuit operates.

Pictorial diagrams are valuable because they help locate a specific component or termination point. **Figure 4-24A** is a pictorial diagram showing the front view of the cover of a control box, and **Figure 4-24B** shows the same control box with the cover opened. You can now see the back side of the control box cover as well as the inside of the control box.

The circuit represented by Figure 4-24B is a forward-reverse control with pushbutton and electrical interlocks. There are also indicator lights to indicate whether power is applied to the circuit,

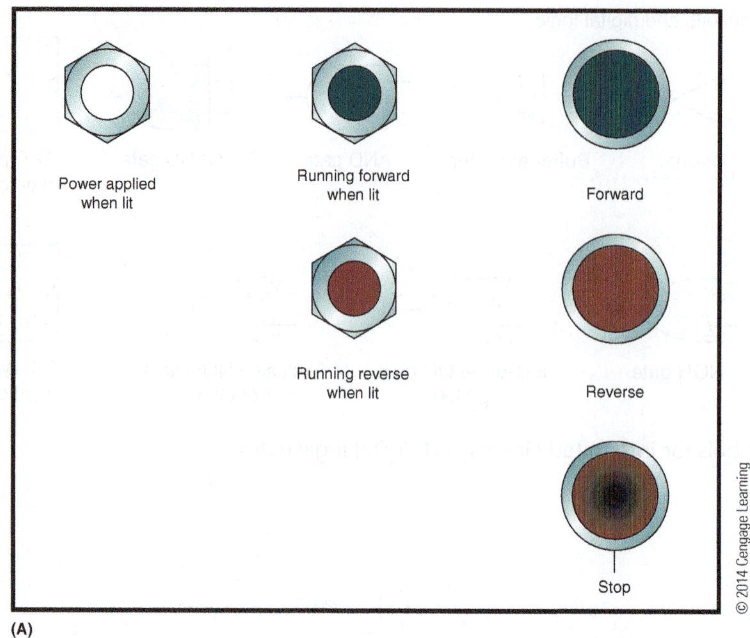

FIGURE 4-24A Pictorial diagram: (A) front view of the cover of a control box.

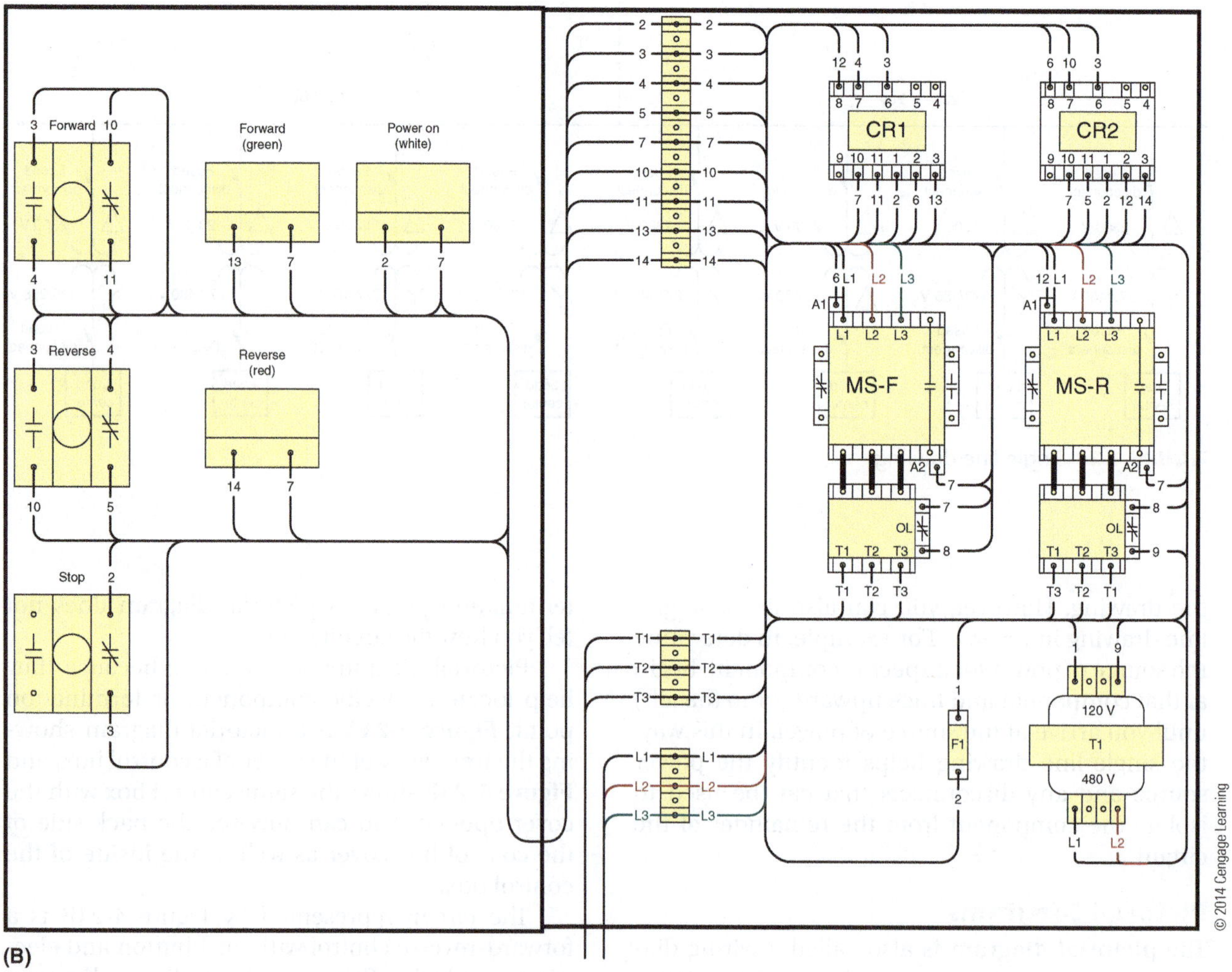

(B)

FIGURE 4-24B (B) Same control box with the cover opened.

and whether the motor is running forward or in reverse. It might be difficult, or impossible, to determine the function of this circuit or of the components by looking at the pictorial diagram. However, the pictorial diagram, for example, would help to locate control relay CR1 within the panel.

Schematic Diagrams

Schematic diagrams show the electrical connections, but not the physical location of components. Schematics are useful in determining how a circuit is designed to operate. Components on a schematic are generally arranged in their approximate location, relative to each other. However, the arrangement on a schematic may be very different from reality.

Figure 4-25 shows a schematic diagram of the forward-reverse control circuit shown in Figure 4-24B. Notice that the power source is located at the upper left side of the drawing. Typically, the power source is located at the top or on the left side. Notice also that the motor (load) is located at the bottom of the drawing. Typically, the load is located at the bottom or on the right side. Schematics help you understand the action of the circuit. By reading a schematic, you can gain an understanding of the function of the various components within the circuit. This information is useful for troubleshooting, modifying, or performing maintenance on the equipment.

Ladder Diagrams

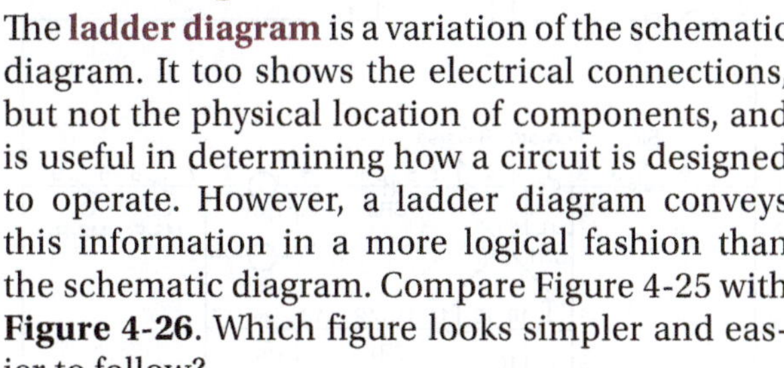

The **ladder diagram** is a variation of the schematic diagram. It too shows the electrical connections, but not the physical location of components, and is useful in determining how a circuit is designed to operate. However, a ladder diagram conveys this information in a more logical fashion than the schematic diagram. Compare Figure 4-25 with **Figure 4-26**. Which figure looks simpler and easier to follow?

Two power rails are located on either side of the diagram in Figure 4-26. The control logic is then placed on rungs between the two power

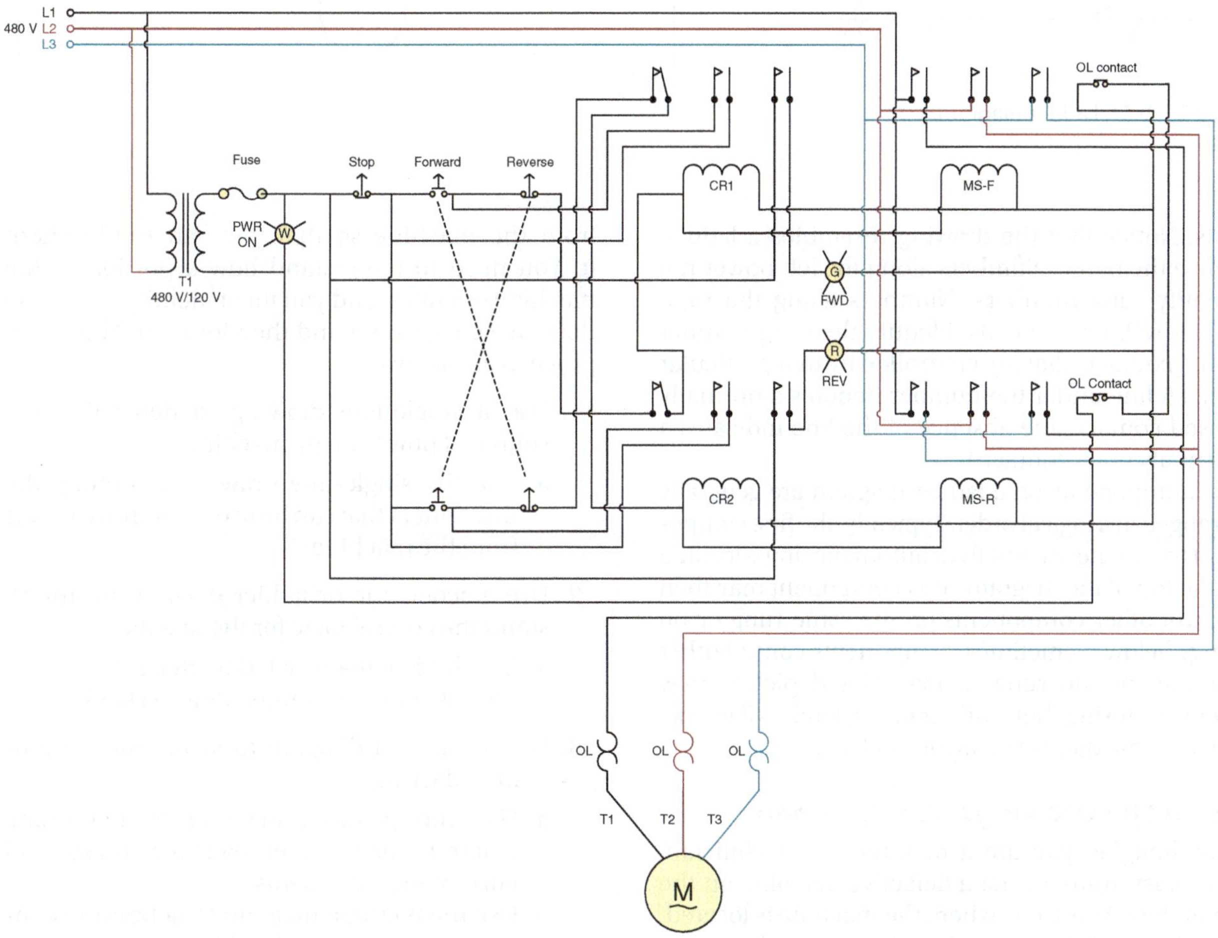

FIGURE 4-25 Schematic diagram of the forward-reverse control circuit shown in Figure 4-24.

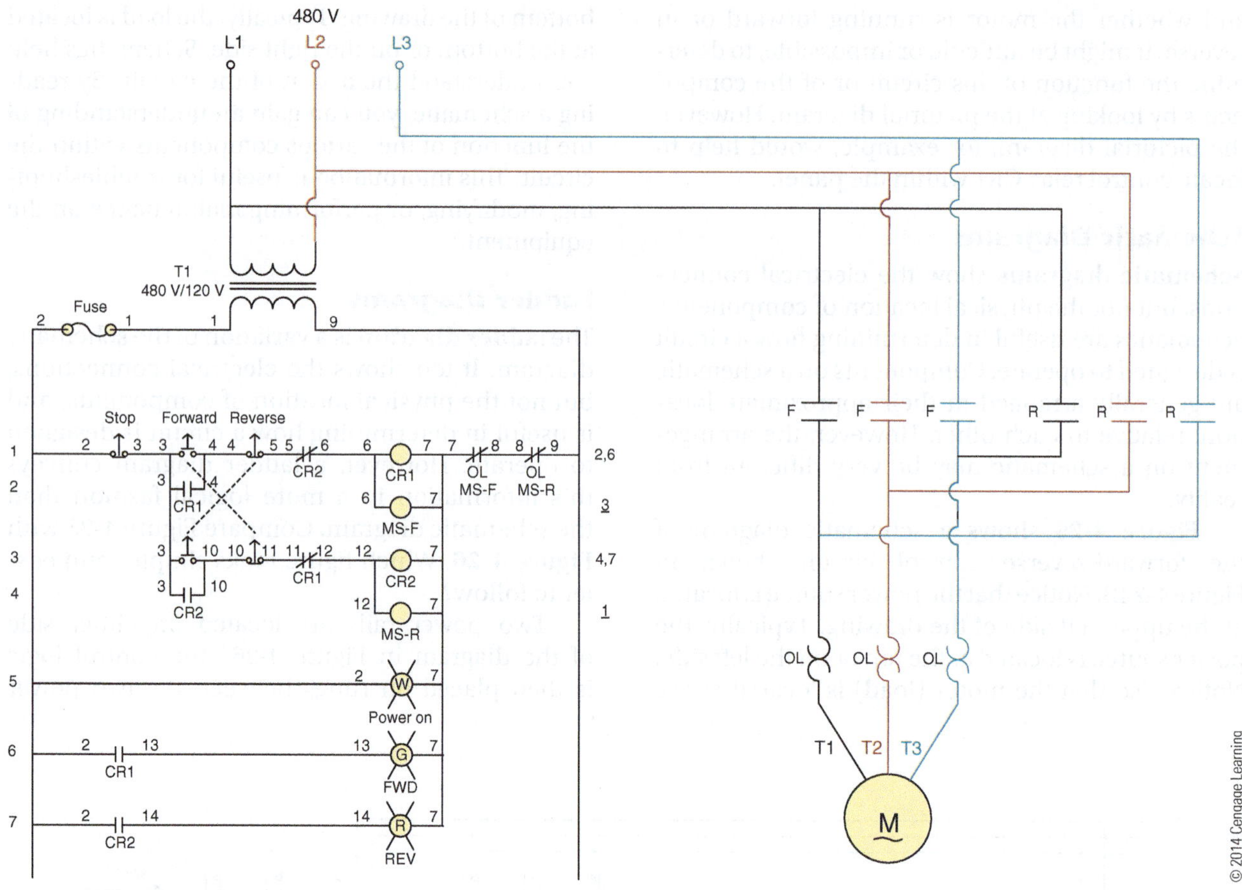

FIGURE 4-26 Ladder diagram.

rails. Notice that the drawing resembles a ladder, hence the name. Numbers along the left power rail identify rung numbers. Numbers along the right power rail, next to coils, identify the rung number of the contacts that are controlled by that particular coil. (A line under the number denotes a normally closed contact. The absence of the line indicates a normally open contact.)

Components on a ladder diagram are generally arranged in a logical order. Typically, the first component to operate (manually or automatically) is located at the top of the diagram. This component may then control other components on the same rung or on rungs below. Sometimes, components control other components on rungs above. This depiction may seem confusing, but, with practice, ladder diagrams can become quite easy and logical to use.

Using the Drawings and Diagrams

Now imagine you are a maintenance technician. You must troubleshoot a defective machine on the plant floor. You know where the machine is located, but you must determine how to remove the power from the machine so that you can troubleshoot it. You need to understand how a portion of the machine operates, and you must identify a possible defective component and then locate it. You could proceed as follows:

1. Use a single-line drawing to determine the source of power to the machine.
   a. Use the single-line drawing to dentify the disconnect that can be used to remove power from the machine.
2. Use a schematic or ladder diagram to understand the control logic for the machine.
   a. Use the schematic or ladder diagram to identify voltages and connections to check.
3. Use a pictorial diagram to locate the component and wiring.
   a. Use the pictorial diagram to determine where to place meter leads for voltage and current measurements.
   b. Use the pictorial diagram to determine connection points that should be checked.

As you can see, you need several types of drawings to help you understand the operating, maintaining, and troubleshooting machinery and equipment at your facility.

4-5 WELDING SYMBOLS AND DRAWINGS

The ANSI has a standard set of symbols used to identify all of the necessary types of welds. **Figure 4-27** shows all of the common types of welds that are used in gas and arc welding and their ANSI symbols.

It is important to know where a weld should go. An angled line and an arrowhead, as shown in **Figure 4-28**, indicate the position of the weld.

At times, symbols may be added to the symbol shown in Figure 4-28 to indicate the method of welding and the contour of the weld.

Another concern identified through the use of symbols may be which side of the joint the weld goes on. See **Figure 4-29**. More concerns may be the dimension of the weld, the finish of the weld,

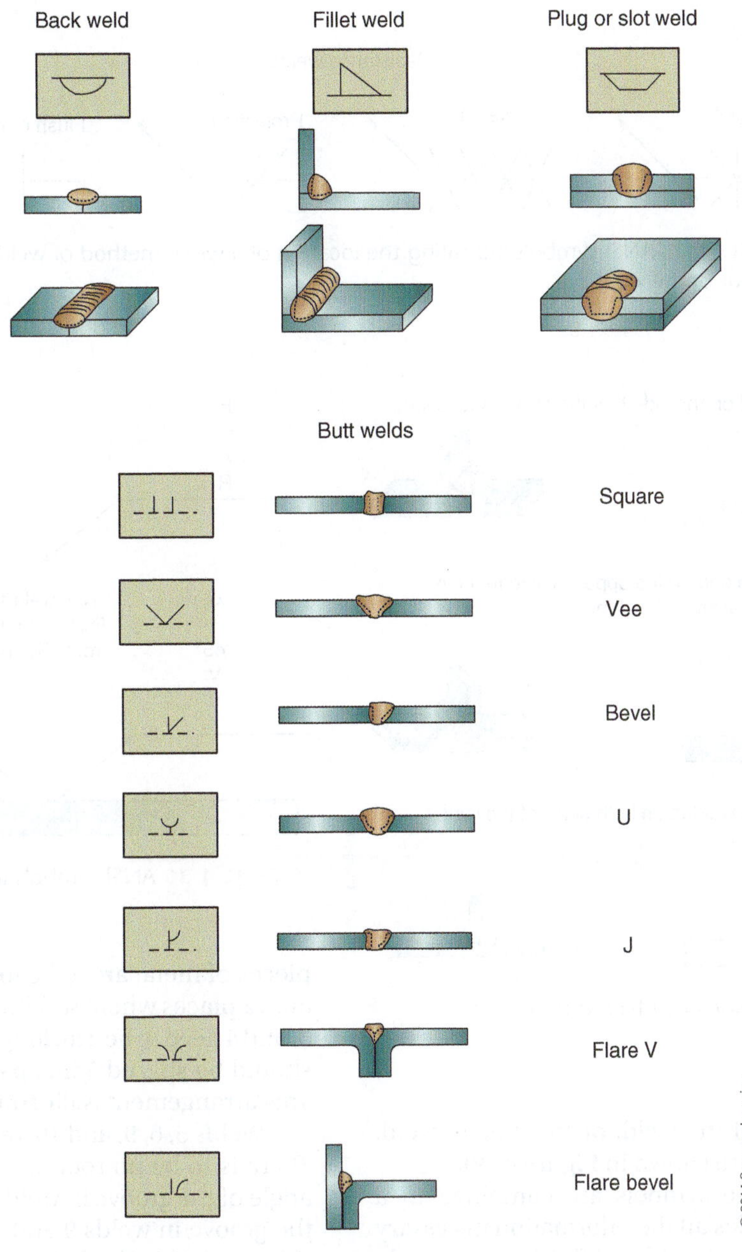

FIGURE 4-27 ANSI welding symbols.

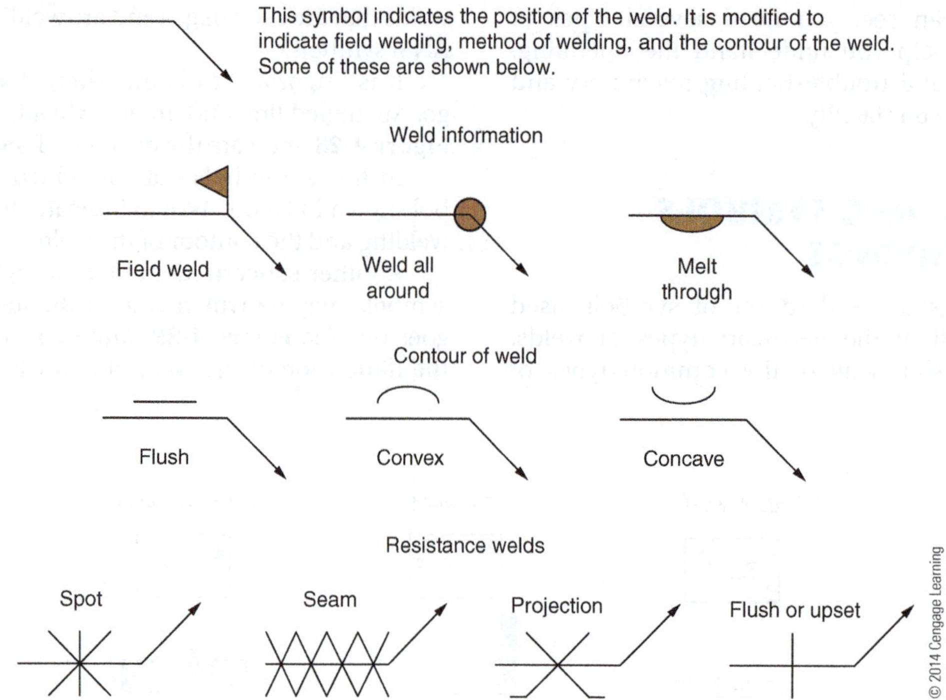

Resistance welds

FIGURE 4-28 ANSI symbols indicating the location of a weld, method of welding, and contour of the weld.

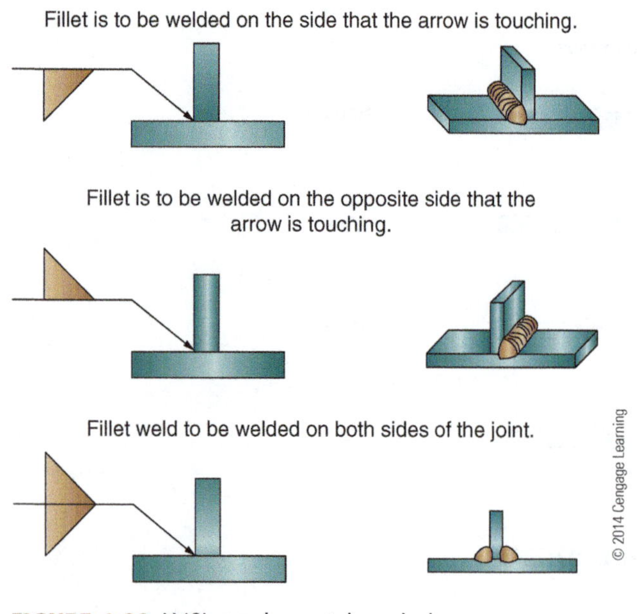

Fillet is to be welded on the side that the arrow is touching.
Fillet is to be welded on the opposite side that the arrow is touching.
Fillet weld to be welded on both sides of the joint.
Projection
Flush or upset
This weld should be ground to a flush finish. The joint is a V-butt joint with no root. The bevels on both sides should be at a 45° angle.
© 2014 Cengage Learning

FIGURE 4-29 ANSI supplemental symbols.

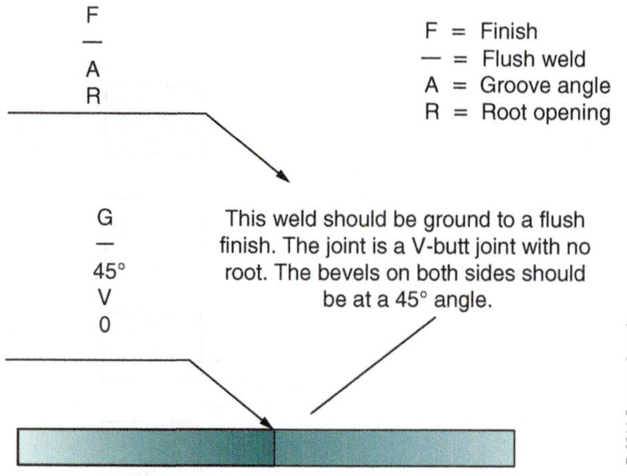

— = Flush weld

FIGURE 4-30 ANSI symbols for finish, groove, and angle.

the groove angle of the weld, or the type of weld. Symbols for these are shown in **Figure 4-30**.

The appropriate symbols are combined in a manner that provides all the information necessary to perform the correct weld and finish. An example of a welding drawing is shown in **Figure 4-31**. Two pieces of metal are to be joined together, and there are 12 places where welding must occur. Welds 1, 2, 3, and 4 need to be 2 in. long in four places. The welds should be spaced 5 in. apart from center to center. This arrangement is illustrated in **Figure 4-32**.

Welds 5, 6, 9, and 10 are to be V-type butt joints. There is to be no root opening in these welds. The angle of the groove in welds 5 and 6 is to be 45°, and the groove in welds 9 and 10 is to be at a 30° angle (**Figure 4-33**). The finish on these welds should be ground to flush.

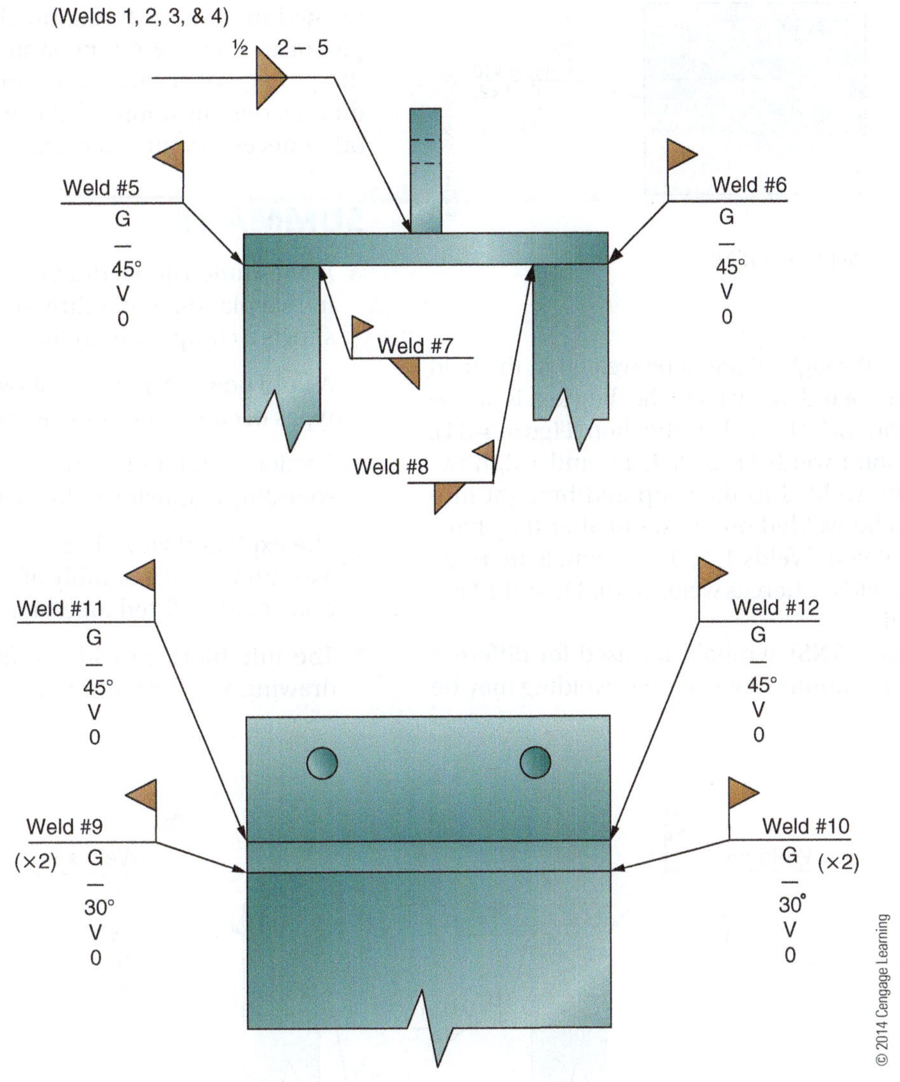

FIGURE 4-31 An example of a simple welding schematic.

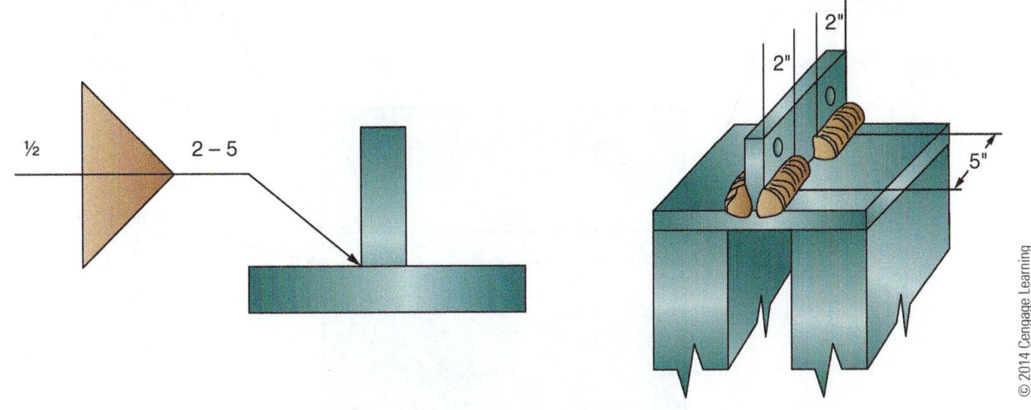

FIGURE 4-32 Welding a fillet: four welds, 2 in. long, 5 in. from center to center.

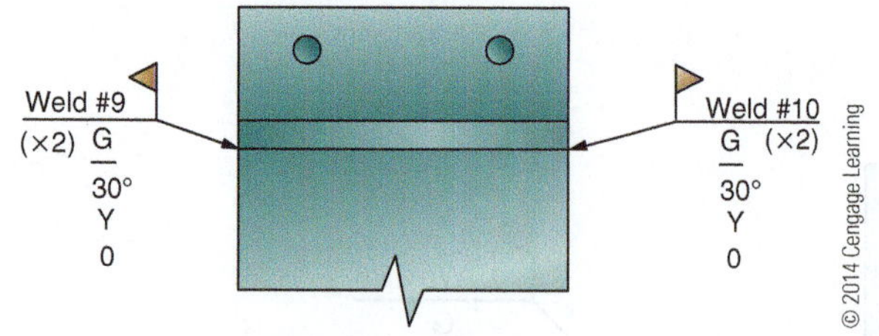

FIGURE 4-33 Welds 9 and 10.

Weds 5 through 10 are to be welded in the field because the stand on which the bracket is to be mounted cannot be brought to the shop (**Figure 4-34**).

The T-joint welds (1, 2, 3, 4, 11, and 12), however, can be welded in the shop and brought into the field to be welded on the stand after they have been completed. Welds 1, 2, 3, 4, 7, and 8 are to be fillet-type welds, whereas welds 9, 10, 11, and 12 are V-butt welds.

The same ANSI symbols are used for different types of applications. For example, welding may be needed to manufacture a bracket, as in the example, or to create a piping system. The drawing for the piping system uses the same symbols to show dimensions, location of the weld, and all of the other necessary information.

SUMMARY

- A two-dimensional drawing shows a component in two planes, and a three-dimensional drawing shows a component in three planes.
- A fluid power schematic shows the locations and types of connections of elements in a system.
- Borders containing numbers and letters surrounding a drawing help locate components.
- The exploded view diagram shows the intended assembly or disassembly of the mechanical parts that are to be fitted together.
- The title block provides information about the drawing and the author of the drawing.

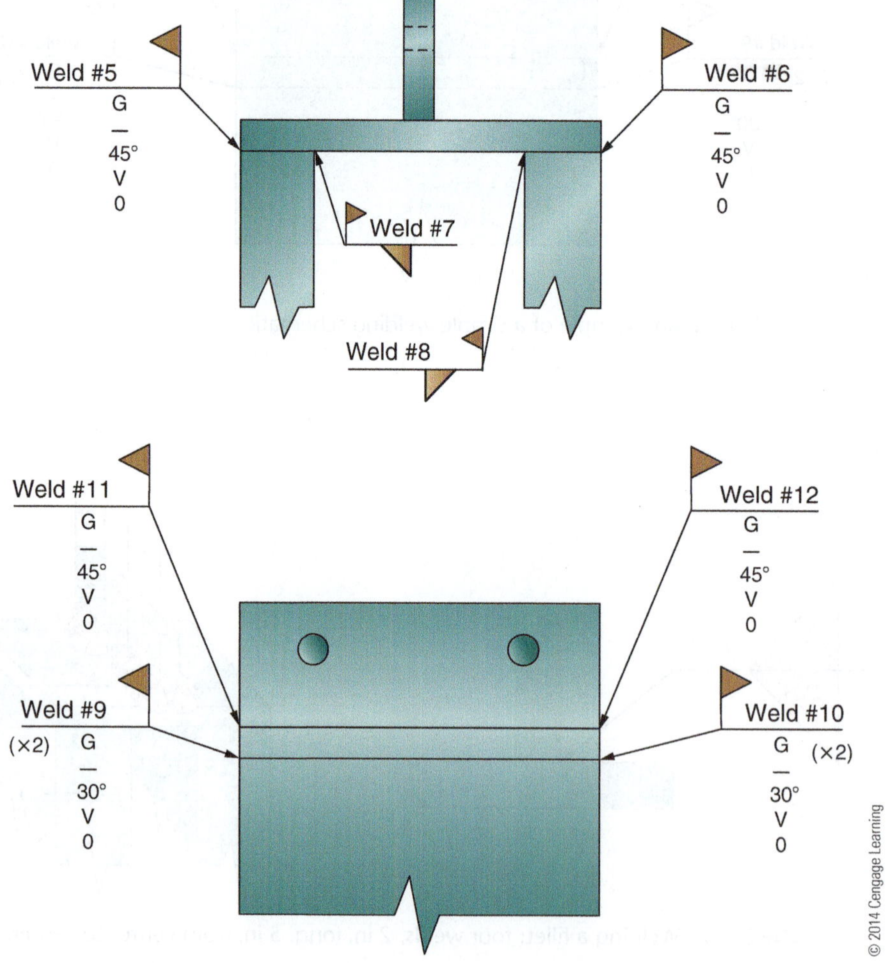

FIGURE 4-34 Field welds.

- Symbols are used to convey meaning, conserving space, and they are combined to form a drawing.
- An electrical drawing provides information about circuit operation and specifications; electrical connections; component function or purpose, location, and specifications; manufacturer; wire gauge and length; motor specifications; and power specifications.
- Electrical drawings are typically drawn with all components de-energized.
- Four types of drawings are used to convey electrical information: the single-line drawing, the pictorial diagram, the schematic diagram, and the ladder diagram.
- The single-line drawing is generally used to convey an overview, but not a lot of detail.
- The pictorial diagram shows the relative physical location of the components, wires, and termination points and is often referred to as a wiring diagram. A pictorial diagram does not tell how the circuit operates.
- Schematic diagrams and ladder diagrams show the electrical connections, but not the location of components.
- American National Standards Institute (ANSI) symbols are used to identify types of welds, where the weld should go, and supplementary information.
- The ANSI symbols are combined in a manner that provides all the information necessary to perform the correct weld and finish.

REVIEW QUESTIONS

1. What is the purpose of a border that has letters and numbers around the outside of a drawing?
2. List five things that may be found in the title block of a drawing.
3. Electrical drawings are typically drawn with all components _____.
4. What is a ladder diagram?
5. Draw the ANSI symbol used to identify a weld that is to be made in the field.

CHAPTER 5

Rigging and Mechanical Installations

Mechanical advantage has been used for thousands of years and has made it possible for mankind to build the world we live in today. Without it, the pyramids would not have been built, nor would we have the great skyscrapers of our age. The Industrial Revolution during the last century would not have occurred had we not been able to move the heavy equipment and machinery that was needed. Often machinery or heavy equipment must be relocated within a plant, or it may be that a machine has to be moved in or out of the facility. To accomplish this task, you must use proper rigging techniques. You must be able to correctly determine the weight of the load, properly plan the lift, select the correct lifting equipment, and determine load balance to execute the lift safely.

OBJECTIVES

After studying this chapter, the student should be able to

- Determine the correct formula to perform weight estimations.
- Discuss the importance of balancing the load.
- Define terminology that is pertinent to the use of slings.
- Identify the common types of synthetic slings that are used in industry.
- Identify the different types of natural fiber rope.
- Demonstrate knot tying using both synthetic and fiber rope.
- Identify the different types of wire rope classifications.
- Identify the different types of chain that are used when rigging.
- Inspect and select the proper equipment needed to make a safe lift.
- Perform prelift planning.

5-1 FORMULAE AND WEIGHT ESTIMATIONS

Units of Measure

There are two modern units of measure in use today, the U.S. Customary System and the International Standard (SI) System. The U.S. Customary system, which is obviously used mostly in the United States, uses units of measure such as the inch, foot, mile, pint, quart, gallon, ounce, and pound. SI is the modern form of the metric system, which is based on multiples of 10 and uses units of measure such as the meter, the liter, and the gram. SI has been adopted across nearly the entire planet, with the United States, Burma, and Liberia being the only countries not yet having adopted it. These two systems can be converted, one to another, and it is often a requirement to do so. For the purpose of this text, the U.S. Customary System is used. Conversion tables are listed in Appendix G. Unit equivalencies and abbreviations (length units only) can be seen in **Table 5-1**.

To execute a safe lift, a close weight estimation of the equipment that is to be lifted must be made. Many formulae are used to help gain this valuable information. Common measurements used in gathering weight estimations are length, area, and volume. Length (distance) uses a single linear

Table 5-1

UNITS OF MEASURE (LENGTH)

U.S. Customary Units		
Unit	*Division*	*SI Equivalent*
1 point (p)		352.8 μm
1 pica (P/)	12 p	4.233 mm
1 inch (in.)	6 P/	2.54 cm
1 foot (ft)	12 in.	0.304 m
1 yard (yd)	3 ft	0.9144 m
1 mile (mi)	5 280 ft	1.609344 km
Surveying Units		
Unit	*Division*	*SI Equivalent*
1 link (li)	33/50 ft or 7.92 in.	2.012 dm
1 (survey) foot (ft)	1200/3937 m	1200/3937 m
1 rod (rd)	25 li or 16.5 ft	5.029210 m
1 chain (ch)	4 rd	2.011684 da m
1 furlong (fur)	10 ch	2.011684 hm
1 survey (or statute) mile (mi)	8 fur	1.609347 km
1 league (lea)	3 mi	4.828042 km
Nautical Units		
Unit	*Division*	*SI Equivalent*
1 fathom (ftm)	2 yd	1.8288 m
1 cable (cb)	120 ftm or 1.09 fur	2.19456 hm
1 nautical mile (NM or nmi)	8.439 cb or 1.151 mi	1.852 km
Internation Standard (SI) Units		
Unit	*Division*	*U.S. Customary Equivalent*
1 micrometer (μm)		0.00003936996 in.
1 millimeter (mm)	1000 μm	0.03936996 in.
1 centimeter (cm)	10 mm	0.3937008 in.
1 metre (recognized as meter in the United States)	100 cm or 1000 mm	39.370 in. or 3.2808 ft
1 kilometer (km)	1000 m	3280.83 ft or 0.6213712 mi

dimension. Area, the measurement of surfaces, uses two linear dimensions. To calculate for volume (solids or space), three linear dimensions must be used. Some or all of these may be needed when estimating the weight for one piece of equipment. Also, remember this rule concerning weight estimations: It is always better to err on the side of more weight than less when rounding or estimating.

Length

Length is a linear measurement used to size or distance. A linear measurement is made when measuring distance in one given dimension; therefore, it is a one-dimensional measurement. The only requirement for executing this type of measurement is being able to correctly use and interpret a measuring device such as a tape measure or ruler. Details for learning how to read these measuring devices can be found in Chapter 2, Section 2-2 of this textbook.

Area

Area is defined as the two-dimensional surface included within a set of lines, space, shape, or boundary. Area is often measured in square inches (in^2) or square feet (ft^2). The three basic shapes that are used when calculating area are the circle, rectangle, and triangle.

The circle is a closed plane curve where every point along the curve is equally distanced from a fixed center point within the curve. To correctly use the formulae related to the circle, some terminologies used when discussing circles must be defined: diameter, radius, and circumference. Each can be seen in **Figure 5-1** below.

The following letters and symbols are used when calculating for the area of a circle:

C = Circumference (the distance around the circle)
r = radius (½ the diameter)
D = Diameter (2 × the radius)
π = 3.1416 (known as *pi*)

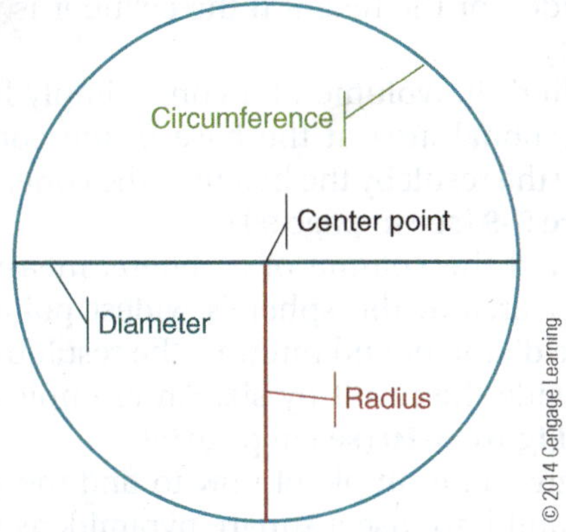

FIGURE 5-1 Properties of the circle.

Use the following formula for calculating the area of a circle when only the radius is known:

$$Area = \pi r^2$$

When the diameter is known, use the following formula, where D = diameter and 0.7854 = π/4:

$$Area = 0.7854 \times D^2$$

When only the circumference is known, find the radius and then apply it to the first formula.

$$Radius = \frac{C}{2\pi}$$

then use,

$$Area = \pi r^2$$

Figure 5-2 shows an example for each of the previous scenarios.

The surface area of a cylinder can be found by multiplying the circumference by the height (or length) of the cylinder. This also can be seen in Figure 5-2.

The area of the rectangle, square, or any other quadrilateral can be found by multiplying the length by the width. A quadrilateral is a polygon that has four sides and four interior angles that always sum to equal 360°. However, the angle of each corner may be greater or smaller than the 90° angles we would find in a rectangle or a square. **Figure 5-3** (see in page 95) shows what each of these shapes looks like.

A trapezoid, which can be seen in **Figure 5-4** (see in page 96), has four sides, two of which are parallel to each other. When trying to calculate the area of a trapezoid, first get the area based on the two largest dimensions; then calculate and subtract the unused triangular areas from the result.

Because the rectangle consists of two triangles as seen in **Figure 5-5** (see in page 97), the area of a triangle is half the area of a rectangle. However, instead of length and width, base (b) and height (h) are used when figuring the area of a triangle. All triangle types can be seen in Figure 5-5 and all triangle types will use this formula:

$$Area = \frac{1}{2} \times b \times h$$

A pyramid can have any polygon shape as its base—a triangle, a quadrilateral, a pentagon, a hexagon, a heptagon, or an octagon. The most common pyramids are the square and triangular pyramids. There are two steps to finding the total surface area of a square pyramid. First, find the surface area of the base, and then add it to the surface area of all the faces on the pyramid. The procedure for calculating total surface area for a square pyramid (as well as the other types of pyramids) can be seen in **Figure 5-6** (see in page 98). When solving area for the triangular, pentagonal, and hexagonal

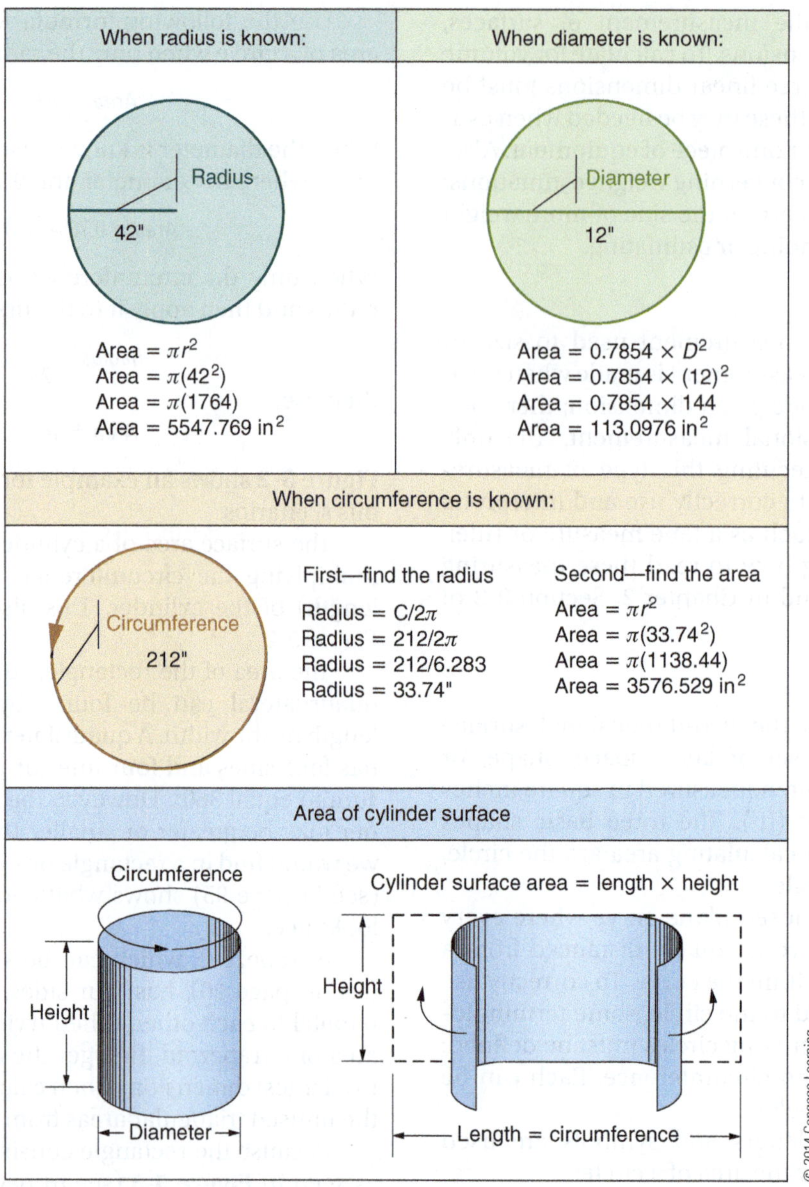

FIGURE 5-2 Area of a circle and cylinder.

pyramids, the apothem length is used. The apothem length is defined as the perpendicular distance from the center of any regular polygon to any of its sides.

Volume

Volume is a three-dimensional measurement that measures the internal size or capacity of an object or container, and its unit of measure is always cubed (*example:* in^3 *or* ft^3). Volume is discussed here for the basic shapes: cubes, cylinders, cones, spheres, and pyramids.

The first shape is the simplest of all the shapes previously listed. The volume of a cube is found by multiplying together the length, width, and height, as seen in **Figure 5-7** (see in page 98).

To find the volume of the cylinder, identify the cross-sectional area of the cylinder, as seen in **Figure 5-8** (see in page 98) and multiply the result by the length if the cylinder is horizontal (lying on its side), or the height if the cylinder is vertical (upright).

To find the volume of a cone, simply find the cross-sectional area at the base of the cone, and multiply the result by the height of the cone, as seen in **Figure 5-9** (see in page 99).

To find the volume of a sphere, measure the diameter area at the sphere's widest point. Then cube the diameter and multiply the result by pi (π). Then divide the result by six. An example can be seen in **Figure 5-10** (see in page 99).

To give an example of how to find the volume of a pyramid, we use a square pyramid, as seen in **Figure 5-11** (see in page 100). First, find the side measurement; then acquire the height (not slant

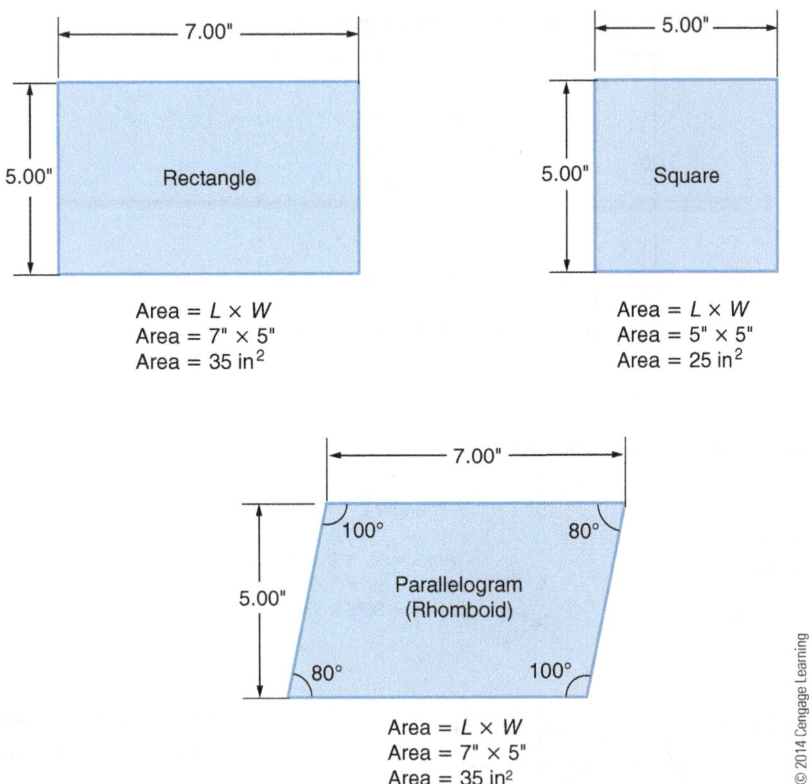

FIGURE 5-3 Area of the rectangle, square, or any other quadrilateral.

height) of the pyramid, place all values into the following formula, and calculate, where b = length of side and h = height:

$$Volume = \left(\frac{1}{3}\right) b^2 h$$

To find the volume of a triangular pyramid, pentagonal pyramid, or a hexagonal pyramid, use the formula given at the bottom of Figure 5-11. A triangular prism can also be seen in the figure as well. To find the volume of this triangular prism, use the following formula, where b = base, h = height, and d = depth:

$$Volume = \left(\frac{1}{2}\right) bhd$$

Irregularly Shaped Objects

Most objects on a machine or pieces of equipment are not shaped like regular solids such as the square cube, rectangular cube, or cylinder. In fact, the shape of most objects is derived from a combination of these shapes in some form or another. This is known as an object having an irregular form. The volume of an irregular form can be calculated by breaking the irregular solid shapes up into regular solid shapes, calculating each solid shape, and then summing the volumes together for total volume. **Figure 5-12** (see in page 100) shows the leveling plate from a machine.

First, look at the leveling plate (Figure 5-12) and identify all of the regular solid shapes that you can without looking at **Figure 5-13** (see in page 101). Hopefully, you recognized that there are five rectangles, one isosceles triangle, one cylinder, and two right triangles. Now, look at Figure 5-13, which shows an exploded view of the outrigger plate. All of the different solid shapes that make up the plate can be seen: five rectangles (A, $D_{(x2)}$, and $E_{(x2)}$), the isosceles triangle (B), the cylinder (C), and the two right triangles ($D_{(x2)}$).

Now, let's go through the all of the steps that are required to calculate the total volume for the entire plate. First, calculate the volume for the cubed rectangles. Use the following formula for the three rectangular pieces (A and $E_{(x2)}$): $Volume = l \times w \times h$

Volume (A)	*Volume* (E_1, E_2)
$Volume = 10'' \times 15'' \times 2''$	$Volume = 1'' \times 1'' \times 10''$
$Volume = 300\ in^3$	$Volume = 10\ in^3$

The volume of the large base plate (A) calculates to 300 in^3 and each piece of 1 in^2 steel bar stock (E) has a calculated volume of 10 in^3. Now, locate the piece that has one rectangle and one right triangle (D). This is actually one piece, but for the purpose of volume calculations, it is looked upon as having separate shapes. Locate the dimension for the rectangular section, and calculate the volume.

Volume (D)

$$Volume = l \times w \times h$$

$$Volume = 1'' \times 1'' \times 2.75''$$

$$Volume = 2.75\ in^3$$

First—find the area using the two largest dimensions

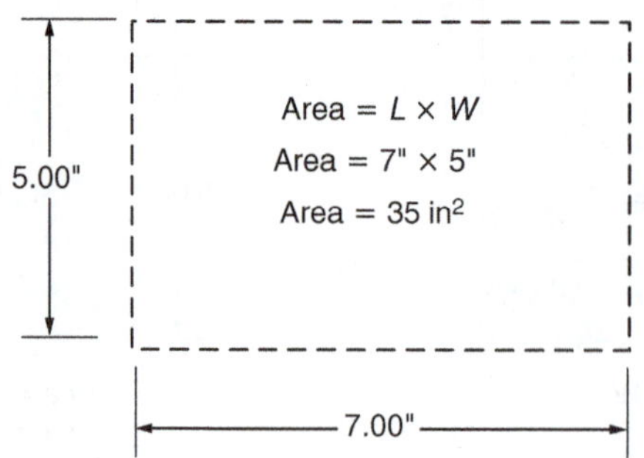

Second—find the area of the unused triangular spaces

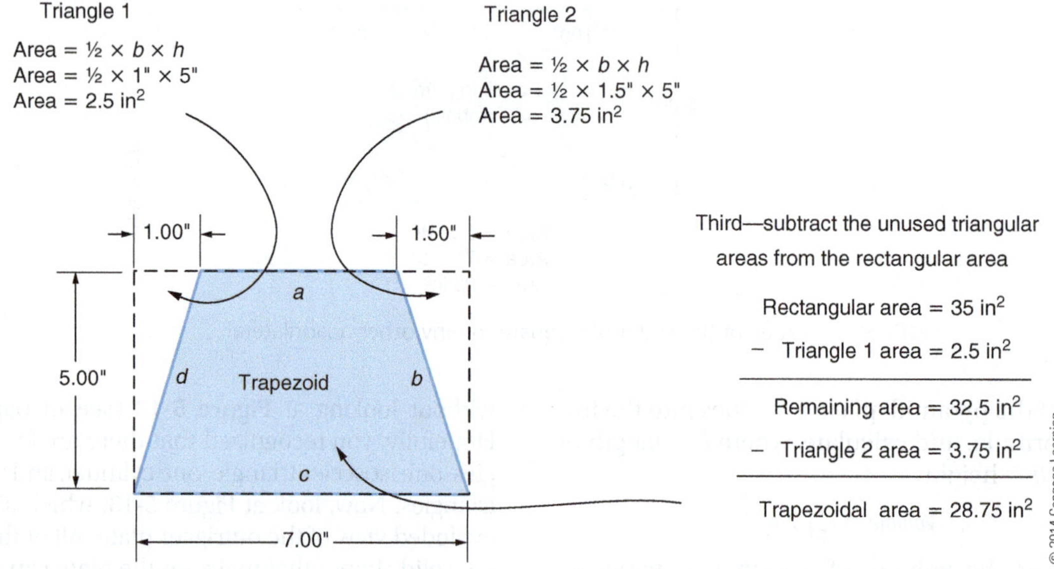

FIGURE 5-4 Area of a trapezoid.

We find that the rectangular section of piece (D) is 2.75 in³. Next, let us solve the right triangle section of piece (D). Because there are three dimensions, this is known as a triangular prism and use the same formula as was used in the previous triangular prism.

$$Volume = \left(\frac{1}{2}\right) bhd$$

$$Volume = \left(\frac{1}{2}\right) \times 2.75'' \times 1'' \times 1''$$

$$Volume = 1.375\ in^3$$

Next, locate the isosceles triangular prism (B) in Figure 5-13. Notice that the peak of the isosceles triangle is rounded. If we needed the exact volume, this portion would be deducted from the total volume. However, because this volume is considered to be negligent with reference to the entire prism, we disregard it and use the prism's full volume. To find the volume of this triangular prism, use the following formula where b = base, h = height, and d = depth:

$$Volume = \left(\frac{1}{2}\right) bhd$$

$$Volume = \left(\frac{1}{2}\right) \times 10'' \times 5'' \times 1''$$

$$Volume = 25\ in^3$$

Now, we calculate the volume for the cylinder (C). Because this cylinder actually represents a hole for the mating pin, we calculate this volume so that it can be deducted from the total isosceles triangle prism (B) volume. As was covered earlier, the area must be found first, using the given diameter of 2 in.

$$Area = 0.7854 \times D^2$$

$$Area = 0.7854 \times 2\ in^2$$

$$Area = 1.5708\ in^2$$

Using the area, calculate for the volume of the cylinder, using a cylinder length of 1 in.

$$Volume = Area \times Length$$

$$Volume = 1.5708\ in^2 \times 1\ in.$$

$$Volume = 1.5708\ in^3$$

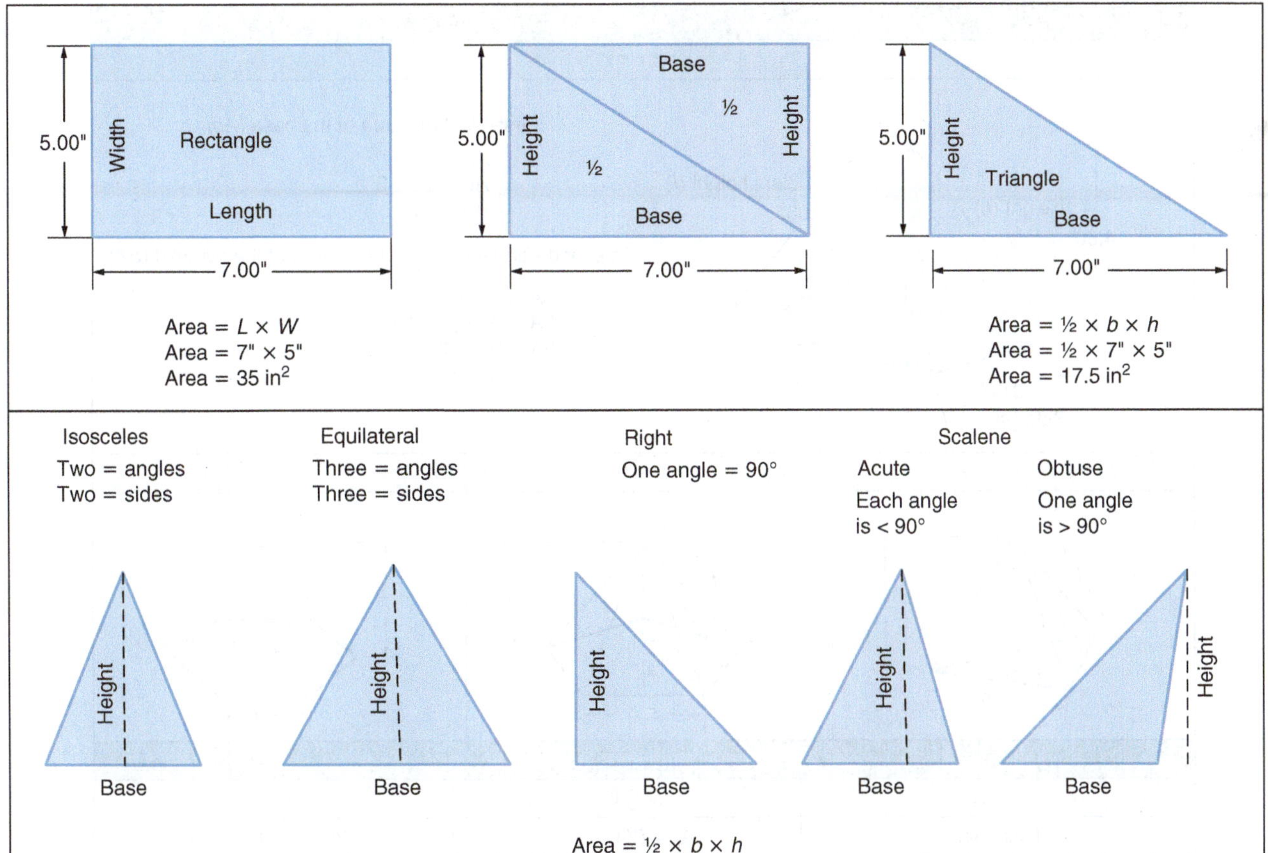

FIGURE 5-5 Area of a triangle.

This volume is then deducted from the total triangular prism volume.

$$\textit{Total Triangular Prism Volume} = 25\text{ in}^3 - 1.5708\text{ in}^3$$
$$\textit{Total Triangular Prism Volume} = 23.4292\text{ in}^3$$

The total volume for the leveling plate can now be calculated by adding the pertinent volumes together, as shown in **Figure 5-14** (see in page 102).

Weight Estimation

Whenever possible, the actual weight of a machine or object should be obtained from the manufacturer's documentation. This could be obtained via Internet sources (.pdf documents), specification data, or shipping documents. However, it is not always possible to get an accurate weight; therefore, you may need to estimate the weight.

Because the bulk weight for most machinery is derived from iron or steel, the weight of the machine can be quickly estimated using 500 lb per cubic-foot (lb/ft^3) as a standard weight. Pure iron is 490.7 lb/ft^3, plain carbon steel is 489.44 lb/ft^3, and stainless steel averages around 495 lb/ft^3. This is why a quick estimation of 500 lb/ft^3 can be used for machines or equipment made of such materials. As mentioned earlier, it is always better to err on the side of more weight rather than less when rounding or estimating. Never make the mistake of underestimating the weight of an object that is to be lifted or moved. This error could result in a failure of the lifting tools or equipment, causing serious injury or death, or at the very least, damage the machine or object that is being moved. If a more accurate calculation is necessary, weight tables should be used. Appendix H contains weight tables for bar stock, steel plate, square and rectangular steel tubing, and standard steel pipe, and **Table 5-2** (see in page 103) contains the weights for other materials such as aggregates, construction material, and wood. You must reference these tables to make accurate weight calculations.

When estimating an object that has an irregular shape, it may be difficult to obtain the exact measurements. In this case, it is generally acceptable to base the weight estimation on the largest dimensions that can be obtained. This gives an estimated weight that is heavier than the actual weight, but remember, it is always better to estimate on the heavy side if the actual weight cannot be obtained.

Figure 5-15 (see in page 104) shows a solid pressure roller that is used in a press machine. The entire roller is made from type 2205 stainless steel and is solid except for where the shaft runs through

Area

Square pyramid

Height (h) 4.00"
Slant height (l) 5.00"
3.00" 3.00"
Side (s)

First—find the area of the base (Ab)

$Ab = s^2$
$Ab = 3" \times 3"$
$Ab = 9 \text{ in}^2$

Second—find the total surface area of the pyramid (SA)

$SA = Ab + 2sl$
$SA = 9 + (2 \times 3 \times 5)$
$SA = 9 + 30$
$SA = 39 \text{ in}^2$

Triangular pyramid	Pentagonal pyramid	Hexagonal pyramid
l, h, a, s, b	l, h, a, s, b	h, l, a, s, b

Other pyramid area formulae

Pyramid type	Area of base	Total surface area
Triangular	$A_b = (1/2)as$	$SA = (1/2)as + (3/2)sl$
Square	$A_b = s^2$	$SA = A_b + 2sl$
Pentagonal	$A_b = (5/2)as$	$SA = (5/2)as + (5/2)sl$
Hexagonal	$A_b = (6/2)as$	$SA = 3as + 3sl$

Where:

A_b = Area of base	s, b = Side	h = Height	V = Volume
SA = Total surface area	a = Apothem length	l = Slant height	

Note: The apothem length is defined as the perpendicular distance from the center of any regular polygon to any of its sides.

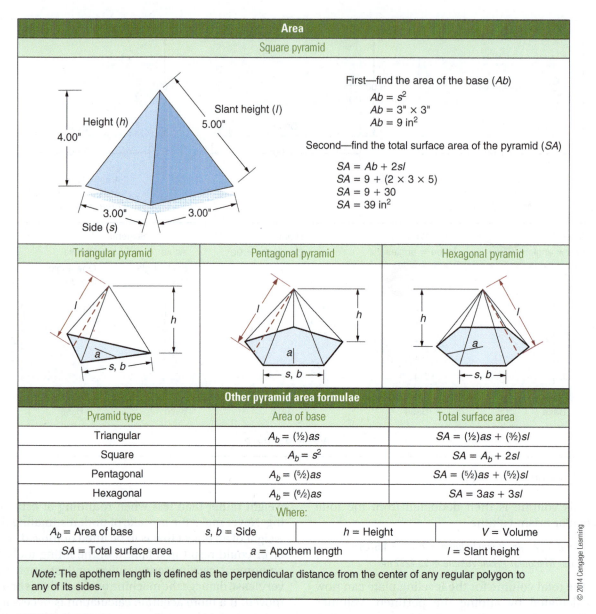

FIGURE 5-6 Surface area of a pyramid.

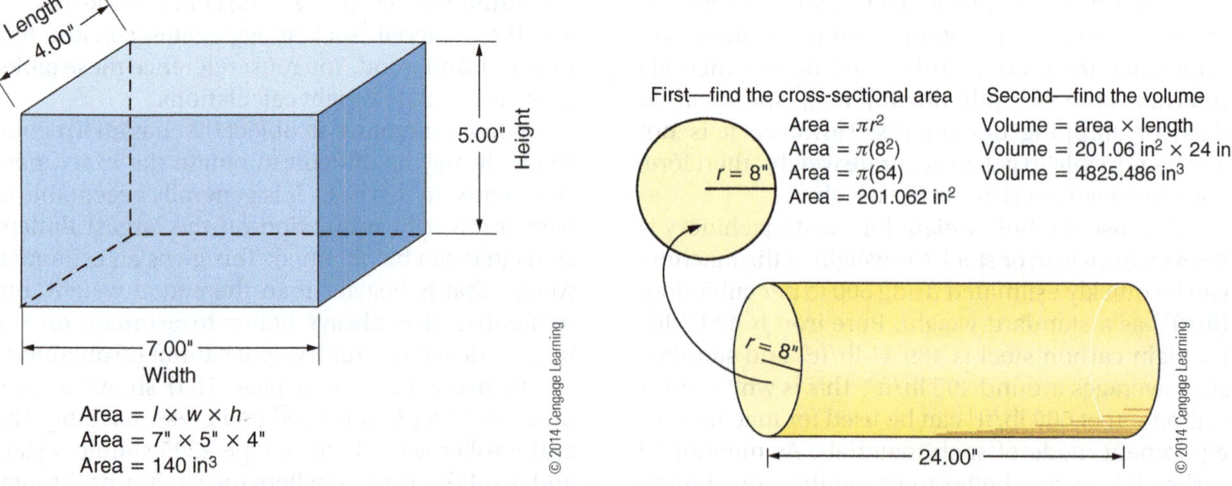

FIGURE 5-7 Volume of a cube.

FIGURE 5-8 Volume of a cylinder.

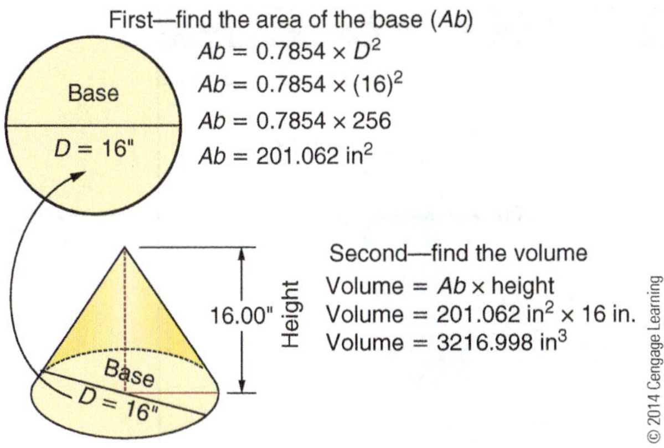

FIGURE 5-9 Volume of a cone.

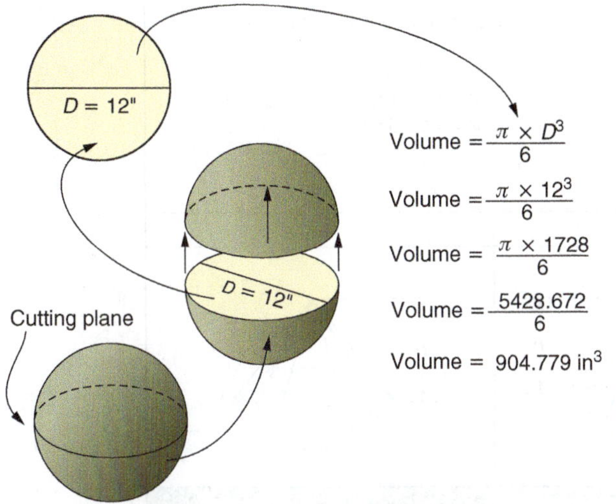

FIGURE 5-10 Volume of a sphere.

its center. The shaft is made from plain carbon steel. The volume of this roller must be figured as if it were made as three separate cylinders; the roller itself, the hole through the center of the roller, and the shaft. Once the volume calculations are complete, weight estimations can be made based upon the density of each type of material that is used.

First, we must calculate the volume of the stainless steel roller. Applying the rules discussed earlier, the area must be found first in order to calculate the volume of the roller. Remembering the formula for area when only the diameter is known ($A = 0.7854 \times D^2$), calculate. *Note:* For now, calculate the volume as if the cylinder were solid, and disregard the hollow space for the shaft in the center of the roller. **Figure 5-16** (see in page 104) shows an example of this.

Now we must calculate the volume of space needed for the shaft. This volume is then deducted from the solid roller volume to give us total roller volume, as seen in **Figure 5-17** (see in page 104).

The total roller volume is now known and is expressed in cubic inches (in^3). For the purpose of weight estimation, we are going to convert cubic inches to cubic feet. This is simply done by dividing the cubic inches by 12^3.

$$\mathit{Volume}\,(\text{cu}-\text{ft}) = \text{in}^3 \div 12^3$$
$$\mathit{Volume}\,(\text{cu}-\text{ft}) = 7916.832\ \text{in}^3 \div 12^3$$
$$\mathit{Volume} = 4.5815\ \text{ft}^3$$

We will use this value later to estimate the weight of this stainless steel roller, but first we must find the volume of the shaft. **Figure 5-18** (see in page 104) shows the solid carbon steel shaft and the calculations for both the cross-sectional area and the volume.

Now we estimate the weight of the entire roller assembly. Because the roller assembly is made of two different types of materials, the weight of the roller and the shaft must be calculated separately and then summed together. The roller is made of 2205 stainless steel, and the shaft is made of plain carbon steel. Refer to Table 5-2 and locate the lb/ft^3 for each type of material.

You should find that the type 2205 stainless steel has a density of 487.25 lb/ft^3. Now, take the total roller volume, which was calculated earlier (4.5815 ft^3), and multiply it by the 487.25 lb/ft^3.

$$\mathit{Total\ Roller\ Weight} = \mathit{Total\ Roller\ Volume} \times \mathit{Material\ Density}$$
$$\mathit{Total\ Roller\ Weight} = 4.5815 \times 487.25$$
$$\mathit{Total\ Roller\ Weight} = 2232.34\ \text{lbs.}$$

From Table 5-1, it is determined that plain carbon steel has a density of 489.44 lb/ft^3. Find the weight of the shaft using the same method as the roller.

$$\mathit{Total\ Shaft\ Weight} = \mathit{Total\ Shaft\ Volume} \times \mathit{Material\ Density}$$
$$\mathit{Total\ Shaft\ Weight} = 0.174533 \times 489.44$$
$$\mathit{Total\ Shaft\ Weight} = 85.42\ \text{lbs.}$$

Now simply add the two weights together to obtain the total weight for the whole roller assembly.

$$\mathit{Total\ Roller\ Assembly\ Weight} = \mathit{Total\ Roller\ Weight} + \mathit{Total\ Shaft\ Weight}$$
$$\mathit{Total\ Roller\ Assembly\ Weight} = 2232.34\ \text{lbs.} + 85.42\ \text{lbs.}$$
$$\mathit{Total\ Roller\ Assembly\ Weight} = 2317.76\ \text{lbs.}$$

Now take a moment to see whether you can estimate the weight of the leveling plate that was used to demonstrate volume calculation for irregularly shaped objects. Refer to Figures 5-11 through 5-15 if you need a reminder. Remember, you can use the tables in Appendix H to help solve this problem.

5-2 LOAD BALANCING

It is imperative to understand the importance of load balancing before making a lift. If an unbalanced load is lifted, the load will most likely shift during the lift, causing damage to the load itself or injury or death to personnel. Two things must be identified when balancing a load: the vertical center

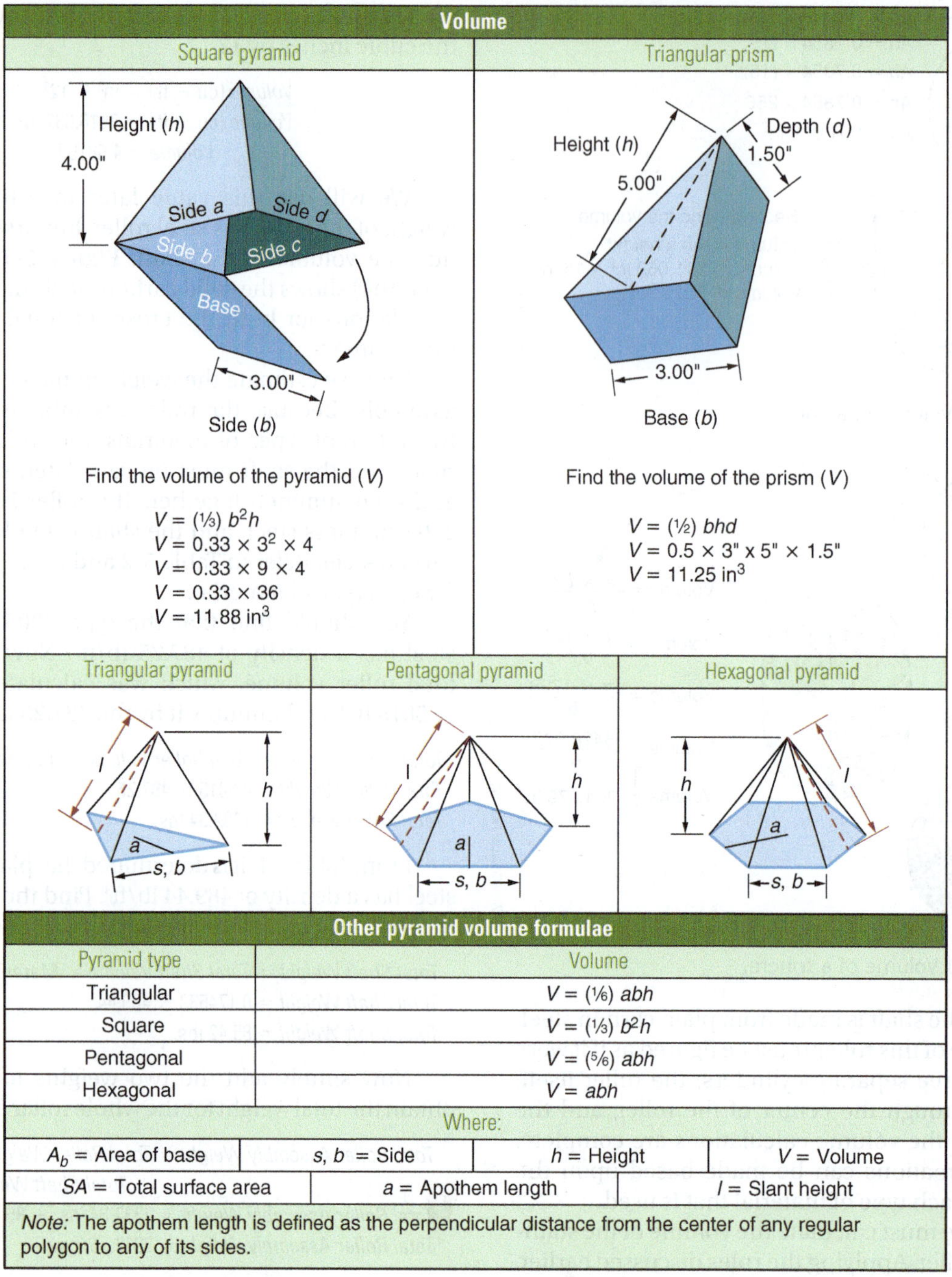

Volume

Square pyramid	Triangular prism
Height (h) 4.00"; Side a; Side d; Side b; Side c; Base; 3.00"; Side (b)	Height (h) 5.00"; Depth (d) 1.50"; 3.00"; Base (b)
Find the volume of the pyramid (V) $V = (⅓)\ b^2h$ $V = 0.33 \times 3^2 \times 4$ $V = 0.33 \times 9 \times 4$ $V = 0.33 \times 36$ $V = 11.88\ in^3$	Find the volume of the prism (V) $V = (½)\ bhd$ $V = 0.5 \times 3" \times 5" \times 1.5"$ $V = 11.25\ in^3$

Triangular pyramid	Pentagonal pyramid	Hexagonal pyramid
l; h; a; s, b	l; h; a; s, b	h; l; a; s, b

Other pyramid volume formulae

Pyramid type	Volume
Triangular	$V = (⅙)\ abh$
Square	$V = (⅓)\ b^2h$
Pentagonal	$V = (⅚)\ abh$
Hexagonal	$V = abh$

Where:

A_b = Area of base	s, b = Side	h = Height	V = Volume
SA = Total surface area	a = Apothem length	l = Slant height	

Note: The apothem length is defined as the perpendicular distance from the center of any regular polygon to any of its sides.

FIGURE 5-11 Volume of a prism.

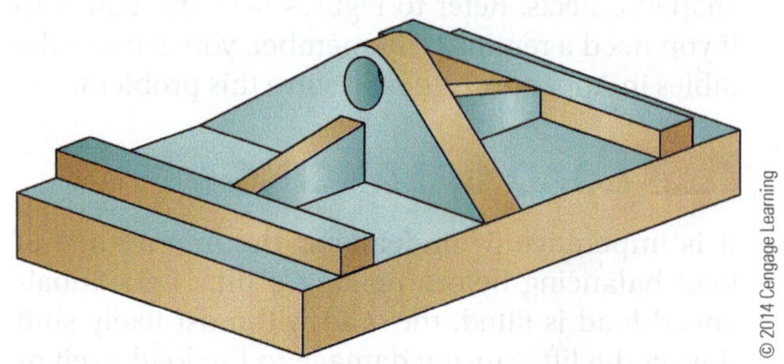

FIGURE 5-12 Leveling plate.

of gravity (vertical balance) and the horizontal center of gravity (center of weight).

Vertical Center of Gravity

Vertical balance must be achieved in order for a safe lift to occur. A safe lift cannot occur if the center of gravity is not correctly determined. For many items that are to be lifted, the center of gravity is easy to determine because the item itself is symmetrical in shape. Let us say for the sake of explanation, an item is cut exactly in half along its center point. If the two resulting pieces were symmetrical, they would be equal in weight

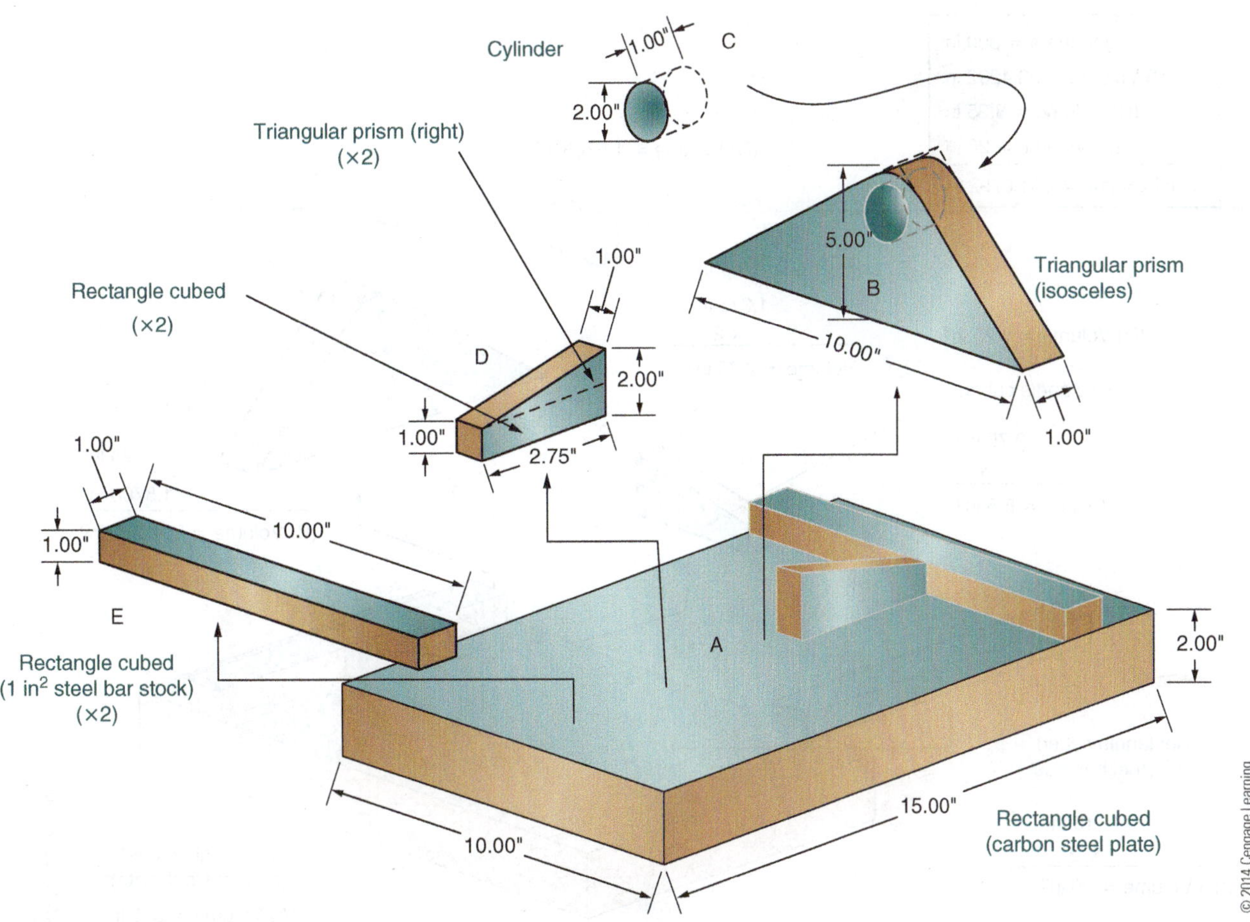

FIGURE 5-13 Identifying the regular shapes and dimensions on the irregularly shaped leveling plate.

because both sides would have an equal amount of material, and they would be mirror images of the opposite side. Often the manufacturer provides lifting points (or eyebolts) at the appropriate locations, which, when connected to properly, are directly over the center of gravity, as seen in **Figure 5-19** (see in page 105).

If the item is asymmetrical, then the two halves are not equal in weight. An example of an asymmetrical load would be a lathe. One end of the lathe is heavier than the other. This is also shown in Figure 5-19. Another thing to consider is palletized loads and liquid loads. The weight of the load may shift if the center of gravity is not correctly identified. When this occurs, the center of gravity moves away from the lifting center point and could cause the rigging equipment to fail or the items being lifted to fall off the pallet that is being lifted. It is extremely important when lifting liquids that the container is secured in a way such that the rigging will not fail if the liquid inside the container moves and causes the center of gravity to shift.

Center of Weight

Toppling a load is to be avoided at all costs. Toppling occurs when the top of the item being lifted weighs more than the bottom, and the lifting equipment is secured to a point below the center of weight. This is referred to as being "top heavy." Whenever possible, the actual center of weight of a machine or object should be obtained from the manufacturer's documentation. This could be obtained via Internet sources (.pdf documents), specification data, or shipping documents. If this information cannot be obtained, then remember this rule of thumb: It is always better to secure the rigging higher upon the load rather than lower. If the center of weight is lower than the point where the lifting equipment is attached, the load should be made safely. If the center of weight is above the point where the lifting equipment is attached, then the load may topple or become unstable during the lift, as seen in **Figure 5-20** (see in page 105).

5-3 SYNTHETIC SLINGS

Rigging equipment comes in many varieties. The most commonly used items to secure or lift a load are chains, slings, straps, and ropes. When considering each of these individually, you find that there are even more choices. For example, ropes come in many different forms. There are wire ropes,

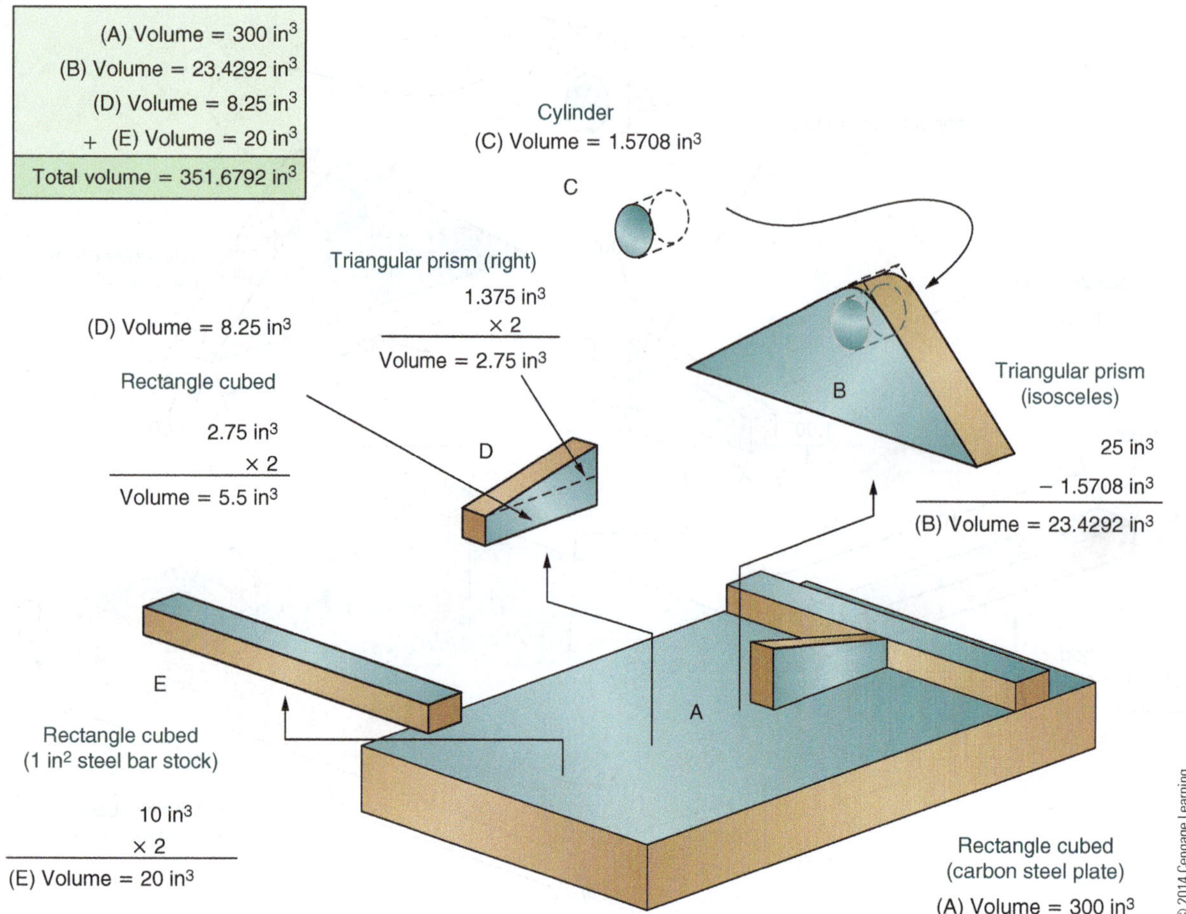

FIGURE 5-14 Calculating the total volume of the leveling plate.

synthetic ropes, and fiber ropes. It is important to know which is right for the type of lift that is being performed. Choosing the wrong equipment could result in death or injury of personnel or, at the least, damage to the load.

Sling Types

Slings are used to secure and lift the load. There is a wide variety of slings that are used in industry today, but only the most commonly ones are mentioned in this chapter. The most common types of slings are the synthetic web sling, the wire trope sling, and the chain sling. OSHA (Standards–29 CFR) Regulation 1910.184 lists the requirements for the proper use and inspection of all slings, and it would be advisable to review this standard prior to using any sling. All synthetic sling types must have sewn onto the sling identification that shows the manufacturer of the sling, the rated load of the sling, and the type of material used to manufacture the sling (if synthetic). If this tag is missing, the sling should not be used and should be removed from service.

Sling Terminology

Some terminology used in this chapter should be understood before moving forward. The **basket hitch** is a sling configuration whereby the sling is passed under the load and has both ends, end attachments, eyes, or handles on the hook or a single master link. A **master link** or **gathering ring** is a forged or welded steel link used to support all members (legs) of an alloy steel chain sling or wire rope sling. **Angle of loading** is the inclination of a leg or branch of a sling measured from the horizontal or vertical plane as shown in **Figure 5-21** (see in page 105).

A **choker hitch** is a sling configuration with one end of the sling passing under the load and through an end attachment, handle, or eye on the other end of the sling. **Proof load** is the load applied in performance of a proof test. A **proof test** is a nondestructive tension test performed by the

Table 5-2

MATERIAL WEIGHTS (LBS./FT³)

Aggregate/Minerals	lbs./ft³
Coal, Anthracite, solid	94
Coal, Bituminous, solid	84
Earth, dry, excavated	78
Earth, moist, excavated	90
Earth, wet, excavated	100
Earth, dense	125
Earth, soft loose mud	108
Earth, packed	95
Granite, solid	168
Granite, broken	103
Gravel, loose, dry	95
Gravel, w/sand, natural	120
Gravel, dry 1/4 to 2 in.	105
Gravel, wet 1/4 to 2 in.	125
Gypsum, solid	174
Gypsum, broken	113
Gypsum, crushed	100
Limestone, solid	163
Limestone, broken	97
Limestone, pulverized	87
Marble, solid	160
Salt, course	50
Salt, fine	75
Sand, wet	120
Sand, wet, packed	130
Sand, dry	100

Construction Material	lbs./ft³
Asphalt, crushed	45
Brick, common red	120
Brick, fire clay	150
Cement, Portland	94
Cement, slurry	90
Concrete, Asphalt	140
Concrete, Gravel	150
Concrete, Limestone	148
Mortar, masonry (wet)	150
Plaster	53
Tar	72

Metals	lbs./ft³
Aluminum, solid	165
Aluminum, oxide	95
Babbitt	454
Brass, cast	534
Brass, rolled	534
Bronze	509
Copper, cast	542
Copper, rolled	556
Gold, pure 24Kt	1204
Iron, cast	450
Iron, pure	490.7
Iron, wrought	483.82
Lead, cast	708
Lead, rolled	711
Lead, red	230
Magnesium, solid	109
Mercury @ 20°C (68°F)	845.63
Nickel, rolled	541
Nickel silver	527
Platinum	1342
Silver	653
Steel, C1020HR	490.06
Steel, carbon AISI-SAE 1020	490.68
Steel, carbon (plain)	489.44
Steel, carbon (tool)	488.19
Steel, cast	490
Steel, cold-drawn	488.81
Steel, rolled	495
Steel, soft (0.06% carbon)	491.31
Steel, stainless (301)	490
Steel, stainless (304)	501.29
Steel, stainless (2205)	487.25
Tin, cast	459
Titanium	281.7
Tungsten	1224
Vanadium	343
Zinc, cast	440
Zinc oxide	25

Other Materials	lbs./ft³
Cork, solid	15
Rubber, caoutchouc	59
Rubber, manufactured	95
Glass	161
Glue, animal (flake)	35
Glue, vegetable (powder)	40
Oil, linseed	58.8
Oil, petroleum	55
Porcelain	150

Water	lbs./ft³
Ice, solid	57.4
Ice, crushed	37
Snow, freshly fallen	10
Snow, compacted	30
Water, pure @ 4°C (39.2°F)	62.4
Water, sea	64.08

Wood/Paper	lbs./ft³
Ash wood, black, dry	34
Ash wood, white, dry	42
Birch wood, yellow	44
Cardboard	43
Cedar, red	24
Cherry wood, dry	35
Chestnut wood, dry	30
Elm, dry	35
Mahogany, Spanish, dry	53
Mahogany, Honduras, dry	34
Maple, dry	44
Oak, live, dry	59
Oak, red	44
Paper, standard	75
Pine, White, dry	26
Pine, Yellow Northern, dry	34
Pine, Yellow Southern, dry	45
Redwood, California, dry	28
Spruce, California, dry	28
Sycamore, dry	37
Walnut, black, dry	38

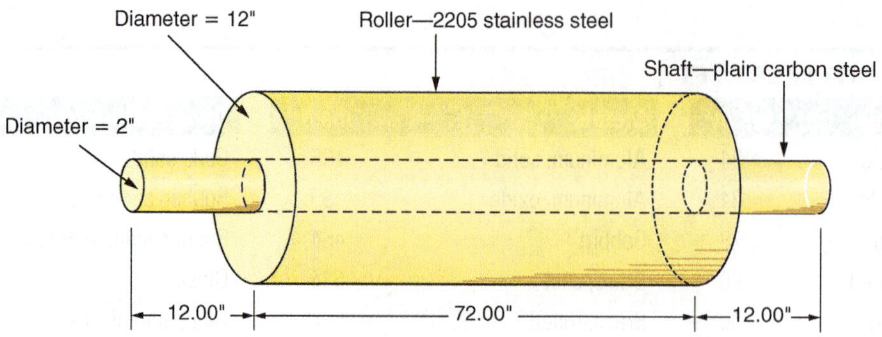
Diameter = 12"
Roller—2205 stainless steel
Shaft—plain carbon steel
Diameter = 2"
12.00"
72.00"
12.00"

FIGURE 5-15 A solid roller used in a pressing machine.

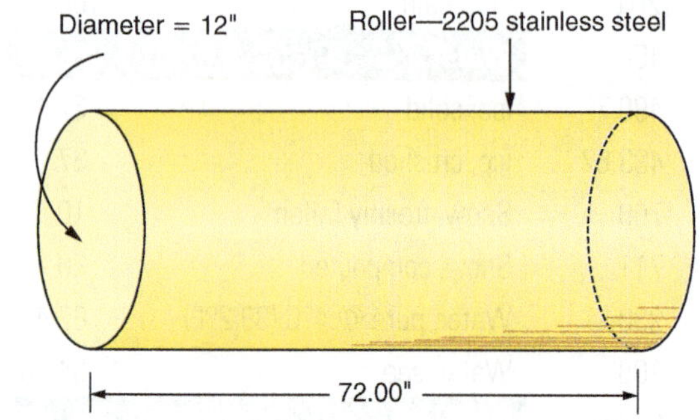
Diameter = 12"
Roller—2205 stainless steel
72.00"

First—find the cross-sectional area

Area = $0.7854 \times D^2$

Area = 0.7854×12^2

Area = 0.7854×144

Area = 113.0976 in^2

Second—find the volume

Volume = area × length

Volume = 113.0976 $in^2 \times$ 72 in.

Volume = 8143.0272 in^3

FIGURE 5-16 Volume of the roller cylinder.

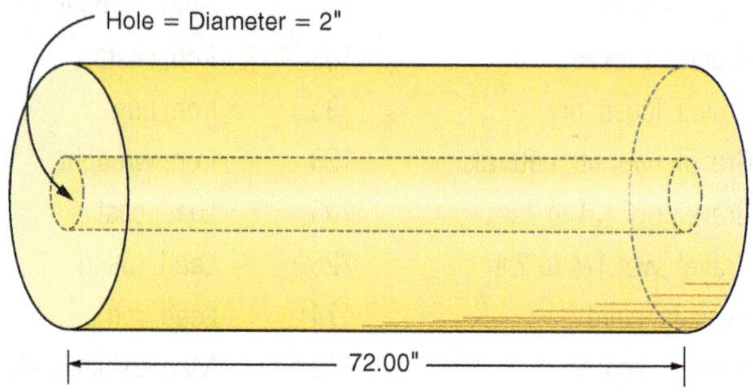
Hole = Diameter = 2"
72.00"

First—find the cross-sectional area

Area = $0.7854 \times D^2$

Area = 0.7854×2^2

Area = 0.7854×4

Area = 3.1416 in^2

Second—find the volume

Volume = area × length

Volume = 3.1416 $in^2 \times$ 72 in.

Volume = 226.1952 in^3

Deduct shaft volume from solid roller volume

Total roller volume = Solid roller volume − shaft volume

Total roller volume = 8143.0272 in^3 − 226.1952 in^3

Total roller volume = 7916.832 in^3

FIGURE 5-17 Subtract the volume where the shaft is inserted.

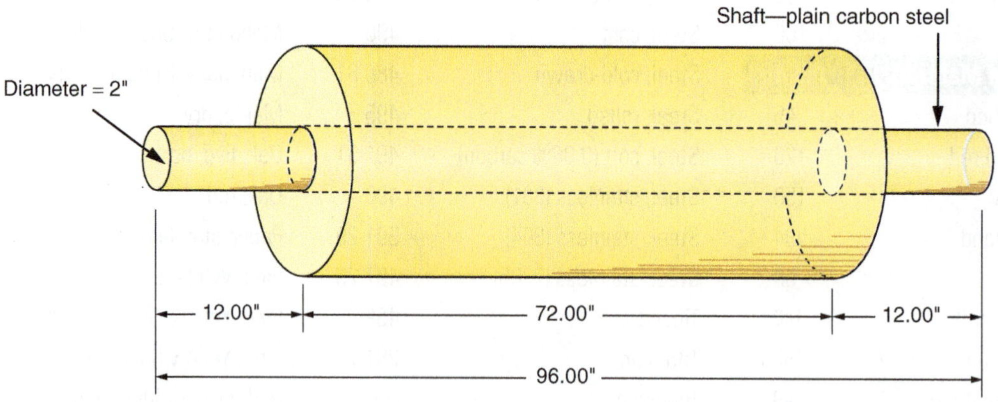
Shaft—plain carbon steel
Diameter = 2"
12.00"
72.00"
12.00"
96.00"

First—find the cross-sectional area

Area = $0.7854 \times D^2$

Area = 0.7854×2^2

Area = 0.7854×4

Area = 3.1416 in^2

Second—find the volume

Volume = area × length

Volume = 3.1416 $in^2 \times$ 96 in.

Volume = 301.5936 in^3

Convert to cu-ft

$$\text{Volume (cu-ft)} = \frac{301.5936\ in^3}{12\ in^3}$$

Volume (cu-ft) = 0.174533 ft^3

FIGURE 5-18 Volume of the shaft.

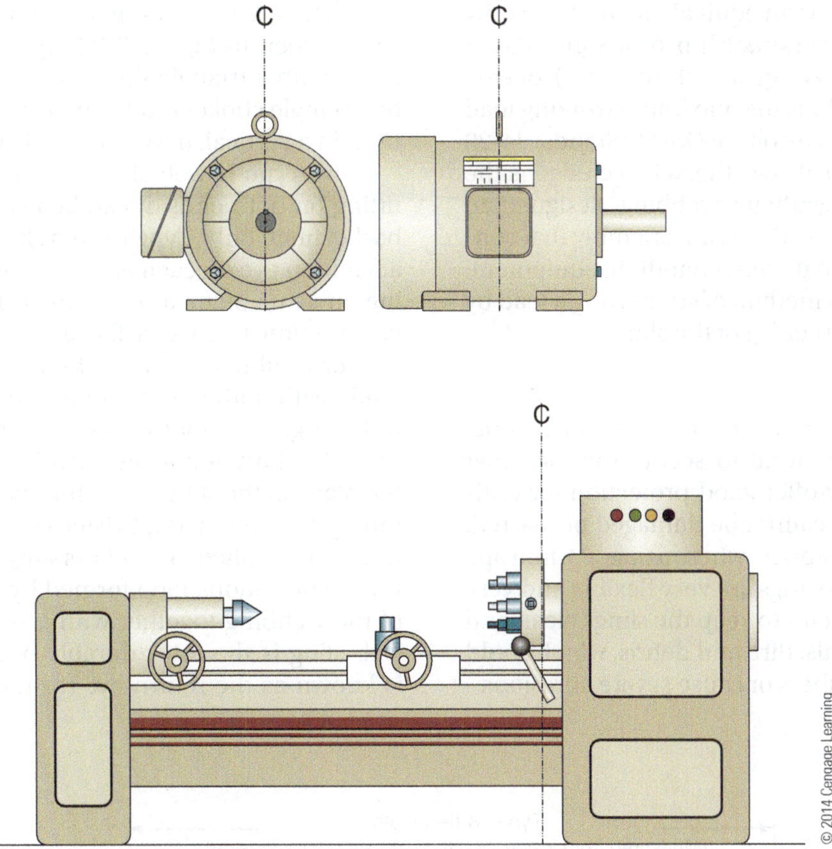

FIGURE 5-19 Center of gravity.

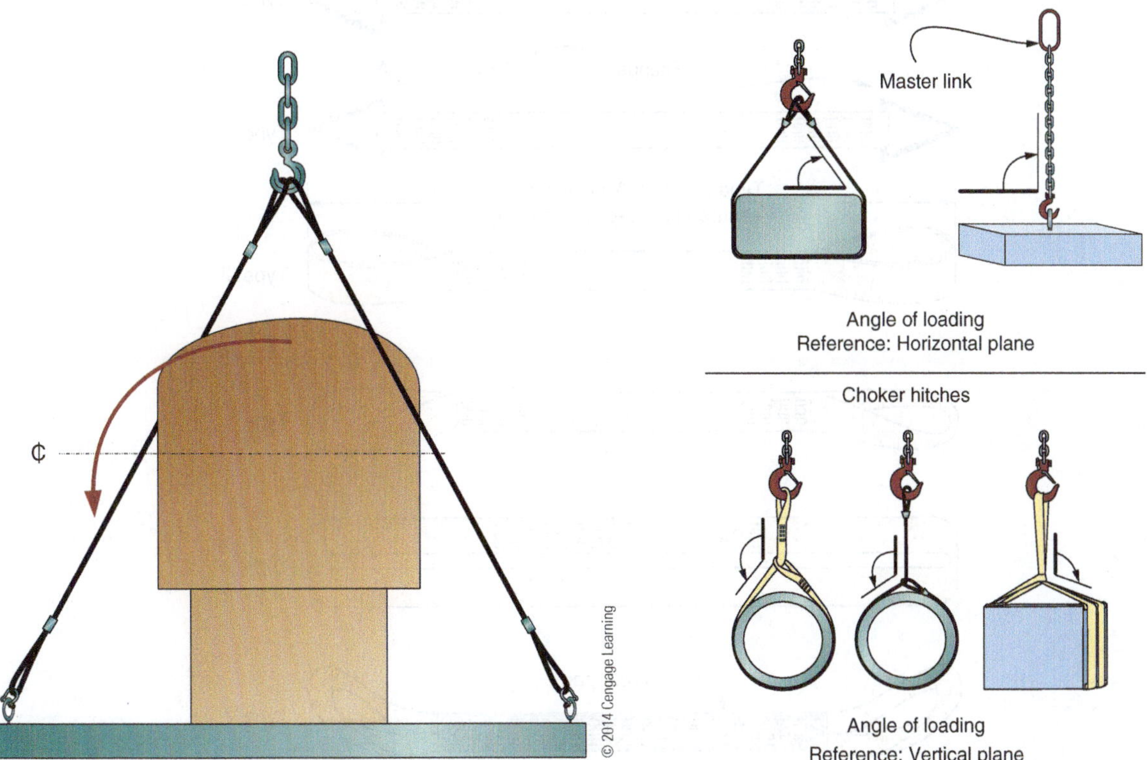

FIGURE 5-20 Center of weight.

FIGURE 5-21 Angle of loading.

sling manufacturer or an equivalent entity to verify construction and workmanship of a sling. **Rated capacity (CAP)**, **working load limit (WLL)**, or **safe working load (SWL)** is the maximum working load permitted by the provisions of OSHA (Standards–29 CFR) Regulation 1910.184. The **selvage edge** is the finished edge of synthetic webbing designed to prevent unraveling. A **sling** is an assembly that connects the load to the material handling equipment. A **vertical hitch** is a method of supporting a load by a single, vertical part or leg of the sling.

Web Slings

Synthetic web slings are used heavily in rigging. Most often they are used to secure the load that is to be lifted. They offer good protection for critical machinery that cannot be damaged or marred. This could easily happen when using a wire rope or chain sling. Web slings are very flexible and very durable. It is important to keep the slings clean and free of oils, chemicals, dirt, and debris, which could cut the individual fibers or cause severe abrasions.

There are six types of web slings, each of which can be seen in **Figure 5-22**. Type I web slings are made with a triangle fitting on one end and a slotted triangle choker fitting on the other end. It can be used in a vertical, basket, or choker hitch.

The Type II web sling is made with a triangle fitting on both ends. It can be used in a vertical or basket hitch only. Type III web slings are made with a flat loop eye on each end with the loop eye opening on same plane as sling body. This type of sling is sometimes called a flat eye-and-eye, eye-and-eye, or double-eye sling. The type IV web sling is made with both loop eyes formed as in the type III web sling, except for the fact that the loop eyes are turned to form a loop eye that is at a right angle to the plane of the sling body. This type of sling is commonly referred to as a twisted-eye sling. The type V web sling is called the endless sling a grommet sling. It is a continuous loop formed by joining the ends of the webbing together with a splice. The type VI web sling is the most durable of all web slings. It is known as the return-eye (or reversed-eye) sling

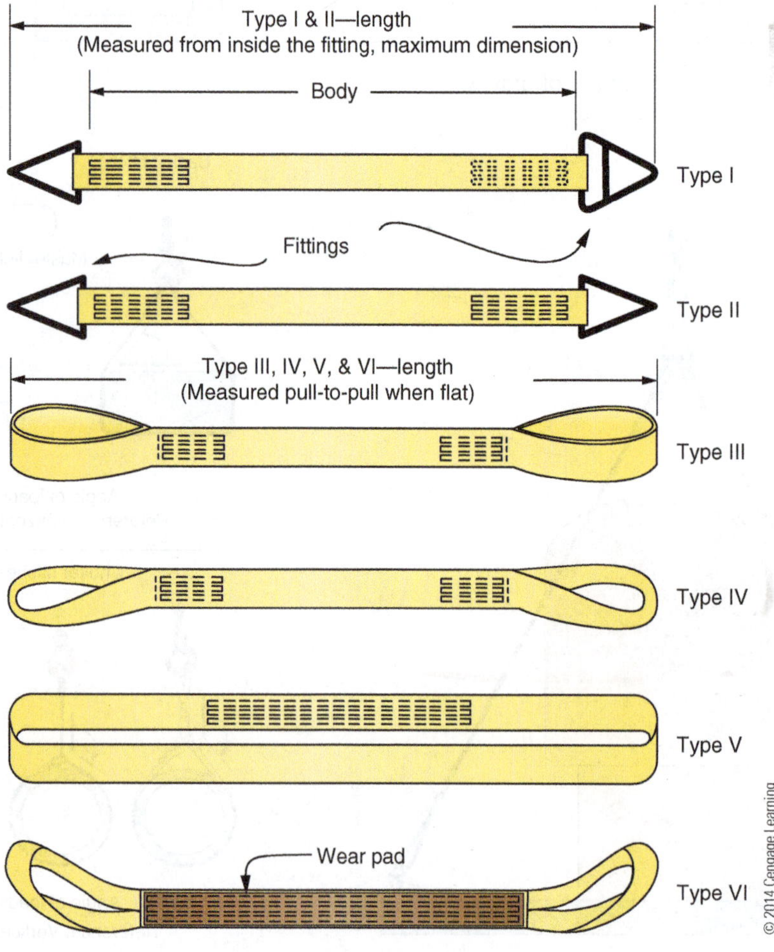

FIGURE 5-22 Web sling types.

and is formed by using multiple widths of webbing held edge to edge. A wear pad is attached on one or both sides of the sling body and on one or both sides of the loop eyes, offering maximum resistance to abrasions. The loop eye at each end of the sling is at a right angle to the plane of the sling body. This eye configuration allows for tight choking and is a good sling to choose for vertical and basket hitches as well.

Web slings are made with internal warning cores that when exposed; mandate the removal of the sling from service. The warning cores are of contrasting and bright coloring so they can indicate excessive wear, cuts, excessive stretching, and severe abrasions. Web slings should be protected from cutting when lifting an object with sharp edges. Slings can be purchased with sewn-on wear pads or separate sliding wear pads can be purchased and installed (slid) onto the sling prior to being place onto the load. Either offers that much-needed protection. Using the slide-on wear pad keeps the sling clean while in service. The sliding wear pad does not add to the load capability of the web sling. Wear pads that are sewn onto a web sling are typically made of leather because of its durability. There are three ways that a load can be lifted using a synthetic sling: the vertical hitch, the choker hitch, and the basket hitch, all three of which can be seen in **Figure 5-23**.

Web Sling Load Capacities

Synthetic web slings are made with a number of plies, which increases the safe load capacity with each ply. A double-ply web sling, for example, may have a capacity range that is 140% to 200% that of a single-ply, depending on the sling type, quality of the sewing, grade of the webbing material, and the method used to hook the sling. A ply is a layer of the web material, and web slings can have up to four of these plies. The sling capacity is sometimes referred to as the working load limit (WLL) or the safe working load (SWL). It is required to be listed on the sling manufacturer's label. Rated capacities are affected by the angle of load for vertical choker lifts (sling to load). When using multilegged slings in a basket lift, the angle of loading is measured from the horizontal plane, as seen in Figure 5-21. To determine the actual sling capacity at a given angle of lift, multiply the original sling rating by the appropriate loss factor determined from the **Table 5-3a** below.

Annex I contains tables with web sling load capacities based on sling type. The capacities given in the tables reflect manufacturers' load capacity ratings when the web sling is used as a choker hitch. The rated loads given in **Table 5-3b** are to be used when a sling has an angle of choke 120° or greater. An angle of choke is formed by the web sling body as it passes through the choking eye. Anytime a web sling is in contact with a sharp edge, you must place protection between the load and the sling, as seen in **Figure 5-24**.

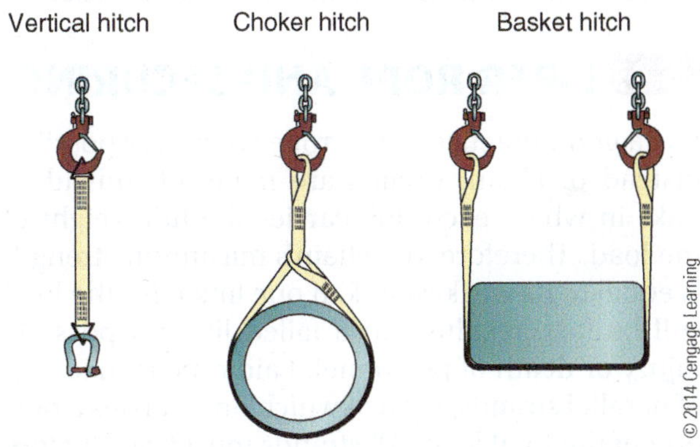

FIGURE 5-23 Types of hitches.

Round Sling

Round slings are excellent for securing almost any object that has to be pulled or lifted. It consists of a continuous (endless) loop of 100% polyester fiber. Its many features include resistance to acids, UV rays, rot, and mildew. Round slings are manufactured to allow minimal elongation, and they are easier to rig than chain or wire rope slings. The round sling is very flexible and lightweight. Unlike the web synthetic sling, the load carrying fiber strands are free to move about inside a protective covering. Also, the load bearing fibers never come in contact with the load, so they do not wear as long as the cover remains intact. These slings are typically endless, meaning that they form an endless loop. They have very limited stretching and weigh less when compared to other types of slings of the same load capacity (i.e., wire rope or chain). The outer cover of a round sling is typically made with nylon, and it should never carry the load. This can only occur if the inner strands have broken, causing the round sling to have stretched. This is a very difficult thing to inspect as the internal strands are completely surrounded by the outer covering. Because of this, new methods of stretch and breakage detection have been put into round slings to indicate when they have stretched beyond acceptable lengths. One is with the use of stretch indicators, which are colored yarns that are sewn through the outer covering so as to be visible from outside the covering. If the round sling is stretched, the colored strands disappear into the covering. If this occurs, the sling must be removed from service. Another method of stretch indication occurs with

Table 5-3a

SYNTHETIC WEB SLING ANGLE LOSS FACTOR TABLE (NOT CHOKER HITCH)

Angle (in degrees)	Loss Factor	Angle (in degrees)	Loss Factor
90	1.000	55	0.819
85	0.996	50	0.0766
80	0.985	45	0.0707
75	0.966	40	0.643
70	0.940	35	0.574
65	0.906	30	0.500
60	0.866	–	–

Table 5-3b

SAFETY FACTOR OF WEB SLINGS WHEN USED AS A CHOKER HITCH

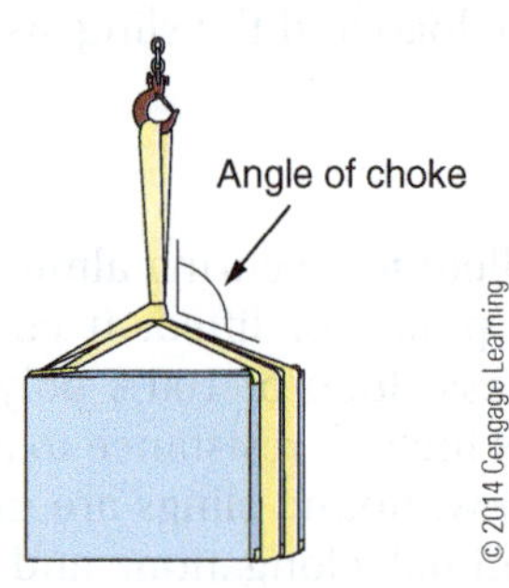

Angle of Choke (in degrees)	Sling Rated Load Factor
120–180	0.750
90–119.9	0.650
60–89.9	0.550
30–59.9	0.400

the use of fiber optics. The construction of this type of sling comprises a fiber-optic signal strand within the inner lifting core. Both ends of the fiber optic strand emerge from under the outer covering near the identification tag. Light shining into one end is visible at the other end if there is no damage to the core by stretching or breaking. Defective continuity (a stretched or damaged core) is indicated when light has been put into the fiber-optic strand and no light can be seen coming out the other end. If this occurs, the sling must be removed from service. **Figure 5-25** shows an example of a round sling that has both types of stretch indicators.

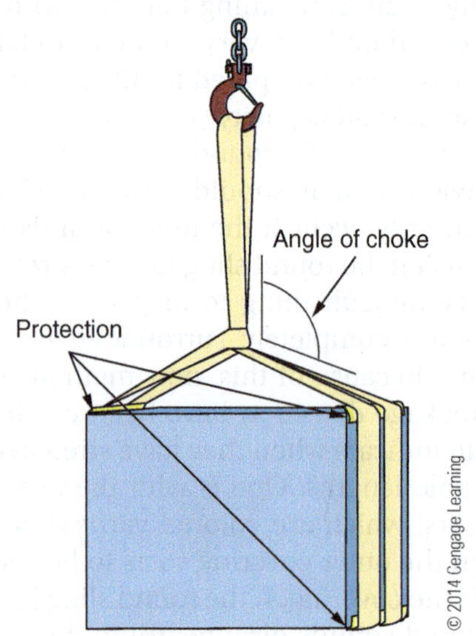

FIGURE 5-24 Protection of web sling.

Round Sling Load Capacities

Most round slings capacities are identified by the color of the outer covering. The colors are listed in **Table 5-4** along with the load capacities for each.

5-4 FIBER ROPE AND SECURING

It is a good practice to use rope whenever possible instead of chain. Chains are made of individual links in which each link carries the full weight of the load. Therefore, the chain's maximum strength is equal to its weakest link. If one link fails, the load will be lost, resulting in a failed lift and possible injury or death of personnel. Laid rope is made up of parallel strands, each of which only carries a portion of the total load. All strands must break before the rope fails.

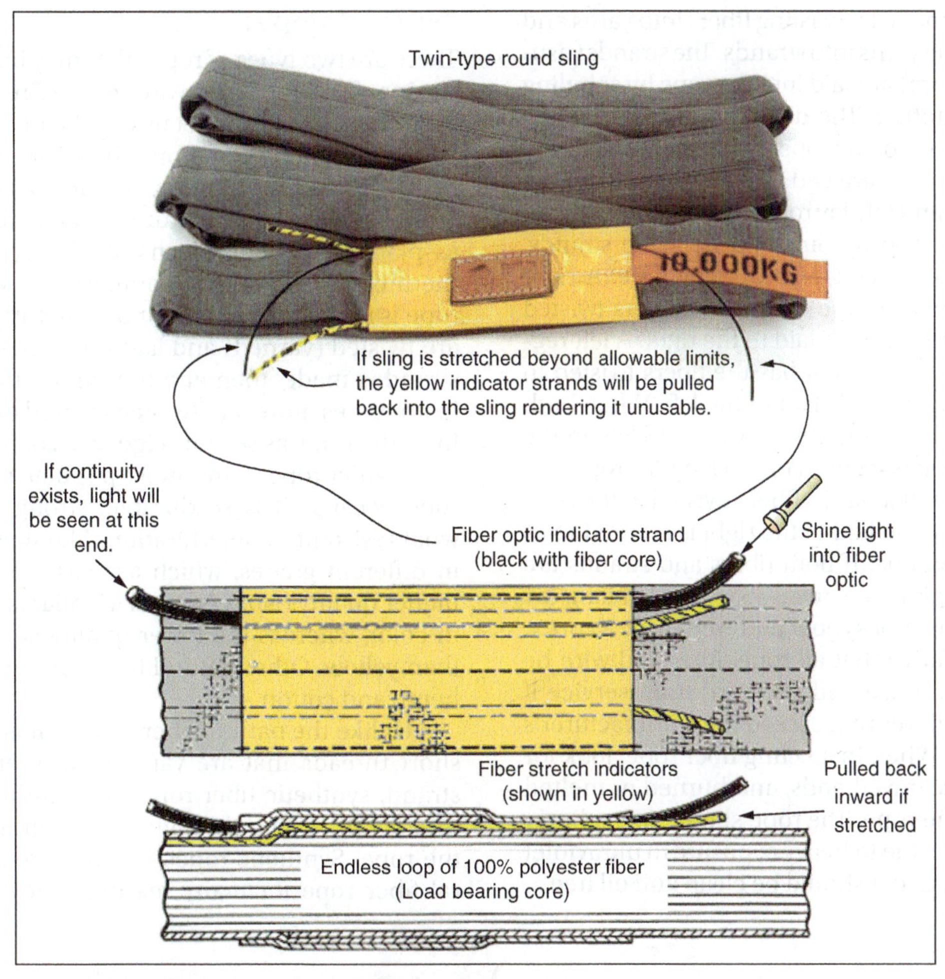

FIGURE 5-25 Round sling safety indicators.

Table 5-4

ROUND SLING LOAD CAPACITIES

Color Code	Weight/ft. (lbs.)	Approx. Diameter (in.)	Vertical (Single Leg)	Choker	Basket Hitch		
					90°	60°	45°
Purple	0.30	0.60	2,600	2,080	5,200	4,500	3,600
Green	0.40	0.80	5,300	4,240	10,600	9,100	7,400
Yellow	0.50	1.00	8,400	6,720	16,800	14,500	11,800
Tan	0.60	1.20	10,600	8,480	21,200	18,300	15,000
Red	0.80	1.30	13,200	10,560	26,400	22,800	18,600
Orange	0.90	1.40	15,900	12,720	31,800	27,500	22,500
Blue	1.20	1.55	21,200	16,960	42,400	36,700	29,900
Orange	1.50	1.75	25,000	20,000	50,000	43,300	35,300
Orange	2.00	1.95	31,000	24,800	62,000	53,600	43,800
Orange	2.80	2.35	40,000	32,000	80,000	69,200	56,500
Orange	3.60	3.15	52,900	43,320	105,800	91,600	74,800
Orange	4.60	3.95	66,100	52,880	132,200	114,400	93,400
Orange	5.80	4.80	90,000	72,000	180,000	155,800	127,200

In no case should the manufacturer's rated capacities listed on the label of the sling be exceed.

Ropes are made by twisting fibers into yarns and then twisting the yarns into strands. The strands (usually three or more) are laid into the rope by spiraling around one another. The direction of the spiraling determines the lay of the rope. Lay refers to the direction that the strands are laid into the rope. There are right-lay ropes and left-lay ropes. When the fibers are twisted in the opposite direction from the strands, it is referred to as being a regular-lay rope. A right regular-lay rope is a rope that has its fibers twisted to the left and the strands laid to the right. A left regular-lay rope is a rope that has the fibers twisted to the right and the strands laid to the left. When both the fibers and the strands are twisted and laid in the same direction, it is referred to as a lang-lay rope. For example, a rope that has its fibers twisted to the right and the strands also laid to the right is referred to as a right lang-lay rope. If both fibers and strands are in the left direction, it is referred to as a left lang-lay rope. Also, some ropes (fiber and wire) have a core.

It is imperative that all rope, fiber and wire be inspected before use and removed from service if the rope has failures that exceed the manufacturer's specifications. When inspecting fiber rope, look for abrasions, cuts, frayed ends, and burned or melted places along the rope. The rope should not splinter into your hands due to heavy exposure to ultraviolet light (the sun), and it should be clean and oil free.

Types of Ropes

There are two types of ropes that may be used when rigging, fiber rope and wire rope. Wire rope is discussed later. For now, let us consider fiber rope and its qualities and limitations. There are two types of fiber rope: natural and synthetic. Natural rope is rope that is made of a natural organic material such as plants. Synthetic means artificially made from the mixing of chemical compounds. Natural fiber rope is named so because it uses natural fibers that are twisted (yarned) and laid into a strand. Once a strand is made, then at a minimum, three strands (sometimes more) are twisted around each other to form the rope, as seen in **Figure 5-26**.

Manila rope is the most popular natural fiber rope because it is so durable, strong, and somewhat resistant to deterioration. Manila rope comes in different grades, which are easy to detect. The higher quality (No. 1 Natural Manila) is light yellow in color, whereas the lower quality is more brown than yellow. Other natural fiber ropes may include hemp and cotton.

Unlike the natural fiber rope that is made from short threads that are yarned together to form a strand, synthetic fiber ropes have individual, continuous threading throughout the entire length of the rope. Synthetic ropes are preferred over natural fiber rope for many reasons. They do not rot,

FIGURE 5-26 Construction of natural rope.

are resistant to mildew, and have more strength than that of natural fiber rope of the same size. However, one undesirable trait of synthetic rope is that all types are susceptible to ultraviolet light, so if left in the sun, they degrade. Synthetic rope is often woven as well as twisted, which gives the rope excellent strength. There are several different types of synthetic rope that may be used in rigging: polypropylene, polyethylene, nylon, and polyester. Polyethylene is the lightest synthetic rope and is not typically used in construction. It is typically limited to home use. Even though it is stronger than natural fiber rope, it only has about 50% the capacity of nylon rope. Polypropylene is also a light-duty synthetic rope, yet it is much stronger than natural fiber rope and polyethylene. It is lightweight, so it floats when put into water and it is resistant to most acids and alkalis. Also, it can withstand shock loading fairly well. Polyester is about twice the strength of manila rope and has an excellent resistance to abrasions, chemicals, and weathering. Nylon is the strongest of the types listed and is the most durable. It is resistant to most alkalis and solvents and can be stored wet, as it does not rot. It also has the greatest shock load of all the synthetics listed.

When cutting a natural fiber rope, you must first tape the rope where it is going to be cut to prevent unraveling. For synthetic ropes, simply cut the rope, and melt the end of the rope to prevent unraveling.

Rope Load Capacities

Any time that rope is used for straight vertical hoisting, a safety factor must be applied to prevent injury to personnel, property, or equipment. A rope's load capability is determined by its breaking strength. The breaking strength is divided by the safety factor, giving the maximum load allowed during the vertical lift. The safety factor for all fiber rope is 5, unless it is used to hoist or support personnel. If this is the case, the safety factor is 10. Safe Working Loads (SWL) are calculated as follows:

$$SWL = \frac{\text{Breaking Strength of the Rope}}{\text{Safety Factor}}$$

Example: If a ½" nylon rope has the breaking strength of 6250 lb then the safe working load is 1250 lb.

$$SWL = \frac{6250\text{ lbs.}}{5} = 1250\text{ lbs.}$$

If the same rope is used to hoist personnel, the safety factor becomes 10.

$$SWL = \frac{6250\text{ lbs.}}{10} = 625\text{ lbs.}$$

Any time that injury to personnel could result while using a rope, the safe working load is reduced significantly so as not to risk failure.

Rope End Preparation

When fiber rope is cut, the ends tend to fray. For synthetic rope, the solution is easy. Simply expose the cut ends to a flame and they melt, causing all of the threading to become solidified once cooled. Be careful not to get burned by the hot melting synthetic material when performing this procedure. Some people use their fingers to "shape" the melted end while it is still in the "plastic" (formable) state. This is not recommended. Use a sheet of metal, a piece of wood, or maybe even a concrete surface to form the hot ends.

Natural fiber rope ends cannot be prepared with a flame. Instead, a method referred to as "whipping" or "seizing" must be performed. To whip or seize a rope end, use a thin twine about a foot long. Make one end into a loop and place it at the end of the rope, as seen in **Figure 5-27**. End A should be fairly short. Wind the longer end of the twine around the rope and the loop, spiraling away and drawing each turn tight. When the whipping is as wide as the diameter of the rope, thread the twine through the end of the loop. Pull end A hard until the loop has disappeared under the whipping. Trim off the two ends.

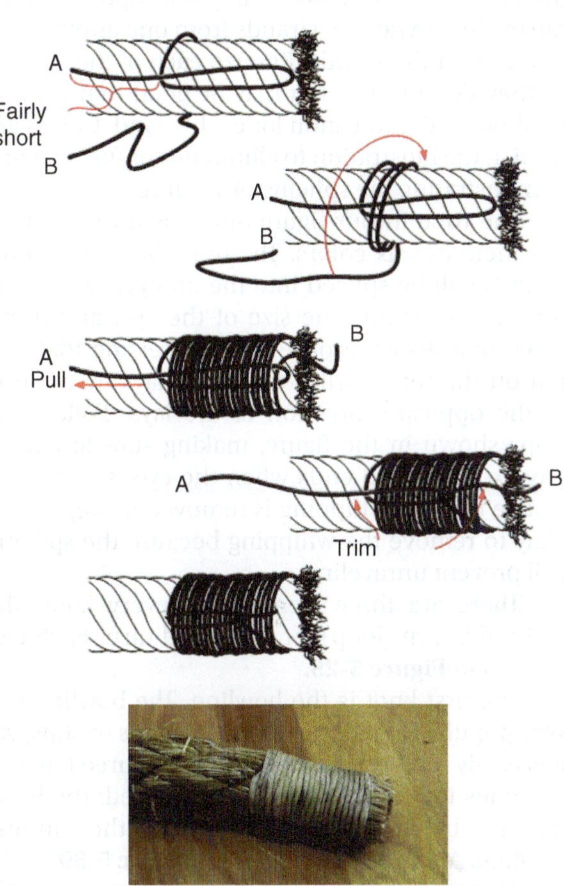

FIGURE 5-27 Whipping.

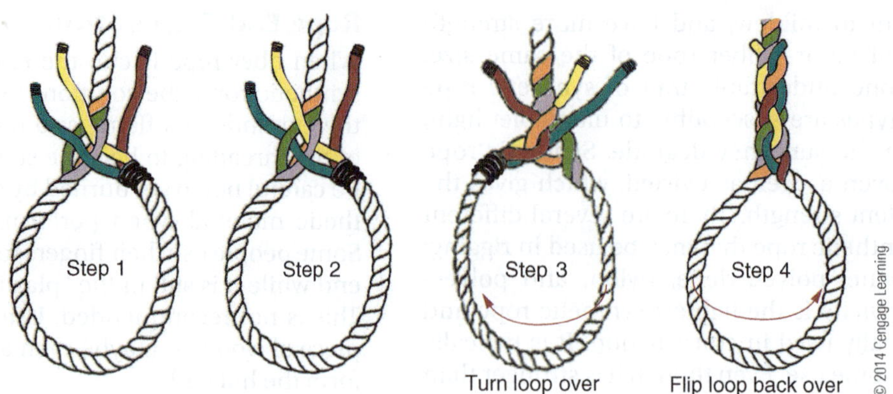

FIGURE 5-28 Making an eye splice in a rope.

Securing Using Synthetic and Fiber Ropes

Knots are used to connect rope to an object, equipment, or to another rope. Rope is often what secures a load as it is being lifted. Hitches and bowlines are the most popular knots when rigging.

Not uncommonly, it is necessary to make an eye from a rope end. **Figure 5-28** shows the procedure. First, tape or seize the rope end approximately six turns from the end of the rope. Once the rope is seized, remove the original seizing and the end of the rope and unlay the rope. To unlay means to unwrap the strands from one another. Do not untwist the stands. Reseize each of the strands so they don't unravel. It may be good to use some method of identification for each strand. Colors are used in the illustration to eliminate confusion while demonstrating the splicing procedure.

The rope in our figure only has three strands, but there are six colors. The red, blue, and yellow strands will be spliced into the orange, green, and purple. Determine the size of the eye, and move twice that distance past the new seizing that you put on the rope earlier. Open the lays by twisting in the opposite direction of the lays. Follow the steps shown in the figure, making sure to cut off any excess strand ends when the eye is complete. Notice that the whipping is removed in step 4. It is okay to remove the whipping because the splicing will prevent unraveling.

There are three elements to every knot: the bight, the turn (loop), or the round turn. Each can be seen in **Figure 5-29.**

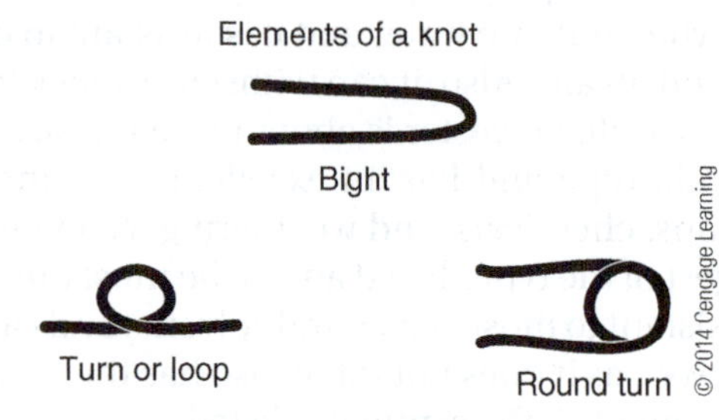

FIGURE 5-29 Elements of a knot.

The first knot is the bowline. The bowline is a very popular knot because it never jams or slips, yet it is easily untied after use. There are three types of bowlines that are most commonly used: the bowline, the bowline on the bright, and the running bowline. All three can be seen in **Figure 5-30**.

The square knot is another very common knot. It is often used when there are two rope ends that have to be tied together but will pull in opposite directions. This knot is fairly easy to make and just as easy to untie when you are done using it. **Figure 5-31** shows the four steps to making a square knot.

The sheep shank knot is used to shorten a rope. There are times that this can be used to take the load off of a weak spot in the rope. The sheepshank has a minimum of three passes, which distributes the load among each pass. Because the passes are in parallel to one another and because all the passes are seized together with a half hitch (which is discussed next), each pass carries approximately one-third the load between the half hitches. The steps for making a sheep shank can be seen in **Figure 5-32.**

The next series of knots are the hitch knots. **Figure 5-33** (see in page 114) shows how to make different types of hitch knots. The most common of all the hitch knots is the half hitch. It is popular because it is quick to make, yet the more it is pulled against, the tighter the connection gets. These types of knots are usually used to secure something to a post or a variation of it is use when lifting pipe, timber, or steel. The different types of hitch knots are the half-hitch, the double hitch, the timber hitch, the cat's claw hitch, and the clove hitch.

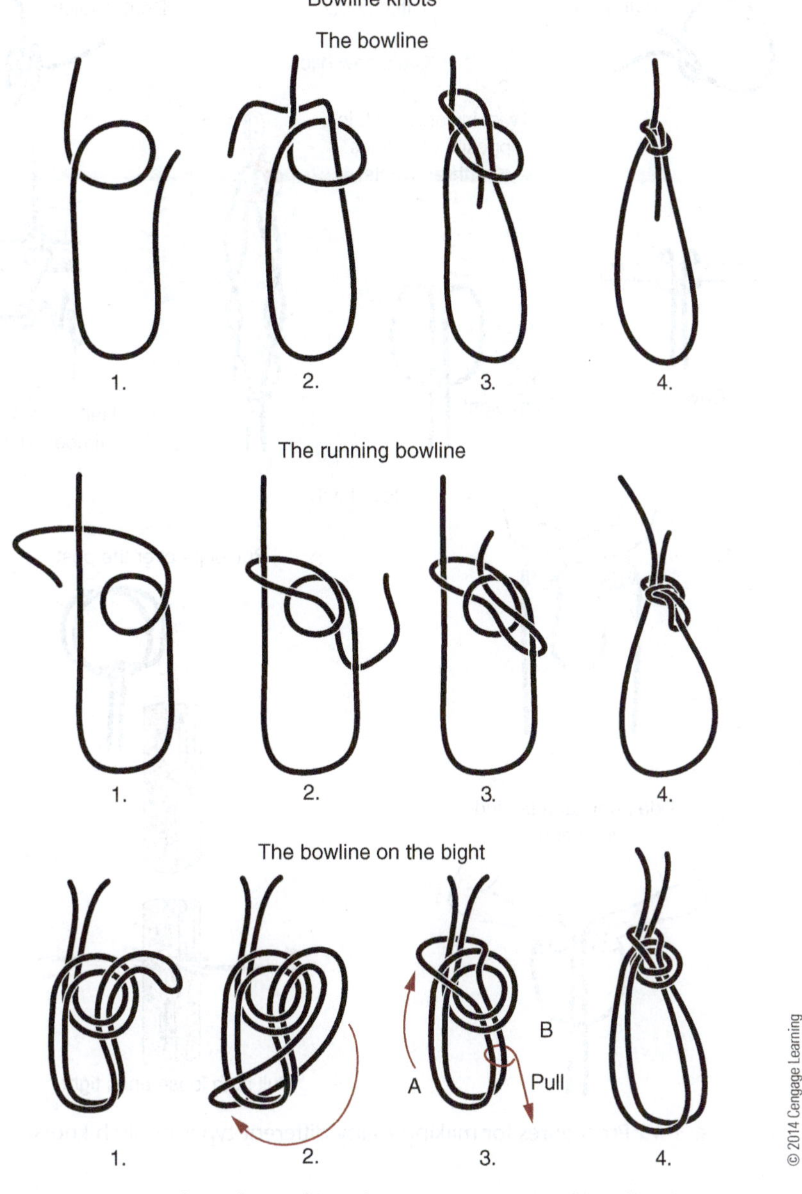

FIGURE 5-30 Procedures for making the common types of bowline knots.

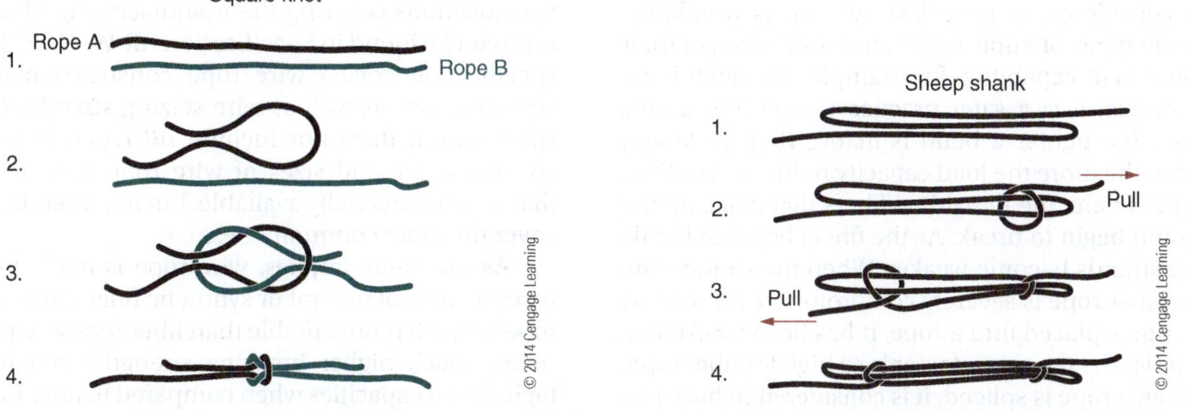

FIGURE 5-31 Making the square knot.

FIGURE 5-32 Making the sheep shank knot.

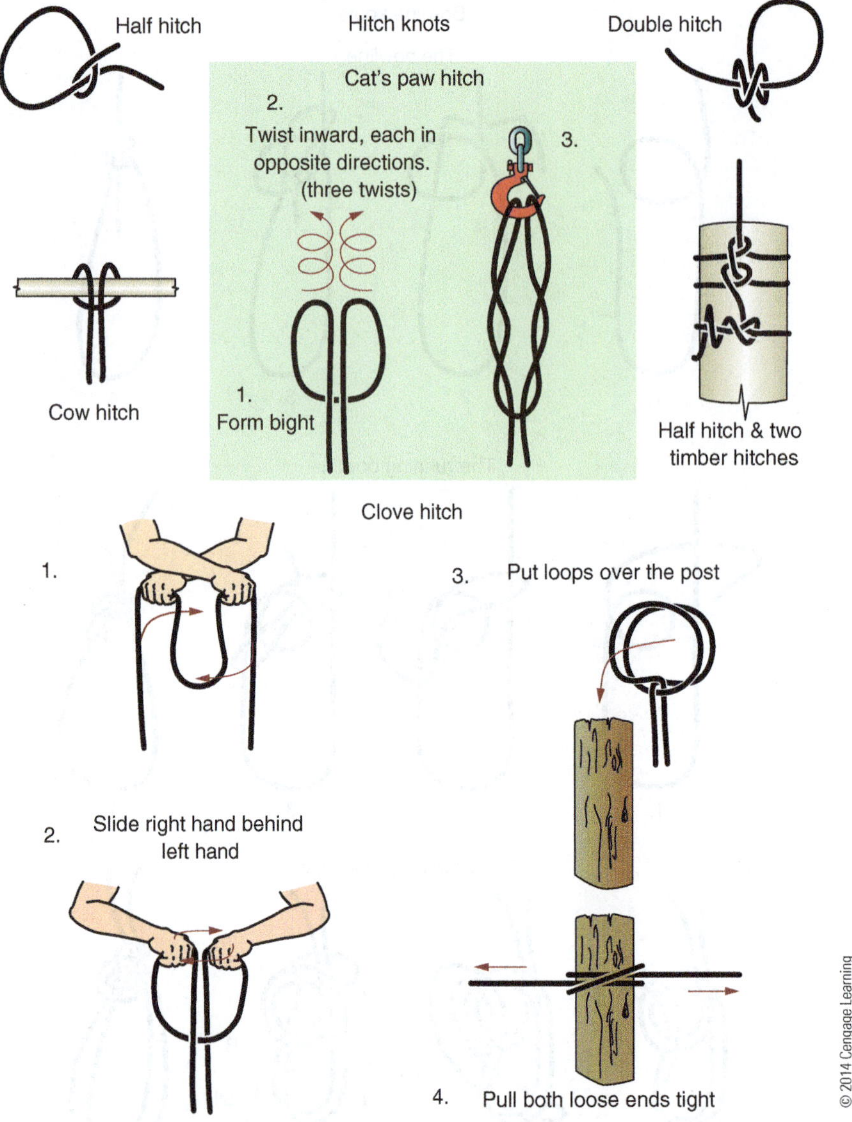

FIGURE 5-33 Procedures for making many different types of hitch knots.

Fiber Rope Strength

Straight lengths of rope can support 100% of their rated load capacity. If a knot is tied in a rope, it is considered to have lost 50% of its capability. Some types of knots only cause a 25% loss of their rated load capability, for example, the hitch knot. However, it is a safer practice to consider a 50% loss. The tighter a bend is made when making a knot, the more the load capacity reduces. As a fiber rope is bent, the individual fibers that make up the strand begin to break. As the fibers begin to break, the strands become weaker. When the strands fail, the fiber rope is severely compromised. As soon as a knot is placed into a rope, it has been weakened. This is why the safety factor is so high for fiber rope. When a rope is spliced, it is considered to have lost 15% of its load capability.

5-5 WIRE ROPE AND WIRE ROPE SLINGS

Specifications covering the manufacturing of wire rope can be found in Specification RR-W-410E. This specification covers wire rope construction and classifications as well as wire seizing strands. This specification does not include *all* types, classes, constructions, and sizes of wire rope and strand that are commercially available, but it is intended to cover the more common ones.

As the name implies, wire rope is made from wires instead of natural or synthetic fibers. The wire rope is much more durable than fiber rope and possesses much higher breaking strengths and thus higher load capacities when compared to fiber rope. Wire rope is made from strands just as fiber ropes are. Wire rope also has a core of some type. The most commonly used wire rope is right regular lay as it offers the most rotation resistance. This means that the rotation forces that are exerted on the strands are opposite those of the individual wires that make up the strand. The load determines the number of wires per strand, the number of strands, the type or the quality of the wire material, and the type of core that is used. The main things that should be considered when determining which type of wire rope to use are loading capabilities, wear, crushing conditions, rotation, conditions that contribute to corrosion, and the amount of tension placed on the wire rope. The wire rope should have enough strands to do the job while offering the greatest amount of flexibility. As would be assumed, when wire rope is constructed from heavier gauge wire, it is naturally less flexible.

As mentioned earlier, wire rope has a core. The core is either fiber or wire. A fiber core is composed of the same materials that are used to make either natural fiber rope or synthetic rope; however, synthetic fiber core would be preferred most often over natural fiber core because it is resistant to rot, corrosion, or deterioration. The fiber core helps cushion the strands while under a load but does not offer much resistance to crushing. Typically, the fiber core is impregnated with lubricant during the manufacturing process. With usage, the lubricant is released over time. Wire core is often referred to as independent wire rope core (IWRC) because it is completely independent from all of the other strands. In fact, the IWRC runs straight through the center, and all the outer strands lie around the core. The IWRC is a "mini" wire rope itself. IWRC is not susceptible to crushing, as is fiber core, but this is at the expense of flexibility. Wire rope also comes in right and left regular- and lang-lays. However, wire rope also has an alternating lay. All of these can be seen in **Figure 5-34**.

Wire Rope Classification, Construction, and Sizes

Classification

Wire rope is classified by the number of wires in its strands. Less flexible wire rope is made of larger wires in each strand and is very durable and resistant to abrasion or wear. In these wire ropes, the strand is usually made up of seven wires. A more flexible wire rope requires smaller gauge wire and may have up to 19 wires. Hoist and traction elevators may use up to 37 wires to give optimum flexibility without compromising strength. **Table 5-5** gives the most common wire rope classifications and the construction of each. The most common classification is a type I (general purpose), Class 1 (6 × 7), or Class 2 (6 × 19) wire rope.

Construction

There are several different types of strand patterns available as shown in **Figure 5-35** (see in page 117): Ordinary, Seale, Warrington, Filler, and Compact. When all of the wires in the strand are the same size and lay, it is referred to as an Ordinary strand pattern. The Seale pattern exists when one layer of wires is laid over a number of small wires with the same direction of lay. A Warrington pattern consists of one layer of wires composed of alternating large and small wires. In a Filler construction, the outer wires are supported by half the number of main inner wires with an equal number of smaller filler wires plus one center wire.

There are cases where the patterns are mixed. For example, you may find a 6 × 19 Seale Warrington. A 6 × 37 wire rope offers many more combinations, such as a Warrington Seale, a Seale Filler wire, a Filler Wire Seale, or even a Seale Warrington Seale.

Compact wire rope has been compressed to remove any air gaps between any wires within the strand. This makes a narrow and very strong cable at the expense of flexibility. **Figure 5-36** (see in page 117) shows a variety of wire rope classifications and pattern types.

Size

Wire rope size is identified by the diameter of the wire rope. The diameter is measured at its major axis using a dial calipers, as seen in **Figure 5-37** (see in page 117). Do not measure the wire rope at the minor axis.

This gives you a reading that is not only inaccurate but also too small. It is important to measure

Right, Alternating-Lay

Left, Lang-Lay

Right, Lang-Lay

Left, Regular-Lay

Right, Regular-Lay

FIGURE 5-34 Wire rope lay types.

Table 5-5

WIRE ROPE CLASSIFICATIONS

Classification	Number of Strands	Wires within the Strand	Core Availability* FC	IWRC	Characteristics Abrasion Resistant	Flexibility	Bending Fatigue	Some Common Uses
1 x 7	1	7	–	–	Excellent	Very Poor	Very Poor	Permanent lightweight securing, no repetitive motion or bending expected—Guy lines
1 x 19	1	19	–	–	Excellent	Very Poor	Very Poor	Permanent medium weight securing, No repetitive motion or bending expected—Guy lines
6 x 7	6	7	√	√	High	Poor	Poor	Dragging and haulage, inclined plane & tramways (cablecars, very little bending expected
6 x 19	6	19	√	√	High	Good	Medium	Haulage, chocker, rotary drilling line
6 x 21	6	21	√	√	Good	Good	Good	Pull rope, load line, back haul ropes, and draglines
6 x 25	6	25	√	√	Good	High	Good	(Most widely used)-crane hoists, skip hoists, haulage, mooring lines, conveyors, etc.
6 x 26	6	26	√	√	Medium	High	Good	Boom hoists & logging
6 x 31	6	31	√	√	Medium	High	High	Overhead and mobile crane hoists
6 x 36	6	36	√	√	Poor	Excellent	Very High	Overhead and mobile crane hoists
7 x 7	7	7	√	√	High	Poor	Poor	Drag line and haulage
7 x 19	7	19	√	√	High	Good	Medium	Chocker haulage
8 x 19	8	19	√	√	High	Good	Medium	Spin resistant—high speed hoists having multiple reeving and single line suspension

*FC, Fiber Core; IWRC, Independent Wire Rope Core.

√ = Available.

the diameter often to see whether the wire rope has been stretched beyond the minimum limits. Stretching and wear can be identified by measuring the outside diameter of the wire rope and comparing it to the values found in **Table 5-6**.

Inspection

Wire rope should be inspected before each use. Problems such as wear, metal fatigue, abrasion, crushing, corrosion, kinking, bird caging, and stretching, all can render a wire rope unsafe for use. These things should be found before a lift is made. Bird caging occurs when the wire rope is twisted in a direction that is opposite the direction of the lay. This causes the cable to open, meaning that the strands are forced apart. The most common cause for this is torsional imbalance caused by sudden stops.

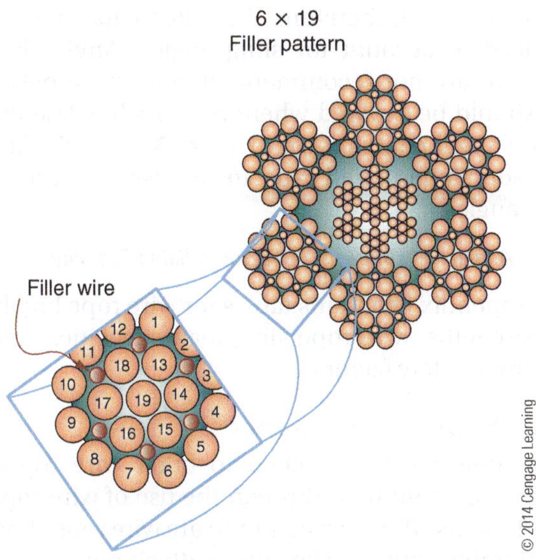

FIGURE 5-35 Construction of filler type wire rope.

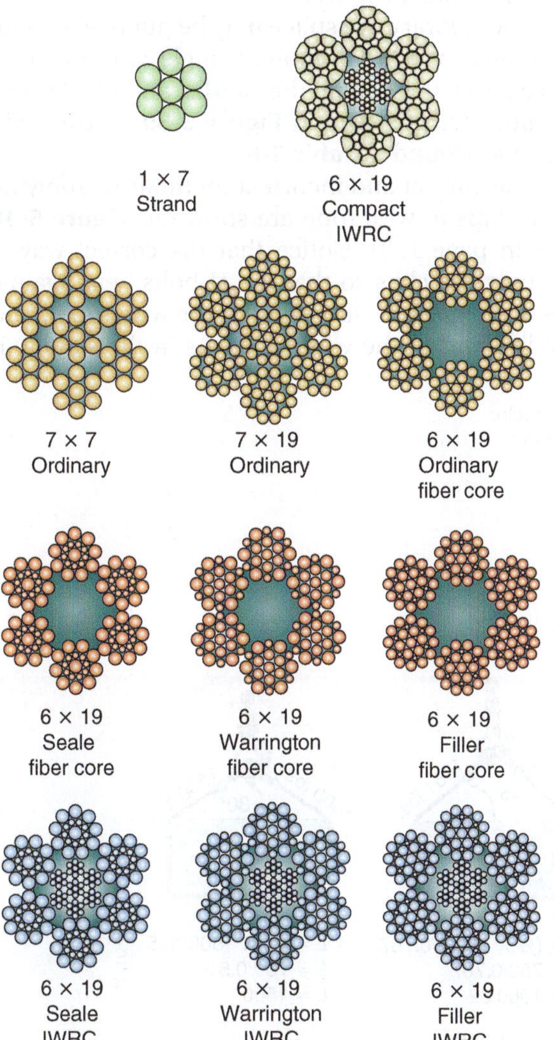

FIGURE 5-36 Classifications and pattern types of the most commonly used wire rope.

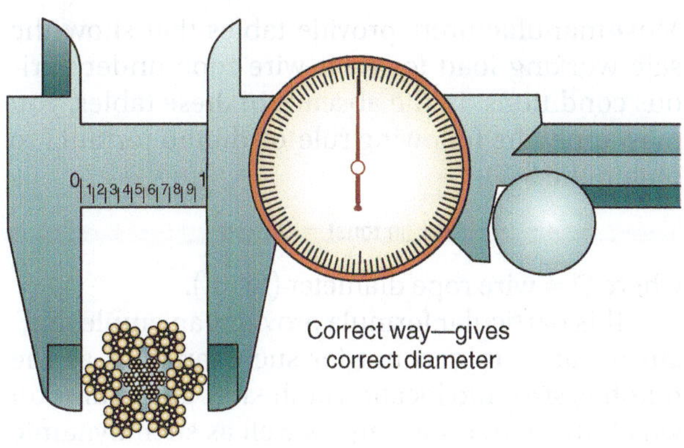

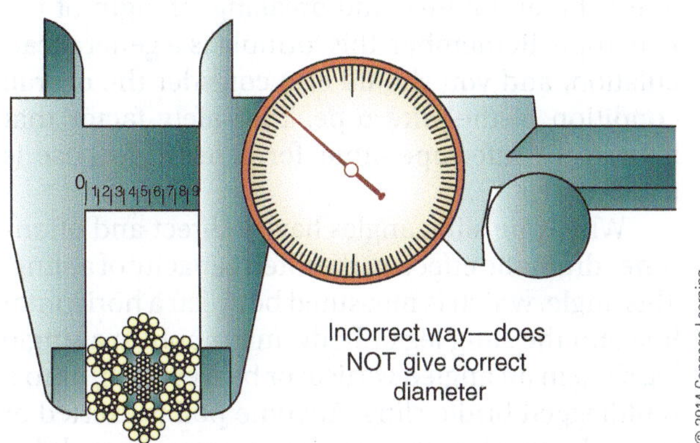

FIGURE 5-37 Proper method for acquiring the diameter of wire rope using a dial indicator.

Table 5-6

MAXIMUM WEAR FOR WIRE ROPE

Nominal Diameter (in.)	Maximum Allowable Wear (in.)
5/16	1/64
3/8–1/2	1/32
9/16–1/2	3/64
7/8–11/8	1/16
1¼–1½	3/32

© 2014 Cengage Learning

Wire Rope Load Capacities

The strength of wire rope is determined by dividing a wire rope's breaking strength by a safety factor. The term *safe working load* (SWL), as used in reference to wire rope, means the load that you can apply and still obtain the most efficient service and also prolong the life of the wire rope.

Most manufacturers provide tables that show the safe working load for their wire rope under various conditions. In the absence of these tables, you may apply the following rule-of-thumb formula to obtain the SWL:

$$SWL \text{ (in tons)} = D^2 \times 8$$

where D = wire rope diameter (in in.).

This particular formula provides an ample margin of safety to account for such variables as the number, size, and location of sheaves and drums on which the wire rope runs, as well as such dynamic stresses as the speed of operation and the acceleration and deceleration of the load, all of which can affect the endurance and breaking strength of the wire rope. Remember, this formula is a general calculation, and you should also consider the overall condition of the wire rope. The safety factor that is used for safe rope limits for general purpose is usually 5:1.

Wire rope sling angles have a direct and oftentimes dramatic effect on the rated capacity of a sling. This angle, which is measured between a horizontal line and the sling leg or body, may apply to a single leg sling in an angled vertical or basket hitch, or to a multilegged bridle sling. Anytime pull is exerted at an angle on a leg, the tension or stress on each leg is increased. To illustrate, each sling leg in a vertical basket hitch absorbs 750 lb of stress from a 1500 lb load. The same load, when lifted in a 60° basket hitch, exerts 866 lb of tension on each leg, as seen in **Figure 5-38**.

It is critical, therefore, that rated capacities be reduced to account for sling angles. Angles less than 45° are not recommended, and those below 30° should be avoided whenever possible. Use the formula and chart shown in **Table 5-7** to calculate the reduction in rated capacities caused by various sling angles.

$$\text{Actual Sling Capacity} = \text{Factor} \times \text{Rated Capacity}$$

Appendix J contains tables for wire rope breaking strengths, wire rope sling load capacities, and wire rope safety factors.

Wire Rope Attachments

Wire rope can be attached to other wire ropes, chains, or equipment through the use of wire rope attachments. The most common are wire rope clips, the thimble, and the clip. These attachments allow the wire rope to be used in a variety of applications and arrangements than would be possible with a more permanent splice.

A temporary eye splice may be put in wire rope by using clips and a thimble. A single clip consists of three parts: the U-bolt, the saddle, and two fastening nuts. Each is seen in **Figure 5-39**. Correct clip spacing is found in **Table 5-8**.

The correct and incorrect methods of applying these clips to wire rope are shown in **Figure 5-40** (see in page 120). Notice that the correct way is to apply the clips so that the U-bolts bear against the dead end (the short end of the wire rope), and the live end of the wire rope sets in the saddle. If

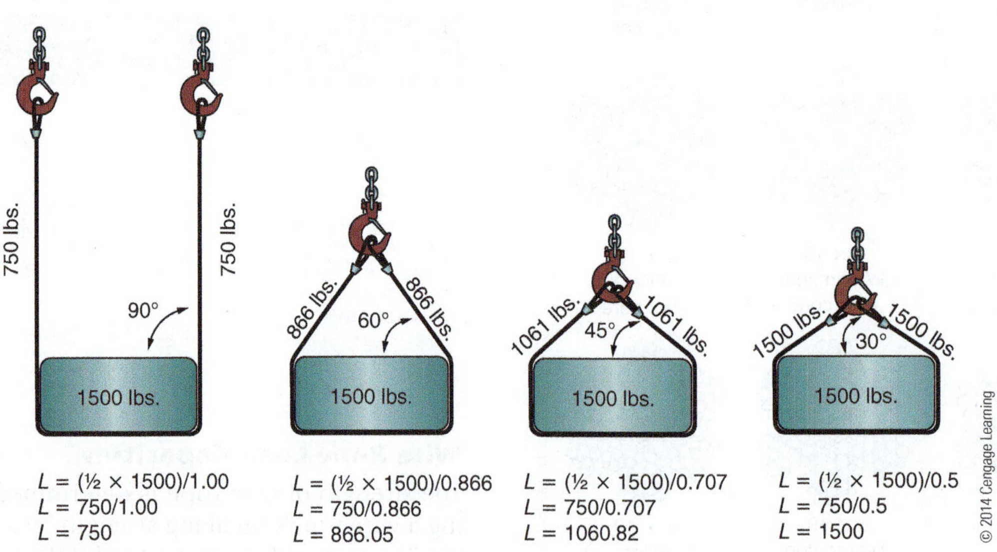

FIGURE 5-38 The load/leg is increased as the angle of the hitch decreases.

Table 5-7

HITCH (SLING) LEG LOADING FACTORS

Sling Angles (in degrees)	Loading Factor	Sling Angles (in degrees)	Loading Factor
15	0.259	55	0.819
20	0.342	60	0.866
25	0.423	65	0.906
30	0.500	70	0.940
35	0.574	75	0.966
40	0.643	80	0.985
45	0.707	85	0.996
50	0.766	90	1.000

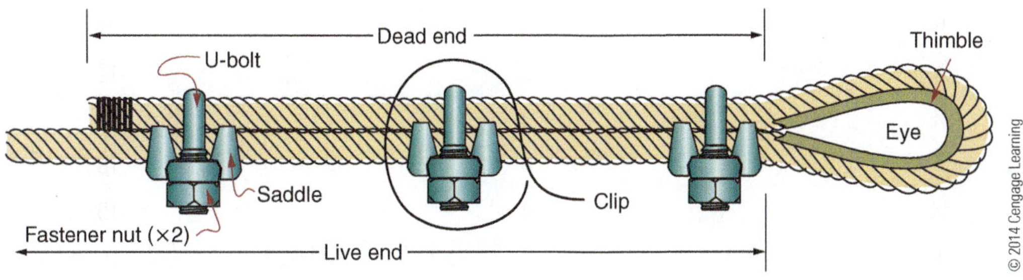

FIGURE 5-39 Correct clip spacing.

the clips are attached incorrectly, the cable could be damaged by distorting or crushing the strands on the live end of the wire rope. The live end, also called the standing end, refers to the end of the rope that supports the load. *Always* place a clip with the U-bolt on the dead end, not on the live end, of the wire rope. A well-known saying to remember when attaching a wire rope clip is "*Never* saddle a dead horse." In other words, never put the saddle of the clip on the dead end of the wire rope. Also, it is a good idea to inspect and tighten the clips after the wire rope assembly has been placed under a load the first time or two.

When an eye is made in a wire rope, a metal fitting, called a thimble, is usually placed in the eye, as seen in Figure 5-39. The thimble protects the inside of the eye against wear. When wire rope eyes have been constructed using thimbles and wire rope clips, the wire rope can only be loaded to approximately 80% of the wire rope's rated load capacity. Once the eye is assembled, attachment can be connected to the eye. Some of the more common attachments that are used in conjunction with the wire rope eye are the eye hook, the master link, or shackle, all of which are discussed in Chain Attachments later in this chapter.

A wedge socket can be seen in **Figure 5-41** (see in page 121). It is used in situations that require frequent changing of the attachments. The wedge socket is constructed in two parts, the socket and the wedge. The socket has a tapered opening for the wire rope, and the wedge is placed in the "eye" of the wire rope, then into the tapered socket. (*Note:* Do not install a thimble!) The socket is always applied to the dead end of the wire rope, and it is good practice to place a clip or two on the dead end to prevent slippage or accidental release of the wedge socket during a load shock. The loop of wire rope must be installed in the wedge socket in a manner that places the load directly in line with the standing part of the wire rope. The wedge socket is designed so that when it is assembled correctly, it tightens as a load is placed on it

Table 5-8

SPACING DIMENSIONS FOR WIRE ROPE CLIPS*

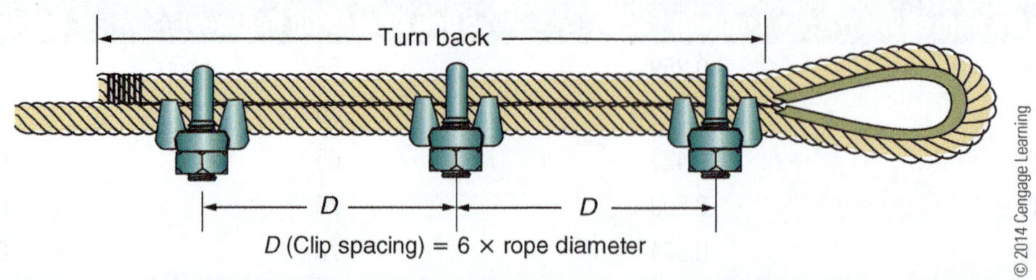

Wire Rope Diameter/Clip Size (in.)	Minimum Number of Clips	Amount of Rope to Turn Back (in.)
1⁄8	2	3¼
3⁄16	2	3¾
¼	2	4¾
5⁄16	2	5¼
3⁄8	2	6½
7⁄16	2	7
½	3	11½
9⁄16	3	12
5⁄8	3	12
¾	4	18
7⁄8	4	19
1	5	26
1⅛	6	34
1¼	7	44
1⅜	8	44
1½	8	54

*Based on 6 x 19 or 6 x 37 wire rope.

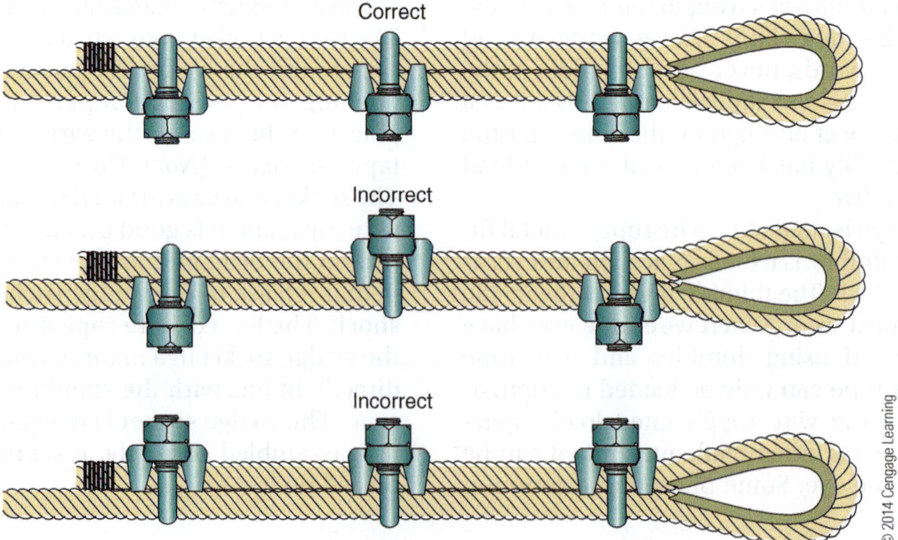

FIGURE 5-40 Proper and improper clip installation.

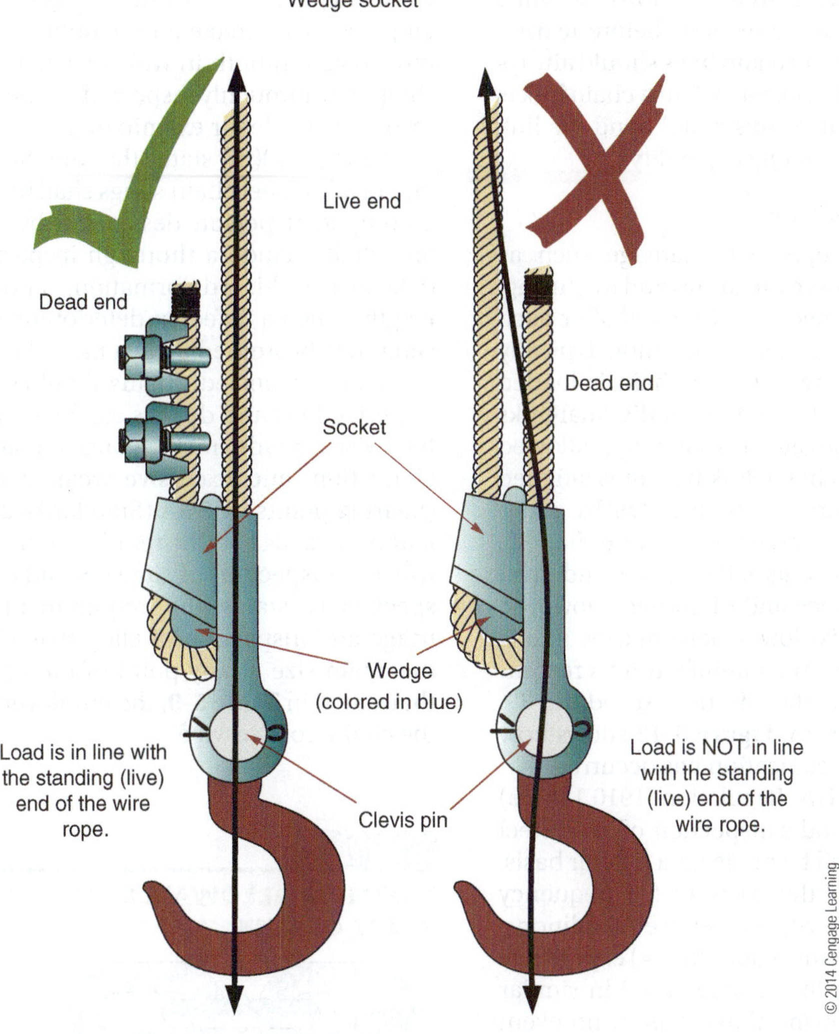

FIGURE 5-41 Proper installation of a wedge socket.

5-6 CHAIN AND CHAIN SLINGS

Chain is very durable and can resist corrosion and sustain misuse or mistreatment better than webbing, fiber rope, or wire rope. It is also extremely flexible and does not lend itself to the problems of kinking and of chafing (abrasions). Chain life is also much higher than wire rope, and chain requires less maintenance. Chain can be used in high-temperature (<600°F) and dirty environments without losing its integrity, and even though chain has no elasticity and is not suitable for heavy shock or impact loading, it can withstand a considerable amount before elongation (stretching) occurs. However, chain is heavy, and when chain is used in a high-temperature environment, the load capability must be derated. This is discussed under Chain Load Capacities later in this chapter.

Several things have to be understood when considering the strength of chain. As stated earlier, chain is made up of individual links. Understanding the strength of the chain comes down to understanding how the individual links are made. For standard chain, each link is made of hardened steel that is then tempered. Heat treating occurs when steel is heated until it is red-hot and then cooled (quenched) very quickly. Once cooled in this manner, the steel becomes very hard and yet brittle. Brittleness is an undesirable property for chain. The tempering process is a heat treat technique that is used after hardening, to reduce some of the hardness (brittleness) created by the heat treating process. When tempered, the links of a chain are reheated to a much lower temperature than was used for hardening, and then they are cooled slowly. This process gives the link hardness with some plasticity. Tempering is used to precisely balance the properties of the link. These properties are shear strength, yield strength, hardness, and tensile strength. **Shear strength** is a term used to describe the amount of strength a material has to resist component failure under a shear load. A **shear load** is a force that that is applied to a material along a plane that is parallel to the direction of force. **Yield strength** is the stress at which a material begins to deform (plasticity). **Tensile strength** is the maximum stress that a material can withstand while being pulled in opposite directions before tearing apart. Also, side loading of chain links should always be avoided. Side loading occurs when a chain link is placed over a sharp edge. This could bend the link and weaken the chain's lifting capability.

Chain Sling Inspection

Chain links are susceptible to damage such as stretching (referred to as elongation) and fracturing. A chain should be inspected before and after each use to ensure that it is safe. Elongation typically occurs before fracturing. When a chain is loaded to more than its yield strength, the individual links begin to stretch or elongate. If stretching is allowed to continue, the metal in the link becomes fatigued and eventually fractures or breaks. OSHA states that if a chain sling has stretched to more than 3% longer than its original length, it is unsafe and must be removed from service and discarded. However, manufacturers may list lower percentages, and it is always best to follow the manufacturer's recommendations as long as they do not exceed the 3% limitation given by OSHA. **Figure 5-42** shows how to determine whether elongation has occurred.

According to OSHA Regulation 1910.184 (e)(3)(i), a thorough periodic inspection of alloy steel chain slings in use shall be made on a regular basis, to be determined on the basis of (1) frequency of sling use, (2) severity of service conditions, (3) the nature of lifts being made, and (4) experience gained on the service life of slings used in similar circumstances. Such inspections shall in no event be at intervals greater than once every 12 months.

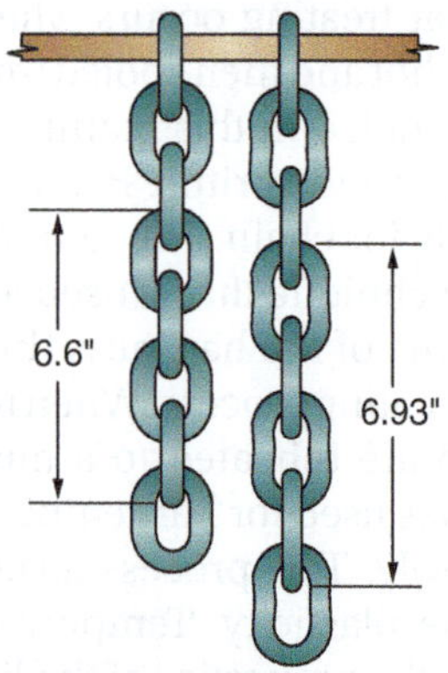
Be sure to pull the chain tight when testing for elongation
6.6"
6.93"
G80 chain: 1 link = 1.65"
(New chain) 4 links = 6.6"
Elongation limit = 3% (6.798")
(Used chain) 4 links = 6.93"
The chain has elongated by 5%
It must be removed from service.

FIGURE 5-42 Testing for chain elongation (stretching).

OSHA Regulation 1910.184 (e)(3)(ii) states that the employer shall make and maintain a record of the most recent month in which each alloy steel chain sling was thoroughly inspected, and shall make such record available for examination. OSHA Regulation 1910.184 (e)(3)(iii) states that the thorough inspection of alloy steel chain slings shall be performed by a competent person designated by the employer, and shall include a thorough inspection for wear, defective welds, deformation, and increase in length. When a defect or deterioration is found, the sling shall be immediately removed from service.

When inspected, chains should cleaned so that imperfections and defects can be easily seen. Look for twists, bends, nicks, gouges, cracks, fractures, elongation, and excessive wear at the link ends (bearing points). OSHA (Standards-29 CFR) Regulation 1910.184 lists the requirements for the proper use and inspection of slings. Standard 1910.184(e) specifically states the requirements for proper usage and inspection of alloy steel chain slings. If the chain size at any point of the link is less than that stated in **Table 5-9**, the employer must remove the chain from service.

Table 5-9

MINIMUM ALLOWABLE CHAIN SIZE AT ANY POINT OF LINK

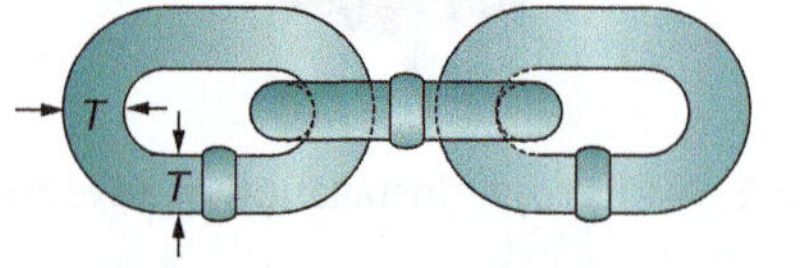
T
T

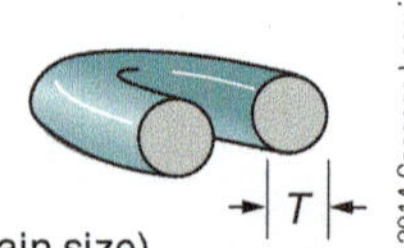
T

T = Material diameter (chain size)

Nominal Chain Size (in.)	Maximum Allowable Size (in.)
1/4	13/64
3/8	19/64
1/2	25/64
5/8	31/64
3/4	19/32
7/8	45/64
1	13/16
1 1/8	29/32
1 1/4	1
1 3/8	1 3/32
1 1/2	1 3/16
1 3/4	1 13/32

Alloy steel chain slings with cracked or deformed master links, coupling links, or other failed components must be removed from service.

Chain Attachments

Shackles, master links, connecting links, swivels, and hooks are the most commonly used attachments when using chain. These devices are used to connect the load to the chain and to connect the chain to other lifting devices and/or equipment. There are two types of shackles, anchor shackles and chain shackles, both of which can be seen in **Figure 5-43**.

A shackle is a U-shaped device that uses a screw pin or a bolt to secure itself to the chain link. Shackles have a loading capacity that should not be exceeded. They are made of alloy steel and are rated in a various load capacities, which should be stamped into the bow section of the shackle. The shackle size is determined by the material diameter of the bow section, not the pin/bolt diameter. Shackle pins/bolts are usually 1⁄16 larger in diameter than the bowed section of the shackle for sizes up to 7⁄16 in. Shackle sizes ½ in. to 1⅝ in. will have a pin/bolt that is ⅛ in. larger than the bow section. For sizes 1¾ in. and up, the pin/bolt will be ¼ in. larger in diameter when compared to the bow section of the shackle. The pin/bolt of the shackle should be hung on the hook, and the slings are usually placed in the body of the shackle. The shackle should be used with spacers on the pin/bolt to keep the load centered on the body of the shackle, as seen in **Figure 5-44**. When the load is allowed to shift off center in the shackle, the rated load capacity goes down, as seen in **Figure 5-45**. Notice the two witness marks that are embossed onto the side of the shackle. This is where the load should reside in order for the shackle to retain its full rated capacity. If the load shifts outside these witness marks, derating must occur.

Shackles			
Screw pin	Slide pin w/cotter pin	Bolt w/cotter pin	
			Anchor
			Chain

FIGURE 5-43 Shackle types.

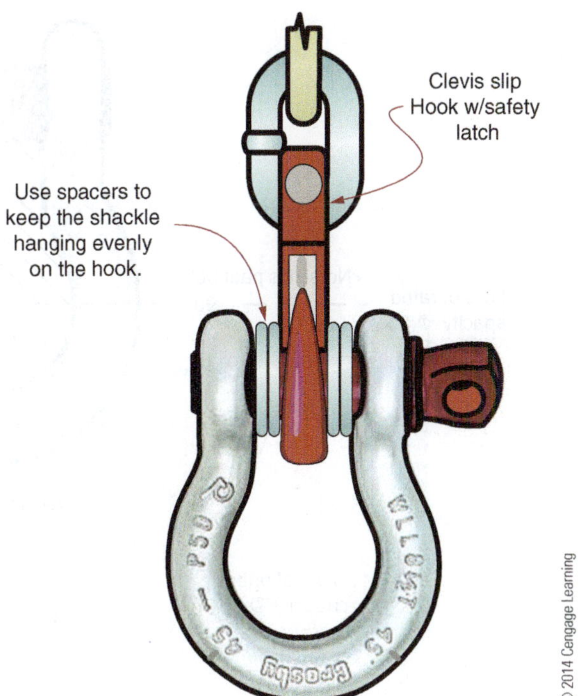
Clevis slip
Hook w/safety
latch
Use spacers to
keep the shackle
hanging evenly
on the hook.

FIGURE 5-44 Centering the shackle.

Master links are easily recognized because their dimensions are noticeably larger than the standard links within the chain, as seen in **Figure 5-46**.

These larger master links can be connected to other lifting attachments and/or equipment more easily than the standard link of a chain. These are also used to create multilegged slings for lifting.

There are many different types of hooks used with chain, but the most commonly used are the clevis hook, the swivel hook, and the eye hook, all of which can be seen in **Figure 5-47**.

The clevis hook has a removable pin, so the hook can be attached directly to a link in the chain. There are two hooks that can only be attached to a chain by means of a connecting link, the eye hook and the swivel hook. A connecting link can also be seen in Figures 5-46 and 5-47. The eye hook has a forged hole at the top of the hook to connect it to the chain, whereas the swivel hook has a rotating connector at the top of the hook assembly. This allows the load to rotate while being lifted.

Each hook, no matter the type, has a loading capacity that should not be exceeded. Typically, the load capacity and the hook size (in in.) are embossed (stamped) into the side of the hook. It would be very dangerous to have everything else sized correctly according to the load that is to be lifted, only to have the lift fail because an undersized hook was used. Hooks should be maintained and inspected thoroughly before each use. It is a

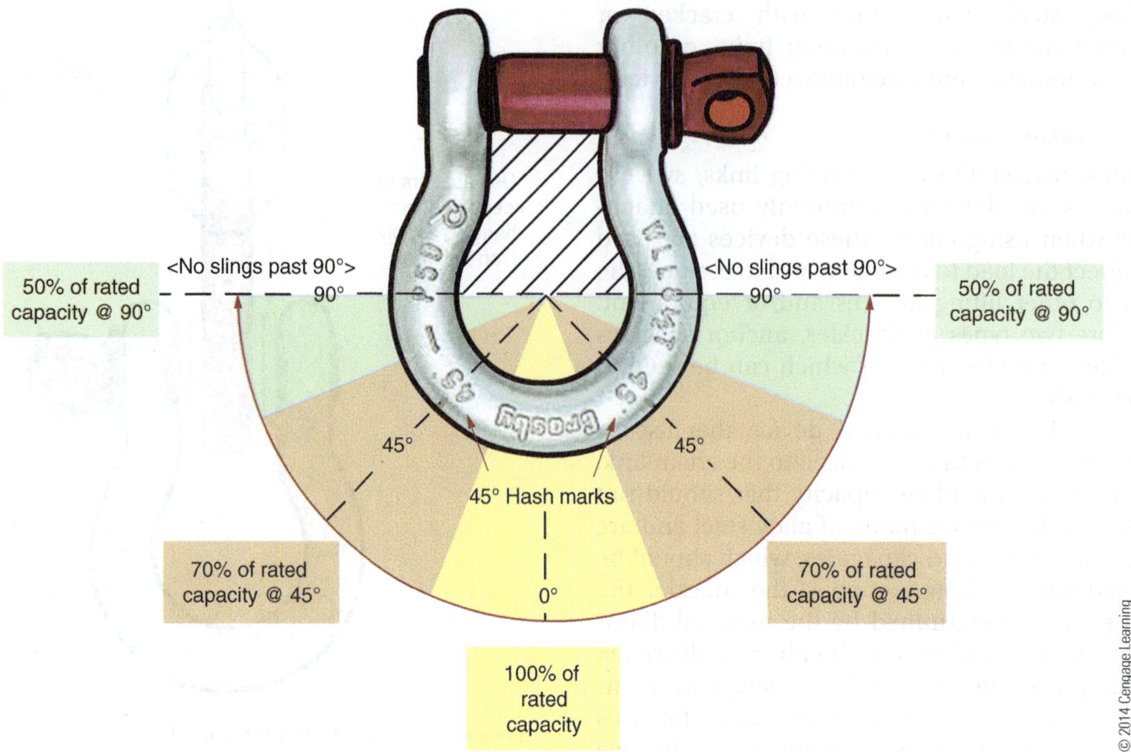

FIGURE 5-45 The load capacity of a shackle is dependent upon the angle of loading.

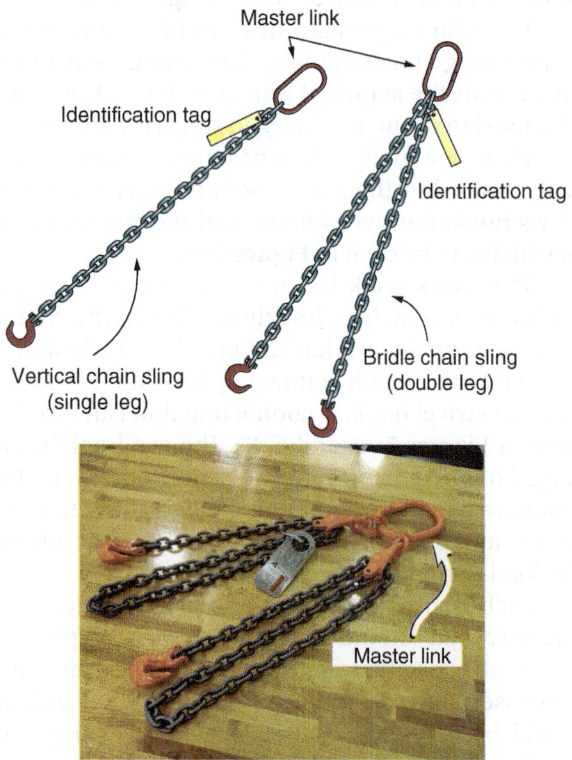

FIGURE 5-46 The chain sling and the master link.

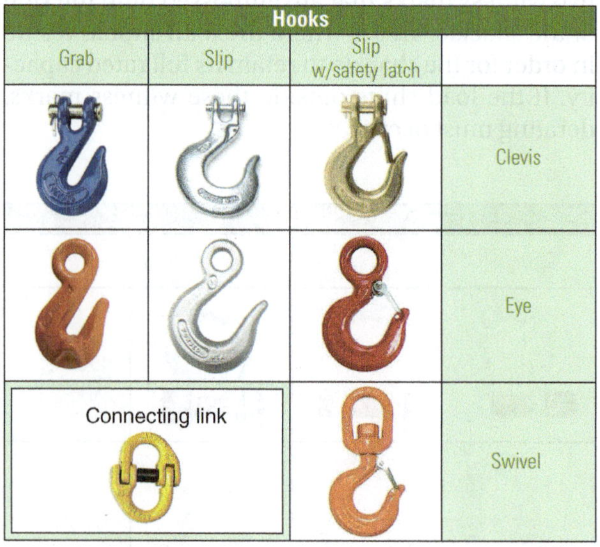

Hooks			
Grab	Slip	Slip w/safety latch	
			Clevis
			Eye
Connecting link			Swivel

FIGURE 5-47 Hook Types.

good idea to use a hook that has the safety catch installed, as this would prevent the sling from slipping out of the hook if it were to shift during the lift.

Chain Grades

The most common chain grades are 30, 43, 70, 80, and 100. These numbers, which manufacturers use to represent the chain's grade, are actually $\frac{1}{10}$ or $\frac{1}{100}$ of the actual grade value and should be hallmarked (stamped onto a link) every 1 to 3 feet (0.3 to 0.9 m) along the chain's length. For example, a chain that has a 43 hallmarked on it is a Grade 430. Sometimes the manufacturer stamps an 8 on a link for a Grade 800 ($\frac{1}{100}$ the actual grade). The grade refers to the tensile strength class of the chain, which is expressed in newtons per square millimeter (a newton is approximately 0.224805 lb).

Grade 300 (G30) is a general-purpose chain of standard commercial quality. It is made from low-carbon steel and is frequently used for towing, logging, and tie down or binding. It is not rated to be used for overhead lifting! Grades 400 (G40) and 430 (G43) chain are primarily used in the boating and trucking industries. Again, these grades are not to be used for overhead lifting or securing! Grade 700 (G70) is a stronger chain that is heavily used to secure transported loads, and it has a higher loading capacity than the previous grades; however, it is not rated for overhead lifting! Grade 800 (G80) is the first chain that was specifically designed for safety and approved by OSHA and other agencies for overhead lifting. Its alloyed, heat-treated steel makes it ideal for making lifting slings and heavy-duty tow chains. Grade 1000 (G100) is a premium alloy chain that has approximately 25% more strength and a higher rated load capacity than that of G80 chain. Grade 1200 (G120) alloy is the strongest chain available and is considered to be the highest-quality chain. Because Grades 800, 1000, and 1200 are alloy chains, they are the only chains that are rated for overhead lifting.

Chain Load Capacities

The load capacity (CAP), most commonly referred to as the working load limit (WLL) or safe working load (SWL), is the maximum load that the chain is rated to handle during a straight pull. The ultimate breaking strength (UBS) is determined by the size and grade of the chain link. It is usually much higher than that of the CAP, WLL, or SWL of the chain. Do you remember that tensile strength is the maximum stress that a material can withstand while being pulled in opposite directions before tearing apart? Therefore, the tensile strength is what determines the ultimate breaking strength (UBS). An example of calculating UBS using $\frac{5}{16}$ in. (8 mm) Grade 800 (G80) is shown in **Figure 5-48**. Once the UBS is calculated, it is typically rounded down to the nearest 1000 lb. In our example, the result is 18,079.89 lb, which would round down to 18,000 lb. Knowing

Example: 5/16" G80 chain

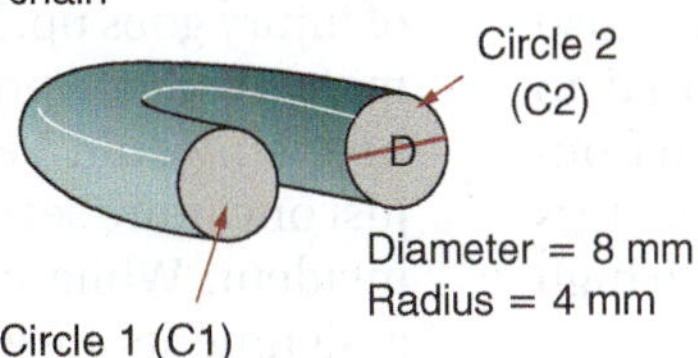
5/16" = 8 mm
Circle 2 (C2)
D
Diameter = 8 mm
Radius = 4 mm
Circle 1 (C1)
1. Calculate for area:
Area = πr^2
Area = $\pi 4^2$
Area = $\pi \times 16$
Area = 50.27 mm^2
2. Calculate for total area:
Total area = C1 area + C2 area
Total area = 50.27 × 2
Total area = 100.53 mm^2
3. Calculate for Ultimate Breaking Strength (UBS):
UBS = Area × grade
UBS = 100.53 mm^2 × 800
UBS = 80,424.77 Newtons
4. Convert ultimate breaking strength into pounds (lbs.):
UBS = Newton × 0.224805
UBS = 80,424.77 × 0.224805
UBS = 18,079.89 lbs.

FIGURE 5-48 An example of calculating UBS using $\frac{5}{16}$" (8 mm) Grade 800 (G80) chain.

that the maximum capacity for 5/16 in. (8 mm) G80 chain is 4500 lb, we can see now that the minimum breaking force is in fact four times the CAP (4 × 4500 lb. = 18,000 lb).

The UBS is three times the CAP for G40 and G 43. G30, G70, G80, and G100 are rated at four times the CAP, whereas G120 is rated at six times the CAP. This is known as the design factor, and it is usually expressed as a ratio (i.e., 4:1 for G100 and 6:1 for G120). Never exceed the CAP/WLL ratings, and never use proof test or minimum breaking force (UBS) values for service or design purposes! Proof testing occurs at two times the CAP (2:1). As stated in OSHA Standard 1910.184(e)(4), proof testing shall be performed in accordance with paragraph 5.2 of the American Society of Testing and Materials (ASTM) Specification A391-65, which is incorporated by reference as specified in Sec. 1910.6 (ANSI G61.1-1968). Also, it is required that the employer shall retain a certificate of the proof test and shall make it available for examination upon request. Appendix K contains the load capacities and chain properties for G30, G40, G43, G70, G80, G100, and G120 chain. Chain sling load capacities are given in Appendix K.

When G80 and G100 alloy chain are used in high-temperature environments, the load capacity must be derated. According to OSHA Standard 1910.184, when alloy chain slings are exposed to service temperatures in excess of 600°F (316°C), employers must reduce the maximum working-load limits permitted by the chain manufacturer, in accordance with the chain or sling manufacturer's recommendations. This standard also requires that all chain slings that are exposed to temperatures over 1000°F (538°C) be removed from service. See Appendix K for the derating factors that must be applied when chain is exposed to high temperatures.

5-7 PRELIFT PLANNING

Once the need to move an item has been identified, planning the lift should begin. Rigging is defined as moving heavy objects through the use of chains, ropes, cables, rollers, and so on, or any of the items listed in the previous section of this text. The goal of any move is that it be done without any injury to personnel and no resulting damage to the item being moved. Planning the lift must be done prior to a lift being executed. Here are a few things that must be considered when planning a load.

Inspect the Load

The item that is to be moved should be thoroughly inspected for anything that might cause a hazard during the move. An example would be a gearbox that might be leaking oil. This could cause the rigging equipment to slip, which would in turn cause the load to shift or fall. Also contaminants could be introduced into the rigging equipment, causing its integrity to weaken. As much as is possible, the load that is to be lifted should be clean and free from sharp edges.

Weight Determination

The weight of the load must be determined. This was covered in great detail at the beginning of this chapter. To execute a safe lift, a close weight estimation of the equipment that is to be lifted must be made. Whenever possible, the actual weight of a machine or object should be obtained from the manufacturer's documentation. However, in the event that it is not possible to obtain an accurate weight from the manufacturer's documentation, it is acceptable to estimate the weight, using the methods prescribed earlier in this chapter.

Needed Personnel

It must also be determined how many people are needed to raise the load, execute the move, and assist with the setting of the load. The planning is usually done by one or two people, a safety officer and/or the qualified (trained) person in charge of executing the lift. Many people may be needed to secure the load, begin the lift, move the load, and set the load down in its new location, but only use as many as are needed—and no more. Each time another person is added to the lift team, the chance of injury goes up. However, be sure not to underestimate needed workforce. The worse thing that could happen would be that someone was injured when just one more set of hands would have prevented the incident. While planning the lift, ensure that duty assignments are written up for each person involved in the lift. A briefing should be held with all who are involved with the lift, prior to execution, so that each knows what his or her duty is during the lift.

Determine Equipment and Inspect

The method of movement must be determined. For example, will the item be moved on rollers or will it be lifted? Once this is identified, the proper equipment must be selected, keeping in mind that all equipment must be adequately rated to handle the load that is to be moved. An inspection of the lifting equipment must be made to verify that it is functional and safe to use. Physical protection of the load and the equipment that is used to secure and lift the load must be considered. When planning the securing of the load, make sure that the rigging equipment is used in a manner that adheres to the manufacturer's specifications. Any deviations from this, "on the fly," could cause equipment failure resulting in possible death or injury, or at the least, damage to the load.

Determine the Path

The path that the load will take during the move must also be planned and cleared. If possible, barricades and warning signs should be put in place to prevent individuals who are not involved with the lift from becoming injured. If this is not possible, plan to post people in strategic areas to warn others of the lift as it occurs. Plan the setting down of the load as well. Have all items at the site, ready to go before the lift is begun. The worst thing is to suspend a load longer than necessary because someone forgot a piece of equipment that is needed to set the load in its proper place.

Planning the lift greatly lessens the odds of an accident during the lift. The paramount goal is the protection of personnel! Secondarily, it is to move the equipment or machinery without causing any damage to it or the surrounding area. If you have an effective plan, and all who are involved with the lift are briefed, the lift should be a successful one.

SUMMARY

In this chapter, we covered the following topics:

- The correct formula to perform weight estimations
- The importance of balancing the load
- Terminology that is pertinent to the use of slings
- The common types of synthetic slings that are used in industry
- The different types of natural fiber rope
- The different methods of knot tying using both synthetic and fiber rope
- The different types of wire rope classifications and their load capabilities
- The different types of chain used when rigging and the load capabilities of each grade
- The importance of inspecting and selecting the proper equipment that is needed to make a safe lift
- The items that are to be considered when performing the prelift plan

REVIEW QUESTIONS

1. At what degree angle to the horizontal is the load on each sling considered to be equal to the weight of the load when lifting a load with two slings?
2. What is the load capacity (WLL) for a blue-covered round sling when it is used as a choker?
3. Which grade of chain is rated for overhead lifting?
4. What is the volume of a triangular prism that has a height of 6.5 in., a depth of 3 in., and a base that is 4.5 in. across?
5. What is the pounds-per-foot (lb/ft) of ¼ in., square, brass bar stock?

SECTION 2

MECHANICAL KNOWLEDGE

Mechanical Power Transmission

Bearings

Coupled Shaft Alignment

Lubrication

Seals and Packing

Pumps and Compressors

Fluid Power

Piping Systems

A TYPICAL DAY IN MAINTENANCE

Check It Out

You arrive at press number 3 at 7:30 A.M. You ask the press operator for additional information and clarification on the operation of the press before it failed. You also ask the operator to try to run the press so that you can verify what you have been told. The press does not run.

You suspect a worn bearing in the hydraulic pump to be the cause of the vibration. This pump is coupled via a rigid coupling to a three-phase motor. After de-energizing and locking out/tagging out the press, on inspection of the pump you determine that the bearing is indeed worn. You also notice the "funny" odor that the operator mentioned. You believe the source of the odor to be the motor, which is coupled to the pump. You make a mental note to investigate this later. You also notice that the pump mounting plate is cracked. This has to be welded. Again, you make a mental note to fix this later.

For now, you direct your efforts to removing and replacing the worn bearing. You begin the replacement process at 7:45 A.M.

Work It Out

1. List the steps that you would follow to replace the worn bearing.
2. What might be the source of the "funny" odor emanating from the motor?
3. What might have caused the pump mounting plate to crack?

CHAPTER

6

Mechanical Power Transmission

Mechanical power transmission is, as the name implies, the means by which mechanical power is transmitted. There are, however, many different methods that are used today to transmit mechanical energy from one point to another. Most of these methods are discussed thoroughly in this chapter, such as belt drives, chain drives, gearboxes, and couplings. Sometimes these methods are used separately, and sometimes they are used together to accomplish the desired power output.

OBJECTIVES

After studying this chapter, the student should be able to

- Discuss the different styles of belts that are used in industry.
- Discuss the benefits of a positive-drive belt.
- Discuss the benefits of a chain drive system.
- Discuss the use of gears and gearboxes.
- Define pitch diameter, circular pitch, pitch line, and gear ratio.
- List the different types of gears.
- Perform speed calculations for belt drives, gear drives, and chain drives.

6-1 BELTS AND SHEAVES

Power transmission through the use of belts is a common occurrence in industry. Today, many different types of belts are in use. It is not within the scope of this textbook to cover all of the different types of belts because there simply are too many. The most commonly used belts, however, are discussed. Specifically two different styles of belts are discussed in this chapter: flat belts and V-belts.

Flat Belts

Flat belts are most often made of leather, canvas, or rubber because these materials offer the most friction and flexibility. Flexibility is important because a belt is constantly flexing at the turning points. The need for friction is self-explanatory. Although not many new installations use flat belts because many new types of V-belts are available, flat belts are still in use today. Many machines that were designed years ago still operate with the flat belt technology; this is therefore why it is important to know about flat belts. V-belts offer the ability to transfer more torque than a flat belt. This comparison is discussed in further detail later.

As stated, flat belts are usually available in rubber, canvas, or leather. This is a little misleading, however, because there may be canvas belts that are coated or impregnated with rubber. These types of belts would be used only in conditions that would harm belts made of rubber alone. The canvas offers reinforcement and resists adverse conditions better than rubber by itself. It is important to know, however, that of all of these types of belts, the leather belt is probably the most commonly used.

Leather belting is specified by thickness and width. There are two thickness classifications, single and double, each of which is further divided into light, medium, and heavy. It is important to use the correct belt for the correct application. If a light belt is used for a heavy application, the belt wears out, often causing downtime. Some of the thickness specifications for first-quality leather belting are

- Medium single—$^{5}/_{32}$ to $^{3}/_{16}$ in.
- Heavy single—$^{3}/_{16}$ to $^{7}/_{32}$ in.
- Light double—$^{15}/_{64}$ to $^{17}/_{64}$ in.
- Medium double—$^{9}/_{32}$ to $^{5}/_{16}$ in.
- Heavy double—$^{21}/_{64}$ to $^{23}/_{64}$ in.

Leather has two different textures. One side is smooth, and the other side is rough but soft. The latter is referred to as the flesh side. Leather belts are cut from the center of the hide of an animal. As the leather cures, the hide dries out. Leather is then treated to give it the needed flexibility. Take the time when installing leather belting to install it correctly. Place the leather with the flesh side on the outside and the smooth side on the inside toward the pulley. A belt that is installed with the flesh side on the inside tends to wear faster. The belt also transmits more power if it is installed correctly.

Belt Speed Calculations

Because flat belts are often used in high-speed situations, you should be careful not to exceed speeds of 5280 ft/min. Centrifugal force tends to throw the belt at extreme speeds. **Belt speed** can be calculated by multiplying pi (π) by the pulley diameter in inches *(D)* and the revolutions per minute (rpm) of the pulley shaft divided by 12. The answer is expressed in feet per minute (ft/min). Following is the formula:

$$Speed = \frac{\pi \times D \times rpm}{12}$$

For example, what is the belt speed when the speed of the motor is 1800 rpm and the pulley is 10 in. in diameter?

$$Speed = \frac{\pi \times D \times rpm}{12}$$

$$Speed = \frac{\pi \times D \times 1800}{12}$$

$$Speed = \frac{56{,}548.67}{12}$$

$$Speed = 4712.39$$

This belt moves at 4712 ft/min.

The length of a belt can be found by using one of two different formulae. One formula is for belts that are crossed over. Belts are crossed over to change the direction of rotation of the driven shaft. To use a belt in a crossed-over configuration, the belt must leave the top of the driver pulley and load onto the driven pulley from the bottom. As the belt leaves the top of the driven pulley, it returns to the bottom of the driver pulley to begin the next revolution. In the open-belt configuration, a belt simply leaves the top of the driver pulley, loads on the top of the driven pulley, leaves the bottom of the driven pulley, and returns to the driver pulley on the bottom. A belt that is used in a crossed-over configuration is longer than a belt that is used in an open fashion. The pulleys of a crossed-over configuration turn in opposite directions, whereas with the open-belt configuration, the pulleys turn in the same direction. If the open-belt configuration is used, the formula is

$$L = \frac{\pi}{2}(D+d) + \left(2\sqrt{C^2+\left(\frac{D-d}{2}\right)^2}\right)$$

L = Belt length
D = Diameter of the large pulley
d = Diameter of the smaller pulley
C = Center distance between the shafts
$\pi = 3.1416$

If the pulleys are of an equal diameter, then the following formula is used:

$$L = \frac{(D+d)\pi}{2} + 2C$$

L = Belt length
D = Diameter of the large pulley
d = Diameter of the smaller pulley
C = Center distance between the shafts
$\pi = 3.1416$

If the belt is crossed and pulleys with different diameters are used, the formula is

$$L = \frac{\pi}{2}(D+d) + \left(2\sqrt{C^2+\left(\frac{D+d}{2}\right)^2}\right)$$

L = Belt length
D = Diameter of the large pulley
d = Diameter of the smaller pulley
C = Center distance between the shafts
$\pi = 3.1416$

Flat Belt Tracking

When working with flat belts, it is important to make sure that the pulleys on which the belts ride are in alignment with each other. This means that the shafts of the pulleys need to be perfectly parallel to each other. If the pulleys are not aligned properly, the belt will roll off the side of a pulley. **Tracking** the belt before the machine is put back into production can prevent this. The first step to tracking a belt is to get the pulleys as closely aligned as possible using a measuring device such as a straightedge, a tape measure, or a laser. A simple way to accomplish pulley alignment is to measure from the center of one pulley shaft to the center of the other pulley shaft on both sides of the pulleys, as shown in **Figure 6-1**. These measurements are known as center-to-center measurements.

If the pulleys are in alignment, these measurements are all equal. When the pulleys are as close as possible to being aligned, put the belt on the pulleys. Once the belt is on the pulleys, mark with chalk where the edge of the belt rides on the pulleys. After you have marked the pulley, run the machine at a low speed. Watch the belt to see whether it is moving over the line, away from the line, or whether it is not moving at all. If the belt is moving away from the line, then the center-to-center measurement of the shafts opposite the mark is less than the center-to-center measurement of the shafts that are on the same side as the mark. This is illustrated in **Figure 6-2**. If the belt is moving over the line, then the center-to-center measurement of the shafts on the same side as the mark is less than the center-to-center measurement of the shafts that are opposite the mark. This is illustrated in **Figure 6-3**.

If either of these scenarios is the case, try to align the shafts again. Once you have given the pulleys another alignment, try tracking the belt again. If the belt did not move off the mark, then speed the machine up in small increments until a speed above the run speed has been achieved. It is critical to have a small delay between adjustments to allow time to monitor the reaction of the belt

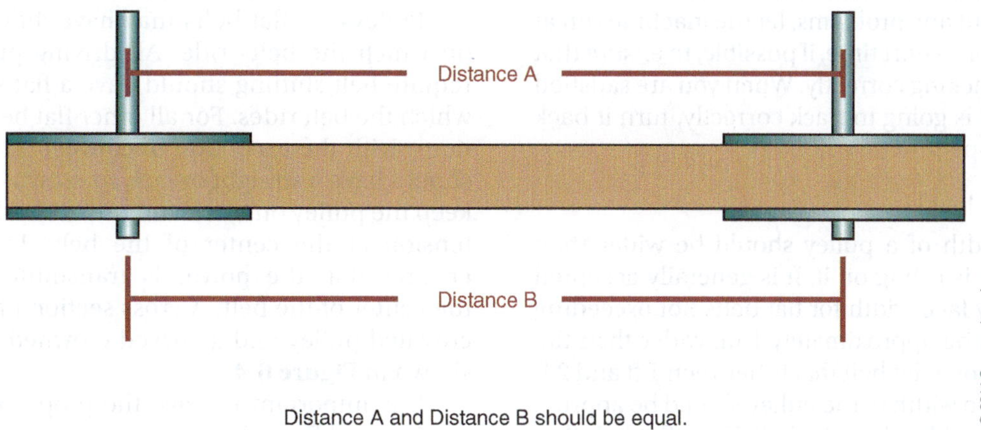

FIGURE 6-1 Flat belt pulley alignment.

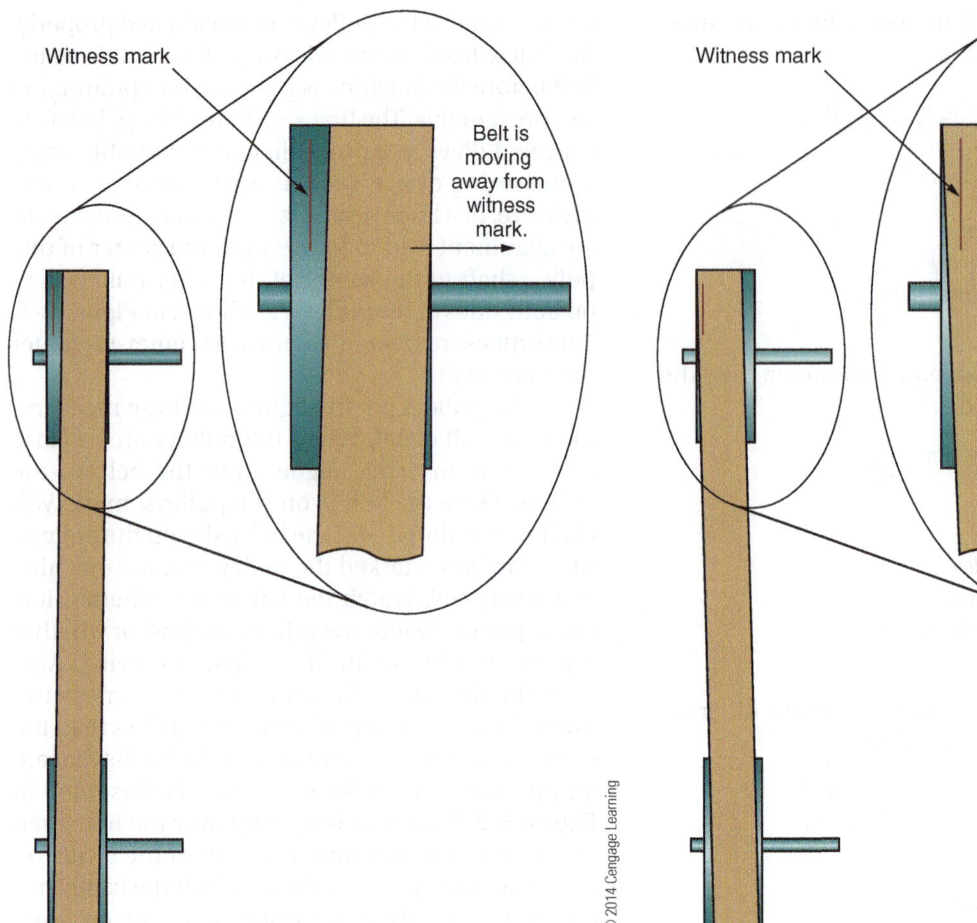
Witness mark
Belt is moving away from witness mark.
© 2014 Cengage Learning
Witness mark
Belt is moving over the witness mark.
© 2014 Cengage Learning

FIGURE 6-2 Flat belt tracking away from witness mark.

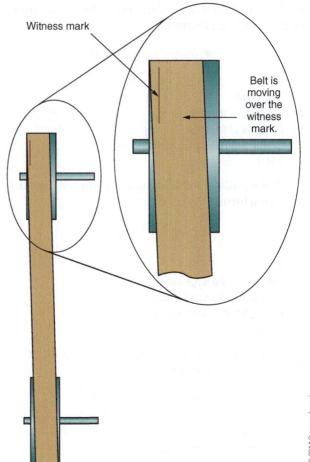

FIGURE 6-3 Flat belt tracking over witness mark.

as well. If the pulleys are just a fraction of an inch off alignment, the belt will begin to ride off the line that you drew once the higher speeds are achieved. The reason that this occurs at higher speeds and not so much at low speeds is that it takes less time for the belt to travel all the way around the pulleys; thus, the higher speeds amplify any misalignment. If you have achieved a speed above the normal-run speed without any problems, let the machine run at that speed for a short time, if possible, to ensure that the belt is tracking correctly. When you are satisfied that the belt is going to track correctly, turn it back over to the operator.

Flat Belt Pulleys

The face width of a pulley should be wider than the belt that is riding on it. It is generally accepted that a pulley face width for flat belts not exceeding 1 ft in width be approximately 1 in. wider than the belt width. For a flat belt that is between 1 ft and 2 ft wide, the face width of the pulley should be approximately 2 in. wider than the belt. For a flat belt that exceeds 2 ft wide, the pulley should be at least 3 in. wider than the belt. This ensures that the belt has room to move on the pulley without rolling off its edge. If you were wondering what kind of flat belt would exceed 2 ft in width, think about a conveyor belt. The major difference is that a conveyor belt, more than likely, is riding on rolls instead of pulleys. This difference does not matter, however, because the rolls have shafts, just as the pulleys do, that can be aligned in the same manner.

Pulleys for flat belts may have different faces on which the belts ride. All driving pulleys that require belt shifting should have a flat surface on which the belt rides. For all other flat belt applications, with the exception of conveyors, the pulley should have a straight or curved crown. This helps keep the pulley on track and provides a little more tension at the center of the belt. This tension ensures that the power is transmitted through the center of the belt. A cross section of a straight crowned pulley and a curved crowned pulley are shown in **Figure 6-4**.

It is important to have the proper amount of tension on a flat belt to prevent slippage. If there is any slippage on a crowned pulley, the belt rides off the crowned surface of the pulley.

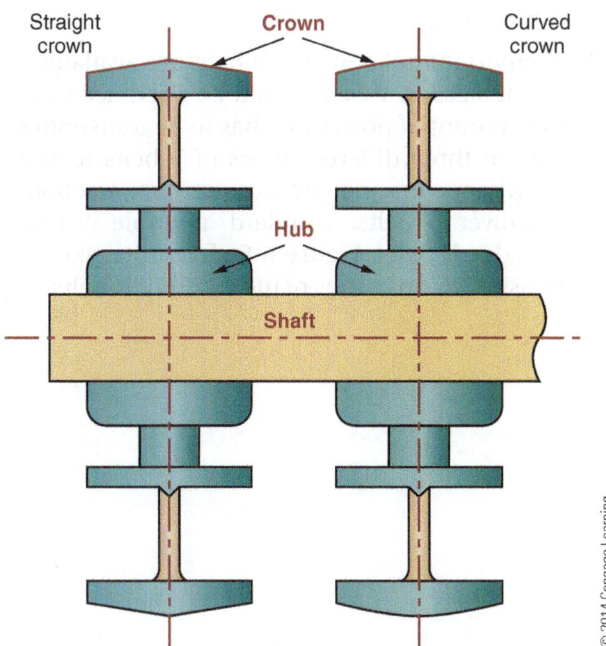

FIGURE 6-4 Cross section of straight crown and curved crown pulleys.

V-Belts

V-belts do not transmit power in the same manner as a flat belt. A V-belt transmits power through the wedging action of its tapered sides when it rides within the pulley groove. This is accomplished because V-belts have a trapezoidal cross section. Side contact between the belt and pulley can be seen in **Figure 6-11**. This is where the power should be transmitted. V-belts are generally made of rubber and a fiber mesh that is used for strengthening. Most of the time, V-belts also have a cloth backing on which an identification code is printed. All V-belts contain fiber or steel cords for reinforcement, referred to as tension members. **Tension members** are the main load-carrying components of V-belts. They also prevent stretching, thus ensuring long belt life. V-belts are resilient, quiet, and strong.

It is important to use the proper procedure for installing a belt onto a sheave (pulley). Never roll a belt onto a sheave because it can cause the tension members to break. Always move one of the pulleys toward the other before placing the belt onto the sheave. By doing this, the distance between the centers of the shafts is reduced, eliminating the need to roll the belts over the edge of the sheave. Once the belt has been placed onto the sheave, using the proper method of installation, the pulley that was moved inward can now be moved back out to tension the belt properly. **Belt deflection** is shown in **Figure 6-5**. Proper belt deflection is approximately 1/64 in. per inch of span between the two shaft centers. A belt deflection tester can test belt deflection and measure deflection force and deflection distance (Figure 6-5).

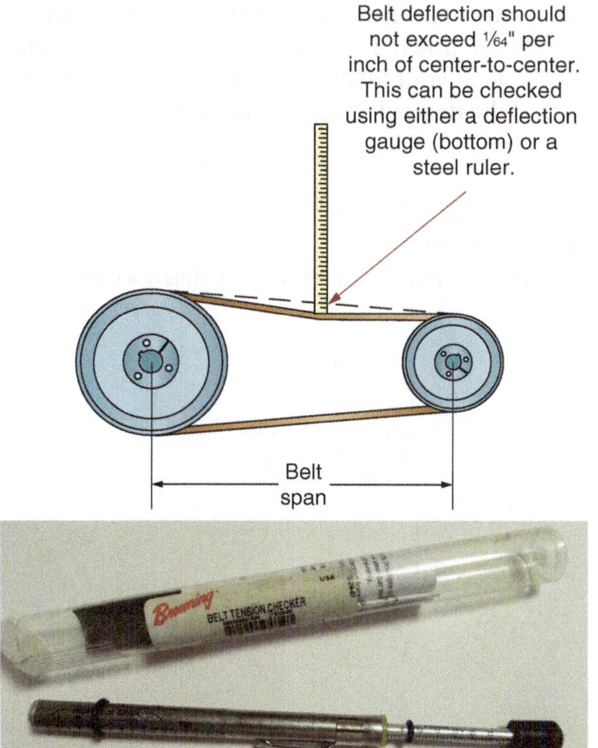

FIGURE 6-5 Belt deflection.

There is a standard procedure that must be used to ensure the proper deflection of a belt: You must first measure the span from the center of one shaft to the center of the other shaft. After the measurement is made, slide the lower O-ring that is on the deflection distance scale to 1/64 in. and the upper O-ring up against the body of the measuring tool. Next, position a straightedge even with the top of the belt. Place the belt deflection guide on the belt in the center of the span, and apply force on the belt deflection guide until the bottom of the O-ring is even with the straightedge. As force is applied to the tool, the upper O-ring moves up the deflection force scale. This marks the amount of force that is needed to achieve the proper deflection. The upper O-ring stays at its measurement even as the tension is released. Once you have removed the belt deflection tester from the belt, locate the force measurement at the topside of the upper O-ring. If the force is less than the minimum deflection force, tighten the belt. If the measured force is greater than the maximum deflection force, loosen the belt. If the measured force is within the minimum and maximum range, tighten the system down to prevent movement of the shafts. If the measured force is not within the range and movement of a pulley is required, follow the same procedure until an adequate measurement is obtained, then tighten the system down. If a deflection tester is not readily available at your facility, use the following formula to find the desired deflection depth:

$$DH = SL \times 0.015625$$

DH = Proper deflection height
SL = Span length
0.015625 = Decimal equivalent of $\frac{1}{64}$ in. (This is a constant.)

V-Belt Selection

It is important to know that V-belts are available in different sizes. A V-belt should be sized according to the amount of power that has to be transmitted. There are three different types of V-belts to consider when choosing the correct one: fractional horsepower V-belts, standard multiple V-belts, and wedge V-belts. **Figure 6-6** shows the size and dimension comparisons of the different V-belts.

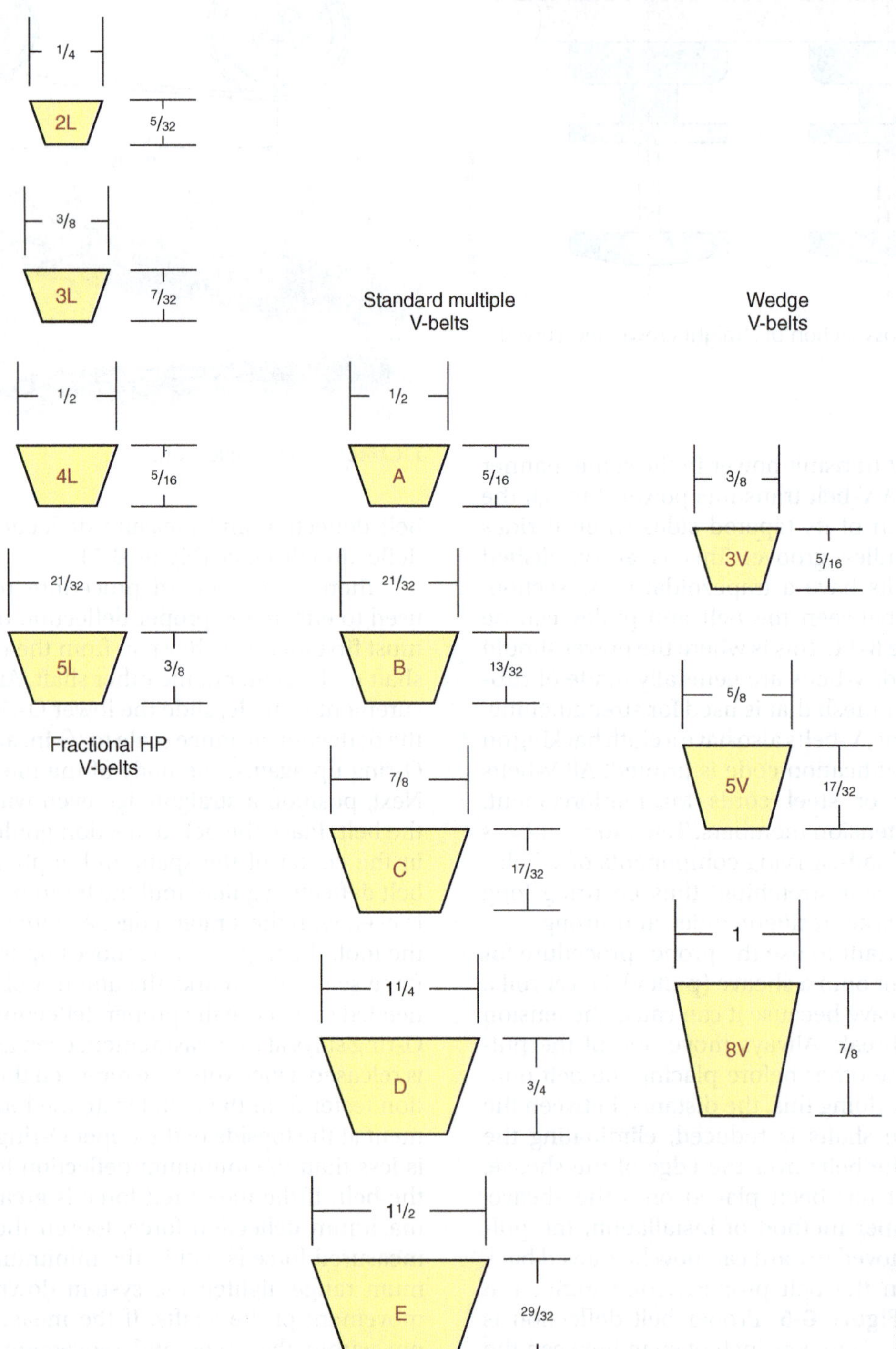

FIGURE 6-6 Dimension comparisons (in inches) of V-belts.

Notice the fractional horsepower belts. These are the smallest V-belts on the left side of Figure 6-6. These belts are typically used to transmit a small amount of power. Also take notice of the wedge belts. The top dimension of the 3V belt is between the A and B belts in the standard column. The same is true for the 5V belt, which is residing between the B and C belts. Finally, you should notice that both dimensions of the 8V belt are between the D and E belts in the standard column. These belts are not used as often as the standard and fractional horsepower belts. Belts have identification markings, for example, B72 48—usually, but not always, the manufacturer's name is between the set of numbers, such as B72 *Manufacturer's Name* 48. Each one of these letters and numbers represents something. The B means that the belt is $^{21}/_{32}$ in. wide across the topside of the belt, as shown in Figure 6-6. The 72 represents the length of the belt. This length is measured at the pitch line, which is at the tension members of the belt. Remember, the tension members are what carry the load. The pitch line of this particular belt is 72 in. long. The 48 represents the nominal size tolerance. Nominal size tolerance is how close a belt is to its rated length. Usually a belt is not exactly the length that is specified. In the example, the belt should be 72 in. long. However, it is known by looking at the 48 on the belt that it is just shy of 72 in. If the belt were exactly 72 in., there would be a 50 in the place of the 48. For every $^{1}/_{10}$ in. over the listed length, 1 is added to the 50. For every $^{1}/_{10}$ in. under the listed length, 1 is subtracted from 50. Therefore, this belt is $^{2}/_{10}$ in. less than 72 in. $^{2}/_{10}$ in. can be simplified to $^{1}/_{5}$ in. Subtracting this from 72 leaves 71$^{4}/_{5}$ in. This is the true length of the belt. If there were a 55 in the place of the 48, then the belt would be $^{5}/_{10}$ in. over 72 in., thus making the belt 72½ in. long ($^{5}/_{10}$ simplifies to ½). If an X were present between the letter and the number—for example, BX72 48—this would indicate that the belt has cogs on its lower side. These cogs are shown in **Figure 6-7.** The cogs have two purposes: to provide more air turbulence at the contact surface of the belt, thus keeping the belt cooler while it is running, and to provide more flexibility.

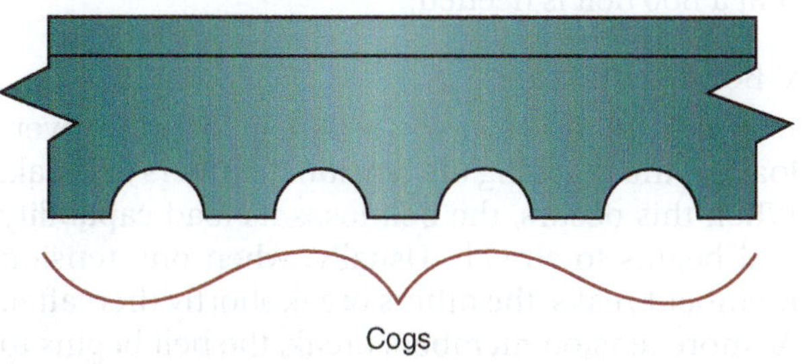

FIGURE 6-7 V-belt cogs.

CAUTION

When belts are used in sets, it is important to match the belts as closely as possible, using the nominal tolerance code. It is generally accepted that all of the belts within a set be within $^{4}/_{10}$ in. ($^{2}/_{5}$ in. simplified) to $^{6}/_{10}$ in. ($^{3}/_{5}$ in. simplified). This ensures that all of the belts in the set share the load. If one of the belts within a set is significantly shorter than the others, it will bear most of the load and thus shorten the life of the belt. Therefore, if one belt in a set goes bad, all of the belts in that set must be changed. If only the bad belt were changed, the new belt would be running with belts that have been stretched, and therefore it would carry most of the load, thus shortening its life.

It is important to know that there is another type of V-belt that is used when one or more pulleys must be driven in the opposite direction of the drive pulley. This belt is most commonly referred to as a serpentine configuration.

A double V-belt, shown in **Figure 6-8**, looks like two standard V-belts placed back to back. It is not, however, made of two separate belts. It is one belt that has a mirrored trapezoidal cross section. The tension members are placed across the centerline of the belt to carry the load of the belt on either side. This can also be seen in Figure 6-8. Because the double V-belt resembles two V-belts back to back, it is given a double-letter code—for example, a BB belt instead of a B belt. **Figure 6-9** shows the size and dimension comparisons of the double V-belts.

Sometimes, in an industrial environment, the information on the belt wears off. If this is the case, the length of the belt may still be determined. There is a simple formula that can be applied to determine the belt size needed for a given set of pulleys. All that is needed is the pitch diameter of the pulleys and the distance between the shafts. Do you remember that the tension members of a belt are what carry the load? The pitch diameter (pitch circle) of a pulley is the point on the pulley where the tension members ride within the pulley. See **Figure 6-10.**

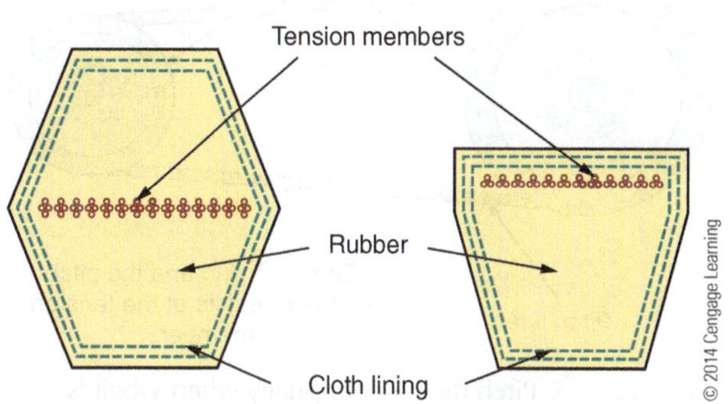

FIGURE 6-8 Tension members of a double V-belt and a standard V-belt.

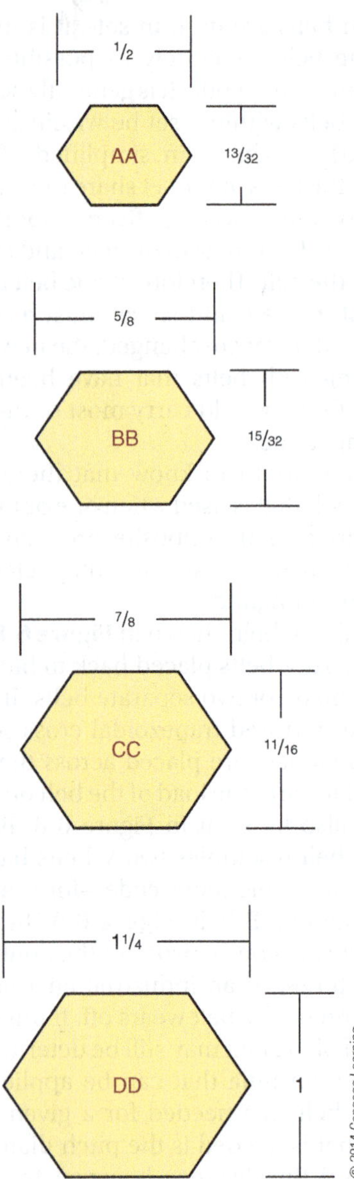

FIGURE 6-9 Dimensions (in inches) of double V-belts.

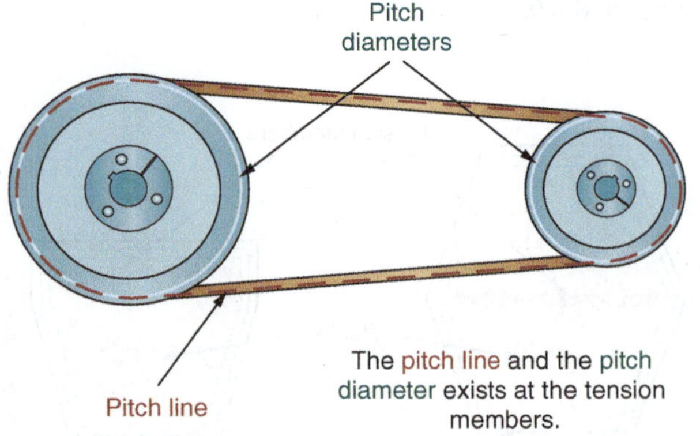

FIGURE 6-10 Pitch diameter of pulley when V-belt is in pulley.

This is important because if you simply measure the diameter of the pulley, your calculations will be off. Usually, the pitch diameter of a pulley is stamped into the face of the pulley itself. For example, a pulley with a 6 in. pitch diameter has a 6.0 stamped on the face somewhere. Also, if this 6 in. pulley were to be measured, the diameter would be more than 6 in. Following is the formula:

$$BPL = 2C + 1.57\,(D + d) + \frac{(D - d)}{4C}$$

BPL = Belt pitch length
C = Shaft distance (this is a center-to-center measurement)
D = Large pulley pitch diameter
d = Small pulley pitch diameter

The 1.57 is derived from 3.14 (π) divided by 2.

Solve the following problem:

The large pulley has a pitch diameter of 9.6 in., and the small pulley has a pitch diameter of 6.4 in. Both pulleys are standard B pulleys. The center-to-center shaft measurement is 18 in. What size belt would be used for this application?

$$BPL = 2C + 1.57\,(D + d) + \frac{(D - d)}{4C}$$

$$BPL = (2 \times 18) + 1.57\,(9.6 + 6.4) + \frac{(9.6 - 6.4)}{(4 \times 18)}$$

$$BPL = 36 + (1.57 \times 16) + \frac{3.2}{72}$$

$$BPL = 36 + 25.12 + 0.044$$

$$BPL = 61.164 \text{ in.}$$

The belt needed for this application is a B60. This is found by looking up the **nominal belt size** in **Table 6-1**. The table shows a portion of nominal belt lengths. The numbers under the letters are the actual calculated pitch lengths. The belt pitch length for the example problem is 61.164 in. A B belt is needed because B pulleys are being used. Look under the B column until you find the calculated pitch. You should always go up to the next closest number if the calculated number is not present. In this case, you would have to go to 61.8. Follow this over to the nominal belt length, and you can see that a B60 belt is needed.

V-Belt Maintenance

It is not uncommon for a belt to become overloaded, thus causing the tension members to break. When this occurs, the belt loses its load capability and begins to stretch. Usually, when one tension member breaks, the others break shortly thereafter. As more tension members break, the belt begins to

Table 6-1

BELT PITCH LENGTHS (IN.) OF STANDARD V-BELTS

Belt Number Indicating Nominal Length	A	B	C	D	E
26	27.3	—	—	—	—
31	32.3	—	—	—	—
33	34.3	—	—	—	—
35	36.3	36.8	—	—	—
38	39.3	39.8	—	—	—
42	43.3	43.8	—	—	—
46	47.3	47.8	—	—	—
48	49.3	49.8	—	—	—
51	52.3	52.8	53.9	—	—
53	54.3	54.8	—	—	—
55	56.3	56.8	—	—	—
60	61.3	61.8	62.9	—	—
62	63.3	63.8	—	—	—
64	65.3	65.8	—	—	—
66	67.3	67.8	—	—	—
68	69.3	69.8	70.9	—	—
71	72.3	72.8	—	—	—
75	76.3	76.8	77.9	—	—
78	79.3	79.8	—	—	—
80	81.3	—	—	—	—
81	—	82.8	83.9	—	—
83	—	84.8	—	—	—
85	86.3	86.8	87.9	—	—
90	91.3	91.8	92.9	—	—
96	97.3	—	98.9	—	—
97	—	98.8	—	—	—
105	106.3	106.8	107.9	—	—
112	113.3	113.8	114.9	—	—
120	121.3	121.8	122.9	123.3	—
128	129.3	129.8	130.9	131.3	—
136	—	137.8	138.9	—	—
144	—	145.8	146.9	147.3	—
158	—	159.8	160.9	161.3	—
162	—	—	164.9	165.3	—
173	—	174.8	175.9	176.3	—
180	—	181.8	182.9	183.3	184.5
195	—	196.8	197.9	198.3	199.5
210	—	211.8	212.9	213.3	214.5
240	—	240.3	240.9	240.8	241.0
270	—	270.3	270.9	270.8	271.0
300	—	300.3	300.9	300.8	301.0
330	—	—	330.9	330.8	331.0
360	—	—	360.9	360.8	361.0
390	—	—	390.9	390.8	391.0
420	—	—	420.9	420.8	421.0
480	—	—	—	480.8	481.0
540	—	—	—	540.8	541.0
600	—	—	—	600.8	601.0
660	—	—	—	660.8	661.0

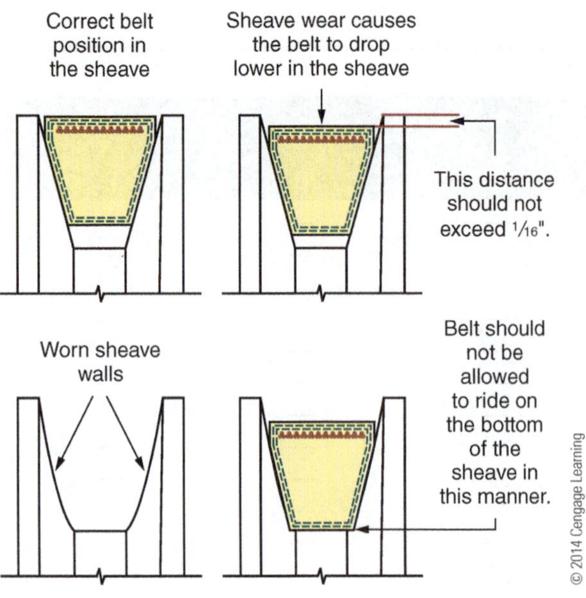
Correct belt position in the sheave
Sheave wear causes the belt to drop lower in the sheave
This distance should not exceed 1/16".
Worn sheave walls
Belt should not be allowed to ride on the bottom of the sheave in this manner.
© 2014 Cengage Learning

FIGURE 6-11 Sheave wear.

ride lower in the pulley groove. This is an improper way to transmit power through the V-belt. As was mentioned earlier, a V-belt should transmit power through the wedging action of the tapered sides of the belt, not the bottom. Eventually, the belt simply begins to slip. This is when it is time to change the belt. Some maintenance mechanics may just tighten the belt and spray it with belt dressing to get it going again. However, this does not solve the problem because the belt eventually breaks if it continues to stretch. Another problem is that, over time, the pulley walls become worn, as shown in **Figure 6-11**. In the first diagram, the belt sits in the pulley in such a way that the top of the belt is flush with the top surface of the pulley wall. This is an ideal setup. As the pulley becomes worn, the belt begins to sit lower in the pulley as shown in the second diagram. **Figure 6-12** shows a belt and sheave wear guide. These come in handy when checking belt or sheave wear.

© 2014 Cengage Learning

FIGURE 6-13 An example of a link belt.

When the pulley has worn enough so as to allow the belt to drop into the pulley $\frac{1}{16}$ in., it is time to replace it. If it is not replaced, wear will continue to the point where the belt is no longer transferring power through the walls and is actually riding on the bottom of the groove. This is not how the manufacturer intended for the belt or pulley to be used. When this occurs, the belt begins to produce a lot of slippage, thus transmitting power ineffectively. Pulley and belt wear can be checked with a pulley gauge. Link belts have a unique "quick connect" belt design that provides for easier and faster belt installation; no tools are required to install this type of belt because of its "twist together" design.

Belts are easily made up to the required length, by hand, in seconds and can be laid into the pulleys and assembled where the two ends meet, with no need to dismantle drive components or change existing pulleys. They also run in industry standard pulley grooves. See **Figure 6-13**.

Positive-Drive Belts

Positive-drive belts, also referred to as timing belts or gear belts, are a combination of a chain drive and a belt drive. A positive-drive belt is shown in **Figure 6-14.**

The benefit of a chain drive is that it is a positive-traction drive. This means that there is no slippage. As discussed earlier, V-belts do slip, which can cause problems in some systems, especially when timing is critical. When using a timing

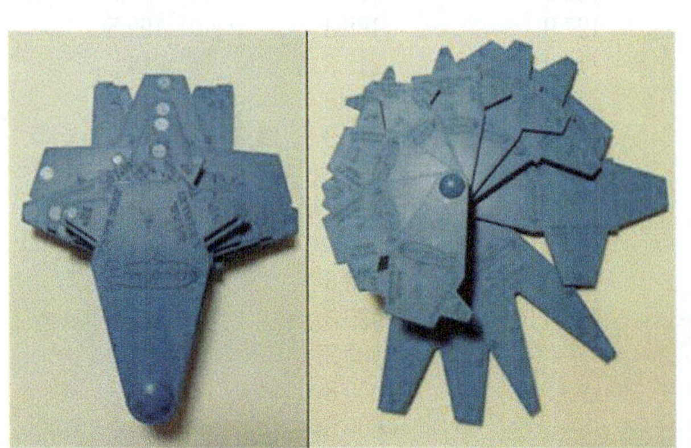
© 2014 Cengage Learning

FIGURE 6-12 Belt and sheave wear guide.

optibelt ZR 270 H
optibelt ZR 270 H
optibelt ZR 270 H
optibelt ZR 270 H
optibelt ZR 270 H
© 2014 Cengage Learning

FIGURE 6-14 Positive-drive belt.

belt, power is transmitted through gearlike teeth because they engage into mating grooves that are in the pulley. Because of this, the timing belt does not depend on friction to transmit power, like V-belts. Something that the timing belt does have in common with the V-belt, however, is there are tension members in the belt. A timing belt is usually made of a neoprene backing, steel tension members, and neoprene teeth with a nylon facing. The pitch line of a positive-drive belt, as with V-belts, is located within the tension members. The pitch diameter (pitch circle) of a positive-drive belt is at its tension members as well. Because the positive-drive belt runs on top of the pulley instead of inside the pulley as with the V-belt, the pitch diameter of a positive-drive belt is always larger than the diameter of the pulley itself. It is important to know that a positive-drive pulley of one pitch cannot be used with a positive-drive belt of a different pitch. The belt must be run with pulleys of the same pitch. The six standard positive-drive belt pitches are as follows:

- Mini-extra-light (MXL) pitch—$^2/_{25}$ in.
- Extra-light (XL) pitch—$^1/_5$ in.
- Light (L) pitch—⅜ in.
- Heavy (H) pitch—½ in.
- Extra-heavy (XH) pitch—$^7/_8$ in.
- Double-extra-heavy (XXH) pitch—¼ in.

It is helpful to know that the circular pitch of a positive-drive belt is measured from the center of one tooth to the center of the next tooth. As with the V-belt, the positive-drive belt also has a numbering system, for example, a 300 H075 positive-drive belt. Each of these numbers also represents something. The 300 represents the pitch length of the belt. This belt has a pitch length of 30 in. Notice that the belt length designation number is the pitch length multiplied by 10. The H represents a heavy-pitch belt. The 075 represents the belt width. The belt width of the example belt is $^3/_4$ in. Notice that the designation number is the belt width multiplied by 100.

Pulleys

Pulleys are sometimes referred to as sheaves. There are typically two types of pulleys available: fixed bore and tapered bore. A fixed-bore pulley has an integral hub cast into the pulley on forging. The hub is bored to fit a certain size shaft with a certain size keyway. The tapered-bore pulley offers a little more flexibility when it comes to shaft sizes. A tapered-bore pulley is shown in **Figure 6-15**. It is a two-piece pulley. The tapered-bore portion of the pulley is known as the hub (sometimes referred to as a bushing). The other piece is known as the flange. The hub can be removed from the center of the flange and be replaced with a hub that has a different shaft bore.

FIGURE 6-15 Tapered-bore pulley.

Why is this helpful? Here is an example: A 6.0 in. pulley is being driven by an electric motor with a 182 frame. The motor has a ⅞ in. shaft with a $^3/_{16}$ in. × $^3/_{32}$ in. keyway. The motor overheats because of clogged air vents, and the windings become shorted. Now it is time to change the motor. On inspection of the warehouse, you find that all you have is a motor with a 182T frame. There are no more motors with a 182 frame in the warehouse. It is going to take 4 days to get a new motor shipped in. However, the machine cannot lose 4 days of production, so you decide to put the motor with the 182T frame in its place. Now the pulley has to fit on a 1⅛ in. shaft with a ¼ in. × ⅛ in. keyway. Because it is a tapered-bore pulley, you can simply change the hub instead of finding a new 6 in. pulley with the correct shaft bore. The tapered-bore assembly becomes as sound as a fixed-bore hub once the two pieces are pulled together and the cap screws are tightened. This is a result of the close fit and extreme pressures that are created from the angled force of the tapered-mating surfaces. It is important when tightening a tapered-bore pulley to tighten the cap screws in an alternate fashion. Always check for alignment as you tighten a tapered-bore pulley. Do not overtighten the cap screws because this may cause the cap screws to shear. Use a torque wrench to tighten the cap screws to their specified torque.

Even though the tapered-bore pulley offers a press fit when it is pulled together, it generally separates with ease because of the tapered fit. If the two pieces do not separate as soon as the cap screws are loosened, a simple, light rap on the flange should cause it to pop off the hub. Do not beat on a flange, because you may bend it. Also, do not lubricate the tapered surface of the hub, because this may cause the hub to break owing to the hydraulic forces that are developed when tightening the cap screws.

6-2 CHAINS AND SPROCKETS

Chain drives are used in industry because of their positive-drive capabilities. They do not slip, nor do they creep. This has made the chain drive system a reliable form of power transmission. The **roller chain** is the most common type of chain that is in use today for power transmission. A roller chain is constructed of different types of links called **roller links** and **pin links**. The roller link consists of two rollers that are mounted on bushings. This configuration allows the roller to roll, thus lowering the amount of friction that is produced on the sprocket. The bushings are pressed into the sidebars. A **sidebar** is a steel plate into which the bushings are pressed on the roller link and the pins are pressed on the pin link. The pin link slides into the bushing in the roller link, as shown in **Figure 6-16**. This mating of roller links and pin links continues in an alternating fashion until the desired length of the chain is met.

A roller chain that has the bushings and pins press fitted is usually premanufactured in given lengths. The maintenance technician usually has to shorten a new chain to a desired length by using a chain splitter. This tool is shown in **Figure 6-17**.

The **chain splitter** pushes against the pin while holding the sidebar. If you do not have access to a chain splitter, the sidebar can be removed by carefully grinding the pins on a pin link. This allows removal of the sidebar, thus allowing the chain to be broken at that point. The chain is then made endless by connecting together both ends—which consist of two roller links—using a master link. This link is sometimes referred to as the connecting link. The master link has the pins press fitted into one

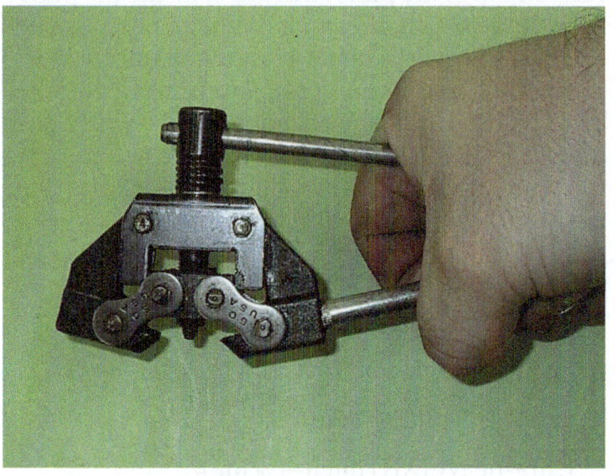

FIGURE 6-17 Chain splitter.

FIGURE 6-16 Various roller chains. Notice the pin link and the roller link at the top.

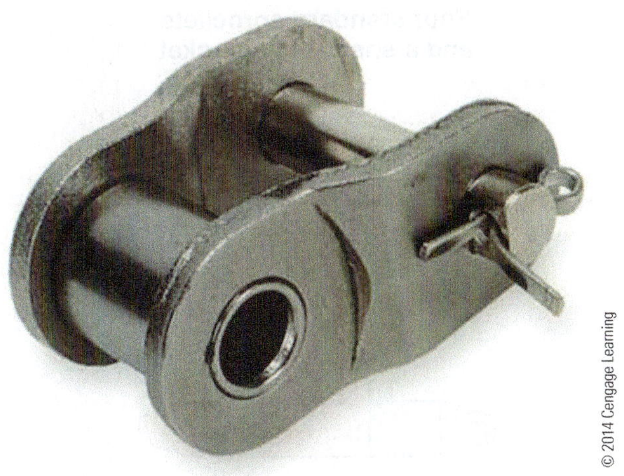

FIGURE 6-18 Offset link.

sidebar. The other sidebar is placed onto the pins after the master link has been placed into the two roller links that make up the end of the chain. After the master link side bar has been placed onto the pins of the master link, a retainer clip is placed on the pins. The pins on a master link have slots cut into them to lock the removable sidebar in place. A roller chain may have cotter pins to hold the sidebars in place. Grinding is not needed to shorten this type of chain. Simply remove the cotter pins and pull off the sidebar. A length of roller chain, before it is made endless, is normally made up of an even number of pitches. Sometimes an odd number of pitches is required in a chain drive system. In this case, an offset link is used. An offset link, which is shown in **Figure 6-18**, has one pin and one bushing roller combination. This can only be used if the chain that is being made endless has a pin link on one end and a roller link on the other. Because the offset link has a sidebar that can be removed, it is usually used as the master link as well.

Roller Chain Identification

Three dimensions are used to identify roller chains: pitch, chain width, and roller diameter. All of the dimensions on a chain are proportional to the pitch dimension. The standard proportions are used to allow for interchangeability. This means that a chain made by one manufacturer can use chain parts from another manufacturer: roller diameter should be approximately ⅝ of the pitch, chain width should be approximately ⅝ of the pitch, pin diameter should be approximately $^5/_{16}$ of the pitch, and the thickness of the sidebars should be approximately ⅛ of the pitch. The standard roller chain numbering system provides a complete identification of the chain by the number that is stamped on the sidebar of each link. This number consists of two or three digits, which may or may not be followed by the letter H. The letter H represents a heavy-duty chain. The number represents two things: the pitch of the chain and the type of chain. The first number indicates the pitch of the chain in eighths of an inch and may contain one or two digits. The pitch of a chain is defined as the distance, center-to-center, from one roller to the next. The second number, which is always one digit, is either a 0, a 1, or a 5.

- 0 indicates a standard roller chain.
- 1 indicates that the chain is a light-duty chain.
- 5 indicates that the chain is a rollerless chain.

On some occasions, a hyphenated number is suffixed to the code. This denotes how many strands the chain has. For example, if a chain has double strands, there is a dash and a two (–2) present just after the last digit of the code. This hyphenated number is not present if the chain has only one strand. Try the problem to see whether you can determine the size and type of chain that is being used.

Describe a chain that has a 60H-3 stamped on its sidebar.

1. The 6 represents $^6/_8$ in. (¾ in.) pitch.
2. The 0 represents a standard roller chain.
3. The H represents a heavy-duty chain.
4. The –3 indicates that this chain happens to be three strands wide.

Solve the following problem:

Describe a chain that has a 25 stamped on its sidebar.

1. The 2 represents $^5/_{16}$ in. (¼ in.) pitch.
2. The 5 indicates that it is a rollerless chain.

Silent Chain

A **silent chain**, sometimes referred to as an inverted-tooth chain, consists of a series of links that are joined together with bushings and pins, like the roller chain; however, this chain does not even closely resemble the roller chain in appearance. A silent chain, which can be seen in **Figure 6-19**, is constructed in a manner that allows the chain to ride on the surface of the sprocket.

A silent chain consists of a series of links that are flat on the top and have teeth on the bottom. This is similar to the timing belt. This chain is very strong and quiet while in operation. It has a guide link (plate) in the center of the chain. This guide link rides in a groove that has been cut in the sprocket, preventing the chain from moving laterally across the face of the sprocket. A silent chain is very

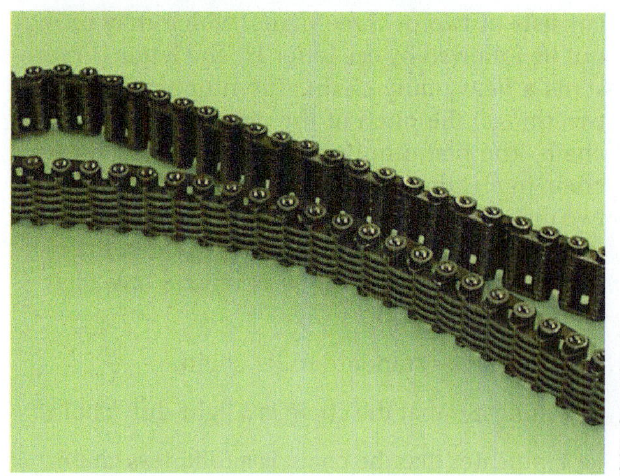

FIGURE 6-19 Silent chain.

expensive in comparison to the roller link chain, but it is more efficient at higher operating speeds. This chain also has a longer life span than the roller chain. The silent chain cannot be used on a roller chain sprocket. It can only be used on a silent chain sprocket.

Ladder Chains

Ladder chains consist of a series of links that are made from precision bent wire with a loop on each end. The loop on each end provides a method for connecting the links together. There are no rollers, pins, or sidebars.

This chain is considered light duty. Most large garage doors that open manually use this type of chain. It is most commonly used for actuating control functions of equipment, such as speed control. As with the silent chain, it is important to use the type of sprocket designed for this chain. Do not use a sprocket from a roller chain.

Sprockets

Sprockets that are used for roller chains are constructed in five different configurations, as shown in **Figure 6-20**.

A type A sprocket is configured in such a manner so it can be fastened directly to what has to be driven. A type A sprocket is a plate sprocket that has no hubs. A type B sprocket has a single hub on one side of the sprocket. The type C sprocket has a double hub, one hub on each side of the sprocket. A type D sprocket has a detachable hub. Finally, the shear-pin type of sprocket has a shear pin in it. This shear pin shears off in the event of an overload, protecting the chain from breaking. If the shear pin were to shear off, the sprocket would be free to rotate on the hub or the shaft, depending on the configuration.

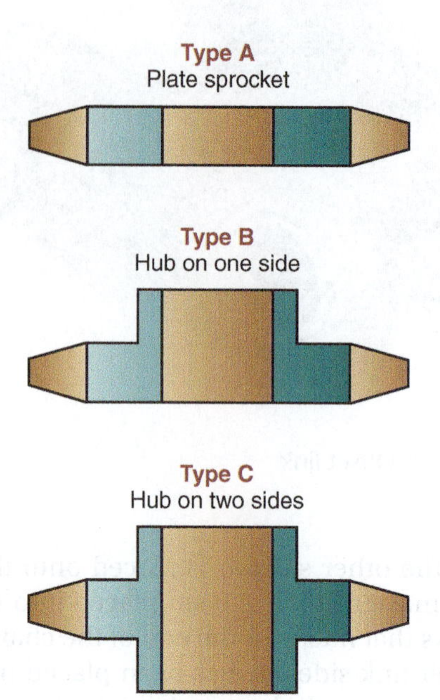

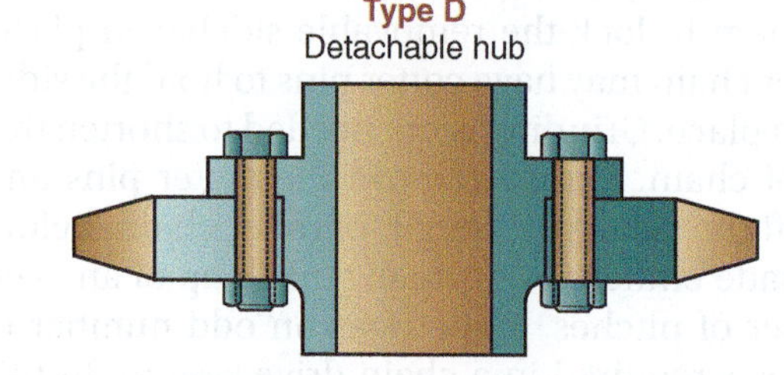

FIGURE 6-20 Five types of sprockets.

6-3 GEARS AND GEARBOXES

A **gear** is a toothed wheel or disc used to transmit mechanical power from one shaft to another through the meshing action of teeth. The teeth surround the entire gear. A gear also has a hole bored in its center so that it can be mounted onto a shaft. There also is a keyway slot cut into the bore so it can receive a key once it is mounted onto the shaft.

The key prevents the gear from spinning on the shaft as it tries to transmit mechanical energy from one shaft to another. When used in drive systems, gears provide positive-drive reaction with the meshing of the teeth. Gears may be used to change direction, speed, or torque. Because gears are used in sets, it is important to distinguish them from one another. The gear that is attached to the driving force (e.g., motor) is referred to as a drive gear, and the gear that is connected to the output shaft is referred to as the driven gear. The smaller gear in a gear set is called a pinion gear. A **pinion gear** may be a drive gear or a driven gear, depending on the use of the gears.

Gear Pitch

It is important to know a few things about gears before discussing the types of gears that are in use today. First, you must know about the pitch diameter, circular pitch, and diametral pitch of a gear. The pitch diameter is the point in which both gears transfer power, as shown in **Figure 6-21**. Notice that the pitch diameter is shown at the point where the two imaginary circles meet. These imaginary circles represent the pitch circle of the two gears. The **pitch circle** is sometimes referred to as the **pitch line**. This point is where the work is done.

Pitch diameter can be calculated by measuring the distance from the root or bottom of a tooth space to the top of the tooth on the opposite side of the gear (180°). This measurement is made across the centerline of the gear, as demonstrated in **Figure 6-22**.

Circular pitch is simply the distance that is measured from one tooth to the exact same point on the next tooth. This measurement is made along the pitch circle.

The **diametral pitch** is the distance along the pitch diameter in which gear teeth will be counted to give the tooth count-to-pitch diameter ratio.

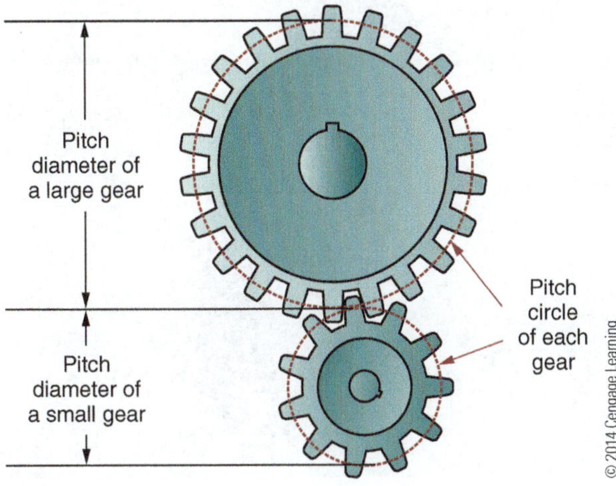

FIGURE 6-21 Pitch diameters of gears.

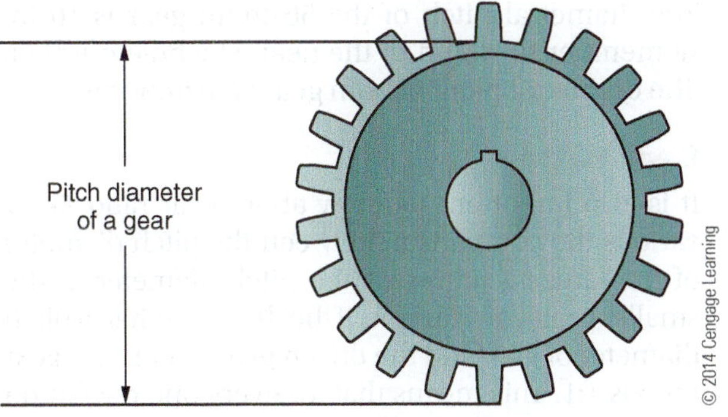

FIGURE 6-22 Measured pitch diameter of a gear.

Solve the following problem for diametral pitch:

The driver gear has forty teeth and a pitch diameter of 5 in. The driven gear has fifty-six teeth and a pitch diameter of 5⅝ in.

First, the diametral pitch must be calculated. The diametral pitch is used because it designates the size and proportion of the gear teeth by specifying the number of teeth in the gear for each inch of the gear's pitch diameter. Diametral pitch numbers are whole numbers; therefore, if a fractional number is present after the calculations, then that number should be rounded to the nearest whole number. In this problem, the 5 in. measurement (taken from the gear with forty teeth) is measured from the bottom of the tooth space to the top of the opposite tooth. This is the pitch diameter. The same is true for the gear with fifty-six teeth. There are 5⅝ in. from the bottom of the tooth space to the top of the opposite tooth. Knowing this, solve the problem by first finding the diametral pitch of each gear.

$$P = \frac{N}{D}$$

P = Diametral pitch
N = Number of teeth
D = Pitch diameter

$$P = \frac{N}{D}$$

$$P = \frac{40}{5}$$

$$P = 8 \text{ in.}$$

The driver gear has a diametral pitch of 8 in. Now find the diametral pitch for the driven gear.

$$P = \frac{N}{D}$$

$$P = \frac{56}{5.625}$$

$$P = 9.96 \text{ in}$$

The diametral pitch of the 56-tooth gear is 10 in. Remember to round to the nearest whole number. The diametral pitch of both gears are now known.

Gear Ratio

It is also important to know about gear ratio. **Gear ratio** is the comparison between the pitch diameter of the larger gear versus the pitch diameter of the smaller gear. For example, if the drive gear had a pitch diameter of 4 in. and the driven gear is 4 in., the gear ratio is 1:1. This means that for every one revolution of the drive gear, there is one revolution of the driven gear as well. If the drive gear were still 4 in. but the driven gear is changed to 6 in., the ratio would now be 1.5:1. This means that the drive gear rotates 1.5 times for every single revolution of the driven gear. Gear ratio can be calculated with the following formula:

$$Ratio = \frac{D}{d}$$

Ratio = Gear ratio
D = Diameter of the larger gear
d = Diameter of the smaller gear

Backlash

Backlash is the amount by which the width of the tooth space exceeds the thickness of the mating gear tooth, as shown in **Figure 6-23**. This space is necessary for several reasons. First, it provides a place for the lubrication (gearbox oil) to flow between the teeth. This helps to protect them from getting worn quickly by providing a lubricant film that acts as a cushion. This also helps keep the gear cooler and prolong its life. Backlash also prevents the gears from binding.

Too much backlash is undesirable when the rotation of the gears is frequently reversed. A large amount of backlash makes the teeth slam into each other with excessive force, causing them to break off over time. Excessive backlash can come from the wear of the teeth during normal operation. It can also come from improper lubrication.

Now that some gear fundamentals have been discussed, it is time to discuss some of the common types of gears that are used in industry today. It is not within the scope of this textbook to discuss every gear in existence; therefore, only the most common types are discussed.

FIGURE 6-23 Backlash.

The Spur Gear

The **spur gear** is the most common type of gear in use today. The teeth on this gear are parallel to the axis of the shaft that the gear is mounted on. A spur gear is shown in **Figure 6-24**.

Notice that its teeth are straight. In order for a set of spur gears to mesh properly, the shafts of the drive gear and the driven gear must be mounted in a parallel configuration. In order for the gears to mesh correctly, the distance between the shaft centers must equal half the sum of the two pitch diameters. Use the following formula:

$$C = \frac{D1 + D2}{2}$$

C = Center distance of the shafts
D1 = Diameter of the first gear

FIGURE 6-24 Spur gear.

$D2$ = Diameter of the second gear
2 = Divides the sum of $D1$ and $D2$ in half

The previous formula can be manipulated to find $D1$ or $D2$.

$$D1 = 2C - D2$$

$$D2 = 2C - D1$$

Spur gears may come in many different configurations. Some of these are the external, internal, and rack gears. External gears are gears that have their teeth on the outside. Internal gear sets have one gear, usually an external gear, meshing on the inside circumference of a larger gear. The large gear with the teeth on its inside circumference is sometimes referred to as a ring gear. This configuration provides a large speed reduction in a limited amount of space. A rack gear is a straight-line gear. It is usually paired with a smaller external spur gear (pinion gear), hence the term *rack and pinion.* A rack and pinion gear set is shown in **Figure 6-25**. This gear setup converts rotary motion into a lateral movement.

Bevel Gears

Bevel gears are very similar to spur gears except that the bevel gear has a conical shape instead of a cylindrical shape and can be helical in design. This type of gear set is used when the axis of the shaft on which one gear is mounted intersects with the axis of another gear's shaft. The gear that is mounted to the shaft the stops short (usually the driver) is referred to as the pinion gear whereas the gear that is mounted to the shaft which is connected to the load (the driven gear) is referred to as the gear wheel. They are usually used in gearboxes where the input shaft and the output shaft are at 90° angles to each other. This does not mean that all gearboxes with bevel gears are at 90° angles to each other, however. Bevel gear sets are sometimes used when the input and output shafts are at more than 90° angles. Because of this, bevel gears are usually manufactured as a pair because of the angle of the taper on the gear. This angle may not be the same as the taper of a different set of bevel gears. For this reason, care must be taken in matching the correct gear sets. Bevel gear sets are available in many different ratios as well. A bevel gear set is shown in **Figure 6-26**.

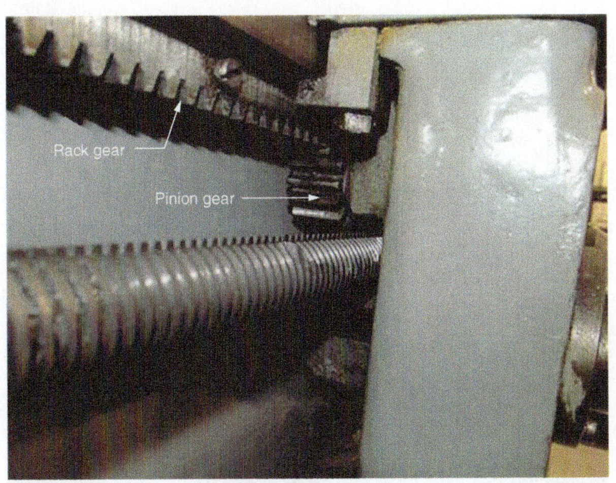

FIGURE 6-25 Rack and pinion gear.

FIGURE 6-26 Bevel gear set.

Miter Gears

Miter gears are very similar to bevel gears, except all miter gear sets have a 1:1 ratio and the axes of the shafts on which they are mounted are always at 90° angles to each other.

Helical Gears

Helical gears look similar to the spur gear with one exception: The teeth of this type of gear are not parallel to the axis shaft. The teeth of the helical gear are slanted and mesh with the teeth of the mating gear. This allows the teeth to stay in contact longer, thus providing a smoother mating action of the teeth. The problem with this configuration, however, is that end thrust is produced during rotation of the gears because the teeth are slanted. If the teeth are slanted to the right, from top to bottom, the end thrust that is produced will be to the left. If the teeth are slanted to the left, the end thrust will be to the right. From this we can conclude that the direction of thrust is dependent on the direction of rotation as well as the direction of tooth slant. A helical gear is shown in **Figure 6-27**.

Herringbone Gears

Herringbone gears eliminate the end thrust problem that exists with helical gears. The herringbone gear is constructed with two sets of teeth that are side by side. One set of teeth is slanted to the right and is on one-half of the gear surface, and another

FIGURE 6-27 Helical gear.

FIGURE 6-28 Herringbone gear.

set of teeth is slanted to the left and is on the other half of the gear surface. This is shown in **Figure 6-28**.

The double slanting cancels out the end thrust that is produced when the teeth are only slanted in one direction. Herringbone gears offer very little noise while operating at high speeds. These types of gears are used for transmitting power between parallel shafts.

Worm Gears

Worm gears are mostly used when there is a need for high-ratio speed reduction. A worm gear is usually meshed with a spur gear or a helical gear, which is called the gear, wheel, or worm wheel. The worm gear is always connected to the driver (motor). As the worm gear rotates, it turns the worm wheel. The worm wheel looks very similar to the spur gear except that the teeth have a slight curve and angle to accommodate the teeth of the worm gear. A worm gear usually has a low number of teeth, whereas the worm wheel may have many. This is how speed reduction occurs. A simple way to count the teeth on a worm gear is to look at the end of the gear, as you are looking into the end of the driveshaft, and count the teeth that are spiraled around the worm gear. It is possible that there may be as little as one tooth around the worm gear. A worm gear is shown in **Figure 6-29**, which has two teeth.

FIGURE 6-29 Worm gear.

Gearboxes

A **gearbox** houses a set of gears. Sometimes there are many gears within a gearbox. Gearboxes are used frequently in industry. They usually have one input shaft and one output shaft. They usually contain oil to keep the gears lubricated and cooled. Gearboxes come in all shapes and sizes. The type of gear set that is in the gearbox determines its shape and size. Gearboxes are used to change the speed or the direction of a particular piece of equipment or machinery. A gearbox is shown in **Figure 6-30**.

FIGURE 6-30 Gearbox.

6-4 SPEED CALCULATIONS

Speed calculations are necessary to prevent overspeeding. Overspeeding can cause damage to the components in a power transmission assembly. By using speed calculations, a desired output speed can be calculated for a given set of gears, belts and pulleys, and chains and sprockets before installation. These calculations can also be useful when the speed of a rotating shaft must be changed, or they can be used to verify the measured speed of a given set of gears, belts and pulleys, and chains and sprockets.

When referring to belt drives, shaft speed and belt speed are not the same thing; therefore, the formula for each is different. Shaft speed is referred to as pulley speed, because the pulley is mounted on the shaft. There are two types of pulley speed: driver pulley speed and driven pulley speed. To find the value of the driven pulley the pitch diameters of both pulleys, driver and driven, and the speed of the driver pulley must be known. The formula that is used to calculate the speed of the driven pulley is

$$S2 = \frac{P1 \times S1}{P2}$$

$S1$ = Speed of the driver pulley
$S2$ = Speed of the driven pulley
$P1$ = Pitch diameter of the driver pulley
$P2$ = Pitch diameter of the driven pulley

This formula can be manipulated to solve for $S1$, $P1$, and $P2$ as well. Following are those formulae:

$$S1 = \frac{P2 \times S2}{P1}$$

$$P1 = \frac{P2 \times S2}{S1}$$

$$P2 = \frac{P1 \times S1}{S2}$$

As you can see, three values must be known to use these formulae.

Solve the following problems:

Refer to **Figure 6-31**. If the driver pulley ($S1$) was rotating clockwise (CW) at 2800 rpm, and the driver and driven pulley pitch diameters were 8 in. and 18 in., respectively, what would the speed and direction of the driven pulley be?

$$S2 = \frac{P1 \times S1}{P2}$$

$$S2 = \frac{8 \times 2800}{18}$$

$$S2 = \frac{22{,}400}{18}$$

$$S2 = 1244.\overline{44} \text{ rpm (CW)}$$

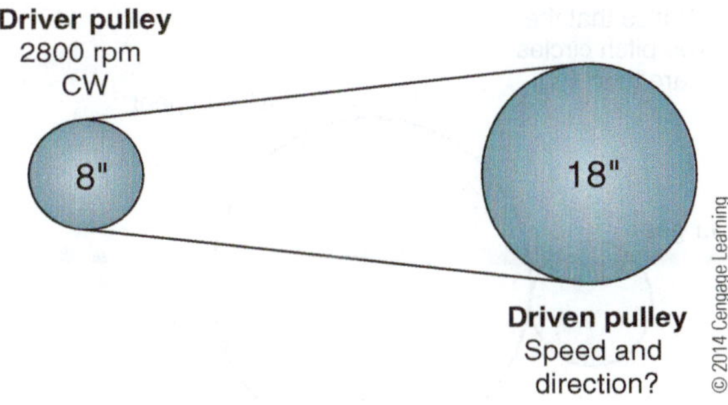

FIGURE 6-31 Speed calculation 1.

A simple way to prove your answer is to use the ratio of the pulley diameters to calculate the speed. The ratio of $P1$ to $P2$ is found by dividing 18 by 8.

$$Ratio = \frac{18}{8}$$

$$Ratio = 2.25{:}1$$

This means that the driver pulley turns 2¼ turns for every one revolution of the driven pulley. Use this ratio to check the speed. Now you know that the driver pulley turns 2.25 times faster than the driven pulley. Because the driver pulley speed was given and the driver pulley is smaller, the driver speed is *divided* by the ratio to get the speed of the driven pulley.

$$S2 = \frac{S1}{Ratio}$$

$$S2 = \frac{2800}{2.25}$$

$$S2 = 1244.\overline{44} \text{ rpm (CW)}$$

It has now been proven that the ratio can also be used to calculate speed.

The same formulae can be used to calculate the speeds on gear and chain drives as well. It is common practice in industry to use the tooth count of a gear and sprocket rather than the pitch diameter for the calculation of speed. This is because the tooth count is directly proportional to the pitch diameter. However, for simplicity's sake, the pitch diameter is used in these examples.

Where the pulleys of a belt drive do not touch each other, gears do, as shown in **Figure 6-32**. Compare Figure 6-32 and Figure 6-31. Notice that there is a span between the pulleys in the belt drive in Figure 6-31. Now notice that there is no space between the gears in Figure 6-32. This is because the circles represent the pitch circles of the two gears. If you remember correctly, the gears mesh at their pitch circles. That is why the circles touch one another. The formulae still apply, however.

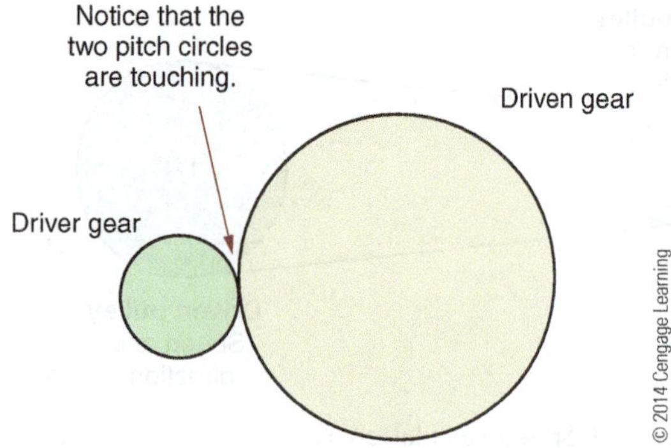

FIGURE 6-32 Notice that the gears are touching.

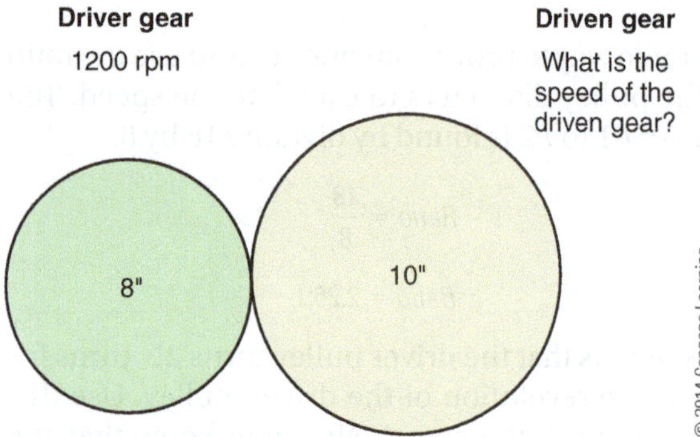

FIGURE 6-33 Speed calculation 2.

Solve the following problem:

Refer to **Figure 6-33**. The driver gear has a pitch diameter of 8 in. and the driven gear has a pitch diameter of 10 in. The driver gear is rotating at 1200 rpm. What is the speed of the driven gear?

The driver gear is 8 in. and the driven gear is 10 in. Find the correct formula, remembering that you want to solve for *S*2.

$$S2 = \frac{P1 \times S1}{P2}$$

$S1$ = Speed of the driver gear
$S2$ = Speed of the driven gear
$P1$ = Pitch diameter of the driver gear
$P2$ = Pitch diameter of the driven gear

$$S2 = \frac{P1 \times S1}{P2}$$

$$S2 = \frac{8 \times 1200}{10}$$

$$S2 = \frac{9600}{10}$$

$$S2 = 960 \text{ rpm}$$

The speed of the driven gear is 960 rpm.

Solve the following problem:

Refer to **Figure 6-34**. The driver gear has a pitch diameter of 4 in. and the driven gear has a pitch diameter of 10 in. The driven gear is rotating at 1200 rpm counterclockwise (CCW). What are the speed and direction of the driver gear?

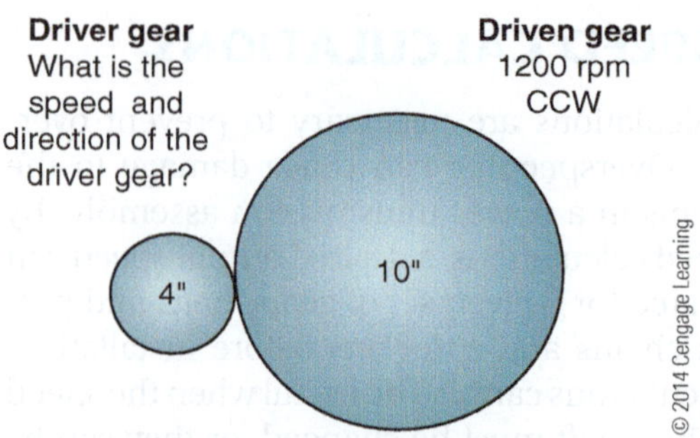

FIGURE 6-34 Speed calculation 3.

$$S1 = \frac{P2 \times S2}{P1}$$

$$S1 = \frac{10 \times 1200}{4}$$

$$S1 = \frac{12{,}000}{4}$$

$$S1 = 3000 \text{ rpm}$$

The driver gear is rotating at 3000 rpm in the clockwise direction.

Unlike the belt drives, the driven gear turns in the opposite direction of the driver gear. This is because the gears are meshed together. Therefore, for all gear sets with an even number of gears (2, 4, 6, etc.), the output gear and every even gear in between turns in the opposite direction (CW) of the driver gear (CCW). This is illustrated in **Figure 6-35**. Knowing this, the driven gear in the problem is turning in the opposite direction of the driver gear (CW).

Solve the same problem using the gear ratio and the speed to check the answer. The ratio of this set of gears is 2.5:1. This means that the driver gear turns 2.5 revolutions in the time that it takes the driven gear to turn one revolution. Being that the driven speed is given and the driven gear is the larger gear, the speed is *multiplied* by the ratio of 2.5.

$$Speed = 1200 \times 2.5$$

$$Speed = 3000$$

The speed of the driver gear is 3000 rpm, just in the first calculation. It should be noticed that a simple way to calculate the speed of a gear is to simply use the gear ratio if it is known. Try another problem using nothing but the gear ratio to calculate the problem.

Now that you have mastered calculating problems using two gears, it is time to learn how to calculate gear sets that use three or more gears in a set.

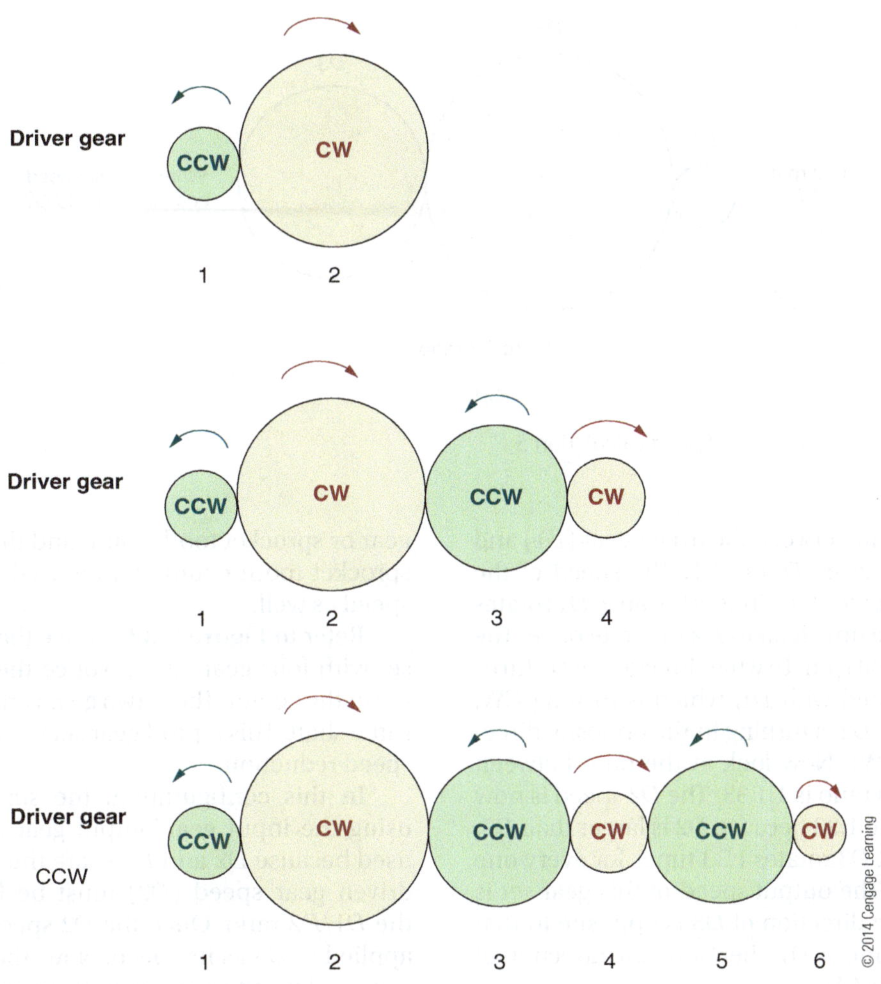

FIGURE 6-35 Gear rotation.

Do not panic! It is not any more difficult than for a set of two gears. The input and output ratios are used if the output speed or the input speed is to be calculated for a gear set of three or more. However, for the following example, work through all three gear combinations first, then work the problem the simple way to prove the answer.

A set of three gears is used to transmit power from one shaft to another. The gear pitch diameters are a 2 in. driver gear (*D*1), an 8 in. driven gear in the center (*D*2), and a 6 in. driven gear (*D*3). *D*3 is the output gear. The speed off the driver gear is 2000 rpm (CW). See **Figure 6-36**. What are the speed and the direction of the output gear (*D*3)?

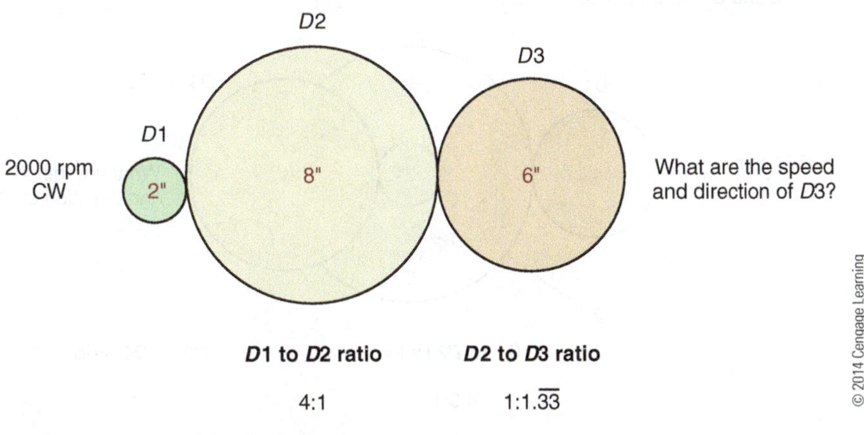

FIGURE 6-36 Speed calculation 4.

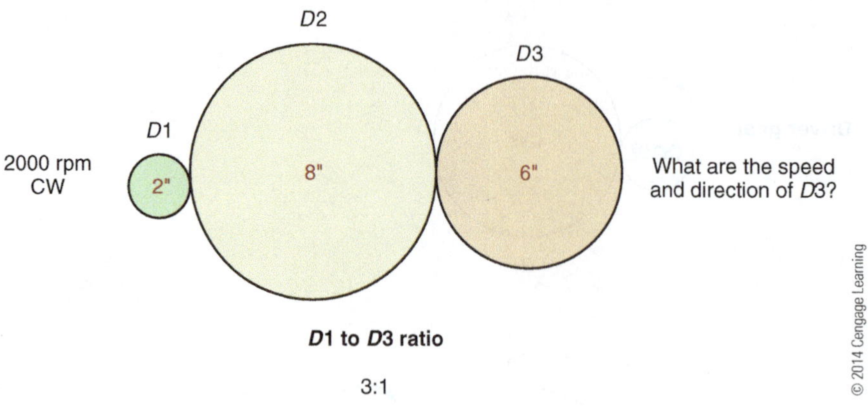

FIGURE 6-37 Speed calculation 5.

The gear ratio between the driver gear (*D*1) and the first driven gear (*D*2) is 4:1. The speed of the driver gear is *divided* by 4. This is because *D*1 rotates four times in the time it takes *D*2 to rotate once. The speed of *D*2 is 500 rpm. In what direction is *D*2 turning? *D*2 is meshed with *D*1, which is turning CW. This means that *D*2 is turning in the opposite direction of *D*1 (CCW). Now look at the ratio between *D*2 and *D*3. This ratio is 1:1.33. The *D*2 speed is now *multiplied* by the 1.33 because *D*2 is larger than *D*3. This means that *D*3 rotates 1.33 times for every one rotation of *D*2. The output speed of this gear set is 666.66 rpm. The direction of *D*3 is opposite to that of *D*2 but the same as *D*1; therefore, the direction of rotation for *D*3 is CW.

Now try it the easy way! Refer to **Figure 6-37**. Find the gear ratio of *D*1 and *D*3. This comes out to be 3:1. Divide 2000 by 3. This is the speed of *D*1 divided by the ratio. *D*1 rotates three times for every one rotation of *D*3. As you can see, this method works! The speed of *D*3 is 666.66 rpm, just as before.

Ratio calculation also works with tooth count. In other words, the ratio derived from the driver gear or sprocket tooth count and the driven gear or sprocket tooth count can be used to calculate the speed as well.

Refer to **Figure 6-38**. Notice that there is a gear set with four gears in it. Notice the two gears that are in the center. These two gears are attached to the same shaft. This type of gear set is used very often in speed reduction.

In this configuration, the simple method of using the input gear/output gear ratio cannot be used because *D*2 and *D*3 share the same shaft. The driven gear speed (*D*2) must be found by using the *D*1/*D*2 ratio. Once the *D*2 speed is found, it is applied to *D*3 as its speed. Now, the *D*3/*D*4 ratio is used to determine the output speed (*D*4).

Speed @ *D*2 = Speed @ *D*3

Mounted on the same shaft.

*D*2 and *D*3 are both being driven by *D*1.

Knowing this, calculate the speeds for this set of gears. The first step is to find the *D*1/*D*2 ratio. Because the driver gear (*D*1) is 4 in. and *D*2 is 18 in.,

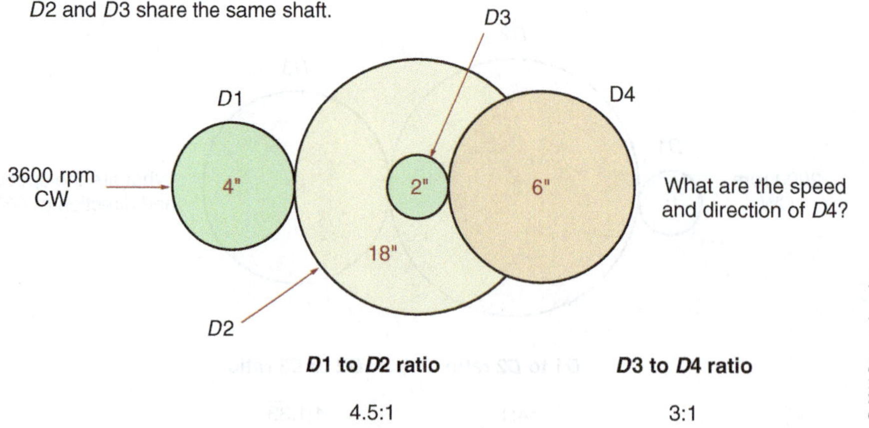

FIGURE 6-38 Stacked gears that share the same shaft.

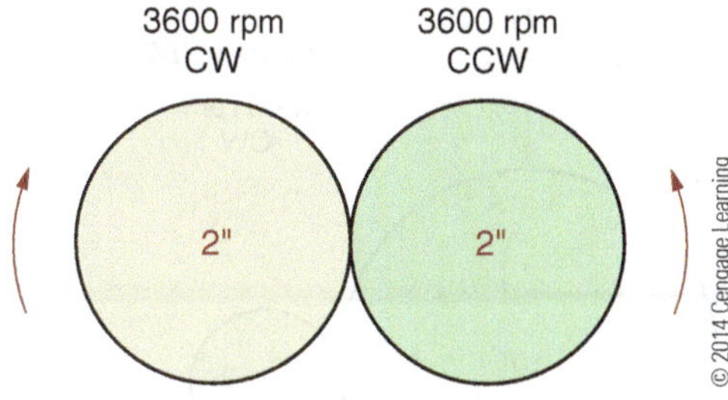

FIGURE 6-39 1:1 gear ratio.

the ratio is 4.5:1. Now divide the *D*1 speed by 4.5. This is because the driver gear (*D*1) rotates 4.5 times during one rotation of the *D*2 gear. You should have 800 rpm (3600 rpm divided by 4.5). Keep track of direction as you go. In Figure 6-38, the direction of rotation for *D*1 is CW. Therefore, the direction of rotation for *D*2 is opposite to that (CCW). Now that the *D*1/*D*2 ratio has been used to calculate the driven gear (*D*2) speed, the *D*3/*D*4 ratio can be used to calculate the *D*4 speed. The ratio is found by dividing 6 by 2, which gives a ratio of 3:1. This means that *D*3 rotates three times in the time it takes *D*4 to rotate once. Now that the ratio is known, and because the 18 in. (*D*2) gear has a speed of 800 rpm and *D*3 (the 2 in. gear) has a speed of 800 rpm, the *D*4 speed can be calculated. This is accomplished by simply dividing 800 (*D*3 Speed) by 3 (*D*3/*D*4 ratio). The output speed of this gear set is 266.67 rpm. Once again, figure out the direction. If *D*2 is rotating CCW, then *D*3 is also rotating CCW. This is because they are mounted on the same shaft. The *D*4 shaft, however, will once again change direction (CW). It is therefore safe to say that this gear set simply provides speed reduction, but not a change in direction.

A gear set with a 1:1 ratio would be used if a change in direction were desired without a change in speed. This is illustrated in **Figure 6-39**.

SUMMARY

- Flat belts and V-belts are the most commonly used belts in industry. Although flat belts are most often made of leather, canvas, or rubber, leather is the most commonly used flat belt. V-belts are able to transfer more torque than a flat belt.
- Belt deflection can be tested with a belt deflection tester, which measures deflection force and deflection distance.
- As opposed to V-belts, which can slip, positive-drive belts do not slip. Positive-drive belts are also referred to as timing belts or gear belts.
- Alignment of a tapered-bore pulley must be checked when tightening it. Overtightening the cap screws may cause the screws to shear.
- Chain drives are used in industry because of their positive-drive capabilities. They do not slip or creep, making them a reliable form of power transmission.
- Three different types of chains discussed are roller chains, silent chains, and ladder chains. Roller chains are the most commonly used chain for power transmission.
- A gear is a toothed wheel or disc used to transmit mechanical power from one shaft to another through the meshing action of teeth.
- Gearboxes house gear sets. They are used frequently in industry and have an input shaft and an output shaft.
- Pitch diameter is the point at which both gears transfer power. Pitch circle is sometimes referred to as the pitch line.
- Circular pitch is the distance that is measured from one tooth to the exact same point on the next tooth.
- The diametral pitch is the number of teeth that are present in 3.14 in. along the pitch circle circumference.
- Gear ratio is the comparison between the pitch diameter of the drive gear and the pitch diameter of the driven gear.
- The different types of gears are spur, bevel, miter, helical, herringbone, and worm. The spur gear is the most commonly used gear.
- Speed calculations are necessary to prevent overspeeding, which can cause damage to the components of a power transmission assembly.

REVIEW QUESTIONS

1. List two things that tension members do in V-belts.
2. Name three different types of misalignment.
3. What are two reasons gears are used for?
4. What are the direction and speed of the driver gear (*D*1) in **Figure 6-40** if the driven gear (*D*2) is a 2 in. gear, *D*1 is a 12 in. gear, and *D*2 is rotating at 1248 rpm (CW)?
5. What are the speed and direction of gear *D*3 in **Figure 6-41**?

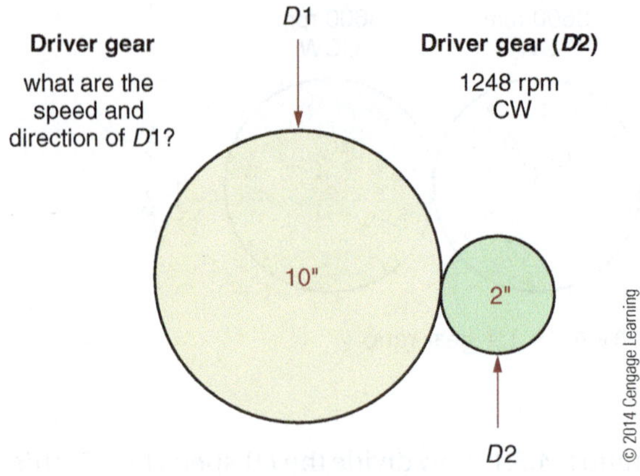

FIGURE 6-40 Speed calculation 6.

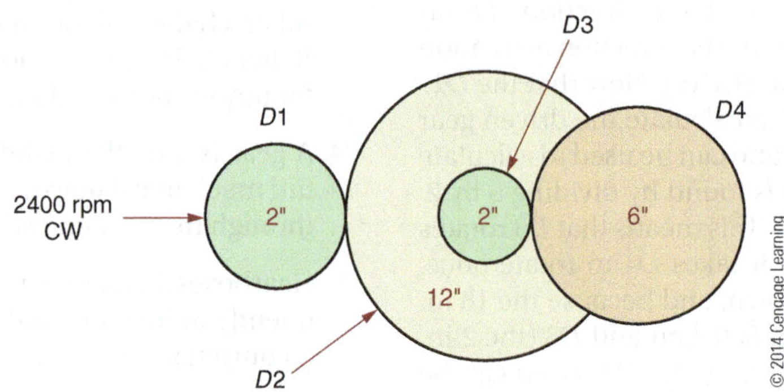

FIGURE 6-41 Speed calculation 7.

CHAPTER 7

Bearings

Bearings are a vital part of mechanical systems. Without them, many of the luxuries that we enjoy today would not be possible. Bearings come in many shapes and sizes. They all, however, are used to reduce friction in machinery and equipment. Because bearing failure is the most common mechanical failure on machinery, it is important to install, maintain, and remove all bearings correctly.

OBJECTIVES

After studying this chapter, the student should be able to

- Define radial, axial, and radial-axial loads.
- List the different parts of a bearing.
- List the different types of antifriction bearings and the different types of plain bearings.
- Correctly install and remove a bearing.
- List several reasons for bearing failure.

7-1 BEARING LOADS

Because bearings are usually used in conjunction with a shaft of some sort, a bearing must be able to support radial, axial, and radial-axial loads. A **radial load** is applied when the pressure from the load is perpendicular to the axis of the shaft. **Figure 7-1** shows the direction of the load and the type of antifrictional bearing that would be used to support that kind of load. An **axial load** is applied when the pressure from the load is parallel to the axis of the shaft. This is shown in Figure 7-1 as well. **Radial-axial loads** exist when pressures from the load are both perpendicular and parallel to the axis of the shaft at the same time. An angular-contact bearing is used in this application, as shown in Figure 7-1. It is important to know this because the type of bearing that is used in a given application is determined by these conditions.

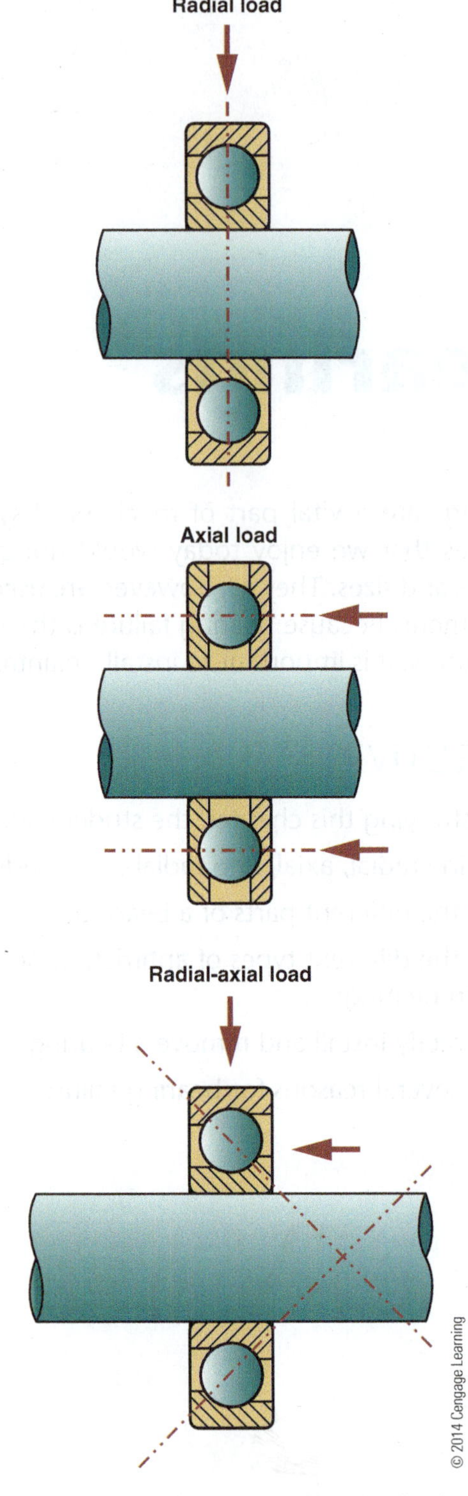

FIGURE 7-1 Three types of bearing loads.

7-2 BEARING CONSTRUCTION

In most antifriction bearings, the **rolling elements** are mounted between an inner ring and an outer ring. These **inner** and **outer rings** are made of hardened steel, as are the rolling elements. Not every antifriction bearing, however, has inner and outer rings. This is discussed later in the needle bearing section. The inner and outer rings are the parts of the bearing that make physical contact with the movable and stationary parts. The rolling elements ride in what is referred to as a **race**. There are two races in an antifriction bearing: the inner race and the outer race. The **inner race** is a shallow groove that is cut into the inner ring in which the bearings ride. The **outer race** is much like the inner race, with the exception that it is cut into the outer ring. The rolling elements reside within the inner race and outer race with a precision fit. The **shoulder** of a bearing is the flat portion of the rings between the rolling element and the face of the bearing. Usually a bearing has a **retainer** around the rolling elements. This retainer is referred to as a separator. Refer to **Figure 7-2**. It holds the rolling elements in place and evenly spaces them around the bearing. Separators are referred to as **cages** when discussing the taper-roller bearing.

The tolerances of the parts that are within an antifriction bearing are very tight. This is to ensure a smooth operation and efficiency of the bearing. A bearing that has a lot of slop in it eventually fails. A bearing with closer tolerances has a longer life expectancy than a less precise bearing.

Some bearings are sealed because the environment they are used in may be dirty. A **sealed bearing** is shown in **Figure 7-3**. These bearings are exactly like the open-faced bearing except that two seals (one metal, the other plastic, vinyl, or rubber) are inset between the inner and outer rings. These seals accomplish two separate tasks. The first is to keep the lubricant in the bearing (bearing grease that is applied during manufacturing); the second

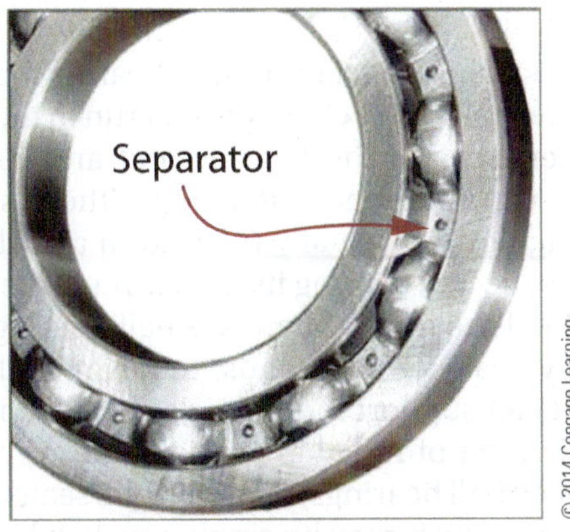

FIGURE 7-2 Bearing retainer or separator.

FIGURE 7-3 Sealed bearing.

is to prevent any contaminants from entering the bearing. Contaminants are a problem because bearings are precision devices and tolerances are usually very close. Any debris or dirt that gets embedded in the bearing may damage the bearing rolling element or the raceway that the rolling element rolls in. This sort of damage shortens the life expectancy of the bearing. Sealed bearings do not have to be lubricated. They are lubricated during the manufacturing process with enough lubrication for the lifetime of the bearing. Removing the seals to add lubricant only shortens the life of the bearing and should not be done.

Bearing nomenclature is derived from several features: the outside diameter of the outer ring, the diameter of the bore, and the width of the bearing.

There are no rolling elements in a plain bearing. A **plain bearing** is usually a sleeve or a bushing that may support radial and axial loads. Unlike the antifriction bearings, the surface of the shaft or journal is part of the plain bearing. Many of these bearings are contained within a device such as a pedestal or **box block**. Most of these bearings possess some sort of hole or port that allows lubrication. Plain bearing material must be durable, resistant to high temperatures, and resistant to corrosion. Plain bearings are constructed from many types of alloys and synthetics.

7-3 SERIES OF BEARINGS

All bearing manufacturers designate the various series as follows:

- Series 100: Extra-Light Series
- Series 200: Light Series
- Series 300: Medium Series
- Series 400: Heavy Series

It is possible for each series to have a bearing with the same bore size and different outside diameters, just as it is possible for each series to have the same outside diameter with four different bore widths. This is illustrated in **Figure 7-4**.

The extra-light series (100) are the narrowest of all the series and should be used only in an extra-light-duty application. This bearing is not designed to support heavy loads. The light series bearing is used in light applications that are too much for the

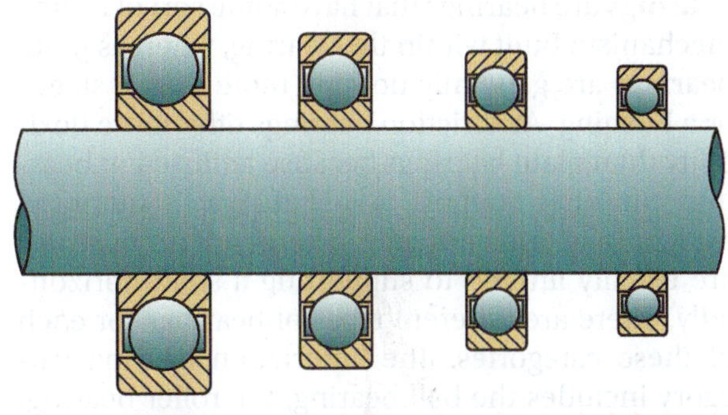

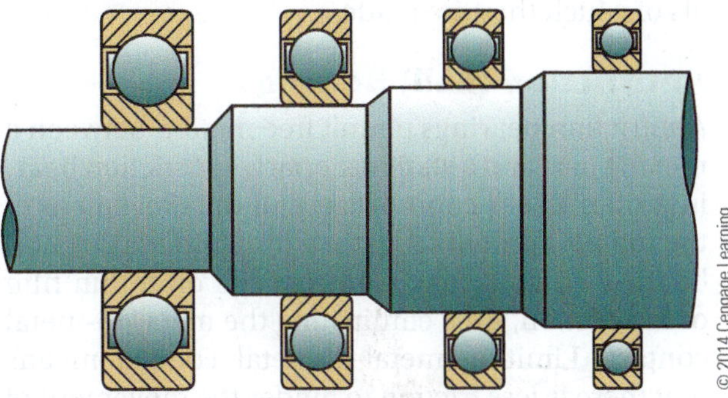

FIGURE 7-4 Inner and outer race diameters.

series 100 bearing. The series 200 and the series 300 bearings are the most commonly used bearings. The series 400 bearing is usually reserved for heavy loads and for heavy-duty applications.

For a long time, all bearing dimensions were in metric units because the first antifriction bearings were made in Europe and standardized in metric dimensions before American bearing manufacturers came into existence. Because bearing manufacturing was already standardized, American manufacturers made their bearings metric to allow for the interchangeability of the bearings. This situation has changed since the 1990s to allow the manufacturing of inch fraction dimensions.

All bearings have the manufacturer's name and an identification number. There are too many identification numbers to name them all. Some bearing manufacturers place proprietary codes on their bearings. The only way to get the dimensions on a bearing with a proprietary code is to contact the manufacturer, to research the number using a catalog, or to simply get the measurements from the old bearing: outer diameter, bore diameter, and width.

7-4 BEARING TYPES

Because there are many types of bearings, only the most common bearings are discussed in this text. In general, there are two categories of bearings: antifriction bearings and plain bearings. **Antifriction bearings** are bearings that have some sort of rolling mechanism built within the bearing, whereas plain bearings are generally nothing more than a sleeve or a bushing. Antifriction bearings offer more flexibility than plain bearings because antifriction bearings can offer support to a shaft that is mounted horizontally and vertically, whereas plain bearings are usually limited to supporting a shaft horizontally. There are different types of bearings for each of these categories. The antifriction bearing category includes the ball bearing, the roller bearing, and the needle bearing. The bearings in the plain bearing category are named for the type of materials of which they are made.

Antifriction (Ball) Bearings

Antifriction bearings permit free motion between a moving part and a stationary part. Antifriction bearings offer less friction because of the movement of the rolling elements that are within the bearing and because these rolling elements roll on a thin film of lubrication, thus eliminating the metal-to-metal contact. Limiting metal-to-metal contact means that there is less friction to hinder the movement of the different parts.

Ball Bearing

The **ball bearing** is an antifriction bearing that contains spherical rolling elements that run in the inner and outer races of the bearing and are confined between the inner and outer rings. These spherical rolling devices, usually referred to as balls, are what give the ball bearing its name. There are many types of ball bearings. There are ball bearings that can carry heavy or light loads; there are ball bearings that can support radial loads, axial loads, and a combination of radial and axial loads; there are open-faced ball bearings; and there are sealed bearings. Finally, there may be single-row ball bearings or double-row ball bearings. The load determines the type of ball bearing used.

The Single-Row Ball Bearing

The **single-row** ball bearing is the most commonly used bearing. It is sometimes referred to as a Conrad bearing. It has one row of balls as its rolling elements. This bearing is of the simplest design and is designed for general usage. This type of bearing can sustain combined radial and thrust loads or thrust loads alone. This bearing operates exceptionally well at extremely high speeds and maintains these characteristics in either direction. A cross section of this bearing is shown in **Figure 7-5**.

The Double-Row Ball Bearing

The double-row bearing is very similar in design to the single-row ball bearing. One exception is that there are two rows of balls instead of just one row. Another difference is that the outer race has a wider groove than the inner race. This difference causes the load to push through the balls onto the outer race. For this reason, this bearing has excellent thrust capabilities. This bearing can be seen in Figure 7-5.

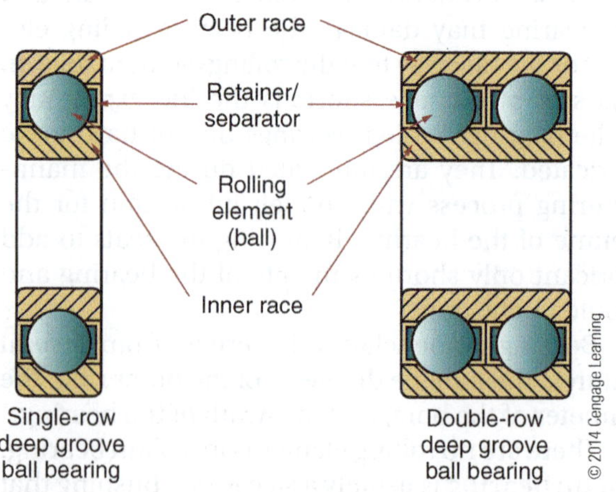

FIGURE 7-5 Single-row and double-row deep groove ball bearings.

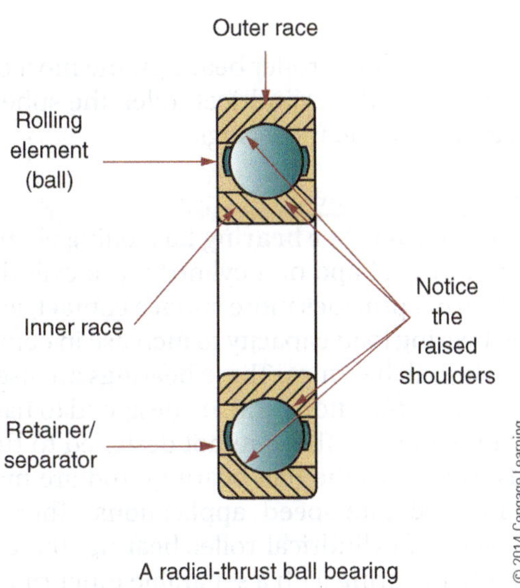

FIGURE 7-6 Radial-thrust ball bearing.

Radial Thrust Ball Bearing

A **radial thrust**, sometimes referred to as an angular-contact, bearing is very similar to the single-row ball bearing, except for two small differences. The shoulders on the inner and outer rings are raised slightly on one side, and these bearings are always used in pairs. The raised shoulders are illustrated in **Figure 7-6**. Raising the shoulder slightly in this manner allows combination loads with a large amount of thrust to be sustained without causing excessive damage to the bearing. This bearing can handle a much greater amount of thrust (in the direction of the raised shoulders) than can the single-row ball bearing of the same size. Because these bearings are used in pairs, they must be placed in a back-to-back or face-to-face orientation. Placing two radial-thrust ball bearings together in a manner in which the high shoulders on the outer ring are toward each other is known as a back-to-back configuration. Placing the high shoulders on the inner ring toward each other is known as the face-to-face configuration. This is illustrated in **Figure 7-7**.

The Ball-Thrust Bearing

Refer to **Figure 7-8**. Notice that this bearing does not have an inner and an outer ring. It has what is commonly referred to as washers. This bearing is designed to carry thrust loads only. This type of bearing cannot handle any radial loads at all. All of the load that is placed on this type of bearing is parallel to the axis of the shaft. In other words, there are no loads perpendicular to the shaft while using this bearing. This type of bearing should not be used in

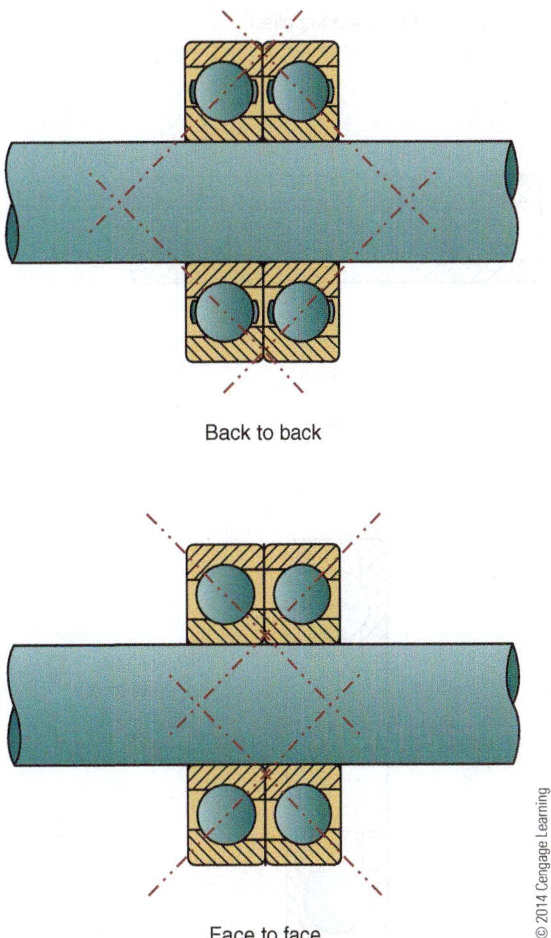

FIGURE 7-7 Back-to-back and face-to-face configurations.

high-speed applications because a high amount of centrifugal force is developed at high speeds. This large amount of centrifugal force would cause excessive loading on the outer edges of the races. Excessive loading causes the races to fail prematurely.

The Self-Aligning Ball Bearing

The self-aligning ball bearing allows a slight amount of angular misalignment of the shaft. It is designed in such a way that the rolling elements are staggered and in a double-row configuration. The separator on this bearing is much more complex than that on the other bearings because the balls are staggered. The outer ring of this bearing does not have a narrow groove for its outer race, as do the other ball bearings. The surface in which the rolling elements ride is curved. This curvature allows the rolling elements to swivel within the outer race instead of restricting the rolling element to a groove. For this reason, it is possible for the inner ring to be in a position of angular misalignment in comparison to the outer ring. This misalignment of the inner and

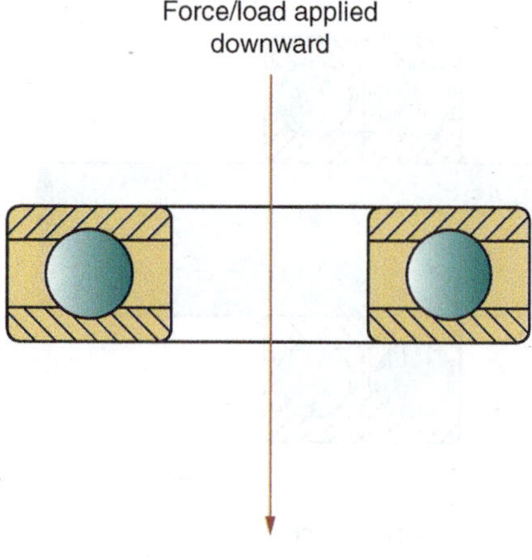

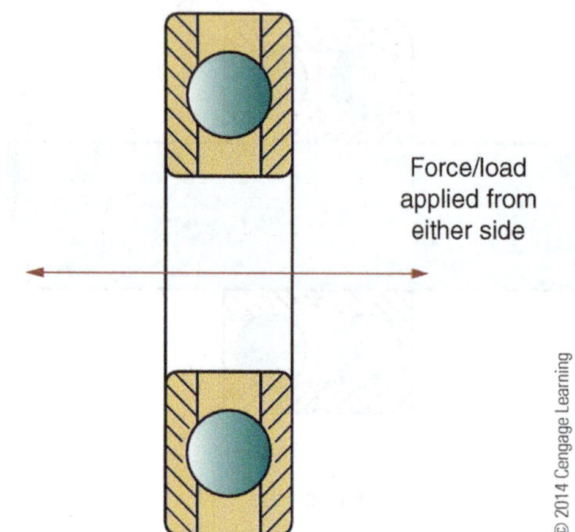

FIGURE 7-8 Ball-thrust bearing.

outer rings compensates for the misalignment of the shaft or bearing housing. A cross section of this bearing is illustrated in **Figure 7-9**. This bearing is designed to handle only moderate radial loads and minimal thrust loads.

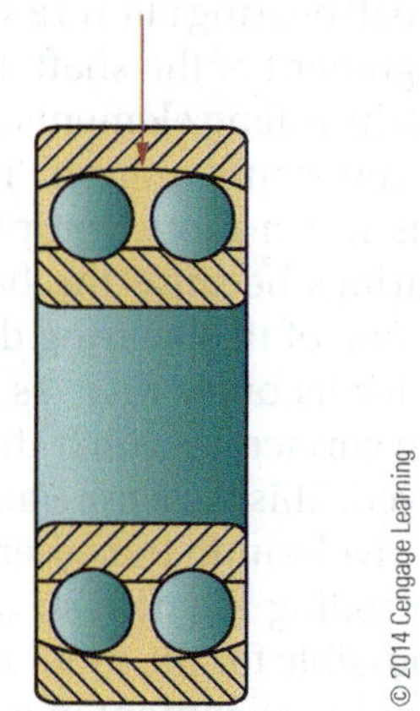

FIGURE 7-9 Self-aligning ball bearing.

Roller Bearings

Of the many types of roller bearings, the most commonly used are the cylindrical roller, the spherical roller, and taper roller bearings.

Cylindrical Roller Bearing

The **cylindrical roller bearing** has rolling elements that are in the shape of a cylinder. The cylindrical roller bearing provides more surface contact, allowing the bearing load capacity to increase in comparison to the ball bearings. These bearings are used in heavy-duty applications and are designed to handle heavy radial loads. They are not designed to run at speeds as high as the ball bearings and are mostly used in moderate-speed applications. There are three types of cylindrical roller bearing: the cylindrical roller bearing with a separable inner ring, the cylindrical roller bearing with a separable outer ring, and the nonseparable cylindrical roller bearing. All three are shown in **Figure 7-10**. The bearings with separable rings allow for axial movement of the shaft in relation to the bearing housing.

Spherical Roller Bearing

The **spherical roller bearing** is a double-row bearing and is referred to as a self-aligning bearing. The rolling elements of this bearing look like swollen cylinders. The midsection of the cylindrical rolling element has a larger diameter at its midpoint than at its ends. As with the self-aligning ball bearing, the surface in which the rolling elements ride is curved. This allows the rolling elements to swivel within the outer race, thus keeping the shaft in alignment. This misalignment of the inner and outer rings compensates

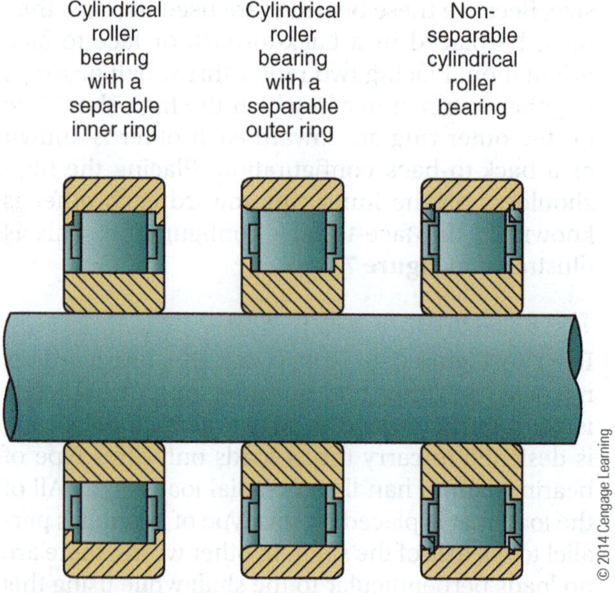

FIGURE 7-10 Three types of cylindrical roller bearings.

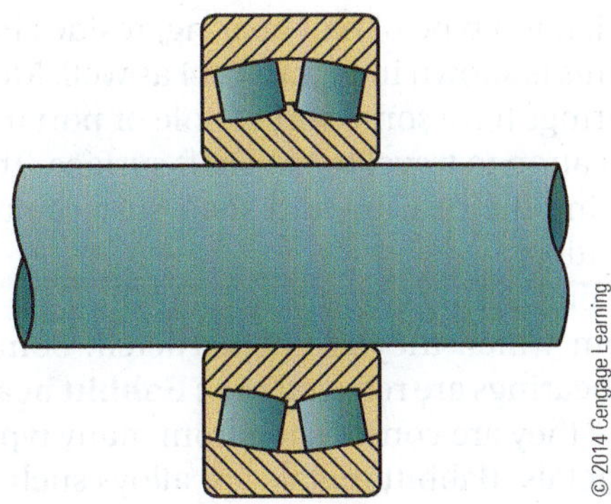

FIGURE 7-11 Double-row spherical roller bearing.

for the misalignment of the shaft or bearing housing. The inner races (keep in mind that this is a double-row bearing) are precisely cut to allow no swiveling action on the inner race. These races are the support races in this type of bearing. In comparison with the self-aligning ball bearing, the spherical roller bearing can support heavy radial loads and heavy thrust loads from both directions. This bearing is considered to be a heavy-duty bearing. A cross section of this bearing is shown in **Figure 7-11**.

The Taper Roller Bearing

The **taper roller** bearing is used where a large amount of thrust load is present, as well as on radial loads. The inner and outer races of this bearing are angled in reference to the bore axis line, as shown in **Figure 7-12**. The outer ring is called a **cup** and the inner ring is called a **cone**. (The outer ring, because of its tapered shape, looks like a cup, and the inner ring, because of its tapered shape, looks like a cone.) The rolling elements of this bearing are very similar to the cylindrical shape of the rolling elements that are in the standard roller bearing except that the rolling element itself is tapered as well. This means that the diameter of the top edge of the rolling element is larger than the bottom edge diameter. The cup, cone, and rolling elements are tapered in such a manner that if straight lines were drawn from the tapered surfaces of each, they would all meet at the same point. Refer to Figure 7-12. The cup, in most cases, is separable. This allows the cup to stay in the housing as the cone and the rolling elements stay on the shaft. These bearings, like the radial thrust ball bearings, are used in pairs. Instead of a back-to-back or face-to-face orientation, taper roller bearings are mounted in a direct or indirect configuration. If direct mounting is used, the bearings are mounted with the cones toward each other. This is sometimes referred to as cone clamped. Mounting the cups toward each other is known as indirect mounting or, sometimes, as cup clamped. Direct mounting is used when maximum stability is not required or desired

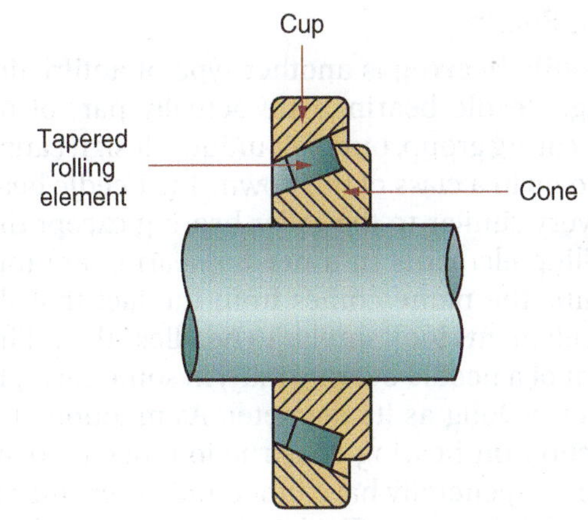

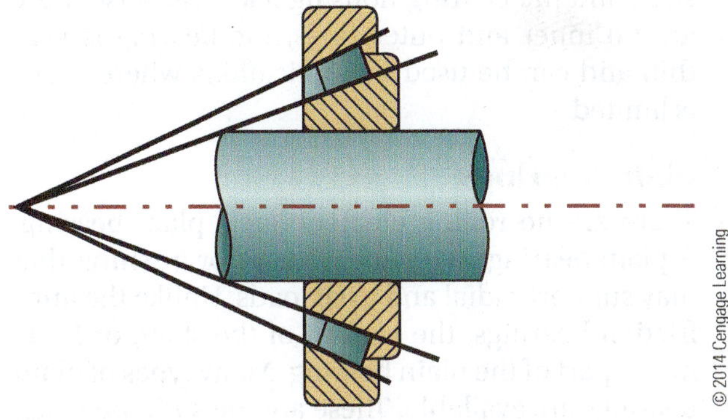

FIGURE 7-12 Taper roller bearing.

and space permits this type of mounting. Indirect mounting is used when maximum stability must be provided in a minimum width.

Taper roller bearings are also available in single- and double-row configurations. The double-row taper roller bearing is one bearing that has the appearance of two single-row taper roller bearings that are cone clamped. There is only one cup and one cone in which the rolling elements ride. This is shown in **Figure 7-13**.

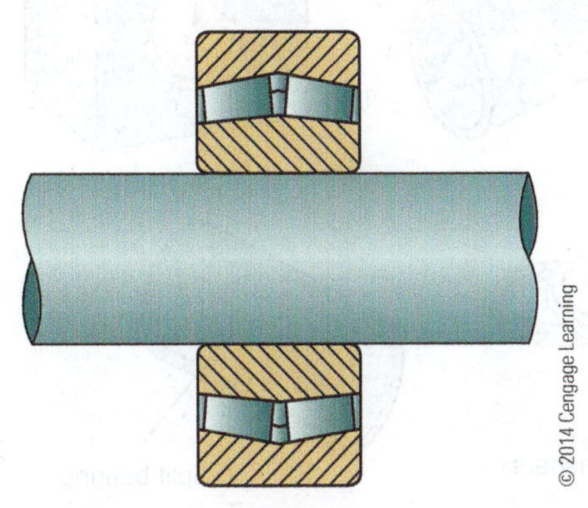

FIGURE 7-13 Double-row tapered-roller bearing.

Needle Bearing

The **needle bearing** is another type of antifriction bearing. Needle bearings are actually part of the roller bearing group, but their unique design causes them to be in a class of their own. The needle bearing is very similar to the roller bearing except that the rolling elements in a needle bearing are long and thin. The name comes from the fact that the rolling elements look similar to needles. The rolling element of a needle bearing may, in some cases, be 10 times as long as its diameter. As mentioned in the section on bearing construction, needle bearings do not generally have inner and outer rings as do the other bearings. The bearings are press-fitted into what is referred to as a case. This causes the rolling elements of the needle bearing to roll on the shaft and the bearing housing itself. Because there are no inner and outer rings, the bearing is very thin and can be used in applications where space is limited.

Plain Bearings

There are no rolling elements in a plain bearing. A plain bearing is usually a sleeve or **bushing** that may support radial and axial loads. Unlike the antifriction bearings, the surface of the shaft, or **journal**, is part of the plain bearing. Many types of plain bearings are available. These are the **solid bearing**, the **split bearing**, the **half bearing**, and the **thrust bearing**. Refer to **Figure 7-14**. Thrust bearings are sometimes referred to as thrust washers. These also can be seen in Figure 7-14.

Many of these bearings are contained within a device such as a pedestal or box block. The stave bearing, which is a type of half bearing, resides in a box block. This is shown in Figure 7-14 as well. Most of these bearings have some sort of hole or port that allows lubrication to travel to the shaft surface. This aids in keeping the bearing and shaft surface cool during operation.

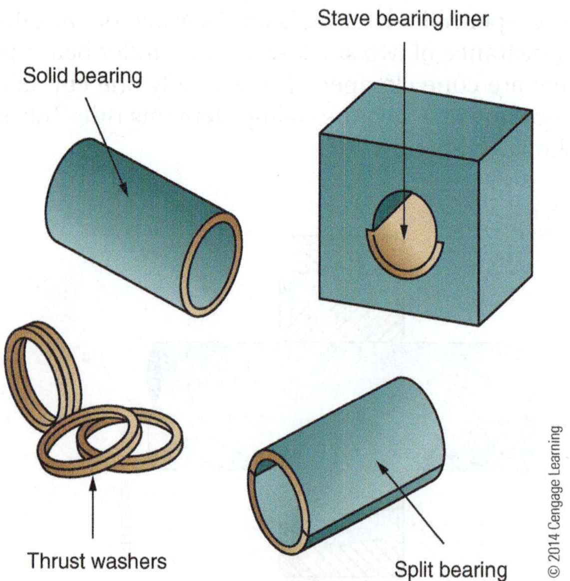

FIGURE 7-14 Types of plain bearings.

Most plain bearings are named for the type of material from which they are constructed. Sometimes plain bearings are referred to as **Babbitt bearings** because they are constructed from many types of Babbitt metals. Babbitt metals are alloys such as copper-lead, bronze, and aluminum. These types of bearings are used for plain bearings requiring increased load-carrying capabilities. The alloy metal is sometimes bonded to a steel support in bearings that are used in heavy applications. The steel gives the softer alloy some rigidity and aids in supporting the load as well. Copper-lead bearings are mostly used for high-temperature environments and high-load applications. Because bronze bearings are porous, they absorb oil when they are lubricated and become impregnated with this oil. After the oil is absorbed into the bronze, it can continually lubricate the shaft during motion. It is important to maintain a continual lubrication schedule for porous bronze bearings because they eventually run out of the oil that they have absorbed during the previous lubrication. If bronze bearings are not lubricated often enough, they begin to run hot and begin to wear as the oil film begins to disappear.

Synthetic Plain Bearings

There are some synthetic bearings available today that were not available just a couple of decades ago. **Synthetic** means something that is produced artificially by chemical means. These newer synthetic bearings are becoming more and more popular for light-duty applications. Some of these synthetic bearings are carbon fiber, nylon, Teflon®, and carbon graphite. These types of bearings are suited for high-temperature applications and are able to withstand a minimal amount of exposure to a chemical environment, whereas plain bearings made of metal would probably corrode and react. The carbon graphite bearings are designed to operate in extremely high temperatures that may even exceed 700°F (371°C).

7-5 BEARING INSTALLATION AND REMOVAL

A bearing must be installed correctly in order to prevent damage. A bearing that is damaged during installation will have a short life and will fail prematurely. Damage may also occur to whatever the bearing is pressed onto or into if it is not installed or removed properly. If the bearing is bad, care need not be taken for the bearing's sake, but the components on which the bearing is mounted must remain unharmed to receive a new bearing.

Bearing Installation

Antifriction bearings are usually gently pressed within a bearing housing after the bearing is pressed onto a shaft or journal. The inner ring generally has to be pressed onto a shaft because the tolerances are tighter on the inner ring than on the outer ring. Because the inner ring has to be pressed onto the shaft and the outer ring simply has a snug fit, the bearing tends to stay on the shaft and pulls out of the bearing housing on disassembly. An exception would be any bearing that has a separable outer ring or cup, or a bearing that has been damaged and has fallen apart. In this case, the outer ring or cup remains in the bearing housing as the inner ring or cone remains on the shaft. Whenever a bearing is pressed onto a shaft or into a part, press only against the mating surfaces. The mating surface of a bearing that is being pressed onto a shaft is the inner ring, whereas the mating surface of a bearing that is being pressed into a part is the outer ring.

It is much easier to install a bearing than it is to remove one. The trick is to install the bearing correctly. It is important to always use the proper tools when installing a bearing. Using the incorrect tools could lead to personal injury and damage to a new bearing. The first step to installing a bearing on a shaft is to make sure that the surface of the shaft in which the bearing will be seated is clean. Clean the surface of the shaft with an extremely fine abrasive pad or sandpaper. While you are cleaning the shaft, it is a good idea to heat the bearing up using a bearing heater. An electric bearing heater is shown in **Figure 7-15**.

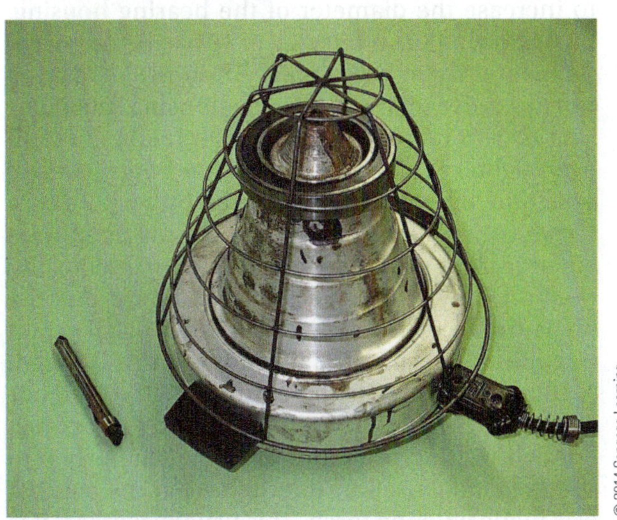

FIGURE 7-15 Electric bearing heater.

This type of bearing heater generally comes with the capability of heating bearings with different bore sizes. There are oil bath heaters that work just as well. It is important to heat, but not overheat, the bearing before placing it on the shaft. The diameter of the bore increases as it is heated. This creates the few extra thousandths of an inch that are required for a smooth bearing installation. When the bearing has reached approximately 200°F (93°C), it is ready to be placed on the shaft. Wear a pair of leather gloves to protect your hands from getting burned as you handle the heated bearing. Also keep in mind that the bearing cools very quickly once it has been removed from the heat source. This is important because if the bearing is not set exactly where it should be when it cools, some adjustments have to be made using a bearing knocker and a ball-peen hammer. If a bearing knocker must be used, it is important to use the correct size. Bearing knockers come in a wide range of diameters. When choosing a bearing knocker, it is wise to choose one that sits on the face of the inner ring or cone. This is only if the bearing is being set on a shaft. If a bearing knocker is being used to set a bearing into a bearing housing, then a bearing knocker that sits on the face of the outer ring or the cup needs to be chosen. It is a good idea to also have all of the tools that are needed at hand before the bearing is removed from the heat source. This is to ensure that no time is wasted looking for tools as the bearing begins to cool.

Another method of installing a bearing onto a shaft is through the use of an arbor press. In this method, the bearing should be shored on the press table, and the shaft should be pressed into the center of the bearing. Care needs to be taken not to insert the shaft into the bearing with too much speed. Slowly press the shaft in small increments into the bearing to ensure that the shaft is going straight into the bore. Remember the safety rules, and always wear a face shield when working around a press. Also make sure to avoid pinch points.

Never hit the rings of a bearing directly with a hammer or a punch. Doing so may cause damage to the bearing, shaft, or bearing housing. If a drift punch is used, it may slip when it is struck, causing damage to the seal, separator, or rolling elements.

Bearing Removal

Because most bearings are pressed onto the shaft or journal, the bearing will, more than likely, have to be pulled off using bearing pullers. There are many types of bearing pullers, including specialty

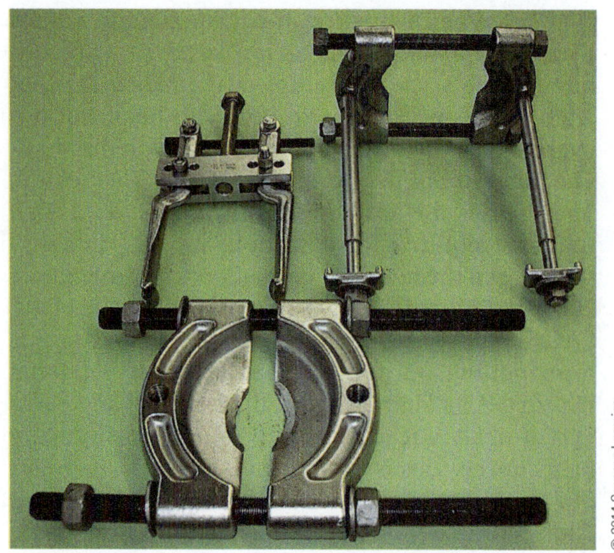

FIGURE 7-16 Three types of bearing pullers.

pullers, because there are so many types of bearings. However, a very common set of bearing pullers is shown in **Figure 7-16**. A common mistake that is made in industry is to mistake gear pullers for bearing pullers and to use gear pullers to remove bearings. There is a difference between gear pullers and bearing pullers. The bearing puller has a foot at the end of each claw that supports the inner and the outer rings as pressure is being applied during removal. A gear puller applies pressure only to the outer ring in most cases. There are many types of bearing pullers available: inside pullers, outside pullers, and hydraulic pullers. The inside pullers are used to remove an outer ring from a bearing housing. The claws of this puller are inserted through the inside of the ring and then opened to allow the claws to pull against the underneath of the ring. The outside pullers are used to remove a bearing that is mounted on a shaft. The hydraulic puller is usually a portable tool that uses hydraulic power to apply the necessary pressure for bearing removal.

If a puller is not readily available, an arbor press may be used. This is accomplished by shoring up the bearing on the table surface of the press, allowing the shaft to pass through the press table. When pressure is applied to the shaft, the shoring supports the bearing and causes the shaft to be pushed out of the center of the bearing. Some safety precautions have to be adhered to while using this method of bearing removal, however. The first is that a safety shield should always be worn while working with a press. Second, try to stand clear in case the bearing disintegrates. Last, make sure that there is something under the shaft that will catch it as it falls out of the bearing. This prevents the shaft from being damaged.

If a set of pullers or a press is not available, the bearing can be removed by manual impact. This is accomplished by shoring up the bearing on the table surface, allowing the shaft to pass through a hole in the table. Place a piece of malleable metal stock on the shaft end, making sure that the stock is of a smaller diameter than that of the bore diameter of the bearing. This prevents the stock from getting jammed in the bearing as the shaft is knocked out. Once the stock is placed on the end of the shaft, squarely hit the malleable metal stock with a heavy ball-peen hammer. This should dislodge the shaft from the bearing after a few hits.

Sometimes when a bearing goes bad, the rolling elements and separator fall out of the bearing during removal. This causes the inner ring or the cone to be the only thing that remains on the shaft, and the outer ring or cup usually remains in the bearing housing. When this occurs, it may become difficult to remove them from the shaft and the bearing housing. First, try to use the bearing pullers to remove the ring or cone. If that does not work, heat the ring or cone with a torch, being careful to keep the flame of the torch on the ring or cone only. This heat causes the metal in the ring or cone to expand, thus making the bore diameter slightly larger by a few thousandths of an inch. This is usually enough to remove the ring or cone. If care is not taken while heating the ring or cone, and the flame is allowed to heat the shaft, the diameter of the shaft increases in diameter, making the fit even tighter. This same method can be applied to an outer ring or cup that may be stuck in a bearing housing. The difference is that now the heat should be placed on the bearing housing instead of on the outer ring or cup. The object is to increase the diameter of the bearing housing, thus allowing the outer ring or cup to be removed. As before, if the heat is placed on the outer ring or cup rather than the bearing housing, the outer ring or cup increases in diameter, making the fit even tighter. If this is tried and the rings just will not come off, then try to split the ring with a cold chisel. Always remember to wear safety glasses when using a cold chisel. If the ring will not split, then the only other alternative is to burn the ring off with an acetylene torch. Care must be taken not to cut into the shaft if this method must be used. A good idea is to use the cutting torch to remove some of the material from the ring but to leave just enough to prevent cutting into the shaft, then to use the cold chisel to split the ring at the cut that was made with the torch. This usually works if it is done correctly.

7-6 BEARING FAILURES

The bearing, if it is properly maintained and not overloaded, should never fail until it simply succumbs to metal fatigue. **Metal fatigue** occurs over time when a metal object is continually placed under a load. Metal that is placed under an intermittent load over time will flex and, in some cases, even distort. Bearing failure is generally caused by a lack of maintenance and overloading. There are some other extraordinary instances that may cause bearing failure. There are signals, once a bearing has failed, that indicate the cause of failure. Some of these signs can be noticed on inspection of the bearing before failure occurs. These signs are clues to the real reason for failure. If a bearing were replaced without correcting the condition that caused the first bearing to fail, chances are that the new bearing would suffer the same fate. Changing the bearing is only fixing the symptom, not the problem. Once the problem is fixed, it is not likely that a replacement bearing will fail for the same reason. Many things can cause premature bearing failure. Some are listed here:

- High temperatures
- Moisture
- Contamination
- Improper lubrication
- Misalignment
- Electric current flow through the bearing
- Overloading or excessive thrust

High Temperatures

High temperatures are indicated by the discoloring of the raceways and the rolling elements. The metal is usually darkened, with a bluish-purple coloring where the overheating occurred. It is not uncommon for the bearing to become deformed because of excessive internal temperatures. High temperatures could exist for many reasons—anything from symptoms of other problems to excessive ambient temperature. It is not uncommon for each one of the following problems to cause a bearing to overheat as well. Another early indication of overheating is the presence of solid or caked lubricant. As a bearing overheats, the oil within the lubricant separates, leaving the thickener to be burned.

Moisture

Rusting surfaces are an indication of moisture. Oxidation occurs when moisture is present and lubrication is lacking. This is commonly referred to as fretting corrosion. If a bearing has a suitable amount of lubrication, oxidation should not occur even when the bearing is in a moist environment. This does not mean that if you cake the bearing with grease or oil, the bearing can be placed in an excessively moist environment; it simply means that a properly maintained bearing can operate in minimal moisture. The problem with rust is that it is not uncommon for small particles of rust to flake off into the bearing, causing damage to the races and the rolling elements. This, in turn, eventually causes high temperatures to be present, causing the bearing to fail prematurely.

Contamination

Contamination occurs any time a foreign particle enters the bearing. This usually occurs when the bearing is operating in a dirty environment. When a foreign particle enters the bearing, it causes damage to the rolling element or to the races of the bearing. The damage is usually in the form of deformity. If the bearing will be operating in a dirty environment, a sealed bearing should used. Contamination causes damage to the races and rolling elements as well. This eventually causes high temperatures to be present, causing the bearing to fail prematurely.

Improper Lubrication

If a bearing is overlubricated or underlubricated, it fails prematurely. Underlubrication is self-explanatory. No lubrication causes friction and overheating within the bearing. Overlubrication can place internal pressure on the bearing because the rolling elements have to move the excessive amount of lubrication within the bearing as well as the load. This could cause the rolling elements to become overloaded. This contributes to the fatigue of the rolling elements and, possibly, to the fatigue of the races as well.

Misalignment

More wear on one side of the bearing than the other indicates misalignment. Also, opposing sides may show signs of misalignment. Uneven wear on the rolling element is another indication.

Electric Current Flow through the Bearing

Current flow is indicated by pitting and fluting. **Electrical pitting** occurs when current flows through the bearing. Voltage is present when there is a difference of potential between two points. Voltage always tries to find a path to ground, even if the path to ground is through a bearing. If one potential exists on the equipment's housing and the opposite potential exists on the shaft of the same piece of equipment, current flows through the bearing. Current flows, for example, from the bearing housing into the outer race, then into the rolling elements as they are rolling, then out of the rolling elements through the inner ring to the shaft. The problem occurs where the contact is made between the outer race and the rolling elements, and the rolling elements and the inner race. Thin lines that are etched into the races make it easy to recognize this condition. This is called **fluting**. Pitting occurs mostly when welding currents pass through the bearing. The simplest way to avoid this problem is to ensure that there is always a path for current to flow through without going through the bearing.

Overloading or Excessive Thrust

Thrust damage is indicated by marks on the shoulder or upper portions of the inner and outer races. There also is discoloration, ranging from slight to heavy galling. **Galling** is a bonding, shearing, and tearing away of material from two contacting, or sliding, metals.

Some other symptoms must be discussed. Spalling and false Brinell damage are some other symptoms that may be recognized. **Spalling** is the flaking away of metal pieces due to metal fatigue. This occurs when the rolling element and the bearing race begin to flex because an excessive load is being applied to the bearing. This flexing (distortion) is momentary and repetitive. If the bearing is on a shaft that is turning at 1800 rpm and is overloaded, the repetitive flexing occurs 2.592 million times in a 24-hour period. This number is then multiplied by the number of rolling elements in the bearing. As the metal begins to fatigue, microscopic fractures begin to appear. This causes the metal to begin flaking. Once spalling occurs, complete bearing failure is imminent. **False Brinell** damage occurs when continual impacting forces (such as vibrations) are passed from one ring to the other through the rolling elements when there is no rotation of the shaft. Indentations are formed on the outer races due to the impacts. This causes the rolling elements, as they begin to rotate, to create heat as they encounter these evenly spaced indentations. This happens quite often to motor and pump bearings because they are usually stored in a warehouse. As heavy forklifts drive by, the bearings absorb the shocks that are transmitted through the floor and shelving units. For this reason, it is a good idea to place rubber absorption pads under the pallets on the floor, which contain motors and pumps, and under the feet of the shelving units, which contain motors and pumps. It is also useful to rotate the shafts of all motors and pumps on a regular basis. This helps eliminate false Brinell damage.

SUMMARY

- Three types of loads that may be placed on a bearing are a radial load, an axial load, or a radial-axial load.
- The inner and outer rings are the parts of the bearing that make physical contact with the movable and stationary parts.
- The inner race is a shallow groove that is cut into the inner ring in which the bearings ride. The outer race is much like the inner race, with the exception that it is cut into the outer ring.
- A separator holds the rolling elements in place and evenly spaces them around the bearing. Separators are referred to as cages when discussing the tapered roller bearing.
- All bearing manufacturers designate the various series as follows: Series 100: Extra-Light Series; Series 200: Light Series; Series 300: Medium Series; Series 400: Heavy Series.
- In the antifriction bearing category are the ball bearing, the roller bearing, and the needle bearing. The different types of bearings in the plain bearing category are named for the type of materials of which they are made.
- There are no rolling elements in a plain bearing. A plain bearing is usually a sleeve or a bushing that may support radial and axial loads.
- It is important to install and remove a bearing correctly. If the bearing is installed incorrectly, it could fail prematurely. If a bearing is removed incorrectly, damage to the mounting surface may occur.
- Bearing failure is generally caused by a lack of maintenance and overloading; however, many other things can cause premature bearing failure as well. Some are high temperatures, moisture, contamination, improper lubrication, misalignment, electric current flow through the bearing, and overloading or excessive thrust.

REVIEW QUESTIONS

1. Define *inner race.*
2. Define *inner ring.*
3. What is a cup on a thrust bearing?
4. What should be done to an antifriction bearing before installing it on a shaft?
5. What is electrical fluting?

CHAPTER

8

Coupled Shaft Alignment

This chapter covers many of the methods that are used to align shafts, which transmit power. It is important to align the shafts because they have some type of mechanism mounted to them that connects them together. These mechanisms could be pulleys, sprockets, or couplings.

OBJECTIVES

After studying this chapter, the student should be able to

- Discuss the fundamentals of shaft coupling alignment.
- Demonstrate how to correctly use the dial indicator to align coupling shafts.
- Demonstrate the use of the reverse dial indicator method to correct coupling shaft misalignments.
- Demonstrate how to use the feeler gauge, taper gauge, and dial caliper to detect and correct coupling shaft misalignments.
- Explain the advantages and disadvantages of using a laser alignment kit to detect and correct coupling shaft misalignments.

8-1 SHAFT ALIGNMENT

Power can be transmitted through the use of pulleys and belts, chains and sprockets, flexible couplings, or mechanical (rigid) couplings. Each type of transmission requires the shafts to be aligned to a very close tolerance in order to preserve the life of the belt, chain, filler, or flange.

Pulley and Sprocket Alignment

Any time that flexible belts and chain drives are used to transmit mechanical power, it is imperative that they be aligned properly. This is true whenever couplings are used to transmit power as well. If this is not done, the belts that are used on the improperly aligned pulleys will fail prematurely. Chains and couplings will also fail prematurely if the drive and driven shafts are not aligned properly. There are three types of misalignment to consider when aligning shafts that are used in pulley and sprocket drives:

- Offset
- Parallel
- Angular

All three are shown in **Figure 8-1**.

Notice in the figure how the misalignments are demonstrated using pulleys. This is important because angular and offset misalignments have different characteristics with reference to coupled shaft alignment. The reason for the difference, for belt and chain drives, is that the drive and driven shaft axes are always parallel to each other, whereas the drive and driven shafts that are to be coupled are always in an end-to-end orientation. This is discussed in greater detail later.

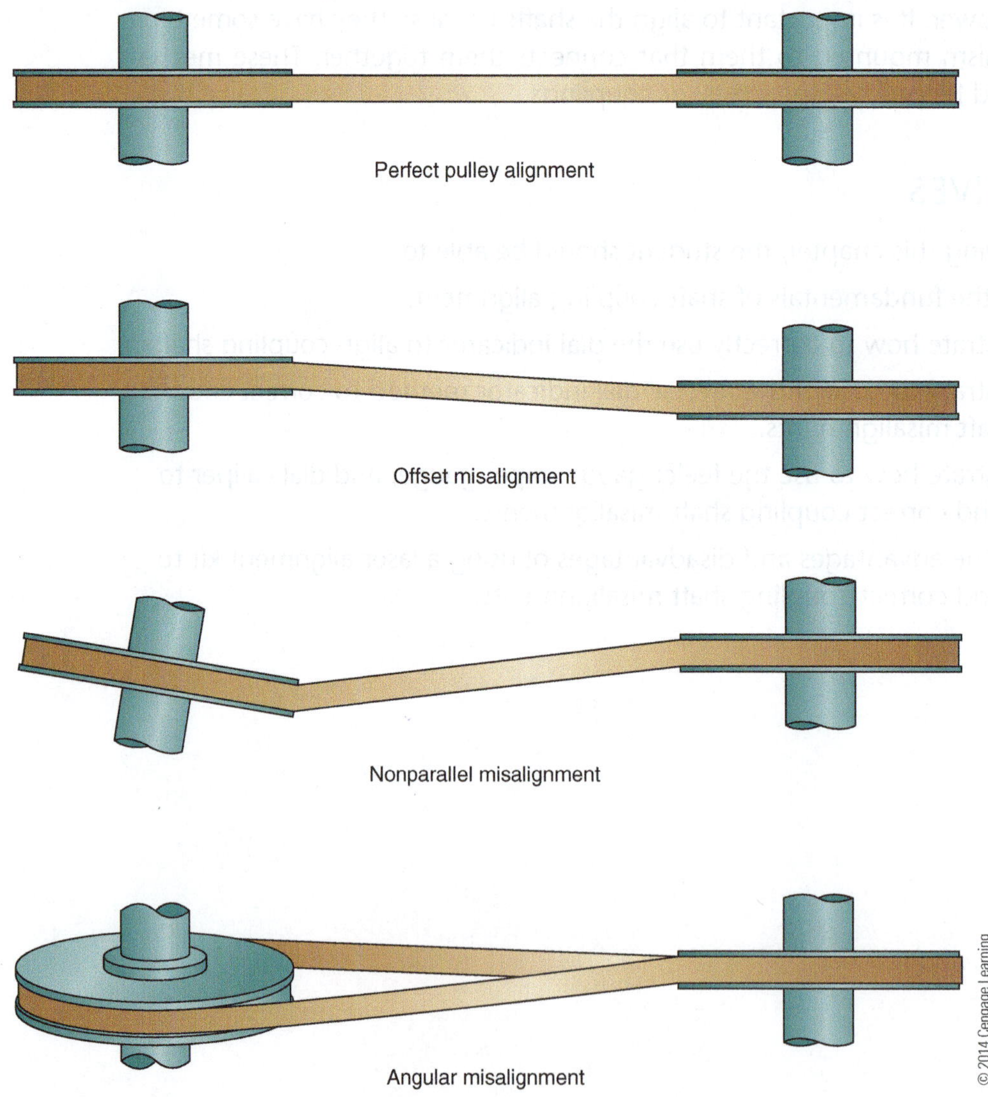

FIGURE 8-1 Three types of pulley misalignment.

In reference to offset alignment, it is usually acceptable to have no more than $\frac{1}{10}$ in. error per foot of span. Remember that span is the distance between the centers of the shafts. Any amount over this requires realignment of the shafts. If not corrected, offset eventually wears out the sides of the belt or the sides of the pulleys, and the belt begins to ride lower in the pulley. If a chain drive is allowed to run with an offset misalignment, the chain or sprockets may become worn prematurely.

Offset Misalignment

Offset misalignment occurs when the two shafts are parallel to each other, but the faces of the pulleys or sprockets are not on the same axis. To correct this problem, simply place a straightedge against the faces of the pulleys or sprockets. If an offset condition is present, only one pulley or sprocket touches the straightedge. Simply slide the pulley or sprocket that is not touching the straightedge toward the straightedge until it touches the straightedge. This should eliminate the offset misalignment.

Angular Misalignment

Angular misalignment can produce the same undesirable outcome as the offset misalignment. Angular misalignment occurs when two shafts are parallel to each other but are at a different angle with the horizontal plane. Placing a level atop the pulleys and leveling both shafts to zero bubble can correct this problem. It is usually acceptable to have not more than half a degree of error.

Nonparallel Misalignment

Nonparallel misalignment exists when two shafts are not parallel to each other. This can be corrected by using a straightedge as well. Reference points must be used to correct this type of misalignment. These reference points exist at each side of the pulley or sprocket, as shown in **Figure 8-2**. As the straightedge is placed against the faces of the pulleys or sprockets, all four reference points should

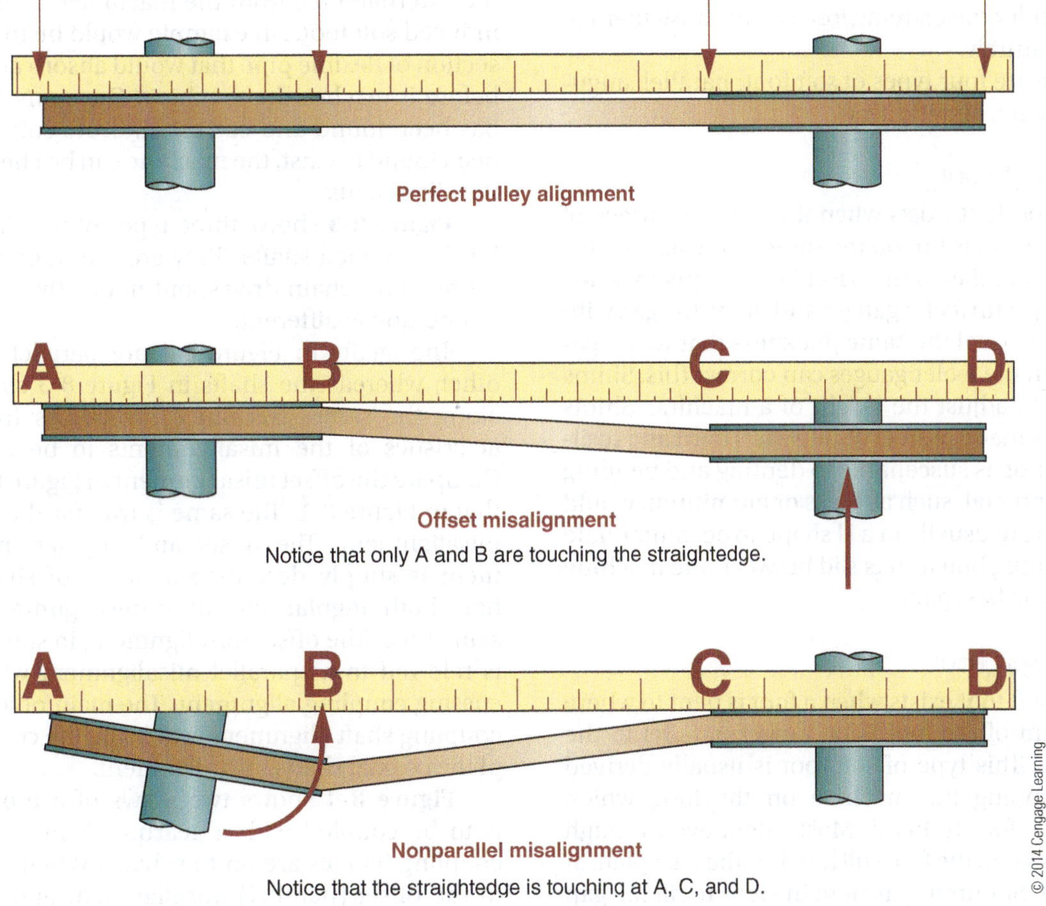

FIGURE 8-2 Pulley face alignment.

be in contact with the straightedge. If a nonparallel misalignment exists, the pulley that is touching the straightedge at one point is rotated toward the straightedge until all four reference points make contact with the straightedge.

Sometimes a set of pulleys or sprockets has all three misalignments that must be corrected. If this is the case, correct one at a time. Start with the offset misalignment, and then check for the nonparallel misalignment. Once that is complete, check to see whether an angular misalignment exists. These steps may have to be repeated two or three times to get the shafts perfectly aligned.

Coupling Alignment

Soft foot must be considered when coupled shaft alignment is performed. Soft foot exists when one or more of the feet on the machine that is being aligned is bent or was poorly manufactured. If soft foot exists, it should be corrected before attempting to correct any misalignments. If not corrected, soft foot causes premature bearing failure due to the frame distortion when it is tightened down to the baseplate. If soft foot is corrected before misalignments, the frame will not distort when it is tightened down, and therefore the bearings are not placed under the extreme forces that cause them to fail prematurely.

There are four types of soft foot: parallel, angular, springing, and induced.

Parallel Soft Foot

Parallel soft foot exists when the bottom surface of one or two feet is not on the same plane as the others, but is parallel to the baseplate. Simply measuring the gap with feeler gauges and filling the gap with shims that equal the same thickness that was measured with the feeler gauges can correct this. Shims are used to adjust the height of a machine. Shims are mostly made of steel because it is hard and resilient. It is not as susceptible to denting and bending as a softer metal, such as brass or aluminum, would be. Shims are usually in a U shape to accommodate the mounting bolt as it is slid between the machine foot and the baseplate.

Angular Soft Foot

Angular soft foot exists when a foot is bent to where the bottom of the foot is no longer parallel to the baseplate. This type of soft foot is usually derived from dropping the machine on the foot, which causes the foot to bend. Most often, even though the foot is bent, the foot still touches the baseplate at one point. Shimming the foot in steps to fill the gap and to support the foot solves this problem. Step shimming is accomplished by taking a measurement at the largest gap and dividing the measurement by 5. This allows the gap to be filled with five shims of equal thickness. As the shims are slid under the foot, they are stepped atop one another.

Springing Soft Foot

Springing soft foot exists when the entire area of the foot does not have contact with the baseplate. This may be caused by dirt, rust, or even a set of poorly manufactured shims. Even grease or oil causes this. If any of these conditions exists and is not corrected, the machine acts as if it were mounted on springs. This type of soft foot can easily be corrected by making sure that the surfaces of the foot and the base are clean and free of debris before mounting the machine. Also make sure to use shims that are not bent or burred when correcting for any misalignments.

Induced Soft Foot

External forces create induced soft foot. These are forces that are absorbed into the machine from such things as pipe vibrations (water hammer), pump vibrations, and bearing failure. Isolating the external force from the machine can eliminate induced soft foot. An example would be to put in a section of flexible pipe that would absorb any shock before it reaches the machine. Once any soft foot has been found and corrected, or no soft foot has been found to exist, the machine can be checked for misalignments.

Figure 8-3 shows three types of misalignment for the coupled shafts. They are the same as those for belt and chain drives, but notice that the shaft orientation is different.

The shafts in Figure 8-1 are parallel to each other, whereas the shafts in Figure 8-3 should be along the same axis line. This causes the characteristics of the misalignments to be different. Compare the offset misalignment in Figure 8-3 with that in Figure 8-1. The same is true for the angular misalignment. The offset and angular misalignment is simply described as a set of shafts that have both angular and offset misalignment at the same time. The offset misalignment, in some cases, is referred to as parallel misalignment when discussing coupling alignment. The main objective of coupling shaft alignment is to bring the centerlines of the coaxial shafts into alignment.

Figure 8-4 shows two views of a motor that is to be coupled with a gearbox. Notice that the coupling flanges are on the shafts. When referring to the offset (parallel) misalignment and angular misalignment, two things have to be considered.

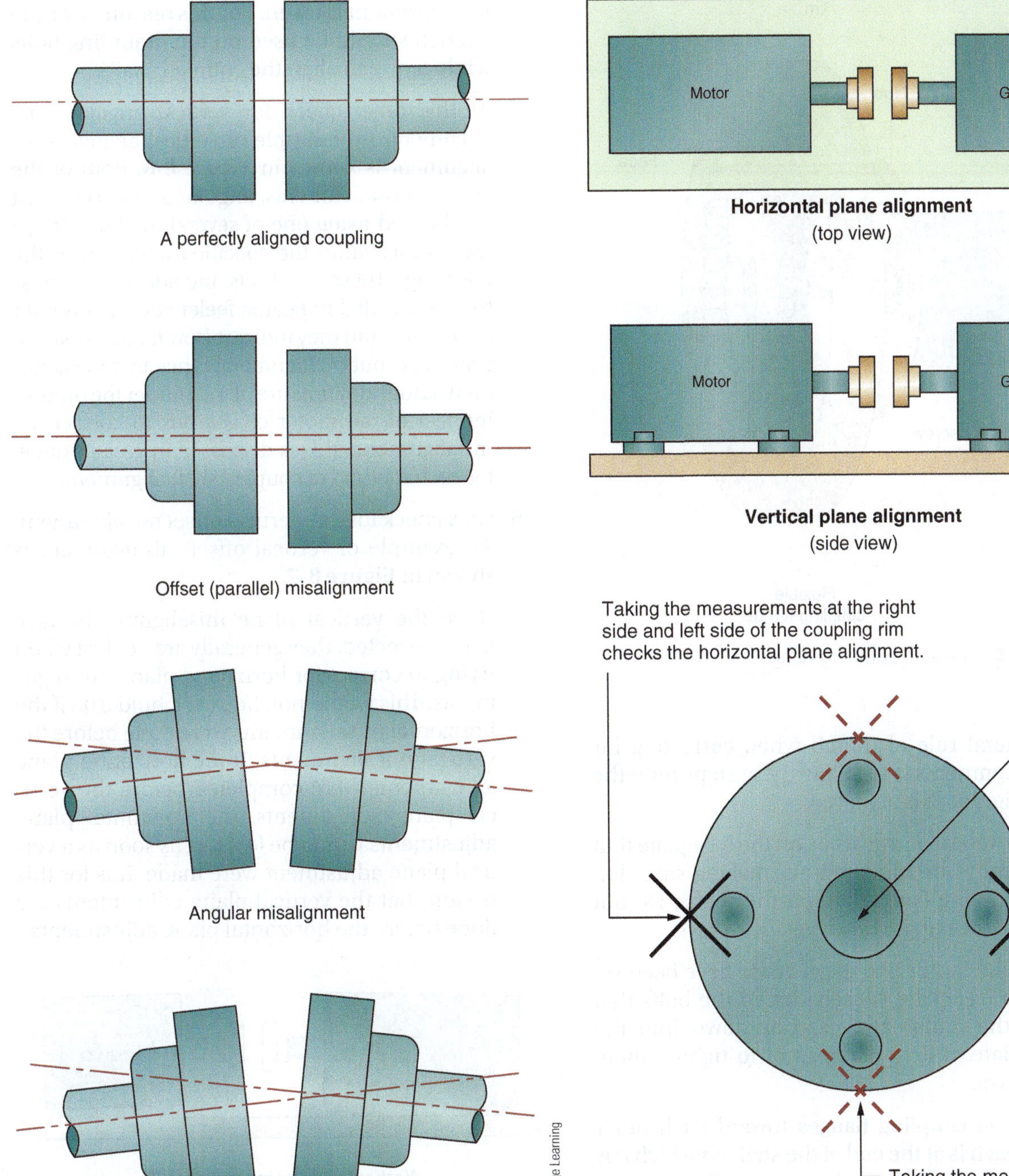
A perfectly aligned coupling
Offset (parallel) misalignment
Angular misalignment
Offset and angular misalignment
© 2014 Cengage Learning
Motor
Gearbox
Horizontal plane alignment
(top view)
Motor
Gearbox
Vertical plane alignment
(side view)
Taking the measurements at the right side and left side of the coupling rim checks the horizontal plane alignment.
Shaft
Taking the measurements at the top and bottom of the coupling rim checks the vertical plane alignment.
© 2014 Cengage Learning

FIGURE 8-3 Coupling misalignments.

FIGURE 8-4 Horizontal and vertical alignments.

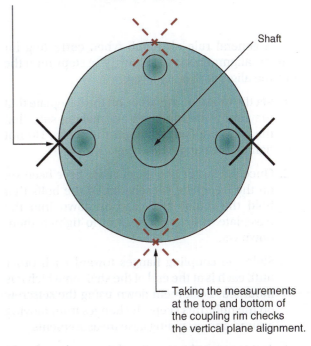

The shafts must be checked for vertical offset and angular alignment as well as for horizontal offset and angular alignment, as illustrated in Figure 8-4. Taking the measurements at the top and bottom of the couplings checks the **vertical alignment**. Taking the measurements at the right side and left side of the couplings checks the **horizontal alignment**.

Notice in **Figure 8-5** that the coupling is labeled. This is important because to properly align a coupling, you have to take measurements from the rim and, in some cases, the face of the coupling.

The coupling flange that is shown in Figure 8-5 is a flexible coupling. No matter what type of coupling is being used (as long as it is not a single-piece coupling), the flange will have a rim and a face.

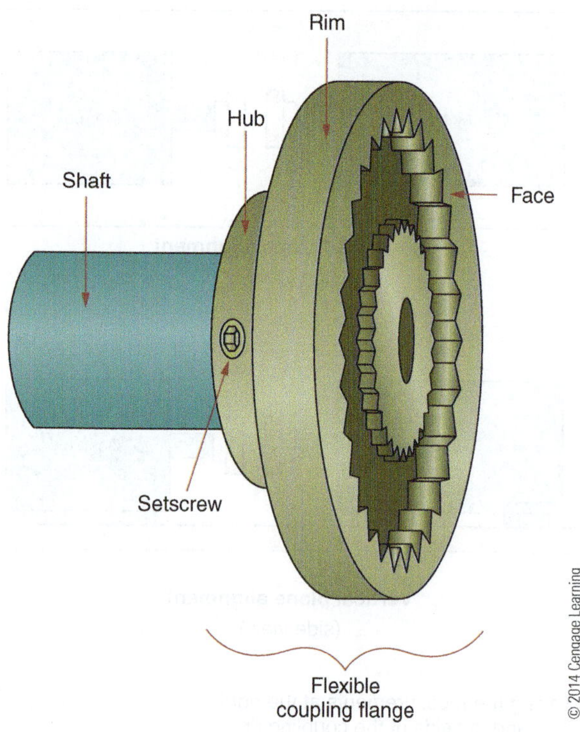

FIGURE 8-5 A flexible coupling flange.

A general rule of thumb when correcting for any misalignments is to follow these steps until the desired alignment is achieved:

1. Set the motor and gearbox on the baseplate that they are to be mounted on, making sure that the coupling flanges are on the shafts. Do not tighten the flanges down yet.
2. Once the motor and the gearbox have been set on the baseplate, loosely thread the bolts that hold the motor and gearbox down into the baseplate. It is important not to tighten them down yet.
3. Slide the coupling flanges toward each other until each is at the end of the shaft on which it is mounted. Lock them down using the setscrew on the hub. This keeps the flanges from moving around and providing false measurements.
4. CAUTION At this point, the motor and the gearbox should be tightened to the baseplate to the required torque specifications. If they are not tightened to the appropriate torque before the misalignment measurements are taken, the true measurements will never be found. It is necessary to loosen the bolts to place shims under the feet as you are trying to correct for misalignment; however, you must not forget to retighten the mounting bolts to the same torque as when the first measurement was taken each time a new measurement is taken. For this reason, a torque wrench should be used on the mounting bolts while trying to align the coupled shafts.
5. At this time, check for vertical angular misalignment. An example of vertical angular misalignment is shown in **Figure 8-6**. Both of the vertical misalignments, angular and offset, must be checked using one of several methods to get the motor within the specified tolerance of the coupling. These methods include dial indicator, reverse dial indicator, feeler gauge, and laser alignment, and they indicate how far the motor or gearbox is out of alignment. Once this measurement is found, shims are placed under the mounting feet of the motor or gearbox to correct the misalignment. This is discussed in greater detail for each method of coupled shaft alignment.
6. Now check for any vertical offset misalignment. An example of vertical offset misalignment is shown in **Figure 8-7**.
7. Once the vertical plane misalignments have been corrected, they generally are not lost when trying to correct for horizontal plane misalignments. This would not, however, hold true if the horizontal misalignments were done before the vertical misalignments. If the horizontal plane adjustments were completed before the vertical plane adjustments, the horizontal plane adjustments would be lost just as soon as a vertical plane adjustment were made. It is for this reason that the vertical plane adjustments are done before the horizontal plane adjustments.

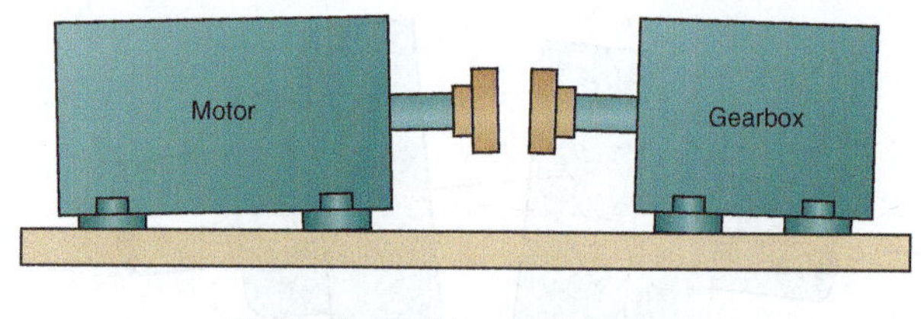

FIGURE 8-6 Vertical angular misalignment.

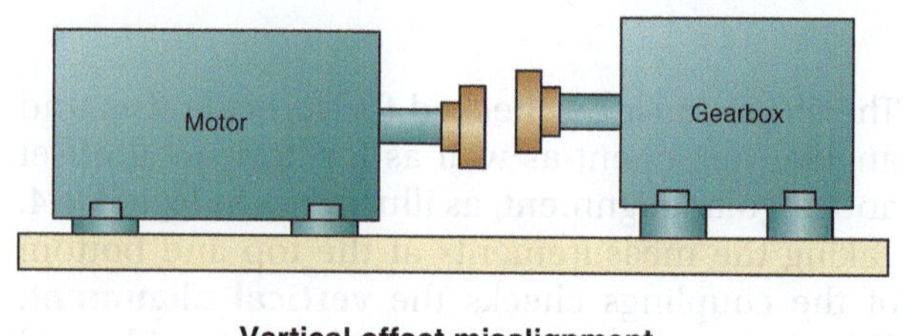

FIGURE 8-7 Vertical offset misalignment.

8. Next check for horizontal angular misalignment. See **Figure 8-8**. A purely horizontal angular misalignment is usually corrected with the use of a straightedge and jackscrews that are

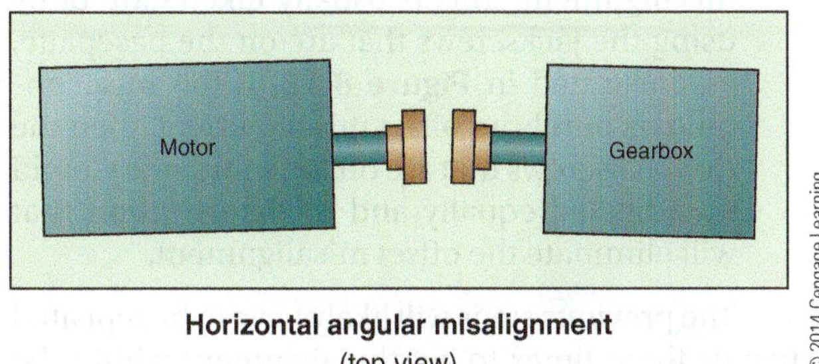

FIGURE 8-8 Horizontal angular misalignment.

present on the baseplate. Horizontal angular misalignment and horizontal offset misalignment can be checked by placing a straightedge against the rim of the couplings. If any offset misalignments are present, the rim of only one coupling will be touching the straightedge. If any angular misalignments are present, the straightedge will sit squarely and completely on one flange but be at an angle on the other. Both of these misalignments are shown in **Figure 8-9**. If horizontal angular misalignment is found, then the two jackscrews that are diagonal to each other should be adjusted to eliminate the angular misalignment. This is illustrated in **Figure 8-10**. If there are no jackscrews to aid in the movement of the machine,

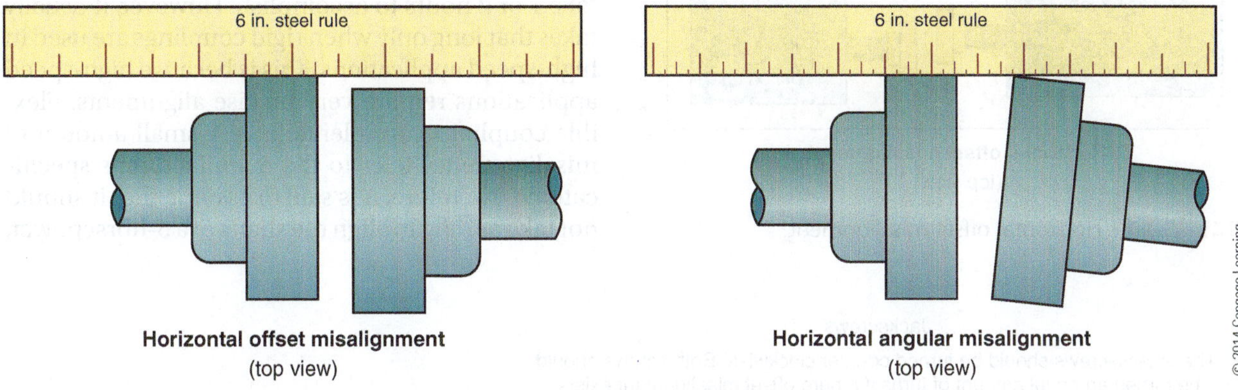

FIGURE 8-9 Using a straightedge to detect horizontal offset and horizontal angular misalignments.

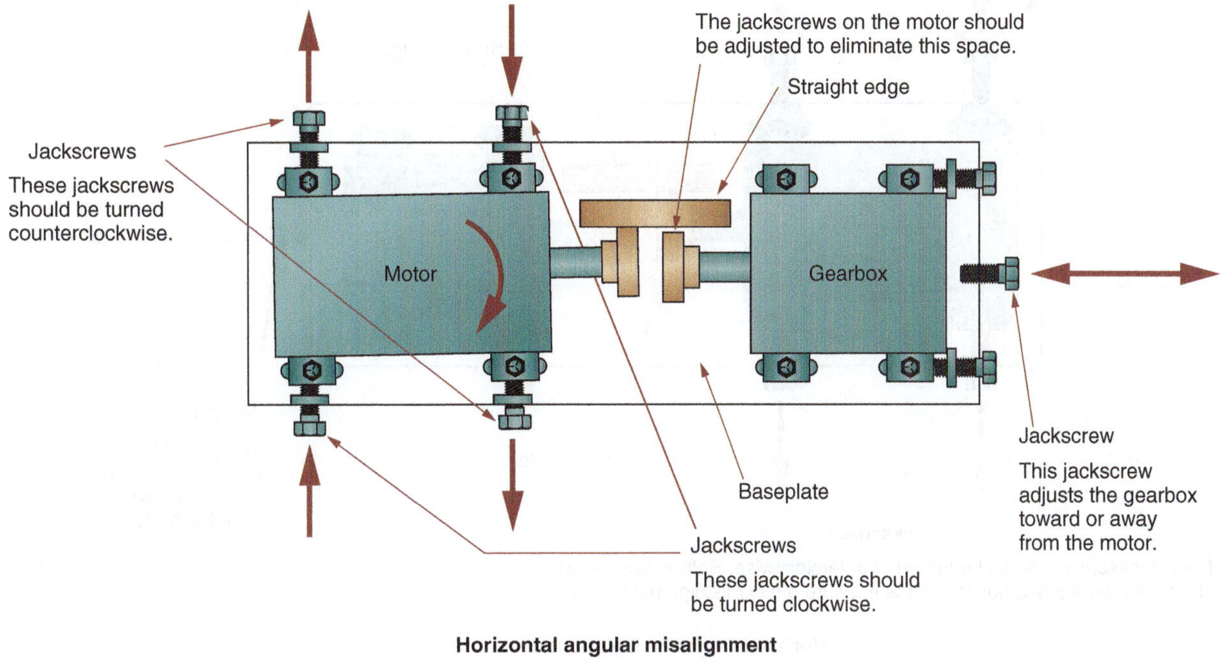

FIGURE 8-10 Use of jackscrews to correct for horizontal angular misalignment.

CAUTION

then try to gently pry against the machine so as to cause very small controlled movements. Another method is to use a dead blow hammer or a soft mallet to very gently tap the machine. This method should be avoided, if at all possible, when using dial indicators or laser alignment devices because the shock from the hammer may cause the devices to become uncalibrated or damaged.

9. Finally, check for horizontal offset misalignment. Horizontal offset misalignment occurs when the axes of the two shafts are not in line with each other, as shown in **Figure 8-11**. This is the easiest misalignment to correct.

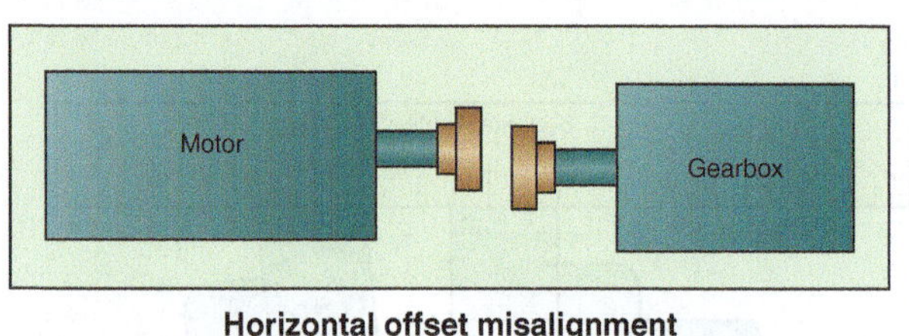
Motor
Gearbox
Horizontal offset misalignment
(top view)
© 2014 Cengage Learning

FIGURE 8-11 Horizontal offset misalignment.

This misalignment is often taken care of by simply placing a straightedge against the rim of the coupling and sliding either the motor or the gearbox over until both flanges touch the straightedge. This should eliminate the offset misalignment. This is usually taken care of by using the jackscrews that are on the baseplate, as illustrated in **Figure 8-12**. If the misalignment is purely an offset misalignment, then the two jackscrews that are on the same side should be adjusted equally and in the direction that will eliminate the offset misalignment.

The previous steps will likely have to be repeated two or three times to get the alignment within the specified tolerance of the coupling. It is very rare to achieve perfect alignment on the first round of checks. It is not unheard of for coupling alignment to take 1 or 2 hours to accomplish. However, it usually takes that long only when rigid couplings are used in high-speed applications. This is because high-speed applications require very precise alignments. Flexible couplings can tolerate a very small amount of misalignment. Refer to the manufacturer's specifications. Therefore, it is safe to assume that it should not take as long to align the shafts of a 5-horsepower,

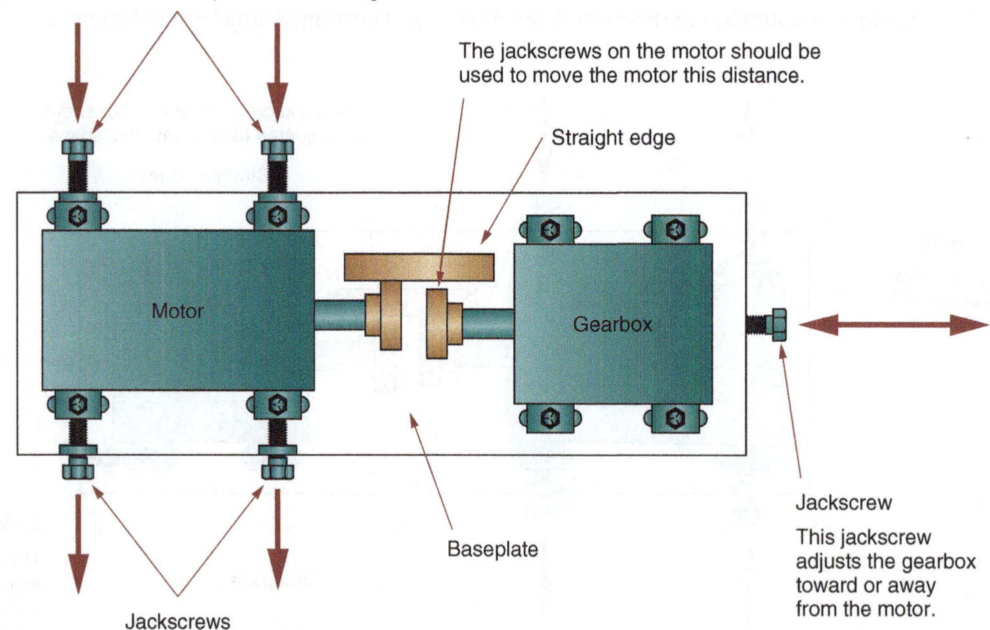
Jackscrews
These jackscrews should be turned counter clockwise. Both screws should be turned an equal amount of turns if a pure offset misalignment exists.
The jackscrews on the motor should be used to move the motor this distance.
Straight edge
Motor
Gearbox
Baseplate
Jackscrew
This jackscrew adjusts the gearbox toward or away from the motor.
Jackscrews
These jackscrews should be turned counterclockwise. Both screws should be turned an equal amount of turns if a pure offset misalignment exists.
Horizontal offset misalignment
(top view)
© 2014 Cengage Learning

FIGURE 8-12 Use of jackscrews to correct for horizontal offset misalignment.

three-phase motor and a 25-gallons-per-minute (gpm) pump that uses a flexible coupling as it would to align the shafts of a steam turbine to a generator that uses a rigid coupling and runs at incredibly high speeds.

8-2 DIAL INDICATOR METHOD

The **dial indicator** is a device that is used to measure very small amounts of variance. It is most commonly used to get measurements that may be as small as a thousandth of an inch (0.001 in.). It is for this reason that a dial indicator is most helpful in aligning coupled shafts. A dial indicator is shown in **Figure 8-13**.

CAUTION

Dial indicators are very precise measuring devices. Any shock or vibration could cause these fragile measuring devices to become inaccurate. Care should be taken when handling dial indicators to avoid damage.

Dial indicators can be mounted by many methods. Two of the most common methods of mounting a dial indicator are mounting the dial indicator on a shaft or on a table. If the dial indicator is mounted directly on a shaft, the supporting rods, which hold the dial indicator in its position, are mounted directly on the shaft and do not allow the supporting rod assembly to slip, which could cause inaccurate readings. If the dial indicator is mounted directly on the surface of a table, a magnetic base is usually used to prevent any slippage, thus eliminating any inaccurate readings.

When one dial indicator is used for aligning purposes, typically one of two methods is used: the rim or the face method. Taking the readings from the rim of the coupling flange is referred to as **rim alignment**. Taking the readings from the face of the flange is referred to as **face alignment**. The two methods are sometimes combined. Taking the readings from the rim and the face of the flange at the same time is referred to as **rim-and-face alignment**. Of course, the rim-and-face method requires the use of two dial indicators. The rim-and-face method is shown in **Figure 8-14**. The rim is used to measure for offset misalignments, and the face is used to measure for angular misalignments.

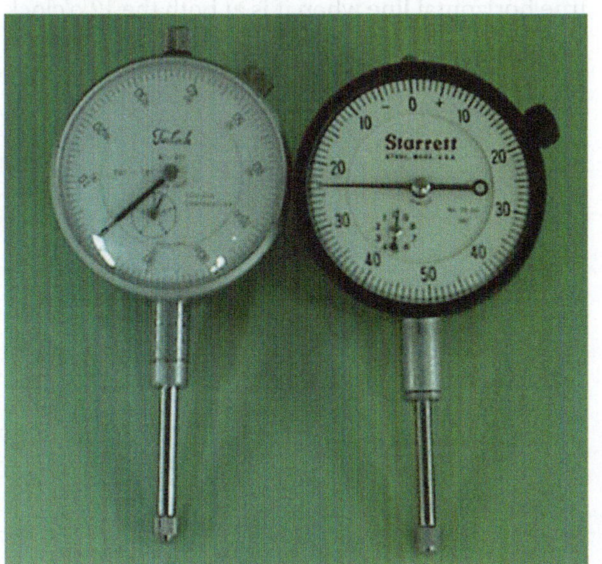

FIGURE 8-13 Dial indicators.

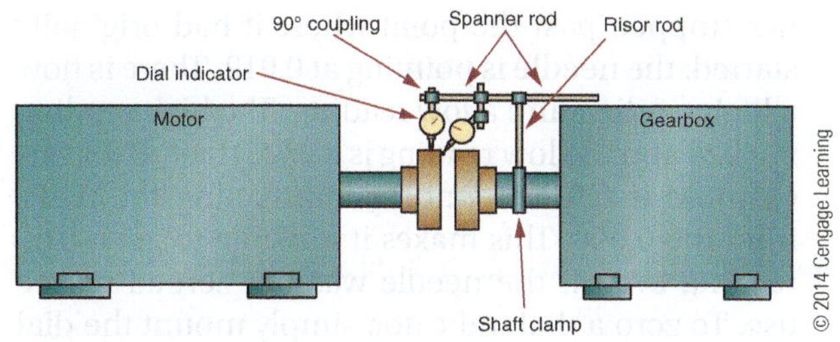

FIGURE 8-14 Rim and face dial indicator setup.

When a dial indicator is mounted on a shaft to measure the amount of misalignment, it must be mounted on the shaft that is opposite of the flange being measured. Notice that this is illustrated in Figure 8-14 as well. It is important to ensure that the mounting rods are solidly connected, to eliminate all bounce or movement. The entire mounting assembly consists of a shaft clamp, the risor rod, two 90° couplings, a spanning rod, and the dial indicator, all of which are shown in Figure 8-14.

The dial indicator has a face that can be rotated 360° so that the needle can be zeroed. The readings that are read from the dial indicator may move above or below the zero point once the plunger of the dial indicator is moved across the surface that is being measured. When the value goes below zero, it is said to be a negative reading. When the value goes above zero, it is said to be a positive reading.

Total Indicator Reading (TIR) is used to find the total amount of needle deflection. Total indicator readings are found by subtracting the lowest reading from the highest reading. For example, assume that a dial indicator is mounted to a table with a magnetic base. The plunger is in contact with the surface of a flat piece of steel. The needle is pointing to 0.022. The dial indicator is not zeroed for this example. As the dial indicator is swept across the surface of the steel, the plunger begins to move inward, causing the needle to move in the positive direction. When the dial indicator is stopped, the needle is pointing at 0.028. The dial indicator is then swept back across the surface of the steel in the other direction. After the dial indicator has stopped past the point where it had originally started, the needle is pointing at 0.012. There is now a high reading and a low reading. The high reading is 0.028 and the low reading is 0.012. The TIR for this example is 0.016, found by subtracting the 0.012 from the 0.028. This makes it possible to get a true reading even if the needle was not zeroed before use. To zero a dial indicator, simply mount the dial indicator with the plunger at about half of its travel, then zero the face of the dial indicator. This allows the dial indicator to travel in the positive and the negative direction, thus giving a positive reading and a negative reading. If only positive values are desired, then the dial indicator should be zeroed only when the plunger of the dial indicator is at the lowest point that is to be measured. This formula is used any time the highest and lowest readings are both positive or both negative. This formula does not work, however, when the high reading is positive and the low reading is negative. If this were the case, the two readings would be summed together as positive numbers to get the TIR. For example, assume that a dial indicator is mounted onto a shaft and was set to zero. As the shaft is turned 10° to the right, the needle begins to move off the zero in the negative direction. It stops at −0.012. The shaft is then rotated back to the zero point. Now the shaft is rotated 10° to the left. As the shaft is turned 10° to the left, the needle begins to move off the zero in the positive direction. It stops on 0.028. There is now a high reading and a low reading. The high reading is 0.028 and the low reading is −0.012. In this case the TIR is 0.040. This is found by adding 0.028 to 0.012. The total amount of indicator reading caused the needle to move 0.040 of an inch. This is why the number is added and not subtracted.

It is important to know that when trying to achieve coupling alignment, the coupling flanges should not be connected. However, they must be turned together to give a true and accurate reading of the shaft misalignments. How do you turn them together without attaching the flanges? Before beginning any shaft rotations, simply take a straightedge and place a mark on the rim of both flanges. These are referred to as witness marks. This gives an indication of how much each shaft has been rotated. The object is to keep the witness marks in line with each other before taking any measurement readings. This is illustrated in **Figure 8-15**.

Once the dial indicator is mounted, the plunger should be resting on the rim at the top of the flange. This is referred to as the 12 o'clock position. It is at this time that the needle should be zeroed. Once this is accomplished, the indicator and the shaft that it is attached to and the flange that the plunger is contacting are both turned simultaneously to the

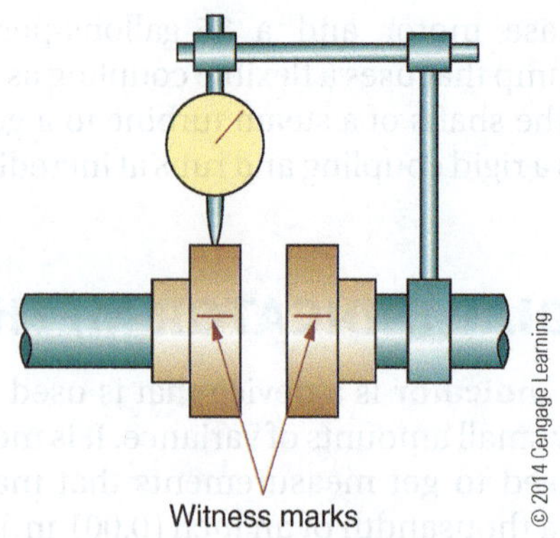

FIGURE 8-15 Witness marks.

next position, which may be 3 o'clock, 6 o'clock, or 9 o'clock. When checking for vertical plane misalignments, readings should be taken at the 12 o'clock and 6 o'clock positions. When checking for horizontal plane misalignments, the readings should be taken from the 3 o'clock and 9 o'clock positions.

Bar sag has to be considered whenever a spanner rod is used. Spanner rod lengths should be kept to a minimum to avoid bar sag. The longer the spanner rod, the more bar sag exists. This is illustrated in **Figure 8-16**. Keep in mind that the bar sag shown in the graphic is somewhat exaggerated to illustrate the point. Bar sag is usually not more than a few thousandths of an inch, depending on the length of the bar. It is important to use a spanner rod that is just long enough to get the dial indicator to the opposite flange. To check bar sag, notice that the entire dial indicator assembly must be mounted on a single shaft that is true and longer than the distance from the clamp to the plunger.

Gravity forces the spanner bar to sag below the true horizontal line when it is at both the 12 o'clock and the 6 o'clock positions. If the needle is zeroed

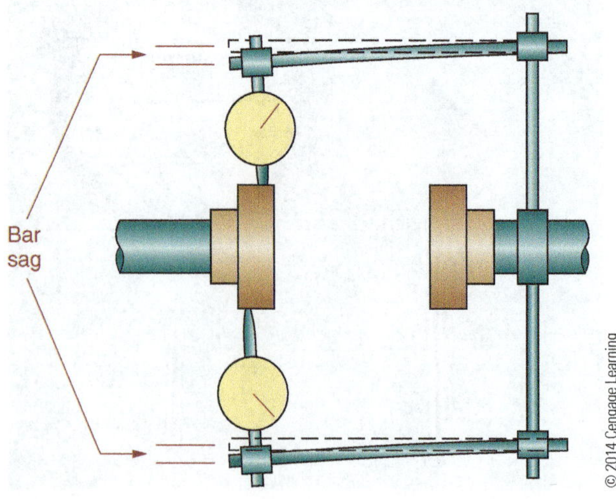

FIGURE 8-16 Bar sag.

while the dial indicator is at the 12 o'clock position and the whole assembly is then moved to the 6 o'clock position, the dial indicates the amount of bar sag that occurred. To find the actual amount of sag at each position, the 12 o'clock position and the 6 o'clock position, the total amount of bar sag indicated at 6 o'clock should be divided by 2. Let's assume that the dial indicator at the 6 o'clock position shows a total deflection of 0.016 in. This means that the actual bar sag at the 12 o'clock position is 0.008 in., and the actual amount of sag at the 6 o'clock position is also 0.008 in. Once the amount of bar sag is found, it is critical that nothing be moved on the assembly except for the height of the riser rod when mounting on the shaft that is to be aligned. Once the actual amount of bar sag is known, it should be subtracted from the vertical offset readings.

When measuring for a vertical angular misalignment, the plunger of the dial indicator should be placed against the face of the opposing flange. The dial indicator should be placed at the top of the flange and zeroed before any measuring is started. Also, before beginning any alignment corrections, ensure that the mounting bolts are tightened to the specified tolerance and that all jackscrews are simply finger tight. Once this is accomplished, both shafts are rotated 180°, making sure to keep the witness marks in line with each other. The dial indicator should show whether there is any angular gap present between the faces of the coupling flanges. If a vertical angular gap is found, the thickness of the shim stock will have to be calculated using this formula:

$$S = \frac{V_A}{2} \div \frac{D_F}{2} \times A$$

S = Shim stock thickness
V_A = Vertical angular gap (in inches)
2 = Constant
D_F = Diameter traveled by the plunger (in inches)
A = Distance between front and back mounting holes (in inches)

If the reading is positive in movement, then the shim, once calculated, should be placed under both of the rear feet. If the reading is negative in movement, the shims should be placed under both of the front feet. It is important to place the shims under the correct machine as well. Place the shims under the machine that is being measured at the face of the flange. This is illustrated in **Figure 8-17**.

Do not forget to retighten the mounting bolts to their specified torque after the shims have been placed under the appropriate feet. Once the shims are placed in the appropriate places and the mounting bolts are securely fastened, place the dial indicator back in the 12 o'clock position and re-zero the dial. Go through the procedure again to see whether all of the vertical angularity is gone. If it is not, continue this process until there is no vertical angular gap between the flange faces. The dial indicator reading 0.000 at the 12 o'clock position and 0.000 at the 6 o'clock position indicates no vertical angular gap. It is very difficult to remove the entire vertical angular gap. That is why you should know what the tolerance is for the particular type of coupling you are aligning for. It is sometimes acceptable to have a small amount of vertical angularity present in some couplings. Do not try too hard to get 0.000; just make sure that the misalignment is within the specified tolerance range.

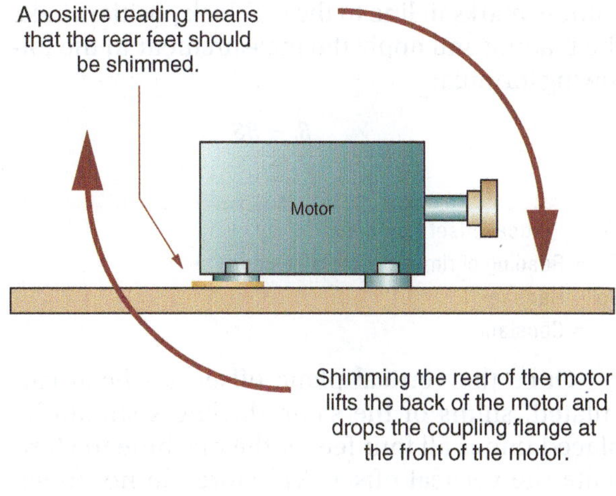
A positive reading means that the rear feet should be shimmed.
Motor
Shimming the rear of the motor lifts the back of the motor and drops the coupling flange at the front of the motor.

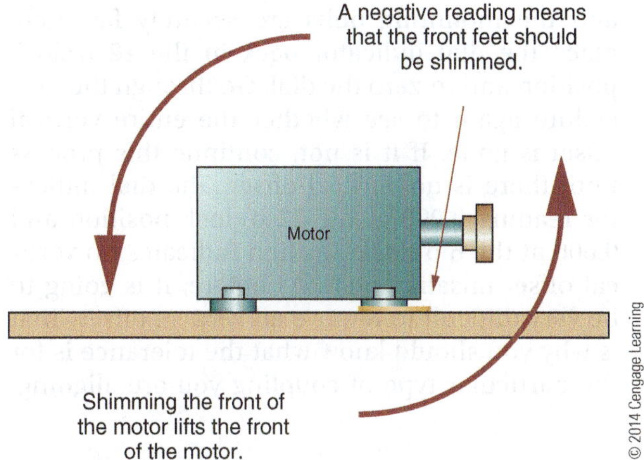
A negative reading means that the front feet should be shimmed.
Motor
Shimming the front of the motor lifts the front of the motor.

FIGURE 8-17 Be sure to shim the correct motor feet.

Once vertical angularity is completed, check for vertical offset misalignment. This is accomplished by checking the rim of the coupling flange at the 12 o'clock and at the 6 o'clock positions. As before, set the dial indicator at the 12 o'clock position and zero the needle, then rotate both shafts, keeping the witness marks in line to the 6 o'clock position. Take the reading and apply the measurement to the following formula:

$$V_0 = \frac{R_6 - BS}{2}$$

V_0 = Vertical offset (in inches)
R_6 = Reading of rim at 6 o'clock position
BS = Bar sag
2 = Constant

Once the vertical plane offset has been calculated, shims of the same thickness should be placed under all four feet of the machine to eliminate the vertical offset. As before, do not forget to retighten the mounting bolts to their specified torque after the shims have been placed under the feet. Once the shims are placed under the feet and the mounting bolts are securely fastened, place the dial indicator back in the 12 o'clock position and re-zero the dial. Go through the procedure again to see whether the entire vertical offset is gone. If it is not, continue this process until there is no vertical offset. The dial indicator reading 0.000 at the 12 o'clock position and 0.000 at the 6 o'clock position indicates no vertical offset misalignment. As before, it is going to be very difficult to remove all vertical offset. That is why you should know what the tolerance is for the particular type of coupling you are aligning. It is sometimes acceptable to have a small amount of vertical offset misalignment present in some couplings. Remember, do not try too hard to get 0.000; just make sure that the misalignment is within the specified tolerance range.

The next misalignment that should be checked is the horizontal angular misalignment. First, ensure that the mounting bolts have been tightened to the specified torque before checking for horizontal angular misalignment. In order to detect this type of misalignment with the dial indicator, it must be set on the face of the coupling at the 3 o'clock position. Once the needle has been zeroed, both shafts should be rotated until the dial indicator is at the 9 o'clock position. Remember to keep the witness marks in line with each other. As was mentioned before, if horizontal angular misalignment exists, the two jackscrews that are diagonal to each other should be adjusted as to eliminate the angular misalignment. This is illustrated in **Figure 8-18**. It is important to know that the measured angularity will be twice the actual angular misalignment. It is for this reason that the motor should have to be rotated only half as much as the TIR by turning the jackscrews. This is also illustrated in Figure 8-18.

The horizontal angular misalignment is easier to correct because it does not require the use of shims. It is simply taken care of by rotating the motor on its center axis to eliminate the angular offset. This is shown in Figure 8-18 as well. Do not forget

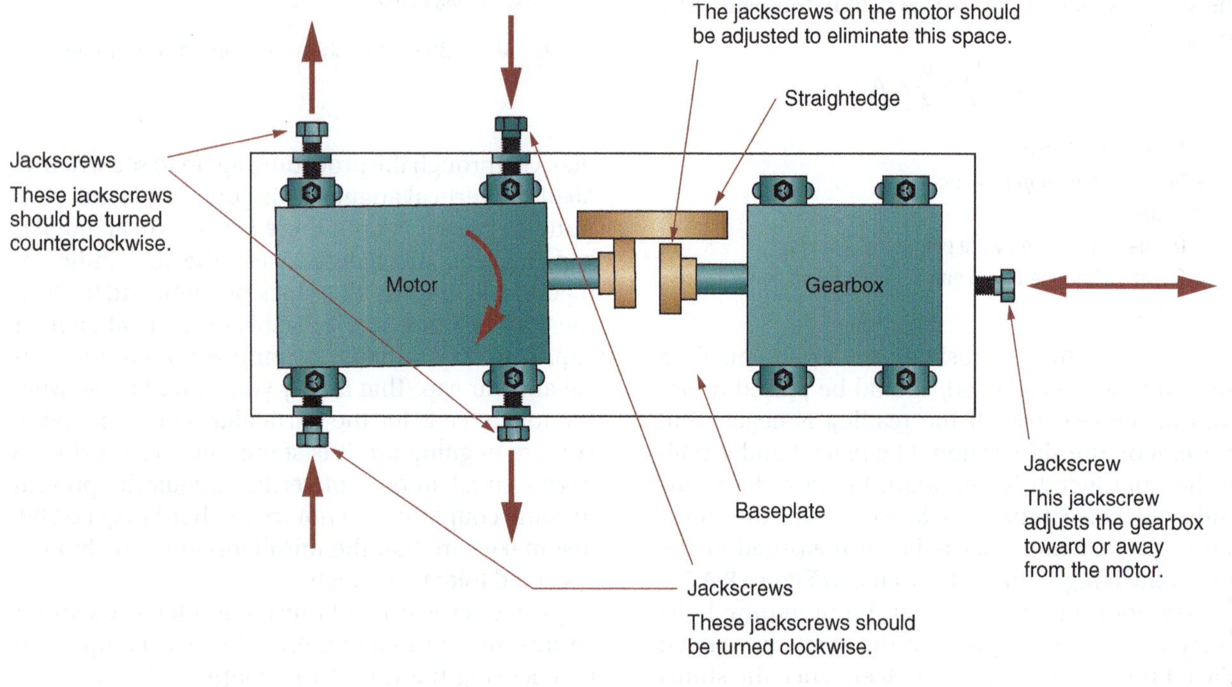
The jackscrews on the motor should be adjusted to eliminate this space.
Straightedge
Jackscrews
These jackscrews should be turned counterclockwise.
Motor
Gearbox
Jackscrew
This jackscrew adjusts the gearbox toward or away from the motor.
Baseplate
Jackscrews
These jackscrews should be turned clockwise.
Horizontal angular misalignment
(top view)
© 2014 Cengage Learning

FIGURE 8-18 Jackscrews can help eliminate the horizontal angular misalignment.

to retighten the mounting bolts to their specified torque after the machine has been adjusted so you can eliminate the angular misalignment. Once the adjustment has been completed and the mounting bolts are retightened, place the dial indicator back in the 3 o'clock position and re-zero the dial. Go through the procedure again to see whether the entire horizontal angular misalignment is gone. If it is not, continue this process until there is no horizontal angular misalignment. A dial indicator reading of 0.000 at the 3 o'clock position and 0.000 at the 9 o'clock position indicates no horizontal angular misalignment. As before, just try to get the angularity misalignment within the specified tolerance. Although eliminating all of the horizontal angularity misalignment would be great, this may prove to be time-consuming, tedious, and frustrating.

Once all of the horizontal angular misalignment has been corrected, the final misalignment that is checked is the horizontal offset misalignment. First, ensure that the mounting bolts are tightened to the specified torque before checking for horizontal offset misalignment. Placing the dial indicator on the rim of the coupling at the 3 o'clock position, zeroing the needle, then rotating both shafts until the dial indicator is at the 9 o'clock position checks for horizontal offset misalignment. Remember to keep the witness marks in line with each other as the shafts are being rotated. Once the dial indicator is at the 9 o'clock position, the offset reading should be read on the dial indicator before any mounting bolts are loosened. It also would be a good idea to scribe the mounting base where the machine is presently mounted before loosening any of the mounting bolts just in case the machine is moved while the mounting bolts are loosened. If a line is scribed and the machine moves while the bolts are being loosened, there is still a reference point by which to measure the adjustment of the machine. The machine should be moved squarely across the centerline of the motor in the direction that eliminates the horizontal offset misalignment. The machine should be moved only one-half the distance of the TIR. It is useful in this case to use two more dial indicators so that they are placed with their plungers on each foot as shown in **Figure 8-19**. Zero the needles before any of the mounting bolts are loosened. This ensures that both sides are moved an equal distance, therefore preventing any horizontal angularity misalignments.

Once the machine is moved so that any horizontal offset misalignment is eliminated, it is extremely important to take care in retightening the mounting bolts because the machine may easily lose some of the adjustment that was just made. Slowly retighten the mounting bolts, watching both of the dial indicators that are on the feet of the machine to ensure that no movement occurs. Once the bolts are securely tightened, set the dial indicator, which is on the coupling flange, back to the 3 o'clock position and re-zero the needle. Run through the procedure again to verify if any offset still exists. Once the horizontal offset has been corrected, it may be necessary to go back and check the horizontal angular alignment to ensure that the last adjustment did not re-create this type of misalignment.

Once all of the misalignments have been corrected and are within the tolerances that are specified by the coupling manufacturer, remove all of the dial indicators from the machine. Loosen the setscrew on the coupling flange, slide the coupling flanges apart, place the insert between the flanges, and then slide the flanges back together so as to sandwich the insert within the two outer flanges.

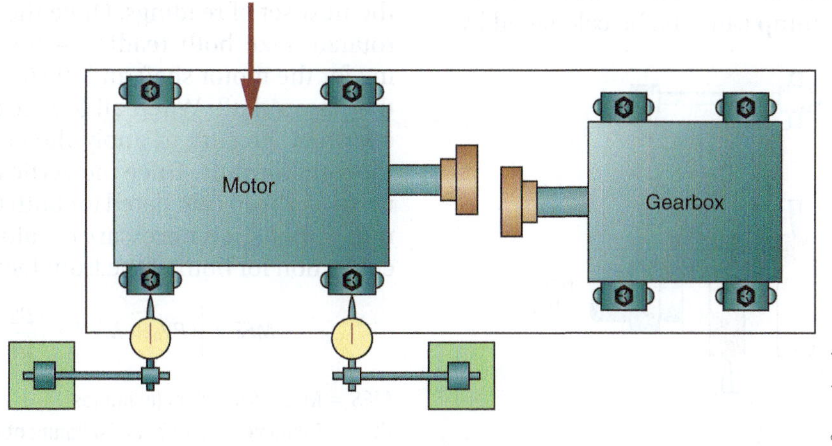
Motor
Gearbox
Watch both indicators to ensure that both ends of the motor travel the same amount of distance as the motor is moved.

FIGURE 8-19 Use of dial indicators to determine the amount of correction to be made.

Once the flanges are fit firmly against the insert, tighten the setscrews. If it is a rigid coupling, there will not be a flexible insert mounted between the flanges. The two flanges are simply slid closer to one another to be fastened together. Once they are fastened together, tighten the setscrews to prevent the flanges from moving on the shafts.

8-3 REVERSE DIAL INDICATOR METHOD

The reverse dial indicator method is the most accurate and easiest method, other than laser alignment, to perform coupled shaft alignment. The object of the reverse dial indicator method is to retrieve all of the data necessary to correct both vertical angular and offset misalignments with one dial indicator reading, and both horizontal angular and offset misalignments with another dial indicator reading. Congruent but oppositional readings allow for a one-step move to correct for each. To acquire these readings, the dial indicators must be set up on both shafts, as shown in **Figure 8-20**.

Notice how the dial indicators are set up in such a manner that they are opposite to each other. It is also important to determine the amount of bar sag before taking any readings. This is because the bar sag must be used in the formulae to calculate for the amount of misalignment. Reverse dial indicator alignment is obtained by applying a procedure that is slightly different from the single dial indicator method.

Both vertical angular and offset misalignments are corrected using the pump and the motor offset readings. Corrections are determined by adjusting the TIR to represent the shaft offset. Take a reading at the 12 o'clock position. Then, rotating both shafts to the 6 o'clock position, take another reading to find the vertical shaft offset. The amount of vertical shaft offset for the pump can then be calculated by applying the readings and bar sag reading to the following formula:

$$PV_O = \frac{PR12 - PR6 - BS}{2}$$

PV_O = Pump vertical offset misalignment (in inches)
$PR12$ = Pump reading at 12 o'clock
$PR6$ = Pump reading at 6 o'clock
BS = Bar sag
2 = Constant

Now the vertical offset can be obtained from the motor shaft. Taking a reading at the 12 o'clock position and then rotating both shafts to the 6 o'clock position and taking another reading accomplishes this. Be sure to get the reading from the correct dial indicator because the wrong reading gives you an undesirable result. The amount of vertical shaft offset for the motor can be calculated by applying the readings and bar sag reading to the following formula:

$$MV_O = \frac{MR12 - MR6 - BS}{2}$$

MV_O = Motor vertical offset misalignment (in inches)
$MR12$ = Motor reading at 12 o'clock
$MR6$ = Motor reading at 6 o'clock
BS = Bar sag
2 = Constant

It is not an uncommon practice to retrieve the TIR from both dial indicators at the same time because both dial indicators have to be rotated to the 12 o'clock and 6 o'clock positions. This can be accomplished by taking the 6 o'clock reading for the motor shaft at the same time that the 12 o'clock reading is being obtained for the pump shaft. When both of the readings have been obtained, rotate both of the shafts together 180° so you can obtain the next set of readings. Once the shafts have been rotated, take both readings—the 12 o'clock reading for the motor shaft and the 6 o'clock reading for the pump shaft. When all four readings have been obtained, be sure to apply the correct readings to the right formula. Once the vertical offset misalignment has been calculated for both the pump and the motor shafts, it is necessary to calculate for the shim correction for both of the front feet of the motor.

$$MFS = \left[(PV_O + MV_O) \times \left(\frac{D2}{D1}\right)\right] - PV_O$$

MFS = Motor front shim (in inches)
PV_O = Pump vertical offset misalignment (in inches)
MV_O = Motor vertical offset misalignment (in inches)
$D1$ = Distance between dial indicators (in inches)
$D2$ = Distance between pump indicator and the front feet on the motor (in inches)

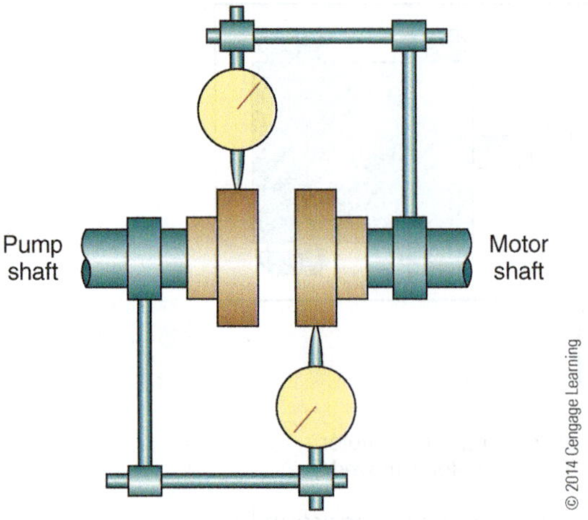

FIGURE 8-20 Reverse dial indicator setup.

Once shim correction for both of the front feet of the motor has been calculated, the shim correction for the back feet has to be calculated. When calculating for the shim thickness needed for the rear feet, it is important to note that a negative value is not possible. The value can only be zero (no shims needed) or positive thickness (shims required).

$$MRS = \left[(PV_O + MV_O) \times \left(\frac{D3}{D1} \right) \right] - PV_O$$

MRS = Motor rear shim (in inches)
PV_O = Pump vertical offset misalignment (in inches)
MV_O = Motor vertical offset misalignment (in inches)
$D1$ = Distance between dial indicators (in inches)
$D3$ = Distance between pump indicator and the rear feet on the motor (in inches)

Taking one set of readings has accomplished all vertical misalignment corrections. Next you must solve for horizontal misalignment.

Both horizontal angular and offset misalignments are corrected using the pump and the motor offset readings. Corrections are determined by adjusting the TIR to represent the shaft offset. Taking a reading at the 3 o'clock position and then rotating both shafts to the 9 o'clock position and taking another reading, find the horizontal shaft offset. The amount of horizontal shaft offset for the motor can then be calculated by applying the readings and bar sag reading to the following formula:

$$MH_O = \frac{PR9 - PR3 - BS}{2}$$

MH_O = Motor horizontal offset misalignment (in inches)
$PR9$ = Pump reading at 9 o'clock
$PR3$ = Pump reading at 3 o'clock
BS = Bar sag
2 = Constant

Now the horizontal offset can be obtained for the pump shaft. Taking a reading at the 3 o'clock position and then rotating both shafts to the 9 o'clock position and taking another reading accomplishes this. Be sure to get the reading from the correct dial indicator because the wrong reading gives you an undesirable result. The amount of horizontal shaft offset for the pump can be calculated by applying the readings and bar sag reading to the following formula:

$$PH_O = \frac{MR9 - MR3 - BS}{2}$$

PH_O = Motor vertical offset misalignment (in inches)
$MR9$ = Motor reading at 9 o'clock
$MR3$ = Motor reading at 3 o'clock
BS = Bar sag
2 = Constant

It is not an uncommon practice to retrieve the TIR from both dial indicators at the same time because both dial indicators have to be rotated to the 9 o'clock and 3 o'clock positions. This can be accomplished by taking the 3 o'clock reading for the motor shaft at the same time that the 9 o'clock reading is being obtained for the pump shaft. Once both of the readings have been obtained, rotate the two shafts together 180° to obtain the next set of readings. Once the shafts are rotated, take both readings—the 9 o'clock reading for the motor shaft and the 3 o'clock reading for the pump shaft. When all four readings have been obtained, be sure to apply the correct readings to the right formula. Horizontal shaft offsets are then used to calculate the side movement of the motor. The horizontal corrective movement of the front feet on the motor is found using this formula:

$$MFM = \left[(MH_O + PH_O) \times \left(\frac{D2}{D1} \right) \right] - MH_O$$

MFM = Motor front feet movement (in inches)
MH_O = Motor horizontal offset misalignment (in inches)
PH_O = Pump horizontal offset misalignment (in inches)
$D1$ = Distance between dial indicators (in inches)
$D2$ = Distance between pump indicator and the front feet on the motor (in inches)

Once the correction for both of the front feet of the motor has been calculated, the correction for the back feet needs to be calculated.

$$MRM = \left[(MH_O + PH_O) \times \left(\frac{D3}{D1} \right) \right] - MH_O$$

MRM = Motor rear feet movement (in inches)
MH_O = Motor horizontal offset misalignment (in inches)
PH_O = Pump horizontal offset misalignment (in inches)
$D1$ = Distance between dial indicators (in inches)
$D3$ = Distance between pump indicator and the rear feet on the motor (in inches)

Taking one set of readings has accomplished all horizontal misalignment corrections.

8-4 FEELER GAUGE METHOD

A **feeler gauge** (also referred to as a thickness gauge) is a thin strip of metal called a leaf, that has a precision thickness and is used along with a straightedge when correcting any misalignment. **Figure 8-21** shows a feeler gauge set, which, when not in use, folds up into the protective covering. The thickness of the feeler gauge can be as thin as 0.001 in. up to as much as 0.25 in. Each feeler gauge increases in thickness by 0.001 in. from the previous feeler gauge. The idea behind the feeler gauge is to identify how far the coupling flanges are out of

FIGURE 8-21 A feeler gauge set.

tolerance. This method of detecting misalignment is, perhaps, the most inaccurate. It is, however, an excellent method for getting the shafts fairly close before another method of alignment is used. Final results may be far out of tolerance on completion of feeler gauge alignment when compared with the other methods that have been discussed.

Before beginning any alignment corrections, ensure that the mounting bolts are tightened to the specified tolerance and that all jackscrews are simply finger tight. The first step is to check for any vertical angular misalignments. This is found by using what is referred to as a taper gauge. The taper gauge is less likely to give false trial-and-error readings than the feeler gauge. A taper gauge is shown in **Figure 8-22**.

A **taper gauge** is a flat, tapered strip of metal with graduations marked along the edge in thousandths of an inch. The taper gauge is lowered vertically between the two flange faces at the 12 o'clock position until a slight resistance is felt. The reading on the taper gauge where the flange faces touch the taper gauge is the measurement of angular misalignment at the 12 o'clock position. Feeler gauges can be used to find the amount of gap at the 12 o'clock position; however, the measurement is not as accurate as the taper gauge because there may be small particles of dirt or gaps of air between the leaves. Once the gap is found at the 12 o'clock position, move to the 6 o'clock position without rotating the shafts and measure the gap. If there is a difference in the two gaps, then vertical angular misalignment exists. If a vertical angular gap is found, the thickness of the shim stock will have to be calculated using this formula:

$$S = \frac{V_A}{2} \div \frac{D}{2} \times A$$

S = Shim stock thickness
V_A = Vertical angular gap (in inches)
2 = Constant
D = Diameter of the coupling (in inches)
A = Distance between front and back mounting holes (in inches)

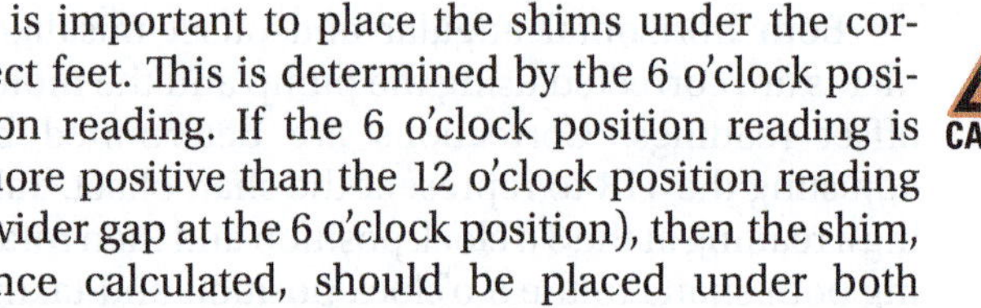

It is important to place the shims under the correct feet. This is determined by the 6 o'clock position reading. If the 6 o'clock position reading is more positive than the 12 o'clock position reading (wider gap at the 6 o'clock position), then the shim, once calculated, should be placed under both of the front feet. If the reading was more negative than the 12 o'clock position reading (wider gap at the 12 o'clock position), the shims should be placed under both of the rear feet. It is important to place the shims under the correct machine as well. It is always common practice to shim the motor because it is not usually connected to anything that is rigid, and it has the flexibility to be moved. Do not forget to retighten the mounting bolts to their specified torque after the shims have been placed under the appropriate feet. Once the shims are placed in the appropriate places and the mounting bolts are securely fastened, check the gap at the 12 o'clock position and 6 o'clock position to see whether all of the vertical angularity is gone. If it is not, continue this process until there is no vertical angular gap present between the flange faces. It is going to

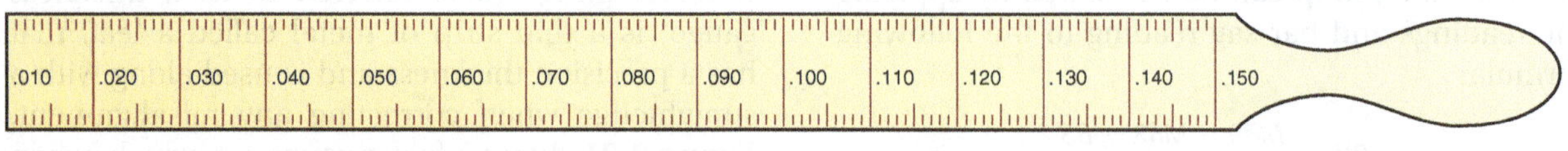

FIGURE 8-22 Taper gauge.

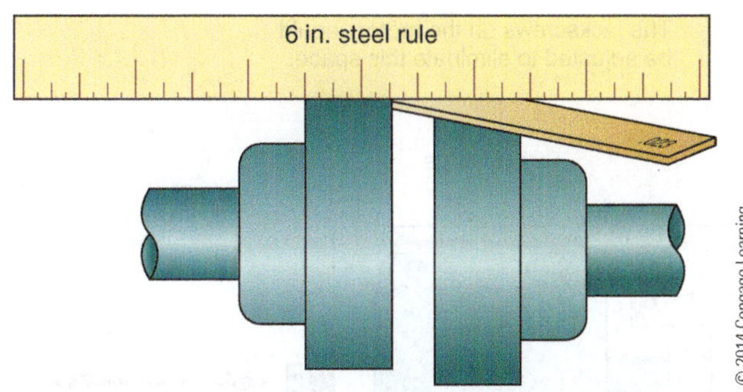

FIGURE 8-23 Insert the feeler gauge between the flange and the straightedge.

be very difficult to remove the entire vertical angular gap using this method. That is why you should know what the tolerance is for the particular type of coupling you are aligning. Keep trying until you are sure that the misalignment is within the specified tolerance range.

Once vertical angularity alignment is completed, check for vertical offset misalignment. To measure for vertical offset misalignment, place a straightedge on the rim of the coupling flange, first at the 12 o'clock position, and then insert the feeler gauge between the flange and the straightedge. This is illustrated in **Figure 8-23**.

Continue to insert feeler gauges until a slight resistance is felt. When this occurs, the measurement can be taken. If more than one feeler gauge is used to fill the gap, then the thicknesses of all the feeler gauges are added together to find the amount of misalignment. For example, if three feeler gauges were used to fill the gap, a 0.003 in., a 0.130 in., and a 0.011 in., the total amount of misalignment would be 0.144 in. Once the vertical plane offset has been found, shims of the same thickness should be placed under all four feet of the appropriate machine so the vertical offset is eliminated. If the coupling flange on the motor were lower than the driven machine coupling, the motor would be shimmed so it is raised to the same vertical level as the driven machine. If the driven machine coupling was lower than the coupling on the motor, the driven machine should be shimmed so as to raise it to the same vertical level as the motor. As before, do not forget to retighten the mounting bolts to their specified torque after the shims have been placed under the feet. Once the shims are placed under the feet and the mounting bolts are securely fastened, go through the procedure again to see whether the entire vertical offset is gone. If it is not, continue this process until there is no vertical offset detected. As before, it is going to be very difficult to remove all vertical offset using this method of correcting misalignment.

The next misalignment that should be checked is the horizontal angular misalignment. First, ensure that the mounting bolts have been tightened to the specified torque before checking for horizontal angular misalignment. In order to detect this type of misalignment, the taper gauge should be inserted horizontally between the two flange faces at the 3 o'clock position until a slight resistance is felt. The reading on the taper gauge where the flange faces touch the taper gauge is the measurement of angular misalignment at the 3 o'clock position. Once again, feeler gauges can be used to find the amount of gap at the 3 o'clock position; however, the measurement is not as accurate as the taper gauge because there may be small particles of dirt or gaps of air between the leaves. Once the gap is found at the 3 o'clock position, move to the 9 o'clock position without rotating the shafts and measure the gap. If there is a difference in the two gaps, then horizontal angular misalignment exists. The amount of angular misalignment is found by subtracting the lower gap measurement from the higher gap measurement. If horizontal angular misalignment is found, the two jackscrews that are diagonal to each other should be adjusted to eliminate the angular misalignment. This is illustrated in **Figure 8-24**.

This type of misalignment is easier to correct because it does not require the use of shims. It is simply taken care of by rotating the motor on its center axis to eliminate the angular offset. Do not forget to retighten the mounting bolts to their specified torque after the machine has been adjusted to eliminate the angular misalignment. Once the adjustment has been completed and the mounting bolts are retightened, go through the procedure again to see whether the entire horizontal angular misalignment is gone. If it is not, continue this process until there is no horizontal angular misalignment. As before, just try to get the angular misalignment within the specified tolerance.

Once all of the horizontal angular misalignment has been corrected, the final misalignment to check is the horizontal offset misalignment. First, ensure that the mounting bolts are tightened to the specified torque before checking for horizontal offset misalignment. To measure for horizontal offset misalignment, place a straightedge on the rim of the coupling flange, first at the 3 o'clock position, and then insert the feeler gauge between the flange and the straightedge as before. Continue to insert feeler gauges until a slight resistance is felt. When this occurs, the measurement can be taken.

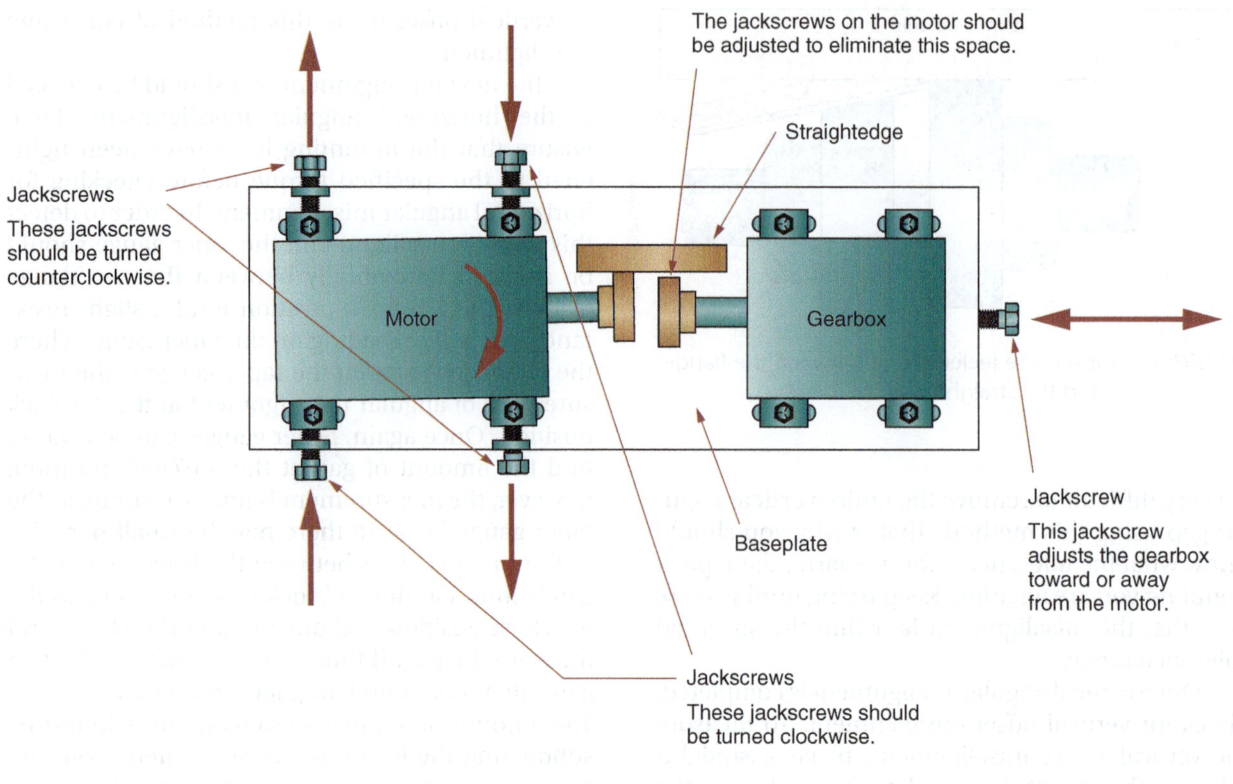
The jackscrews on the motor should be adjusted to eliminate this space.
Straightedge
Jackscrews
These jackscrews should be turned counterclockwise.
Motor
Gearbox
Jackscrew
This jackscrew adjusts the gearbox toward or away from the motor.
Baseplate
Jackscrews
These jackscrews should be turned clockwise.
Horizontal angular misalignment
(top view)
© 2014 Cengage Learning

FIGURE 8-24 Jackscrews can help eliminate the horizontal angular misalignment.

If more than one feeler gauge is used to fill the gap, the thicknesses of all the feeler gauges are added together to find the amount of misalignment. Once the horizontal plane offset has been found, scribe the mounting base where the machine is presently mounted, and move the machine squarely across the centerline of the motor in the direction that will eliminate the horizontal offset misalignment. This is accomplished by adjusting the two jackscrews on the same side as the distance that was measured by the feeler gauges. It is important that both of the jackscrews be turned an equal amount. If they are not, an angularity condition will arise. This would be undesirable because the horizontal angularity has already been corrected. Once the machine is moved to eliminate any horizontal offset misalignments, it is extremely important to carefully retighten the mounting bolts because the machine may easily lose some of the adjustment that was just made. Slowly retighten the mounting bolts, watching the feet of the machine to ensure that no movement occurs. Once the bolts are securely tightened, run through the procedure again to verify that no offset conditions exist. If a slight offset continues to exist, repeat the procedure until the entire offset is eliminated. Once the horizontal offset has been corrected, it may be necessary to go back to check the horizontal angular misalignment again to ensure that the last adjustment did not re-create this type of misalignment.

Another device that may be used in place of feeler gauges and taper gauges may be the dial caliper. A **dial caliper** is shown in **Figure 8-25.** It is a device that can measure internal values, external values, and depth values.

The dial caliper, like the dial indicator, has a face that can rotate so you can zero the needle before measuring. Follow the same procedures that were given for the feeler gauges and taper gauges when using the dial caliper to correct for any misalignments. It is always important, however, to ensure that the dial caliper is being used properly to ensure that the readings are accurate. It is important for the calipers to fit snugly and squarely on the surfaces that are being tested, in this case, the flange faces. If the calipers are not square, an inaccurate reading may be obtained, and therefore an improper adjustment will be made.

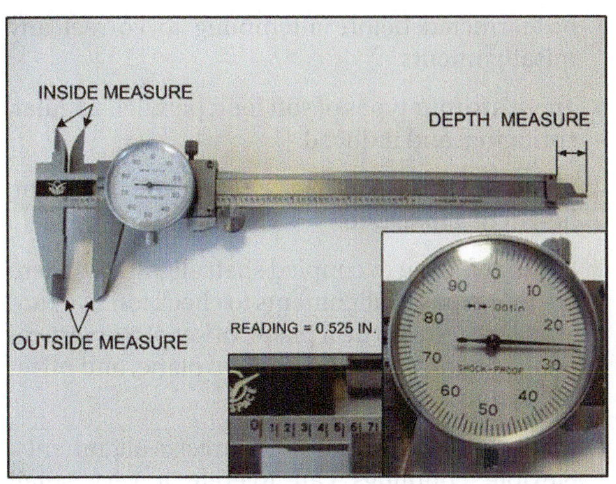

© 2014 Cengage Learning

FIGURE 8-25 Dial caliper.

© 2014 Cengage Learning

FIGURE 8-26 A laser alignment kit in use.

8-5 LASER SHAFT ALIGNMENT

Laser shaft alignment is the most accurate and probably the easiest to accomplish. Even though the initial cost of the laser alignment kit is higher than the devices used in the other methods, the expense is offset by the accuracy of the measurements taken and the ease of the procedure. There are many different types of laser alignment kits available today. They are increasing in variety and accuracy each day; therefore, it is not within the scope of this text to go over a specific model or brand. However, the basic operation of the laser alignment kit is discussed.

CAUTION

The laser alignment kit is very fragile and therefore should not be dropped or bumped, which cause the equipment to give inaccurate readings and lose alignment integrity. Also, because this is electronic equipment, it should not be kept in a harmful environment, such as in places that may contain high levels of moisture, dirt, or heat. These conditions may have an adverse effect on the laser beam. It is also important to note that one should never look into the laser beam because injury to the eyes may result.

The laser alignment kit usually shows indications of all misalignments with one measurement taken by mounting a laser transmitter on one shaft and a target receiver on the opposite shaft. The laser alignment kit often does not even use the coupling flanges to align the shafts, as can be seen in **Figure 8-26**. Once both of the devices are mounted, the laser alignment kit should be able to check for soft foot and determine any existing misalignments. It should also give the shim sizes and shim placements that are necessary to correct for any misalignments. Some models even have a liquid crystal display (LCD) that shows the actual machine displacements graphically. Another positive characteristic of the laser alignment kit is that it is not affected by bar sag, as is the dial indicator assembly.

The laser alignment kit works by using a microprocessor, which processes and calculates the misalignments as the laser hits the target receiver. Offset and angles are accurately detected when the target receives the laser beam at the 12 o'clock position, then it is slowly rotated to the 6 o'clock position. As the shafts are rotated together, the laser beam moves across the target face as misalignments occur, causing data to be collected. The data are then processed and the amount of offset is determined. The processor measuring for any right triangle existence as the laser hits the target also determines the amount of angularity. See **Figure 8-27**.

Once any misalignments are found and the recommendations are followed to correct the misalignments, a recheck is performed. This usually is accomplished in a matter of seconds. If there are no misalignments, the laser beam will be shot directly into the center of the target with no presence of triangular calculation possible.

Laser alignment should be used for applications that require extreme precision alignment. Laser alignment is becoming widely accepted as the primary method of alignment because of its accuracy and ease of operation.

Offset

The laser has missed the center target; therefore, the amount of offset can be calculated by determining the distance between where the laser hit the target and the center of the target. Horizontal and vertical offsets are calculated at the same time.

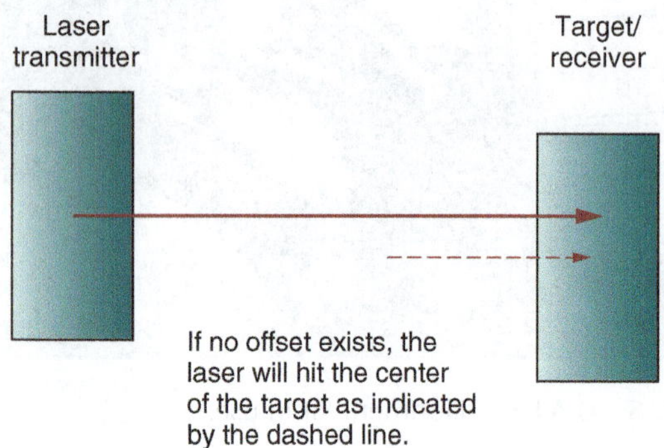
Laser
transmitter
Target/
receiver
If no offset exists, the
laser will hit the center
of the target as indicated
by the dashed line.

Angular

The computer can calculate the amount of total angularity using right triangle trigonometry. As with the offset misalignment, horizontal and vertical angular offsets are calculated at the same time.

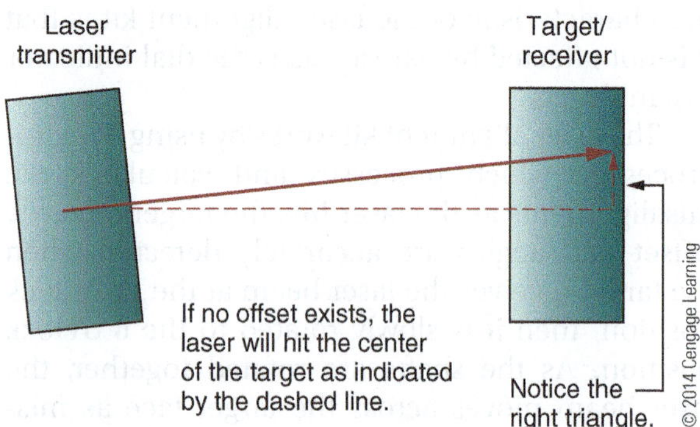
Laser
transmitter
Target/
receiver
If no offset exists, the
laser will hit the center
of the target as indicated
by the dashed line.
Notice the
right triangle.
© 2014 Cengage Learning

FIGURE 8-27 Laser targeting is used to determine offset and angularity misalignments.

SUMMARY

- Pulley and sprocket alignment is not the same as coupled shaft alignment. For belt and chain drives, the drive and driven shaft axes are always going to be parallel to each other, whereas the drive and driven shafts that are to be coupled are always in an end-to-end orientation.
- There are three types of misalignment to consider when aligning shafts that are used in pulley and sprocket drives: offset, angular, and parallel.
- Soft foot exists when one or more of the feet on the machine that is being aligned is bent or was poorly manufactured. If soft foot exists, it should be corrected before attempting to correct any misalignments.
- There are four types of soft foot: parallel, angular, springing, and induced.
- Shims are used to correct soft foot and adjust the height of a machine.
- When referring to coupled shaft alignment, there are four types of alignments to check for. They are angular of the vertical plane, offset of the vertical plane, angular of the horizontal plane, and offset of the horizontal plane.
- Rigid couplings require very precise alignments. Flexible couplings can tolerate a very small amount of misalignment. (Refer to the manufacturer's specifications.)
- Two of the most common methods of mounting a dial indicator are mounting the dial indicator on a shaft or on a table.
- There are typically two types of alignment methods that are used when using one dial indicator for aligning purposes: rim alignment and face alignment. Rim alignment is used to detect offsets, and face alignment is used to check for angularity.
- When a dial indicator is mounted on a shaft to measure the amount of misalignment, it must be mounted on the shaft that is opposite the flange being measured.
- The entire mounting assembly consists of a shaft clamp, the risor rod, two 90° couplings, a spanning rod, and the dial indicator. It is important to ensure that the mounting rods are solidly connected to eliminate all bouncing or movement.
- Total indicator reading (TIR) is used to find the total amount of needle deflection. Total indicator readings are found by subtracting the lowest reading from the highest reading. This formula is used any time the highest and lowest readings are both positive or both negative. This formula does not work, however, when the high reading is positive and the low reading is negative. If this were the case, the two readings would be added together as positive numbers to get the TIR.
- Before beginning any shaft rotations, simply take a straightedge and place a mark on the rim of both flanges. These are referred to as witness marks, and they indicate how much each shaft has been rotated. The object is to keep the witness marks in line with each other before taking any measurement readings.

- When checking for vertical plane misalignments, readings should be taken at the 12 o'clock and 6 o'clock positions. When checking for horizontal plane misalignments, the readings should be taken from the 3 o'clock and 9 o'clock positions.
- Bar sag needs to be considered whenever a spanner rod is used. Once the actual amount of bar sag is known, it should be subtracted from the vertical offset readings.
- The reverse dial indicator is more accurate than the single dial indicator method.
- The feeler gauge method of detecting misalignment is, perhaps, the most inaccurate.
- The laser alignment method is the most accurate of all the alignment methods.

REVIEW QUESTIONS

1. Name the three types of misalignments for a pulley drive.
2. What is soft foot?
3. Which is the more difficult to correct, rigid coupling alignment or flexible coupling alignment?
4. What is TIR?
5. What is the most inaccurate of all the coupled shaft alignment methods?

CHAPTER 9

Lubrication

Facilities are in business to make money. If their machinery is not properly lubricated, the facilities will have excessive downtime. Excessive downtime means no product. No product means money is not being made. For this reason, an aggressive lubrication schedule must be adhered to. Lubricants can prevent excessive downtime by providing a smooth layer of film between parts. This smooth layer of film reduces friction. Lubrication also cools components while they are in use.

OBJECTIVES

After studying this chapter, the student should be able to

- Explain some basic terms that are used when referring to lubrication.
- Discuss the necessity for lubrication.
- Discuss the different forms of lubrication.
- List the types of lubricants used in industry.
- Discuss applications of lubrication.
- Discuss the importance of a lubrication schedule.

9-1 GENERAL TERMS

There are a few terms that need to be known before discussing lubrication. This is to ensure that you have a complete understanding of what is being discussed later in this chapter. Here are a few terms that you should know:

- **Lubricant**: A substance that reduces friction by providing a smooth surface of film over parts that move against each other.
- **Viscosity**: The internal friction of a lubricant caused by molecular attraction.
- **Hydrocarbon**: Any compound containing mostly hydrogen and carbon. A pure hydrocarbon contains only hydrogen and carbon.
- **Semisolid**: A gel-like substance that has the characteristics of both a solid and a liquid.

9-2 TYPES OF LUBRICANTS

Lubricants come in one of four forms: solid, semisolid, liquid, and gas. A hydrocarbon, petroleum, is the basis for most of the lubricants in use today. Petroleum is composed of 12% hydrogen and 85% carbon. The missing 3% consists of sulfur, oxygen, nitrogen, and other miscellaneous elements. Petroleum resists oxidation, and petroleum-based lubricants are available in all four forms.

Another type of lubricant is made from animal oils or vegetable oils. These oils are primarily used in the food industry because of the toxic nature of the petroleum-based lubricants. It is not uncommon to find a lubricant that has a combination of petroleum, animal oils, and vegetable oils. The animal and vegetable oils are much slipperier than the petroleum lubricants, but they are not resistant to bacteria as is petroleum. Another problem with animal and vegetable oils is that they are very acidic and tend to form more acid during use. These oils also do not prevent oxidation, and they tend to break down because of their organic nature. Consequently, systems that use animal and vegetable oils need to be monitored very closely to ensure that the lubrication is always present.

Additives are used to improve the properties of lubricants. An example of an additive may be an oxidation inhibitor. Some other examples may be these:

- Graphite
- Fatty materials
- Demulsifiers
- Viscosity improvers

Graphite is used to prevent galling. Galling is a bonding, shearing, and tearing away of material from two contacting, or sliding, metals. Fatty materials are used to improve film strength. **Demulsifiers** are used to remove water from within the lubrication. Viscosity improvers are used to aid in machine startup when the lubrication is cold.

Viscosity is a very important factor to consider when dealing with lubrication. Basically, viscosity is the measurement of the resistance of a fluid's molecules as they move past each other. The fluid's ability to flow is inversely proportional to its viscosity. A fluid with a higher viscosity has a lower flow rate. Lubricants with higher viscosity offer greater film thickness than those of a lower viscosity; however, the trade-off is a lower flow rate. Grease has a higher viscosity than oil. It is important to realize that viscosity is affected by temperature. Higher temperatures and pressures can break down a fluid's viscosity, thus lowering its ability to provide the protective film that is needed to reduce friction. Synthetic lubricants are usually made from a petroleum base; however, they are made to resist changes in viscosity caused by increased temperatures and pressures. It is for this reason that they usually cost more.

It is important upon changing old lubricant that it is not mixed with a different type of lubricant. The reason for this is that some synthetic lubricants when mixed with petroleum lubricants may cause rapid deterioration of the seals that are in contact with the lubricant. A semisolid lubricant is one with the consistency of grease. Grease is neither a solid nor a liquid. This is useful when the lubricant needs to be able to get into small areas and not run off like liquid. This is accomplished by combining low-viscosity oils with thickeners such as clay, lead, graphite, and soap. Graphite is a very good lubricant that is produced from coal. It can be in the form of a solid, semisolid, or powder. Graphite has a very low coefficient of friction, which makes it desirable for many applications, such as lubricating the tumblers of a lock. The coefficient of friction is the measure of the frictional force between two surfaces that are in contact with each other. Because of graphite's low coefficient of friction, it is often mixed with other lubricants. Graphite lubricants must have moisture to be effective. Silicone is another common type of lubricant used in industry. Silicone is made from a silicon compound, and it is extremely resistant to heat, moisture, and water.

Film Lubrication

Film lubrication is accomplished when a lubricant creates a film that is thick enough to completely separate the two surfaces of metal. In the absence of lubrication, the metal surfaces would rub together, causing heat and wear due to friction. **Figure 9-1** shows two pieces of metal without film lubrication. The surfaces of both pieces of metal are rough and jagged. Without film lubrication, these rough surfaces would drag across each other, each having to overcome the other's imperfections. Film lubrication allows the two pieces of metal to move without ever coming into contact with each other's imperfections, thus lowering the friction and heat. This is represented in **Figure 9-2**.

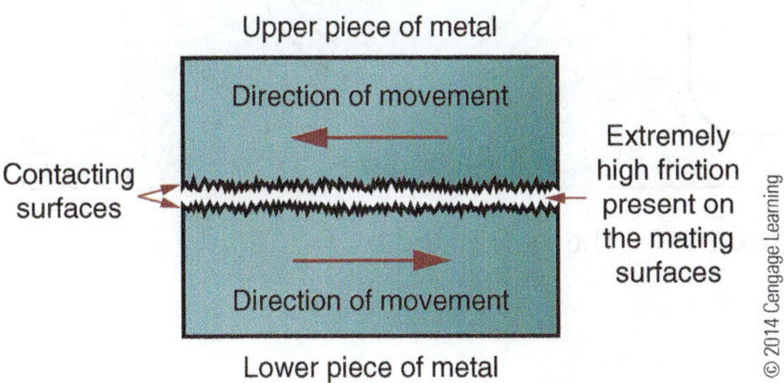

FIGURE 9-1 Metal to metal with no lubrication.

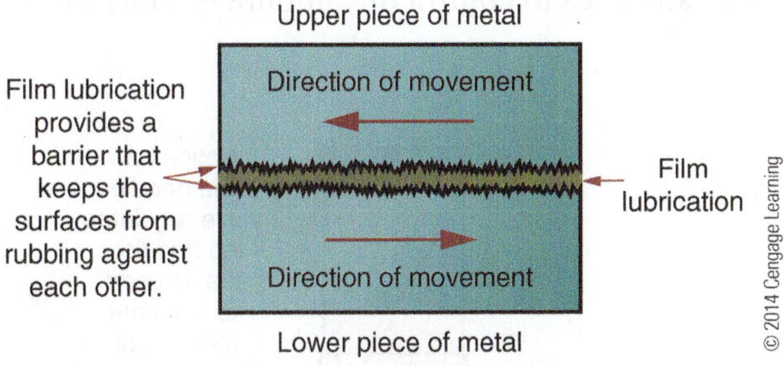

FIGURE 9-2 Metal to metal with film lubrication.

9-3 METHODS OF APPLICATION

Oil is applied by several methods. Because oil runs off the part that is being lubricated, there must be some method of replenishing the oil in order to ensure proper lubrication. One method used to accomplish this task is to make sure that the part needing lubrication is submersed in the oil continually or cyclically.

Submersion

If submersion is used to lubricate bearings, it is important to ensure that the proper oil level in the oil reservoir is maintained. When at the correct level, the oil covers half of the lowest rolling element of the bearing. If the oil level is higher, the bearing begins churning the oil. Submersion is most commonly used to lubricate gears within a gearbox. This is done by filling the gearbox so the gears are immersed in the oil to a depth that is equal to twice the height of one tooth. Two other methods of submersion, ring and chain, are shown in **Figure 9-3**.

The Splash Method

A slinger is used in the splash method. Refer to **Figure 9-4**. Notice how the slinger enters the oil reservoir and slings the oil in a cyclic manner.

The Wick Method

The wick method is used to transfer oil from a reservoir through a wick to the part being lubricated. To understand how wick lubrication works, think of an oil lamp. The fuel from the reservoir is pulled into the wick through capillary action. This brings the level of the fuel to the burn chamber of the lamp. Capillary action is when the surface of a liquid is elevated on a material because of its relative molecular

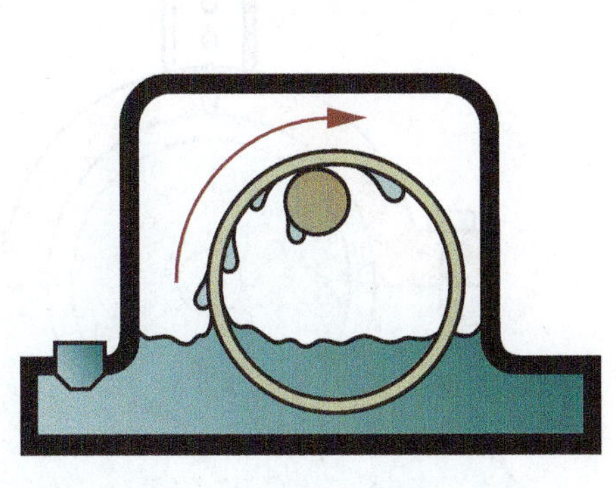

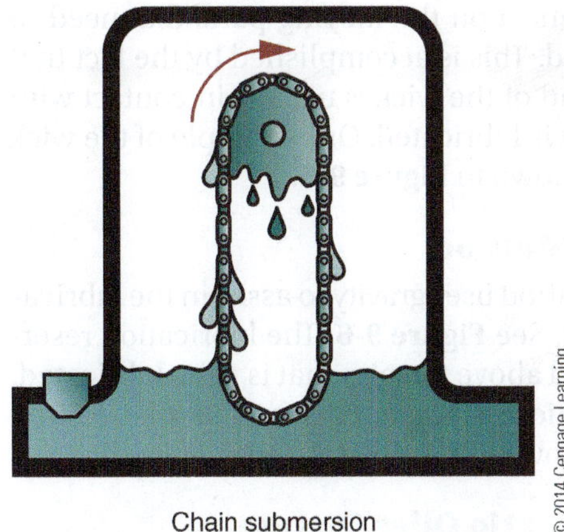

FIGURE 9-3 Two methods of submersion.

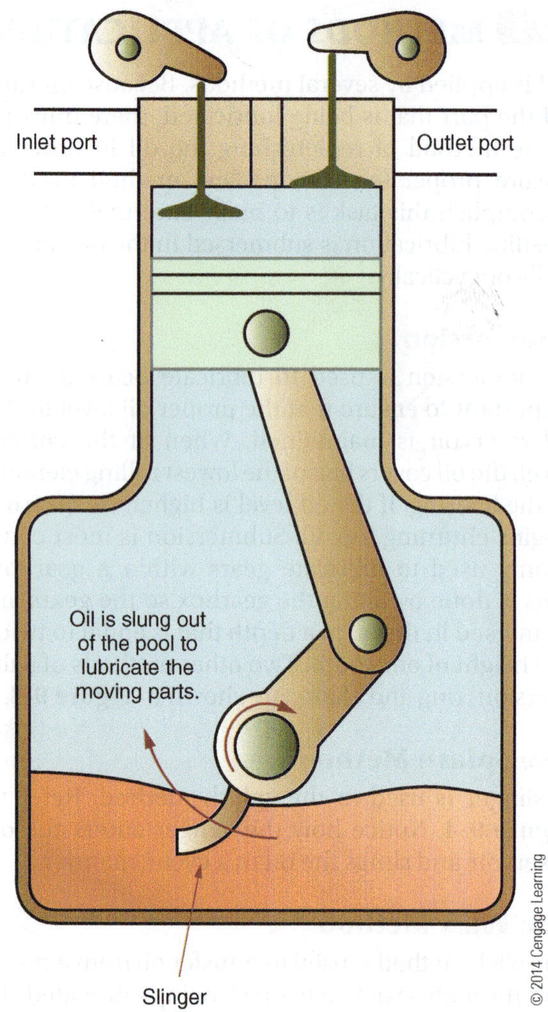

FIGURE 9-4 Slinger lubricator.

attraction. In layman's terms, this is nothing more than absorption. Much like the oil lamp, oil in a lubrication system is pulled into a wick because one end of the wick resides in the oil reservoir. The oil is then pulled up into the wick by capillary action and is dispersed on the moving parts that need to be lubricated. This is accomplished by the fact that the other end of the wick is usually in contact with the part that is lubricated. One example of the wick method is shown in **Figure 9-5**.

The Drip Method

The drip method uses gravity to assist in the lubrication of parts. See **Figure 9-6**. The lubrication reservoir is placed above the part that is to be lubricated. The lubrication is then regulated to the desired amount of flow with the use of a needle valve.

The Automatic Oiler System

Many facilities use an automatic oiler to lubricate their components. The oiler system controls the amount of lubrication through electronic means. An electronic time delay is usually used to control a pneumatic valve. Sometimes electronic counters are used in tandem with proximity switches or reed switches to control the amount of lubrication.

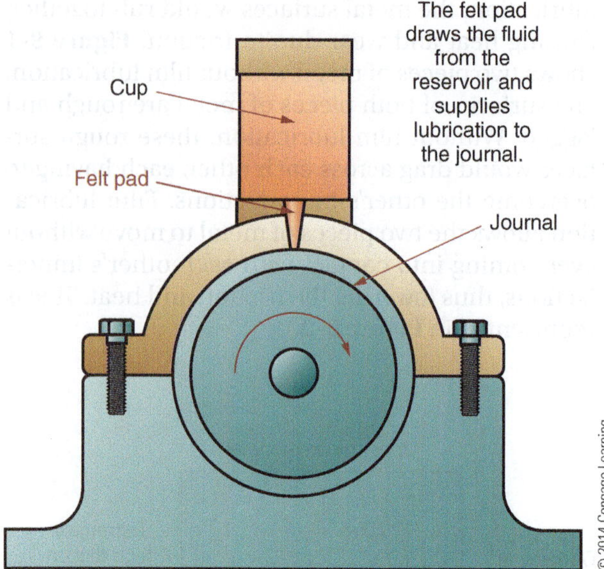

FIGURE 9-5 The wick method.

Notice how the needle valve regulates the drip flow that is used to lubricate the journal.
© 2014 Cengage Learning

FIGURE 9-6 The drip method.

This method ensures that the rate of lubrication increases as the speed increases. The proximity switch counts the keyway every time a revolution is made. If the revolutions per minute are increased, the desired count will be achieved sooner, thus oiling the component sooner. If the revolutions per minute decrease, the desired count will come later, thus oiling the component later. This system is desirable because components have to be lubricated more often when they are running at higher speeds.

The automatic lubricator, which can be seen in **Figure 9-7**, comes in different sizes and lubricating methods, and it will supply lubrication at a desired rate.

Bearings that are not lubricated sufficiently fail long before they have reached their life expectancy. These lubricators guarantee a consistent supply of lubricant in the pre-adjusted proportioning. Shutdown times are decreased and costs in comparison to time-consuming lubrication by hand are clearly reduced.

These automatic lubricators typically thread directly into the point where the grease fitting is installed and can be installed in seconds. They are powered by a gas-producing dry cell that builds pressure behind the piston within the lubricator cylinder, which dispenses the lubricant automatically and evenly in the lubrication point.

The amount of lubricant can easily be adjusted by means of an Allen key. Filled with oil, these automatic oilers manage the automatic lubrication of chains, open gears, or guide rails, for example.

FIGURE 9-7 Simalube lubricator.

Methods of Grease Application

Grease is usually applied using one of four methods: the grease gun, the grease cup, the centralized system, or the grease block application. The grease gun is a small hand-operated pump that will pump grease out of a tube and into a grease fitting. It is important to state that in order to ensure that no contaminants enter the grease chamber, the grease fitting should be cleaned off before attaching the grease gun to it. Once the fitting is cleaned and the grease gun is attached, the handle of the pump is slowly pulled downward, thus applying pressure to the grease that is in the grease tube and causing the grease to flow out of the grease gun into the fitting. Once the grease flows through the fitting, it will flow into the grease chamber, where it will be used to lubricate the necessary components. Some grease guns do not have a hand pump or even a tube of grease. This type of grease gun uses compressed air to pump grease out of a 5-gallon bucket.

The grease cup uses a receptacle that contains the grease that will be used to lubricate the bearings. The grease cup is usually mounted just above the part that is to be lubricated. Pressure is developed within the cup when the cap is put on the reservoir. This causes the grease to be pushed into the bearing. A frequent check on this type of system is necessary to ensure proper operation.

A centralized system is often used in industry to lubricate machinery. A centralized system uses a reservoir, compressed air, and tubing to transfer grease to many locations. The grease is placed into the reservoir. The reservoir is then pressurized with positive air pressure, which causes the grease to be forced out of the reservoir to the parts that need lubrication. This type of system needs to be thoroughly inspected frequently to ensure proper operation.

The use of a grease block is usually reserved for large journals. A grease block is nothing more than grease that has a heavy amount of thickener in it, allowing the grease to be shaped into a block. It is not uncommon for a large journal to have a diameter of 6 in. or more. These shafts are usually supporting an extremely large amount of weight. It is not uncommon to find these journals just sitting on a plain bearing. See **Figure 9-8**. In this configuration, the grease block is placed atop the journal. As the temperature rises in the journal because of friction between the journal and the plain bearing, the thickener in the grease block will begin to melt, thus allowing the lubricant to lubricate and cool the journal, as it is needed. This type of application has several drawbacks.

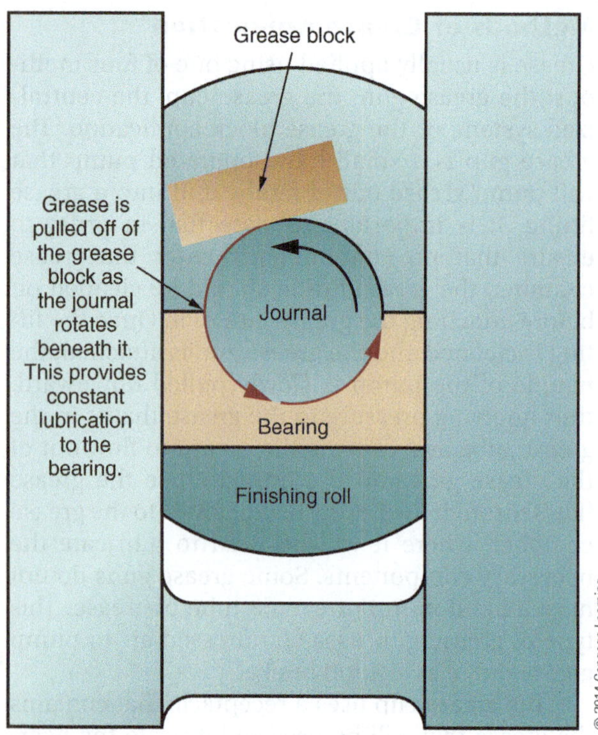

FIGURE 9-8 Grease block application.

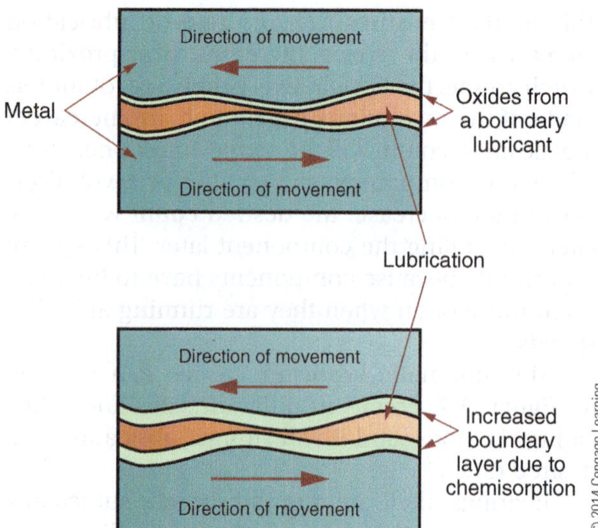

FIGURE 9-9 Boundary lubrication.

One is that the lubrication is exposed to contaminants. The second is that once the thickener has begun to melt, allowing the lubrication to flow, the excess grease that flows off the journal and bearing must be contained. This requires much attention to ensure that the grease trap does not overflow.

Boundary Lubrication

Boundary lubrication is the condition of lubrication in which the friction between two surfaces, which are in motion, is determined by the properties of the surfaces and the properties of the lubrication that is used, excluding viscosity. Boundary lubrication is a thin layer of lubrication that has been made up of properties of both the lubrication and the metal, and it resides between the metal and the lubrication. When lubrication is added to the surface of a metal, the lubricant and the metal begin to share molecules by exchanging electrons. Boundary lubrication is attributable to **chemisorption**, which occurs when weak chemical bonds are formed between liquid or gas molecules and a solid surface. This provides a secondary boundary that further protects the metal in the event that the first boundary (film) is lost. However, the second boundary should not be relied on as the sole means of lubrication. A graphical representation of boundary lubrication is shown in **Figure 9-9**.

9-4 LUBRICATION SCHEDULES

A rigorous lubrication schedule is recommended for each machine in the facility. It is common practice for a facility to have a lubrication technician. This is usually a mechanic who is very knowledgeable in the subject of lubrication. The lubrication technician works closely with the safety administrator of the facility to ensure that all Material Safety Data Sheets (MSDS) are in the proper location and are kept up to date. The lubrication technician uses a wide variety of chemicals and must be aware of the procedures that must be followed when in contact with these chemicals.

The lubrication technician should have a regular maintenance schedule for each machine in the facility and should be aware of the preventive maintenance (PM) scheduling for each machine. This is so the lubrication PM can be coordinated at the same time as the mechanical or electrical PM. The lubrication technician must be ready to inspect everything on a machine that requires lubrication when PMs are being performed on a given machine. On inspection, the lubrication technician may find that some lubricants may need to be replenished or replaced. It is at this time that all bearings should be inspected for overheating. If caking up of lubricant is noticed, a closer inspection of the bearing may be required. This is the time for the lubrication technician to notify the PM mechanic of a bearing that is overheating. Failure to replace a bearing during a scheduled shutdown will more than likely cause an unscheduled shutdown later. It is best to change the bearing while the machine is already down.

An efficient facility will have an aggressive lubrication schedule. Depending on the size of the facility, there may be two or even three lubrication technicians who maintain different portions of the plant. Having more than one lubrication technician saves the facility money in the long run by preventing downtime caused from lubrication problems.

The lubrication technician should have a list of every machine that needs lubricating. Some machines must be lubricated more often than others because they will consume lubricant more quickly. The lubrication schedule should be made according to these demands. For example, a machine that uses grease for lubrication probably would not have to be inspected as frequently as one that uses oil, because grease is not consumed as fast as oil. And machines that run at high speeds need to be inspected more often than those that run at low speeds because the lubricant tends to be consumed or evaporate due to the heat that is produced while running machines at a higher speed.

It should also be the responsibility of the lubrication technician to inspect all centralized lubrication systems on a regular basis to ensure their proper operation. This is probably going to be one of the most important systems the lubrication technician will have to monitor. The centralized system for a single-shift facility should be inspected once a day. For a plant or facility that runs 24 hours a day, 7 days a week (better known as 24/7), the centralized system should be checked at the beginning of every shift.

Most facilities do not put a strong emphasis on lubrication. A strong and aggressive lubrication program along with an efficient lubrication schedule put into effect by a facility would almost guarantee an increase in profit because there would be a decrease in downtime caused by bearing failure.

It is important for the maintenance technician to keep a log of the MSDS for each type of lubricant that is used by the facility. The MSDS log should be readily accessible in the event of an emergency.

SUMMARY

- Here are a few terms that should be known:
  a. Lubricant: A substance that reduces friction by providing a smooth surface of film over parts that move against each other.
  b. Viscosity: The internal friction of a lubricant, caused by the molecular attraction.
  c. Hydrocarbon: Any compound containing mostly hydrogen and carbon. A pure hydrocarbon contains only hydrogen and carbon.
  d. Semisolid: A gel-like substance that has the characteristics of both a solid and a liquid.
- Lubrication is used to provide a smooth layer of film between parts that move against each other, thus reducing friction. This allows the moving parts to stay cooler as they move against each other. Most components that have friction overheat if they are not lubricated.
- Lubricants come in four forms: solid, semisolid, liquid, and gas.
- Lubricants are developed from a hydrocarbon base, animal oils, or vegetable oils.
- Animal oils or vegetable oils are primarily used in the food industry because of the toxic nature of the petroleum-based lubricants.
- Additives are used to improve the properties of lubricants.
- Viscosity is the measurement of the resistance of a fluid's molecules while moving past each other.
- A semisolid lubricant is one with the consistency of grease; it is neither a solid nor a liquid.
- Oil is applied by several methods: submersion, the splash method, the wick method, the drip method, an automatic oiler, a grease gun, a grease cup, a centralized system, and grease block application.
- A rigorous lubrication schedule is recommended for each machine in the facility.
- Ensure that all MSDS are in the proper location and are kept up to date.
- It should also be the responsibility of the lubrication technician to inspect all centralized lubrication systems on a daily basis for a single-shift facility and at the beginning of every shift for a plant or facility that runs 24/7 to ensure proper operation.
- A facility that puts a strong and aggressive emphasis on lubrication, a lubrication program, and an effective lubrication schedule will see a decrease in downtime and an increase in profit.

REVIEW QUESTIONS

1. Define viscosity.
2. What is the purpose of lubrication?
3. What is an additive when referring to lubrication?
4. Which has the higher viscosity, grease or oil?
5. List at least eight methods used to apply lubrication.

CHAPTER 10

Seals and Packing

Seals prevent the escape of fluid from a cavity when a shaft penetrates the wall of the cavity. Most commonly, the shaft has a rotary or linear movement. If a seal is not installed correctly or maintained properly, it may fail, causing the loss of fluid. The two main functions of a seal are to keep the fluid in while keeping dirt and debris out.

OBJECTIVES

After studying this chapter, the student should be able to

- List three types of seals.
- List the different materials that packing materials are made of.
- Explain the purpose of impregnating a packing material with lubricants.
- Name four types of packing materials.
- Properly install all of the components of a stuffing box.

10-1 PACKING SEALS

Seals are used to contain a fluid in a bodied vessel or cavity at any point where the wall of the vessel or cavity is compromised by something that passes through it. This is usually a shaft or rod of some sort. An example of this is the input and output shafts on a gearbox. The seals that are used in a gearbox allow the shafts to pass through the wall without the loss of the oil that is so vital to keeping the gears cool. There are many other applications for seals. In this chapter, a few of these applications are covered as well as the types of seals that are used in industry today.

Packing Material

The packing material that is used in packing seals is sometimes referred to as **compression packing**. Most compression packing is impregnated with a variety of lubricants to make the packing material resistant to the absorption of the fluid that it is trying to contain. Most compression packing is made of various types of fibers, including vegetable, mineral, animal, and synthetic fibers. The most common mineral fiber, asbestos, is slowly being removed, through attrition, because of its health hazard to human lungs. This type of packing was popular because of the strength that asbestos mineral offers; however, it has been determined that asbestosis (a lung disease) can develop from heavy exposure to fibrous asbestos. Now most packing is made of resilient natural or synthetic fibers that are twisted or braided. Sometimes these twists or braids are covered in graphite. This is referred to as being graphited. There are four different types of packing:

- Twisted fiber packing
- Square-braided packing
- Braid-over-braid packing
- Interlocking packing

The twisted fiber is the most widely used type of packing. It is usually made with twisted strands of cotton that have been lubricated with mineral oil and graphite. Because the twisted type of packing is simply twisted and not interlocked, it is not as strong as the other types of packing.

The square-braided packing can be made of cotton, plastic, or leather, and it is manufactured to have a square-shaped cross section. It is not uncommon for square-braided packing to have metal wires within the strands of packing material to add strength and to give the packing better shape-holding characteristics. Because this type of packing is usually used in heavier applications than the twisted-type packing, square-braided packing is impregnated with oil or grease. This packing is generally stronger than the twisted type of packing because the fibers are braided. The braiding also limits the unraveling of the fiber, which may occur with the twisted-type packing.

The braid-over-braid type of packing has a circular cross section. The center is sometimes composed of lead wires, which give strength to the packing and give the packing better shape-holding characteristics. The lead wires are covered with a braided jacket, which in turn is covered with another braided jacket. It is for this reason that the braid-over-braid type of packing is sometimes referred to as a jacket-over-jacket. This type of packing is also impregnated with lubricants.

The interlocking type of packing is the strongest of all the packing types. It has interlocking fibers that are resistant to fraying or unraveling. This type of packing has a square cross section that is impregnated with a lubricant as well. This type of packing is used for heavy-duty applications where it would not be desirable to use the other types of packing.

All of these packing materials are usually sold in rolls of different lengths. The packing material has to be cut to the appropriate length. The proper procedure for cutting the packing material is stated in the installation section of this chapter.

Packing Seal Parts and Assemblies

The most common type of seal that is used is the **stuffing box seal**. This is one of the first types of seals to be used. The stuffing box seal is made up of three parts:

- Packing chamber
- Packing rings
- Stuffing gland

The principle is fairly simple. A sealing action is accomplished by compressing the packing material between two surfaces. Leakage is prevented once the packing material is compressed against the packing chamber, which is commonly referred to as a box, and the shaft. It is compressed by the stuffing gland, which is sometimes referred to as the follower.

On some occasions, it is desirable to have a small amount of leakage through the packing material. This keeps the surface of the shaft or rod lubricated as it rubs or rotates against the packing material.

FIGURE 10-1 A lantern ring allows the packing material to be lubricated by external means.

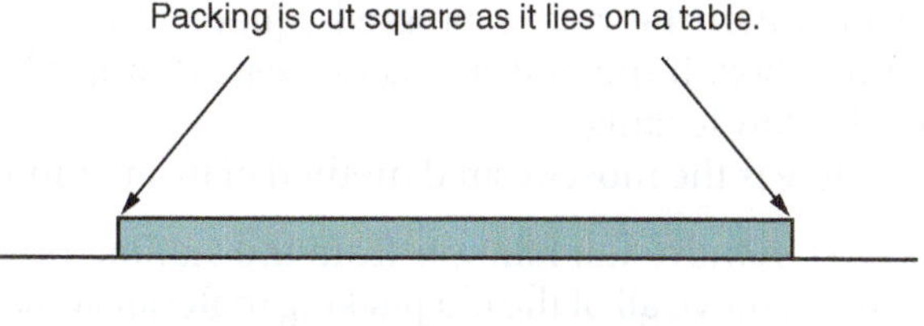

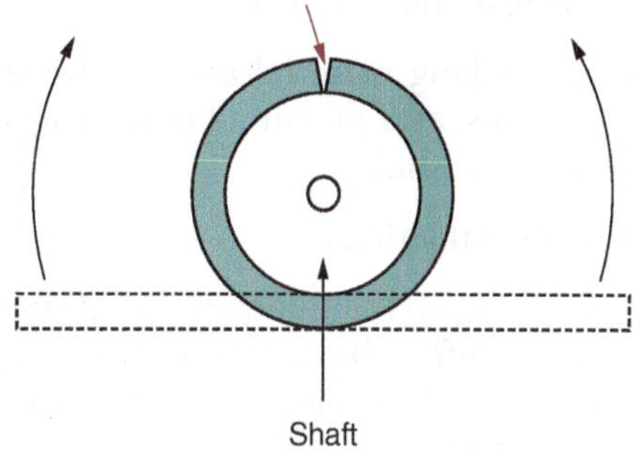

FIGURE 10-2 An improper method of cutting packing material.

There is usually some wear that occurs on the packing material and the shaft or rod during normal operation because there is friction present between the packing material and the shaft or rod. Because there is friction, material will wear off on both packing and shaft material. This has to be compensated for by making frequent adjustments to the follower. It is not recommended, when you are tightening the follower, to tighten it as far as you can. This would prevent any leakage at all. Allowing a small amount of leakage through the packing material slows the wear of both the packing and shaft material down to a minimum. Most packing seals consist of five rings. This ensures the proper operation of the seal. A lantern ring may be present in the stuffing box.

A **lantern ring**, which is shown in **Figure 10-1**, is used to allow the packing material to be lubricated by external means. The lantern ring, through its design, provides spacing between the third and fourth packing rings. This void can be filled with lubricant through a lubrication port, which should be in line with the lantern ring. It is important for the lantern ring to be installed properly and that it not be left out of the stuffing box. The lantern ring should be placed in the stuffing box in such a manner that its leading edge does not pass the lubrication port. If the lantern ring's leading edge passes the lubrication port, when the follower is tightened up, the entire lantern ring will move forward and pass the lubrication port. This does not allow the proper lubrication of the packing material. If the lantern ring is installed properly, it will be directly under the lubrication port after the follower has been tightened to its operating pressure.

Installation

As was mentioned before, the packing rings are cut into small rings and pressed into a stuffing box. The most common mistake that most mechanics make is during the cutting of the packing material. Most mechanics unroll the packing material onto a table, cut it to the desired length, wrap it around the shaft, and pack it into the stuffing box, overlooking a serious problem that is created when the packing material is cut in this manner. By laying the packing material on the table and cutting it perpendicularly across its center, a V-shaped gap is created when the packing is wrapped around the shaft. This is illustrated in **Figure 10-2.**

When the packing material is laid on the table and cut perpendicularly across its center, the length of the top side and the length of the bottom side are exactly the same. This creates a problem because the outside diameter of a ring is always larger than the inside diameter. This means that the top side should be longer than the bottom side. If the packing material is cut improperly, the gap will provide a place for fluid to leak through the packing material. The proper method for cutting the packing material to the desired length is to wrap the packing around the shaft that it will be used on or around one of the same diameter. Cutting the packing material after it has been wrapped around the shaft prevents the creation of a V-shaped gap.

Another common mistake is when the packing material is tamped into the stuffing box with anything other than a tamping tool. The tamping tool ensures that the ring is pressed squarely into the stuffing box. If the first ring is not seated properly, it will allow leaking.

Here is the most desired method of installation:

- First, remove the follower from the stuffing box, and remove all of the old packing material. Make sure to clean all surfaces (including the follower) before trying to install any new packing material.
- Wrap the packing material around the shaft at least five times, and precut all of the rings without scarring the shaft.
- Separate all of the rings.
- Set the first ring on the shaft, and gently tamp the ring into the stuffing box. Make sure the ring goes into the stuffing box squarely. This is to ensure that the packing does not bind or roll.
- Repeat the process for the next two rings, making sure to alternate the cuts. Be sure to firmly tamp each ring into position against the preceding ring. Alternating the cuts ensures that the seal will not leak as it would if all of the cuts were placed in line with each other. This is shown in **Figure 10-3.**
- Place the lantern ring (if one is used) on the shaft and gently slide it into position, making sure not to allow the front edge to pass the lubrication port.
- Install the final two rings, making sure to firmly tamp each ring into position.
- Install the follower. If everything is done correctly, the follower should barely enter the stuffing box.
- Tighten the stuffing box, alternating from side to side, until the follower has compressed the packing material to the point where it is about one-third of the way into the stuffing box. This allows the packing and lantern ring to occupy about two-thirds of the stuffing box. Remember, when tightening, it is important to keep an eye on the lantern ring through the lubrication port. Do not let it pass the port.
- Fill the vessel or cavity with the fluid to be used during operation. Monitor the seal. If there is an excessive amount of leakage, slowly tighten the follower until only the desired amount of leakage is present. If there is no leakage at all within a few minutes of filling the vessel or cavity, loosen the follower slightly until the desired amount of leakage is present.
- Allow the unit to run for a few hours, then check the leakage. The follower may need to be adjusted slightly after the unit has been run.

10-2 MECHANICAL SEALS

A mechanical seal eliminates excess leakage, at least to the naked eye. Some mechanical seals must have a small amount of leakage to ensure proper operation of the seal. Some leakage will still occur with a mechanical seal, but it will be of such a small amount that it most likely will not be visible. This is accomplished through the machining of the faces that are used in the sealing action to a very smooth finish. The mechanical seal is primarily used for rotary sealing and consists of three sealing points. **Mating rings** are used to accomplish this task. Each mating ring, one rotating and one stationary, has a face that is machined to extremely high tolerances. The rotating mating ring is made to be static with respect to the shaft, whereas the stationary mating ring is made static with respect to the housing. An inside seal design is illustrated in **Figure 10-4.**

Because pressure is present within the chamber, the rotary mating seal is held in place by the pressure of the fluid. This pressurized fluid applies a force to the rotating mating ring so that it presses against the machined surfaces of the mating rings, causing the sealing action. There are some occasions, however, when the use of springs is necessary because of low pressures. The springs in this case help provide the necessary force to cause the sealing action at the machined surfaces of the mating rings.

There are generally three types of mechanical seals: the inside seal, the outside seal, and the double seal. The **inside seal** is a seal in which the

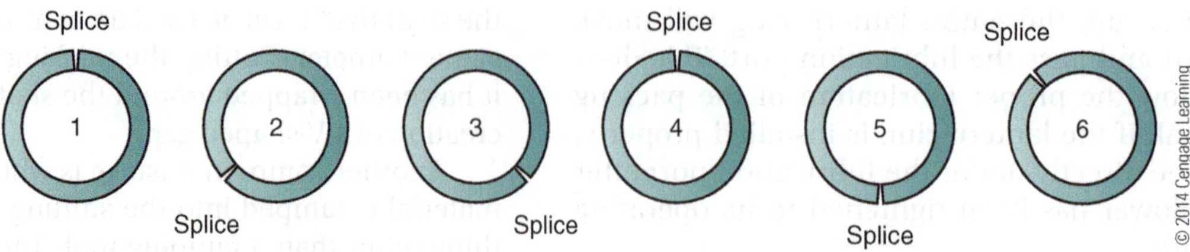

FIGURE 10-3 Alternate the splices.

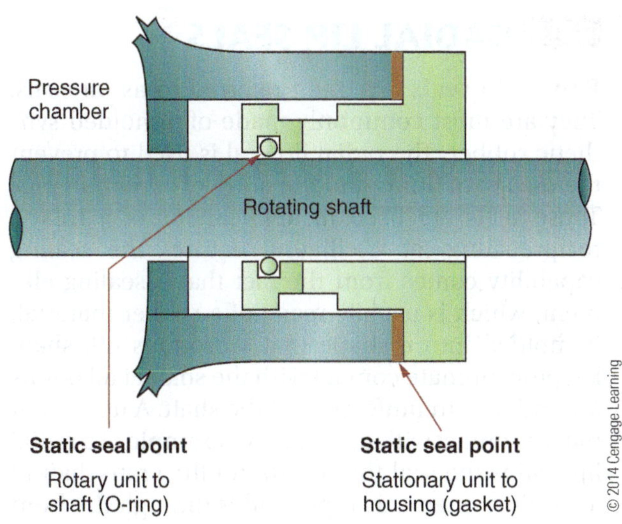

FIGURE 10-4 An inside mechanical seal.

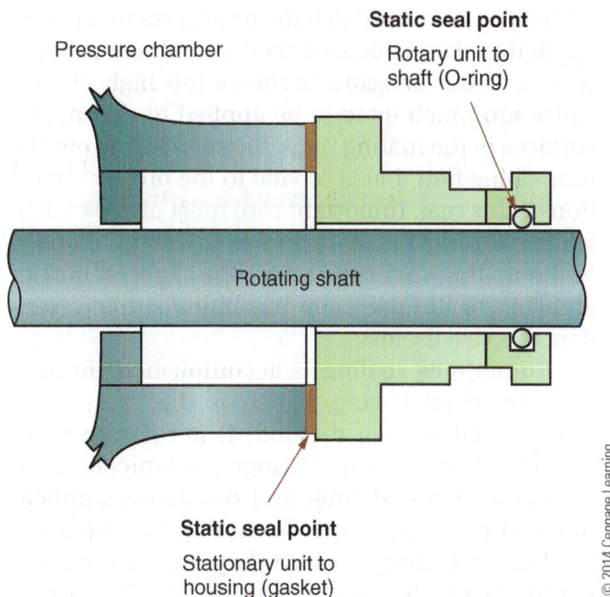

FIGURE 10-5 An outside mechanical seal.

rotating mating ring is internally mounted in the stuffing box. This is the most difficult configuration to install or assemble; however, because the pressure in the fluid chamber aids in the sealing action, it is a better seal in comparison to the outside seal. The **outside seal** is a mechanical seal in which the rotating mating surface is mounted externally from the stuffing box. This configuration makes for an easy installation, but the trade-off is that this seal is used on moderate to low pressures because the pressurized fluid is now working against the sealing surfaces. This configuration is shown in **Figure 10-5**.

A **double seal** is basically two inside seals that are in a back-to-back orientation. This seal, which is shown in **Figure 10-6**, is typically used for sealing hazardous liquids. This seal provides a high degree of safety because of its sealing operation. It uses a nonhazardous liquid in the stuffing box that is at a higher pressure than that of the liquid it is sealing. This ensures that any leakage will be directed inward, toward the fluid that is being sealed, thus preventing a hazardous condition.

It is important that mechanical seals are never run while they are dry. This causes damage to the precision-machined surfaces of the mating faces, which would reduce their sealing capabilities.

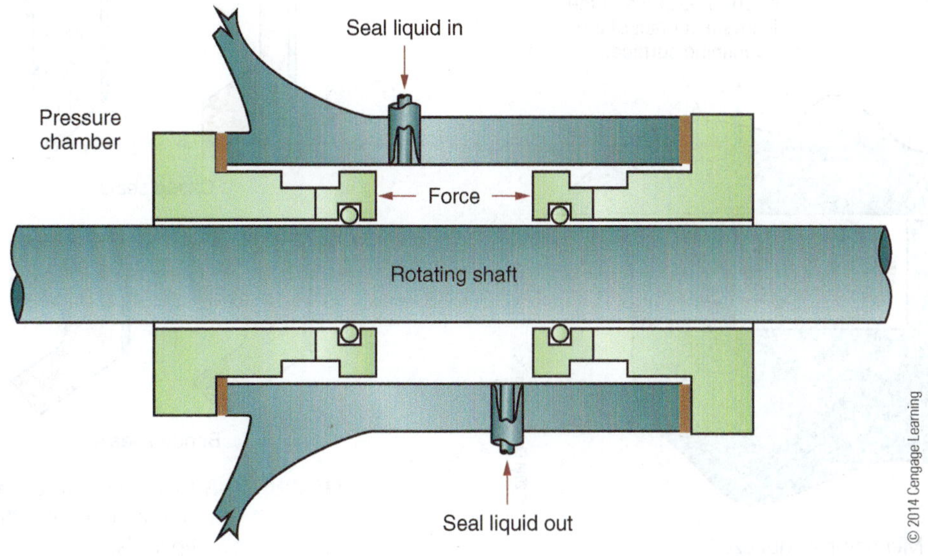

FIGURE 10-6 A double mechanical seal.

It is just as important that too much pressure not be applied to the inside seal via the fluid that is being sealed. If the pressure becomes too high, it may cause too much force to be applied to the mating surfaces of the mating rings, thus squeezing out the lubricating film that is so vital to the proper operation of this seal. Important care must also be taken while handling the mating rings so that no damage to the mating surfaces occurs. The slightest nick or abrasion to the precision-machined surfaces renders this seal useless.

Sometimes sealing is accomplished through the use of an O-ring, which is the most commonly used sealing method. It is referred to as squeeze-type packing. O-rings are typically used in rotary, reciprocating, and oscillating applications. O-rings are slowly becoming the preferred method of sealing because they are inexpensive, readily available, and easy to install. The O-ring is made of a synthetic rubber that is very flexible. The O-ring gets its name from its basic shape. O-rings, however, may come in different shapes and sizes. The principal operation of the O-ring can be described as controlled deformation. This deformation occurs because all O-rings are manufactured to allow an initial squeeze of approximately 10% of their cross-sectional dimension. This basically means that the O-ring is manufactured to be larger than necessary, causing it to roll and distort when pressure is applied. The deformation squeeze flattens the O-ring into intimate contact with the confining surfaces. This contact with the confining surfaces provides the sealing action. Mechanical squeeze and deformation is represented in **Figure 10-7**.

10-3 RADIAL LIP SEALS

Radial lip seals are often referred to as oil seals. They are most commonly made of a molded synthetic rubber. The radial lip seal is used to prevent the escape of fluids and the entry of contaminants. These seals are used in applications of a rotary, reciprocating, or oscillating nature. The sealing capability comes from the fact that a sealing element, which is usually made of a rubber material, is molded into a shape that surrounds the shaft, keeping intimate contact with the shaft at all points around the circumference of the shaft. A lip is what makes contact with the shaft. An example of a radial lip seal is the seal that is around the crankshaft of a gasoline engine as it protrudes through the front of the engine block. Another example is the rear main seal.

Three broad categories of radial lip seals are the cased seals, the bonded seals, and the dual-element seals. The sealing element of the cased seal resides within a case that holds the seal itself. The bonded seal holds the seal because the sealing element is chemically bonded to the metal casing. Refer to **Figure 10-8**. The dual-element seal is a seal that, in some form or another, has two lips that remain in contact with the shaft at all times. As was mentioned earlier, seals are used to prevent the escape of fluids and the entry of contaminants. A seal that is used to prevent an escape of fluid is said to be used for retention, whereas a seal that is used to prevent contaminants from entering the fluid cavity is said to be used for exclusion. A radial lip seal that has only one lip can only be used either for retention or for exclusion, but not both. If the seal is installed

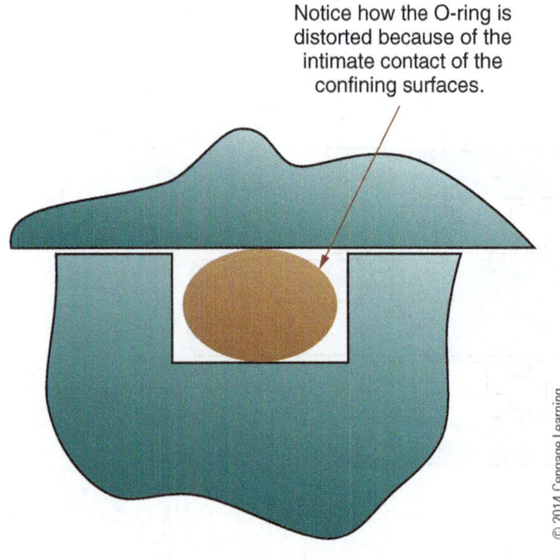

Notice how the O-ring is distorted because of the intimate contact of the confining surfaces.
© 2014 Cengage Learning

FIGURE 10-7 Mechanical squeeze.

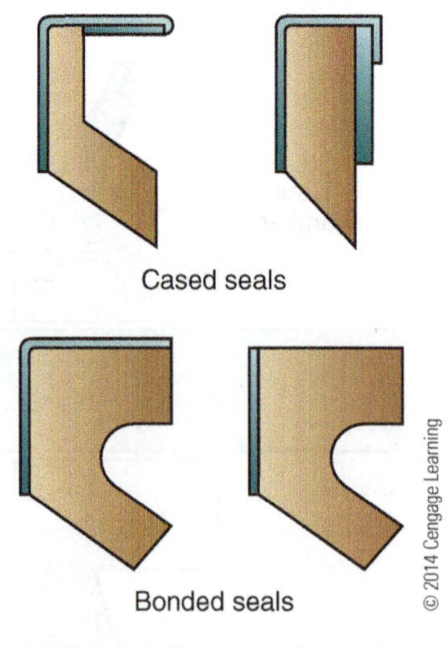

FIGURE 10-8 Cross sections of two single-element radial lip seals.

in a manner that makes the lip face inward, toward the fluid, it is being used for retention. If the lip is facing outward, the seal is being used for exclusion. A dual-element seal, however, does both retention and exclusion. If a radial lip seal is being used for exclusion, the fluid must not be under a lot of pressure. This is because the seal lip is facing outward, away from the fluid. It is quite possible that if the fluid builds enough pressure, it will overcome the tension of the radial lip seal. The seal will then lose contact with the shaft, causing leakage.

It is important that the surface of the shaft be smooth and free of debris before attempting to install a new radial lip seal. Any dirt or scars that may be on the shaft may cause damage to the lip of the seal. Most radial lip seals have a pressed fit. This means that they usually have to be tapped into the housing in which they reside. It is critical that this is done correctly to avoid damaging the seal. If the seal is not carefully knocked into the housing, it may become warped or bent. If this occurs, the seal will probably leak on the side with the dent or bend. Use a device such as a thimble or a sleeve that has a large enough diameter to set on the face of the casing. If a thimble or a sleeve cannot be found, a bearing knocker, a bushing, or a block of hard wood may substitute just as well. Be careful not to use an oversized or undersized device to drive the seal into the housing. As you begin to drive the seal into the housing, gently tap with a mallet squarely on the installation tool that is being used. If you do not tap squarely, the seal may become cocked in the housing, causing damage to the seal. Frequently remove the installation tool and check to ensure that the seal is going into the housing squarely. If it is, continue to gently tap the installation tool until the seal is firmly set in the housing. It is easy to know when the seal is completely installed. As the mallet strikes the installation tool, it feels like a solid hit. If the seal is not completely in the housing, there is a springy feeling when the mallet strikes the installation tool. Never strike the seal directly with a hammer or mallet. Also, never apply pressure to the soft portion of the seal because this may cause damage to the sealing element.

If installed correctly, the radial lip seal will not leak at all. Radial lip seals do however, begin to show signs of stress after a long period of time if they are not maintained, and especially if they are exposed to a lot of heat. When this occurs, the radial lip seal begins to leak. The seal should be checked and cleaned regularly to prolong its life. The seal should be replaced if leaking is noticed upon inspection, if it looks dry and brittle, if it is hard to the touch and has lost its flexibility, and/or if it has numerous nicks and cuts. Nicks and cuts occur because of failure to keep dirt and debris off the seal. If dirt and debris are allowed to accumulate on the surface of the seal, it will more than likely fail prematurely.

SUMMARY

- Seals are used to contain a fluid in a bodied vessel or cavity at any point where the wall of the vessel or cavity is compromised by something that passes through it.
- The three different types of seals are the packing seal, the mechanical seal, and the radial lip seal.
- Most compression packing is made of vegetable, mineral, animal, and synthetic fibers.
- Most packing is impregnated with a variety of lubricants to ensure that the packing material is resistant to the absorption of the fluid that it is trying to contain.
- There are four different types of packing: twisted fiber packing, square-braided packing, braid-over-braid packing, and interlocking packing.
- The stuffing box seal is made up of three parts: the packing chamber, the packing rings, and the stuffing gland.
- On some occasions, it is desirable to have a small amount of leakage through the packing seal. This keeps the surface of the shaft or rod lubricated as the packing material rubs or rotates against the packing material.
- A mechanical seal eliminates excess leakage, at least to the naked eye. The mechanical seal is primarily used for rotary sealing and consists of three sealing points.
- There are generally three types of mechanical seals: the inside seal, the outside seal, and the double seal.
- It is important that mechanical seals are never run while they are dry. Running a mechanical seal while it is dry causes damage to the precision-machined surfaces of the mating faces, therefore reducing their sealing capabilities.
- An O-ring is a squeeze-type packing, and it is typically used in rotary, reciprocating, and oscillating applications. O-rings are slowly becoming the preferred method of sealing because they are inexpensive, readily available, and easy to install. The O-ring is made of a synthetic rubber that is very flexible.

- The radial lip seal is most commonly made of a molded synthetic rubber. The sealing capability comes from the fact that a sealing element is molded into a shape that surrounds the shaft while keeping intimate contact with the shaft at all points around the circumference of the shaft. A lip is what makes contact with the shaft, hence the name radial lip.
- Three broad categories of radial lip seals are the cased seals, the bonded seals, and the dual-element seals.

REVIEW QUESTIONS

1. List two functions of a seal.
2. Which fiber is slowly being eliminated through attrition, and why?
3. What is meant by the term *graphited*?
4. What is a lantern ring? State its purpose.
5. In reference to the O-ring, what is deformation and what is the general amount of deformation that occurs during the initial squeeze?

CHAPTER 11

Pumps and Compressors

Pumps and compressors are used throughout industry as well as in much of the world. It is important to be able to produce the flow and pressure of a fluid. In this example, fluid is described as a liquid or a gas. A pump is used to move a liquid or a gas in large volumes, amounts that would not be possible with atmospheric pressure alone. Compressors are used to pressurize air to levels that are above atmospheric pressure. Once the flow of a fluid is produced or air has been pressurized, the fluid can be used in a process either as an ingredient or to produce work.

OBJECTIVES

After studying this chapter, the student should be able to

- List the types of pumps that are in use today.
- Explain and calculate volumetric efficiency.
- Explain and calculate the delivery of a pump.
- Discuss some of the common problems that may be encountered while troubleshooting a pump.
- List the most common types of compressors that are used in industry, and explain their operation.
- Discuss some of the common problems that may be encountered while troubleshooting a compressor.

11-1 HYDRAULIC PUMPS

A **pump** is a mechanical device that changes mechanical power into fluid power. A **positive displacement pump** is a pump that allows very little leakage through its internal components. This is because the components of the pump fit so closely together. Because of the tight fit of these components, a positive displacement pump can produce very high pressures in the fluid that is being pumped. The ability to pressurize the fluid to higher pressures depends on the tolerance of the components within the pump—the closer the tolerance, the higher the pressure capabilities. The word **displacement** refers to how much fluid a pump can move in a single rotation. The displacement of a pump is usually expressed in cubic inches per revolution (CIR). The flow volume of a pump is rated in gallons per minute (gpm) at certain revolutions per minute (rpm); therefore, the flow volume is proportional to the shaft speed of the pump.

There are generally three types of positive displacement pumps in use today: the vane pump, the gear pump, and the piston pump. Of these, the piston pump offers the best tolerances among its internal components. Therefore, it is generally considered to carry a higher pressure rating than the others. There are other types of pumps that are not positive displacement pumps. These are the impeller and the centrifugal types. These two types of pumps have very poor tolerances among the internal components, thus allowing too much fluid slippage to be considered positive displacement pumps. These pumps produce a pressure that is too low for most hydraulic systems.

Vane Pumps

A vane pump has vanes in the rotor of the pump. These vanes slide in and out of their slots as the rotor rotates, causing fluid to be moved from the input port to the output port of the pump.

Fixed Displacement Vane Pump

Figure 11-1 shows the internal construction of a fixed displacement vane pump. The vanes are pressed against a cam surface either by centrifugal force or by springs that are placed beneath the vanes themselves. While operating at low speeds, vane pumps that rely on centrifugal force alone do not perform as well as vanes that are spring loaded. In a fixed displacement vane pump, the cam is machined into the housing of the pump. This cam is eccentric (off center) to the centerline of the rotor shaft. Because the cam is machined in an eccentric fashion, the vanes move in and out as they are rotated around.

To understand the principle of operation, the internal portion of the pump must be broken into four different sections: bottom dead center, the inlet side, top dead center, and the outlet side. Beginning with the bottom dead center, all of the vanes that are in contact with the cam at bottom dead center are pressed completely into their respective slots. The rotor, which houses the vanes, is against the cam itself. Because the rotor is against the cam and the vanes do not protrude from the rotor, there is no cavity in which fluid can be transferred. As the rotor turns off bottom dead center (in a clockwise motion), the top surface of the vanes remains in contact with the cam. However, because the rotor is of a smaller diameter than the cam and the rotor is eccentric to the cam, a void begins to appear between the vanes. A vacuum is also produced as the cavities between the vanes get larger. The vanes are now off bottom dead center and are at the inlet side of the pump. Fluid is pulled into the opening cavities due to the vacuum, thus they fill to their maximum capacity as the rotor continues to turn. Once the rotor has rotated to top dead center, the vanes are extended to their maximum potential and fluid is trapped in the cavities. As the rotor continues to turn, the vanes leave top dead center and enter the outlet side of the pump. The fluid is forced out of the cavities by the vanes beginning to retract back into their slots. As fluid is forced out of the cavities, it is pushed out the outlet side of the pump. The rotor continues to rotate and the vanes return back to bottom dead center. At this time the pump has made one complete revolution.

FIGURE 11-1 A fixed displacement vane pump.

Variable Displacement Vane Pump

A pressure-compensated variable displacement vane pump, shown in **Figure 11-2**, is very similar

FIGURE 11-2 A variable displacement vane pump.

to the fixed displacement vane pump with several exceptions.

- The cam is not machined directly into the housing. It is a movable ring that is placed within the housing and held in its eccentric location with a compensator spring.
- The tension on the compensator spring can be adjusted.
- There is a maximum volume adjustment that is mounted on the housing. This maximum adjustment changes the area of the cavities, making them smaller for minimum displacement and larger for maximum displacement.

When the cam ring is in the far left position, the pump is at its maximum displacement and produces maximum flow. When hydraulic pressure is developed, it pushes against the cam ring, causing the cam ring to move toward the right. If enough pressure is present to make the cam ring move into a concentric position with respect to the rotor, the pump output will be reduced to zero. If the compensating spring is adjusted inward (tightened), it will take more system pressure to overcome the spring. If the compensating spring is adjusted outward (loosened), it will take less system pressure to overcome the spring. The pressure at which the spring is overcome is known as the firing pressure. The maximum is used when the driving motor or engine is not large enough, in horsepower, to run the pump at full displacement.

Balanced Vane Pump

A balanced vane pump is shown in **Figure 11-3**. The vane pumps that have been discussed up until now have been pumps in which the vanes extend and retract only once per revolution. In the balanced vane pump, the vanes extend and retract twice per revolution.

FIGURE 11-3 A balanced vane pump.

Where the cams were circular but placed in an eccentric position in the other types of vane pumps, the cam surface on a balanced vane pump is somewhat elliptical, allowing cavities on both sides of the rotor instead of on one side only. This is why the vanes extend and retract twice instead of once. The balanced vane pump has some benefits that the others do not have; for example, the bearing load on this type of pump from internal pressure is minimal (almost nonexistent) because the load produced by one pressure chamber is offset by the load from the opposite pressure chamber. Thus they are balanced. Because of their balanced construction, balanced vane pumps do not usually offer variable displacement or pressure compensation.

Gear Pumps

Gear pumps are widely used because of their cost and durability. Because gears are solid in nature and cannot change their shape or size, gear pump displacement is fixed. The only thing that can change while using a gear pump is the output flow. This can be accomplished by changing the speed of the pump. There are four types of gear pumps that are discussed in this chapter: the external gear pump (sometimes referred to as a gear-on-gear), the lobe pump, the internal gear pump, and the gerotor pump.

External Gear Pump

The external gear pump, which is shown in **Figure 11-4**, has two gears that mesh externally. This means that the each gear is external to each

FIGURE 11-4 An external gear, or gear-on-gear, pump.

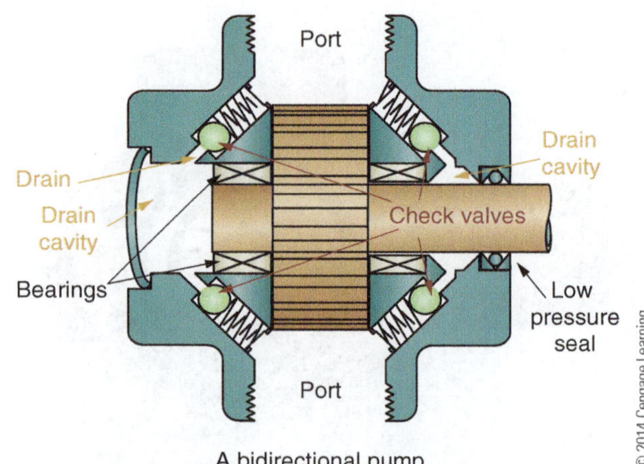

FIGURE 11-5 A bidirectional gear pump.

other, unlike in an internal gear pump, which has one gear inside the other.

The external gear pump is usually constructed of a set of spur gears, helical gears, or herringbone gears. Figure 11-4 shows a gear pump with a set of spur gears. The herringbone gear pump offers less noise and vibration than that of the spur gear configuration. The helical gear configuration offers less noise than the spur gear as well but is not as quiet as the herringbone gear. Most gear pumps can only be run in one direction. They should be marked by the manufacturer as to whether they were designed for use in a clockwise (CW) or counterclockwise (CCW) direction when looking into the shaft end of the pump. Running a gear pump in the wrong direction can cause the shaft seal to blow out, thus damaging the pump and possibly causing bodily injury. This is because the output side of the pump is usually the high-pressure side of the pump. If the direction of rotation were changed, the inlet port would become the outlet port. The shaft seal is usually not a high-pressure seal because it is on the inlet side. Because of this, it would probably blow out if the direction of the pump were backwards. There are, however, some gear pumps that are manufactured to be bidirectional. These pumps are probably manufactured with four internal check valves that would prevent the seal from blowing out because of high pressures, as shown in **Figure 11-5**.

SAFETY

Each of the gears in a gear pump has a name. The gear that is connected to the drive shaft of the motor is called the drive gear, whereas the gear that is meshing with the drive gear is called the driven gear. The teeth on the gears mesh, making a tight, sliding fit as they turn in the housing. This meshing of the teeth separates the inlet port from the outlet port. As the gears rotate, suction is created at the inlet port as the rotating gears carry fluid away. The fluid then becomes trapped between the teeth of both gears and is forced around the outside of the gears. After the fluid is forced around the outside of the gears, it is forced through the outlet port of the pump by the fluid that is coming in behind it.

The lobe pump works much like the gear pump with several differences:

- The lobe pump offers more volume (displacement) than that of a gear pump of the same size because the fluid chambers are larger in the lobe pump.
- The lobe pump does not seal the inlet from the outlet as effectively as the gear pump does.
- The lobe pump has more slippage and leakage than the gear pump.

The use of this pump is usually reserved for high-volume applications when pump size is an issue.

Internal Gear Pump

The internal gear pump, has one gear inside another, as shown in **Figure 11-6**.

The inner gear is usually the drive gear. The inner gear usually has one less tooth than the outer gear. As the drive gear rotates, it meshes with and carries the outer gear. This pump has a crescent-shaped divider that separates the inlet side from the outlet side. As the drive gear turns, it creates a vacuum at the inlet port, causing fluid to be pulled into the pump from the reservoir. The fluid is then carried in the cavities between the teeth of the gears past the crescent divider to the outlet port of the pump. The fluid is forced out of the pump from the teeth that are meshing back together. This type of

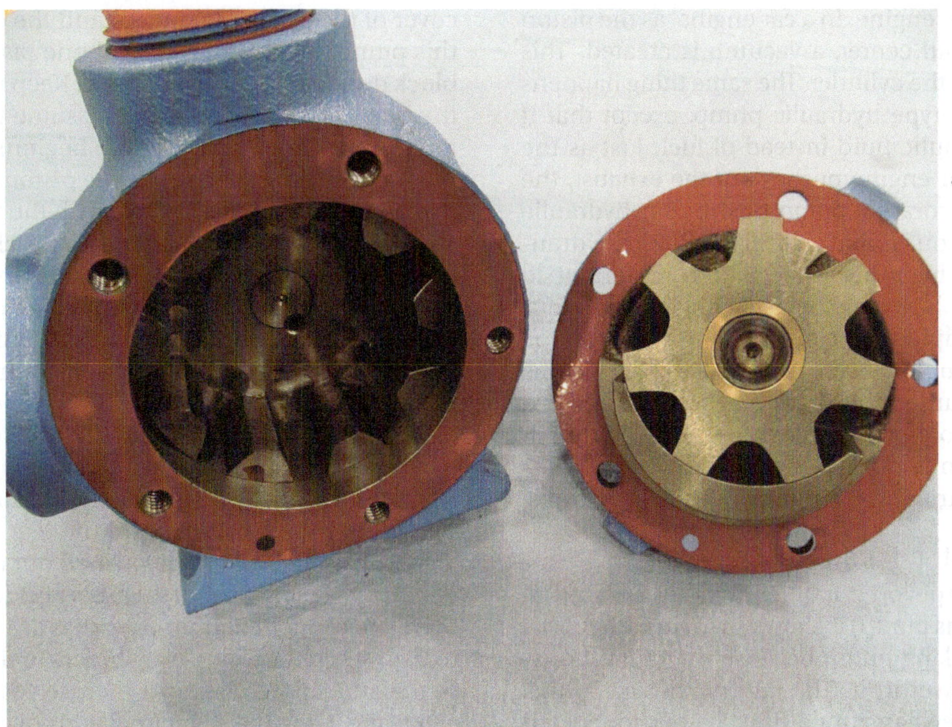

FIGURE 11-6 An internal gear pump.

pump forces more gallons per minute (gpm) and pressure compared with an external gear pump of the same size and produces a constant flow of fluid that has no pulsations. The internal gear pump is more expensive than the external gear pump.

Gerotor Gear Pump

The gerotor-type gear pump is very similar to the internal gear pump. It is shown in **Figure 11-7**. Notice that the shaft is connected to the inner element.

FIGURE 11-7 A gerotor gear, or gear-in-gear, pump.

As the inner element turns, the outer element is turned within the pump housing. As with the internal gear pump, the inner element of the gerotor has one less tooth than the outer element. Notice that the drive element (inner) is eccentrically positioned in the outer element. As the inner element is rotated, each tooth maintains contact with the outer element at all times. However, because there is one less tooth on the inner element, a void is produced as the tooth reaches bottom dead center. This void is what carries the fluid from the inlet port to the outlet port. As with the other pumps, a vacuum is created at the inlet port because of the opening of a cavity as the tooth rotates. Fluid is forced out through the outlet port because of the meshing of the teeth of the elements. These pumps are very smooth in operation and very sensitive to dirty fluid.

Piston Pumps

Even though there are several types of piston pumps, they are manufactured to offer more pressure capabilities as compared with the vane and gear pumps. This is because the piston pumps have closer internal tolerances. These tight tolerances allow the piston pumps to operate at higher efficiencies than the vane and gear pumps. All piston pumps are manufactured with an odd number of pistons to reduce the amount of pressure ripple in the output flow. Piston pumps work in a manner similar to a car engine. In a car engine, as the piston leaves top dead center, a vacuum is created. This pulls fuel into the cylinder. The same thing happens with a piston-type hydraulic pump, except that it pulls in hydraulic fluid instead of fuel. Just as the piston in a car engine pushes out the exhaust, the piston-type hydraulic pump pushes out hydraulic fluid. This action is what creates the flow of hydraulic fluid in a hydraulic system. It is important to keep in mind, however, that a car engine takes four strokes to complete a cycle, whereas the piston-type hydraulic pump only takes two strokes. The pistons of a piston pump move back and forth in a block. The orientation of this block gives the different types of piston pumps their names. There are bent-axis pumps, straight-axis pumps, and radial pumps.

Bent-Axis Piston Pump

The bent-axis piston pump (sometimes referred to as an offset axis pump) is shown in **Figure 11-8**. This is an odd-looking pump because the pump housing has an offset in it. The internal design of this pump is somewhat simple. The shaft is connected to a thrust plate. The pistons have connecting rods that connect them to the thrust plate. The pistons reside in a block, which is held in place with what is referred to as a rear cover. This is like the head on a car engine. As the valves reside in the head in a car engine, the inlet and outlet ports reside in the rear cover of the pump. To understand the operation of this pump, follow the action of one piston and the block during one complete cycle. Keep in mind that the block is tilted at an angle. Assume that the piston is at top dead center before beginning to rotate the pump. As the shaft turns, the pistons are moved in and out as the block rotates with the thrust plate. Top dead center is when the piston is closest to the rear cover. As rotation begins, the pistons are pulled in an outward motion, away from the bottom of the cylinder. A vacuum is created and fluid is pulled into the cylinder. After the thrust plate and the block have rotated 180°, the piston has reached its maximum backward stroke. This is referred to as bottom dead center. At this point the cylinder is full of fluid. As the thrust plate and the block continue to rotate, the piston moves back toward the rear cover plate. This causes the fluid to be pushed out of the cylinder, thus causing flow. It is important to know that there are semicircular grooves that are cut into the rear cover connecting the cylinders to the inlet port or the outlet port. The ports themselves are located at 90° and 270°. The semicircular groove that is connected to the inlet port is cut into the rear cover (from approximately 5° to 175°) to allow fluid to be pulled into the cylinder as the piston moves from top dead center to bottom dead center. As the piston passes bottom dead center, the cylinder is not connected to either the inlet port or the outlet port. This prevents slippage. As with the inlet port, the semicircular groove that is connected to the outlet port is cut into the rear cover (from approximately 185° to 355°) as well. This allows fluid to be pushed out of the cylinder as it moves from bottom dead center to top dead center. When the piston passes top dead center, the cylinder is not connected to either the outlet port or the inlet port. It is important to know that the volume of flow is directly proportional to the rpm, the number of pistons, the bore of the cylinder, and the length of the stroke of the pistons. If this were not a variable displacement pump, the only way to vary the flow volume would be to vary the rpm of the pump. A variable displacement bent-axis piston pump has a volume control adjustment that changes the angle of the offset, usually ranging from 0° to 20°, thus changing the maximum allowable volume within the cylinder. As the angle is increased, the flow volume increases. As the angle is decreased, the flow volume decreases. If a variable displacement bent-axis pump were set at 0°, there would be no flow at all.

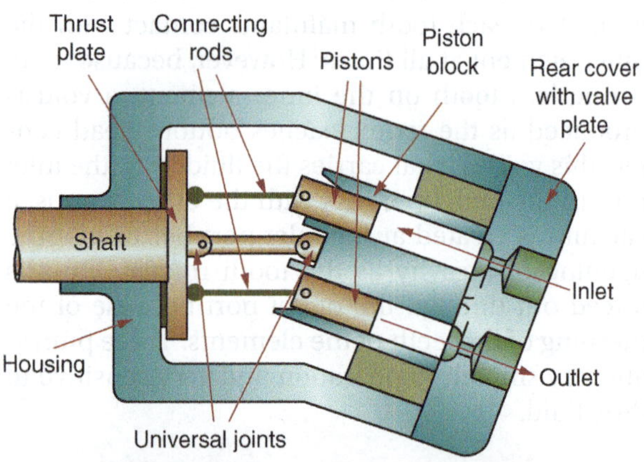

FIGURE 11-8 A bent-axis piston pump.

Straight-Axis Piston Pump

The straight-axis piston pump is very similar to the bent-axis piston pump. There are some differences, however. The pistons in this type of pump

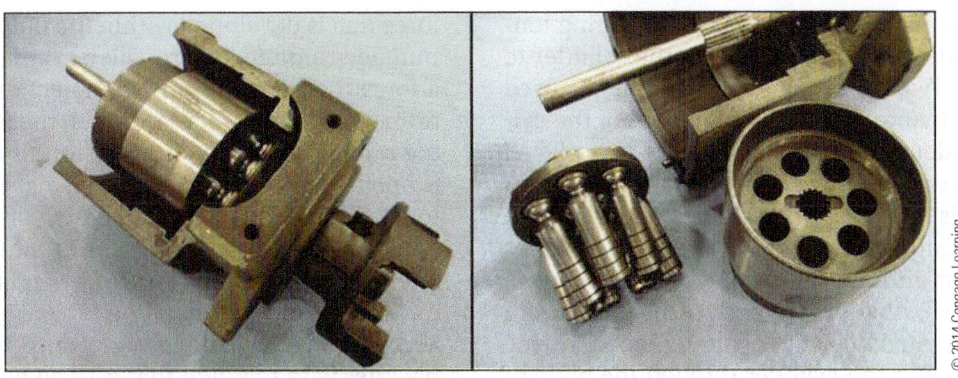

FIGURE 11-9 A straight-axis piston pump.

are arranged to be parallel to the shaft axis. There is no offset of the block on this type of pump. This type of pump uses a swash plate to cause the pistons to move in and out. The swash plate does not rotate with the block and pistons. Instead of placing the block at an angle, as in the bent-axis pump, the swash plate is placed in this pump in such a way as to provide the angle that is necessary to produce the reciprocating motion of the pistons. A straight-axis piston pump with a swash plate is shown in **Figure 11-9**.

As the shaft rotates, the block rotates. The pistons that are in the rotating block are connected to the swash plate by means of a piston shoe. As the block rotates, a reciprocating motion is produced in the cylinder. As with the bent-axis pump, this reciprocating motion causes fluid to be pulled into and out of the pump, thus causing flow. The variable displacement type of swash plate piston pump has an adjustment that changes the angle of the swash plate. As before, when the angle is increased, the flow volume increases. As the angle is decreased, the flow volume decreases. If a variable displacement swash plate were set at 0°, there would be no flow at all.

Another type of straight-axis piston pump is the wobble-plate pump. This type of pump is shown in **Figure 11-10**.

In this type of pump, the drive shaft is not connected to the block and the pistons. It is connected to a **wobble plate**, which is sometimes referred to as a rotating cam plate. The block is stationary in this type of pump. The pistons are spring loaded so as to press against the wobble plate. The wobble plate, shaped like a cam, causes the piston to reciprocate. As the piston shoe is at the narrower portion of the wobble plate, the piston has moved backwards because it is spring loaded. This creates a vacuum that is necessary to fill the cylinder with fluid. As the wobble plate continues to rotate,

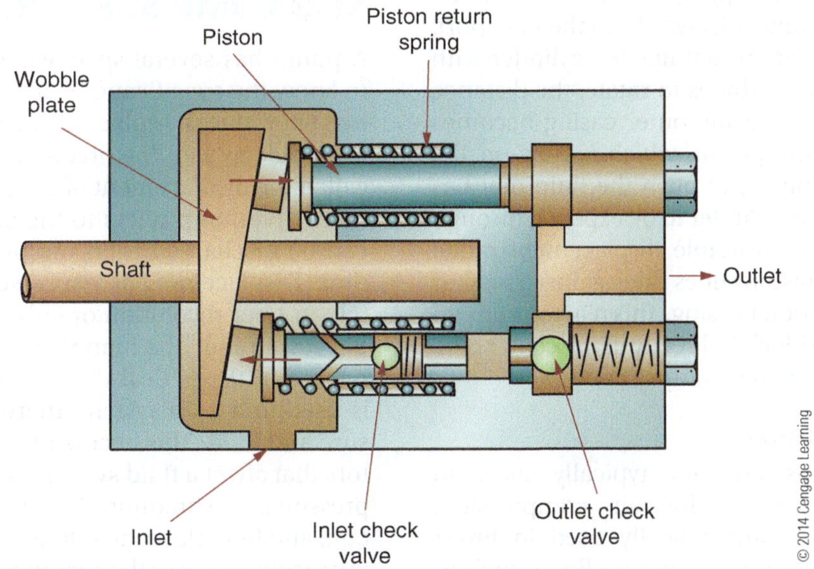

FIGURE 11-10 A wobble-plate pump.

it presses the piston inward against the spring tension. This causes any fluid that is in the cylinder to be expelled. Two check valves per cylinder are used to valve the hydraulic fluid into and out of the cylinder. The inlet port on this type of pump is placed to allow the wobble-plate cavity to fill with hydraulic fluid. This helps keep the surface of the wobble plate and pistons lubricated, and it also provides a small reservoir for each piston to pull fluid from. The first check valve is in the body of the piston. As a vacuum is created, the check valve opens, allowing fluid to flow into the cylinder through the body of the piston. As the piston starts pressing the fluid out of the cylinder, the check valve in the body of the piston closes and a check valve at the outlet manifold opens, causing the fluid to be expelled from the cylinder. Variable displacement, straight-axis wobble-plate pumps are available. These types of pumps are exactly like the wobble-plate pump that was just described except that there is an adjustment that will restrict how far a piston may retract, thus creating less area in the cylinder. This causes the amount of flow to be restricted.

Radial Piston Pump

The radial piston pump is very similar to the unbalanced vane pump. A radial piston pump has the cylinder block attached to the shaft. As the shaft rotates, the block rotates. The pistons, either because of centrifugal force or spring tension, are forced outward until the piston shoe makes contact with the outer casing. The shaft and the block are oriented in the pump in an eccentric position. This is what causes the reciprocating motion of the pistons. The inlet and outlet ports are located in the center of the block. The two ports are separated by a pintle. The pintle is what prevents slippage. As the block rotates, the pistons move outward, away from the inlet port. This causes the vacuum that fills the cylinder with fluid. As the block continues to rotate, the distance between the block and the outer casing becomes less. This causes the piston, which is now on the other side of the pintle, to push the fluid that has been pulled into the cylinder to be expelled through the outlet port. In a variable displacement radial piston pump, the piston shoes ride against a reactor ring instead of the outer casing. This reactor ring can be adjusted so the block is directly in the center of it. In this position, there is no output flow.

Centrifugal Pumps

Centrifugal pumps are not typically used in a hydraulic system or for air compression. **Centrifugal pumps** are typically used in lower pressure systems that move thinner fluids such as water. Centrifugal pumps rely on the centrifugal force that is developed within the fluid as it is being pumped through the propeller. **Centrifugal force** is a force that tends to pull an object outward, away from the center of rotation, when the object is rotating rapidly around a center point; thus, a spinning propeller enclosed within a housing causes suction to be created at the input (center) of the propeller and discharge of the fluid at the output of the pump. These types of pumps typically need to be primed in order to begin operation. **Priming** a pump is accomplished by filling the pump and the supply lines with the fluid that is to be pumped before the pump is started. This is usually done right after the pump has been installed. Priming a pump is not required again unless it loses its prime. The centrifugal pump is simple in construction when compared with the other pumps that have been discussed in this chapter. It has a disk, mounted to the shaft of a motor, and fins that have been machined onto the disk. The fins are curved in design to throw the fluid outward. The space between the pump housing and the propeller is usually very close in tolerance. However, it is nowhere as close in tolerance as the previous pumps. One of the most common problems with the centrifugal pump is that the propeller often wears out prematurely because of cavitation. Remember that cavitation occurs when air becomes entrained into the fluid as it is pumped. **Cavitation** can be defined as vacuum void within the fluid. This can be avoided by ensuring that all of the air has been removed from the system before putting the pump into service. Repair of this pump is fairly easy and more cost-effective than buying a new pump. This is because there are fewer moving parts in this pump than in the others.

11-2 PUMP SPECIFICATIONS

A pump has several specifications. It is important to know the specifications of a pump when installing a new one or replacing one that may have gone bad while in use. The first of these specifications is called the displacement of the pump. The displacement of a pump refers to the amount of fluid that can be discharged from the pump in one revolution. Displacement is expressed in cubic inches (cu. in.) per revolution or cubic centimeters (ccm) per revolution. The manufacturer's rated displacement is a theoretical displacement. As the pump is used in a fluid system, there are losses of pressure and fluid. This occurs due to the different factors that affect a fluid system, such as the operating pressures, the amount of restrictions in a fluid circuit, and how close the tolerances are in the pumping mechanisms as they are in use. Any one of these factors can change the rated output flow of a pump;

therefore, it is safe to say that there are two types of displacement: theoretical and actual. The efficiency of a pump can be found using this relationship of displacement:

$$VE = \frac{AD}{TD} \times 100$$

VE = Volumetric efficiency
AD = Actual displacement
TD = Theoretical displacement

The **volumetric efficiency** of a pump can be found by dividing the theoretical displacement (manufacturer's rating) into the actual displacement. Because efficiency is expressed as a percentage, multiply the answer by 100 to get the percentage of efficiency. It is important to size a pump correctly in a system. Keep in mind, however, that over time the pump will begin to show some signs of aging. When this occurs, the actual displacement decreases in volume. It is for this reason that a variable displacement pump is handy. Over time, as the pump wears and the hydraulic system is changed, the actual displacement of a pump can be changed to keep up with the demand of the system by simply adjusting the displacement of the pump.

Solve the following problem using the formula that was just learned.

What is the volumetric efficiency of a pump that is rated at 15 CIR when it is only supplying 12.5 cu. in.?

$$VE = \frac{12.5}{1.5} \times 100$$

$$VE = 0.833 \times 100$$

$$VE = 83\%$$

Notice that this particular pump is running at 83% efficiency.

Another specification is the **delivery** of a pump. The delivery of a pump is the rate at which the hydraulic fluid is supplied to a system from the pump in a specified amount of time. The delivery is expressed in gpm. The pump may have metric ratings on it as well. In this case, the delivery would be expressed as liters per minute (lpm). It is important to know that the delivery of a pump is determined by the displacement of the pump and the motor speed that is driving the pump. Knowing this, the delivery of a pump can be determined by using the following formula:

$$Delivery = \frac{D\,(CIR) \times MS\,(rpm)}{231}$$

$Delivery$ = Gallons per minute or liters per minute
D = Displacement (CIR or ccm)
MS = Motor speed (rpm)
231 = Cubic inches per gallon

As can be seen in the formula, there are 231 cu. in. in 1 gallon (cu. in./gal). If a pump with metric displacement values were being used, the 231 would be replaced with 1000 because there are 1000 cubic centimeters in 1 liter (cu. cm/l). The theoretical displacement would also be changed to cubic centimeters.

Solve the following problem.

What is the delivery capability of a pump that has a displacement of 15 CIR and a motor speed of 1800 rpm?

$$Delivery = \frac{15 \times 1800}{231}$$

$$Delivery = \frac{27{,}000}{231}$$

$$Delivery = 116.88 \text{ gpm}$$

Notice that this particular pump has the capability of delivering 116.88 gpm.

A pump is rated according to the amount of power it produces. **Power** is the equivalent to how much work is being produced in a certain amount of time. Work can be defined, in this case, as inch-pounds (in.-lb). Inch-pound refers to how far (in inches) a given amount of weight (pounds) moves. Power can be determined by using the following formula.

$$Power = \frac{W}{T}$$

$Power$ = Inch-pounds per second (in.-lb/sec)
W = How far (in inches) a given amount of weight (pounds) moves
T = How much time (in seconds) it takes to perform the work

This formula does not give the horsepower rating of a pump. It only determines the power that is required in a system or the power that is output in a system.

Solve the following problem.

How much power is needed to move 8000 pounds 2.5 ft in 2 minutes?

$$Power = \frac{W\,(in. \times lb)}{T\,(sec)}$$

$$Power = \frac{30 \times 8000}{120}$$

$$Power = \frac{240{,}000}{120}$$

$$Power = 2000$$

Notice that 2000 in.-lb per second worth of power is required to move this 8000-pound load 2.5 ft in 2 minutes. In this example, 2000 in.-lb/sec can be converted to 166.66 ft-lb/sec (2000 divided by 12) or 10,000 ft-lb/min (166.66 multiplied by 60).

Horsepower is a different rating than power. **Horsepower (hp)** is a unit of power in the English system of units, just as are the joule, watt, calorie, and BTU. One horsepower (hp) is equal to 33,000 ft-lb per minute or 550 ft-lb per second. One foot-pound is the work done when a weight of 1 lb is moved a distance of 1 ft. James Watt, an English inventor, originated the term *horsepower*. He is known for developing the first practical steam engine. He also determined, through an experiment, that a draft horse could do 550 ft-lb of work (force) per second in drawing coal from a coal pit; therefore, if a mechanical machine moves 550 ft-lb per second, it is said to have produced 1 mechanical horsepower. If 33,000 ft-lb/min equals 1 horsepower, and 550 ft-lb/sec (33,000 divided by 60) equals 1 horsepower, then 6600 in.-lb/sec (550 times 12) equals 1 mechanical horsepower as well. The formula for finding mechanical horsepower is as follows:

$$hp = \frac{W(ft\text{-}lb)}{T(\text{sec}) \times 550} \quad \text{or} \quad hp = \frac{W(ft\text{-}lb)}{T(\text{min}) \times 33{,}000} \quad \text{or}$$

$$hp = \frac{W(in.\text{-}lb)}{T(\text{sec}) \times 6000}$$

hp = Horsepower
W = How far (in inches or feet) a given amount of weight (pounds) moves
T = How much time (in seconds or minutes) it takes to perform the work
Standard unit of power = 550 (ft-lb/sec) or 33,000 (ft-lb/min) or 6,600 (in.-lb/sec)

Solve the following problems to prove the formulae.

How much horsepower is required to move 8000 lb 2.5 ft in 2 minutes?

$$hp = \frac{W(ft\text{-}lb)}{T(\text{sec}) \times 550}$$

$$hp = \frac{2.5\,(ft) \times 8000(lb)}{120\,(\text{sec}) \times 550}$$

$$hp = \frac{20{,}000\,(ft\text{-}lb)}{120\,(\text{sec}) \times 550}$$

$$hp = \frac{20{,}000\,(ft\text{-}lb)}{66{,}000}$$

$$hp = 0.3$$

Notice that a ⅓-horsepower pump could move the load in the previous problem.

Try another that has a much larger load.

How much horsepower is required to move 200 tons 10 ft in 2.5 minutes?

$$hp = \frac{W(ft\text{-}lb)}{T(\text{min}) \times 33{,}000}$$

$$hp = \frac{4{,}000{,}000}{2.5 \times 33{,}000}$$

$$hp = \frac{4{,}000{,}000}{82{,}500}$$

$$hp = 48.48$$

Notice that a much higher horsepower is required to move 200 tons 10 ft compared to the horsepower that is required to move 4 tons 2.5 ft in the same amount of time; therefore, a pump with a much higher horsepower rating is needed to move the heavier load.

It is also important to mention that a hydraulic pump can never put out more horsepower than that applied to the circuit by the motor, which drives the pump. In other words, if a 10-horsepower motor is driving a 25-horsepower pump, the output horsepower of the pump will never exceed 10 horsepower. Also, if there is a 20-horsepower motor driving a 20 horsepower pump, it is safe to say that the output horsepower of the pump will be slightly lower than 20 horsepower because of losses in the motor. It will not be because of the hydraulic circuit.

11-3 PUMP TROUBLESHOOTING

As a pump fails, many common problems may be discovered as you begin to troubleshoot the reason for failure. One of the most common problems with pumps is bearing failure. Bearing failure is most common with the gear-type pumps because gears produce a large amount of side load. The type of gear pump that is likely to have the most side load is the helical gear pump. Side load shortens the life of the bearings. If the bearings that are in a pump fail, they can usually be replaced. This would be more cost-effective than buying a new pump. If you have to change a bearing because of failure, it is always a good idea to change all of the bearings that are in the pump even though the others appear to be good. If this is not done, it is possible that a different bearing may fail shortly after the repair of the first one, meaning that the pump will have to be pulled again, thus causing more downtime.

Another common problem is that seals sometimes begin to leak. There may be several types of seals on a pump. Shaft seals are the most common seals to go bad. Shaft seals are located anywhere that a shaft penetrates the housing of a pump. These seals are designed to keep fluid in and air out. Most shaft seal problems arise from improper alignment or excessive operating pressures. When a seal begins to fail, fluid starts to leak from the seal.

As the fluid begins to leak, it begins to spill onto the floor, and the fluid level in the reservoir begins to drop. This may cause several problems. As the fluid leaks onto the floor, a slip hazard may arise. A preventive action that can be taken to prevent this hazard is simply placing a basin under the pump to catch and contain any fluid that may leak from a seal.

The second problem is that if the fluid reservoir is not checked on a regular basis, the drop in fluid may go unnoticed. If the fluid levels drop too low on the inlet side, a vacuum may be formed and the pump may begin to suck air. It is not good for air to be present in any part of the hydraulic system. If air is sucked into the pump through a bad seal, air can become entrained in the fluid, possibly causing cavitation.

Common problems that occur with vane pumps are that the vanes tend to wear out over time, as do the cam ring and housing. These components may be obtained from the pump manufacturer. It will more likely be cost-effective to repair the vanes versus buying a new pump. The vanes wear because of the constant contact with the housing or cam ring. This constant friction causes wear over time. The hydraulic fluid itself provides lubrication; however, it is common for the fluid to have contaminants in it. Contaminants cause the components to wear faster than usual.

It is also common for most pumps to have an inlet strainer, and it is not uncommon for this strainer to become clogged. If this occurs, the flow of oil will be restricted because the pump cannot pull the fluid through the strainer. This causes damage to the pump, because a pump is not designed to run dry.

Sometimes vane springs may fail. These are the springs that keep pressure on the vane or piston so it stays in constant contact with the surface of the housing or cylinder. These springs sometimes wear out over time. As the spring is constantly flexed, stress begins to produce heat. This heat causes fatigue. Fatigue starts to degrade the integrity of the spring, thus causing the spring to fail. If a spring fails in a vane pump, the vane will rely on centrifugal force to maintain contact with the surface of the housing. This is fine if the pump is operating at lower flow and lower pressures; however, the centrifugal force is not enough to prevent leakage under high-pressure applications. If a spring fails in a piston pump, the piston with the broken spring will not produce the full amount of flow. The broken spring may also cause the cylinder wall to become scratched or scarred. If the scarring is too deep, the pump may be irreparable.

Another point to keep in mind is the method that is used to link the pump to the prime mover, which is usually a motor. The motor is usually directly coupled to the pump via a coupling. It is rare to see a hydraulic pump connected to a motor via a V-belt. If a hydraulic pump is connected to the motor with a belt, it will probably be a positive-drive belt. This is a belt that has teeth or cogs on it. Pumps that use belts to connect to the motor are usually of a lower horsepower rating. When a belt breaks, the pump does not produce the output flow that is necessary to make the hydraulic system work. There are generally two types of couplings that are used to connect the pump to the motor: flexible and rigid. If a rigid coupling is used, it is important to perform a coupling shaft alignment to ensure that the shafts of both the motor and the pump are in true alignment along the axis of their respective shafts. If this is not done, or if it is done incorrectly, the rigid coupling will become fatigued and fail. If this is not corrected when the coupling is replaced, the new coupling will fail as well. This will continue until the shafts are aligned correctly or the shaft becomes worn from misalignment. If this occurs, it is possible that the pump, the motor, or both, will have to be replaced. Flexible couplings have more tolerance than rigid couplings when it comes to a shaft being out of alignment; however, this is no reason not to perform the coupling shaft alignment. If the alignment is not within the tolerances specified by the flexible coupling manufacturer, it too will fail.

Another common problem is that the shaft key or keyway sometimes fails. This usually occurs only after a setscrew has loosened and gone unnoticed for some time. As the setscrew becomes loose, all of the turning force is placed on the shaft key and the keyway. This eventually causes the key to fail. When this occurs, the coupling will, more than likely, slip on the shaft, causing a reduction in output flow of the pump and damage to the shaft.

Most pumps can be rebuilt fairly inexpensively instead of replacing them with new ones. However, it would be a good idea to have a replacement pump in stock. This way the machine can be repaired quickly, and the damaged pump can be repaired more leisurely. Once the pump has been repaired, it can be used to replace the next pump that goes out. If a bad pump is not repaired as soon as possible, you may find that you are overstocked on bad pumps and understocked on good pumps. This is why the bad pump should be repaired as soon as possible.

11-4 AIR COMPRESSION

It is important to realize, when referring to pneumatic fluid (air) versus hydraulic oil (liquid), that air is compressible, whereas oil is only slightly compressible. Although hydraulic oil is slightly compressible, it is generally considered to be noncompressible. Because of its compressibility and availability, air is used in many industrial applications. **Compression** occurs when air is forced into a smaller space than it originally occupied. This compression is what creates the pressure that is used in a pneumatic system. Compressed air is usually reserved for use at lower operating pressures. It is important to know the laws that apply when compressing air. When a space that contains a volume of air is changed, the pressure and temperature of the air that is trapped in that space changes according to these three laws:

- Pascal's law—"Pressure set up in a confined body of fluid acts equally in all directions, and always at right angles to the containing surfaces."
- Boyle's law—"At constant temperature, the *absolute* pressure of a gas is inversely proportional to its volume."
- Charles's law—"At constant pressure, the volume of gas is proportional to its *absolute* temperature."

What is volume? **Volume** is the size of an object using all three dimensions—height, width, and depth. The volume of liquids and solids is definite and does not vary significantly when they are compressed, whereas gases, such as air, have an indefinite volume that varies significantly when compressed. As noted in Boyle's law and Charles's law, temperature is an important factor when compressing air. As air is compressed, the temperature of the air begins to rise because the air molecules are being forced to rub against each other as they are being compressed. This is called friction, and where there is friction, there is heat. The heat that is produced during the compression stage causes the air molecules to try to push farther apart, causing the hot air to try to occupy more space in the system. Because the air is contained, it cannot expand; therefore, the pressure rises equally in all directions against the container's surfaces. As the air begins to cool after compression, more air can be stored in the system. Because air is compressible, it reacts much like a spring. For this reason, it is difficult to position a device accurately using compressed air. It would be more suitable to use a hydraulic system for these critical applications because of its lack of compressibility.

On occasion, air has to be compressed to a very high pressure. If this is the case, the air would have to be compressed in stages. A two-stage compressor is used to accomplish this task. This type of compressor allows the compressed air to cool in-between stages. If the air were continually compressed instead of compressing in stages to reach a high pressure, the temperature of the air would rise to dangerous levels. This would be a problem only if a hydrocarbon material, a petroleum product such as oil or grease, or some other temperature-sensitive combustible should come in contact with the overheated air.

Refer to **Figure 11-11**. Notice that a cylinder is shown containing atmospheric pressure. As the piston is pushed inward, the volume of air is reduced. As this happens, the pressure increases proportionally to the amount of travel. For example, if the piston were to be pushed inward to one-quarter of its possible travel, the air within the cylinder would be three-quarters of its original volume and the pressure would have increased from 30 to 45 psig. If the piston were pushed inward to half of its potential travel, the volume of the air trapped in the cylinder would be half of its original volume and the pressure would now have increased to 60 psig. Review Figure 11-11 to see the reaction of the air as it continues to compress. Notice how the pressure has increased to 240 psig when it has been compressed to one-eighth of its original volume. The pressures that are shown are referred to as gauge pressures. This is what a pressure would read if it were attached to the cylinder.

It is important to keep in mind that compressed air can be corrosive. **Corrosion** is often referred to as oxygenation and occurs when oxygen and water are present. When air is compressed, water tends to form because of condensation, which means the hot compressed air makes contact with a cool container. As the compressed air is distributed throughout the pneumatic system, the water often is transferred along with the compressed air. This causes components to seize up over time if the system is not installed correctly so as to catch the water before it can do any harm to the components. This is accomplished by using an accumulator or filter. Water also accumulates in the bottom of the receiver tank. It is critical that the water be drained out of the bottom of this storage tank periodically to prevent the water from going down the system.

It is not uncommon for some compressors to get quite large. Consequently, the drive motor for these compressors is also quite large. In this case, it would not be cost-effective to turn the drive motor on and off when the pressure levels fall and rise. This is undesirable because the inrush currents of these monster motors (some as large as a couple hundred horsepower) are in some cases up to 600% of their full-load current. It is better to leave them running and simply unload the compressor until the system pressure has dropped below the pressure limit. It would be acceptable, however, to allow a smaller compressor to start and stop because these motors have lower current demands and would not be too

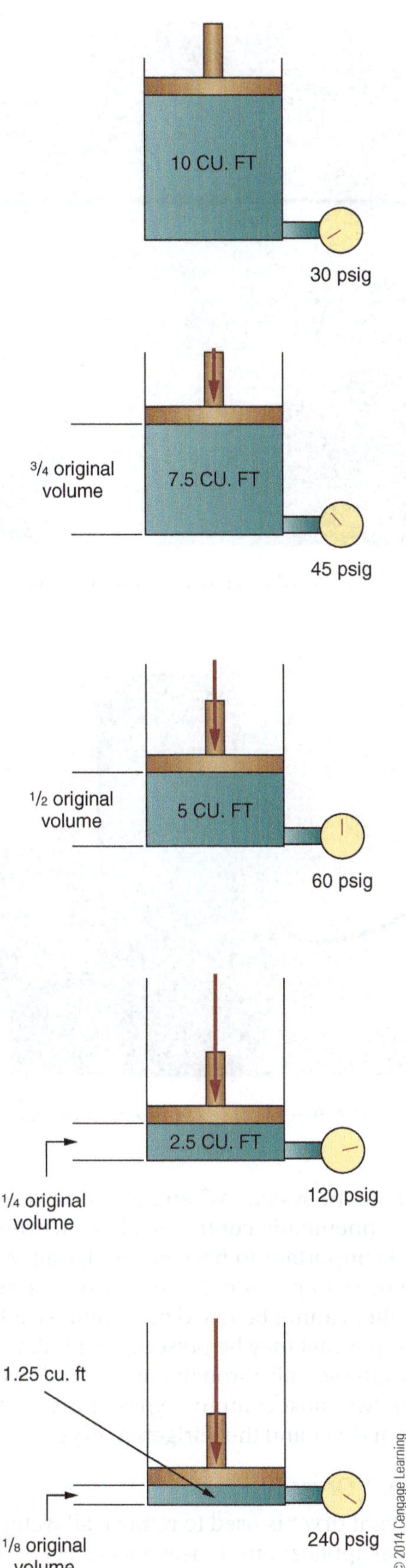

FIGURE 11-11 PSIG increases as the piston moves downward.

cost prohibitive to start and stop on a regular basis. When referring to **unloading**, this means that once the pressure limit has been met, an unloading valve simply vents the air that is compressed into the atmosphere during each cycle of the compressor instead of turning off the motor. When the pressure drops because of plant use, the unloading valve closes, allowing the compressed air to be forced back into the system.

Finally, it is important to know that the rate at which a compressor transfers air from the inlet port to the outlet port is known as the compressor's **delivery rate** and it is usually expressed in standard cubic feet per minute (scfm) or standard cubic meters per minute (scmm).

11-5 COMPRESSOR TYPES

There are many types of compressors used in industry today. An air compressor is a device that takes a form of mechanical input, usually an electric motor, and delivers compressed air that is under pressure through a pneumatic system for use. The design and operation of the different types of air compressors are very similar to those of the hydraulic pumps that were discussed earlier in this chapter. Just as before, there are gear, vane, axial piston, and radial piston-type compressors, although the internal and external gear compressors are not used very often. This is because they do not have the high-volume intakes that are required to be of sufficient capacity to afford a high quantity of air transfer. All of these compressors, however, work exactly the same as the hydraulic pump except that they transfer air, instead of oil, from the inlet port to the outlet port. Of these, the vane and the lobe compressors are the most commonly used. They are not, however, the most commonly used compressors when considering all of the different types of air compressors. One of them is a piston-style compressor. These compressors may have one or two cylinders. A single-cylinder compressor is shown in **Figure 11-12**. Notice that this type of compressor has a motor connected to the piston connecting rod via a belt. A direct drive system would not use a belt to connect the piston rod to the motor. The piston rod of a direct drive compressor would be connected directly to the motor shaft.

A compressor that has two cylinders (a twin-piston compressor) will most likely be in a "V" configuration. It looks like a motorcycle engine. In this particular configuration, a crankshaft is used to move the pistons. The connecting rod is what connects the crankshaft to the piston. This design is very similar to that of the gasoline engine. The difference is that the compressor needs a prime mover to operate,

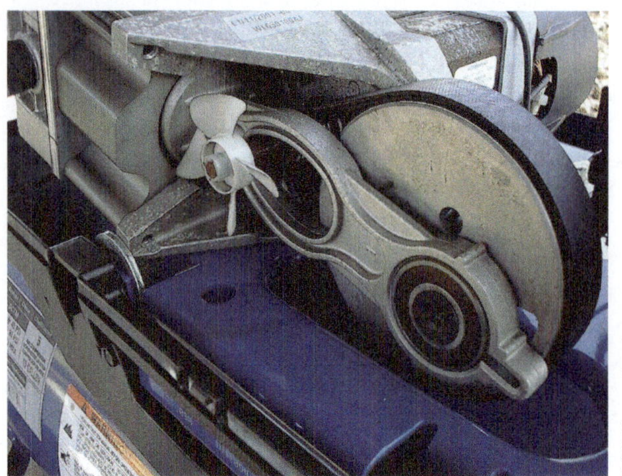

FIGURE 11-12 A single-cylinder compressor.

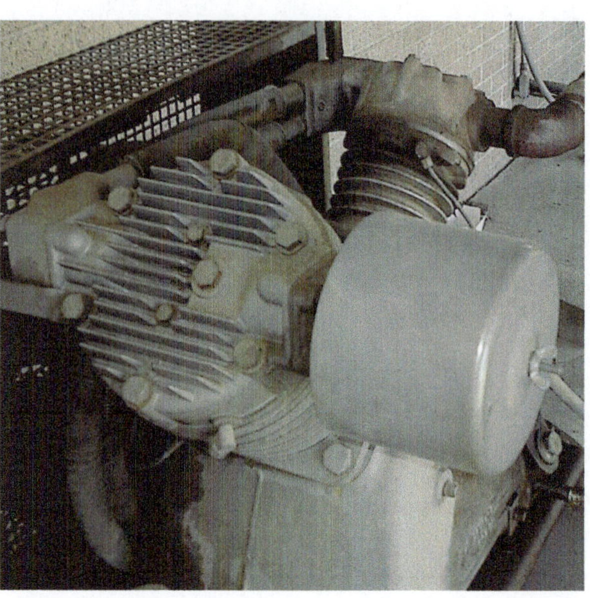

FIGURE 11-13 A twin-cylinder, two-stage compressor.

FIGURE 11-14 A vertical, screw-type compressor.

whereas the gasoline engine itself is a prime mover. An electric motor is usually attached to a flywheel via a belt. The flywheel is mounted onto the crankshaft and provides momentum once the compressor is up to speed. This helps the driver motor and keeps operating costs down. Some of the larger piston compressors have two belts to transfer more torque. As the motor turns, power is transferred through the belt to the crankshaft of the compressor. The crankshaft then alternately moves the two pistons in and out. Not all twin-cylinder compressors, however, have a "V" configuration. Some have the cylinders placed at 180° from each other. This type of compressor is usually a direct-drive compressor. A twin-piston compressor is usually used for staging. Staging occurs when the compressed air is taken from the outlet port of the first cylinder and is fed into the inlet port of the second cylinder. This allows the compressed air from the first cylinder to cool a little before being further compressed a second time. A two-stage, twin-piston compressor is shown in **Figure 11-13**.

Another type of compressor is the screw-type compressor. This is a very quiet compressor in comparison to the other types of compressors. The screw type usually supplies cooler compressed air with no pulsations. This is because the screw-type compressor mixes cooler air with the compressed air. This type of compressor is becoming widespread because it has the most up-to-date compression technology. It is a more expensive unit to purchase than other types of compressors. A vertical screw-type compressor is shown in **Figure 11-14**.

Air Dryers

Whenever air is compressed, the air becomes extremely hot. **Condensation** occurs when the air cools. This introduces water and water vapor into the pneumatic system. When the compressed air is used for pneumatic control, such as instrumentation, it is important to have clean, dry air. Air dryers are used to provide the dry air that is needed. An air filter cannot be relied on to remove all of the water vapor that may be present. An air dryer dries the compressed air through cooling and condensing. The two most common types of dryers are the desiccant dryer and the refrigerant dryer.

Desiccant Dryer

A desiccant dryer is used to remove all water vapor in critical applications. It uses a silica gel or an aluminizing material to absorb the water vapor. This process removes 99.9% of the water vapor. The desiccant material becomes saturated as it absorbs the water vapor. When this occurs, the desiccant becomes ineffective and must be regenerated in order to be used again. Two drying cylinders are used to continue the drying process while the saturated desiccant is being regenerated. Once the compressed air is routed to the cylinder with the dry desiccant through the use of a valve, the saturated desiccant can then be heated to cause the evaporation and regeneration of the desiccant material.

Refrigerant Dryer

The refrigerant dryer works similar to an air conditioner. The hotter, moist air is run through a heat exchanger, then through a refrigeration system. This lowers the temperature of the compressed air to approximately 35°F (1.6°C). Cooling the compressed air causes the water to condensate by lowering the relative humidity and dew point of the air. This water is collected at the lowest point of the system. The dry air is then sent back out to the facility for use. It is not recommended to run lubricated air through the dryer, because this lessens the drying effect that the dryer has on the air. If the air should be lubricated, lubricate it after it has been dried.

11-6 COMPRESSOR TROUBLESHOOTING

There are many things that can go wrong with an air compressor, one of which is the same problem that may be encountered with hydraulic pumps. This is bearing failure. Bearing failure in an air compressor, as with the hydraulic pump, can be repaired in the shop if the damage is not too severe.

Many other problems may arise. One of these is excessive system leakage. Any time that compressed air is allowed to leak into the atmosphere, the compressor cycles more often, causing unnecessary wear and tear on the compressor. An ideal pneumatic system has no leaks in it. When this is accomplished, the system is very efficient. Although this is the ideal running condition, it is not very realistic. It is important, though, to get air leaks to a minimum. Therefore, it is good practice to check the entire system for leaks on a regular basis.

Another problem is that the intake filters get clogged. Most compressors have some sort of filtration system on the intake of the compressor to prevent the compressor from pulling foreign debris into its internal parts. If a filter is clogged, the compressor does not supply the rated delivery and therefore may not keep the system pressure at an adequate level to operate. Clogged filters also cause a compressor to overheat. This should be avoided to prevent damage to the compressor.

Another common problem is belt slippage. When a motor drives a compressor using belts, it is important that the belts be properly maintained. As a belt is used, it begins to stretch. This causes the overall size of the belt to increase. When this occurs, slippage may occur. There are times that a maintenance technician simply coats the belt with belt dressing. This stops the belts from slipping for a short time. However, the belt begins to slip again simply because the symptom, not the problem, was fixed. The symptom is the belt's slipping; the problem is that the belt has probably stretched. If the belt is slipping because of its stretching, try to tighten it by sliding the motor base away from the compressor. Most motors have some sort of adjustment capability just for this reason. When the belt stretches to the point that the tension members are broken, it should be replaced. If a belt is worn because of misalignment, stop the compressor and align the motor before putting new belts on.

As with the hydraulic pump, it is a good idea to have a spare compressor on hand. This allows for rapid repair of the pneumatic system while repair of the compressor could be done at leisure. It is not always going to be cost-effective to have a spare compressor on hand. In this case, it would be good to at least have a repair kit and replacement parts readily available for the compressor type so that the compressor can be fixed as quickly as possible.

SUMMARY

- A positive displacement pump is a pump that allows very little leakage through its internal components, because the internal components of the pump fit so closely together that they can produce very high pressures in the fluid that is being pumped.
- There are generally three types of positive displacement pumps that are in use today: the vane pump, the gear pump, and the piston pump.
- A variable displacement adjustment changes the area of the cavities in a pump, making it smaller for minimum displacement and larger for maximum displacement.
- The word *displacement* refers to how much fluid a pump can move in a single rotation. The displacement of a pump is usually expressed in cubic inches per revolution (CIR). The flow volume of a pump is rated in gallons per minute (gpm) at a certain revolutions per minute (rpm); therefore, the flow volume is proportional to the shaft speed of the pump.

- The volumetric efficiency of a pump can be found by dividing the theoretical displacement (manufacturer's rating) by the actual displacement.
- The delivery of a pump is the rate at which the hydraulic fluid is supplied to a system from the pump in a specified amount of time. The delivery of a pump is determined by the displacement of the pump multiplied by the motor speed, divided by the standard unit of 231 cubic inches. If a pump with metric displacement values were being used, the 231 would be replaced with 1000 because there are 1000 cubic centimeters in 1 liter (cu. cm/l). The theoretical displacement would also be changed to cubic centimeters.
- Power is the equivalent of how much work is being produced in a certain amount of time. Work can be defined, in this case, as inch-pounds. Inch-pound refers to how far (in inches) a given amount of weight (pounds) moves. Power can be determined by dividing the amount of work by the time that it took to do the work.
- One horsepower (hp) is equal to 33,000 foot-pounds (ft-lb) per minute, 550 foot-pounds (ft-lb) per second, or 6600 inch-pounds (in.-lb) per second. There are three different formulae that could be used to find hp.

$$hp = \frac{W(ft\text{-}lb)}{T(sec) \times 550} \quad \text{or} \quad hp = \frac{W(ft\text{-}lb)}{T(min) \times 33{,}000} \quad \text{or}$$

$$hp = \frac{W(in.\text{-}lb)}{T(sec) \times 6000}$$

- One of the most common problems with pumps is bearing failure. It is always a good idea to change all of the bearings that are in the pump even though the others appear to be good to prevent another bearing from failing shortly after the repair of the first one. Another common problem is that seals sometimes begin to leak. Most shaft seal problems arise from improper alignment or excessive operating pressures. Reservoir fluid levels may drop too low and the pump may begin to suck air. If this happens, air bubbles form and is carried along with the fluid stream. This is known as entrained air. Entrained air can cause cavitation. If there is a bad seal on the inlet side of the pump, air may be sucked into the pump as well, causing entrained air. The vanes in a vane pump tend to wear out over time, as do the cam ring, housing, and vane springs. The vanes wear out because of the constant contact with the housing or cam ring. This constant friction eventually causes wear. It is not uncommon to find a clogged strainer. This causes damage to the pump because a pump is not designed to run dry. It is also important to keep a check on the belt or coupling that is used to turn the pump. Belts wear and couplings tend to break, especially if the shafts are not aligned correctly.
- As air is compressed, the temperature of the air begins to rise because the air molecules are being forced to rub against each other as they are being compressed. The heat that is produced during the compression stage causes the air molecules to try to push farther apart, causing the hot air to try to occupy more space in the system. Because air is contained, it cannot expand; therefore, the pressure rises equally in all directions against the container's surfaces. As the air begins to cool after compression, more air can be stored in the system.
- Pascal's law—"Pressure set up in a confined body of fluid acts equally in all directions, and always at right angles to the containing surfaces."
- Boyle's law—"At constant temperature, the *absolute* pressure of a gas is inversely proportional to its volume."
- Charles's law—"At constant pressure, the volume of gas is proportional to its *absolute* temperature."
- Volume is the size of an object, using all three dimensions—height, width, and depth. The volume of liquids and solids is definite and does not vary significantly when they are compressed, whereas gases, such as air, have an indefinite volume that varies significantly when compressed.
- Unloading occurs once the pressure limit has been met. An unloading valve simply vents the air that has been compressed into the atmosphere.
- There are gear, vane, lobe, axial piston, and radial piston-type compressors. Keep in mind that the internal and external gear compressors are not used very often because of a lack of the high-volume intake that is required to supply a sufficient amount of compressed air. The vane and the lobe compressors are the most commonly used. There are also several types of piston-style compressors. These compressors may have one or two cylinders. Another type of compressor is the screw-type compressor, the quietest with the smoothest output flow of compressed air.

- Bearing failure, as with the hydraulic pump, is the most common problem of compressors. Excessive system leakage causes the compressor to cycle more often, causing unnecessary wear and tear on the compressor. The intake filters may become clogged. If a filter is clogged, the compressor will not supply the rated delivery and therefore may not keep the system pressure at an adequate level to operate. Clogged filters also cause a compressor to overheat. Another common problem with compressors is belt slippage. If the full amount of torque is not transferred from the motor to the compressor, the compressor will not run as efficiently as possible. This causes a lack of compressed air to the system.

REVIEW QUESTIONS

1. What is displacement, and how is it expressed?
2. What two types of pumps have the worst internal tolerances?
3. What is volumetric efficiency usually expressed in?
4. What is the most common failure in both pumps and compressors?
5. Whose law states "At constant pressure, the volume of gas is proportional to its *absolute* temperature"?

CHAPTER 12

Fluid Power

Fluid power is widely used in industry to transmit power from one location to another. There are two methods of transmitting power through a fluid: hydraulic and pneumatic. The word **hydraulic** refers to something that is to be operated by the movement and pressure of a liquid. An example of this would be a hydraulic cylinder. This cylinder is operated by a moving liquid that is under pressure. The word **pneumatic** refers to a fluid power system that uses compressed air or a vacuum. Although air is a gas, it is referred to as a fluid; however, it is not a liquid as is the oil in a hydraulic system.

There is a vast amount of information that can be covered when referring to fluid power; however, this chapter discusses only the basics of fluid power systems and some components that are used in a fluid power system, and shows some simple applications of fluid power.

OBJECTIVES

After studying this chapter, the student should be able to

- Discuss some fluid power fundamentals.
- Explain psi, psig, psia, and inches of mercury ("Hg).
- Understand how force is transmitted through a hydraulic system.
- Understand the effects of compressing air for a fluid power system.
- Learn how to recognize the different valves that may be used in a fluid power system.
- List and explain at least eight methods of valve actuation.
- Discuss the operation of different actuators.

12-1 FLUID POWER FUNDAMENTALS

Pumps are needed to produce the flow that moves a fluid. A **pump** is a mechanical device that changes mechanical power into fluid power. A pump, however, loses some power; therefore, the fluid power that is produced may be slightly lower than the mechanical power. It is important to know that a pump's main purpose is not to produce pressure, but to produce the flow of a fluid. **Pressure** is only present when there is a restriction against the flow of the fluid. The amount of pressure that is present is just enough to overcome the amount of restriction. In case of a blockage (deadheaded), the pump continues to build pressure until something in the system breaks or bursts. Because of this, it is important to unload the pump when the flow is restricted or when the fluid is not moving. To unload a pump, the fluid must be allowed to flow back to the reservoir (hydraulic) or be exhausted (pneumatic), thus reducing the amount of pressure in the system. This is accomplished through the use of a pressure-relief valve. This type of valve only opens when the pressure in a circuit has risen to its rated pressure. This is the device that unloads the pump in the event of a blockage, thus preventing damage.

Fluid power has several advantages over mechanical power. A few of them are listed here:

- Ease of control
- Accuracy
- Ability to multiply force
- Constant force
- Constant torque
- Instantly reversible

Ease of control is gained through the use of simple levers or pushbuttons. By using these devices, the operator can easily start, stop, and change the speed and position of different processes that may require extremely high horsepower. The *accuracy* is usually very precise, controlling within a fraction of an inch.

Force can be *multiplied* very efficiently through the use of a fluid power system. Think about trying to stop your car without the use of its power brake system. Because the automobile uses hydraulics to control the brakes, a very small amount of force applied to the brake pedal creates enough force to stop the automobile. This is because the force that is applied to the pedal is multiplied through the hydraulic system and then delivered to the brakes of the car. If that is not enough, think about the ability to lift up a car with a hydraulic jack. A relatively small amount of force on the input piston can create enough force to push the output piston upward, lifting the vehicle off the ground. This is explained in greater detail later.

Only a fluid power system can provide a *constant force* and a *constant torque* regardless of speed changes. Fluid can also be *reversed instantly* on command. Because the fluid is controlled through the use of valves, the direction of flow can be changed as quickly as one can operate the valve that is controlling the direction. This is done without any spin down time—a necessity for many applications.

A **gauge** reads the pressure that is in the system. The amount of pressure in a system is totally responsive to the amount of restrictions present in that system. The total pressure is just the amount that is necessary to overcome the restrictions that are present in the system.

It is important to know the units of measurement for a fluid power system. Pressures in both the hydraulic and pneumatic systems are measured in pounds per square inch (psi). A gauge is generally used to show the pressure level above the reference base of the surrounding atmosphere. **Pressure gauge readings** are a measure of intensity of the force or torque that is produced by the fluid system. When pressures are read from a gauge, the reading is referred to as psig. This is so that the gauge pressure can be distinguished from **absolute pressure**, which is indicated by psia. The psig is always 14.7 psi less than psia because psia is a measure of the pressure above the absolute base of a perfect vacuum. The psia has to be measured with barometric-type instruments. Atmospheric pressure is psia. Under standard sea level conditions, atmospheric pressure is 14.7 psi higher than a perfect vacuum. A vacuum can also be measured in psi but it is more commonly measured in inches of mercury ("Hg). Thirty in. of mercury is equal to 14.7 psi. A bar is a metric unit of measurement for fluid pressure. This is known as the International Standard (SI) System. One bar is equal to 14.5 psi. **Table 12-1** shows the relationship among absolute bars, psig, and psia.

12-2 HYDRAULICS

Hydraulics is the science of transmitting power through a liquid. The liquid that is used is most commonly referred to as hydraulic oil. Oil has been the preferred fluid for hydraulic systems because of its desirable characteristics, such as its lubricating quality, its low specific gravity, and oil film.

Table 12-1
COMPARISON AMONG ABSOLUTE BARS, PSIG, AND PSIA

Bars (Gauge Pressure)	Bars (Absolute Pressure)	PSIG	PSIA
0	1	0	14.5
1	2	14.5	29.0
2	3	29.0	43.5
3	4	43.5	58.0
4	5	58.0	72.5
5	6	72.5	87.0
6	7	87.0	101.5
7	8	101.5	116.0
8	9	116.0	130.5

Oil Temperature

As oil is pumped, it creates a flow through the hydraulic system. If there are restrictions in that system, the pressure will begin to rise. As the pressure begins to rise, the temperature of the oil begins to rise. If allowed to continue, this will cause damage to the pump or hydraulic motor. It also causes many other malfunctions throughout the hydraulic system, such as valve damage and seal damage. Always check the manufacturer's specifications to see what the leveling-off temperature of the oil should be. The temperature should be read after the pump has run for a few minutes. Monitor the temperature of the oil and verify whether it is within the manufacturer's specifications. If it is too high, a heat exchanger may have to be installed to lower the temperature of the oil. Never work on a hydraulic system when the oil is hot. Allow time for the oil to cool to avoid getting burned by hot oil before working on a hydraulic system.

Static Head Pressure

Static head pressure is defined as the amount of pressure that is developed for every inch of rise in elevation above the point of measurement. Water has a specific gravity of 1.0 and weighs 0.0361 pound per cubic inch. For every inch of elevation, this weight increases by 0.0361 psi. If a 1 cu. in. column of water were to rise 4 in., the static head pressure would increase from 0.0361 to 0.1444 psi. Refer to **Figure 12-1**. Notice that the two tanks of water are the same height, approximately 34 ft tall. This would cause a static head pressure of 14.7 psig.

The pressure developed by both tanks, because of head pressure, is the same even though one tank has more water in it than the other. The gauge does not see how much water is in the tank. Each gauge sees 1 in.^{2} of water that is 16 in. high.

Hydraulic oil has a specific gravity of 0.9. Therefore, the static head pressure for oil is 0.9 × 0.0361 for every inch of rise in elevation. The first cubic inch of oil weighs 0.0325 psi and increases by 0.0325 psi for every inch of rise in elevation thereafter. This has to be considered because static head pressure can be either helpful or detrimental to a fluid power system. When pumping oil to a lower elevation, head pressure is produced, which adds to the pump pressure as each 1 in. drop in elevation occurs. Hydraulic pistons operating at lower elevations than the pump may become overpressured because of the static head pressure that is developed in the system. When pumping oil to a higher elevation, static head pressure is produced for each 1 in. rise in elevation that exists. The pump therefore has to create an equal amount of pressure to offset the static head pressure that is developed from each 1 in. rise in elevation in order for the static head pressure to be nonexistent. Head pressure is also important when discussing hydraulic pumps. A hydraulic pump pulls a vacuum when it is running. This vacuum pulls the oil from the reservoir. If the pump is placed more than 3 ft above the reservoir, the elevation will cause a rise in vacuum. If the vacuum is too much, the pump

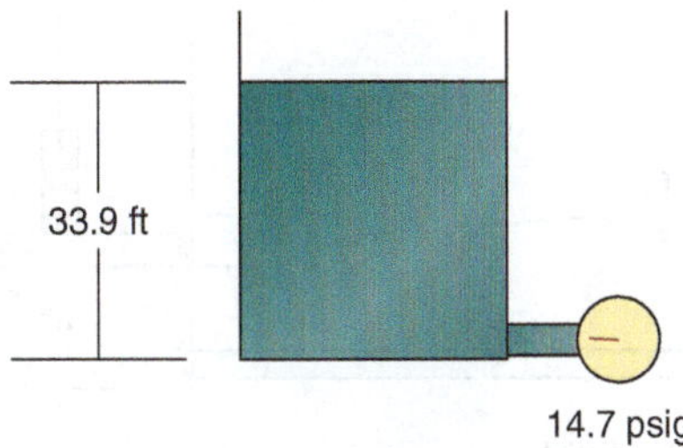

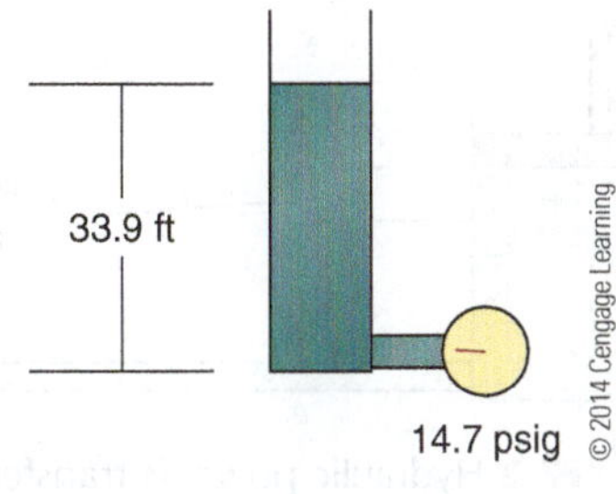

FIGURE 12-1 Static head pressure.

may begin cavitation. This causes excessive wear to the pump.

Hydraulic Operation

The principle behind fluid power is to contain the fluid within the system and apply pressure to the fluid. When pressure is applied to the fluid, the pressure is transmitted through the fluid, as shown in **Figure 12-2**.

Notice how the fluid is contained within the system. On the left, there is a 4 in. cylinder, which contains a piston. Also notice that a pipe is transmitting the fluid to another piston, on the right, of the same diameter. If 10 lb of force is applied to the piston on the right, causing a downward movement of the piston, 10 lb of pressure will then be delivered to the cylinder walls and the surface of the piston on the right, causing the piston on the right to rise out of its cylinder as far as the piston on the left was pushed into its cylinder. As can be seen, work has been accomplished from a distance through the use of fluid power. Notice also that power in equals power out in this example. This is not always the case because of losses in a circuit, such as mechanical friction and flow friction.

Notice in the last example that 10 lb of force was present at the input and the output pistons. As mentioned before, force can be multiplied. Notice in **Figure 12-3** that two different size cylinders are being used for this example.

The one on the right is still 4 in. in diameter, whereas the one on the left is now 40 in. in diameter. This causes an increase in force at the output shaft of cylinder B by 100 times as much. If 10 lb of force is applied to the input shaft of cylinder A, then the output force of cylinder B will be 1000 lb. Notice how the force was multiplied 100 times just by increasing the diameter of the second piston 10 times. However, the rate of travel of the output piston will only be one-hundredth as far as the input piston and only one-hundredth of the speed. Notice how an increase of force was gained at the expense of speed and distance. This simply means that you cannot get something for nothing. The reason for the increase in force and the decrease in speed and distance is because the power was never changed. The power delivered by the output piston is the same as the power delivered by the input piston except for any power losses that may exist because of mechanical and flow friction. The work applied to the left piston (force × distance moved) is transmitted to the piston on the right as the system output. The amount of power (horsepower) transmitted through the fluid depends on how fast the pistons move. A fast movement transfers a greater horsepower than a slower movement because horsepower is defined as the rate at which work is done over a given amount of time.

It is important to know that the force of a piston can be calculated by multiplying the pressure by the area ($F = P \times A$). According to Pascal's law, this force is present against all internal surfaces that are in contact with the fluid, including the cylinder walls, piping, piston surface, pump surfaces, and control valves.

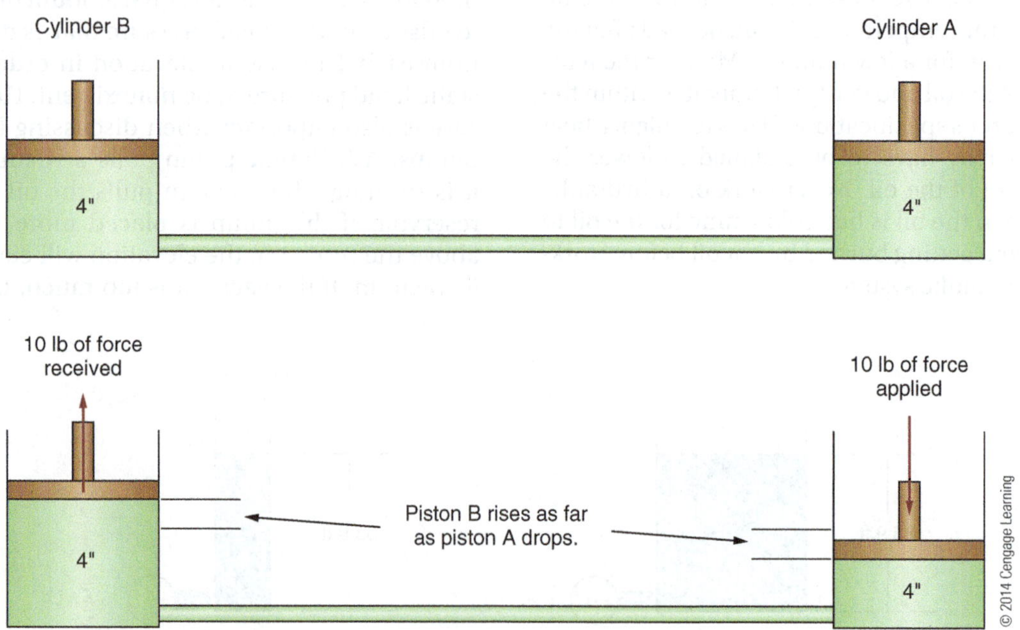

FIGURE 12-2 Hydraulic power is transferred through the fluid.

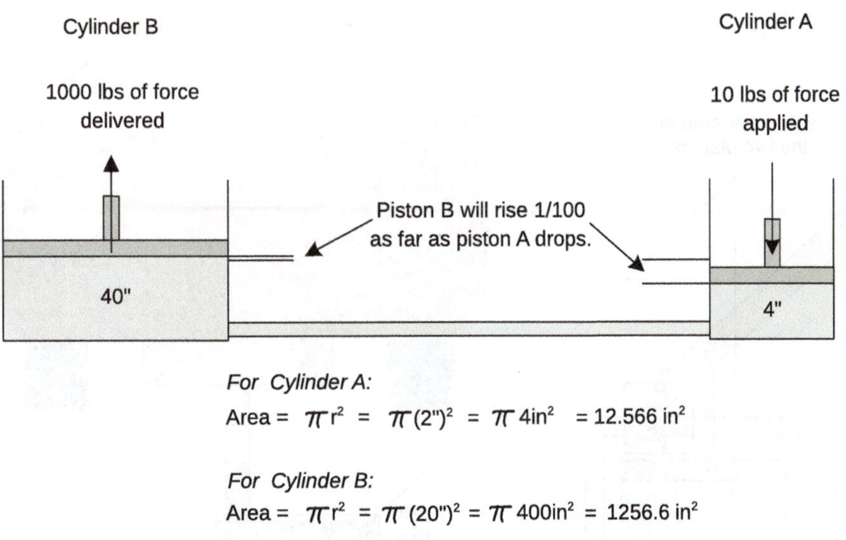

Cylinder B
1000 lbs of force delivered
Cylinder A
10 lbs of force applied
Piston B will rise 1/100 as far as piston A drops.
40"
4"
For Cylinder A:
Area = $\pi r^2 = \pi(2")^2 = \pi 4in^2 = 12.566 in^2$
For Cylinder B:
Area = $\pi r^2 = \pi(20")^2 = \pi 400in^2 = 1256.6 in^2$
The surface area of the piston in cylinder B is 100 x greater than that of cylinder A; therefore, the force will be multiplied by 100.
Cylinder A = 10lbs of Force applied
Cylinder B = 10lbs of Force X 100 = 1000 lbs. of Force delivered.
© 2014 Cengage Learning

FIGURE 12-3 Power is still transferred even when different size pistons are used.

The formula can be manipulated to find the pressure if the force and area are known or the area of a piston if the force and pressure are known.

$$P = \frac{F}{A}$$

$$A = \frac{F}{P}$$

This is illustrated in **Figure 12-4**.

Notice that there is a force of 1500 lb applied to the top of the piston. The pressure gauge is reading 75 psi. If you were to calculate for the area of the piston, you would find that it is 20 in.^{2}

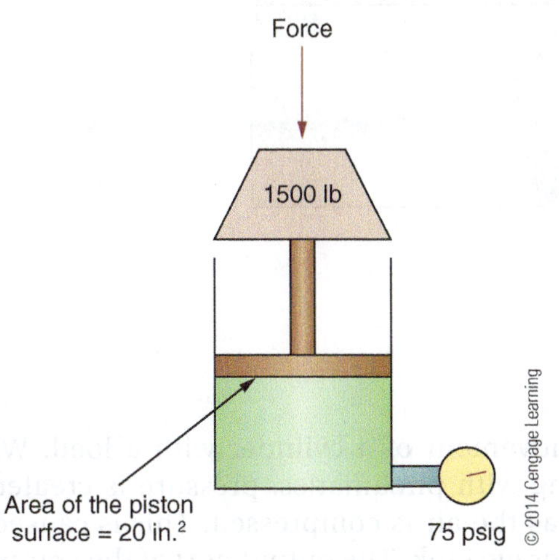

Force
1500 lb
Area of the piston surface = 20 in.^{2}
75 psig
© 2014 Cengage Learning

FIGURE 12-4 Force = Pressure × Area.

$$A = \frac{F}{P}$$

$$A = \frac{1500}{75}$$

$$A = 20 \text{ in.}^2$$

Turn the formula around and solve for pressure.

$$P = \frac{F}{A}$$

$$P = \frac{1500}{20}$$

$$P = 75 \text{ psi}$$

As you can see, the values are all proportional to each other.

Refer back to the hydraulic jack. How is it possible that a small amount of force can create enough force to lift a 2000 lb car? It is accomplished by a small amount of force being applied to a piston with less surface area. Because the second piston has a load placed on it (the car), pressure begins to build as more force is applied to the smaller piston. The pressure eventually builds to a level that creates enough force to lift the car off the ground. This is accomplished simply because there is more surface area on the second piston, thus multiplying the force that is applied to the car. This is illustrated in **Figure 12-5**.

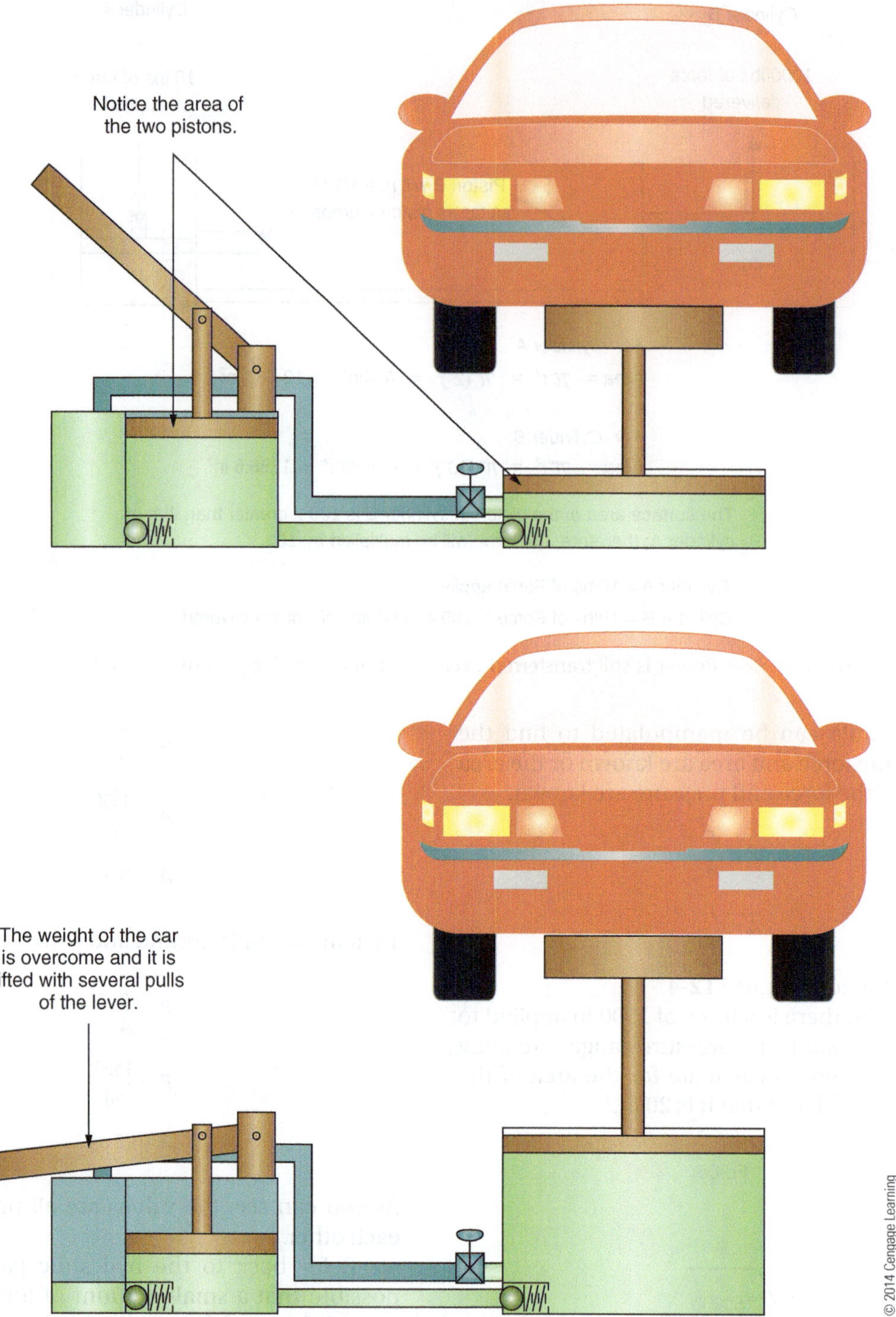
Notice the area of the two pistons.
The weight of the car is overcome and it is lifted with several pulls of the lever.
© 2014 Cengage Learning

FIGURE 12-5 An example of hydraulic leverage.

12-3 PNEUMATICS

There is a big difference between pneumatics and hydraulics. With hydraulics, pressure is created because fluid is restricted. The restriction derives from the work that is being done, that is, the movement of a cylinder with a load. When dealing with pneumatics, pressure is created as soon as the air is compressed. This is caused by the storage tank. The output port of the compressor should be directly connected to the storage tank unless a dryer or lubricator is used in the system. For simplicity's sake, assume that the output port is connected straight to the storage tank. The major difference between air and oil is that a liquid has a finite capacity to be compressed. Air can be compressed, then cooled in its compressed state, then compressed again. This is referred to as staging. It is for this reason that air is considered to have a higher compressibility than oil. Pneumatic energy is often referred to in the field as "air." When air is mentioned, it usually refers to the compressed plant air. This compressed air is usually at a pressure somewhere between 75 and 125 psi. A pneumatic system has a method of storing compressed air. A hydraulic system does not store energy in this manner. When the compressor has compressed the air to the desired pressure level, the compressor turns off (unless the compressor is cycled). The compressed air that is in the storage tank remains there until it is used. A gauge should be placed on the tank to visually verify the pressure level in the tank. If the pneumatic system is efficient and there are no leaks, the pressure should not drop from the cutoff pressure level until the air is used in some manner.

The energy in the stored air can now be used for different applications. When the air has to be used, it must be controlled through the use of plumbing and valving. A valve controls the flow (on/off) and direction of the air. The main goal of compressed air is to get back to atmospheric pressure. An example of this theory would be if a tank has compressed air in it and then develops a small hole, the compressed air will leave the tank through the hole and the pressure in the tank will eventually equalize to atmospheric pressure. The object of pneumatics is to control the escaping air and to use the stored energy to do work. As the compressed air is used, the pressure begins to drop in the storage tank. When this happens, the compressor begins compressing air again.

Fluid Conditioners

Fluid conditioners are separate devices that are used to clean, purify, and lubricate compressed air. The devices that are used to accomplish these tasks are the filter, the dryer, the regulator, and the lubricator. The air filter is self-explanatory. It removes contaminants from the air before they reach pneumatic components, such as valves, and actuators, such as a cylinder. This is to prevent the components from becoming corroded. The pressure regulator is used to maintain constant reduced pressures in specified locations of a pneumatic system. The regulator contains an adjustable upper spring, which allows the valve to hold a desired pressure on the downstream side. The pressure of the pressure regulator is set for the desired downstream pressure. The fluid, if below the pressure setting of the regulator, flows freely through the regulator until the pressure increases to a point that exceeds the internal spring tension of the regulator. At this time, the pushrod is allowed to move up, and the spring-loaded valve at the bottom of the regulator begins to close to throttle the air supply to the controlled pressure side. The air lubricator is used to lubricate the components in a pneumatic system. This is accomplished by mixing small droplets of oil with the moving compressed air to form a mist. The mist is then carried downstream to the components to lubricate the moving parts of the components. It is not uncommon for all three of these devices to be installed in a pneumatic system side by side, forming what is referred to as a **trio assembly**, or **filter, lubricator, regulator (FLR)**. The air dryer is used to remove excess moisture from the compressed air. If too much moisture is present in a pneumatic system, the components may corrode. This occurs because the moisture, if not removed, is often sent downstream as it mixes with the compressed air to form a mist. This mist eventually forms into water droplets within the components of the system, often stripping the components of their protective coating of oil. This exposes all of the metal within the components, which begins to oxidize. Most air dryers are preassembled at the factory. When they are shipped to a facility for use, they usually only require the external hookups, such as the piping and the electrical supply.

A Vacuum

Vacuum fluid power may be described as power that derives its force from the weight of the atmosphere. Remember that atmospheric pressure at sea level is 14.7 psi. Also remember that a vacuum is measured in inches of mercury ("Hg). Refer to **Figure 12-6**. Notice the glass tube in the center of the bowl.

The bowl is full of mercury. As the glass tube is submersed into the mercury, it is tipped to the side to let all of the trapped air escape. When the tube fills completely up to the top with mercury, the tube is raised out of the mercury pool by the closed end. The open end of the glass tube remains submerged but is not resting on the bottom of the bowl. As the tube is pulled out of the mercury, the atmospheric pressure causes the mercury to rise to 30 in. This is known as a **perfect vacuum**. Because the maximum atmospheric pressure at sea level is 14.7 psi, this is the maximum pressure that is available in a vacuum system. A vacuum pump cannot

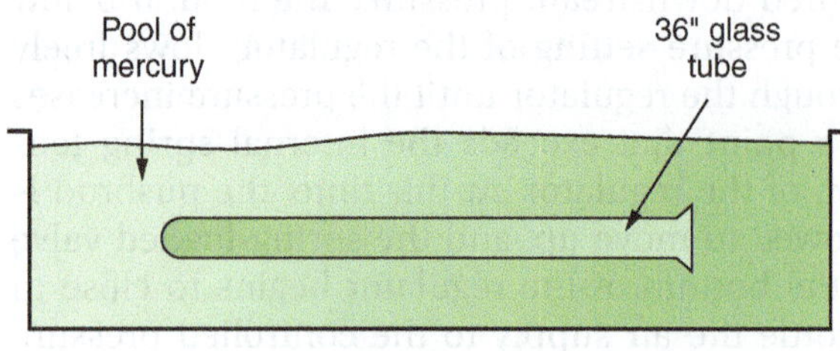
Fully submerge the tube to fill it completely with mercury.
Pool of mercury
36" glass tube

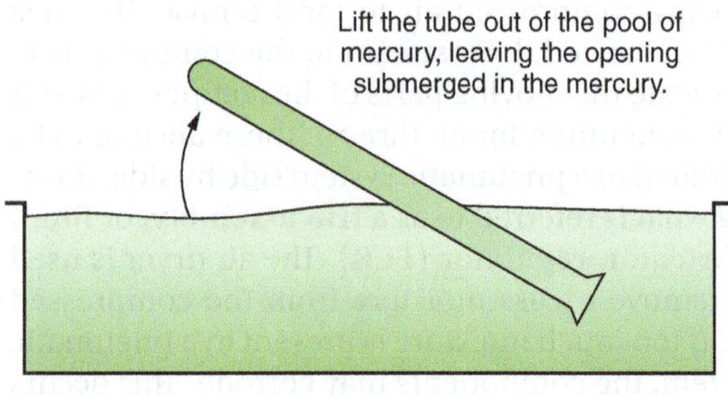
Lift the tube out of the pool of mercury, leaving the opening submerged in the mercury.

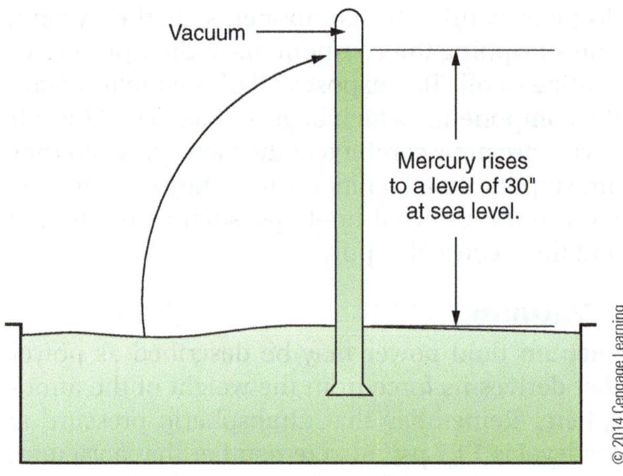
Vacuum
Mercury rises to a level of 30" at sea level.

FIGURE 12-6 A mercury barometer.

pull a perfect vacuum. A vacuum pump is therefore rated according to how much vacuum it can pull in comparison to a perfect vacuum. An example would be a vacuum pump that is capable of producing 27 in. of mercury at sea level and is said to be a pump with a 90% rating (27 divided by 30 = 0.90, or 90%).

Table 12-2 shows the relationship between a vacuum and atmospheric pressure. Notice in the table that for every 1 in. of mercury, the pressure increases 0.49 psi.

Any time a vacuum system is in operation, the atmosphere exerts force on all outside surfaces of the system containing the vacuum because of the lower pressure within the system. Just as with compressed air, a storage tank can store a vacuum. Just as the compressed air is always trying to equalize with the atmospheric pressure, so is the vacuum. A vacuum is a space with a pressure lower than atmospheric pressure; therefore, instead of the air trying to get out of the storage tank in a compressed air system, air from outside the vacuum system tries to get into the system to equalize the pressure to the atmospheric pressure. This means that if a hole were to develop in the tank, the outside air would enter the tank through the hole, and the pressure in the tank would eventually equalize to atmospheric pressure.

A vacuum system is sometimes used for controlling systems such as heating, ventilation, and air conditioning, and has some use in instrumentation. Vacuums are also used to pick up a load through the use of suction cups. A vacuum is created at the center of the cup, thus causing the product to be sucked up into the cup as the lip of the cup begins to seal on the product. It is not unusual to lift extremely heavy loads with a vacuum. This is shown in **Figure 12-7**. Sometimes a vacuum is used to hold something down as well. This is accomplished by drilling small

Table 12-2

CONVERSION OF INCHES OF MERCURY ("HG) TO ATMOSPHERIC PRESSURE

Vacuum ("Hg)	PSI
30	14.7
28	13.75
26	12.77
24	11.79
22	10.81
20	9.82
18	8.84
16	7.86
14	6.88
12	5.89
10	4.91
8	3.93
6	2.95
4	1.97
2	.983
0	0

1 psi = 0.049125"Hg

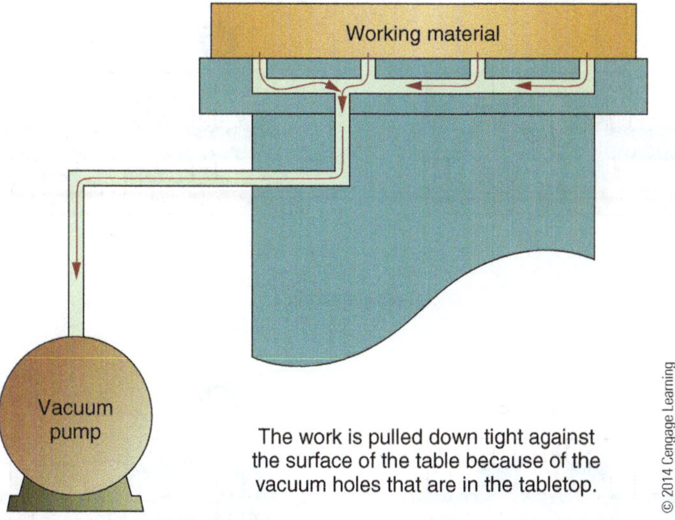
Working material
Vacuum pump
The work is pulled down tight against the surface of the table because of the vacuum holes that are in the tabletop.
© 2014 Cengage Learning

FIGURE 12-7 An example of using a vacuum to do work.

holes in the table surface in which a vacuum is pulled through. As a piece of material is laid across the holes, the vacuum causes the material to be pulled down onto the table with great force. This is also shown in Figure 12-7.

12-4 VALVES

This section explains the operation of each component and shows the American Standards Institute (ANSI) symbol while explaining the component; Chapter 4 (Industrial Print Reading), Section 4-3 (Hydraulic/Pneumatic Symbols & Drawings) lists some of the most common symbols that are used in hydraulic and pneumatic systems. Valves can be used for several purposes in a fluid power system. These are directional control, flow control, and pressure control. A valve is identified by its internal components, the number of ports, and the method used to actuate the valve. Another parameter for valve description may be normally closed or normally open. A normally open valve means that fluid can flow through the valve in its normal state until the valve is actuated. When a valve is actuated, the spool shifts, causing a change in the flow or direction of the fluid. A valve may be actuated either automatically or manually. A normally closed valve does not have a path for fluid to flow until the valve is actuated. It is a common practice to identify a valve by how many ports it has. For example, a valve that has two ports, an input port and an output port, is referred to as a two-way valve. A three-way valve is a valve that has three ports. There are also four-way and five-way valves. These valves are shown in **Figure 12-8** and are discussed later. Never use a valve unless all ports are correctly connected.

CAUTION

As can be seen in Figure 12-8, each valve has a different path for the fluid to flow. Each valve has its own purpose. Before each one is discussed in greater detail, you must be able to identify the components that are found in a valve and identify each port. Before the parts of a valve are labeled, you must know that there are many types of valves available. There are spool valves, which are shown in Figure 12-8, poppet valves, gate valves, ball valves, and needle valves. It is not within the scope of this text to include every type of valve that is in existence. Only the most common types of valves that are used in a fluid power system are discussed.

The Spool Valve

The spool valve, sometimes referred to as the sliding spool valve, is the most preferred type of valve for controlling the direction of a fluid in a hydraulic system. The spool moves back and forth in an enclosed body in which tunnels are machined. The tunnels, shown in Figure 12-8, provide a path for fluid to flow. The spool determines which tunnel the fluid flows through. The back-and-forth sliding motion of the spool within the valve body accomplishes this.

The Poppet Valve

The poppet valve, as shown in **Figure 12-9**, is a simple two-way valve that is most commonly used to stop and start fluid flow. It is not uncommon to find a three-way poppet valve.

The poppet valve can be either normally open or normally closed. The main problem with the poppet valve is that if the pressure within the fluid system rises to a level greater than the spring pressure within the valve, the valve may change state

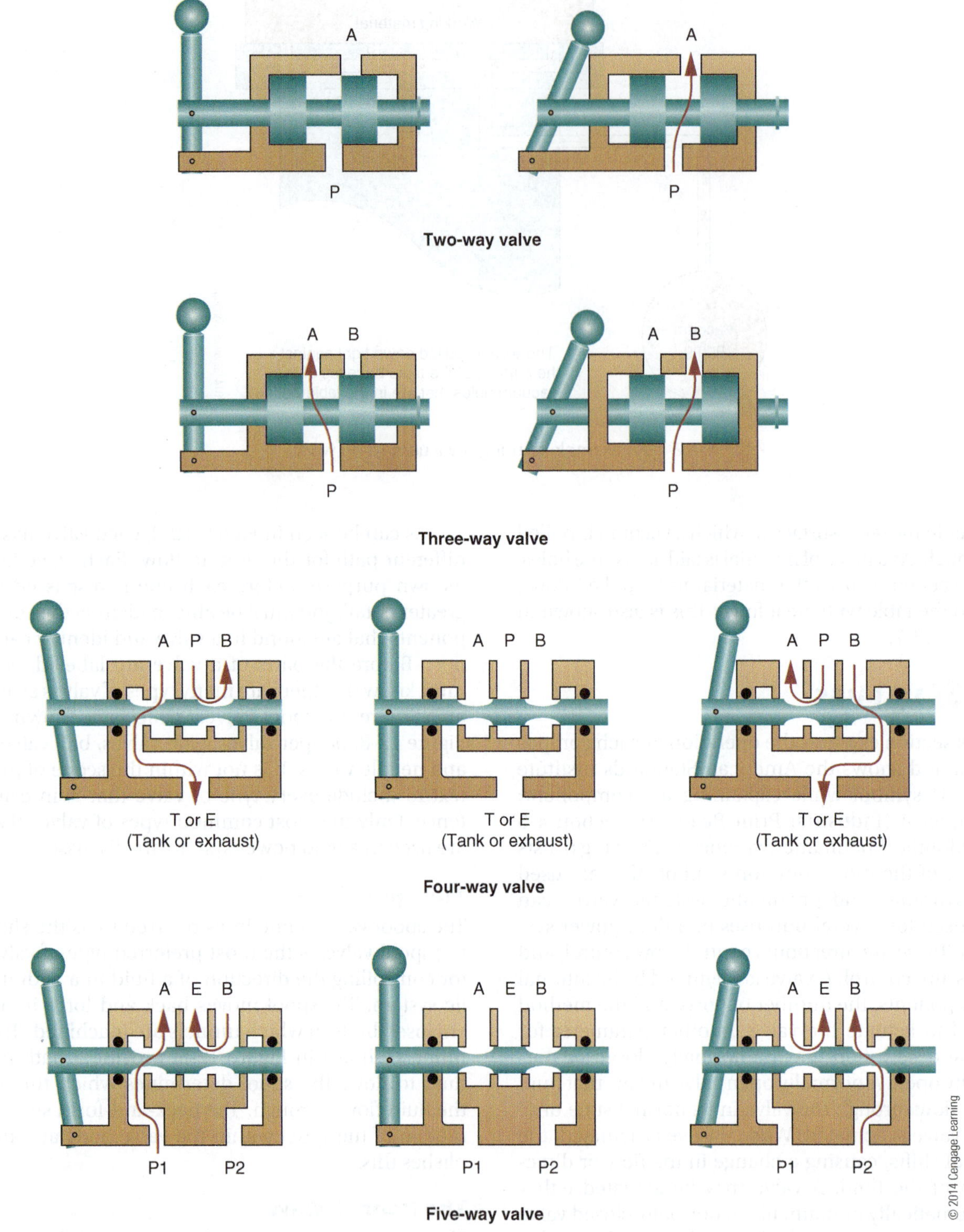

FIGURE 12-8 Various spool valves.

automatically without being actuated and without warning. This type of valve is usually reserved for use on a low-pressure fluid system for this reason. It is important not to exceed the pressure rating of a poppet valve.

The Gate Valve

A gate valve uses a wedge-shaped gate to choke off the flow of fluid as it is closed. The gate valve usually has a hand wheel at the top of the valve body. Gate valves are usually very large in comparison to the

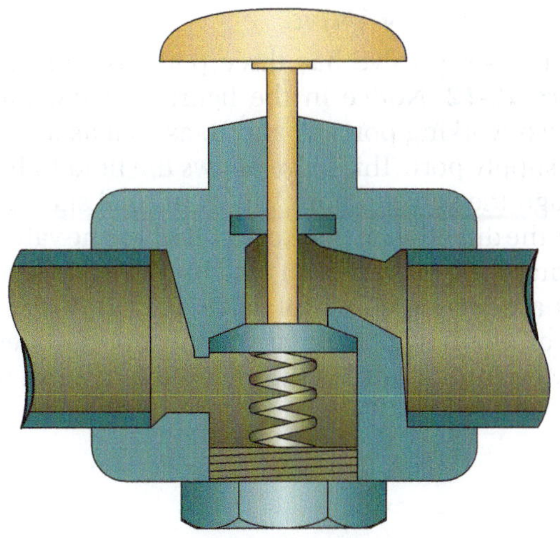

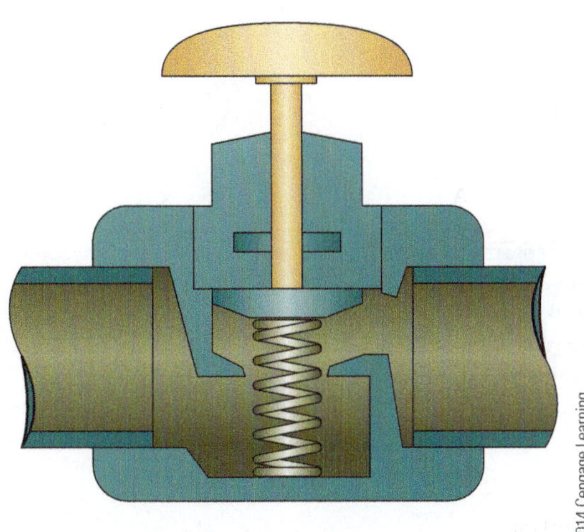

FIGURE 12-9 Normally closed and normally open poppet valves.

other valves. These types of valves are used mostly for flow control. This gate offers very little resistance to flow and therefore causes little to no turbulence in the fluid as it flows through the valve.

The Ball Valve

The ball valve is another very common type of valve. It is also used for flow control. A ball valve, which is shown in **Figure 12-10**, has a ball in the center of the valve body. The ball fits so tightly in the valve body that it does not permit any leakage when the ball valve is in the closed position. The ball has a hole bored through its center that is the same diameter as the inside diameter of the piping in which the fluid is flowing. When the handle is turned so it is at a right angle to the valve body, the passageway through the valve is blocked. When the handle is turned so it is in line with the valve body, the fluid

When the handle of the ball valve is turned parallel to the valve body, fluid is unrestricted and can flow through the valve.

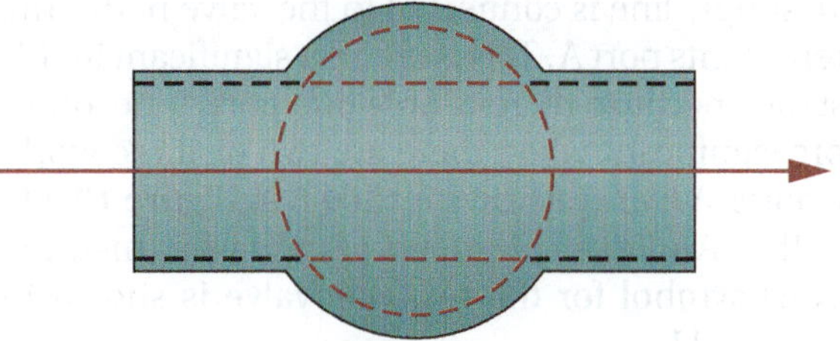

When the handle is turned perpendicular with the valve body, fluid is blocked and cannot flow through the valve.

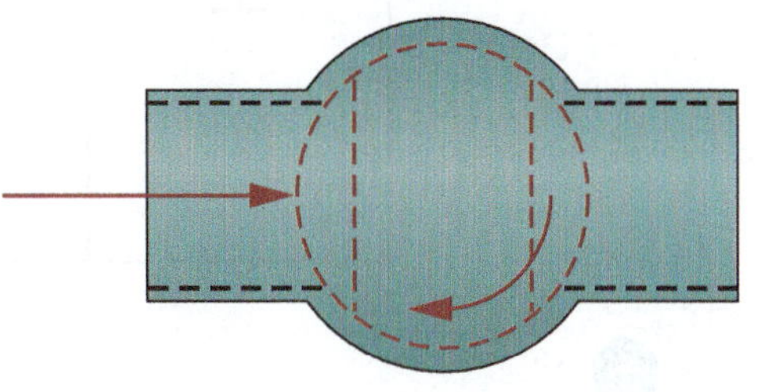

FIGURE 12-10 The internal operation of a ball valve.

can flow through the hole in the center of the ball. This is achieved because the hole in the ball is in line with the handle of the valve. The handle only need be turned one-quarter of a full turn to go from completely open to completely closed.

The Needle Valve

Needle valves are used for flow control as well. These types of valves are used to throttle the fluid to a desired flow rate. These can be used for shut-off purposes as well as flow control. The needle valve is a delicate valve in the sense that if too much tension is applied to the needle as it is being tightened, it may become scored. If the needle becomes scored, leakage may begin to occur. The needle valve is used mostly for high-pressure circuits (up to 15,000 psi) with low flow. It is important to pay close attention to the direction of flow on the needle valve. The flow of fluid should be toward the bottom of the needle. This keeps the extremely high pressures off the seal at the top of the stem. If the flow were fed into the needle valve incorrectly, the seal could conceivably blow out if the valve were closed because of a buildup of pressure.

The Two-Way Valve

The two-way valve is usually used for flow control only. This is because when the spool is in one position, the fluid cannot flow, and when the spool is shifted, the fluid can then flow. This is a simple on-off valve. The ports on this valve are labeled A and P. The P stands for pressure. This is where the pressure supply line is connected to the valve body. The A represents port A. This is not that significant in this instance because there is only one port. It becomes more significant when there are two or three working ports. A two-way valve is shown in **Figure 12-11**.

The American National Standards Institute (ANSI) symbol for the two-way valve is shown in Figure 12-11.

The Three-Way Valve

The three-way valve has three ports as shown in **Figure 12-12**. Notice in the figure that the valve has two working ports, A and B, as well as a pressure supply port. This valve allows the fluid to flow through the valve to port A in its normal state and shifts the direction of flow to port B when the valve is actuated. This is the reason that this valve is referred to as a directional control valve. It is not used to shut off the flow of fluid as are most of the previous valves. This valve is simply used to control the

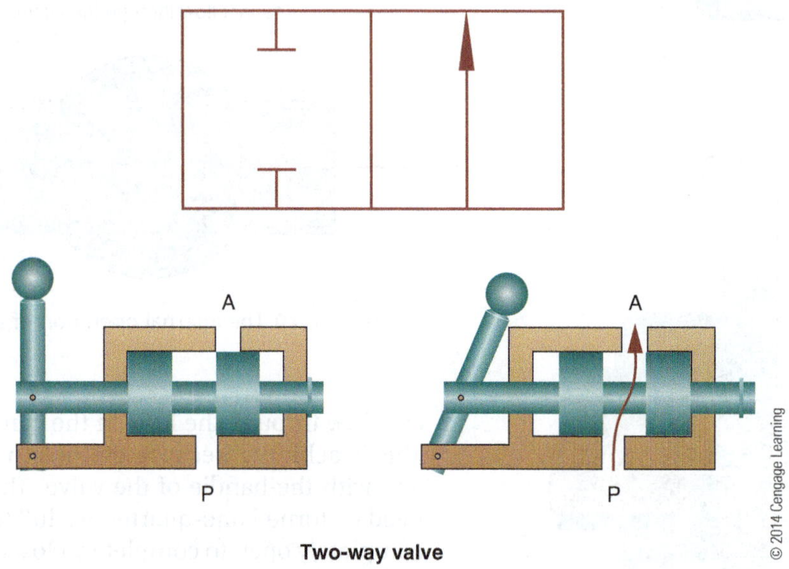

FIGURE 12-11 A two-way spool valve with ANSI symbol.

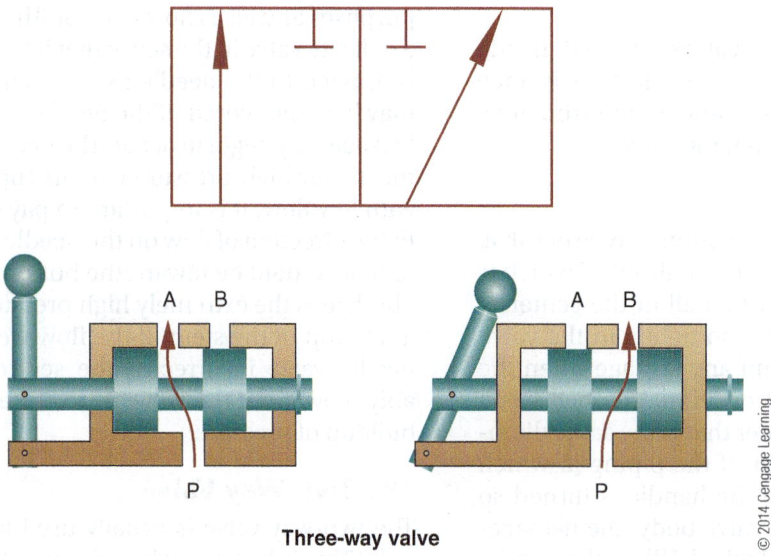

FIGURE 12-12 A three-way spool valve with ANSI symbol.

direction of flow. This valve is most commonly used on hydraulic and pneumatic cylinders or motors.

The ANSI symbol for the three-way valve is shown in Figure 12-12.

The Four-Way Valve

The four-way valve is shown in **Figure 12-13.** Notice that there are four ports labeled A, B, P, and T or E. You already know that the A and B ports are the working ports and that the P port is the pressure supply port. This valve not only controls the direction of flow, as does the three-way valve, but it has shut-off capabilities as well. Notice the center valve in Figure 12-13.

When the valve is not actuated in one direction or the other, there is no flow through the valve. It is also important to notice that the spool has two separate grooves that are used to control both the A port and the B port. These two grooves are used to supply pressure to one port as it is exhausting the opposite port. This only occurs when the spool has been shifted one way or the other. Notice that when the spool is shifted to the left (left valve in Figure 12-13), the pressure port is connected to port B and port A is connected to the exhaust or drain port. The letter T represents Tank when this valve is used in a hydraulic system, and the E represents Exhaust when this valve is used in a pneumatic system. Any time a four-way valve is used in a hydraulic system, the fluid must be drained back to the tank or reservoir. If this valve is being used on a pneumatic system, the compressed air can simply be exhausted into the atmosphere. It is common practice, however, to use a muffler on the exhaust port of the four-way valve when used in a pneumatic system. The ANSI symbol for the four-way valve is shown in Figure 12-13.

The Five-Way Valve

Look at **Figure 12-14.** Notice that the port that was labeled P on the four-way valve is now labeled E, and there are two new types of ports, P1 and P2. This type of valve offers the flexibility of using two separate pressures that are applied to ports A and B.

Notice that when the spool is in the center, there is no flow of fluid in any direction. However, when the spool is shifted to the left, the spool allows the pressure at port P1 to flow through to port A, and the pressure in port B is exhausted to the atmosphere (pneumatic use). The pressure at P2 is static. However, if the spool is shifted to the right, the pressure at P2 is no longer static and is in fact flowing through the valve to port B, while port A is being exhausted to the atmosphere (pneumatic use). The ANSI symbol for the five-way valve is shown in Figure 12-14.

Methods of Actuation

Valves can be actuated manually or automatically by manual lever, pushbutton, foot pedal, cam or mechanical operation, or solenoid or can be pilot operated. The most common automatic method of

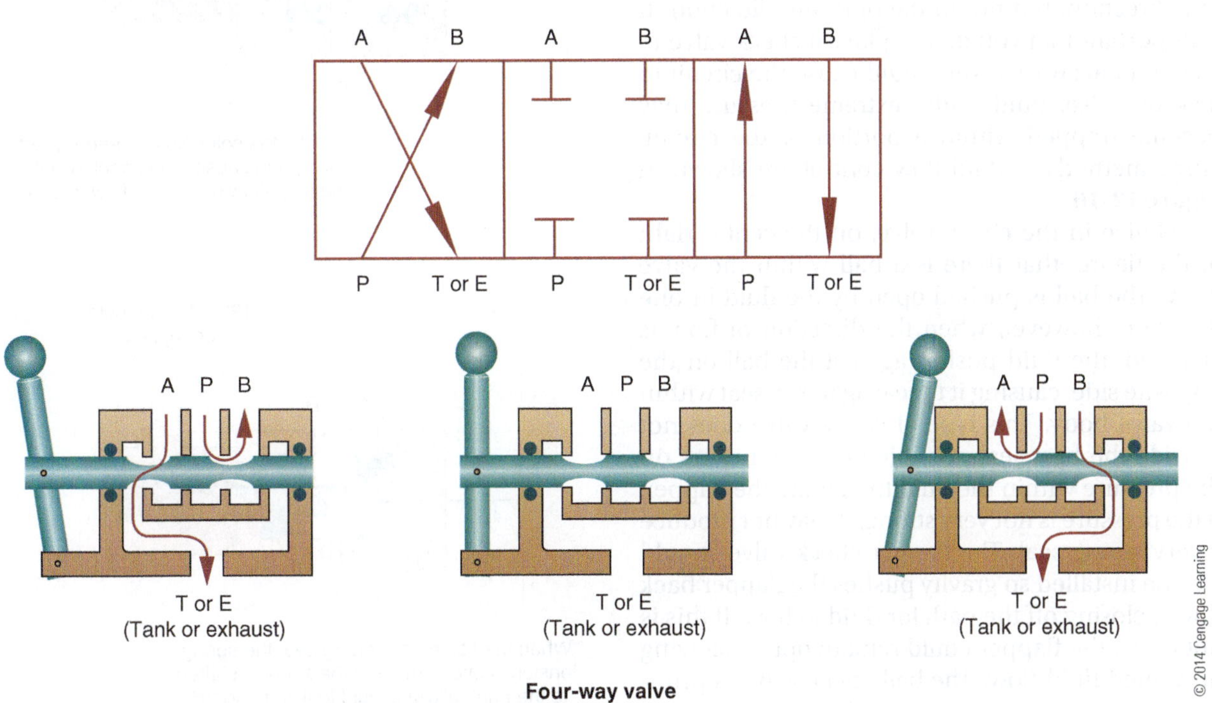

FIGURE 12-13 A four-way spool valve with ANSI symbol.

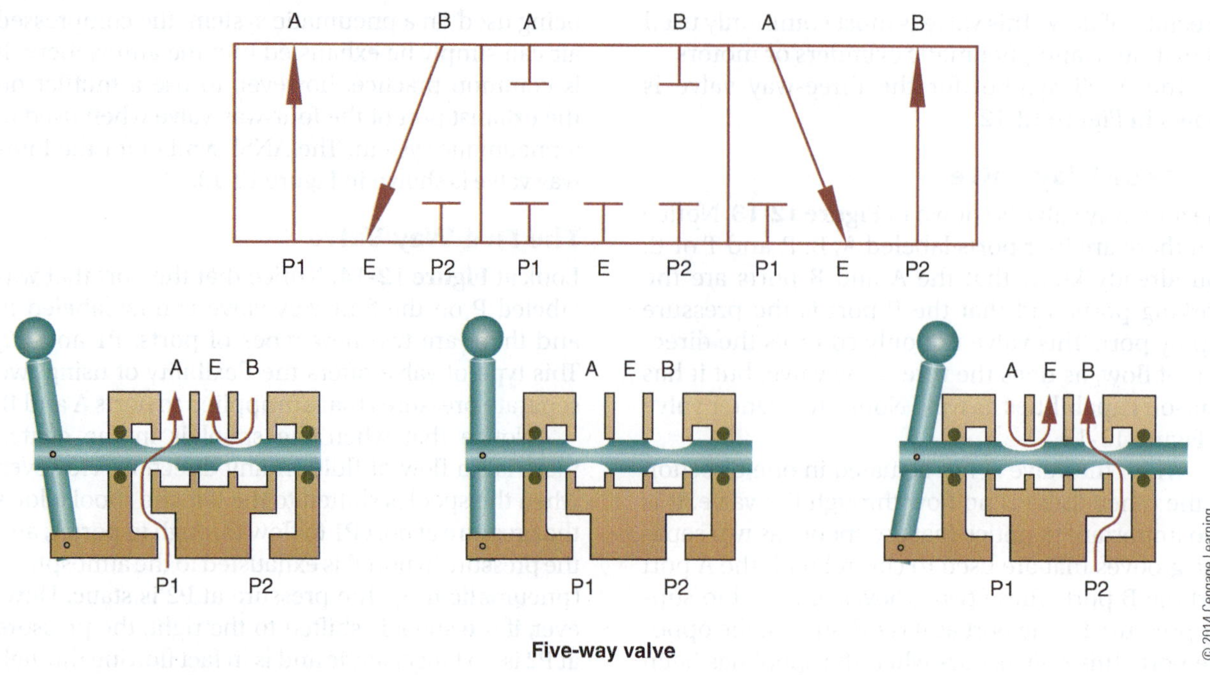

FIGURE 12-14 A five-way spool valve with ANSI symbol.

actuation is usually through the use of an electric solenoid. A solenoid valve is shown in **Figure 12-15.**

The ANSI symbols for all of the actuating methods are shown in **Figure 4-10**, located in Chapter 4.

The Check Valve

The check valve is used to allow fluid to flow in one direction, but not in the opposite direction. It is important that you do not place a check valve in a circuit backwards. This could cause the circuit to lock up. Also, fluid under extreme pressure may become trapped within a portion of the circuit. Three methods of fluid flow control are shown in **Figure 12-16.**

CAUTION

Notice in the check valve, on the center right of the figure, that there is a ball within the valve body. The ball is pushed open by the fluid in one direction; however, when the direction of flow is changed, the fluid pushes against the ball on the opposite side, causing it to seal against a seat within the valve body. This type of check valve does not provide the best seal possible because it relies on the pressure within the fluid to operate the flapper. If the pressure is not very strong, it may not produce a very strong seal. The flapper check valve should also be installed so gravity pushes the flapper back down, closing off the path for fluid to flow. If this is not done, the flapper could remain open, allowing unwanted fluid flow. The ball check valve is probably the most common check valve that is in use. It has a ball in the center that is spring loaded in order

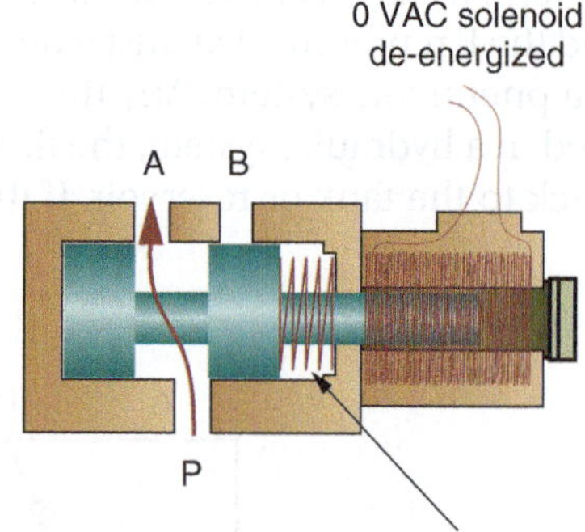

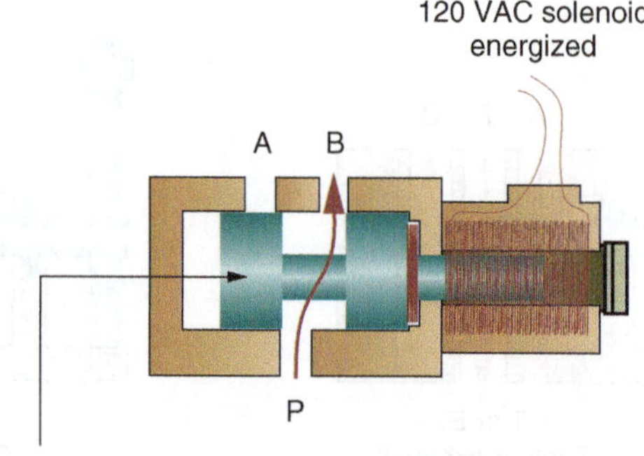

FIGURE 12-15 A solenoid valve.

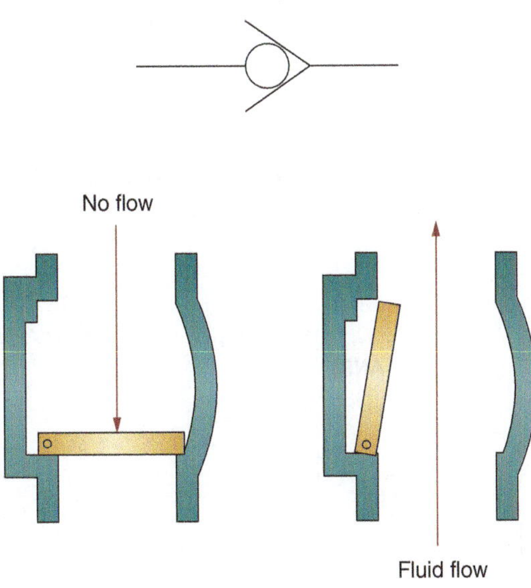

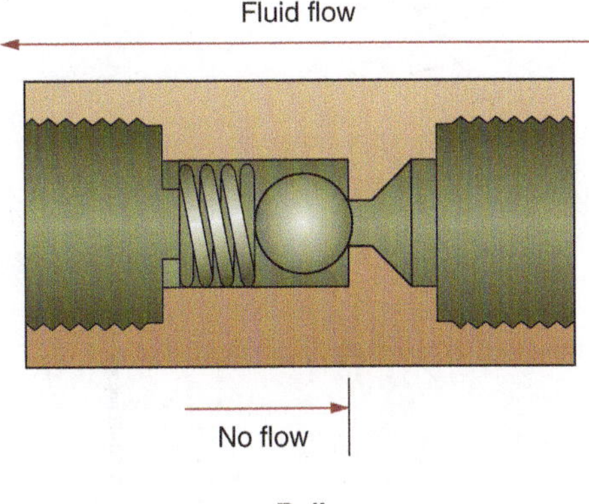

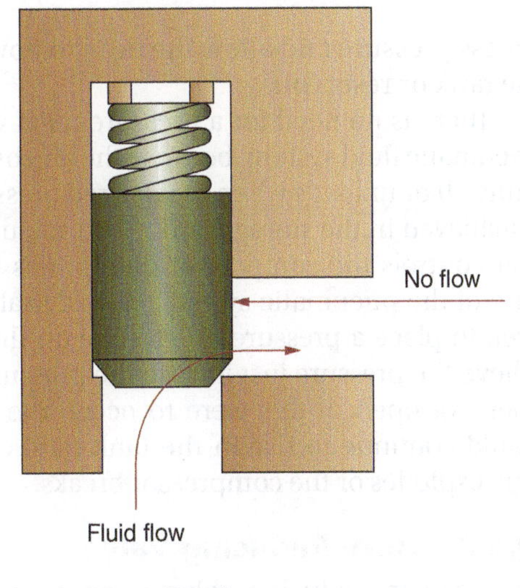

FIGURE 12-16 Three types of check valves.

to remain sealed in one direction, but it allows fluid flow in the opposite direction when the pressure in the fluid rises to a level high enough to overcome the spring tension.

The final check valve shown in the figure is a barrel check valve. It operates very similarly to the ball check valve except that it does not have a ball in it. It has a barrel that provides the seal. Like the ball check valve, it is spring loaded in order to prevent fluid flow in one direction, but it allows it in the opposite direction.

The Pressure-Relief Valve

The pressure-relief valve is similar to the check valve in the sense that it too has an internal spring. The pressure relief valve is constructed in a manner that there is no flow through the valve until a certain pressure within the system has been obtained. The pressure that is achieved when the pressure-relief valve actuates is the pressure rating of the relief valve. The pressure-relief valve is a pressure control valve and not a directional or flow control device. The pressure-relief valve is used in both hydraulic and pneumatic systems. It is a safety device that protects both the individual and property. Never set the pressure-relief valve higher than the lowest rated operating pressure in the circuit. Failure to do this may cause a component to explode or an unexpected release of fluid at the point that has the lowest rating. A poppet-type pressure-relief valve is shown in **Figure 12-17**.

Some pressure-relief valves can be adjusted by simply adjusting the tension on the spring that holds the poppet against its seat as shown in Figure 12-17. This is not the case for all pressure-relief valves, however. The basic operation of the pressure-relief valve were that if the pressure remains below the rated pressure of the valve, the flow will continue in the system as though the valve were not even present in the system. However, when the pressure rises to a level in which the valve is rated, the spring tension is overcome, and the valve is opened to allow a path for fluid to flow back to the tank. This relieves the pressure in the system. The pressure-relief valve remains open as long as the pressure is at or above the pressure rating of the valve. As soon as the pressure drops below the actuating pressure level, the valve closes again, shutting off the flow of fluid back to the tank. If the pressure rises again, the valve again relieves the pressure by bleeding the excess pressure off to the tank. This description of the operation of a pressure-relief valve is for a direct-acting, hydraulic fluid power system.

There are also pilot-operated pressure-relief valves that are used for hydraulic systems. They are usually used in hydraulic systems that are of

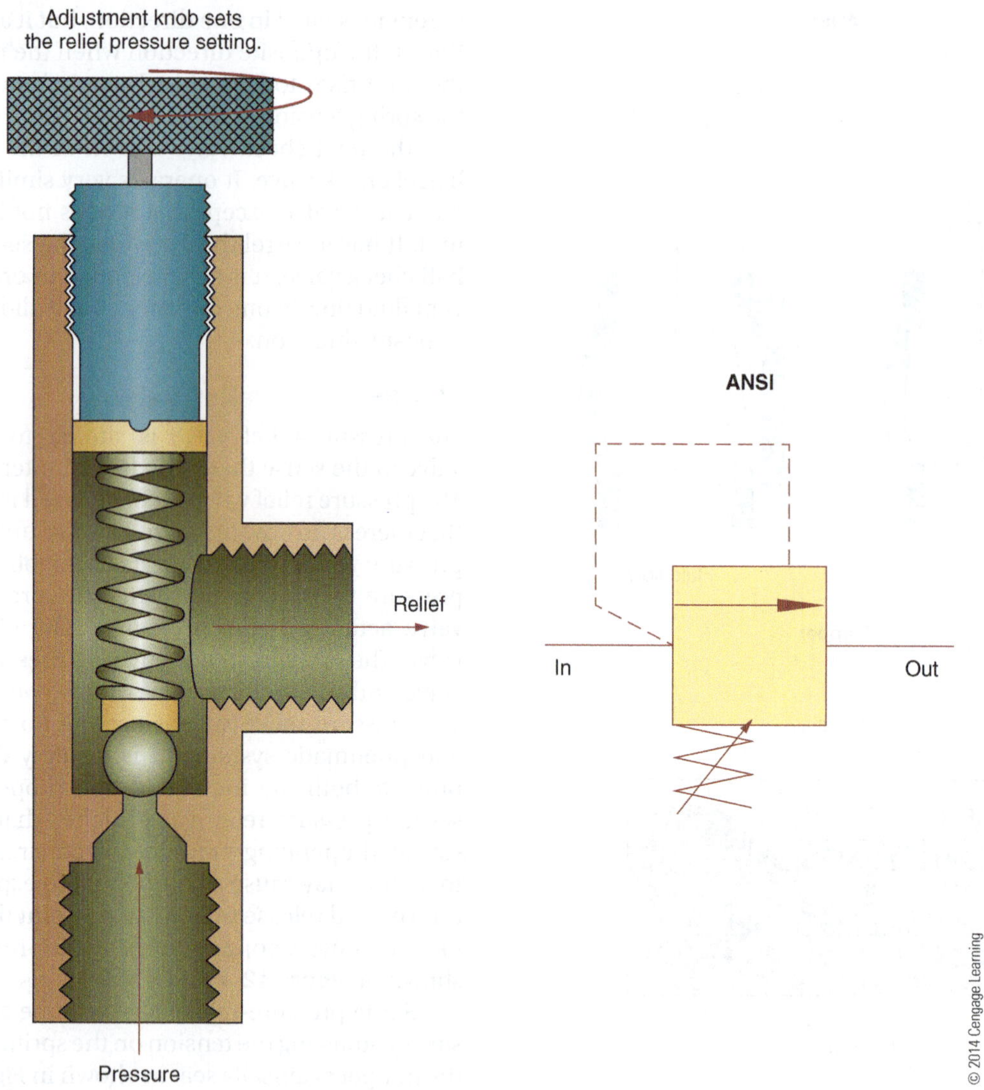

FIGURE 12-17 A pressure-relief valve (poppet type).

a higher pressure and flow. Refer to **Figure 12-18.** Notice the small orifice in the center of the main poppet.

The purpose of the small hole is to allow the pressurized fluid to flow from the main poppet to the pilot poppet. The small orifice equalizes the pressure between the top and bottom sides of the main poppet. The spring that is holding the main poppet in place is a light spring that is rated at a cracking pressure of 25 psi to 75 psi. Cracking pressure is the pressure at which the valve poppet unseats, causing the pressure to be relieved. When the pressure rises to a level that is high enough to make the pilot poppet open, fluid begins to flow back to the tank through the orifice in the main poppet. Once this occurs, a difference in pressure is present between the top and the bottom of the main poppet. This causes the main poppet to open just enough to alleviate the excess pressure and allow the fluid to flow back to the tank or reservoir.

There is no need for a pressure relief valve in a pneumatic fluid system, because the air compressor shuts off or unloads when the desired pressure level is achieved in the storage tank. The pressure switch that controls the compressor determines the pressure of the pneumatic system. It is advisable, however, to place a pressure-relief valve on the tank to relieve the pressure just in case the pressure switch does not open. If this were to occur, the pressure would continue to rise in the tank until either the tank explodes or the compressor breaks.

The Pressure-Reducing Valve

The pressure-reducing valve is another pressure control valve. This valve makes it possible to have more than one operating pressure within a system. The output pressure of the pump, any restrictions

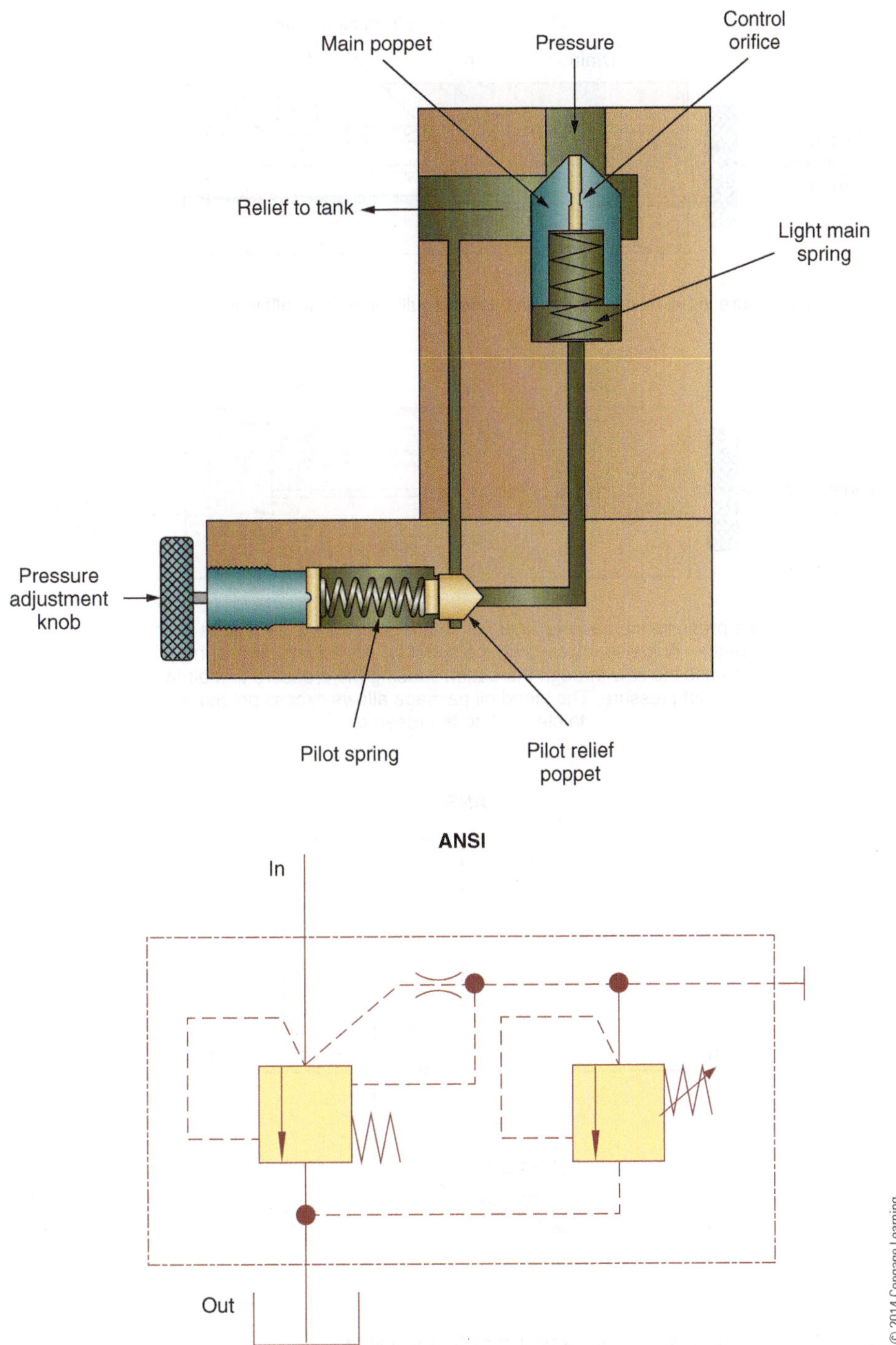

FIGURE 12-18 A pressure-relief valve (pilot type).

that may be present in a system, and the rating of the pressure-relief valve determine the highest pressure that is available in a hydraulic fluid system; the pressure-reducing valve does not. The cutoff pressure of the compressor determines the maximum pressure of a pneumatic system. This type of valve is used to maintain reduced pressures in specified locations of a hydraulic system. There are many types of pressure-reducing valves in industry today. The most common are the direct-acting pressure-reducing valve and the pilot-operated pressure-reducing valves. The direct-acting pressure-reducing valve operates with a pilot pressure that is taken from the outlet side of the valve, not from the inlet side as occurs with the pressure-relief valve. The valve is a normally open valve that tends to close when

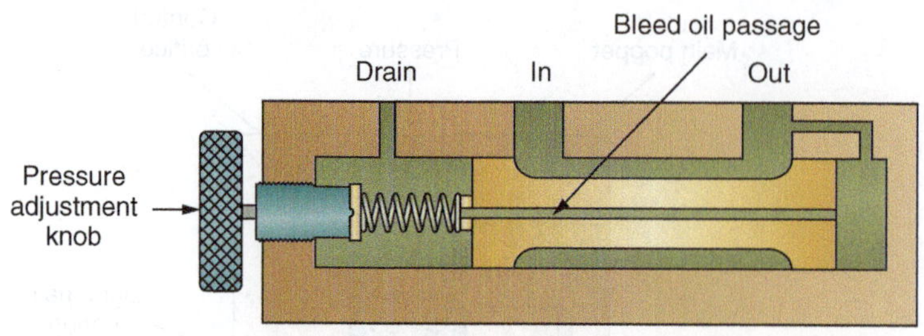

If pressure is below the setting, the valve is fully opened to allow maximum flow.

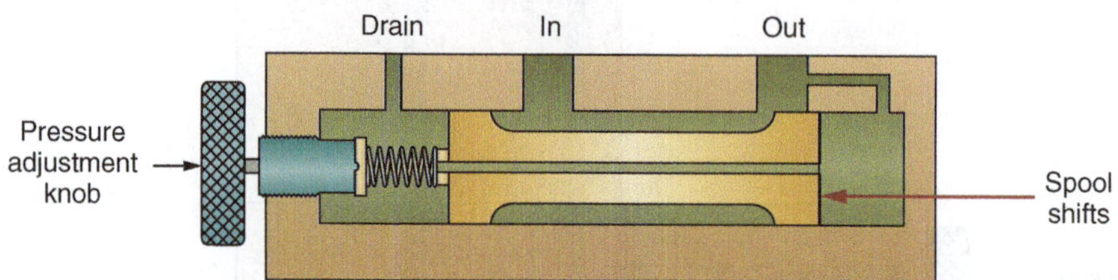

As pressure increases above the setting, pressure is sensed at the bottom of the spool, causing the spool to shift against the spring. This limits the flow through the outlet, causing the pressure to maintain the set pressure. The bleed oil passage allows excess pressure to bleed off to the reservoir.

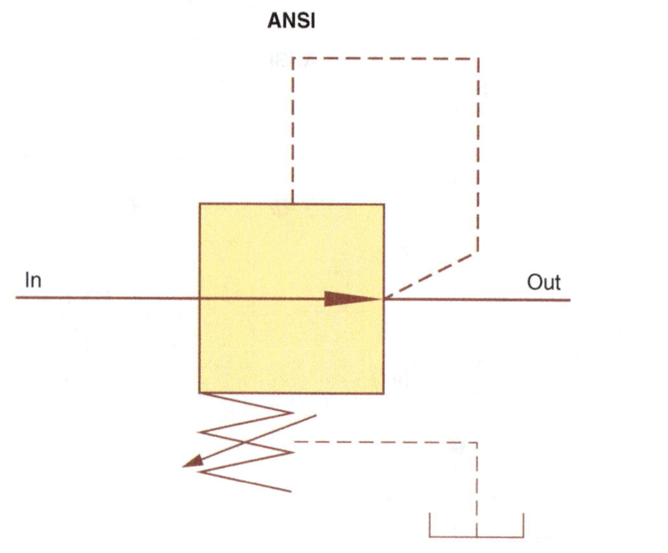

FIGURE 12-19 A direct-acting pressure-reducing valve.

the downstream pressure reaches the valve setting. A direct-acting pressure-reducing valve is shown in **Figure 12-19**.

Notice that the spool is spring-loaded. If the downstream pressure is lower than the valve setting, the fluid flows freely from the inlet port through the valve to the outlet port. Now, notice the pilot passageway through the center of the spool. When the downstream pressure increases to the valve setting, the spool moves to partially cover the outlet port. The spool is moved just enough to allow enough fluid to flow through the outlet port to maintain the pressure equal to the setting of the valve.

12-5 LINEAR ACTUATORS

Cylinders are often referred to as **linear actuators**. Cylinders and pistons are easy to understand. **Figure 12-20** shows a graphic representation of a cylinder. The term *cylinder* usually refers to all of

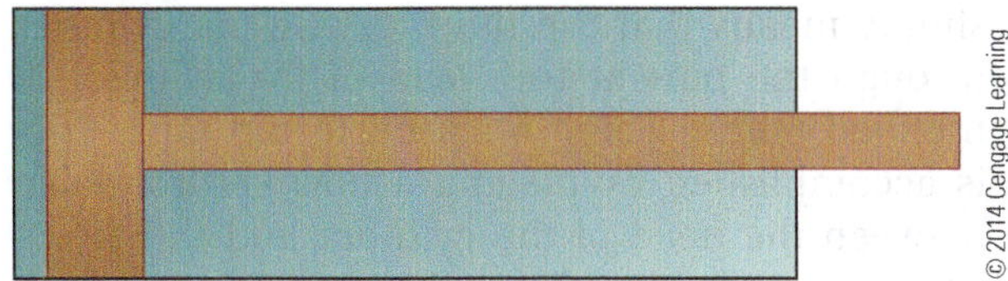

FIGURE 12-20 The ANSI symbol for a cylinder.

the components that are housed within the cylinder, including the piston. The piston is the part of the cylinder that does the work because the shaft of the piston is connected to the load. It fits within the cylinder with tolerances that are so close that a seal is formed against the cylinder wall. It is also not uncommon for a rubber, fiber, or leather seal to be set into the piston so a seal is formed against the cylinder wall. This is what makes it possible to create a difference of pressure within the cylinder. Movement of the piston is accomplished by creating a higher pressure on one side of the piston than on the other side. When a cylinder is used in a system, it is referred to as a linear system.

The Single-Acting Cylinder

Notice in **Figure 12-21** that the cylinder contains a spring, which is pressing against the piston. This is referred to as a single-acting cylinder. Look at the fluid system in Figure 12-21. Notice how the three-way valve is controlling the directional movement of the piston. As the three-way valve is moved, the fluid flows from the pressure supply port to port A. The fluid travels into the back of the cylinder, causing the piston to extend out of the cylinder. The valve is then moved in the opposite direction, allowing the fluid to flow from the cylinder, back through the valve, and out of the valve through the tank or exhaust port.

The Double-Acting Cylinder

Refer to **Figure 12-22**. Notice that there is no spring present within the cylinder and that a four-way valve is being used.

The operation of a double-acting cylinder is simple. The cylinder is extended in the same manner as the single-acting cylinder. The valve is shifted, allowing fluid to flow from the pressure supply port to port A. The fluid then travels into the back of the cylinder, causing the piston to extend out of the cylinder. When the valve is centered, the cylinder holds its extended state. It remains this way until the four-way valve is shifted in the opposite direction.

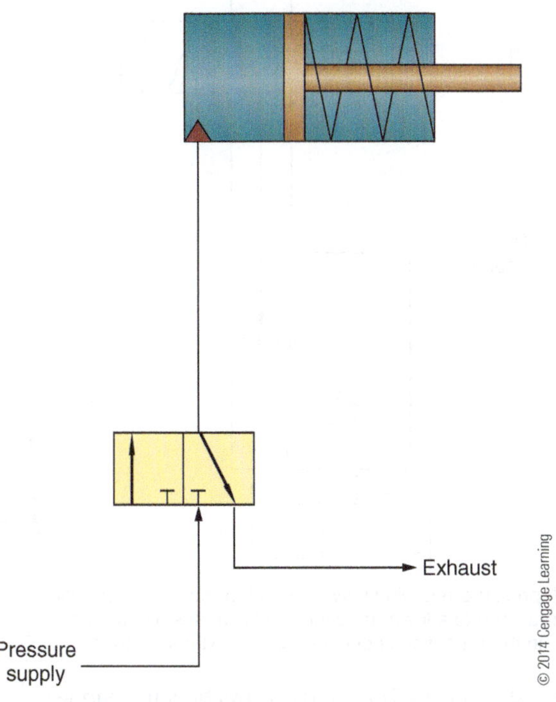

FIGURE 12-21 The single-acting cylinder principle.

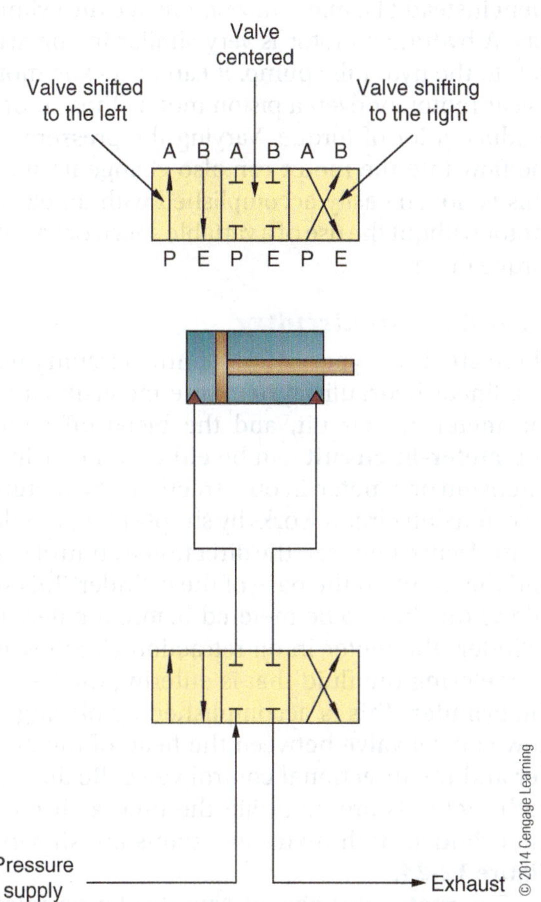

FIGURE 12-22 The double-acting cylinder principle.

Moving the valve to this position accomplishes two things. First, it allows the pressurized fluid from the pressure port to flow out of the valve through port B. This makes the fluid flow into the other side of the piston, causing the piston to retract. Second, it provides a path from port A to the E or T port. This allows the fluid that caused the piston to extend to be evacuated from the cylinder and be exhausted or sent back to the tank. When the valve provided a path for the fluid to be released from the cylinder, the pressure on that side of the cylinder dropped in comparison to the pressure being applied to the other side of the piston. This difference in potential is what causes the piston to retract. The retraction of a double-acting cylinder in most cases is faster than that of a single-acting cylinder. This is because the pressure that is applied to the retraction side of the piston is more than that of a spring. With the use of a five-way valve, the double-acting cylinder can make the extension speed and the retraction speed of the double-acting cylinder different. This is shown in **Figure 12-23**.

Rotary Actuators

The hydraulic and pneumatic motor is referred to as a **rotary actuator**. This device has a rotary movement instead of a linear movement like the cylinder has. A hydraulic motor is very similar in construction to the hydraulic pump. It can be a vane motor, a gear motor, or even a piston motor. These motors produce a lot of torque. Varying the pressure and the flow into the motor can also change its speed. This is not so easily accomplished with an electric motor without the use of a variable speed or variable torque drive.

Flow Control Circuitry

There are three types of flow control circuitry used in a linear hydraulic system: the meter-in circuit, the meter-out circuit, and the bleed-off circuit. The **meter-in circuit** can be either a meter-in on extension or a meter-in on retraction. The meter-in on extension circuit works by simply placing a flow control valve between the directional control valve and the input on the base of the cylinder. This will allows the fluid to be metered before it enters the cylinder. The meter-in on retraction circuit works by metering the fluid that is entering the head of the cylinder. This is accomplished by placing the flow control valve between the head of the cylinder and the directional control valve. Realize that both methods are metering the fluid as it enters the cylinder. Both meter-in circuits are shown in **Figure 12-24**.

The **meter-out circuit** can also be on retraction and on extension. The meter-out on extension simply means that the fluid leaving the cylinder through the port at the head of the cylinder is metered before being released to the tank. This is accomplished by placing a flow control valve between the head of the cylinder and the directional control valve. The meter-out on retraction has the flow control valve connected between the base port of the cylinder and the directional control valve. Notice that in both cases, the fluid is being metered as it leaves the cylinder and returns

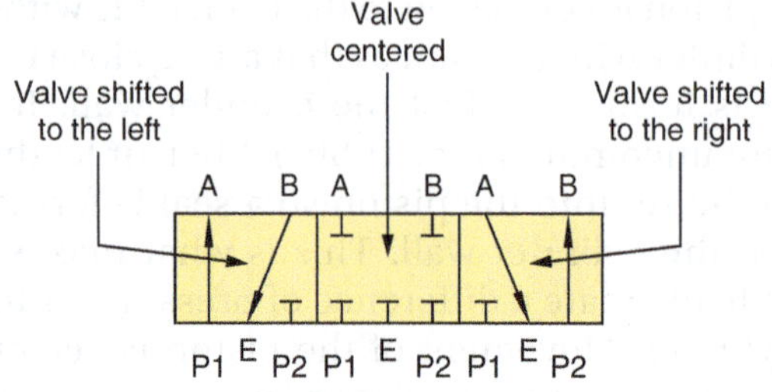

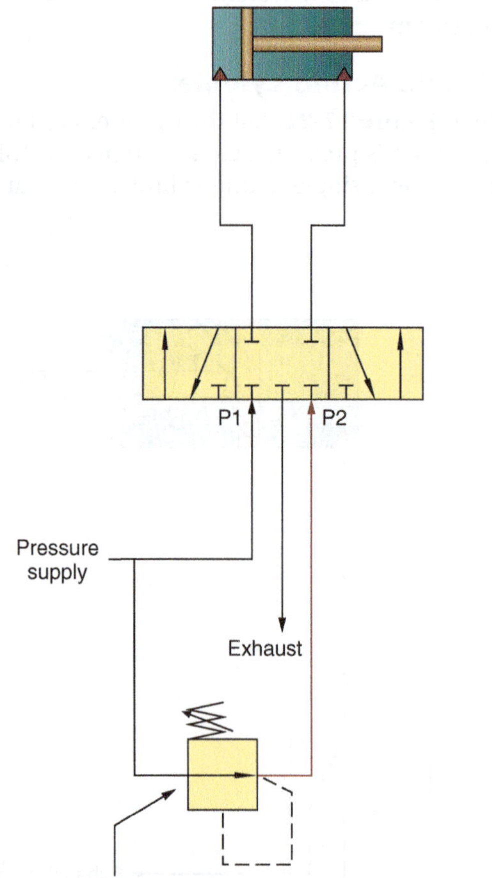

FIGURE 12-23 This circuit shows how to change the retraction speed.

Meter-in on extension
Notice that the flow control valve is on the base of the cylinder. Also notice that the check valve provides a path for the cylinder to move unrestricted during retraction.
Pressure supply
Meter-in on retraction
Notice that the flow control valve is on the head of the cylinder. Also notice that the check valve provides a path for the cylinder to move unrestricted during extension.
Pressure supply
© 2014 Cengage Learning

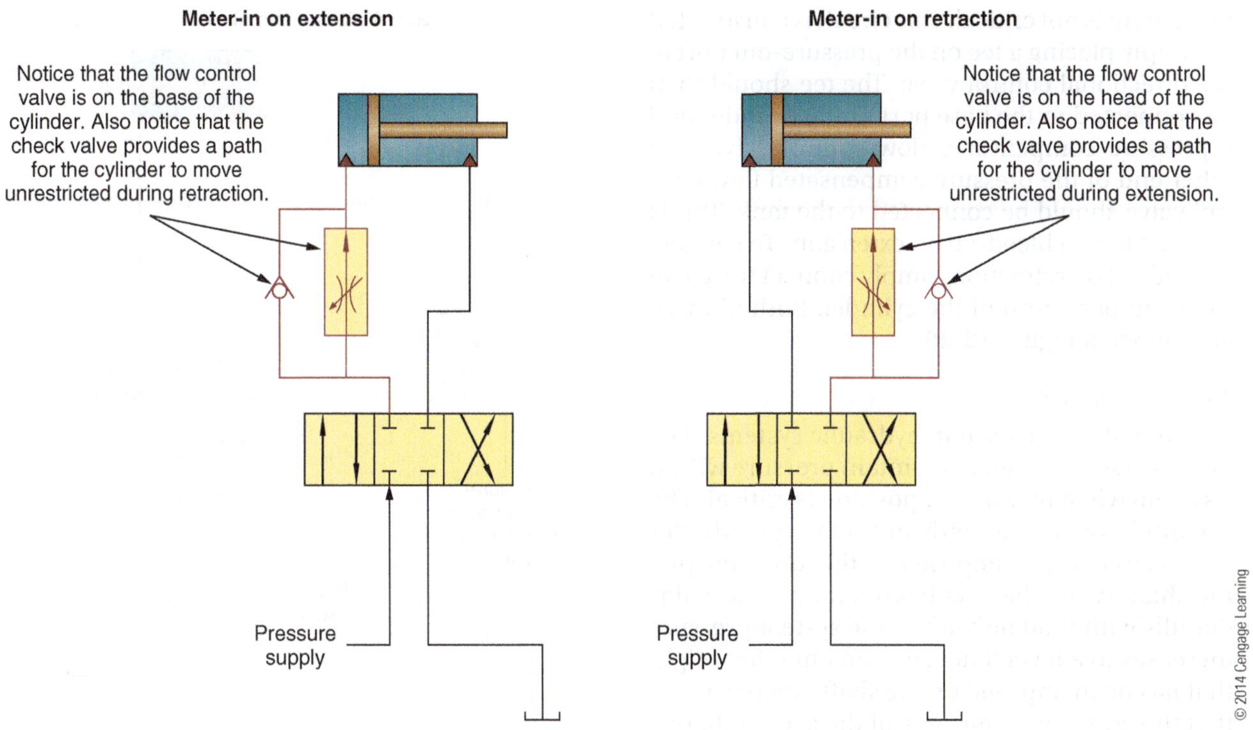

FIGURE 12-24 Two types of meter-in circuitry.

to the tank. This method is most desirable when a cylinder is mounted vertically and gravity assists the stroke. An example would be an elevator. The retraction of the elevator must be metered to prevent a rapid descent. Both meter-out circuits are shown in **Figure 12-25.**

The **bleed-off circuit** is used when speed control is required but the actual rate of cylinder

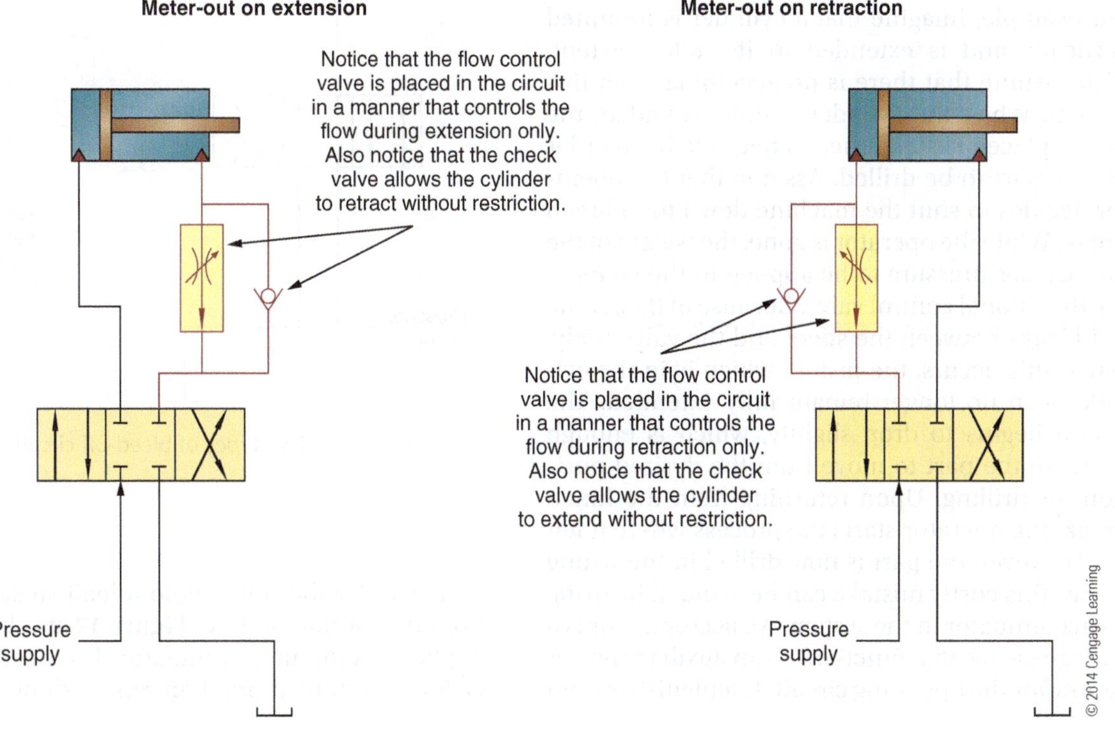
Meter-out on extension
Notice that the flow control valve is placed in the circuit in a manner that controls the flow during extension only. Also notice that the check valve allows the cylinder to retract without restriction.
Pressure supply
Meter-out on retraction
Notice that the flow control valve is placed in the circuit in a manner that controls the flow during retraction only. Also notice that the check valve allows the cylinder to extend without restriction.
Pressure supply
© 2014 Cengage Learning

FIGURE 12-25 Two types of meter-out circuitry.

movement is not critical. This circuit is constructed by simply placing a tee on the pressure-out port of the directional control valve. The tee should then be connected to the base port of the cylinder and a pressure compensated flow control valve. The other end of the pressure compensated flow control valve should be connected to the tank. This is referred to as a bleed-off on extension. To connect a bleed-off on retraction, simply connect the tee to the head port (top) of the cylinder. Both of these are shown in **Figure 12-26**.

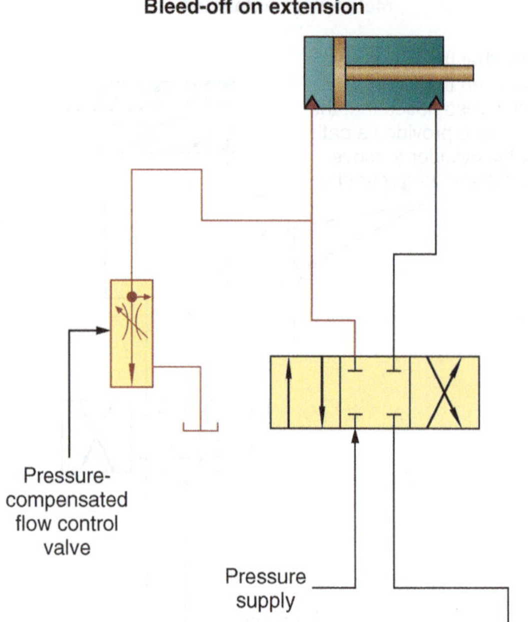

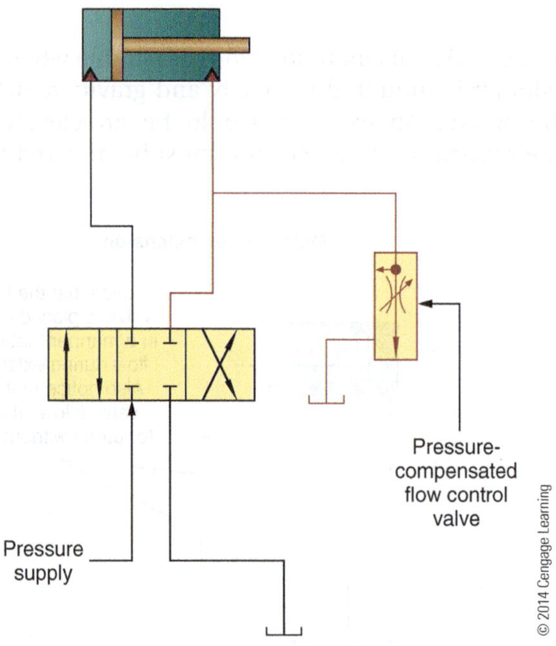

FIGURE 12-26 Two types of bleed-off circuitry.

Accumulators

Accumulators are used in hydraulic systems. The **accumulator** is used to maintain pressure within a system when pressure or position is critical. The accumulator acts similarly to a ram-type cylinder with a large weight imposed on the top of the piston shaft. When the circuit is operating, the cylinder fills with fluid only when the system pressure increases to a level that can overcome the weight that has been imposed on the shaft. Keep in mind that the cross-sectional area of the accumulator's piston does affect the pressure at which the accumulator begins to fill (piston begins to rise).

Most sliding spool valves have a small amount of leakage between the body of the valve and the spool when they are in the closed position. Over time, this could cause a drop in pressure within the system if the load is opposing the cylinder. A drop in the system pressure can cause the cylinder or motor to move from the desired position. For example, imagine that a cylinder is mounted vertically and is extended to its fullest extent. Also assume that there is no accumulator in this system. When the cylinder is fully extended, the part is placed exactly where it needs to be in order for the part to be drilled. Assume that the operator decides to shut the machine down for a lunch break. While the operator is gone, the weight of the part causes pressure to be applied to the spool of the directional control valve. Because of this, some fluid leaks between the spool and the valve body. When this occurs, the piston, which is in the cylinder, can no longer remain fully extended. The piston begins to drop slightly, which is enough to cause the part to move from the desired position for drilling. Upon returning from the lunch break, the operator starts the process where it left off. However, the part is now drilled in the wrong place. This costly mistake can be avoided by using an accumulator in the system. An accumulator is a storage device that functions as an auxiliary power supply for the operating circuit. It replenishes fluid to a cylinder that must hold a load steady for an extended period of time. **Figure 12-27** shows one application for an accumulator. This accumulator is doing nothing more than replenishing the fluid

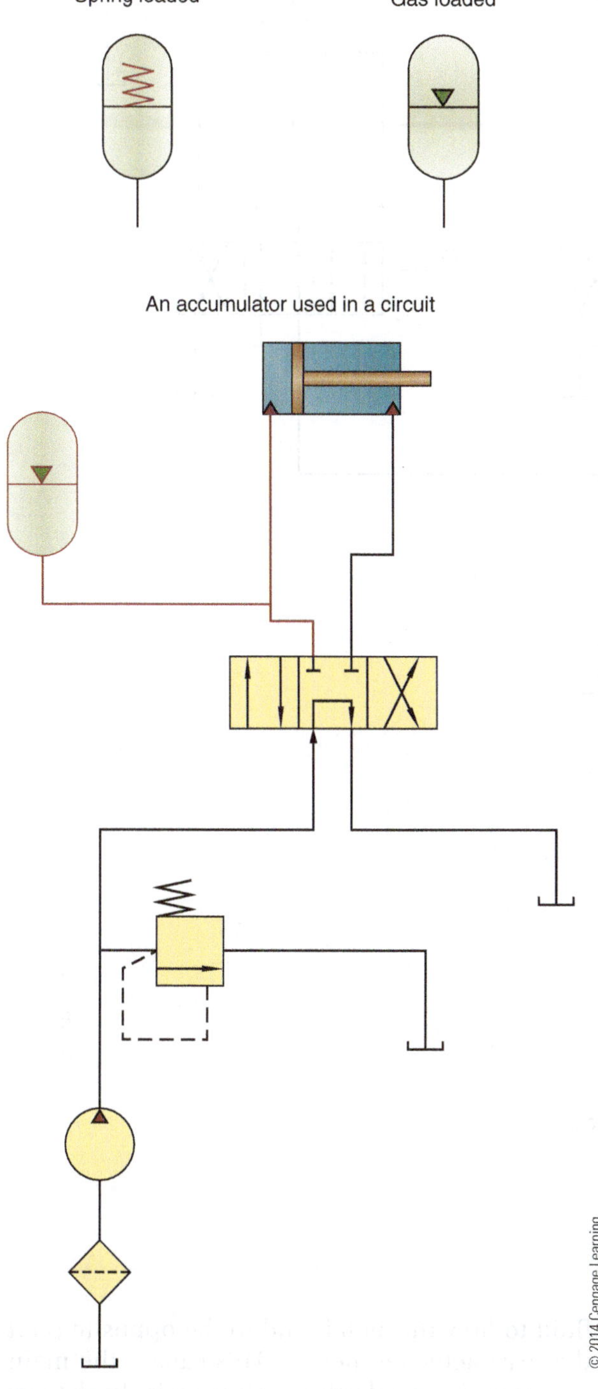

FIGURE 12-27 An accumulator replenishing circuit.

that is lost because of leakage. This ANSI symbol is also shown in Figure 12-27.

Notice that the accumulator in the circuit charges when the cylinder extends. This causes the accumulator to charge to the same pressure as the cylinder. Replenishing circuits can be used as safety devices as well. Used in this condition, the capacity of the accumulator is generally sized to be able to fully extend and retract completely. If a motor is used, the accumulator should be sized to allow one complete rotation of the motor. This type of system is critical in the event of a malfunction or when the potential supply has been shut off. This system is shown in **Figure 12-28**.

This circuit can also be automatic, as shown in **Figure 12-29**. Notice in Figure 12-29 that the replenishing circuit can be made automatic by using a normally open, two-position, one-way directional control valve that is pilot activated from the main circuit. When the pressure drops in the circuit because of pump failure, the pilot-operated valve shifts and causes the accumulator to discharge into the circuit. This allows the cylinder to be shifted into either position to complete the cylinder cycle.

Accumulators are also used to control surging and pulsations in the hydraulic system, acting somewhat like a shock absorber.

The two most common types of accumulators that are in use today are the spring-type accumulator and the gas-type accumulator. A cross-sectional view of a spring-type and two types of gas accumulators are shown in **Figure 12-30**. **Figure 12-31** (see in page 248) shows a bladder-type, gas-charged accumulator.

Pneumatic Directional Control Circuitry

The most common pneumatic control circuit is referred to as the **flip-flop circuit**. It uses a double-end rod, double-acting cylinder, which is sometimes referred to as an **oscillator**. The schematic is shown in **Figure 12-32** (see in page 248).

Notice the two limit switches in the schematic. These switches cause the oscillation of the cylinder automatically. The supply pressure of LS1 is connected to the extend right (ER) output of the main directional control valve. Also notice that the supply pressure for LS2 is connected to the extend left (EL) output of the main directional control valve. To explain the operation of the flip-flop circuit, first assume that the piston is in the center as shown in Figure 12-32. As supply pressure is allowed to flow through the directional control valve, the fluid enters the left side of the cylinder. This causes the piston to extend to the right (ER). When the cylinder is fully extended, the cam on the end of the rod strikes the limit roller, causing it to open. This allows

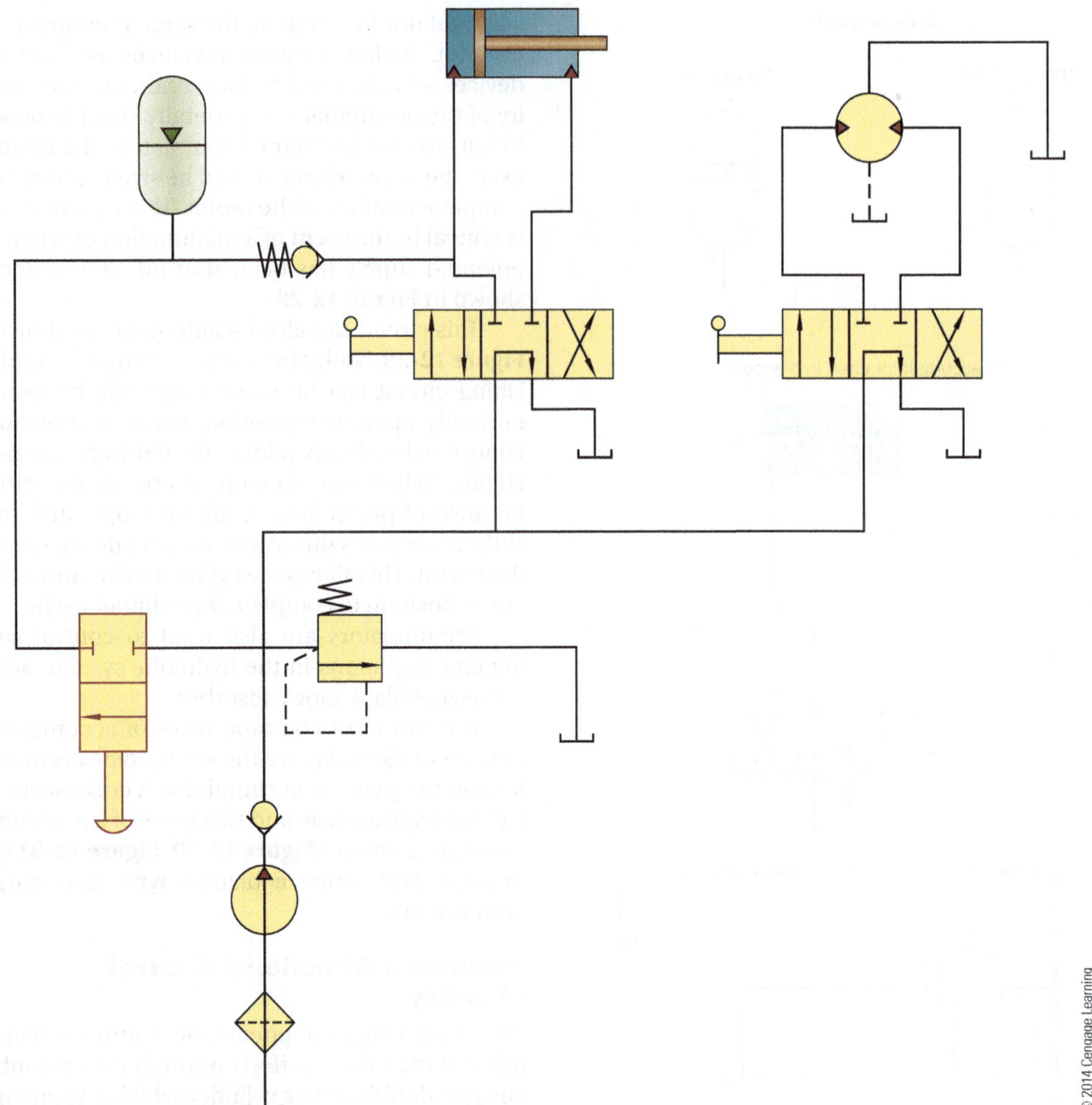

FIGURE 12-28 A replenishing circuit used as a safety device.

the fluid to pass through to the pilot of the main directional control valve. This, in turn, causes the main directional control valve to shift. When the main directional control valve shifts, the fluid enters the cylinder on the right side, causing it to extend to the left (EL). As the piston begins to move to the left, the cam moves off LS2, thus removing the continual pilot pressure from the main directional control valve. As the piston extends to the far left, it strikes LS1, causing the fluid to flow through it and to the opposite pilot that was activated before. This causes the main directional control valve to shift again, back to its original position, thus causing the piston in the cylinder to extend to the right once again. This continues until the pressure supply is removed from the circuit. This flow control valve can be added to the pilot circuits to create a time-delay effect on the shifting of the main directional control valve.

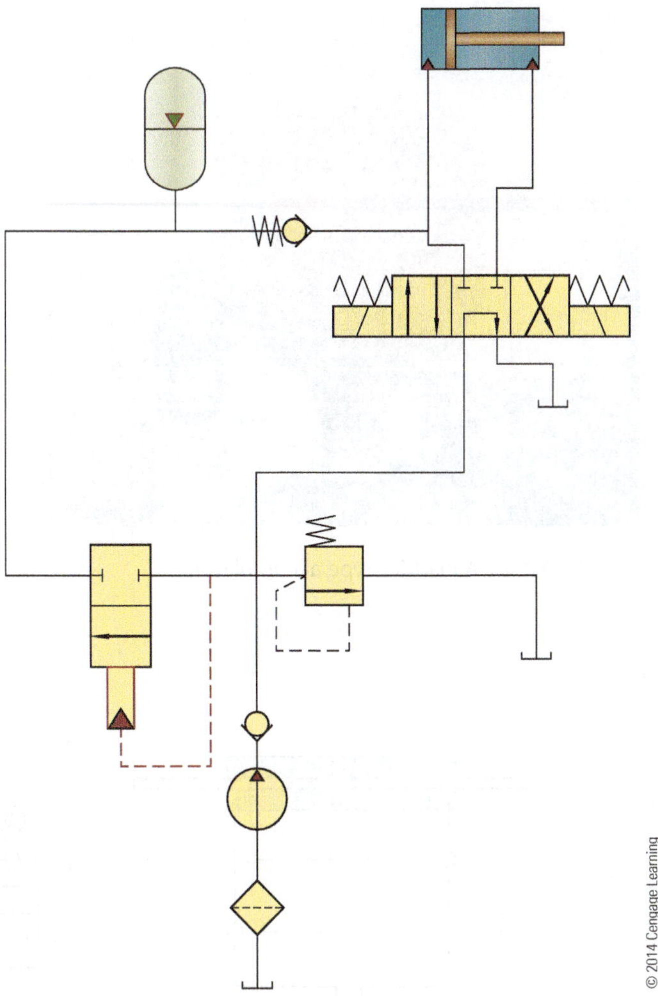

FIGURE 12-29 An automatic accumulator replenishing circuit.

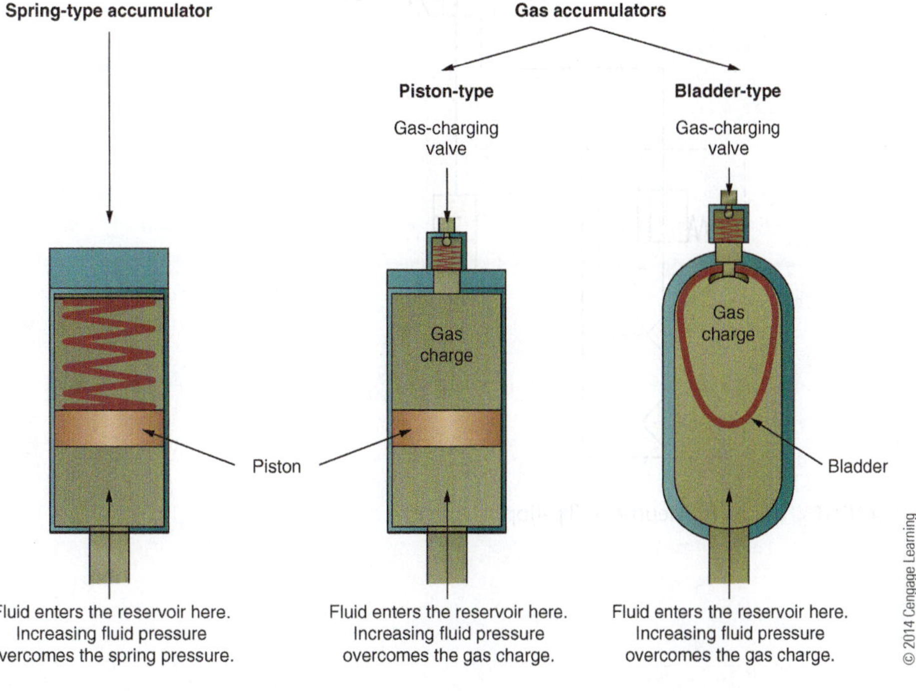

FIGURE 12-30 Two common types of accumulators.

FIGURE 12-31 A bladder-type accumulator.

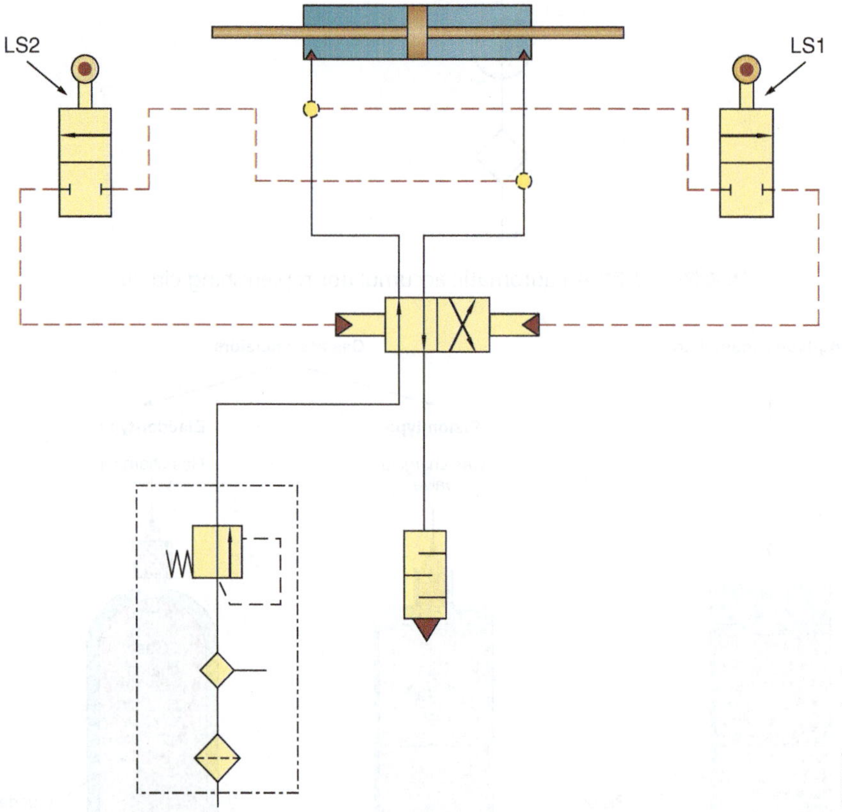

FIGURE 12-32 A pneumatic flip-flop circuit.

SUMMARY

- There are two methods of transmitting power through a fluid, hydraulic and pneumatic.
- The word *hydraulic* refers to something that is operated by the movement and pressure of a liquid.
- The word *pneumatic* refers to a fluid power system that uses compressed air or a vacuum.
- Although air is a gas, it is referred to as a fluid; however, it is not a liquid like the oil in a hydraulic system.
- Pressure is only present when there is a restriction against the flow of the fluid.
- Fluid power has several advantages over mechanical power. These include ease of control, accuracy, force multiplication, constant force, constant torque, and rapid reversal capabilities.
- Pressures in both the hydraulic and pneumatic systems are measured in pounds per square inch (psi).
- When pressures are read from a gauge, the reading is referred to as psig.
- Atmospheric pressure is psia.
- A vacuum can also be measured in psi but is more commonly measured in inches of mercury ("Hg).
- Static head pressure is defined as the amount of pressure that is developed for every inch of rise in elevation above the point of measurement.
- When pressure is applied to the fluid, the pressure is transmitted through the fluid.
- The force of a piston can be calculated by multiplying the pressure by the area ($F = P \times A$).
- When dealing with pneumatics, pressure is created as soon as the air is compressed.
- A hydraulic system does not store energy as does the pneumatic system.
- The main goal of compressed air is to get back to atmospheric pressure.
- Fluid conditioners are separate devices that are used to clean, purify, dry, and lubricate compressed air.
- Vacuum fluid power may be described as power that derives its force from the weight of the atmosphere.
- A perfect vacuum is 30 in. of mercury.
- A vacuum pump is rated according to how much vacuum it can pull in comparison to a perfect vacuum. For example, a vacuum pump that is capable of producing 27 in. of mercury at sea level is said to be a pump with a 90% rating ($27 \div 30 = 0.90$, or 90%).
- A valve is used for directional control, flow control, and pressure control.
- A normally open valve allows fluid to flow through the valve in its normal state until the valve is actuated.
- A normally closed valve does not have a path for fluid to flow in its normal state until the valve is actuated.
- When a valve is actuated, it changes state internally.
- The two-way valve has two ports and is usually used for flow control only.
- The three-way valve has three ports and is a directional control valve.
- The four-way valve not only controls the direction of flow, as does the three-way valve, but it also has shut-off capabilities and four ports.
- The five-way valve has five ports and offers the flexibility of using two separate pressures that are applied to ports A and B.
- Valves can be actuated by manual lever, pushbutton, foot pedal, cam or mechanical operation, or solenoid, or can be pilot operated.
- The check valve is used to allow fluid to flow in one direction, but not in the opposite direction.
- The pressure-relief valve is constructed in such a manner that there is no flow through the valve until a certain pressure within the system has been obtained. The pressure that is achieved when the pressure-relief valve actuates is the pressure rating of the relief valve. The pressure-relief valve is a pressure control valve instead of a directional or flow control device.
- The pressure-reducing valve is another pressure control valve. This valve makes it possible to have more than one operating pressure within a system.
- The piston is the part of the cylinder that does the work because the shaft of the piston is connected to the load.
- The single-acting cylinder uses a spring to return the piston back to the retracted position.

- The double-acting cylinder can have fluid pressure applied to both sides of the piston, unlike the single-acting cylinder.
- The single-acting and double-acting cylinders are referred to as linear motion actuators.
- A hydraulic motor produces rotary movement instead of linear movement, which the cylinder produces.
- Flow control circuitry is often used to control the speed of extension and retraction. Three types of flow control are the meter-in circuit, the meter-out circuit, and the bleed-off circuit.
- Accumulators are a form of auxiliary pressure or fluid supply.
- The flip-flop circuit is used in pneumatics and is commonly used in industry. The flip-flop circuit is a circuit in which a cylinder cycles back and forth automatically.

REVIEW QUESTIONS

1. What type of fluid is used in a pneumatic system?
2. What causes pressure in a hydraulic system?
3. List at least three advantages that fluid power has over mechanical power.
4. What is a perfect vacuum?
5. What is the purpose of the check valve?

CHAPTER

13

Piping Systems

Many types of piping systems are used to deliver fluids from one place to another. This chapter discusses a number of the piping systems that are in use today and covers some of the most common tools that are used when installing piping systems.

OBJECTIVES

After studying this chapter, the student should be able to

- List the necessary tools that may be used in connecting piping systems together.
- List and explain, in detail, the types of piping systems that are in use today.
- List, describe, and explain the use of various types of fittings in piping systems.
- Be able to interpret a piping sketch and correctly size, cut, and assemble pipes according to the dimensions laid out in the piping sketch.
- Explain the differences between center-to-center, end-to-center, and end-to-end measurements.
- Correctly deduce the necessary fitting allowance from the center-to-center measurements.
- Demonstrate the different methods that are used to connect piping systems together.

13-1 PIPING TOOLS

Piping is used to contain the movement of a fluid or gas and direct it to a desired location. To do this, pipes and fittings are assembled together to provide the path in which the fluid or gas flows. To assemble a piping system, several tools may be needed. Every tool that is mentioned may not be needed; however, it may be wise to have the tool close by in case it is needed. End preparation tools, consisting of the hacksaw and the pipe cutter, are discussed first.

Hacksaw

The hacksaw has two major components to its construction: the frame and the blade. Most hacksaws are adjustable in order to accept blades from 8 in. to 16 in. in length. Hacksaw blades are hardened and tempered. This makes the cutting teeth of the hacksaw blade harder than the material that is to be cut. The hacksaw blade has between fourteen and thirty-two teeth per inch. Be sure to use the proper hacksaw blade for the task at hand. Use a blade with 14 teeth per inch when cutting a large section of mild material. Use 18 teeth per inch when cutting on a large section of hard material. Use 24 teeth per inch when cutting brass, thick wall pipe, angle iron, and copper. Use 32 teeth per inch when cutting thin wall tubing. For the proper use of the hacksaw, refer to Chapter 2. Care must be taken when cutting a pipe for end preparation. Do not allow the hacksaw blade to angle as the cut is being made. If this occurs, the cut will be at an angle and will pose a problem upon starting the threads.

Pipe Cutter

The pipe cutter does not rely on a back-and-forth motion to cut the pipe, and it produces a much smoother cut than the hacksaw. It cuts the pipe in a rotary motion. Pipe cutters may have from one to four cutting wheels. Most, however, have only one. A pipe cutter that has four cutting wheels does not have to be rotated around the pipe as far as a pipe cutter that has only one wheel because the pipe is being cut in four places simultaneously rather than being cut in only one place. The pipe cutter with four wheels is used in very tight places where there is limited space around the pipe, which hampers the ability to rotate a single-wheel cutter a full 360° around the pipe. Place the cutting wheel on the pipe where the cut has to be made and tighten the T-handle until the cutting wheel is pressed firmly against the piece of pipe. Make one full rotation, if a single-wheel pipe cutter is being used, around the pipe. After every one rotation of 360°, tighten the T-handle ¼ turn. This presses the cutting wheel farther into the pipe. Make another rotation around the pipe. Continue this operation until the pipe has been cut all the way through.

Reamer

Once the pipe has been cut, whether by a hacksaw or a pipe cutter, the inside of the pipe has to be reamed. This is accomplished by using a reamer. A reamer is shown in **Figure 13-1**.

The sharp edges on the reamer remove any sharp edges that were made when the pipe was cut with a pipe cutter, or deburr a cut that was made with a hacksaw. It is important to ream the pipe that has just been cut to remove any burrs or sharp edges that may become an obstruction to the fluid flow. Many types of reamers are in use today, but the most common one is the ratcheting type. This allows the user to apply continuous pressure while deburring the pipe.

Pipe Threader

Once the end of the pipe has been cut and reamed, the pipe may have to be threaded. A pipe threader is used to put threads on a pipe. The specifications of the threads are discussed later in this chapter. However, it is important that you know what a pipe threader is and how it works. There are two types of pipe threader: the manual pipe threader, sometimes referred to as a hand threader, and the power threader. The hand threader, which is shown in **Figure 13-2**, is the ratcheting type.

This type of threader has the capability of swapping the stock to allow a larger diameter pipe to be threaded. The stock is the part of the threader that holds the cutting die. The cutting die is what cuts the threads into the surface of the pipe as it is rotated around the pipe. Although this threader is referred to as a hand threader, it can be used in conjunction

FIGURE 13-1 A reamer.

FIGURE 13-2 A hand threader.

with a power vise or power drive, which cuts the threads into the surface of the pipe automatically instead of manually. A pipe-threading machine, shown in **Figure 13-3**, is usually used when a large quantity of pipe is to be threaded.

The power threader usually has removable dies that allow the operator to thread pipes from $^3/_8$ in. up to 4 in. or more. The power threader also has a cutter and a reamer, which, when not in use, are flipped out of the way to allow the pipe to be threaded. The power threader has a built-in vise that is used to hold the pipe in place while it is being cut, reamed, or threaded. The die and reamer are both driven into the end of the pipe, which is held by the vise, by a rotating lever that is mounted on the side of the power threader. This lever turns a rack and pinion gear set that causes the floating head to drive the die or reamer into the pipe. When the pipe is being threaded, the die is pressed against the pipe end. Once the die catches and begins to cut the threads, the lever can be released, and the die begins to travel on its own from that point on. It is very important to ensure that the end of the pipe is lubricated continually, before and during the threading procedure. Most power threaders have an oil lubricator built into the threader to pump oil any time the threader is turned on. When the cutting die is not in use, it is flipped out of the way. This shuts off the flow of oil and moves the lubricating stem out of the way during the cutting and reaming procedures. Once the threads are cut to the desired length, a release on the top of the die can be opened to remove the cutting die from the surface of the pipe. If the release is difficult to operate, the power threader can be shut off with the control switch and reversed to a point where the tension is released. Running the threader backwards with the die still engaged cleans and hones the threads as they travel back across the threads that have been cut.

Other Piping Tools

Whatever method is used to cut the threads on a pipe, the pipe has to be held down. This can be accomplished through the use of a pipe vise. The three types of vises are the power vise, chain vise, and yoke vise.

Wrenches are needed to put the pipe together when all of the threading has been completed. It may be necessary to have an adjustable wrench, a pipe wrench, a chain wrench, or a strap wrench to accomplish this task. The type of wrench that is used depends on the task at hand.

It is also necessary to have a steel tape. A 25 ft (7.6 m) steel tape should be sufficient for most tasks. If it is not, then a 100 ft (30.4 m) tape should be close at hand just in case it is needed.

A level is also used in leveling or grading a piping system. A longer level provides a more accurate reading than a shorter one.

The framing square is a useful tool when you are welding a piping system or making square cuts on a pipe.

FIGURE 13-3 A threading machine.

The plumb bob can show you where a vertical reference point may be on a floor. If a point on a ceiling has to be exactly above a reference point on the floor, the plumb bob helps in locating the point on the ceiling. This is accomplished by hanging the plumb bob from the ceiling with a string and adjusting it to where the plumb bob is directly centered over the target on the floor.

A chalk line may be helpful in different applications. It is used to make a straight line on a horizontal or vertical surface. This is accomplished by unrolling a string coated with a bright colored chalk and pulling the string taut between two reference points. Once the string is pulled taut, it is lifted at the center and popped against the surface on which it is placed. This causes the chalk that is on the string to mark the line on the surface, leaving a reference line to follow.

13-2 PIPING SYSTEMS

Now that the tools that are used to install piping systems have been discussed, it is important to discuss the types of piping systems.

Gas Piping

The most common type of material that is used for gas piping is threaded steel pipe. Most authorities accept threaded steel piping as a safe means of conveying a combustible gas to its final destination. Many local guidelines must be followed when piping in a gas system, so you have to contact your local gas authorities to ensure the proper installation of the gas piping system before any work is begun. The most important thing that must be achieved when piping in a gas system is to get the gas to the desired destination with no leaks. All gas piping must be reamed before installation. The threads also should be threaded to a specified length depending on the pipe diameter. **Table 13-1** shows the suggested thread lengths.

Table 13-1

SUGGESTED GAS PIPE THREAD LENGTHS

Nominal Pipe Size	Thread Length
½	¾
¾	¾
1	⅞
1¼	1
1½	1
2	1
2½	1½
3	1½
4	1⅝

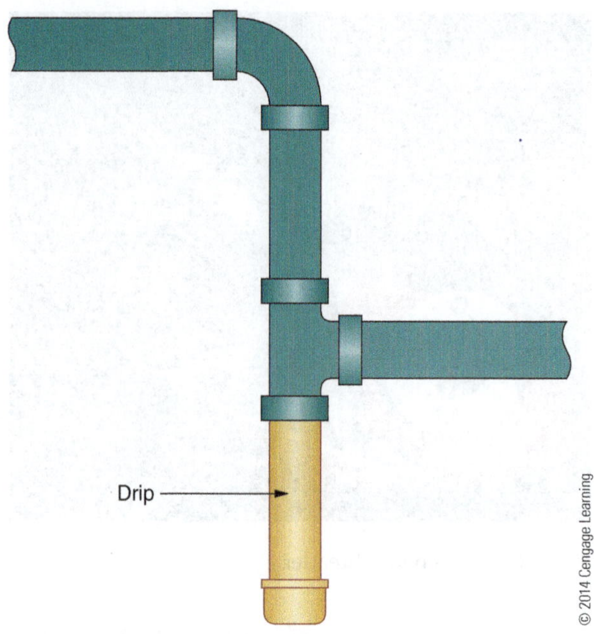

FIGURE 13-4 An example of a drip. Moisture collects in the cap.

Most authorities mandate the use of a joint compound when making up a gas pipe joint. This helps ensure that there will be no gas leaks in case the threads have been nicked or chipped. Wrenches must be used when making up gas piping to ensure a very tight fit of the joint, thus preventing any leakage of the gas. It is also very important, as the system is being piped in, that a **drip** is installed to catch any moisture that may exist in the piping system that is caused by condensation. A drip is shown in **Figure 13-4**. As pipe is installed in vertical runs, it must be supported at every floor. For horizontal runs, the pipe is supported at intervals, according to its diameter, as shown in **Table 13-2**.

Waste Disposal

It is important to provide a path into which soiled water can flow, uninhibited, to its place of disposal. This may be a public sewer or a private septic system. The lowest pipe in a waste system is considered the **drainpipe**. It is part of the piping system that handles all the wastewater and sewage. This type of

Table 13-2

GAS PIPING SUPPORT INTERVALS

Nominal Pipe Size	Spacing of Supports
½	Every 6 ft
¾ or 1	Every 8 ft
1¼ or larger	Every 10 ft

system used to be constructed with cast iron piping, but because of advances in technology, more manageable piping systems are available today. Plastic piping is the preferred piping for this type of system because of its light weight and durability. It is also easier to work with plastic piping than with cast iron. The drainage system must all collect to one main drain line. This drain line is then extended out of the building (underground) to the waste disposal system. Once the piping is 5 ft outside the structure, it is no longer referred to as a drainage system; it is now a sewage system. The main drainpipe under the building or structure should be constructed to have a ¼ in. drop for every 1 ft of horizontal distance. The slant should be in the direction that allows the wastewater and the sewage to flow toward the sewage disposal system. The angle of the slant should produce a flow so wastewater travels at 260 ft per minute. This speed is desirable because as the wastewater runs through the piping system at this speed, it scours the inside of the pipe, keeping it clean. If the improper slant is put on a drainage system, waste begins to collect and the system eventually and continually becomes clogged. It is important as the waste system is being piped in to include stack vents. The stack vents provide two functions:

- They vent any gases that may come from the septic or sewage system into the atmosphere.
- They provide the air that is necessary to prevent siphoning and backpressure.

The waste piping system must also include traps below every fixture that is to be drained in the structure. The trap provides a seal that prevents harmful methane gases from entering the structure through the drain piping system. Refer to **Figure 13-5**. Notice the trap that is in the figure. The liquid that is present in the trap is what provides the seal and prevents any gases from flowing through the trap.

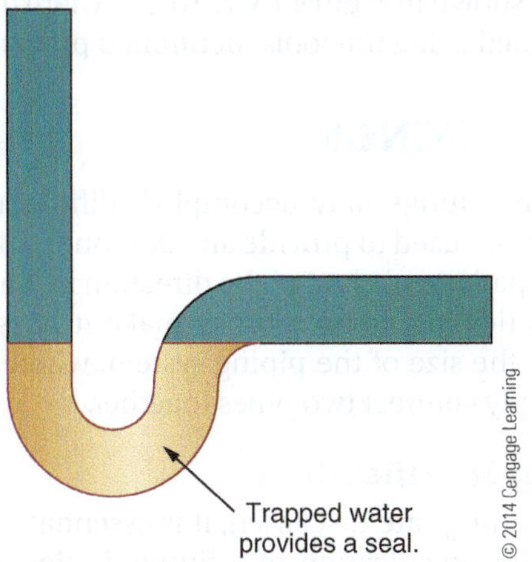
Trapped water provides a seal.
© 2014 Cengage Learning

FIGURE 13-5 An example of a P-trap.

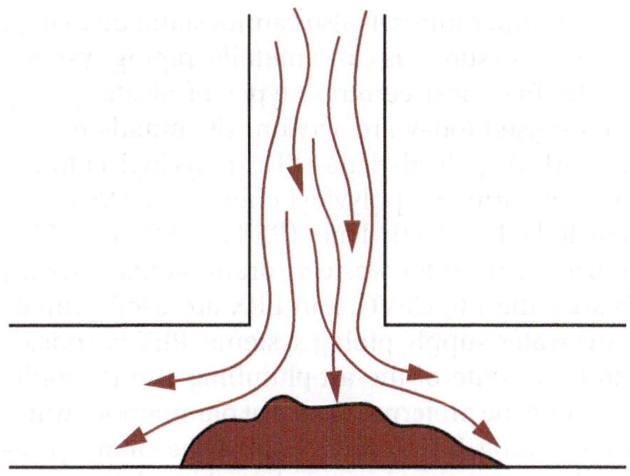
Waste collects at the bottom of the tee.

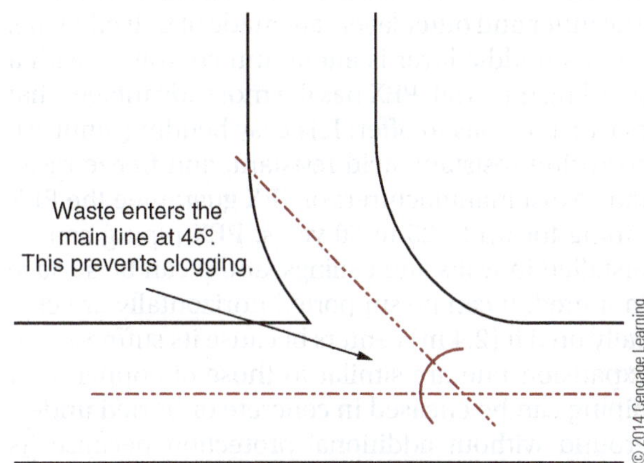
Waste enters the main line at 45°. This prevents clogging.
© 2014 Cengage Learning

FIGURE 13-6 Enter a main drain line at no more than 45°.

When a drainpipe is placed under a structure, it is important that it be supported in a manner that does not allow the drainpipe to sag. This is accomplished through the use of piers. Piers should be placed under every stack vent and at intervals not to exceed 5 ft, for the best results. If the ground level is in excess of 2 ft, strapping hangers to the floor joist in intervals not to exceed 5 ft should support the piping.

It is also a good idea, when using fittings that have a turn or an angle in a drainage system, to use fittings that have angles that are not greater than 45°. This is illustrated in **Figure 13-6**.

Plastic Piping

Plastic piping has a lot of advantages that metal pipes do not. For example, it is lightweight and is usually less expensive than steel pipe. A plastic piping system usually can be installed in less time than a metallic piping system. In addition, plastic does not rust or corrode. There are some disadvantages to plastic piping, however. The biggest problem with plastic piping is that it tends to sag when exposed to higher temperatures. It also cannot stand up to high internal pressures, as can a metallic piping system.

The five most common types of plastic piping that are used today are acrylonitrile butadiene styrene (ABS), polyethylene (PE), polyvinyl chloride (PVC), chlorinated polyvinyl chloride (CPVC), and cross-linked polyethylene (PEX). ABS and PVC are usually used for waste or drain piping systems, whereas the PE, CPVC, and PEX are used primarily for water supply piping systems. PEX is considered to be state-of-the-art plumbing and is rapidly becoming the preferred choice of plumbers for water supply piping. This new technology has many positive characteristics that are desirable when piping a water supply system. It is manufactured by cross-linking electron beams, and it is made in three layers. The inner and outer layers are made of polyethylene, and its middle layer is aluminum combined with a bonding material. PEX has the most advantages that plastic pipe has to offer: It is easy bending, antirust, corrosion resistant, acid resistant, and freeze resistant. Most manufacturers of PEX guarantee the PEX piping for up to 25 to 50 years. PEX piping can be installed in walls and ceilings, and it can be surface mounted. It can be supported horizontally or vertically on 8 ft (2.4 m) centers because its stiffness and expansion rate are similar to those of copper. PEX piping can be encased in concrete or buried underground without additional protection because its tough outer plastic layer resists chemicals and corrosive activity. The inner aluminum core also provides a 100% permeation barrier against ground source contaminants such as fertilizer and pesticide.

Water Supply Systems

Copper piping is mostly used for water supply systems. It is rapidly being replaced with new plastic piping in new installations. However, copper piping has a lot of desirable characteristics and is therefore still being used today. The biggest factor that is causing copper to be replaced in new installations is that copper is expensive in comparison to the new plastic piping. Copper resists rust and is, in its soft form, soft enough to be bent around obstructions and still maintain a metallic rigidity.

Copper expands and contracts about twice the distance of iron or steel. Sufficient space should be allowed upon installation to allow for this high degree of expansion and contraction. If room for expansion and contraction is not considered, the copper may fatigue and rupture.

Copper pipe also has a tendency to split open when the water inside it freezes. This occurs because as the water turns into ice, it expands. This puts an extreme amount of pressure on the inside surface of the pipe, causing it to rupture. If copper piping is going to be exposed to freezing temperatures, it should be wrapped with a pipe insulation to prevent the water inside from freezing. The joints and fittings of a copper piping system are usually attached by sweating (soldering) or flaring.

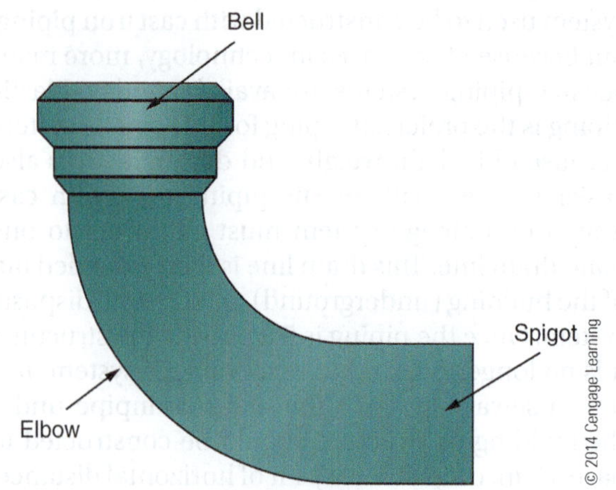

FIGURE 13-7 A bell and spigot.

Clay and Cast Iron Piping

Clay piping, much like cast iron piping, is outdated because of advances in technology. This does not mean that clay and cast iron are not used. These types of piping systems are so reliable that there are clay and cast iron piping systems that have been in use for hundreds of years. Some clay piping systems may even be thousands of years old, because the Egyptians, Greeks, and Romans first used clay piping to supply many of their cities with water. These types of piping systems are not the precision piping system that we are accustomed to today. Of the two types of piping, cast iron piping is used more. The cast iron piping system requires special tools to assemble, such as ladles, yarning irons, packing irons, and caulking irons. The common joint of the cast iron pipe is referred to as a **bell and spigot**, shown in **Figure 13-7**. This is where the joint was sealed using the tools mentioned previously.

13-3 FITTINGS

Different fittings may accomplish different tasks. **Fittings** are used to provide an additional path from a main path, or to change the direction in which the fluid is flowing. Some fittings make it possible to change the size of the piping system, whereas others simply connect two pipes together.

Fitting Specifications

Before fittings are discussed, it is essential that you know the specification of a fitting. Fitting specifications are needed any time a request for a given fitting is made. The fitting specification is usually given in this order:

1. Quantity of fittings that are needed
2. Diameter of the pipes to which they will be joined
3. Degrees of the angle, if applicable
4. Type of fitting
5. Material of which the fitting is made
6. Method to be used to connect the fitting to the piping system
7. Weight of the fitting

   An example of this would be:
   4¾ in., 45°, elbow, brass, sweat, standard
   An abbreviation would be:
   4¾", 45°, ell., br., swt., std

Types of Fittings

Elbow Fitting

The first type of fitting that is discussed is the elbow fitting. The elbow, which is shown in **Figure 13-8**, is used to change the direction of the piping system, therefore changing the direction of fluid flow, or it is used to change the direction of the piping system to avoid an obstruction. The elbow is available in 90° and 45° bends. The 45° angle elbow offers less flow resistance than does the 90° angle elbow. Elbows are usually made of malleable iron and are used for piping compressed air, lubrication flow, and hand railings. Galvanized iron is used on water supply systems or systems that are exposed to the weather. The galvanized coating protects the iron from rusting caused by water or the weather. Two types of elbows are service and return. A service elbow is one that has a set of female threads and a set of male threads, each at opposite ends of the elbow. A return elbow is an elbow that turns a full 180°. This elbow is usually reserved for use on pipe coils.

Wye Fitting

The wye is a fitting that allows a branch to be extended from a main run. The wye portion of the fitting is at a 45° angle from the main run in order to reduce the restriction of flow. The wye is often written as a Y fitting.

T Fitting

T fittings come in three different configurations: the straight tee, the short turn T, and the long turn T. The T fitting, much like the Y, provides a place where a branch is to be extended from the main run. The difference is that the Y branch is at a 45° angle from the main run, whereas the T is at a 90° angle from the main run. The T has a run and a branch. The run of a T is not on the branch. When you are looking into an opening of the T and you can see straight through the T as if you were looking through a pipe, you are looking through the run of the T.

The YT is a fitting that has characteristics of both the Y and the T fittings. This fitting is shown in **Figure 13-9**.

Other Types of Fittings

A coupling is a fitting that simply joins two pieces of pipe together. It is usually straight in orientation and receives the same size diameter pipe into each end.

A bushing is a fitting that is used to reduce an outlet to a smaller pipe diameter. An example would be a fixture outlet that has a ¾ in. outlet connected to a ½ in. piping system. Many people get this device confused with a reducer.

FIGURE 13-8 A 90° angle elbow.

FIGURE 13-9 A YT fitting.

A reducer changes the diameter of the piping system to a smaller size. An eccentric reducer is used in steam lines as the reduced output is offset to allow any condensation to be drained through the smaller pipe. In order for this reducer to work properly, the eccentric output of the reducer should be at the lowest point in the system.

An adapter is a fitting that connects two different piping systems together, for example, when a galvanized piping system is connected to a copper piping system. Another example is changing an older cast iron piping system to a newer PVC piping system.

A union allows a piping system to be connected to a fixed, unmovable object. All pipe threads are right-hand threads and therefore must be turned clockwise to be tightened. When a pipe must be connected to a fixed apparatus, such as a compressed air storage tank, it cannot be tightened into the apparatus and the last fitting at the same time without the use of a union. The union also provides a place where the piping system can be broken without the pipe being cut. This is possible because of the three-part configuration of the union. There are two hubs, with machined faces, that form a seal when pulled together by the third part, the collar. The hubs mount on two separate pieces of pipe. When the union is wrench tight, it provides an unobstructed path for fluid to flow. A union is shown in **Figure 13-10**.

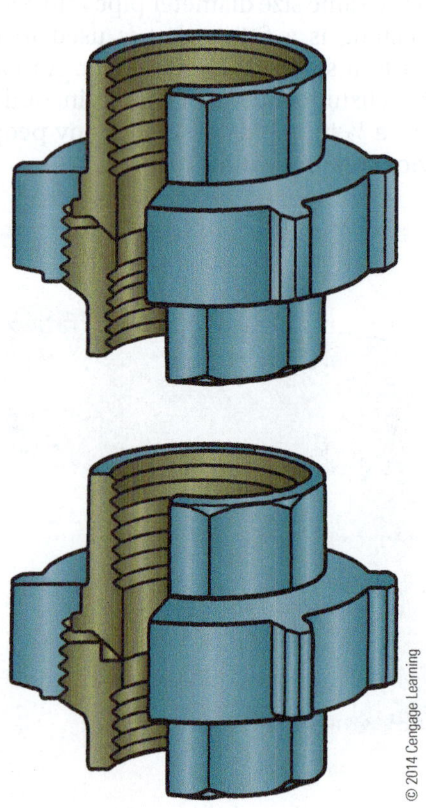

FIGURE 13-10 A union.

Pipe Nipple

A pipe nipple is considered to be a fitting because it is often used in conjunction with another type of fitting. A pipe nipple is a piece of pipe that is 12 in. or less in length, with threads at both ends of the nipple. Nipples are usually made of black iron, galvanized pipe, steel, or brass. Three common nipples are used in piping systems: the close nipple, the shoulder nipple, and the long nipple. The close nipple is approximately 2 in. in total length. There are threads on both ends of the nipple. The threads are opposite each other at the center because both are right-hand threads that have been cut from opposite ends of the pipe. This type of nipple does not have a shoulder. The shoulder nipple usually has a small shoulder between the two sets of threads. This nipple is usually 2 in. to 4 in. in length. The long nipple is a longer version of the shoulder nipple. They are usually ordered in graduations of ½ in. in length, up to 12 in. long.

The last two fittings are plugs and caps. A plug and a cap are used for the same purpose—to close off the end of a piping system but still allow for future expansion—or to allow access to the internal portion of the piping system. A plug has male threads and is threaded into a fitting, whereas the cap has female threads and is threaded onto a fitting.

13-4 FITTING ALLOWANCES

When sketching a piping system, it is always going to be necessary to obtain the end-to-end measurements. The **end-to-end (E-to-E) measurement** is the actual length of the pipe after it is cut and threaded without any fittings. As a fitting is threaded onto each end of the pipe, the overall length of the pipe becomes longer. Look at **Figure 13-11**A. A wall space of 22 in. is available to install a pipe. If the pipe were cut to 22 in. E to E, it would fit nicely into the available space. The problem, however, is that the pipe is not connected to anything to carry a fluid through it. In order to do this, fittings must be used. If two elbow fittings were installed on each end of a pipe with a 22 in. E-to-E measurement, the entire assembly would be too long to fit into the space (Figure 13-11) because the threads of the pipe do not absorb the entire length of the fittings. Because the entire assembly must be able to fit into the available space, the E-to-E measurement must be reduced to accommodate the increase in length caused by the added fittings (see Figure 13-11C). The reduction of the overall E-to-E measurement is accomplished by finding the fitting allowance of a given fitting and subtracting that known value from the E-to-E measurement.

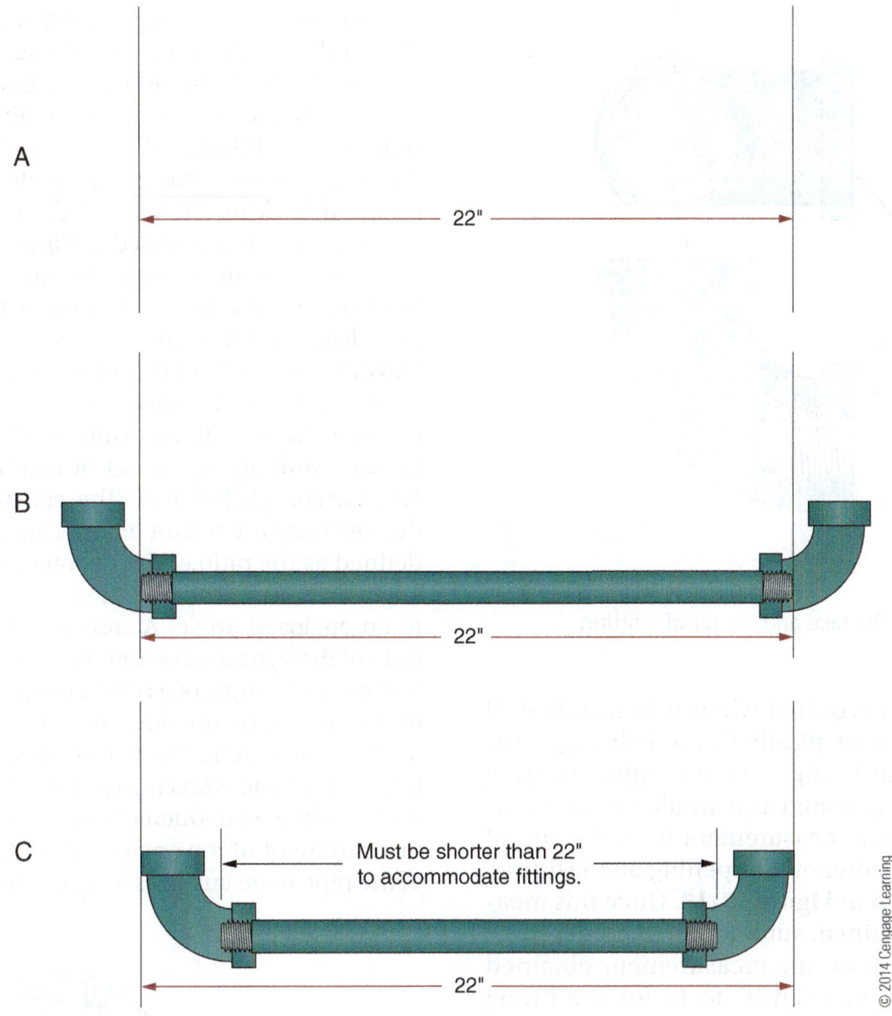

FIGURE 13-11 A fitting allowance.

This is discussed in greater detail later. Before going on, you must know about two other types of measurements: the end-to-center (E-to-C) and the center-to-center (C-to-C) measurements. An **end-to-center (E-to-C) measurement** is taken from the end of a piece of pipe to the center of a fitting. The center of the fitting is defined as the point that the centerlines of the pipe would cross if they were continued on into the fitting—*not the physical center*. This is shown in **Figure 13-12**. The fitting that is being referred to in this statement may be an elbow, a T, a Y, or a union. A **center-to-center (C-to-C) measurement** is taken from the center of one fitting, down the length of the pipe, to the center of the other fitting. Both the E-to-C and the C-to-C measurements are shown in **Figure 4-9**, page 72, along with the E-to-E measurement. These measurements are bolded so they can be easily recognized.

The fitting allowance is defined as the space between the end of a pipe, as it is threaded into the fitting, to the center of the opening on the fitting. This can be found by checking the manufacturer's specification sheet, which is commonly referred to as a *take-off chart*. Most manufacturers supply take-off charts for their fittings upon request. Take-off charts provide all of the fitting allowances for a given fitting. If the supplier cannot supply you with take-off charts, the fitting allowance can be calculated. It is important not to become dependent on charts; you should understand the fitting allowance process in the event that take-off charts are not available. Fitting allowance can be calculated by simply placing the desired fitting on a short piece of threaded pipe, such as a scrap piece. The length of the pipe should be known and recorded before placing the fitting on it. It does not matter what fitting is used as long as they are all from the same manufacturer. This is because a manufacturer uses the same fitting allowance for all of its fittings. The fitting, if it is a threaded type, should be tightened

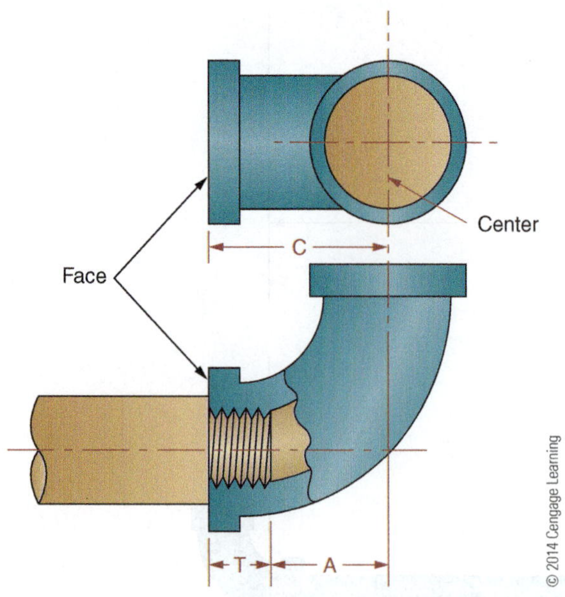

FIGURE 13-12 The face and center of a fitting.

to the tightness required when it is installed. If a sweated fitting or plastic fitting is being used, simply slide the fitting onto the pipe, ensuring that it is properly seated as it would be on installation; then take a measurement from the end of the pipe to the center of the opening on the fitting. This is illustrated in **Figure 13-13**. Once this measurement is obtained, subtract the recorded pipe length (E to E) from the measurement obtained with the fitting attached (C to E) for the fitting allowance. Here is the formula:

$$A = C - E$$

A = Fitting allowance
C = Center-to-end (C-to-E) measurement
E = End-to-end (E-to-E) measurement

An easy way to remember this is to remember the word *ACE*. Do not forget, though, that the measurements should be subtracted, not added!

The **offset** occurs when a piping system is routed to a different direction with the use of a 45° angle elbow, travels for a distance, then is rerouted back to the original direction that it was traveling. This is illustrated in **Figure 13-14**.

There are three parts to an offset: the set, the run, and the travel. The **set** is the distance of the offset. The **run** is equal in distance to the set. The **travel** is the length of the angled pipe. The distance between the two elbows (C-to-C measurement) can be calculated if the distance of the set or run is known. Multiplying the set or run by the constant 1.414 accomplishes this. The constant of 1.414 is derived from the secant or cosecant of 45°. *Secant* is defined as the ratio of the hypotenuse of a right triangle to either of the other two sides with reference to an enclosed angle. A *cosecant* is defined as the ratio of the hypotenuse and the side that is opposite a given acute angle of a right triangle. The end result of the calculation provides the C-to-C measurement of the travel length. The C-to-C measurement must have the fitting *allowances* (plural because there are two elbows) deducted from it to give the E-to-E measurement of the angled pipe. This is the length of the pipe to be cut. This is shown in **Figure 13-15**.

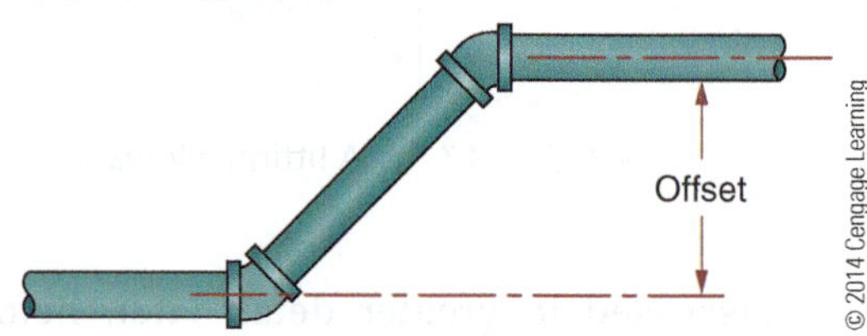

FIGURE 13-14 An offset in a piping system.

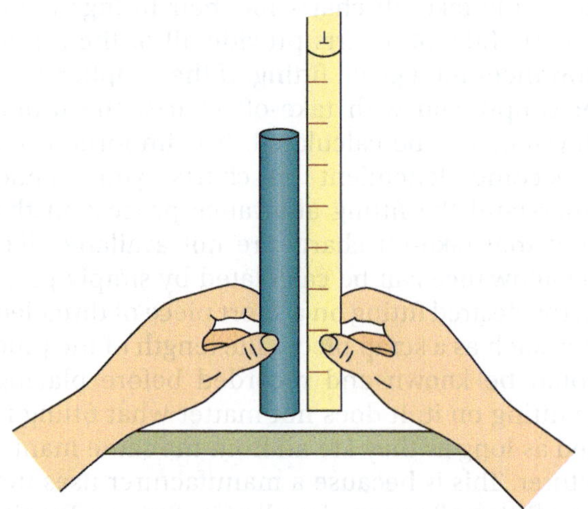
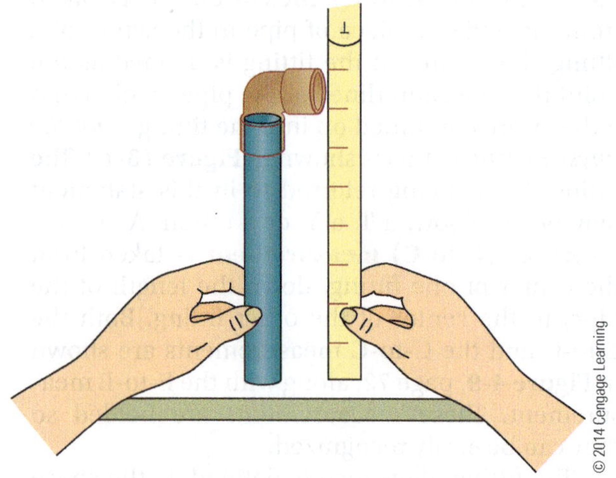

FIGURE 13-13 How to measure for fitting allowance using a ruler.

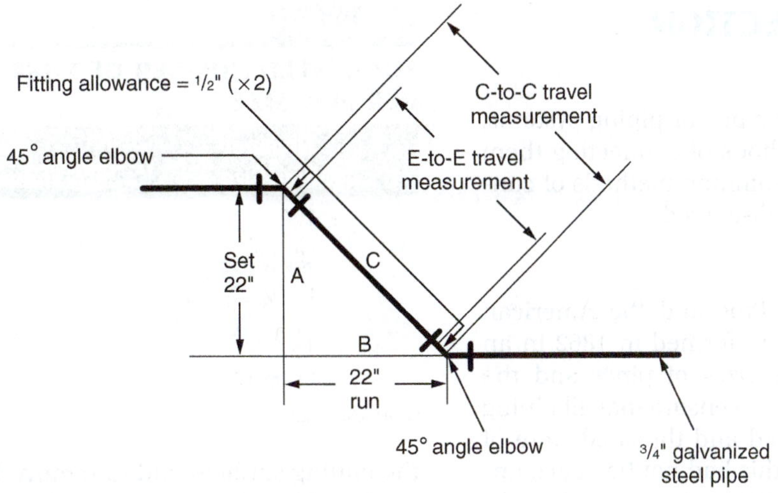

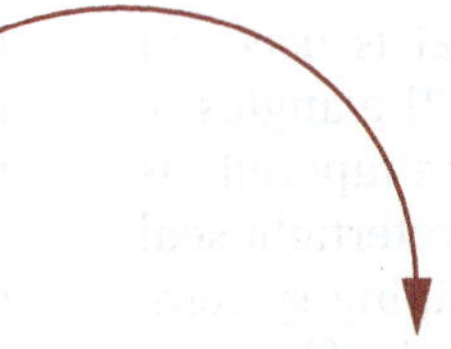

$$C = A \times \sqrt{1 + \frac{A^2}{B^2}}$$

$$\text{Travel} = \text{set} \times \sqrt{1 + \frac{\text{Set}^2}{\text{Run}^2}}$$

$$\text{Travel} = 22 \times \sqrt{1 + \frac{22^2}{22^2}}$$

$$\text{Travel} = 22 \times \sqrt{1 + \frac{484}{484}}$$

$$\text{Travel} = 22 \times \sqrt{1 + 1}$$

$$\text{Travel} = 22 \times \sqrt{2}$$

$$\text{Travel} = 22 \times 1.414$$

Travel = 31.11" C to C

The C-to-C measurement can be checked using the Pythagorean theorem.

$$C = \sqrt{A^2 + B^2}$$

$$C = \sqrt{22^2 + 22^2}$$

$$C = \sqrt{484 + 484}$$

$$C = \sqrt{968}$$

C = 31.11"

Now calculate the fitting allowance.

E-to-E measurement = C-to-C measurement – FA (×2)

E-to-E measurement = 31.11" – 1/2" (×2)

E-to-E Measurement = 31.11" – 1"

E-to-E measurement = 30.11"

FIGURE 13-15 A fitting allowance calculation.

13-5 PIPE CONNECTION METHODS

Because there are many types of piping systems, there are also many methods of connecting them together. Only the most common methods of connecting pipe together are discussed.

Threaded Pipe

The American Standard Pipe and the American Standard Pipe Thread were formed in 1862 in an effort to standardize the sizes of pipes and the threads on pipes. This was to ensure that all piping systems that were installed and threaded were of the same dimensions. If this had not been accomplished, there would be different types of threads, which would cause the fitting of one manufacturer and the fitting from another not to be interchangeable. The standard thread that is used on piping today is known as a V-thread. The angles of the threads are at 60°. The threads are tapered, as shown in **Figure 13-16**, to provide a watertight seal to keep fluid contained within the piping system. The taper of the threads is 1⁄32 in. per inch of length.

The seven threads closest to the end of the pipe should be perfect threads. The crest and the root of the thread should be clean, with a definite sharpness. The taper begins on the next two or three threads. For these two or three threads, the crest of the thread is slightly rounded, whereas the root of the thread still remains sharp. The remaining threads are referred to as the starting threads and therefore do not provide any sealing capabilities at all. There are a suggested number of threads for each size of pipe. This is shown in **Table 13-3**.

Cutting Dies

It is very important to provide lubrication to the cutting dies before and during their use. This accomplishes two things: to provide lubrication to the cutting surfaces and to remove the heat that is produced by friction as the cutting die cuts into the metal. It is also very important to install the cutting dies in the proper order. Cutting dies usually come in sets of four and are numbered to be installed in a sequence. This causes the cutting edges of the dies to be staggered to make a continual cut at four sides of the pipe at the same time. It is also important to have the dies set to the correct length. Each die of the set has a line that indicates the depth to which it is to be set. If the sets of dies are not installed to their indicator lines, the cutting surfaces of the dies will be too close, causing the thread to be cut too deep. This causes the fitting to be too loose, resulting in fluid leaks. If the dies are installed past the indicator lines, the cut will be too shallow. The fitting will not screw far enough onto the fitting to make a good seal, and the fluid or gas will probably leak.

Most dies that are used in threading have a space for chips of metal to be removed into as they are torn away from the surface of the pipe. It is important to ensure that these chip spaces are not clogged to prevent the cutting edges of the die from breaking. Threads are cut to their standard length when the face of the cutting die is flush with the end of the pipe. It is a common practice,

Table 13-3

SUGGESTED NUMBER OF THREADS PER PIPE SIZE

Nominal Pipe Size	Suggested Threads Per Inch
1⅛"	27
¼", ⅜"	18
½", ¾"	14
1", 2"	11½
2½ to 12"	8

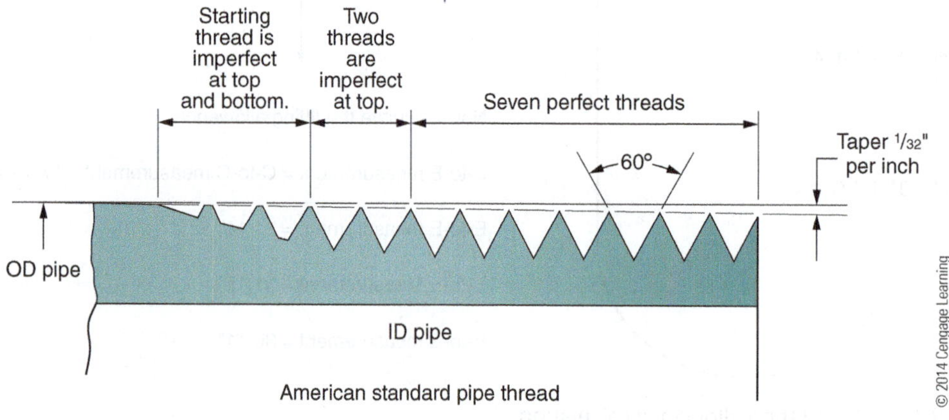

FIGURE 13-16 Thread taper.

however, for most pipe fitters to continue threading until there is one complete thread past the face of the die. You should not exceed this cutting length because if the threads are cut too long, the pipe may become jammed into the rear threads of the fittings. If the pipe is threaded correctly, the starter threads should both start to engage at the same time.

Cast Iron Caulking

Sealing the bell and spigot is a tedious task at best. Oakum is packed tightly into the joint. Oakum is continually packed into the joint until it is about 1 in. from the top of the bell. Once the oakum is packed, the bell is then poured with molten lead. After a lead bead is poured, the lead seal is then caulked. See **Figure 13-17**.

When the cast iron pipe is installed horizontally, a problem is encountered when pouring the lead joint. A lead runner must be used to contain the lead until it cools if the piping is horizontal. This prevents the molten lead from pouring out of the joint as it is being poured into the joint. Because lead affects the nervous system, this form of sealing a bell and spigot is rarely used. When repairs have to be made to a cast iron piping system today, they are accomplished through the use of neoprene gaskets, which have taken the place of oakum and lead.

Cast iron is difficult to cut at best. It takes a special knack to get it right on the first try without cracking the pipe. A chain cutter is used any time cast iron piping must be cut. The chain cutter has very small cutting wheels on the links of the chain. The chain is wrapped around the cast iron pipe and then tightened to a snug fit. If the chain is tightened too much, the pipe will crack. The chain cutter is then rolled around the pipe until the pipe snaps apart at the cut. The cutter rarely cuts all the way through the pipe before it separates at the cut.

Because there are only a few plumbers left who are extremely experienced with cast iron piping, any damaged portions of the cast iron piping system are being replaced with PVC piping with the use of an adapter.

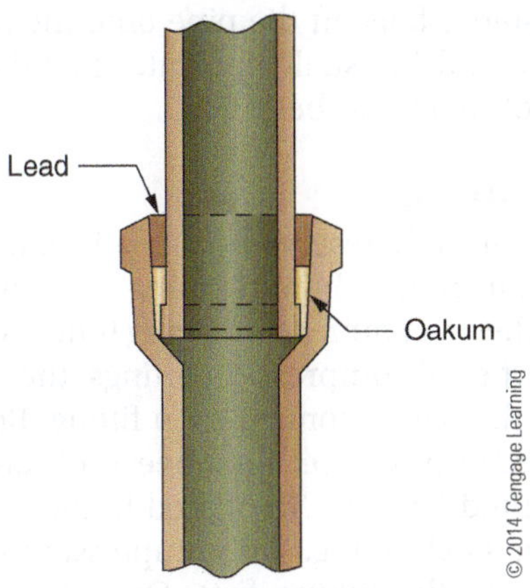

FIGURE 13-17 Caulking a joint.

Plastic Pipe

The method that is used to connect the fittings to each of these types of pipes is different, with the exception of the PVC and the CPVC piping. The PE piping uses barbed fittings that are held into place with hose clamps. The ABS, PVC, and CPVC all use a solvent, or glue, to bond the fittings to the pipe. The solvent that is used causes a chemical reaction in the plastic as it is applied. This chemical reaction, in most cases, does not last long; therefore, it is critical to have the fitting or pipe close by and ready to be installed. When the fitting is placed in contact with the solvent, the plastic on the fitting begins to melt as well. The melting plastic from the pipe and the melting plastic from the fitting bond together to form an unbreakable seal. This is why it is critical to be aware of the placement of the fitting as it is being installed because there is not much time to make corrections. There is something that must be mentioned at this point: The PVC and CPVC should have the ends prepared with a primer before using the solvent, whereas the ABS does not need a primer. It should also be noted that the solvent for ABS piping should not be used on PVC or CPVC, nor should the PVC/CPVC solvent be used on ABS piping. Fittings are attached to the PEX tubing by means of crimping.

It is not unheard of to weld plastic pipe. A heat gun, sometimes referred to as a torch, is used in welding plastic piping. The torch heats the ambient air as it is pulled from the atmosphere through the torch and out the discharge end of the torch. The forced, heated air is usually at a temperature that brings the plastic to its melting point. A filler rod is then added to the area that is heated. The filler rod also reaches its melting temperature and begins to melt with the plastic of the pipe. The area that is to be welded should be dry and clean before attempting to weld the plastic. This can be accomplished with a solvent. The solvent removes the finished surface of the pipe and provide a rough surface for the weld to be applied. The filler rod that is used should be of the same type of plastic as the one that is being welded. This ensures that the weld goes into the plastic. If a filler rod of a different type is used, the plastics may reject each other and not bond properly. Once the area to be welded is heated to its melting point, the heat is removed, and the filler rod is then gently pushed into the seam between the fitting and the pipe. This causes the plastic on the fitting and the plastic from the pipe to bond together as one continuous flow of plastic. If done properly, this is an effective means of preventing leakage.

Sweating the Fittings

Sweating is sometimes referred to as soldering and is usually performed on copper and brass because these piping systems usually do not have threads. Because there are no threads, the fittings must be slid onto the pipe and made waterproof by soldering the joint. Before any soldering can begin, however, many things have to be done. First, gather all of the necessary tools that are needed to cut, ream, clean, and solder. The cutting of the pipe is usually accomplished with a pipe cutter. Once the pipe has been cut, the inside diameter of the pipe must be reamed to remove all obstructions that may prevent a smooth flow of the fluid.

Step 1: Cleaning the Surfaces to Be Sweated

Once the reaming is complete, all surfaces that need to be soldered have to be cleaned thoroughly. These are the outside of the pipe and the inside of the fitting. It is important to clean the outside of the pipe to a length that exceeds where the fitting will be placed. The inside of the fitting should be cleaned to the bottom of the socket that receives the pipe. Steel wool or an emery cloth prepares the surfaces of the copper or brass, leaving them bright and clean. Care should be taken not to touch any of the cleaned surfaces with your fingers, to prevent depositing oil on the cleaned surface. The presence of oil prevents the solder from sticking to the surface that was touched. Always be sure to wear the proper safety apparel when soldering. Also avoid touching the heated pipe until it has had time to cool.

Step 2: Applying Flux When Sweating

Once the cleaning is complete, flux should be applied to the surfaces that will be soldered. The **flux** further cleans the surface and prevents the copper or brass from oxidizing when heat is applied. It is important to use the correct amount of flux. An excessive amount of flux may cause it to be pushed into the pipe as the fitting is put onto the pipe, causing the inside of the piping to corrode. A lack of flux will not prevent the copper or brass from oxidizing. This causes the solder not to stick to the surface of the copper or brass. This condition can be recognized by the surface of the metal turning black once heat is applied. If this occurs, the joint has to be disassembled, cleaned, recoated with flux, and reassembled. Slide the fitting onto the end of the pipe after the appropriate amount of flux has been applied. The fitting should be twisted to smear the flux evenly throughout the joint. Do not allow flux to contact the skin. If it does, it may cause irritation. Do not allow flux to get into your eyes.

Step 3: Applying the Heat

Heat should then be applied to the fitting—*not to the solder or the pipe.* The heat is applied to the fitting because the fitting is usually thicker than the pipe. This means that it will require more heat. If the piping system is ½ in. or smaller in diameter, the pipe will not need to be heated because the heat will be transferred to the pipe through the heating of the fitting. If the pipe is over ½ in. but less than 3 in. in diameter, it may be necessary to heat the fitting first, then apply small amounts of heat to the pipe as you begin to solder. For pipes that are over 3 in. in diameter, two torches should be used because there is more metal to keep hot. After the heat has been applied to the ½ in. fitting, the temperature needs to be checked to see whether it is hot enough to apply the solder. This is done by touching the fitting with the solder on the opposite side of the flame. If the solder melts, it is the right temperature to begin soldering.

Step 4: Applying the Solder

Solder with a lead/tin composition was used in the past to sweat pipes. This type of solder should not be used because lead affects the nervous system. If lead is used, the water may become contaminated as it comes in contact with the lead, possibly causing low-level amounts of lead poisoning. It is for this reason that a mixture of 95% tin and 5% antimony is used today. This type of solder is referred to as 95-5 solder. When the solder is applied to the face of the fitting, the heat should be removed. If it has been heated correctly, the solder will be pulled up between the fitting and the pipe, filling the space completely with solder. It is important to get enough, but not too much, solder into the joint. If enough solder is not fed into the joint, it may fail over time and begin to leak. If too much solder is fed into the joint, the solder may form a ball on the inside of the pipe. This ball of solder may break away when the fluid flows through the system and damage components. If the ball does not break away, it will obstruct fluid flow within the pipe. It is common practice to feed solder into a joint until one drop of solder falls off the pipe onto the floor. The forming of a drip usually indicates that the proper amount of solder has been used.

Flared Fittings

Flaring is usually reserved for use in most tubing applications rather than in piping systems. Flaring is used when a compression fitting is needed. There are two types of compression fittings: the flared fitting and the barrel compression fitting. Both work off the same principle: to squeeze off any possibility of fluid leakage. The flared fitting is used on copper or steel tubing. The compression nut must be slid onto the tubing first. Once this is done, the end of the tubing is flared with a flaring tool.

The flared end of the tubing is then set onto the male portion of the fitting, and the compression nut is brought down to the male threads on the fitting. As the compression nut is threaded onto the fitting, the tubing is compressed between the compression nut and the fitting, causing a watertight seal. The barrel compression fitting works without flaring the end of the tubing. Instead, it works with brass components that accomplish the same task. As before, the compression nut should be slid onto the tubing first, and then a compression ring is slid onto the tubing. Once the compression nut and the compression ring have been placed onto the outside of the tubing, a barrel is slid into the end of the tubing. The compression ring is then slid down, and the compression nut is slid over the top of the compression ring. As the compression nut is threaded onto the fitting, it causes the compression ring to become wedged between the compression nut and the barrel. This, in turn, causes the tubing to become crimped between the compression ring and the barrel, sealing the fitting.

Welding

This method of pipefitting is usually reserved for large piping systems and requires a skilled welder. The two methods of welding piping systems are the gas welding method and the electric arc welding method. The most common types of welds that are used on piping systems are usually the fillet weld and the butt weld, both of which are explained in greater detail in Section 4 (Welding) of this text. As with all of the other methods of connecting pipe, the ends of the pipe must be prepared before any joining is done. The end is prepared according to the type of weld that is being used. A fillet weld is usually used on a piping system that uses a socket-type fitting. This is when the fitting has an internal diameter that is just large enough to accept a pipe end. This is referred to as a socket. If a fillet weld is to be used, the pipe end can be prepared with a square end. The pipe end should be pushed all the way into the socket, and a mark should be made on the pipe at the face of the fitting. The pipe should then be pulled out ⅛ in. to allow room for expansion. Expansion will occur once welding is started on the pipe. Because the socket type of fitting is self-aligning, no alignment exercises need be performed. Moving the pipe out gives the pipe room for this growth. The fillet weld should be run in a consistent bead all the way around the pipe. This is where the skill of a welder is tested. An experienced welder will have no problems with this type of welding. On the other hand, a beginner should not work on any critical systems until more experience is gained because most pipe welds will be tested in the field either by ultrasound or X-ray. The butt weld is used when the fitting does not have a socket. This means that the fitting and the pipe have the same diameter. In this case, the pipe and fitting ends should be prepared to have a 45° angle bevel. The ends are then placed close to each other, leaving a ⅛ in. gap, and tacked in place on the top surface. The ⅛ in. gap allows room for expansion and allows the weld to penetrate all the way to the inside of the pipe and fitting. Before the joint is welded, the pipe needs to be aligned with whatever it is being welded to. If the pipe is to be aligned with another pipe, the pipe ends are placed close together, leaving a ⅛ in. gap, and then are squared with the use of two framing squares. The framing squares are set atop the two pipes, as shown in **Figure 13-18**. If the pipe is aligned correctly, the numbers on the two squares will match. When this condition is met, tack the top of the pipe. Next, align the bottom and sides, tacking as you go. Once all four sides have been tacked, begin the weld.

To align a 90° angle elbow with the pipe, place the end of the fitting close to the end of the pipe, allowing the ⅛ in. gap. Center one square on top of the pipe and the other on top of the elbow's alternate face. This is illustrated in **Figure 13-19**. The fitting should then be adjusted until it is centered

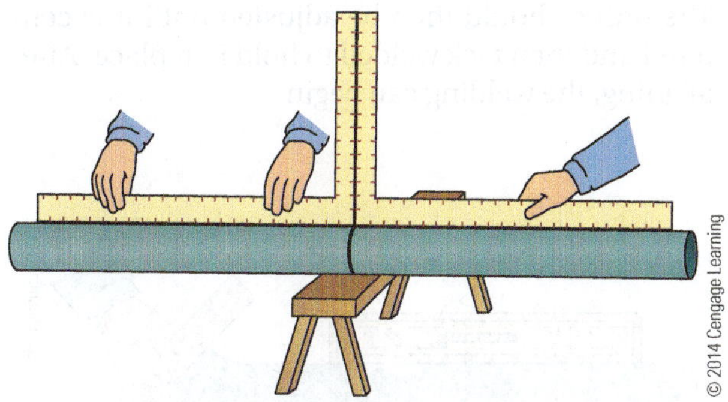

FIGURE 13-18 Aligning two pipes for welding.

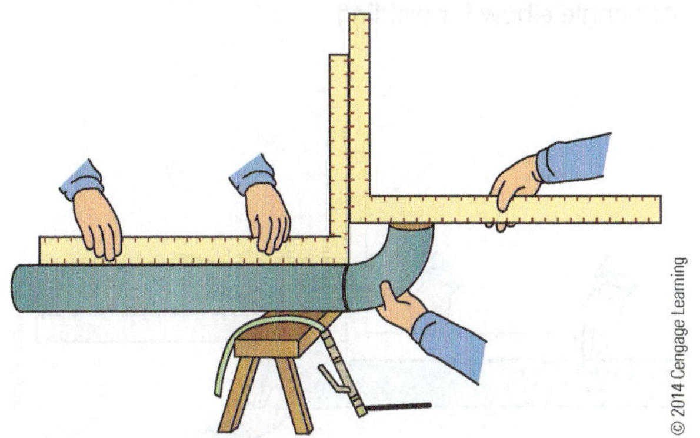

FIGURE 13-19 Aligning a pipe and a 90° angle elbow for welding.

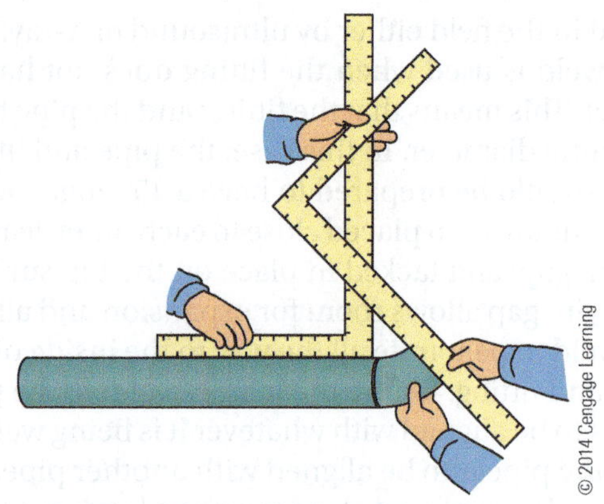

FIGURE 13-20 Aligning a pipe and a 45° angle elbow for welding.

then tack welded to hold it in place. After aligning, the welding can begin.

A 45° angle elbow is aligned similarly to the 90° angle elbow except that the squares will be crossed, as shown in **Figure 13-20**. A set of torpedo levels can also be used to align the 45° angle elbow as long as one of them has a 45° angle bubble. This is shown in **Figure 13-21**.

The tee is aligned by centering one square on top of the pipe and the other on top of the face of the branch outlet. This is illustrated in **Figure 13-22**. The fitting should then be adjusted until it is centered and then tack welded to hold it in place. After aligning, the welding can begin.

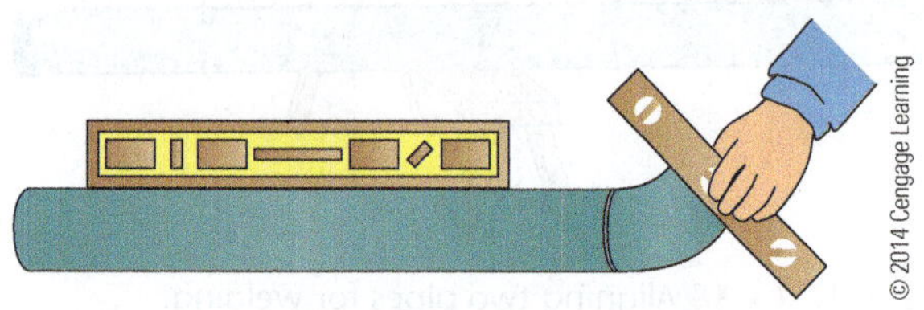

FIGURE 13-21 Using levels to align a pipe and a 45° angle elbow for welding.

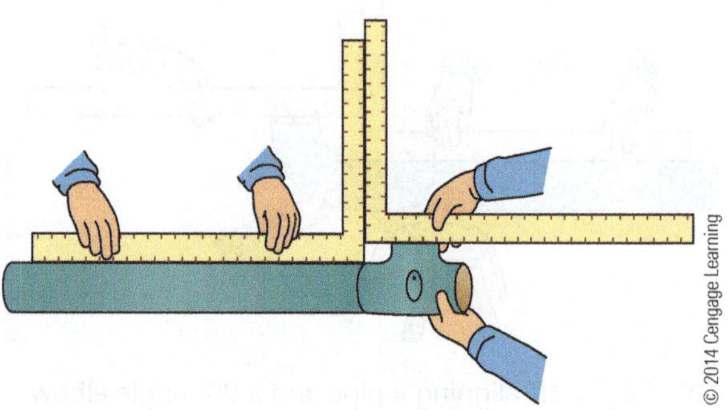

FIGURE 13-22 Aligning a pipe and a tee for welding.

SUMMARY

- To assemble a piping system, several tools may be needed: a hacksaw or pipe cutter; a pipe reamer; a pipe threader; a power vise, chain vise, or yoke vise; several types of wrenches; a steel tape; a torpedo level; a framing square; a plumb bob; and a chalk line.
- Some types of piping systems may include gas piping; waste disposal; plastic piping; copper piping, which is mostly used for water supply systems; clay piping; and cast iron piping.
- Different types of fittings may accomplish different tasks. Some provide an additional path from a main path or change the direction in which the fluid is flowing. Some fittings make it possible to change the size of the piping system, whereas others simply connect two pipes together.
- It is essential to know the specification of a fitting.
- The 45° angle elbow offers less flow resistance than the 90° angle elbow.
- The wye fitting allows a branch to be extended from a main run.
- The T fitting, much like the Y, provides a place in which a branch is extended from the main run. The difference is that the Y branch is a 45° angle from the main run, whereas the T is at a 90° angle from the main run.
- The YT is a fitting that has characteristics of both the Y and the T fittings.
- A coupling is a fitting that simply joins two pieces of pipe together.
- A bushing is a fitting that is used to reduce an outlet to a smaller pipe diameter.
- A reducer changes the diameter of the piping system to a smaller size.
- An adapter is a fitting that connects two different types of piping systems together.
- A union allows a piping system to be connected to a fixed, unmovable object.
- Pipe nipples are considered to be fittings because they are often used in conjunction with another type of fitting.
- A plug and a cap are used for the same purpose—to close off the end of a piping system but still allow for future expansion—or to allow access to the internal portion of the piping system. A plug has male threads and is threaded into a fitting, whereas the cap has female threads and is threaded onto a fitting.

- A sketch is a drawing that is made with minimal detail. It is used to convey information from the person who drew the sketch to the person who will interpret the sketch.
- There are two types of single-line drawings that are used in piping sketches: orthographic and isometric.
- The orthographic drawing is made by using two separate views. One view is from the top of the piping system (looking down), showing all of the horizontal runs that are made. This is referred to as a plan view. The other orthographic view is from the side, to show all of the vertical runs. This is referred to as an elevated view.
- The isometric drawing has both the horizontal and the vertical runs of the piping system included on one drawing.
- The end-to-end (E-to-E) measurement is the actual length of the pipe after it is cut and threaded without any fittings.
- An E-to-C measurement refers to a measurement that is taken from the end of a piece of pipe to the center of a fitting.
- A C-to-C measurement is a measurement that is taken from the center of one fitting, down the length of the pipe, to the center of the other fitting.
- The fitting allowance is defined as the space between the end of a pipe, as it is threaded into the fitting, to the center of the opening on the fitting.
- Subtract the recorded pipe length (E to E) from the measurement obtained with the fitting attached (C to E) for the fitting allowance.
- The offset occurs when a piping system is routed to a different direction with the use of a 45° angle elbow, travels for a distance, then is rerouted back in the original direction that it was traveling.
- There are three parts to an offset: the set, the run, and the travel.
- Piping systems may be connected by means of threading.
- Cast iron piping systems are caulked.
- Plastic pipes may be connected using different methods such as barbed fittings; a solvent, or glue; crimping; and welded plastic.
- Sweating is used to connect the components of a copper or brass piping system.
- Flaring is used for most tubing.
- Welding may be the method of connecting a steel piping system.

REVIEW QUESTIONS

1. What type of waste disposal piping system would you expect to find in a really old structure?
2. There are several reasons that plastic piping systems are preferred over copper. Name a few of these reasons.
3. State the main difference between the T and the Y fittings.
4. What is the E-to-E measurement of the angled piece of pipe that is in the travel of the offset in **Figure 13-23**?
5. List at least six methods of connecting piping systems.

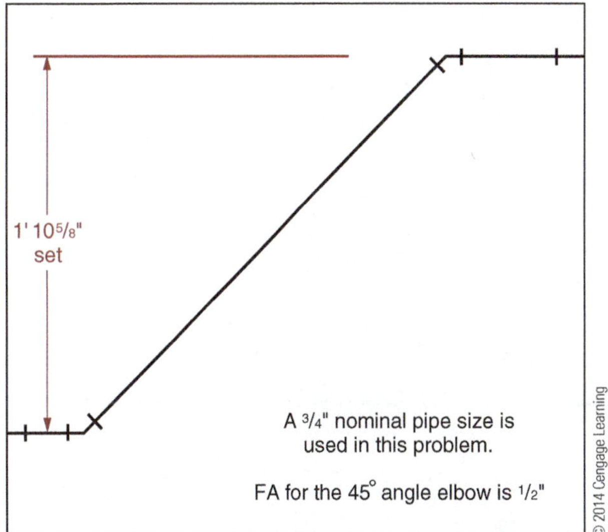
1' 10 5/8"
set
A 3/4" nominal pipe size is used in this problem.
FA for the 45° angle elbow is 1/2"
© 2014 Cengage Learning

FIGURE 13-23 Solve for the E-to-E measurement of the travel.

SECTION 3

ELECTRICAL KNOWLEDGE

- Electrical Fundamentals
- Test Equipment
- Basic Resistive Electrical Circuits
- Reactive Circuits and Power Factor
- Wiring Methods
- Transformers
- Electrical Machinery
- Control and Controlled Devices
- Motor Control Circuits
- Basic Industrial Electronics
- Electronic Variable Speed Drives
- Programmable Logic Controllers
- Lighting

A TYPICAL DAY IN MAINTENANCE

Check It Out

You finished the bearing replacement at 10:00 A.M. You now focus your attention on the source of the "funny" odor.

Upon investigation, you determine that the motor windings have overheated. Apparently, when the bearing failed, the motor shaft was placed in a bind. Normally, the overload relay for the motor should have tripped. However, upon inspection, you discover that the heaters have been improperly sized.

You determine and install the correct heaters needed for this particular motor. You also begin the process of replacing the defective motor and preparing the required paperwork to have the defective motor sent to a rewind shop for repair.

The motor replacement is completed at 11:30 A.M. Because it is now lunchtime, you secure your tools and equipment and break for lunch.

Work It Out

1. How do you determine the proper size heaters for a motor?
2. List the tools that you would need to remove and replace the motor.
3. List the steps that you would follow to remove and replace the motor.

CHAPTER 14

Electrical Fundamentals

This chapter introduces you to electricity. It is important to have an understanding of electricity because practically everything uses electricity either directly or in the process of being manufactured. Maintenance technicians must understand electricity if they are to possess the knowledge and skills necessary to be considered multicrafted individuals. The study of electricity covers many areas. In this chapter, you will learn the foundations of electricity. It is on these foundations that future electrical knowledge will build. It is very important that you fully understand and grasp this material before continuing on to the next chapter.

OBJECTIVES

After studying this chapter, the student should be able to

- Describe the structure of matter.
- Define electricity.
- Describe current, voltage, resistance, and power.
- Explain resistor color codes.
- Identify the three basic Ohm's law formulae and three basic power law formulae.
- Identify the six derived Ohm's law and power law formulae.
- Solve problems for an unknown value of resistance, voltage, current, or power, using the correct formulae when two values are known.

14-1 ATOMIC STRUCTURE

Before you can begin to understand complex electrical circuits and devices, you need to have a good understanding of the basics of electricity. In order to understand the basics of electricity, you have to understand the structure of **matter**. Matter is any material or object that occupies space and has mass. Some examples of matter are plastic, metal, water, gasoline, air, and hydrogen. As you can see from these examples, matter can be found in several forms. In fact, matter can be found as a solid (the plastic and the metal), a liquid (the water and the gasoline), or a gas (the air and the hydrogen).

Matter consists of very small particles. When matter is broken down into smaller parts, you will find that matter is made up of **molecules**. A molecule is the smallest particle within matter that retains the same chemical properties of the matter. Look at water as an example. If you filled a cup with water, you would have a cup of water. However, that cup of water actually has millions of individual molecules of water, which, when combined together, fill the cup with water.

A water molecule can be reduced into simpler or smaller parts. The smallest part of a molecule is called an **atom**. An atom is an **element** that, when combined with other elements, forms a molecule. A water molecule consists of two atoms of hydrogen gas and one atom of oxygen gas. You may be familiar with the chemical symbol for water, H_2O. Notice that hydrogen by itself is a gas and is unlike water. Notice too that oxygen by itself is also a gas and is unlike water. However, when these two gases are combined in the correct amounts, they form a molecule of water, a liquid. This is the definition of an atom. Elements or atoms by themselves do not possess the same chemical properties of the matter that they help to form.

Currently, there are over 100 known elements, shown in **Figure 14-1**. These elements can be naturally occurring or man-made. This study of electricity begins by focusing on one of these elements—copper. **Figure 14-2** shows the atomic structure of a copper atom. At the center of the atom is a structure called the **nucleus**. The nucleus contains two components called **protons** and **neutrons**. Protons have a positive charge, whereas neutrons have no charge, or are said to be neutral. Within the nucleus of a copper atom are twenty-nine protons and thirty-five neutrons. The effect of this is that the nucleus has an overall net positive charge.

Again referring to Figure 14-2, notice the components that are in orbit around the nucleus. These components are called **electrons**. Electrons have a negative charge. In a copper atom, there are twenty-nine electrons. Notice that a copper atom has the same number of electrons as protons. Actually, all atoms in their natural state have an equal number of electrons and protons.

Notice also that the electrons are located at different distances from the nucleus. These are called orbits, or shells. The shell closest to the nucleus is called the first shell. This shell can hold only two electrons. The second shell, moving outward from the nucleus, can hold a maximum of eight electrons. The third shell, moving outward, can hold a maximum of eighteen electrons. The fourth shell can hold no more than thirty-two electrons. The fifth and sixth shells can hold up to eighteen electrons each. Finally, the seventh shell can hold a maximum of only two electrons. Some atoms have the shell that is farthest from the nucleus completely filled with electrons. Other atoms have the farthest shell partly filled. It is important to observe that a copper atom has only one electron in its outermost shell.

As mentioned earlier, the nucleus has a net positive charge, and the orbiting electrons have a negative charge. You may recall learning elsewhere about something called the **law of charges**. The law of charges states that opposite charges attract and like charges repel. This means that in the copper atom example, the electrons (negatively charged) should be attracted to the nucleus (positively charged). However, the electrons remain in orbit about the nucleus. There must be some other force in effect that prevents the electrons and the nucleus from colliding with each other. Actually, there are two forces at play here: centripetal and centrifugal.

The force of attraction between the negatively charged electron and the positively charged nucleus is called **centripetal** force. This is an inward force that tries to draw the electron toward the nucleus. The force that offsets the centripetal force is called **centrifugal** force. This is an outward force that tries to push the electron away from the nucleus. If these forces are balanced, the electron remains in orbit at a specific distance from the nucleus.

As mentioned earlier, a copper atom has only one electron in its outermost shell. Refer to **Figure 14-3**. This electron is called a **valence electron**. The outermost electrons in any atom are called valence electrons. Because the valence electron is located the farthest from the nucleus, the forces that hold it in orbit are weaker. If some external force were applied to the copper atom, the valence electron could be freed from its orbit.

1	2	3	4	5	6	7	8	9	10	11	12	13	14	15	16	17	18
hydrogen 1 **H**																	helium 2 **He**
lithium 3 **Li**	beryllium 4 **Be**											boron 5 **B**	carbon 6 **C**	nitrogen 7 **N**	oxygen 8 **O**	fluorine 9 **F**	neon 10 **Ne**
sodium 11 **Na**	magnesium 12 **Mg**											aluminum 13 **Al**	silicon 14 **Si**	phosphorus 15 **P**	sulfur 16 **S**	chlorine 17 **Cl**	argon 18 **Ar**
potassium 19 **K**	calcium 20 **Ca**	scandium 21 **Sc**	titanium 22 **Ti**	vanadium 23 **V**	chromium 24 **Cr**	manganese 25 **Mn**	iron 26 **Fe**	cobalt 27 **Co**	nickel 28 **Ni**	copper 29 **Cu**	zinc 30 **Zn**	gallium 31 **Ga**	germanium 32 **Ge**	arsenic 33 **As**	selenium 34 **Se**	bromine 35 **Br**	krypton 36 **Kr**
rubidium 37 **Rb**	strontium 38 **Sr**	yttrium 39 **Y**	zirconium 40 **Zr**	niobium 41 **Nb**	molybdenum 42 **Mo**	technetium 43 **Tc**	ruthenium 44 **Ru**	rhodium 45 **Rh**	palladium 46 **Pd**	silver 47 **Ag**	cadmium 48 **Cd**	indium 49 **In**	tin 50 **Sn**	antimony 51 **Sb**	tellurium 52 **Te**	iodine 53 **I**	xenon 54 **Xe**
caesium 55 **Cs**	barium 56 **Ba**	lanthanum 57 **La***	hafnium 72 **Hf**	tantalum 73 **Ta**	tungsten 74 **W**	rhenium 75 **Re**	osmium 76 **Os**	indium 77 **Ir**	platinum 78 **Pt**	gold 79 **Au**	mercury 80 **Hg**	thallium 81 **Tl**	lead 82 **Pb**	bismuth 83 **Bi**	polonium 84 **Po**	astatine 85 **At**	radon 86 **Rn**
francium 87 **Fr**	radium 88 **Ra**	actinium 89 **Ac#**	rutherfordium 104 **Rf**	dubnium 105 **Db**	seaborgium 106 **Sg**	bohrium 107 **Bh**	hassium 108 **Hs**	meitnerium 109 **Mt**	darm-stadium 110 **Ds**	roentgenium 111 **Rg**	Ununbium 112 **Uub**	Ununtrium 113 **Uut**	Ununquadium 114 **Uuq**	Ununpentium 115 **Uup**	Ununhexium 116 **Uuh**	Ununseptium 117 **Uus**	Ununoctium 118 **Uuo**

- Alkali Metals
- Alkali Earth Metals
- Transition Metals
- Lanthanides
- Actinides
- Non Metals
- Metalloid
- Halogens
- Noble Gases

***lanthanoids**	cerium 58 **Ce**	praseo-dymium 59 **Pr**	neodymium 60 **Nd**	promethium 61 **Pm**	samarium 62 **Sm**	europium 63 **Eu**	gadolinium 64 **Gd**	terbium 65 **Tb**	dysprosium 66 **Dy**	holmium 67 **Ho**	erbium 68 **Er**	thulium 69 **Tm**	ytterbium 70 **Yb**	lutetium 71 **Lu**
#actinoids	thorium 90 **Th**	protactinium 90 **Pa**	uranium 92 **U**	neptunium 93 **Np**	plutonium 94 **Pu**	americium 95 **Am**	curium 96 **Cm**	berkelium 97 **Bk**	californium 98 **Cf**	einsteinium 99 **Es**	fermium 100 **Fm**	mendelevium 101 **Md**	nobelium 102 **No**	lawrencium 103 **Lr**

FIGURE 14-1 The periodic table of the elements.

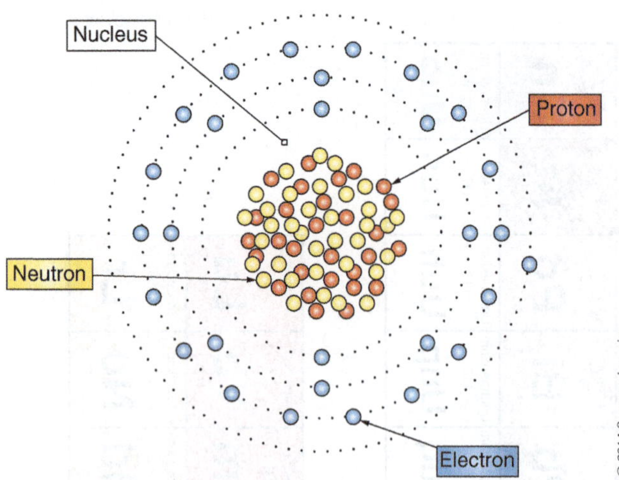

FIGURE 14-2 The atomic structure of a copper atom.

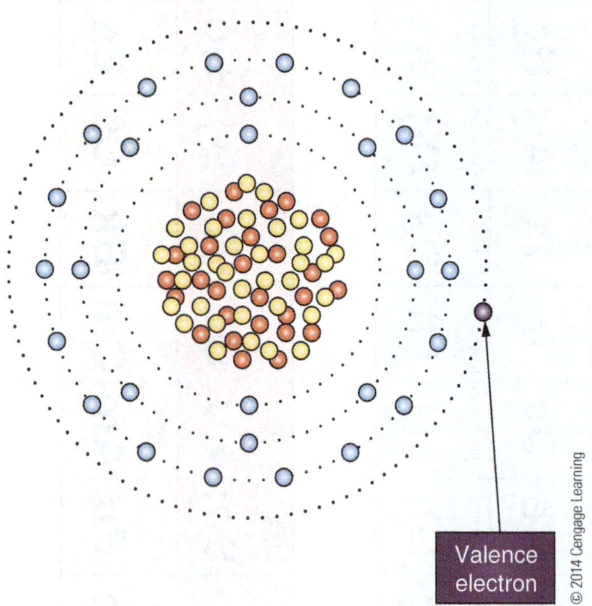

FIGURE 14-3 A valence electron.

This free electron would then wander from atom to atom. Should this free electron collide with a valence electron of another copper atom, the free electron could knock the valence electron out of its orbit. The free electron would be captured by the copper atom and the former valence electron would become the free electron. Refer to **Figure 14-4**.

Elements that have between one and two valence electrons are considered good conductors of electricity. This is because the valence electrons are loosely held in place and can be quite readily freed. Elements that have seven or eight valence electrons are considered poor conductors of electricity, because the valence electrons are very tightly held and are not so readily freed. These materials are called **insulators**. Examples of insulators are rubber, plastic, glass, and fiberglass. Some elements have three, four, or five valence electrons. These elements are neither good conductors nor good insulators. They are called **semiconductors**. Semiconductors are discussed in Chapter 23.

14-2 CURRENT

Electric **current** is the movement of electrons. As the free electrons travel from one atom to another, an electrical current is produced. If you could look at a material through which electrons were moving and count the number of electrons traveling by, you could determine the amount of electrical current. In electricity, the rate of flow of electrical current is measured in units called amperes, or amps. The **ampere (amp)** is the standard unit of measurement of electrical current. One ampere is equal to 6.25×10^{18} electrons (6,250,000,000,000,000,000) traveling past a given point in one second of time. The letter A is the abbreviation for amperes, or amps. In formulas, the letter I represents the intensity of the current. It might help you to think of current as similar to the flow of liquid in a plumbing system. Refer to **Figure 14-5**. In a plumbing system, the rate of flow of the liquid is measured in gallons per minute.

14-3 VOLTAGE

In a plumbing system, pressure is needed to make the liquid flow. Refer to **Figure 14-6**. Typically, this pressure is measured in pounds per square inch. Without this pressure, liquid could not flow. This is also true in an electrical system. Electrical current could not flow if there is no pressure behind it. The pressure in an electrical system is called **voltage**. Voltage is measured in units called **volts** and is abbreviated with the letter *V*. The letter *E* will be used to represent voltage in formulas. The letter *E* represents **electromotive force**, which is another term for voltage. Because voltage is the force that causes electrons to move, voltage is sometimes referred to as electromotive force. Voltage may also be called **difference of potential or potential difference**. These two terms mean the same thing. Imagine water behind a dam. The water

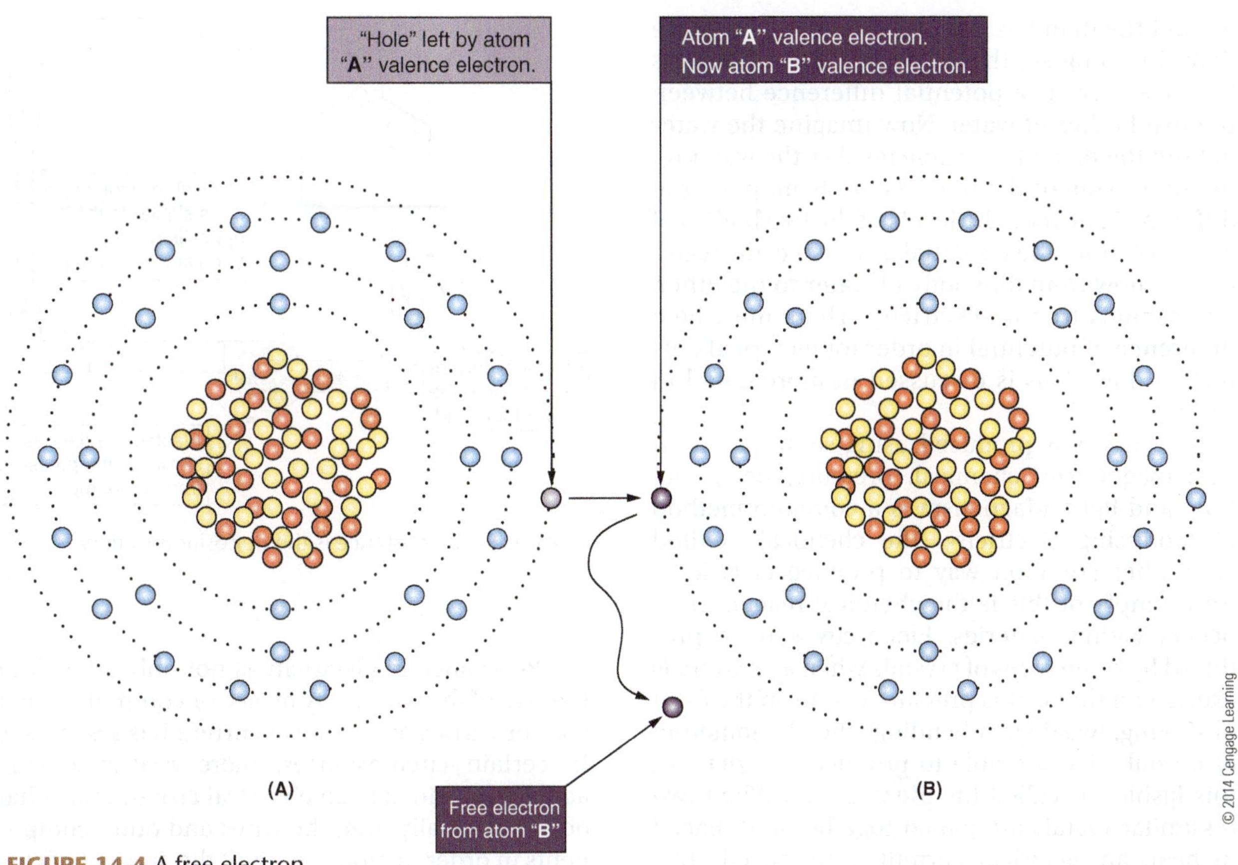

"Hole" left by atom "A" valence electron.
Atom "A" valence electron. Now atom "B" valence electron.
(A)
(B)
Free electron from atom "B"
© 2014 Cengage Learning

FIGURE 14-4 A free electron.

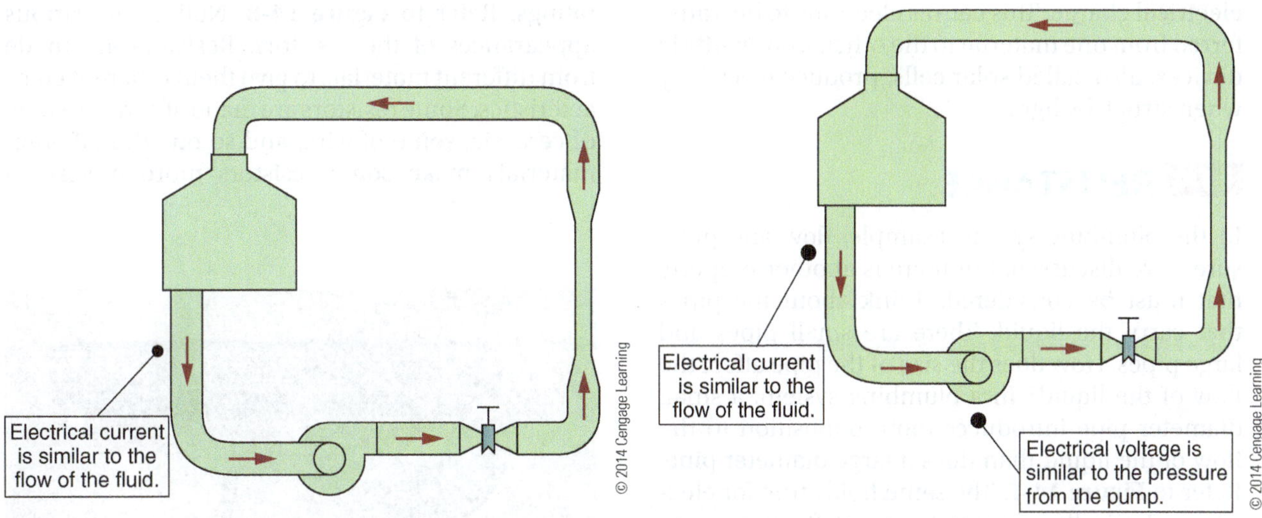

Electrical current is similar to the flow of the fluid.
© 2014 Cengage Learning
Electrical current is similar to the flow of the fluid.
Electrical voltage is similar to the pressure from the pump.
© 2014 Cengage Learning

FIGURE 14-5 Current is analogous to water flow.

FIGURE 14-6 Voltage is analogous to pressure.

behind the dam has the potential to fall over the dam. In doing so, the water would flow. This is because there is a potential difference between the two bodies of water. Now imagine the water behind the dam at the same level as the water on the other side of the dam. There is no potential difference between the levels of the two bodies of water. Without the potential difference, the water cannot flow from one body of water to the other. The same is true in electricity. There must be a difference of potential in order for electrical current to flow. This is discussed in more detail in Chapter 16.

Electricity is produced using six basic methods: magnetism, chemical, pressure, heat, friction, and light. Magnetism is a common method of producing electricity. The chemical method is another common way to produce electricity. An example of this is the chemical reaction that occurs within batteries. Electricity can be produced by some types of crystals when a pressure is exerted on them. This pressure can be in the form of striking, twisting, or bending. The phenomenon of crystals that are able to produce electricity in this fashion is called the **piezo effect**. When two dissimilar metals are joined together and placed in heat, an electrical current is produced. This device is called a thermocouple. The phenomenon of producing electricity from heat is called the **Seebeck effect**. The friction between certain materials when rubbed together produces a static electrical charge. This causes electrons to be transferred from one material to the other. **Photovoltaic** devices, also called solar cells, produce electricity when struck by light.

14-4 RESISTANCE

In the plumbing system example, flow and pressure were discussed, but there is another property that must be considered. Think about the pipes that carry the liquid. There are small pipes and large pipes. How does the size of the pipe affect the flow of the liquid? In a plumbing system, a small diameter pipe introduces more opposition to the flow of the liquid than does a large diameter pipe. Refer to **Figure 14-7**. The same holds true for electricity. Typically, electrical current flows through wires. There are small wires and large wires. A small diameter wire introduces more opposition to the flow of electrical current than does a large diameter wire. This opposition to electrical current is called **resistance**. Resistance is measured in units called **ohms**. The Greek letter for *omega* (Ω) represents ohms. In formulae, the letter *R* represents resistance.

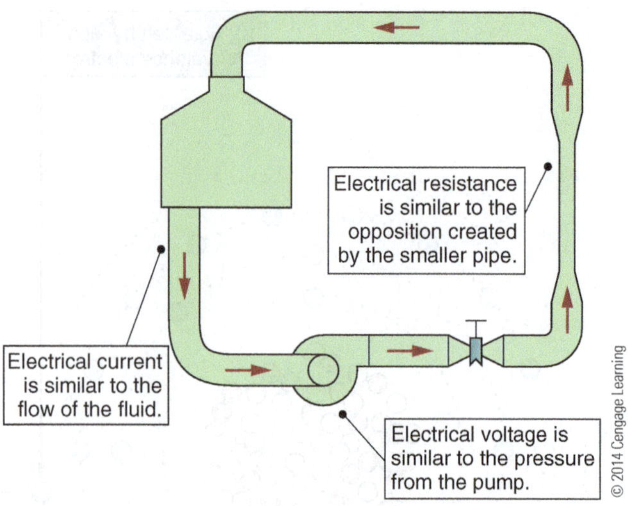
Electrical resistance is similar to the opposition created by the smaller pipe.
Electrical current is similar to the flow of the fluid.
Electrical voltage is similar to the pressure from the pump.
© 2014 Cengage Learning

FIGURE 14-7 Resistance is the opposition to flow.

Resistance in electricity is not only a result of the size of the wires. Any device or component that uses or carries an electrical current has resistance. In certain circumstances, more resistance may actually be added to an electrical **circuit** than what occurs normally from the wires and other components in order to reduce or limit the amount of current that is flowing in a circuit. To reduce or limit the current, one or more devices called **resistors** are added to the circuit.

Resistors come in different shapes, sizes, and ratings. Refer to **Figure 14-8**. Notice the various appearances of the resistors. Resistors are made from different materials to give them different characteristics. Some resistors are made of carbon, some of ceramic, some of wire, and so on. The different materials make some resistors more precise in

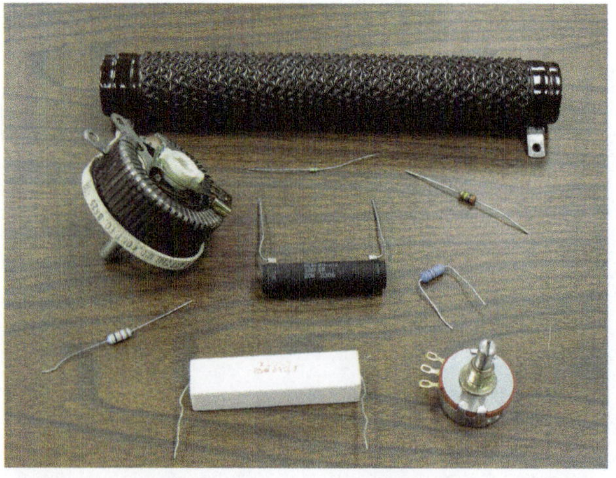
© 2014 Cengage Learning

FIGURE 14-8 Various types of resistors.

their values. Other materials make the resistors capable of operating at higher temperatures. Notice also the different physical sizes of the resistors in Figure 14-8. Typically, the physical size is an indicator of how much heat the resistors can handle before becoming damaged. Finally, notice the various color bands that encircle the resistors. Some resistors have four color bands, whereas others have five color bands. These color bands represent a code that indicates the amount of resistance a particular resistor offers.

Four-Band Resistor

Refer to **Figure 14-9**. The first thing that you must do to decode the color code of a resistor is to position it properly. Notice that the entire group of color bands is shifted slightly toward one end of the resistor. You must position the resistor so that the group of color bands is pointed toward your left side. Figure 14-9 shows the resistor in the correct orientation. Refer to **Figure 14-10**. With the resistor positioned in this fashion, the first color band to your left represents the first significant digit in the value of the resistor. The next color band to its right represents the second significant digit. The next color band (third band from the left) represents the multiplier value. Finally, the last color band (fourth band from the left) represents the tolerance.

The colors in each of the bands have values assigned to them. **Table 14-1** lists the colors in the bands and the values assigned to those colors. Refer to **Figure 14-11**. In this example, the first band is brown, the second band is black, the third band is red, and the fourth band is silver. To decode the color-code of this resistor, look at the color of the first band. It is brown. A brown-colored first band has a value of 1. Now look at the second band. Its color is black. A black-colored second band has a value of 0. The third band is red. A red-colored third band has a multiplier value of ×100. Ignore the fourth band for the moment. Now put the values of the first three bands together. You will have 1, 0, ×100 or 10 ×100, or 1000. Therefore, the value of a brown, a black, and a red resistor is 1000 ohms.

The fourth band is silver. A silver-colored fourth band has a tolerance of ±10%. Therefore, the final value of the resistor is 1000 ohms ±10%. This means that if you were to measure the resistance of this brown, black, red, and silver resistor, the measured value would be between 900 ohms and 1100 ohms. The color-code value of the resistor (1000 ohms) can vary up or down by as much as 10% and still be considered to be within tolerance. Because 10% of 1000 is 100, the value of the

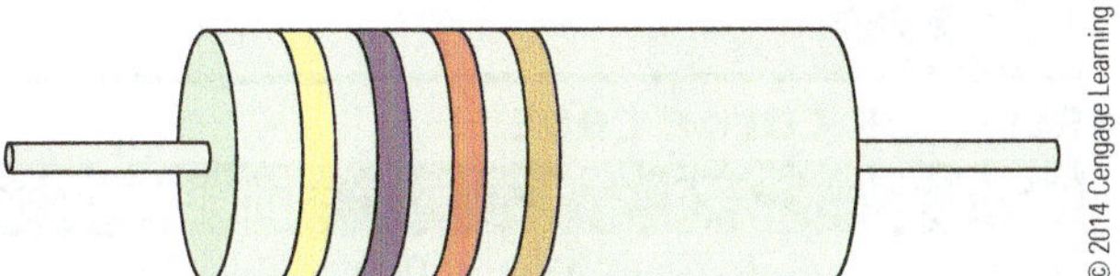

FIGURE 14-9 Four-band resistor.

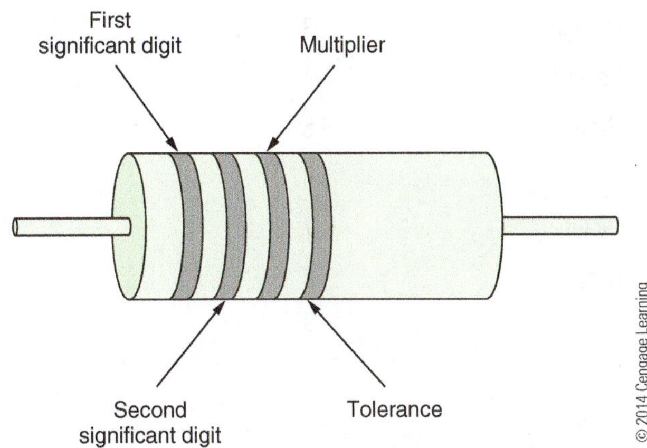

FIGURE 14-10 The correct orientation for determining color-code value.

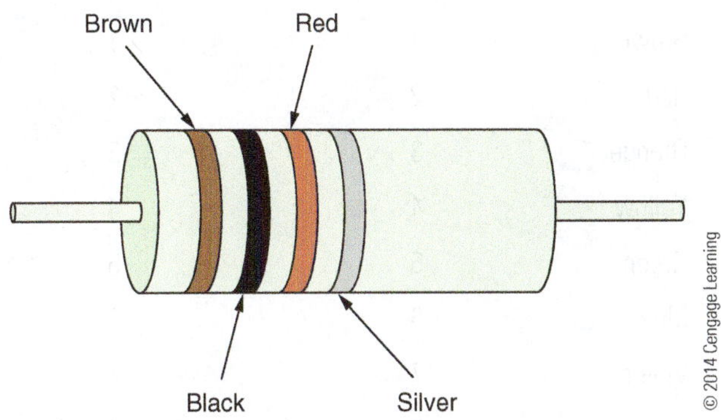

FIGURE 14-11 A four-band resistor color-code example.

resistor can be 100 less than 1000, or 900 ohms; 100 more than 1000, or 1100 ohms; or anywhere in between.

Five-Band Resistor

Refer to **Figure 14-12**. In this example, the first band is brown, the second band is black, the third band is black, the fourth band is brown, and the fifth band is brown. To decode the color code of this resistor, look at the color of the first band. It is brown. A brown first band has a value of 1. Now look at the second band. It is black. A black second band has a value of 0. The third band is black.

Table 14-1

RESISTOR COLOR CODE

Four-Band Resistor Color-Code Chart				
Color	*1st Significant Digit*	*2nd Significant Digit*	*Multiplier*	*Tolerance*
Black		0	×1	
Brown	1	1	×10	±1%
Red	2	2	×100	±2%
Orange	3	3	×1,000	
Yellow	4	4	×10,000	
Green	5	5	×100,000	
Blue	6	6	×1,000,000	
Violet	7	7	×10,000,000	
Grey	8	8		
White	9	9		
Gold			×0.1	±5%
Silver			×0.01	±10%
None				±20%

Five-Band Resistor Color-Code Chart					
Color	*1st Significant Digit*	*2nd Significant Digit*	*3rd Significant Digit*	*Multiplier*	*Tolerance*
Black		0	0	×1	
Brown	1	1	1	×10	±1%
Red	2	2	2	×100	±2%
Orange	3	3	3	×1,000	
Yellow	4	4	4	×10,000	
Green	5	5	5	×100,000	±0.5%
Blue	6	6	6	×1,000,000	±0.25%
Violet	7	7	7	×10,000,000	±0.1%
Grey	8	8	8		
White	9	9	9		
Gold				×0.1	
Silver				×0.01	

A black-colored band has a value of 0. The fourth band is brown. A brown fourth band has a multiplier value of ×10. Ignore the fifth band for the moment. Now put the values of the first four bands together. You will have 1, 0, 0, ×10 or 100 ×10, or 1000. Therefore, the value of a brown, black, black, and brown resistor is 1000 ohms.

Now let us look at the fifth band. Five-band resistors are more precise than four-band resistors. Therefore, the tolerances on five-band resistors are tighter than those on four-band resistors. The fifth band is brown. A brown fifth band has a tolerance of ±1%. Therefore, the final value of the resistor is 1000 ohms ±1%. This means that if you were to measure the resistance of this brown, black, black, brown, and brown resistor, the measured value could be between 990 ohms and 1010 ohms. The color-code value of the resistor (1000 ohms) can vary up or down by as much as 1% and still be considered to be within tolerance. Because 1% of

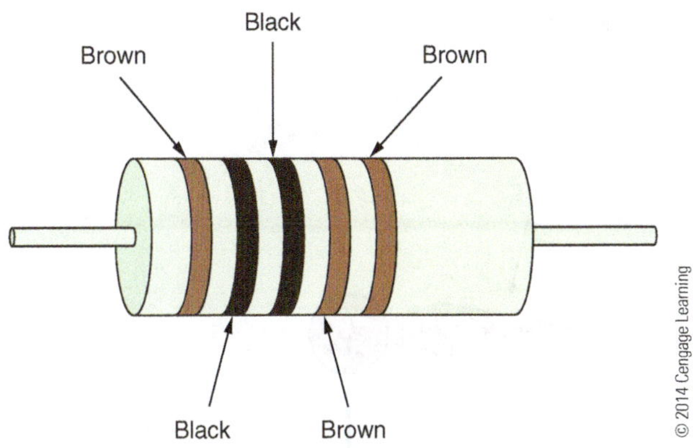

FIGURE 14-12 A five-band resistor color-code example.

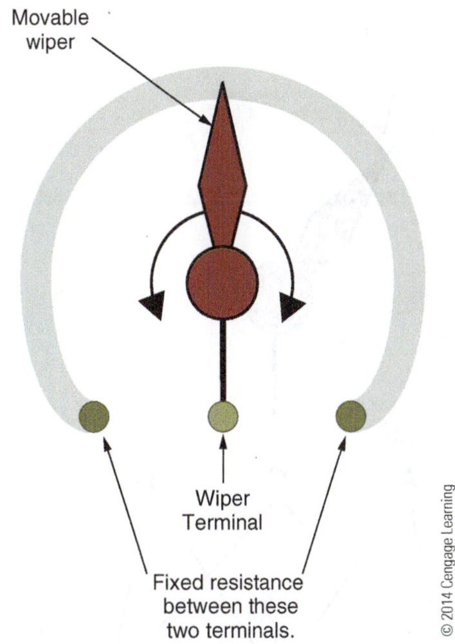

FIGURE 14-13 Variable resistor construction.

1000 is 10, the value of the resistor can be 10 less than 1000, or 990 ohms; 10 more than 1000, or 1010 ohms; or anywhere in between. As you can see, a variation of ±10 ohms is a smaller variation than ±100 ohms, as was seen in the four-band resistor example.

Four-band resistors and five-band resistors belong to a family of resistors known as *fixed resistors*. This is because their values are determined during the manufacturing process and cannot be changed.

Variable Resistor

Another family of resistors with which you must become familiar is called **variable resistors**. **Figure 14-13** shows the construction of a variable resistor. Variable resistors are constructed by creating a fixed resistance with a connection terminal at each end. A movable wiper that also has a connection terminal is then added. The wiper can be moved from one end of the resistor to the other and stopped anywhere in between. This allows adjustment of the resistance between the wiper and an end terminal to a desired value. Some wipers are adjusted by using a rotary motion. This is similar to the volume control found on most radios. Other variable resistors use a linear slide adjustment. This is similar to the adjustments found on most audio equalizer controls. **Figure 14-14** shows examples of rotary and linear slide-type variable resistors.

Refer to **Figure 14-15** for an example. Notice that the total resistance of this variable resistor is 600 ohms. This means that the resistance measured from one end of the element, point *A*, to the other end of the element, point *B*, is 600 ohms. Now notice the position of the wiper, point *C*. The wiper is not located at the exact midpoint of the element between points *A* and *B*. In fact, the wiper is located approximately one-third of the distance from point *A* to point *B*. This means that the resistance measured between point *A* and the wiper is approximately 200 ohms (⅓ of the total resistance), and the resistance measured between point *B* and the wiper is approximately 400 ohms (⅔ of the total resistance). The sum of these two measurements must equal the total resistance of the element, 600 ohms.

FIGURE 14-14 Rotary and linear resistors.

Figure 14-16 shows the wiper repositioned to a point approximately one-fifth of the distance from point *A*. The resistance from point *A* to the wiper now measures approximately 120 ohms. The resistance from point *B* to the wiper now measures approximately 480 ohms. The sum of these two measurements still equals the total resistance of the element, 600 ohms. However, if a circuit were connected between point *A* and the wiper, the

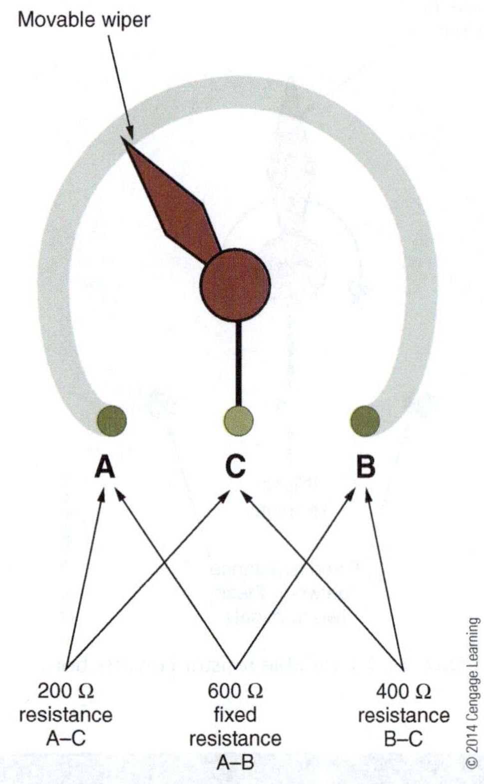

FIGURE 14-15 A variable resistor—⅓ rotation.

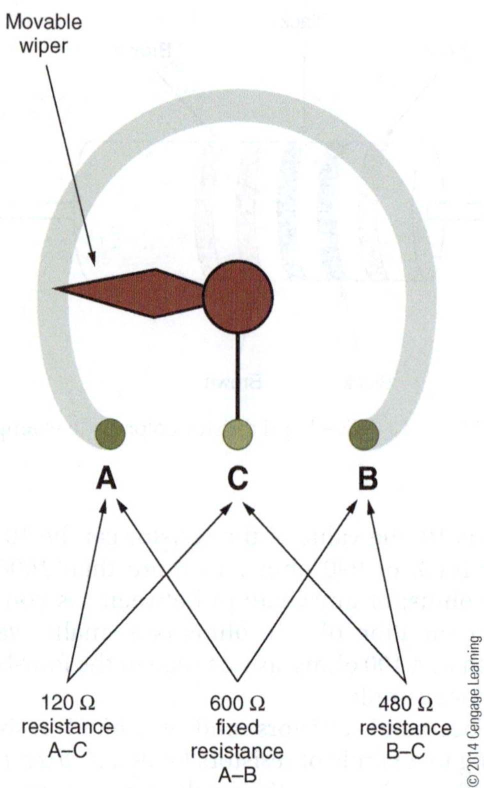

FIGURE 14-16 A variable resistor— ⅕ rotation.

resistance would be adjusted from 200 ohms to 120 ohms. Using a variable resistor allows adjustment of the resistance in a circuit without having to replace a component.

In addition to varying the amount of resistance, a variable resistor can be used to vary voltage or current. This is a function of how the variable resistor is connected in the circuit. If the variable resistor is connected to vary the voltage, the variable resistor is referred to as a **potentiometer**, or **pot** for short. The word *potentiometer* refers to potential, or voltage. If the variable resistor is connected to vary the current, the variable resistor is referred to as a **rheostat**.

14-5 RESISTOR WATTAGE RATINGS AND POWER

An electrical current under pressure from an electrical voltage flowing through a resistance produces heat. This heat is a result of the friction produced by the electrons moving through the conductors (wires, resistors, and others). The heat must be dissipated by the wires and any resistors. When using resistors, the amount of heat that the resistor must be capable of handling must be considered. This is known as the **wattage rating** of the resistor. Resistors are available in many different wattage ratings. Usually an indicator of the wattage rating of a resistor is its physical size and the type of material used to construct it. Notice the different sizes of the resistors shown in **Figure 14-17**. The smallest resistor shown has a wattage rating of ⅛ watt. The largest resistor shown has a wattage rating of 2 watts. Typically, resistors with a wattage rating higher than 2 watts will have the actual wattage rating marked on the body of the resistor. Wattage ratings as high as several hundred watts are available.

To determine the wattage rating needed for a particular resistor in a particular circuit, you must perform some simple calculations. Following is an example:

A circuit contains a 100-ohm resistor, which is connected to 120 volts. How much heat must the resistor dissipate? To find the amount of heat that the resistor must dissipate, you need to know the following formula:

$$P = \frac{E^2}{R}$$

P = Power (measured in watts)
E = Voltage (measured in volts)
R = Resistance (measured in ohms)

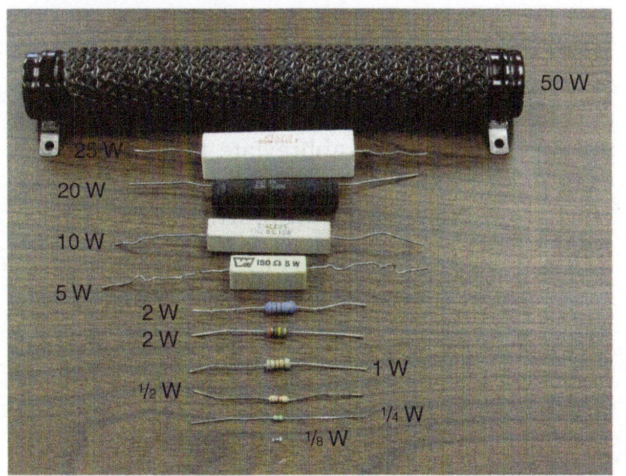

FIGURE 14-17 Various resistors and wattage ratings.

Now insert the known values into the formula and solve for *P*:

$$P = \frac{E^2}{R} = \frac{120\ V^2}{100\ \Omega} = \frac{14400\ V}{100\ \Omega} = 144\ W$$

This means that the resistor in this circuit must be capable of dissipating 144 watts of heat. Following is another example:

A circuit contains a 50-ohm resistor that has 5 amperes of current flowing through it. How much heat must this resistor dissipate? The formula that was used in the first example cannot be used because the voltage in this circuit is not known. Fortunately, another formula can be used when the resistance and current are known:

$$P = I^2R$$

P = Power (measured in watts)
I = Current (measured in amps)
R = Resistance (measured in ohms)

Now insert the known values into the formula and solve for *P*:

$$P = I^2R = 5\ A^2 \times 50\ \Omega = 25\ A \times 50\ \Omega = 1250\ W$$

This means that the resistor in this circuit must be capable of dissipating 1250 watts of heat. Following is one more example:

A circuit with a resistor is supplied with 277 volts and has 2 amperes of current flowing in the circuit. How much heat must this resistor dissipate? The formula from the first example cannot be used because the value of the resistor is not known. Likewise, the formula used in the second example will not work because it also requires the value of the resistor. Fortunately, another formula can be used when the value of the resistor is unknown:

$$P = IE$$

P = Power (measured in watts)
I = Current (measured in amps)
E = Voltage (measured in volts)

Now insert the known values into the formula and solve for *P*:

$$P = IE = 2\ A \times 277\ V = 554\ W$$

This means that the resistor in this circuit must be capable of dissipating 554 watts of heat.

When determining the wattage rating of a resistor for a new circuit, it is good practice to double the wattage rating of the resistor to provide a margin of error.

Power in electrical circuits is also a measure of the rate at which **work** is done. Work is the overcoming of resistance through a distance. Work is generally measured in **foot-pounds** (ft-lb). Foot-pounds are defined as the work performed when a force of 1 pound acts through a distance of 1 ft. Power is a measurement of the rate at which work is done. If Work = Force × Distance, then Power = Work/Time (measured in minutes).

Perhaps the most familiar unit of measurement of mechanical power is *horsepower* (hp). The formula used to find horsepower is:

$$hp = \frac{ft\text{-}lb/min}{33{,}000}$$

Look at the following example:

A crane can lift a pallet of 55-gallon drums weighing 10,000 pounds onto a flatbed railroad car that is 8 ft high. It takes the crane 3 minutes to lift the load. How much horsepower does the crane deliver?

First, find the ft-lb.

$$ft\text{-}lb = ft \times lb = 8\ ft \times 10{,}000\ lb = 80{,}000\ ft\text{-}lb$$

Now find the ft-lb/min.

$$ft\text{-}lb/min = \frac{ft\text{-}lb}{min} = \frac{80{,}000\ ft\text{-}lb}{3\ min} = 26{,}666.67\ ft\text{-}lb/min$$

Finally, find the hp.

$$hp = \frac{ft\text{-}lb/min}{33{,}000} = \frac{26{,}666.67\ ft\text{-}lb/min}{33{,}000} = 0.81\ hp$$

In electrical circuits, power is measured in watts. To convert from mechanical power to electrical power, use the following conversion:

$$1\ hp = 746\ W$$

In the previous example of the crane, it was determined that the crane developed 0.81 hp. How much electrical power was this? Use the conversion 1 hp = 746 W to determine the equivalent electrical power. From this, a new formula can be derived:

$$W = hp \times 746\ W/hp$$

Now insert the known values and solve for W.

$$W = hp \times 746\ W/hp = 0.81\ hp \times 746\ \text{W/hp} = 604.26\ \text{W}$$

Therefore, the crane develops 0.81 hp or 604.26 W of power.

Electrical energy is the product of power and time. The unit of measurement for electrical energy is the watt-hour (Wh) or perhaps more commonly, the kilowatt-hour (kWh). A 100-watt lightbulb that is operated for 10 hours uses 1000 watt-hours (Wh) of electricity, or 1 kWh (100 W × 10 hours = 1000 Wh or 1 kWh). Most consumers are familiar with the kilowatt-hour. The kilowatt-hour is the unit of measurement used by the electric company to determine your electric bill. The electric utility multiplies your electric consumption (in kWh) by the current rate ($/kWh) to arrive at your bill amount. Look at the following example:

An electric utility charges $0.08/kWh for its electricity. In a one-month period, a homeowner uses 1150 kWh of electricity. How much is the homeowner's electric bill for this month? Simply multiply the electric usage by the rate to determine the amount of the bill.

$$cost = usage \times rate = 1150\ \text{kWh} \times \$0.08/\text{kWh} = \$92.00$$

Therefore, this homeowner will receive a bill for $92 to cover the electricity that was used for the month.

14-6 SCIENTIFIC NOTATION

When dealing with measurements and quantities in the field of electricity, it is not uncommon to encounter values that are quite large as well as quite small. An example would be a current of 10,000 amperes or a voltage of 0.000006 volt. It can be seen that these numbers can be somewhat difficult to handle accurately. How many zeros appear to the right of the decimal point in the voltage example? It might be necessary to double-check to be certain that there are actually five.

Fortunately, there is an easier and more accurate method of writing and even speaking very large and very small numbers like these. The method is known as **scientific notation**. Scientific notation is a type of shorthand that is used to make large and small numbers more easily manageable. Powers of 10 are used to abbreviate the number. For example, 10,000 amperes can be rewritten as 1.0×10^4. Because 10 raised to the fourth power is 10,000, 1.0×10^4 is $1 \times 10{,}000$ or 10,000.

Here is another example. The number 3,404,655 can be rewritten using scientific notation. The number becomes 3.404655×10^6. 10 raised to the sixth power is 1,000,000. Therefore, $3.404655 \times 1{,}000{,}000$ is 3,404,655.

The same process is used for very small numbers. The only difference is that a negative (−) sign is placed in front of the power of 10 exponent to indicate that the zeros are to the right of the decimal point. The voltage example used earlier will illustrate this. The voltage was 0.000006 volt. Rewritten in scientific notation, the new number becomes 6.0×10^{-6}. 10 raised to the −6 power is 0.000001. Therefore, 6.0×0.000001 becomes 0.000006.

Here is another example: 0.0234. Using scientific notation, 0.0234 becomes 2.34×10^{-2}. 10 raised to the −2 power is 0.01. Therefore, 2.34×0.01 becomes 0.0234. Note that this example can also be written as 23.4×10^{-3} or 234×10^{-4}. These are all correct; however, it is preferable to rewrite a number in scientific notation with one whole number to the left of the decimal point. Therefore, 2.34×10^{-2} is the preferred method.

14-7 ENGINEERING NOTATION

A variation on scientific notation is **engineering notation**. Engineering notation works in a similar manner to scientific notation except that engineering notation moves in steps of 1000, not 10.

The first example of 10,000 amperes will illustrate this. Using engineering notation, 10,000 amperes can be rewritten as 10.0×10^3. 10 raised to the third power is 1000. Therefore, 10.0×1000 becomes 10,000. However, this is taken one step further in engineering notation. Words are substituted for the power of 10. For instance, 10^3 becomes **kilo**, which means "times 1000." So 10,000 amperes can be rewritten as 10 kiloamperes. This is the same as saying 10×10^3 amperes. The number can be shortened even further by abbreviating the word *kilo* with a k and writing it as 10 k amperes. Because this is new, it might seem confusing, but actually this can be more accurate, because it is not necessary to get the correct number of zeros in the number.

Remember the number 3,404,655? If it is rewritten in engineering notation, 3,404,655 becomes $3.404{,}655 \times 10^6$. This is the same as the scientific notation example. The reason is that engineering notation moves in steps of thousands. From one thousand, the next step is one thousand thousand, or one million. This means that the power of 10 increases by threes, from 10^3 to 10^6, to 10^9 and so on.

Remember that in engineering notation, words are used to replace the powers of 10. 10^3 was called kilo or k. 10^6 is called mega or M. **Mega** means "times 1,000,000." Now $3.404{,}655 \times 10^6$ can be rewritten as 3.404,655 megavolts or 3.404,655 M volts.

Table 14-2 shows the standard units of engineering notation as well as the associated scientific notation values. Refer to Table 14-2 in examining the next example.

In an earlier example, the voltage was 0.000006 volt. How would this value be expressed in engineering notation? 0.000006 becomes 6.0×10^{-6} volts. This is the same as the scientific notation example. Engineering notation moves in steps of thousandths as well. From one one-thousandth, the next step is one thousand one-thousandths, or one millionth. This means that the power of 10 increases by threes, from 10^{-3} to 10^{-6}, to 10^{-9} and so on. Remember that in engineering notation, words are used to replace the powers of 10. 10^{-3} is called milli or m. **Milli** means "one-thousandth." 10^{-6} is called micro or μ (Greek letter called mu). **Micro** means "one-millionth." Now 6.0×10^{-6} can be rewritten as 6.0 microvolts or 6.0 μ volts.

Here is one final example: 0.0234. Using engineering notation, 0.0234 becomes 23.4×10^{-3}, or 23.4 milli (23.4 m). 10 raised to the −3 power is 0.001. Therefore, 23.4×0.001 becomes 0.0234.

Notice that unlike the example in scientific notation, this is the only way to write 0.0234 in engineering notation. This number could be written as $23{,}400 \times 10^{-6}$, or 23,400 μ, but notice how large the number becomes. Likewise, 0.0234 could be written as 0.0000234×10^3, but look at all those zeros! 23.4 μ is the easiest number to use and understand.

14-8 OHM'S LAW

A German scientist and mathematician, Georg S. Ohm, observed a definite relationship among current, voltage, and resistance in an electrical circuit. Ohm observed that current and resistance were inversely proportional, whereas current and voltage were directly proportional. Ohm also realized that these relationships could be expressed and demonstrated mathematically. All individuals involved in the electrical field need a solid understanding of Ohm's law and the relationships among current, voltage, and resistance. An understanding of Ohm's law will allow you to determine an unknown quantity in an electrical circuit when two other variables are known.

Ohm developed a formula to express the relationships among current, voltage, and resistance in an electrical circuit. The formula that he developed is now referred to as Ohm's law in his honor, just

Table 14-2

STANDARD UNITS OF SCIENTIFIC AND ENGINEERING NOTATIONS

Scientific Notation	Engineering Notation	Symbol	Multiply By:
1×10^{12}	Tera	T	1,000,000,000,000
1×10^{9}	Giga	G	1,000,000,000
1×10^{6}	Mega	M	1,000,000
1×10^{3}	Kilo	K	1,000
1×10^{0}	Base Unit	—	1
1×10^{-3}	milli	m	0.001
1×10^{-6}	micro	μ	0.000,001
1×10^{-9}	nano	n	0.000,000,001
1×10^{-12}	pico	p	0.000,000,000,001

as the unit of resistance, the ohm, also honors the man. Ohm's law states:

$$I = \frac{E}{R}$$

I = Current flowing in the circuit measured in amperes (A)
E = Voltage applied to the circuit measured in volts (V)
R = Resistance of the circuit measured in ohms (Ω)

Solve the following problems to better understand a circuit according to Ohm's law.

Suppose that a circuit consists of a 100-ohm resistor with 115 volts applied to the circuit. How much current is flowing in this circuit? Use Ohm's law to determine the amount of current flow:

What is known?

$E = 115\ \text{V}$
$R = 100\ \Omega$

What is not known?

$I = ?$

What formula can be used?

$$I = \frac{E}{R}$$

Substitute the known values and solve for *I*:

$$I = \frac{E}{R} = \frac{115\ \text{V}}{100\ \Omega} = 1.15\ \text{A}$$

According to Ohm's law, 1.15 amperes of current is flowing in this circuit.

Here is another example:

A circuit is energized with 48 volts and contains 75 ohms of resistance. How much current is flowing in this circuit?

What is known?

$E = 48\ \text{V}$
$R = 75\ \Omega$

What is not known?

$I = ?$

What formula can be used?

$$I = \frac{E}{R}$$

Substitute the known values and solve for *I*:

$$I = \frac{E}{R} = \frac{48\ \text{V}}{75\ \Omega} = 0.64\ \text{A, or 640 mA}$$

According to Ohm's Law, 0.64 ampere, or 640 milliamperes, of current is flowing in this circuit.

However, what is Ohm's law really trying to tell? Look at a few more examples. Begin with a circuit that contains 50 ohms of resistance and is energized with 200 volts. Find the current flowing in this circuit.

What is known?

$E = 200\ \text{V}$
$R = 50\ \Omega$

What is not known?

$I = ?$

What formula can be used?

$$I = \frac{E}{R}$$

Substitute the known values and solve for *I*:

$$I = \frac{E}{R} = \frac{200\ \text{V}}{50\ \Omega} = 4\ \text{A}$$

Therefore, the circuit will have 4 amperes of current flowing through it. Now leave the voltage at the same level (200 volts) but increase the resistance to 100 ohms. What will happen to the current?

What is known?

$E = 200\ \text{V}$
$R = 100\ \Omega$

What is not known?

$I = ?$

What formula can be used?

$$I = \frac{E}{R}$$

Substitute the known values and solve for *I*:

$$I = \frac{E}{R} = \frac{200\ \text{V}}{100\ \Omega} = 2\ \text{A}$$

Now the circuit current is 2 amperes instead of the 4 amperes in the previous example. Notice that doubling the resistance while keeping the voltage constant has halved the current. In other words, with the voltage unchanged, increasing the resistance decreases the current.

Look at another example. Again, the voltage will be kept constant, but the resistance will be lowered to 25 ohms from the original value of 50 ohms.

What is known?

$E = 200\ \text{V}$
$R = 25\ \Omega$

What is not known?

$I = ?$

What formula can be used?

$$I = \frac{E}{R}$$

Substitute the known values and solve for *I*:

$$I = \frac{E}{R} = \frac{200\ \text{V}}{25\ \Omega} = 8\ \text{A}$$

Now the circuit current is 8 amperes instead of the original 4 amperes. Notice that halving the resistance and keeping the voltage constant have doubled the current. In other words, with the voltage unchanged, decreasing the resistance increases the current. This is what Ohm meant when he stated that if the voltage of a circuit is kept constant, current and resistance are inversely proportional.

What happens when the resistance of a circuit is held constant and the voltage is varied? Following are three examples to find the answer. For the first example, use the original problem of a circuit consisting of 50 ohms energized with 200 volts. Recall that the current flowing in this circuit was 4 amperes. For the second example, the voltage will be increased to 400 volts, leaving the resistance at 50 ohms. What will the effect on the current be?

What is known?

$$E = 400\text{ V}$$
$$R = 50\ \Omega$$

What is not known?

$$I = ?$$

What formula can be used?

$$I = \frac{E}{R}$$

Substitute the known values and solve for *I*:

$$I = \frac{E}{R} = \frac{400\text{ V}}{50\ \Omega} = 8\text{ A}$$

Now the circuit current is 8 amperes instead of the 4 amperes in the original example. Notice that doubling the voltage while keeping the resistance constant has doubled the current. In other words, with the resistance unchanged, increasing the voltage increases the current.

For the third example, decrease the voltage to 100 volts, leaving the resistance at 50 ohms. What will the effect on the current be?

What is known?

$$E = 100\text{ V}$$
$$R = 50\ \Omega$$

What is not known?

$$I = ?$$

What formula can be used?

$$I = \frac{E}{R}$$

Substitute the known values and solve for *I*:

$$I = \frac{E}{R} = \frac{100\text{ V}}{50\ \Omega} = 2\text{ A}$$

Now the circuit current is 2 amperes instead of the 4 amperes in the original example. Notice that halving the voltage while keeping the resistance constant has halved the current. In other words, with the resistance unchanged, decreasing the voltage decreases the current. This is what Ohm meant when he stated that if the resistance of a circuit is kept constant, current and voltage are proportional.

The previous examples work very well for circuits where the voltage and resistance are known and you are trying to determine the amount of current flow in a circuit. However, suppose you knew the amount of current flowing in the circuit as well as the amount of voltage applied to the circuit, but you did not know the amount of resistance in the circuit. What then? On the other hand, suppose you knew the amount of current flowing through a certain amount of circuit resistance, but you did not know how much voltage was applied to the circuit. Could this be determined? The answer to both of these questions is yes. The Ohm's law formula can be manipulated for finding current so that you can find resistance if the voltage and current are known, or find voltage if the current and resistance are known. Here is how. First, manipulate the Ohm's law formula used to find current so that you can find resistance.

Ohm's Law formula to find current:

$$I = \frac{E}{R}$$

Multiply both sides of the equation by *R*.

$$R \times I = \frac{E}{R} \times R$$

Cancel like terms.

$$R \times I = E \text{ or } E = IR$$

Therefore, $E = IR$ is the formula that can be used to find voltage (E) when current (I) and resistance (R) are known.

Now find the formula to find resistance when current and voltage are known. Again, begin with the original Ohm's law formula that was used to find current.

Ohm's Law formula to find current:

$$I = \frac{E}{R}$$

Multiply both sides of the equation by *R*.

$$R \times I = \frac{E}{R} \times R$$

Cancel like terms.

$$R \times I = E$$

Divide both sides of the equation by *I*.

$$\frac{R \times I}{I} = \frac{E}{I}$$

Cancel like terms.

$$R = \frac{E}{I}$$

Therefore, $R = \frac{E}{I}$ is the formula that can be used to find resistance (*R*) when voltage (*E*) and current (*I*) are known.

Now use these two new formulae to solve some examples. Imagine that a circuit has 5 amperes of current flowing through a 2000-ohm resistor. How much voltage is applied to the circuit?

What is known?

$$I = 5\text{ A}$$
$$R = 2000\ \Omega$$

What is not known?

$$E = ?$$

What formula can be used?

$$E = IR$$

Substitute the known values and solve for *E*:

$$E = IR$$
$$= 5\text{ A} \times 2000\ \Omega = 10{,}000\text{ V, or } 10\text{ kV}$$

Therefore, according to Ohm's law, 10 kilovolts must be applied to a circuit containing 2000 ohms of resistance in order to produce a current flow of 5 amperes.

Solve the following problem in which you need to determine the amount of resistance that allows 150 milliamperes of current flow in a circuit energized with 240 volts.

What is known?

$$E = 240\text{ V}$$
$$I = 150\text{ mA}$$

What is not known?

$$R = ?$$

What formula can be used?

$$R = \frac{E}{I}$$

Substitute the known values and solve for *R*:

$$R = \frac{E}{I} = \frac{240\text{ V}}{150\text{ mA}} = 1600\ \Omega\text{, or } 1.6\text{ k}\Omega$$

Therefore, according to Ohm's law a resistance of 1.6 kilohms will allow 150 milliamperes of current flow when 240 volts is applied to this circuit.

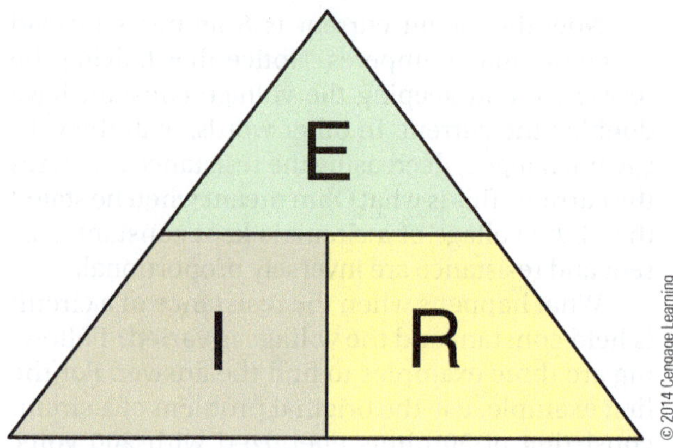

FIGURE 14-18 The Ohm's law triangle.

If you have difficulty manipulating these three Ohm's law formulae, a simple device can help. This device is called the **Ohm's law triangle**. Refer to **Figure 14-18**. Notice the triangle and how it is divided. Notice also that the letter *E* is located in the top compartment, the letter *I* is located in the lower left compartment, and the letter *R* is located in the lower right compartment. To use the Ohm's law triangle, you simply cover up the compartment of the variable that you are trying to find. The remaining compartments show you the formula that you should use. For example, if you need to find the current, cover the *I*. This leaves the *E* and the *R* uncovered. However, notice the arrangement of the *E* and the *R*. They appear as the formula $\frac{E}{R}$. Therefore, the triangle shows you the formula for finding current, $I = \frac{E}{R}$.

Suppose you need to find the voltage when the current and the resistance are known. Simply cover the *E* because you are trying to find the voltage, and you have the formula $I \times R$. Therefore, the triangle shows you the formula for finding voltage, $E = I \times R$. Finally, if you need to determine the resistance when you know the voltage and current, cover the *R*. The remaining uncovered variables give you the formula $\frac{E}{I}$. Therefore, the triangle shows you the formula for finding resistance, $R = \frac{E}{I}$.

14-9 POWER LAW

As you have seen, the relationships among current, voltage, and resistance can be demonstrated mathematically. The same is true for power. The amount of power used by a circuit depends on the amount of voltage present and the amount of current flow in the circuit. This is seen mathematically by using the following formula:

$$P = I \times E$$

P = Power used by the circuit measured in watts (W)
I = Current flowing in the circuit measured in amperes (A)
E = Voltage applied to the circuit measured in volts (V)

This relationship is known as the **power law**. Work an example to better understand what the power law states about a circuit.

Suppose a circuit has 10 amperes of current flow with 115 volts applied to it. How much power does this circuit consume? Use the power law to determine the amount of power consumed. Here is how:

What is known?

$$I = 10\text{ A}$$
$$E = 115\text{ V}$$

What is not known?

$$P = ?$$

What formula can be used?

$$P = I \times E$$

Substitute the known values and solve for P:

$$P = I \times E = 10\text{ A} \times 115\text{ V} = 1150\text{ W}$$

Therefore, according to the power law this circuit consumes 1150 watts of power.

The previous example works very well for circuits where the voltage and current are known and you are trying to determine the amount of power consumed by a circuit. However, suppose you knew the amount of current flowing in the circuit as well as the amount of power consumed by the circuit, but you did not know the amount of voltage applied to the circuit. What then? On the other hand, suppose you knew the amount of voltage applied to the circuit as well as the amount of power consumed by the circuit, but you did not know how much current was flowing in the circuit. Could this be determined? The answer is yes to both of these questions. The power law formula can be manipulated just like the Ohm's law formula.

Original power law formula:

$$P = I \times E$$

To solve for I, divide both sides of the equation by E.

$$\frac{P}{E} = \frac{I \times E}{E}$$

Cancel like terms.

$$\frac{P}{E} = I \quad \text{or} \quad I = \frac{P}{E}$$

Therefore, $I = \frac{P}{E}$ is the formula that can be used to find current (I) when power (P) and voltage (E) are known.

Now determine the formula to find voltage when power and current are known. Again, begin with the original power law formula.

Original power law formula:

$$P = I \times E$$

To solve for E, divide both sides of the equation by I.

$$\frac{P}{I} = \frac{I \times E}{I}$$

Cancel like terms.

$$\frac{P}{I} = E \quad \text{or} \quad E = \frac{P}{I}$$

Therefore, $E = \frac{P}{I}$ is the formula that can be used to find voltage (E) when power (P) and current (I) are known.

Now work two problems using the last two formulae that you have learned. Given a circuit where the power consumed by a circuit operating from 120 volts is 75 watts, how much current is flowing in this circuit?

What is known?

$$E = 120\text{ V}$$
$$P = 75\text{ W}$$

What is not known?

$$I = ?$$

What formula can be used?

$$I = \frac{P}{E}$$

Substitute the known values and solve for I:

$$I = \frac{P}{E} = \frac{75\text{ W}}{120\text{ V}} = 0.625\text{ A, or } 625\text{ mA}$$

Therefore, according to the power law this circuit has a current of 625 milliamperes flowing through it.

Now suppose you have a circuit where a current flow of 3.75 amperes causes the circuit to consume 500 watts. How much voltage must be applied to this circuit for this to happen?

What is known?

$$I = 3.75\text{ A}$$
$$P = 500\text{ W}$$

What is not known?

$$E = ?$$

What formula can be used?

$$E = \frac{P}{I}$$

Substitute the known values and solve for E:

$$E = \frac{P}{I} = \frac{500\text{ W}}{3.75\text{ A}} = 133.33\text{ V}$$

Therefore, according to the power law you will need to apply 133.33 volts to this circuit.

As was the situation with Ohm's law, if you have difficulty manipulating these three power law formulae, a simple device can help. This device is called the **power law triangle**. Refer to **Figure 14-19**. Notice the triangle and how it is divided. Notice also that the letter *P* is located in the top compartment, the letter *I* is located in the lower left compartment, and the letter *E* is located in the lower right compartment. To use the power law triangle, simply cover up the compartment of the variable that you are trying to find. The remaining compartments show you the formula that you should use. For example, if you need to find the current, cover the *I*. This leaves the *E* and the *P* uncovered. But notice the arrangement of the *E* and the *P*. They appear as the formula $\frac{P}{E}$. Therefore, the triangle shows you the formula for finding current, $I = \frac{P}{E}$.

Suppose you need to find the voltage when the current and the power are known. Simply cover the *E*, because you are trying to find the voltage, and you have the formula $\frac{P}{I}$. Therefore, the triangle shows you the formula for finding voltage, $E = \frac{P}{I}$. Finally, if you need to determine the power when you know the voltage and current, cover the *P*. The remaining uncovered variables give you the formula $I \times E$. Therefore, the triangle shows you the formula for finding power, $P = I \times E$.

You now know six formulae for determining an unknown value when two values are known. However, suppose you had a circuit in which you knew the voltage and the power but needed to find the resistance. You do not have a formula that allows you to find *R* when you know *E* and *P*. Likewise, suppose you wanted to find the circuit current but you only knew the circuit power and resistance. Again, you do not have a formula for finding *I* when you know *P* and *R*. There are six more formulae that

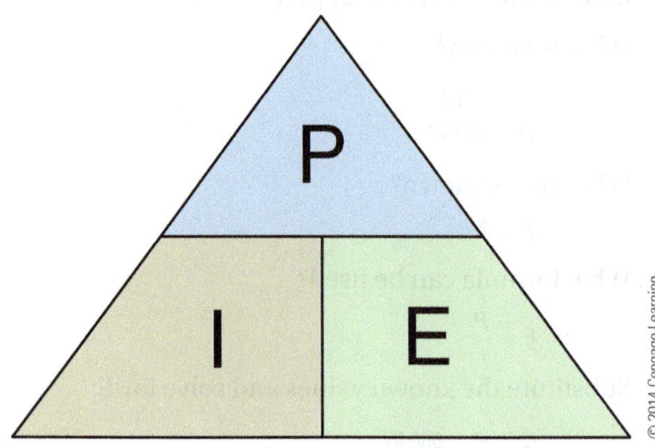

FIGURE 14-19 The power law triangle.

will enable you to solve for any unknown when two items are known. Look at the formulae first, and then you will see how they were developed.

1. $P = \frac{E^2}{R}$
2. $P = I^2 \times R$
3. $R = \frac{E^2}{P}$
4. $R = \frac{P}{I^2}$
5. $E = \sqrt{P \times R}$
6. $I = \sqrt{\frac{P}{R}}$

Begin by looking at formula 1, $P = \frac{E^2}{R}$. This formula is a combination of two formulae. A: $P = I \times E$ and B: $I = \frac{E}{R}$. Here is how these two formulae were combined.

Begin with formula A:

$$P = I \times E$$

From formula B it is known that:

$$I = \frac{E}{R}$$

Substitute formula B for the variable *I* in formula A. Formula A now becomes

$$P = \frac{E \times E}{R}$$

Combining like terms gives

$$P = \frac{E^2}{R}$$

This is formula 1.

Now look at formula 2. Formula 2 is also a combination of two formulae. It is a combination of A: $P = I \times E$ and B: $E = I \times R$. Here is how these two formulae were combined.

Begin with formula A:

$$P = I \times E$$

From formula B it is known that

$$E = I \times R$$

Substitute formula B for the variable *E* in formula A. Formula A now becomes

$$P = I \times I \times R$$

Combining like terms gives

$$P = I^2 \times R$$

This is formula 2.

Formula 3, $R = \frac{E^2}{P}$, is found by rearranging formula 1. Here is how.

Formula 1:

$$P = \frac{E^2}{R}$$

Multiply both sides by R:

$$R \times P = \frac{E^2}{R} \times R$$

Cancel like terms:

$$R \times P = E^2$$

Now divide both sides of the equation by P:

$$\frac{R \times P}{P} = \frac{E^2}{P}$$

Cancel like terms:

$$R = \frac{E^2}{P}$$

This is formula 3.

Now derive formula 4. Formula 4 is derived from formula 2. Here is how it is done.

Formula 2:

$$P = I^2 \times R$$

Divide both sides of the equation by I^2:

$$\frac{P}{I^2} = \frac{I^2 \times R}{I^2}$$

Cancel like terms:

$$\frac{P}{I^2} = R \quad \text{or} \quad R = \frac{P}{I^2}$$

This is formula 4.

Formula 5, $E = \sqrt{P \times R}$, is also found by rearranging formula 1. Here is how.

Formula 1:

$$P = \frac{E^2}{R}$$

Multiply both sides by R:

$$R \times P = \frac{E^2}{R} \times R$$

Cancel like terms:

$$R \times P = E^2$$

Take the square root of both sides of the equation:

$$\sqrt{R \times P} = \sqrt{E^2}$$

Simplify both sides of the equation:

$$\sqrt{R \times P} = E \quad \text{or} \quad E = \sqrt{R \times P} \quad \text{or} \quad E = \sqrt{P \times R}$$

This is formula 5.

Formula 6 is also derived from formula 2. Here is how.

Formula 2:

$$P = I^2 \times R$$

Divide both sides of the equation by R:

$$\frac{P}{R} = \frac{I^2 \times R}{R}$$

Cancel like terms:

$$\frac{P}{R} = I^2$$

Take the square root of both sides of the equation:

$$\sqrt{\frac{P}{R}} = \sqrt{I^2}$$

Simplify both sides of the equation:

$$\sqrt{\frac{P}{R}} = I \quad \text{or} \quad I = \sqrt{\frac{P}{R}}$$

This is equation 6.

Again, a device may help you remember all 12 formulae. The device is called a **PIRE wheel**. The acronym *PIRE* is from the letters used to represent (P)ower, (I)ntensity, (R)esistance, and (E)lectromotive force. **Figure 14-20** shows a PIRE wheel. Notice that all 12 formulae are shown on the wheel. Notice also that the formulae are grouped by the variable that you are trying to find. The wheel is divided into four quadrants, and each quadrant contains the three formulae that can be used to find the same unknown variable. For instance, in the upper left quadrant are the three formulae that can be used to find power. You simply choose the formula that contains the known variables that you have available to you.

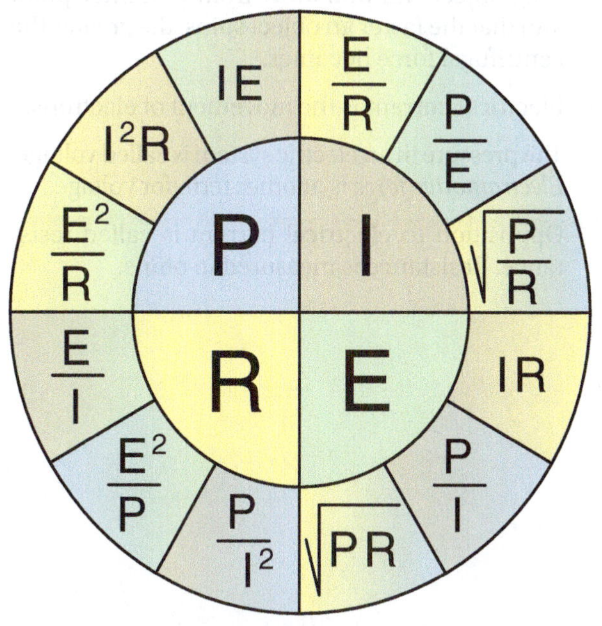

FIGURE 14-20 The PIRE wheel.

14-10 MAGNETISM

When studying electricity, a familiarity with magnetism is essential. Without magnetism, most of the electrical power produced in the world would not exist.

You may remember playing with magnets when you were younger. At that time, you may have observed a fundamental law of magnetism; that is that like poles repel and unlike poles attract.

In your study of electricity, you will also learn that whenever an electrical current flows through a conductor, a magnetic field is formed around that conductor. The magnetic field will only be present while the current is flowing. When the current is removed, the magnetic field collapses.

You will learn more about magnetism in later chapters.

SUMMARY

- Matter is made up of molecules. The smallest part of the molecule is called an atom.
- The nucleus is in the center of an atom and contains protons and neutrons.
- Protons have a positive charge, whereas neutrons have no charge and are said to be neutral.
- Electrons are components that orbit around the nucleus.
- The law of charges states that opposite charges attract and like charges repel.
- The law of centrifugal force states that a spinning object will pull away from its center point and that the faster an object spins, the greater the centrifugal force becomes.
- Electrical current is the movement of electrons.
- The pressure in an electric system is called voltage. *Electromotive force* is another term for voltage.
- Opposition to electrical current is called resistance. Resistance is measured in ohms.
- Resistors are added to a circuit to reduce or limit the current. they are made from different materials to give them different characteristics. Resistors are made of carbon, ceramic, wire, and other materials.
- Color bands encircle some resistors. They represent a code that indicates the amount of resistance a particular resistor offers.
- The colors in each of the bands have values assigned to them.
- Georg S. Ohm developed a formula to express the relationship among current, voltage, and resistance. This formula is referred to as Ohm's law.
- The amount of power used by a circuit depends on the amount of voltage present and the amount of current flow in the circuit. The formula $P = I \times E$ is referred to as power law.

REVIEW QUESTIONS

1. Electric current is the movement of _____.
2. A four-band resistor is color-coded yellow, violet, brown, and gold. What is the value of this resistor?
3. You need an 820-ohm ± 2% five-band resistor. What is the color code for this resistor?
4. A 2-watt, 220-ohm resistor has 0.2 ampere of current flowing through it. Can this resistor handle this much current?
5. 3 amperes of current is flowing through a 120-V soldering iron. What is the resistance of the heating element of this iron?
6. A resistor measures 470 ohms and is connected in a circuit. There are 350 milliamperes of current flowing through the resistor. How much voltage is measured across the resistor?
7. How much power does an electric water heater that operates from 240 volts consume? The resistance of the heating element is 14.4 ohms.

CHAPTER 15

Test Equipment

This chapter discusses the four most common pieces of test equipment used in industry today: the digital multimeter, the clamp-on ammeter, the megohmmeter, and the oscilloscope. It is very important that you gain an understanding of these pieces of test equipment and know how to use them properly and safely. You must also understand their limitations. These pieces of test equipment provide valuable information that you need in maintaining and troubleshooting various pieces of electrical equipment.

OBJECTIVES

After studying this chapter, the student should be able to

- Correctly set up and use a digital multimeter to measure voltage, current, and resistance.
- Correctly set up and use a clamp-on ammeter.
- Correctly set up and use a megohmmeter to measure high resistances.
- Correctly set up and use an oscilloscope to measure voltage, time, and frequency.

15-1 DIGITAL MULTIMETER

The most common multimeter in use today is the **digital multimeter (DMM)**, shown in **Figure 15-1**. The DMM is so named because the measured value is displayed in a digital readout. This is an improvement over the older analog style of display in which you had to match a moving indicator (vane or needle) to a fixed scale. These analog indicators were prone to inaccuracies and difficulties in obtaining correct measurements. The digital display has made using a multimeter much easier.

The DMM may be bench or panel mounted, or more commonly in maintenance, handheld. The operation of DMMs is essentially the same. The major differences are in their functions and accuracies. Most bench- or panel-type DMMs have more ranges, or more specific ranges with higher accuracies. However, this is not a problem for the maintenance technician who needs a portable, handheld unit. The handheld DMMs have the function and accuracy necessary to get the job done.

Premeasurement Inspection

SAFETY

Before using the multimeter, you should visually inspect the meter and test leads thoroughly. Take your time and fully examine both the red and black test leads from the probe end to the plug end. Look for cracks or breaks in the probe, insulation, or plug. If you find any damage to either test lead, *replace the set of leads!* Test leads are very inexpensive, and it is far safer to replace a damaged test lead set than to try to repair it with electrical tape. Remember, the voltage or current that you are measuring must be sensed by the meter by way of the test leads. The insulation quality of these leads is all that is between your hands and the voltage or current that you are measuring. If your leads are damaged, *replace them* with a new set. The life you save may be your own!

Also thoroughly inspect the multimeter itself. If you find any cracks or missing pieces from the case, have the meter repaired. *Do not use it!* The case protects you from the voltages and currents that you will be measuring. Cracked or missing pieces of the case provide an opportunity for you to come into contact with dangerous levels of voltage or current.

If the meter and test leads check out visually, you should perform an operational test. When preparing to measure resistance, you should *short* the leads together to verify that the meter is responding properly when set to measure resistance as shown in **Figure 15-2**. When preparing to measure voltage or current, you should verify the operation of the meter on a known good source of voltage or current. You should test the meter on the function and range of its expected use. Many technicians have been injured or killed as a result of trusting a faulty meter.

Imagine that you are preparing to work on a circuit. You need to know if the circuit is energized or not. You connect the meter into the circuit to measure voltage, and you read 0 volt. You assume that the circuit is de-energized and begin to work on it. Upon coming into contact with the circuit, you receive an electrical shock! How could this happen if the meter indicated 0 volt? The meter is defective and cannot measure voltage.

Before measuring the circuit voltage, you should have tested the meter on a known good source of voltage. You should have set the meter

FIGURE 15-1 A digital multimeter.

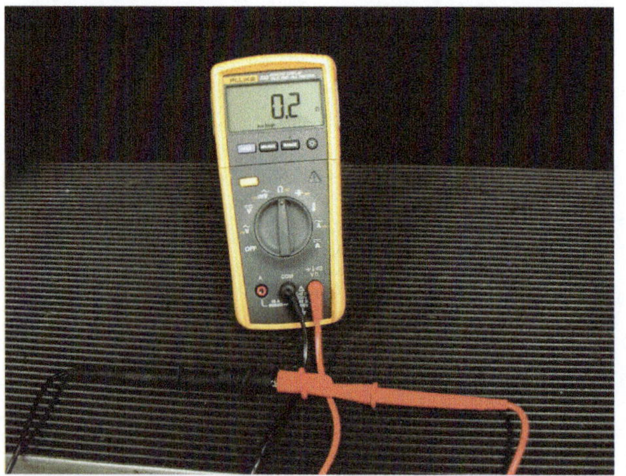

FIGURE 15-2 Ohmmeter check—leads are shorted.

to the correct voltage function and range and then measured the voltage of a known energized circuit. If the meter had indicated the proper type and amount of voltage, it would have been safe to use it on the circuit that you are troubleshooting.

Premeasurement Setup

Before connecting the meter into a circuit to be tested, you must set up the meter for the measurement that you expect to make. For example, if you are going to measure the voltage of a circuit, you must connect the test leads into the appropriate jacks. You must then set the function and/or range switch to measure the correct type of voltage and the proper range for the expected amount. After you double-check your setup, you are ready to connect the meter to the circuit to be tested.

Following are the correct procedures for setting up and using the meter for measuring voltage, current, and resistance. Keep in mind that there are different manufacturers and models of digital multimeters. The information presented here is generic in nature. You should familiarize yourself with the operating manual for the particular meter that you will be using. Check with your supervisor or the manufacturer of the meter that you are using if you have any questions or need additional help.

Measuring Voltage

Refer to **Figure 15-3**.

1. Perform a visual inspection of the meter and test leads.
2. Determine the amount of voltage that you expect to measure and verify that the voltage rating of the meter and test leads will not be exceeded.

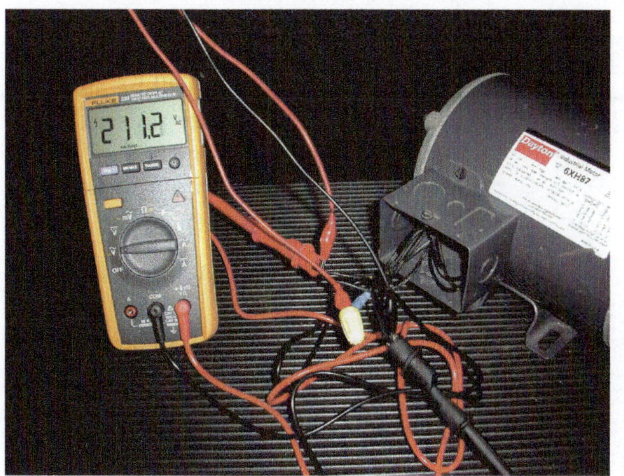

FIGURE 15-3 Checking the voltage applied to a motor.

3. Set up the meter to measure voltage as follows:
   a. Insert the black test lead into the COM test lead jack.
      i. Make certain that the test lead is fully and securely inserted into the COM jack.
   b. Insert the red test lead into the Volts or V test lead jack.
      i. Make certain that the test lead is fully and securely inserted into the Volts or V jack.
   c. Determine the type of voltage (ac or dc), and set the voltage selection switch (if so equipped) to the proper voltage type.
4. Determine the approximate amount of voltage to be measured, and set the range switch to the proper voltage range. Again, verify that the voltage rating of the meter and test leads will not be exceeded.
   a. If you are unsure of the amount of voltage that you will be measuring, *set the range switch to the highest voltage range.* If you have an auto ranging meter, you will not need to set the voltage range. The meter will determine the correct range automatically!

5. Double-check the meter setup.
6. Connect the red and black test probes to a circuit with a similar known amount of voltage present.
   a. When measuring voltage, the test leads will be connected across, or in parallel, with the portion of the circuit that you are testing.
      i. The safest method for making the meter connections is to de-energize the circuit to be tested, connect the meter into the circuit, and then re-energize the circuit. Unfortunately, in reality, this is impractical. Therefore, exercise extreme caution when connecting the meter to an energized circuit. Be sure to wear appropriate personal protective equipment required by your facility.
7. Read the indicated voltage from the digital display of the meter.
   a. If the indicated voltage is correct for the amount of voltage that should be present in the circuit, the meter is functioning normally and you may proceed with your test. Disconnect the test probes from the test circuit and proceed to step 8.
   b. If the indicated voltage is not correct for the amount of voltage that should be present in the circuit, the meter is faulty and should not be used until repaired. Disconnect the test probes from the test circuit.

8. Connect the red and black test probes to the circuit that you are troubleshooting.
   a. Remember that when measuring voltage, the test leads will be connected across, or in parallel, with the portion of the circuit that you are testing.
9. Read the indicated voltage from the digital display of the meter.
10. Carefully disconnect the meter from the circuit that you are testing.
11. Reconnect the meter to the circuit with the similar known amount of voltage present that you had used previously. This will verify that the meter is still functioning properly.

Measuring Current

Refer to **Figure 15-4**.

1. Perform a visual inspection of the meter and test leads.
2. Determine the amount of current that you expect to measure and verify that the current rating of the meter and test leads will not be exceeded.
3. Set up the meter to measure current as follows:
   a. Insert the black test lead into the COM test lead jack.
      i. Make certain that the test lead is fully and securely inserted into the COM jack.
   b. Insert the red test lead into the mA or A test lead jack, depending on the amount of current that you expect to measure. If you are uncertain of the amount of current that you will be measuring, insert the red test lead into the A jack. You can move it to the mA jack later if you discover that the current is less than 1 ampere.
      i. Make certain that the test lead is fully and securely inserted into the mA or A jack.
   c. Determine the type of current (ac or dc) and set the current selection switch, if so equipped, to the proper current type.
   d. Determine the approximate amount of current to be measured, and set the range switch to the proper current range. If you are unsure of the amount of current that you will be measuring, *set the range switch to the highest current range.* If you have an auto ranging meter, you will not need to set the current range. The meter will determine the correct range automatically!

4. Double-check the meter setup.
5. Connect the red and black test probes to a circuit with a known amount of current present.
   a. When measuring current, the test leads will be connected in series with the portion of the circuit that you are testing.
      i. Remove power from the circuit.
      ii. Break or open the circuit at the point where you will measure the current.
      iii. Insert the meter in series with the break so that one test lead is connected to one side of the break and the other test lead is connected to the remaining side of the break. Now reapply power. Connecting the meter in this fashion causes the circuit current to flow from the circuit into the meter, through the meter, and then back to the circuit. This can be very difficult to do in an operating facility. Therefore, the measurement of current in this manner is not widely performed. However, the correct procedure to follow is discussed so that you will be aware of it should the opportunity present itself. Be sure to wear appropriate personal protective equipment as required by your facility.
6. Read the indicated current from the digital display of your meter.
   a. If the indicated current is correct for the amount of current that should be present in the circuit, the meter is functioning normally and you may proceed with the test. Remove power from the circuit and disconnect the test probes from the test circuit. Proceed to step 7. Do not forget

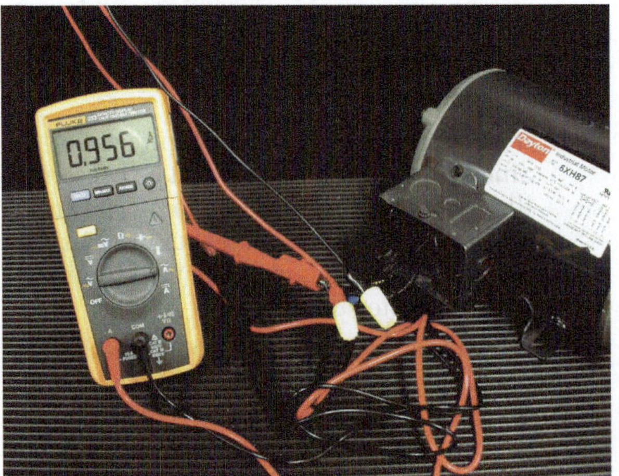

FIGURE 15-4 Checking motor current.

to reconnect the circuit at the break and reapply power.

b. If the indicated current is not correct for the amount of current that should be present in the circuit, the meter is faulty and should not be used until repaired. Remove power from the circuit and disconnect the test probes from the test circuit. Do not forget to reconnect the circuit at the break and reapply power.

7. Connect the red and black test probes to the circuit that you are troubleshooting.
   a. Remember that when measuring current, you will need to remove power from the circuit. Break or open the circuit at the point where you will measure the current. Insert the meter in series with the break so that one test lead is connected to one side of the break and the other test lead is connected to the remaining side of the break. Reapply power.
8. Read the indicated current from the digital display of the meter.
9. Remove power from the circuit.
10. Carefully disconnect the meter from the circuit that you are testing. Remember to reconnect the circuit at the break and reapply power.
11. Reconnect the meter to the circuit with the similar known amount of current present that you had used previously. This will verify that the meter is still functioning properly.

Measuring Resistance

Refer to **Figure 15-5**.

FIGURE 15-5 Checking motor winding resistance.

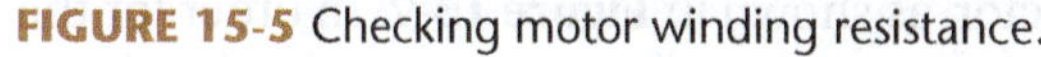

1. Perform a visual inspection of the meter and test leads.
2. Set up the meter to measure resistance as follows:
   a. Insert the black test lead into the COM test lead jack.
      i. Make certain that the test lead is fully and securely inserted into the COM jack.
   b. Insert the red test lead into the Ohms or Ω test lead jack.
      i. Make certain that the test lead is fully and securely inserted into the Ohms or Ω jack.

   c. *Verify that all voltage has been removed from the circuit or component that you will be testing.*
   d. Determine the approximate amount of resistance to be measured, and set the range switch to the proper resistance range. If you are unsure of the amount of resistance that you will be measuring, *set the range switch to the highest resistance range.* If you have an auto ranging meter, you will not need to set the resistance range. The meter will determine the correct range automatically!

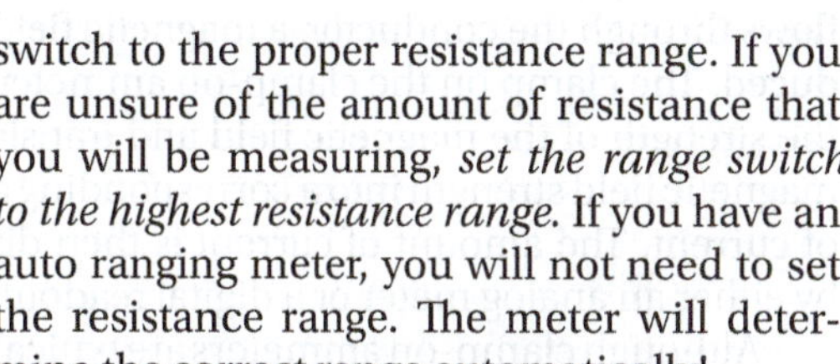

   e. Touch or short the red and black probe tips together. This verifies that the meter is properly set up to measure resistance, the test leads are making good connection, and the meter is in working order. In addition, this verifies that the internal battery of the meter is capable of supplying enough energy to perform the resistance measurement.
      i. You should see a reading of almost 0 ohm of resistance on the meter's display.
      ii. If the reading is correct, proceed to step 3.
      iii. If the reading is incorrect, the meter is faulty and should not be used until repaired. This may be as simple as replacing the internal battery.
3. Double-check the meter setup, and *double-check that all power has been removed from the circuit or component that you are testing.*

4. Connect the red and black test probes to the circuit or component that you are testing.
   a. When measuring resistance, the test leads will be connected across, or in parallel, with the component or portion of the circuit that you are testing.
5. Read the indicated resistance from the digital display of the meter.
6. Disconnect the meter from the circuit or component that you are testing.

15-2 CLAMP-ON AMMETER

When measuring current with a multimeter, you must remove power, break the circuit, insert the meter into the break, and reapply power. This is almost always impractical to do in an operating manufacturing facility. Power cannot be removed easily without disrupting other processes. The **clamp-on ammeter** allows the measurement of current without the need to de-energize the circuit.

Figure 15-6 shows a clamp-on ammeter. Notice the lack of test leads. Notice also the clamp at the top of the meter. The clamp is opened by depressing a button located on the side of the meter housing. Opening the clamp allows it to be placed around a conductor, as shown in **Figure 15-7**. In order for the clamp-on ammeter to function properly, the clamp must be fully closed around the conductor. As current flows through the conductor, a magnetic field is produced. The clamp on the clamp-on ammeter senses the strength of the magnetic field and translates the magnetic field strength into a corresponding amount of current. The amount of current is then displayed by either an analog meter or a digital readout.

Although clamp-on ammeters are typically used to measure significant amounts of current, they can be used to measure smaller amounts of current as well. There is, however, a technique that is used to aid in measuring smaller currents. **Figure 15-8** shows the technique that is used to measure small values of current with a clamp-on ammeter.

Notice that the conductor has been wound around one of the jaws of the clamp. In fact, the conductor has been wound around the jaw three times. This has the effect of tripling the strength

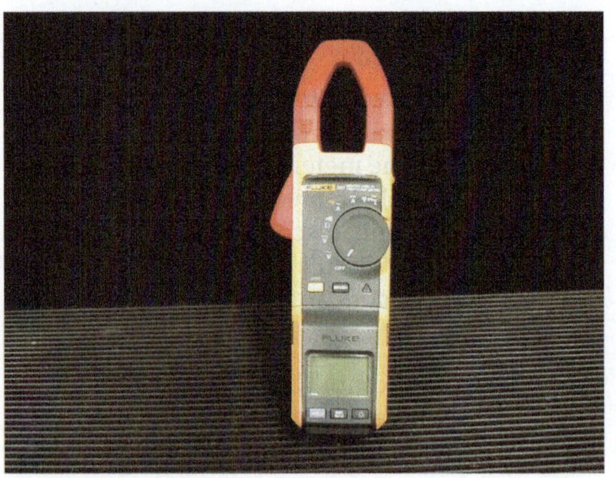

FIGURE 15-6 A clamp-on ammeter.

FIGURE 15-7 Using a clamp-on ammeter to measure conductor current.

FIGURE 15-8 Technique for measuring small amounts of current.

of the magnetic field created by the current flowing through the conductor. The ammeter will now display a current reading. However, the amount of current displayed is artificially three times higher than the actual value because of the winding of the conductor around the jaw. Now simply read the indicated amount of current and divide by 3. You now have a fairly accurate indication of the amount of current flow in the conductor. You will need to experiment to determine how many times to wind the conductor around the jaw to obtain a measurable amount of current. Just remember to divide the indicated current by the number of times the conductor is wound around the jaw.

15-3 MEGOHMMETER

Figure 15-9 shows a photograph of a **megohmmeter**. The megohmmeter, more commonly referred to as a *Megger®*, allows the measurement of very high resistance values. (*Megger* is a registered trademark of AVO International Limited.) This is useful in determining the quality of insulation of wires, cables, transformers, motors, and generators. There are many different types, makes, and models of megohmmeters on the market today.

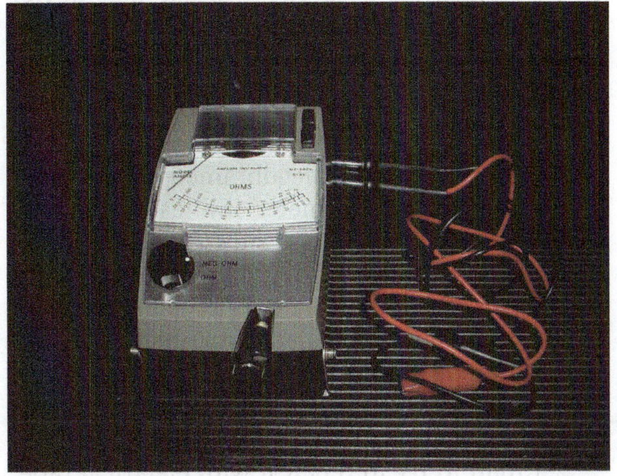

FIGURE 15-9 A megohmmeter.

Some models simply measure high resistance values. Other models incorporate additional features, which allow them to measure continuity and voltage as well. Some models use a hand crank to operate an internal DC generator. This DC generator is what produces the high voltages needed to test high resistance values. Other megohmmeters use internal electronic circuitry to develop the high test voltages and therefore do not have a hand crank. In this textbook, the measurement of high resistance values using a hand crank model is discussed.

Following are the correct procedures for setting up and using the megohmmeter to measure high resistance values. The information presented here is generic in nature. You should familiarize yourself with the operating manual for the particular megohmmeter that you will be using. Should you have any questions or need additional help, check with your supervisor or the manufacturer of the megohmmeter that you are using.

In order to measure high values of resistance, a megohmmeter must deliver high test voltages to the circuit that is being tested. It is not uncommon for a megohmmeter to operate with 500 volts or even 1000 volts at its test terminals! Therefore, before using the megohmmeter, some safety precautions *must* be observed.

Safety Precautions

1. Do not use the megohmmeter unless you have been trained properly.
2. Wear your facility's approved and appropriate safety equipment when performing tests with a megohmmeter.
3. You must de-energize the circuit that you will be testing. Be absolutely certain that power is removed and the circuit is isolated before performing any measurements with the megohmmeter.

4. Perform a visual inspection of the megohmmeter. Before using the megohmmeter, you should visually inspect the meter and test leads thoroughly. Take your time and fully examine the red, black, and guard or ground test leads from the clip end to the plug end. Look for cracks or breaks in the clip, insulation, or plug. If you find any damage to either test lead, *replace the set of leads!* Test leads are very inexpensive, and it is far safer to replace a damaged test lead set than to try to repair it with electrical tape. Remember, the megohmmeter operates with very high test voltages. If your leads are damaged, *replace them* with a new set. The life you save may be your own! Also, thoroughly inspect the megohmmeter itself. If you find any cracks or missing pieces from the case, have the meter repaired. *Do not use it!* Cracked or missing pieces of the case may provide an opportunity for you to come into contact with dangerous levels of voltage. The case is your protection from the voltages and currents that you will be measuring.

5. *While the test is in process, you must not touch or come into contact with the megohmmeter connections to the circuit that you are testing. High voltages will be present!*
6. *Discharge the circuit that has been tested before disconnecting the megohmmeter leads.* This not only applies to circuits containing capacitors, but also to circuits that become capacitive because of long lengths of cable. Most megohmmeters contain an internal discharge circuit to perform a circuit discharge automatically. However, be certain that the megohmmeter is equipped with this circuit and that it is operating properly. An automatic discharge circuit should not be taken for granted and viewed as a substitute for safe working practices.
7. *Never have someone hold the test leads while operating the megohmmeter! This is very dangerous and can cause injury or death!*

Preparing the Megohmmeter for Resistance Tests

1. Place the megohmmeter on a flat, level, solid surface.
2. Before connecting the test leads, set the test voltage selector switch to the highest test voltage position (500 volts or 1000 volts depending on your unit).
3. Hold down the test button while turning the generator hand crank.
   a. The crank should be turned at a constant rate according to the megohmmeter's manufacturer's specifications.
   b. The meter movement indicator should remain at infinity (∞).
   c. This test verifies that the megohmmeter does not have any internal leakage.
4. Stop turning the generator hand crank.
5. Release the test button.
   a. When the test button is released, any voltage stored by the circuit will be indicated on the voltage scale of the meter movement. Wait a few seconds for the voltage to completely discharge to 0 volt before proceeding further.
6. Insert the red test lead into the + terminal, and insert the black test lead into the − terminal. Verify that the clip ends of the test leads are not in contact with anything.

7. Hold down the test button while turning the generator hand crank.
   a. The crank should be turned at a constant rate according to the megohmmeter's manufacturer's specifications.
   b. The meter movement indicator should remain at infinity (∞).
   c. This verifies the quality of the insulation of the test leads.
8. Stop turning the generator hand crank.
9. Release the test button.
   a. When the test button is released, any voltage stored by the circuit will be indicated on the voltage scale of the meter movement. Wait a few seconds for the voltage to completely discharge to 0 volt before handling the test leads.
10. Connect the clip ends of the test leads together.
11. Hold down the test button while turning the generator hand crank.
    a. The crank should be turned at a constant rate according to the megohmmeter's manufacturer's specifications.
    b. The meter movement indicator should now read approximately 0 ohm.
    c. A high or infinity reading means that one or both test leads are open and will need replacement. This may also indicate a problem with the megohmmeter. If the same reading is obtained after repeating the test with replacement leads, the megohmmeter is faulty and must be repaired.
12. Stop turning the generator hand crank.
13. Release the test button.
    a. When the test button is released, any voltage stored by the circuit will be indicated on the voltage scale of the meter movement. Wait a few seconds for the voltage to completely discharge to 0 volt before handling or disconnecting the test leads.

Resistance Tests to Ground

Refer to **Figure 15-10**.

1. Place the test voltage selector switch to the required test voltage position.
2. Connect the red (+) test lead to a good ground or to the metal frame of the equipment under test.
3. Connect the black (–) test lead to the portion of the circuit that is being tested.

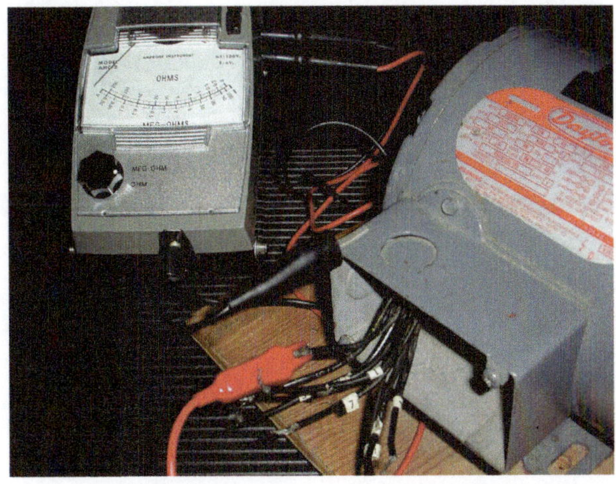

FIGURE 15-10 Using a megohmmeter to measure winding resistance to ground.

4. Hold down the test button while turning the generator hand crank.
   a. The crank should be turned at a constant rate according to the megohmmeter's manufacturer's specifications.
   b. The meter movement indicator will indicate the amount of insulation resistance on the megaohm scale.
5. Stop turning the generator hand crank.
6. Release the test button.
   a. When the test button is released, any voltage stored by the circuit will be indicated on the voltage scale of the meter movement. Wait a few seconds for the voltage to completely discharge to 0 volt before handling or disconnecting the test leads.

Insulation Test Between Two Insulated Wires

Refer to **Figure 15-11**.

1. Place the test voltage selector switch to the required test voltage position.
2. Connect the red (+) test lead to the conductor of one of the wires.
3. Connect the black (–) test lead to the conductor of the remaining wire.
4. Hold down the test button while turning the generator hand crank.
   a. The crank should be turned at a constant rate according to the megohmmeter's manufacturer's specifications.

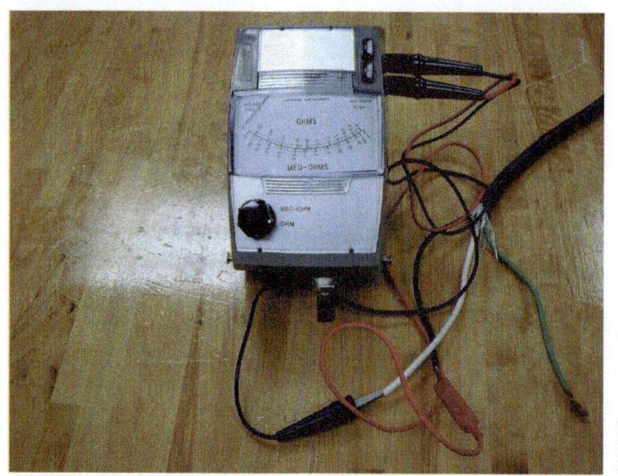

FIGURE 15-11 Using a megohmmeter to measure the insulation quality between two insulated conductors.

   b. The meter movement indicator will indicate the amount of insulation resistance on the megaohm scale.

5. Stop turning the generator hand crank.
6. Release the test button.
   a. When the test button is released, any voltage stored by the circuit will be indicated on the voltage scale of the meter movement. Wait a few seconds for the voltage to completely discharge to 0 volt before handling or disconnecting the test leads.
7. If a reading of infinity (∞) is obtained, the opposite ends (not the ends connected to the megohmmeter) of the conductors being tested should be connected together.
8. Hold down the test button while turning the generator hand crank.
   a. The crank should be turned at a constant rate according to the megohmmeter's manufacturer's specifications.
9. The meter movement should indicate approximately 0 ohm.
   a. This verifies the measurement of the insulation as infinity and that the test leads are not disconnected or broken.
10. Stop turning the generator hand crank.
11. Release the test button.
   a. When the test button is released, any voltage stored by the circuit will be indicated on the voltage scale of the meter movement. Wait a few seconds for the voltage to completely discharge to 0 volt before handling or disconnecting the test leads.

15-4 OSCILLOSCOPE

As industrial controls and devices become more intelligent and advanced, the technician will need more advanced test equipment as well. One piece of test equipment that is becoming more and more commonplace in the industrial environment is the **oscilloscope** (or **o-scope** or **scope**). The oscilloscope, as shown in **Figure 15-12**, is being used more frequently in the maintenance field. The oscilloscope not only measures voltage, it allows the user to see a representation of the voltage as well. This is quite helpful when diagnosing problems caused by voltage spikes or *dirty* power or simply as a way to find failed components more rapidly.

At first glance, an oscilloscope may seem rather intimidating. There are many knobs and adjustments on the front of the scope. Rather than a digital readout or a moving pointer, an oscilloscope displays an image of the voltage that you are measuring. This is what makes the scope so attractive for troubleshooting. Now you will not only be able to measure the voltage, but you will be able to see it as well! So that you will have a basic understanding of the setup, operation, and interpretation of an oscilloscope, the function of the different controls found on a typical scope is discussed next. Once a scope is properly set up, you will learn how to make and interpret the scope display. Be aware that there are many manufacturers of oscilloscopes. Each manufacturer may produce different models. It is therefore impossible to present all of the possible features and functions in this text. The most common controls and their function are presented here. You will need to familiarize yourself with your facility's oscilloscope.

The Display Section

Refer to **Figure 15-13**. The section of an oscilloscope on which measurements are made may be either a **cathode ray tube (CRT)** or a liquid crystal display (LCD). The CRT or LCD is part of the display section. These controls are called **INTENSITY**, **TRACE ROTATION**, **BEAM FIND**, and **FOCUS**.

Intensity

The intensity control allows you to adjust the brightness, or intensity, of the trace in varying lighting conditions. For example, if you were to use the scope in a brightly lit area, you may need to increase the intensity of the trace so that the ambient light does not wash out the display. On the other hand, if the scope is used in a dimly lit area, the intensity of the trace would be set to a lower level so that the trace does not *bloom*, or *blossom*. Blooming, or blossoming, causes inaccuracies in

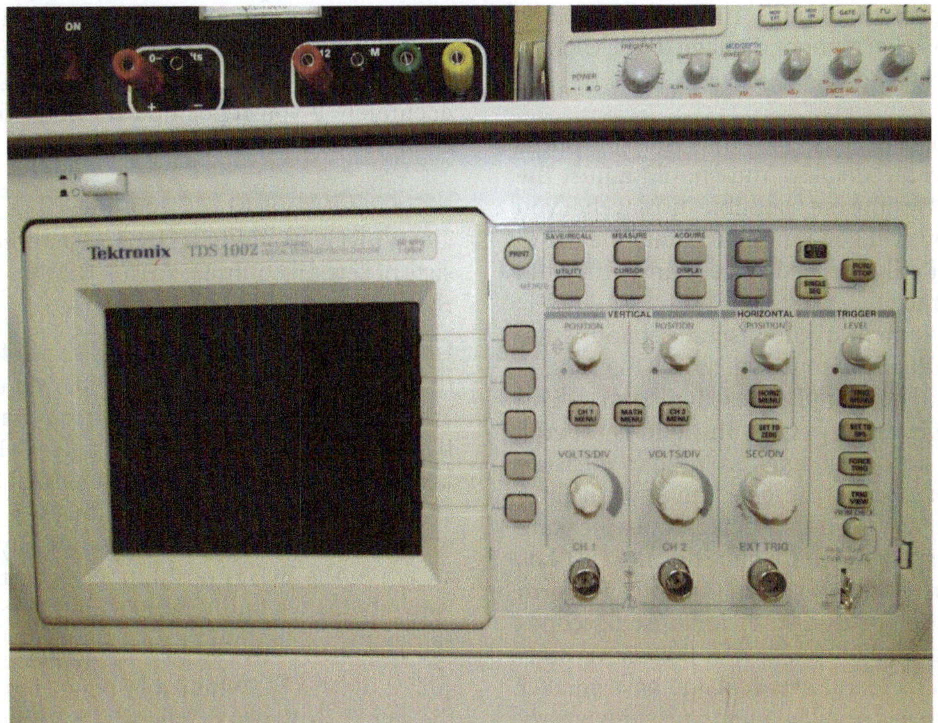

FIGURE 15-12 An oscilloscope.

your measurements. You should always adjust the INTENSITY control for the *minimum* brightness of the trace for a comfortable measurement. It is common practice to leave a scope turned on when not making measurements. This allows the scope to be ready for use at a moment's notice. However, leaving a trace displayed on the CRT for extended periods of time will damage the CRT permanently. This will lead to a very costly repair. It is therefore good practice to turn the intensity of the trace down to a point where the trace is no longer visible. You can still leave the scope on; just turn down the intensity. Now when you need to take a measurement, simply turn up the intensity to the required level. This avoids damage to the CRT.

CAUTION

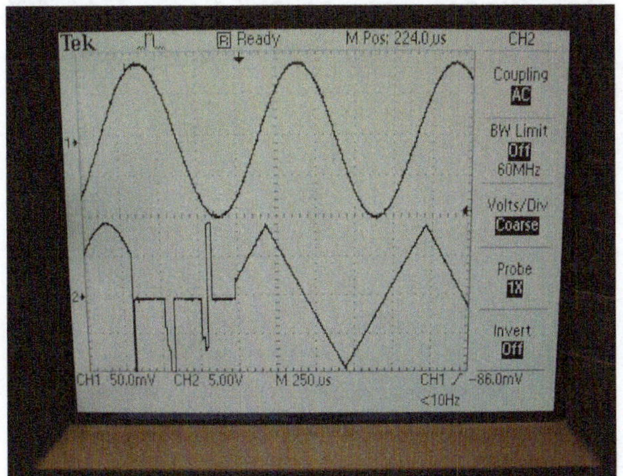

FIGURE 15-13 The display section of an oscilloscope.

Trace Rotation

The CRT display is divided into grids by vertical and horizontal lines. For accurate measurements, it is important that the trace be perfectly horizontal. That is, the trace must run parallel to the horizontal grid lines. Moving a scope from place to place will cause the trace to *tilt*; the trace is no longer parallel to a horizontal grid line but is now at an angle. This is a result of the variations in the Earth's magnetic field as the scope is repositioned. To compensate for these variations, simply take a small screwdriver and adjust the TRACE ROTATION control until the display trace is aligned horizontally with the horizontal grid lines.

Beam Find

Sometimes when you turn on your scope and wait for the display to appear, you may be disappointed. This may be caused by a previous user's adjustments, which resulted in the trace being off-screen. To quickly locate the trace, press the BEAM FIND button. You will then see an intensified spot on the CRT display. The spot will appear in one of the four quadrants on the CRT. This helps you adjust the POSITION controls to quickly bring the trace back on-screen. Do not depress the BEAM FIND button for extended periods of time. The intensity of the spot could cause permanent damage to the CRT.

Focus

The **FOCUS control** is used to adjust the sharpness of the trace on the display. You should adjust the FOCUS control for the best possible display that you can obtain. You will get a more accurate measurement using a finely focused trace than you would using a fuzzy trace.

The final element of the display section is the CRT itself. As already mentioned, the CRT is divided by vertical and horizontal lines that form a grid pattern on the face of the CRT. This grid is called a **graticule**. The horizontal lines are used to measure the amount of voltage present. The vertical lines are used to measure the time. Oscilloscopes measure voltage with respect to time. Notice that the CRT has 8 horizontal lines and 10 vertical lines. These form the *major* divisions of the display. Most oscilloscopes use an 8 × 10 grid. Notice also that the center horizontal line and the center vertical line have smaller graduations. These are called *minor divisions* or *subdivisions*. There are four minor divisions between two major divisions. These major and minor divisions are used to measure voltage and time.

The Vertical Section

Refer to **Figure 15-14**. One of the reasons that the oscilloscope seems so intimidating is because of the number of knobs or controls that are found on the front of the scope. The vertical section is partly to blame for this. Most oscilloscopes are *two-channel*, or *dual-trace*, scopes. This means that the scope has the capability to display two different waveforms at the same time. To have control over each of these displays, you must have two separate sets of controls.

When you look at the vertical section of the scope, you see that some of the controls have been duplicated. This accounts for many of the knobs or controls that make the scope intimidating. But remember, they are duplicate controls. If you understand how one set of controls works, you will understand the operation of the other set.

The vertical section consists of the **VERTICAL POSITION**, **VERTICAL OPERATING MODE**, **VARIABLE VOLTS/DIV**, **INPUT SENSITIVITY**, and **INPUT COUPLING** controls. In addition, the **INPUT JACKS** for the **test probes** are a part of this section as well.

Vertical Position

The VERTICAL POSITION control allows you to move the trace displayed on the CRT up or down. In other words, if the displayed waveform is near the bottom of the CRT, you can use the VERTICAL POSITION control to reposition the trace to the center or top of the CRT. Do not be afraid to do this. You will make measurements more easily and more accurately if you reposition the trace for the optimum viewing position.

Vertical Operating Mode

The VERTICAL OPERATING MODE control consists of several controls: **CH 1 - BOTH - CH 2**, **CH 2 INVERT**, and **ADD - ALT - CHOP**.

CH 1 – BOTH – CH 2 This control allows you to determine which trace is displayed on the CRT. You may want to view the trace from channel 1 only. If so, simply place this switch in the CH 1 position. Likewise, to view the trace for channel 2, place the switch in the CH 2 position. However, if you want to view both channels simultaneously, place the switch in the BOTH position. Now you are able to view channel 1 and channel 2 on the CRT at the same time.

CH 2 INVERT When this control is depressed, the signal displayed on channel 2 is inverted. Using this switch in conjunction with the ADD position of the ADD - ALT - CHOP switch allows you to make **differential measurements**. For example, assume

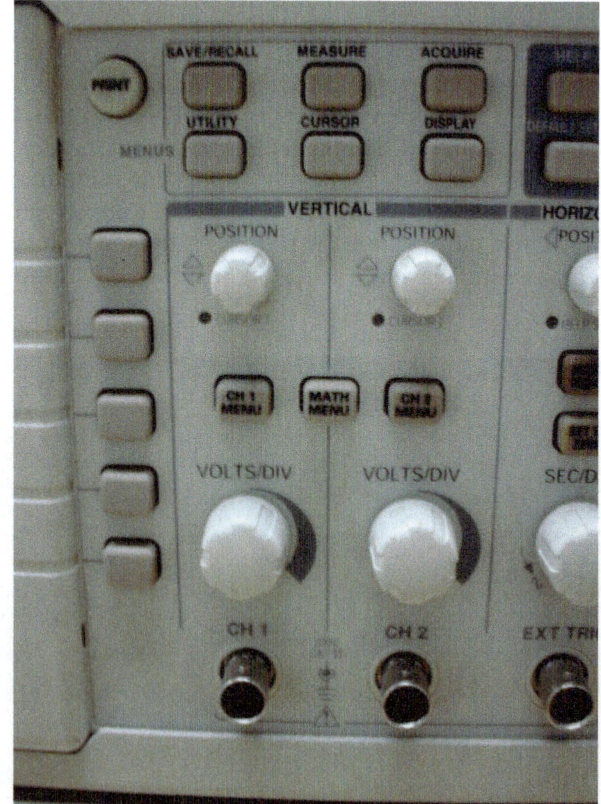

FIGURE 15-14 The vertical section of an oscilloscope.

that the input signal on channel 1 is 20 volts peak to peak, and the input signal on channel 2 is 7 volts peak to peak. To display the difference between these two input signals, place the ADD - ALT - CHOP switch in the ADD position and depress the CH 2 INVERT switch. The signal for channel 2 is now inverted and when added to the signal from channel 1 will yield the difference of the two input signals. Unless you are making differential measurements, always verify that the CH 2 INVERT switch is in the normal position.

ADD – ALT – CHOP As shown previously, the ADD position allows adding the signals from channel 1 and channel 2 together algebraically. This gives the sum of the two input signals or the difference between the two input signals, depending on the setting of the CH 2 INVERT switch. The ALT, or alternate, position allows the scope to show the channel 1 signal, then the channel 2 signal, then the channel 1 signal, and so on. When the scope is set to display higher frequencies, the ALT position allows you to see both input signals at the same time. The scope alternates between channel 1 and channel 2 at such a high speed that your eyes do not see the switching, and the display appears to show both input signals simultaneously. When the scope is set to display lower frequencies, you must use the CHOP position.

Variable Volts/Div

A dual-trace scope has two of these controls, one for channel 1 and one for channel 2. This is a variable control that allows adjusting the vertical height of the displayed signal for channel 1 or channel 2 independently. This is helpful for making quick comparisons of signals. For example, imagine that you only wanted to see whether a signal was amplified, but you really did not care by how much. You measure the input signal to the circuit. While displaying the input signal, adjust the VARIABLE VOLTS/DIV control until you fit the displayed waveform within the horizontal major division graticule lines. Now without adjusting the setting, move your probe to the output of the circuit. If the displayed waveform extends beyond the graticule lines that you used for the input, the signal has been amplified. If the displayed waveform does not fill in the area between the graticule lines that you used for the input, the signal has not been amplified and is in fact smaller than the input signal. Notice that this control has a curved arrow and the word CAL on it. On this particular scope, you turn this control fully clockwise. There is actually a detent in the switch that clicks when the control is turned fully clockwise. The VARIABLE VOLTS/DIV control must be in the detent or CAL position in order make use of the values on the VOLTS/DIV switch behind.

Input Sensitivity

The INPUT SENSITIVITY control is actually labeled CH 1 VOLTS/DIV or CH 2 VOLTS/DIV. Think of this as a range switch. You adjust this control until the displayed waveform is the largest it can be, vertically, and still fit within the graticule of the CRT. This provides the highest accuracy for your measurements. Notice that there are two windows on this control. The **1X window** is located at the 10 o'clock position, and the **10X window** is located at the 2 o'clock position. These windows work in conjunction with the type of oscilloscope probes that you use. If you use a 1X scope probe, you will make your measurements using the 1X window. If you use a 10X scope probe, you will make your measurements using the 10X window. The numbers on this control represent the volts per division for the graticule. For example, suppose you are using a 1X probe and displaying a waveform that is three major divisions high. If the VOLTS/DIV switch is set to 0.2 under the 1X window, the waveform measures 0.6 V p-p, because 3 divisions times 0.2 volt per division equals 0.6 V p-p.

Input Coupling

There are two INPUT COUPLING switches, one for each channel. Each switch has three positions: AC, GND, and DC. The DC position represents **direct coupled**. This allows the scope to display all components of the input signal. The ac position inserts a capacitor into the input circuit. This blocks any dc component from the signal that you are measuring. The scope will display only the ac component of the signal. The GND position disconnects the input signal from the scope and connects the scope input to the chassis ground of the scope. This prevents unwanted signals from being displayed while the trace is adjusted for a reference on the CRT. For example, by placing the input coupling switch in the GND position, the trace can be adjusted vertically, using the VERTICAL POSITION control, until it lines up exactly with the horizontal grid line that is second from the bottom of the graticule. Now the input coupling switch is placed in the DC position. When a measurement is made, the trace moves vertically upward to the fourth horizontal graticule line from the bottom. Now you know that a dc component is present because the trace is now at some positive voltage above ground.

The Horizontal Section

Refer to **Figure 15-15**. The horizontal section consists of the **HORIZONTAL POSITION**, **VARIABLE**

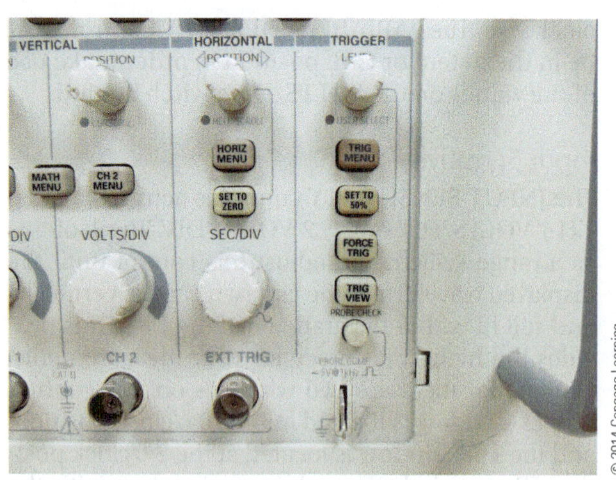

FIGURE 15-15 The horizontal section of an oscilloscope.

SEC/DIV, **HORIZONTAL MAGNIFICATION**, and **SWEEP SPEED** controls.

Horizontal Position

The HORIZONTAL POSITION control allows you to move the trace that is displayed on the CRT to the left or to the right. In other words, if the displayed waveform is near the left side of the CRT, you can use the HORIZONTAL POSITION control to reposition the trace to the center or right side of the CRT. Do not be afraid to do this. You will make measurements more easily and more accurately if you reposition the trace for the optimum viewing position.

Variable Sec/Div

The VARIABLE SEC/DIV control allows you to independently adjust the horizontal width of the displayed signal for channel 1 and channel 2. This is helpful for making quick comparisons of signals. For example, imagine that you only wanted to compare the period of one waveform with another. You display one input signal. While displaying the input signal, adjust the VARIABLE SEC/DIV control until you fit one complete cycle of the displayed waveform within the vertical major division graticule lines. Without adjusting the setting, move your probe to the signal which you wish to compare with. If the one complete cycle of the displayed waveform extends beyond the graticule lines that you used for the input, the signal has a longer period. If the displayed waveform does not fill in the area between the graticule lines that you used for the input, the signal has a shorter period. Notice that this control has a curved arrow and the word CAL on it. On this particular scope, you turn this control fully clockwise. There is actually a detent in the switch that clicks when the control is turned fully clockwise. The VARIABLE SEC/DIV control must be in the detent or CAL position in order to make use of the values on the SEC/DIV switch behind.

Horizontal Magnification

Pulling the VARIABLE SEC/DIV knob gently enables you to magnify the displayed waveform by a factor of 10. This allows you to examine a waveform more closely. The HORIZONTAL MAGNIFICATION control must be depressed in order to make use of the SEC/DIV switch behind.

Sweep Speed

Think of the SWEEP SPEED as a range switch. You adjust this control until the displayed waveform is the largest it can be horizontally and still fit within the graticule of the CRT. This provides the highest accuracy for your measurements. Notice that the settings are divided into S (for seconds), mS (for milliseconds), and μs (for microseconds). After adjusting this control, count the number of major and minor divisions horizontally on the graticule for one complete cycle of the measured signal. For example, imagine a signal that is displayed on the CRT. From the beginning to the end of one complete cycle you count five major divisions. You look at the SEC/DIV control and find it set on 5 ms. This means that there is 5 milliseconds per division. Therefore, the time of one cycle is 25 milliseconds, because 5 major divisions times 5 milliseconds per division equals 25 milliseconds.

The Trigger System

Refer to **Figure 15-16**. The **trigger system** is the portion of the scope that controls when the waveform is displayed on the CRT. If the trigger system controls are not adjusted properly, the waveform may drift slowly across the screen, or it may be a jumble of images. In any event, the display will be impossible to read. Proper adjustment of the trigger system controls is critical in displaying a usable image on the CRT.

Typical controls found in the trigger system are: **VARIABLE TRIGGER HOLDOFF**, **TRIGGER OPERATING MODES**, **TRIGGER SLOPE**, **TRIGGER LEVEL CONTROL**, and **TRIGGER COUPLING**.

- VARIABLE TRIGGER HOLDOFF: This control adjusts the delay time before the waveform is drawn on the CRT. For example, imagine that you are trying to display a waveform that is rather complex. The scope has difficulty displaying a stable waveform because there are many points on the complex wave that cause the scope to

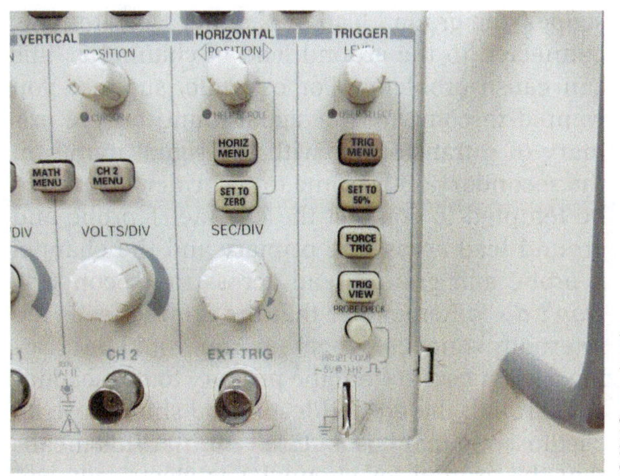

FIGURE 15-16 The trigger section of an oscilloscope.

begin drawing the pattern. By adjusting the *VARIABLE TRIGGER HOLDOFF*, you cause the scope to wait for a particular portion of the waveform to be present before the waveform is drawn. This provides a stable display. Typically, this control is set to the *NORM* position, but do not be afraid to adjust this control if the display is not stable.

- TRIGGER OPERATING MODES: These modes consist of several controls: **NORMAL**, **P-P AUTO**, and **TELEVISION** modes and **TRIGGER SOURCE** switches.
- NORMAL: This control allows the scope to be triggered when a waveform is detected. When a signal is not present, the scope is not triggered and the CRT is blank.
- P-P AUTO: This control allows the scope to be triggered when a waveform is detected. When a signal is not present, a trigger signal is internally sent to the scope, causing a bright baseline to appear on the CRT. The scope is typically set for *P-P AUTO* mode.
- TELEVISION: This control allows the scope to trigger on TV fields or lines. This function is not applicable in the industrial maintenance field.
- TRIGGER SOURCE SWITCHES: These are composed of two switches, SOURCE and INT, that work in conjunction with one another.
    1. SOURCE: The *SOURCE* switch allows the user to choose the trigger source. *INT*, or internal, means that the scope triggers on the signal applied to either channel 1 or channel 2. *LINE* means that the scope uses the frequency of the ac power to the scope for the trigger signal. EXT, or external, means that the scope triggers on a separate trigger signal that is applied to the EXT INPUT jack. Usually, the *SOURCE* switch is set to the *INT* position.
    2. INT: In the *CH* 1 position, the scope\uses the signal from CH 1 for triggering. In the *CH* 2 position, the scope uses the signal from CH 2 as the trigger signal. Setting this switch to the VERT MODE position tells the scope to use whichever signal is present. This switch is used when the *SOURCE* switch is set to the INT position. Typically, the *INT* switch is set to CH 1; however, many individuals find it more convenient to set the *INT* switch to *VERT MODE.*
- TRIGGER SLOPE: This controls whether the scope triggers on the rising or falling edge of a signal. Usually, the *TRIGGER SLOPE* switch is set for a rising slope.
- TRIGGER LEVEL CONTROL: This controls where the trigger point occurs on the signal. For most measurements you set this control to its midposition. However, do not be afraid to adjust this control if your waveform is unstable.
- TRIGGER COUPLING: This switch is used when the *SOURCE* switch is set to the *EXT* position. Trigger coupling functions similarly to the coupling switches for the vertical section. This switch has three positions: AC, DC, and DC÷10. The *AC* position inserts a capacitor into the external trigger input circuit. This blocks any DC component from the signal on which you are triggering. The scope will trigger on the ac component of the signal only. The *DC* position represents direct coupled. This allows the scope to trigger on all components of the input signal. The *DC÷10* position also uses direct coupling, but attenuates the signal by a factor of 10. This is useful if the external trigger signal is too large.

Oscilloscope Probes

Refer to **Figure 15-17**. In order for you to make any measurements with an oscilloscope, you must connect the circuit to be tested to the channel 1 or channel 2 input of the scope. Although it is possible to use an ordinary set of wires to do the job, you will get better, more accurate measurements by using oscilloscope probes.

There are three basic types of oscilloscope probes. The first type is called a **direct measurement**, or **1X, probe**. This probe supplies an input signal that is of the same amplitude as the signal being measured. The second type of probe is called a **10X probe**. This probe provides an input signal that is

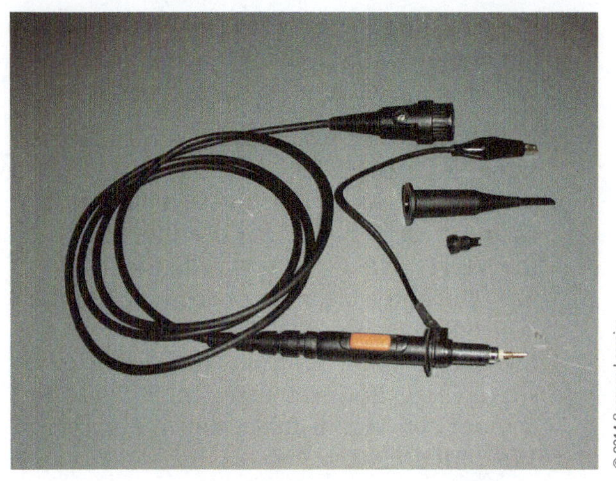

FIGURE 15-17 A ×1, GND, and ×10 switchable oscilloscope probe with ground lead.

attenuated by a factor of 10. In addition, because of its design, a 10X probe introduces less loading effect to the circuit being measured. For this reason, most users of oscilloscopes use a 10X probe at all times. The third type of oscilloscope probe combines the first two types. This probe has a small switch that allows the user to switch between the 1X setting and the 10X setting. This is the most convenient type of probe to use. When using this type of probe, always double-check the position of the switch so that your measurements are not off by a factor of 10.

The 10X probe must be matched to the oscilloscope input being used. This is called **compensating the probe**. To compensate the probe, connect the probe tip to the **probe adjust** test point on the oscilloscope. The probe adjust test point provides a test signal for compensating the probe. Adjust the scope for a stable display. You should see a square wave. Some probes are adjusted by rotating a collar near the probe tip. Other probes are compensated by adjusting a small screw with a nonmetallic screwdriver. The adjustment may be located in the probe or at the connector end. While adjusting the probe, look at the square wave on the CRT. Adjust the probe so that the tops and bottoms of the square wave are flat and the corners are sharp. Once the probe is compensated, you do not have to readjust it unless you move the probe to a different input, or use it on a different scope. The probe is always compensated to the input that you are using.

Oscilloscope probes also contain a ground lead or clip. When making measurements, the probe tip is attached to the point of the circuit where the measurement is taken. The ground lead is attached to the reference point for the circuit. Be very careful when using the ground lead. You should be aware that on many dual-channel scopes, the ground lead of channel 1 is internally connected to the ground lead of channel 2. This can cause problems. For example, suppose you wanted to compare the signal found on the primary of a transformer with the signal found on the secondary of the same transformer. You might be tempted to connect the channel 1 probe and ground lead across the primary and the channel 2 probe and ground lead across the secondary. *Don't do it!* Remember that the ground leads are internally connected together. You would be creating a short circuit from the primary to the secondary through the ground leads. You should use an oscilloscope that has **isolated inputs**. This means that the ground for channel 1 is electrically isolated from the ground for channel 2.

Often there is another precaution that you must observe when using ground leads: The ground leads may also be connected to the scope ground, which is the grounding conductor of the three-wire ac plug. This means that when you make a measurement, it is possible to connect the reference point in the circuit to the X ground through the ground lead and ground prong of the scope x plug. To prevent this from occurring, some individuals may attach a three-prong to two-prong adapter to the X plug of the scope. This is not the safest procedure to follow, because it defeats the equipment ground for the scope and may create a shock hazard. The ideal fix is to use a scope with isolated inputs.

Making Measurements

Initializing the Scope

Before you begin to make your measurement, you should initialize the scope and check your setup. Do the following to initialize the scope:

1. Display System Controls
   a. Set the INTENSITY control to the midrange position.
   b. Turn the FOCUS control to the midrange position.
2. Vertical System Controls
   a. Turn the channel 1 POSITION control to the midrange position.
   b. Turn the channel 2 POSITION control to the midrange position.
   c. Place the VERTICAL MODE switch in the CH 1 position, if you are using CH 1 only; the BOTH position, if you will be using CH 1 and CH 2 simultaneously; or CH 2 position, if you are using CH 2 only.

d. Turn both CH 1 and CH 2 VOLTS/DIV switches fully to their highest setting.
e. Verify that the CH 1 and CH 2 VARIABLE VOLTS/DIV switch is in the CAL position or detent.
f. Place the CH 1 and CH 2 INPUT COUPLING switches to DC, if you want to view the DC and AC components of the signal; AC, if you only want to view the AC component of the signal; or GND, if you want to establish a reference.

3. Horizontal System Controls
   a. Turn the horizontal POSITION control to the midrange position.
   b. Set the SEC/DIV switch to the 0.5 ms position.
   c. Verify that the VARIABLE SEC/DIV switch is in the CAL position or detent.
   d. Verify that HORIZONTAL MAGNIFICATION is not selected by pushing the VARIABLE SEC/DIV switch.
4. Trigger System Controls
   a. Set the VAR HOLDOFF control to the NORM position.
   b. Set the TRIGGER OPERATING MODE switches as follows:
      i. P-P AUTO on
      ii. SOURCE to INT
      iii. INT to VERT MODE
   c. Set the TRIGGER SLOPE switch to positive, if you want to trigger on the leading edge of the signal, or negative, if you want to trigger on the falling edge of the signal.
   d. Set the TRIGGER LEVEL control to the midrange position.

Your scope is now initialized. Plug your scope into a properly grounded outlet and power up the scope. Connect an oscilloscope probe to the CH 1 input jack or CH 2 if you are using channel 2 for your measurements, or connect a probe to both CH 1 and CH 2 if you will be measuring two different input signals.

5. Compensate the probe(s).
6. Verify that the ground of the circuit you will be testing is at the same potential as the oscilloscope ground.
   a. Touch the probe tip to the circuit ground. If no difference of potential is detected, you may connect the oscilloscope ground lead to the circuit ground.

Making Voltage Measurements

Refer to **Figure 15-18**. Voltage is measured vertically on the oscilloscope graticule. This means that when a waveform is displayed on the CRT, you will measure the height or amplitude of the waveform to determine the amount of voltage present. Following are the steps you should take to make a voltage measurement:

1. Connect the ground lead of the oscilloscope probe to the circuit ground.
2. Connect the probe tip of the oscilloscope probe to the point of the circuit where the voltage is to be measured.
3. Adjust the SEC/DIV switch until a minimum of one complete cycle of the waveform is displayed horizontally across the CRT.
4. Adjust the TRIGGER LEVEL control as necessary to obtain a stable display.
5. Adjust the VERTICAL POSITION control until the waveform is centered on the CRT graticule.
6. Adjust the VOLTS/DIV switch until the displayed waveform fills the CRT from top to bottom. If the waveform extends beyond the top or bottom of the CRT, you need to use the next larger VOLTS/DIV setting. If you cannot adjust the waveform to fit vertically within the CRT, you will need to use a 10X probe instead of a 1X probe.
7. Use the VERTICAL POSITION control to reposition the displayed waveform so that the bottom peak of the waveform is just touching the bottommost horizontal line (major division) on the CRT graticule.

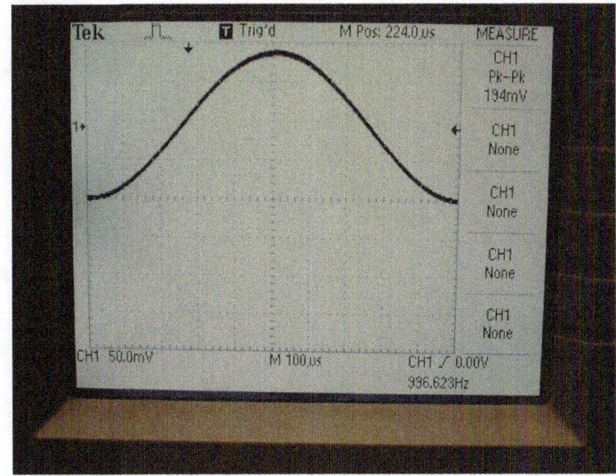

FIGURE 15-18 Making a voltage measurement.

8. Use the HORIZONTAL POSITION control to reposition the displayed waveform so that the top peak of the waveform is centered over the center vertical graticule line, which is the one with the minor divisions.
9. Count the major and minor divisions from the bottom of the negative peak to the top of the positive peak.
10. Multiply this number by the VOLTS/DIV setting. Be sure to use the correct value of VOLTS/DIV for the type probe that you are using. The product is the amount of peak-to-peak voltage for the displayed waveform.

Look at Figure 15-18 for an example. This figure shows an oscilloscope with the display properly adjusted for voltage measurement. If you count the number of major and minor divisions from the bottom of the negative peak to the top of the positive peak, you will have 5.2 divisions. Now multiply the number of divisions, 5.2, by the VOLTS/DIV setting of 0.5 volt and the product is 2.6 volts peak to peak.

Making Time Measurements

Refer to **Figure 15-19**. Time is measured horizontally on the oscilloscope graticule. This means that when a waveform is displayed on the CRT, you will measure the width of one complete cycle of the waveform to determine the amount of time of one complete cycle. Following are the steps you should take to make time measurements:

1. Connect the ground lead of the oscilloscope probe to the circuit ground.
2. Connect the probe tip of the oscilloscope probe to the point of the circuit where the time is to be measured.
3. Adjust the SEC/DIV switch until a minimum of one complete cycle of the waveform is displayed horizontally across the CRT.
4. Adjust the TRIGGER LEVEL control as necessary to obtain a stable display.
5. Adjust the VERTICAL POSITION control until the waveform is centered on the CRT graticule.
6. Adjust the VOLTS/DIV switch until the displayed waveform fills the CRT from top to bottom. If the waveform extends beyond the top or bottom of the CRT, you need to use the next larger VOLTS/DIV setting. If you cannot adjust the waveform to fit vertically within the CRT, use a 10X probe instead of a 1X probe.
7. Use the HORIZONTAL POSITION control to reposition the displayed waveform so that an easily identified key portion of the waveform is just touching the leftmost vertical line (major division) of the CRT graticule.
8. Use the VERTICAL POSITION control to reposition the displayed waveform so that the key portion of the waveform is located on the center horizontal graticule line, which is the one with the minor divisions.
9. Count the major and minor divisions from the key portion of the left side of the waveform to the same key portion at the right side of the waveform where the cycle repeats.
10. Multiply this number by the SEC/DIV setting. The product is the amount of time in seconds, milliseconds, or microseconds of one complete cycle of the displayed waveform.

For example, Figure 15-19 shows an oscilloscope with the display properly adjusted for time measurement. If you count the number of major and minor divisions from the key portion of the left side of the waveform to the same key portion of the right side of the waveform where the cycle repeats, you will have 5.1 divisions. Now multiply the number of divisions, 5.1, by the SEC/DIV setting of 2 milliseconds and the product is 10.2 milliseconds.

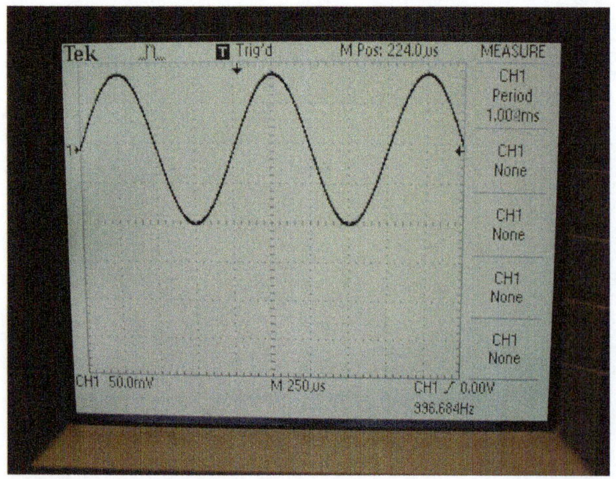

FIGURE 15-19 Making a time measurement.

Making Frequency Measurements

Once the time it takes to complete one cycle is known, it is possible to calculate the frequency of the waveform. The formula used to determine frequency when time is known is:

$$F = \frac{1}{T}$$

F = Frequency in Hz
T = Time for one complete cycle in seconds

In the next example, use the values from the time measurement exercise above:

$$T = 10.2 \text{ mS}$$

$$F = \frac{1}{T} = \frac{1}{10.2 \text{ ms}} = 98.04 \text{ Hz}$$

SUMMARY

- The most common multimeter used today is the digital multimeter. It is so named because the measured value is displayed in a digital readout.
- The digital multimeter may be bench or panel mounted, but the one most commonly used in maintenance is the handheld type.
- The operation of digital multimeters is essentially the same. The major differences are in their functions and accuracies.
- Before using the multimeter, the meter and test leads should be visually inspected. Replace the set of leads even if only one test lead is damaged. It is far safer to replace them than to try to repair a damaged lead with electrical tape.
- Thoroughly inspect the meter itself. If there are any cracks or missing pieces from the case, the meter must be repaired. The case protects you from the voltages and currents that you will be measuring.
- If the meter and test leads check out visually, you should perform an operational test.
- Before connecting the meter into a circuit to be tested, you must set up the meter for the measurement that you expect to make.
- The clamp-on ammeter allows the measurement of current without the need to de-energize the circuit.
- For the clamp-on ammeter to function properly, the clamp must be fully closed around the conductor.
- Clamp-on ammeters are typically used to measure significant amounts of current, but they can be used to measure smaller amounts of current as well.
- The megohmmeter allows the measurement of very high resistance values.
- There are many types of megohmmeters on the market today. Some models measure high resistance values, others measure continuity and voltage, and some models use a hand crank to operate an internal dc generator.
- The oscilloscope is used not only to measure voltage, but it also allows the user to see a representation of the voltage. This is quite helpful when diagnosing problems caused by voltage spikes or dirty power or simply as a way to find failed components more rapidly.

REVIEW QUESTIONS

1. List the steps that you would follow in preparing to make a voltage measurement with a digital multimeter.
2. How is a digital multimeter connected in a circuit when measuring current?
3. What precaution must you observe when measuring the resistance of a circuit with a digital multimeter?
4. Describe the technique that you would use to measure small amounts of current with a clamp-on ammeter.
5. When would you use a megohmmeter?
6. The vertical sensitivity adjustment of an oscilloscope is set to 5 volts/Div, and the horizontal sensitivity adjustment is set to 2 mS/Div. A displayed waveform is 6.2 divisions from peak to peak. What is the peak-to-peak voltage of the displayed waveform?
7. One cycle of the waveform in question 6 spans 4.7 divisions. What is the frequency of this waveform?

CHAPTER 16

Basic Resistive Electrical Circuits

Electrical circuits are the building blocks that comprise all electrical devices, no matter how simple or complex. Technicians must gain a solid understanding of the basics if they ever hope to understand more complicated devices. Regardless of the complexity, all electrical circuits consist of series, parallel, or combination circuits. This chapter introduces you to these fundamental circuits upon which more complex circuits are built.

OBJECTIVES

After studying this chapter, the student should be able to

- Identify a series resistive circuit.
- Identify a parallel resistive circuit.
- Identify a combination resistive circuit.
- Perform all necessary calculations to analyze a resistive electrical circuit.

16-1 SERIES CIRCUITS

Before beginning the studies of series, parallel, and combination circuits, you must first understand what comprises a circuit. A *circuit* must contain a minimum of three elements: a **power source**, a **complete path** for current flow, and a **load**. Some examples of a power source are a battery, a power supply, and a generator. Conductors are used to provide a path for the current flow. Conductors are generally in the form of wires, which connect the power source to the load. Conductors may also be the copper traces, or tracks, on a printed circuit board. A load is any device that draws current from the power source. A load could be a resistor, a lightbulb, a motor, or other such devices.

In addition to these three items, some circuits also contain a form of **control**. A control device is used to switch the current flow on or off. Some examples of control devices include a switch, a thermostat, a relay, and others. **Figure 16-1** shows a complete circuit. Notice the power source (the battery), the conductors (wires), the load (lightbulb), and the control (switch).

A **series circuit** is shown in **Figure 16-2**. Notice that this circuit is simple. The circuit consists of a power source (indicated by the AC sine wave), conductors, loads (the three lightbulbs), and a control device (the switch). When you look at this circuit, imagine yourself as the current flowing from the power source. As you leave the power source, you flow through the conductor to the switch. In Figure 16-2, the switch is open. This means that the current flow cannot continue to the loads. There is no complete path because of the open switch. Therefore, the lightbulbs do not light.

Now look at **Figure 16-3**. The switch has been closed. The current can now flow from the power source, through the conductor to the switch, through the now closed switch, through the conductor to

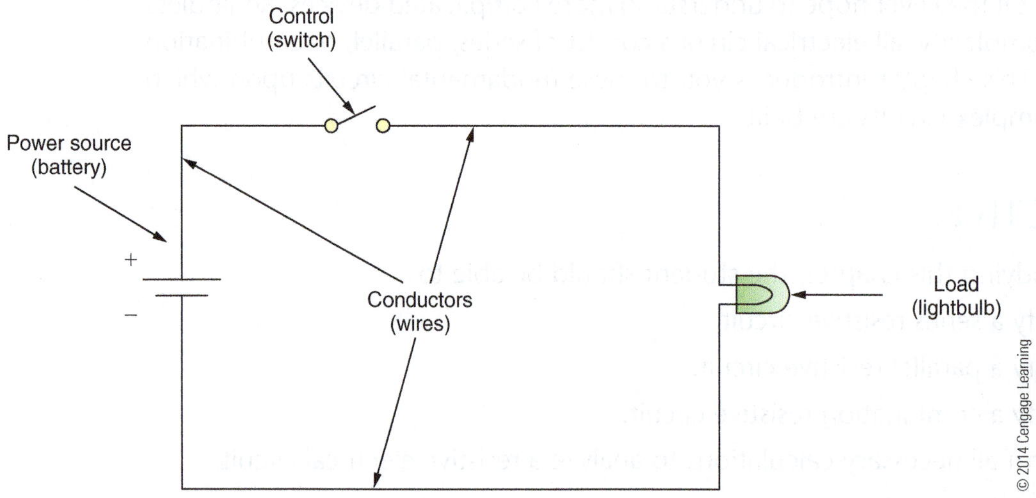

FIGURE 16-1 A complete circuit.

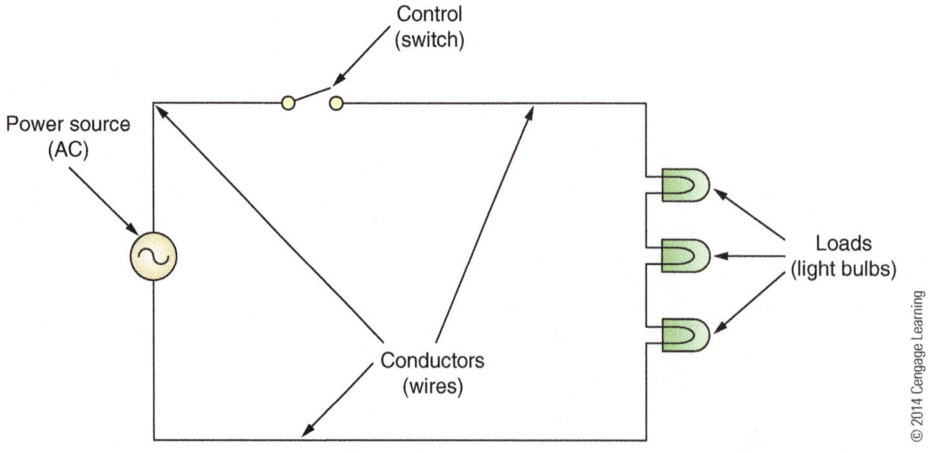

FIGURE 16-2 A series circuit—switch open.

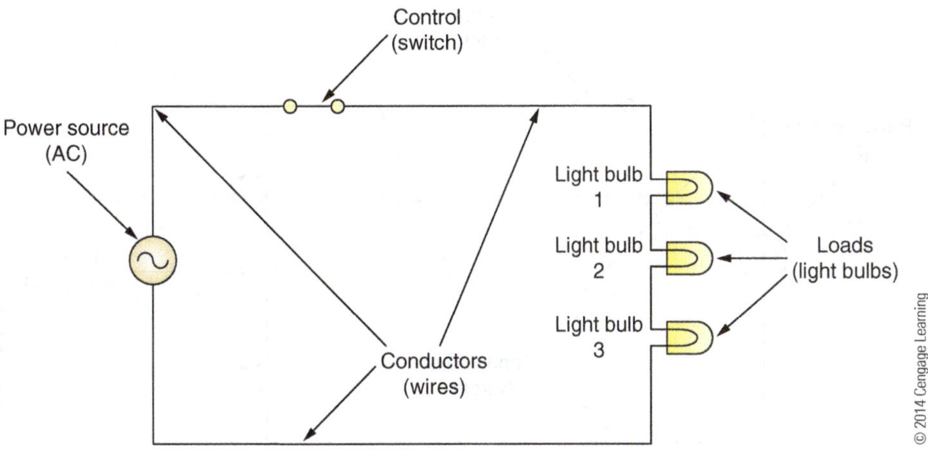

FIGURE 16-3 A series circuit—switch closed.

lightbulb 1, through lightbulb 1, through the conductor to lightbulb 2, through lightbulb 2, through the conductor to lightbulb 3, through lightbulb 3, and through the conductor back to the power source. Because there is a complete path for current flow, the lightbulbs light. Notice that the current could also flow in the opposite direction; that is, from the power source, through the conductor to lightbulb 3, through lightbulb 3, through the conductor to lightbulb 2, through lightbulb 2, through the conductor to lightbulb 1, through lightbulb 1, through the conductor to the switch, through the closed switch, and through the conductor back to the power source. Regardless of the direction of the current flow, there is no other path that the current can travel. This is an important characteristic of a series circuit. *A series circuit only has one path for current flow.*

Now imagine that the filament of lightbulb 2 has burned open, as shown in **Figure 16-4**. It is obvious that lightbulb 2 will not light, but what about lightbulbs 1 and 3? Again, imagine yourself as the current flowing from the power source. As you leave the power source, you flow through the conductor to the closed switch. The current flows through the closed switch through the conductor to lightbulb 1, through lightbulb 1, and through the conductor to lightbulb 2. Because the filament of lightbulb 2 has burned open, the current flow stops at this point, and because there is no complete path for current flow, *none of the lightbulbs light.* If one lightbulb burns out in a circuit with lightbulbs connected in series, all of the lightbulbs will fail to light.

Figure 16-5 shows a circuit in which the lightbulbs have been replaced with three resistors, R_1, R_2, and R_3. Is this a series circuit? Follow the path for current flow. Is there only one path? Current flows from the power source, through the conductor, through the closed switch, through the conductor to R_1, through R_1, through the conductor to R_2, through R_2, through the conductor to R_3, through R_3, and through the conductor back to the power

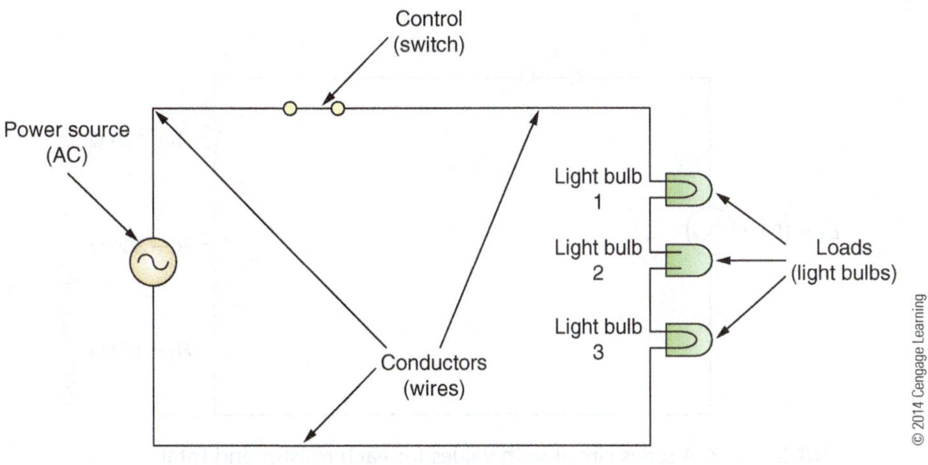

FIGURE 16-4 Open filament in lightbulb 2.

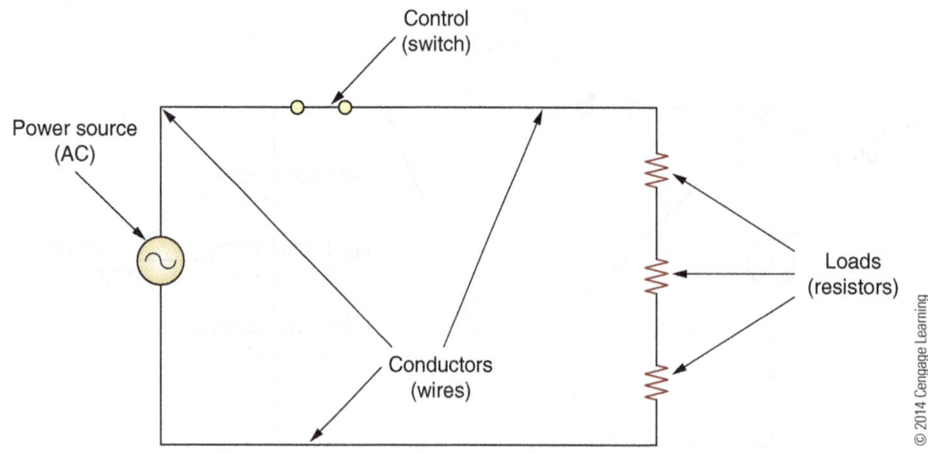

FIGURE 16-5 A three-resistor series circuit.

source. This is the only path that current can follow. The current could flow in the opposite direction, but the current must still follow the same path. There is only one current path in this circuit. Therefore, this is also a series circuit. In fact, resistors R_1, R_2, and R_3 are said to be connected in series.

Once the circuit has been identified as being a series circuit, other useful information about the circuit can be determined. Refer to **Figure 16-6**. Figure 16-6 is the same circuit as shown in Figure 16-5, except that values have been added for the amount of voltage present as well as the values of R_1, R_2, and R_3. Because the amount of voltage present is known, can the amount of current flowing in the circuit be determined? In Chapter 14 you learned that by using Ohm's law, the amount of current can be determined if the amount of voltage and resistance are known. Recall the formula:

$$I = \frac{E}{R}$$

I = Current in amperes
E = Voltage in volts
R = Resistance in ohms

Trying to determine the amount of current flowing in this circuit is difficult. What should be used for the value of R? Should the value of R_1, the value of R_2, or the value of R_3 be used? Actually, the combined values of $R_1 + R_2 + R_3$ are used. To find the total circuit current, the total circuit resistance as well as the total circuit voltage must be used. This is another important characteristic about a series circuit. *The total resistance of a series circuit is equal to the sum of the individual resistances.* Therefore, the formula for finding the total resistance of a series circuit is

$$R_T = R_1 + R_2 + R_3 + \ldots$$

What is the total resistance of the circuit shown in Figure 16-6?
What values are known?

$$R_1 = 100\ \Omega$$

$$R_2 = 330\ \Omega$$

$$R_3 = 470\ \Omega$$

What value is not known?

$$R_T = ?$$

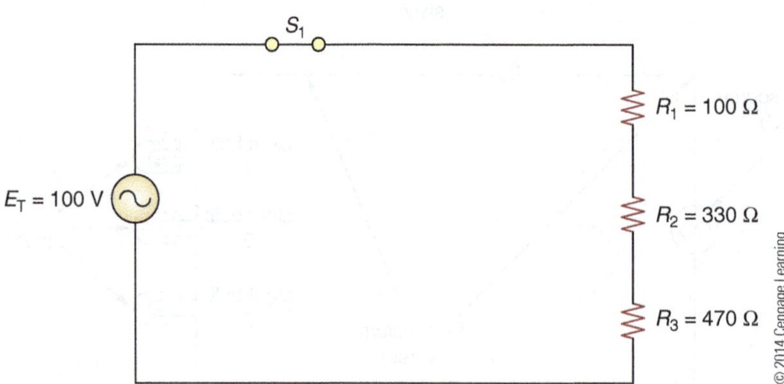

FIGURE 16-6 A series circuit with values for each resistor and total circuit voltage.

What formula can be used?

$$R_T = R_1 + R_2 + R_3 + \ldots$$

Substitute the known values and solve for R_T:

$$R_T = R_1 + R_2 + R_3 + \ldots$$

$$R_T = 100\ \Omega + 330\ \Omega + 470\ \Omega$$

$$R_T = 900\ \Omega$$

Therefore, the total resistance of the circuit shown in Figure 16-6 is 900 ohms. Now that the total resistance of the circuit and the total applied voltage to the circuit are known, the total circuit current can be determined.

What values are known?

$$E_T = 100\ \text{V}$$

$$R_T = 900\ \Omega$$

What value is not known?

$$I_T = ?$$

What formula can be used?

$$I_T = \frac{E_T}{R_T}$$

Substitute the known values and solve for I_T:

$$I_T = \frac{E_T}{R_T}$$

$$I_T = \frac{100\ \text{V}}{900\ \Omega}$$

$$I_T = 0.111\ \text{A or}$$

$$I_T = 111.11\ \text{mA}$$

Therefore, the total current flowing in the circuit in Figure 16-6 is 111.11 milliamperes.

If the total current flowing in the circuit of Figure 16-6 is 111.11 milliamperes, how much current is flowing through R_1? The answer is 111.11 milliamperes. How much current is flowing through R_2? Again, the answer is 111.11 milliamperes. The current flowing through R_3 is also 111.11 milliamperes. Because there is only one path for current flow in a series circuit, the current must be the same throughout a series circuit. This means that no matter where you measure the current in a series circuit, the amount of current is always equal to the total circuit current. This is another important characteristic of a series circuit. *In a series circuit, the current is the same value throughout the entire circuit. The value of the current flowing through each individual component is equal to the total circuit current.* This can be expressed in the formula:

$$I_T = I_1 = I_2 = I_3$$

Now turn your attention to the voltage. If a digital voltmeter were placed across resistor R_1, how much voltage would be measured? How much voltage would be measured across resistor R_2? What about resistor R_3? Before answering these questions, analyze this: 100 volts is applied to the circuit shown in Figure 16-6. There are 111.11 milliamperes of current flowing in this circuit. The power source is supplying 100 volts of pressure to push 111.11 milliamperes of current through R_1, R_2, and R_3. What happens to the pressure when the current reaches R_1? There is a pressure drop. How much does the pressure drop? The amount of pressure, or voltage, drop can be determined by using an Ohm's law formula. Recall the formula for finding voltage when current and resistance are known:

$$E = I \times R$$

To know the voltage drop across resistor R_1, use the current flowing through R_1 and the resistance value of R_1.

What values are known?

$$I_1 = 111.11\ \text{mA}$$

$$R_1 = 100\ \Omega$$

What value is not known?

$$E_1 = ?$$

What formula can be used?

$$E_1 = I_1 \times R_1$$

Substitute the known values and solve for E_1:

$$E_1 = I_1 \times R_1$$

$$E_1 = 111.11\ \text{mA} \times 100\ \Omega$$

$$E_1 = 11.11\ \text{V}$$

Therefore, the voltage drop across R_1 is 11.11 volts. How much voltage is dropped across resistor R_2? Use the same value for current, but you must use the resistance value for R_2.

What values are known?

$$I_2 = 111.11\ \text{mA}$$

$$R_2 = 330\ \Omega$$

What value is not known?

$$E_2 = ?$$

What formula can be used?

$$E_2 = I_2 \times R_2$$

Substitute the known values and solve for E_2:

$$E_2 = I_2 \times R_2$$

$$E_2 = 111.11\ \text{mA} \times 330\ \Omega$$

$$E_2 = 36.67\ \text{V}$$

Therefore, the voltage drop across R_2 is 36.67 volts. How much voltage is dropped across resistor R_3? Use the same value for current, but you must use the resistance value for R_3.

What values are known?

$I_3 = 111.11\ \text{mA}$

$R_3 = 470\ \Omega$

What is not known?

$E_3 = ?$

What formula can be used?

$E_3 = I_3 \times R_3$

Substitute the known values and solve for E_3:

$E_3 = I_3 \times R_3$

$E_3 = 111.11\ \text{mA} \times 470\ \Omega$

$E_3 = 52.22\ \text{V}$

Therefore, the voltage drop across R_3 is 52.22 volts. Now notice something interesting: If you take the voltage drop across R_1, add it to the voltage drop across R_2, and add the voltage drop across R_3, you find the total applied voltage.

$E_1 = 11.11\ \text{V}$

$E_2 = 36.67\ \text{V}$

$E_3 = 52.22\ \text{V}$

$E_T = E_1 + E_2 + E_3$

$E_T = 11.11\ \text{V} + 36.67\ \text{V} + 52.22\ \text{V}$

$E_T = 100\ \text{V}$

This is another important characteristic of a series circuit: *The sum of all individual voltage drops equals the total applied voltage.* You should notice something else about the voltage drops in this circuit. Compare the voltage drop across resistor R_1 with the voltage drop across resistor R_2 and with the voltage drop across resistor R_3. Which voltage drop is larger? The 52.22-volt voltage drop across R_3 is larger. Which resistor is larger? R_3 at 470 ohms is larger than R_1 at 100 ohms or R_2 at 330 ohms. *The larger resistor has the largest voltage drop.*

There is one other parameter of a circuit to be concerned about: the amount of power consumed by the circuit and the individual components of the circuit. Circuit power is measured in watts (W). With the individual components, watts are a measure of the amount of power (in the form of heat) that the component must dissipate. Refer to Figure 16-6 again. The total circuit voltage, the total circuit current, and the total circuit resistance are known, but the total circuit power is not. The voltage drops across R_1, R_2, and R_3 are also known, as well as the current flowing through each resistor and the resistive value of each resistor. However, the power that each resistor must dissipate is not known. Begin by finding the total power, P_T. Recall from Chapter 14 the three formulae for finding power:

$$P = I \times E$$

$$P = \frac{E^2}{R}$$

$$P = I^2 \times R$$

Any of the three formulae can be used to determine the total circuit power. The values of I_T and E_T are known; therefore, the first formula can be used. The values of E_T and R_T are known; therefore, the second formula can be used. Finally, the values of I_T and R_T are known; therefore, the third formula can be used. For this example, use the second formula. You may want to try to solve the example using either one or both of the remaining formulae. Your results should be the same. Now find P_T.

What values are known?

$E_T = 100\ \text{V}$

$R_T = 900\ \Omega$

What value is not known?

$P_T = ?$

What formula can be used?

$$P_T = \frac{E^2}{R}$$

Substitute the known values and solve for P_T:

$$P_T = \frac{E^2}{R}$$

$$P_T = \frac{100\ \text{V}^2}{900\ \Omega}$$

$P_T = 11.11\ \text{W}$

Therefore, this circuit consumes 11.11 watts of power. Now turn your attention to the individual resistors, R_1, R_2, and R_3. How much power must each resistor dissipate? Begin with R_1.

What values are known?

$E_1 = 11.11\ \text{V}$

$I_1 = 111.11\ \text{mA}$

$R_1 = 100\ \Omega$

What value is not known?

$P_1 = ?$

What formula can be used? Use any one of the following.

$$P_1 = I_1 \times E_1$$

$$P_3 = \frac{E_1^2}{R_3}$$

$$P_1 = I_1^2 \times R_1$$

Substitute the known values and solve for P_1. Use the first equation.

$$P_1 = I_1 \times E_1$$

$$P_1 = 111.11 \text{ mA} \times 11.11 \text{ V}$$

$$P_1 = 1.23 \text{ W}$$

Therefore, resistor R_1 must dissipate 1.23 watts of heat. (In actual circuit design, the minimum wattage rating for resistor R_1 should be 2.46 watts. A good rule of thumb is that the resistor wattage rating should be twice the actual wattage that must be handled.) Now determine how much power resistor R_2 must dissipate.

What values are known?

$$E_2 = 36.67 \text{ V}$$

$$I_2 = 111.11 \text{ mA}$$

$$R_2 = 330 \ \Omega$$

What value is not known?

$$P_2 = ?$$

What formula can be used? Use any one of the following.

$$P_2 = I_2 \times E_2$$

$$P_2 = \frac{E_2^2}{R_2}$$

$$P_2 = I_2^2 \times R_2$$

Substitute the known values and solve for P_2. Use the third equation.

$$P_2 = I_2^2 \times R_2$$

$$P_2 = 111.11 \text{ mA}^2 \times 330 \ \Omega$$

$$P_2 = 4.07 \text{ W}$$

Therefore, resistor R_2 must dissipate 4.07 watts of heat. Now determine how much power resistor R_3 must dissipate.

What values are known?

$$E_3 = 52.22 \text{ V}$$

$$I_3 = 111.11 \text{ mA}$$

$$R_3 = 470 \ \Omega$$

What value is not known?

$$P_3 = ?$$

What formula can be used? Use any one of the following.

$$P_3 = I_3 \times E_3$$

$$P_3 = \frac{E_3^2}{R_3}$$

$$P_3 = I_3^2 \times R_3$$

Substitute the known values and solve for P_3. Use the second equation.

$$P_3 = \frac{E_3^2}{R_3}$$

$$P_3 = \frac{52.22 \text{ V}^2}{470 \ \Omega}$$

$$P_3 = 5.80 \text{ W}$$

Therefore, resistor R_3 must dissipate 5.80 watts of heat.

Now notice something interesting. If you take the actual wattage dissipated by R_1, add it to the actual wattage dissipated by R_2, and add the actual wattage dissipated by R_3, you find the total wattage consumed by the circuit.

$$P_1 = 1.23 \text{ W}$$

$$P_2 = 4.07 \text{ W}$$

$$P_3 = 5.80 \text{ W}$$

$$P_T = P_1 + P_2 + P_3 + \ldots$$

$$P_T = 1.23 \text{ W} + 4.07 \text{ W} + 5.80 \text{ W}$$

$$P_T = 11.10 \text{ W}$$

Do not be concerned with the 0.01-watt difference between the two P_T answers. This is a result of the rounding off of numbers to two decimal places during calculations. This is another important characteristic of a series circuit: *The sum of all individual wattages equals the total circuit wattage.* You should notice something else about the wattage in this circuit. Compare the power dissipated by the three resistors, R_1, R_2, and R_3. Which wattage is larger? The 5.80 watts of R_3 is larger. Which resistor is larger? R_3 at 470 ohms is larger than R_1 at 100 ohms or R_2 at 330 ohms. *The largest resistor must dissipate the most wattage.*

16-2 PARALLEL CIRCUITS

A **parallel circuit** is shown in **Figure 16-7**. The circuit consists of a power source (indicated by the AC sine wave), conductors, three loads (the lightbulbs), and a control device (the switch). When you look at this circuit, imagine yourself as the current flowing from the power source. As you leave the power source, you flow through the conductor to the switch. The switch is open in Figure 16-7. This means that the current flow cannot continue to the load. There is no complete path because of the open switch. Therefore, the lightbulbs do not light.

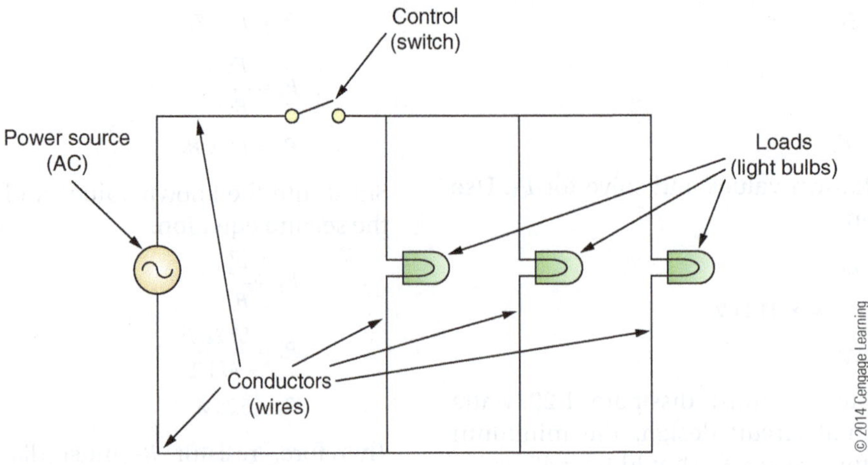

FIGURE 16-7 A parallel circuit—switch open.

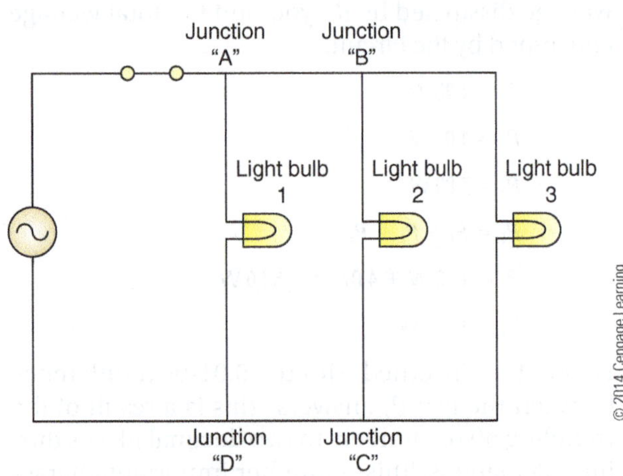

FIGURE 16-8 A parallel circuit—switch closed.

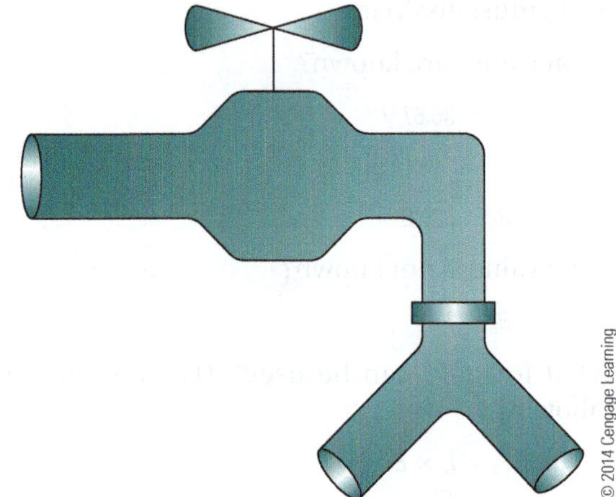

FIGURE 16-9 Current divides in a parallel circuit as water divides in a Y pipe.

Now look at **Figure 16-8**. The switch has been closed. The current can now flow from the power source, through the conductor to the switch, through the now closed switch, and through the conductor to junction A. At junction A, the current can take one of two paths. The current can flow through lightbulb 1 or continue toward lightbulb 2 and lightbulb 3. Actually, the current does both. The current *divides*, with some of the current flowing through lightbulb 1 and some of the current flowing toward lightbulb 2 and lightbulb 3. If you have difficulty understanding this, think of connecting two garden hoses to a single faucet with a "Y" connector, as shown in **Figure 16-9**. Some of the water flows through one hose, whereas some of the water flows through the second hose. The same principle applies to the circuit. Now return to Figure 16-8. Leave the current flowing through lightbulb 1 for a moment, and turn your attention to the current flowing toward lightbulb 2 and lightbulb 3. At junction B, the current again divides. Some of the current flows through lightbulb 2, whereas some of the current flows through lightbulb 3. The current flowing through the lightbulbs must return to the power source. Therefore, after the current flows through lightbulb 3, it combines at junction C with the current flowing through lightbulb 2. The combined currents (from lightbulbs 2 and 3) unite at junction D with the current flowing through lightbulb 1. The total recombined current then flows through the conductor back to the power source. Because there is a complete path for current flow, the lightbulbs light. Notice that the current could also flow in the opposite direction, that is, from the power source through the conductor to junction D, where the current divides. Some of the current flows through lightbulb 1 and some of the current flows toward lightbulb 2 and lightbulb 3.

At junction C, the current divides. Some of the current flows through lightbulb 2, and some of the current flows through lightbulb 3. The current from lightbulb 3 combines with the current from lightbulb 2 at junction B. This current then combines at junction A with the current from lightbulb 1. The total recombined current then flows through the conductor to the closed switch, through the switch, and back to the power source. Notice in this circuit that there is more than one path for current flow. Current can flow from the source through lightbulb 1, or current can flow from the source through lightbulb 2, or current can flow from the source through lightbulb 3. In fact, should the filament of lightbulb 2 burn open, lightbulb 1 and lightbulb 3 would still light. Likewise, if lightbulb 1 burned out, lightbulb 2 and lightbulb 3 would still light. Finally, if lightbulb 3 burned out, lightbulb 1 and lightbulb 2 would still light. This means that there are three paths for current flow in this circuit.

Now look at **Figure 16-10**. This circuit contains five lightbulbs. Notice that each lightbulb works independently of the others. This circuit also has more than one path for current flow. In fact, there are five current paths in this circuit. Therefore, this circuit is also a parallel circuit. This is an important characteristic of a parallel circuit. *A parallel circuit has two or more paths for current flow.*

Figure 16-11 shows a circuit in which the lightbulbs have been replaced with three resistors, R_1, R_2, and R_3. Is this a parallel circuit? Follow the path for current flow. Is there only one path, or are there two or more paths? Current flows from the power source, through the conductor, through the closed switch, to junction A. At junction A, the current divides. Some of the current flows through R_1 and some of the current flows toward R_2 and R_3. At junction B, the current divides again. Some of the current flows through R_2 and some of the current flows through R_3. The currents from R_2 and R_3 recombine

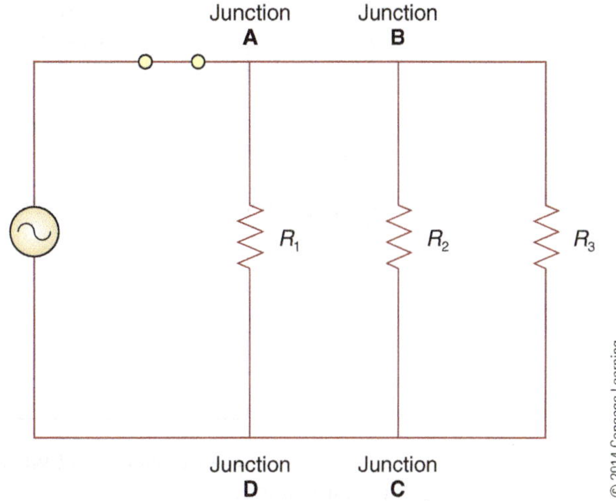

FIGURE 16-11 A three-resistor parallel circuit.

at junction C. This current then recombines with the current for resistor R_1 at junction A and then flows through the conductor back to the power source. How many paths exist for current flow? There are three paths. Therefore, this is a parallel circuit. In fact, resistors R_1, R_2, and R_3 are said to be connected in parallel.

Once the circuit has been identified as being a parallel circuit, other useful information about the circuit can be determined. Refer to **Figure 16-12**. Figure 16-12 is the same circuit as the one shown in Figure 16-11, except that now values have been added for the amount of voltage present as well as the values of R_1, R_2, and R_3. The same values that were used in the previous example of a series circuit are used again so that you can see the effects of connecting components in series or parallel. Because the amount of voltage present is known, can the amount of current flowing in the circuit be determined?

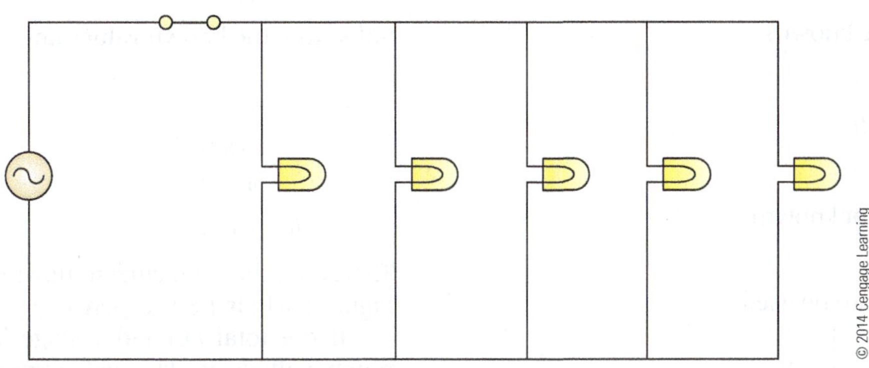

FIGURE 16-10 A five-branch parallel circuit.

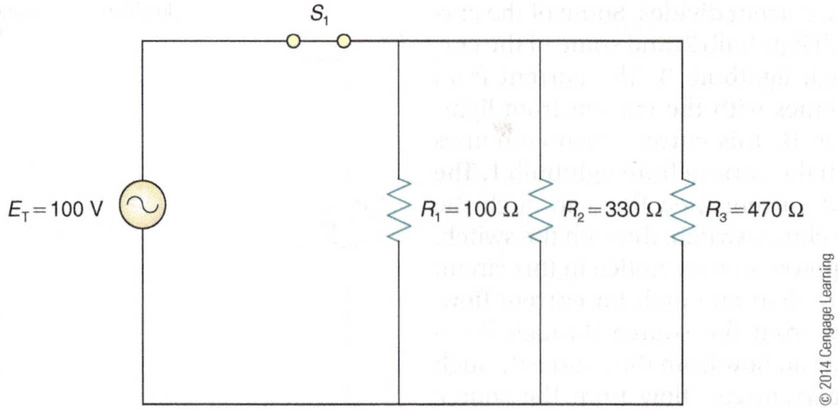

S_1

$E_T = 100$ V

$R_1 = 100\ \Omega$ $R_2 = 330\ \Omega$ $R_3 = 470\ \Omega$

FIGURE 16-12 A parallel circuit with values for each resistor and total circuit voltage.

Trying to determine the amount of current flowing in this circuit is difficult. What should be used for the value of R? Should the value of R_1, R_2, or R_3 be used? Actually, the total circuit resistance is *less* than either individual resistor. How can this be? Think about a single entrance to an amusement park, with a row of turnstiles, only one of which is open. Imagine there are 5000 people trying to enter the park through one turnstile. There is a lot of opposition, right? Now imagine if five more turnstiles were opened. What effect would that have on the crowd? Would it make entering the park easier? Would there be less opposition? The same applies to a parallel circuit. If more paths are provided for current flow, then the overall opposition to current flow must be less. *In a parallel circuit, the total circuit resistance is smaller than the smallest branch resistance.*

To determine the opposition to current flow in a parallel circuit, use the following formula:

$$R_T = \frac{1}{\frac{1}{R_1} + \frac{1}{R_2} + \frac{1}{R_3} + \ldots}$$

The total resistance of a parallel circuit is equal to the reciprocal of the sum of the reciprocals of the individual resistances. What is the total resistance of the circuit shown in Figure 16-12?

What values are known?

$$R_1 = 100\ \Omega$$

$$R_2 = 330\ \Omega$$

$$R_3 = 470\ \Omega$$

What value is not known?

$$R_T = ?$$

What formula can be used?

$$R_T = \frac{1}{\frac{1}{R_1} + \frac{1}{R_2} + \frac{1}{R_3} + \ldots}$$

Substitute the known values and solve for R_T:

$$R_T = \frac{1}{\frac{1}{R_1} + \frac{1}{R_2} + \frac{1}{R_3} + \ldots}$$

$$R_T = \frac{1}{\frac{1}{100\ \Omega} + \frac{1}{330\ \Omega} + \frac{1}{470\ \Omega}}$$

$$R_T = 65.97\ \Omega$$

Therefore, the total resistance of the circuit shown in Figure 16-12 is 65.97 ohms. Notice that the total resistance, 65.97 ohms, is smaller than the smallest branch resistance of 100 ohms. Now that the total resistance of the circuit and the total applied voltage to the circuit are known, the total circuit current can be determined.

What values are known?

$$E_T = 100\ \text{V}$$

$$R_T = 65.97\ \Omega$$

What value is not known?

$$I_T = ?$$

What formula can be used?

$$I_T = \frac{E_T}{R_T}$$

Substitute the known values and solve for I_T:

$$I_T = \frac{E_T}{R_T}$$

$$I_T = \frac{100\ \text{V}}{65.97\ \Omega}$$

$$I_T = 1.52\ \text{A}$$

Therefore, the total current flowing in the circuit in Figure 16-12 is 1.52 amperes.

If the total applied voltage is 100 volts, how much voltage is dropped across R_1, R_2, and R_3? Would you believe 100 volts? Look carefully at

Figure 16-12. Imagine placing one probe of your voltmeter at the top of R_1. Now imagine placing the other probe of your voltmeter at the bottom of R_1. This is how you would measure the voltage drop across R_1. Now look at what you are actually measuring. Follow the connection from the top of R_1 to the left to the top of the power source. Now follow the connection from the bottom of R_1 to the left to the bottom of the power source. You have actually placed your voltmeter across the power source. This means that you actually measure the voltage of the power source, 100 volts. Therefore, the voltage drop across R_1 is the same as the voltage of the power source, 100 volts. Now determine the voltage drop across R_2. It is also 100 volts. Likewise, the voltage drop across R_3 is 100 volts. *In a parallel circuit, the voltage across the individual branch circuits is the same.* Expressed as a formula:

$$E_T = E_1 = E_2 = E_3 = \ldots$$

If the total current flowing in the circuit of Figure 16-12 is 1.52 amperes, how much current is flowing through R_1? Because the voltage drop across R_1 is known, 100 volts, and because the resistive value of R_1 is known, 100 ohms, the current flowing through R_1 can be determined. This is how:

What values are known?

$$E_1 = 100\text{ V}$$

$$R_1 = 100\ \Omega$$

What value is not known?

$$I_1 = ?$$

What formula can be used?

$$I_1 = \frac{E_1}{R_1}$$

Substitute the known values and solve for I_1:

$$I_1 = \frac{E_1}{R_1}$$

$$I_1 = \frac{100\text{ V}}{100\ \Omega}$$

$$I_1 = 1\text{ A}$$

Therefore, there is 1 ampere of current flowing through R_1. How much current is flowing through R_2?

What values are known?

$$E_2 = 100\text{ V}$$

$$R_2 = 330\ \Omega$$

What value is not known?

$$I_2 = ?$$

What formula can be used?

$$I_2 = \frac{E_2}{R_2}$$

Substitute the known values and solve for I_2:

$$I_2 = \frac{E_2}{R_2}$$

$$I_2 = \frac{100\text{ V}}{330\ \Omega}$$

$$I_2 = 303.03\text{ mA}$$

Therefore, there is 303.03 milliamperes of current flowing through resistor R_2. How much current is flowing through R_3?

What values are known?

$$E_3 = 100\text{ V}$$

$$R_3 = 470\ \Omega$$

What value is not known?

$$I_3 = ?$$

What formula can be used?

$$I_3 = \frac{E_3}{R_3}$$

Substitute the known values and solve for I_3:

$$I_3 = \frac{E_3}{R_3}$$

$$I_3 = \frac{100\text{ V}}{470\ \Omega}$$

$$I_3 = 212.77\text{ mA}$$

Recall from the tracing of the current flow in the parallel circuit that the current divided at junctions A and B and recombined at junctions C and D. This means that the total circuit current of 1.52 amperes divided into three branch currents of 1 ampere, 303.03 milliamperes, and 212.77 milliamperes. These three branch currents must recombine into the total circuit current. This can be expressed as a formula:

$$I_T = I_1 + I_2 + I_3 + \ldots$$

To see whether this is true:

What values are known?

$$I_1 = 1\text{ A}$$

$$I_2 = 303.03\text{ mA}$$

$$I_3 = 212.77\text{ mA}$$

What value is not known?

$$I_T = ?$$

What formula can be used?

$$I_T = I_1 + I_2 + I_3 + \ldots$$

Substitute the known values and solve for I_T:

$$I_T = I_1 + I_2 + I_3 + \ldots$$

$$I_T = 1\text{ A} + 303.03\text{ mA} + 212.77\text{ mA}$$

$$I_T = 1.52\text{ A}$$

As you can see, this is the same value of current that was calculated earlier. The current leaving the power source divides through the individual branch circuits. The branch-circuit currents then recombine and return to the power source. Therefore, the current that leaves the power source must equal the current that returns to the power source. Expressed another way, *in a parallel circuit, the sum of the individual branch currents equals the total circuit current.*

There is one other parameter of a circuit to be concerned about: the amount of power consumed by the circuit and the individual components of the circuit. Refer to Figure 16-12 again. The total circuit voltage, the total circuit current, and the total circuit resistance are known, but the total circuit power is not. Also known are the voltage drops across R_1, R_2, and R_3, the current flowing through each resistor, and the resistive value of each resistor. However, the power that each resistor must dissipate is not known. Begin by finding the total power, P_T. Recall again the three formulae for finding power:

$$P = I \times E$$

$$P = \frac{E^2}{R}$$

$$P = I^2 \times R$$

Use any of the above three formulae to determine the total circuit power. The values of I_T and E_T are known; therefore, you can use the first formula. The values of E_T and R_T are known; therefore, you can use the second formula. Finally, the values of I_T and R_T are known; therefore, you can use the third formula. For this example, use the second formula. You may want to try to solve the example using either one or both of the remaining formulae. Your results should be the same. Now find P_T.

What values are known?

$$E_T = 100\text{ V}$$

$$R_T = 65.97\ \Omega$$

What value is not known?

$$P_T = ?$$

What formula can be used?

$$P_T = \frac{E^2}{R}$$

Substitute the known values and solve for P_T:

$$P_T = \frac{E^2}{R}$$

$$P_T = \frac{100\text{ V}^2}{65.97\ \Omega}$$

$$P_T = 151.58\text{ W}$$

Therefore, this circuit consumes 151.58 watts of power. Now turn your attention to the individual resistors, R_1, R_2, and R_3. How much power must each resistor dissipate? Begin with R_1.

What values are known?

$$E_1 = 100\text{ V}$$

$$I_1 = 1\text{ A}$$

$$R_1 = 100\ \Omega$$

What value is not known?

$$P_1 = ?$$

What formula can be used? (Use any one of the following.)

$$P_1 = I_1 \times E_1$$

$$P_1 = \frac{E_1^2}{R_1}$$

$$P_1 = I_1^2 \times R_1$$

Substitute the known values and solve for P_1. Use the first equation.

$$P_1 = I_1 \times E_1$$

$$P_1 = 1\text{ A} \times 100\text{ V}$$

$$P_1 = 100\text{ W}$$

Therefore, resistor R_1 must dissipate 100 watts of heat. Now determine how much power resistor R_2 must dissipate.

What values are known?

$$E_2 = 100\text{ V}$$

$$I_2 = 303.03\text{ mA}$$

$$R_2 = 330\ \Omega$$

What value is not known?

$$P_2 = ?$$

What formula can be used? Use any one of the following.

$$P_2 = I_2 \times E_2$$

$$P_2 = \frac{E_2^2}{R_2}$$

$$P_2 = I_2^2 \times R_2$$

Substitute the known values and solve for P_2. Use the third equation.

$$P_2 = I_2^2 \times R_2$$

$$P_2 = 303.03\text{ mA}^2 \times 330\ \Omega$$

$$P_2 = 30.30\text{ W}$$

Therefore, resistor R_2 must dissipate 30.30 watts of heat. Now determine how much power resistor R_3 must dissipate.

What values are known?

$E_3 = 100\text{ V}$

$I_3 = 212.77\text{ mA}$

$R_3 = 470\ \Omega$

What value is not known?

$P_3 = ?$

What formula can be used? Use any one of the following.

$P_3 = I_3 \times E_3$

$P_3 = \frac{E_3^2}{R_3}$

$P_3 = I_3^2 \times R_3$

Substitute the known values and solve for P_3. Use the second equation.

$P_3 = \frac{E_3^2}{R_3}$

$P_3 = \frac{100\text{ V}^2}{470\ \Omega}$

$P_3 = 21.28\text{ W}$

Therefore, resistor R_3 must dissipate 21.28 watts of heat.

Now notice something interesting. If you take the wattage dissipated by R_1 and add it to the wattage dissipated by R_2 and R_3, you find the total wattage consumed by the circuit.

$P_1 = 100\text{ W}$

$P_2 = 30.30\text{ W}$

$P_3 = 21.28\text{ W}$

$P_T = P_1 + P_2 + P_3 + \ldots$

$P_T = 100\text{ W} + 30.30\text{ W} + 21.28\text{ W}$

$P_T = 151.58\text{ W}$

As you have learned earlier in the discussion of series circuits, the same is true for a parallel circuit: *The sum of all individual wattages equals the total circuit wattage.*

16-3 COMBINATION CIRCUITS

As the name implies, a **combination circuit** consists of components that are connected in both series and parallel. **Figure 16-13** and **Figure 16-14** show two types of combination circuits. In Figure 16-13, resistor R_1 is connected in series with the parallel combination of resistors R_2 and R_3. In Figure 16-14, resistors R_1 and R_2 are connected in series. This series combination is then paralleled by resistor R_3. Look at each of these circuits in more detail so that you may better understand how to analyze them.

The circuit in Figure 16-13 consists of a power source (indicated by the AC sine wave), conductors, three loads (R_1, R_2, and R_3), and a control device (the switch). When you look at this circuit, imagine yourself as the current flowing from the power source. As you leave the power source, you flow through the conductor to the switch. In Figure 16-13, the switch is closed. The current can now flow through the closed switch, through the conductor to resistor R_1, and through resistor R_1 to the junction at A. At junction A, the current can take one of two paths. The current can flow through resistor R_2 or continue toward resistor R_3. Actually, the current does both. The current *divides*, with some of the current flowing through R_2 and some of the current flowing through R_3. The current flowing through R_2 and R_3

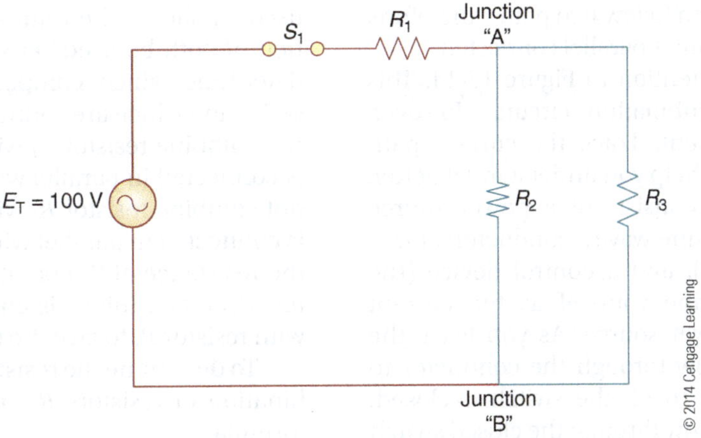

FIGURE 16-13 One form of a combination circuit. Resistor R_1 is connected in series with the parallel combination of resistors R_2 and R_3.

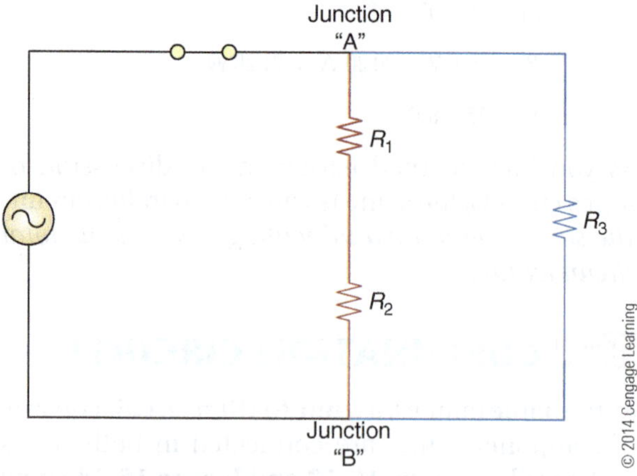

FIGURE 16-14 Another form of a combination circuit. Resistors R_1 and R_2 are connected in series then paralleled by resistor R_3.

must return to the power source. Therefore, after the current flows through R_3, it combines at junction B with the current flowing through R_2. The total recombined current then flow through the conductor back to the power source. Notice that the current could also flow in the opposite direction; that is, from the power source through the conductor to junction B, where the current divides. Some of the current flows through R_2, and some of the current flows through R_3. The current from R_3 combines with the current from R_2 at junction A. This current then flows through R_1, through the conductor to the closed switch, through the switch, and back to the power source. Notice in this circuit that there is a portion of the circuit where all of the circuit current must flow, between the power source and junction A and between the power source and junction B. The current has the same value at either one of these two points. Because this is the only path through which current can flow, these portions of the circuit are series connected. At junctions A and B, the current can follow two paths; therefore, this portion of the circuit is parallel connected.

Now turn your attention to Figure 16-14. This circuit is also a combination circuit. However, it looks a little different. Trace the current path through this circuit to help you understand it better.

This circuit also consists of a power source (indicated by the AC sine wave), conductors, three loads (R_1, R_2, and R_3), and a control device (the switch). Again, imagine yourself as the current flowing from the power source. As you leave the power source, you flow through the conductor to the switch. In Figure 16-14, the switch is closed. The current can now flow through the closed switch through the conductor to junction A. At junction A, the current can take one of two paths: Some of the current flows through resistor R_1, through resistor R_2, to the junction at B; some of the current flows from junction A, through resistor R_3, to junction B. The current flowing through resistor R_1 and resistor R_2 has only one path to follow. This means that resistors R_1 and R_2 must be connected in series. This also means that the current flowing through R_1 must be the same value as the current flowing through R_2. Because the current divided at junction A, resistor R_3 is connected in parallel with the series combination of R_1 and R_2. The current flowing through the R_1, R_2 branch and the current flowing through the R_3 branch must return to the power source. These two branch currents combine at junction B. The total combined current then flows through the conductor back to the power source. Notice that the current could also flow in the opposite direction, that is, from the power source through the conductor to junction B, where the current divides. Some of the current flows through the R_1, R_2 series branch, and some of the current flows through the R_3 branch. The current from R_3 combines with the current from the R_1, R_2 series branch at junction A. This current then flows through the conductor to the closed switch, through the switch, and back to the power source.

Once the circuit has been identified as being a combination circuit, other useful information about the circuit can be determined. Refer to **Figure 16-15**. Figure 16-15 is the same circuit shown in Figure 16-13, except that values have been added for the amount of voltage present as well as the values of R_1, R_2, and R_3. The same values that were used in the previous examples are used again so that you can see the effects of connecting components in a combination circuit. Because the amount of voltage present is known, can the amount of current flowing in the circuit be determined?

Trying to determine the amount of current flowing in this circuit is difficult. What should be used for the value of R_T? Should the value of R_1 be used, or should the value of R_2 or R_3 or a combination of both be used? First, look at the circuit and determine which components are connected in series and which are connected in parallel. You cannot combine resistor R_1 with R_2 because resistor R_3 is connected in parallel with R_2. Likewise, you cannot combine resistor R_1 with R_3 because resistor R_2 is connected in parallel with R_3. First, you must find the resistance of the parallel combination of resistors R_2 and R_3. This resistance can then be combined with resistor R_1 to give the total circuit resistance.

To determine the resistance of the parallel combination of resistors R_2 and R_3, use the following formula:

$$R_A = \frac{1}{\frac{1}{R_2} + \frac{1}{R_3}}$$

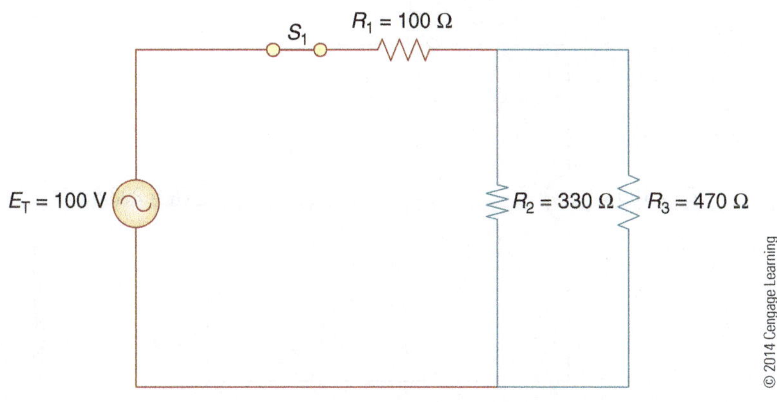

FIGURE 16-15 A combination circuit with values for each resistor and total circuit voltage.

Use R_A to represent the parallel combination of R_2 and R_3.

What values are known?

$R_2 = 330\ \Omega$

$R_3 = 470\ \Omega$

What value is not known?

$R_A = ?$

What formula can be used?

$$R_A = \frac{1}{\frac{1}{R_2} + \frac{1}{R_3}}$$

Substitute the known values and solve for R_T:

$$R_A = \frac{1}{\frac{1}{R_2} + \frac{1}{R_3}}$$

$$R_A = \frac{1}{\frac{1}{330\ \Omega} + \frac{1}{470\ \Omega}}$$

$$R_A = 193.88\ \Omega$$

Therefore, the resistance of the parallel combination of R_2 and R_3 is 193.88 ohms. Now redraw the circuit shown in Figure 16-15 to reflect the combination of R_2 and R_3. The redrawn circuit appears in **Figure 16-16**. Notice that resistor R_1 is connected in series with R_A, the parallel combination of R_2 and R_3. To find the total circuit resistance, R_T, simply add the value of R_1 to the value of R_A.

What values are known?

$R_1 = 100\ \Omega$

$R_A = 193.88\ \Omega$

What value is not known?

$R_T = ?$

What formula can be used?

$R_T = R_1 + R_A$

Substitute the known values and solve for R_T:

$R_T = R_1 + R_A$

$R_T = 100\ \Omega + 193.88\ \Omega$

$R_T = 293.88\ \Omega$

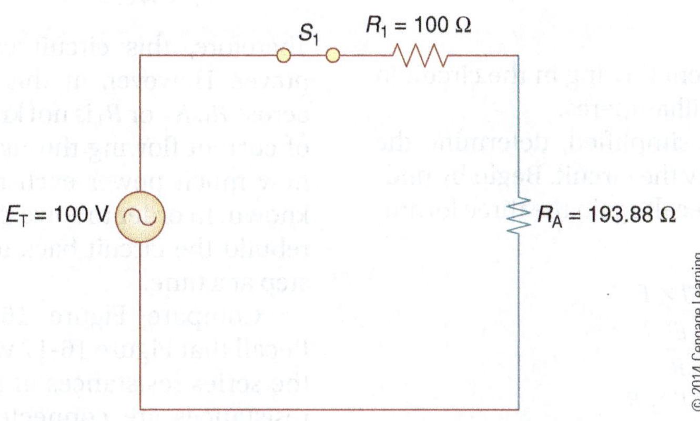

FIGURE 16-16 Parallel resistors R_2 and R_3 have been combined into R_A.

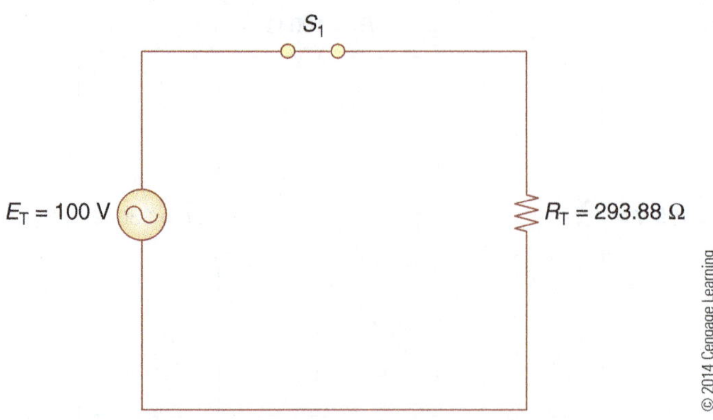

FIGURE 16-17 Series resistances R_1 and R_A have been combined into R_T.

Therefore, the total circuit resistance in Figure 16-15 is 293.88 ohms. Now redraw the circuit shown in Figure 16-16 to reflect the combination of R_1 and R_A. The redrawn circuit appears in **Figure 16-17**. Notice that this circuit is very simple. The circuit now consists of a power source, a switch, conductors, and one resistor, R_T.

With the circuit simplified in this manner, the total circuit current can be determined.

What values are known?

$$E_T = 100\text{ V}$$

$$R_T = 293.88\ \Omega$$

What value is not known?

$$I_T = ?$$

What formula can be used?

$$I_T = \frac{E_T}{R_T}$$

Substitute the known values and solve for I_T:

$$I_T = \frac{E_T}{R_T}$$

$$I_T = \frac{100\text{ V}}{293.88\ \Omega}$$

$$I_T = 340.27\text{ mA}$$

Therefore, the total current flowing in the circuit in Figure 16-17 is 340.27 milliamperes.

While the circuit is simplified, determine the total power consumed by the circuit. Begin by finding the total power, P_T. Recall again the three formulae for finding power:

$$P = I \times E$$

$$P = \frac{E^2}{R}$$

$$P = I^2 \times R$$

You can use any of the three formulae to determine the total circuit power. The values of I_T and E_T are known; therefore, you can use the first formula. The values of E_T and R_T are known; therefore, you can use the second formula. Finally, the values of I_T and R_T are known; therefore, you can use the third formula. For this example, use the second formula. You may want to try to solve the example using either one or both of the remaining formulae. Your results should be the same. Now find P_T.

What values are known?

$$E_T = 100\text{ V}$$

$$R_T = 293.88\ \Omega$$

What value is not known?

$$P_T = ?$$

What formula can be used?

$$P_T = \frac{E^2}{R}$$

Substitute the known values and solve for P_T:

$$P_T = \frac{E^2}{R}$$

$$P_T = \frac{100\text{ V}^2}{293.88\ \Omega}$$

$$P_T = 34.03\text{ W}$$

Therefore, this circuit consumes 34.03 watts of power. However, at this point, the voltage drop across R_1, R_2, or R_3 is not known. Neither the amount of current flowing through the three resistors, nor how much power each resistor must dissipate is known. In order to solve these unknowns, you must rebuild the circuit back to the original circuit one step at a time.

Compare Figure 16-17 with Figure 16-16. Recall that Figure 16-17 was created by combining the series resistances of R_1 and R_A. Because these resistances are connected in series, the current flowing through each resistance must be the same. This means there must be 340.27 milliamperes of

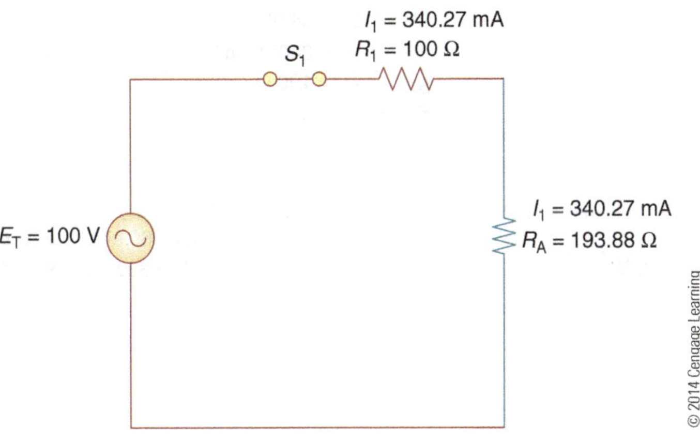

FIGURE 16-18 The current in a series circuit is the same value throughout.

current flowing through R_1 and 340.27 milliamperes of current flowing through R_A. This is shown in **Figure 16-18.**

When you look at Figure 16-18, you can see that two things are known about each resistor: the value of the resistance and the amount of current flowing through each resistance. This means that the amount of voltage drop across each resistance can now be determined.

What values are known?

$I_1 = 340.27 \text{ mA}$

$R_1 = 100\ \Omega$

What value is not known?

$E_1 = ?$

What formula can be used?

$E_1 = I_1 \times R_1$

Substitute the known values and solve for E_1:

$E_1 = I_1 \times R_1$

$E_1 = 340.27 \text{ mA} \times 100\ \Omega$

$E_1 = 34.03 \text{ V}$

Therefore, the voltage drop across R_1 is 34.03 volts. How much voltage is dropped across resistor R_A? Use the same value for current, but you must use the resistance value for R_A.

What values are known?

$I_A = 340.27 \text{ mA}$

$R_A = 193.88\ \Omega$

What value is not known?

$E_A = ?$

What formula can be used?

$E_A = I_A \times R_A$

Substitute the known values and solve for E_A:

$E_A = I_A \times R_A$

$E_A = 340.27 \text{ mA} \times 193.88\ \Omega$

$E_A = 65.97 \text{ V}$

Therefore, the voltage drop across R_A is 65.97 volts. Now check your work. Recall that the sum of the individual voltage drops in a series circuit must total the applied voltage. Is this true in this example?

$E_1 = 34.03 \text{ V}$

$E_A = 65.97 \text{ V}$

$E_T = E_1 + E_A$

$E_T = 34.03 \text{ V} + 65.97 \text{ V}$

$E_T = 100 \text{ V}$

The sum of the voltage drops is equal to the applied voltage. This means that you are on the right track. Now turn your attention to resistor R_A. Recall that the value of this resistor was a result of the parallel combination of resistors R_2 and R_3. Now rebuild the circuit back to the original circuit by separating resistor R_A back into resistors R_2 and R_3.

Because resistors R_2 and R_3 are connected in parallel, the voltage drop across R_2 and R_3 must be the same value. This is the value of the voltage that is dropped across resistor R_A. In other words, there are 65.97 volts dropped across resistor R_2, and 65.97 volts dropped across resistor R_3.

Figure 16-19 shows the circuit as it originally appeared, only now all of the known values have been added. If you turn your attention to resistors R_2 and R_3, you see that two things about each of these resistors are now known: the voltage drop across each resistor and the amount of resistance of each resistor. You are therefore able to determine the

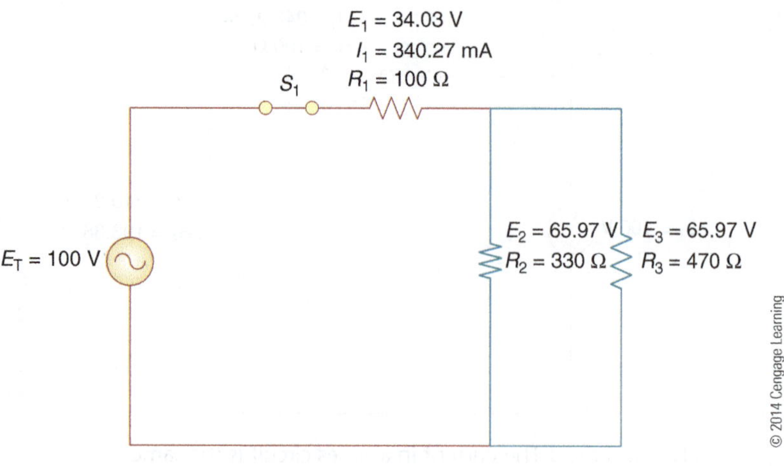

FIGURE 16-19 The voltage is the same in a parallel circuit.

amount of current flowing through each resistor. Begin with resistor R_2:

What values are known?

$$E_2 = 65.97\text{ V}$$

$$R_2 = 330\ \Omega$$

What value is not known?

$$I_2 = ?$$

What formula can be used?

$$I_2 = \frac{E_2}{R_2}$$

Substitute the known values and solve for I_2:

$$I_2 = \frac{E_2}{R_2}$$

$$I_2 = \frac{65.97\text{ V}}{330\ \Omega}$$

$$I_2 = 199.91\text{ mA}$$

Therefore, the current flowing through resistor R_2 in the circuit in Figure 16-19 is 199.91 milliamperes. Now determine the amount of current flowing through resistor R_3:

What values are known?

$$E_3 = 65.97\text{ V}$$

$$R_3 = 470\ \Omega$$

What value is not known?

$$I_3 = ?$$

What formula can be used?

$$I_3 = \frac{E_3}{R_3}$$

Substitute the known values and solve for I_3:

$$I_3 = \frac{E_3}{R_3}$$

$$I_3 = \frac{65.97\text{ V}}{470\ \Omega}$$

$$I_3 = 140.36\text{ mA}$$

Therefore, the current flowing through resistor R_3 in the circuit in Figure 16-19 is 140.36 milliamperes.

Now notice something interesting. Follow the current from the power source, through the switch and the resistors, and back to the power source. Refer to **Figure 16-20**.

When the switch is closed, 340.27 milliamperes of current flows from the power source through the switch and through resistor R_1 to junction A. At junction A, this current divides: 199.91 milliamperes of current flows through resistor R_2 toward junction B and 140.36 milliamperes of current flows from junction A through resistor R_3 toward junction B. At junction B, the 199.91 milliamperes of current from R_2 combines with the 140.36 milliamperes of current from R_3. This produces a total of 340.27 milliamperes of current flow back to the power source. *The amount of current that left the power source is the amount of current that returned to the power source. The amount of current that flowed into junction A is the amount of current that flowed out of junction B.*

There are three final calculations to make. Earlier you calculated the total circuit power. You now need to determine the amount of power that each resistor must dissipate. Begin with R_1:

What values are known?

$$E_1 = 34.03\text{ V}$$

$$I_1 = 340.27\text{ mA}$$

$$R_1 = 100\ \Omega$$

What value is not known?

$$P_1 = ?$$

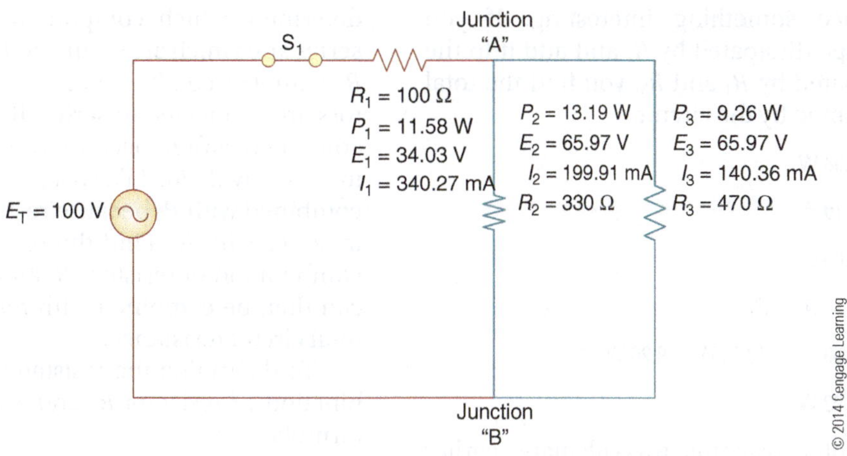

FIGURE 16-20 The sum of the branch currents equals the total circuit current.

What formula can be used? Use any one of the following.

$$P_1 = I_1 \times E_1$$

$$P_1 = \frac{E_1^2}{R_1}$$

$$P_1 = I_1^2 \times R_1$$

Substitute the known values and solve for P_1. Use the first equation.

$$P_1 = I_1 \times E_1$$

$$P_1 = 340.27 \text{ mA} \times 34.03 \text{ V}$$

$$P_1 = 11.58 \text{ W}$$

Therefore, resistor R_1 must dissipate 11.58 watts of heat. Recall that a good rule of thumb is for the resistor wattage rating to be twice the actual wattage that must be handled. Therefore, R_1 should be rated at 23.16 watts. Now determine how much power resistor R_2 must dissipate.

What values are known?

$$E_2 = 65.97 \text{ V}$$

$$I_2 = 199.91 \text{ mA}$$

$$R_2 = 330\ \Omega$$

What value is not known?

$$P_2 = ?$$

What formula can be used? Use any one of the following.

$$P_2 = I_2 \times E_2$$

$$P_2 = \frac{E_2^2}{R_2}$$

$$P_2 = I_2^2 \times R_2$$

Substitute the known values and solve for P_2. Use the third equation.

$$P_2 = I_2^2 \times R_2$$

$$P_2 = 199.91 \text{ mA}^2 \times 330\ \Omega$$

$$P_2 = 13.19 \text{ W}$$

Therefore, resistor R_2 must dissipate 13.19 watts of heat. Now determine how much power resistor R_3 must dissipate.

What values are known?

$$E_3 = 65.97 \text{ V}$$

$$I_3 = 140.36 \text{ mA}$$

$$R_3 = 470\ \Omega$$

What value is not known?

$$P_3 = ?$$

What formula can be used? Use any one of the following.

$$P_3 = I_3 \times E_3$$

$$P_3 = \frac{E_3^2}{R_3}$$

$$P_3 = I_3^2 \times R_3$$

Substitute the known values and solve for P_3. Use the second equation.

$$P_3 = \frac{E_3^2}{R_3}$$

$$P_3 = \frac{65.97 \text{ V}^2}{470\ \Omega}$$

$$P_3 = 9.26 \text{ W}$$

Therefore, resistor R_3 must dissipate 9.26 watts of heat.

Now notice something interesting. If you take the wattage dissipated by R_1 and add it to the wattage dissipated by R_2 and R_3, you find the total wattage consumed by the circuit.

$$P_1 = 11.58 \text{ W}$$

$$P_2 = 13.19 \text{ W}$$

$$P_3 = 9.26 \text{ W}$$

$$P_T = P_1 + P_2 + P_3$$

$$P_T = 11.58 \text{ W} + 13.19 \text{ W} + 9.26 \text{ W}$$

$$P_T = 34.03 \text{ W}$$

Compare the total power that was calculated earlier (34.03 watts) with the total power found by adding the individual powers (34.03 watts). This proves that in a combination circuit, the sum of all individual wattages equals the total circuit wattage.

Now turn your attention to **Figure 16-21**. Figure 16-21 is the same circuit shown earlier in Figure 16-14. You have already determined that this circuit is a combination circuit: Resistors R_1 and R_2 are connected in series and resistor R_3 is connected in parallel with the series combination of R_1 and R_2. Values have been added for the amount of voltage present as well as the values of R_1, R_2, and R_3. The same values that were used in the previous examples are used again so that you can see the effects of connecting components in a combination circuit of different designs. Because the amount of voltage present is known, can the amount of current flowing in the circuit be determined?

Trying to determine the amount of current flowing in this circuit is difficult. What should be used for the value of R_T? Should the value of R_1 be used, or should the value of R_2 or R_3 or a combination be used? First, look at the circuit and determine which components are connected in series and which are connected in parallel. Resistor R_1 cannot be combined with R_2 because these resistors are connected in series. Resistor R_1 cannot be combined with R_3 because resistor R_2 is connected in series with R_1. Likewise, resistor R_2 cannot be combined with R_3 because resistor R_1 is connected in series with R_2. Find the resistance of the series combination of resistors R_1 and R_2. This resistance can then be combined with resistor R_3 to give the total circuit resistance.

To determine the resistance of the series com bination of resistors R_1 and R_2, use the following formula:

$$R_A = R_1 + R_2$$

Use R_A to represent the series combination of R_1 and R_2.

What values are known?

$$R_1 = 100\ \Omega$$

$$R_2 = 330\ \Omega$$

What value is not known?

$$R_A = ?$$

What formula can be used?

$$R_A = R_1 + R_2$$

Substitute the known values and solve for R_A:

$$R_A = R_1 + R_2$$

$$R_A = 100\ \Omega + 330\ \Omega$$

$$R_A = 430\ \Omega$$

Therefore, the resistance of the series combination of R_1 and R_2 is 430 ohms. Now redraw the circuit shown in Figure 16-21 to reflect the combination of R_1 and R_2.

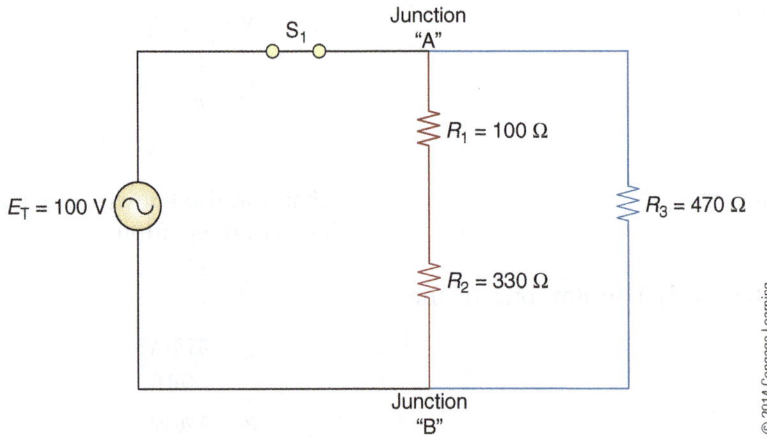

FIGURE 16-21 A combination circuit with values for each resistor and total circuit voltage.

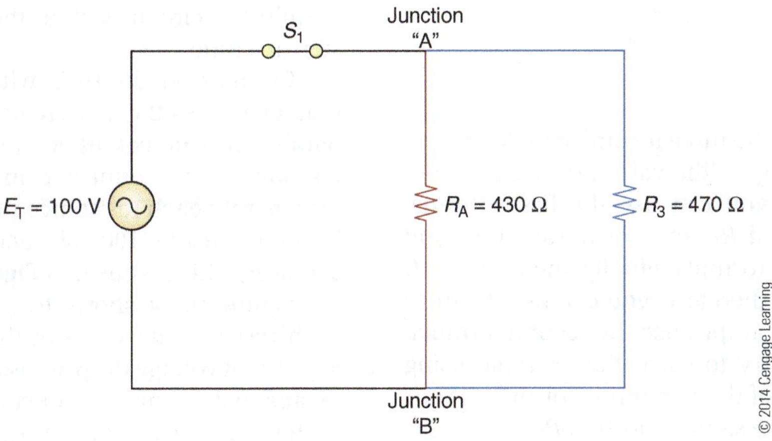

FIGURE 16-22 Series resistors R_1 and R_2 have been combined into R_A.

The redrawn circuit appears in **Figure 16-22**. Notice that resistor R_3 is connected in parallel with R_A, which is the series combination of R_1 and R_2. Now determine the total circuit resistance, R_T.

What values are known?

$R_A = 430\ \Omega$

$R_3 = 470\ \Omega$

What value is not known?

$R_T = ?$

What formula can be used?

$$R_T = \frac{1}{\frac{1}{R_A} + \frac{1}{R_3}}$$

Substitute the known values and solve for R_T:

$$R_T = \frac{1}{\frac{1}{R_A} + \frac{1}{R_3}}$$

$$R_T = \frac{1}{\frac{1}{430\ \Omega} + \frac{1}{470\ \Omega}}$$

$R_T = 224.56\ \Omega$

Therefore, the total circuit resistance in Figure 16-22 is 224.56 ohms. Now redraw the circuit shown in Figure 16-22 to reflect the combination of R_A and R_3. The redrawn circuit appears in **Figure 16-23**. Notice that this circuit is very simple. The circuit now consists of a power source, a switch, conductors, and one resistor, R_T.

With the circuit simplified in this manner, the total circuit current can now be determined.

What values are known?

$E_T = 100\ V$

$R_T = 224.56\ \Omega$

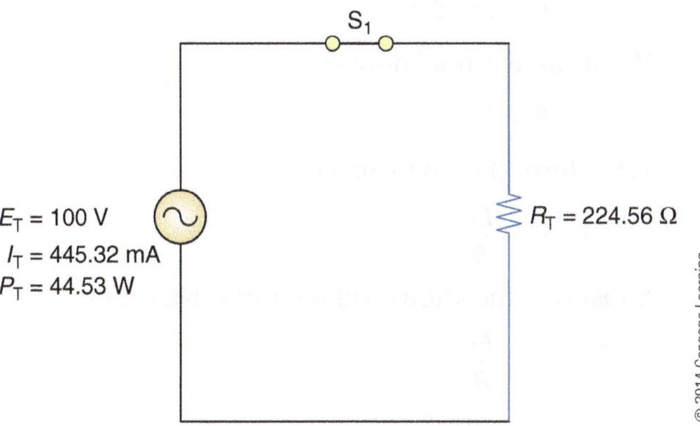

FIGURE 16-23 Parallel resistances R_3 and R_A have been combined into R_T.

What value is not known?

$I_T = ?$

What formula can be used?

$$I_T = \frac{E_T}{R_T}$$

Substitute the known values and solve for I_T:

$$I_T = \frac{E_T}{R_T}$$

$$I_T = \frac{100\ V}{224.56\ \Omega}$$

$I_T = 445.32\ mA$

Therefore, the total current flowing in the circuit in Figure 16-23 is 445.32 milliamperes.

While the circuit is simplified, you can determine the total power consumed by the circuit. Begin by finding the total power, P_T. Recall again the three formulae for finding power:

$P = I \times E$

$$P = \frac{E^2}{R}$$

$$P = I^2 \times R$$

You can use any of the three formulae to determine the total circuit power. The values of I_T and E_T are known; therefore, you can use the first formula. The values of E_T and R_T are known; therefore, you can use the second formula. Finally, the values of I_T and R_T are known; therefore, you can use the third formula. For this example, use the second formula. You may want to try to solve the example using either one or both of the remaining formulae. Your results should be the same. Now find P_T.

What values are known?

$$E_T = 100\ \text{V}$$

$$R_T = 224.56\ \Omega$$

What value is not known?

$$P_T = ?$$

What formula can be used?

$$P_T = \frac{E^2}{R}$$

Substitute the known values and solve for P_T:

$$P_T = \frac{E^2}{R}$$

$$P_T = \frac{100\ \text{V}^2}{224.56\ \Omega}$$

$$P_T = 44.53\ \text{W}$$

Therefore, this circuit consumes 44.53 watts of power. However, at this point, the amount of voltage drop across R_1, R_2, or R_3, the amount of current flowing through the three resistors, or the amount of power that each resistor must dissipate are not known. In order to solve these unknowns, you must rebuild the circuit back to the original circuit one step at a time.

Compare Figure 16-23 with Figure 16-22. Recall that Figure 16-23 was created by combining the parallel resistances of R_A and R_3. Because these resistances are connected in parallel, the voltage drop across each resistance must be the same; that is, there must be 100 volts across R_A, and 100 volts across R_3. This is shown in **Figure 16-24**.

Figure 16-24 shows two things known about each resistor: the value of the resistance and the amount of voltage drop across each resistance. This means that the amount of current flowing through each resistance can now be determined.

What values are known?

$$E_A = 100\ \text{V}$$

$$R_A = 430\ \Omega$$

What value is not known?

$$I_A = ?$$

What formula can be used?

$$I_A = \frac{E_A}{R_A}$$

Substitute the known values and solve for I_A:

$$I_A = \frac{E_A}{R_A}$$

$$I_A = \frac{100\ \text{V}}{430\ \Omega}$$

$$I_A = 232.56\ \text{mA}$$

Therefore, the current flowing through R_A is 232.56 milliamperes. How much current is flowing through resistor R_3? Use the same value for voltage, but you must use the resistance value for R_3.

What values are known?

$$E_3 = 100\ \text{V}$$

$$R_3 = 470\ \Omega$$

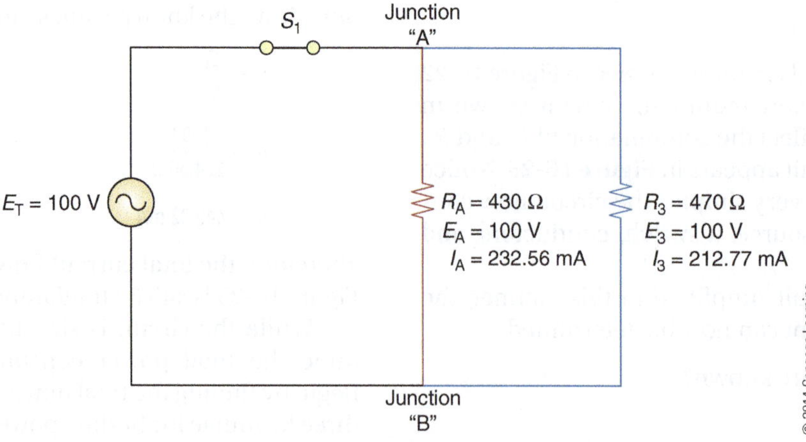

FIGURE 16-24 The voltage is the same in a parallel circuit.

What value is not known?

$I_3 = ?$

What formula can be used?

$$I_3 = \frac{E_3}{R_3}$$

Substitute the known values and solve for I_3:

$$I_3 = \frac{E_3}{R_3}$$

$$I_3 = \frac{100\ \text{V}}{470\ \Omega}$$

$$I_3 = 212.77\ \text{mA}$$

Therefore, the current flowing through R_3 is 212.77 milliamperes. Now check your work. Recall that the sum of the individual branch currents in a parallel circuit must equal the total circuit current. Is this true in this example?

$I_A = 232.56$ mA

$I_3 = 212.77$ mA

$I_T = I_A + I_3$

$I_T = 232.56\ \text{mA} + 212.77\ \text{mA}$

$I_T = 445.33$ mA

The sum of the individual branch-circuit currents is equal to the total circuit current. This means that you are on the right track. Do not be concerned with the 0.01-milliampere difference between the two answers. This is a result of the rounding off of numbers to two decimal places during calculations.

Turn your attention to resistor R_A. Recall that the value of this resistor was a result of the series combination of resistors R_1 and R_2. Now rebuild the circuit back to the original circuit by separating resistor R_A back into resistors R_1 and R_2.

Because resistors R_1 and R_2 are connected in series, the current flowing through R_1 and R_2 must be the same value. This is the value of the current flowing through resistor R_A. In other words, there are 232.56 milliamperes of current flowing through resistor R_1 and 232.56 milliamperes of current flowing through resistor R_2.

Figure 16-25 shows the circuit as it originally appeared, but now all of the known values have been added. If you turn your attention to resistors R_1 and R_2, you see that two things about each of these resistors are now known: the amount of current flowing through each resistor and the amount of resistance of each resistor. Therefore the amount of voltage drop across each resistor can be determined. Begin with resistor R_1:

What values are known?

$I_1 = 232.56$ mA

$R_1 = 100\ \Omega$

What value is not known?

$E_1 = ?$

What formula can be used?

$E_1 = I_1 \times R_1$

Substitute the known values and solve for E_1:

$E_1 = I_1 \times R_1$

$E_1 = 232.56\ \text{mA} \times 100\ \Omega$

$E_1 = 23.26$ V

Therefore, the voltage drop across resistor R_1 in the circuit in Figure 16-25 is 23.26 volts. Now determine the amount of voltage drop across resistor R_2:

What values are known?

$I_2 = 232.56$ mA

$R_2 = 330\ \Omega$

What value is not known?

$E_2 = ?$

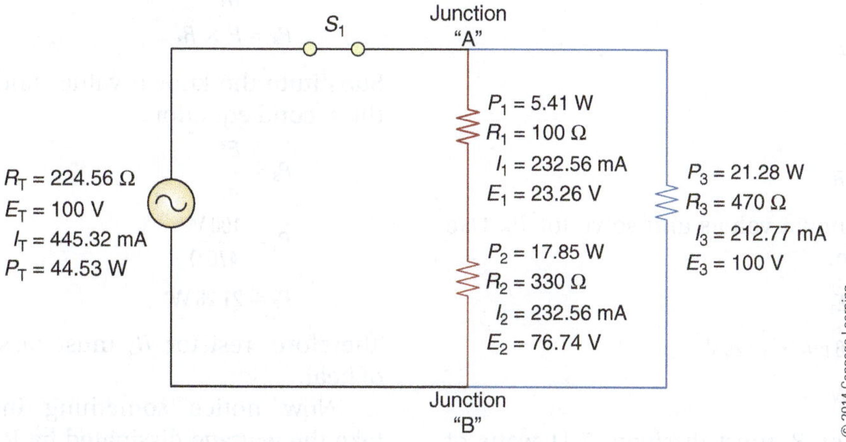

FIGURE 16-25 The current in a series circuit is the same value throughout.

What formula can be used?

$$E_2 = I_2 \times R_2$$

Substitute the known values and solve for E_2:

$$E_2 = I_2 \times R_2$$

$$E_2 = 232.56 \text{ mA} \times 330\ \Omega$$

$$E_2 = 76.74 \text{ V}$$

Therefore, the voltage drop across resistor R_2 in the circuit in Figure 16-25 is 76.74 volts.

Now notice something interesting. This circuit is supplied with 100 volts. Imagine placing one probe of your voltmeter at the top of R_1 and then placing the other probe of your voltmeter at the bottom of R_2. Now look at what you are actually measuring. Follow the connection from the top of R_1 to the left to the top of the power source. Now follow the connection from the bottom of R_2 to the left to the bottom of the power source. You have actually placed your voltmeter across the power source. This means that you actually measure the voltage of the power source, 100 volts. Now notice the voltage drops across R_1 and R_2. R_1 has a voltage drop of 23.26 volts. R_2 has a voltage drop of 76.74 volts. Your meter is measuring the total voltage across resistors R_1 and R_2, or 23.26 volts + 76.74 volts or 100 volts!

Three final calculations must be made. Earlier the total circuit power was calculated. You now need to determine the amount of power that each resistor must dissipate. Begin with R_1:

What values are known?

$$E_1 = 23.26 \text{ V}$$

$$I_1 = 232.56 \text{ mA}$$

$$R_1 = 100\ \Omega$$

What value is not known?

$$P_1 = ?$$

What formula can be used? Use any one of the following.

$$P_1 = I_1 \times E_1$$

$$P_1 = \frac{E_1^2}{R_1}$$

$$P_1 = I_1^2 \times R_1$$

Substitute the known values and solve for P_1. Use the first equation.

$$P_1 = I_1 \times E_1$$

$$P_1 = 232.56 \text{ mA} \times 23.26 \text{ V}$$

$$P_1 = 5.41 \text{ W}$$

Therefore, resistor R_1 must dissipate 5.41 watts of heat. Now determine how much power resistor R_2 must dissipate.

What values are known?

$$E_2 = 76.74 \text{ V}$$

$$I_2 = 232.56 \text{ mA}$$

$$R_2 = 330\ \Omega$$

What value is not known?

$$P_2 = ?$$

What formula can be used? Use any one of the following.

$$P_2 = I_2 \times E_2$$

$$P_2 = \frac{E_2^2}{R_2}$$

$$P_2 = I_2^2 \times R_2$$

Substitute the known values and solve for P_2. Use the third equation.

$$P_2 = I_2^2 \times R_2$$

$$P_2 = 232.56 \text{ mA}^2 \times 330\ \Omega$$

$$P_2 = 17.85 \text{ W}$$

Therefore, resistor R_2 must dissipate 17.85 watts of heat. Now determine how much power resistor R_3 must dissipate.

What values are known?

$$E_3 = 100 \text{ V}$$

$$I_3 = 212.77 \text{ mA}$$

$$R_3 = 470\ \Omega$$

What value is not known?

$$P_3 = ?$$

What formula can be used? Use any one of the following.

$$P_3 = I_3 \times E_3$$

$$P_3 = \frac{E_3^2}{R_3}$$

$$P_3 = I_3^2 \times R_3$$

Substitute the known values and solve for P_3. Use the second equation.

$$P_3 = \frac{E_3^2}{R_3}$$

$$P_3 = \frac{100 \text{ V}^2}{470\ \Omega}$$

$$P_3 = 21.28 \text{ W}$$

Therefore, resistor R_3 must dissipate 21.28 watts of heat.

Now notice something interesting. If you take the wattage dissipated by R_1 and add it to the wattage dissipated by R_2 and R_3, you find the total wattage consumed by the circuit.

$P_1 = 5.41\text{ W}$

$P_2 = 17.85\text{ W}$

$P_3 = 21.28\text{ W}$

$P_T = P_1 + P_2 + P_3$

$P_T = 5.41\text{ W} + 17.85\text{ W} + 21.28\text{ W}$

$P_T = 44.54\text{ W}$

Compare the total power that was calculated earlier (44.53 watts) with the total power found by adding the individual powers (44.54 watts). This proves that in a combination circuit, the sum of all individual wattages equals the total circuit wattage. Again, the difference is caused by the rounding off of numbers during calculations.

SUMMARY

- A circuit contains a minimum of three elements: a power source, a complete path for current flow, and a load.
- A power source can be a battery, a power supply, or a generator.
- Conductors are used to provide a path for the current flow. They generally are in the form of wires, which connect the power source to the load. Conductors may be the copper traces, or tracks, on a printed circuit board.
- A load is any device that draws current from the power source. A load could be a resistor, a lightbulb, or a motor.
- A series circuit has only one path for current flow.
- The total resistance of a series circuit is equal to the sum of the individual resistances.
- In a series circuit, the current is the same value throughout the entire circuit. The value of the current flowing through each individual component is equal to the total circuit current.
- A parallel circuit has two or more paths for current flow.
- In a parallel circuit, the total circuit resistance is smaller than the smallest branch resistance.
- The total resistance of a parallel circuit is equal to the reciprocal of the sum of the reciprocals of the individual resistances.
- In a parallel circuit, the voltage across the individual branch circuits is the same.
- In a parallel circuit, the sum of the individual branch currents equals the total circuit current.
- Combination circuits consist of components that are connected in both series and parallel.

REVIEW QUESTIONS

1. State a rule about current, voltage, resistance, and power as it applies to series circuits.
   a. Current: ______________________
   b. Voltage: ______________________
   c. Resistance: ______________________
   d. Power: ______________________
2. State a rule about current, voltage, resistance, and power as it applies to parallel circuits.
   a. Current: ______________________
   b. Voltage: ______________________
   c. Resistance: ______________________
   d. Power: ______________________
3. Given the circuit in **Figure 16-26**, find the unknown values.
4. Given the circuit in **Figure 16-27**, find the unknown values.
5. Given the circuit in **Figure 16-28**, find the unknown values.

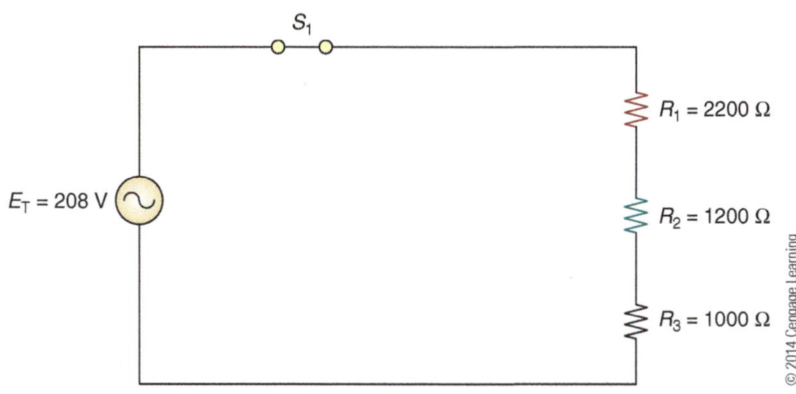

FIGURE 16-26

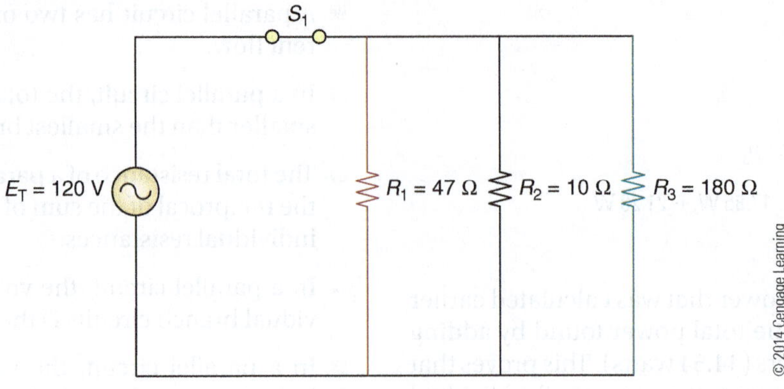

FIGURE 16-27

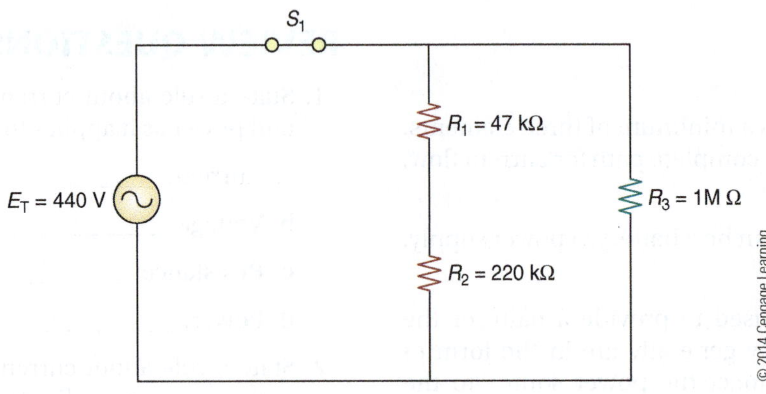

FIGURE 16-28

CHAPTER

17

Reactive Circuits and Power Factor

Very few electrical circuits are purely resistive in nature. Practically all electrical circuits contain resistance and reactive components. This chapter takes you deeper into the most common types of electrical circuits. You learn about inductors and capacitors and how they affect current, voltage, and power in a circuit. In addition, you are introduced to a procedure called *power factor correction*, which enables motors to operate more efficiently.

OBJECTIVES

After studying this chapter, the student should be able to

- Identify inductance and R-L circuits.
- Identify capacitance and R-C circuits.
- Identify series and parallel R-L-C circuits.
- Perform all necessary calculations to analyze a resistive-reactive electrical circuit.
- Correct the power factor of single-phase and three-phase motors.

17-1 INDUCTANCE AND R-L CIRCUITS

Whenever a current flows through a conductor, a **magnetic field** is formed around the conductor. This is shown in **Figure 17-1**. If the current flowing through the conductor increases in intensity, the magnetic field becomes stronger. Likewise, if the current flow decreases in intensity, the magnetic field becomes weaker. When an alternating current flows through a conductor, the current increases and decreases at a rate equal to double the frequency of the applied AC. This means that the magnetic field strengthens (or expands) and weakens (or collapses) at a similar rate in response to the current alterations.

When the magnetic field weakens or collapses, the magnetic lines of force cut the conductor, inducing a voltage into it. The amount of induced voltage caused by the changing current is called **inductance.** Inductance is represented in formulae by the letter *L* and is measured in **henrys**.

The induced voltage opposes the applied voltage, thus providing an opposition to the current flow. This opposition is called **inductive reactance**, which is represented in formulae by X_L and is measured in ohms.

The amount of inductance, and thereby the amount of inductive reactance, can be increased by forming a conductor into a coil. This is a result of increased concentration of the magnetic field caused by forming the conductor into the coil. With a more concentrated magnetic field, there is more induced voltage, causing more opposition to current flow. This is shown in the following formula for determining inductive reactance:

$$X_L = 2\pi fL$$

2 = a constant
π = 3.14 (a constant)
f = frequency, measured in hertz (Hz)
L = inductance, measured in henrys (H)

The formula shows that two factors directly influence the amount of inductive reactance of a coil: the frequency of the applied AC voltage and the amount of inductance. In other words, if the inductance is increased, the inductive reactance increases proportionately. If the frequency is increased, the inductive reactance also increases proportionately.

An interesting effect is seen when a circuit consists of an AC source and an inductor. For the purposes of this discussion assume the inductor to be perfect; that is, there is no resistance in the wire that is used to create the inductor. Recall that the magnetic field expands and contracts as the alternating current increases and decreases. The expanding and collapsing magnetic field induces a voltage that *opposes* the applied voltage. In looking at this in more detail, you see that after the applied voltage reaches its peak or maximum value, the current flowing through the inductor is at its minimum value. This is a result of the maximum value of inductive reactance caused by the magnetic field's expansion to its maximum value.

Look at this in another way. Imagine that the voltage applied across the inductor has increased to its maximum or peak value. Therefore, the magnetic field around the conductors has also increased in strength. As the magnetic field increases in strength, more opposition to current flow is created. Therefore, the current flowing through the inductor does not reach its peak or maximum value at the same time that the applied voltage does.

As the applied voltage decreases toward the zero level, the magnetic field collapses. As the magnetic field collapses, the opposition to current flow decreases. The decrease in opposition allows current flow to increase. Current flow is at its maximum or peak value when its opposition is at its minimum value. The minimum opposition occurs when the applied voltage is also at the minimum value. Therefore, when the applied voltage is at its minimum, the current is at its maximum, and when the applied voltage is at its maximum, the current is at its minimum.

This means that the current and voltage are no longer *in phase* because they are in an

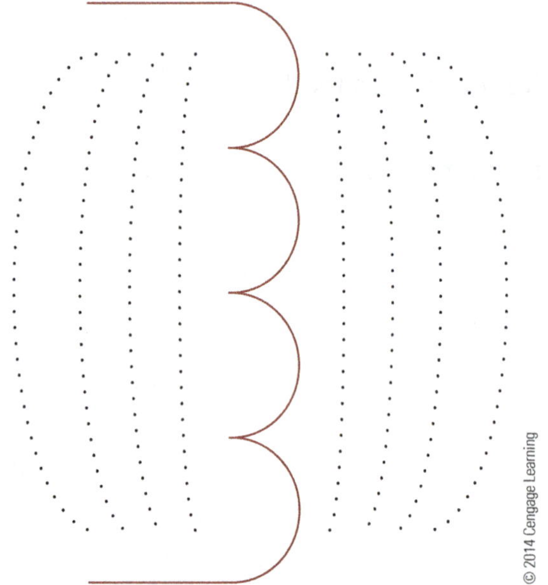

FIGURE 17-1 Whenever a current flows through a conductor, a magnetic field is formed around the conductor.

inductive circuit. Recall that in a resistive circuit, when the applied voltage is at its maximum value, the current is also at its maximum value. When the applied voltage is at its minimum value, the current is also at its minimum value. The current and voltage are therefore in phase because the current and voltage peak at the same time.

In a circuit containing only pure inductance, the current and voltage do not peak at the same time. In fact, the current reaches its peak value *after* the voltage has peaked, which in a purely inductive circuit occurs 90° after the voltage peaks. Therefore, *in a purely inductive circuit, the current lags the voltage by 90°*. To help you remember when the current reaches its peak, memorize the phrase *ELI the ICE man*. The letters *E*, *L*, and *I* in the word *ELI* are used to represent *voltage*, *inductance*, and *current*, respectively. Notice that the letter *I* appears after the letter *E*. This should help you remember that *current lags, or follows, the voltage* in an inductor. The word *ICE* is discussed later in this chapter.

The out-of-phase relationship between the current and voltage in an inductor creates another interesting occurrence. Recall one of the formulae for finding power $P = I \times E$. Recall also that current and voltage are in phase in a purely resistive circuit. If the value of the current is positive, and the value of the voltage is also positive, then the amount of power will be a positive value. If the current and voltage are both negative in value, the amount of power will be positive in value. **Figure 17-2** shows the voltage and current waveforms for a purely resistive circuit.

Notice that within the first 180°, the voltage and current are both positive in value. Therefore, the power is also positive ($P = I \times E$). In the second 180°, the voltage and current are both negative in value, but the value of the power is still positive ($P = -I \times -E$). Positive power means that power is taken from the circuit. If power is used, work is done.

Figure 17-3 shows a circuit that contains pure inductance. Notice that during the first 90°, the voltage is positive in value, but the current is negative in value. Therefore, the power is negative in value ($-P = -I \times E$). During the second 90°, the voltage and current are positive in value, so the power is positive in value as well ($P = I \times E$). During the third 90°, the voltage is negative in value, but the current is positive in value. Therefore, the power is negative in value ($-P = I \times -E$). Finally, during the fourth 90°, both the voltage and the current are negative in value; therefore, the power is again positive in value ($P = -I \times -E$).

Notice that positive power is produced during the second and fourth 90°. During the first and third 90°, the circuit is producing **negative power**. This is called **reactive power** and is measured in units called **volt-amps-reactive (VARs)**.

During the 360° of an AC sine wave, an inductor uses power from the circuit during the second and

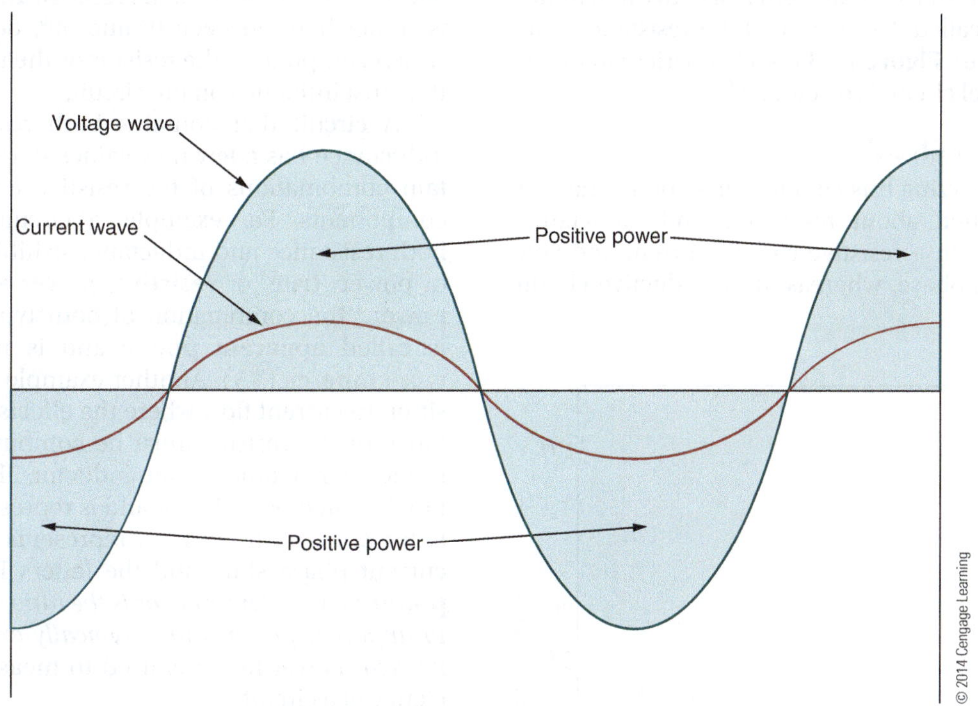

FIGURE 17-2 In a purely resistive circuit, the current and voltage are in phase.

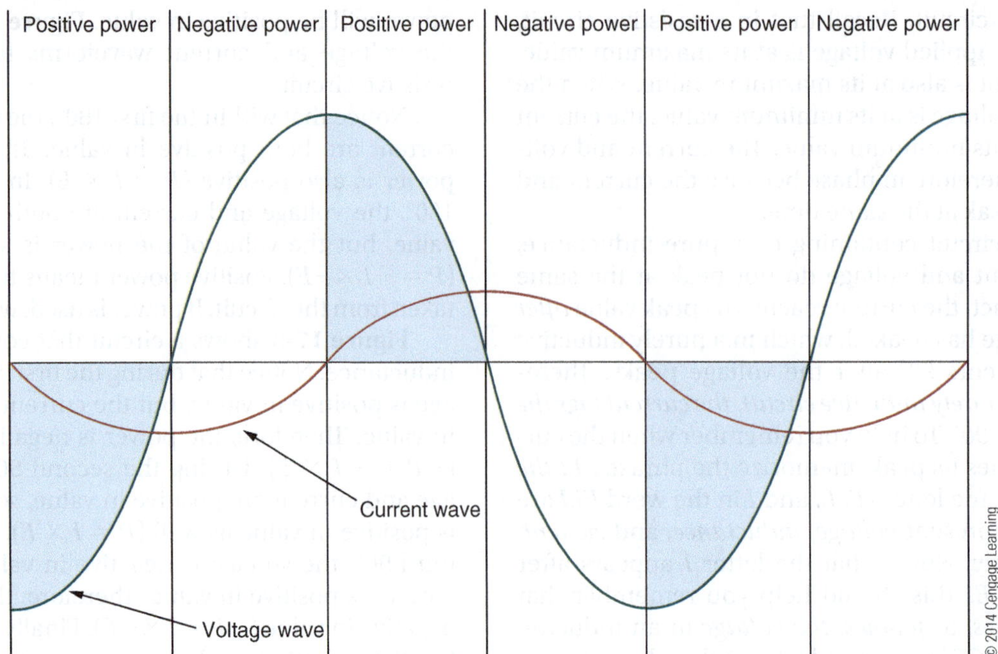

FIGURE 17-3 In a purely inductive circuit, the current lags the voltage by 90°. Positive power is produced during the second and fourth 90°. During the first and third 90°, the circuit is producing negative power.

fourth 90°. As a result of the collapsing magnetic field, an inductor returns power to the circuit during the first and third 90°. The net effect of this is that no power is used. If no power is used, no work is done. Therefore, the energy is wasted.

Circuits that contain both resistance and inductance are called **R-L circuits** (R for resistance, L for inductance). **Figure 17-4** is an R-L series circuit. An R-L parallel circuit is discussed later.

R-L Series Circuit

Before analyzing this circuit, think about what you have learned about resistance and inductance. Recall that in a resistive circuit, current and voltage are in phase, whereas in an inductive circuit, current and voltage are out of phase by 90°. When a resistance and an inductance are combined in one circuit, the current and voltage are not in phase because of the effects of the inductor. Likewise, the current and voltage are not out of phase by 90° because of the effects of the resistor. The phase shift is somewhere between 0° and 90°, depending on which component, the resistor or the inductor, has the most influence on the circuit.

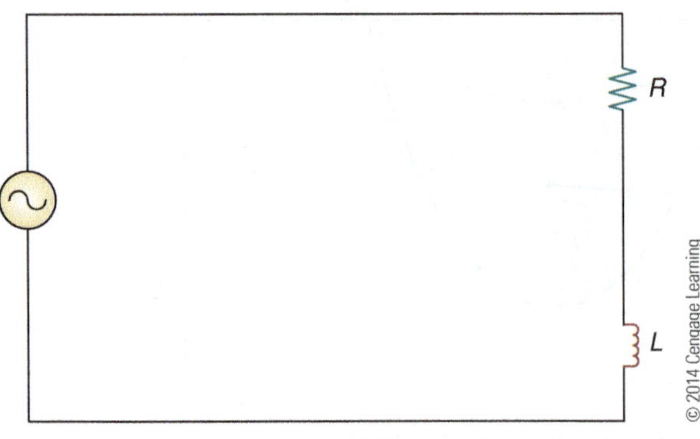

FIGURE 17-4 An R-L series circuit.

A circuit that contains both resistance and inductance has a few new values to represent certain combinations of the resistive and inductive components. For example, a circuit containing both resistance and inductance exhibits two types of power: **true**, or **resistive**, **power** and **reactive power**. The combination of both types of power is called **apparent power** and is measured in **volt-amperes (VA)**. Another example is the opposition to current flow where the effects of the resistance of the resistor must be combined with the inductive reactance of the inductor. The combination is called **impedance** and is represented by the letter Z. **Angle theta ($\angle\theta$)** represents the voltage-current phase shift, and the letters PF represent **power factor**. *Power factor is the ratio of true power to apparent power and is typically expressed as a percent.* Power factor is used to measure the efficiency of a circuit.

To help you better understand this, solve the following example problem. **Figure 17-5** shows an

FIGURE 17-5 An R-L series circuit with values.

R-L series circuit with values. Solve for the unknown values.

What values are known?

$E_T = 120\text{ V}$

$f = 60\text{ Hz}$

$R = 15\ \Omega$

$L = 20\text{ mH}$

What values are not known?

$I_T = ?$

$Z_T = ?$

$VA = ?$

$PF = ?$

$\angle\theta = ?$

$E_R = ?$

$I_R = ?$

$P_R = ?$

$E_L = ?$

$I_L = ?$

$X_L = ?$

$VARs_L = ?$

Begin by solving for X_L.

What values are known?

$f = 60\text{ Hz}$

$L = 20\text{ mH}$

What value is not known?

$X_L = ?$

What formula can be used

$X_L = 2\pi fL$

Substitute the known values and solve for X_L.

$X_L = 2\pi fL$

$X_L = 2 \times \pi \times 60\text{ Hz} \times 20\text{ mH}$

$X_L = 7.54\ \Omega$

Therefore, the inductor presents 7.54 ohms of opposition to the current flowing in this circuit. Now that the opposition presented by the inductor as well as by the resistor is known, you can determine the total opposition (impedance) of the circuit. Find Z_T.

What values are known?

$R = 15\ \Omega$

$X_L = 7.54\ \Omega$

What value is not known?

$Z_T = ?$

As a result of the phase shift introduced by the inductor, the resistance of the resistor cannot be added to the inductive reactance of the inductor. A different formula that takes into account the effects of the inductor must be used.

What formula can be used?

$Z_T = \sqrt{R^2 + X_L^2}$

Substitute the known values and solve for Z_T.

$Z_T = \sqrt{R^2 + X_L^2}$

$Z_T = \sqrt{15\ \Omega^2 + 7.54\ \Omega^2}$

$Z_T = \sqrt{225\ \Omega + 56.85\ \Omega}$

$Z_T = \sqrt{281.85\ \Omega}$

$Z_T = 16.79\ \Omega$

Thus, the combined opposition to the current of the circuit is 16.79 ohms. Now that you know the total opposition and the total applied voltage, you can determine the total circuit current.

What values are known?

$E_T = 120\ V$

$Z_T = 16.79\ \Omega$

What value is not known?

$I_T = ?$

What formula can be used?

$$I_T = \frac{E_T}{Z_T}$$

Notice that this formula looks very similar to the Ohm's law formula, $I_T = \frac{E_T}{R_T}$. Because this circuit contains both resistance and inductance, use a formula that represents the combination of resistance and inductive reactance. The Ohm's law formula is changed slightly to reflect the total opposition (Z) to current flow.

Substitute the known values and solve for I_T.

$$I_T = \frac{E_T}{Z_T}$$

$$I_T = \frac{120\ V}{16.79\ \Omega}$$

$I_T = 7.15\ A$

Therefore, there are 7.15 amperes of current flowing in this R-L series circuit. As you have learned previously, this is the same amount of current flowing throughout this circuit. This means that there are 7.15 amperes of current flowing through the resistor, and there are 7.15 amperes of current flowing through the inductor. Therefore, you now know that

$I_R = 7.15\ A$

$I_L = 7.15\ A$

Because in a series circuit,

$I_T = I_R = I_L$

Now focus your attention on the resistor. You know the value of the resistor (15 ohms) and the amount of current flowing through the resistor (7.15 amperes). Determine the amount of voltage dropped across the resistor, E_R.

What values are known?

$R = 15\ \Omega$

$I_R = 7.15\ A$

What value is not known?

$E_R = ?$

What formula can be used?

$E_R = I_R \times R$

Substitute the known values and solve for E_R.

$E_R = I_R \times R$

$E_R = 7.15\ A \times 15\ \Omega$

$E_R = 107.25\ V$

Therefore, there are 107.25 volts dropped across the resistor. The last unknown value for the resistor is the amount of power that the resistor must handle.

What values are known?

$R = 15\ \Omega$

$I_R = 7.15\ A$

What value is not known?

$P_R = ?$

What formula can be used?

$P_R = I_R^2 \times R$

Substitute the known values and solve for P_R.

$P_R = I_R^2 \times R$

$P_R = 7.15\ A^2 \times 15\ \Omega$

$P_R = 51.12\ A \times 15\ \Omega$

$P_R = 766.84\ W$

Therefore, the resistor consumes 766.84 watts of power.

Now turn your attention to the inductor. So far, the values of L, X_L, and I_L are known. Find the amount of voltage dropped across the inductor, E_L.

What values are known?

$I_L = 7.15\ A$

$X_L = 7.54\ \Omega$

What value is not known?

$E_L = ?$

What formula can be used?

$E_L = I_L \times X_L$

Notice that this formula looks very similar to the Ohm's law formula, $E = I \times R$. Because this circuit contains inductive reactance, you must use a formula that represents the opposition presented by the inductor. The Ohm's law formula is changed slightly to reflect this opposition (X_L) to current flow.

Now substitute the known values and solve for E_L.

$E_L = I_L \times X_L$

$E_L = 7.15\ A \times 7.54\ \Omega$

$E_L = 53.91\ V$

Therefore, there are 53.91 volts dropped across the inductor. Now notice something interesting. Recall the rule about voltage drops in the study of series resistive circuits, which states that the sum of all voltage drops equals the total applied voltage. Is this true in this R-L series circuit? To find out, solve the following problem:

What values are known?

$$E_T = 120\text{ V}$$

$$E_R = 107.25\text{ V}$$

$$E_L = 53.9\text{ V}$$

Does $E_T = E_R + E_L$?

$$E_T = E_R + E_L$$

$$E_T = 107.25\text{ V} + 53.91\text{ V}$$

$$E_T = 161.16\text{ V}$$

This appears to be wrong, but it is not. Remember, the inductor introduces a phase shift in the circuit. As a result of this phase shift, you cannot simply add the voltage drops together. You must calculate the sum of the voltage drops differently. Here is a new formula to use:

$$E_T = \sqrt{E_R^2 + E_L^2}$$

Now substitute the known values and solve for E_T.

$$E_T = \sqrt{E_R^2 + E_L^2}$$

$$E_T = \sqrt{107.25\text{ V}^2 + 53.91\text{ V}^2}$$

$$E_T = \sqrt{11502.56\text{ V} + 2906.29\text{ V}}$$

$$E_T = \sqrt{14408.85\text{ V}}$$

$$E_T = 120.04\text{ V}$$

The calculation was correct! You simply cannot ignore the phase shift introduced by the inductor in the circuit. Do not be concerned with the 0.04-volt difference between this answer and the given value. This is a result of the rounding off of numbers to two decimal places during calculations. Now determine the reactive power of the inductor. Recall that reactive power is measured in VARs. To determine the amount of VARs, multiply the reactive volts by the reactive current. This is how:

What values are known?

$$E_L = 53.91\text{ V}$$

$$I_L = 7.15\text{ A}$$

What value is not known?

$$VARs_L = ?$$

What formula can be used?

$$VARs_L = E_L \times I_L$$

Substitute the known values and solve for $VARs_L$.

$$VARs_L = E_L \times I_L$$

$$VARs_L = 53.91\text{ V} \times 7.15\text{ A}$$

$$VARs_L = 385.46\text{ VARs}_L$$

Therefore, the inductive reactive power of the circuit is 385.46 VARs_L.

In a circuit containing resistance and reactance, there are several types of power to be concerned about. Resistive, or true, power, which is measured in watts (W), has already been discussed, as has reactive power, which is measured in VARs. There is a third type of power called apparent power, which is measured in volt-amperes (VA). The apparent power of a circuit is the combination of the true power and the reactive power. This is better understood by finding the apparent power in the following example:

What values are known?

$$P_R = 766.84\text{ W}$$

$$VARs_L = 385.46\text{ VARs}_L$$

What value is not known?

$$VA = ?$$

What formula can be used?

$$VA = \sqrt{P_R^2 + VARs_L^2}$$

Substitute the known values and solve for VA.

$$VA = \sqrt{P_R^2 + VARs_L^2}$$

$$VA = \sqrt{766.84\text{ W}^2 + 385.46\text{ VARs}_L^2}$$

$$VA = \sqrt{588043.59\text{ W} + 148579.41\text{ VARs}_L}$$

$$VA = \sqrt{736623}$$

$$VA = 858.27\text{ VA}$$

Therefore, the total apparent power of this circuit is 858.27 volt-amperes.

Before performing the final calculations on this circuit, a detailed analysis of some of the results needs to be done. In particular, compare the amount of opposition offered by the resistor with that of the inductor. The resistor offers 15 ohms of opposition, whereas the inductor offers 7.54 ohms. The resistor has more influence on the current flow than does the inductor. Now compare the amount of voltage dropped across the resistor with the amount of voltage dropped across the inductor. The resistor voltage is 107.25 volts, whereas the inductor voltage is 53.91 volts. The resistor has the larger voltage drop. Finally, compare the amount of resistive power with the amount of reactive power. The resistive power is 766.84 watts, whereas the reactive power is 385.46 $VARs_L$. The resistive power is larger. When all of this is considered, it should be apparent that the resistor has more influence on the circuit than does the inductor. You know that if the circuit were purely resistive, the current and voltage would be in phase, and the phase angle difference between the current and voltage would be 0°. If the circuit were purely inductive, the current and voltage would be out of phase, and the phase angle difference would be 90° lagging. This circuit is a combination of resistance and inductive reactance. This means that the phase angle must fall somewhere between 0° and 90°. When comparing the influence of the resistor with that of the inductor in this circuit, it shows that the resistor has more influence. This means the phase angle difference must fall somewhere between 0° and 45°. To see whether this assumption is correct, first calculate the power factor of the circuit using the formula $PF = \frac{P}{VA} \times 100$.

What values are known?

$$P_R = 766.84 \text{ W}$$

$$VA = 858.27 \text{ VA}$$

What value is not known?

$$PF = ?$$

What formula can be used?

$$PF = \frac{P}{VA} \times 100$$

Substitute the known values and solve for *PF*.

$$PF = \frac{P}{VA} \times 100$$

$$PF = \frac{766.84 \text{ W}}{858.27 \text{ VA}} \times 100$$

$$PF = 0.8935 \times 100$$

$$PF = 89.35\%$$

Therefore, this circuit has a power factor of 89.35%. Now determine the phase angle difference, $\angle\theta$.

What value is known?

$PF = 0.8935$ (For this calculation, the power factor is not expressed as a percent.)

What value is not known?

$$\angle\theta = ?$$

What formula can be used?

$$\angle\theta = COS^{-1}\,PF$$

Substitute the known values and solve for $\angle\theta$.

$$\angle\theta = COS^{-1}\,PF$$

$$\angle\theta = COS^{-1}0.8935$$

$$\angle\theta = 26.68°$$

Therefore, in this circuit the current lags the voltage by 26.68°.

Parallel Circuit

Figure 17-6 shows an R-L parallel circuit. So that you can compare the effects of a series circuit with those of a parallel circuit, use the same values from the R-L series circuit.

What values are known?

$$E_T = 120 \text{ V}$$

$$f = 60 \text{ Hz}$$

$$R = 15\ \Omega$$

$$L = 20 \text{ mH}$$

What values are not known?

$$I_T = ?$$

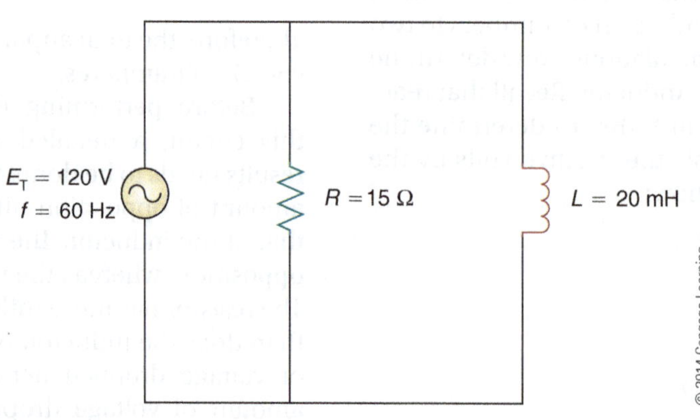

FIGURE 17-6 An R-L parallel circuit with values.

$Z_T = ?$

$VA = ?$

$PF = ?$

$\angle\theta = ?$

$E_R = ?$

$I_R = ?$

$P_R = ?$

$E_L = ?$

$I_L = ?$

$X_L = ?$

$VARs_L = ?$

Begin by solving for X_L.

What values are known?

$f = 60\text{ Hz}$

$L = 20\text{ mH}$

What value is not known?

$X_L = ?$

What formula can be used?

$X_L = 2\pi fL$

Substitute the known values and solve for X_L.

$X_L = 2\pi fL$

$X_L = 2 \times \pi \times 60\text{ Hz} \times 20\text{ mH}$

$X_L = 7.54\ \Omega$

Therefore, the inductor presents 7.54 ohms of opposition to the current flowing in this circuit. Now that you know the opposition presented by the inductor and the resistor, you can determine the total opposition (impedance) of the circuit. Now find Z_T.

What values are known?

$R = 15\ \Omega$

$X_L = 7.54\ \Omega$

What value is not known?

$Z_T = ?$

As a result of the phase shift introduced by the inductor, you cannot use the reciprocal of the sum of the reciprocals formula that you learned when you studied resistive parallel circuits. You must consider the effects of the inductor and the resulting phase shift between the current and the voltage. You must use a different formula that takes into account the effects of the inductor.

What formula can be used?

$$Z_T = \frac{1}{\sqrt{\left(\frac{1}{R}\right)^2 + \left(\frac{1}{X_L}\right)^2}}$$

Substitute the known values and solve for Z_T.

$$Z_T = \frac{1}{\sqrt{\left(\frac{1}{R}\right)^2 + \left(\frac{1}{X_L}\right)^2}}$$

$$Z_T = \frac{1}{\sqrt{\left(\frac{1}{15\ \Omega}\right)^2 + \left(\frac{1}{7.54\ \Omega}\right)^2}}$$

$$Z_T = \frac{1}{\sqrt{(0.067)^2 + (0.133)^2}}$$

$$Z_T = \frac{1}{\sqrt{0.0044 + 0.018}}$$

$$Z_T = \frac{1}{\sqrt{0.0224}}$$

$$Z_T = \frac{1}{0.1497}$$

$Z_T = 6.68\ \Omega$

Thus, the combined opposition to the current of the circuit is 6.68 ohms. Now that you know the total opposition and the total applied voltage, you can determine the total circuit current.

What values are known?

$E_T = 120\text{ V}$

$Z_T = 6.68\ \Omega$

What value is not known?

$I_T = ?$

What formula can be used?

$$I_T = \frac{E_T}{Z_T}$$

Now substitute the known values and solve for I_T.

$$I_T = \frac{E_T}{Z_T}$$

$$I_T = \frac{120\text{ V}}{6.68\ \Omega}$$

$I_T = 17.96\text{ A}$

Therefore, there are 17.96 amperes of current flowing in this R-L parallel circuit.

As you have learned previously, the voltage in a parallel circuit remains the same across each of the branch circuits, because in a parallel circuit,

$E_T = E_R = E_L$

$E_T = 120\text{ V}$

$E_R = 120\text{ V}$

$E_L = 120\text{ V}$

Now focus your attention on the resistor. You know the value of the resistor (15 ohms) and the amount of voltage dropped across the resistor (120 volts). Determine the amount of current flowing through the resistor, I_R.

What values are known?

$$R = 15\ \Omega$$

$$E_R = 120\ \text{V}$$

What value is not known?

$$I_R = ?$$

What formula can be used?

$$I_R = \frac{E_R}{R}$$

Substitute the known values and solve for I_R.

$$I_R = \frac{E_R}{R}$$

$$I_R = \frac{120\ \text{V}}{15\ \Omega}$$

$$I_R = 8.0\ \text{A}$$

Therefore, there are 8.0 amperes of current flowing through the resistor. The last unknown value for the resistor is the amount of power that the resistor must handle.

What values are known?

$$I_R = 8.0\ \text{A}$$

$$R = 15\ \Omega$$

What value is not known?

$$P_R = ?$$

What formula can be used?

$$P_R = I_R^2 \times R$$

Substitute the known values and solve for P_R.

$$P_R = I_R^2 \times R$$

$$P_R = 8.0\ \text{A}^2 \times 15\ \Omega$$

$$P_R = 64.0\ \text{A} \times 15\ \Omega$$

$$P_R = 960.0\ \text{W}$$

Therefore, the resistor consumes 960.0 watts of power. Now turn your attention to the inductor. So far, you know the values of L, X_L, and E_L. Find the amount of current flowing through the inductor, I_L.

What values are known?

$$E_L = 120\ \text{V}$$

$$X_L = 7.54\ \Omega$$

What value is not known?

$$I_L = ?$$

What formula can be used?

$$I_L = \frac{E_L}{X_L}$$

Now substitute the known values and solve for I_L.

$$I_L = \frac{E_L}{X_L}$$

$$I_L = \frac{120\ \text{V}}{7.54\ \Omega}$$

$$I_L = 15.92\ \text{A}$$

Therefore, there are 15.92 amperes of current flowing through the inductor. Now notice something interesting. Recall the rule about branch currents in the study of parallel resistive circuits, which states that the sum of all branch currents equals the total circuit current. Is this true in this R-L parallel circuit? To find out, solve the following problem:

What values are known?

$$I_T = 17.96\ \text{A}$$

$$I_R = 8.0\ \text{A}$$

$$I_L = 15.92\ \text{A}$$

Does $I_T = I_R + I_L$?

$$I_T = I_R + I_L$$

$$I_T = 8.0\ \text{A} + 15.92\ \text{A}$$

$$I_T = 23.92\ \text{A}$$

This appears to be wrong, but it is not. Remember, the inductor introduces a phase shift in the circuit. As a result of this phase shift, you cannot simply add the branch currents together. You must calculate the sum of the branch currents differently. Here is a new formula to use:

$$I_T = \sqrt{I_R^2 + I_L^2}$$

Substitute the known values and solve for I_T.

$$I_T = \sqrt{I_R^2 + I_L^2}$$

$$I_T = \sqrt{8.0\ \text{A}^2 + 15.92\ \text{A}^2}$$

$$I_T = \sqrt{64.0\ \text{A} + 253.45\ \text{A}}$$

$$I_T = \sqrt{317.45\ \text{A}}$$

$$I_T = 17.82\ \text{A}$$

The calculation was correct! You simply cannot ignore the phase shift introduced by the inductor in the circuit. Do not be concerned with the 0.14-ampere difference between this answer and the given value. This is a result of the rounding off of numbers to two decimal places during calculations. Now determine the reactive power of the inductor.

What values are known?

$$E_L = 120\ V$$

$$I_L = 15.92\ A$$

What value is not known?

$$VARs_L = ?$$

What formula can be used?

$$VARs_L = E_L \times I_L$$

Substitute the known values and solve for $VARs_L$.

$$VARs_L = E_L \times I_L$$

$$VARs_L = 120\ V \times 15.92\ A$$

$$VARs_L = 1910.4\ VARs_L$$

Therefore, the inductive reactive power of the circuit is 1910.4 $VARs_L$. Now determine the circuit's apparent power.

What values are known?

$$P_R = 960\ W$$

$$VARs_L = 1910.4\ VARs_L$$

What value is not known?

$$VA = ?$$

What formula can be used?

$$VA = \sqrt{P^2 + VARs_L^{\,2}}$$

Substitute the known values and solve for VA.

$$VA = \sqrt{P^2 + VARs_L^{\,2}}$$

$$VA = \sqrt{960\ W^2 + 1910.4\ VARs_L^{\,2}}$$

$$VA = \sqrt{921600\ W + 3649628.16\ VARs_L}$$

$$VA = \sqrt{4571228.16}$$

$$VA = 2138.04\ VA$$

Therefore, the total apparent power of this circuit is 2138.04 volt-amperes.

Before performing the final calculations on this circuit, a detailed analysis of some of the results needs to be done. In particular, compare the amount of opposition offered by the resistor with that of the inductor. The resistor offers 15 ohms of opposition, whereas the inductor offers 7.54 ohms. The inductor has more influence on the current flow than does the resistor. Now compare the amount of current flow through the resistor with the amount of current flow through the inductor. The resistor current is 8.0 amperes, whereas the inductor current is 15.92 amperes. The inductor has the larger current flow. Finally, compare the amount of resistive power with the amount of reactive power. The resistive power is 960 watts, whereas the reactive power is 1910.4 $VARs_L$. The reactive power is larger. When all of this is considered, it should be apparent that the inductor has more influence on the circuit than does the resistor. You know that if the circuit were purely resistive, the current and voltage would be in phase, and the phase angle difference between the current and voltage would be 0°. If the circuit were purely inductive, the current and voltage would be out of phase and the phase angle difference would be 90° lagging. This circuit is a combination of resistance and inductive reactance. This means that the phase angle must fall somewhere between 0° and 90°. By comparing the influence of the resistor with that of the inductor in this circuit, it was determined that the inductor has more influence. This means the phase angle difference must fall somewhere between 45° and 90°. To see if this assumption is correct, first calculate the power factor of the circuit.

What values are known?

$$P_R = 960\ W$$

$$VA = 2138.04\ VA$$

What value is not known?

$$PF = ?$$

What formula can be used?

$$PF = \frac{P}{VA} \times 100$$

Substitute the known values and solve for PF.

$$PF = \frac{P}{VA} \times 100$$

$$PF = \frac{960\ W}{2138.04\ VA} \times 100$$

$$PF = 0.4490 \times 100$$

$$PF = 44.9\%$$

Therefore, this circuit has a power factor of 44.9%. Now determine the phase angle difference, $\angle\theta$.

What value is known?

$PF = 0.4490$ (For this calculation, the power factor is not expressed as a percent.)

What value is not known?

$$\angle\theta = ?$$

What formula can be used?

$$\angle\theta = COS^{-1}\ PF$$

Substitute the known values and solve for $\angle\theta$.

$$\angle\theta = COS^{-1}\ PF$$

$$\angle\theta = COS^{-1} 0.4490$$

$$\angle\theta = 63.32°$$

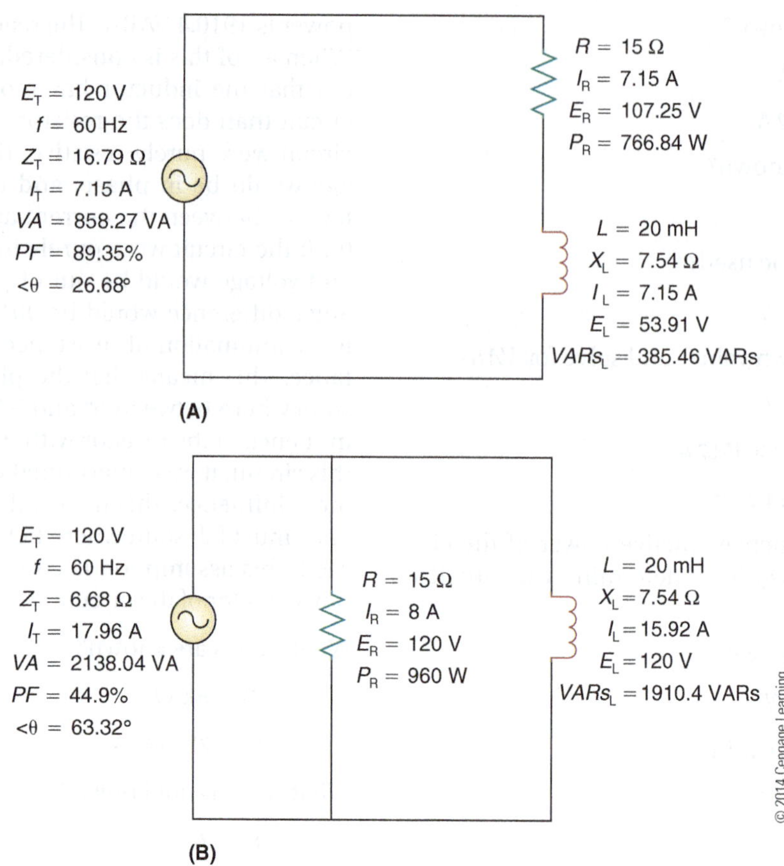

FIGURE 17-7 (A) A fully analyzed R-L series circuit; and (B) a fully analyzed R-L parallel circuit.

Therefore, in this circuit the current lags the voltage by 63.32°. Because the inductor has more influence in the circuit, the circuit is more inductive and the phase angle approaches 90°. Compare the remaining parameters of the two circuits shown in **Figure 17-7A** and **Figure 17-7B**. Notice the changes caused by connecting the components in either series or parallel.

17-2 CAPACITANCE AND R-C CIRCUITS

Two conductors in close proximity that are separated by some type of insulating material form a device called a **capacitor**. A capacitor is a device that is capable of storing an electrical charge. **Figure 17-8** shows a picture of several types of capacitors. **Figure 17-9** shows the schematic symbols used to represent a capacitor. When dealing with capacitors, the two conductors are called **plates**, and the insulating material is called a **dielectric**.

Figure 17-10 shows a capacitor connected to a DC source. Notice switch S_1. In position A, switch S_1 connects the capacitor across the power source. In position B, switch S_1 connects the capacitor across load resistor R_1. The capacitor is connected across the power source by placing switch S_1 in position A as shown in Figure 17-10. Notice that the plate that is connected to the negative terminal of the power source is labeled "−", whereas the plate that is connected to the positive terminal of the power source is labeled "+".

Electrons accumulate or build up on the negative terminal of the capacitor. At the same time, electrons are forced from the positive plate, causing a deficiency of electrons. This causes a charge to be built up across the capacitor. The charge across the capacitor continues to build up until the amount of charge across the capacitor is equal to the amount of the applied charge of the DC source. *Notice that current is not flowing through the capacitor from one plate to the other.* However, there is a current flow between each plate of the capacitor and the power source. When the charge on the capacitor is equal to the value of the DC source, the current flow stops.

At this point, the capacitor is said to be fully charged. The charge remains on the capacitor even if the capacitor is removed from the circuit. *This can be very dangerous! Never come into contact with a charged capacitor. The capacitor could discharge into*

FIGURE 17-8 Various types and sizes of capacitors.

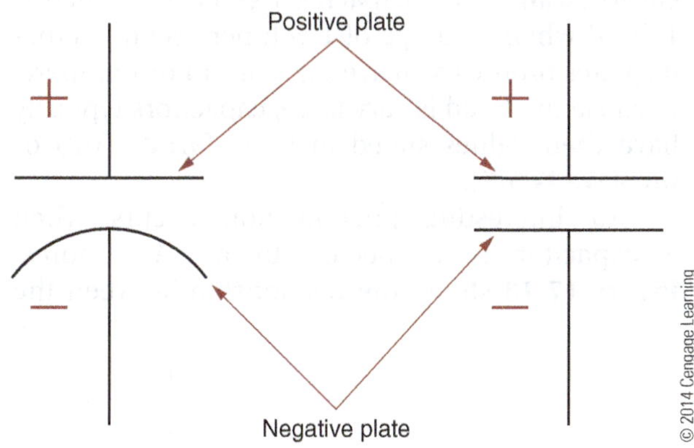

FIGURE 17-9 Alternative schematic symbols to represent a capacitor.

your body and cause burns or even interrupt the beating of your heart! Always treat a capacitor as having a full charge until you can prove that no charge exists. Simply place a voltmeter across the terminals of the capacitor to check for the presence of voltage. If a voltage is present, the capacitor must be discharged with an approved shorting device before proceeding.

Now refer to **Figure 17-11**. Switch S_1 has been placed in the B position. This disconnects the charged capacitor from the power source and connects the load resistor R_1 across the capacitor terminals. The capacitor discharges across resistor R_1. The electrons that have built up on the negative terminal flow through the resistor to the positive terminal. This flow of electrons continues until the

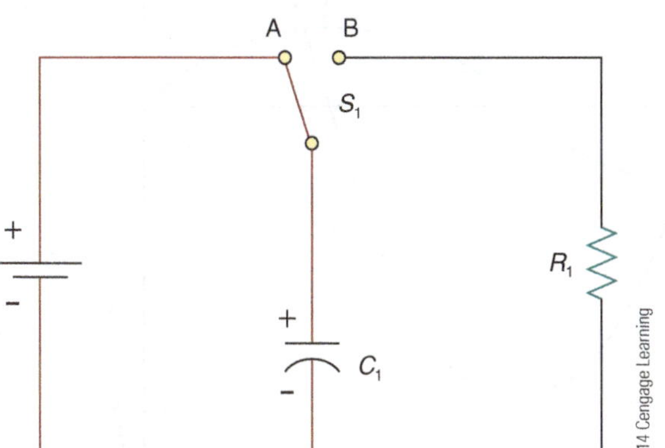

FIGURE 17-10 Charging a capacitor.

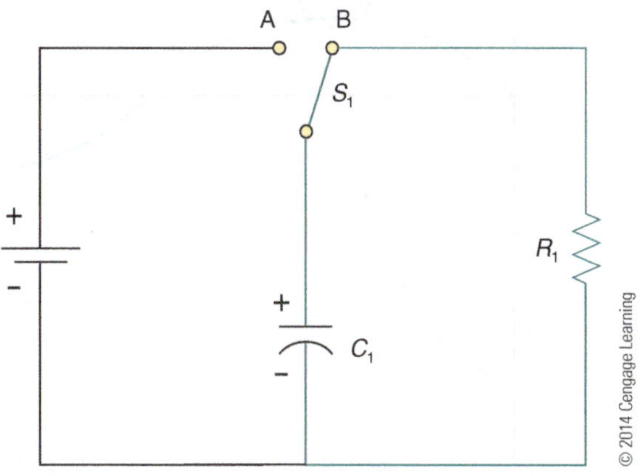

FIGURE 17-11 Discharging a capacitor.

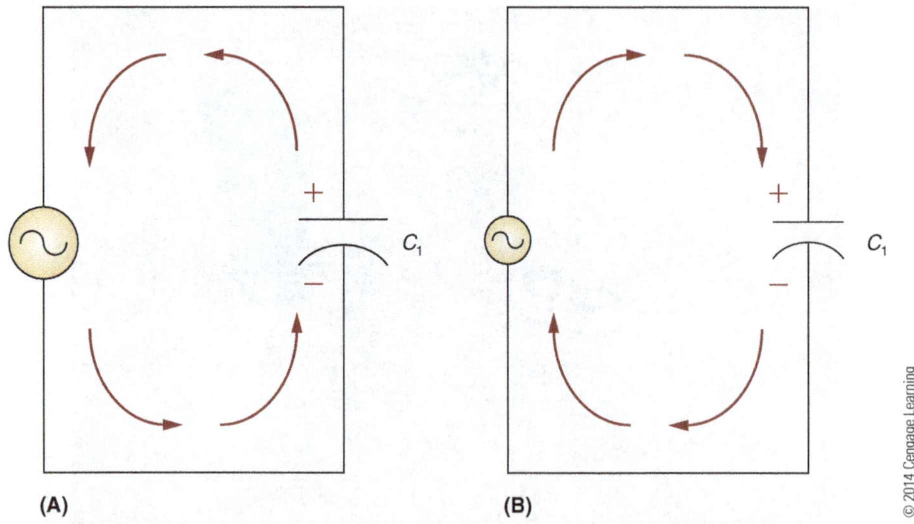

FIGURE 17-12 (A) A capacitor current flow during half of an AC cycle; and (B) a capacitor current flow during the other half of an AC cycle.

number of electrons on the negative plate is equal to the number of electrons on the positive plate. At this point, the capacitor is said to be discharged.

Figure 17-12A and **Figure 17-12B** show a capacitor connected to an AC source. Recall that alternating current reverses its direction at some frequency. When a capacitor is placed across an AC source, the current flows back and forth continually. The current does not flow through the capacitor, but it flows alternately in one direction and then in the other direction as the capacitor charges (Figure 17-12A) and discharges (Figure 17-12B).

Capacitance is a measure of a capacitor's ability to store an electrical charge. The amount of capacitance that a capacitor can have is measured in units called **farads (F)**. A capacitor has a capacitance of 1 farad when a charge of 1 volt per second across its plates produces an average current of 1 ampere. Because the farad is very large, capacitors typically have their values stated in **microfarads (μF)** or **picofarads (ρF)**.

An interesting phenomenon occurs when a capacitor is connected to an AC source. **Figure 17-13** shows the relationship between the

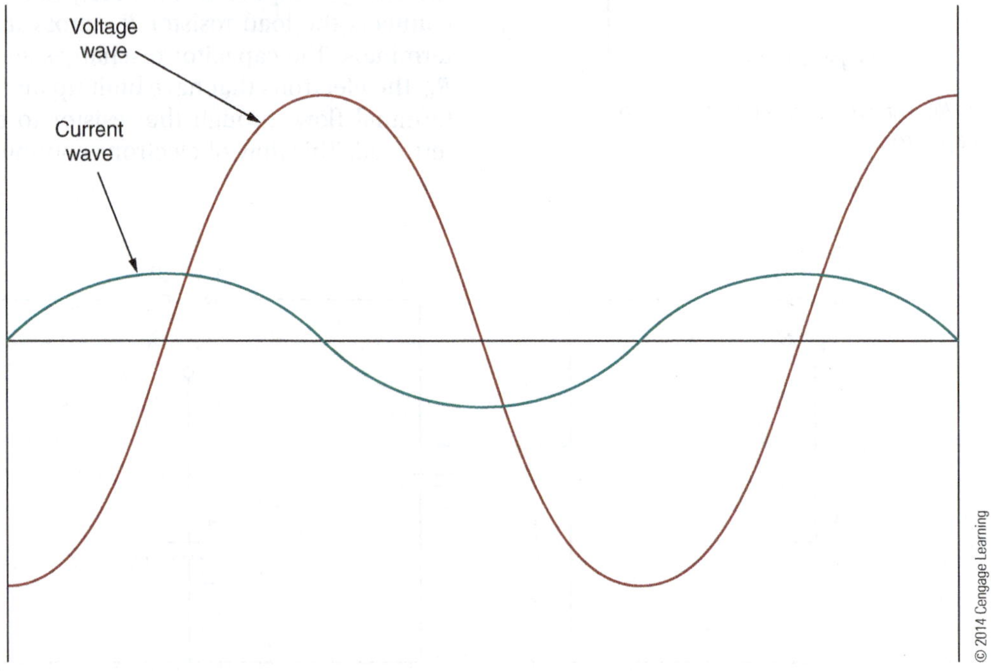

FIGURE 17-13 In a purely capacitive circuit, the current leads the voltage by 90°.

voltage and the current for a capacitor connected to an AC source. Notice that when the voltage waveform is at zero (capacitor discharged), the current is at its peak positive value. As the voltage increases in the positive direction (capacitor charging), the current decreases toward zero. When the voltage is at its peak positive value (capacitor fully charged), the current is at zero. As the voltage decreases toward zero (capacitor discharging), the current increases in the negative direction. When the voltage has reached zero (capacitor fully discharged), the current is at its peak negative value. This continues through the negative half cycle of the applied AC. You should notice that the current and voltage are out of phase as occurs with inductors. However, look closely at the relationship between the current and voltage. Notice that the current reaches its peak value *before* the voltage, which is opposite to what occurs with inductors, where the current reaches its peak value after the voltage. To help you remember this relationship, recall the phrase *ELI the ICE man.* You learned that ELI represented the voltage (E) leading the current (I) in an inductor (L). Now focus on the word *ICE*. The letters *I, C,* and *E* represent *current, capacitance,* and *voltage,* respectively. Notice that the letter *I* appears before the letter *E.* This should help you remember that *current leads the voltage in a capacitor.* Notice that in a circuit containing only capacitance (purely capacitive), the current and voltage are 90° out of phase. Therefore, *in a purely capacitive circuit, the current leads the voltage by 90°.*

As a result of the phase shift between the current and the voltage in a capacitive circuit, the capacitor presents opposition to alternating current flow. This opposition is called **capacitive reactance**, represented in formulae by X_C and measured in ohms. The formula for finding capacitive reactance is

$$X_C = \frac{1}{2\pi fC}$$

2 = a constant
π = 3.14 (a constant)
f = frequency, measured in hertz (Hz)
C = capacitance, measured in farads (F)

By analyzing the above formula, you can see that two factors inversely influence the amount of capacitive reactance of a capacitor: the frequency of the applied AC voltage and the amount of capacitance. In other words, if the capacitance is increased, the capacitive reactance decreases proportionately, and if the frequency is increased, the capacitive reactance also decreases proportionately.

As was seen earlier in the study of inductors, the out-of-phase relationship between the current and the voltage in a capacitor creates another interesting occurrence. Recall one of the formulae for finding power: $P = I \times E$. Recall also that current and voltage are in phase in a purely resistive circuit. If the value of the current is positive, and the value of the voltage is positive, then the amount of power is a positive value. If the current and voltage are both negative in value, the amount of power is positive in value. Refer back to Figure 17-2, which shows the voltage and current waveforms for a purely resistive circuit.

Now look at **Figure 17-14**, which shows a circuit that contains pure capacitance. Notice that during the first 90°, both the voltage and the current are positive in value. Therefore, the power is positive in value ($P = I \times E$). During the second 90°, the voltage is positive in value, but the current is negative in value; therefore, power is negative in value ($-P = -I \times E$). During the third 90°, both the voltage and the current are negative in value. Therefore, the power is positive in value ($P = -I \times -E$). Finally, during the fourth 90°, the voltage is negative, but the current is positive. The power is again negative in value ($P = I \times -E$).

Notice that positive power is produced during the first and third 90°. During the second and fourth 90°, the circuit is producing *negative power.* This is called *reactive power* and is measured in units called *VARs.*

During the 360° of an AC sine wave, a capacitor uses power from the circuit during the first and third 90°. As a result of the discharging, a capacitor returns power to the circuit during the second and fourth 90°. The net effect of this is that no power is used. If no power is used, no work is done. Therefore, the energy is wasted.

Circuits that contain both resistance and capacitance are called **R-C circuits** (R for resistance, C for capacitance). An R-C series circuit is shown in **Figure 17-15**. An R-C parallel circuit is discussed later.

R-C Series Circuit

Before analyzing this circuit, think about what you have learned about resistance and capacitance. Recall that in a resistive circuit, current and voltage are in phase, whereas in a capacitive circuit, current and voltage are out of phase by 90°. When a resistance and a capacitance are combined in one circuit, the current and voltage are not in phase because of the effects of the capacitor. Likewise, the current and voltage are not out of phase by 90° because of the effects of the resistor. The phase shift is somewhere between 0° and 90°, depending on which component, the resistor or the capacitor, has the most influence on the circuit.

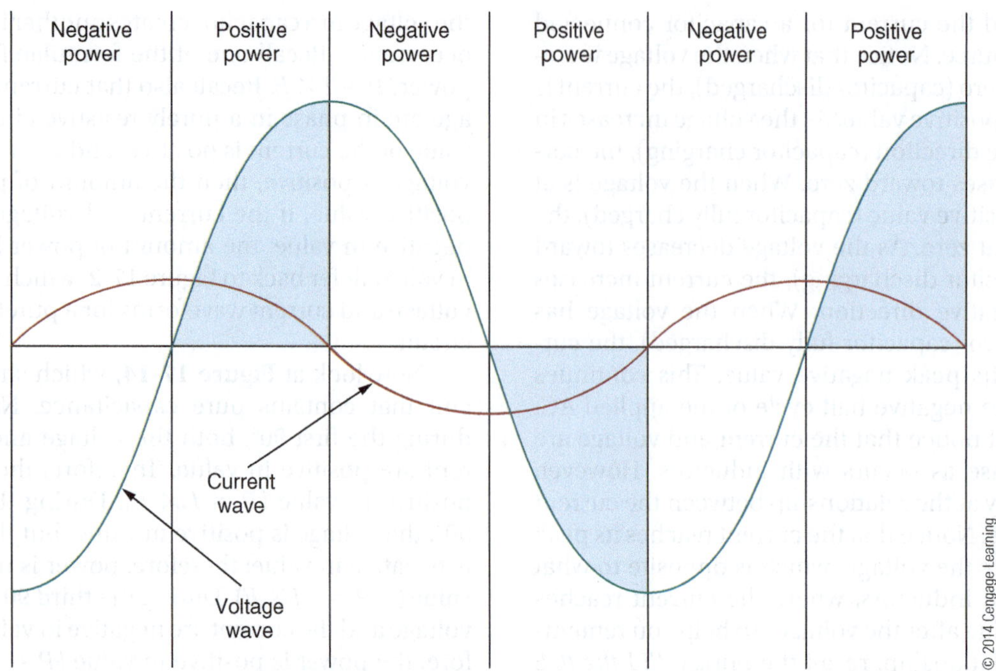

FIGURE 17-14 Positive power is produced during the first and third 90°. During the second and fourth 90°, the circuit is producing negative power.

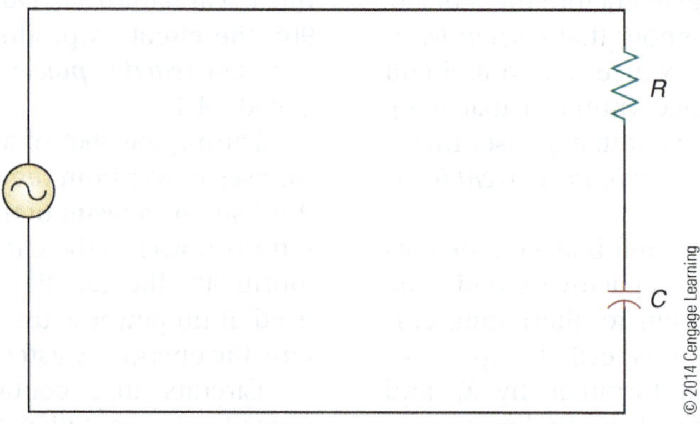

FIGURE 17-15 An R-C series circuit.

A circuit that contains both resistance and capacitance has a few new values to represent certain combinations of the resistive and capacitive components. For example, a circuit containing both resistance and capacitance exhibits two types of power: true, or resistive, power and reactive power. The combination of both types of power is called *apparent power* and is measured in volt-amperes (VA). Another example is the opposition to current flow, where the effects of the resistance of the resistor must be combined with the capacitive reactance of the capacitor. The combination is called *impedance* and is represented by the letter Z. Angle theta ($\angle\theta$) represents the voltage-current phase shift, and the letters PF represent power factor. Power factor is the ratio of true power to apparent power and is typically expressed as a percent. Power factor is used to measure the efficiency of a circuit.

To help you better understand this, solve the following problem. **Figure 17-16** shows an R-C series circuit with values. Solve for the unknown values.

What values are known?

$E_T = 120$ V

$f = 60$ Hz

$R = 15\ \Omega$

$C = 100\ \mu$F

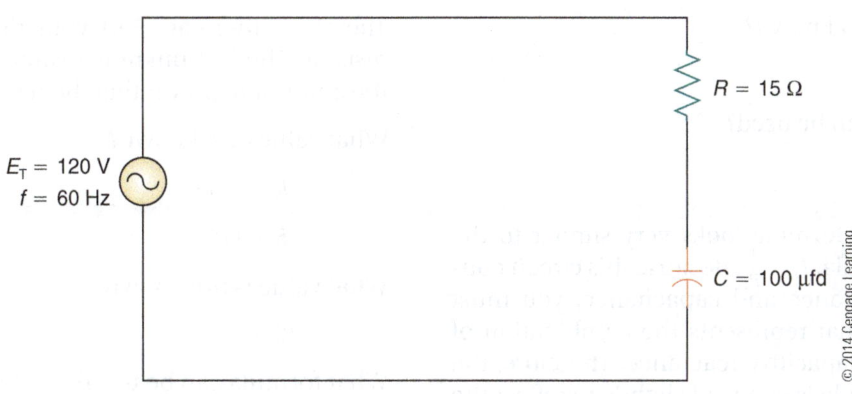

FIGURE 17-16 An R-C series circuit with values.

What values are not known?

$I_T = ?$

$Z_T = ?$

$VA = ?$

$PF = ?$

$\angle\theta = ?$

$E_R = ?$

$I_R = ?$

$P_R = ?$

$E_C = ?$

$I_C = ?$

$X_C = ?$

$VARs_C = ?$

Begin by solving for X_C.

What values are known?

$f = 60\text{ Hz}$

$C = 100\ \mu\text{F}$

What value is not known?

$X_C = ?$

What formula can be used?

$$X_C = \frac{1}{2\pi fC}$$

Substitute the known values and solve for X_C.

$$X_C = \frac{1}{2\pi fC}$$

$$X_C = \frac{1}{2 \times \pi \times 60\text{ Hz} \times 100\ \mu\text{F}}$$

$$X_C = 26.54\ \Omega$$

Therefore, the capacitor presents 26.54 ohms of opposition to the current flowing in this circuit. Now that you know the opposition presented by the capacitor as well as that by the resistor, you can determine the total opposition (impedance) of the circuit. Now find Z_T.

What values are known?

$R = 15\ \Omega$

$X_C = 26.54\ \Omega$

What value is not known?

$Z_T = ?$

As a result of the phase shift introduced by the capacitor, you cannot simply add the resistance of the resistor to the capacitive reactance of the capacitor. A different formula that takes into account the effects of the capacitor must be used.

What formula can be used?

$$Z_T = \sqrt{R^2 + X_C^2}$$

Substitute the known values and solve for Z_T.

$$Z_T = \sqrt{R^2 + X_C^2}$$

$$Z_T = \sqrt{15\ \Omega^2 + 26.54\ \Omega^2}$$

$$Z_T = \sqrt{225\ \Omega + 704.37\ \Omega}$$

$$Z_T = \sqrt{929.37\ \Omega}$$

$$Z_T = 30.49\ \Omega$$

Thus, the combined opposition to the current of the circuit is 30.49 ohms. Because you now know the total opposition and the total applied voltage, you can determine the total circuit current.

What values are known?

$E_T = 120\text{ V}$

$Z_T = 30.49\ \Omega$

What value is not known?

$$I_T = ?$$

What formula can be used?

$$I_T = \frac{E_T}{Z_T}$$

Notice that this formula looks very similar to the Ohm's law formula, $I_T = \frac{E_T}{R_T}$. Because this circuit contains both resistance and capacitance, you must use a formula that represents the combination of resistance and capacitive reactance. Therefore, the Ohm's law formula is changed slightly to reflect the total opposition (Z) to current flow.

Now substitute the known values and solve for I_T.

$$I_T = \frac{E_T}{Z_T}$$

$$I_T = \frac{120\text{ V}}{30.49\ \Omega}$$

$$I_T = 3.94\text{ A}$$

Therefore, there are 3.94 amperes of current flowing in this R-C series circuit. This is the same amount of current flowing throughout this circuit. This means that there are 3.94 amperes of resistive current, and there are 3.94 amperes of capacitive current. Therefore, you now know that

$$I_R = 3.94\text{ A}$$

$$I_C = 3.94\text{ A}$$

because in a series circuit,

$$I_T = I_R = I_C$$

Now focus your attention on the resistor. You know the value of the resistor (15 ohms) and the amount of current flowing through the resistor (3.94 amperes). Determine the amount of voltage dropped across the resistor, E_R.

What values are known?

$$R = 15\ \Omega$$

$$I_R = 3.94\text{ A}$$

What value is not known?

$$E_R = ?$$

What formula can be used?

$$E_R = I_R \times R$$

Substitute the known values and solve for E_R.

$$E_R = I_R \times R$$

$$E_R = 3.94\text{ A} \times 15\ \Omega$$

$$E_R = 59.1\text{ V}$$

Therefore, there are 59.1 volts dropped across the resistor. The last unknown value for the resistor is the amount of power that the resistor must handle.

What values are known?

$$I_R = 3.94\text{ A}$$

$$R = 15\ \Omega$$

What value is not known?

$$P_R = ?$$

What formula can be used?

$$P_R = I_R^2 \times R$$

Substitute the known values and solve for P_R.

$$P_R = I_R^2 \times R$$

$$P_R = 3.94\text{ A}^2 \times 15\ \Omega$$

$$P_R = 15.12\text{ A} \times 15\ \Omega$$

$$P_R = 232.85\text{ W}$$

Therefore, the resistor consumes 232.8 watts of power. Now turn your attention to the capacitor. So far, you know the values of C, X_C, and I_C. Find the amount of voltage dropped across the capacitor, E_C.

What values are known?

$$I_C = 3.94\text{ A}$$

$$X_C = 26.54\ \Omega$$

What value is not known?

$$E_C = ?$$

What formula can be used?

$$E_C = I_C \times X_C$$

Notice that this formula looks very similar to the Ohm's law formula, $E = I \times R$. Because this circuit contains inductive reactance, you must use a formula that represents the opposition presented by the capacitor. Therefore, the Ohm's law formula is changed slightly to reflect this opposition (X_C) to current flow.

Now substitute the known values and solve for E_C.

$$E_C = I_C \times X_C$$

$$E_C = 3.94\text{ A} \times 26.54\ \Omega$$

$$E_C = 104.57\text{ V}$$

Therefore, there are 104.57 volts dropped across the capacitor. Now notice something interesting. Recall the rule about voltage drops in the study of series resistive circuits, which states that the sum of all voltage drops equals the total applied voltage. Is this true in this R-C series circuit? To find out, solve the following problem:

What values are known?

$$E_T = 120 \text{ V}$$

$$E_R = 59.1 \text{ V}$$

$$E_C = 104.57 \text{ V}$$

Does $E_T = E_R + E_C$?

$$E_T = E_R + E_C$$

$$E_T = 59.1 \text{ V} + 104.57 \text{ V}$$

$$E_T = 163.67 \text{ V}$$

This appears to be wrong, but it is not. Remember, the capacitor introduces a phase shift in the circuit. As a result of this phase shift, you cannot simply add the voltage drops together. You must calculate the sum of the voltage drops differently. Here is a new formula to use:

$$E_T = \sqrt{E_R^2 + E_C^2}$$

Now substitute the known values and solve for E_T.

$$E_T = \sqrt{E_R^2 + E_C^2}$$

$$E_T = \sqrt{59.1 \text{ V}^2 + 104.57 \text{ V}^2}$$

$$E_T = \sqrt{3492.81 \text{ V} + 10934.88 \text{ V}}$$

$$E_T = \sqrt{14427.69 \text{ V}}$$

$$E_T = 120.12 \text{ V}$$

The calculation was correct! You simply cannot ignore the phase shift introduced by the capacitor in the circuit. Do not be concerned with the 0.12-volt difference between this answer and the given value. This is a result of the rounding off of numbers to two decimal places during calculations. Now determine the reactive power of the capacitor. Recall that reactive power is measured in VARs. To determine the amount of VARs, multiply the reactive volts by the reactive current. This is how:

What values are known?

$$E_C = 104.57 \text{ V}$$

$$I_C = 3.94 \text{ A}$$

What value is not known?

$$VARs_C = ?$$

What formula can be used?

$$VARs_C = E_C \times I_C$$

Substitute the known values and solve for $VARs_C$.

$$VARs_C = E_C \times I_C$$

$$VARs_C = 104.57 \text{ V} \times 3.94 \text{ A}$$

$$VARs_C = 412.01 \text{ VARs}_C$$

Therefore, the capacitive reactive power of the circuit is 412.01 VARs_C.

In a circuit containing resistance and reactance, there are several types of power to be concerned about. Resistive, or true power, which is measured in watts, has already been discussed, as has reactive power, which is measured in VARs. There is a third type of power called apparent power, which is measured in volt-amperes (VAs). The apparent power of a circuit is the combination of the true power and the reactive power. This is better understood by finding the apparent power in the following example. What values are known?

$$P_R = 232.85 \text{ W}$$

$$VARs_C = 412.01 \text{ VARs}_C$$

What value is not known?

$$VA = ?$$

What formula can be used?

$$VA = \sqrt{P^2 + VARs_C^2}$$

Substitute the known values and solve for VA.

$$VA = \sqrt{P^2 + VARs_C^2}$$

$$VA = \sqrt{232.85 \text{ W}^2 + 412.01 \text{ VARs}_C^2}$$

$$VA = \sqrt{54219.12 \text{ W} + 169752.24 \text{ VARs}_C}$$

$$VA = \sqrt{223971.36}$$

$$VA = 473.26 \text{ VA}$$

Therefore, the total apparent power of this circuit is 473.26 volt-amperes.

Before performing the final calculations on this circuit, a detailed analysis of some of the results has to be done. In particular, compare the amount of opposition offered by the resistor with that of the capacitor. The resistor offers 15 ohms of opposition, whereas the capacitor offers 26.54 ohms. The capacitor has more influence on the current flow than does the resistor. Now compare the amount of voltage dropped across the resistor with the amount of voltage dropped across the capacitor. The resistor voltage is 59.1 volts, whereas the capacitor voltage is 104.57 volts. The capacitor has the larger voltage drop. Finally, compare the amount of resistive power with the amount of reactive power. The resistive power is 232.85 watts, whereas the reactive power is 412.01 VARs_C. The reactive power is larger. When all of this is considered, it should be apparent that the capacitor has more influence on the circuit than does the resistor. You know that if the circuit were purely resistive, the current and voltage would be in phase, and the phase angle difference between the current and voltage would be 0°. If the circuit were purely capacitive, the current and voltage would be out of phase, and the phase angle difference would be 90° leading. This circuit is a combination of resistance and capacitive reactance. This means that the phase angle must fall somewhere between 0° and 90°. When comparing the influence of the resistor with that of the capacitor in this circuit, it shows that the capacitor has more influence. This means the phase angle difference must fall somewhere between 45° and 90°. To see whether this assumption is correct, first calculate the power factor of the circuit using the formula $PF = \frac{P}{VA} \times 100$.

What values are known?

$P_R = 232.85\ W$

$VA = 473.26\ VA$

What value is not known?

$PF = ?$

What formula can be used?

$$PF = \frac{P}{VA} \times 100$$

Substitute the known values and solve for PF.

$$PF = \frac{P}{VA} \times 100$$

$$PF = \frac{232.85\ W}{473.26\ VA} \times 100$$

$$PF = 0.4920 \times 100$$

$$PF = 49.2\%$$

Therefore, this circuit has a power factor of 49.2%. Now determine the phase angle difference, $\angle\theta$.

What value is known?

$PF = 0.4920$ (For this calculation, the power factor is not expressed as a percent.)

What value is not known?

$\angle\theta = ?$

What formula can be used?

$$\angle\theta = COS^{-1}\ PF$$

Substitute the known values and solve for $\angle\theta$.

$$\angle\theta = COS^{-1}\ PF$$

$$\angle\theta = COS^{-1} 0.4920$$

$$\angle\theta = 60.53°$$

Therefore, in this circuit the current leads the voltage by 60.53°.

R-C Parallel Circuit

Figure 17-17 shows an R-C parallel circuit. So that you can compare the effects of a series circuit with those of a parallel circuit, use the same values from the R-C series circuit.

What values are known?

$E_T = 120\ V$

$f = 60\ Hz$

$R = 15\ \Omega$

$C = 100\ \mu F$

What values are not known?

$I_T = ?$

$Z_T = ?$

$VA = ?$

$PF = ?$

$\angle\theta = ?$

$E_R = ?$

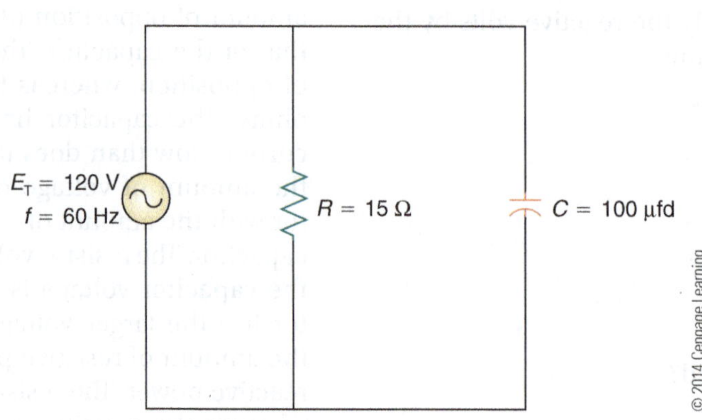

FIGURE 17-17 An R-C parallel circuit with values.

$I_R = ?$

$P_R = ?$

$E_C = ?$

$I_C = ?$

$X_C = ?$

$VARs_C = ?$

Begin by solving for X_C.

What values are known?

$f = 60 \text{ Hz}$

$C = 100 \text{ µF}$

What value is not known?

$X_C = ?$

What formula can be used?

$$X_C = \frac{1}{2\pi fC}$$

Substitute the known values and solve for X_C.

$$X_C = \frac{1}{2\pi fC}$$

$$X_C = \frac{1}{2 \times \pi \times 60 \text{ Hz} \times 100 \text{ µF}}$$

$$X_C = 26.54\ \Omega$$

Therefore, the capacitor presents 26.54 ohms of opposition to the current flowing in this circuit. Now that you know the opposition presented by the capacitor as well as that by the resistor, you can determine the total opposition (impedance) of the circuit. Find Z_T.

What values are known?

$R = 15\ \Omega$

$X_C = 26.54\ \Omega$

What value is not known?

$Z_T = ?$

As a result of the phase shift introduced by the capacitor, you cannot use the reciprocal of the sum of the reciprocals formula that you learned when you studied resistive parallel circuits. You must consider the effects of the capacitor and the resulting phase shift between the current and the voltage. A different formula that takes into account the effects of the capacitor must be used.

What formula can be used?

$$Z_T = \frac{1}{\sqrt{\left(\frac{1}{R}\right)^2 + \left(\frac{1}{X_C}\right)^2}}$$

Substitute the known values and solve for Z_T.

$$Z_T = \frac{1}{\sqrt{\left(\frac{1}{R}\right)^2 + \left(\frac{1}{X_C}\right)^2}}$$

$$Z_T = \frac{1}{\sqrt{\left(\frac{1}{15\ \Omega}\right)^2 + \left(\frac{1}{26.54\ \Omega}\right)^2}}$$

$$Z_T = \frac{1}{\sqrt{(0.067)^2 + (0.038)^2}}$$

$$Z_T = \frac{1}{\sqrt{0.0044 + 0.0014}}$$

$$Z_T = \frac{1}{\sqrt{0.0058}}$$

$$Z_T = \frac{1}{0.076}$$

$$Z_T = 13.16\ \Omega$$

Thus, the combined opposition to the current of the circuit is 13.16 ohms. Because you now know the total opposition and the total applied voltage, you can determine the total circuit current.

What values are known?

$E_T = 120 \text{ V}$

$Z_T = 13.16\ \Omega$

What value is not known?

$I_T = ?$

What formula can be used?

$$I_T = \frac{E_T}{Z_T}$$

Now substitute the known values and solve for I_T.

$$I_T = \frac{E_T}{Z_T}$$

$$I_T = \frac{120 \text{ V}}{13.16\ \Omega}$$

$$I_T = 9.19 \text{ A}$$

Therefore, there are 9.19 amperes of current flowing in this R-C parallel circuit.

As you have learned previously, the voltage in a parallel circuit remains the same across each of the branch circuits, because in a parallel circuit,

$E_T = E_R = E_C$

$E_T = 120 \text{ V}$

$E_R = 120 \text{ V}$

$E_C = 120 \text{ V}$

Now focus your attention on the resistor. You know the value of the resistor (15 ohms) and the amount of voltage dropped across the resistor (120 volts).

Now determine the amount of current flowing through the resistor, I_R.

What values are known?

$$R = 15\ \Omega$$

$$E_R = 120\ V$$

What value is not known?

$$I_R = ?$$

What formula can be used?

$$I_R = \frac{E_R}{R}$$

Substitute the known values and solve for I_R.

$$I_R = \frac{E_R}{R}$$

$$I_R = \frac{120\ V}{15\ \Omega}$$

$$I_R = 8.0\ A$$

Therefore, there are 8.0 amperes of current flowing through the resistor. The last unknown value for the resistor is the amount of power that the resistor must handle.

What values are known?

$$I_R = 8.0\ A$$

$$R = 15\ \Omega$$

What value is not known?

$$P_R = ?$$

What formula can be used?

$$P_R = I_R^2 \times R$$

Substitute the known values and solve for P_R.

$$P_R = I_R^2 \times R$$

$$P_R = 8.0\ A^2 \times 15\ \Omega$$

$$P_R = 64.0\ A \times 15\ \Omega$$

$$P_R = 960.0\ W$$

Therefore, the resistor consumes 960.0 watts of power. Now turn your attention to the capacitor. So far, you know values of C, X_C, and E_C. Find the amount of capacitive current, I_C, flowing in this circuit.

What values are known?

$$E_C = 120\ V$$

$$X_C = 26.54\ \Omega$$

What value is not known?

$$I_C = ?$$

What formula can be used?

$$I_C = \frac{E_C}{X_C}$$

Now substitute the known values and solve for I_C.

$$I_C = \frac{E_C}{X_C}$$

$$I_C = \frac{120\ V}{26.54\ \Omega}$$

$$I_C = 4.52\ A$$

Therefore, there are 4.52 amperes of capacitive current. Now notice something interesting. Recall the rule about branch currents in the study of parallel resistive circuits, which states that the sum of all branch currents is equal to the total circuit current. Is this true in this R-L parallel circuit? To find out, solve the following problem:

What values are known?

$$I_T = 9.19\ A$$

$$I_R = 8.0\ A$$

$$I_C = 4.52\ A$$

Does $I_T = I_R + I_C$?

$$I_T = I_R + I_C$$

$$I_T = 8.0\ A + 4.52\ A$$

$$I_T = 12.52\ A$$

This appears to be wrong, but it is not. Remember, the capacitor introduces a phase shift in the circuit. As a result of this phase shift, you cannot simply add the branch currents together. You must calculate the sum of the branch currents differently. Here is a new formula to use:

$$I_T = \sqrt{I_R^2 + I_C^2}$$

Now substitute the known values and solve for I_T.

$$I_T = \sqrt{I_R^2 + I_C^2}$$

$$I_T = \sqrt{8.0\ A^2 + 4.52\ A^2}$$

$$I_T = \sqrt{64.0\ A + 20.43\ A}$$

$$I_T = \sqrt{84.43\ A}$$

$$I_T = 9.19\ A$$

The calculation was correct! You simply cannot ignore the phase shift introduced by the capacitor in the circuit. Now determine the reactive power of the capacitor.

What values are known?

$$E_C = 120\ V$$

$$I_C = 4.52\ A$$

What value is not known?

$$VARs_C = ?$$

What formula can be used?

$$VARs_C = E_C \times I_C$$

Substitute the known values and solve for $VARs_C$.

$$VARs_C = E_C \times I_C$$

$$VARs_C = 120\ \text{V} \times 4.52\ \text{A}$$

$$VARs_C = 542.40\ \text{VARs}_C$$

Therefore, the capacitive reactive power of the circuit is 542.40 VARs_C. Now determine the circuit's apparent power.

What values are known?

$$P_R = 960\ \text{W}$$

$$VARs_C = 542.40\ \text{VARs}_C$$

What value is not known?

$$VA = ?$$

What formula can be used?

$$VA = \sqrt{P^2 + VARs_C^2}$$

Substitute the known values and solve for VA.

$$VA = \sqrt{P^2 + VARs_C^2}$$

$$VA = \sqrt{960\ \text{W}^2 + 542.40\ \text{VARs}_C^2}$$

$$VA = \sqrt{921600\ \text{W} + 294197.76\ \text{VARs}_C}$$

$$VA = \sqrt{1215797.76}$$

$$VA = 1102.63\ \text{VA}$$

Therefore, the total apparent power of this circuit is 1102.63 volt-amperes.

Before performing the final calculations on this circuit, a detailed analysis of some of the results has to be done. In particular, compare the amount of opposition offered by the resistor with that of the capacitor. The resistor offers 15 ohms of opposition, whereas the capacitor offers 26.54 ohms. The resistor has more influence on the current flow than does the capacitor. Now compare the amount of current flow through the resistor with the amount of current flow through the capacitor. The resistor current is 8.0 amperes, whereas the capacitor current is 4.52 amperes. The resistor has the larger current flow. Finally, compare the amount of resistive power with the amount of reactive power. The resistive power is 960 watts, whereas the reactive power is 542.40 VARs_C. The resistive power is larger. When all of this is considered, it should be apparent that the resistor has more influence on the circuit than does the capacitor. You know that if the circuit were purely resistive, the current and voltage would be in phase, and the phase angle difference between the current and voltage would be 0°. If the circuit were purely capacitive, the current and voltage would be out of phase, and the phase angle difference would be 90° leading. This circuit is a combination of resistance and capacitive reactance. This means that the phase angle must fall somewhere between 0° and 90°. When comparing the influence of the resistor with that of the capacitor in this circuit, it shows that the resistor has more influence. This means the phase angle difference must fall somewhere between 0° and 45°. To see whether this assumption is correct, first calculate the power factor of the circuit.

What values are known?

$$P_R = 960\ \text{W}$$

$$VA = 1102.63\ \text{VA}$$

What value is not known?

$$PF = ?$$

What formula can be used?

$$PF = \frac{P}{VA} \times 100$$

Substitute the known values and solve for PF.

$$PF = \frac{P}{VA} \times 100$$

$$PF = \frac{960\ \text{W}}{1102.63\ \text{VA}} \times 100$$

$$PF = 0.8706 \times 100$$

$$PF = 87.06\%$$

Therefore, this circuit has a power factor of 87.06%. Now determine the phase angle difference, $\angle\theta$.

What value is known?

$PF = 0.8706$ (For this calculation, the power factor is not expressed as a percent.)

What value is not known?

$$\angle\theta = ?$$

What formula can be used?

$$\angle\theta = COS^{-1}PF$$

Substitute the known values and solve for $\angle\theta$.

$$\angle\theta = COS^{-1}PF$$

$$\angle\theta = COS^{-1}0.8706$$

$$\angle\theta = 29.47°$$

Therefore, in this circuit the current leads the voltage by 29.47°. Because the resistor has more influence in the circuit, the circuit is more resistive and the phase angle approaches 0°. Compare the remaining parameters of the two circuits shown in **Figure 17-18A** and **Figure 17-18B**. Notice the changes caused by connecting the components in either series or parallel.

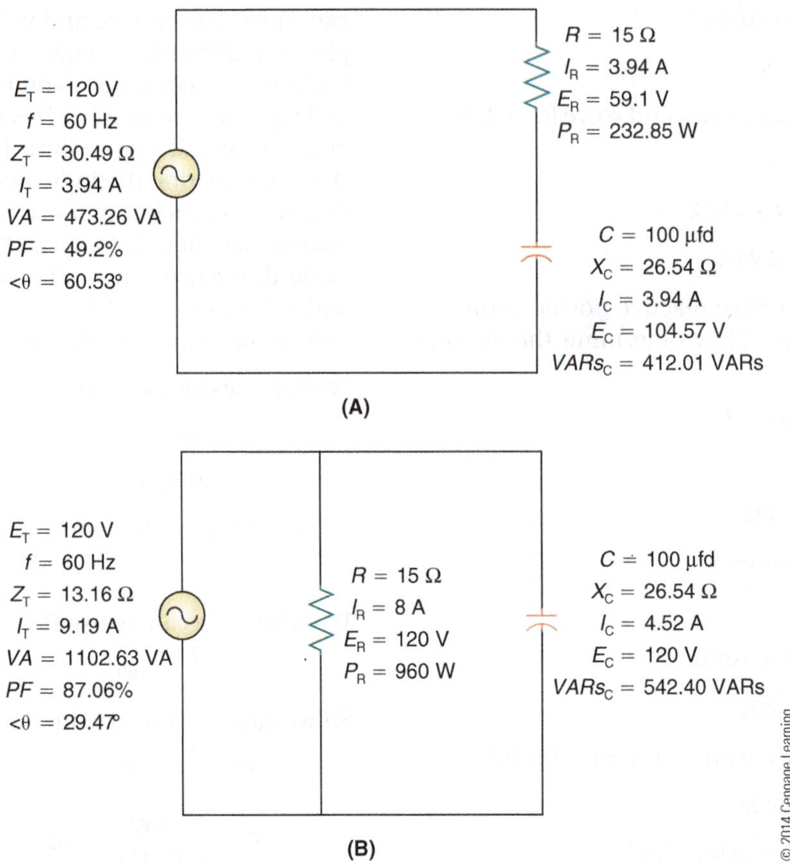

FIGURE 17-18 (A) A fully analyzed R-C series circuit; and (B) a fully analyzed R-C parallel circuit.

17-3 R-L-C SERIES AND PARALLEL CIRCUITS

Now we analyze a circuit that can be either series or parallel and contain resistance, inductance, and capacitance, otherwise known as **R-L-C circuits.** In analyzing an R-L-C circuit, you must keep in mind the effects of the inductance and capacitance in the circuit. Recall that in a purely inductive circuit, the current lags the voltage by 90°, whereas in a purely capacitive circuit the current leads the voltage by 90°. When a circuit contains both inductance and capacitance, the two reactive components have an effect on each other. The smaller reactance is cancelled out by the larger reactance. You will better understand this concept by working through two examples: one of an R-L-C series circuit and the other of an R-L-C parallel circuit.

R-L-C Series Circuit

Figure 17-19 shows an R-L-C series circuit with values. Solve for the unknown values, using the same values for resistance, inductance, and capacitance as those in previous examples. This provides you with a better understanding of the effects of combining the two reactive components.

What values are known?

$E_T = 120$ V

$f = 60$ Hz

$R = 15\ \Omega$

$L = 20$ mH

$C = 100\ \mu$F

What values are not known?

$I_T = ?$

$Z_T = ?$

$VA = ?$

$PF = ?$

$\angle\theta = ?$

$E_R = ?$

$I_R = ?$

$P_R = ?$

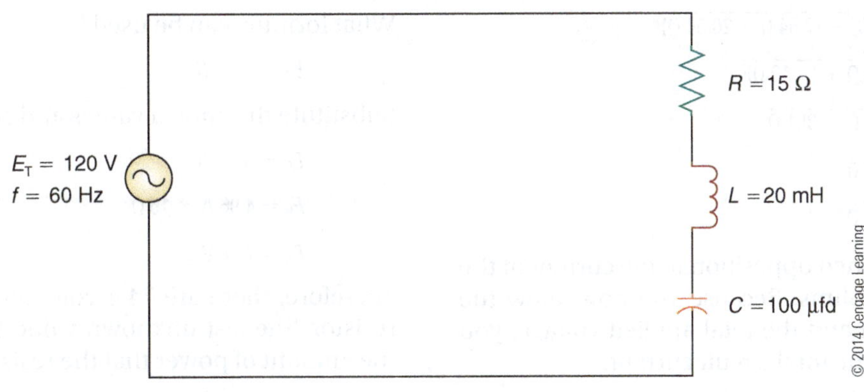

FIGURE 17-19 An R-L-C series circuit with values.

$E_L = ?$

$I_L = ?$

$X_L = ?$

$VARs_L = ?$

$E_C = ?$

$I_C = ?$

$X_C = ?$

$VARs_C = ?$

Begin by solving for X_L.

What values are known?

$f = 60\text{ Hz}$

$L = 20\text{ mH}$

What value is not known?

$X_L = ?$

What formula can be used?

$$X_L = 2\pi fL$$

Substitute the known values and solve for X_L.

$$X_L = 2\pi fL$$

$$X_L = 2 \times \pi \times 60\text{ Hz} \times 20\text{ mH}$$

$$X_L = 7.54\ \Omega$$

Therefore, the inductor presents 7.54 ohms of opposition to the current flowing in this circuit. Next, we solve for X_C.

What values are known?

$f = 60\text{ Hz}$

$C = 100\ \mu\text{F}$

What value is not known?

$X_C = ?$

What formula can be used?

$$X_C = \frac{1}{2\pi fC}$$

Substitute the known values and solve for X_C.

$$X_C = \frac{1}{2\pi fC}$$

$$X_C = \frac{1}{2 \times \pi \times 60\text{ Hz} \times 100\ \mu\text{F}}$$

$$X_C = 26.54\ \Omega$$

Therefore, the capacitor presents 26.54 ohms of opposition to the current flowing in this circuit. Now that you know the opposition presented by the inductor, the opposition presented by the capacitor, and the opposition presented by the resistor, you can determine the total opposition (impedance) of the circuit. Find Z_T.

What values are known?

$R = 15\ \Omega$

$X_C = 26.54\ \Omega$

$X_L = 7.54\ \Omega$

What value is not known?

$Z_T = ?$

As a result of the lagging phase shift introduced by the inductor and the leading phase shift introduced by the capacitor, you cannot simply add the resistance of the resistor to the inductive reactance and the capacitive reactance. You must use a different formula that takes into account the effects of the inductor and capacitor.

What formula can be used?

$$Z_T = \sqrt{R^2 + (X_L - X_C)^2}$$

Substitute the known values and solve for Z_T.

$$Z_T = \sqrt{R^2 + (X_L - X_C)^2}$$

$Z_T = \sqrt{15\ \Omega^2 + (7.54\ \Omega - 26.54\ \Omega)^2}$

$Z_T = \sqrt{225\ \Omega + (-19\ \Omega)^2}$

$Z_T = \sqrt{225\ \Omega + 361\ \Omega}$

$Z_T = \sqrt{586\ \Omega}$

$Z_T = 24.21\ \Omega$

Thus, the combined opposition to the current of the circuit is 24.21 ohms. Because you now know the total opposition and the total applied voltage, you can determine the total circuit current.

What values are known?

$E_T = 120\ V$

$Z_T = 24.21\ \Omega$

What value is not known?

$I_T = ?$

What formula can be used?

$I_T = \frac{E_T}{Z_T}$

Now substitute the known values and solve for I_T.

$I_T = \frac{E_T}{Z_T}$

$I_T = \frac{120\ V}{24.21\ \Omega}$

$I_T = 4.96\ A$

Therefore, there are 4.96 amperes of current flowing in this R-L-C series circuit. This is the same amount of current flowing throughout this circuit. This means that there are 4.96 amperes of resistive current, 4.96 amperes of inductive current, and 4.96 amperes of capacitive current. Therefore, you now know that:

$I_R = 4.96\ A$

$I_L = 4.96\ A$

$I_C = 4.96\ A$

because in a series circuit,

$$I_T = I_R = I_L = I_C$$

Now focus your attention on the resistor. You know the value of the resistor (15 ohms) and the amount of current flowing through the resistor (4.96 amperes). Determine the amount of voltage dropped across the resistor, E_R.

What values are known?

$R = 15\ \Omega$

$I_R = 4.96\ A$

What value is not known?

$E_R = ?$

What formula can be used?

$E_R = I_R \times R$

Substitute the known values and solve for E_R.

$E_R = I_R \times R$

$E_R = 4.96\ A \times 15\ \Omega$

$E_R = 74.4\ V$

Therefore, there are 74.4 volts dropped across the resistor. The last unknown value for the resistor is the amount of power that the resistor must handle.

What values are known?

$I_R = 4.96\ A$

$R = 15\ \Omega$

What value is not known?

$P_R = ?$

What formula can be used?

$P_R = I_R{}^2 \times R$

Substitute the known values and solve for P_R.

$P_R = I_R{}^2 \times R$

$P_R = 4.96\ A^2 \times 15\ \Omega$

$P_R = 24.6\ A \times 15\ \Omega$

$P_R = 369.02\ W$

Therefore, the resistor consumes 369.02 watts of power. Now turn your attention to the inductor. So far, you know the values for L, X_L, and I_L. Find the amount of voltage dropped across the inductor, E_L.

What values are known?

$I_L = 4.96\ A$

$X_L = 7.54\ \Omega$

What value is not known?

$E_L = ?$

What formula can be used?

$E_L = I_L \times X_L$

Now substitute the known values and solve for E_L.

$E_L = I_L \times X_L$

$E_L = 4.96\ A \times 7.54\ \Omega$

$E_L = 37.4\ V$

Therefore, there are 37.4 volts dropped across the inductor. Now find the amount of voltage dropped across the capacitor, E_C.

What values are known?

$I_C = 4.96\ A$

$X_C = 26.54\ \Omega$

What value is not known?

$E_C = ?$

What formula can be used?

$E_C = I_C \times X_C$

Now substitute the known values and solve for E_C.

$E_C = I_C \times X_C$

$E_C = 4.96\text{ A} \times 26.54\ \Omega$

$E_C = 131.64\text{ V}$

Therefore, there are 131.64 volts dropped across the capacitor.

Now notice something interesting. Recall the rule about voltage drops in the study of series circuits, which states that the sum of all voltage drops equals the total applied voltage. Is this true in this R-L-C series circuit? To find out, solve the following problem:

What values are known?

$E_T = 120\text{ V}$

$E_R = 74.4\text{ V}$

$E_L = 37.4\text{ V}$

$E_C = 131.64\text{ V}$

Does $E_T = E_R + E_L + E_C$?

$E_T = E_R + E_L + E_C$

$E_T = 74.4\text{ V} + 37.4\text{ V} + 131.64\text{ V}$

$E_T = 243.44\text{ V}$

This appears to be wrong, but it it not. Remember, the inductor and capacitor introduce a phase shift in the circuit. As a result of this phase shift, you cannot simply add the voltage drops together. You must calculate the sum of the voltage drops differently. Here is a new formula to use:

$$E_T = \sqrt{E_R^2 + (E_L - E_C)^2}$$

Now substitute the known values and solve for E_T.

$E_T = \sqrt{E_R^2 + (E_L - E_C)^2}$

$E_T = \sqrt{74.4\text{ V}^2 + (37.4\text{ V} - 131.64\text{ V})^2}$

$E_T = \sqrt{74.4\text{ V}^2 + (-94.24\text{ V})^2}$

$E_T = \sqrt{5535.36\text{ V} + 8881.18\text{ V}}$

$E_T = \sqrt{14416.54\text{ V}}$

$E_T = 120.07\text{ V}$

The calculation was correct! You simply cannot ignore the phase shift introduced by the inductor and the capacitor in the circuit. Do not be concerned with the 0.07 volt difference between this answer and the given value. This is a result of the rounding off of numbers to two decimal places during calculations. Now determine the reactive power of the inductor.

What values are known?

$E_L = 37.4\text{ V}$

$I_L = 4.96\text{ A}$

What value is not known?

$VARs_L = ?$

What formula can be used?

$VARs_L = E_L \times I_L$

Substitute the known values and solve for $VARs_L$.

$VARs_L = E_L \times I_L$

$VARs_L = 37.4\text{ V} \times 4.96\text{ A}$

$VARs_L = 185.5\text{ VARs}_L$

Therefore, the inductive reactive power of the circuit is 185.5 VARs_L. Now determine the reactive power of the capacitor.

What values are known?

$E_C = 131.64\text{ V}$

$I_C = 4.96\text{ A}$

What value is not known?

$VARs_C = ?$

What formula can be used?

$VARs_C = E_C \times I_C$

Substitute the known values and solve for $VARs_C$.

$VARs_C = E_C \times I_C$

$VARs_C = 131.64\text{ V} \times 4.96\text{ A}$

$VARs_C = 652.93\text{ VARs}_C$

Therefore, the capacitive reactive power of the circuit is 652.93 VARs_C.

Next, calculate the apparent power of the circuit. Again, you must consider the effects of the inductor and those of the capacitor in the circuit. Here is how you can calculate the apparent power of the R-L-C series circuit.

What values are known?

$P_R = 369.02\text{ W}$

$VARs_L = 185.5\text{ VARs}_L$

$VARs_C = 652.93\text{ VARs}_C$

What value is not known?

$VA = ?$

What formula can be used?

$$VA = \sqrt{P_R^2 + (VARs_L - VARs_C)^2}$$

Substitute the known values and solve for *VA*.

$$VA = \sqrt{P_R^2 + (VARs_L - VARs_C)^2}$$

$$VA = \sqrt{369.02\ W^2 + (185.5\ VARs_L - 652.93\ VARs_C)^2}$$

$$VA = \sqrt{369.02\ W^2 + (-467.43\ VARs_C)^2}$$

$$VA = \sqrt{136175.76\ W + 218490.8\ VARs_C}$$

$$VA = \sqrt{354666.56}$$

$$VA = 595.54\ VA$$

Therefore, the total apparent power of this circuit is 595.54 volt-amperes.

Before performing the final calculations on this circuit, a detailed analysis of some of the results needs to be done. In particular, compare the amount of opposition offered by the inductor with that of the capacitor. The inductor offers 7.54 ohms of opposition, whereas the capacitor offers 26.54 ohms. The capacitor has more influence on the current flow than does the inductor. Now compare the amount of voltage dropped across the inductor with the amount of voltage dropped across the capacitor. The inductor voltage is 37.54 volts, whereas the capacitor voltage is 131.64 volts. The capacitor has the larger voltage drop. Finally, compare the amount of inductive reactive power with the amount of capacitive reactive power. The inductive reactive power is 185.5 $VARs_L$, whereas the capacitive reactive power is 652.93 $VARs_C$. The capacitive reactive power is larger. When all of this is considered, it should be apparent that the capacitor has more influence on the circuit than does the inductor. You know that if the circuit were purely resistive, the current and voltage would be in phase, and the phase angle difference between the current and voltage would be 0°. If the circuit were purely inductive, the current and voltage would be out of phase, and the phase angle difference would be 90° lagging. If the circuit were purely capacitive, the current and voltage would be out of phase, and the phase angle difference would be 90° leading. This circuit is a combination of resistance, inductance, and capacitance. This means that the phase angle must fall somewhere between 0° and 90° and could be lagging or leading. When comparing the influence of the inductor and that of the capacitor in this circuit, it shows that the capacitor has more influence. This means the phase angle difference must fall somewhere between 0° and 90° *leading*. Now compare the resistor with the capacitor. The resistor offers 15 ohms of opposition, whereas the capacitor offers 19 ohms (7.54 ohms of capacitive reactance was cancelled by the 7.54 ohms of inductive reactance). Therefore, 19 ohms of capacitive reactance remains in the circuit. The capacitor has more influence on the current flow than does the resistor. Now compare the amount of voltage dropped across the resistor with the amount of voltage dropped across the capacitor. The resistor voltage is 74.4 volts, whereas the capacitor voltage is 94.24 volts (37.4 volts of capacitive voltage was cancelled by the 37.4 volts of inductive voltage). The capacitor has the larger voltage drop. Finally, compare the amount of resistive power with the amount of capacitive reactive power. The resistive power is 369.02 watts, whereas the capacitive reactive power is 467.43 $VARs_C$ (185.5 $VARs_C$ of capacitive reactive power was cancelled by the 185.5 $VARs_L$ of inductive reactive power). The capacitive reactive power is larger. When all of this is considered, it should be apparent that the capacitor has more influence on the circuit than does the resistor. You know that if the circuit were purely resistive, the current and voltage would be in phase, and the phase angle difference between the current and voltage would be 0°. If the circuit were purely capacitive, the current and voltage would be out of phase, and the phase angle difference would be 90° leading. This circuit is a combination of resistance and capacitive reactance; the inductive influence has been cancelled by the capacitor. This means that the phase angle must fall somewhere between 0° and 90°. When comparing the influence of the resistor and that of the capacitor in this circuit, it shows that the capacitor has more influence. This means the phase angle difference must fall somewhere between 45° and 90°. To see if this assumption is correct, first calculate the power factor of the circuit using the formula $PF = \frac{P}{VA} \times 100$.

What values are known?

$$P_R = 369.02\ W$$

$$VA = 595.54\ VA$$

What value is not known?

$$PF = ?$$

What formula can be used?

$$PF = \frac{P}{VA} \times 100$$

Substitute the known values and solve for *PF*.

$$PF = \frac{P}{VA} \times 100$$

$$PF = \frac{369.02\ W}{595.54\ VA} \times 100$$

$$PF = 0.6196 \times 100$$

$$PF = 61.96\%$$

Therefore, this circuit has a power factor of 61.96%. Now determine the phase angle difference, $\angle\theta$.

What value is known?

$PF = 0.6196$ (For this calculation, the power factor is not expressed as a percent.)

What value is not known?

$\angle\theta = ?$

What formula can be used?

$\angle\theta = COS^{-1}\ PF$

Substitute the known values and solve for $\angle\theta$.

$\angle\theta = COS^{-1}\ PF$

$\angle\theta = COS^{-1}\ 0.6196$

$\angle\theta = 51.71°$

Therefore, in this circuit the current leads the voltage by 51.71°.

R-L-C Parallel Circuit

Figure 17-20 shows an R-L-C parallel circuit. So that you can compare the effects of a series circuit with those of a parallel circuit, use the same values from the R-L-C series circuit.

What values are known?

$E_T = 120\ V$

$f = 60\ Hz$

$R = 15\ \Omega$

$L = 20\ mH$

$C = 100\ \mu F$

What values are not known?

$I_T = ?$

$Z_T = ?$

$VA = ?$

$PF = ?$

$\angle\theta = ?$

$E_R = ?$

$I_R = ?$

$P_R = ?$

$E_L = ?$

$I_L = ?$

$X_L = ?$

$VARs_L = ?$

$E_C = ?$

$I_C = ?$

$X_C = ?$

$VARs_C = ?$

Begin by solving for X_L.

What values are known?

$f = 60\ Hz$

$L = 20\ mH$

What value is not known?

$X_L = ?$

What formula can be used?

$X_L = 2\pi fL$

Substitute the known values and solve for X_L.

$X_L = 2\pi fL$

$X_L = 2 \times \pi \times 60\ Hz \times 20\ mH$

$X_L = 7.54\ \Omega$

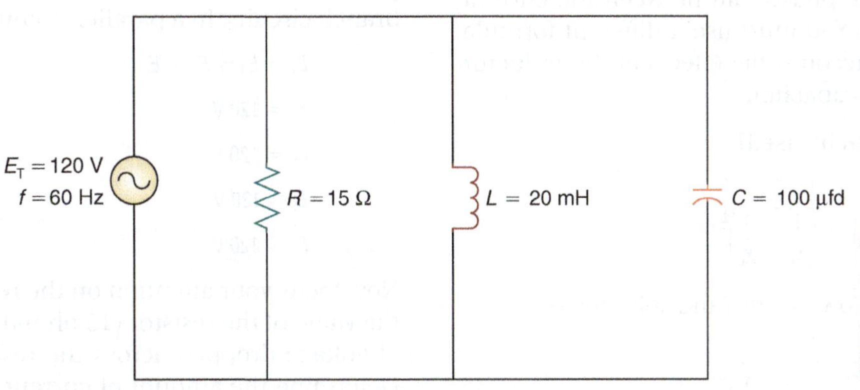

FIGURE 17-20 An R-L-C parallel circuit with values.

Therefore, the inductor presents 7.54 ohms of opposition to the current flowing in this circuit. Now calculate the capacitive reactance, X_C.

What values are known?

$$f = 60\text{ Hz}$$

$$C = 100\ \mu\text{F}$$

What value is not known?

$$X_C = ?$$

What formula can be used?

$$X_C = \frac{1}{2\pi fC}$$

Substitute the known values and solve for X_C.

$$X_C = \frac{1}{2\pi fC}$$

$$X_C = \frac{1}{2 \times \pi \times 60\text{ Hz} \times 100\ \mu\text{F}}$$

$$X_C = 26.54\ \Omega$$

Therefore, the capacitor presents 26.54 ohms of opposition to the current flowing in this circuit. Now that you know the opposition presented by the inductor, the capacitor, and the resistor, you can determine the total opposition (impedance) of the circuit. Now find Z_T.

What values are known?

$$R = 15\ \Omega$$

$$X_L = 7.54\ \Omega$$

$$X_C = 26.54\ \Omega$$

What value is not known?

$$Z_T = ?$$

As a result of the phase shift introduced by the inductor, you cannot use the reciprocal of the sum of the reciprocals formula that you learned when you studied resistive parallel circuits. You must consider the effects of the inductor and the capacitor and the resulting phase shift between the current and the voltage. You must use a different formula that takes into account the effects of the inductor and those of the capacitor.

What formula can be used?

$$Z_T = \frac{1}{\sqrt{\left(\frac{1}{R}\right)^2 + \left(\frac{1}{X_L} - \frac{1}{X_C}\right)^2}}$$

Substitute the known values and solve for Z_T.

$$Z_T = \frac{1}{\sqrt{\left(\frac{1}{R}\right)^2 + \left(\frac{1}{X_L} - \frac{1}{X_C}\right)^2}}$$

$$Z_T = \frac{1}{\sqrt{\left(\frac{1}{15\ \Omega}\right)^2 + \left(\frac{1}{7.54\ \Omega} - \frac{1}{26.54\ \Omega}\right)^2}}$$

$$Z_T = \frac{1}{\sqrt{(0.067\ \Omega)^2 + (0.133\ \Omega - 0.038\ \Omega)^2}}$$

$$Z_T = \frac{1}{\sqrt{(0.067\ \Omega)^2 + (0.095\ \Omega)^2}}$$

$$Z_T = \frac{1}{\sqrt{0.0045 + 0.009}}$$

$$Z_T = \frac{1}{\sqrt{0.014}}$$

$$Z_T = \frac{1}{0.118}$$

$$Z_T = 8.47\ \Omega$$

Thus, the combined opposition to the current of the circuit is 8.47 ohms. Because you now know the total opposition and the total applied voltage, you can determine the total circuit current.

What values are known?

$$E_T = 120\text{ V}$$

$$Z_T = 8.47\ \Omega$$

What value is not known?

$$I_T = ?$$

What formula can be used?

$$I_T = \frac{E_T}{Z_T}$$

Now substitute the known values and solve for I_T.

$$I_T = \frac{E_T}{Z_T}$$

$$I_T = \frac{120\text{ V}}{8.47\ \Omega}$$

$$I_T = 14.17\text{ A}$$

Therefore, there are 14.17 amperes of current flowing in this R-L-C parallel circuit.

As you have learned previously, the voltage in a parallel circuit remains the same across each of the branch circuits. In a parallel circuit,

$$E_T = E_R = E_L = E_C$$

$$E_T = 120\text{ V}$$

$$E_R = 120\text{ V}$$

$$E_L = 120\text{ V}$$

$$E_C = 120\text{ V}$$

Now focus your attention on the resistor. You know the value of the resistor (15 ohms) and the amount of voltage dropped across the resistor (120 volts). Determine the amount of current flowing through the resistor, I_R.

What values are known?

$R = 15\ \Omega$

$E_R = 120\ V$

What value is not known?

$I_R = ?$

What formula can be used?

$$I_R = \frac{E_R}{R}$$

Substitute the known values and solve for I_R.

$$I_R = \frac{E_R}{R}$$

$$I_R = \frac{120\ V}{15\ \Omega}$$

$I_R = 8.0\ A$

Therefore, there are 8.0 amperes of current flowing through the resistor. The last unknown value for the resistor is the amount of power that the resistor must handle.

What values are known?

$I_R = 8.0\ A$

$R = 15\ \Omega$

What value is not known?

$P_R = ?$

What formula can be used?

$P_R = I_R^2 \times R$

Substitute the known values and solve for P_R.

$P_R = I_R^2 \times R$

$P_R = 8.0\ A^2 \times 15\ \Omega$

$P_R = 64.0\ A \times 15\ \Omega$

$P_R = 960.0\ W$

Therefore, the resistor consumes 960.0 watts of power. Now turn your attention to the inductor. So far, you know the values of L, X_L, and E_L. Find the amount of current flowing through the inductor, I_L.

What values are known?

$E_L = 120\ V$

$X_L = 7.54\ \Omega$

What value is not known?

$I_L = ?$

What formula can be used?

$$I_L = \frac{E_L}{X_L}$$

Now substitute the known values and solve for I_L.

$$I_L = \frac{E_L}{X_L}$$

$$I_L = \frac{120\ V}{7.54\ \Omega}$$

$I_L = 15.92\ A$

Therefore, there are 15.92 amperes of current flowing through the inductor. Next turn your attention to the capacitor. So far, you know the values of C, X_C, and E_C. Find the amount of current flowing in the capacitor circuit, I_C.

What values are known?

$E_C = 120\ V$

$X_C = 26.54\ \Omega$

What value is not known?

$I_C = ?$

What formula can be used?

$$I_C = \frac{E_C}{X_C}$$

Now substitute the known values and solve for I_C.

$$I_C = \frac{E_C}{X_C}$$

$$I_C = \frac{120\ V}{26.54\ \Omega}$$

$I_C = 4.52\ A$

Therefore, there are 4.52 amperes of current flowing in the capacitor circuit. Now notice something interesting. Recall the rule about branch currents in the study of parallel resistive circuits, which states that the sum of all branch currents equals the total circuit current. Is this true in this R-L-C parallel circuit? To find out, solve the following problem:

What values are known?

$I_T = 14.17\ A$

$I_R = 8.0\ A$

$I_L = 15.92\ A$

$I_C = 4.52\ A$

Does $I_T = I_R + I_L + I_C$?

$I_T = I_R + I_L + I_C$

$I_T = 8.0\ A + 15.92\ A + 4.52\ A$

$I_T = 28.44\ A$

This appears to be wrong, but it is not. Remember, the inductor and capacitor introduce a phase shift in the circuit. As a result of this phase shift, you cannot simply add the branch currents together.

You must calculate the sum of the branch currents differently. Here is a new formula to use:

$$I_T = \sqrt{I_R^2 + (I_L - I_C)^2}$$

Now substitute the known values and solve for I_T.

$$I_T = \sqrt{I_R^2 + (I_L - I_C)^2}$$

$$I_T = \sqrt{8.0\ A^2 + (15.92\ A - 4.52\ A)^2}$$

$$I_T = \sqrt{8.0\ A^2 + 11.4\ A^2}$$

$$I_T = \sqrt{64.0\ A + 129.96\ A}$$

$$I_T = \sqrt{193.96\ A}$$

$$I_T = 13.93\ A$$

The calculation was correct! You simply cannot ignore the phase shift introduced by the inductor in the circuit. Do not be concerned with the 0.24-ampere difference between this answer and the given value. This is a result of the rounding off of numbers to two decimal places during calculations. Now determine the reactive power of the inductor.

What values are known?

$$E_L = 120\ V$$

$$I_L = 15.92\ A$$

What value is not known?

$$VARs_L = ?$$

What formula can be used?

$$VARs_L = E_L \times I_L$$

Substitute the known values and solve for $VARs_L$.

$$VARs_L = E_L \times I_L$$

$$VARs_L = 120\ V \times 15.92\ A$$

$$VARs_L = 1910.4\ VARs_L$$

Therefore, the inductive reactive power of the circuit is 1910.4 $VARs_L$. Next, calculate the reactive power of the capacitor.

What values are known?

$$E_C = 120\ V$$

$$I_C = 4.52\ A$$

What value is not known?

$$VARs_C = ?$$

What formula can be used?

$$VARs_C = E_C \times I_C$$

Substitute the known values and solve for $VARs_C$.

$$VARs_C = E_C \times I_C$$

$$VARs_C = 120\ V \times 4.52\ A$$

$$VARs_C = 542.4\ VARs_C$$

Therefore, the capacitive reactive power of the circuit is 542.4 $VARs_C$. Now determine the circuit's apparent power.

What values are known?

$$P_R = 960\ W$$

$$VARs_L = 1910.4\ VARs_L$$

$$VARs_C = 542.4\ VARs_C$$

What value is not known?

$$VA = ?$$

What formula can be used?

$$VA = \sqrt{P_R^2 + (VARs_L - VARs_C)^2}$$

Substitute the known values and solve for VA.

$$VA = \sqrt{P_R^2 + (VARs_L - VARs_C)^2}$$

$$VA = \sqrt{960\ W^2 + (1910.4\ VARs_L - 542.4\ VARs_C)^2}$$

$$VA = \sqrt{960\ W^2 + 1368\ VARs^2}$$

$$VA = \sqrt{921600\ W + 1871424\ VARs}$$

$$VA = \sqrt{2793024}$$

$$VA = 1671.23\ VA$$

Therefore, the total apparent power of this circuit is 1671.23 volt-amperes.

Before performing the final calculations on this circuit, a detailed analysis of some of the results must be done. In particular, compare the amount of opposition offered by the inductor with that of the capacitor. The inductor offers 7.54 ohms of opposition, whereas the capacitor offers 26.54 ohms. The inductor has more influence on the current flow than does the capacitor. Now compare the amount of inductive current with the amount of capacitive current. The inductive current is 15.92 amperes, whereas the capacitive current is 4.52 amperes. The inductive current is larger. Finally, compare the amount of inductive reactive power with the amount of capacitive reactive power. The inductive reactive power is 1910.4 $VARs_L$, whereas the capacitive reactive power is 542.4 $VARs_C$. The inductive reactive power is larger. When all of this is considered, it should be apparent that the inductor has more influence on the circuit than does the capacitor. You know that if the circuit were purely resistive, the current and voltage would be in phase, and the phase angle difference between the current and voltage would be 0°. If the circuit were purely inductive, the current and voltage would be out of phase, and the phase angle difference would be 90° lagging. If the circuit were purely capacitive, the current and voltage would be out of phase, and the phase angle difference would be 90° leading. This circuit is a combination of resistance, inductance, and capacitance. This means that the phase angle must fall somewhere between 0° and 90° and could be lagging or leading. By comparing the influence of the inductor with that of the capacitor in this circuit, it was determined that the inductor has more influence. This means the phase angle difference must fall somewhere between 0° and 90° *lagging*. Now compare the resistor and the inductor. The resistor offers 15 ohms of opposition, whereas the inductor offers 7.54 ohms. The inductor has more influence on the current flow than does the resistor. Now compare the amount of resistive current with the amount of inductive current. The resistive current is 8.0 amperes, whereas the inductive current is 15.92 amperes. The inductive current is larger. Finally, compare the amount of resistive power with the amount of inductive reactive power. The resistive power is 960 watts, whereas the inductive reactive power is 1910.4 $VARs_L$. The inductive reactive power is larger. When all of this is considered, it should be apparent that the inductor has more influence on the circuit than does the resistor. We know that if the circuit were purely resistive, the current and voltage would be in phase, and the phase angle difference between the current and voltage would be 0°. If the circuit were purely inductive, the current and voltage would be out of phase, and the phase angle difference would be 90° lagging. This circuit is a combination of resistance and inductive reactance; the capacitive influence has been cancelled by the inductor. This means that the phase angle must fall somewhere between 0° and 90°. By comparing the influence of the resistor with that of the inductor in this circuit, it was determined that the inductor has more influence. This means the phase angle difference must fall somewhere between 45° and 90° lagging. To see whether this assumption is correct, first calculate the power factor of the circuit.

What values are known?

$$P_R = 960 \text{ W}$$

$$VA = 1671.23 \text{ VA}$$

What value is not known?

$$PF = ?$$

What formula can be used?

$$PF = \frac{P}{VA} \times 100$$

Substitute the known values and solve for *PF*.

$$PF = \frac{P}{VA} \times 100$$

$$PF = \frac{960 \text{ W}}{1671.23 \text{ VA}} \times 100$$

$$PF = 0.5744 \times 100$$

$$PF = 57.44\%$$

Therefore, this circuit has a power factor of 57.44%. Now determine the phase angle difference, $\angle\theta$.

What value is known?

$PF = 0.5744$ (For this calculation, the power factor is not expressed as a percent.)

What value is not known?

$$\angle\theta = ?$$

What formula can be used?

$$\angle\theta = COS^{-1}PF$$

Substitute the known values and solve for $\angle\theta$.

$$\angle\theta = COS^{-1}PF$$

$$\angle\theta = COS^{-1}0.5744$$

$$\angle\theta = 54.94°$$

Therefore, in this circuit the current lags the voltage by 54.94°. Because the inductor has more influence in the circuit, the circuit is more inductive and the phase angle approaches 90°. Compare the remaining parameters of the two circuits shown in **Figure 17-21A** and **Figure 17-21B**. Notice the changes caused by connecting the components either in series or in parallel.

17-4 POWER FACTOR CORRECTION

Now that you have learned how to analyze a circuit that contains resistive, inductive, and capacitive components, this information is used to improve the operating efficiency of an AC motor. This process is called **power factor correction**. Because an AC motor is constructed of wire formed into coils or windings, it is considered an inductive device. In addition, the windings have a resistance to the flow of current. Therefore, the AC motor presents both a resistive and an inductive component to the circuit. **Figure 17-22** shows an AC motor circuit where the motor is represented by the inductor, and the resistance of the windings is represented by the resistor. Notice that this circuit appears as an R-L series circuit. To calculate the power factor and phase angle for this circuit, begin by determining the apparent power of the circuit.

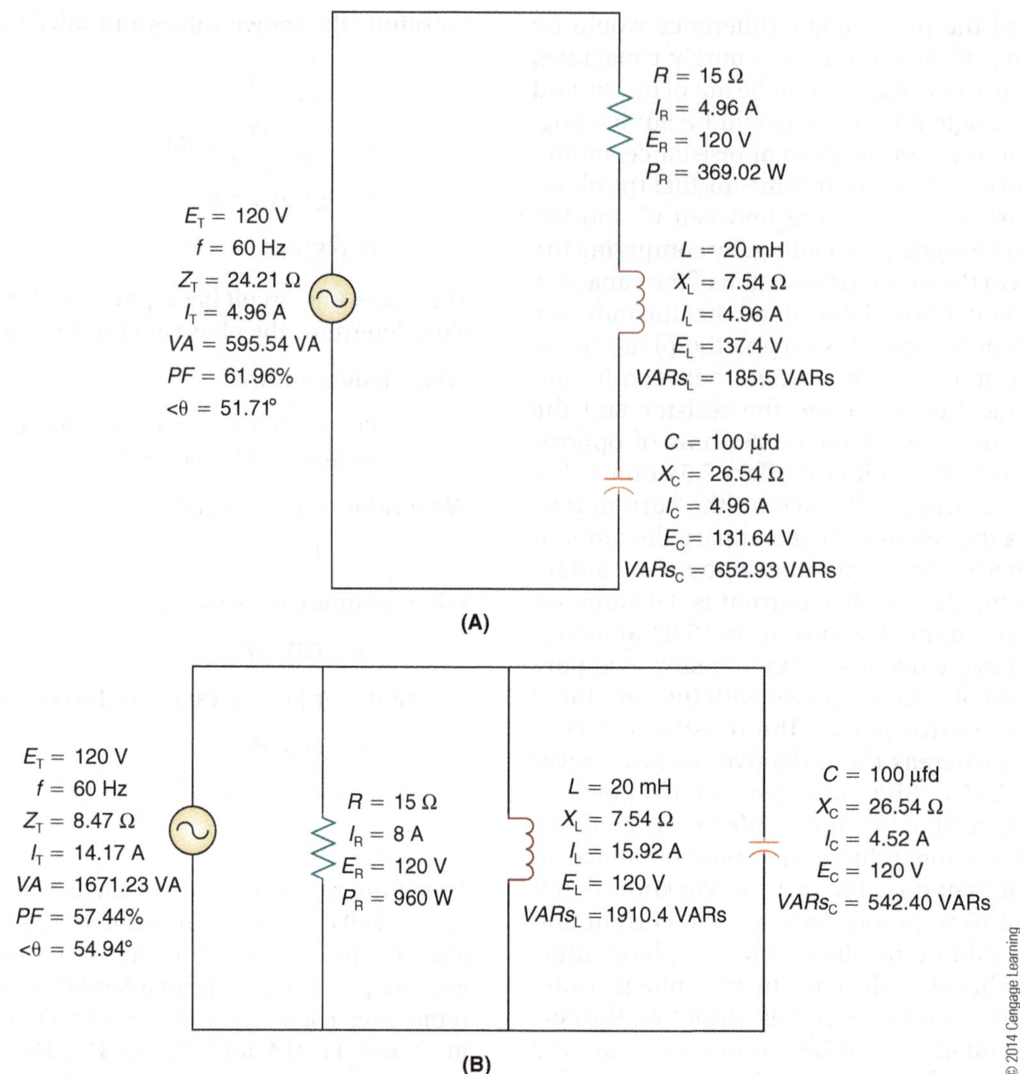

FIGURE 17-21 (A) A fully analyzed R-L-C series circuit; and (B) a fully analyzed R-L-C parallel circuit.

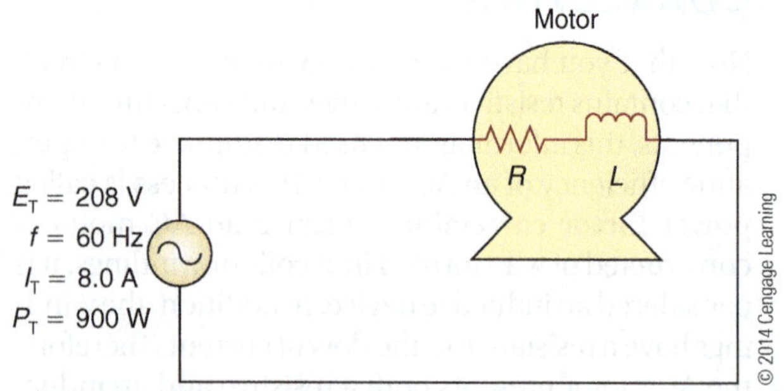

FIGURE 17-22 The AC motor presents both a resistive component and an inductive component to the circuit.

What values are known?

$E = 208\text{ V}$

$I = 8\text{ A}$

What value is not known?

$VA = ?$

What formula can be used?

$VA = E \times I$

Substitute the known values and solve for VA.

$VA = E \times I$

$VA = 208\text{ V} \times 8\text{ A}$

$VA = 1664\text{ VA}$

Therefore, this circuit has an apparent power of 1664 volt-amperes. Now determine the power factor of the motor.

What values are known?

$$P = 900\text{ W}$$

$$VA = 1664\text{ VA}$$

What value is not known?

$$PF = ?$$

What formula can be used?

$$PF = \frac{P}{VA} \times 100$$

Substitute the known values and solve for *PF*.

$$PF = \frac{P}{VA} \times 100$$

$$PF = \frac{900\text{ W}}{1664\text{ VA}} \times 100$$

$$PF = 0.5409 \times 100$$

$$PF = 54.09\%$$

Therefore, the power factor of this motor is 54.09%. Now determine the phase angle difference, $\angle\theta$.

What value is known?

$PF = 0.5409$ (For this calculation, the power factor is not expressed as a percent.)

What value is not known?

$$\angle\theta = ?$$

What formula can be used?

$$\angle\theta = COS^{-1}PF$$

Substitute the known values and solve for $\angle\theta$.

$$\angle\theta = COS^{-1}PF$$

$$\angle\theta = COS^{-1}0.5409$$

$$\angle\theta = 57.26°$$

Therefore, in this circuit the current lags the voltage by 57.26°.

This motor is not running very efficiently. Essentially, this means that for every dollar spent to operate this motor, only $0.54 worth of work is received in return. The inductive nature of the motor produces a phase shift between the current and voltage of the circuit. The phase shift causes reactive power to be produced. The reactive power is responsible for the wasted energy in this circuit.

The efficiency of this motor circuit needs improvement. If the current and voltage could be brought back in phase, the power factor of the circuit would be 100%, and the circuit would be operating at maximum efficiency. Quite often, it is not practical to operate at a power factor of 100%. An acceptable power factor is approximately 85%. How can the power factor of this circuit be improved from 54% to 85%? What can be done to reduce the phase angle difference between the current and voltage?

Recall that a capacitor, unlike an inductor, creates an opposite effect on the phase relationship between the circuit current and voltage. Recall also ELI the ICE man. In an inductive circuit, the current lags the voltage, whereas in a capacitive circuit the current leads the voltage. If you were to add a capacitor to the motor circuit, the effect of the capacitor would offset the effect of the inductor. In essence, the capacitor would cancel the effects of the inductor, and the only component affecting the circuit current and voltage would be the resistive element. This would result in the current and voltage being in phase. The power factor of the circuit would be 100%. As mentioned earlier, it is more practical to improve the power factor to approximately 85%. To do this, you must determine the proper amount of capacitance to add to the circuit. First, determine the amount of inductive reactive power that is produced in the circuit.

What values are known?

$$VA = 1664\text{ VA}$$

$$P = 900\text{ W}$$

What value is not known?

$$VARs_L = ?$$

What formula can be used?

$$VARs_L = \sqrt{VA^2 - P^2}$$

Substitute the known values and solve for $VARs_L$.

$$VARs_L = \sqrt{VA^2 - P^2}$$

$$VARs_L = \sqrt{1664\text{ VA}^2 - 900\text{ W}^2}$$

$$VARs_L = \sqrt{2768896 - 810000}$$

$$VARs_L = \sqrt{1958896}$$

$$VARs_L = 1399.61\text{ VARs}_L$$

Therefore, the inductive portion of this circuit is producing 1399.61 $VARs_L$. To correct the power factor to approximately 85%, you must determine the amount of apparent power needed for an 85% power factor.

What values are known?

$$P = 900\text{ W}$$

$$PF = 85\%$$

What value is not known?

$$VA = ?$$

What formula can be used?

$$VA = \frac{P}{PF}$$

Substitute the known values and solve for *VA*.

$$VA = \frac{P}{PF}$$

$$VA = \frac{900\text{ W}}{0.85}$$

$$VA = 1058.82\text{ VA}$$

Therefore, 1058.82 volt-amperes is needed to produce a power factor of 85%. Next, determine the amount of inductive VARs required to produce this amount of apparent power.

What values are known?

$$VA = 1058.82\text{ VA}$$

$$P = 900\text{ W}$$

What value is not known?

$$VARs_L = ?$$

What formula can be used?

$$VARs_L = \sqrt{VA^2 - P^2}$$

Substitute the known values and solve for $VARs_L$.

$$VARs_L = \sqrt{VA^2 - P^2}$$

$$VARs_L = \sqrt{1058.82\text{ VA}^2 - 900\text{ W}^2}$$

$$VARs_L = \sqrt{1121099.79 - 810000}$$

$$VARs_L = \sqrt{311099.79}$$

$$VARs_L = 556.87\text{ VARs}_L$$

Therefore, the amount of induction required is 556.87 VARs_L. Originally, the circuit produced 1399.61 VARs_L. To correct the power factor to 85%, the circuit must only produce 556.87 VARs_L. This means that a capacitor that is capable of supplying an amount of capacitive VARs equal to the difference is needed. Now determine the amount of capacitive VARs needed.

What values are known?

$$Original\ VARs_L = 1399.61\text{ VARs}_L$$

$$85\%\ PF\ VARs_L = 556.87\text{ VARs}_L$$

What value is not known?

$$VARs_C = ?$$

What formula can be used?

$$VARs_C = Original\ VARs_L - 85\%\ PF\ VARs_L$$

Substitute the known values and solve for $VARs_C$:

$$VARs_C = Original\ VARs_L - 85\%\ PF\ VARs_L$$

$$VARs_C = 1399.61\text{ VARs}_L - 556.87\text{ VARs}_L$$

$$VARs_C = 842.74\text{ VARs}_C$$

Now you know that you need a capacitor that is capable of supplying 842.72 VARs_C to the circuit. Next, determine the amount of capacitive reactance needed to produce this amount of VARs_C.

What values are known?

$$E = 208\text{ V}$$

$$VARs_C = 842.74\text{ VARs}_C$$

What value is not known?

$$X_C = ?$$

What formula can be used?

$$X_C = \frac{E^2}{VARs_C}$$

Substitute the known values and solve for X_C.

$$X_C = \frac{E^2}{VARs_C}$$

$$X_C = \frac{208\text{ V}^2}{842.74\text{ VARs}_C}$$

$$X_C = \frac{43264}{842.74\text{ VARs}_C}$$

$$X_C = 51.34\ \Omega$$

Therefore, a capacitor with 51.34 ohms of capacitive reactance is needed. Now determine the amount of capacitance required to produce 51.34 ohms of capacitive reactance at 60 hertz.

What values are known?

$$f = 60\text{ Hz}$$

$$X_C = 51.34\ \Omega$$

What value is not known?

$$C = ?$$

What formula can be used?

$$C = \frac{1}{2\pi f X_C}$$

Substitute the known values and solve for *C*.

$$C = \frac{1}{2\pi f X_C}$$

$$C = \frac{1}{2 \times 3.14 \times 60\text{ Hz} \times 51.34\ \Omega}$$

$$C = \frac{1}{19.34}$$

$$C = 51.69\ \mu\text{f}$$

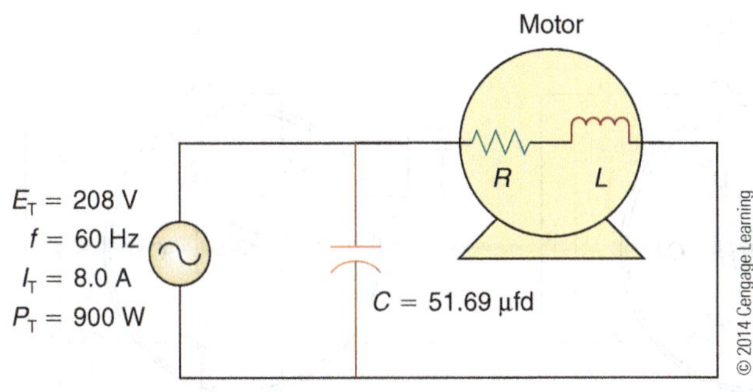

FIGURE 17-23 A motor power factor corrected to 85%.

This means that if you were to add a 51.69 microfarad capacitor in parallel with this motor, as shown in **Figure 17-23**, the power factor is corrected to 85%. Now when $1.00 is paid to operate this motor, $0.85 worth of work is received. The motor is now operating more efficiently.

17-5 THREE-PHASE CIRCUITS

The circuits in the previous examples used a type of electricity referred to as **single-phase AC**. Most electrical power used in industry is **three-phase AC**. **Figure 17-24** shows a three-phase alternator. The armature rotates within the field due to an external source of mechanical energy such as wind, water, steam, etc. Refer to **Figure 17-25**. As the armature revolves, the magnetic field produced in the armature by the DC current cuts the windings of one of the stationary fields, inducing a current in field winding, A. This induced current increases and decreases in intensity as the armature moves closer and then farther away from the field winding. As the armature rotates away from winding A, it rotates toward winding B, inducing a current in this field winding. This induced current increases and decreases in intensity as the armature moves closer and then farther away from field winding B. As the armature rotates away from winding B, it rotates toward winding C, inducing a current in this field winding. This induced current increases and decreases in intensity as the armature moves closer and then farther away from field winding C. As the armature rotates away from winding C, it again rotates toward winding A, inducing a current in this field winding.

Notice the waveform, shown in **Figure 17-26**, that is produced by the redesigned alternator. The waveform consists of three single-phase AC sine waves 120° apart. This is referred to as three-phase

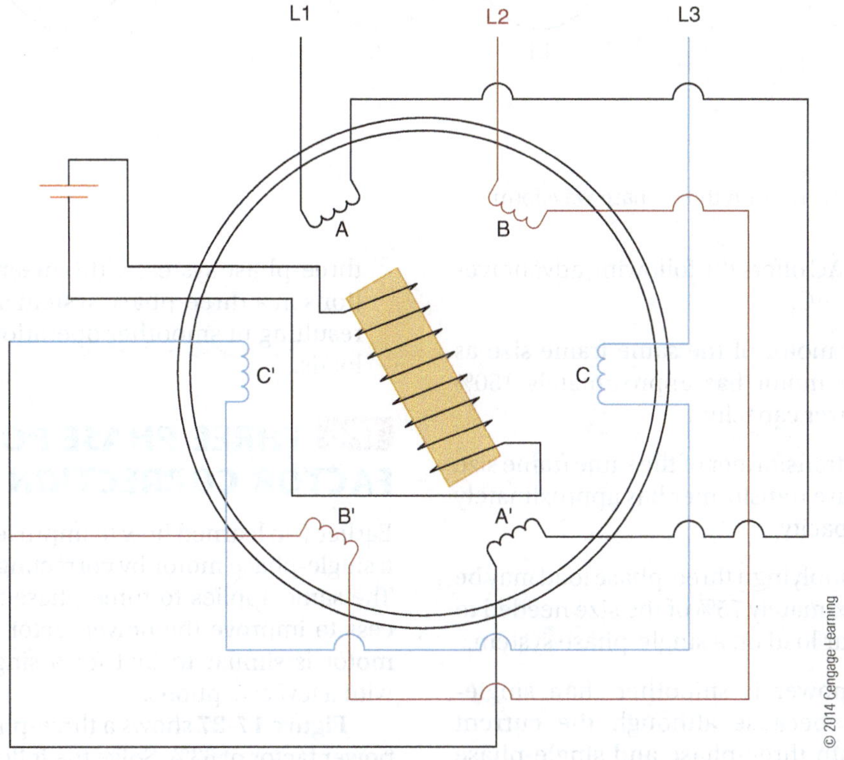

FIGURE 17-24 A three-phase alternator.

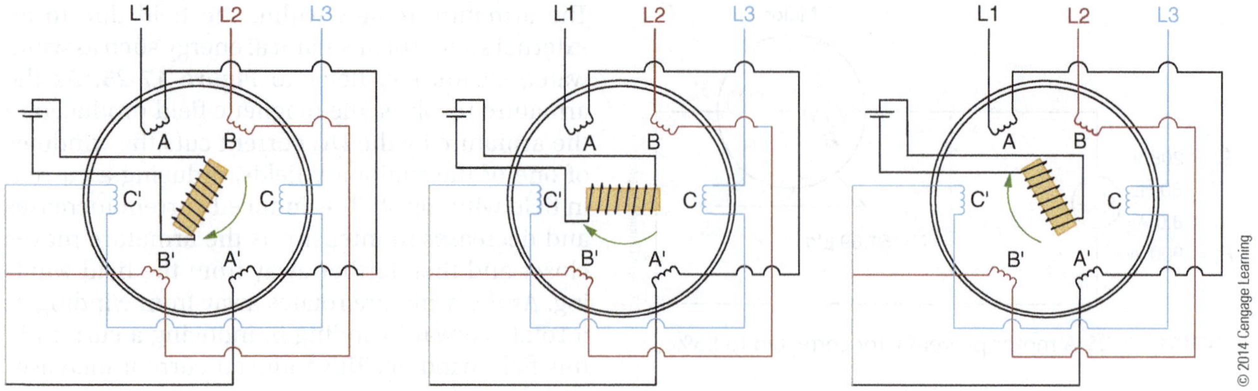

FIGURE 17-25 A rotating armature.

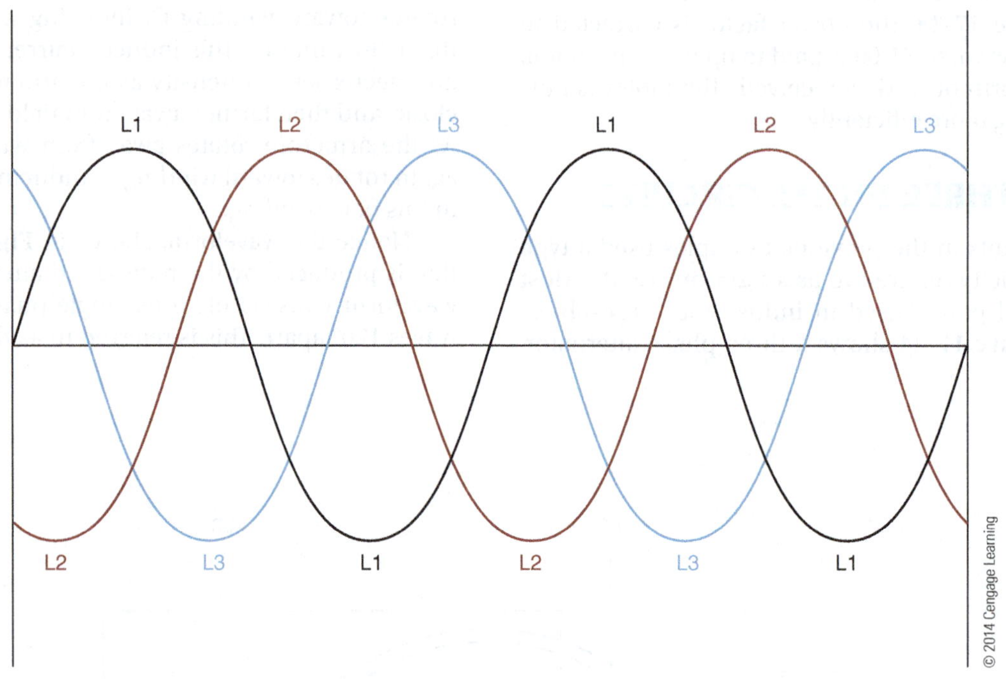

FIGURE 17-26 A three-phase waveform.

AC. Three-phase AC offers the following advantages over single-phase AC:

- A three-phase motor of the same frame size as a single-phase motor has approximately 150% more horsepower capacity.
- A three-phase transformer of the same frame size as a single-phase transformer has approximately 150% more capacity.
- Conductors supplying a three-phase load may be sized at approximately 75% of the size needed to supply a similar load on a single-phase system.
- Three-phase power is smoother than single-phase power, because although the current pulsates in both three-phase and single-phase systems, the current never falls to the 0 level in three-phase systems. This means that the pulsations in a three-phase system are not as severe, resulting in smoother operation of three-phase loads.

17-6 THREE-PHASE POWER FACTOR CORRECTION

Earlier you learned how to improve the efficiency of a single-phase motor by correcting its power factor. The same applies to three-phase motors. The process to improve the power factor of a three-phase motor is similar to that for a single-phase motor, with a few exceptions.

Figure 17-27 shows a three-phase motor with a power factor of 63%. Solve the following problem to improve the power factor of this motor to 87%.

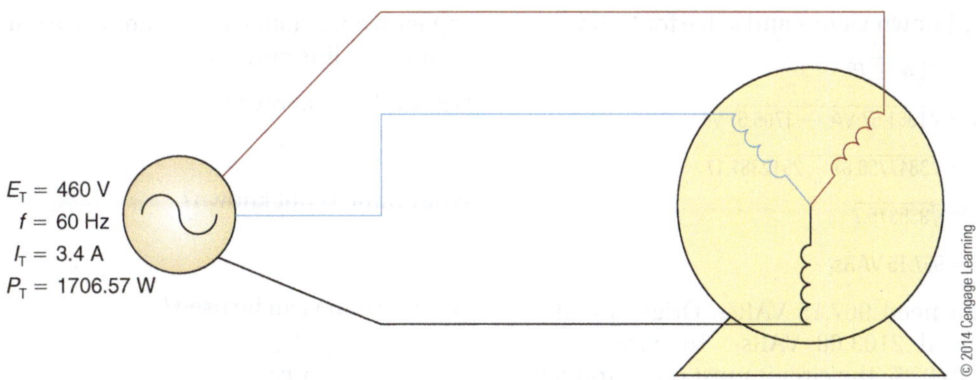

FIGURE 17-27 A three-phase motor with 63% power.

First, find the circuit apparent power. What values are known?

$$E = 460 \text{ V}$$
$$I = 3.4 \text{ A}$$

What value is not known?

$$VA = ?$$

What formula can be used?

$VA = E \times I \times 1.732$ (1.732, or $\sqrt{3}$, is a constant used when calculating three-phase values.)

Substitute the known values and solve for VA:

$$VA = E \times I \times 1.732$$
$$VA = 460 \text{ V} \times 3.4 \text{ A} \times 1.732$$
$$VA = 2708.85 \text{ VA}$$

Therefore, the circuit has an apparent power of 2708.85 VA. Now determine the inductive VARs of the circuit:

What values are known?

$$VA = 2708.85 \text{ VA}$$
$$P = 1706.57 \text{ W}$$

What value is not known?

$$VARs_L = ?$$

What formula can be used?

$$VARs_L = VA2 - P2$$

Substitute the known values and solve for $VARs_L$:

$$VARs_L = \sqrt{VA^2 - P^2}$$
$$VARs_L = \sqrt{2708.85 \text{ VA}^2 - 1706.57 \text{ W}^2}$$
$$VARs_L = \sqrt{7337868.32 - 2912381.17}$$
$$VARs_L = \sqrt{4425487.16}$$
$$VARs_L = 2103.68 \text{ VARs}_L$$

Therefore the inductive portion of this circuit is producing 2103.68 $VARs_L$. To correct the power factor to approximately 87%, you must determine the amount of apparent power needed for an 87% power factor.

What values are known?

$$P = 1706.57 \text{ W}$$
$$PF = 87\%$$

What value is not known?

$$VA = ?$$

What formula can be used?

$$VA = \frac{P}{PF}$$

Substitute the known values and solve for VA:

$$VA = \frac{P}{PF}$$
$$VA = \frac{1706.57 \text{ W}}{0.87}$$
$$VA = 1961.57 \text{ VA}$$

Therefore, 1961.57 VAs are needed to produce a power factor of 87%. Next, you need to determine the amount of inductive VARs required to produce this amount of apparent power.

What values are known?

$$VA = 1961.57 \text{ VA}$$
$$P = 1706.57 \text{ W}$$

What value is not know?

$$VARs_L = ?$$

What formula can be used?

$$VARs_L = \sqrt{VA^2 - P^2}$$

Substitute the known values and solve for $VARs_L$:

$$VARs_L = \sqrt{VA^2 - P^2}$$

$$VARs_L = \sqrt{1961.57\ VA^2 - 1706.57\ W^2}$$

$$VARs_L = \sqrt{3847756.87 - 2912381.17}$$

$$VARs_L = \sqrt{935375.7}$$

$$VARs_L = 967.15\ VARs_L$$

Therefore, you need 967.15 $VARs_L$. Originally, the circuit produced 2103.68 $VARs_L$. To correct the power factor to 87%, the circuit must only produce 967.15 $VARs_L$. This means that a capacitor that is capable of supplying an amount of capacitive VARs equal to the difference is needed. Now determine the amount of capacitive VARs needed.

What values are known?

$$\text{Original } VARs_L = 2103.68\ VARs_L$$

$$87\%\ PF\ VARs_L = 967.15\ VARs_L$$

What value is not known?

$$VARs_C = ?$$

What formula can be used?

$$VARs_C = \text{Original } VARs_L - 87\%\ PF\ VARs_L$$

Substitute the known values and solve for $VARs_C$.

$$VARs_C = \text{Original } VARs_L - 87\%\ PF\ VARs_L$$

$$VARs_C = 2103.68\ VARs_L - 967.15\ VARs_L$$

$$VARs_C = 1136.53\ VARs_C$$

Next determine the amount of line current supplying the capacitor.

What values are known?

$$VARs_C = 1136.53\ VARs_C$$

$$E_T = 460\ V$$

What value is not known?

$$I_{Line} = ?$$

What formula can be used?

$$I_{Line} = \frac{VARs_C}{E_T \times 1.732}$$

Substitute the known values and solve for I_{Line}.

$$I_{Line} = \frac{VARs_C}{E_T \times 1.732}$$

$$I_{Line} = \frac{1136.53\ VARs_C}{460\ V \times 1.732}$$

$$I_{Line} = 1.43\ A$$

Since the three-phase capacitor bank is connected in a delta configuration, you must determine the capacitive reactance using the phase current value for a 1.43 A line current.

What value is known?

$$I_{Line} = 1.43\ A$$

What value is not known?

$$I_{Phase} = ?$$

What formula can be used?

$$I_{Phase} = \frac{I_{Line}}{1.732}$$

Substitute the known values and solve for I_{Phase}.

$$I_{Phase} = \frac{I_{Line}}{1.732}$$

$$I_{Phase} = \frac{1.43\ A}{1.732}$$

$$I_{Phase} = 0.83\ A$$

Now determine the capacitive reactance. Recall that in a delta configuration the line voltage is equal to the phase voltage.

What values are known?

$$E_{Phase} = 460\ V$$

$$I_{Phase} = 0.83\ A$$

What value is not known?

$$X_C = ?$$

What formula can be used?

$$X_C = \frac{E_{Phase}}{I_{Phase}}$$

Substitute the known values and solve for X_C.

$$X_C = \frac{E_{Phase}}{I_{Phase}}$$

$$X_C = \frac{460\ V}{0.83\ A}$$

$$X_C = 554.22\ \Omega$$

Therefore, a capacitor with 554.22 Ω of capacitive reactance is needed. Now determine the amount of capacitance required to produce 554.22 Ω of capacitive reactance at 60 Hz.

What values are known?

$$f = 60\ Hz$$

$$X_C = 554.22\ \Omega$$

What value is not known?

$$C = ?$$

What formula can be used?

$$X_C = \frac{1}{2\pi f X_c}$$

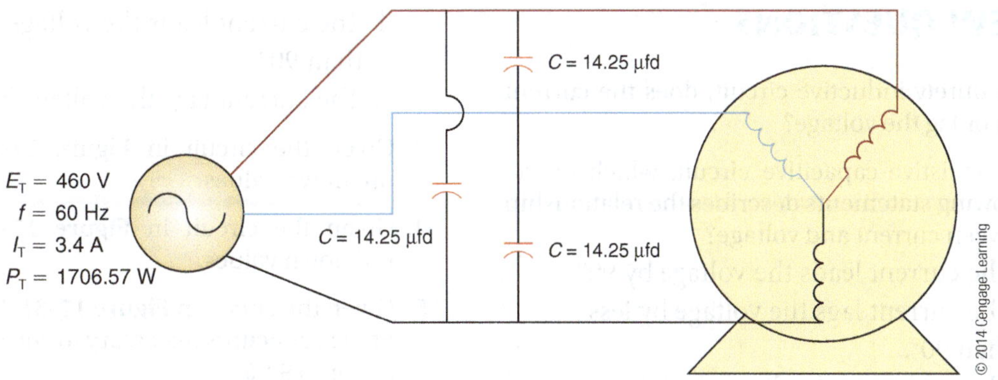

FIGURE 17-28 A three-phase motor with power factor corrected to 87%.

Substitute the known values and solve for X_C.

$$X_C = \frac{1}{2\pi f X_C}$$

$$X_C = \frac{1}{2 \times 3.14 \times 60 \text{ Hz} \times 554.22\ \Omega}$$

$$X_C = 4.79\ \mu\text{F}$$

This means that if you were to add one 4.79 µF capacitor from L1 to L2, a second 4.79 µF capacitor from L2 to L3, and a third 4.79 µF capacitor from L1 to L3, as shown in **Figure 17-28**, the power factor will be corrected to approximately 87%. Now when $1.00 is paid to operate this motor, $0.87 worth of work will be received. The motor is now operating more efficiently.

SUMMARY

- Whenever a current flows through a conductor, a magnetic field is formed around the conductor.
- If the current flowing through the conductor increases in intensity, the magnetic field becomes stronger, and if the current flow decreases in intensity, the magnetic field becomes weaker.
- When an alternating current flows through a conductor, the current increases and decreases at a rate equal to double the frequency of the applied AC.
- The amount of induced voltage caused by the changing current is called inductance. It is represented in formulae by the letter L and is measured in henrys.
- The induced voltage opposes the applied voltage, thus providing an opposition to current flow. This opposition is called inductive reactance, represented in formulae by the letter X_L and is measured in ohms.
- In a purely inductive circuit, the current lags the voltage by 90°.
- A circuit that contains both resistance and inductance exhibits two types of power: true, or resistive, power and reactive power. The combination of both types of power is called apparent power and is measured in volt-amperes.
- Power factor is the ratio of true power to apparent power and is expressed as a percent. Power factor is used to measure the efficiency of a circuit.
- Two conductors in close proximity that are separated by some type of insulating material form a device called a capacitor. A capacitor is capable of storing an electrical charge.
- When dealing with capacitors, the two conductors are called plates, and the insulating material is called a dielectric.
- Capacitance is the measure of a capacitor's ability to store an electrical charge.
- The amount of capacitance is measured in farads.
- In a purely capacitive circuit, the current leads the voltage by 90°.
- Capacitive reactance is the opposition to alternating current flow caused by a capacitor, resulting from the phase shift between the current and the voltage in a capacitive circuit.
- A circuit that contains both resistance and capacitance is called an R-C circuit.
- There are two types of R-C circuits: an R-C series circuit and an R-C parallel circuit.
- An R-L-C circuit can be either series or parallel and contains resistance, inductance, and capacitance.
- Power factor correction is the process to improve the operating efficiency of an AC motor.
- Most electrical power used in industry is three-phase AC.

REVIEW QUESTIONS

1. In a purely inductive circuit, does the current lead or lag the voltage?
2. In a resistive-capacitive circuit, which of the following statements describes the relationship between current and voltage?
   a. The current leads the voltage by 90°.
   b. The current lags the voltage by less than 90°.
   c. The current and voltage are in phase.
   d. The current leads the voltage by less than 90°.
   e. The current lags the voltage by 90°.
3. Given the circuit in **Figure 17-29**, find the unknown values.
4. Given the circuit in **Figure 17-30**, find the unknown values.
5. Given the circuit in **Figure 17-31**, find the value of the capacitors necessary to correct the power factor to 93%.

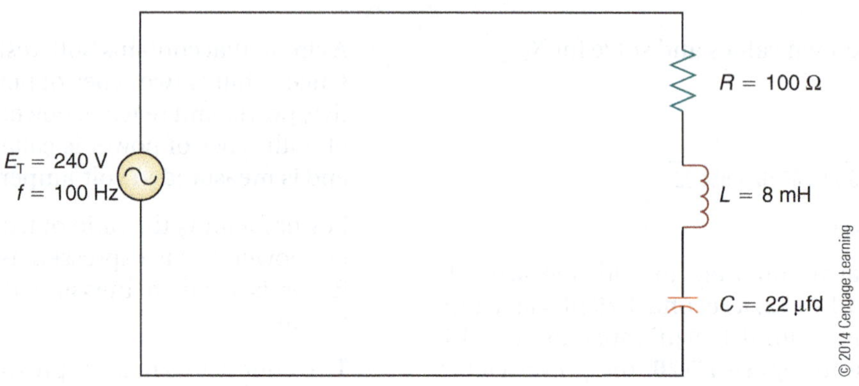

FIGURE 17-29

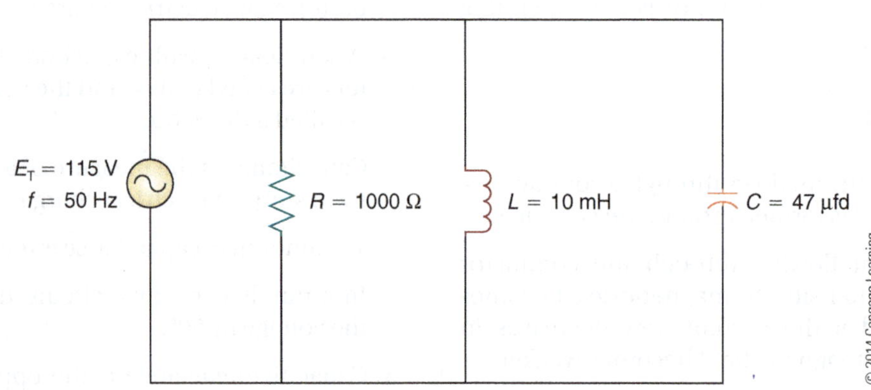

FIGURE 17-30

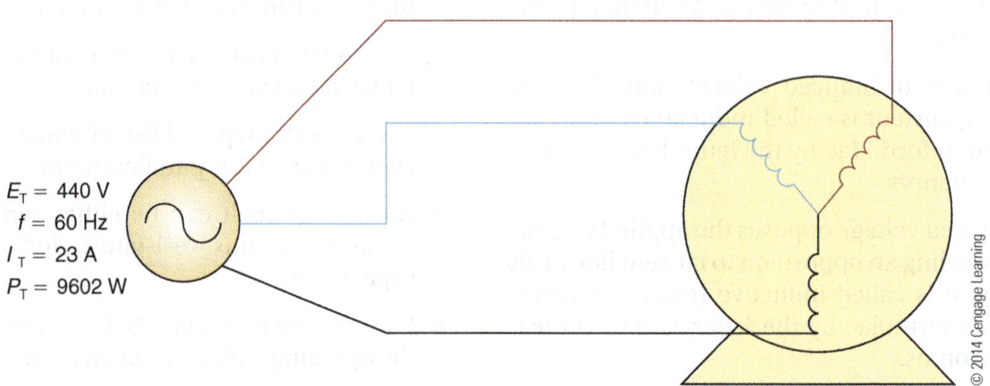

FIGURE 17-31

CHAPTER 18

Wiring Methods

The *National Electrical Code* (*NEC*)is considered by many to be the standard by which all wiring and associated components of an electrical system are designed, specified, and installed. By this virtue, the *NEC* is very comprehensive and challenging to read and understand. This chapter presents a very small portion of the *NEC* relating to sizing conductors and conduits. The sizing of conductors is crucial in ensuring a safe and efficient electrical system. In addition, protecting those conductors by encasing them in various types of conduits is equally important.

OBJECTIVES

After studying this chapter, the student should be able to

- Use the correct tables from the *National Electrical Code* to properly select and size conductors.
- Determine conductor voltage rating by the conductor insulation color.
- Use the correct tables from the *National Electrical Code* to properly select and size conduits.

18-1 CONDUCTOR SELECTION AND SIZING

Conductors used in industrial environments are constructed from three materials and are of several types: copper and nickel or nickel-coated copper, aluminum, or copper-clad aluminum. These three materials have different conduction characteristics that must be considered when sizing conductors for given loads. It is important that a conductor be properly sized for a given load. If the conductor is sized too small, there is an excessive amount of heat buildup when current flows through the conductor. This excessive heat damages the conductor insulation, resulting in a short circuit, a grounded circuit, or an electrical fire. Conductors may certainly be sized larger than required, allowing for future expansion. How ever, this may be taken to extremes and could result in unacceptable increases in cost.

Most facilities use the *NEC* as the standard by which conductors are sized. Your facility may use a different standard. Check with your supervisor to determine the standard that you should use. In this textbook, the *NEC* is used to properly size conductors.

NEC Article 310 covers *Conductors for General Wiring*. Within this article, you find *Table 310.15(B)(16)*, *Table 310.15(B)(17)*, *Table 310.15(B)(18)*, and *Table 310.15(B)(19)*. You will also find these tables in Appendix A of this textbook. Refer to *Table 310.15(B)(16)*.

Table 310.15(B)(16) covers *Allowable Ampacities of Insulated Conductors Rated Up to and Including 2000 Volts, 60°C Through 90°C (140°F Through 194°F) Not More Than Three Current-Carrying Conductors in Raceway, Cable, or Earth (Directly Buried), Based on Ambient Tem perature of 30°C (86°F)**. There is a lot of information in the heading for this table. What does it all mean? Following is a breakdown of this table into smaller bits of information.

Allowable Ampacities:

> *This is the maximum current, in amperes, that a conductor can carry continuously under the conditions of use without exceeding its temperature rating* (*NEC Article 100*).

Insulated Conductors

> *A conductor encased within material of composition and thickness that is recognized by this Code as electrical insulation* (*NEC Article 100*).

Raceway

> *An enclosed channel of metal or nonmetallic materials designed expressly for holding wires, cables, or busbars, with additional functions as permitted in this Code. Raceways include, but are not limited to, rigid metal conduit, rigid nonmetallic conduit, intermediate metal conduit, liquidtight flexible conduit, flexible metallic tubing, flexible metal conduit, electrical nonmetallic tubing, electrical metallic tubing, underfloor raceways, cellular concrete floor race ways, cellular metal floor raceways, surface race ways, wire ways, and busways* (*NEC Article 100*).

Therefore, *Table 310.15(B)(16)* is to be used when you have no more than three insulated current-carrying conductors contained within a raceway, a cable, or directly buried in the earth. These conductors must be rated up to and including 2000 volts. The temperature rating of the conductors is from 60°C (140°F) through 90°C (194°F), and the ambient temperature is 30°C (86°F).

Now compare this with *Table 310.15(B)(17)*. Notice the difference between the headings of the two tables. *Table 310.15(B)(17)* is to be used for *Single-Insulated Con ductors in Free Air*. This means that the conductors are not part of a cable assembly, nor are the conductors contained within a raceway or directly buried in the earth. *Table 310.15(B)(17)* is to be used to determine the ampacity rating of single, individual-insulated conductors.

Table 310.15(B)(18) is to be used when you have ... *Insulated Conductors, Rated Up to and Including 2000 Volts, 150°C Through 250°C (302°F Through 482°F), Not More Than Three Current-Carrying Conductors in Raceway or Cable, Based on Ambient Air Temperature of 40°C (104°F)**. This is similar to *Table 310.15(B)(16)* in that *Table 310.15(B)(18)* is used when there are no more than three insulated current-carrying conductors contained within a raceway or a cable. However, the temperature rating of the conductors is higher, and the ambient air temperature is also higher.

Finally, refer to *Table 310.15(B)(19)*. How is this table different? According to the heading for *Table 310.15(B)(19)*, this table is used for ... *Single-Insulated Conductors, Rated Up to and Including 2000 Volts, 150°C Through 250°C (302°F Through 482°F), in Free Air, Based on Ambient Air Temperature of 40°C (104°F)**. This is similar to *Table 310.15(B)(17)* in that *Table 310.15(B)(19)* is used for single-insulated conductors in free air. However, the temperature rating of the conductors is higher, and the ambient air temperature is also higher.

Each of these tables is further divided into columns. The columns have the following headings, Size (AWG or kcmil) and Temperature Rating of Conductor (see *Table 310.104(A)*), and Size. The Size (AWG or kcmil) column on the *left* is to be used for sizing

copper conductors. The Size (AWG or kcmil) column on the *right* is to be used when sizing **aluminum** or **copper-clad aluminum conductors**. Notice the line that runs vertically through the center of the table. This line divides the table into two sections: copper conductors on the left and aluminum or copper-clad aluminum conductors on the right. Notice also that each section is further divided by insulation type.

It is important to use the proper table when sizing conductors. It is also important to use the proper section of the table. Using the wrong table or the wrong section within a table could result in undersizing the conductors, which could lead to an electrical fire. Following are a few examples to learn how to use these tables.

What is the allowable ampacity of a single 6 AWG Type THWN copper wire in free air, based on an ambient air temperature of 30°C (86°F)?

1. Determine the proper table to use. Because this is a single-insulated conductor in free air, use *Table 310.15(B)(17)*. Do not use *Table 310.15(B)(16)* or *Table 310.15(B)(18)* because there is only a single conductor in free air. Do not use *Table 310.15(B)(19)* because the ambient air temperature is 30°C, and THWN insulation is not listed in it.
2. Because this is a copper conductor, use the Size column on the *left* side of *Table 310.15(B)(17)*. Move down the Size column until you see 6 AWG. Rest your finger at this point.
3. Now move your finger to the *right* until it rests under the column for THWN insulation for a copper conductor. This will be the 75°C (167°F) column.
4. Now read the ampacity of the conductor. You should read 95.

Now repeat the same problem with an aluminum conductor.

What is the allowable ampacity of a single 6 AWG Type THWN aluminum wire in free air based on an ambient air temperature of 30°C (86°F)?

1. Determine the proper table to use. Because this is a single-insulated conductor in free air, use *Table 310.15(B)(17)*. Do not use *Table 310.15(B)(16)* or *Table 310.15(B)(18)* because there is only a single conductor in free air. Do not use *Table 310.15(B)(19)* because the ambient air temperature is 30°C, and THWN insulation is not listed in it.
2. Because this is an aluminum conductor, use the Size column on the *right* side of *Table 310.15(B)(17)*. Move down the Size column until you see 6 AWG. Rest your finger at this point.
3. Now move your finger to the *left* until it rests under the column for THWN insulation for an aluminum conductor. This will be the 75°C (167°F) column.
4. Now read the ampacity of the conductor. You should read 75.

Notice the difference! The same size wire (6 AWG) has a different ampacity depending on whether the conductor is copper or aluminum. This is why it is so important that you use the correct tables and sections within the tables. Try another example.

What is the allowable ampacity of a nonmetallic-sheathed cable containing three 12 AWG Type THW-2 copper wires based on an ambient air temperature of 30°C (86°F)?

1. Determine the proper table to use. Because this is cable containing three conductors, use *Table 310.15(B)(16)*. Do not use *Table 310.15(B)(17)* or *Table 310.15(B)(19)* because there are three conductors in a cable. Do not use *Table 310.15(B)(18)* because the ambient air temperature is 30°C, and THW-2 insulation is not listed in it.
2. Because these are copper conductors, use the Size column on the *left* side of *Table 310.15(B)(16)*. Move down the Size column until you see 12 AWG. Rest your finger at this point.
3. Now move your finger to the *right* until it rests under the column for THW-2 insulation for a copper conductor. This will be the 90°C (194°F) column.
4. Now read the ampacity of the conductor. You should read 30.

Now try something a bit different in the next example.

A load draws 120 amperes of current. What is the minimum size of aluminum conductor with Type THHN insulation in free air that is capable of supplying this load based on an ambient air temperature of 30°C (86°F)?

1. Determine the proper table to use. Because this is a single-insulated conductor in free air, use *Table 310.15(B)(17)*. Do not use *Table 310.15(B)(16)* or *Table 310.15(B)(18)* because there is only a single conductor in free air. Do not use *Table 310.15(B)(19)* because the ambient air temperature is 30°C and THHN insulation is not listed in it.
2. Because this is an aluminum conductor, use the Insulation Type columns on the *right* side of *Table 310.15(B)(17)*. Find the column with Type THHN insulation. This will be the 90°C (194°F) column.

3. Move down the THHN insulation column until you see an ampacity rating of 120 A or more. Because you need to find the minimum size, rest your finger at the 130 A. You cannot use the 115 A value because it is too small. You could use the 150 A value, but you are looking for the minimum value.

4. Now move your finger to the right until it rests on 3 AWG. This is the minimum size of aluminum conductor Type THHN insulation in free air for a 120-ampere load.

Refer again to *Table 310.15(B)(17)*. Recall that this table is to be used for an ambient air temperature of 30°C (86°F). Assume that the ambient air temperature is 49°C. What is the allowable ampacity of a single 6 AWG Type THWN copper wire in free air?

1. Because this is a copper conductor, use the Size column on the *left* side of *Table 310.15(B)(17)*. Move down the Size column until you see 6 AWG. Rest your finger at this point.

2. Now move your finger to the *right* until it rests under the column for THWN insulation for a copper conductor. This will be the 75°C (167°F) column.

3. Read the ampacity of the conductor. You should read 95.

4. Look at the note with one* located below *Table 310.15(B)(17)*. This note refers you to *310.15(B)(2)* for the ampacity correction factors where the ambient temperature is other than 30°C (86°F). Notice that there are actually two ampacity correction factor tables; *310.15(B)(2)(a)* and *310.15(B)(2)(b)*. *Table 310.15(B)(2)(a)* is used when the ambient temperature is based on 30°C (86°F). *Table 310.15(B)(2)(b)* is used when the ambient temperature is based on 40°C (104°F). Because the temperature is based on 30°C, we use *Table 310.15(B)(2)(a)*. You are now directed to *Table 310.15(B)(2)* for the ampacity correction factors where the ambient temperature is other than 30°C (86°F). (*Table 310.15(B)(2)* can be found in Appendix B of this textbook.) Notice the three columns in the middle of the table. These columns represent the temperature rating of the conductor as 60°C, 75°C, or 90°C. Notice also the extreme left and right columns. The left column is the *Ambient Temperature (°C)* and the right column is the *Ambient Temperature (°F)*.

5. Place your finger at the ambient air temperature of 46-50°C (115-122°F; the actual temperature of 49°C [120°C] falls within this range) in the extreme *left* column (ninth row down). Move your finger to the *right* until your finger is under the column for 75°C (the conductor temperature rating). You should find a correction factor of 0.75.

6. Now take the ampacity of 95 A found in step 3 and multiply it by the correction factor of 0.75: 95 A × 0.75 = 71.25 A. Therefore, the new ampacity for this conductor is 71.25 A.

Because the ambient temperature is higher (49°C vs. 30°C) [120°F vs. 86°F], the conductor cannot have as much ampacity. This is a result of the heat produced by a current flowing through the conductor. With a higher ambient temperature, the conductor cannot dissipate heat as readily. There fore, the amount of current that the conductor can safely carry must be reduced. The Ambient Temperature Correction Factors tables *(310.15(B)(2)(a)* and *310.15(B)(2)(b))* compensate for the effects of changes in the ambient temperature.

Refer again to *Table 310.15(B)(16)*. Recall that this table is to be used when there are ... *Not More Than Three Current-Carrying Conductors in Race way, Cable, or Earth (Directly Buried)*. What is the allowable ampacity of a nonmetallic-sheathed cable containing six 12 AWG Type THW-2 copper wires, based on an ambient air temperature of 30°C (86°F)?

1. Because these are copper conductors, use the Size column on the *left* side of *Table 310.15(B)(16)*. Move down the Size column until you see 12 AWG. Rest your finger at this point.

2. Move your finger to the *right* until it rests under the column for THW-2 insulation for a copper conductor. This will be the 90°C (194°F) column.

3. Read the ampacity of the conductor. You should read 30.

4. Look at *Table 310.15(B)(3)(a), Adjustment Factors for More Than Three Current-Carrying Conductors in a Raceway or Cable*, found in Appendix B of this textbook.

5. This example has six current-carrying conductors in the cable. Find the column labeled *Number of Conductors* and place your finger at the 4-6 listing. Move your finger to the *right* until your finger is resting on 80. This is the *Percent of Values in Tables 310.15(B)(16)* through *Table 310.15(B)(19) as Adjusted for Ambient Temperature if Necessary*.

6. Now take the ampacity of 30 A found in step 3 and multiply it by the *Percent of Values*, 80, found in step 5: 30 A × 0.80 = 24 A. There fore, the new ampacity for these conductors is 24 A.

Because there are more than three current-carrying conductors within the same cable, the conductors cannot have as much ampacity. This is a result of the heat produced by a current flowing through the conductors. With more than three current-carrying conductors within the same cable, the conductors cannot dissipate heat as readily. Therefore, the amount of current that the conductors can carry safely must be reduced. *Table 310.15(B)(3)(a), Adjustment Factors for More Than Three Current-Carrying Conductors in a Raceway or Cable* provides the adjustment factor to use, depending on the number of current-carrying conductors.

Imagine a raceway or cable containing more than three current-carrying conductors located in an area in which the ambient temperature was different from 30°C (86°F) for *Table 310.15(B)(16)* or 40°C (104°F) for *Table 310.15(B)(18)*. You would first correct the ampacity rating of the conductor for the ambient temperature, then you would correct the new ampacity rating based on the number of current-carrying conductors. Here is an example.

What is the allowable ampacity of 12 single-insulated 10 AWG Type-Z aluminum wires in a raceway, based on an ambient air temperature of 65°C (149°F)?

1. Because these are aluminum conductors Type-Z insulation in the raceway, use *Table 310.15(B)(18)*. Move down the Size column on the *right* until you see 10 AWG. Rest your finger at this point.
2. Move your finger to the *left* until it rests under the column for Type-Z insulation for an aluminum conductor. This will be the 150°C (302°F) column.
3. Now read the ampacity of the conductor. You should read 44.
4. Look at *Table 310.15(B)(2)(b) Ambient Temperature Correction Factors Based on 40°C (104°F)*.
5. Place your finger at the ambient air temperature range of 61–65°C (142–149°F), which includes 65°C (149°F) in the extreme left column (12th row down). Move your finger to the right until your finger is under the 150°C column for Type-Z insulation with aluminum conductors. You should find a correction factor of 0.88.
6. Now take the ampacity of 44 A found in step 3 and multiply it by the correction factor of 0.88: 44 A × 0.88 = 38.72 A. Therefore, the new ampacity for this conductor is 38.72 A based on the difference in the ambient air temperature.
7. Correct the ampacity of the conductors based on the fact that there are more than three current-carrying conductors in the raceway.
8. Refer to *Table 310.15(B)(3)(a), Adjustment Factors for More Than Three Current-Carrying Conductors in a Raceway or Cable.*
9. This example has twelve current-carrying conductors in the raceway. Find the column labeled *Number of Conductors* and place your finger at the 10–20 listing. Move your finger to the right until your finger is resting on 50. This is the *Percent of Values in Tables 310.15(B)(16).* Through *310.15(B)(19) as Adjusted for Ambient Temperature if Necessary.*
10. Now take the ampacity of 38.72 A found in step 6 and multiply it by the *Percent of Values* found in step 9, 50: 38.72 A × 0.50 = 19.36 A. Therefore, the new ampacity for these conductors is 19.36 A.

In the preceding examples, you have determined the ampacity rating of the conductors based on the number of conductors, the conductor insulation, and the ambient temperature in which the conductor will operate. A further consideration must be given to the temperature rating of the termination or device to which the conductor is connected. *Article 110.14(C)* states, *The temperature rating associated with the ampacity of a conductor shall be selected and coordinated so as not to exceed the lowest temperature rating of any connected termination, conductor, or device. Conductors with temperature ratings higher than specified for terminations shall be permitted to be used for ampacity adjustment, correction, or both.*

There are many other articles and tables within *Chapter 3* of the *NEC* that apply to *Wiring Methods and Materials.* In addition, *Chapter 4* of the *NEC* contains *Article 400*, which deals with *Flexible Cords and Cables.* Also, *Article 402* deals with *Fixture Wires.* It is of tremendous benefit to maintenance technicians to become familiar with these and all other aspects of the *NEC* as well as their facility's policies and procedures.

18-2 CONDUCTOR COLOR CODE

To assist in providing safety for maintenance personnel, it is good practice to identify the voltage and purpose of the conductors of an electrical distribution system. Conductors are typically identified by the color of their insulation or by other types of color coding such as tape, paint, and so on.

Article 200.6 of the *NEC* is titled, *Means of Identifying Grounded Conductors. NEC 200.6(A), Sizes 6 AWG or Smaller,* states:

> *An insulated grounded conductor of AWG or smaller shall be identified by a continuous*

white or natural gray outer finish or by one of the following means:

1. *A continuous white outer finish.*
2. *A continuous gray outer finish.*
3. *Three continuous white stipes along the conductor's entire length on other than green insulation.*
4. *Wires that have their outer covering finished to show a white or gray color but have colored tracer threads in the braid identifying the source of manufacturer shall be considered as meeting the provisions of this section.*
5. *The grounded conductor of a mineral-insulated, metal-sheathed cable shall be identified at the time of installation by distinctive marking at its termination.*
6. *A single-conductor, sunlight-resistant, outdoor-rated cable used as a grounded conductor in photovoltaic power systems, as permitted by 690.31, shall be identified at the time of installation by distinctive white marking at all terminations.*
7. *Fixture wire shall comply with the requirements for grounded conductor identification as specified in 402.8.*
8. *For aerial cable, the identification shall be as above, or by means of a ridge located on the exterior of the cable so as to identify it.*

NEC 200.6(B), Sizes 4 AWG or Larger, states:

An insulated grounded conductor 4 AWG or larger shall be identified by one of the following means:

1. *A continuous white outer finish.*
2. *A continuous gray outer finish.*
3. *Three continuous white stripes along its entire length on other than green insulation.*
4. *At the time of installation, by a distinctive white or gray marking at its terminations. This marking shall encircle the conductor or insulation.*

Therefore, the general rule is that the grounded conductor is identified by either a white or natural gray marking. This is further defined when conductors of different systems are contained within the same raceway, cable, box, auxiliary gutter, or other type of enclosure. *NEC 200.6(D), Grounded Conductors of Different Systems*, states:

Where grounded conductors of different systems are installed in the same raceway, cable, box, auxiliary gutter, or other type of enclosure, each grounded conductor shall be identified by system. Identification that distinguishes each system grounded conductor shall be permitted by one of the following means:

1. *One system grounded conductor shall have an outer covering conforming to 200.6(A) or (B).*
2. *The grounded conductor(s) of other systems shall have a different outer covering conforming to 200.6(A) or 200.6(B) or by an outer covering of white or gray with a readily distinguishable colored stripe other than green running along the insulation.*
3. *Other and different means of identification as allowed by 200.6(A) or (B) that will distinguish each system grounded conductor.*

The means of identification shall be documented in a manner that is readily available or shall be permanently posted where the conductors of different systems originate.

NEC Article 310.110 further states requirements for *Conductor Identification. NEC 310.110(C), Ungrounded Conductors*, states:

Conductors that are intended for use as ungrounded conductors, whether used as a single conductor or in multiconductor cables, shall be finished to be clearly distinguishable from grounded and grounding conductors. Distinguishing markings shall not conflict in any manner with the surface markings required by 310.120(B)(1). Branch-circuit ungrounded conductors shall be identified in accordance with 210.5(C). Feeders shall be identified in accordance with 215.12.

Article 250.119 of the *NEC* addresses the *Identification of Equipment Grounding Conductors*. This article states:

Unless required elsewhere in this Code, equipment grounding conductors shall be permitted to be bare, covered, or insulated. Individually covered or insulated equipment grounding conductors shall have a continuous outer finish that is either green or green with one or more yellow stripes except as permitted in this section.

There are additional sections to *Article 250.119*. How ever, the broad meaning is that the equipment grounding conductor will be either a bare conductor or insulated with a green insulation or an insulation that is green with yellow stripes.

Often a transformer secondary is wired in a delta configuration with the midpoint of one phase winding grounded. This is done to supply lighting and other similar loads. The result is that one phase conductor will have a higher voltage to ground. This phase conductor is referred to as the *high leg* or *bastard leg*. The *NEC* requires this *high leg* to be identified with an orange color. This is addressed in three separate locations within the *NEC*. Under *Requirements For Electrical Installations, NEC Article 110.15, High-Leg Marking* states:

> *On a 4-wire, delta-connected system where the midpoint of one phase winding is grounded, only the conductor or busbar having the higher phase voltage to ground shall be durably and permanently marked by an outer finish that is orange in color or by other effective means. Such identification shall be placed at each point on the system where a connection is made if the grounded conductor is also present.*

NEC Article 230.56, Service Conductor with the Higher Voltage to Ground, is used for service conductors. This article states:

> *On a 4-wire, delta-connected service where the midpoint of one phase winding is grounded, the service conductor having the higher phase voltage to ground shall be durably and permanently marked by an outer finish that is orange in color, or by other effective means, at each termination or junction point.*

Finally, when there is more than one nominal voltage system within a building, the *NEC* requires each ungrounded conductor of a multiwire branch circuit to be identified by phase and system. *Article 210.5(C), Identification of Ungrounded Conductors*, states:

> *Identification of Ungrounded Conductors. Ungrounded conductors shall be identified in accordance with 210.5(C)(1), (2), and (3).*
>
> 1. *Application. Where the premises wiring system has branch circuits supplied from more than one nominal voltage system, each ungrounded conductor of a brandch circuit shall be identified by phase or line and system at all termination, connection, and splice points.*
> 2. *Means of Identification. The means of identification shall be permitted to be by separate color coding, marking tape, tagging, or other approved means.*
> 3. *Posting of Identification Means. The method utilized for conductors originating within each branch-circuit panelboard or similar branch-circuit distribution equipment shall be documented in a manner that is readily available or shall be permanently posted at each branch-circuit panel-board or similar branch-circuit distribution equipment.*

Notice that the *NEC* does not specify the colors that are to be used. Following is a list of the most common colors and their meaning:

Three-phase greater than 600 volts
Phase X(A) = Black
Phase Y(B) = Red
Phase Z(C) = Blue

Three-phase 480/277 volts
Phase X(A) = Brown
Phase Y(B) = Orange
Phase Z(C) = Yellow

Three-phase 208/120 volts
Phase X(A) = Black
Phase Y(B) = Red
Phase Z(C) = Blue
Neutral = White
Ground = Green

Single-phase 240/120 volts
Phase X(A) = Black
Phase Y(B) = Red
Neutral = White
Ground = Green

Other facilities may use the following method to determine the color codes and their meaning:

First System (4-wire)	Black, Red, Blue, and White
Second System (4-wire)	Brown, Orange, Yellow, and White
First System (3-wire)	Black, Red, and White
Second System (3-wire)	Brown, Orange, and White

Because there is no national standard for these colors, you may find different colors and meanings at your facility. This is the reason why the *NEC* requires the color code and meaning to be permanently posted at each panelboard.

18-3 RACEWAY SELECTION

The *NEC* defines a raceway as

> *An enclosed channel of metal or nonmetallic materials designed expressly for holding wires, cables, or busbars, with additional functions as permitted in this Code. Raceways include, but are not limited to, rigid metal conduit, rigid nonmetallic conduit, intermediate metal conduit, liquidtight flexible conduit, flexible metallic tubing, flexible metal conduit, electrical*

nonmetallic tubing, electrical metallic tubing, underfloor raceways, cellular concrete floor raceways, cellular metal floor raceways, surface raceways, wireways, and busways.

For the purposes of this text, the focus is on the type of raceways more commonly referred to as conduit. This includes electrical nonmetallic tubing, inter mediate metal conduit, rigid metal conduit, rigid nonmetallic conduit, liquidtight flexible conduit, flexible metallic tubing, flexible metal conduit, and electrical metallic tubing.

Article 362 of the *NEC* addresses electrical nonmetallic tubing (ENT). ENT is a pliable corrugated raceway of circular cross section that is composed of a material that is resistant to moisture and chemical atmospheres. It is also flame retardant. Because ENT is a pliable raceway, it can be bent by hand with relative ease. *NEC Article 362.10* explains the permitted uses of ENT whereas *Article 362.12* explains the uses or areas for which ENT is not approved. The installation of ENT is covered by *Article 362.10* through *Article 362.60.*

Intermediate metal conduit (IMC) is covered in *NEC Article 342.* IMC is a listed steel raceway of circular cross section. It has a thicker wall construction than EMT but a thinner wall construction than Rigid Metal Conduit (RMC). IMC is permitted to be threaded. *NEC Article 342.10* lists the permitted uses of IMC. *Article 342.10* through *Article 342.60* covers its installation requirements.

NEC Article 344 addresses rigid metal conduit (RMC). RMC is a listed metal raceway of circular cross-sectional area. RMC has a thicker wall construction than EMT and IMC. RMC is also permitted to be threaded, and it is generally made of steel with protective coatings or aluminum. Red brass and stainless steel may be used for special circumstances. *Article 344.10* explains the permitted uses of RMC, and *Article 344.10* through *Article 344.60* specify its installation requirements.

NEC Article 358 addresses electrical metallic tubing (EMT). EMT is a listed metallic tubing of circular cross section approved for the installation of electrical conductors when joined together with listed fittings. EMT has the thinnest wall construction when compared with IMC and RMC. *Article 358.10* explains the permitted uses of EMT, whereas *Article 358.12* identifies the uses that are not permitted. *Article 358.10* through *Article 358.60* specify the installation requirements for EMT.

Flexible metallic tubing (FMT) is covered in *NEC Article 360.* FMT is a listed tubing that is circular in cross section, flexible, metallic, and liquidtight without a nonmetallic jacket. *Article 360.10* lists the permitted uses of FMT, whereas *Article 360.12* lists the uses where FMT is not permitted. The installation requirements for FMT are listed in *Article 360.10* through *Article 360.60.*

NEC Article 348 addresses flexible metal conduit (FMC). FMC is a raceway of circular cross section made of helically wound, formed, interlocked metal strips. *Article 348.10* explains the permitted uses of FMC, whereas *Article 348.12* identifies the uses that are not permitted. *Article 348.10* through *Article 348.60* specify the installation requirements for FMC.

Liquidtight flexible metal conduit (LFMC) is covered in *NEC Article 350. Article 350.2* defines LFMC as a listed raceway of circular cross section that has an outer liquidtight, nonmetallic, sunlight-resistant jacket over an inner flexible metal core with associated couplings, connectors, and fittings approved for the installation of electric conductors. The permitted uses are covered in *Article 350.10. Article 350.12* identifies the uses not permitted.

Article 356 defines liquidtight flexible nonmetallic conduit (LFNC) as a listed raceway of circular cross section of various types. These types include:

1. A smooth, seamless inner core and cover bonded together and having one or more reinforcement layers between the core and cover designated as Type LFNC-A
2. A smooth inner surface with integral reinforcement within the conduit wall designated as Type LFNC-B
3. A corrugated internal and external surface without integral reinforcement within the conduit wall designated as Type LFNC-C

The permitted uses for LFNC are covered in *Article 356.10*, whereas *Article 356.12* covers the uses for LFNC which are not permitted. The installation requirements for LFNC are covered by *Article 356.10* through *Article 356.60.*

As you can see, there are several types of conduit from which to choose. You must make your selection based on the environment and level of protection required for the conductors that will be enclosed by the conduit.

18-4 RACEWAY SIZING

One final factor that will have an impact on your selection of conduit will be the size required.

It would be ideal for all of the conductors within a conduit to be the same size and have the same type of insulation. This makes sizing a conduit very easy. *Note 1* of *Chapter 9* of the *NEC* states that you should *See Informative Appendix C for the maximum number of conductors and fixture wires, all of the same size (total cross-sectional area including insulation) permitted in trade sizes of the applicable conduit or tubing. NEC Table C.1* through *Table C.12(A)* show the *Conduit and Tubing Fill Tables for Conductors and Fixture Wires of the Same Size.* These tables are found in Appendix C of this textbook.

Following are some examples to work through. You will be installing twelve 14 AWG Type THHN insulated conductors within some EMT conduit. What is the minimum trade size diameter of EMT conduit needed?

To find the answer, refer to *Table C.1*. You must use this table because you are using EMT. Find the section that lists THHN insulation. Next, look under the Conductor Size (AWG/kcmil) column. Find 14 AWG. Now slide to the *right* until you find the first number that is equal to or greater than the number of conductors that will be installed in this conduit. You should rest your finger on the 12 because twelve conductors are being installed. Now move up the table until you see the Metric Designator (Trade Size) required. For this example, you will need Metric Designator 16 (or ½") EMT. Notice that this size does not allow for any future installation of conductors. In this example, ½-in. EMT is permitted to have no more than twelve 14 AWG Type THHN conductors installed. Running additional conductors in this conduit in the future will be in violation of the Code. You would have to replace this conduit with one that has a larger diameter or install another separate conduit to accommodate the additional conductors. The best solution may be to install the next larger conduit initially; that is, install a 21 (or ¾"). EMT now instead of 16 (or ½") This will allow for up to ten additional 14 AWG Type THHN conductors to be installed at a later date.

Try another example. You will be installing five 2/0 AWG Type TW conductors in FMC. What is the minimum trade size diameter of FMC required?

First, identify the correct table. Because you will be using FMC, you must refer to *Table C.3*. Find the section that lists TW insulation. Next, look under the Conductor Size (AWG/kcmil) column. Find 2/0 AWG. Now slide to the *right* until you find the first number that is equal to or greater than the number of conductors that will be installed in this conduit. You should rest your finger on the 5 because you are installing five conductors. Now move up the table until you see the Trade Size required. For this example, you will need 53 (or 2") FMC. Notice that there is no allowance for any future installation of conductors. In this example, 53 (or 2") FMC is permitted to have no more than five 2/0 AWG Type TW conductors installed. Running additional conductors in this conduit in the future will be in violation of the Code. You would have to replace this conduit with one that has a larger diameter or install another separate conduit to accommodate the additional conductors. The best solution may be to install the next larger conduit initially; that is, instead of installing 53 (or 2") FMC now, install 63 (or 2 ½"). This will allow for up to two additional 2/0 AWG Type TW conductors to be installed at a later date.

Unfortunately, it is rare to have all of the conductors the same size with the same insulation within the same conduit. The next example shows what to do when there are different sizes and insulation types within the same conduit.

What is the minimum trade diameter of RMC required to contain the following conductors: four 4/0 AWG Type THHW, three 10 AWG Type XFF, and one 4 AWG bare copper equipment grounding conductor?

To find the answer to this question, you must use different tables within the *NEC*. These tables will allow you to calculate the cross-sectional area that each of these conductors occupies. All of the following tables are found in *Chapter 9* of the *NEC*. The tables have been reprinted in the Appendix of this textbook for your convenience. Appendix D of this textbook contains *Table 5* of the *NEC*. *Table 5* lists *Dimensions of Insulated Conductors and Fixture Wires. Table 8*, found in Appendix E of this textbook, lists *Conductor Properties*. You will need to find the total of the cross-sectional areas and determine the size of conduit that can accommodate these conductors. *Table 4* of the *NEC* lists *Dimensions and Percent Area of Conduit and Tubing. Table 4* is found in Appendix F of this textbook.

1. Find the cross-sectional area of four 4/0 AWG Type THHW conductors.
    a. From *Table 5*, find the section containing THHW insulation.
    b. Now find the section containing 4/0 AWG.
    c. The *Approximate Area* of one 4/0 AWG Type THHW conductor is 239.9 mm^2 or 0.3718 in^2.
    d. Multiply this value by the number of 4/0 AWG Type THHW conductors.
        (i) Four 4/0 AWG conductors × 239.9 mm^2 per conductor = 959.6 mm^2 (Four 4/0 AWG conductors × 0.3718 in^2 per conductor = 1.4872 in^2.)
2. Find the cross-sectional area of three 10 AWG Type XFF conductors.
    a. From *Table 5*, find the section containing XFF insulation.
    b. Now find the section containing 10 AWG.

c. The Approximate Area of one 10 AWG Type XFF conductor is 21.48 mm^2 or 0.0333 in^2.

d. Multiply this value by the number of 10 AWG Type XFF conductors.

(i) Three 10 AWG conductors × 21.48 mm^2, or 0.0333 in^2 per conductor = 64.44 mm^2. (Three 10 AWG conductors × 0.0333 in^2 per conductor = 0.0999 in^2.)

3. Find the cross-sectional area of one 4 AWG bare copper conductor.

a. From *Table 8*, find the column with the heading, Size (AWG or kcmil). Move down the column until you find 4 AWG.

b. Move your finger to the right until it rests under the column with the headings Con ductors, Overall, Area mm^2 (in^2.).

(i) The area in square millimeters (inches) of a bare 4 AWG conductor is 27.19 mm^2 (0.042 in^2).

4. Now add the total area found in step 1. d. (i) to the total area found in step 2. d. (i) to the total area found in step 3. b. (i).

a. 959.6 mm^2 + 64.44 mm^2 + 27.19 mm^2 = 1051.23 mm^2 (1.4872 in^2 + 0.0999 in^2 + 0.042 in^2 = 1.6291 in^2).

5. This means that the total area occupied by all of the conductors contained within this conduit is 1051.23 mm^2 (1.6291 in^2).

6. You must now determine the minimum trade diameter of RMC to contain these conductors. *Table 4* of the *NEC* shows the *Dimensions and Percent Area of Conduit and Tubing (Areas of Conduit or Tubing for the Combina tions of Wires Permitted in Table 1, Chapter 9).*

a. From *Table 4*, find the section titled *Article 344—Rigid Metal Conduit (RMC).*

b. Find the column with the heading, Over 2 Wires 40% mm^2 (in^2.).

c. Move your finger down the column until you arrive at a number that is equal to 1051.23 mm^2 (1.6291 in^2) or more.

d. 1051.23 mm^2 (1.6291 in^2) does not appear in this column. You will see the number 879 mm^2 (1.363 in^2) and 1255 mm^2 (1.946 in^2). Because 879 mm^2 (1.363 in^2) is smaller than 1051.23 mm^2 (1.6291 in^2), you must use the conduit that is the next larger size—1255 mm^2 (1.946 in^2).

e. Place your finger at the 1255 mm^2 (1.946 in^2) location and move your finger to the *left* until it rests under the column with the heading, Metric Designator (Trade Size). Your finger should be resting on the number 63 (2½ in.)

Therefore, you must use 63 (2½ in.) RMC.

As you can see, sizing conduit is not that difficult. However, it is very important that conduit not be overfilled with conductors. This creates a fire hazard, because heat builds up from the flow of current through the conductors. Without sufficient air space for the heat to dissipate, the insulation will break down and a short circuit or a grounded circuit can result. This can lead to further damage and increased maintenance costs.

SUMMARY

- The *National Electrical Code* is considered by many to be the standard by which all wiring and associated components of an electrical system are designed, specified, and installed.
- Conductors used in industrial environments are constructed from three materials: copper and nickel or nickel-coated copper, aluminum, and copper-clad aluminum.
- These three materials have different conduction characteristics that must be considered when sizing conductors for given loads.
- It is important that a conductor be properly sized for a given load.
- Conductors are typically identified by the color of their insulation or by other types of color coding such as tape, paint, etc.
- Electrical nonmetallic tubing (ENT) is a pliable corrugated raceway of circular cross section that is composed of a material that is resistant to moisture and chemical atmospheres. It is flame retardant and can be bent by hand easily.
- Intermediate metal conduit (IMC) is a listed steel raceway of circular cross section. IMC is permitted to be threaded.
- Rigid metal conduit (RMC) is a listed metal raceway of circular cross-sectional area. It is also permitted to be threaded and may be constructed of galvanized steel or aluminum.
- Electrical metallic tubing (EMT) has a circular cross section approved for installation of electrical conductors when joined together with listed fittings.

- Flexible metallic tubing (FMT) is circular in cross section, flexible, metallic, and liquidtight without a nonmetallic jacket.
- Flexible metal conduit (FMC) is a raceway of circular cross section made of helically wound, formed, interlocked metal strips.
- Liquidtight flexible metal conduit (LFMC) is a raceway of circular cross section that has an outer liquidtight, nonmetallic, sunlight resistant jacket over an inner flexible metal core with associated couplings, connectors, and fittings approved for the installation of electrical conductors.
- Liquidtight flexible nonmetallic conduit (LFNC) is a raceway of circular cross section of different types: LFNC-a, LFNC-b, and LFNC-C.

REVIEW QUESTIONS

1. What is the allowable ampacity of a single 10 AWG Type THW copper wire in free air based on an ambient air temperature of 30°C (86°F)?
2. What is the allowable ampacity of a single 1/0 AWG Type THWN aluminum wire in free air based on an ambient air temperature of 30°C (86°F)?
3. A load draws 97 amperes of current. What is the minimum size aluminum conductor with Type THHN insulation in free air that is capable of supplying this load based on an ambient air temperature of 30°C (86°F)?
4. What is the minimum trade size diameter of EMT conduit needed for the installation of eight 10 AWG Type THHN insulated conductors?
5. What is the minimum trade diameter of RMC required to contain the following conductors: six 4 AWG Type THHW, two 12 AWG Type XFF, and one 2 AWG bare copper equipment grounding conductor?

CHAPTER 19

Transformers

Transformers are a vital and essential component of practically all electrical systems in use today. Whether discussing the power distribution grid from the electric utility's generating station to homes, the power distribution within a plant, or even the electronic circuitry within a television set, transformers perform essential functions necessary for the operation of electrical equipment.

OBJECTIVES

After studying this chapter, the student should be able to

- Explain the operation of a basic transformer.
- Identify the different types of single-phase transformers.
- Identify the different connection methods of three-phase transformers.
- Perform transformer calculations.
- Discuss power distribution systems.

19-1 BASIC TRANSFORMERS

One of the most efficient electrical machines is the **transformer**. Transformers typically have an efficiency of approximately 93% to 98%. In addition, because transformers have no moving parts, they are efficient and reliable. Transformers are used to perform several functions such as changing values of voltage, current, or impedance and providing **electrical isolation** between circuits. However, transformers cannot be used to change frequency.

When used to change the voltage, a transformer can increase or decrease the applied voltage. A transformer can also be used to increase or decrease current or to increase or decrease a circuit's impedance without altering the voltage or current. Finally, when a transformer is said to provide electrical isolation between circuits, this means that it prevents unwanted electrical noise from passing between circuits without altering the voltage, current, or impedance.

The basic transformer is very simple in construction. Refer to **Figure 19-1**. To make a transformer, take a length of wire and form it into a coil. Make a second coil from another length of wire. Place the two coils of wire near, but not touching, each other. Pass a small AC current through one coil of wire in order to measure a small AC voltage across the other coil. **Figure 19-1A** shows the schematic symbol for an air core transformer. This is a transformer in which the coils are wound around a cardboard tube or simply formed into a coil with nothing but air in the center. **Figure 19-1B** shows the schematic for an iron core transformer. This is the most common type of transformer that you will find in industry. The iron core transformer has the windings wound around an iron frame, as can be seen in **Figure 19-2**. This provides for higher efficiency and better heat handling capabilities.

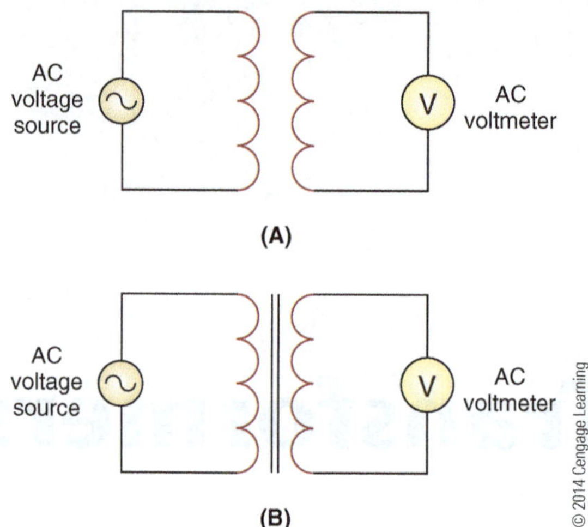

FIGURE 19-1 Figure A shows an air core transformer while figure B shows an iron core transformer.

A transformer works by **mutual induction**. Examine the basic transformer in Figure 19-2 to better understand how a transformer works. Notice that the winding on the left side of the drawing is labeled *primary*. The **primary winding** of the transformer is where the electrical power is applied. The winding on the right is labeled *secondary*. The **secondary winding** is where the load is connected. Notice that the primary and secondary windings show the same number of "loops" of wire. Notice also that both the primary and secondary windings are wound around an iron frame or **core**. This is different from the transformer shown in Figure 19-1.

An AC current is applied to the primary winding of the transformer in **Figure 19-3A**. Recall that whenever a current flows through a conductor, a magnetic field is produced around the conductor. Figure 19-3A shows the AC current sine wave rising from the zero crossing point toward the peak positive value. As the current increases toward the positive peak, the magnetic field around the conductor increases as well.

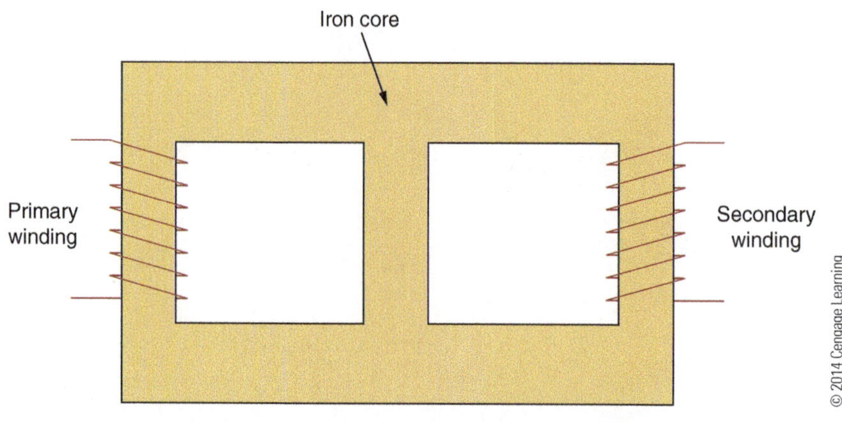

FIGURE 19-2 Construction of a transformer.

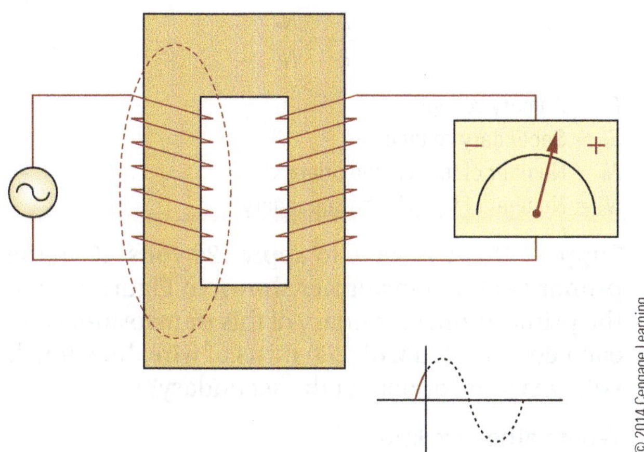

FIGURE 19-3A A small increase in magnetic field strength.

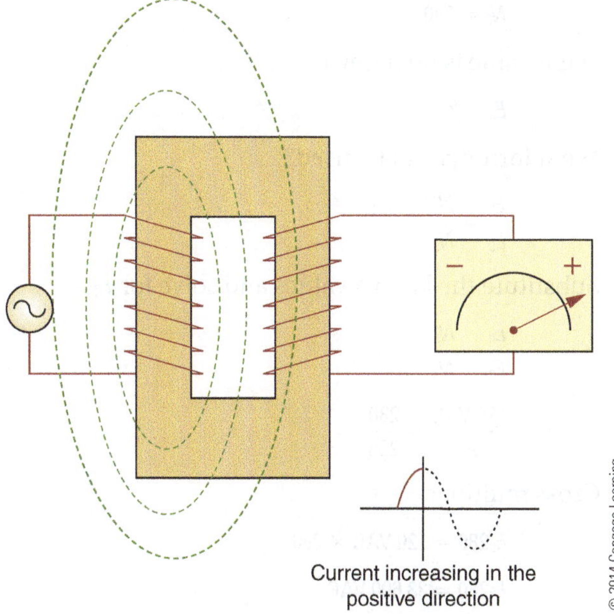

FIGURE 19-3B Maximum magnetic field strength.

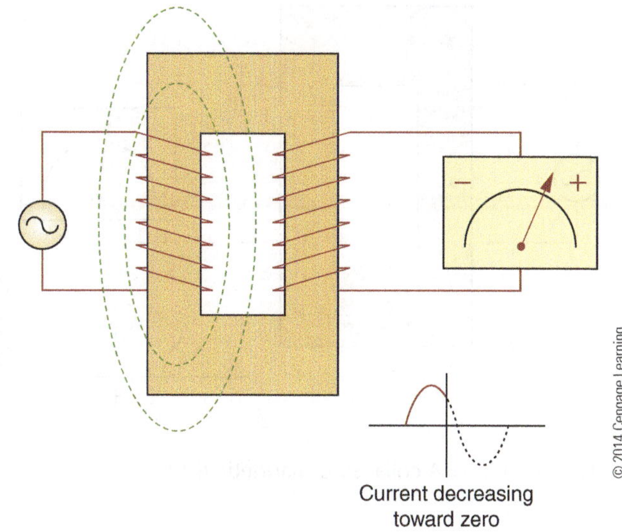

FIGURE 19-3C A collapsing magnetic field.

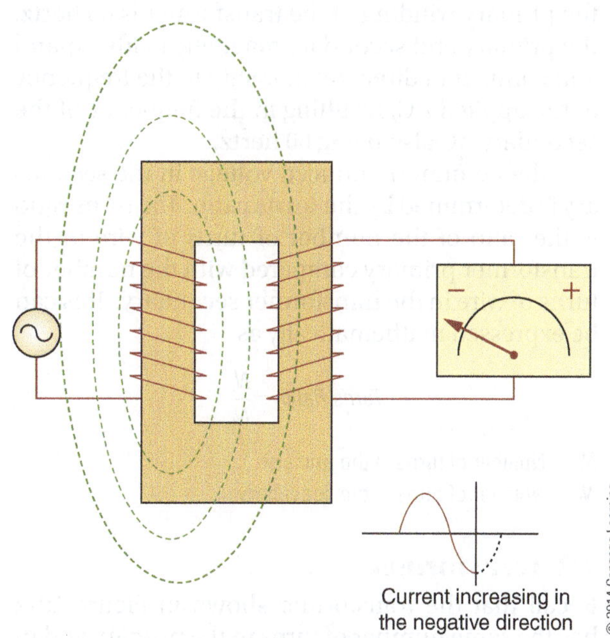

FIGURE 19-3D Maximum magnetic field strength.

As the magnetic field increases in strength, it expands outward from the primary winding. This causes the magnetic lines of force to cut through the conductor that forms the secondary of the transformer, as shown in **Figure 19-3B**. As the magnetic lines of force cut through the secondary winding of the transformer, a voltage is induced into the secondary.

After the applied AC current reaches the positive peak value, the applied AC current decreases toward the zero crossing point, as shown in **Figure 19-3C**. This causes the expanding magnetic lines of force, which no longer cut through the secondary winding of the transformer, to collapse. This causes the induced secondary voltage to decrease as well.

After the applied AC current reaches the zero crossing point, the AC current increases toward the negative peak value. As the current increases toward the negative peak, the magnetic field around the conductor increases as well. As the magnetic field increases in strength, the field expands outward from the primary winding. This causes the magnetic lines of force to cut through the conductor that forms the secondary of the transformer, as shown in **Figure 19-3D**. As the magnetic lines of force cut through the secondary winding of the transformer, a voltage is induced into the secondary.

After the applied AC current reaches the negative peak value, the applied AC current decreases toward the zero crossing point, as shown in **Figure 19-3E**. This causes the expanding magnetic lines of force, which no longer cut through the secondary winding of the transformer, to collapse. This causes the induced secondary voltage to decrease as well.

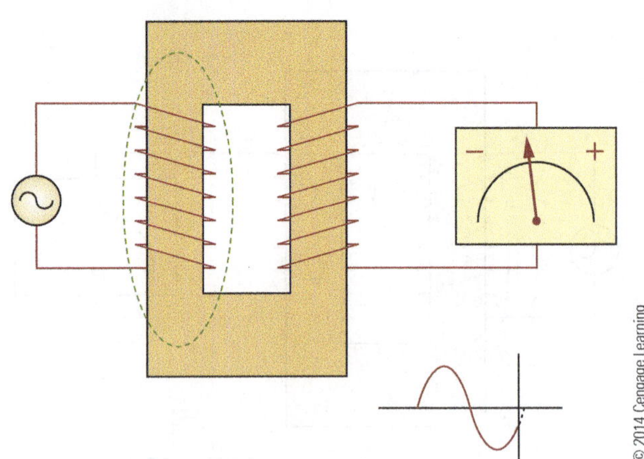

FIGURE 19-3E A collapsing magnetic field.

If the frequency of the applied AC current to the primary winding of the transformer is 60 hertz, the primary and secondary magnetic fields expand and contract in direct relationship to the frequency of the applied AC, resulting in the frequency of the secondary AC also being 60 hertz.

The amount of induced voltage in the secondary is determined by the **turns ratio**. The turns ratio is the ratio of the number of turns of wire in the transformer primary compared with the number of turns of wire in the transformer secondary. This can be expressed mathematically as

$$Turns\ Ratio = \frac{N_P}{N_S}$$

N_P = Number of turns in the primary
N_S = Number of turns in the secondary

1:1 Transformer

Recall that the transformer shown in Figure 19-2 has the same number of turns in the primary and in the secondary. This is a **1:1 transformer** because it has a 1:1 turns ratio. Whether there are 50 turns in the primary and 50 turns in the secondary, or 150 turns in the primary and 150 turns in the secondary, or even 1270 turns in the primary and 1270 turns in the secondary, does not matter because the ratio is still 1:1.

There are transformers in which the turns ratio of the primary and secondary circuits is different. For example, a transformer may have a primary that has *fewer* turns than the secondary, or it may have a primary that has *more* turns than the secondary. These types of transformers are discussed later.

As a result of the 1:1 turns ratio of the transformer shown in Figure 19-2, the secondary voltage can be determined if the primary voltage is known by using the following formula:

$$\frac{E_P}{E_S} = \frac{N_P}{N_S}$$

E_P = Primary voltage
E_S = Secondary voltage
N_P = Number of turns in the primary
N_S = Number of turns in the secondary

Suppose that you were to apply 120 volts AC to the primary of the transformer shown in Figure 19-2. If the primary and secondary of this transformer were each constructed with 280 turns of wire, how much voltage would appear at the secondary?

What values are known?

$E_P = 120\ VAC$

$N_P = 280$

$N_S = 280$

What value is not known?

$E_S = ?$

What formula can be used?

$$\frac{E_P}{E_S} = \frac{N_P}{N_S}$$

Substitute the known values and solve for E_S.

$$\frac{E_P}{E_S} = \frac{N_P}{N_S}$$

$$\frac{120\ VAC}{E_S} = \frac{280}{280}$$

Cross multiply:

$$E_S 280 = 120\ VAC \times 280$$

$$E_S 280 = 33{,}600\ VAC$$

Divide both sides by 280:

$$\frac{E_S 280}{280} = \frac{33{,}600\ VAC}{280}$$

$$E_S = 120\ VAC$$

Therefore, the secondary voltage is also 120 volts AC. Because this is a 1:1 transformer, the secondary voltage has the same value as the primary voltage.

The 1:1 transformer is used to provide isolation between electrical circuits. Notice that there is no electrical connection between the primary and the secondary of the transformer shown in Figure 19-2. This means that if this transformer is placed between two electrical circuits, the voltage and current is coupled by the transformer, but there is not a direct electrical connection between the two circuits. This is done to prevent or eliminate unwanted electrical noise or interference passing between the two circuits.

Recall that an iron core was used in the construction of the transformer shown in Figure 19-2. An iron core is used in practically all industrial transformers because the iron allows the magnetic lines of force to travel between the windings more easily than if they were to travel through air alone. However, although this helps the efficiency of the transformer, the iron core also introduces a problem called **eddy currents**. Eddy currents form when the magnetic lines of force travel through the iron core. These currents circulate around the metal and produce heat. To minimize eddy currents, the iron core can be constructed of thin sheets of iron fastened together to form a **laminated iron core**. The laminations form layers of iron oxide, which acts as an insulator between the laminations, thus preventing the formation of eddy currents.

Step-Up Transformer

Refer to **Figure 19-4**. Notice that there are fewer turns of wire in the primary circuit than there are in the secondary circuit. Assume that there are 50 turns of wire in the primary circuit and 100 turns of wire in the secondary circuit. This means that for every turn of wire in the primary circuit, there are two turns of wire in the secondary circuit. The turns ratio of this transformer is 1:2:

$$\frac{N_P}{N_S} = \frac{50}{100} = \frac{1}{2} \text{ or } 1{:}2$$

Because the secondary has twice as many turns of wire as the primary, there are twice as much voltage induced into the secondary. This can be shown mathematically. Assume that there is 120 volts AC applied to the primary of this transformer.

What values are known?

$N_P = 50$

$N_S = 100$

$E_P = 120 \text{ VAC}$

What value is not known?

$E_S = ?$

What formula can be used?

$$\frac{E_P}{E_S} = \frac{N_P}{N_S}$$

Substitute the known values and solve for E_S.

$$\frac{E_P}{E_S} = \frac{N_P}{N_S}$$

$$\frac{120 \text{ VAC}}{E_S} = \frac{50}{100}$$

Cross multiply:

$$E_S 50 = 12{,}000 \text{ VAC}$$

Divide both sides by 50:

$$\frac{E_S 50}{50} = \frac{12{,}000 \text{ VAC}}{50}$$

$$E_S = 240 \text{ VAC}$$

Therefore, with 120 volts AC applied to the primary of this 1:2 transformer, the secondary voltage is 240 volts AC. This is called a **step-up transformer**, because the primary voltage has been increased from 120 volts AC to 240 volts AC.

Up to this point, the discussion has been focused on voltage. Before proceeding further, the relationship between the turns ratio and current must be examined. Use the transformer in Figure 19-4, except that now a 10-ohm load has been added to the secondary side, as shown in **Figure 19-5**.

How much secondary current flows through the 10-ohm load? This can be determined by using Ohm's law as follows:

What values are known?

$E_S = 240 \text{ VAC}$

$R_L = 10\ \Omega$

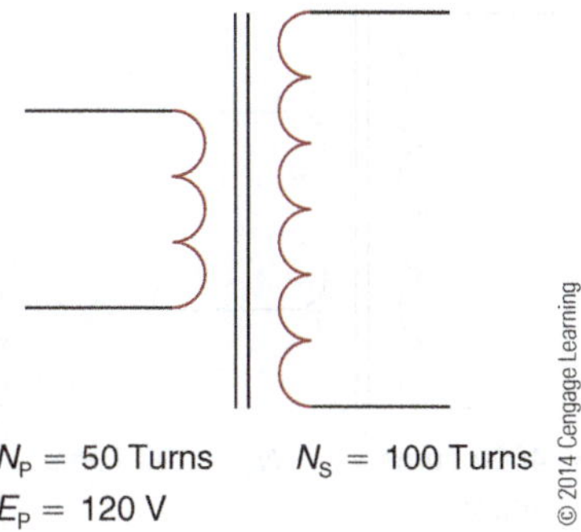

FIGURE 19-4 A step-up transformer.

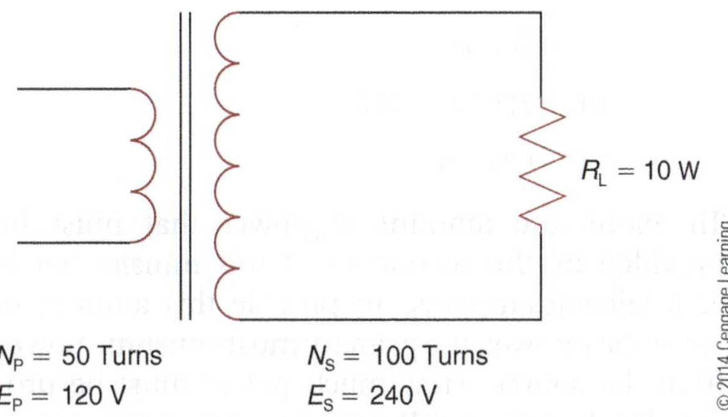

FIGURE 19-5 A step-up transformer with a load.

What value is not known?

$$I_S = ?$$

What formula can be used?

$$I_S = \frac{E_S}{R_L}$$

Substitute the known values and solve for I_S.

$$I_S = \frac{E_S}{R_L}$$

$$I_S = \frac{240 \text{ VAC}}{10\ \Omega}$$

$$I_S = 24 \text{ A}$$

Therefore, there are 24 amperes of current flowing in the secondary circuit of this transformer. In order for there to be secondary current, there must be primary current. How much primary current is flowing in this transformer?

Before answering this question, you need to know an important rule when working with transformers. Recall that transformer efficiency is nearly 100% (95% to 98%). For this chapter, these transformers are considered to be 100% efficient. This means that the power applied to a transformer must equal the power taken from it. Stated another way, *power in must equal power out*. When dealing with transformers, power is measured in VA. For large transformers, power is measured in kilovolt-amperes (kVA). In the above example, the secondary current was 24 amperes for the 240-volt secondary circuit. How much power is this?

What values are known?

$$E_S = 240 \text{ VAC}$$

$$I_S = 24 \text{ A}$$

What value is not known?

$$VA_S = ?$$

What formula can be used?

$$VA_S = E_S \times I_S$$

Substitute the known values and solve for VA_S.

$$VA_S = E_S \times I_S$$

$$VA_S = 240 \text{ VAC} \times 24 \text{ A}$$

$$VA_S = 5.76 \text{ kVA}$$

Therefore, the amount of power that must be provided by the secondary of this transformer is 5.76 kilovolt-amperes. To provide this amount of secondary power, the primary must consume power from the source. How much power must be provided to the primary? Remember that power in must equal power out. This means that if the secondary is to deliver 5.76 kilovolt-amperes of power, the primary must consume 5.76 kilovolt-amperes of power from the source.

If the primary operating at 120 volts AC, consumes 5.76 kilovolt-amperes of power, how much primary current flows through the primary winding? While you are determining this mathematically, you will discover something interesting as well.

What values are known?

$$E_P = 120 \text{ VAC}$$

$$VA_P = 5.76 \text{ kVA}$$

What value is not known?

$$I_P = ?$$

What formula can be used?

$$I_P = \frac{VA_P}{E_P}$$

Substitute the known values and solve for I_P.

$$I_P = \frac{VA_P}{E_P}$$

$$I_P = \frac{5.76 \text{ kVA}}{120 \text{ VAC}}$$

$$I_P = 48 \text{ A}$$

Therefore, the primary current must be 48 amperes. Notice what has happened. While the primary voltage was stepped up from 120 volts AC to 240 volts AC, the primary current was stepped down from 48 amperes to 24 amperes. Notice that while the voltage was stepped up by a 1:2 ratio, the current was stepped down by a 2:1 ratio. Voltage and current must change inversely to one another in order for the power in to equal the power out. Remember that *if you step up the voltage, you step down the current.*

Step-Down Transformer

Refer to **Figure 19-6**. Notice that there are more turns of wire in the primary circuit than there are

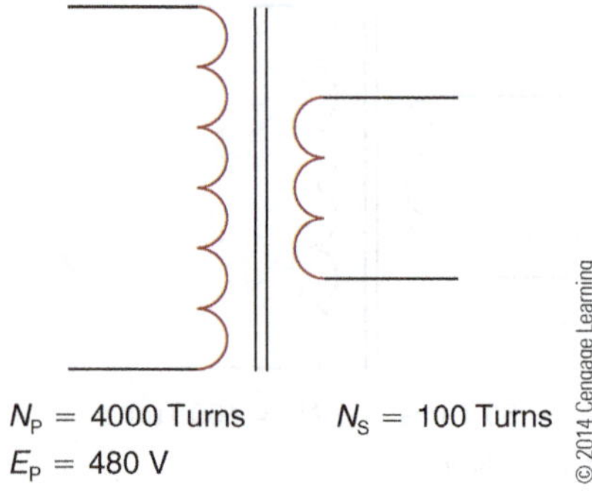

FIGURE 19-6 A step-down transformer.

in the secondary circuit. Assume that there are 4000 turns of wire in the primary circuit and 100 turns of wire in the secondary circuit. This means that for every 40 turns of wire in the primary circuit, there is 1 turn of wire in the secondary circuit. The turns ratio of this transformer is 40:1.

$$\frac{N_P}{N_S} = \frac{4000}{100} = \frac{40}{1} \text{ or } 40{:}1$$

Because the primary has 40 times as many turns of wire as the secondary, there is 1/40th as much voltage induced into the secondary. This can be shown mathematically. Assume that there are 480 volts AC applied to the primary of this transformer.

What values are known?

$N_P = 4000$

$N_S = 100$

$E_P = 480$ VAC

What value is not known?

$E_S = ?$

What formula can be used?

$$\frac{E_P}{E_S} = \frac{N_P}{N_S}$$

Substitute the known values and solve for E_S.

$$\frac{E_P}{E_S} = \frac{N_P}{N_S}$$

$$\frac{480 \text{ VAC}}{E_S} = \frac{4000}{100}$$

Cross multiply:

$E_S 4000 = 48{,}000$ VAC

Divide both sides by 4000:

$$\frac{E_S 4000}{4000} = \frac{48{,}000 \text{ VAC}}{4000}$$

$E_S = 12$ VAC

Therefore, with 480 volts AC applied to the primary of this 40:1 transformer, the secondary voltage is 12 volts AC. This is called a **step-down transformer**, because the primary voltage has been decreased from 480 volts AC to 12 volts AC.

Figure 19-7 shows this same transformer with the addition of an 8-ohm load in the secondary circuit. You need to determine the amount of current that flows in both the primary and secondary circuits. Begin with the secondary.

What values are known?

$E_S = 12$ VAC

$R_L = 8\ \Omega$

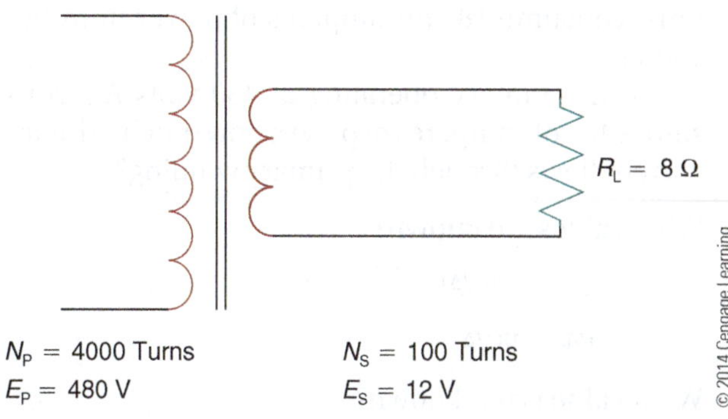

FIGURE 19-7 A step-down transformer with a load.

What value is not known?

$I_S = ?$

What formula can be used?

$$I_S = \frac{E_S}{R_L}$$

Substitute the known values and solve for I_S.

$$I_S = \frac{E_S}{R_L}$$

$$I_S = \frac{12 \text{ VAC}}{8\ \Omega}$$

$I_S = 1.5$ A

Therefore, there are 1.5 amperes of current flowing in the secondary circuit of this transformer. How much power is provided by this transformer?

What values are known?

$E_S = 12$ VAC

$I_S = 1.5$ A

What value is not known?

$VA_S = ?$

What formula can be used?

$VA_S = E_S \times I_S$

Substitute the known values and solve for VA_S.

$VA_S = E_S \times I_S$

$VA_S = 12 \text{ VAC} \times 1.5$ A

$VA_S = 18$ VA

Therefore, the amount of power that must be provided by the secondary of this transformer is 18 volt-amperes. To provide this amount of secondary power, the primary must consume power from the source. How much power must be provided to the primary? Remember that power in must equal power out. This means that if the secondary is to deliver 18 volt-amperes of power, the primary must consume 18 volt-amperes of power from the source.

If the primary operating at 480 volts AC consumes 18 volt-amperes of power, how much primary current flows through the primary winding?

What values are known?

$E_P = 480$ VAC

$VA_P = 18$ VA

What value is not known?

$I_P = ?$

What formula can be used?

$$I_P = \frac{VA_P}{E_P}$$

Substitute the known values and solve for I_P.

$$I_P = \frac{VA_P}{E_P}$$

$$I_P = \frac{18 \text{ VA}}{480 \text{ VAC}}$$

$$I_P = 37.5 \text{ mA}$$

Therefore, the primary current must be 37.5 milliamperes. Again, notice what has happened. Whereas the primary voltage was stepped down from 480 volts AC to 12 volts AC, the primary current was stepped up from 37.5 milliamperes to 1.5 amperes. Notice that whereas the voltage was stepped down by a 40:1 ratio, the current was stepped up by a 1:40 ratio. Again, voltage and current must change inversely to one another in order for the power in to equal the power out. Remember that *if you step down the voltage, you step up the current.*

Transformer with a Center-Tapped Secondary

Figure 19-8 shows a transformer with a **center-tapped secondary**. Notice that there is a third connection to the secondary winding of this transformer. This connection is connected to the electrical center of the secondary winding. Imagine having a transformer with a 240-volt AC secondary. If this transformer had a center tap, you would have the full-winding voltage of 240 volts AC and the capability of using each half of the full winding, giving two windings rated at 120-volt AC each, as shown in **Figure 19-9**. This type of transformer is similar to the transformer used by the electric utility to reduce the high voltage that is transmitted from the generating station to the low voltage used in homes. The primary on the electric utility's transformer may be 7200 volts AC, 13,200 volts AC, or something else. The secondary side, which supplies power to residences, is 120/240 volts AC. This transformer uses a center-tapped secondary winding with a full-winding voltage of 240 volts AC. The voltage from the center tap to each leg of the winding is 120 volts AC.

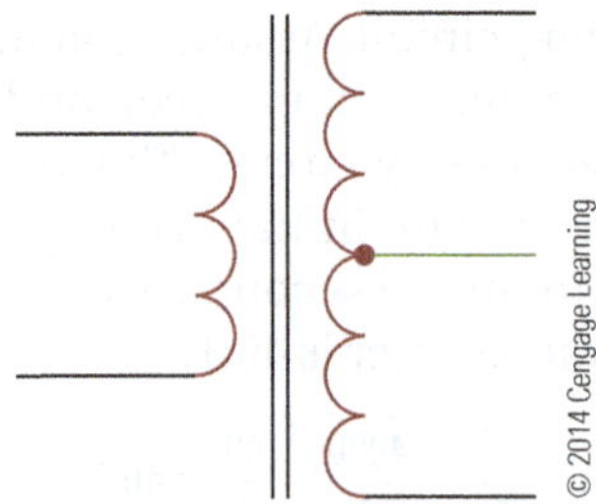

FIGURE 19-8 A step-up transformer with a center-tapped secondary.

Transformer with Multiple-Tapped Secondary

The transformer shown in **Figure 19-10** also has a **tapped secondary**. However, this transformer has a secondary that has **multiple taps**. These taps are not located at the electrical center of the secondary winding. Rather they are located at a point on the secondary winding to yield the desired voltage.

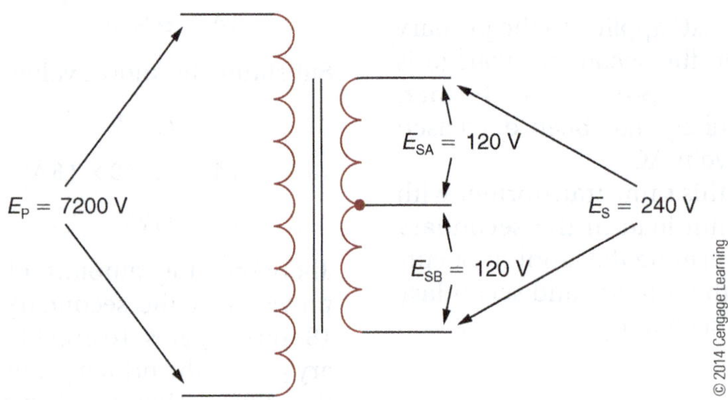

FIGURE 19-9 A step-down transformer with a center-tapped secondary and voltages.

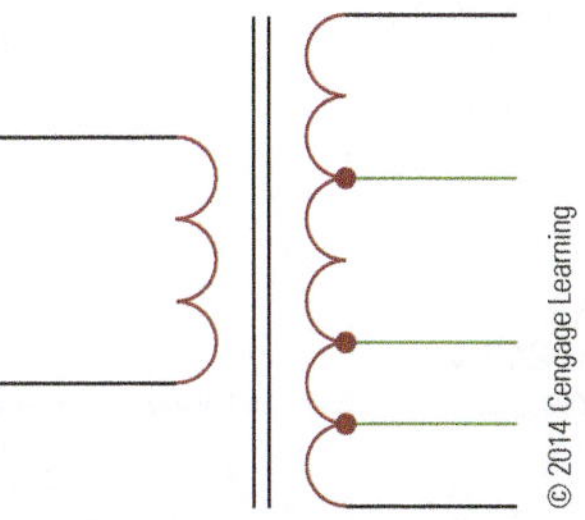

FIGURE 19-10 A step-up transformer with multiple secondary taps.

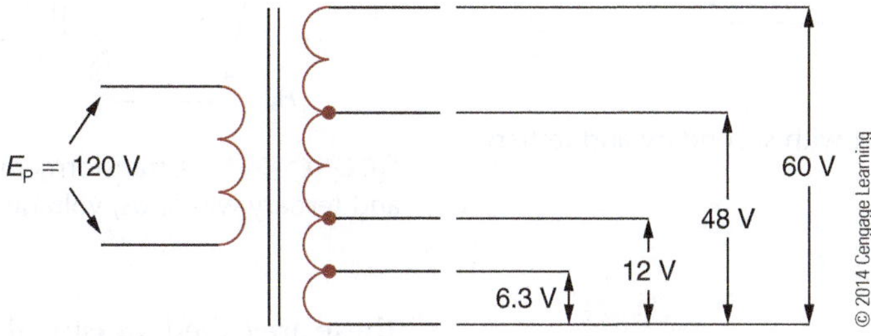

FIGURE 19-11 A step-down transformer with multiple secondary taps and voltages.

Notice the transformer shown in **Figure 19-11**. This transformer has a 120-volt AC primary. The full voltage rating of the secondary winding is 60 volts AC. In addition, there are taps that provide 48 volts AC, 12 volts AC, and 6.3 volts AC. As you can see, it is possible to obtain different voltages from this one transformer.

Transformer with Multiple Windings

Figure 19-12 shows another transformer that also produces different voltages. However, this transformer uses **multiple windings** to accomplish this feat. Refer to **Figure 19-13**. Notice that the primary voltage is 480 volts AC. This transformer uses a *secondary winding* and a **tertiary winding**. The voltage rating of the secondary is 240 volts AC, and the voltage rating of the tertiary is 120 volts AC. The main difference between a transformer with multiple windings and a transformer with multiple taps is that the transformer with multiple windings has two or more totally isolated windings, whereas the transformer with multiple taps must share a connection or a **common**.

Now look at **Figure 19-14**. Do you notice anything different in the way the secondary windings are drawn? Notice how the wires wrap around the iron core at the top and bottom of the windings. Focusing on the top of the windings, Figure 19-14 shows

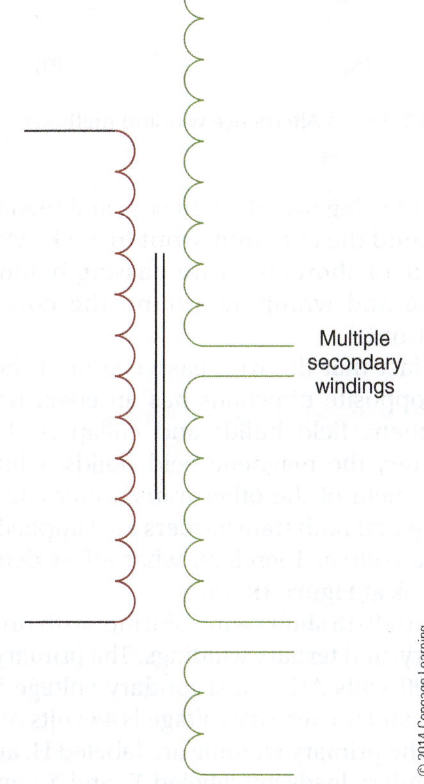

FIGURE 19-12 A transformer with multiple secondary windings.

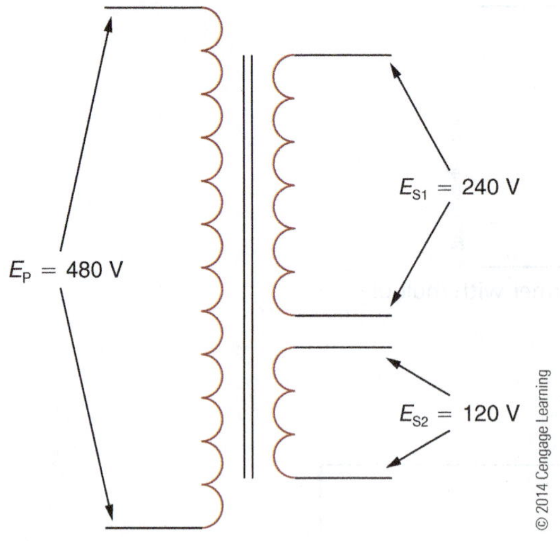

FIGURE 19-13 A transformer with secondary and tertiary windings and voltages.

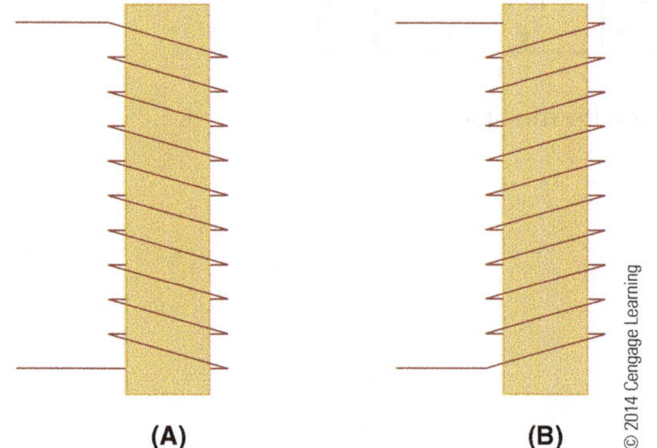

FIGURE 19-14 Alternative winding methods.

the wire crossing over the iron core and then wrapping around the core from front to back, whereas Figure 19-14 shows the wire passing behind the iron core and wrapping around the core from back to front.

The fact that the wire passes around the iron core in opposite directions has an effect on how the magnetic field builds and collapses. In one transformer, the magnetic field builds, while the magnetic field of the other transformer collapses, assuming that both transformers are supplied from the same source. Therefore, what effect does this have? Look at **Figure 19-15**.

Figure 19-15 shows a transformer with primary, secondary, and tertiary windings. The primary voltage is 480 volts AC; the secondary voltage is 120 volts AC; and the tertiary voltage is 48 volts AC. The leads of the primary winding are labeled H_1 and H_2; the secondary leads are labeled X_1 and X_2; and the tertiary leads are labeled X_3 and X_4. Notice the small black dots that have been added to the windings.

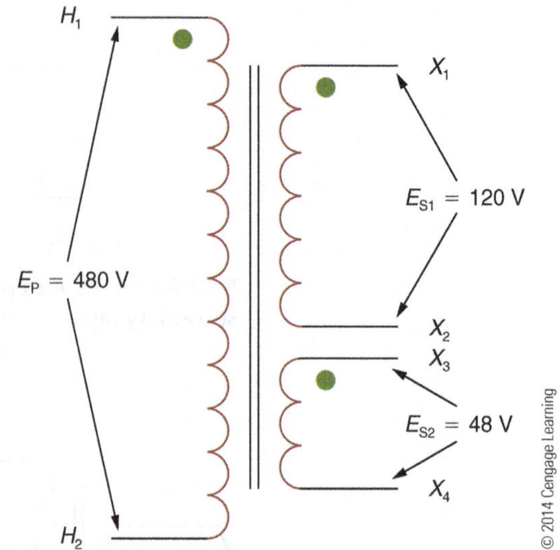

FIGURE 19-15 A transformer with primary, secondary, and tertiary windings, voltages, and phasing dots.

These are called **phasing dots**, and they are used to indicate the phase relationship between the windings. Notice that the phasing dots are located at the tops of the primary (H_1), secondary (X_1), and tertiary (X_3) windings. This means that at any point, the tops of the primary, secondary, and tertiary windings EW all in phase.

Here is why this is important to know. **Figure 19-16** shows the same transformer shown in Figure 19-15. The only difference is that the secondary winding, X_2, has been connected to the tertiary winding, X_3. In effect, there is now one large secondary winding because the two separate windings have been connected in series. Because the windings have been connected in series with a non-phasing

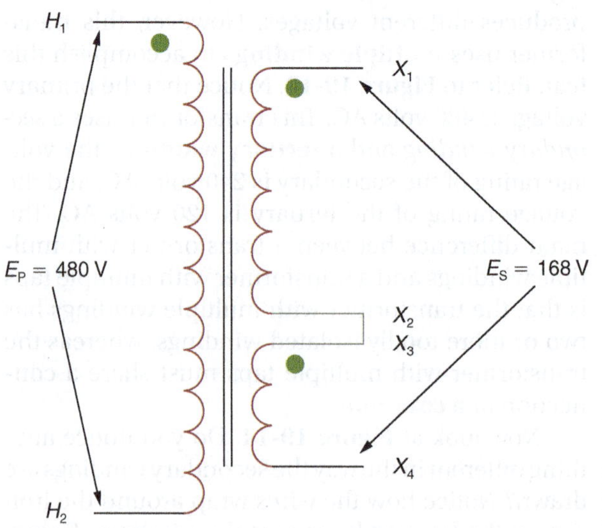

FIGURE 19-16 Additive polarity.

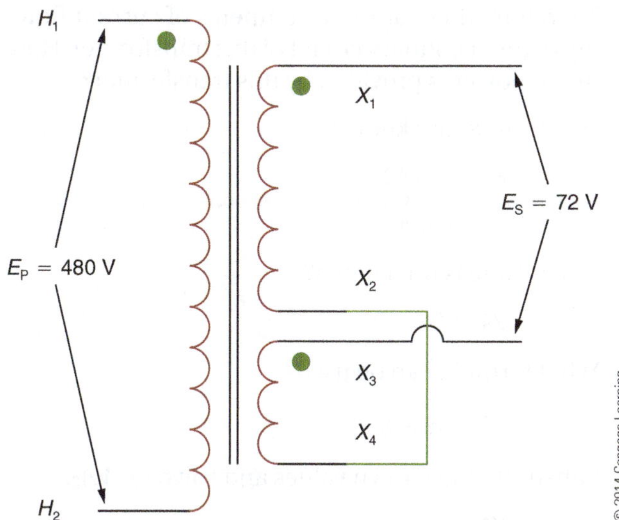

FIGURE 19-17 Subtractive polarity.

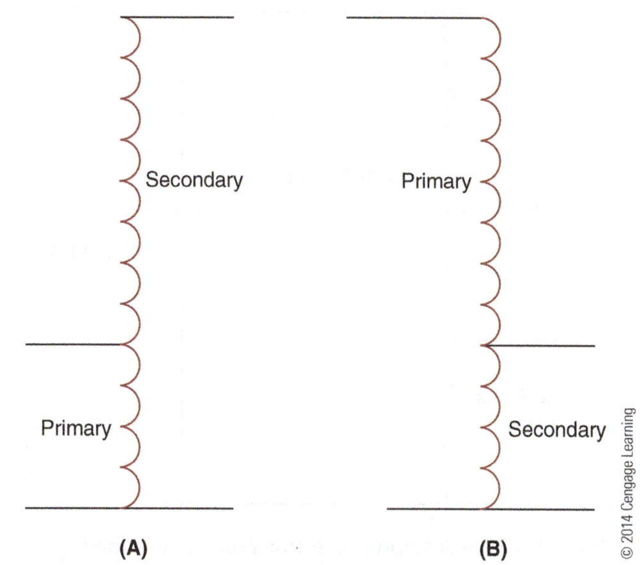

FIGURE 19-18 Autotransformers: (A) step-up, (B) step-down.

dot lead connected to a phasing dot lead, the voltages add or aid each other (series aiding). In other words, the voltage measured across the full winding is the sum of the two individual winding voltages; 120 VAC + 48 VAC = 168 VAC.

Now look at **Figure 19-17**. At first glance, this figure may look identical to Figure 19-16, but look closely. Notice that the connections to the tertiary winding have been reversed. The secondary and tertiary windings are still connected in series; however, the two non-phasing dot leads are now connected together. With these connections, the voltages do not add; instead they subtract (series opposing). The result is that the voltage across the full winding is the difference between the two voltages; 120 VAC − 48 VAC = 72 VAC.

19-2 AUTOTRANSFORMERS

Some transformers are constructed with only one winding. These transformers are called **autotransformers**, or **autoformers**. Autotransformers have the advantage of being less expensive than the transformers studied previously, because the single windings are less expensive to manufacture than two or more separate windings. There is a disadvantage to autotransformers, however. Because there is only one winding, there is no electrical isolation between the primary and secondary windings. This means that electrical noise or interference may pass from the primary circuit to the secondary circuit. This could present problems for the electrical circuitry on the secondary side of the autotransformer.

Two types of autotransformers are shown in **Figure 19-18**. Figure 19-18A shows a step-up autotransformer, whereas Figure 19-18B shows a step-down autotransformer.

Step-Up Autotransformer

In Figure 19-18A, the portion of the winding labeled *primary* has fewer turns of wire than the portion of the winding labeled *secondary*. This means that this is a step-up autotransformer. When an alternating current is applied to the primary, the expanding and collapsing magnetic field cuts through the full winding, which is the secondary. This causes a voltage to be induced into the full winding of the secondary. Because the total number of turns of wire in the primary is less than the total number of turns in the secondary, the voltage is increased, or stepped up. The same formula that was used to determine the amount of voltage for a transformer with two separate windings can be used for an autotransformer. For example, imagine that the primary of the autotransformer in **Figure 19-19** has 50 turns, and the secondary has 183⅓ turns. The primary is supplied with 120 volts AC and the secondary load is 30 ohms. Determine the amount of secondary voltage, secondary current, primary current, and the power rating for this transformer.

What values are known?

$N_P = 50$

$N_S = 183.33$

$E_P = 120$ VAC

$R_L = 30\ \Omega$

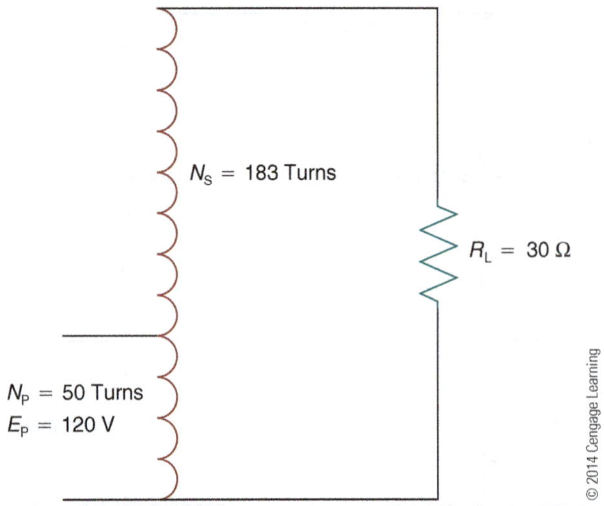

N_S = 183 Turns

R_L = 30 Ω

N_P = 50 Turns
E_P = 120 V

FIGURE 19-19 A step-up autotransformer with load.

What values are not known?

$E_S = ?$

$I_S = ?$

$I_P = ?$

$VA = ?$

Begin by finding the secondary voltage, E_S.

What formula can be used?

$$\frac{E_P}{E_S} = \frac{N_P}{N_S}$$

Substitute the known values and solve for E_S.

$$\frac{E_P}{E_S} = \frac{N_P}{N_S}$$

$$\frac{120\text{ VAC}}{E_S} = \frac{50}{183.33}$$

Cross multiply:

$$E_S 50 = 22{,}000\text{ VAC}$$

Divide both sides by 50:

$$\frac{E_S 50}{50} = \frac{22000\text{ VAC}}{50}$$

$$E_S = 440\text{ VAC}$$

Therefore, the secondary voltage is 440 volts AC. Now determine the secondary current.

What formula can be used?

$$I_S = \frac{E_S}{R_L}$$

Substitute the known values and solve for I_S.

$$I_S = \frac{E_S}{R_L}$$

$$I_S = \frac{440\text{ VAC}}{30\ \Omega}$$

$$I_S = 14.67\text{ A}$$

Therefore, there are 14.67 amperes of current flowing in the secondary circuit of this transformer. How much power is provided by this transformer?

What values are known?

$E_S = 440$ VAC

$I_S = 14.67$ A

What value is not known?

$VA_S = ?$

What formula can be used?

$$VA_S = E_S \times I_S$$

Substitute the known values and solve for VA_S.

$$VA_S = E_S \times I_S$$

$$VA_S = 440\text{ VAC} \times 14.67\text{ A}$$

$$VA_S = 6.45\text{ kVA}$$

Therefore, the amount of power that must be provided by the secondary of this transformer is 6.45 kilovolt-amperes. Remember a rule about transformers: Power in must equal power out. This same rule applies to autotransformers as well. This means that if the secondary delivers 6.45 kilovolt-amperes, the primary must use 6.45 kilovolt-amperes of power from the source. To supply this power, the primary must draw current from the source. Now determine the amount of primary current that this autotransformer draws.

What values are known?

$E_P = 120$ VAC

$VA_P = 6.45$ kVA

What value is not known?

$I_P = ?$

What formula can be used?

$$I_P = \frac{VA_P}{E_P}$$

Substitute the known values and solve for I_P.

$$I_P = \frac{VA_P}{E_P}$$

$$I_P = \frac{6.45\text{ kVA}}{120\text{ VAC}}$$

$$I_P = 53.75\text{ A}$$

Therefore, the primary current must be 53.75 amperes.

As in the study of step-up transformers, if you step up the voltage, you step down the current. In this example of a step-up autotransformer, the primary voltage of 120 volts AC was stepped up to a secondary voltage of 440 volts AC. The primary current of 53.75 amperes was stepped down to a secondary current of 14.67 amperes.

Step-Down Autotransformer

A step-down autotransformer was shown previously in Figure 19-18B. **Figure 19-20** shows the same step-down autotransformer with load. Notice that the primary winding has more turns of wire (3500 turns) than the secondary winding (404 turns). The primary voltage is 208 volts AC, and there is a 20-ohm load connected to the secondary winding of this autotransformer. Determine the amount of secondary voltage, secondary current, primary current, and the power rating for this transformer.

What values are known?

$$N_P = 3500$$

$$N_S = 404$$

$$E_P = 208 \text{ VAC}$$

$$R_L = 20\ \Omega$$

What values are not known?

$$E_S = ?$$

$$I_S = ?$$

$$I_P = ?$$

$$VA = ?$$

Begin by finding the secondary voltage, E_S.

What formula can be used?

$$\frac{E_P}{E_S} = \frac{N_P}{N_S}$$

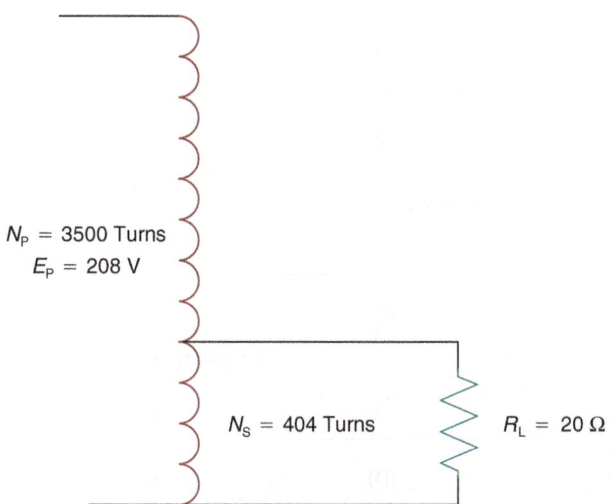

FIGURE 19-20 A step-down autotransformer with load.

Substitute the known values and solve for E_S.

$$\frac{E_P}{E_S} = \frac{N_P}{N_S}$$

$$\frac{208 \text{ VAC}}{E_S} = \frac{3500}{404}$$

Cross multiply:

$$E_S 3500 = 84032 \text{ VAC}$$

Divide both sides by 3500:

$$\frac{E_S 3500}{3500} = \frac{84032 \text{ VAC}}{3500}$$

$$E_S = 24 \text{ VAC}$$

Therefore, the secondary voltage is 24 volts AC. Now determine the secondary current.

What formula can be used?

$$I_S = \frac{E_S}{R_L}$$

Substitute the known values and solve for I_S.

$$I_S = \frac{E_S}{R_L}$$

$$I_S = \frac{24 \text{ VAC}}{20\ \Omega}$$

$$I_S = 1.2 \text{ A}$$

Therefore, there are 1.2 amperes of current flowing in the secondary circuit of this transformer. How much power is provided by this transformer?

What values are known?

$$E_S = 24 \text{ VAC}$$

$$I_S = 1.2 \text{ A}$$

What value is not known?

$$VA_S = ?$$

What formula can be used?

$$VA_S = E_S \times I_S$$

Substitute the known values and solve for VA_S.

$$VA_S = E_S \times I_S$$

$$VA_S = 24 \text{ VAC} \times 1.2 \text{ A}$$

$$VA_S = 28.8 \text{ VA}$$

Therefore, the amount of power that must be provided by the secondary of this transformer is 28.8 volt-amperes. Remember, power in must equal power out. If the secondary delivers 28.8 volt-amperes, the primary must use 28.8 volt-amperes of power from the source. Now determine the amount of primary current that this autotransformer draws.

What values are known?

$$E_P = 208 \text{ VAC}$$

$$VA_P = 28.8 \text{ VA}$$

What value is not known?

$$I_P = ?$$

What formula can be used?

$$I_P = \frac{VA_P}{E_P}$$

Substitute the known values and solve for I_P.

$$I_P = \frac{VA_P}{E_P}$$

$$I_P = \frac{28.8 \text{ VA}}{208 \text{ VAC}}$$

$$I_P = 136.46 \text{ mA}$$

Therefore, the primary current must be 138.46 milli-amperes.

Again, if you step down the voltage, you step up the current. In this example of a step-down autotransformer, the primary voltage of 208 volts AC was stepped down to a secondary voltage of 24 volts AC. The primary current of 138.46 milli-amperes was stepped up to a secondary current of 1.2 amperes.

Autotransformers with Multiple Taps

Autotransformers are also available with multiple taps, as shown in **Figure 19-21**. Figure 19-21A shows

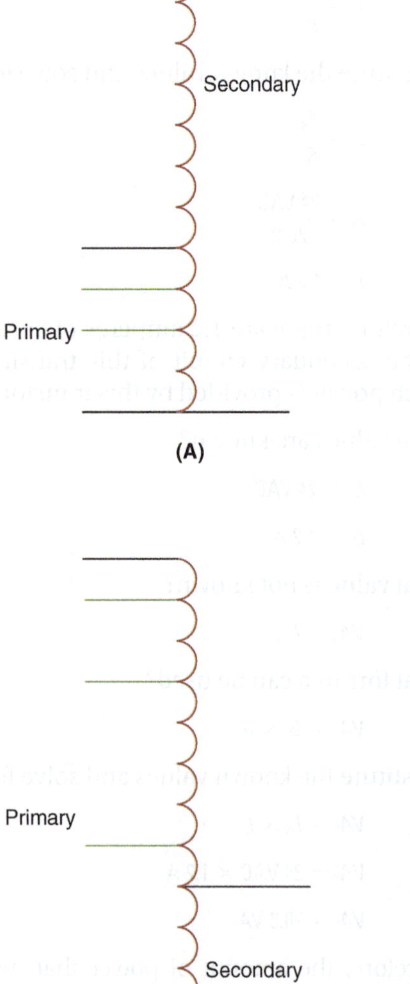

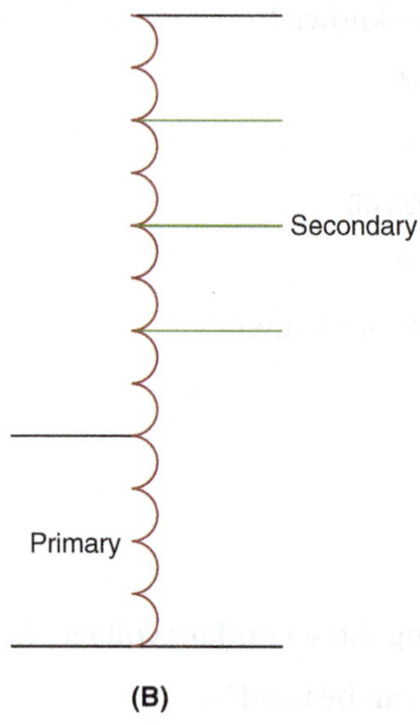

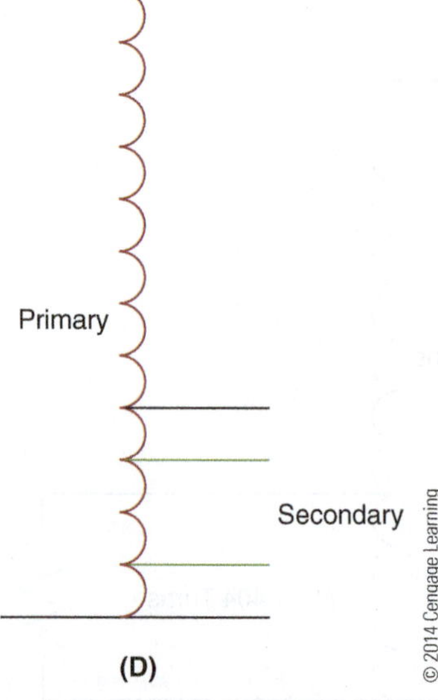

FIGURE 19-21 Various types of autotransformers: (A) step-up with multiple primary taps; (B) step-up with multiple secondary taps; (C) step-down with multiple primary taps; and (D) step-down with multiple secondary taps.

a step-up autotransformer with multiple primary taps. Figure 19-21B shows a step-up autotransformer with multiple secondary taps. A step-down autotransformer with multiple primary taps is shown in Figure 19-21C, and Figure 19-21D shows one with multiple secondary taps.

19-3 CURRENT TRANSFORMERS

Often, in industrial settings large amounts of AC current must be measured and monitored. This typically occurs with the AC power supplying a distribution transformer, panel, or load. To measure this current, a special type of transformer, called a current transformer, or CT, is used. **Figure 19-22** shows a current transformer, or CT, of the type typically found in an electrical cabinet.

A typical CT is of a toroidal design. A toroid is defined as a donut-shaped object; therefore, a toroidal transformer is a donut-shaped core around which conductors are wound, as seen in **Figure 19-23**.

CTs are constructed so that the conductor in which you wish to measure the current serves as the primary of the CT. The winding in the CT itself serves as the secondary. The industry standard rating for the secondary of a CT is 5 amperes. CT ratios begin at 100:5 and can be found with ratios of several thousand to 5. Because the secondary is rated at 5 amperes, a standard 5-amp AC current meter is connected across the secondary of the CT. In this fashion, a full-scale deflection of 5 amps can occur.

FIGURE 19-22 A current transformer or CT.

The conductor in which you wish to measure the current is inserted through the hole in the CT, as seen in **Figure 19-24**. A 5-amp full-scale AC ammeter is connected across the secondary terminals of the CT. The current measured on the ammeter is proportional to the ratio of the CT. For example, if the CT has a 600:5 ratio, then should 400 amperes of current flow through the power conductor, the ammeter indicates 3.33 amps of current.

FIGURE 19-23 A current transformer. Note the toroidal shape.

FIGURE 19-24 Note the conductor passing through the "hole" of the current transformer.

If the opening in the CT is large enough, the power conductor can be looped back through the opening of the CT. This will cause the primary to effectively have two turns of wire instead of the original one turn. Now a current of 300 amperes flowing through the power conductor produces a full scale reading on the ammeter. To increase the sensitivity of the CT, the power conductor can be looped through the opening as often as the opening in the CT allows. Just remember that looping the power conductor through two times reduces the current needed for full-scale reading to 300 amperes; looping it three times reduces the current needed to 200 amperes; looping it four times reduces the current needed to 150 amperes; and so on.

19-4 THREE-PHASE TRANSFORMERS

Up to this point, you have been learning about single-phase transformers. However, three-phase power is very much the standard in industry. There fore, you must be able to transform three-phase power as well as single-phase power. This can be accomplished by using **three-phase transformers**. A three-phase transformer may take the form of a single transformer, or it may be created from three single-phase transformers.

There are two ways that the primary or secondary of a three-phase transformer can be connected, as shown in **Figure 19-25**. Figure 19-25(A) shows a three-phase transformer with a **delta connection**. Both the primary and secondary windings are connected in a delta configuration, usually referred to as a **delta-delta** connection. Figure 19-25B shows a three-phase transformer with a **wye**, or **star**, **connection**. Both the primary and secondary windings are connected in a wye, or star, configuration, usually referred to as a **wye-wye**, or **star-star**, connection.

Figure 19-26 shows two other methods of connecting a three-phase transformer. Figure 19-26A shows a three-phase transformer with the primary connected in a delta connection and the secondary connected in a wye, or star, connection, usually referred to as a **delta-wye**, or **delta-star, configuration**. Figure 19-26B shows a three-phase transformer with the primary connected in a wye, or star, connection and the secondary connected in a delta connection, usually referred to as a **wye-delta**, or **star-delta**, **configuration**.

Figure 19-27, **Figure 19-28** (see in page 411), **Figure 19-29** (see in page 411), and **Figure 19-30** (see in page 412) show the same connections as those in Figure 19-25A and B, and Figure 19-26A and B, except the transformers used are not single three-phase transformers, but rather three single-phase transformers. Figure 19-27 shows a delta-delta connection, Figure 19-28 shows a wye-wye connection, Figure 19-29 shows a delta-wye connection, and Figure 19-30 shows a wye-delta connection.

Also show CAUTION and SAFETY when connecting three single-phase transformers to form a delta secondary connection. Extreme care must be taken to ensure that the delta windings are connected in the correct fashion because of the

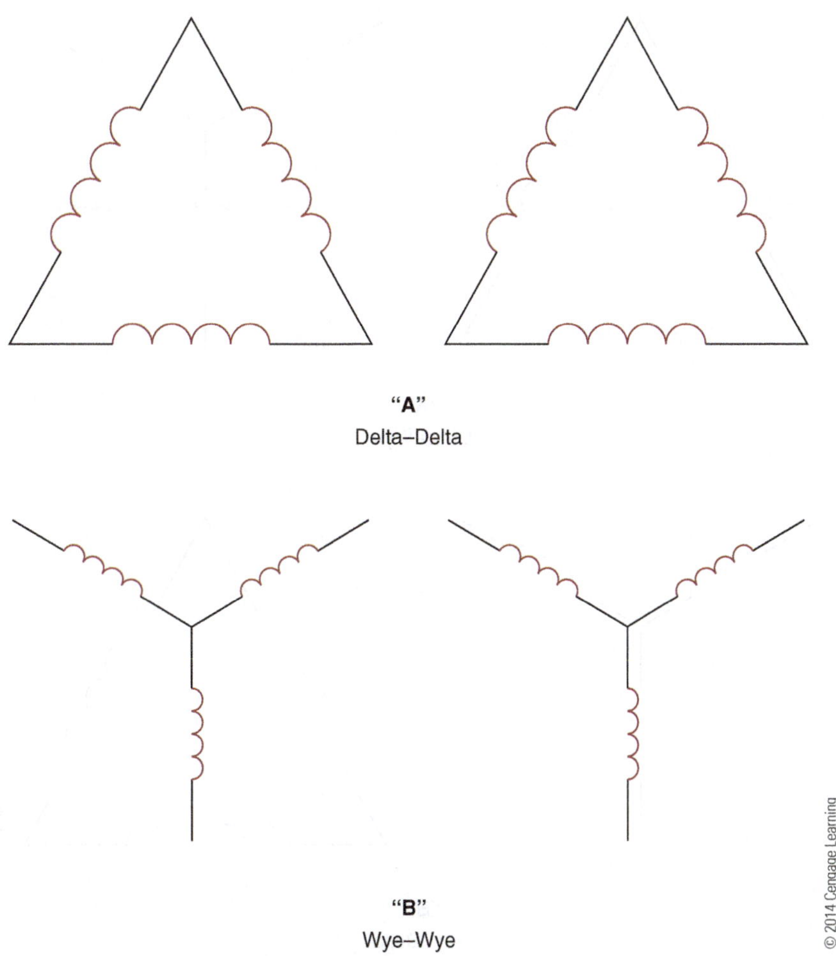

FIGURE 19-25 Three-phase transformer connections: (A) delta primary–delta secondary, and (B) wye primary–wye secondary.

polarities of the windings. To verify the correct connections, connect two of the delta windings in their usual fashion. Next, use a voltmeter to measure the voltage across the two open ends, as shown in Figure 19-31. If the windings are connected correctly, the measured voltage will be the same as the output of each transformer. If one of the windings is connected with a reverse polarity, the measured voltage will be 1.732 times the output voltage of one transformer. *Do not close the connection with the transformer polarity reversed!* Switch the connections to one of the two existing transformer windings, and then re-check the voltage across the open circuit. The measured voltage should now be equal to the same value as the output of one of the transformers. You can now close the delta by connecting the third transformer winding.

When working with delta wound transformers, care must be taken when making the connection that closes the delta. Recall that transformer windings have a polarity. When connecting the windings in a delta configuration, the polarity of the windings must be observed. To check this, measure the voltage across the open terminals of two of the three windings connected in delta. Refer to **Figure 19-31** (see in page 412). The voltmeter should indicate a low voltage when power is applied. If a significant voltage is measured, reverse the connections to one of the two windings and measure the voltage again. Whichever connection produced the lowest voltage reading is the correct configuration. *Never close a delta connection without measuring the open delta voltage. Closing the delta with incorrect connections can produce short-circuit magnitude currents resulting in damage to the transformers.*

Before examining the different three-phase transformer configurations, you need to understand new terminology. Because you are dealing with three-phase connections, there are two different voltage and current measurements. One measurement is called the **line value** and the other measurement is called the **phase value**. **Figure 19-32** (see in page 412) shows three windings connected in a wye configuration. Notice that each winding is

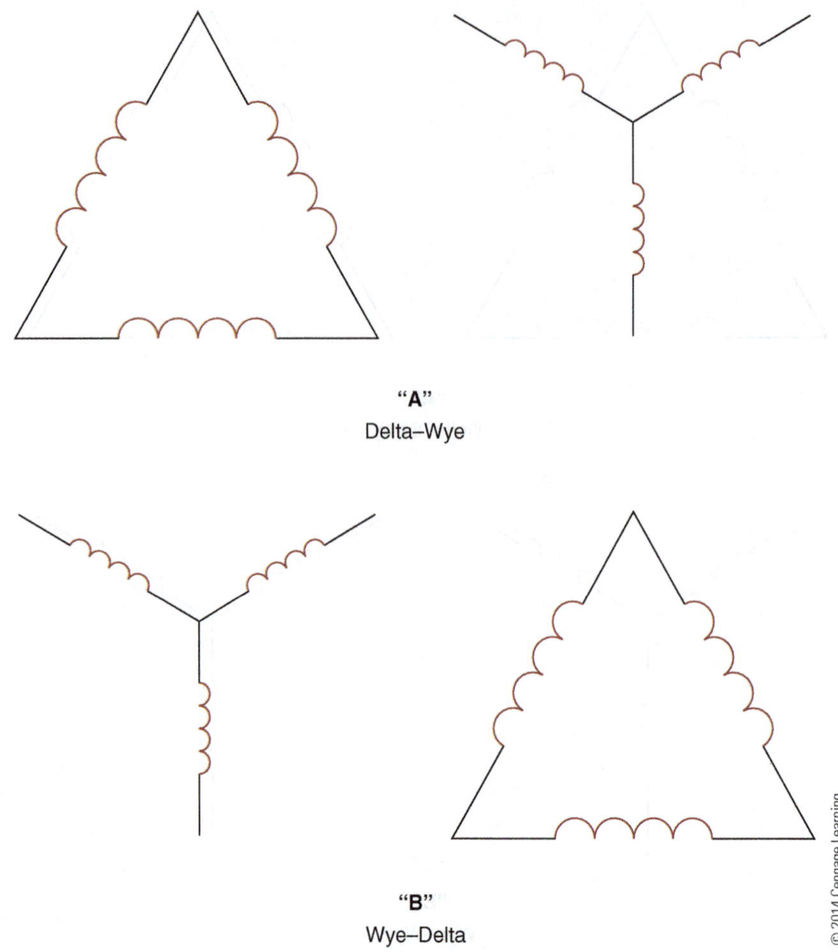

FIGURE 19-26 Other three-phase transformer connections: (A) delta primary–wye secondary, and (B) wye primary–delta secondary.

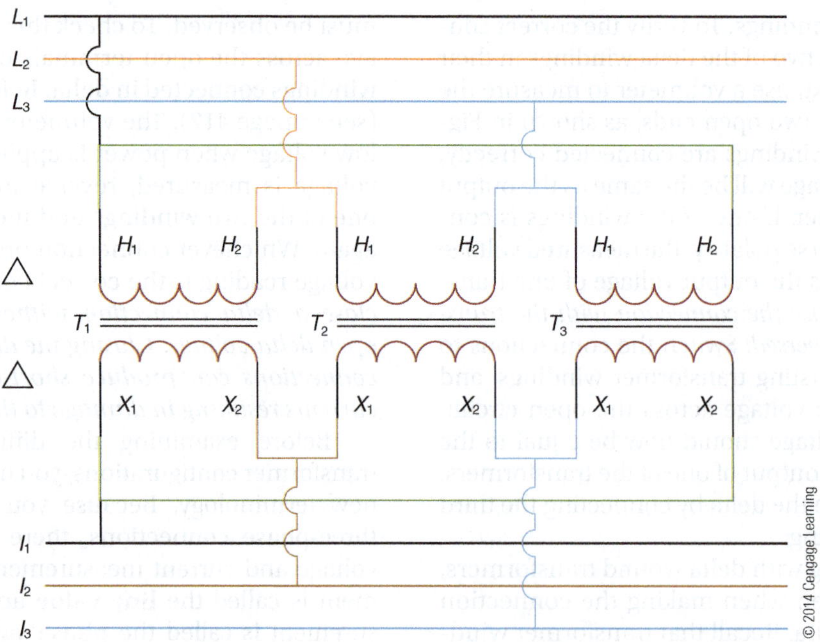

FIGURE 19-27 Delta–delta connections with three single-phase transformers.

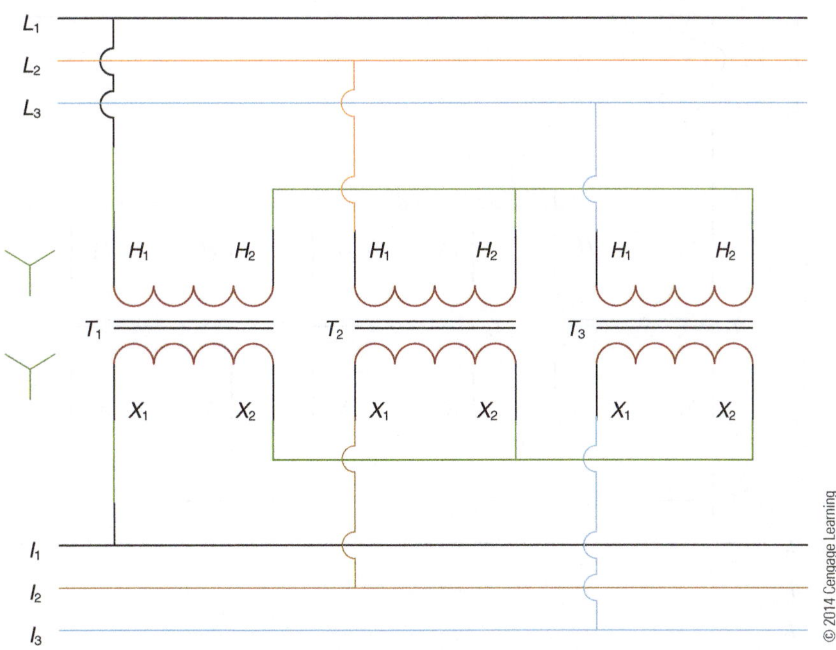

FIGURE 19-28 Wye–wye connections with three single-phase transformers.

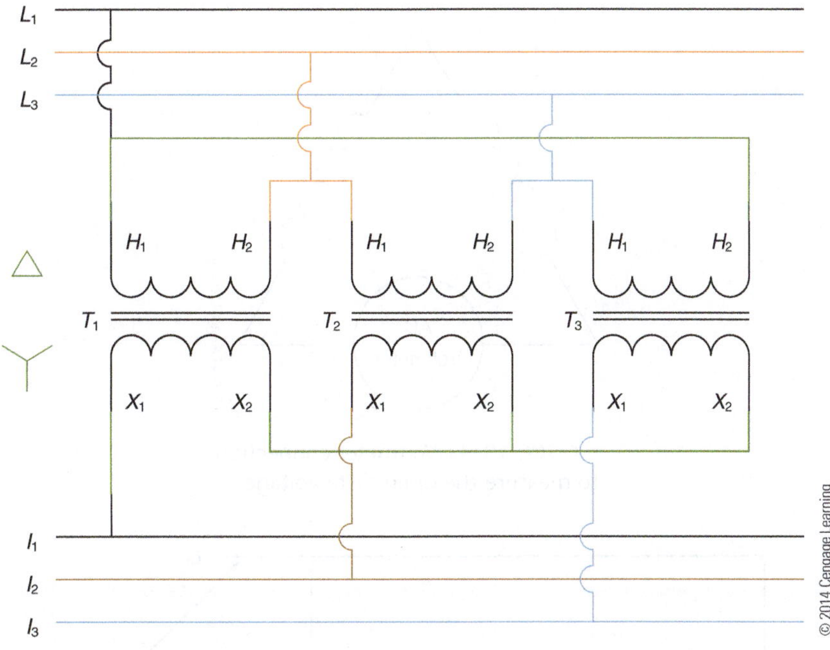

FIGURE 19-29 Delta–wye connections with three single-phase transformers.

connected to one of the three-phase power lines, L_1, L_2, and L_3. The voltage measured between L_1 and L_2, or between L_2 and L_3, or between L_1 and L_3 is called the **line voltage** because this is the amount of voltage measured from *line to line.* On the other hand, if you were to measure the voltage from L_1 to the common point where all three windings are connected, you would be measuring the **phase voltage**. The phase voltage is the voltage measured across each phase winding. This is also true if you measure the voltage from L_2 to common or L_3 to common. *In a wye-connected system, the line voltage is not equal to the phase voltage.* In fact, *the line voltage is 1.732 times greater than the phase voltage.*

In Figure 19-32, if you imagine yourself as an electron flowing from L_1 to L_2, you would take the path that takes you through windings A and B. Essentially, you only have one path to follow. Hence,

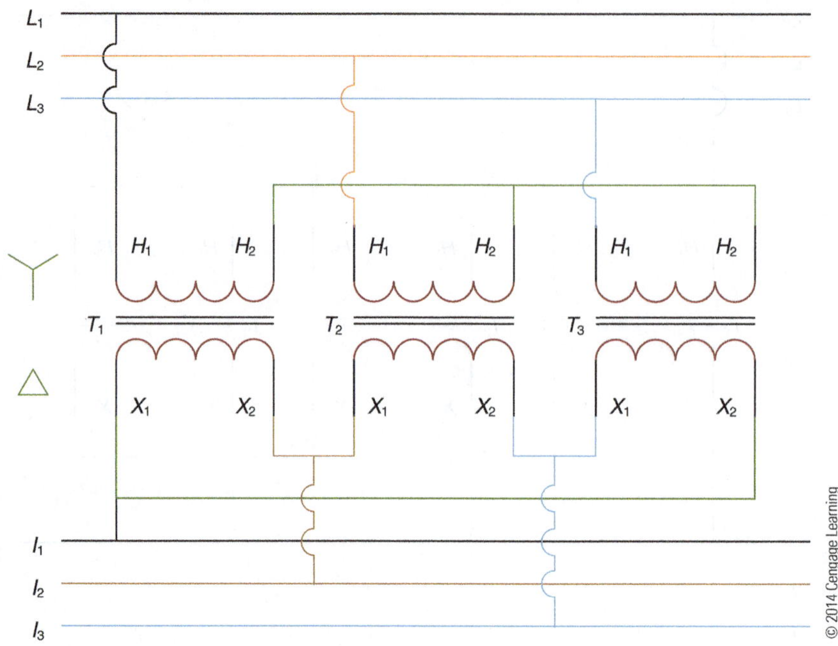

FIGURE 19-30 Wye–delta connections with three single-phase transformers.

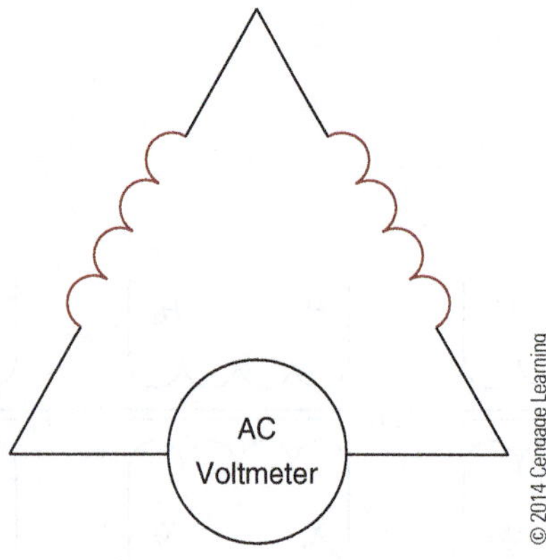

FIGURE 19-31 Voltmeter connection to measure the open delta voltage.

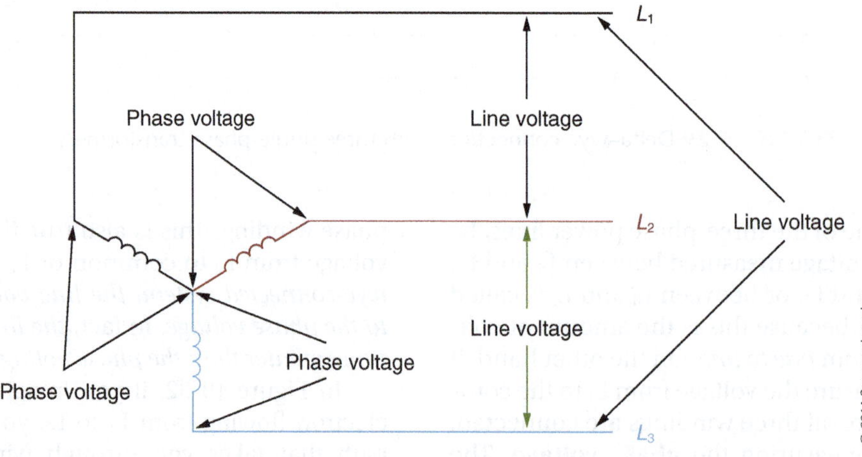

FIGURE 19-32 Line voltages and phase voltages for wye-connected transformers.

it appears as though windings A and B are connected in series. The same can be said for current flowing from L_2 to L_3. In this instance, the current can only flow through windings B and C. Likewise, current flowing from L_1 to L_3 can only flow through windings A and C. In each example, there is only one path for current flow. Therefore, the line current must equal the phase current in a wye-connected circuit.

Figure 19-33 shows three windings connected in a delta configuration. Notice that one end of winding A (A_2) is connected to one end of winding B (B_1). The other end of winding B (B_2) is connected to one end of winding C (C_1). The remaining end of winding C (C_2) is connected to the remaining end of winding A (A_1). The three-phase power connections are made at the junctions of two windings. As seen in the wye connection, the voltage measured between L_1 and L_2, or between L_2 and L_3, or between L_1 and L_3 is called the line voltage, because this is the amount of voltage measured from *line to line.* Likewise, if you were to measure the voltage across one winding, you would be measuring the *phase voltage. In a delta-connected system, the line voltage equals the phase voltage.* Imagine placing the leads of a voltmeter across the line connections. Notice that the voltmeter is also placed across one of the windings at the same time. This is why the phase voltage and the line voltage are equal in a delta-connected system.

If you imagine yourself as an electron flowing from L_1 to L_2, you would take the path that takes you through winding B, but you could also take the path that takes you through windings A and C. Essentially, you have two paths to follow. Hence, it appears as though winding B is in parallel with the series combination of windings A and C. The same can be said for current flowing from L_2 to L_3. In this instance, the current can flow through winding C and through the series combination of windings A and B. Likewise, current flowing from L_1 to L_3 can flow through winding A and through the series combination of windings B and C. In each example, there are two paths for current flow. Therefore, the line current must be larger than the phase current in a delta-connected circuit. This is because the line current must divide through the phase windings, then the phase current recombines to become the line current. In fact, *the line current is 1.732 times greater than the phase current.*

The following variables is used in the analysis of three-phase transformer circuits:

$E_{P(Phase)}$ = Primary phase voltage
$E_{P(Line)}$ = Primary line voltage
$E_{S(Phase)}$ = Secondary phase voltage
$E_{S(Line)}$ = Secondary line voltage
$E_{L(Phase)}$ = Load phase voltage
$E_{L(Line)}$ = Load line voltage
$I_{P(Phase)}$ = Primary phase current
$I_{P(Line)}$ = Primary line current
$I_{S(Phase)}$ = Secondary phase current
$I_{S(Line)}$ = Secondary line current
$I_{L(Phase)}$ = Load phase current
$I_{L(Line)}$ = Load line current

Delta–Delta-Configured Transformer

Figure 19-34 shows a delta-delta-connected transformer. The primary line voltage, $E_{P(Line)}$ is 480 volts AC. The load (a wye-connected motor) is designed to operate from 208-volt AC power, $E_{L(Line)}$. Each motor

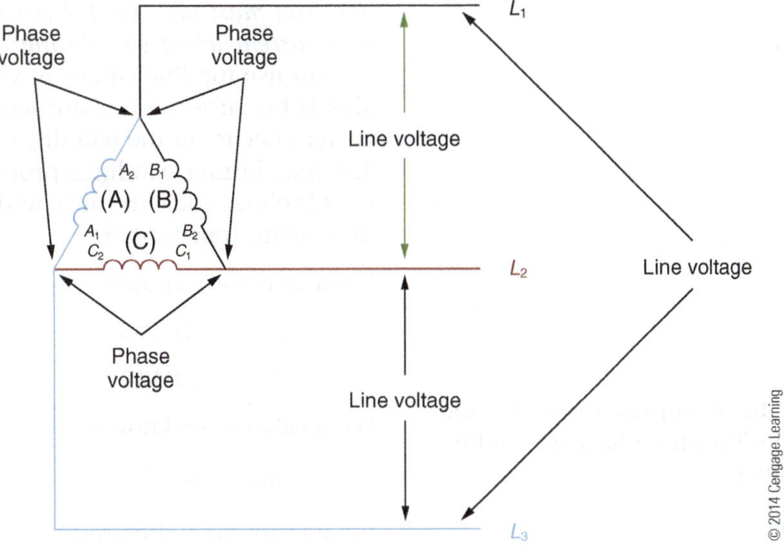

FIGURE 19-33 Line voltages and phase voltages for delta-connected transformers.

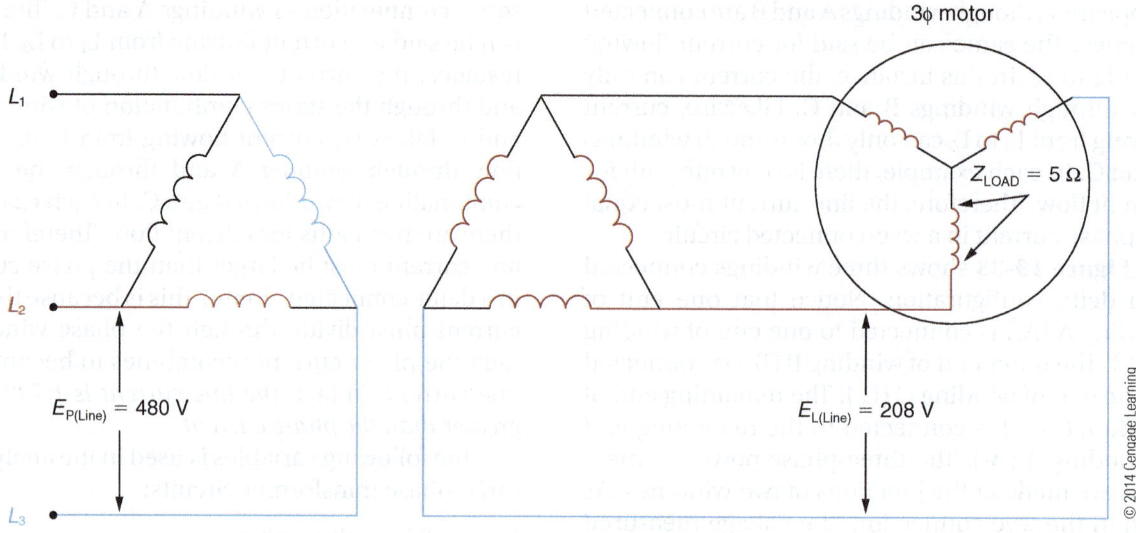

FIGURE 19-34 A delta–delta-connected transformer with a wye-connected load.

winding has an impedance of 5 ohms. Deter mine the following:

What values are known?

$E_{P(Line)} = 480$ VAC

$E_{L(Line)} = 208$ VAC

$k = 1.732$

$Z_{Load} = 5\ \Omega$

What values are not known?

$E_{P(Phase)} = ?$

$E_{S(Phase)} = ?$

$E_{S(Line)} = ?$

$E_{L(Phase)} = ?$

$I_{P(Phase)} = ?$

$I_{P(Line)} = ?$

$I_{S(Phase)} = ?$

$I_{S(Line)} = ?$

$I_{L(Phase)} = ?$

$I_{L(Line)} = ?$

Turns Ratio = ?

Begin by recognizing that the primary is connected in a delta configuration. The line voltage is therefore equal to the phase voltage.

$E_{P(Line)} = 480$ VAC

$E_{P(Line)} = E_{P(Phase)}$

$E_{P(Phase)} = 480$ VAC

Notice that the secondary is also connected in a delta configuration, and that the load operates from a 208 volt AC line voltage. This means that the secondary line voltage must be 208 volts AC as well. If the secondary line voltage is 208 volts AC and the secondary is connected in a delta configuration, the secondary phase voltage must also be equal to 208 volts AC.

$E_{L(Line)} = 208$ VAC

$E_{S(Line)} = 208$ VAC

$E_{S(Line)} = E_{S(Phase)}$

$E_{S(Phase)} = 208$ VAC

Before going further, there is an important rule that you must know when calculating transformer values: *You must use only the phase values of voltage and current when calculating transformer values.* Do not use the line values of voltage and current. This is because the transformation of voltage and current occurs in the windings of the transformer. The transformer windings produce the phase values of voltage and current. Now determine the turns ratio of this transformer.

What values are known?

$E_{P(Phase)} = 480$ VAC

$E_{S(Phase)} = 208$ VAC

What value is not known?

Turns Ratio = ?

What formula can be used?

$$Turns\ Ratio = \frac{E_{P(Phase)}}{E_{S(Phase)}}$$

Substitute the known values and solve for the *Turns Ratio*.

$$\text{Turns Ratio} = \frac{E_{P(Phase)}}{E_{S(Phase)}}$$

$$\text{Turns Ratio} = \frac{480 \text{ VAC}}{208 \text{ VAC}}$$

$$\text{Turns Ratio} = \frac{2.31}{1} \text{ or } 2.31{:}1$$

Therefore, the turns ratio of this transformer is 2.31:1.

Recall that the load is connected in a wye configuration. As a result of this, the load line voltage, $E_{L(Line)}$, is 1.732 times greater than the load phase voltage $E_{L(Phase)}$. Recall that the load line voltage, $E_{L(Line)}$, is equal to 208 volts AC. Now determine the load phase voltage, $E_{L(Phase)}$.

What values are known?

$$E_{L(Line)} = 208 \text{ VAC}$$

$$k = 1.732$$

What value is not known?

$$E_{L(Phase)} = ?$$

What formula can be used?

$$E_{L(Phase)} = \frac{E_{L(Line)}}{k}$$

Now substitute the known values and solve for $E_{L(Phase)}$.

$$E_{L(Phase)} = \frac{E_{L(Line)}}{k}$$

$$E_{L(Phase)} = \frac{208 \text{ VAC}}{1.732}$$

$$E_{L(Phase)} = 120.09 \text{ VAC}$$

Therefore, the load phase voltage is 120.09 volts AC. Now that you know that each phase of the load consists of a 5-ohm impedance, and each load phase voltage is equal to 120.09 volts AC, you can determine the amount of load phase current, $I_{L(Phase)}$, by using Ohm's law.

What values are known?

$$E_{L(Phase)} = 120.09 \text{ VAC}$$

$$Z_{Load} = 5\ \Omega$$

What value is not known?

$$I_{L(Phase)} = ?$$

What formula can be used?

$$I_{L(Phase)} = \frac{E_{L(Phase)}}{Z_{Load}}$$

Now substitute the known values and solve for $I_{L(Phase)}$.

$$I_{L(Phase)} = \frac{E_{L(Phase)}}{Z_{Load}}$$

$$I_{L(Phase)} = \frac{120.09 \text{ VAC}}{5\ \Omega}$$

$$I_{L(Phase)} = 24.02 \text{ A}$$

Therefore, the load phase current, $I_{L(Phase)}$, is equal to 24.02 amperes. Next, determine the load line current, $I_{L(Line)}$. Recall that the load is connected in a wye configuration. In a wye configuration, the phase current and the line current are equal.

$$I_{L(Phase)} = 24.02 \text{ A}$$

$$I_{L(Phase)} = I_{L(Line)}$$

$$I_{L(Line)} = 24.02 \text{ A}$$

So in the example, the load line current, $I_{L(Line)}$, is equal to 24.02 amperes. The load line current must be equal to the secondary line current, because the secondary is the source of the line current that is supplied to the load.

$$I_{L(Line)} = 24.02 \text{ A}$$

$$I_{L(Line)} = I_{S(Line)}$$

$$I_{S(Line)} = 24.02 \text{ A}$$

Therefore, the secondary line current, $I_{S(Line)}$, is equal to 24.02 amperes. Now determine the secondary phase current, $I_{S(Phase)}$. Recall that the secondary of the transformer is connected in a delta configuration; therefore, the line current is 1.732 times greater than the phase current.

What values are known?

$$I_{S(Line)} = 24.02 \text{ A}$$

$$k = 1.732$$

What value is not known?

$$I_{S(Phase)} = ?$$

What formula can be used?

$$I_{S(Phase)} = \frac{I_{S(Line)}}{k}$$

Substitute the known values and solve for $I_{S(Phase)}$.

$$I_{S(Phase)} = \frac{I_{S(Line)}}{k}$$

$$I_{S(Phase)} = \frac{24.02 \text{ A}}{1.732}$$

$$I_{S(Phase)} = 13.87 \text{ A}$$

Therefore, the secondary phase current, $I_{S(Phase)}$, equals 13.87 amperes. Now that you know the secondary phase current, you can determine the primary phase current, $I_{P(Phase)}$, by using the transformer ratio calculated earlier. Recall that when the primary voltage is stepped down, the current is stepped up by the same ratio. Likewise, when the primary voltage is stepped up, the current is stepped down by the same ratio. The transformer in this example is a step-down transformer. This means that if the primary voltage is stepped down, the current must be stepped up. Because you are looking at this transformer from the secondary side, it is easy to get confused on this issue.

Before determining the primary phase current, $I_{P(Phase)}$, you must realize that the primary phase current must be smaller than the secondary phase current. Remember to use the phase value of current when performing the calculations.

What values are known?

$$I_{S(Phase)} = 13.87\ A$$

$$Turns\ Ratio = 2.31{:}1$$

What value is not known?

$$I_{P(Phase)} = ?$$

What formula can be used?

$$I_{P(Phase)} = \frac{I_{S(Phase)}}{Turns\ Ratio}$$

Substitute the known values and solve for $I_{P(Phase)}$.

$$I_{P(Phase)} = \frac{I_{S(Phase)}}{Turns\ Ratio}$$

$$I_{P(Phase)} = \frac{13.87\ A}{2.31}$$

$$I_{P(Phase)} = 6.0\ A$$

Therefore, the primary phase current of the transformer is 6.0 amperes. Finally, you can determine the primary line current, $I_{P(Line)}$. Recall that the primary of the transformer is connected in a delta configuration; therefore, the line current is 1.732 times greater than the phase current.

What values are known?

$$I_{P(Phase)} = 6.0\ A$$

$$k = 1.732$$

What value is not known?

$$I_{P(Line)} = ?$$

What formula can be used?

$$I_{P(Line)} = I_{P(Phase)} \times k$$

Substitute the known values and solve for $I_{P(Line)}$.

$$I_{P(Line)} = I_{P(Phase)} \times k$$

$$I_{P(Line)} = 6.0\ A \times 1.732$$

$$I_{P(Line)} = 10.39\ A$$

Therefore, the primary line current, $I_{P(Line)}$, is equal to 10.39 amperes. **Figure 19-35** shows the same circuit with all of the values entered.

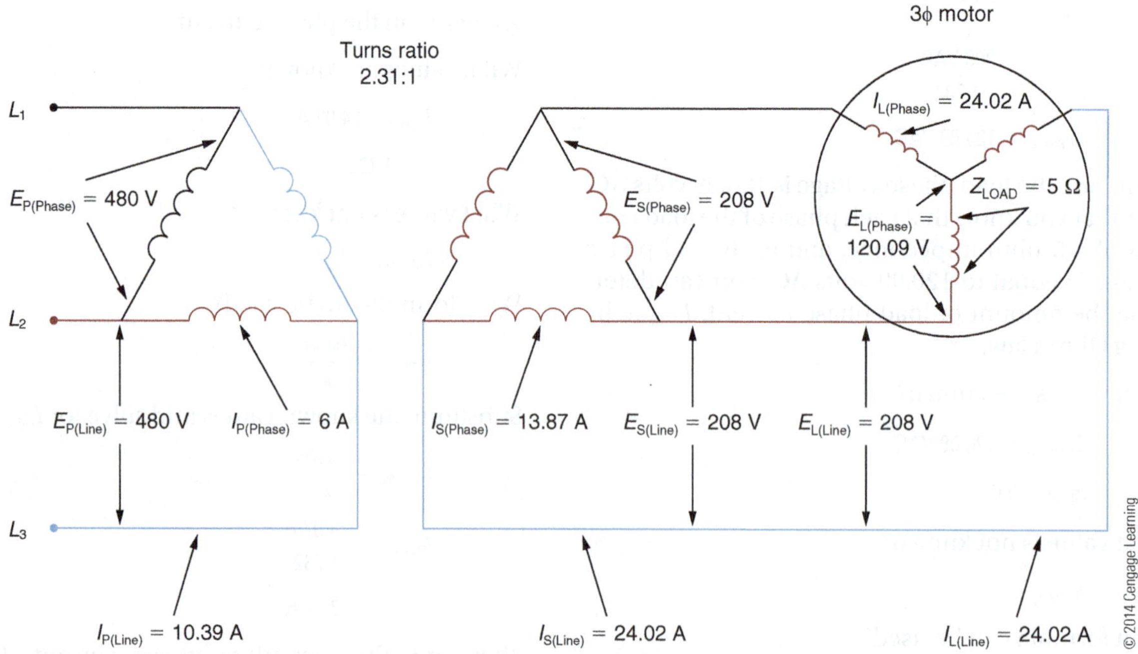

FIGURE 19-35 A delta–delta-connected transformer with a wye-connected load with all values of voltage and current shown.

Wye–Wye-Configured Transformer

We now examine a wye–wye configured transformer. **Figure 19-36** shows a transformer with a wye–wye connection. So that you can compare the effects of a wye–wye connection, the same values for $E_{P(Line)}$, $E_{L(Line)}$, and Z_{Load} are used. Therefore, the primary line voltage, $E_{P(Line)}$, is 480 volts AC, the load (a wye-connected motor) operates from 208 volts AC power, $E_{L(Line)}$, and each motor winding has an impedance of 5 ohms. Determine the following:

What values are known?

$$E_{P(Line)} = 480 \text{ VAC}$$

$$E_{L(Line)} = 208 \text{ VAC}$$

$$k = 1.732$$

$$Z_{Load} = 5\ \Omega$$

What values are not known?

$$E_{P(Phase)} = ?$$

$$E_{S(Phase)} = ?$$

$$E_{S(Line)} = ?$$

$$E_{L(Phase)} = ?$$

$$I_{P(Phase)} = ?$$

$$I_{P(Line)} = ?$$

$$I_{S(Phase)} = ?$$

$$I_{S(Line)} = ?$$

$$I_{L(Phase)} = ?$$

$$I_{L(Line)} = ?$$

$$Turns\ Ratio = ?$$

Begin by recognizing that the primary is connected in a wye configuration. The line voltage is therefore 1.732 times greater than the phase voltage.

$$E_{P(Line)} = 480 \text{ VAC}$$

$$E_{P(Phase)} = \frac{E_{P(Line)}}{1.732}$$

$$E_{P(Phase)} = \frac{480 \text{ VAC}}{1.732}$$

$$E_{P(Phase)} = 277.14 \text{ VAC}$$

Notice that the secondary is also connected in a wye configuration, and that the load operates from a 208-volt AC line voltage. This means that the secondary line voltage must be 208 volts AC as well. If the secondary line voltage is 208 volts AC and the secondary is connected in a wye configuration, the secondary phase voltage must be less than the secondary line voltage by a factor of 1.732.

$$E_{L(Line)} = 208 \text{ VAC}$$

$$E_{S(Line)} = 208 \text{ VAC}$$

$$E_{S(Phase)} = \frac{E_{S(Line)}}{1.732}$$

$$E_{S(Phase)} = \frac{208 \text{ VAC}}{1.732}$$

$$E_{S(Phase)} = 120.09 \text{ VAC}$$

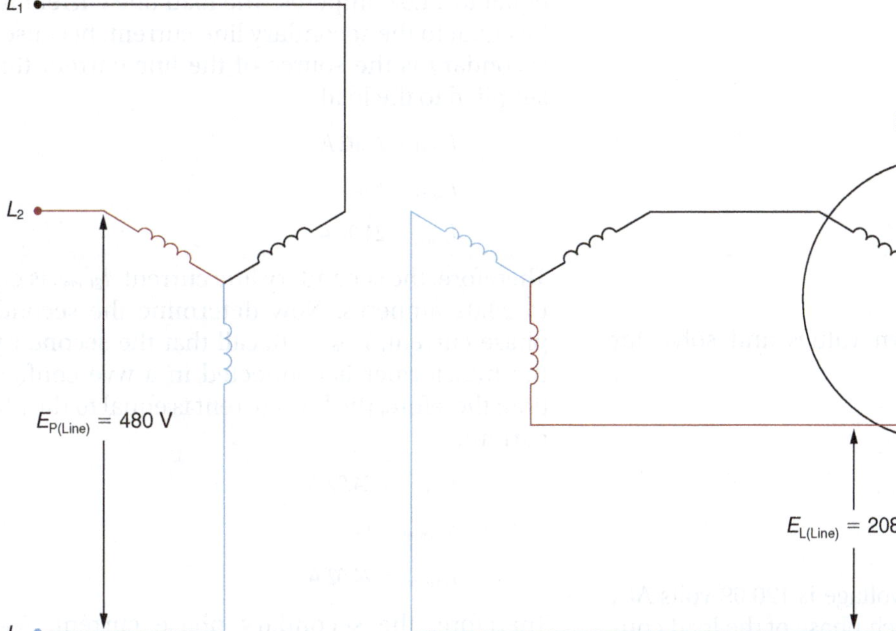

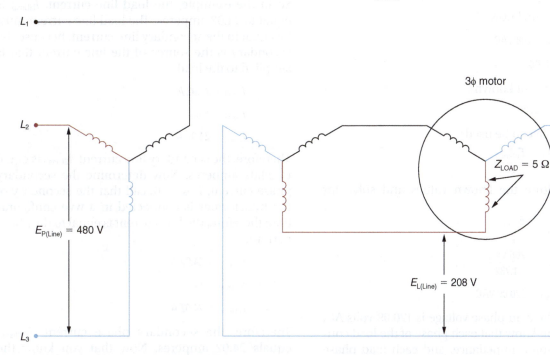

FIGURE 19-36 A wye–wye-connected transformer with a wye-connected load.

Recall the rule when calculating transformer values. *You must use only the phase values of voltage and current when calculating transformer values.* Do not use the line values of voltage and current. Now determine the turns ratio of this transformer.

What values are known?

$$E_{P(Phase)} = 277.14 \text{ VAC}$$

$$E_{S(Phase)} = 120.09 \text{ VAC}$$

What value is not known?

$$\textit{Turns Ratio} = ?$$

What formula can be used?

$$\textit{Turns Ratio} = \frac{E_{P(Phase)}}{E_{S(Phase)}}$$

Substitute the known values and solve for the *Turns Ratio*.

$$\textit{Turns Ratio} = \frac{E_{P(Phase)}}{E_{S(Phase)}}$$

$$\textit{Turns Ratio} = \frac{277.14 \text{ VAC}}{120.09 \text{ VAC}}$$

$$\textit{Turns Ratio} = \frac{2.31}{1} \text{ or } 2.31{:}1$$

Therefore, the turns ratio of this transformer is 2.31:1.

Recall that the load is connected in a wye configuration. As a result of this, the load line voltage, $E_{L(Line)}$, is 1.732 times greater than the load phase voltage, $E_{L(Phase)}$. Recall that the load line voltage, $E_{L(Line)}$, is equal to 208 volts AC. Now determine the load phase voltage, $E_{L(Phase)}$.

What values are known?

$$E_{L(Line)} = 208 \text{ VAC}$$

$$k = 1.732$$

What value is not known?

$$E_{L(Phase)} = ?$$

What formula can be used?

$$E_{L(Phase)} = \frac{E_{L(Line)}}{k}$$

Now substitute the known values and solve for $E_{L(Phase)}$.

$$E_{L(Phase)} = \frac{E_{L(Line)}}{k}$$

$$E_{L(Phase)} = \frac{208 \text{ VAC}}{1.732}$$

$$E_{L(Phase)} = 120.09 \text{ VAC}$$

Therefore, the load phase voltage is 120.09 volts AC. Now that you know that each phase of the load consists of a 5-ohm impedance, and each load phase voltage is equal to 120.09 volts AC, you can determine the amount of load phase current, $I_{L(Phase)}$, by using Ohm's law.

What values are known?

$$E_{L(Phase)} = 120.09 \text{ VAC}$$

$$Z_{Load} = 5\ \Omega$$

What value is not known?

$$I_{L(Phase)} = ?$$

What formula can be used?

$$I_{L(Phase)} = \frac{E_{L(Phase)}}{Z_{Load}}$$

Now substitute the known values and solve for $I_{L(Phase)}$.

$$I_{L(Phase)} = \frac{E_{L(Phase)}}{Z_{Load}}$$

$$I_{L(Phase)} = \frac{120.09 \text{ VAC}}{5\ \Omega}$$

$$I_{L(Phase)} = 24.02 \text{ A}$$

Therefore, the load phase current, $I_{L(Phase)}$, is equal to 24.02 amperes. Next, determine the load line current, $I_{L(Line)}$. Recall that the load is connected in a wye configuration; therefore, the phase current and the line current are equal.

$$I_{L(Phase)} = 24.02 \text{ A}$$

$$I_{L(Phase)} = I_{L(Line)}$$

$$I_{L(Line)} = 24.02 \text{ A}$$

So in the example, the load line current, $I_{L(Line)}$, is equal to 24.02 amperes. The load line current must be equal to the secondary line current, because the secondary is the source of the line current that is supplied to the load.

$$I_{L(Line)} = 24.02 \text{ A}$$

$$I_{L(Line)} = I_{S(Line)}$$

$$I_{S(Line)} = 24.02 \text{ A}$$

Therefore, the secondary line current, $I_{S(Line)}$, is equal to 24.02 amperes. Now determine the secondary phase current, $I_{S(Phase)}$. Recall that the secondary of the transformer is connected in a wye configuration; therefore, the line current is equal to the phase current.

$$I_{L(Line)} = 24.02 \text{ A}$$

$$I_{S(Phase)} = I_{S(Line)}$$

$$I_{S(Phase)} = 24.02 \text{ A}$$

Therefore, the secondary phase current, $I_{S(Phase)}$, equals 24.02 amperes. Now that you know the secondary phase current, you can determine the primary phase current, $I_{P(Phase)}$, by using the transformer ratio calculated earlier. Recall that when the primary voltage is stepped down, the current is stepped up by the same ratio. Likewise, when the primary voltage is stepped up, the current is stepped down by the same ratio. The transformer in this example is a step-down transformer. This means that if the primary voltage is stepped down, the current must be stepped up. Because you are looking at this transformer from the secondary side, it is easy to get confused on this issue. Before determining the primary phase current, $I_{P(Phase)}$, you must realize that the primary phase current must be smaller than the secondary phase current. Remember to use the phase value of current when performing the calculations.

What values are known?

$$I_{S(Phase)} = 24.02 \text{ A}$$

$$\textit{Turns Ratio} = 2.31{:}1$$

What value is not known?

$$I_{P(Phase)} = ?$$

What formula can be used?

$$I_{P(Phase)} = \frac{I_{S(Phase)}}{\textit{Turns Ratio}}$$

Substitute the known values and solve for $I_{P(Phase)}$.

$$I_{P(Phase)} = \frac{I_{S(Phase)}}{\textit{Turns Ratio}}$$

$$I_{P(Phase)} = \frac{24.02 \text{ A}}{2.31}$$

$$I_{P(Phase)} = 10.40 \text{ A}$$

Therefore, the primary phase current of the transformer is 10.40 amperes. Finally, you can determine the primary line current, $I_{P(Line)}$. Recall that the primary of the transformer is connected in a wye configuration; therefore, the line current is equal to the phase current.

$$I_{P(Phase)} = 10.40 \text{ A}$$

$$I_{P(Line)} = I_{P(Phase)}$$

$$I_{P(Line)} = 10.40 \text{ A}$$

Therefore, the primary line current, $I_{P(Line)}$, is equal to 10.40 amperes. **Figure 19-37** shows the same circuit with all of the values entered.

Compare the values shown in Figure 19-35 with those shown in Figure 19-37. Notice that the primary line current values are essentially the same. However, the primary phase current is larger in the wye-connected transformer of Figure 19-37. Now

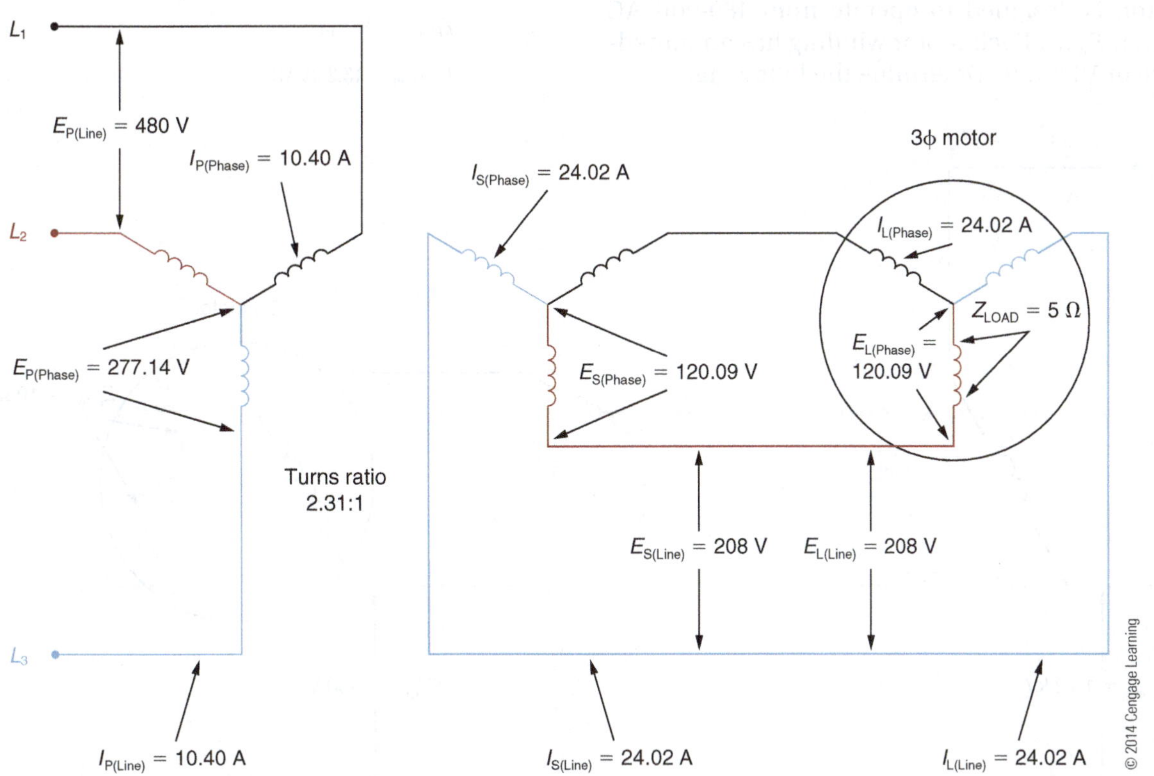

FIGURE 19-37 A wye–wye-connected transformer with a wye-connected load with all values of voltage and current shown.

look at the secondary currents. The line currents of the wye- and delta-connected secondaries are essentially the same value. However, the secondary phase current is larger in the wye-connected transformer of Figure 19-37.

The primary phase voltage of the wye-connected transformer of Figure 19-37 is lower than the primary phase voltage of the delta-connected transformer of Figure 19-35. Likewise, the secondary phase voltage of the wye-connected transformer is lower than the secondary phase voltage of the delta-connected transformer.

The current and voltage values for the load did not change regardless of the type of transformer connection used. This means that should you decide to use a delta-delta configuration, the transformer could be constructed with smaller gauge wire for the primary and secondary phase windings. This could result in a smaller physical size and lower cost. On the other hand, should there be a need to develop voltages other than 480 volts AC and 208 volts AC, you may opt to use the wye-wye configuration. This configuration produces 480 volts AC, 277 volts AC, 208 volts AC, and 120 volts AC.

Delta-Wye-Configured Transformer

Figure 19-38 shows a delta-wye-connected transformer. The primary line voltage, $E_{P(Line)}$, is 13.2 kilovolts AC. The load, a delta-connected motor, is designed to operate from 480-volt AC power, $E_{L(Line)}$. Each motor winding has an impedance of 10 ohms. Determine the following:

What values are known?

$E_{P(Line)} = 13.2$ kVAC

$E_{L(Line)} = 480$ VAC

$k = 1.732$

$Z_{Load} = 10\ \Omega$

What values are not known?

$E_{P(Phase)} = ?$

$E_{S(Phase)} = ?$

$E_{S(Line)} = ?$

$E_{L(Phase)} = ?$

$I_{P(Phase)} = ?$

$I_{P(Line)} = ?$

$I_{S(Phase)} = ?$

$I_{S(Line)} = ?$

$I_{L(Phase)} = ?$

$I_{L(Line)} = ?$

Turns Ratio = ?

Begin by recognizing that the primary is connected in a delta configuration. The line voltage is therefore equal to the phase voltage.

$E_{P(Line)} = 13.2$ kVAC

$E_{P(Line)} = E_{P(Phase)}$

$E_{P(Phase)} = 13.2$ kVAC

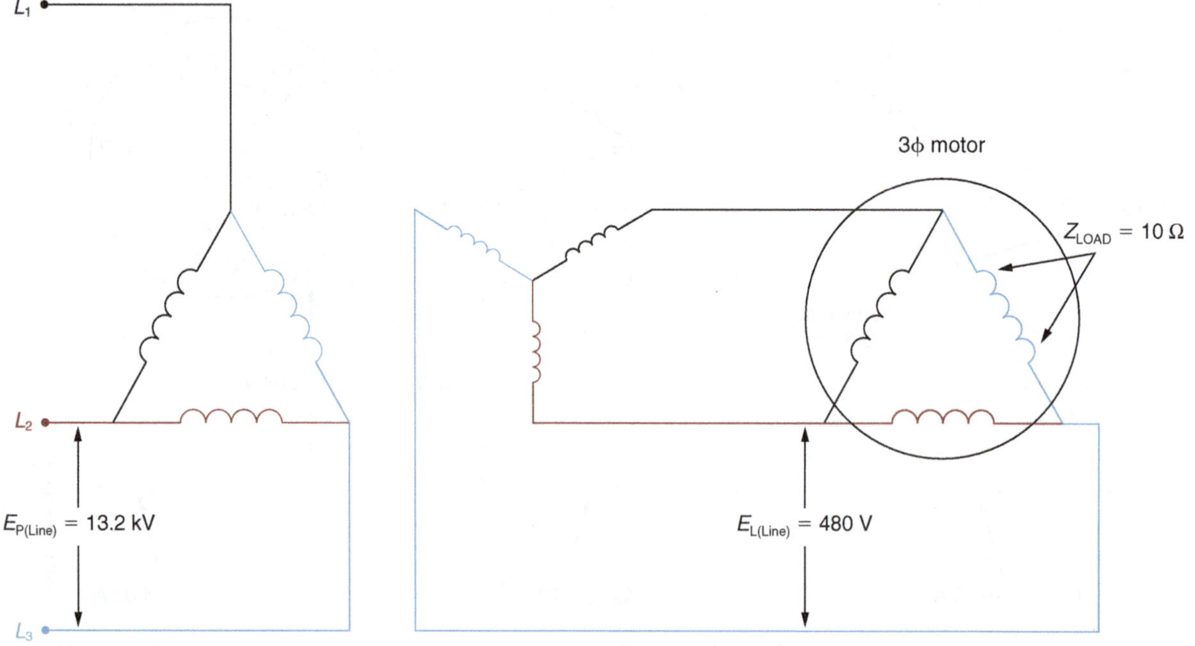

FIGURE 19-38 A delta–wye-connected transformer with a delta-connected load.

Notice that the secondary is connected in a wye configuration, and that the load operates from a 480-volt AC line voltage. This means that the secondary line voltage must be 480 volts AC as well. If the secondary line voltage is 480 volts AC, and the secondary is connected in a wye configuration, the secondary phase voltage must be less than the secondary line voltage by a factor of 1.732.

$$E_{L(Line)} = 480 \text{ VAC}$$

$$E_{S(Line)} = 480 \text{ VAC}$$

$$E_{S(Phase)} = \frac{E_{S(Line)}}{1.732}$$

$$E_{S(Phase)} = \frac{480 \text{ VAC}}{1.732}$$

$$E_{S(Phase)} = 277.14 \text{ VAC}$$

Now determine the turns ratio of this transformer.

What values are known?

$$E_{P(Phase)} = 13.2 \text{ kVAC}$$

$$E_{S(Phase)} = 277.14 \text{ VAC}$$

What value is not known?

$$Turns\ Ratio = ?$$

What formula can be used?

$$Turns\ Ratio = \frac{E_{P(Phase)}}{E_{S(Phase)}}$$

Substitute the known values and solve for the *Turns Ratio*.

$$Turns\ Ratio = \frac{E_{P(Phase)}}{E_{S(Phase)}}$$

$$Turns\ Ratio = \frac{13.2 \text{ kVAC}}{277.14 \text{ VAC}}$$

$$Turns\ Ratio = \frac{47.63}{1} \text{ or } 47.63{:}1$$

Therefore, the turns ratio of this transformer is 47.63:1.

Recall that the load is connected in a delta configuration. As a result of the delta configuration, the load line voltage, $E_{L(Line)}$, is equal to the load phase voltage $E_{L(Phase)}$. Recall that the load line voltage, $E_{L(Line)}$, is equal to 480 volts AC. Now determine the load phase voltage, $E_{L(Phase)}$.

$$E_{L(Line)} = 480 \text{ VAC}$$

$$E_{L(Phase)} = E_{L(Line)}$$

$$E_{L(Phase)} = 480 \text{ VAC}$$

Therefore, the load phase voltage is 480 volts AC. Now that you know that each phase of the load consists of a 10-ohm impedance and each load phase voltage is equal to 480 volts AC, you can determine the amount of load phase current, $I_{L(Phase)}$, by using Ohm's law.

What values are known?

$$E_{L(Phase)} = 480 \text{ VAC}$$

$$Z_{Load} = 10\ \Omega$$

What value is not known?

$$I_{L(Phase)} = ?$$

What formula can be used?

$$I_{L(Phase)} = \frac{E_{L(Phase)}}{Z_{Load}}$$

Now substitute the known values and solve for $I_{L(Phase)}$.

$$I_{L(Phase)} = \frac{E_{L(Phase)}}{Z_{Load}}$$

$$I_{L(Phase)} = \frac{480 \text{ VAC}}{10\ \Omega}$$

$$I_{L(Phase)} = 48 \text{ A}$$

Therefore, the load phase current, $I_{L(Phase)}$, is equal to 48 amperes. Next, determine the load line current, $I_{L(Line)}$. Recall that the load is connected in a delta configuration; therefore, the line current is 1.732 times greater than the phase current.

$$I_{L(Phase)} = 48 \text{ A}$$

$$I_{L(Line)} = I_{L(Phase)} \times 1.732$$

$$I_{L(Line)} = 48 \text{ A} \times 1.732$$

$$I_{L(Line)} = 83.14 \text{ A}$$

So in the example, the load line current, $I_{L(Line)}$, is equal to 83.14 amperes. The load line current must be equal to the secondary line current, because the secondary is the source of the line current that is supplied to the load.

$$I_{L(Line)} = 83.14 \text{ A}$$

$$I_{S(Line)} = I_{L(Line)}$$

$$I_{S(Line)} = 83.14 \text{ A}$$

Therefore, the secondary line current, $I_{S(Line)}$, is equal to 83.14 amperes. Now you can determine the secondary phase current, $I_{S(Phase)}$. Recall that the secondary of the transformer is connected in a wye configuration; therefore, the line current is equal to the phase current.

$$I_{S(Line)} = 83.14 \text{ A}$$

$$I_{S(Phase)} = I_{S(Line)}$$

$$I_{S(Phase)} = 83.14 \text{ A}$$

Therefore, the secondary phase current, $I_{S(Phase)}$, equals 83.14 amperes. Now that you know the secondary phase current, you can determine the primary phase current, $I_{P(Phase)}$.

What values are known?

$$I_{S(Phase)} = 83.14\text{ A}$$

$$\textit{Turns Ratio} = 47.63{:}1$$

What value is not known?

$$I_{P(Phase)} = ?$$

What formula can be used?

$$I_{P(Phase)} = \frac{I_{S(Phase)}}{\textit{Turns Ratio}}$$

Substitute the known values and solve for $I_{P(Phase)}$.

$$I_{P(Phase)} = \frac{I_{S(Phase)}}{\textit{Turns Ratio}}$$

$$I_{P(Phase)} = \frac{83.14\text{ A}}{47.63}$$

$$I_{P(Phase)} = 1.75\text{ A}$$

Therefore, the primary phase current of the transformer is 1.75 amperes. Finally, you can determine the primary line current, $I_{P(Line)}$. Recall that the primary of the transformer is connected in a delta configuration; therefore, the line current is 1.732 times greater than the phase current.

What values are known?

$$I_{P(Phase)} = 1.75\text{ A}$$

$$k = 1.732$$

What value is not known?

$$I_{P(Line)} = ?$$

What formula can be used?

$$I_{P(Line)} = I_{P(Phase)} \times k$$

Substitute the known values and solve for $I_{P(Line)}$.

$$I_{P(Line)} = I_{P(Phase)} \times k$$

$$I_{P(Line)} = 1.75\text{ A} \times 1.732$$

$$I_{P(Line)} = 3.03\text{ A}$$

Therefore, the primary line current, $I_{P(Line)}$, is equal to 3.03 amperes. **Figure 19-39** shows the same circuit with all of the values entered.

Wye–Delta-Configured Transformer

Figure 19-40 shows a wye-delta-connected transformer. The primary line voltage, $E_{P(Line)}$, is 120 volts AC. The load, a delta-connected motor, is designed to operate from 440-volt AC power, $E_{L(Line)}$. Each motor

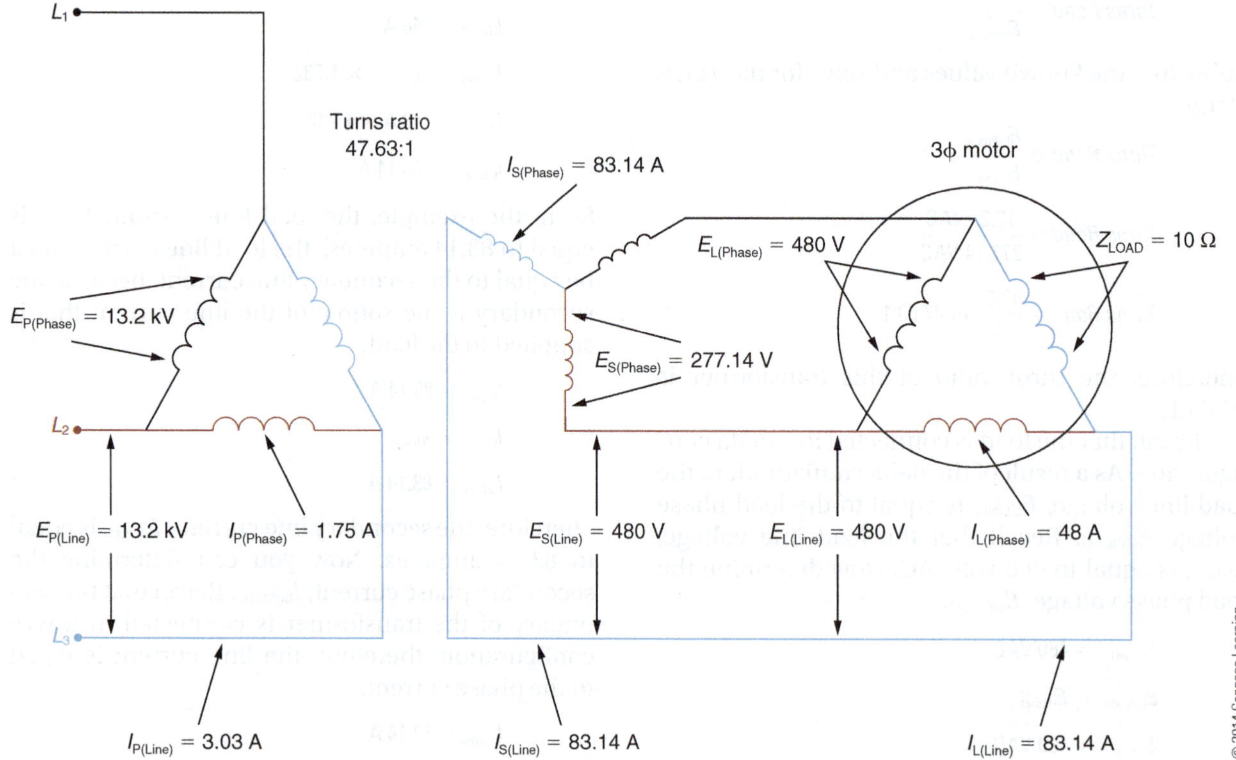

FIGURE 19-39 A delta–wye-connected transformer with a delta-connected load with all values of voltage and current shown.

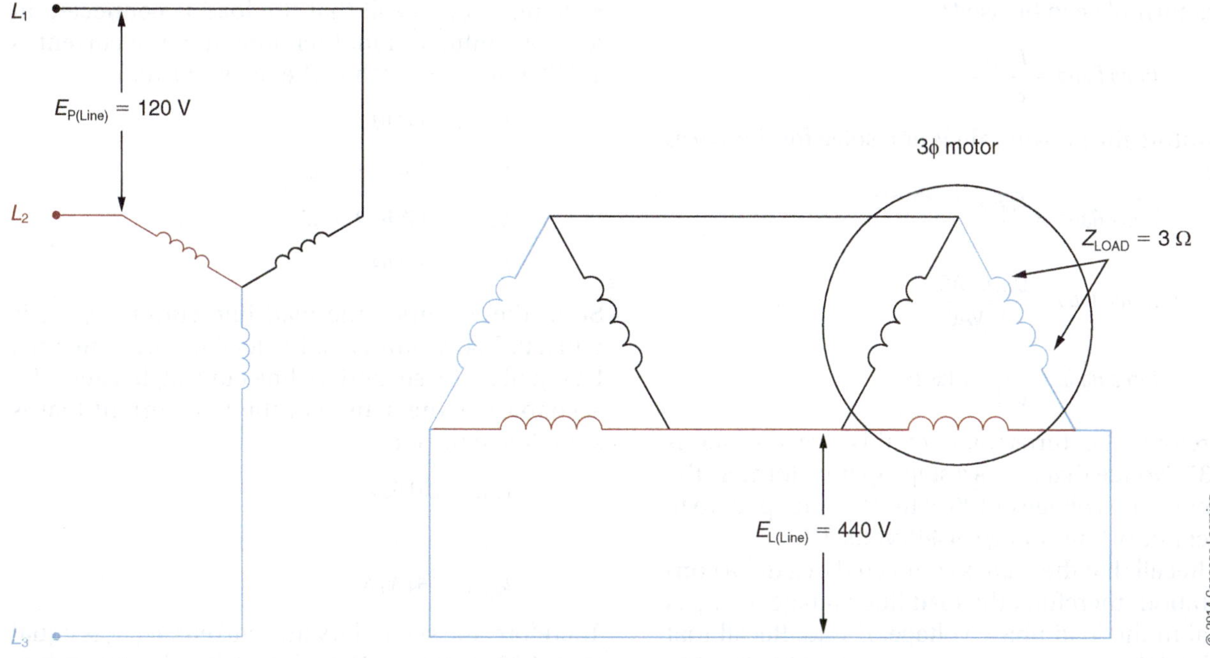

FIGURE 19-40 A wye–delta-connected transformer with a delta-connected load.

winding has an impedance of 3 ohms. Determine the following:

What values are known?

$E_{P(Line)} = 120$ VAC

$E_{L(Line)} = 440$ VAC

$k = 1.732$

$Z_{Load} = 3\ \Omega$

What values are not known?

$E_{P(Phase)} = ?$

$E_{S(Phase)} = ?$

$E_{S(Line)} = ?$

$E_{L(Phase)} = ?$

$I_{P(Phase)} = ?$

$I_{P(Line)} = ?$

$I_{S(Phase)} = ?$

$I_{S(Line)} = ?$

$I_{L(Phase)} = ?$

$I_{L(Line)} = ?$

Turns Ratio = ?

Begin by recognizing that the primary is connected in a wye configuration. The line voltage is therefore 1.732 times greater than the phase voltage.

$E_{P(Line)} = 120$ VAC

$$E_{P(Phase)} = \frac{E_{P(Line)}}{1.732}$$

$$E_{P(Phase)} = \frac{120\text{ VAC}}{1.732}$$

$E_{P(Phase)} = 69.28$ VAC

Notice that the secondary is connected in a delta configuration and that the load operates from a 440-volt AC line voltage. This means that the secondary line voltage must be 440 volts AC as well. If the secondary line voltage is 440 volts AC and the secondary is connected in a delta configuration, the secondary phase voltage must also be equal to 440 volts AC.

$E_{L(Line)} = 440$ VAC

$E_{S(Line)} = 440$ VAC

$E_{S(Line)} = E_{S(Phase)}$

$E_{S(Phase)} = 440$ VAC

Now determine the turns ratio of this transformer.

What values are known?

$E_{P(Phase)} = 69.28$ VAC

$E_{S(Phase)} = 440$ VAC

What value is not known?

Turns Ratio = ?

What formula can be used?

$$Turns\ Ratio = \frac{E_{P(Phase)}}{E_{S(Phase)}}$$

Substitute the known values and solve for the *Turns Ratio*.

$$Turns\ Ratio = \frac{E_{P(Phase)}}{E_{S(Phase)}}$$

$$Turns\ Ratio = \frac{69.28\ VAC}{440\ VAC}$$

$$Turns\ Ratio = \frac{1}{6.35}\ \text{or}\ 1{:}6.35$$

Therefore, the turns ratio of this transformer is 1:6.35. Notice that this is a step-up transformer. The primary line voltage of 120 volts AC is stepped up to a secondary line voltage of 440 volts AC.

Recall that the load is connected in a delta configuration; therefore, the load line voltage, $E_{L(Line)}$, is equal to the load phase voltage, $E_{L(Phase)}$. Recall that the load line voltage, $E_{L(Line)}$, is equal to 440 volts AC. Now determine the load phase voltage, $E_{L(Phase)}$.

$$E_{L(Line)} = 440\ VAC$$

$$E_{L(Phase)} = E_{L(Line)}$$

$$E_{L(Phase)} = 440\ VAC$$

Therefore, the load phase voltage is 440 volts AC. Now that you know that each phase of the load consists of a 3-ohm impedance and each load phase voltage is equal to 440 volts AC, you can determine the amount of load phase current, $I_{L(Phase)}$, by using Ohm's law.

What values are known?

$$E_{L(Phase)} = 440\ VAC$$

$$Z_{Load} = 3\ \Omega$$

What value is not known?

$$I_{L(Phase)} = ?$$

What formula can be used?

$$I_{L(Phase)} = \frac{E_{L(Phase)}}{Z_{Load}}$$

Now substitute the known values and solve for $I_{L(Phase)}$.

$$I_{L(Phase)} = \frac{E_{L(Phase)}}{Z_{Load}}$$

$$I_{L(Phase)} = \frac{440\ VAC}{3\ \Omega}$$

$$I_{L(Phase)} = 146.67\ A$$

Therefore, the load phase current, $I_{L(Phase)}$, is equal to 146.67 amperes. Next, determine the load line current, $I_{L(Line)}$. Recall that the load is connected in a delta configuration; therefore, the line current is 1.732 times greater than the phase current.

$$I_{L(Phase)} = 146.67\ A$$

$$I_{L(Line)} = I_{L(Phase)} \times 1.732$$

$$I_{L(Line)} = 146.67\ A \times 1.732$$

$$I_{L(Line)} = 254.03\ A$$

So in the example, the load line current, $I_{L(Line)}$, is equal to 254.03 amperes. The load line current must be equal to the secondary line current, because the secondary is the source of the line current that is supplied to the load.

$$I_{L(Line)} = 254.03\ A$$

$$I_{S(Line)} = I_{L(Line)}$$

$$I_{S(Line)} = 254.03\ A$$

Therefore, the secondary line current, $I_{S(Line)}$, is equal to 254.03 amperes. Now determine the secondary phase current, $I_{S(Phase)}$. Recall that the secondary of the transformer is connected in a delta configuration; therefore, the line current is 1.732 times greater than the phase current.

What values are known?

$$I_{S(Line)} = 254.03\ A$$

$$k = 1.732$$

What value is not known?

$$I_{S(Phase)} = ?$$

What formula can be used?

$$I_{S(Phase)} = \frac{I_{S(Line)}}{k}$$

Substitute the known values and solve for $I_{S(Phase)}$.

$$I_{S(Phase)} = \frac{I_{S(Line)}}{k}$$

$$I_{S(Phase)} = \frac{254.03\ A}{1.732}$$

$$I_{S(Phase)} = 146.67\ A$$

Therefore, the secondary phase current, $I_{S(Phase)}$, equals 146.67 amperes. Now that you know the secondary phase current, you can determine the primary phase current, $I_{P(Phase)}$.

What values are known?

$$I_{S(Phase)} = 146.67\ A$$

$$Turns\ Ratio = 1{:}6.35$$

What value is not known?

$$I_{P(Phase)} = ?$$

What formula can be used? (Remember, this is a step-up transformer.)

$$I_{P(Phase)} = I_{S(Phase)} \times \textit{Turns Ratio}$$

Substitute the known values and solve for $I_{P(Phase)}$.

$$I_{P(Phase)} = I_{S(Phase)} \times \textit{Turns Ratio}$$

$$I_{P(Phase)} = 146.67\text{ A} \times 6.35$$

$$I_{P(Phase)} = 931.35\text{ A}$$

Therefore, the primary phase current of the transformer is 931.35 amperes. Finally, you can determine the primary line current, $I_{P(Line)}$. Recall that the primary of the transformer is connected in a wye configuration; therefore, the line current is equal to the phase current.

$$I_{P(Phase)} = 931.35\text{ A}$$

$$I_{P(Line)} = I_{P(Phase)}$$

$$I_{P(Line)} = 931.35\text{ A}$$

Therefore, the primary line current, $I_{P(Line)}$, for the example is equal to 931.35 amperes. **Figure 19-41** shows the same circuit with all of the values entered.

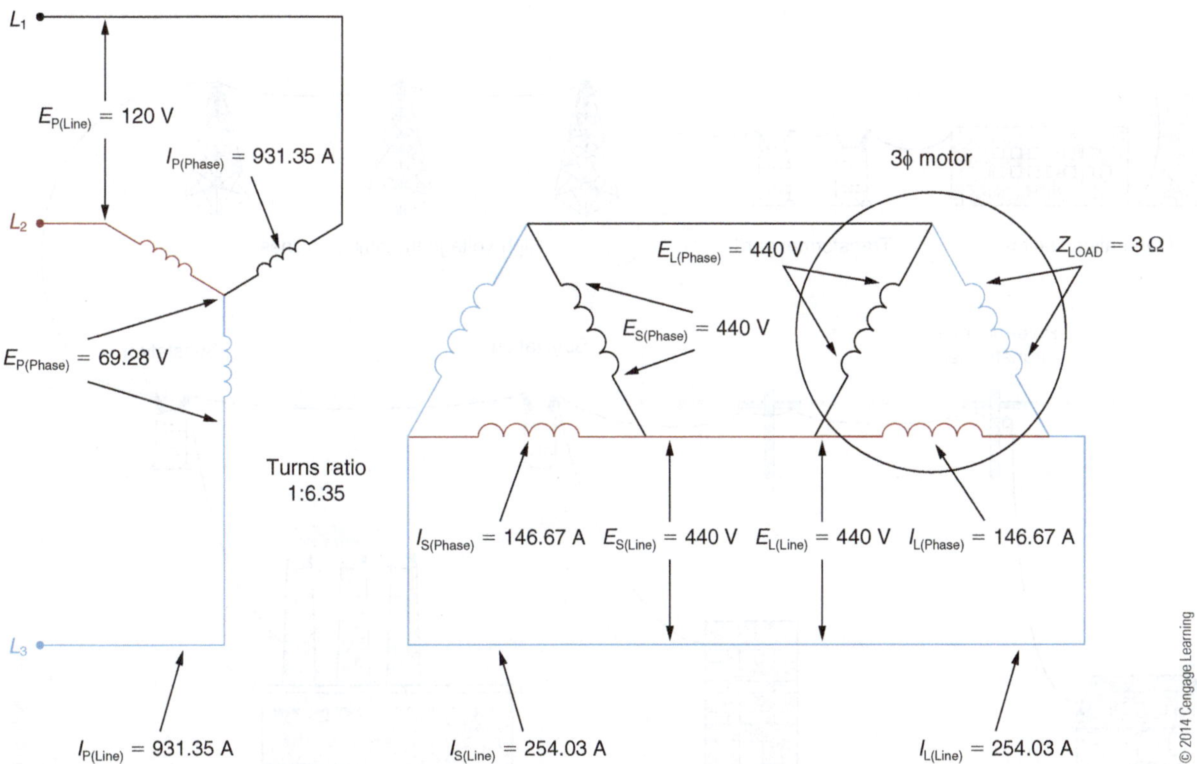

FIGURE 19-41 A wye–delta-connected transformer with a delta-connected load with all values of voltage and current shown.

Open-Delta or V Connection

It is possible to obtain three-phase power from only two transformers. This is known as an open-delta or V connection as seen in **Figure 19-42**. The open delta connection can be used to save money as only two transformers are needed to produce three-phase power. Also, should one of the transformers in a three-transformer delta connection fail, the transformer bank can be re-wired into an open-delta configuration to restore three-phase power quickly while a replacement transformer is obtained. There is, however, a down side to using an open-delta connection. The power supplied by the open-delta transformer will be approximately 58%

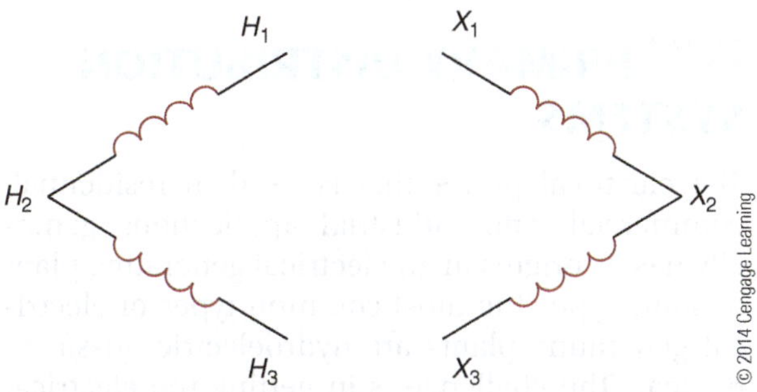

FIGURE 19-42 Open delta primary and secondary configuration.

of the closed, three-transformer delta connection. For example, if three 50 kVA transformers were connected in a closed delta bank, the total power capacity of the bank would be 50 kVA + 50 kVA + 50 kVA for a total of 150 kVA. If the bank were re-wired as a two transformer open-delta bank, the total capacity would now decrease to approximately 58% or 150 kVA × 0.58 = 87 kVA. If only two transformers are installed initially as a cost saving move, the total capacity is found by taking the sum of the kVA ratings of both transformers and multiplying the total by 86.5%. For example, two 50 kVA transformers connected in an open-delta configuration produces a total capacity of 50 kVA + 50 kVA = 100 kVA × 0.865 = 87 kVA (when rounded off).

19-5 PRIMARY DISTRIBUTION SYSTEMS

The electrical power that is used in residential, commercial, and industrial applications generally has its origins in an electrical generating plant of some type. The most common types of electrical generating plants are hydroelectric, fossil, or nuclear. The challenge is in getting the electrical power generated at these plants to the residence, commercial site, or industrial facility. The mechanism used to accomplish this is called the **primary distribution system**. It consists of the wiring and associated transformers and substations, as shown in **Figure 19-43**.

The voltage generated at the generating station undergoes several changes before it can be used by the consumer. **Figure 19-44** shows a typical primary distribution system's voltages. Notice that the electrical generating station generates 69 kilovolts. This is the voltage produced by the generators at the generating plant. The 69 kilovolts is fed to a transformer yard located a short distance away. Here the 69 kilovolts is stepped up to 100 kilovolts for transmission to the community.

The voltage is stepped up to 100 kilovolts for several reasons. By stepping up the voltage, the current is stepped down. This means that smaller conductors can be used, thus lowering the cost of the transmission wires. Also, by transmitting higher voltage, an allowance can be made for voltage losses along the way.

Again referring to Figure 19-44, notice that the 100-kilovolt transmission line feeds a substation. The purpose of the substation is to step down the 100-kilovolt transmission voltage to lower voltage levels for use by the residential, commercial, and industrial consumers. Typically, the 100 kilovolts is stepped down to voltages of 13.2 kilovolts, 4160 volts, and so on.

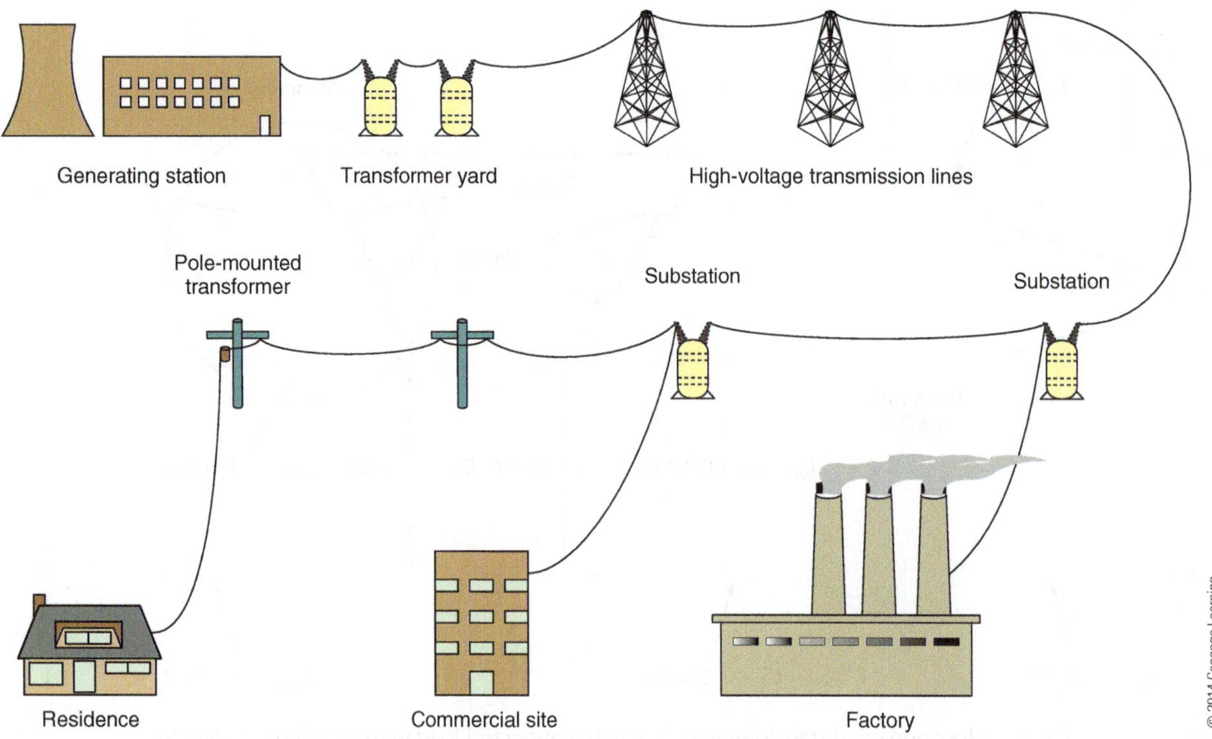

FIGURE 19-43 The primary distribution system.

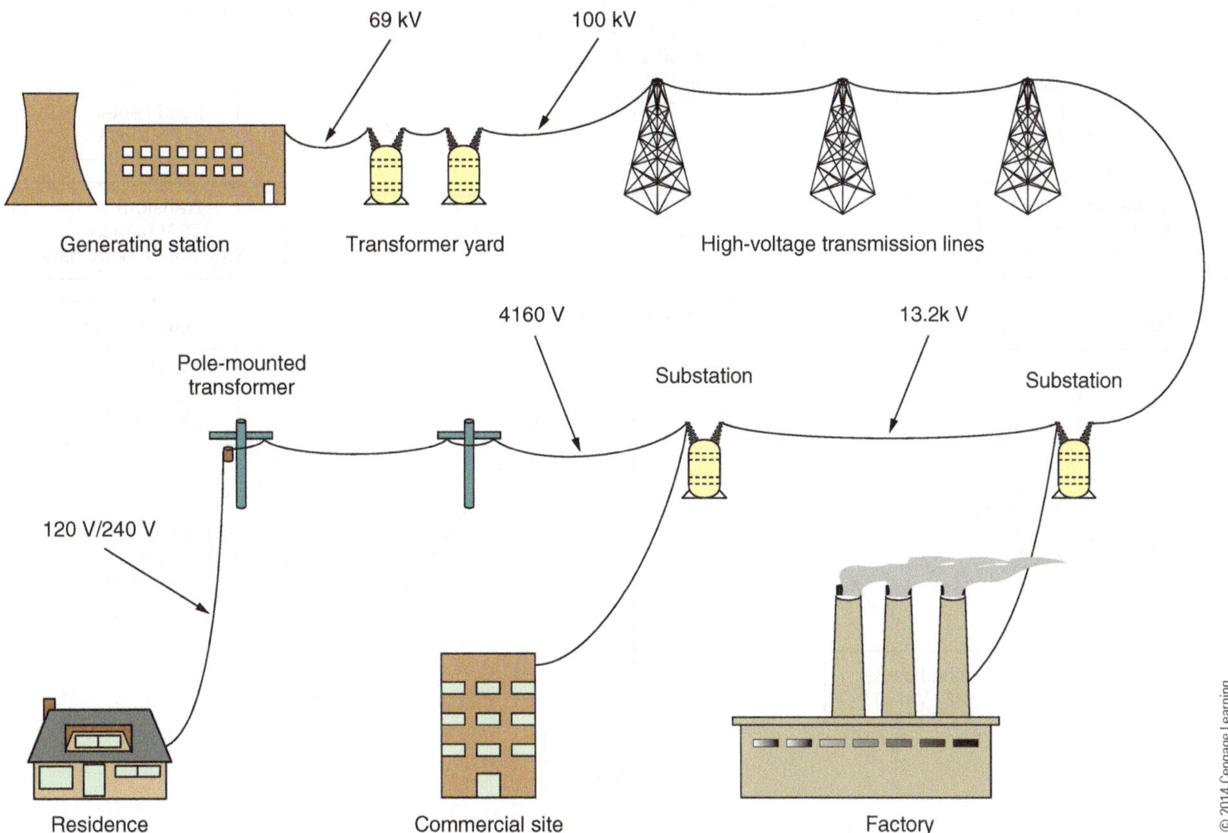

FIGURE 19-44 Typical primary distribution system voltages.

It should be noted here that different parts of the country use different voltages. This is determined by the power generated by the local electric utility and the transformer configurations used.

After the substation, the power is again transmitted through the community. At this point, the voltage is again stepped down to meet the local demand. This may involve another substation located on the premises of a factory, or a pole or padmounted transformer near a residence.

19-6 CONSUMER DISTRIBUTION SYSTEM

A consumer distribution system is used to distribute the electrical power within a residence, commercial, or industrial environment. The residential distribution system is fairly simple and straightforward. There are two forms of consumer distribution systems: the **radial system** and the **loop system**. **Figure 19-45** shows a radial system. Notice that in the radial system, the service appears as the hub in the spokes of a wheel. **Figure 19-46** shows a loop system. Notice that the distribution lines form a complete loop, starting and ending at the service. With a loop system, the loads are supplied from both ends, and sections can be isolated in the event of trouble.

Figure 19-47 shows a simple residential distribution system. The power supplied to a residence is typically called single-phase, three-wire, 120/240 volts. The 240 volts is used to supply loads such as an electric range, electric water heater, electric dryer, and so on. The 120 volts is used to supply general lighting loads and branch circuits.

The commercial and industrial distribution systems can be similar in complexity, depending on the various voltages required. **Figure 19-48** shows an industrial distribution system. Notice the different voltages that are derived to satisfy the needs of specific machines or processes. Notice also how the different areas can be isolated for maintenance and troubleshooting.

19-7 GROUNDING SYSTEM

The grounding system is necessary to provide protection from lightning surges and momentary power surges and for protection of personnel from accidental contact with energized components. System grounding electrodes must be designed and installed to provide a low impedance path for undesirable

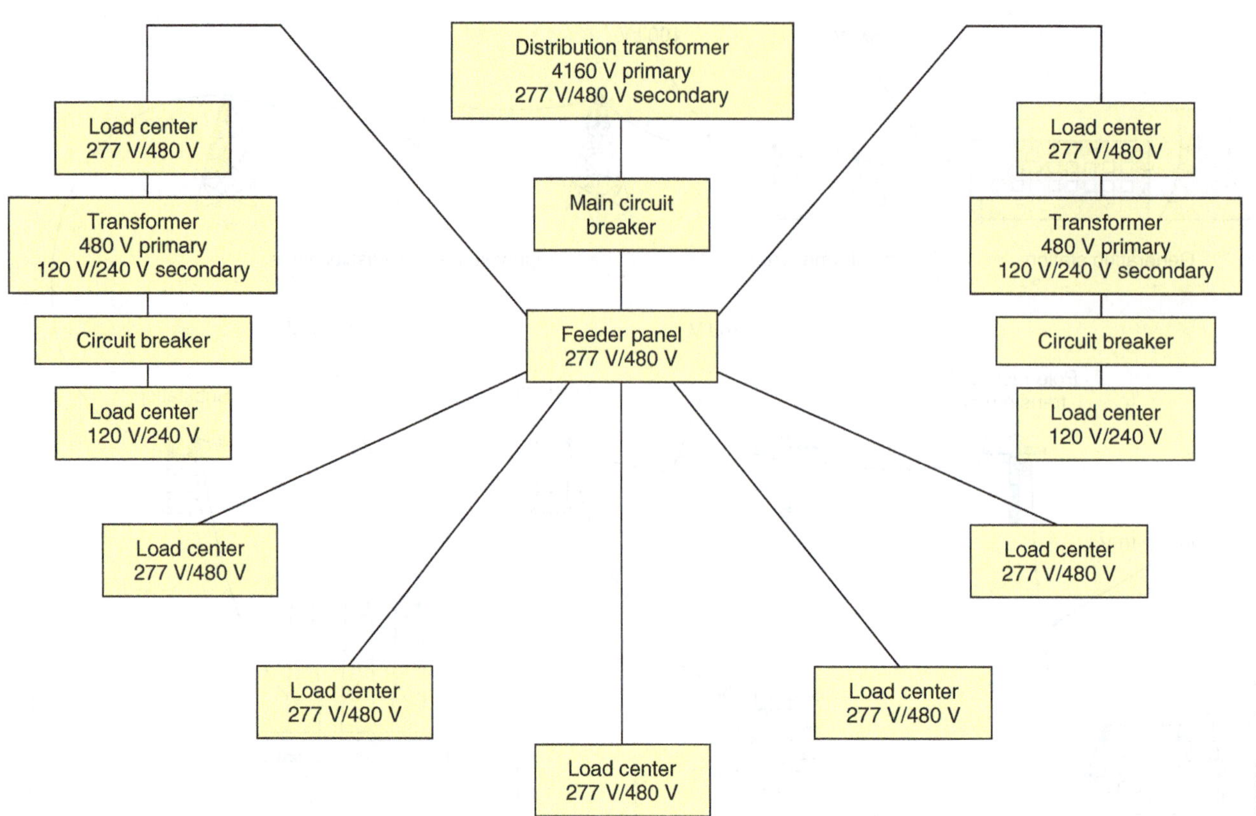

FIGURE 19-45 A radial-style consumer distribution system.

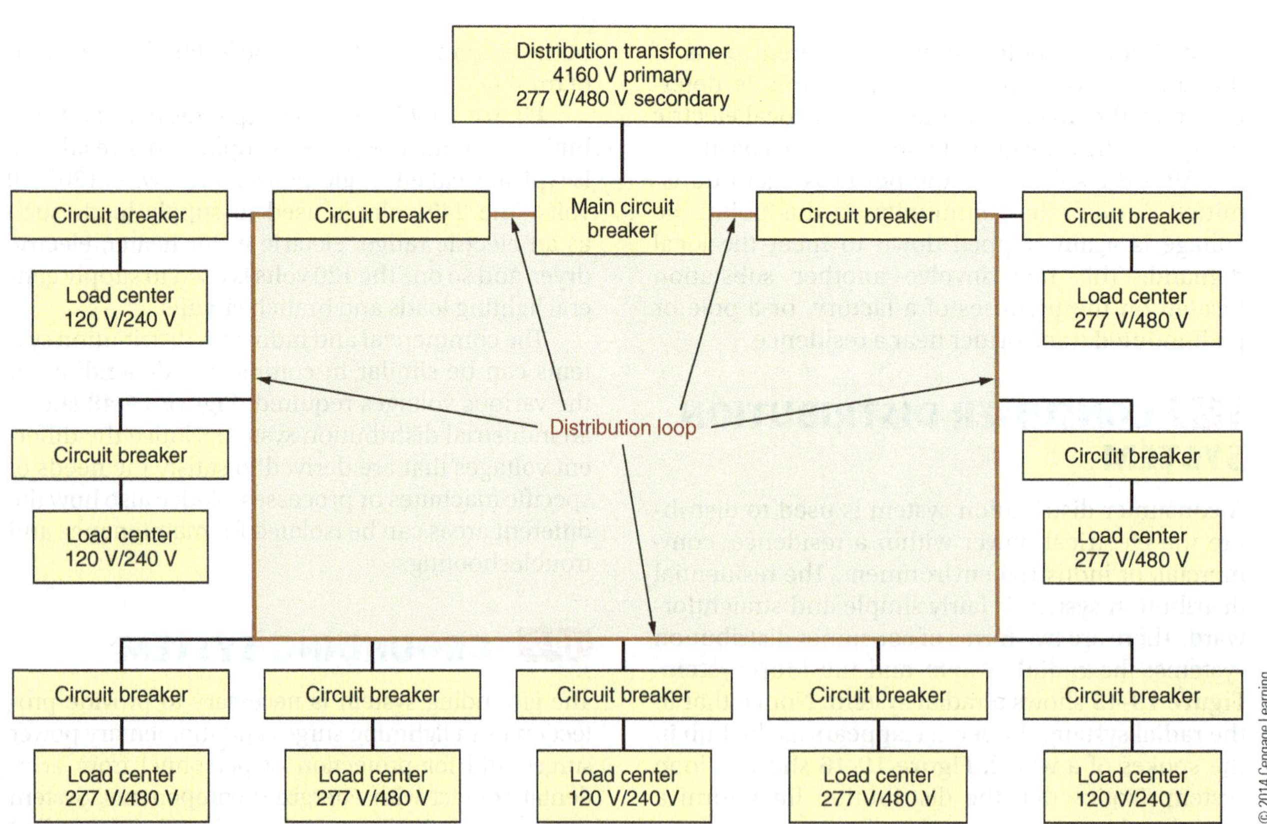

FIGURE 19-46 A loop-style consumer distribution system.

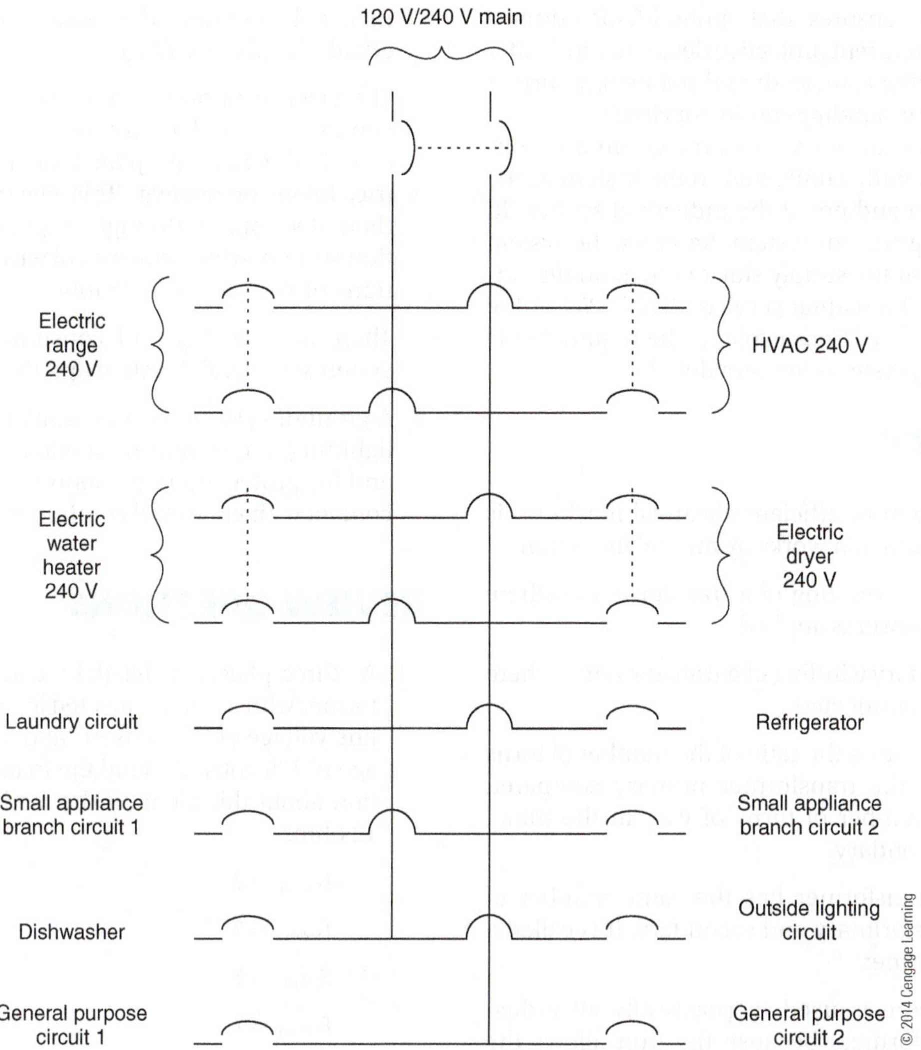

FIGURE 19-47 A residential distribution system.

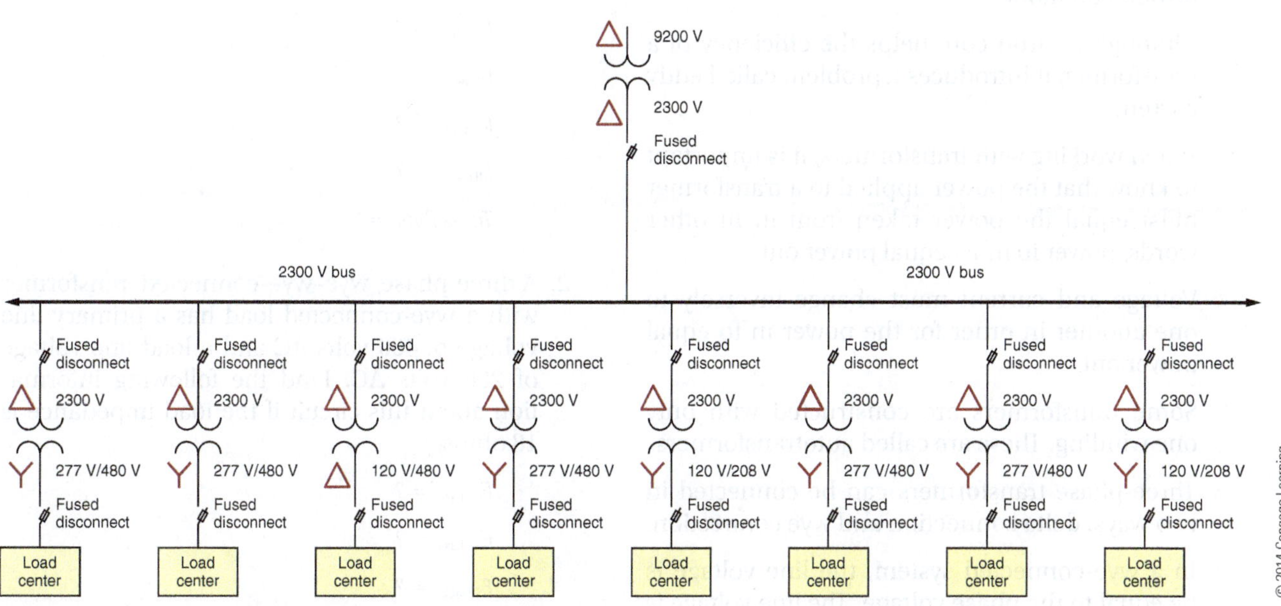

FIGURE 19-48 An industrial distribution system.

currents. This ensures that ground-fault currents cause the overcurrent protective device to trip. Under normal conditions, there should not be any current flowing in the grounding conductor circuit.

In a direct current system, the grounding conductor is generally connected to the system at the supply station and not at the individual service. In an alternating current system, however, the system is grounded on the supply side of the main disconnect at every individual service. *Article 250* of the *National Electrical Code* explains the requirements for grounding systems in more detail.

SUMMARY

- One of the most efficient electrical machines is the transformer. It works by mutual induction.
- The primary winding of a transformer is where electrical power is applied.
- The secondary winding of a transformer is where the load is connected.
- The turns ratio is the ratio of the number of turns of wire in the transformer primary compared with the number of turns of wire in the transformer secondary.
- When a transformer has the same number of turns in its primary and secondary, it is called a 1:1 transformer.
- An iron core is used in practically all industrial transformers because the iron allows the magnetic lines of force to travel between the windings more easily than if they were to travel through air alone.
- Although an iron core helps the efficiency of a transformer, it introduces a problem called eddy currents.
- When working with transformers, it is important to know that the power applied to a transformer must equal the power taken from it. In other words, power in must equal power out.
- Voltage and current must change inversely to one another in order for the power in to equal power out.
- Some transformers are constructed with only one winding. These are called autotransformers.
- Three-phase transformers can be connected in two ways: delta connection and wye connection.
- In a wye-connected system, the line voltage is be equal to the phase voltage. The line voltage is 1.732 times greater than the phase voltage.
- In a delta-connected system, the line voltage equals the phase voltage.
- The electrical power that is used in residential, commercial, and industrial sites originates in electrical generating plants such as hydroelectric, fossil, or nuclear. This electrical power is then distributed through a primary distribution system, which consists of wiring, associated transformers, and substations.
- There are two types of consumer distribution systems: the radial system and the loop system.
- A grounding system is necessary to protect from lightning surges and momentary power surges and for protection of personnel from accidental contact with energized components.

REVIEW QUESTIONS

1. A three-phase, delta-delta-connected transformer with a wye-connected load has a primary line voltage of 277 volts AC and a load line voltage of 120 volts AC. Find the following information about this circuit if the load impedance is 12 ohms.

   $E_{P(Phase)} = ?$

   $E_{S(Phase)} = ?$

   $E_{S(Line)} = ?$

   $E_{L(Phase)} = ?$

   $I_{P(Phase)} = ?$

   $I_{P(Line)} = ?$

   $I_{S(Phase)} = ?$

   $I_{S(Line)} = ?$

   $I_{L(Phase)} = ?$

   $I_{L(Line)} = ?$

   $Turns\ Ratio = ?$

2. A three-phase, wye-wye-connected transformer with a wye-connected load has a primary line voltage of 460 volts AC and a load line voltage of 208 volts AC. Find the following information about this circuit if the load impedance is 18 ohms.

   $E_{P(Phase)} = ?$

   $E_{S(Phase)} = ?$

   $E_{S(Line)} = ?$

   $E_{L(Phase)} = ?$

$I_{P(Phase)} = ?$

$I_{P(Line)} = ?$

$I_{S(Phase)} = ?$

$I_{S(Line)} = ?$

$I_{L(Phase)} = ?$

$I_{L(Line)} = ?$

Turns Ratio = ?

3. A three-phase, delta-wye-connected transformer with a delta-connected load has a primary line voltage of 560 volts AC and a load line voltage of 277 volts AC. Find the following information about this circuit if the load impedance is 7 ohms.

$E_{P(Phase)} = ?$

$E_{S(Phase)} = ?$

$E_{S(Line)} = ?$

$E_{L(Phase)} = ?$

$I_{P(Phase)} = ?$

$I_{P(Line)} = ?$

$I_{S(Phase)} = ?$

$I_{S(Line)} = ?$

$I_{L(Phase)} = ?$

$I_{L(Line)} = ?$

Turns Ratio = ?

4. A three-phase, wye-delta-connected transformer with a delta-connected load has a primary line voltage of 208 volts AC and a load line voltage of 460 volts AC. Find the following information about this circuit if the load impedance is 15 ohms.

$E_{P(Phase)} = ?$

$E_{S(Phase)} = ?$

$E_{S(Line)} = ?$

$E_{L(Phase)} = ?$

$I_{P(Phase)} = ?$

$I_{P(Line)} = ?$

$I_{S(Phase)} = ?$

$I_{S(Line)} = ?$

$I_{L(Phase)} = ?$

$I_{L(Line)} = ?$

Turns Ratio = ?

CHAPTER

20

Electrical Machinery

Electrical motors are used regardless of the type of facility or product manufactured. There are many types of electrical motors. Some operate on DC, whereas others operate on AC. AC motors may be single phase or three phase. In addition to motors, many facilities find it economical and necessary to generate their own electrical power. For this reason, a basic understanding of DC and AC generators is necessary. This chapter introduces you to the different types of DC and AC generators as well as the different DC, AC single-phase, and AC three-phase motors.

OBJECTIVES

After studying this chapter, the student should be able to

- Discuss the different types of DC generators.
- Identify the three basic types of DC motors.
- Discuss an AC alternator.
- Explain the theory of operation of three-phase motors.
- Identify different types of single-phase motors.

20-1 DC GENERATORS

The typical **DC generator** consists of four main parts: the **armature**, the **commutator**, the **brush assembly**, and the **field windings**. The armature typically has loops of wire mounted on a rotating **shaft**. Some mechanical energy is used to rotate the shaft. The mechanical energy may be in the form of water, steam, or air turning a turbine attached to the shaft. Other methods may also be used. The ends of the armature loops are connected to the commutator. The commutator provides a surface on which the brush assembly makes electrical contact, allowing current to be drawn through the assembly from the revolving armature windings. The field windings, which are connected to a source of **excitation current**, do not revolve; instead they remain stationary. **Figure 20-1** shows the parts of a DC generator.

The excitation current flowing through the field winding creates a magnetic field. As the armature is rotated by the mechanical energy, the loops of wire in the armature rotate through the magnetic field of the field windings. This induces a current and voltage in the armature windings. The amount of generated voltage is determined by the strength of the magnetic field and the speed of rotation of the armature. A stronger magnetic field caused by increasing the excitation current will produce a higher output voltage. Likewise, rotating the armature at a faster rate cuts more lines of force of the magnetic field, resulting in a higher output voltage. Therefore, varying either the excitation current or the speed of rotation will have a direct effect on the generated voltage.

The load is connected to the armature windings through the brushes. Therefore, the current drawn by the load must pass through the brush assembly.

The purpose of the commutator is to maintain the polarity of the DC voltage generated by the armature. Notice in **Figure 20-2** that the field is connected to a source of DC excitation current. Observe the polarity of the field windings. Also, notice that the connections to the field do not change. The field will always have the same magnetic polarity. Now watch what happens as the armature revolves. When mechanical energy is applied to the armature in Figure 20-2, the magnetic field of the field windings causes a current to be induced into the armature winding. Now look at **Figure 20-3**, **Figure 20-4**, and **Figure 20-5**. The polarity of the DC voltage that is produced within the armature is maintained by the commutator, causing the load to continue to receive the same polarity of voltage.

Sometimes excessive arcing will be noticed at the point where the brushes contact the commutator. This may be a result of the brush assembly being out of proper position. The brushes must be set to a position known as the **neutral plane**. To set the

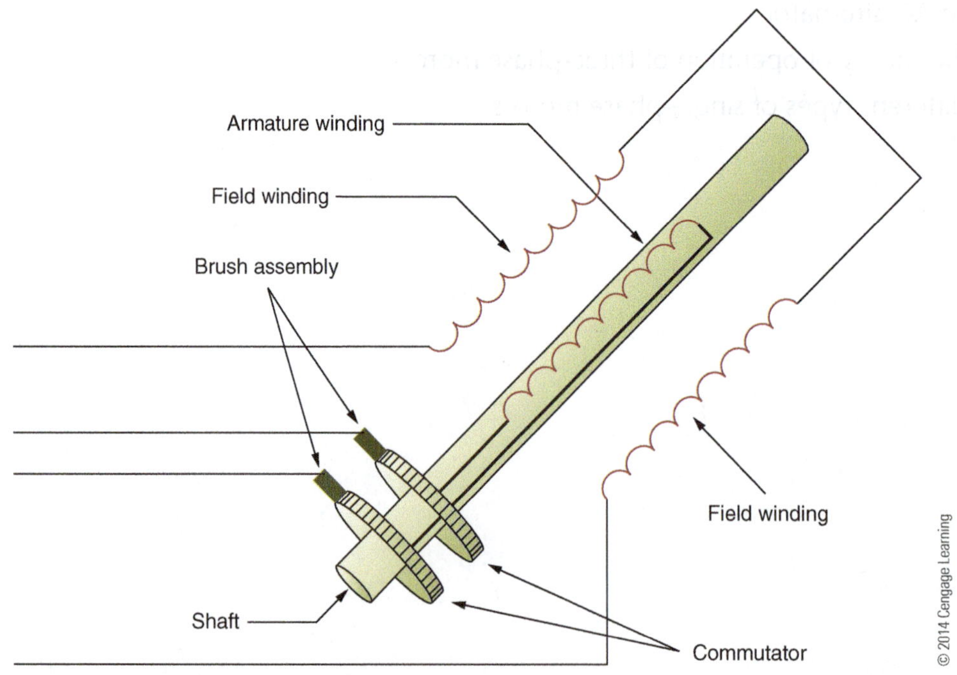

FIGURE 20-1 The parts of a DC generator.

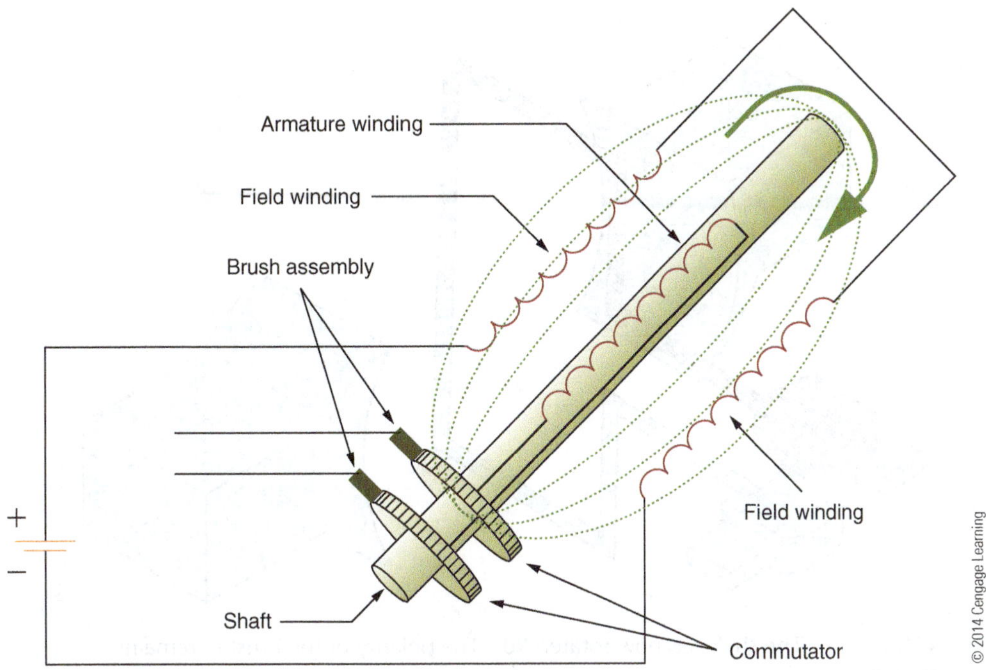

FIGURE 20-2 A magnetic field is created by applying DC excitation current.

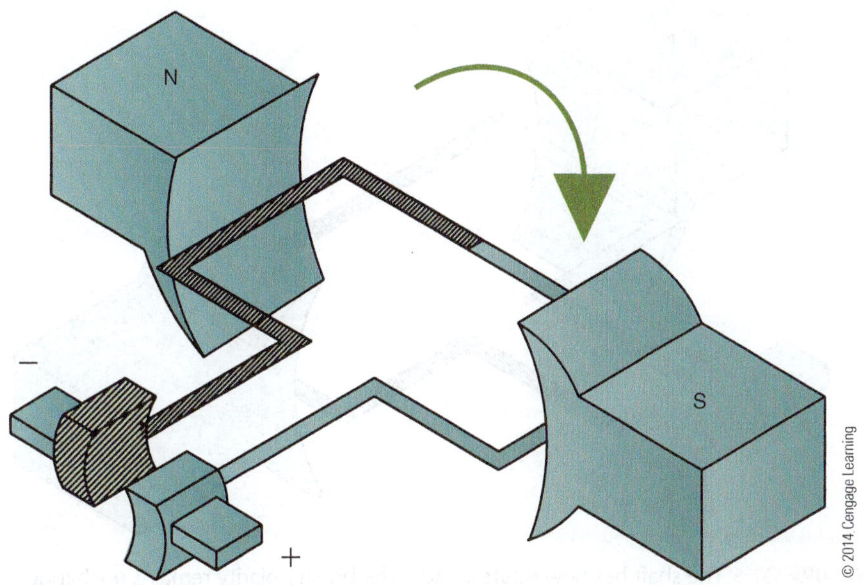

FIGURE 20-3 Mechanical energy causes the shaft to rotate. The magnets represent the magnetic field.

brushes to the neutral plane, refer to **Figure 20-6** and follow these steps:

1. Connect an AC voltmeter across the shunt field winding.
2. Obtain a low-voltage AC source.
3. Connect the AC source across the armature.
4. Carefully move the brush assembly back and forth, noting the change in voltage as indicated on the AC voltmeter.
5. Place the brush assembly in the location that produces the lowest indicated voltage on the AC voltmeter.

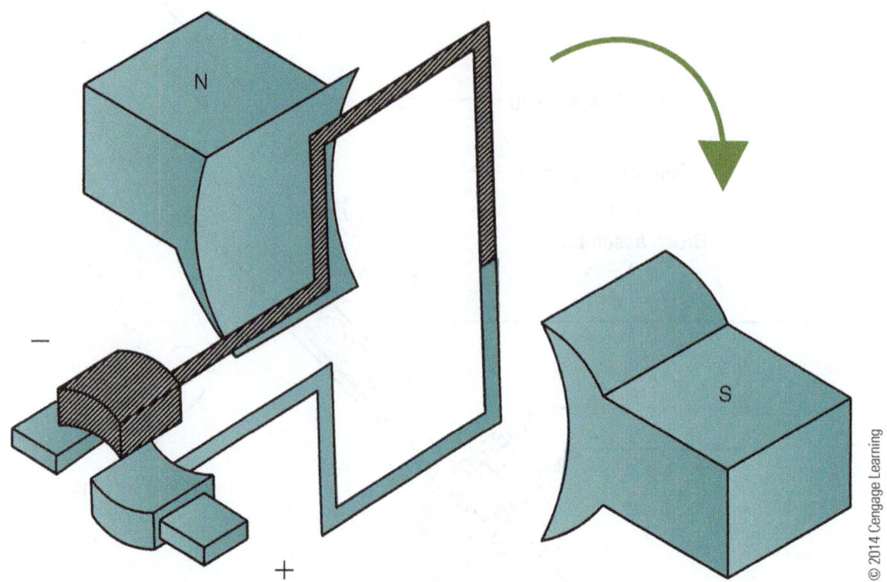

FIGURE 20-4 The shaft has now rotated 90°. The polarity of the brushes remains constant.

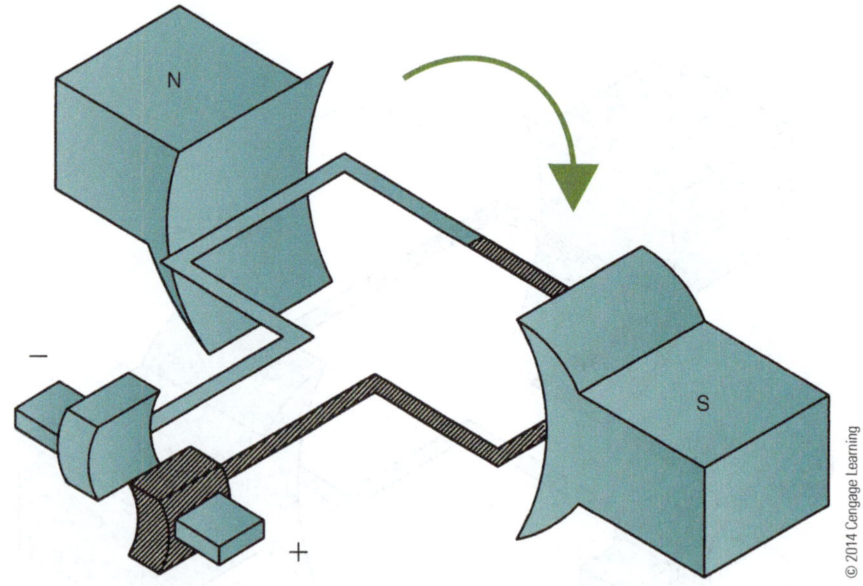

FIGURE 20-5 The shaft has now rotated 180°. The brush polarity remains unchanged.

DC generators can be connected in one of four ways: series wound, self-excited shunt wound, separately excited shunt wound, and compound wound.

Series-Wound DC Generator

Figure 20-7 shows a **series-wound DC generator** because the field is connected in series with the armature. In a series-wound generator, the series field excitation current is the same one that is supplied to the load. A high-resistance load draws a minimal amount of current from the generator, causing a minimal amount of voltage to be produced. Should the load become less resistive, the current will increase, causing an increase in generated voltage. The net result is that the generated output voltage of a series-wound generator varies with the load. Therefore, the series-wound generator is considered to have very poor voltage regulation. Series-wound generators are rarely used today.

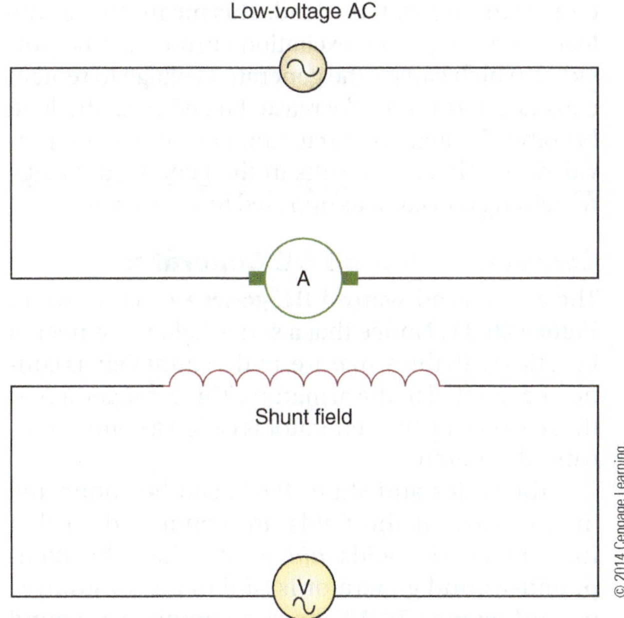

FIGURE 20-6 Set-up for finding the brush neutral position.

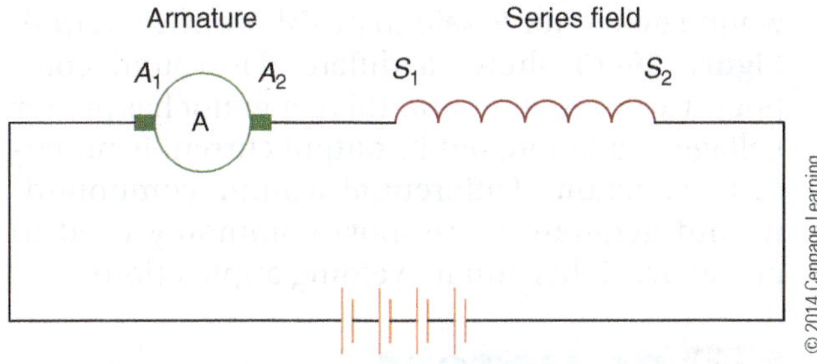

FIGURE 20-7 A series-wound DC generator.

Self-Excited Shunt-Wound DC Generator

A second way to connect a DC generator is to connect the field windings across or parallel to the armature. This configuration, known as a **shunt-wound DC generator**, is shown in **Figure 20-8**. A shunt-wound DC generator can be *self-excited* or *separately excited*. Figure 20-8 shows a **self-excited shunt-wound DC generator**. In this circuit, the shunt winding receives its excitation current from the **residual magnetism** within the stator poles of the DC generator. Usually enough magnetism remains to induce a small excitation current in the field windings. The amount of excitation current is sufficient to cause the generator to produce power when the armature is rotated. As the armature rotates, more voltage is produced, which induces more current, thus producing more voltage, and so on. Because the excitation current of a self-excited shunt-wound generator is not the same as that of the load current, its voltage regulation is better than a series-wound generator.

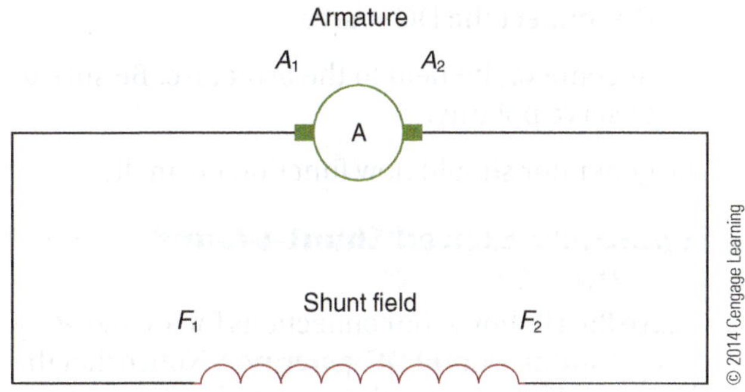

FIGURE 20-8 A self-excited shunt-wound DC generator.

Sometimes a self-excited shunt-wound generator fails to produce a voltage. This is usually because the residual magnetism within the pole pieces has been lost. The problem can be easily fixed through a procedure called **flashing the field**. **Figure 20-9** shows the connections to flash the field and restore the residual magnetism. Follow these steps to restore the residual magnetism:

1. Disconnect the field from the armature. Be sure to note the polarity of the field.
2. Obtain a low-voltage DC source. A 12-volt automotive-type battery is sufficient for a 120-volt DC to 600-volt DC generator.
3. Connect the DC source across the field winding. Be sure to connect the DC source in the same polarity as noted in step 1.
4. Allow the DC source to remain connected for approximately 10 minutes.

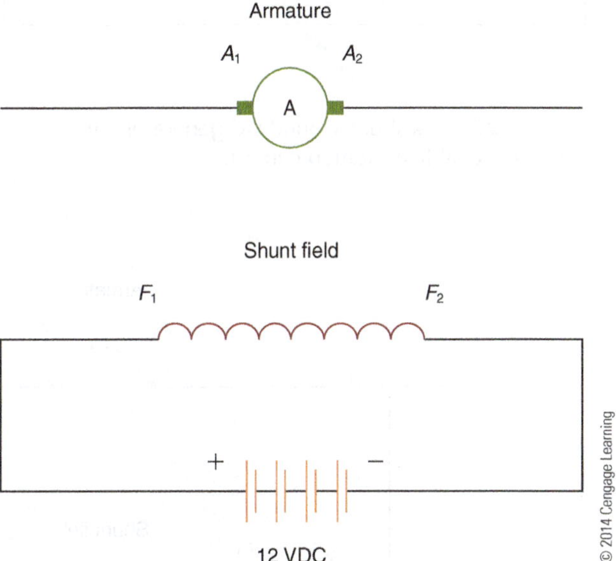

FIGURE 20-9 Flashing the field of a shunt-wound DC generator.

5. Disconnect the DC source.
6. Reconnect the field to the armature. Be sure to observe polarity.

The generator should now function normally.

Separately Excited Shunt-Wound DC Generator

Figure 20-10 shows the connections for a **separately excited shunt-wound DC generator**. Notice that the shunt winding now receives power from a separate DC source. Most DC generators are connected in this manner. The separately-excited shunt-wound generator offers better voltage regulation than does the self-excited generator. However, the separately-excited shunt-wound generator is more costly because of the additional source of DC required to provide the field excitation current. Having a separate source of excitation current allows the excitation current to vary with the load. This means that as the load increases, more excitation current can be provided, which causes the generated voltage to remain constant and not to decrease. Likewise, as the load becomes lighter, less excitation current can be provided, which again results in the generated voltage remaining constant, as opposed to increasing.

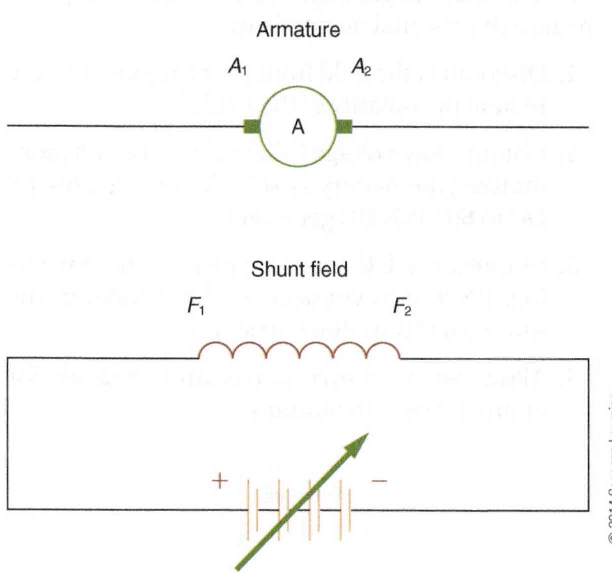

FIGURE 20-10 A shunt-wound DC generator with separate variable excitation current.

Compound-Wound DC Generator

The **compound-wound DC generator** is shown in **Figure 20-11**. Notice that a series field is connected in series with the armature, and a shunt field is connected parallel to the armature. This arrangement is the reason why the generator is called a compound-wound generator.

The series and shunt fields can be connected in two ways. If the fields are connected so that their magnetic fields *aid* each other, the compound-wound generator is said to be **cumulative wound**. **Figure 20-12** shows a cumulative-wound compound-wound generator. This generator has excellent voltage regulation.

If the fields are connected so that their magnetic fields *oppose* each other, the compound-wound generator is said to be **differential wound**. **Figure 20-13** shows a differential-wound compound-wound generator. This generator has poorer voltage regulation, but its output current is practically constant. Differential-wound compound-wound generators are most commonly used to power arc lights and in welding applications.

20-2 DC MOTORS

The typical DC motor consists of four main parts: the armature, the commutator, the brush assembly, and the field windings. The armature typically has loops of wire mounted on a rotating shaft. The ends of these loops are connected to the commutator. The commutator provides a surface on which the brush assembly makes electrical contact, allowing

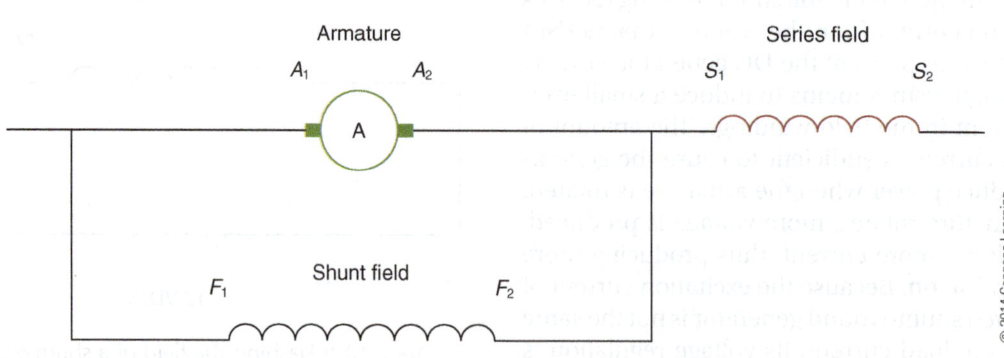

FIGURE 20-11 A compound-wound DC generator.

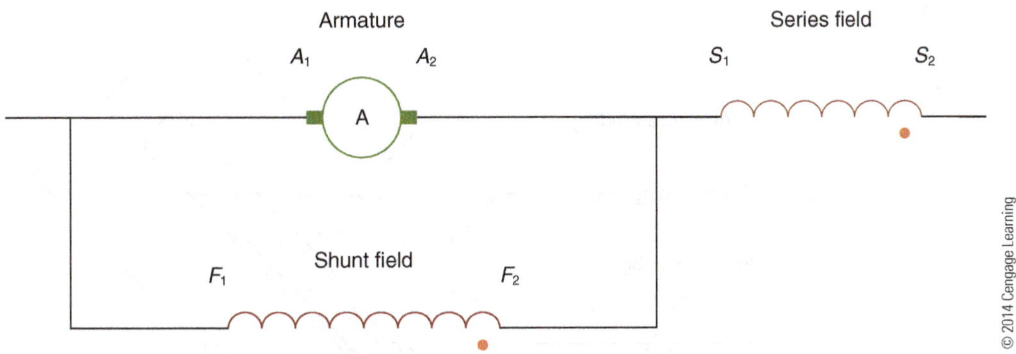

FIGURE 20-12 A cumulative-wound compound-wound DC generator.

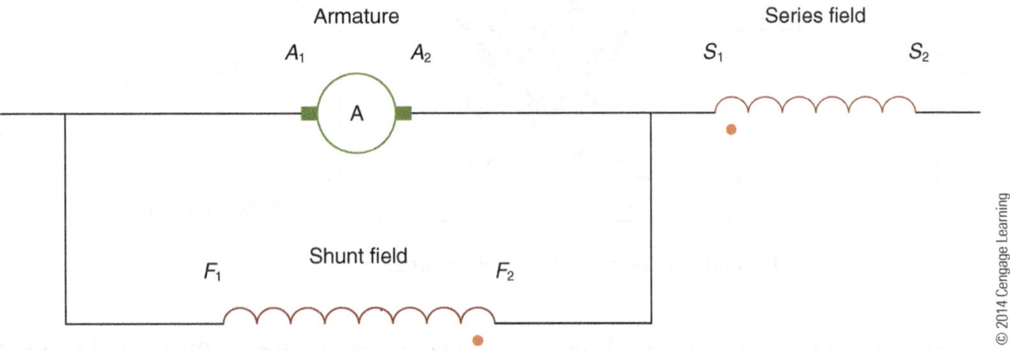

FIGURE 20-13 A differential-wound compound-wound DC generator.

current to be applied through the assembly to the revolving armature windings. The field windings, which are also connected to a source of electrical power, do not revolve but instead remain stationary. **Figure 20-14** shows the parts of a DC motor.

The purpose of the commutator is to reverse the polarity of the DC voltage applied to the armature. This causes the armature to turn in only one direction because of the attraction between the magnetic fields in the field and armature. Notice in **Figure 20-15**

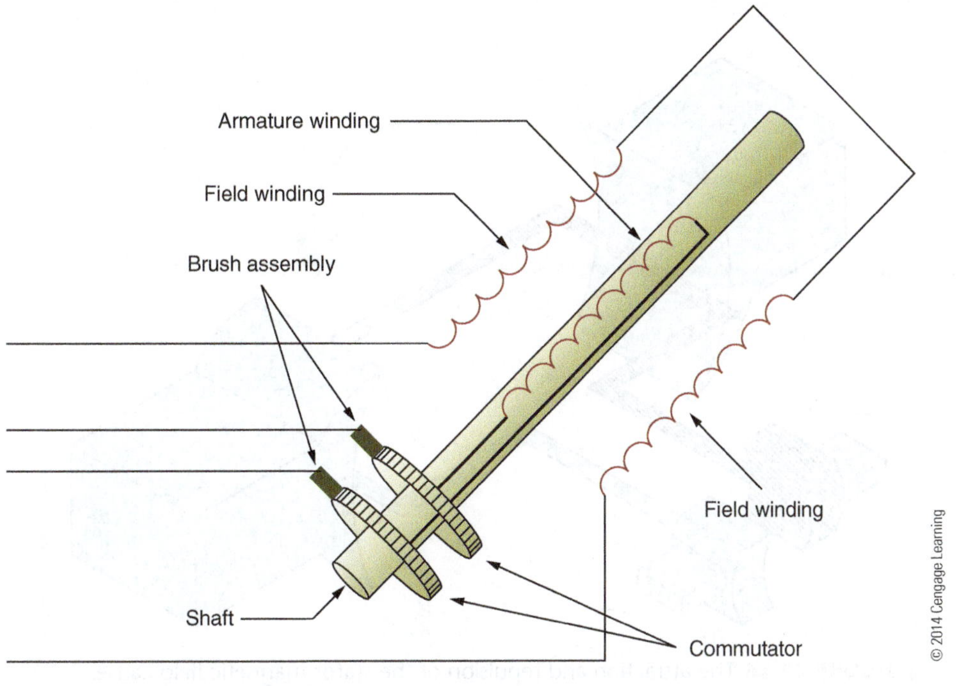

FIGURE 20-14 Parts of a DC motor.

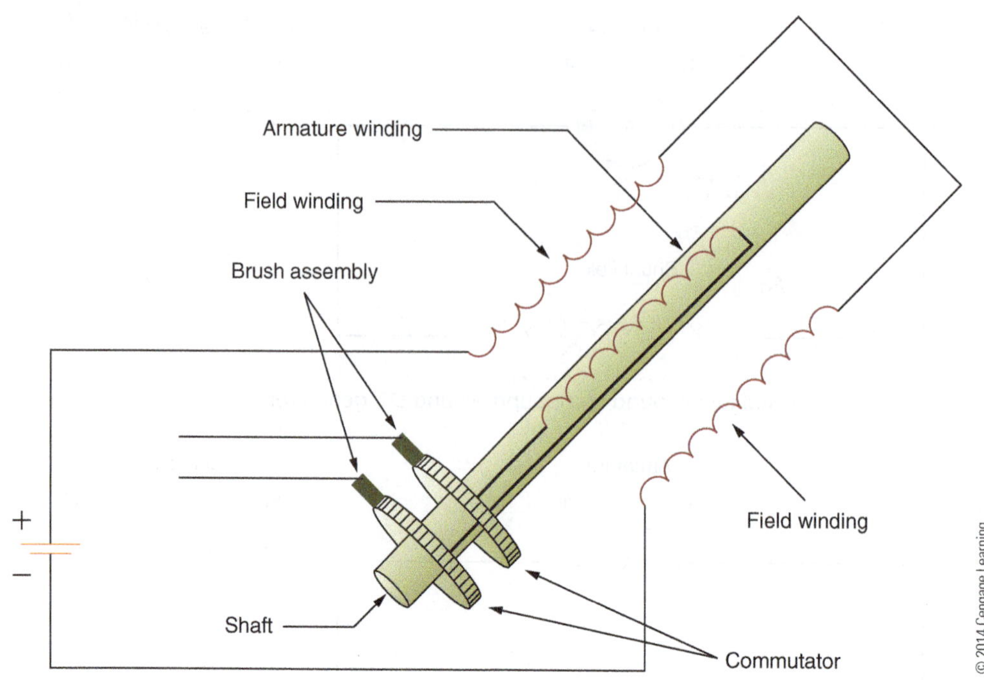

FIGURE 20-15 The stator is connected to a DC source.

that the field is connected to a DC source. Observe the polarity of the field windings. Also, notice that the connections to the field do not change. The field will always have the same magnetic polarity. When DC is applied to the armature, causing it to revolve, as shown in Figure 20-15, the magnetic field of the armature causes it to be attracted to the field winding with the opposite magnetic polarity. The armature thus turns in the direction of field windings. Now look at **Figure 20-16**, **Figure 20-17**, and **Figure 20-18**. The polarity of the DC voltage that is applied to the armature is reversed by the commutator, causing the armature to continue to rotate in the same direction. The speed of the armature is controlled by the strength of its magnetic field, usually by means of a variable resistor in the armature circuit. To change the direction of the armature's rotation, you can reverse the DC connections either to the armature or to the field.

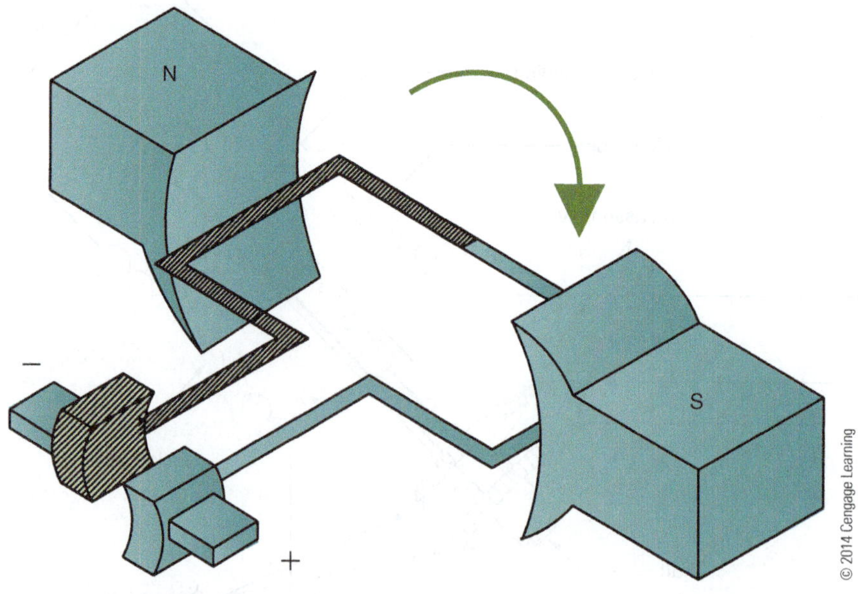

FIGURE 20-16 The attraction and repulsion of the stator magnetic field causes the armature to rotate.

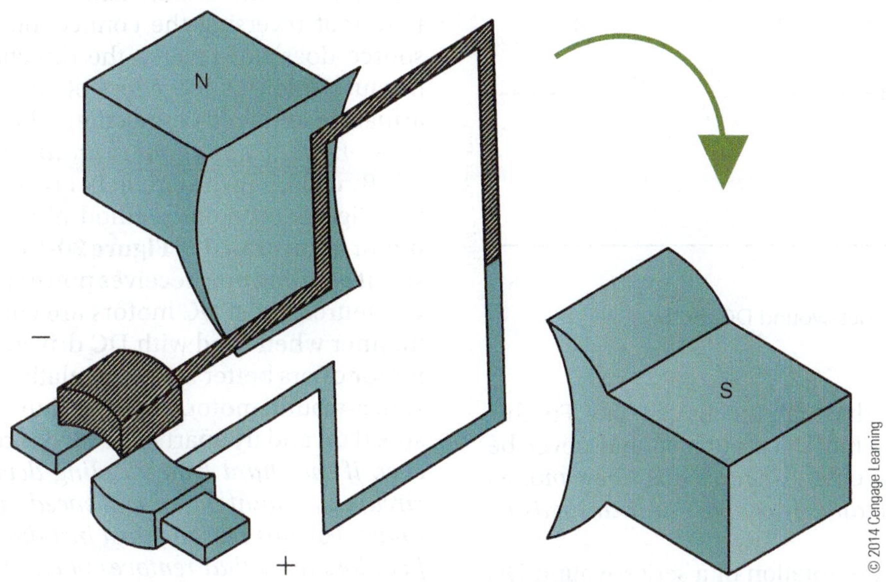

FIGURE 20-17 As the armature rotates, the current flow through the armature windings is reversed.

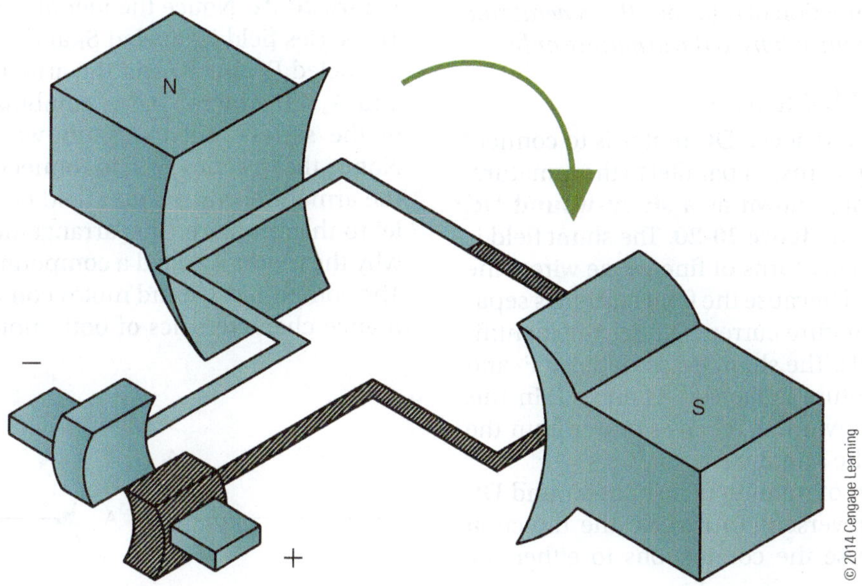

FIGURE 20-18 The armature has rotated 180°. Notice the polarity of the armature winding.

DC motors can be connected in three ways: series wound, shunt wound, and compound wound.

Series-Wound DC Motor

A motor connected as shown in **Figure 20-19** is called a **series-wound DC motor** because the series field is connected in series with the armature. The series field is constructed of few turns of heavy gauge wire. Heavy gauge wire is required because the armature current must also flow through the series-connected series field. Notice the identification of the leads. The series field is labeled S_1 and S_2, and the armature is labeled A_1 and A_2. The series-wound motor produces high torque at low speeds. Unfortunately, it does not possess good speed regulation. In fact, a loss of load creates a serious problem in the series-wound DC motor. If the load is removed, the series-wound DC motor speeds up. The motor's speed can increase to such a high value that the motor will actually fly apart because of the

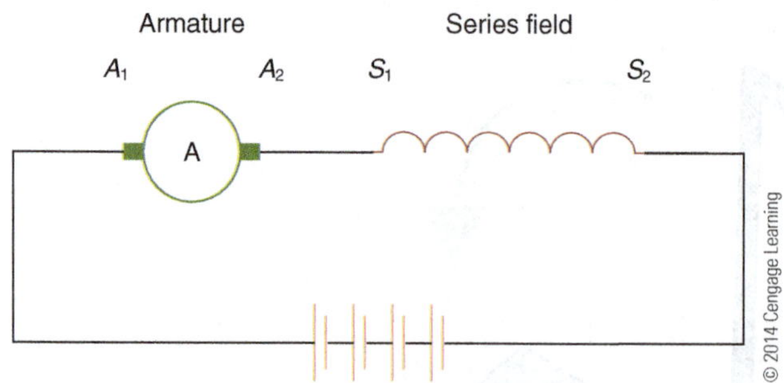

FIGURE 20-19 A series-wound DC motor.

centrifugal forces that high speeds generate. For this reason, series-wound DC motors should never be used with belt- or chain-driven loads. *These motors should always be directly coupled to their loads by shafts.*

The direction of rotation of a series-wound DC motor may be reversed. To reverse the direction of rotation, reverse the connections to either the armature *or* the series field, *but not both.* Also, note that reversing the connections of the power source does not reverse the direction of rotation. *The general rule is to reverse the connections to the armature only.*

Shunt-Wound DC Motor

A second way to connect a DC motor is to connect the field windings across or parallel to the armature. This configuration, known as a **shunt-wound DC motor**, is shown in **Figure 20-20**. The shunt field is constructed of many turns of fine gauge wire. Fine gauge wire is used because the field current is separate from the armature current. Notice the identification of the leads. The shunt field is labeled F_1 and F_2, and the armature is labeled A_1 and A_2. In this circuit, the shunt winding receives power from the same source as the armature.

The direction of rotation of a shunt-wound DC motor may be reversed. To reverse the direction of rotation, reverse the connections to either the armature *or* the shunt field, *but not both.* Also, note that reversing the connections of the power source does not reverse the direction of rotation because this has the effect of reversing both the armature and field connections. *The general rule is to reverse the connections to the armature only.*

By changing the circuit in Figure 20-20 slightly, the slightly different method of connecting a DC motor is illustrated in **Figure 20-21**. Notice that the shunt winding now receives power from a separate DC source. Most DC motors are connected in this manner when used with DC drives. The DC shunt motor offers better speed regulation than does the series-wound motor. Also, the shunt motor will not speed up and fly apart if the load is removed. *However, if the shunt field winding develops an open circuit, the shunt motor will speed up and could fly apart. For this reason, most installations include a field loss relay that removes power from the motor when it detects an open field condition.*

Compound-Wound DC Motor

The **compound-wound DC motor** is shown in **Figure 20-22**. Notice the identification of the leads. The series field is labeled S_1 and S_2, the shunt field is labeled F_1 and F_2, and the armature is labeled A_1 and A_2. This motor has a combination of features of the series- and the shunt-wound DC motors. Notice that a series field is connected in series with the armature, and a shunt field is connected parallel to the armature. This arrangement is the reason why the motor is called a compound-wound motor. The compound-wound motor combines the performance characteristics of both motor types as well.

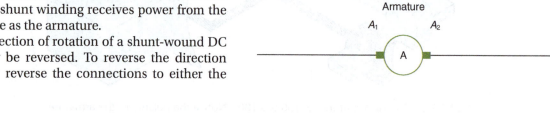

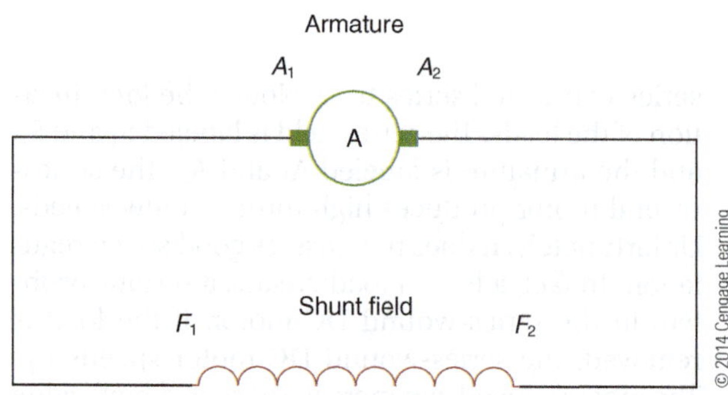

FIGURE 20-20 A shunt-wound DC motor.

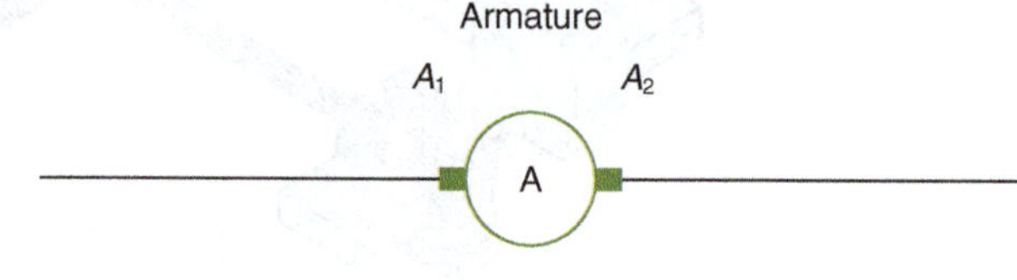
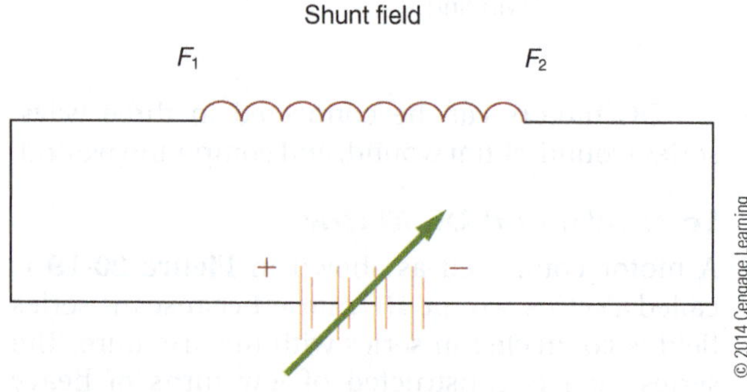

FIGURE 20-21 A shunt-wound DC motor with variable field excitation.

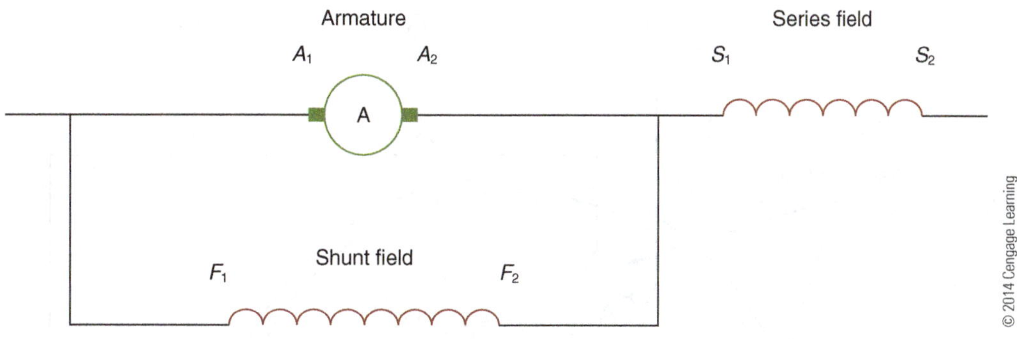

FIGURE 20-22 A compound-wound DC motor.

Its speed regulation is not as good as that of the shunt-wound motor, but it is better than that of the series-wound motor. The starting torque is better than that of the shunt-wound motor, but it is not as good as that of the series-wound motor. To reverse the direction of rotation of a compound-wound DC motor, *the general rule is to reverse the connections to the armature only.*

20-3 ALTERNATORS

The typical alternator consists of four main parts: the armature, **slip rings**, brush assembly, and field windings. The armature typically has loops of wire mounted on a rotating shaft. Some mechanical energy is used to rotate the shaft. The mechanical energy may be in the form of water, steam, or air turning a turbine attached to the shaft. Other methods may also be used. The ends of the armature loops are connected to the slip rings. The slip rings provide a surface on which the brush assembly makes electrical contact, allowing excitation current to flow through the assembly to the revolving armature windings. The field windings, from which the generated current is drawn, do not revolve, but instead remain stationary. **Figure 20-23** shows the parts of an alternator.

20-4 THREE-PHASE MOTORS

Three-phase motors are the most widely used motors in industry. Therefore, it is important that you recognize and understand the operation of the following types of three-phase motors: the squirrel-cage induction motor, the wound-rotor induction motor, and the synchronous motor.

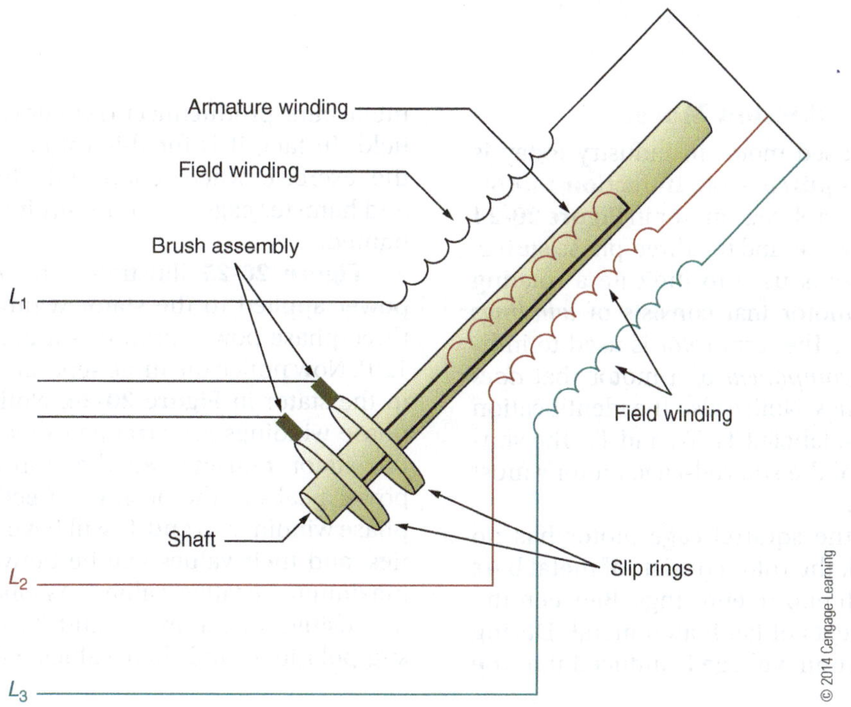

FIGURE 20-23 Parts of an alternator.

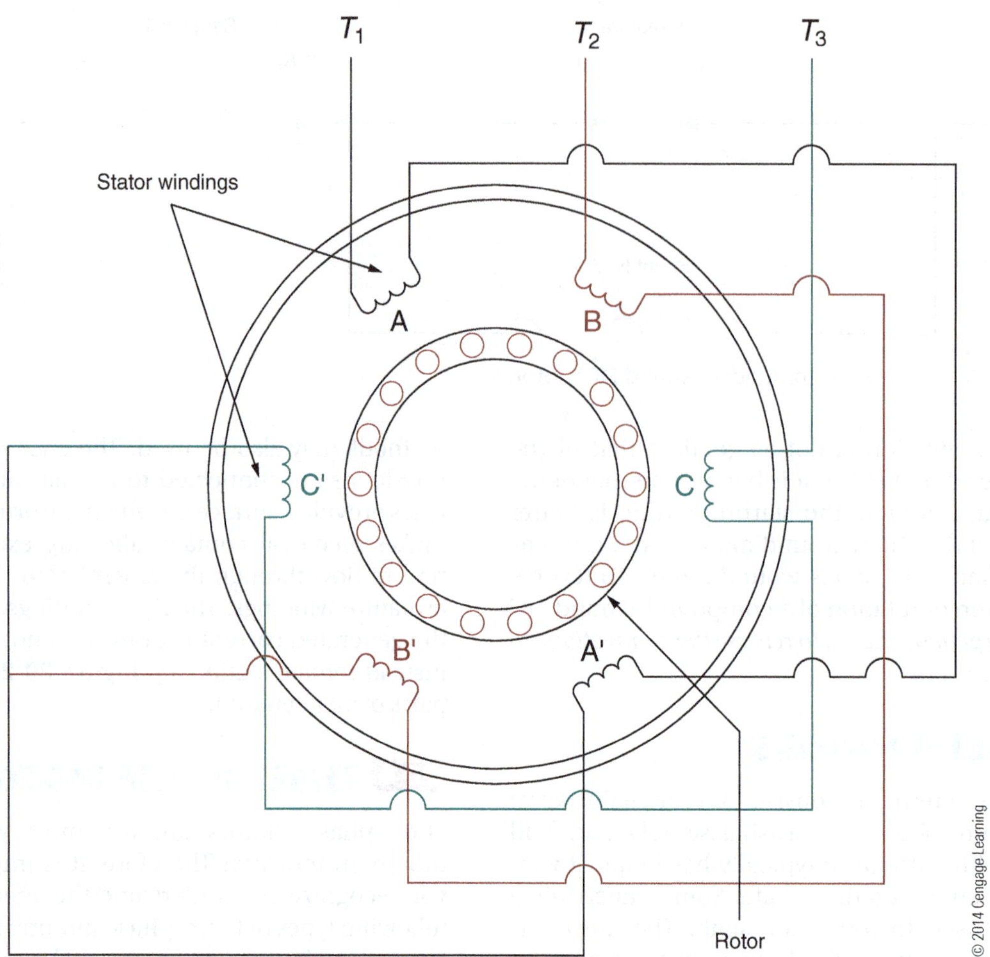

FIGURE 20-24 The squirrel-cage induction motor.

Squirrel-Cage Induction Motor

The most widely used motor in industry today is the **three-phase squirrel-cage induction motor**. Notice that the squirrel-cage motor in **Figure 20-24** has two parts: the **rotor** and the three-phase **stator**. The term *armature* is used to indicate a rotating component of a motor that consists of *windings* and a *commutator*. The term *rotor* is used to indicate the *rotating component* of a motor that *does not contain windings*. Notice also the identification of the motor leads, labeled T_1, T_2, and T_3. The simple design is one of the squirrel-cage motor's most attractive features.

The rotor of the squirrel-cage motor has no windings. Instead, the rotor consists of metal bars connected at each end to end rings. Between the metal bars are sheets of laminated metal. During the motor's operation, voltage is induced into the metal bars, producing current flow and a magnetic field. In fact, it is for this rotor, which resembles the exercise wheel commonly found in mouse and hamster cages, that the squirrel-cage motor is named.

Figure 20-25 illustrates the three-phase AC power applied to the stator windings. Recall that three-phase power produces a peak voltage every 120°. Now notice the three separate phase windings in the stator in **Figure 20-26**. Notice also that the phase windings are arranged in sequence around the stator housing. As the applied three-phase power peaks in the positive direction in phase T_1, phase windings T_2 and T_3 will have opposite polarities, and their values will be between 0 and their maximum negative values. As phase T_2 peaks in its positive direction, T_1 and T_3 will have opposite polarities, and their values will be between 0

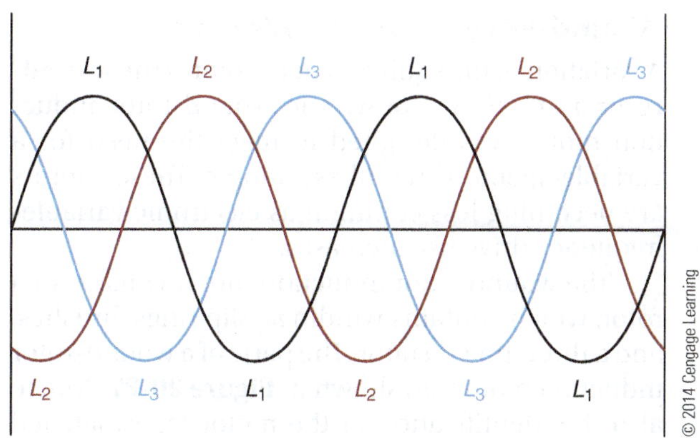

FIGURE 20-25 The three-phase waveform.

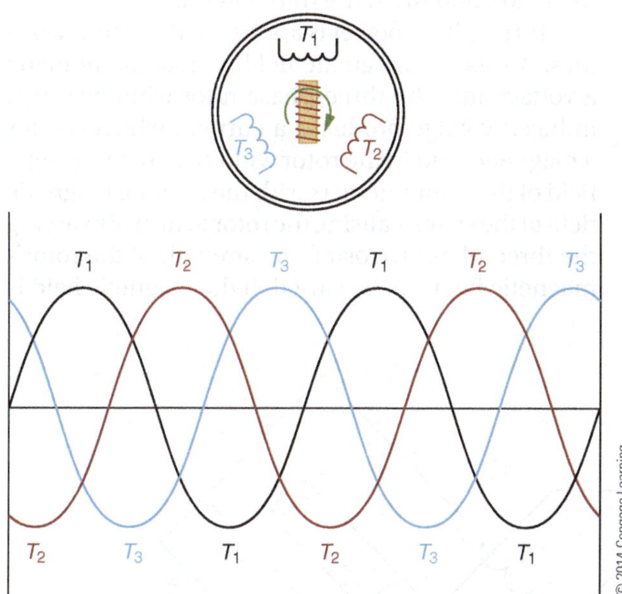

FIGURE 20-26 The phase rotation produces a rotating magnetic field in the stator.

and their maximum negative values. This pattern continues through the entire 360° rotation of the applied AC sine wave.

If you concentrate on the stator windings, you will see that a rotating magnetic field appears around the stator. The speed at which the rotating magnetic speed revolves is called the **synchronous speed**. Synchronous speed is affected by the number of stator poles and the frequency of the applied AC. This relationship can be shown by using the following mathematical formula:

$$S = \frac{120 \times f}{P}$$

S = Synchronous speed in rpm
120 = A constant value
f = Frequency of the applied AC in hertz (Hz)
P = Number of stator poles

For example, what is the synchronous speed of a four-pole squirrel-cage motor operating from 60-hertz AC power?

What values are known?

$$f = 60 \text{ Hz}$$

$$P = 4$$

What value is not known?

$$S = ?$$

What formula can be used?

$$S = \frac{120 \times f}{P}$$

Substitute the known values and solve for *S*.

$$S = \frac{120 \times f}{P}$$

$$S = \frac{120 \times 60 \text{ Hz}}{4}$$

$$S = \frac{7200}{4}$$

$$S = 1800 \text{ rpm}$$

Therefore, the synchronous speed of this squirrel cage motor is 1800 revolutions per minute.

The rotating magnetic field of the stator induces a voltage into the rotor bars, causing a current flow. The current flow in the rotor bars produces another magnetic field that is attracted to the revolving magnetic field in the stator. This attraction causes the rotor to turn and produce a torque. However, the rotor will not rotate at the same speed as the rotating field of the stator. Friction in the bearings and wind resistance (**windage**) will cause the rotor to rotate at a slightly slower speed. This is known as **rotor slip**. Rotor slip is the difference in speed between the synchronous speed and the rotor speed.

Until recently, the squirrel-cage motor was considered a fixed-speed motor. As can be seen from the above formula for synchronous speed, the only way to vary the speed of the motor would be to change the frequency of the applied AC voltage or change the number of poles in the motor. The power companies in the United States accurately maintain the power line frequency at 60 hertz. Therefore, changing the frequency of the applied AC power was not an option. It is also not possible to change the number of poles in the motor. The motor would have to be totally rebuilt. Today, however, with **electronic variable-speed drives**, the frequency of the applied AC power can be varied, and, consequently, so can the speed of the squirrel-cage motor. Using solid-state components, electronic variable-speed drives accomplish this by converting the applied AC power to DC, chopping the DC into pulses whose frequency can be varied, and converting those pulses into an artificial AC, which is then applied to the motor. Now the frequency of the applied AC power can be varied, thus allowing for variable-speed operation of the motor.

To reverse the direction of rotation of a squirrel-cage motor, you simply interchange the connections of any two of the three stator leads. As a rule, T_1 and T_3 are interchanged. This will cause the rotating magnetic field in the stator to revolve in the opposite direction. The rotor will then turn in the new direction of the rotating magnetic field, which will be opposite from the original.

Wound-Rotor Induction Motor

A variation of the squirrel-cage motor is the **wound-rotor induction motor**. The wound-rotor induction motor was designed to meet the need for a variable-speed, three-phase motor. These motors are becoming less common as electronic variable-frequency drive use increases.

The wound-rotor induction motor consists of a rotor, which contains windings, slip rings, brushes, and a three-phase stator. The parts of a wound-rotor induction motor are shown in **Figure 20-27**. Notice also the identification of the motor leads, labeled T_1, T_2, and T_3 for the stator connections, and M_1, M_2, and M_3 for the wound-rotor connections. Figure 20-27 also shows a three-phase rheostat connected to M_1, M_2, and M_3 in the rotor circuit.

Three-phase power is applied to the stator windings. A rotating magnetic field is created, inducing a voltage into the three-phase rotor windings. This induced voltage produces a current, which creates a magnetic field in the rotor windings. The magnetic field of the rotor interacts with the rotating magnetic field of the stator, causing the rotor to turn. By varying the three-phase rheostat, the strength of the rotor's magnetic field can be varied. If the magnetic field is

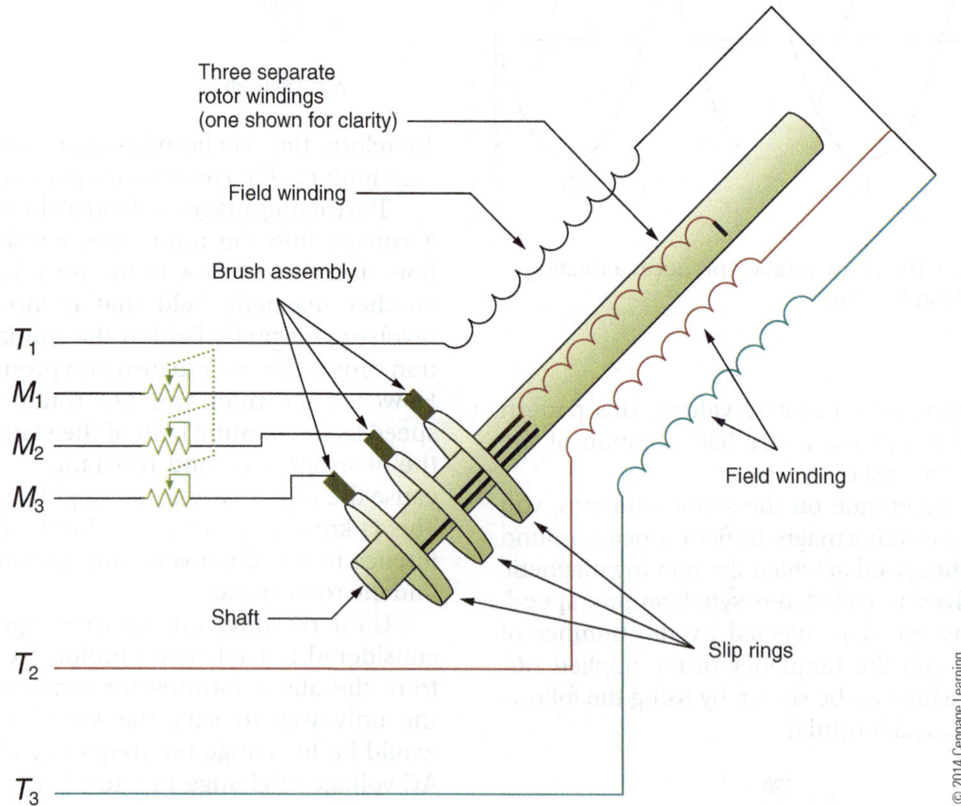

FIGURE 20-27 Parts of a wound-rotor induction motor. Only one winding is shown on the rotor shaft for clarity.

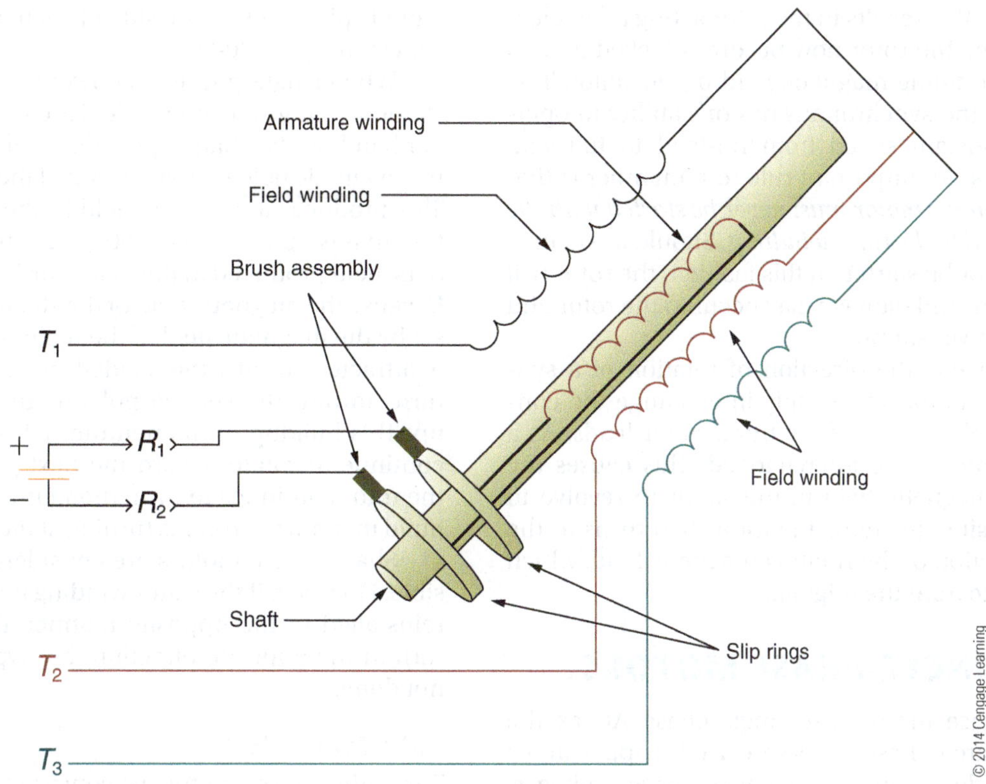

FIGURE 20-28 Parts of a synchronous motor. Note the DC applied to the rotor winding.

lessened, the rotor speed decreases. If the magnetic field is strengthened, the rotor speed increases.

This control over the rotor magnetic field can be used to vary the speed of the wound-rotor induction motor when driving a constant load. In addition, the rotor magnetic field can be varied by an automated process in response to varying load conditions. In this manner, the speed of the wound-rotor induction motor can be held fairly constant.

To reverse the direction of rotation of a wound-rotor induction motor, you simply interchange the connections of any two of the three stator leads. As a rule, T_1 and T_3 are interchanged. This causes the rotating magnetic field in the stator to revolve in the opposite direction. The rotor then turns in the new direction of the rotating magnetic field, which is opposite from the original.

The wound-rotor induction motor has disadvantages: It is more expensive than a similar size squirrel-cage motor; it requires more maintenance because of brush and slip ring wear; and although it has the benefit of providing variable speed, this benefit has become less attractive because of the use of drives.

Synchronous Motor

Figure 20-28 shows the parts of a **synchronous motor**. Notice that a synchronous motor has a three-phase stator, a wound rotor, slip rings, and brushes. Notice, however, that the rotor has a single winding as compared to the three windings of the wound-rotor induction motor. In addition, the rotor also contains shorting bars similar to the rotor of the squirrel-cage motor. The synchronous motor leads are labeled T_1, T_2, and T_3 for the stator connections, and R_1 and R_2 for the rotor connections. In addition, a DC source is used to supply power to the rotor.

A synchronous motor is started by applying three-phase power to the stator. *However, power is not applied to the rotor at this time.* When the three-phase power is applied to the stator, a rotating magnetic field is created. The rotating magnetic field induces a voltage into the shorting bars of the rotor. This induced voltage produces a current, which creates a magnetic field in the rotor. The rotor is then attracted to the rotating magnetic field of the stator, causing the rotor to turn.

When the rotor is up to speed, the DC source is energized, allowing DC to be applied to the rotor windings. This results in the rotor acting as an electromagnet. The rotor now becomes locked in step with the rotating magnetic field of the stator. This results in the synchronous motor's ability to operate at constant speed from no-load to full-load conditions. An important rule to remember is that *a synchronous motor must never be started with the DC applied to the rotor windings.* Should a synchronous motor be started in this fashion, the rotor will fail to turn, and damage may occur to the rotor and the DC power supply.

To reverse the direction of rotation of a synchronous motor, you simply interchange the connections of any two of the three stator leads. As a rule, T_1 and T_3 are interchanged. This causes the rotating magnetic field in the stator to revolve in the opposite direction. The rotor then turns in the new direction of the rotating magnetic field, which is opposite from the original.

20-5 SINGLE-PHASE MOTORS

Single-phase motors use single-phase AC as the power source. These motors are used in applications requiring less horsepower and/or lower voltages. Single-phase motors are found in the fractional horsepower to several horsepower in size. They generally operate from 120 volts AC or 240 volts AC.

Whereas all three-phase motors operate from a rotating magnetic field, the same is not true for single-phase motors. Some use a rotating magnetic field, whereas others use different methods. Following are several types of single-phase motors and their theory of operation.

Shaded-Pole Motor

As its name implies, the **shaded-pole motor** uses a **shaded pole** to cause its rotor to turn. **Figure 20-29** shows a diagram of a shaded pole. A shaded pole is constructed from a loop of copper wire or copper banding joined together to form a ring. This copper band is placed over one side of each pole piece, as shown in Figure 20-29.

When single-phase AC power is applied to the stator winding, a magnetic field is created. The copper band on the shaded pole delays the creation of the magnetic field in that portion of the stator poles. This produces a magnetic field in the shaded portion that is approximately 90° apart from the magnetic field produced in the main portion of the pole. Because the magnetic field of the shaded pole is offset by the magnetic field of the main pole, the rotor is attracted toward the shaded pole. As the rotor turns toward the shaded pole, momentum builds up. This buildup of momentum helps the rotor to continue to rotate toward the next pole piece. As the rotor continues to turn, momentum and speed build up until the rotor is turning at the rated speed.

Shaded-pole motors are considered nonreversible. However, if the stator winding is removed and reinstalled in the opposite manner, the rotor will turn in the opposite direction, but typically this is not done.

Split-Phase Motor

The **split-phase motor** is constructed with two stator windings: the **start winding** and the **run winding**. **Figure 20-30** shows the construction of a split-phase motor. Notice the start and run windings. Notice also the switch contact located in the start winding circuit. This switch contact is operated by a **centrifugal switch**, which is shown in the photograph in **Figure 20-31**. The centrifugal switch is located on the shaft of the split-phase motor. When the motor is at rest, the centrifugal switch contacts are closed. This connects the start winding in the circuit.

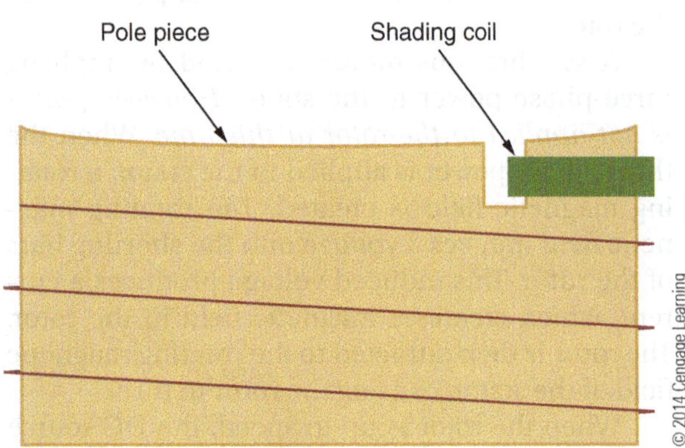

FIGURE 20-29 A shading coil.

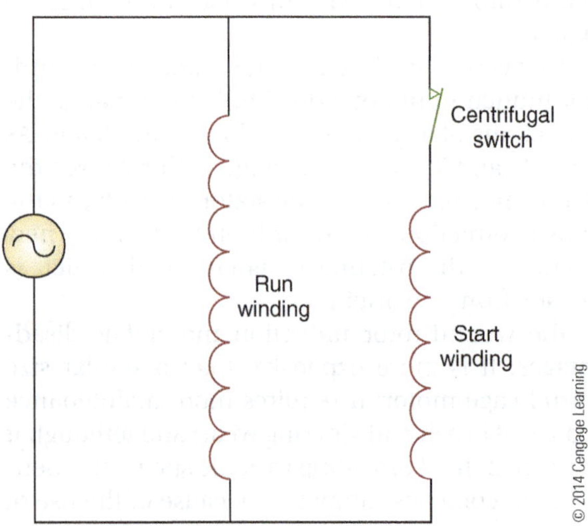

FIGURE 20-30 The split-phase motor.

FIGURE 20-31 A centrifugal switch.

The start winding is constructed of a few turns of fine gauge wire, whereas the run winding is constructed of many turns of heavy gauge wire. This gives the two windings different electrical properties. The start winding has more resistance than does the run winding. The run winding has more reactance than does the start winding. This creates a phase shift between the two windings. This phase shift produces a rotating magnetic field, which causes the rotor to turn.

The turning rotor builds up speed. When a preset speed is reached, somewhere around 75% of rated speed, the centrifugal switch contacts open, disconnecting the start winding from the circuit. The rotor continues to speed up until the rated speed of the split-phase motor is reached, operating with only the run winding connected to the power source.

To reverse the direction of rotation of a split-phase motor, the connections to the start or run windings are reversed. Reversing the connections to both the start and run windings does not reverse the direction of rotation. As a rule, the connections to the start winding are the ones that are reversed.

Capacitor-Start Motor

The **capacitor-start motor** is similar in construction to the split-phase motor. The capacitor-start motor has two stator windings: the start winding and the run winding. In addition, a capacitor and a normally closed contact operated by a centrifugal switch are connected in series with the start winding, as shown in **Figure 20-32**.

As in the split-phase motor, the start winding is constructed of a few turns of fine gauge wire, and the run winding is constructed of many turns of heavy gauge wire. The start winding has more resistance than does the run winding. The run winding

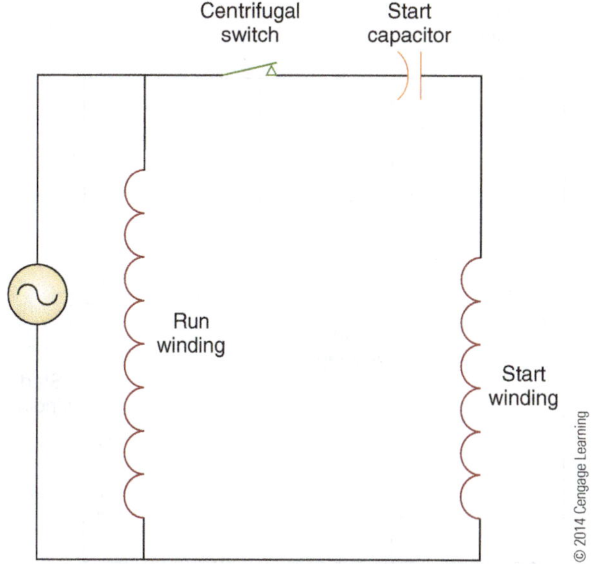

FIGURE 20-32 The capacitor-start motor.

has more reactance than does the start winding. This creates a phase shift between the two windings. This phase shift produces a rotating magnetic field, which causes the rotor to turn. The start capacitor is used to produce a higher starting torque than that found in the split-phase motor.

The turning rotor builds up speed. When a preset speed is reached, somewhere around 75% of rated speed, the centrifugal switch contacts open, disconnecting the start winding and start capacitor from the circuit. The rotor continues to speed up until the rated speed of the capacitor-start motor is reached, operating with only the run winding connected to the power source.

To reverse the direction of rotation of a capacitor-start motor, the connections to the start or run windings are reversed. Reversing the connections to both the start and run windings does not reverse the direction of rotation. As a rule, the connections to the start winding are the ones that are reversed.

Capacitor-Run Motor

The **capacitor-run motor** is similar in construction to the capacitor-start motor. However, the capacitor-run motor allows the start winding and capacitor to remain connected at all times. There is no centrifugal switch in a capacitor-run motor. **Figure 20-33** shows the construction of a capacitor-run motor.

Because of the run capacitor remaining in the circuit, the full-load speed of the capacitor-run motor is reduced. However, by adding the run capacitor to the circuit, the capacitor-run motor produces higher running torque than does the capacitor-start or split-phase motors.

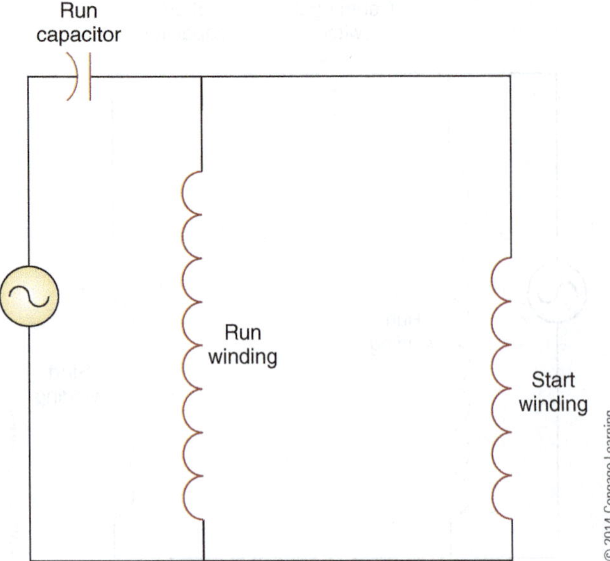

FIGURE 20-33 The capacitor-run motor.

To reverse the direction of rotation of a capacitor-run motor, the connections to the start or run windings are reversed. Reversing the connections to both the start and run windings will not reverse the direction of rotation. As a rule, the connections to the start winding are the ones that are reversed.

Capacitor-Start/Capacitor-Run Motor

One of the most widely used single-phase motors is the **capacitor-start/capacitor-run motor** shown in **Figure 20-34**. Notice the various parts of the capacitor-start/capacitor-run motor. This motor has a rotor, two capacitors (a start capacitor and a run capacitor), and a stator that consists of a start winding and a run winding. *The start capacitor is a larger value capacitor than the run capacitor.*

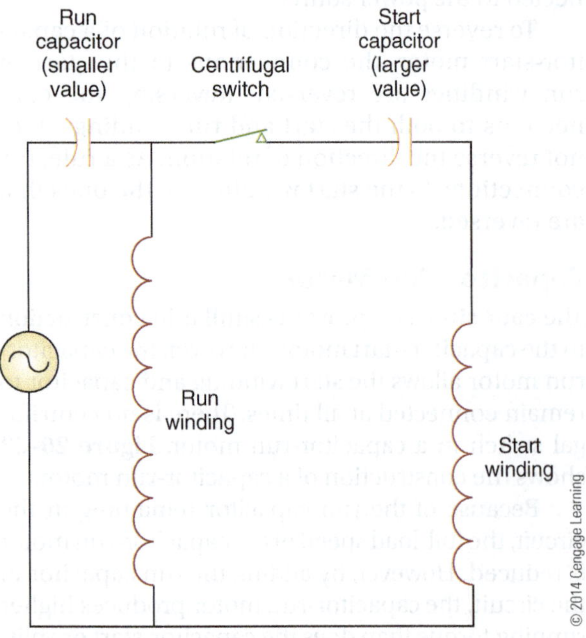

FIGURE 20-34 The capacitor-start/capacitor-run motor.

Notice that the start capacitor is connected in series with the start winding by means of a centrifugal switch. The combination of the capacitor and the start winding causes the starting current to undergo a phase shift that typically approaches 90°. Because of the high phase shift, the capacitor-start/capacitor-run motor has a high starting torque. To allow the motor to maintain a high running torque, the run capacitor remains in the circuit after the motor has attained a preset speed.

To reverse the direction of rotation of a capacitor-start/capacitor-run motor, the connections to the start or run windings are reversed. Reversing the connections to both the start and run windings does not reverse the direction of rotation. As a rule, the connections to the start winding are the ones that are reversed.

Permanent-Magnet Motors

A **permanent-magnet motor** is a motor in which the main field flux is produced by permanent magnets. An electromagnet is used for the secondary field or armature flux.

Because of the constant field flux, the standard permanent-magnet motor has many of the same characteristics of the shunt motor. Variations in design, however, can change these characteristics considerably.

These motors are frequently used for servomotors, torque motors, and industrial drive motors. They are used on machines requiring exact positioning of an object or component, where high starting and operating torque are required and where a constant torque is required. Some examples are the opening of a valve under pressure, precise positioning of dampers and three-way valves, and other specific operations in various control systems.

The stationary permanent-magnet motor consists of permanent magnets mounted on a frame with a rotating armature placed between them. Electrical energy is supplied to the rotor by way of a commutator and brushes. This is the conventional construction, and it is suitable for many uses. **Figure 20-35** shows a motor of this type.

The revolving permanent-magnet and the wound-stator types of motors have several advantages over the wound-rotor type of motor. Because the windings are on the stator, as seen in **Figure 20-36**, they can dissipate heat more rapidly. In addition, there is less stress on the windings because of the lack of centrifugal force. Another advantage is easy access to the windings. This permits easier and more frequent

FIGURE 20-35 Permanent magnet motor. Note the commutator and the permanent magnets just visible inside the motor housing.

FIGURE 20-36 A disassembled permanent magnet motor. Note the two magnets inside the motor housing.

insulation checks, monitoring of temperature, and use of static control.

The method used for commutation (reversing the current flow in the stator windings) is quite different from the methods for motors discussed previously. A specially designed commutator and slip rings are mounted inside the stator. Rolling contacts are mounted on the rotor. As the rotor revolves, one roller makes contact with a slip ring. Half of the rollers contact one slip ring for one polarity; the other half contact the other slip ring for the opposite polarity. The rollers are positioned so as to energize the stator at the correct instant and with the proper polarity to maintain a constant torque.

The development of the rotating magnet motor has led to improvements in the magnetic materials used. One of the materials that have received wide acclaim is the cobalt-rare earth magnet. These magnets are made thin in the direction of magnetization to allow a large number of poles to be installed in the rotor, where they will funnel maximum flux into the air gap. This greater flux density results in a higher torque.

This type of construction yields a more efficient motor than those previously used. The new design has approximately the same dimensions as the conventional one. However, for the same physical size, it provides 50% higher continuous rating. Its peak torque is 55% greater than that of the conventional design. Its accelerating ability is almost twice as good as that of the standard machine.

Brushless DC Motors

A brushless DC motor is a DC motor that does not contain or need brushes. The brushless DC motor is a permanent magnet motor that uses an electronic circuit to perform the commutation of the applied DC.

Figure 20-37 shows the simplified construction of a brushless DC motor. Notice that the rotor is constructed of permanent magnets, whereas the field is constructed of windings. Because the rotor is a permanent magnet, there is no need for an electrical connection to the rotor. Therefore, brushes and a commutator are no longer required.

To provide commutation, a special circuit must be employed. This can take the form of one of two types of circuits. One type of circuit is known as an optical encoder. This type of encoder uses an electronic device known as a phototransistor. Phototransistors will be studied in Chapter 23. The other type of circuit is known as a magnetic encoder. This type of circuit uses an electronic device known as a Hall-effect device.

Figure 20-38 shows a simplified drawing of a brushless DC motor using an optical encoder for commutation. Notice the shutter that blocks phototransistors Q_2 and Q_3. This means that only Q_1 is conducting, because Q_1 is the only phototransistor to be exposed to light. Because Q_1 is conducting, current flows into the base of transistor Q_4, causing Q_4 to conduct also. With Q_4 conducting, current will flow through field winding S_1. This causes the permanent magnet rotor to rotate into alignment with field winding S_1.

As the permanent magnet rotor revolves in a clockwise fashion, the shutter on the end of the motor shaft rotates as well. This causes the light to be blocked from phototransistor Q_1, and simultaneously allows the light to strike phototransistor Q_2. Because light is blocked from phototransistor Q_1, Q_1 is turned off. This causes transistor Q_4 to turn off as well, stopping any current from flowing through

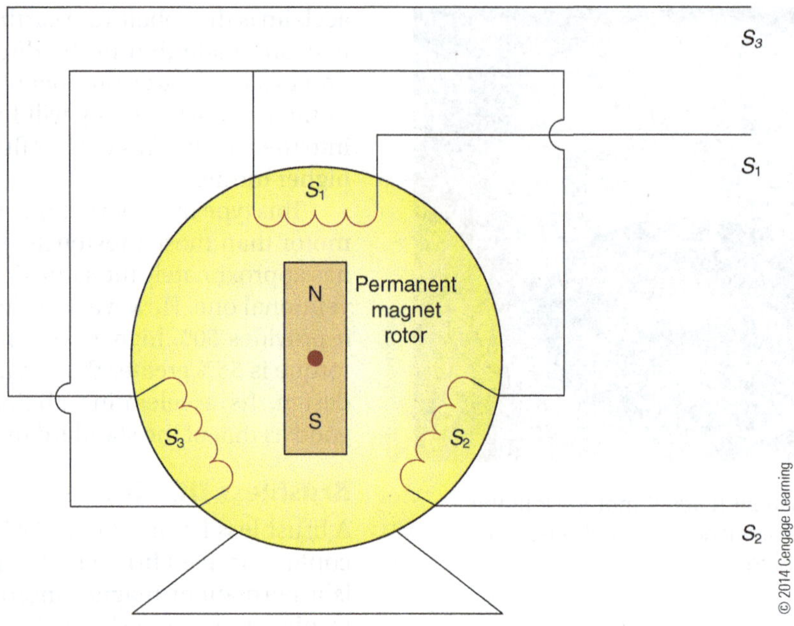

FIGURE 20-37 Simplified construction of a brushless DC motor.

field winding S_1. However, because phototransistor Q_2 is now conducting, transistor Q_5 is now turned on. With Q_5 turned on, current flows through field winding S_2. This causes the permanent magnet rotor to once again turn in a clockwise fashion and align itself with field winding S_2.

As the permanent magnet rotor aligns with field winding S_2, the shutter on the end of the motor shaft rotates as well. This causes the light to be blocked from phototransistor Q_2 and simultaneously allows the light to strike phototransistor Q_3. Because light is blocked from phototransistor Q_2, Q_2 is turned off. This causes transistor Q_5 to turn off as well, stopping any current from flowing through field winding S_2. However, since phototransistor Q_3 is now conducting, transistor Q_6 will now be turned on. With Q_6 turned on, current flows through field winding S_3. This causes the permanent magnet rotor to once again turn in a clockwise fashion and align itself with field winding S_3.

A magnetic encoder type brushless DC motor is shown in **Figure 20-39**. Notice that this type of encoder does not require a shutter on the end of the motor shaft. This encoder uses two Hall-effect devices to sense the magnetic field of the permanent magnet rotor.

Notice in Figure 20-39 that the south pole of the permanent magnet rotor is positioned under Hall-effect sensor H_1. Hall-effect sensor H_1 detects the presence of a magnetic field and supplies a voltage to the base of transistor Q_1. This voltage causes Q_1 to conduct a current, which flows through field winding S_1. The magnetic field created by this current in field winding S_1 repels the permanent magnet rotor in a clockwise direction. As the permanent magnet rotor rotates, a magnetic field is no longer sensed by Hall-effect device H_1. This turns off transistor Q_1, removing the current from field winding S_1.

As a result of the clockwise rotation of the permanent magnet rotor, the south pole is now aligned with Hall-effect device H_2. Hall-effect sensor H_2 detects the presence of a magnetic field and supplies a voltage to the base of transistor Q_2. This voltage causes Q_2 to conduct a current, which flows through field winding S_2. The magnetic field created by this current in field winding S_2 repels the permanent magnet rotor in a clockwise direction. As the permanent magnet rotor rotates, a magnetic field is no longer sensed by Hall-effect device H_2. This turns off transistor Q_2, removing the current from field winding S_2.

Hall-effect sensor H_1 detects the presence of a magnetic field from the north pole of the permanent magnet rotor and supplies a voltage to the base of transistor Q_3. Because the north pole of the permanent magnet rotor is aligned with Hall-effect device H_1, the voltage produced is seen at the second terminal of H_1. This voltage causes Q_3 to conduct a current, which flows through field winding S_3. The magnetic field created by this current in field winding S_3 repels the permanent magnet rotor in a clockwise direction. As the permanent magnet rotor rotates, a magnetic field is

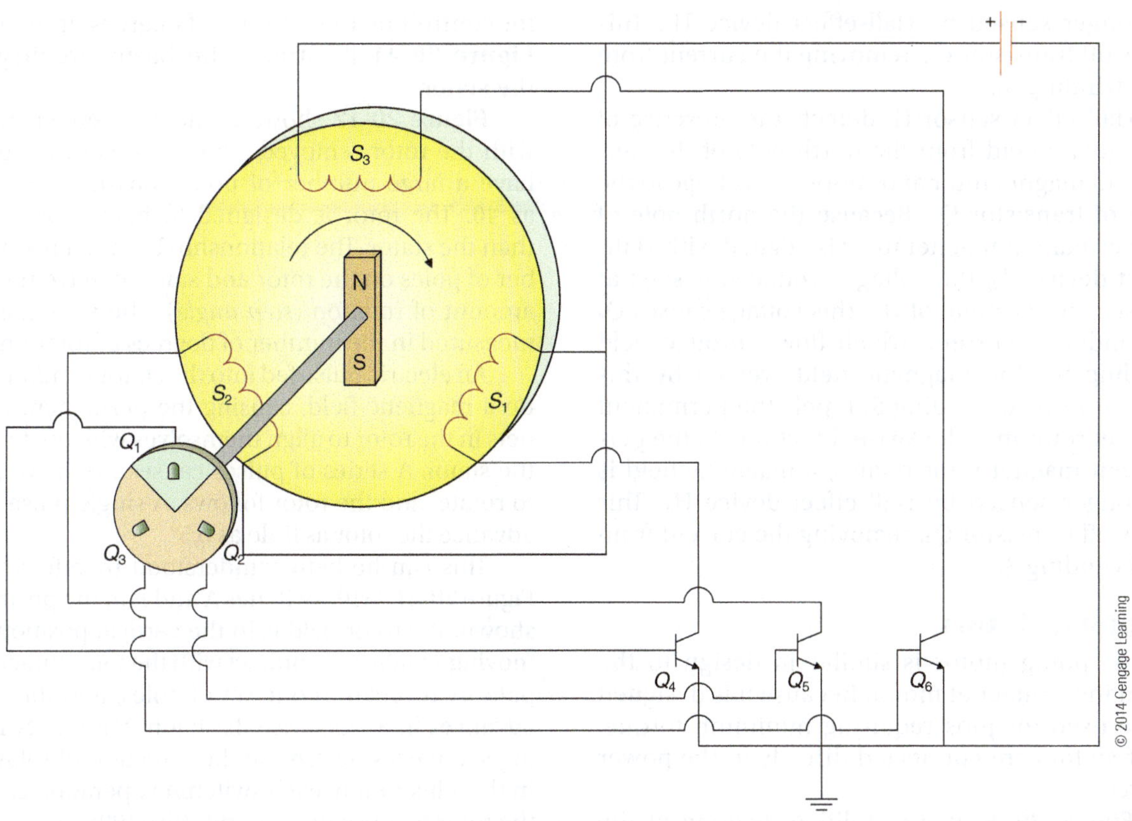

FIGURE 20-38 Brushless DC motor with optical encoder for commutation.

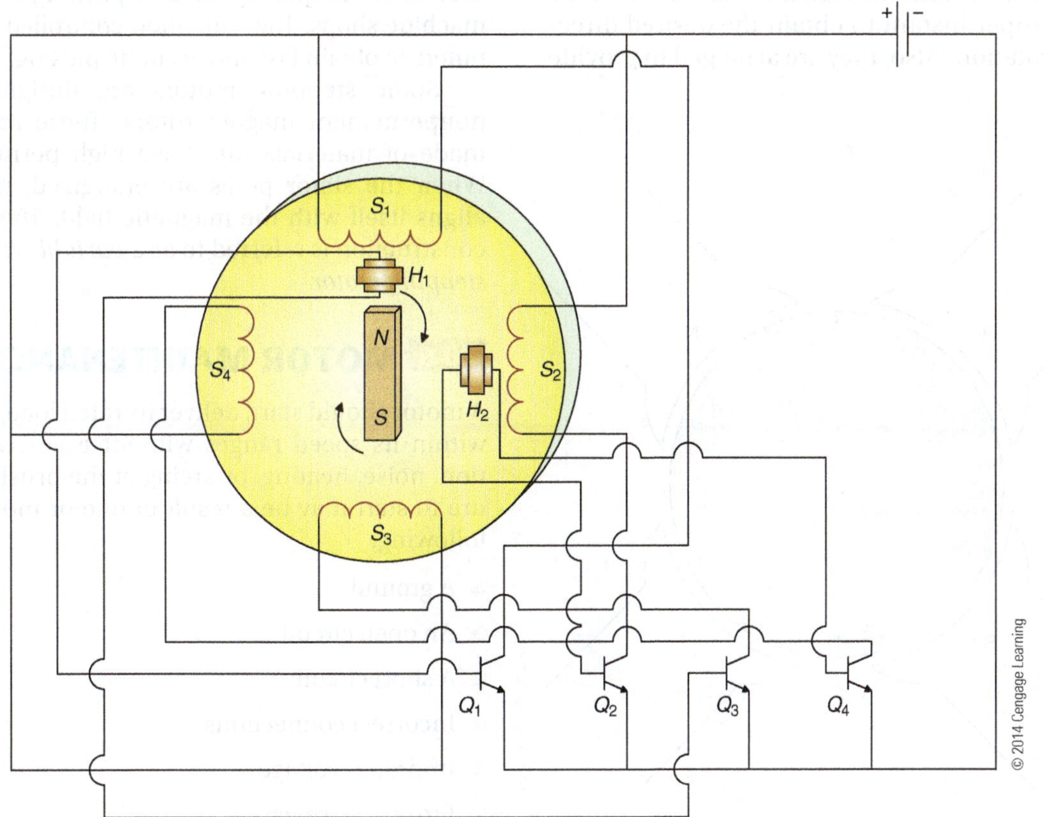

FIGURE 20-39 Brushless DC motor with magnetic encoder for commutation.

no longer sensed by Hall-effect device H_1. This turns off transistor Q_3, removing the current from field winding S_3.

Hall-effect sensor H_2 detects the presence of a magnetic field from the north pole of the permanent magnet rotor and supplies a voltage to the base of transistor Q_4. Because the north pole of the permanent magnet rotor is aligned with Hall-effect device H_2, the voltage produced is seen at the second terminal of H_2. This voltage causes Q_4 to conduct a current, which flows through field winding S_4. The magnetic field created by this current in field winding S_4 repels the permanent magnet rotor in a clockwise direction. As the permanent magnet rotor rotates, a magnetic field is no longer sensed by Hall-effect device H_4. This turns off transistor Q_4, removing the current from field winding S_4.

Stepping Motors

The stepping motor is similar in design to the permanent-magnet motor. Because it is designed to be used for jobs requiring minimum torque, their stators are connected directly to the power source.

Figure 20-40 is a simplified diagram of the magnetic polarity arrangement of the stator and rotor. The stator windings are arranged so that the direction of current through them can be reversed at the proper instant to obtain the desired direction of rotation. Also, they are arranged to provide for controlling the number of degrees of rotation. **Figure 20-41** illustrates the basic circuitry for the stator.

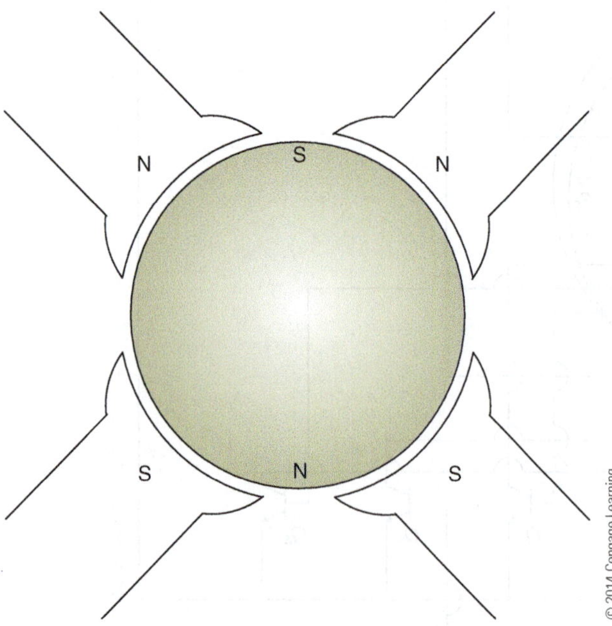

FIGURE 20-40 Stepper motor magnetic polarity arrangement of the stator and rotor.

Figure 20-42 shows an actual stepper motor with the rotor removed. The stator is designed to have a large number of poles, possibly as many as 40. The rotor is designed to have more poles than the stator. The relationship between the number of poles on the rotor and stator determines the amount of rotation (*step angle*). The step angle is measured in the number of degrees of rotation.

An electric pulse fed into the stator winding sets up a magnetic field, causing the permanent magnets in the rotor to align themselves with the field of the stator. A series of pulses causes the stator field to rotate, and the rotor follows. A single pulse may advance the rotor as little as 0.5°.

This can be better understood by referring to Figure 20-41. With switches A and B in the positions shown, the rotor field is in the vertical position. By moving switch A to connect with the top contact, the polarity reverses in one set of poles, and the rotor advances in a clockwise direction. Various switching sequences determine the direction of rotation. In this illustration, each switching operation causes the rotor to move one-quarter turn (90°).

The ability to control the rotation to less than 1° allows for many applications in industry. Stepping motors are frequently used in printing shops and machine shops. They are often controlled by computers to obtain fast and accurate pulsing.

Some stepping motors are designed with nonpermanent magnet rotors. These rotors are made of materials that have high permeability. When the stator poles are energized, the rotor aligns itself with the magnetic field. This type of construction is referred to as a *variable reluctance stepping motor.*

20-6 MOTOR MAINTENANCE

A motor should start, deliver its rated load, and run within its speed ranges without excessive vibration, noise, heating, or arcing at the brushes. Failure to start may be a result of one or more of the following:

- A ground
- An open circuit
- A short circuit
- Incorrect connections
- Improper voltage
- Frozen bearings
- An overload

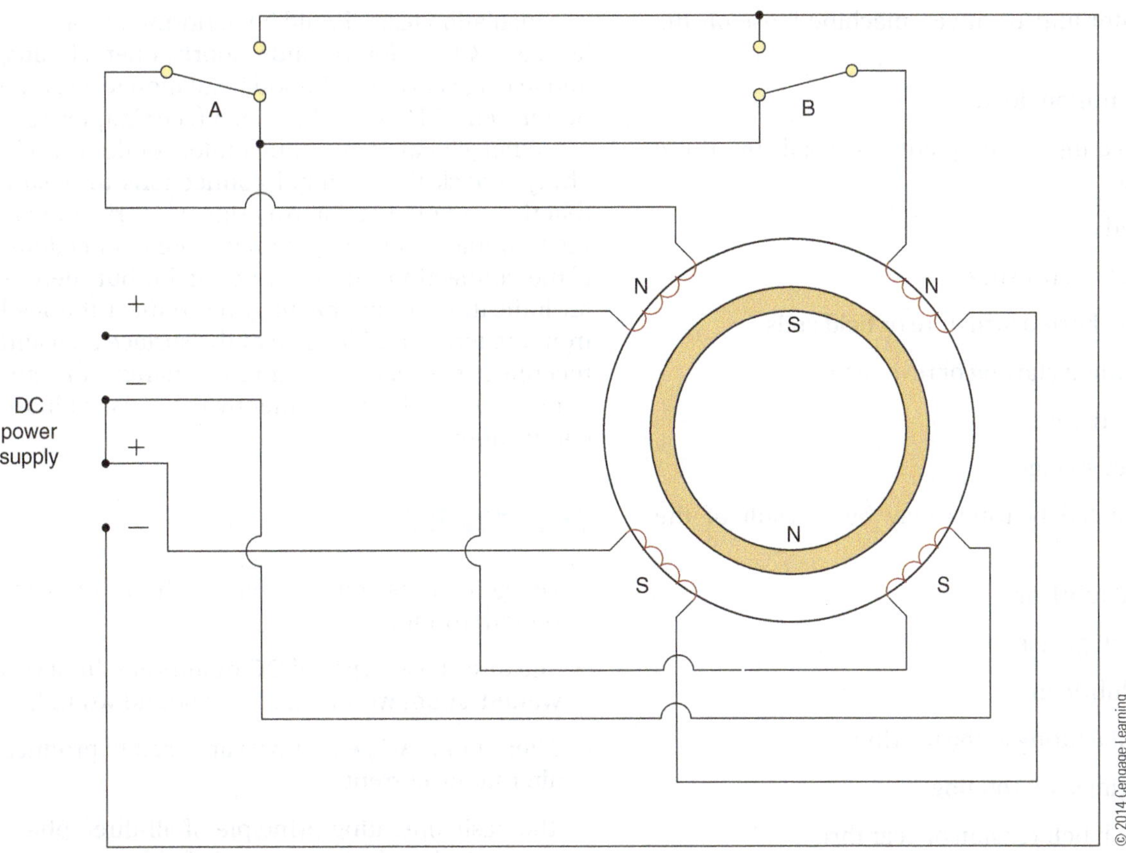

FIGURE 20-41 Simplified drawing of a stepper motor's stator winding arrangement.

FIGURE 20-42 Actual stepper motor with the rotor removed.

Arcing at the brushes may be traced to the brushes, commutator, armature, or an overload. Arcing may be the result of any of the following:

- Insufficient brush contact
- Worn or improper brushes
- Not enough spring tension
- An incorrect brush setting
- A dirty commutator
- High mica
- A rough or pitted commutator surface
- Eccentricity (commutator off center)
- An opening in the armature coil
- A short in the armature or field coils

The following may cause vibration and pounding noises:

- Worn bearings
- Loose parts
- Rotating parts hitting stationary parts
- Armature unbalance
- Improper alignment of the motor with the driven machine
- Loose coupling
- Insufficient end play

- The motor and/or driven machine loose on the base
- An intermittent load

Overheating is frequently caused by these conditions:

- Overload
- Arcing at the brushes
- A wet or shorted armature or field coils
- Too frequent starting or reversing
- Poor ventilation
- Incorrect voltage

Overheated bearings may be a result of the following:

- A lack of lubricant
- Too much grease
- A dirty lubricant
- Too tight a fitting at the bearings
- The oil rings not rotating
- Too much belt tension or gear thrust
- Insufficient end play
- A rough or bent shaft
- A shaft out of round

To ensure good operating conditions, all motors should be checked periodically. Keep a permanent record of the results, replacement of parts, and general maintenance tasks. The log should indicate the date, time of day, ambient temperature, humidity, and other pertinent data that may affect the test results.

The interior and exterior of the motor should be kept clean and dry. Depending upon the atmospheric conditions, periodically disassemble the motor and clean the interior thoroughly. All loose dirt should be vacuumed away. Clean the commutator and contacts with a nontoxic, nonabrasive cleaner. Bearings should be lubricated as needed. (Do not over-lubricate.) Replace worn bearings immediately.

Insulation tests should be performed on a regular basis. Check for grounds, shorts, open circuits, and leakage currents. A good insulation test can be performed with a megohmmeter (see Chapter 15).

When cleaning or testing motors or doing both, always check the terminal connections to ensure that they are made tight. Any sign of overheating at the terminals is an indication of poor connections. If the connections appear to be tight, but there is an indication of overheating, disconnect the lead from the terminal. Clean all of the surface area and reconnect the lead. Vibration or changes in temperature, or both, are frequently the cause of loose connections.

SUMMARY

- DC generators may be series, shunt, or compound wound.
- The three basic types of DC motors are the series wound, shunt wound, and compound wound.
- Alternators (AC generators) are used to produce alternating current.
- The basic operating principle of all three-phase motors is the rotating magnetic field.
- Single-phase AC motors are the shaded-pole, the split-phase, the capacitor-start, the capacitor-run, and the capacitor-start/capacitor-run motors.

REVIEW QUESTIONS

1. Which type of DC generator has the best speed regulation?
2. Explain what happens to a shunt-wound DC motor that is operating under load if the shunt field winding opens.
3. Explain why it is not a good idea to couple a series-wound DC motor to a load with a belt.
4. Describe the operation of a three-phase squirrel-cage motor.
5. Explain the purpose of the start capacitor in a capacitor-start motor.

CHAPTER

21

Control and Controlled Devices

Motor control concerns controlling the motor; that is, starting, stopping, braking, changing direction, varying or maintaining speed, and so on. There are many types of devices that are used to control a motor. These are called control devices. Control devices can be pushbuttons, thermostats, switches, or others. Controlled devices can be relays, motor starters, alarms, or others. This chapter explains various types of control and controlled devices in use today.

OBJECTIVES

After studying this chapter, the student should be able to

- Discuss different types of control devices.
- Identify the proper usage of different types of control devices.
- Discuss different types of controlled devices.
- Identify the proper usage of different types of controlled devices.

21-1 PUSHBUTTONS

Perhaps the most common type of control device used in motor control is the **pushbutton**. **Figure 21-1** shows two types of common pushbuttons. Pushbuttons are **manual** devices. This means that some type of human intervention is needed in order to operate them.

Pushbuttons are typically used to control a motor by performing the start, stop, and emergency stop functions. In addition, pushbuttons can be used to alter the operation of a running motor, such as causing a motor to run forward, run reverse, jog, jog forward, jog reverse, run slow, and run fast.

Pushbuttons come in many shapes and styles. However, they all perform similar functions: they *open, close,* or *open and close one or more circuits simultaneously.*

Pushbuttons may operate on a **momentary** basis. A momentary pushbutton is spring loaded and remains in its operated state for as long as the button is depressed. The contacts return to their normal state when the button is released.

Pushbuttons may also be **mechanically held**. This type of pushbutton contains a mechanical latching mechanism and remains in its operated state after being depressed. Typically, another pushbutton that is mechanically linked to the first pushbutton is depressed to release the first pushbutton to its original state. Sometimes a mechanically held pushbutton is designed to push on and push off. When this type of pushbutton is depressed, its contacts change state and remain in that condition until the pushbutton is depressed again. This returns the contacts to their original condition. Another type of mechanically held pushbutton uses a locking head design. This is typically found on emergency stop pushbuttons. When this pushbutton is depressed, the button head is mechanically locked in the depressed position. The button head returns to the normal position by either pulling it straight back or by twisting and pulling. This is a safety feature intended to keep the button and equipment in the off state when operated during an emergency shutdown.

Normally Closed and Normally Open Contacts

Probably the most common pushbutton is a momentary pushbutton that contains a **normally closed (NC)** and **normally open (NO)** set of contacts, as shown in **Figure 21-2**. A design feature of most pushbuttons allows additional contact blocks to be added to the switch body. This allows additional NO and/or NC contacts to be added, depending on the circuit requirements. In addition, there are different operator styles available. Pushbuttons typically have a large surface on which to push. Again refer to Figure 21-2. This is especially helpful when operating a pushbutton while wearing work gloves. Other pushbuttons may have large, mushroom-shaped heads, as shown in **Figure 21-3**. These are typically found on pushbuttons used for emergency stop functions. The large, mushroom-shaped head can be slapped with the palm of the hand, allowing easy operation of the pushbutton. Quick and sure operation is essential in an emergency situation.

The frequency of the operation and the circuit operating parameters determine the rating of the pushbutton to be used. Pushbuttons are rated as **standard duty** and **heavy duty**. Standard-duty-rated

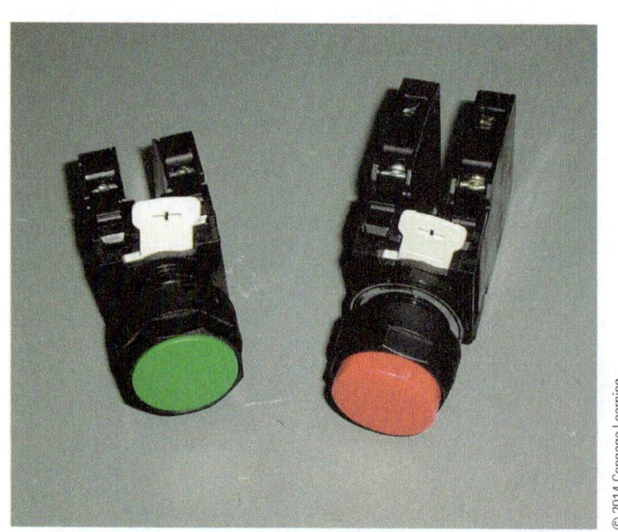

FIGURE 21-1 Two types of pushbuttons.

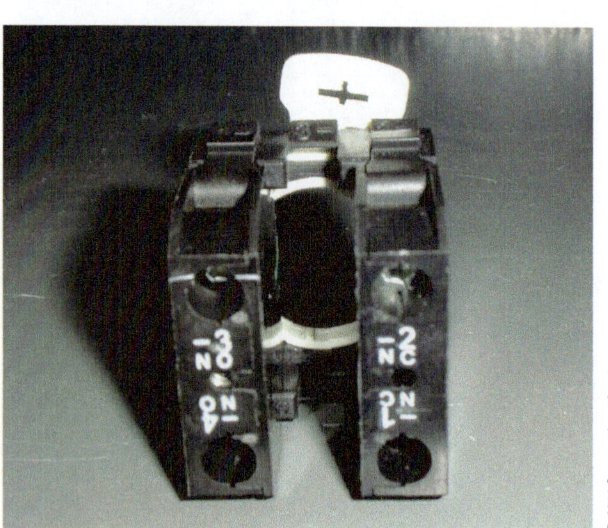

FIGURE 21-2 Normally open (NO) and normally closed (NC) contact blocks.

FIGURE 21-3 A mushroom head (E-stop) pushbutton.

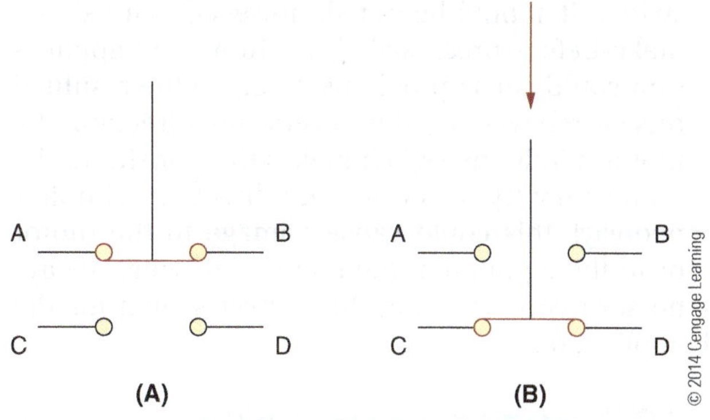

FIGURE 21-4 A break-before-make contact: (A) before the pushbutton is depressed; (B) as the pushbutton is depressed, the movable contact breaks the connection between A and B before making the connection with C and D.

switches are designed for **pilot duty**. This means that the switch operates low-voltage, low-current circuits. A standard-duty switch is also operated on a regular, but somewhat infrequent, basis. Heavy-duty-rated switches are designed to switch higher voltages and currents. These switches also are operated very frequently and are therefore subject to hard and rough usage.

Break-before-Make and Make-before-Break Contacts

In addition to NO and NC contacts, there also are **break-before-make contacts** (**Figure 21-4**) and **make-before-break contacts** (**Figure 21-5**). Figure 21-4A shows a break-before-make contact before the pushbutton is depressed. As the pushbutton is depressed, the movable contact breaks the connection between terminals A and B before making the connection with terminals C and D, as shown in Figure 21-4B. For a brief period, the circuit will be open as the movable contact makes its transition from the top to the bottom set of contacts. Usually this is not a problem because the transition occurs fairly quickly.

Figure 21-5A shows a make-before-break contact before the pushbutton is depressed. As the pushbutton is depressed, the lower movable contact makes the connection between terminals C and D before the upper movable contact breaks the connection between terminals A and B, as shown in Figure 21-5B. As the pushbutton is depressed fully, the upper movable contact breaks the connection between terminals A and B, as shown in Figure 21-5C. This type of switch is used in circuits that cannot tolerate the brief open-circuit condition created by the break-before-make

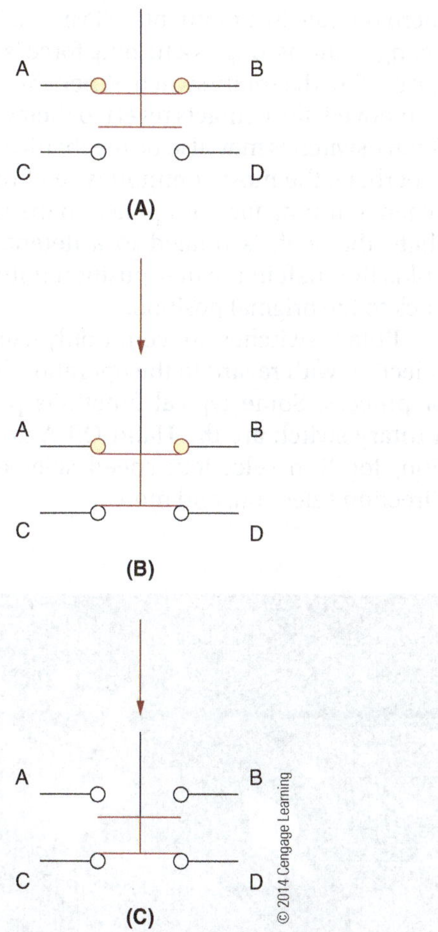

FIGURE 21-5 A make-before-break contact: (A) before the pushbutton is depressed; (B) as the pushbutton is depressed, the lower movable contact makes the connection with C and D before the upper movable contact breaks the connection between A and B; (C) as the pushbutton is depressed fully, the upper movable contact breaks the connection between A and B.

switch. It should be noted, however, that using a make-before-break switch in the wrong application could cause problems. Imagine the results if this switch were used to reverse the direction of a motor. While the switch makes the transition, the motor may try to run in both directions simultaneously! This could cause damage to the motor or to the equipment the motor is driving. Always be sure you are using the correct switch for the application.

CAUTION

21-2 ROTARY SWITCHES

Rotary switches are manually operated and require a turning motion to operate. They are sometimes known as **selector switches**. **Figure 21-6** shows different types of rotary switches. A rotary switch is used to open, close, or open and close one or more circuits simultaneously.

Rotary switches may be spring loaded, making them operate in a momentary fashion. The contacts change state as long as a turning force is held against the shaft of the rotary switch. When the turning force is removed, the contacts revert to their original state. Rotary switches may also be mechanically held. This is perhaps the most common type of rotary switch. When a turning force is applied to the rotary switch shaft, the shaft is rotated to a detent. The detent holds the shaft in the new position until it is rotated back to the original position.

Rotary switches are commonly used to make a selection with regard to the operation of a machine or process. Some typical functions performed by a rotary switch are the Hand-Off-Automatic selection, Jog-Run selection, speed selection, rotation direction selection, and more.

FIGURE 21-6 Types of rotary switches.

Normally Closed and Normally Open Contacts

One of the biggest benefits of using a rotary switch is the almost limitless arrangement of contact configurations. The most basic rotary switch consists of a shaft, the switch deck or wafer, and a set of contacts (NO or NC), as shown in **Figure 21-7**. However, it is possible to add additional switch decks to the same shaft, as shown in **Figure 21-8**, which increases the number of contact sets available. In addition, the rotary switch can be configured for 2 to 12 positions. Often the rotary switch is equipped with an adjustable stop. This allows the user to take a 12-position switch and adjust the stop so that the switch has however many positions are desired between 2 and 12.

FIGURE 21-7 A rotary switch with one NO and one NC set of contacts.

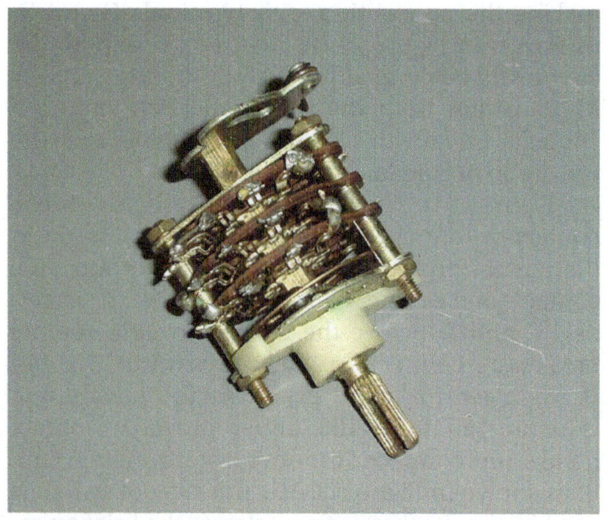

FIGURE 21-8 A multideck rotary switch.

Break-before-Make and Make-before Break Contacts

As in the pushbutton switch, the rotary switch may also have a break-before-make (**Figure 21-9**) or a make-before-break (**Figure 21-10**) contact. Figure 21-9A shows a rotary switch break-before-make contact before the shaft is turned. Notice the width of the pole (movable contact). As the shaft is turned, the movable contact breaks the connection from terminal A before making the connection with terminal B, as shown in Figure 21-9B. For a brief period, the circuit will be open as the movable contact makes the transition from terminal A to terminal B. Usually this is not a problem because the transition occurs fairly quickly.

Figure 21-10A shows a rotary switch make-before-break contact before the shaft is turned. Again, notice the width of the pole piece. The pole piece is wider in the make-before-break-type contact. As the shaft is turned, the movable contact makes the connection with terminal B before breaking the connection with terminal A, as shown in Figure 21-10B. As the shaft is rotated further, the movable contact breaks the connection with terminal A while maintaining the connection with terminal B, as shown in Figure 21-10c.

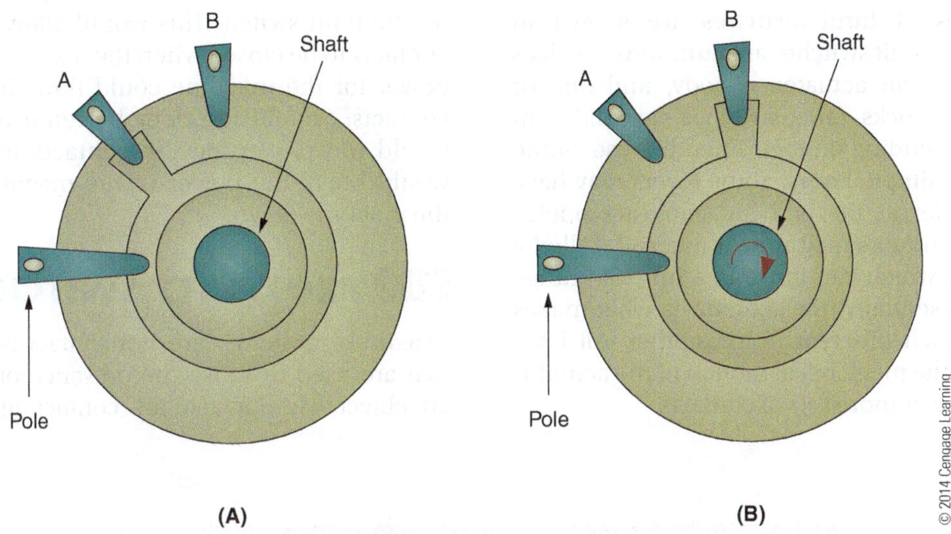

FIGURE 21-9 A break-before-make rotary switch: (A) before the shaft is turned; (B) as the shaft is turned, the movable contact breaks the connection from terminal A before making the connection with terminal B.

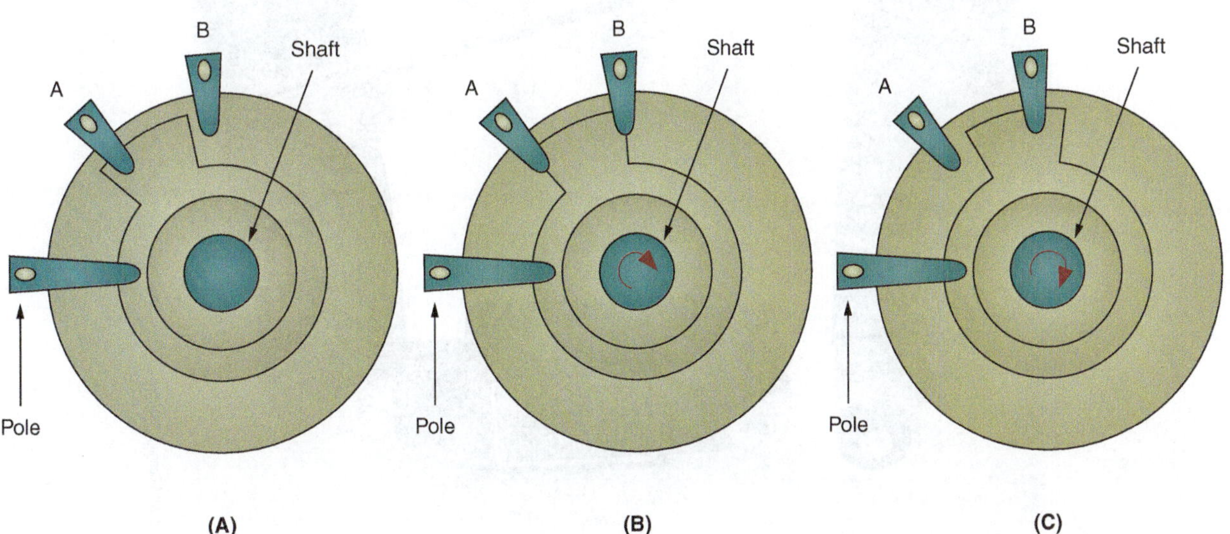

FIGURE 21-10 A make-before-break rotary switch: (A) before the shaft is turned; (B) as the shaft is turned, the movable contact makes the connection with terminal B before breaking the connection with terminal A; (C) as the shaft is rotated further, the movable contact breaks the connection with terminal A while maintaining connection with terminal B.

This type of switch is used in circuits that cannot tolerate the brief open-circuit condition created by the break-before-make switch. It should be noted, however, that using a make-before-break switch in the wrong application could cause problems. Imagine the results if this switch were used between two power sources. While the switch makes the transition, the two power sources would be connected together, possibly causing damage to one or both of the power sources. Always be sure you are using the correct switch for the application.

21-3 LIMIT SWITCHES

Several types of **limit switches** are shown in **Figure 21-11**. Limit switches are **automatic** devices that consist of an actuator, a body, and one or more contact blocks. Limit switches have different actuators, depending on their intended use. Some actuators are simple levers. Some levers may have a smaller roller at the end for smoother operation. Other actuators may appear as a rod, called a wobble stick, which can move in a full 360° range of motion. Essentially, the actuator is what makes one limit switch different from another. All limit switches use the mechanical motion of the actuator to operate one or more sets of contacts.

Limit switches are used to sense the presence or absence of an item or to verify that a component of a machine is in the correct position. A box moving along a conveyor could be sensed with a limit switch, which would be located on the conveyor in such a fashion that the box would press against the lever of the limit switch. This causes the switch contacts within the limit switch body to change state, indicating the presence of the box. Another example would be an access door to a piece of equipment. Because of the existence of dangerous voltages behind the access door, the machine must automatically shut down if the door is removed. A limit switch could be mounted behind the door so that the presence of the door depresses the lever on the limit switch. This would allow a set of NO contacts to be closed when the door is in place. The power for the machine could flow through these contacts. Should the door be removed, the lever would move, causing the contacts to open. This would break the power to the machine, shutting the machine down.

21-4 PROXIMITY SWITCHES

Proximity switches are noncontact-type switches that are used to sense the presence or absence of an object without actually contacting the object.

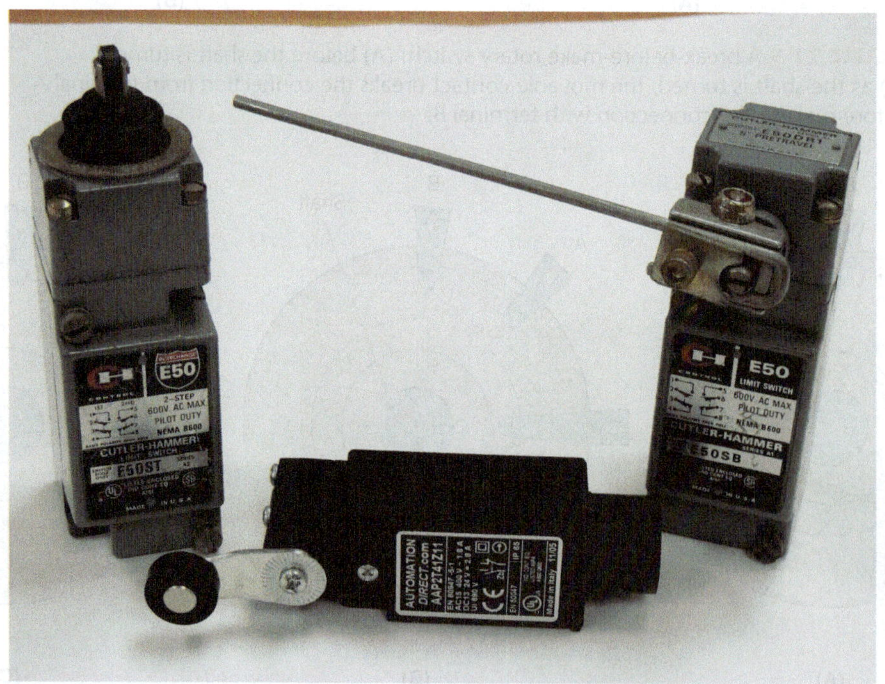

FIGURE 21-11 Limit switches.

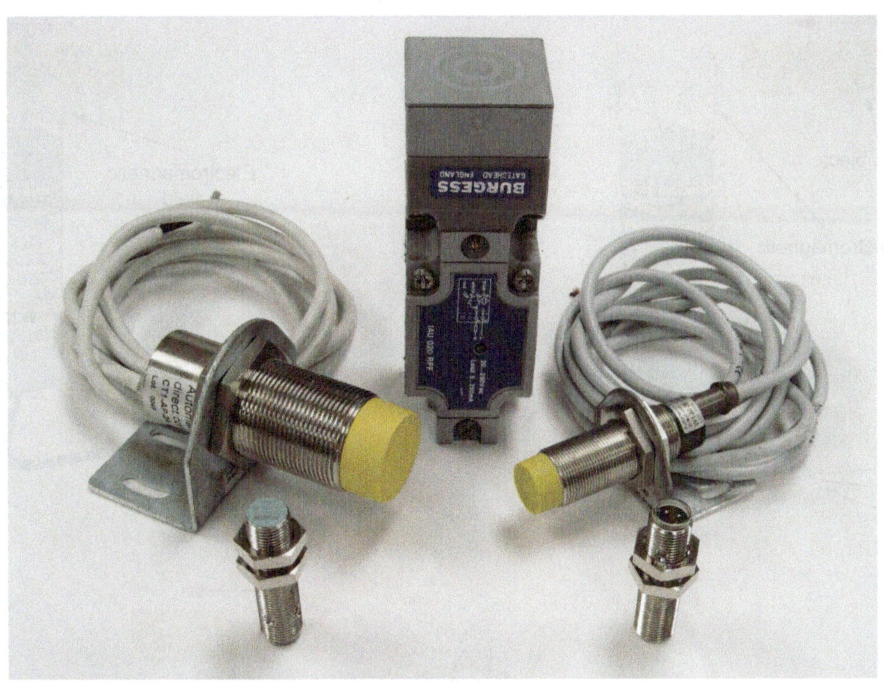

FIGURE 21-12 Proximity switches: inductive proximity switches are on the left and rear; a capacitive proximity switch is on the right.

Proximity switches combine high-speed switching with small physical size. **Figure 21-12** shows three different proximity switches. There are two basic types of proximity switches: inductive and capacitive.

Inductive Proximity Switches

Inductive proximity switches contain a coil and an oscillator circuit. The oscillator circuit produces an electromagnetic field, which surrounds the coil. The electromagnetic field extends outward from the target area of the inductive proximity switch, as shown in **Figure 21-13A**. The intensity of the electromagnetic field decreases when a metal object enters it. The weakened magnetic field is sensed by the electronic circuit contained within the inductive proximity switch. This causes the output of the inductive proximity switch to change state, as shown in **Figure 21-13B**. The field returns to its original intensity when the metal object moves out. This increase in electromagnetic field strength is sensed by the electronic circuit within the inductive proximity switch. The output of the inductive proximity switch will then revert to its original state, as shown in **Figure 21-13C**. Inductive proximity switches only detect conductive materials. This makes them ideal for applications where the presence of metal must be detected.

There are two versions of the inductive proximity switch: **inductive shielded** and **inductive unshielded**. The shielded version is either cylindrical in shape or shaped like a limit switch. It is designed to be flush mounted. The shielded type can detect ferrous and nonferrous metal. Its maximum sensing distance is 0.4 in. The unshielded version may be cylindrical in shape or shaped like a limit switch, a small block, or a flat rectangle. It must have clearance around the target end; therefore, it cannot be flush mounted. As a result, the body of the switch will protrude. This means that the sensor is subject to physical damage. In addition, because the switch protrudes and is unshielded, the maximum sensing distance is 0.7 in. The unshielded version can detect ferrous and nonferrous metal.

Capacitive Proximity Switches

Unlike an inductive proximity switch, a **capacitive proximity switch** senses conductive and nonconductive material. The capacitive proximity switch contains an oscillator circuit, which produces an **electrostatic field**. The electrostatic field extends from the target area of the capacitive proximity switch, as shown in **Figure 21-14A**. When an object enters the electrostatic field, the capacitance of the proximity switch's circuit alters. The capacitance of the circuit increases when an object is within the target area and decreases when no object is detected. This change in capacitance

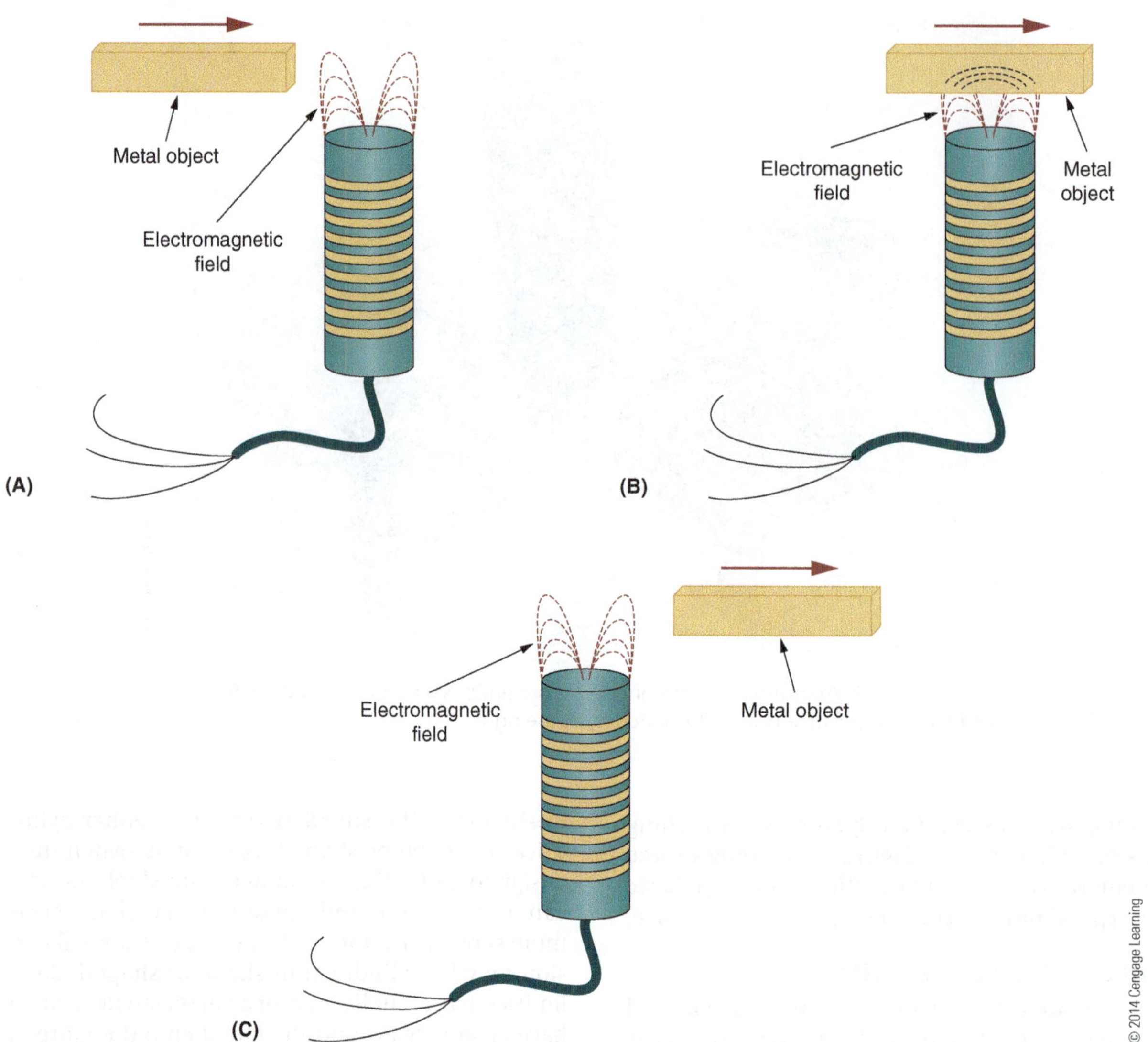

FIGURES 21-13A, B, and C An inductive proximity switch with an electromagnetic field: (A) the metal object is not sensed; (B) the metal object is sensed; and (C) the metal object is not sensed.

causes the output of the capacitive proximity switch to change state, as shown in **Figure 21-14B** and **Figure 21-14C**.

21-5 PHOTOELECTRIC SWITCHES

Photoelectric switches are noncontact-type sensors. This means that a photoelectric switch can be used to sense the presence or absence of an item without having to make physical contact with the item. **Figure 21-15** shows several types of photoelectric switches. Photoelectric switches consist of three basic components: the **transmitter**, the **receiver**, and the **switching device**.

Transmitter

The transmitter may also be called the *source*, *light source*, or *emitter*. Transmitters use incandescent light or **light-emitting diodes (LEDs)** to provide the light source. Incandescent light is identified by the visible white light that is emitted by the transmitter. Unfortunately, incandescent lights have a short life span and high failure rate. They do not operate well in areas of high vibration. Quite frequently, the repair costs of an incandescent-type transmitter exceed the cost of a new transmitter. For these reasons, incandescent transmitters are practically obsolete today.

Most transmitters today use LED technology for the light source. LEDs offer several advantages. LED transmitters are less prone to false activation because of ambient light pollution. They are very inexpensive and have a long life expectancy. Because an LED is a solid-state device, it can withstand operation in areas of high vibration. LEDs do present some problems, however. Heat is the

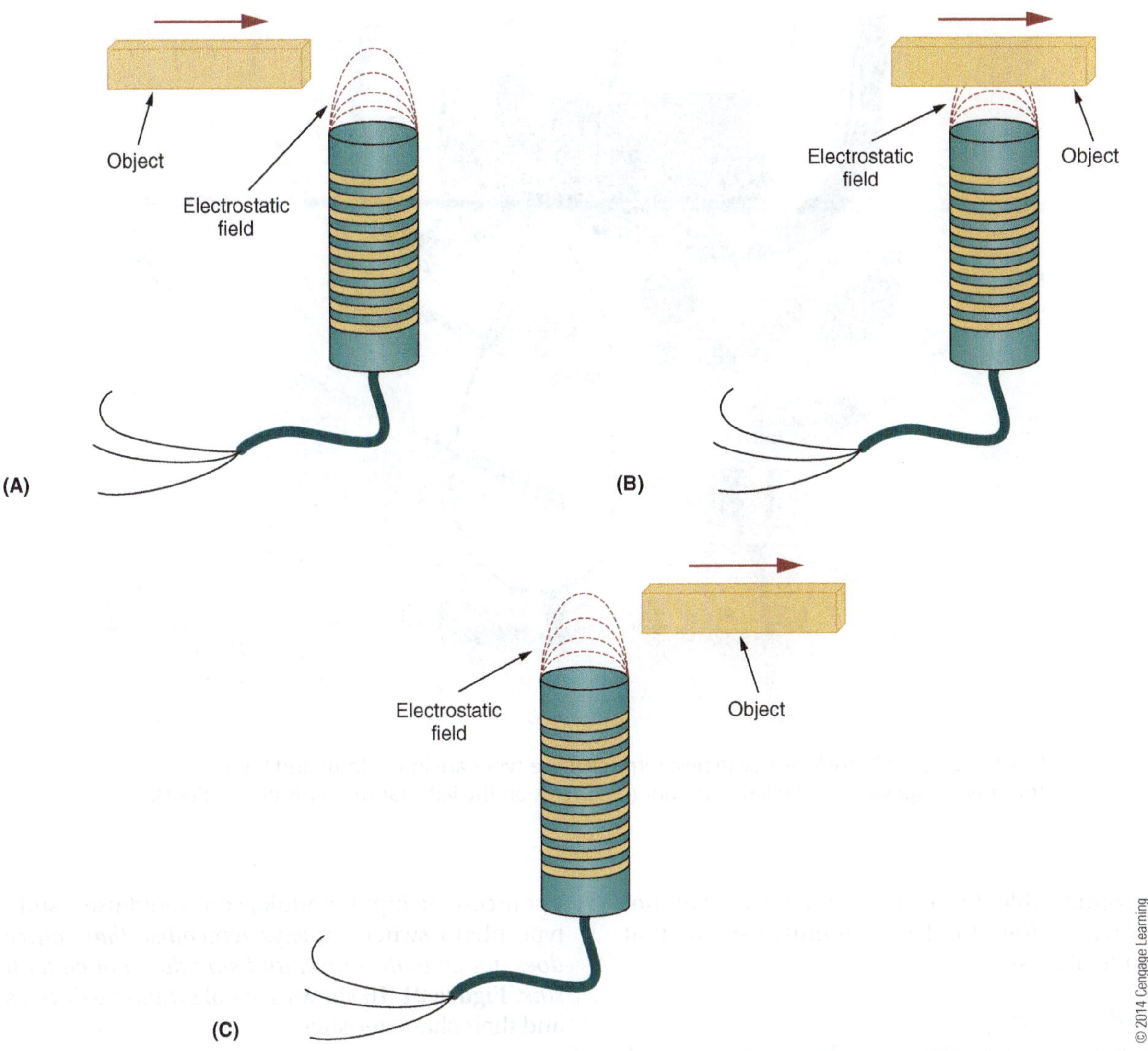

FIGURES 21-14A, B, and C A capacitive proximity switch with an electrostatic field: (A) an object is not sensed; (B) an object is sensed; and (C) an object is not sensed.

primary cause of failure of an LED. To minimize heating, the LED is rapidly cycled on and off. Because of this cycling, the transmitter must be matched to a specific receiver, which must be synchronized to the switching frequency of the transmitter in order to "see" the transmitted signal. This prevents one manufacturer's transmitter from being used with a different manufacturer's receiver.

A benefit of the switching frequency is that one transmitter-receiver pair will not typically "see" the signal from another transmitter, because the second transmitter is switching at a different frequency. This allows transmitter-receiver pairs to be located in close proximity to one another. However, if the transmitters are too close to one another, interference or cross talk may occur. The transmitters will then need to be repositioned to eliminate the interference.

There are two types of LED transmitters: one type produces visible light, whereas the other emits an **infrared light**. Infrared light cannot be seen with the naked eye. This can make troubleshooting difficult because you cannot look at the transmitter and see the light. However, there are devices available from local electronic supply stores that resemble a small card. The card is made of a material that is sensitive to infrared light. Placing the card in the infrared light beam will cause the material to glow. If you do not have one of these cards or are unable to locate one, you can use a camcorder that uses **charge-coupled device (CCD)** technology. Most digital cameras also use CCD devices. Pointing an operating camera at an operating infrared LED will allow you to "see" the infrared light in the camera's viewfinder. A drawback to the use of infrared LEDs occurs when using

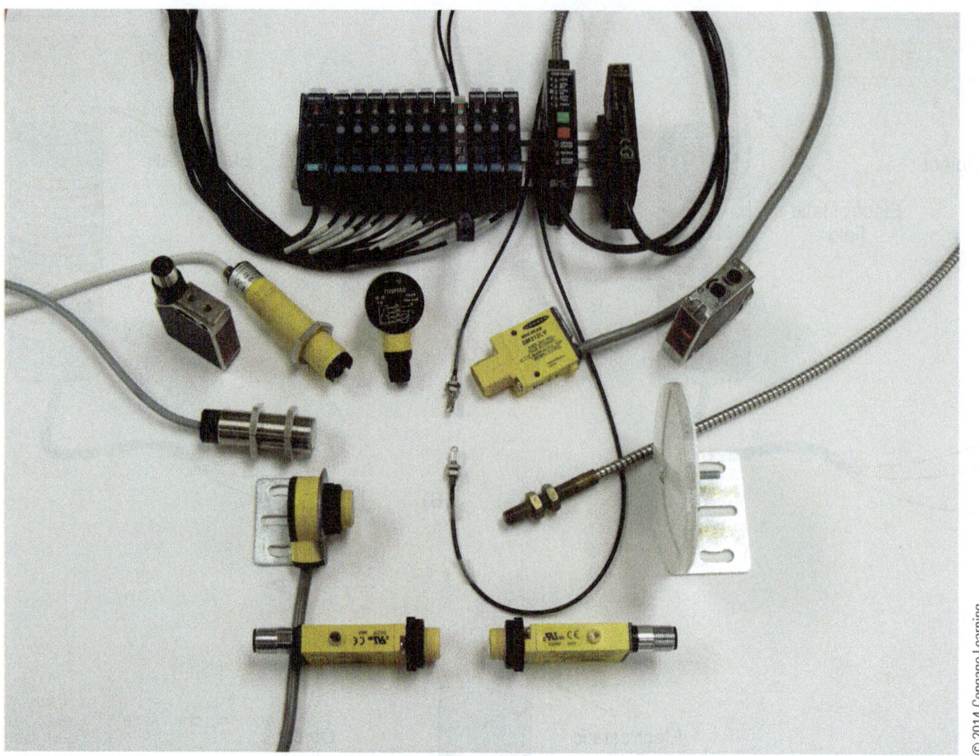

FIGURE 21-15 Photoelectric switches: retroreflective types are in the front and back; thru-beam types are on the left and right (receiver is on the left, transmitter is on the right).

fiber-optic cable. Plastic fiber-optic cable will not work with infrared LEDs—you must use one that is made of glass.

The Receiver

The receiver may also be called the *photocell, photodiode*, or *photodetector*. The receiver detects the presence or absence of the light transmitted by the light source and then activates or deactivates the switching device.

The Switching Device

Different methods are used to provide the switching function. The main difference among them is the **response time**. The response time is the difference between the time when the light is or is not detected and when a change in the state of the switching device occurs. Another difference among switching devices is in their classification as either a **source** or a **sink device**. A source device requires current to flow from the positive (+) through the output and the load to the negative (−). A sink device requires current to flow from the positive (+) through the load and the output to the negative (−). Source and sink devices must always be paired with their opposite partner. For example, when using a programmable logic controller (PLC) with a source-type input module, you would use a sink-type photo switch. *Always remember that source does not go with source, and sink does not go with sink.* **Figure 21-16** shows several switching devices and their characteristics.

Operating Methods

There are three basic methods of operating a photoelectric switch: **thru-beam**, **retroreflective**, and **diffuse**.

Thru-Beam

The thru-beam mode is also known as *transmitted beam* or *opposed pair sensing*. With this form of sensing, the transmitter is housed in one unit and the receiver is housed in another unit. The transmitter and receiver are located some distance apart. The item to be sensed is allowed to pass between the transmitter and receiver, thus breaking the light beam, as shown in **Figure 21-17**. An alternative approach is to have the object constantly block the light beam. Should the object be removed, the beam is allowed to pass from the transmitter to the receiver, changing the state of the switching device.

Thru-beam sensing may be used over distances of up to 700 ft. However, environmental conditions may limit the distance to considerably less. If the

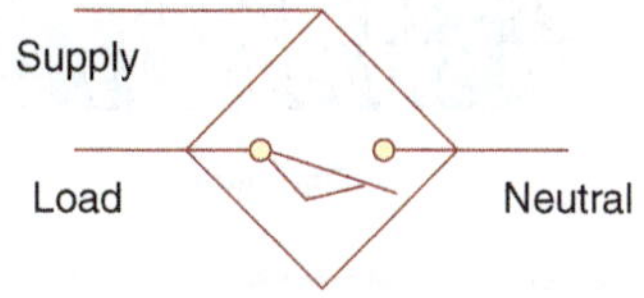

Supply

Load

Neutral

NPN output
(Sink or open-collector output)

Response time:	500 μSec
Leakage current:	10 μA
Maximum current:	100 μA
Voltage:	DC only

Notes:
- Fastest transition of DC outputs
- Ideal for PLCs
- Ideal for high-speed counting
- Parallel connected only

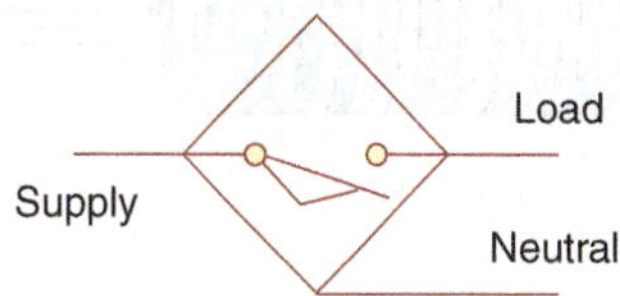

PNP output
(Source or open-emitter output)

Response time:	500 μSec
Leakage current:	10 μA
Maximum current:	100 μA
Voltage:	DC only

Notes:
- Fastest transition of DC outputs
- Ideal for PLCs
- Ideal for high-speed counting
- Series or parallel connected

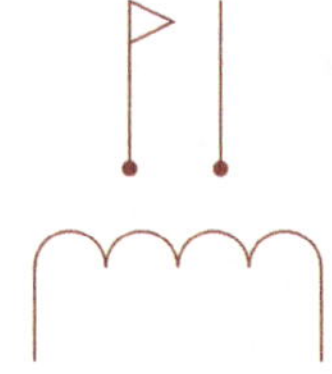

Relay output
(Electromechanical relay)

Response time:	10–20 mSec
Leakage current:	None
Maximum current:	1 A@120 V
Voltage:	AC or DC

Notes:
- Slowest transition time
- Mechanically noisy
- Shortest life expectancy
- Series or parallel connected

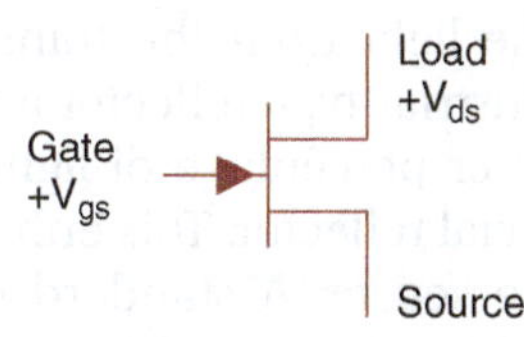

FET
(Field-effect transistor)

Response time:	1 mSec
Leakage current:	Low
Maximum current:	30 mA
Voltage:	AC or DC

Notes:
- Faster than the TRIAC
- Used with PLCs
- Used with solid-state controls
- Parallel connected only

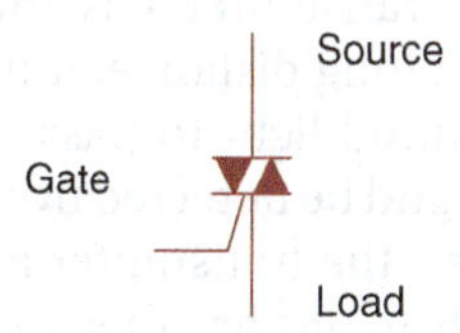

TRIAC
(Triode AC switch)

Response time:	8 mSec
Leakage current:	Some
Maximum current:	0.75 A@120 V
Voltage:	AC only

Notes:
- Faster than the relay output
- Longer life expectancy
- Series or parallel connected

FIGURE 21-16 Different switching devices and their characteristics.

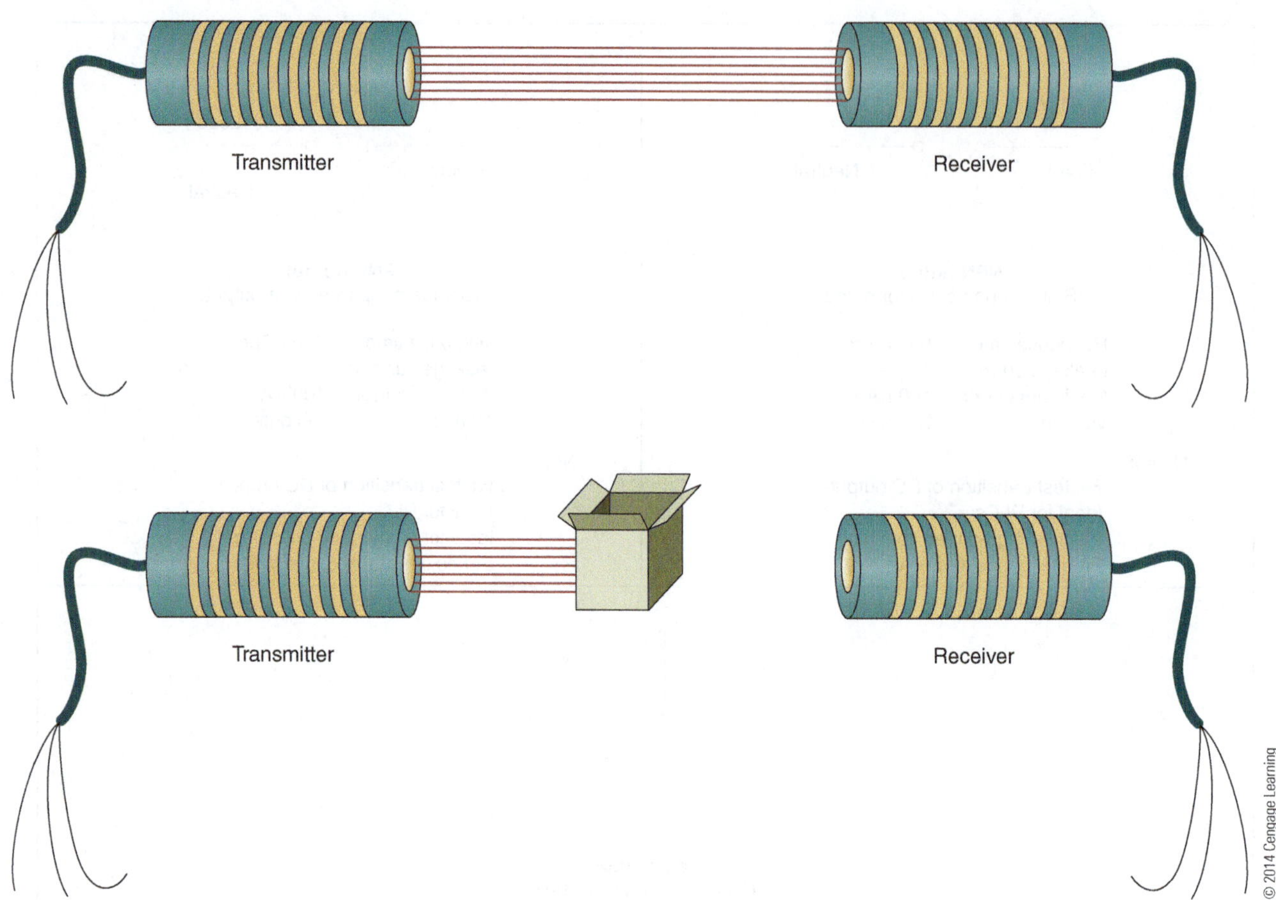

FIGURE 21-17 A thru-beam–type photoelectric switch.

surrounding air is very dirty or has a high moisture content, the distance between the transmitter and receiver will be restricted. Because the potential to cover distances of up to 700 ft exists, another problem can occur when the environment is relatively clean. Because the transmitter has the power to send the light over a long distance, it may be possible for the transmitted light to pass through the object being sensed and be detected by the receiver. This may occur when the transmitter and receiver are in close proximity to one another and the object being sensed is somewhat opaque. On the other hand, this may be a desired effect. You may want to determine whether a product has been placed within a container. The light beam is able to pass through an empty container, but it is blocked by a full container.

Retroreflective

Retroreflective devices are also called *reflective sensors* or *reflex senors.* A retroreflective sensor houses the transmitter and receiver in a single unit. In essence, a retroreflective device transmits the light to a reflector and detects the reflected light at its receiver. See **Figure 21-18**. There are two basic types of retroreflective sensors: **standard retroreflective** and **polarized retroreflective**.

Standard Retroreflective Sensor The standard retro reflective sensor uses an industrial reflector to return the light from the transmitter to the receiver. An industrial-type reflector is used because it reflects a higher percentage of light than does a typical commercial reflector. This enhances the performance of the sensor. A standard retroreflective sensor has trouble when used in an environment with high moisture content in the air because the moisture scatters the light, allowing less light to be returned to the receiver. This may result in the sensor becoming blind.

Polarized Retroreflective Sensor The polarized retroreflective sensor is also called an *antiglare sensor.* A polarized lens is placed over the receiver and allows only light beams that are oriented properly to pass through to the receiver. This is similar to the effect of polarized sunglasses. The benefit of polarized retroreflective sensing is that shiny objects can be reliably sensed.

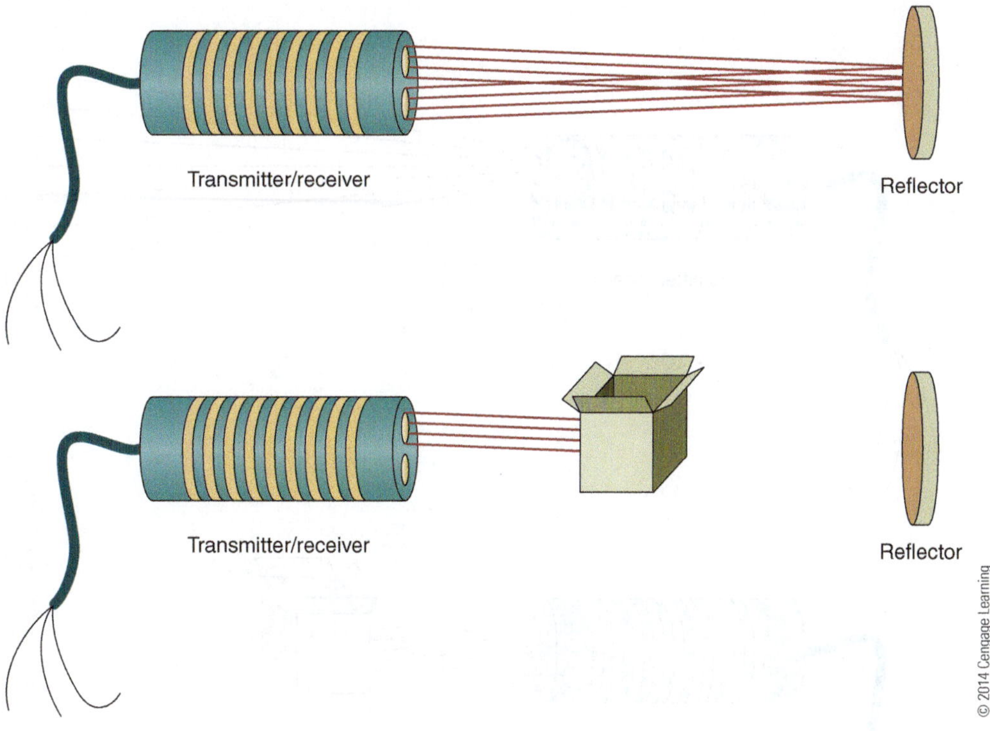

FIGURE 21-18 A retroreflective photoelectric switch.

Diffuse

Diffuse sensors use the reflectivity of the object being sensed to return the light from the transmitter to the receiver, as shown in **Figure 21-19**. There are several types of diffuse sensors: the **standard diffuse sensor**, the **long-range diffuse sensor**, the **fixed-focus diffuse sensor**, the **wide-angle diffuse sensor**, and the **background suppression sensor**.

Standard Diffuse Sensor Diffuse sensors sense the reflectivity of an object. This can create a problem when the object is no longer present, because diffuse sensors may trigger on the reflectivity of the background that was behind the object being sensed. This results in a false indication that the object is still present. To minimize the possibility of this occurring, the distance among the sensor, the object being sensed, and the background must be kept proportional. The recommended proportion is 1:3. This means that if the object to be sensed is located 1 ft away from the sensor, the background should be 3 ft away.

Another challenge for sensing objects with a diffuse sensor is the physical appearance or characteristics of the object to be detected. Texture and color determine the accuracy and range of detection, so a smooth, light-colored object is detected more readily than a rough, dark-colored one.

Long-Range Diffuse Sensor As its name implies, a long-range diffuse sensor is capable of detecting objects over a longer distance than the standard diffuse sensor. The long-range diffuse sensor works best when there is no background present or when the background is a considerable distance away from the object being detected.

Fixed-Focus Diffuse Sensor The fixed-focus diffuse sensor uses a focusing lens to create a focal point. Light is transmitted from the transmitter, reflects off the object to be detected, and is sensed by the receiver. If the object is too close or too far away, the reflected light is out of focus and the object is not sensed by the sensor.

Wide-Angle Diffuse Sensor The wide-angle diffuse sensor also uses a special lens. This lens, however, widens the field of view of the sensor, allowing the sensor to sense objects in a wider area. Although this may be desired, it can also present new problems. Insects, dust, and other airborne objects may cause false detections to occur.

Background Suppression Sensor As its name implies, a background suppression sensor senses an object but ignores the background. This type of sensor works best when sensing dark objects against a shiny background. Although background suppression offers improved sensing reliability, it reduces sensing range. Typical background suppression sensors are only effective when sensing objects that are located between 3 in. and 18 in. from the sensor.

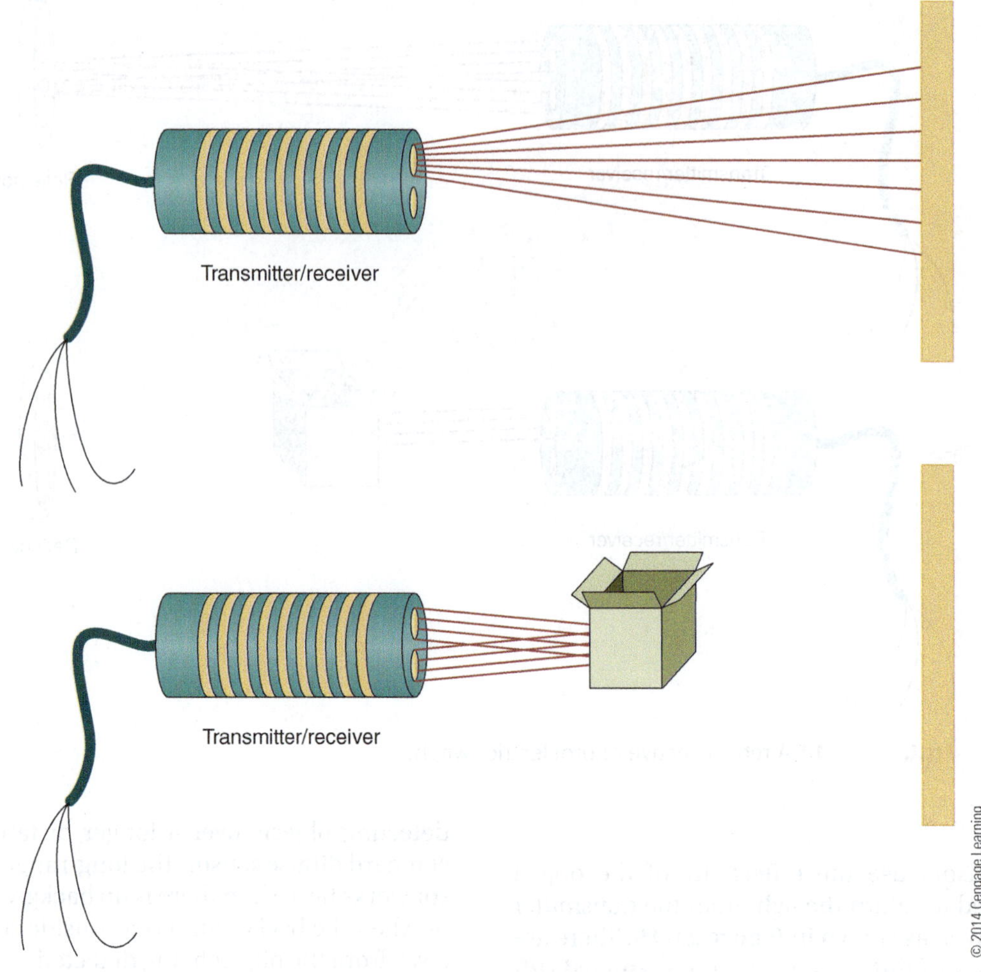

FIGURE 21-19 A diffuse sensor.

Table 21-1 shows a comparison of different types of photoelectric switches and their advantages and disadvantages.

21-6 RELAYS

A relay is an **electromechanical switch. Figure 21-20** shows several types of relays. The construction of a typical relay is shown in **Figure 21-21**. Notice that a relay contains a coil of wire wound around an iron core. Attached to the frame of the relay is a set of contacts. One contact is spring loaded and movable (the pole), and the other contact is fixed in place. The contacts are electrically isolated from the coil.

Relays are manufactured to operate from a DC or an AC source. This is noted on the relay. It is important to use the correct relay for the appropriate voltage source. When a relay is energized, a magnetic field is created around the coil of the relay. The magnetic field attracts the movable contact, causing the contact to move. When the contact moves, the circuit between the movable contact and the fixed contact is closed. Essentially, this is an electrically operated switch. More precisely, it is an electromagnetically operated switch. **Figure 21-22** shows the relay in Figure 21-21 in its energized state. When the energizing current is removed from the relay, the spring causes the movable contact to move, opening the circuit with the fixed contact.

Relays are used in different circuits to perform different functions. Perhaps the most widely used function of a relay is allowing one type or magnitude of voltage to be switched with a different type or magnitude of voltage. For example, a relay rated at 12 volts DC may be used to switch 240 volts AC. Recall that the contacts of a relay are electrically isolated from the coil. This allows 12 volts DC to be applied to the coil without interfering with the voltage switched by the contacts (240 volts AC), as shown in **Figure 21-23**.

Table 21-1

TYPES OF PHOTOELECTRIC SWITCHES AND THEIR BENEFITS AND WEAKNESSES

		Retroreflective		Diffuse				
	Thru-beam	*Standard*	*Polarized*	*Standard*	*Long Range*	*Fixed Focus*	*Wide Angle*	*Background Suppression*
Benefits	Senses over great distances	Costs less than thru-beam	Senses shiny objects reliably	Senses shiny objects reliably	Can sense over long distances	Senses small objects in precise location	Good sensor to detect absence of object	Can sense various colors and textures at dif ferent distances
	Senses through most dirt and moisture	Aligns easily	Aligns easily	Works best without a background	Can sense multiple-colored objects at close range	Aligns easily	Senses small objects easily	Does not sense close items
	Can sense through some objects	Works well on vibrating equipment		Works best with reflective objects and dark backgrounds	Senses shiny objects easily		Ignores most backgrounds	Does not falsely sense shiny backgrounds
	Most reliable	Senses large items on conveyors easily			Works best without a background		Immune to vibration	
Weaknesses	Unwanted sensing through objects	Cannot sense over long distances	Requires special reflectors	Sensing distance is affected by surface character istics of object	Can falsely sense surrounding objects and backgrounds at great distances	Can sense various colors and textures at different distances	Can sense various colors and textures at different distances	Limited sensing range
	Expensive to buy and install	Difficulty in sensing shiny objects	Slightly less sensing distance than the standard retroreflective	Can falsely sense surrounding objects and background		Short sensing range	Senses everything within its sensing area	Complicated installation
	Difficult to align at great distances	Does not work well in moist or dirty environments		Shiny backgrounds may cause false sensing		Sensing distance is not adjustable	Does not work in moist or dirty environments	
		Reflector quality and condition af fect performance		Shorter sensing distance than retroreflective		False sensing from vibration		

FIGURE 21-20 Various types of relays.

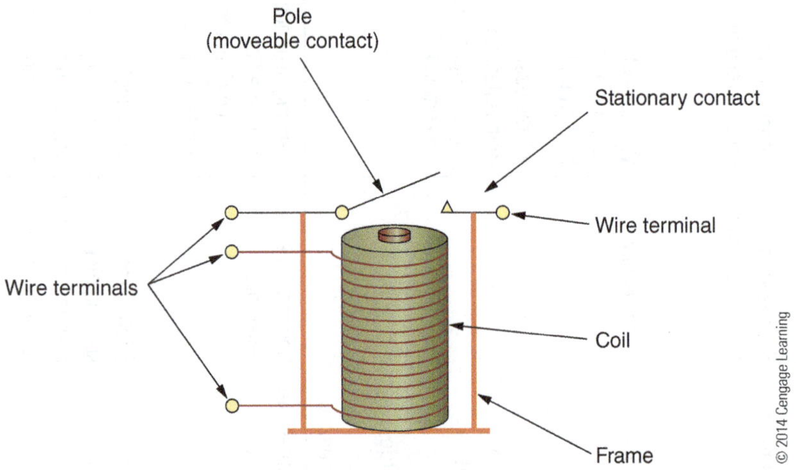

FIGURE 21-21 A de-energized relay construction.

Another common use for relays is extending the number of contacts for a device. For example, a pushbutton switch may only have one NO contact, but the circuit operation requires four circuits to be closed when the pushbutton is pressed. A relay will solve the dilemma. The single NO contact from the pushbutton is used to energize a relay that has four sets of NO contacts. When the pushbutton is pressed, the relay energizes, closing all four contacts simultaneously, as shown in **Figure 21-24**.

There are many varieties of relays. There are multipole relays, which have two or more sets of contacts: NO, NC, or a combination of both. In addition, the contacts may be **single-throw** or **double-throw**.

Time-delay relays are another example of a different variety of relay. There are **on-delay** time-delay relays and **off-delay** time-delay relays. On-delay relays provide an adjustable time delay between the time the relay is energized and when the contacts change state. Off-delay relay contacts change state immediately when the relay is energized. However, there is an adjustable time delay between the time the relay is de-energized and when the contacts change to their initial state.

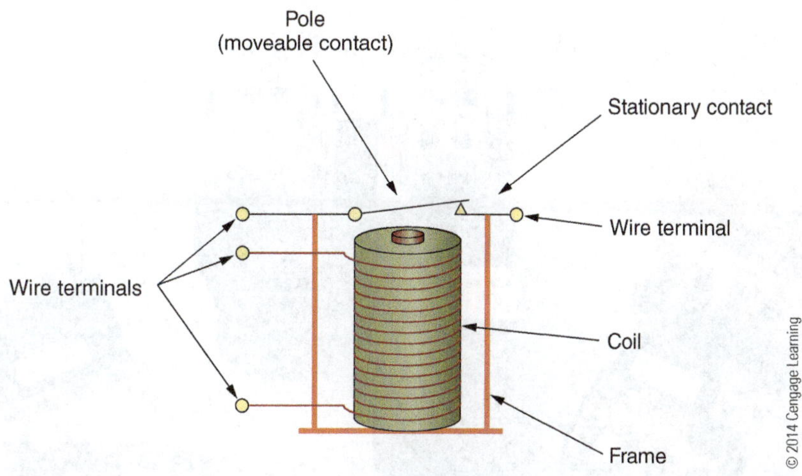

FIGURE 21-22 An energized relay construction.

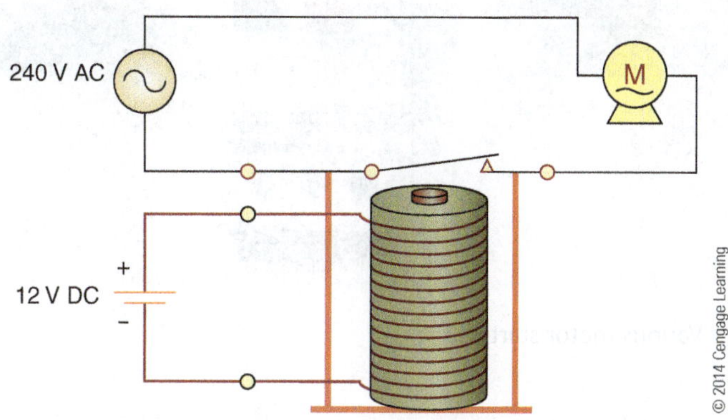

FIGURE 21-23 Low-voltage control for high-voltage switching.

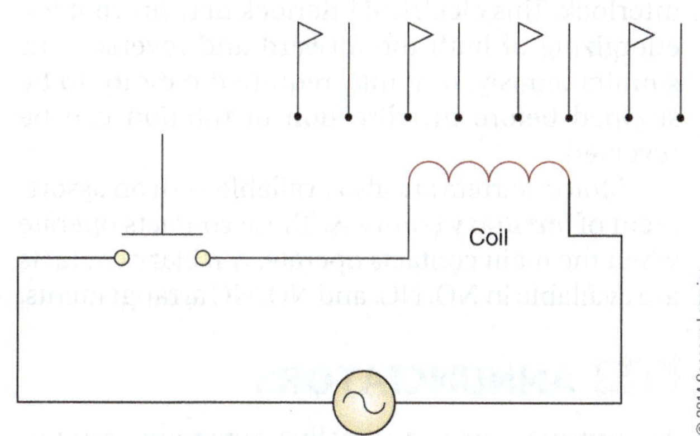

FIGURE 21-24 A four-pole relay.

Relays may also be of the **latching** type. These relays consist of two separate coils and a mechanical linkage. One relay coil is called the latch, or **set, coil** and the other is called the unlatch, or **reset, coil.** When the latch coil is energized, the contacts change state. The contacts remain in this condition even after the relay is de-energized. This is a result of the mechanical linkage. To change the state of the contacts back to their original condition, the unlatch coil is energized. The contacts remain in their original condition after the unlatch coil is de-energized. Again, the mechanical linkage maintains the contacts in their last state.

Some relays may have heavy-duty contacts. These contacts are physically larger and therefore can handle higher currents. A relay with heavy-duty contacts is called a **contactor**.

21-7 MOTOR STARTERS

A motor starter is an electromechanical switch similar to a relay. A motor starter also contains heavy-duty contacts similar to a contactor. *The difference between a motor starter and a relay or contactor is that a motor starter also contains overload protection for the motor.* **Figure 21-25** shows different types of motor starters.

Motor starters are available for operation from AC or DC voltage. Different coil voltage ratings are

FIGURE 21-25 Various motor starters.

available to match the motor starter to the control circuit voltage. In addition, motor starters are available in different sizes as determined by the motor that they are controlling. **Table 21-2** shows an example of a motor starter selection chart.

There are essentially two different types of motor starters: the *standard motor starter* and the **reversing motor starter**. The standard motor starter allows a control voltage to be used to energize the motor starter and subsequently cause the motor to run. Removing the control voltage from the motor starter causes the motor to stop. The motor is allowed to run in one direction only. The reversing motor starter, as its name implies, allows the direction of rotation of a motor to be changed. This motor starter consists of two coils: a forward coil and a reverse coil. When control voltage is applied to the forward coil, the forward set of contacts closes, causing the motor to start and run in the forward direction. When the control voltage is applied to the reverse coil, the reverse set of contacts closes, causing the motor to start and run in the reverse direction. Typically, reversing motor starters contain some type of mechanical interlock, which prevents both the forward and reverse sets of contacts from closing simultaneously. In addition, and as a further precaution, most forward and reverse control circuits use an electrical interlock. This electrical interlock may prevent the energizing of both the forward and reverse coils simultaneously, or it may require the motor to be stopped before the direction of rotation can be reversed.

Motor starters are also available with an assortment of **auxiliary contacts**. These contacts operate when the main contacts operate. Auxiliary contacts are available in NO, NC, and NO/NC arrangements.

21-8 ANNUNCIATORS

An **annunciator** is a signaling apparatus. Annunciators may be audible or visible. Examples of an *audible* annunciator are a *bell*, a *klaxon*, a *chime*, a *horn*, a *loudspeaker*, a *siren*, and so on. Examples of a *visible* annunciator are an *indicator light*, a *strobe light*, a *rotating beacon*, and so forth. Annunciators are often used as warning devices to alert personnel of an event or change in status. Annunciators may also be used to indicate the current condition of a machine or process.

Table 21-2

MOTOR STARTER SIZE CHART

NEMA Size	Continuous Current Rating	Motor Voltage	Maximum Horsepower	Coil Voltage
00	9 A	200 V	1½	208 V
		230 V	1½	240 V
		460 V	2	480 V
		575 V	2	600 V
0	18 A	200 V	3	208 V
		230 V	3	240 V
		460 V	5	480 V
		575 V	5	600 V
1	27 A	200 V	7½	208 V
		230 V	7½	240 V
		460 V	10	480 V
		575 V	10	600 V
2	45 A	200 V	10	208 V
		230 V	15	240 V
		460 V	25	480 V
		575 V	25	600 V
3	90 A	200 V	25	208 V
		230 V	30	240 V
		460 V	50	480 V
		575 V	50	600 V
4	135 A	200 V	40	208 V
		230 V	50	240 V
		460 V	100	480 V
		575 V	100	600 V
5	270 A	200 V	75	208 V
		230 V	100	240 V
		460 V	200	480 V
		575 V	200	600 V
6	540 A	200 V	150	208 V
		230 V	200	240 V
		460 V	400	480 V
		575 V	400	600 V
7	810 A	200 V	—	208 V
		230 V	300	240 V
		460 V	600	480 V
		575 V	600	600 V

A, amperes; NEMA, National Electrical Manufacturers Association; V, volts.

SUMMARY

- The most common type of control device used in motor control is the pushbutton. Pushbuttons are manual devices that are typically used to control a motor by performing start, stop, and emergency stop functions.
- Pushbuttons open, close, or open and close one or more circuits simultaneously.
- The most common pushbutton is a momentary pushbutton that contains a normally closed and normally open set of contacts.
- Pushbuttons are rated as standard duty and heavy duty. Standard-duty-rated switches are designed for pilot duty. This means that the switch operates low-voltage, low-current circuits.
- There also are break-before-make and make-before-break contacts.
- Rotary switches are operated manually and require a turning motion to operate.
- Proximity switches are noncontact-type switches. They are used to sense the presence or absence of an object without actually contacting the object.
- Photoelectric switches are noncontact-type sensors. They are used to sense the presence or absence of an item without having to make physical contact with the item.
- There are three methods of operating a photoelectric switch: thru-beam, retroreflective, and diffuse.
- A relay is an electromechanical switch. It contains a coil of wire wound around an iron core.
- A motor starter is an electromechanical switch similar to a relay. The difference between them is that a motor starter also contains overload protection for the motor.
- An annunciator is a signaling apparatus. It may be audible or visible. Annunciators are often used to alert personnel of an event or a change in status but may also be used to indicate the current condition of a machine or process.

REVIEW QUESTIONS

1. Name three types of control devices.
2. Describe the difference between a break-before-make contact and a make-before-break contact.
3. What is the difference in operation between a thru-beam and a retroreflective photoelectric switch?
4. What are the differences among a relay, a contactor, and a motor starter?
5. List three audible and three visible annunciators.

CHAPTER

22

Motor Control Circuits

The design possibilities to develop circuits that start, control the speed, reverse, brake, and stop motors are endless. This chapter presents you with some basic designs of motor control circuits. These are by no means the only circuits that perform these functions. Likewise, there are other functions of motor control that are not covered in this chapter. Each of the following sections presents one or more circuits and a brief description of the circuit operation. Remember, there are other possible designs that accomplish the same task. The circuits presented here are perhaps the most common.

OBJECTIVES

After studying this chapter, the student should be able to

- Describe different methods of starting, controlling speed, reversing, braking, and stopping motors.
- Develop control circuits to perform the functions of starting, controlling speed, reversing, braking, and stopping motors.

22-1 TWO-WIRE CONTROLS

A **two-wire control** is a method of motor control that uses manual devices to start and stop the motor. As a result, only two wires are needed to connect the control device into the circuit. Because the control device is mechanically held in the on position, the motor will restart automatically when power is restored after a power failure.

SAFETY

Figure 22-1A shows a two-wire control using a snap-action toggle switch. Closing the toggle switch causes the motor to run; opening the toggle switch causes the motor to stop. **Figure 22-1B** shows a two-wire control using two pushbuttons that are mechanically linked. Pressing the start pushbutton causes the motor to run. The start button is mechanically held in the depressed position, while the stop pushbutton mechanically moves to the extended position. Depressing the stop pushbutton stops the motor, causing the start pushbutton to move to the extended position. The stop pushbutton is mechanically held in the depressed position.

22-2 THREE-WIRE CONTROLS

As the name implies, **three-wire controls** need three wires to connect the control device into the circuit. Refer to **Figure 22-2A**. Three-wire controls also use some type of electromechanical switch to provide a latching, seal-in, memory, or holding function. This function is what causes the motor to continue to run even after the start pushbutton is released. Following is the step-by-step operation of this circuit:

1. The circuit is shown in its stopped mode.
2. Pressing the start pushbutton energizes the motor starter coil, MS.
   a. This causes the holding contacts, MS-1, to close, maintaining a current path to the MS coil.
   b. This also causes the main motor contacts, MS-A, MS-B, and MS-C, to close.
3. The motor runs.
4. Releasing the start pushbutton has no effect on the circuit because MS-1 maintains current flow to MS.

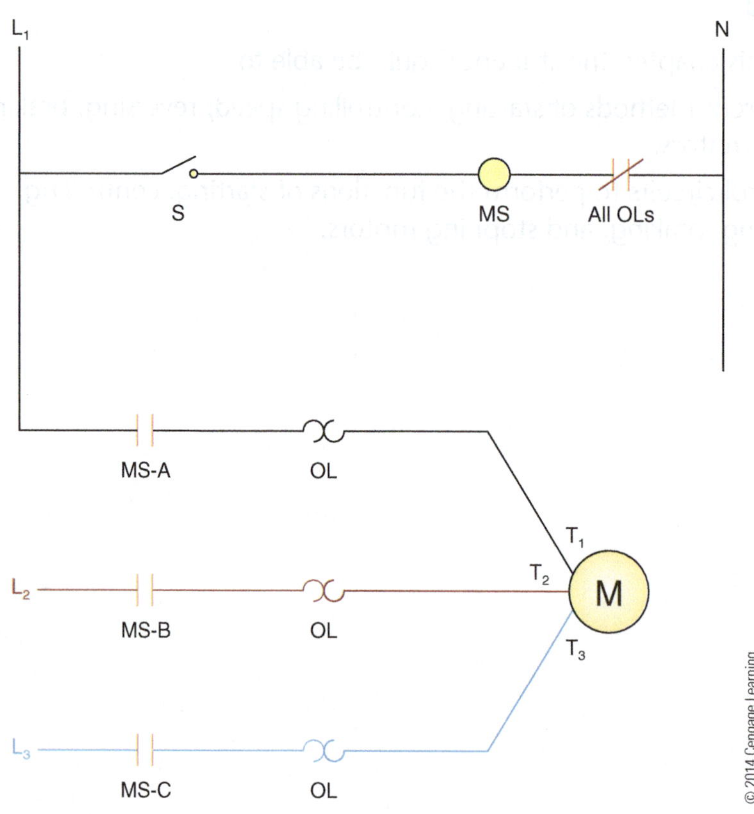

FIGURE 22-1A A two-wire control.

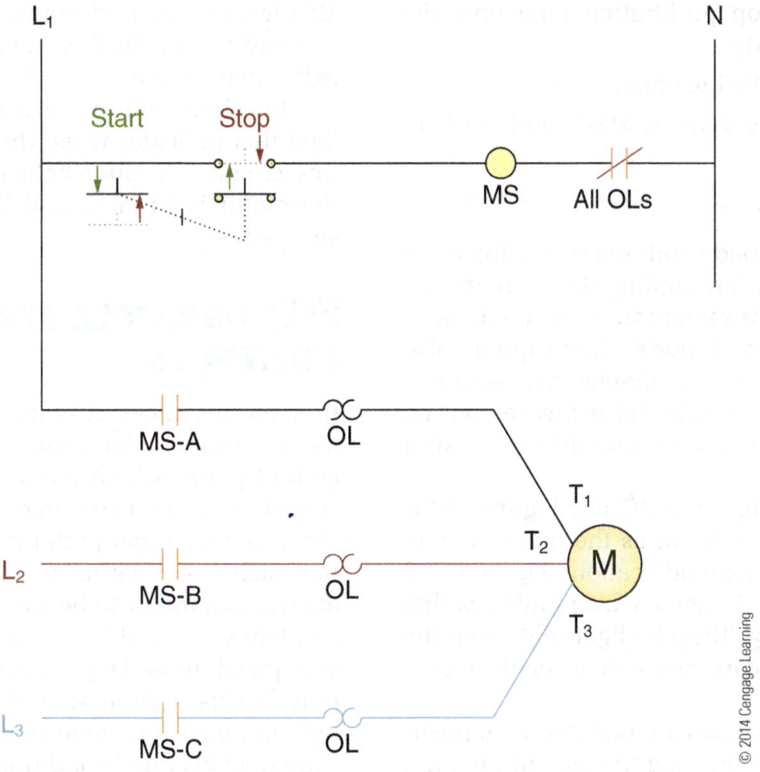

FIGURE 22-1B A two-wire control with two mechanically linked pushbuttons.

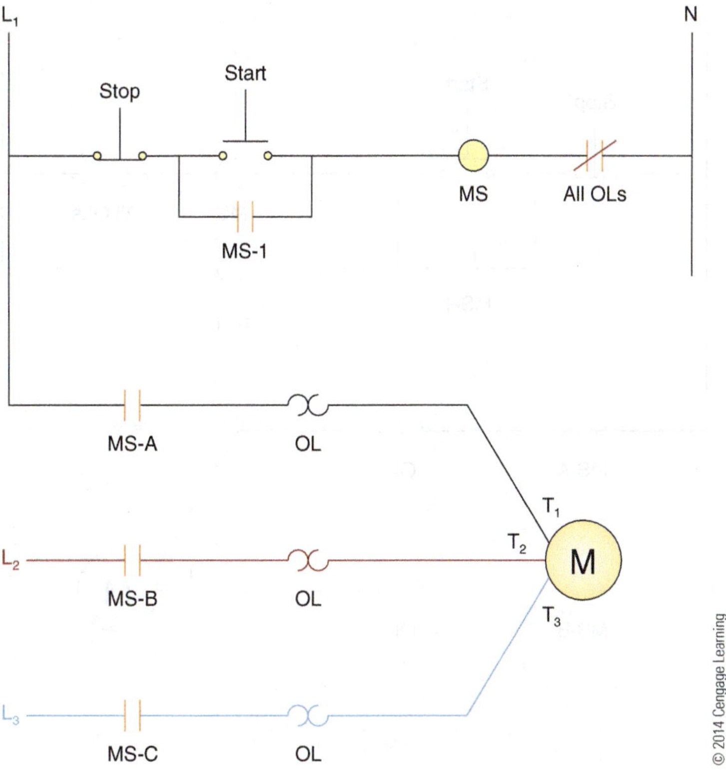

FIGURE 22-2A A three-wire control.

5. Pressing the stop pushbutton interrupts the current flow to MS.
   a. This causes MS-1 to open.
   b. This also causes MS-A, MS-B, and MS-C to open.
6. The motor stops.

Should an overload condition occur, the overload contacts open, interrupting the current path of MS. This causes the motor starter to de-energize, stopping the motor. A power interruption also causes the motor starter to de-energize, stopping the motor. However, unlike the two-wire control, when power is restored, the motor does not restart automatically.

Figure 22-2B, **Figure 22-2C**, and **Figure 22-2D** show the same basic circuit as the one in Figure 22-2A with some minor modifications. Figure 22-2B has a pilot light (PL1) added, which indicates that the motor is running. The pilot light is lit when the motor is running and extinguished when the motor is stopped.

Figure 22-2C includes a motor running, push-to-test pilot light. Depressing the lens of the pilot light causes the light to illuminate if it is working. This test can be performed whether the motor is running or not. In this manner, a defective pilot light can be found.

The circuit found in Figure 22-2D shows a pilot light that indicates when the motor is stopped. In this circuit, the pilot light is extinguished while the motor is running and lit when the motor is stopped.

22-3 MULTIPLE START/STOP CONTROLS

In some instances, it is necessary to control two motors from one location. An example might be a control panel, which requires two separate start/stop stations and two separate motors. **Figure 22-3** shows a circuit that performs this function. Notice that each motor has its own start/stop station. This allows each motor to be started and stopped independently of the other. However, notice the master stop pushbutton. Depressing it interrupts the current flowing to both start/stop circuits. This stops both motors simultaneously. Notice, however, that should an overload condition occur on one motor, the other motor is not affected.

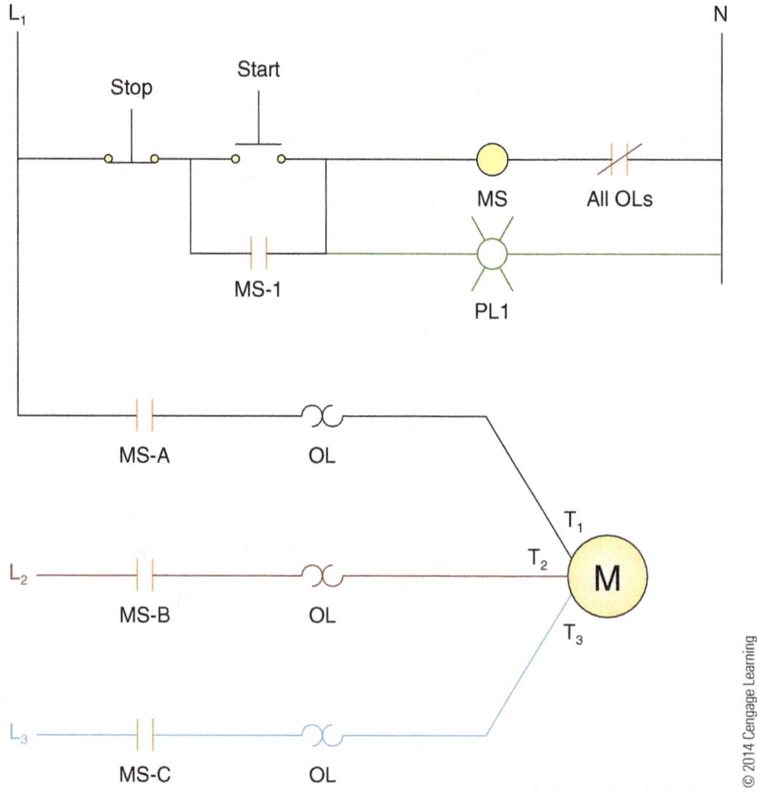

FIGURE 22-2B A three-wire control with a motor running indicator light.

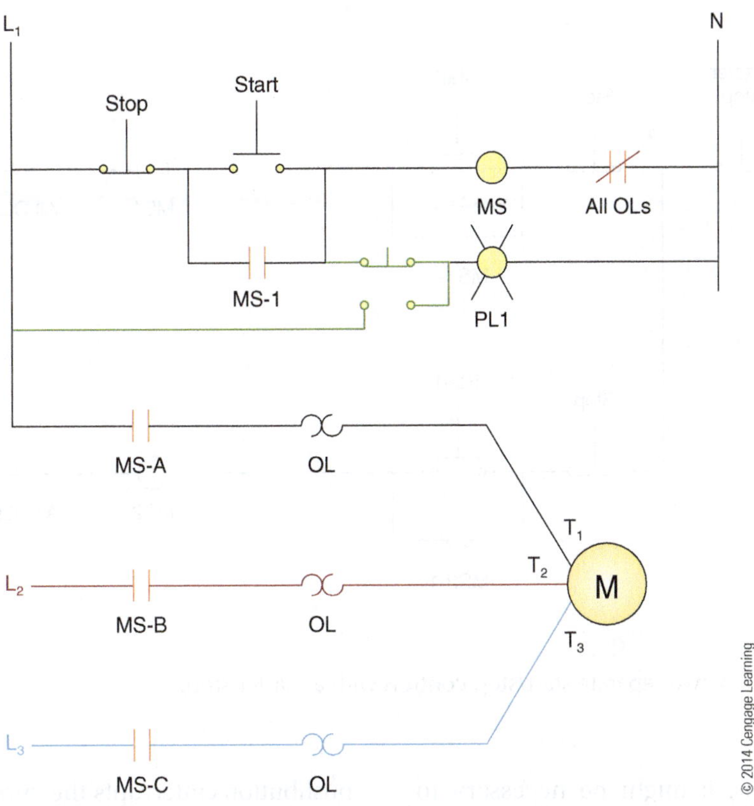

FIGURE 22-2C A three-wire control with a push-to-test motor running indicator light.

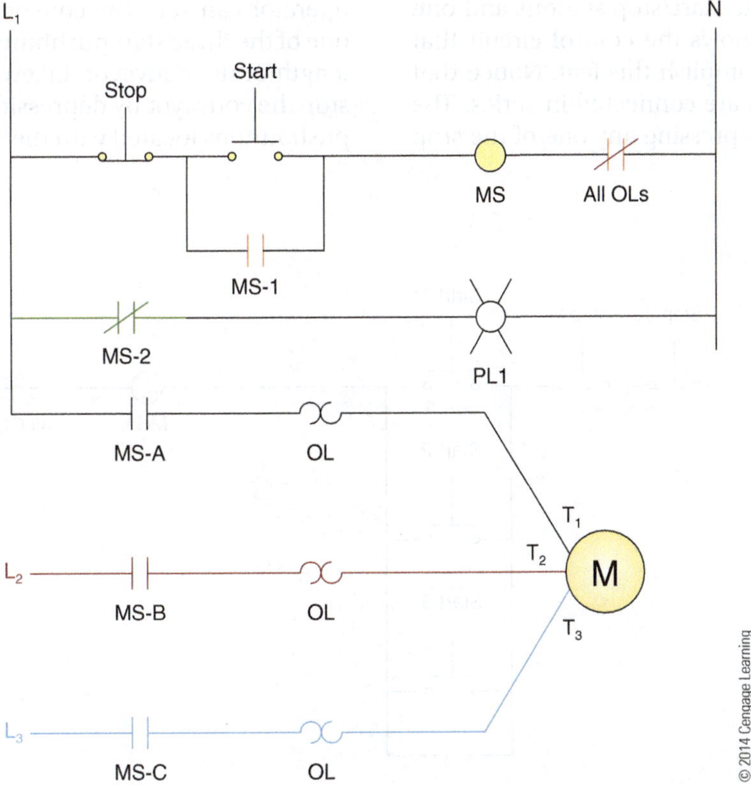

FIGURE 22-2D A three-wire control with a motor stopped indicator light.

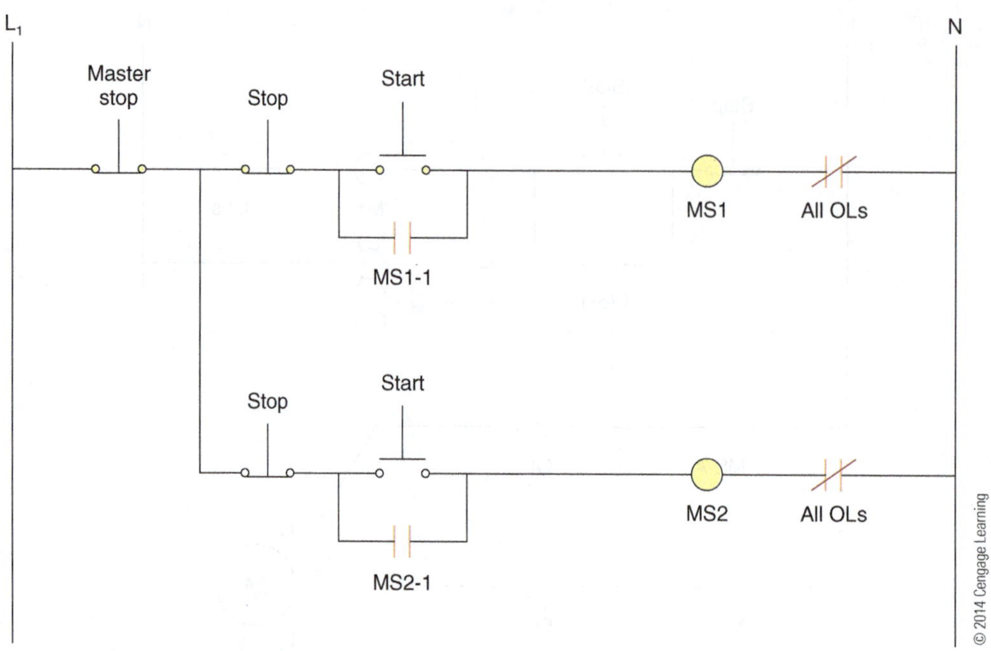

FIGURE 22-3 Two separate start/stop controls with a master stop.

In some instances, it might be necessary to control one motor from several different locations. For example, imagine a long conveyor belt. It may be desirable to be able to start and stop the conveyor at either end and in the middle. This requires three separate start/stop stations and one motor. **Figure 22-4** shows the control circuit that could be used to accomplish this feat. Notice that the stop pushbuttons are connected in series. The result of this is that depressing any one of the stop pushbuttons interrupts the current to the remainder of the control circuit. Notice also that the start pushbuttons are wired in parallel. The result of this is that depressing any one of the start pushbuttons causes current to flow to MS1. This means that an operator can start the conveyor by depressing any one of the three start pushbuttons located along the length of the conveyor. Likewise, the operator can stop the conveyor by depressing any one of the stop pushbuttons located with the start pushbuttons.

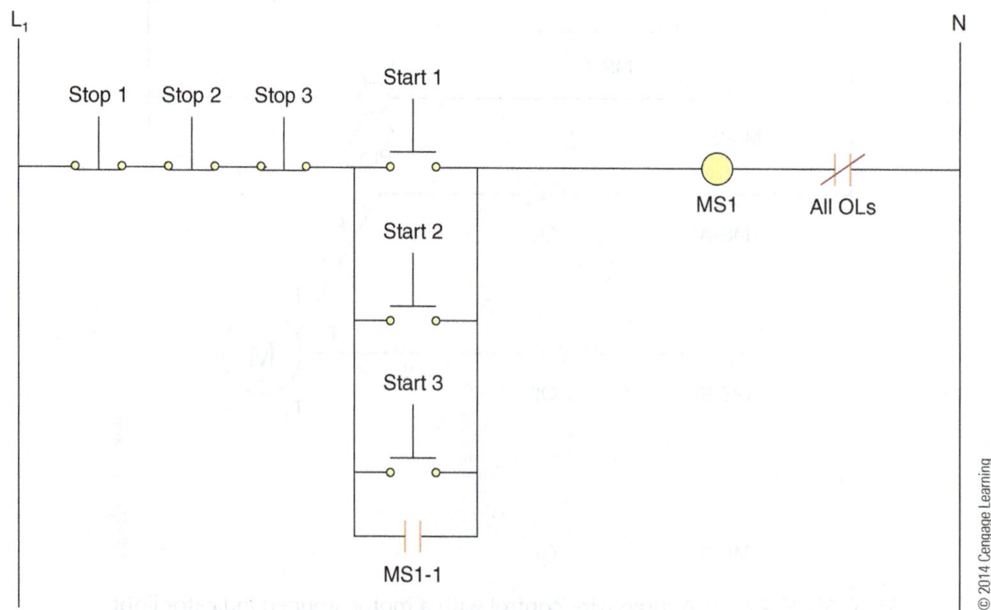

FIGURE 22-4 Multiple start/stop controls controlling a single motor.

22-4 FORWARD/REVERSE CONTROLS

Often it is necessary to change the direction of rotation of a motor. **Figure 22-5A** shows a circuit that allows a motor to run in either the forward or reverse direction. Following is a step-by-step operation of this circuit:

1. The circuit is shown in the stopped mode.
   a. Notice that there are two sets of contacts, one NO and one NC, associated with the forward pushbutton. These contacts are mechanically linked (indicated by the dotted line).
   b. Notice that there are two sets of contacts, one NO and one NC, associated with the reverse pushbutton. These contacts are mechanically linked (indicated by the dotted line).
2. The forward pushbutton is depressed.
   a. This energizes the forward motor starter, FWD.
      i. This causes the FWD holding contacts to close.
      ii. This causes the FWD NC contacts in front of the reverse coil, REV, to open.
3. The motor is now running in the forward direction.
4. Depressing the reverse pushbutton interrupts the current flow to the FWD coil.
   a. This de-energizes the FWD coil.
      i. This causes the FWD holding contacts to open.
      ii. This causes the FWD NC contacts in front of the reverse coil, REV, to reclose.
   b. This allows the REV coil to energize.
      i. This causes the REV holding contacts to close.
      ii. This causes the REV NC contacts in front of the forward coil, FWD, to open.
5. The motor is now running in the reverse direction.
6. Depressing the stop pushbutton interrupts the current to both the FWD and REV circuits, causing the motor to stop.

Notice that it was not necessary to stop the motor before changing the direction of rotation. This may be a desirable characteristic because changing direction without stopping the motor allows for a rapid change of direction. However, this may be hard on the motor and may shorten its life expectancy. The requirements of the application dictate whether the motor should be designed for this type of operation or whether a circuit that requires stopping should be used.

Figure 22-5B and **Figure 22-5C** show the same essential circuit as the one in Figure 22-5A, but with some slight modifications. Figure 22-5B shows a forward/reverse circuit with a pilot light that indicates the motor is running. The pilot light does not indicate in which direction the motor is running, however.

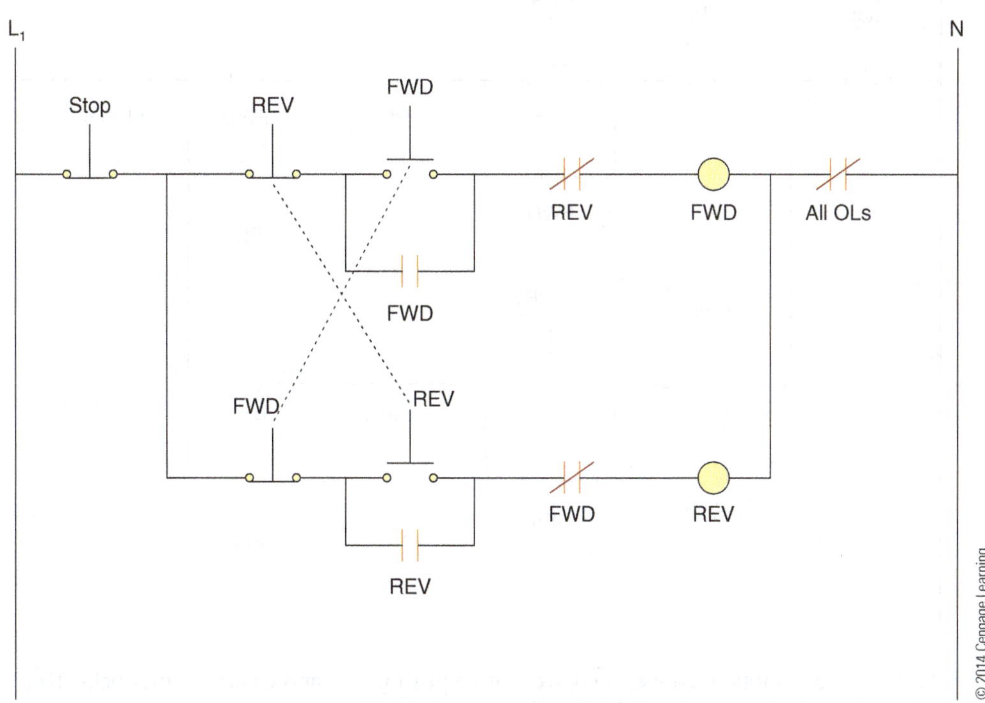

FIGURE 22-5A A forward/reverse control with a pushbutton and electrical interlocks.

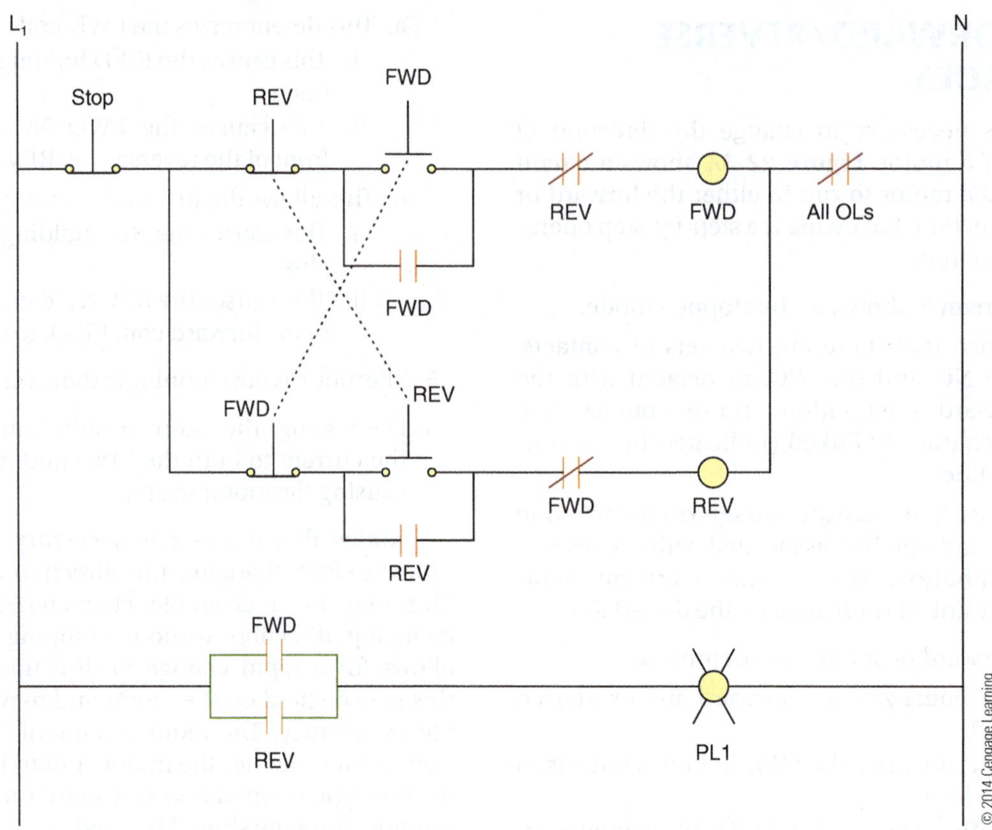

FIGURE 22-5B A forward–reverse control with a pushbutton and electrical interlocks. This circuit also includes a motor running indicator light.

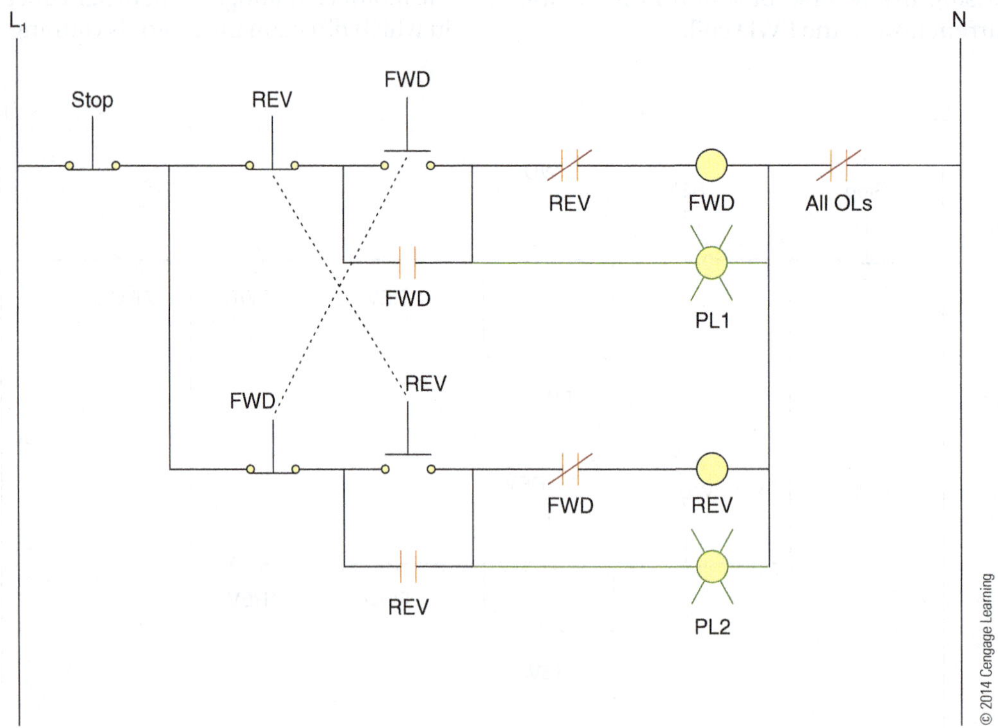

FIGURE 22-5C A forward–reverse control with a pushbutton and electrical interlocks. This circuit also includes indicator lights for direction of rotation.

Figure 22-5C shows a circuit where the direction of rotation is indicated. This circuit uses two pilot lights. When illuminated, one pilot light, PL1, indicates that the motor is running in the forward direction, whereas the other pilot light, PL2, indicates that the motor is running in reverse.

Another type of direction control circuit is shown in **Figure 22-5D**. This circuit uses a control device known as a **drum switch**. A drum switch is a mechanically held device that allows the direction of rotation of a motor to be changed. Notice the switching arrangement as shown in the small diagrams at the bottom of the drawing in Figure 22-5D. Essentially, when the drum switch is placed in the forward position, L_1 is connected to T_1, L_2 is connected to T_2, and L_3 is connected to T_3. Placing the drum switch in the reverse position connects L_1 to T_3, L_2 to T_2, and L_3 to T_1. You may recall that to reverse the direction of rotation of a three-phase motor, you must interchange the connections to two of the three motor leads. The drum switch accomplishes this function by switching the connections to T_1 and T_3.

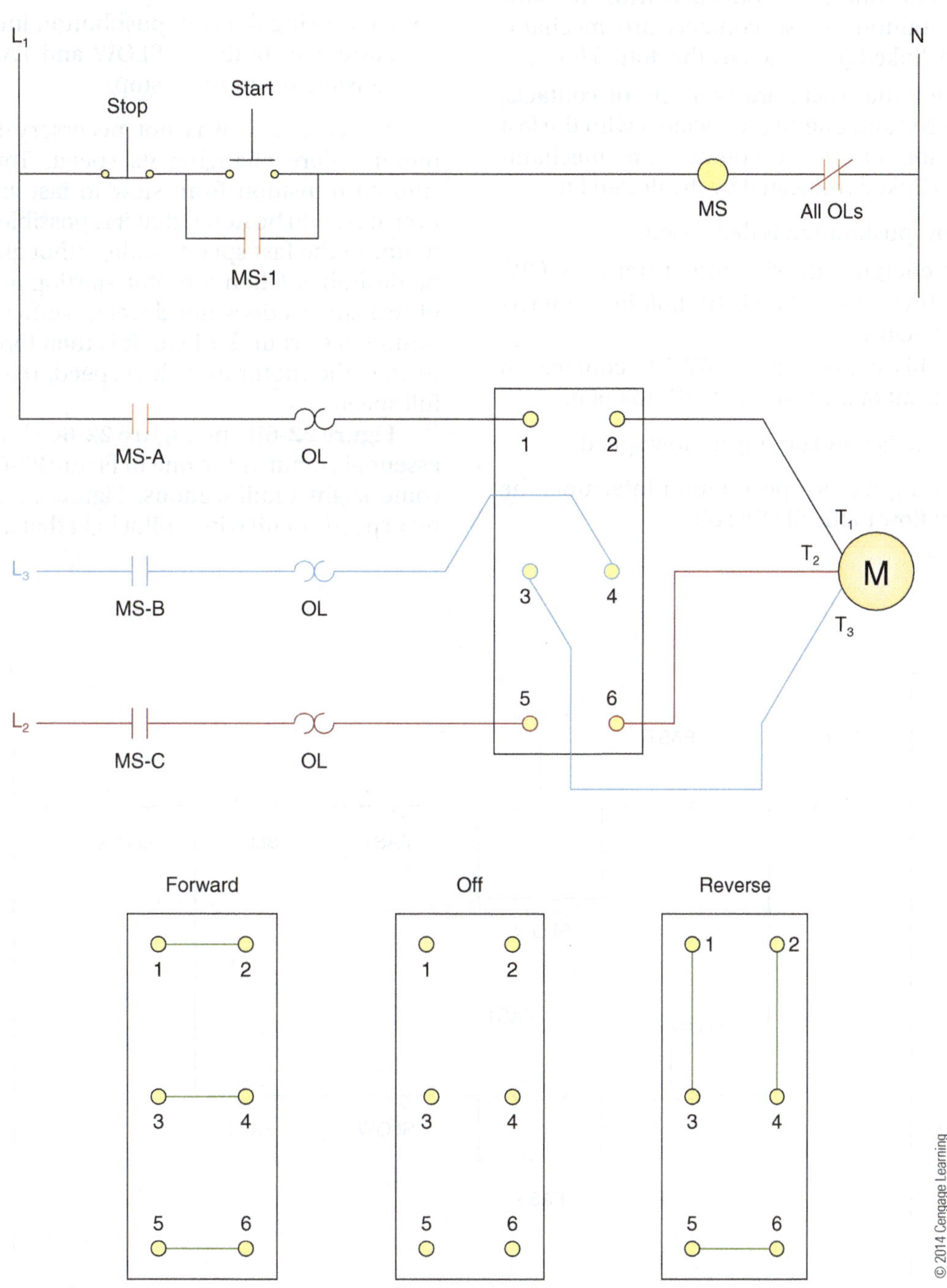

FIGURE 22-5D A drum switch control.

22-5 SPEED CONTROL

Another function required of motor control circuits is to control the speed of the motor. This can be accomplished by a circuit that resembles the forward–reverse control shown in Figure 22-5A. **Figure 22-6A** shows a basic two-speed control circuit. Following is the sequence of operation of this circuit:

1. The circuit is shown in the stopped mode.
   a. Notice that there are two sets of contacts, one NO and one NC, associated with the slow pushbutton. These contacts are mechanically linked (indicated by the dotted line).
   b. Notice that there are two sets of contacts, one NO and one NC, associated with the fast pushbutton. These contacts are mechanically linked (indicated by the dotted line).
2. The slow pushbutton is depressed.
   a. This energizes the slow motor starter, SLOW.
      i. This causes the SLOW holding contacts to close.
      ii. This causes the SLOW NC contacts in front of the fast coil, FAST, to open.
3. The motor is now running at slow speed.
4. Depressing the fast pushbutton interrupts the current flow to the SLOW coil.
   a. This de-energizes the SLOW coil.
      i. This causes the SLOW holding contacts to open.
      ii. This causes the SLOW NC contacts in front of the fast coil, FAST, to re-close.
   b. This allows the FAST coil to energize.
      i. This causes the FAST holding contacts to close.
      ii. This causes the FAST NC contacts in front of the slow coil, SLOW, to open.
5. The motor is now running at fast speed.
6. Depressing the stop pushbutton interrupts the current to both the SLOW and FAST circuits, causing the motor to stop.

Notice that it was not necessary to stop the motor before changing its speed. This allows a smooth transition from slow to fast speed. However, it should be noted that it is possible to start the motor in the fast speed mode, although it may not be desirable. Often a motor starting at the higher of two speeds does not develop sufficient starting torque to start under load. It is therefore necessary to start the motor at a slow speed, then switch to full speed.

Figure 22-6B and **Figure 22-6C** show the same essential circuit as the one in Figure 22-6A, but with some slight modifications. Figure 22-6B shows a two-speed circuit with a pilot light that indicates the

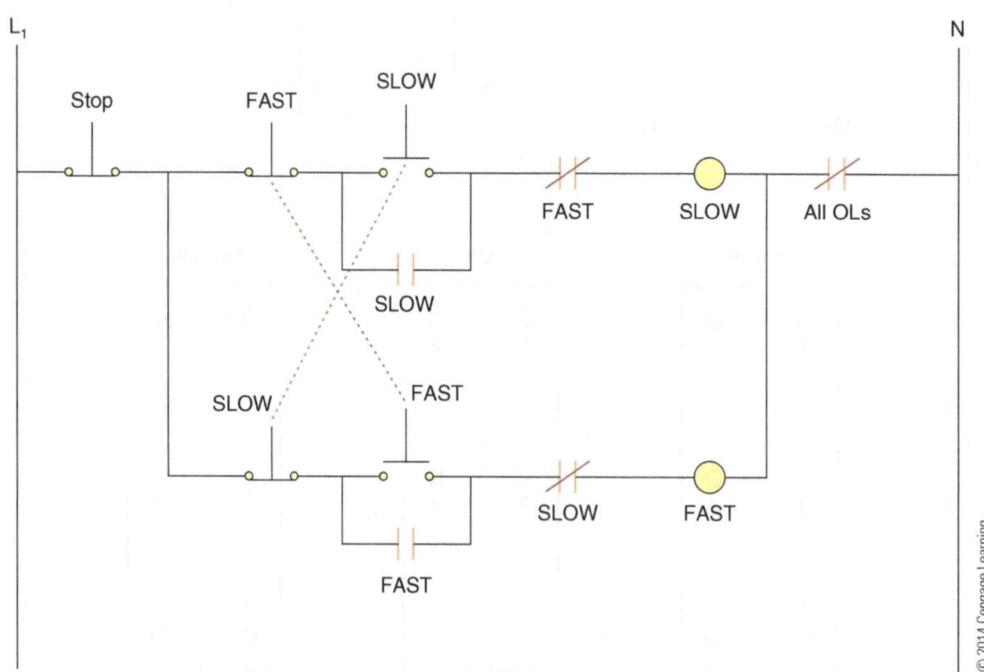

FIGURE 22-6A A two-speed control with a pushbutton and electrical interlocks.

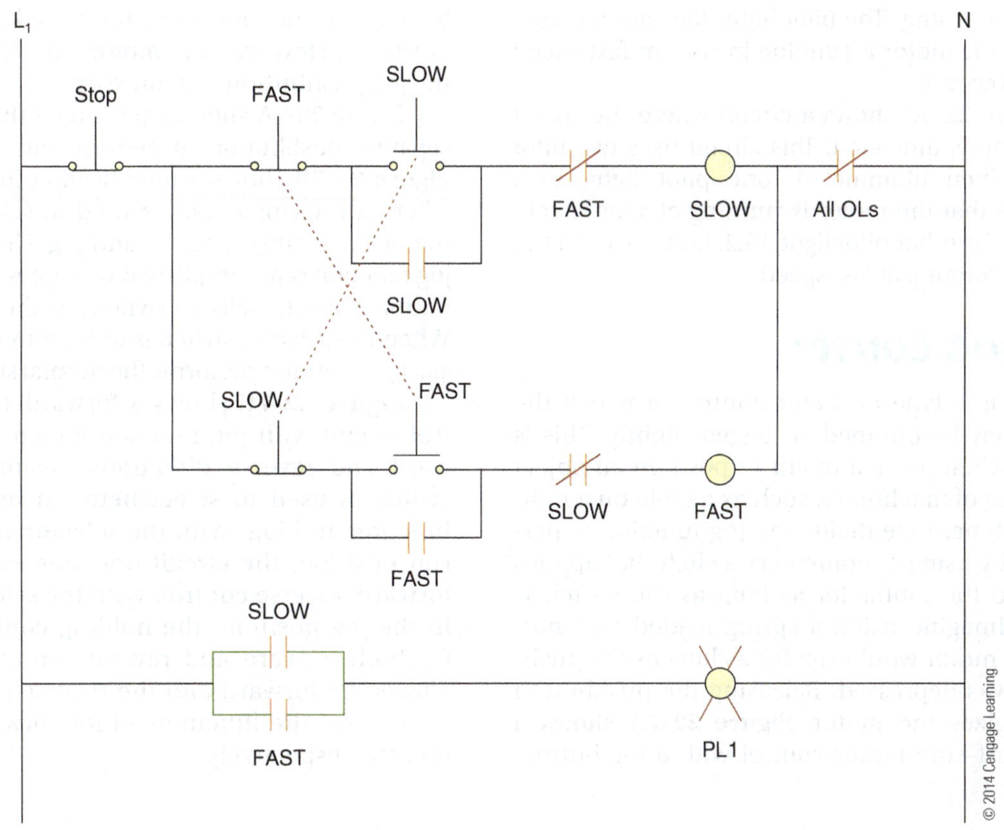

FIGURE 22-6B A two-speed control with a pushbutton and electrical interlocks. This circuit also includes a motor running indicator light.

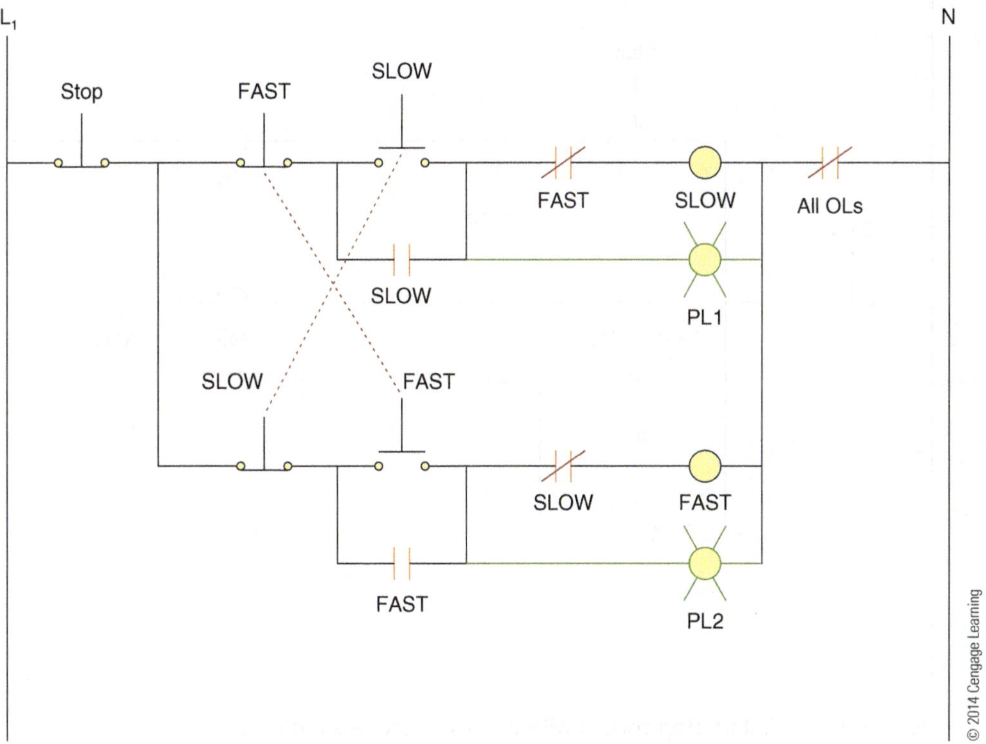

FIGURE 22-6C A two-speed control with a pushbutton and electrical interlocks. This circuit also includes indicator lights for operating speed.

motor is running. The pilot light does not indicate whether the motor is running in slow or fast speed mode, however.

Figure 22-6C shows a circuit where the speed of rotation is indicated. This circuit uses two pilot lights. When illuminated, one pilot light, PL1, indicates that the motor is running at slow speed, whereas the other pilot light, PL2, indicates that the motor is running at fast speed.

22-6 JOG CONTOL

Jogging is a type of motor control in which the motor may be bumped or jogged slightly. This is helpful when using a motor to position an object or a piece of machinery, such as a table on a milling machine. Essentially, the jog function is performed by using a momentary switch that applies power to the motor for as long as the switch is closed. Imagine using a spring-loaded pushbutton. The motor would run for as long as the pushbutton was depressed. Releasing the pushbutton de-energizes the motor. **Figure 22-7A** shows a basic start–stop motor control with a jog button. Notice that the motor can be started and stopped normally. However, the motor runs for as long as the jog pushbutton is depressed.

Figure 22-7A shows a jog circuit that requires a separate pushbutton to perform the jog function. **Figure 22-7B** shows a modification to this circuit, where a selector switch is used to select between the normal (run) function and jog. Notice that the jog function is accomplished by depressing the start button when the selector switch is in the jog position. When the selector switch is in the run position, the start pushbutton performs the normal start function.

Figure 22-7C shows a forward–reverse control circuit with jog function added to both forward and reverse directions. Again a selector switch is used to select between normal (run) function and jog. With the selector switch in the run position, the circuit operates as a standard forward–reverse control. With the selector switch in the jog position, the holding contact circuits for both forward and reverse are opened. This allows the forward and the reverse pushbuttons to perform the functions of jog forward and jog reverse, respectively.

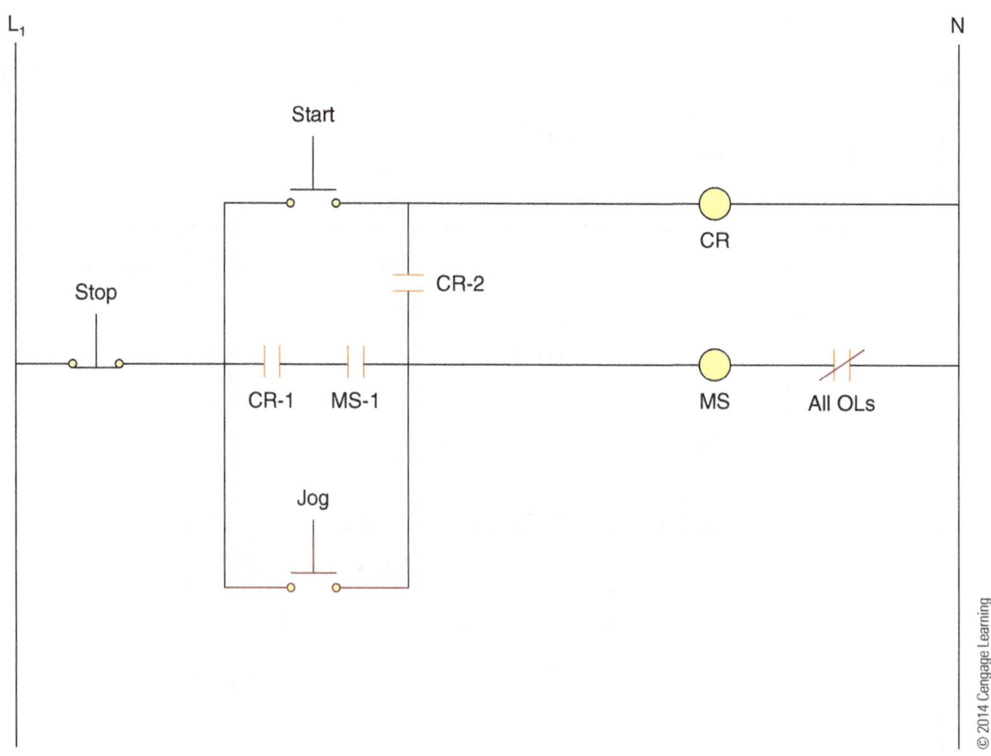

FIGURE 22-7A A start–stop control with a separate jog pushbutton.

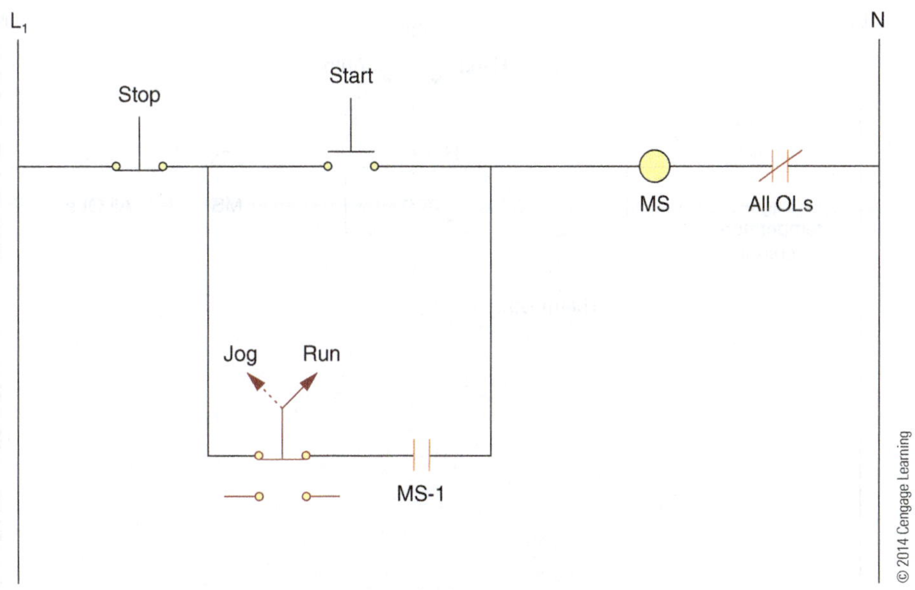

FIGURE 22-7B A jog–run selector switch control.

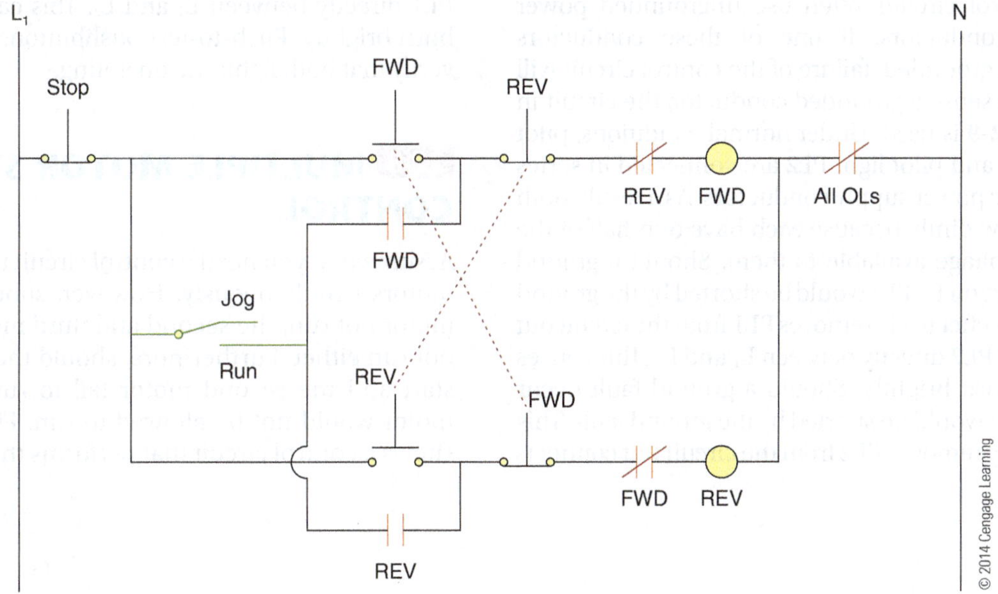

FIGURE 22-7C A forward–reverse control with jog forward and jog reverse features.

22-7 MISCELLANEOUS CONTROL

Figure 22-8 also uses a selector switch; however, this circuit does not perform a jog function. The purpose of the circuit shown in Figure 22-8 is known as a **hand-off-automatic (HOA) control**. HOA control allows a circuit to function under the control of either an automatic device or manually. Notice the thermostat control. This is an automatic device. With the HOA switch in the Automatic position, and depending on the temperature sensed, the motor will turn on or off automatically. However, if the HOA switch is placed in the Hand position, the motor turns on, regardless of the state of the thermostat. The high-temperature cutout switch prevents the motor from operating, regardless of the position of the HOA switch, if the temperature rises too high.

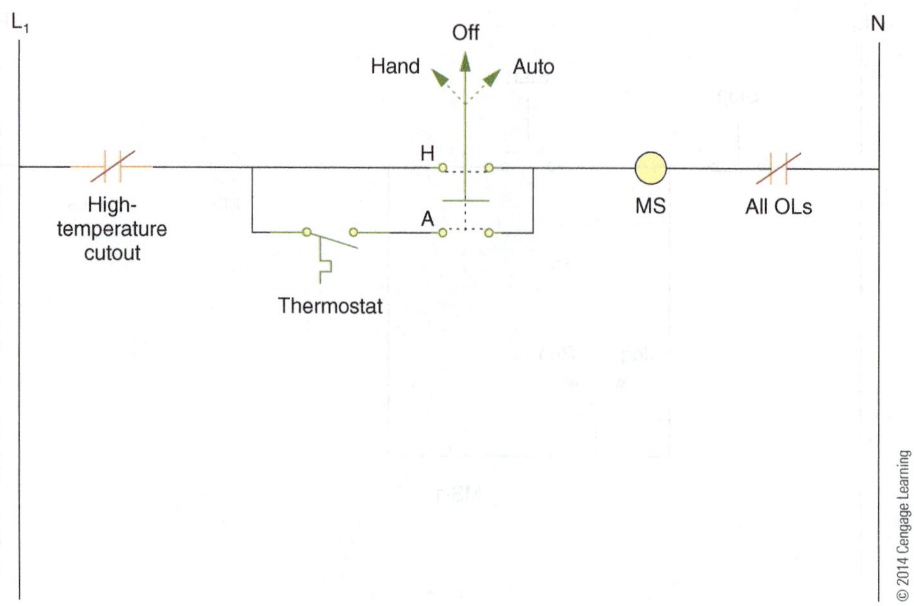

FIGURE 22-8 A hand-off-automatic (HOA) control.

Control circuits often use ungrounded power supply conductors. If one of these conductors becomes grounded, failure of the control circuit will occur. To sense a grounded conductor, the circuit in **Figure 22-9** is used. Under normal conditions, pilot light PL1 and pilot light PL2 are connected in series across the power supply conductors. As a result, both lights glow dimly because each have one-half of the supply voltage available to them. Should a ground fault occur on L_1, PL1 would be shorted by the ground fault. This effectively removes PL1 from the circuit but connects PL2 directly between L_1 and L_2. This causes PL2 to burn brightly. Should a ground fault occur on L_2, PL2 would be shorted by the ground fault. This effectively removes PL2 from the circuit but connects PL1 directly between L_1 and L_2. This causes PL1 to burn brightly. Push-to-test pushbuttons are used to verify that both lights are operating.

22-8 MULTIPLE MOTOR STARTER CONTROL

Assume that you need a control circuit to start three motors simultaneously. However, should the first motor not run, the second and third motors would not run either. Furthermore, should the first motor start and the second motor fail to run, the third motor would not be allowed to run. **Figure 22-10** shows a control circuit that performs this function.

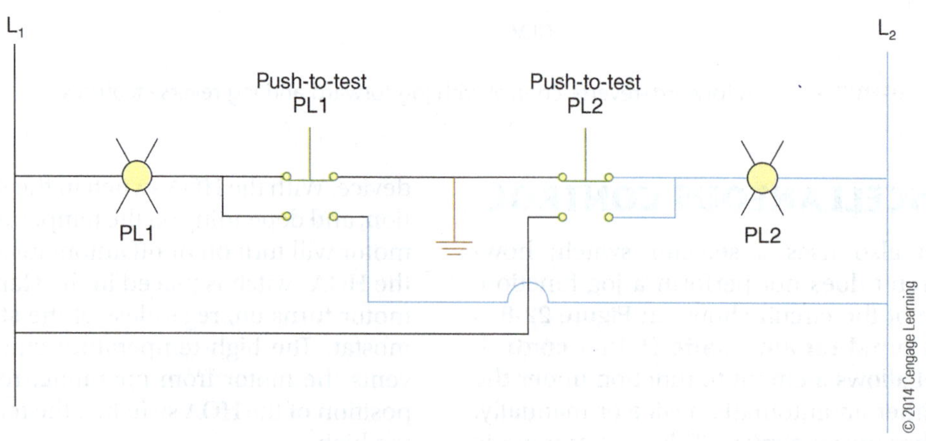

FIGURE 22-9 A ground-fault detection circuit.

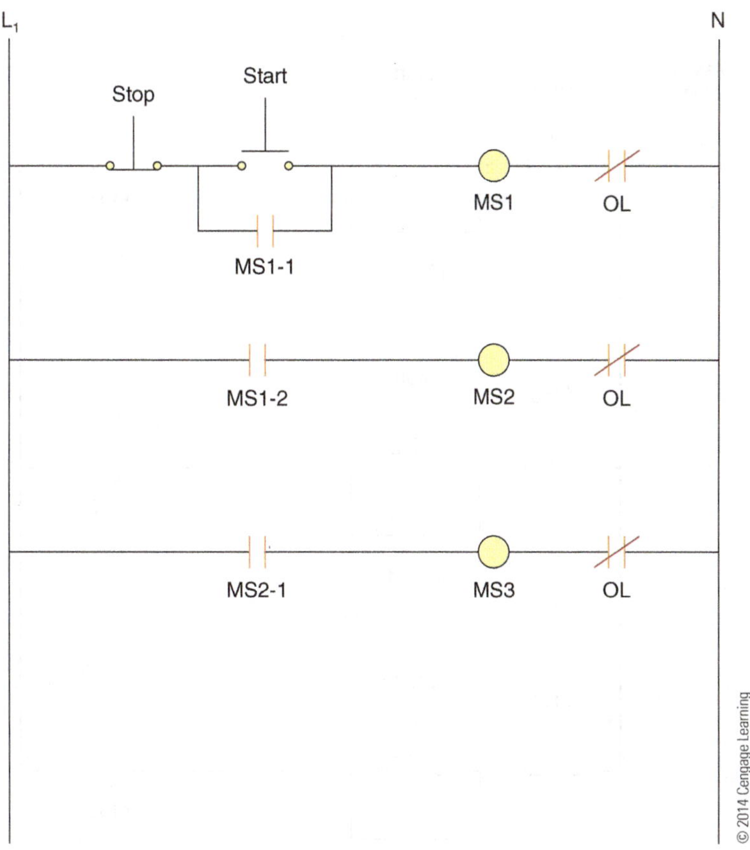

FIGURE 22-10 Simultaneous starting of three motors.

Notice that a set of NO contacts from MS1 controls current to the second motor starter, MS2. Notice also that a set of NO contacts from MS2 controls current to the third motor starter, MS3. This type of circuit is known as a sequential starting circuit. MS1 must be energized before the MS1-1 NO contact can close to provide current to the MS2 motor starter. Likewise, MS2 must be energized before the MS2-1 NO contact can close to provide current to the MS3 motor starter.

Should motor starter MS1 not energize, motor starters MS2 and MS3 do not energize. Should motor starter MS1 energize and motor starter MS2 not energize, motor starter MS3 do not energize.

An overload on MS3 shuts down motor starter MS3 only. An overload on MS2 shuts down motor starters MS2 and MS3, but not MS1. An overload on motor starter MS1 shuts down all motor starters.

Figure 22-11 also shows the control of three separate motors. This circuit provides independent control of the three motors, however. Notice that each motor has its own start-stop station. This means each motor can be started and stopped independently of the others. However, there is a master stop pushbutton that stops all motors simultaneously. In addition, notice that all overloads are located on the first rung. This means that if an overload condition occurs on any of the motors, all motors is shut down.

22-9 SEQUENTIAL STARTING CONTROL

Under certain conditions, it may be desirable to start one motor, and then, after a short time delay, start a second motor. An example might be a cutting tool. Initially, a coolant pump is started to pump cutting fluid to the cutting tool. After the coolant starts pumping, the cutting tool motor is allowed to run. This ensures that cutting fluid is available before the cutting operation can begin. **Figure 22-12** shows a

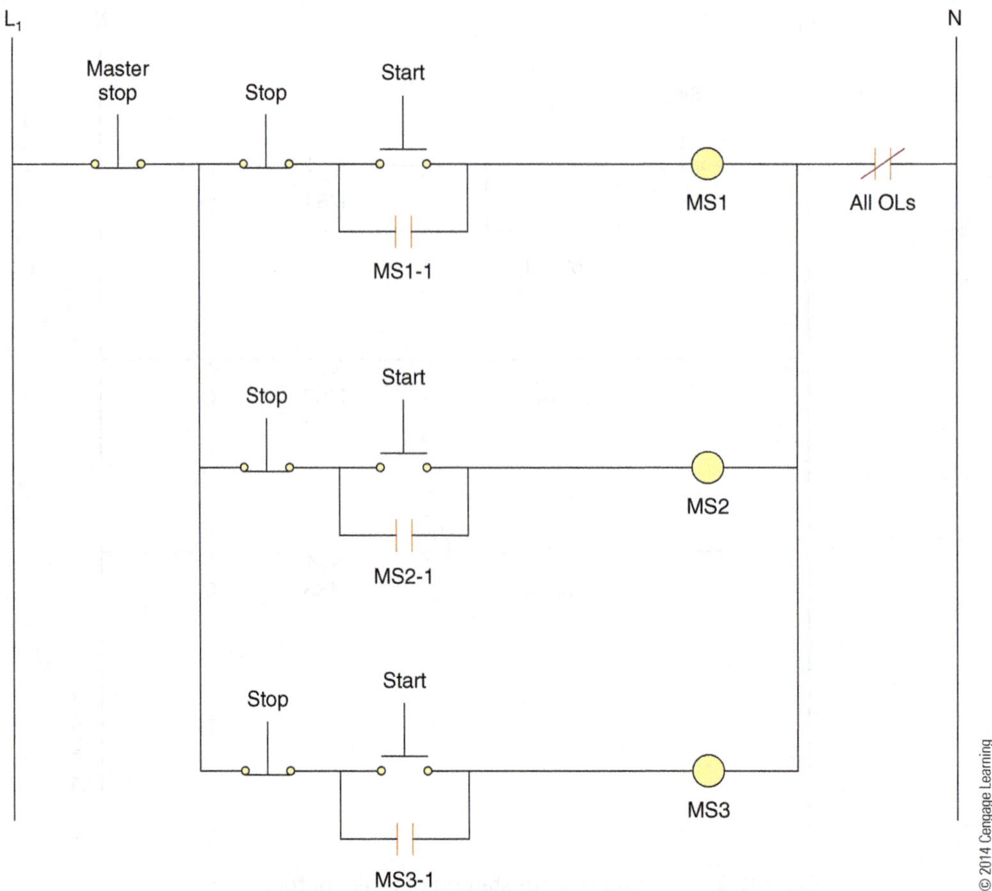

FIGURE 22-11 Three separate start–stop controls with one master stop.

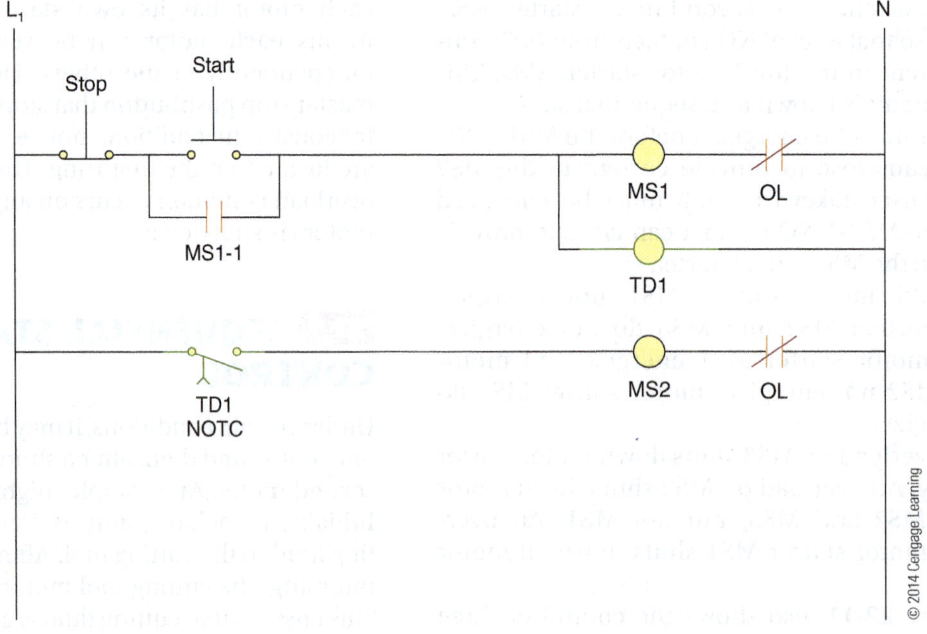

FIGURE 22-12 Time-delayed sequential start.

circuit that could perform this action. Following is how this circuit works:

1. The circuit is shown in the initial de-energized state.
2. The start pushbutton is depressed.
   a. This energizes the MS1 coil.
      i. This causes the MS1-1 holding contacts to close.
   b. TD1, the time-delay relay coil, also energizes.
3. After a delay, the TD1 normally open timed close (NOTC) contacts close.
   a. This energizes the MS2 coil.
4. Pressing the stop pushbutton stops the complete circuit.

It should be noted that if an overload should occur on the MS2 motor, only this motor stops. The MS1 motor continues to operate. However, should an overload occur on the MS1 motor, the entire circuit would shut down.

Figure 22-13A shows a different type of sequential control. In this circuit, depressing the start pushbutton, which is connected to MS2, has no effect until MS1 is started. This circuit could be used for the same application as the one used in the circuit in Figure 22-11; however, the second motor must be started manually, not automatically.

A sequential control is also shown in **Figure 22-13B**. This circuit also uses automatic controls. However, one automatic control is in the form of a pressure switch, whereas the other is a time-delay relay. Following is the operation of this circuit:

1. The circuit is shown in the de-energized state.
2. The start pushbutton is depressed.
   a. This energizes the MS1 coil and PL1.
      i. This causes the MS1-1 holding contacts to close.
      ii. PL1 indicates that MS1 is energized.
3. If MS1 were controlling a pump, pressure switch PS1 would close after MS1 started and built up pressure.
   a. This would cause MS2, PL2, and TD1 to energize.
   b. PL2 indicates that MS2 is energized.
4. After a time delay, the NOTC contact of TD1 closes.
   a. This energizes MS3 and PL3.
   b. PL3 indicates that MS3 is energized.
5. Pressing the stop pushbutton stops the entire process.

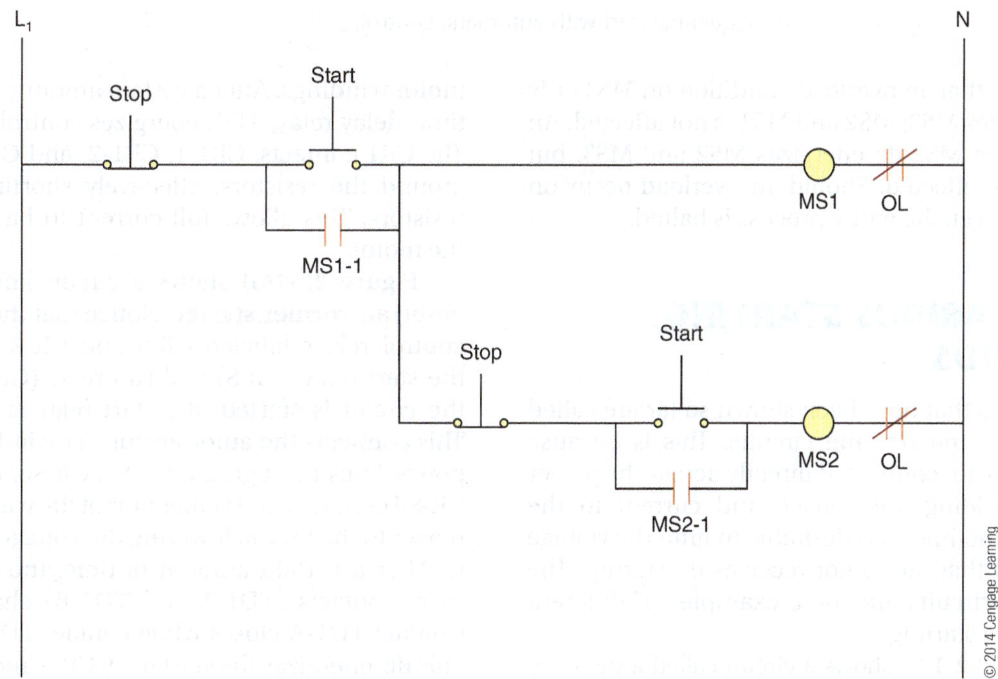

FIGURE 22-13A Sequential start. MS1 must be started before MS2 can be started.

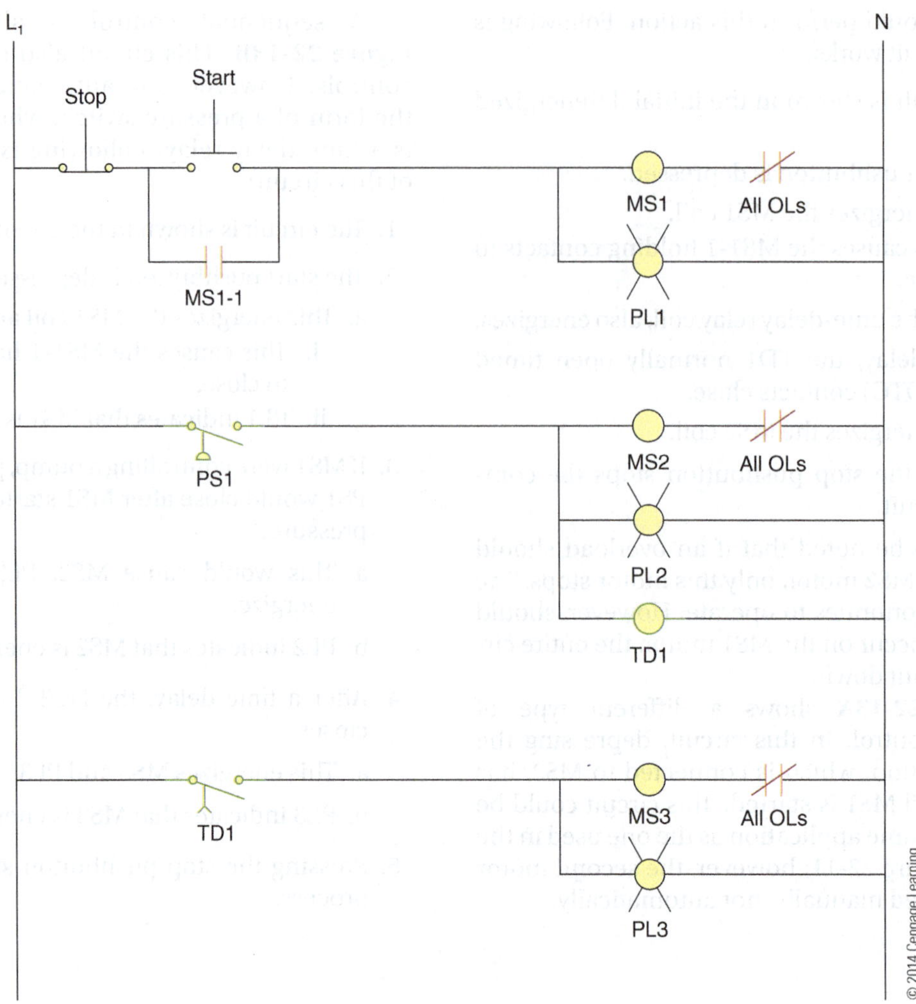

FIGURE 22-13B Sequential start with automatic control.

Notice that an overload condition on MS3 only de-energizes MS3; MS2 and MS1 is not affected. An overload on MS2 de-energizes MS2 and MS3, but MS1 is not affected. Should an overload occur on MS1, however, the entire process is halted.

22-10 VARIOUS STARTING METHODS

The circuits that have been shown so far are called **across-the-line starting** circuits. This is because the motors are connected directly across the power lines, providing full voltage and current to the motor. Sometimes it is desirable to limit the voltage or current that the motor receives on startup. The following circuits are some examples of different methods of starting.

Figure 22-14A shows a circuit called a **primary resistor starter**. When the motor is initially started, resistors are connected in series with the motor. These resistors limit the inrush of current to the motor windings. After a certain amount of time, the time-delay relay, TD1, energizes control relay CR1. The CR1 contacts, CR1-1, CR1-2, and CR1-3, close around the resistors, effectively shorting out the resistors. This allows full current to be applied to the motor.

Figure 22-14B shows a circuit known as an **autotransformer starter**. Notice that there are two control relays labeled CR-S and CR-R. These are the start relay (CR-S) and run relay (CR-R). When the circuit is started, the start relay is energized. This connects the autotransformer windings to the power lines through the CR-Sa, CR-Sb, CR-Sc, and CR-Sd contacts. The center taps of the windings feed power to the motor, lowering the voltage applied to it. After a certain amount of time, the time-delay relay contacts (TD1-A and TD1-B) change state. Contact TD1-A closes while contact TD1-B opens. This de-energizes the start relay CR-S and energizes the run relay CR-R, opening the start relay contacts CR-Sa, CR-Sb, CR-Sc, and CR-Sd. At the same time, the run relay contacts CR-Ra and CR-Rb close.

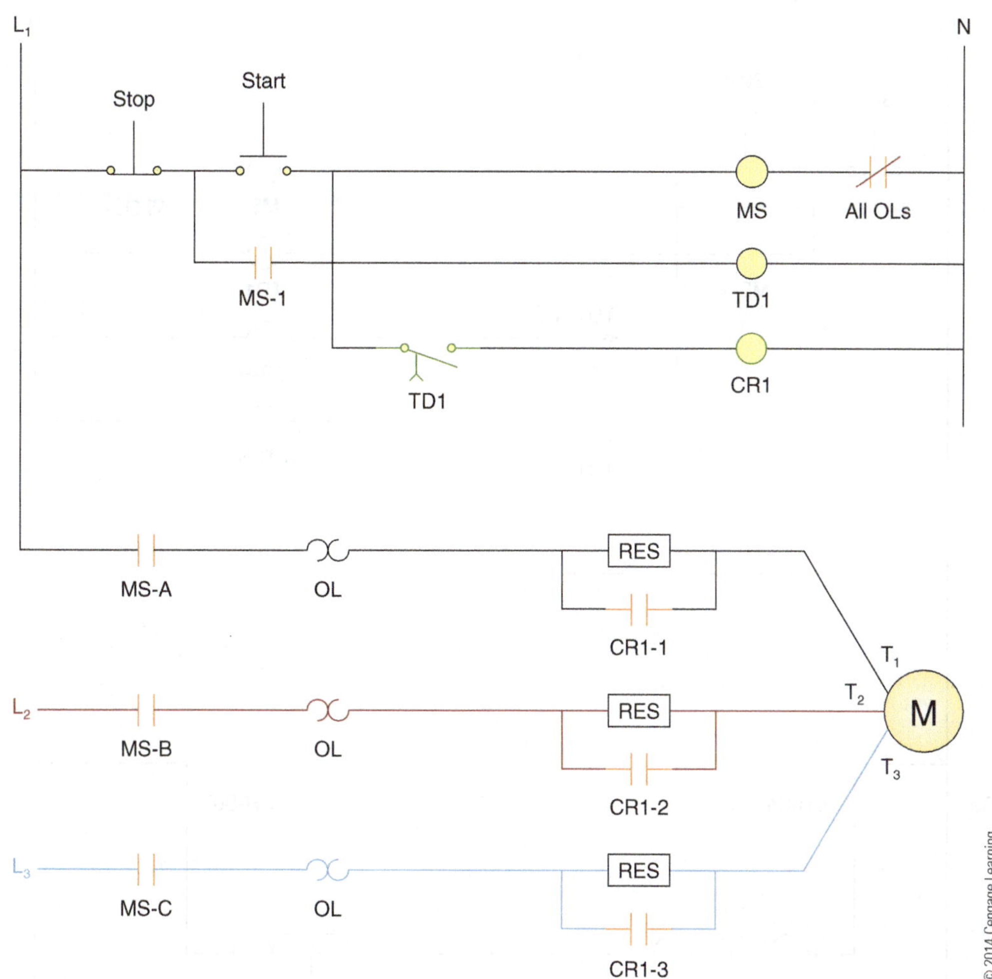

FIGURE 22-14A A primary resistor starter.

This results in the autotransformer being electrically removed from the circuit and the motor being connected directly to the power lines. This applies full voltage and current to the motor.

The circuit in **Figure 22-14C** uses a *nine-lead motor*. This is necessary because this circuit uses one winding of the motor for starting and parallels a second winding of the motor for running. At startup, MS1 is energized, closing contacts MS1-A, MS1-B, and MS1-C. This allows the motor to start on windings T_1, T_2, and T_3. After a preset time delay, the NOTC contact of TD1 closes. This energizes MS2, closing contacts MS2-A, MS2-B, and MS2-C. Windings T_7, T_8, and T_9 are now connected in parallel with T_1, T_2, and T_3. This circuit is known as a two-step part-winding starting circuit.

The circuit shown in **Figure 22-14D** is known as a **wye–delta starting** circuit. A wye–delta motor is required for this type of circuit. This is a motor that contains two sets of windings. However, the connections for each end of the winding must extend to the motor terminal box. This is known as a *12-lead motor*. Following is the operation of this circuit:

1. The circuit is shown in its de-energized state.
2. The start pushbutton is depressed.
   a. This energizes C1, CR, MS1, and TD and keeps C2 from energizing.
      i. C1 and C2 are contactors; CR is a control relay.
      ii. Contacts C1-1, C1-2, and C1-3 will close.
         1) Contact C1-1 provides part of the holding circuit for the start pushbutton.
         2) Contacts C1-2 and C1-3 connect one end of each of the motor windings together to form the middle (common point) of the wye circuit.
      iii. Contact MS-1 forms the remaining part of the holding circuits for the start pushbutton.

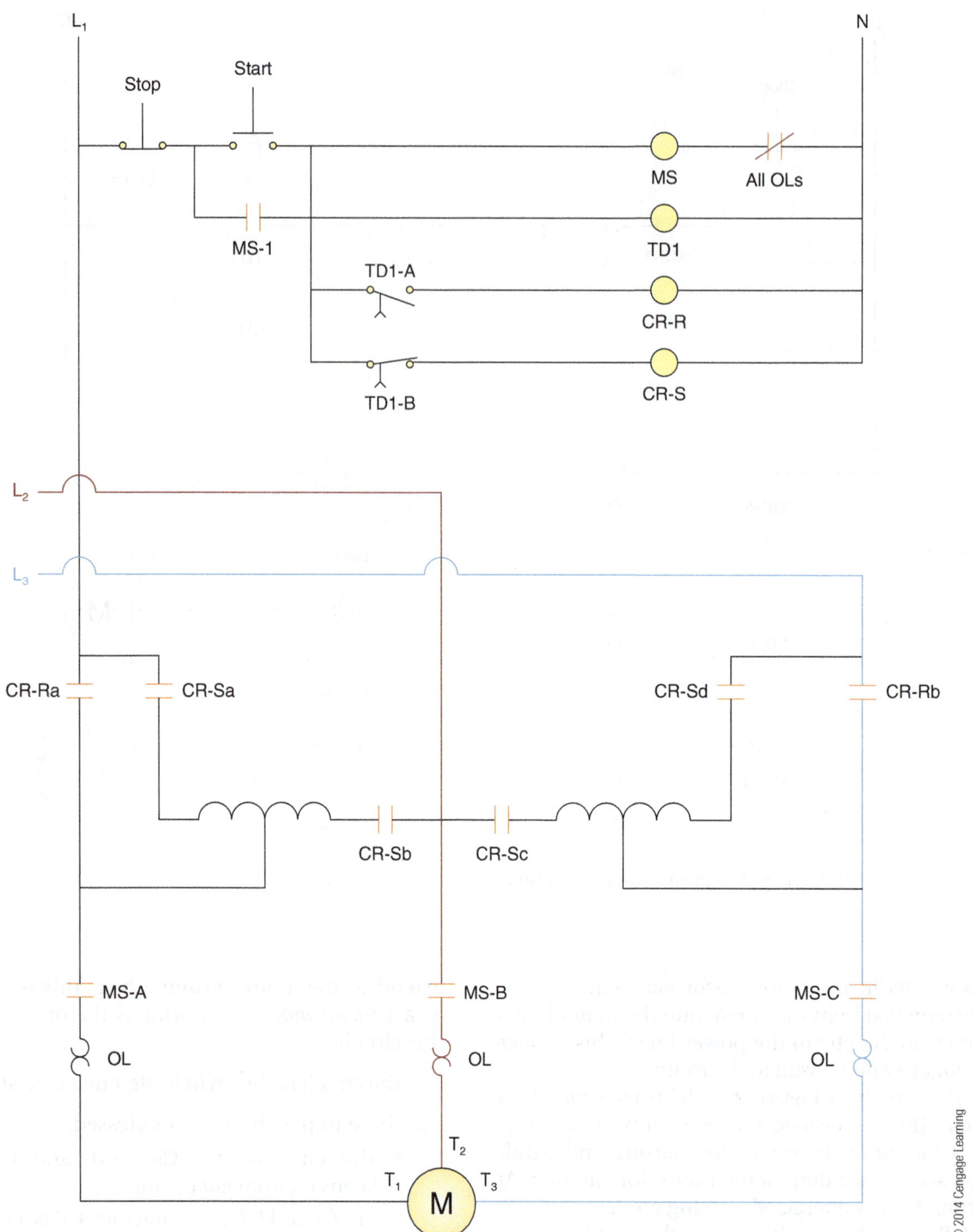

FIGURE 22-14B An autotransformer starter.

iv. Contacts MS1-A, MS1-B, and MS1-C will close.

1) These contacts connect motor terminals T_1, T_2, and T_3 to the power lines.

b. The motor has now started as a wye-connected motor.

c. After a time delay, the NOTC contact TD1-1 will open.

i. This de-energizes C1 and CR.

1) Contact C1-1 opens. The holding circuit is still maintained by the MS-1 contact.

2) Contacts C1-2 and C1-3 open. This removes the wye connection from the motor windings.

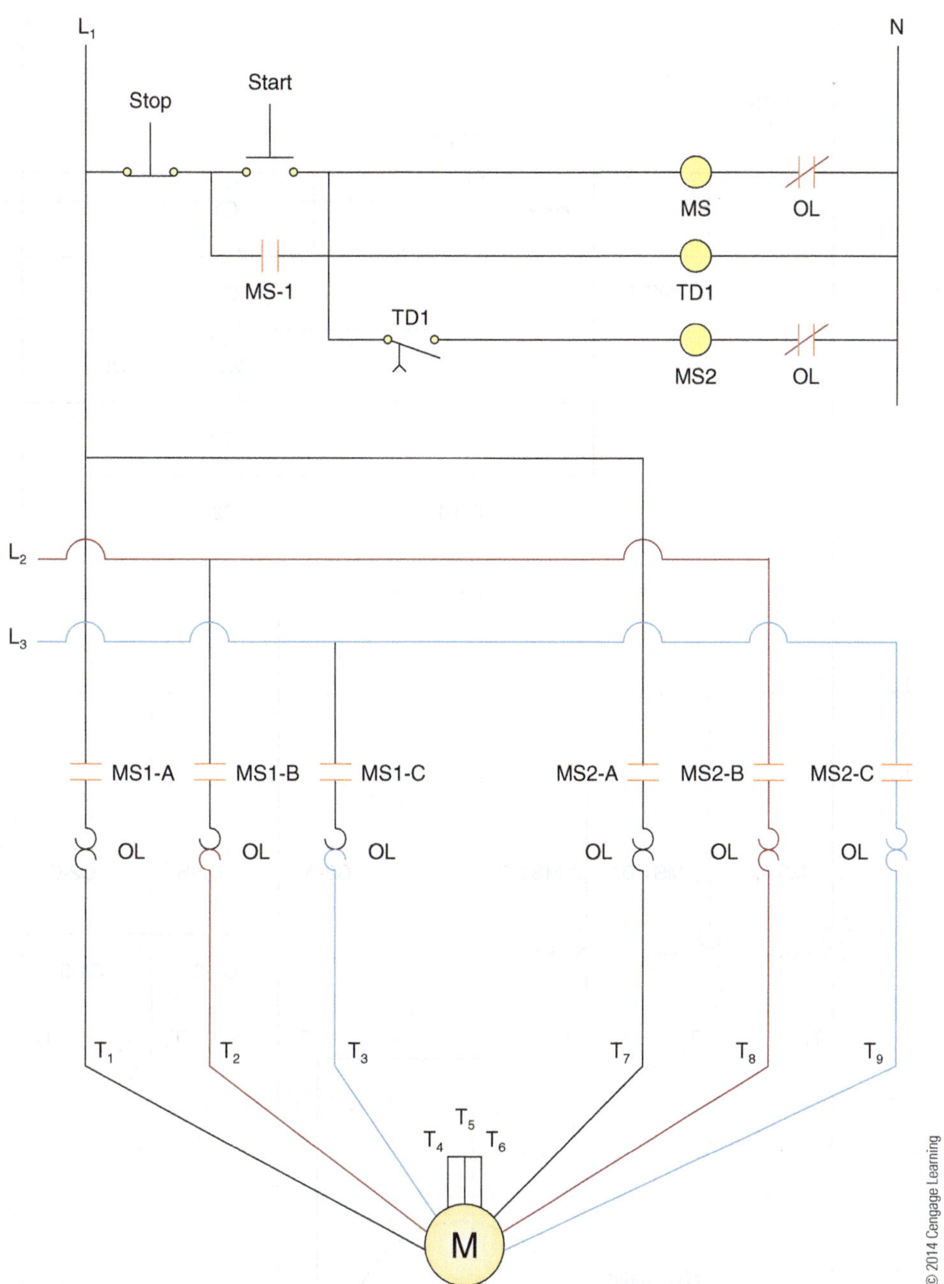

FIGURE 22-14C A two-step part-winding starter.

3) Contact CR-1 closes, energizing C2.
   a) This causes contacts C2-A, C2-B, and C2-C to close.
   b) These contacts connect the motor windings in a delta configuration.

d. The motor is now running as a delta-connected motor.

Figure 22-14E shows a circuit that provides **automatic three-point starting** for a wound-rotor induction motor. Following is how this circuit operates:

1. The circuit is shown in its de-energized state.
2. Pressing the start pushbutton energizes MS1 and TD1.
   a. This causes the MS-1 contact to close.
      i. This provides a holding contact for the start pushbutton.
   b. Contacts MS1-A, MS1-B, and MS1-C close.
      i. These contacts connect motor terminals T_1, T_2, and T_3 to the power lines.

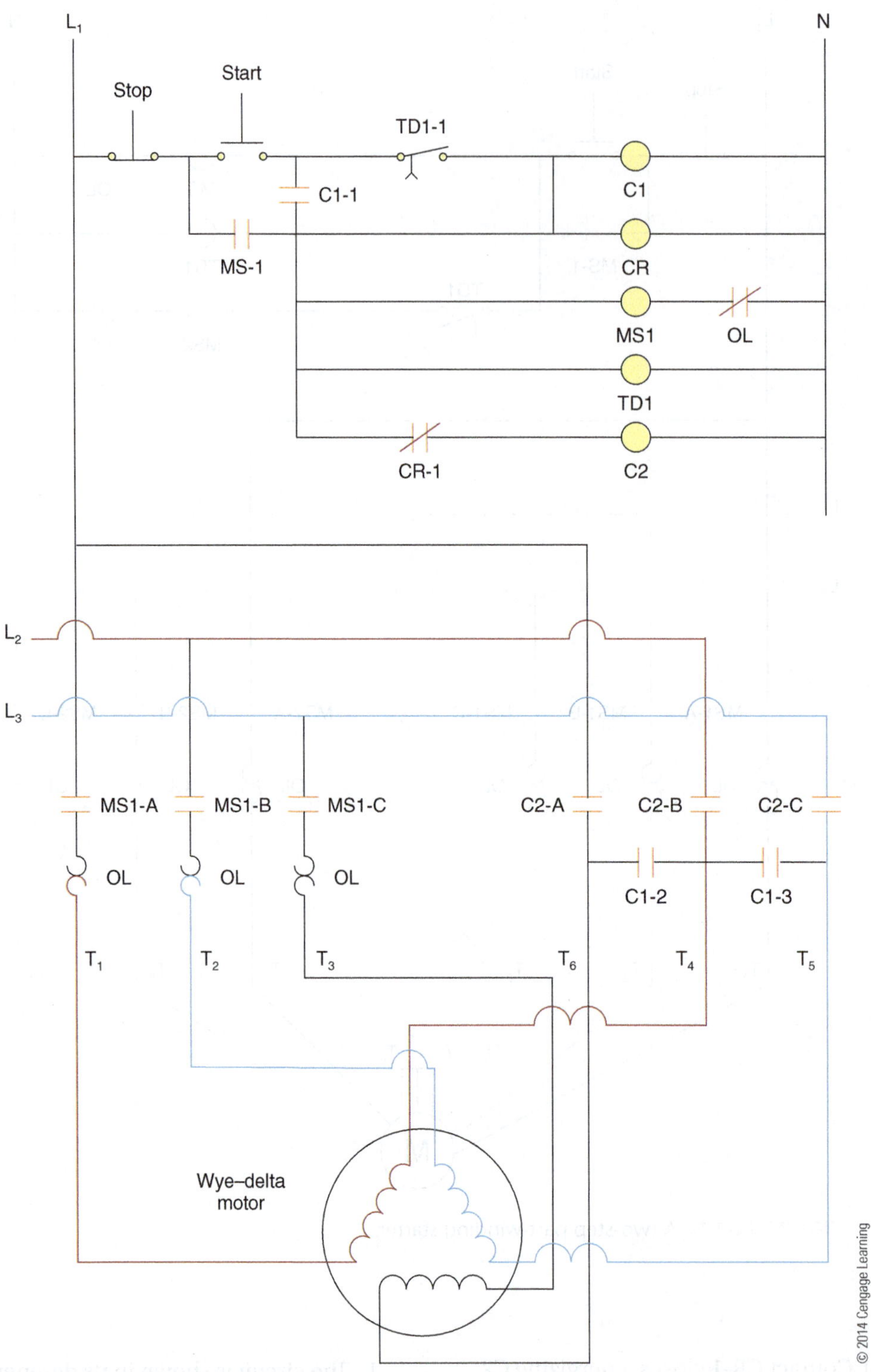

FIGURE 22-14D A wye–delta starter.

ii. This causes the motor to begin to rotate.
   1) All six resistors are now connected to the rotor circuit.
   2) This greatly limits the amount of rotor current.

c. After a delay, NOTC timer contact TD1-1 closes.

i. This energizes contactor C1 and time-delay relay TD2.
   1) Contacts C1-1 and C1-2 close.
   2) This electrically shorts out the right-most set of resistors in the rotor circuit.
   3) This allows an increase in rotor current, resulting in an increase in rotor speed.

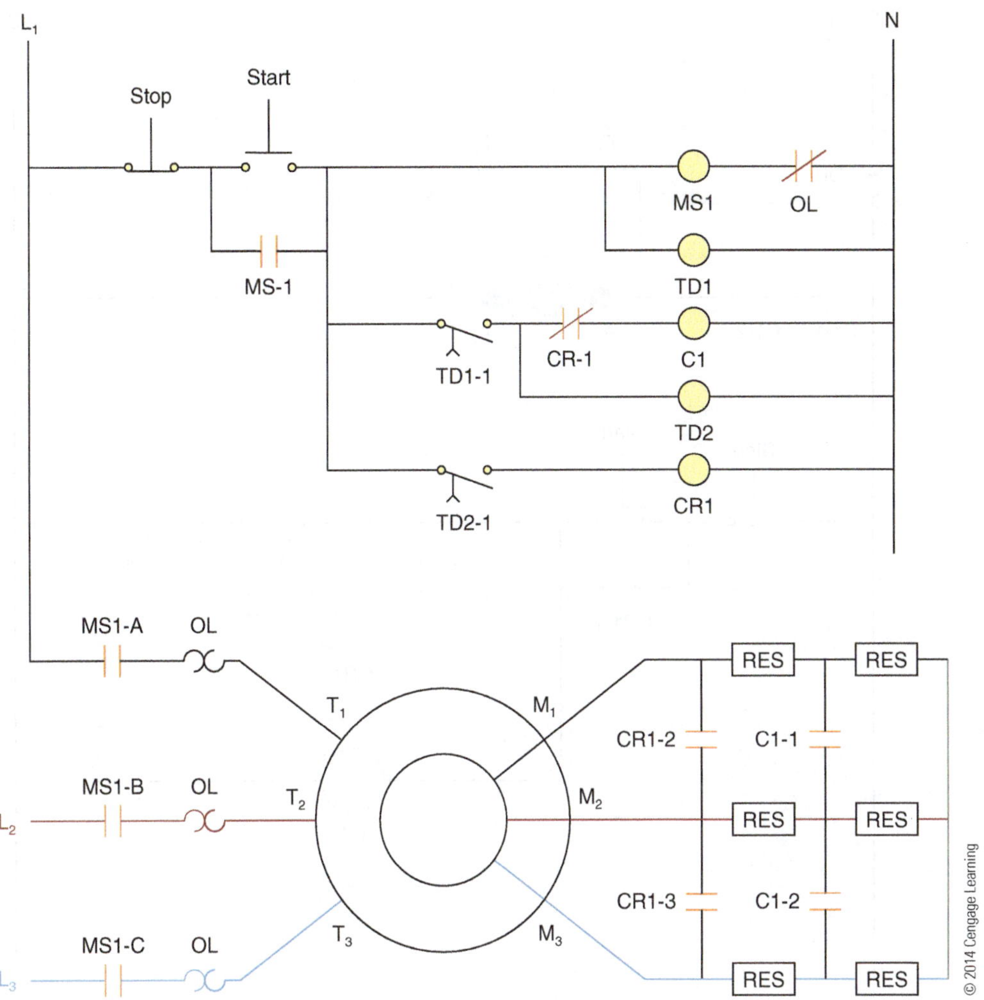

FIGURE 22-14E Three-point starting of a wound-rotor induction motor.

d. After a delay, NOTC timer contact TD2-1 closes.
   i. This energizes control relay CR1.
      1) This causes the NC CR1-1 contact to open.
         a) This de-energizes contactor C1.
            i) This causes contacts C1-1 and C1-2 to open.
      2) This causes contacts CR1-2 and CR1-3 to close.
         a) This shorts out all of the resistors electrically in the rotor circuit.
         b) Full current is now applied to the rotor, and the motor accelerates to full speed.

Figure 22-15 shows a circuit that can be used to provide across-the-line starting for a DC motor. Essentially, this is the same circuit shown for an AC motor with only a few minor modifications. Pressing the start pushbutton energizes the control relay CR1 and the contactor C1. The control relay contact CR1-1 provides the holding contact for the start pushbutton. The energized contactor, C1, causes the three contacts, C1-1, C1-2, and C1-3, to close. This connects the armature and the shunt field directly across the DC supply. The motor is now running. Pressing the stop pushbutton de-energizes control relay CR1 and contactor C1, causing the motor to stop.

You may notice that the control relay appears slightly different in this drawing than in ones shown previously. This is because this is a DC control relay. These relays have two internal coils and one internal contact. One coil is much larger than the other coil. This large coil is the start winding and is used only when the coil is first energized. It assists the other coil in moving the contacts. Once energized, the internal contacts open, removing the start winding from the circuit. Now only the smaller, holding coil remains energized to maintain the contacts in

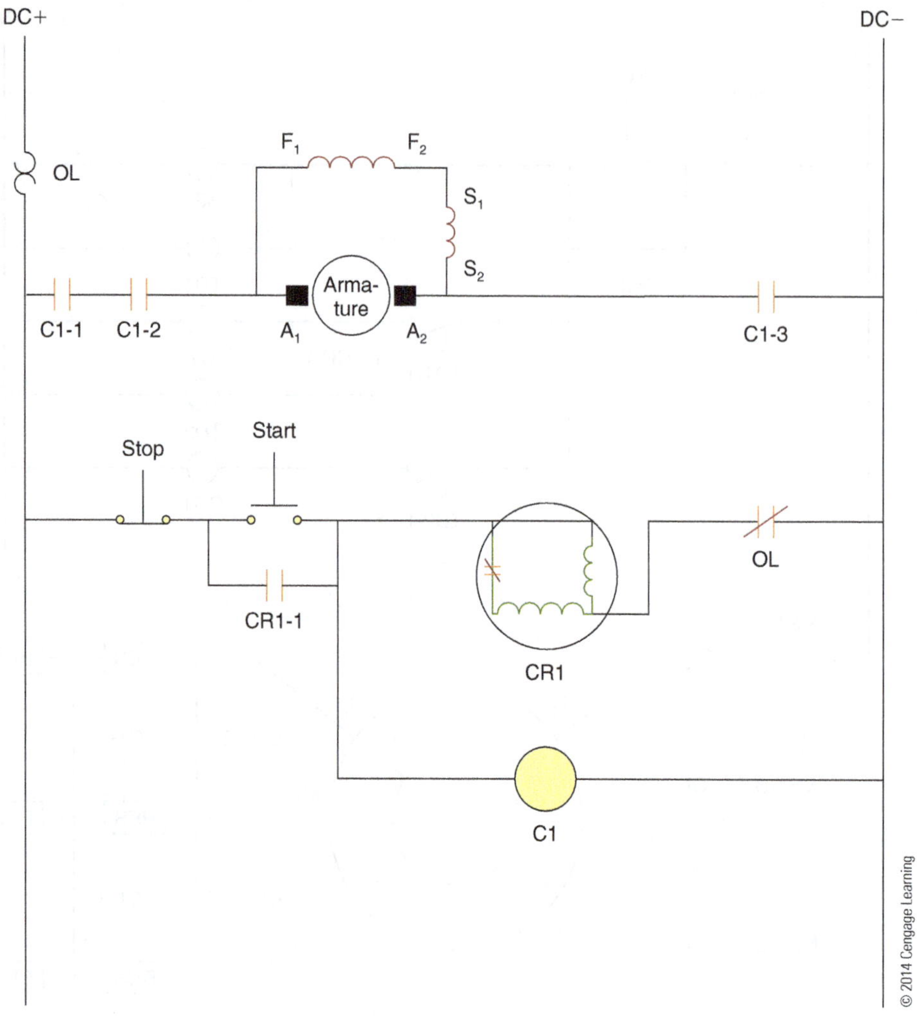

FIGURE 22-15 Across-the-line starting of a DC motor.

their energized state. Contactor contacts C1-1 and C1-2, connected in series as shown in this circuit, is necessary to suppress the arcing that occurs when switching DC circuits.

22-11 BRAKING

Plugging is a method of braking a motor. A motor is brought to a rapid stop by forcing it to rotate in the opposite direction from the direction in which it was running. Imagine driving your car at 70 miles per hour down the interstate and throwing the transmission into reverse! You would come to a quick halt, but you would possibly do some damage to your vehicle. The same principle is applied to a motor. Plugging reverses the direction of a running motor. The braking action is very quick. Plugging should be reserved for emergency stop applications only. Repeated plugging of a motor causes damage to it.

Figure 22-16A shows a plugging circuit. Following is how this circuit works:

1. The circuit is shown in its de-energized state.
   a. Notice timer contact TD1-1. This is a normally open timed-open (NOTO) contact. Although the contact is drawn in the open state, the contact is closed when the circuit is started.
2. The run pushbutton is pressed.
   a. This energizes control relay CR1, time-delay relay TD1, and the FWD motor starter.
      i. CR1-1 contacts close to provide a holding function around the run pushbutton.
      ii. CR1-2 contacts open.
      iii. Timer contact TD1-1 closes.
   b. The motor is now running.
3. The emergency stop pushbutton is pressed.
   a. This de-energizes control relay CR1, time-delay relay TD1, and the FWD motor starter.

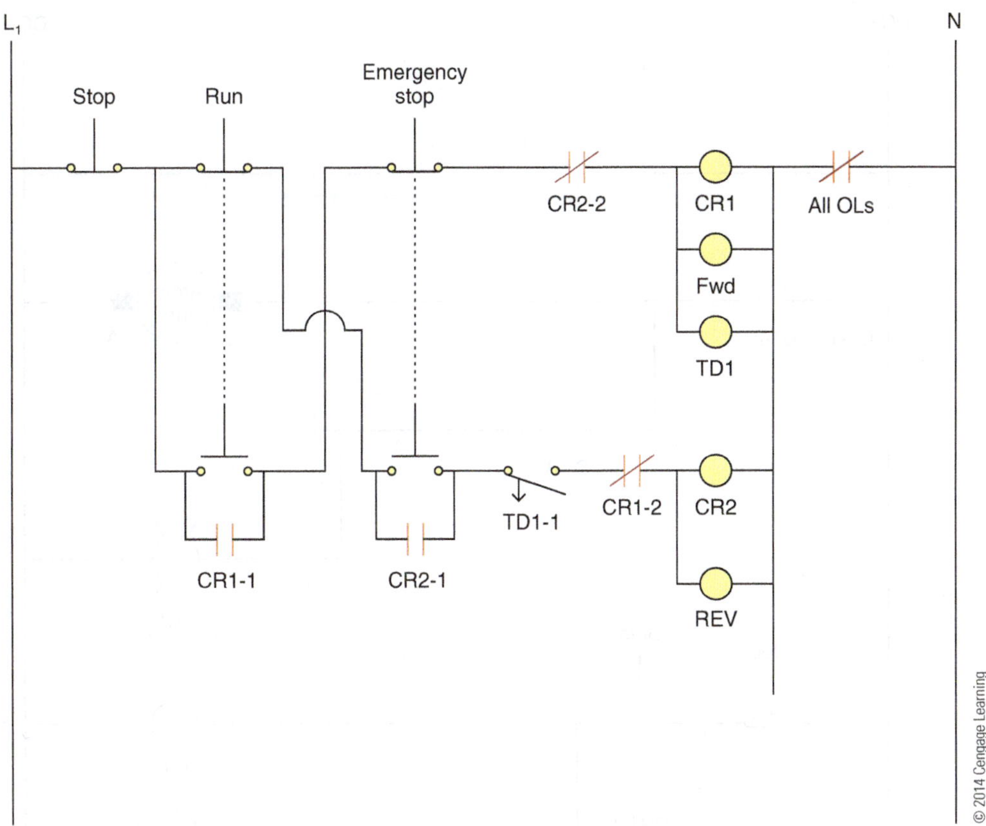

FIGURE 22-16A A plugging circuit.

i. The holding contacts, CR1-1, around the run pushbutton open.
ii. CR1-2 contacts reclose.
iii. Time-delay relay TD1 begins timing.
iv. Control relay CR2 and the REV motor starter energize.

b. The motor reverses its direction of rotation.

4. Time-delay relay TD1 times out.
   a. Timer contacts TD1-1 reopen.
      i. This de-energizes CR2 and the REV motor starter.
   b. The motor has now stopped.

It should be noted that the setting of the time delay of timer TD1 is critical. If the time is set too short, the motor does not come to a full stop. If the time is set too long, the motor continues to run in the opposite direction. The setting of the time delay is also dependent on the load that the motor is driving. A setting made with the motor lightly loaded would not be correct should the load increase. Likewise, a setting made with a heavy load would not be correct should the load decrease.

Figure 22-16B shows a braking circuit for a DC motor. This braking circuit uses a form of braking known as **dynamic braking**. When power is removed from an operating motor, the motor coasts to a stop. While coasting, the motor acts like a generator, because there is a residual magnetic field remaining in the iron of the motor. The coasting motor provides the relative motion of the motor windings, causing the motor to generate a small voltage. This voltage opposes the applied voltage to the motor. If this voltage could be reapplied to the motor, the motor would try to rotate in the opposite direction. This would have the effect of braking the motor. In the circuit shown in Figure 22-16B, resistors are used to provide a path for the generated voltage back to the motor. Following is how this circuit works:

1. The circuit is shown in its de-energized state.
2. The start pushbutton is pressed.
   a. This causes contactor C1, control relay CR1, and pilot light PL2 to energize.
      i. Contactor contacts C1-1 and C1-2 close.
         1) The motor is now running.
      ii. Contactor contact C1-3 opens.
      iii. Control relay contact CR1-1 closes to provide a holding function for the start pushbutton.
      iv. Pilot light PL2 indicates that the motor is running.

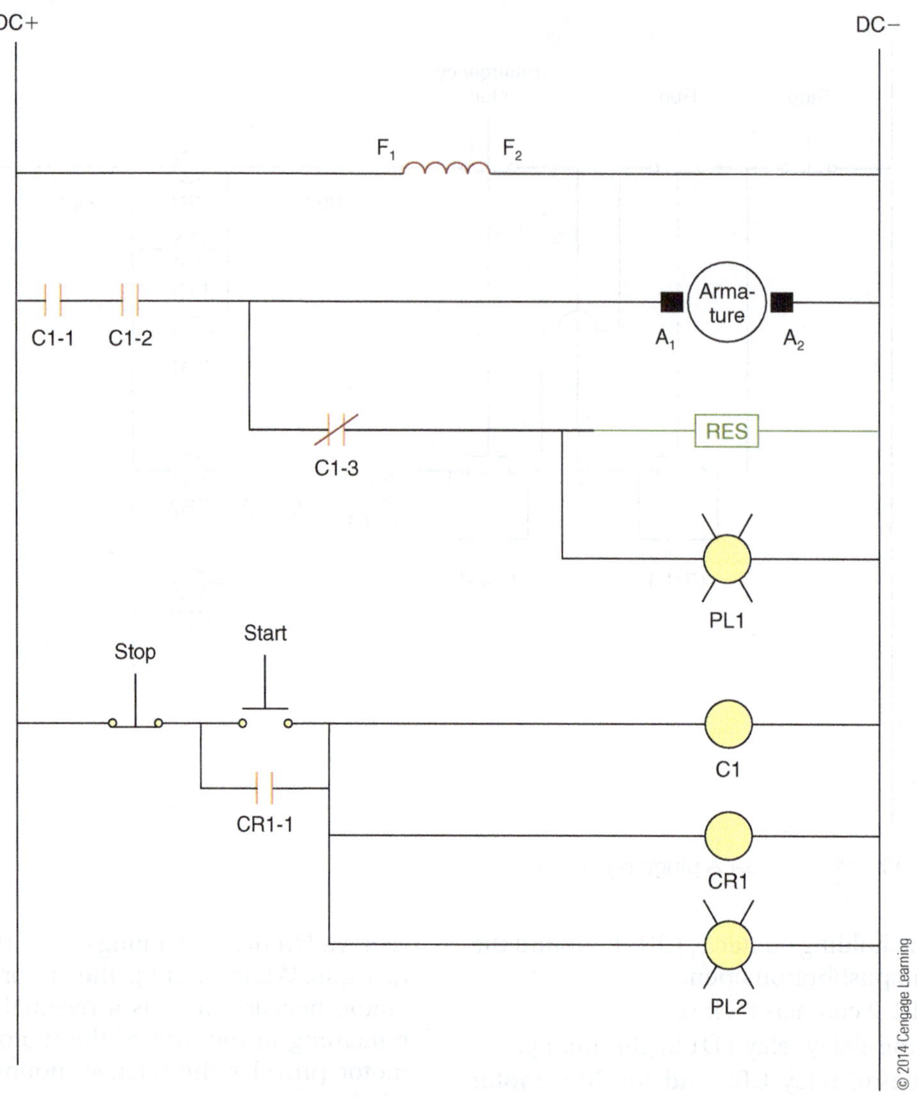

FIGURE 22-16B Dynamic braking of a DC motor.

3. The stop pushbutton is pressed.
   a. This causes contactor C1, control relay CR1, and pilot light PL2 to de-energize.
      i. Contactor contacts C1-1 and C1-2 reopen.
      ii. The motor is now coasting.
      iii. Contactor contact C1-3 recloses.
         1) This connects the dynamic braking resistor and pilot light PL1 across the armature of the motor.
            a) This circuit provides the braking effect to the motor.
            b) Pilot light PL1 will initially be bright and gradually grow dim as the braking energy from the motor bleeds off.
   b. The motor is now stopped.

One disadvantage of dynamic braking is that the braking effect diminishes as the motor slows down. It is, therefore, necessary to provide some means of auxiliary braking if a complete stop is required. This is usually accomplished by the addition of a mechanical brake.

SUMMARY

- A two-wire control is a method of motor control that uses manual devices to start and stop a motor. Only two wires are therefore needed to connect the control device into the circuit.
- Three-wire controls need three wires to connect the control device into the circuit. Three-wire controls also use some type of electromechanical switch to provide a latching, seal-in, memory, or holding function.

- It is sometimes necessary to control two motors from one location, thus the need for multiple start–stop controls.
- Often it is necessary to be able to change the direction of rotation of a motor.
- A drum switch is a mechanically held device that allows the direction of rotation of a motor to be changed.
- Jogging is a type of motor control in which the motor may be bumped or jogged slightly. This is helpful when using a motor to position an object or a piece of machinery.
- Plugging is a method of braking a motor. It brings a motor to a rapid stop by forcing it to rotate in the opposite direction from which it was running.

REVIEW QUESTIONS

1. What is meant by the term *two-wire control*?
2. Explain how a three-wire control differs from a two-wire control.
3. Describe the wiring method used on the stop and start pushbuttons when using multiple start–stop stations.
4. What is the purpose of interlocking the forward and reverse controls of a forward–reverse circuit?
5. Explain how dynamic braking is accomplished.

CHAPTER

23

Basic Industrial Electronics

This chapter introduces you to some of the more common electronic devices that are found in industrial applications today. Not too long ago, maintenance technicians pulled wire and ran conduit. The electronic technicians were the individuals who were responsible for understanding, installing, and troubleshooting devices containing electronic components. With the advent of programmable logic controllers (PLCs) and electronic variable speed/frequency drives, the field of electronics has entered the world of the maintenance technician. It is therefore important that you achieve a basic understanding of electronic devices and their applications.

OBJECTIVES

After studying this chapter, the student should be able to

- Identify the common symbols used in solid-state devices.
- Explain the operation of common solid-state devices.
- Identify common operational amplifier circuits.
- Identify common symbols used in digital logic.
- Explain the operation of common digital logic gates.
- Perform basic go–no go tests on some common devices.

23-1 DIODES

The **diode** is a two-terminal device. Several styles of diodes are shown in **Figure 23-1**. The schematic symbol and a physical drawing of a diode are shown in **Figure 23-2**. One terminal is called the **anode**, represented by the *arrowhead* symbol, and the other terminal is called the **cathode**, represented by the *T*-shaped symbol.

Diodes are placed in a circuit in such a fashion that the cathode has a more negative voltage applied to it with respect to the anode. When a diode is connected in this fashion, it conducts an electrical current and is said to be **forward biased**, as shown in **Figure 23-3**. If the polarities of the diode are reversed; that is, if the cathode is more positive with respect to the anode, the diode will block the current and is now said to be **reverse biased**, as shown in **Figure 23-4**.

When a diode is reverse biased, it could break down and conduct a current in the reverse direction. Diodes have a specification called the **peak inverse voltage (PIV)** rating. This rating is a measure of how much peak voltage a diode can withstand in the reverse direction. For example, if you have a diode that is rated at 400 volts PIV, this means that it can withstand a peak reverse voltage of up to 400 volts before it breaks down and conducts in the reverse direction. At this point, the diode is destroyed and must be replaced. It is important to realize that when replacing a defective diode, you must always equal or exceed its PIV rating. For example, if the defective diode in your circuit has a PIV rating of 200 volts, you can replace it with one that is rated at 200 volts PIV or 400 volts PIV or more, but not with one rated at 100 volts PIV or less, which would probably be destroyed when the circuit operated.

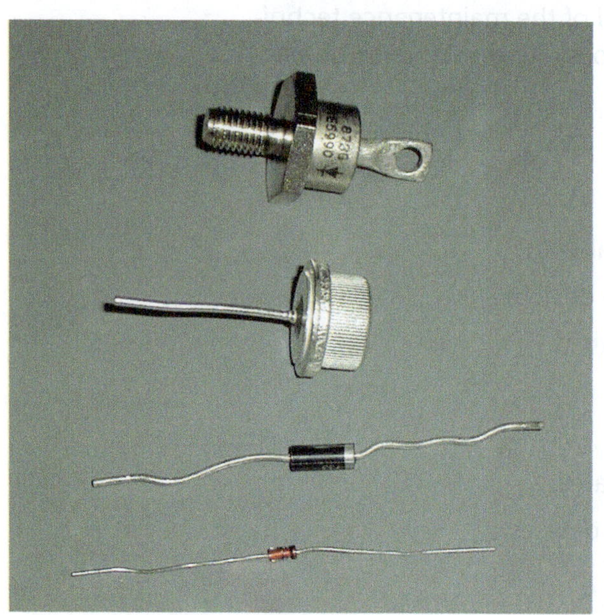

FIGURE 23-1 Various diodes.

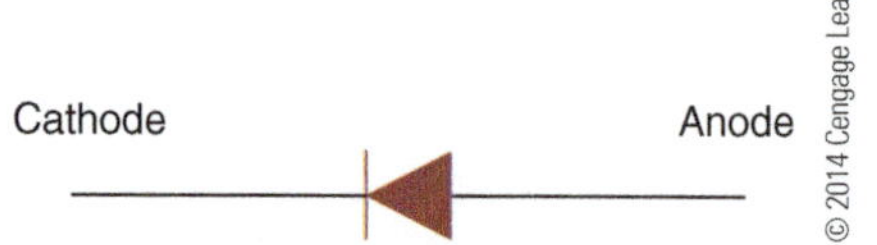

FIGURE 23-2 A schematic symbol of a diode. Note the cathode and anode. The anode is represented by the arrowhead symbol, and the cathode is represented by the T-shaped symbol.

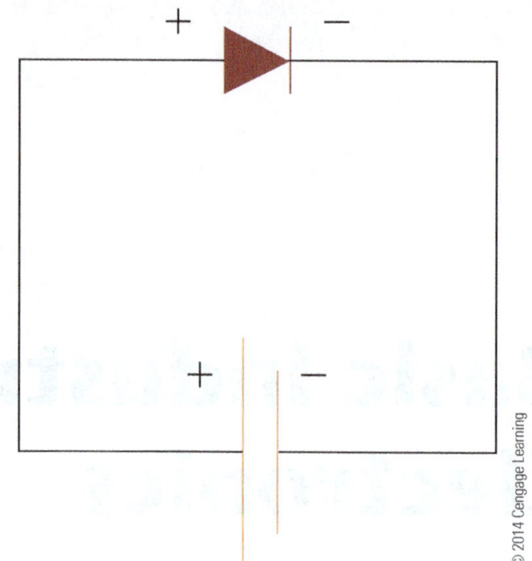

FIGURE 23-3 Forward-biasing a diode.

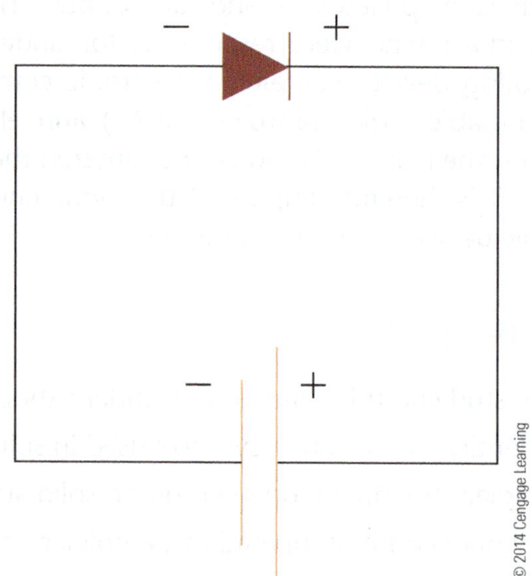

FIGURE 23-4 Reverse-biasing a diode.

Another factor to consider when replacing a defective diode is its current rating. For example, a diode may be rated at 1 ampere of current. This means that the maximum current that can flow through this diode when it conducts in the forward direction is 1 ampere. If the forward current flow exceeds 1 ampere, the diode will be destroyed. The replacement diode must, therefore, have a current rating equal to 1 ampere or greater.

Rectifier Diodes

A diode only conducts current when forward biased. If a diode is placed in a circuit that has an alternating current applied to it, it only conducts for half of the 360° of the AC sine wave. **Figure 23-5** shows a circuit with a diode and an applied AC voltage. Following is a step-by-step analysis of this circuit.

Assume that L1 is positive and L2 is negative. Current flows from L2 through the load and the diode and return to L1. Notice that the diode is forward biased, allowing current to flow. This causes the output voltage to appear for this half of the AC sine wave input. Refer to **Figure 23-6**. Now imagine that the AC sine wave input voltage has reversed its polarity, so L1 is now negative and L2 is positive. Current is blocked because the diode is now reverse biased, and there can be no voltage across the load for this half of the AC sine wave input.

Depending on how the diode is placed in the circuit, the output voltage is either all positive or all negative. The output voltage is therefore DC. The diode has been used to convert the AC input voltage into a DC output voltage. This conversion is called **rectification**. When a diode is used to perform rectification, it is called a **rectifier diode**, or simply, a **rectifier**. Because the output voltage appears for only half of the AC sine wave input voltage, this circuit is called a **half-wave rectifier**.

Assume that a DC voltage is needed to be present for both half-cycles of the AC sine wave input voltage. If you use two diodes and a center-tapped transformer with an AC sine wave input voltage, the output voltage appears during both half-cycles of the AC input. This full-wave rectifier circuit appears in **Figure 23-7**. Following is a step-by-step analysis of this circuit.

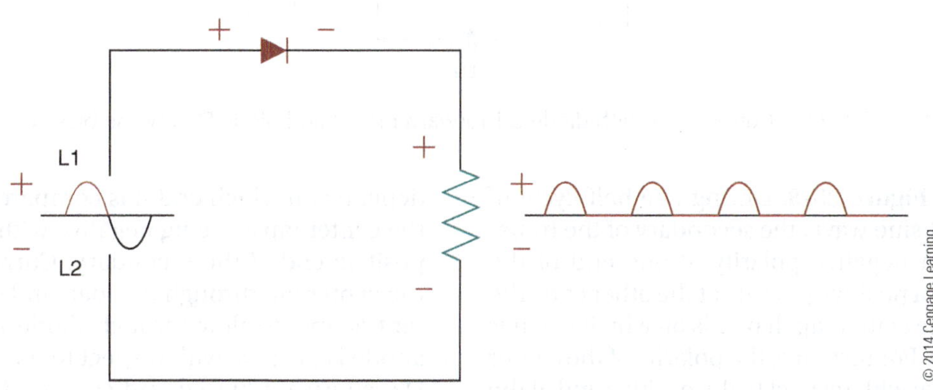

FIGURE 23-5 Rectification—the diode is forward biased.

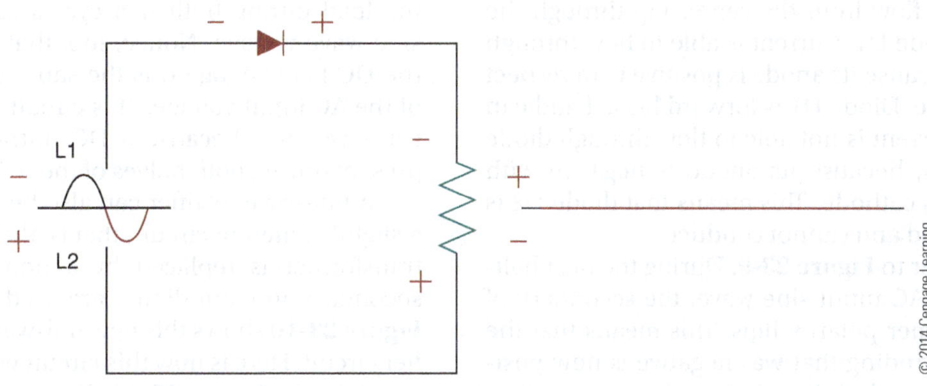

FIGURE 23-6 Rectification—the diode is reverse biased.

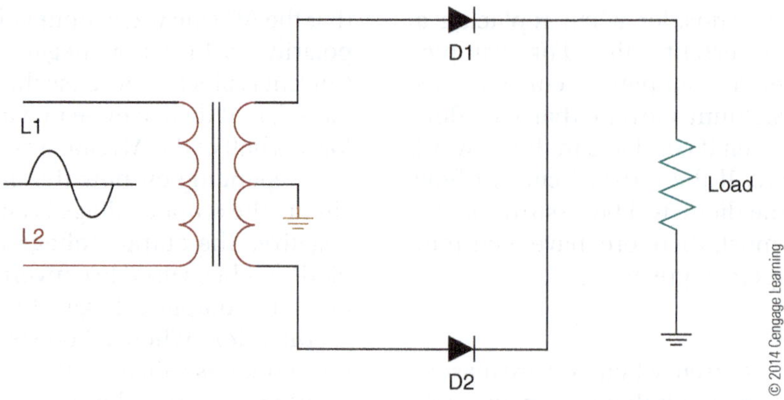

FIGURE 23-7 A full-wave rectifier circuit with a center-tapped transformer.

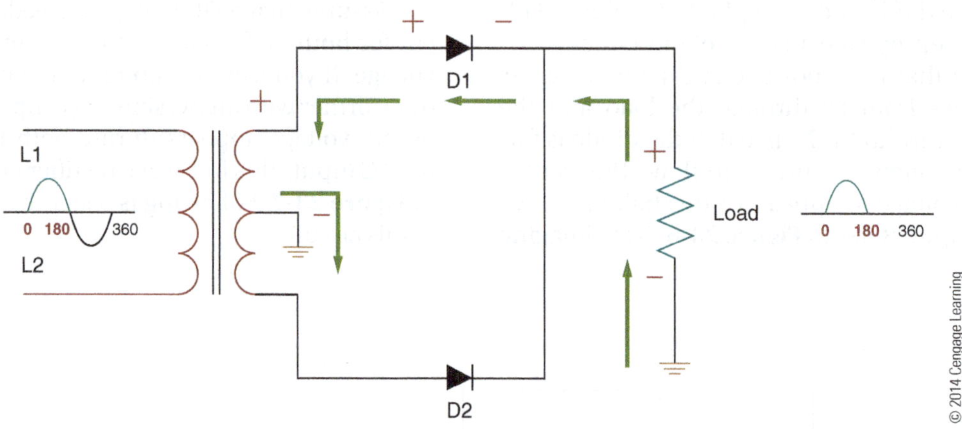

FIGURE 23-8 Current flow with diode D1 forward biased and diode D2 reverse biased.

Refer to **Figure 23-8**. During one half-cycle of the AC input sine wave, the secondary of the transformer has a negative polarity at one end of the winding and a positive polarity at the other end. The polarity of the center tap depends on which end it is compared to. For instance, the polarity of the center tap is negative with respect to the positive end of the secondary. However, the polarity of the center tap is positive with respect to the negative end of the secondary. Consider the center tap as being negative with respect to the positive end of the secondary. Current will flow from the center tap through the load and diode D1. Current is able to flow through diode D1 because its anode is positive with respect to its cathode. Diode D1 is forward biased and can conduct. Current is not able to flow through diode D2, however, because its anode is negative with respect to its cathode. This means that diode D2 is reverse biased and cannot conduct.

Now refer to **Figure 23-9**. During the next half-cycle of the AC input sine wave, the secondary of the transformer polarity flips. This means that the end of the winding that was negative is now positive, and the end of the winding that was positive is now negative. The polarity of the center tap still depends on which end it is compared to. Consider the center tap as being negative with respect to the positive end of the secondary. Current flows from the center tap through the load and diode D2. Current is able to flow through diode D2 because its anode is positive with respect to its cathode. Diode D2 is forward biased and can conduct. Current is not able to flow through diode D1 because its anode is negative with respect to its cathode. This means that diode D1 is reverse biased and cannot conduct.

Notice that a DC voltage was present across the load during both half-cycles of the AC input sine wave voltage. Notice, too, that the polarity of the DC load voltage was the same for both halves of the AC input voltage. This circuit is called a **full-wave rectifier** because a DC output voltage was present during both halves of the AC input voltage.

A full-wave rectifier can also be fashioned with a slightly different circuit; that is, the center-tapped transformer is replaced by a non-center-tapped secondary and four diodes are used instead of two. **Figure 23-10** shows this new full-wave bridge rectifier circuit. Here is how this circuit works.

Refer to **Figure 23-11**. During one half-cycle of the AC input sine wave, the secondary of the

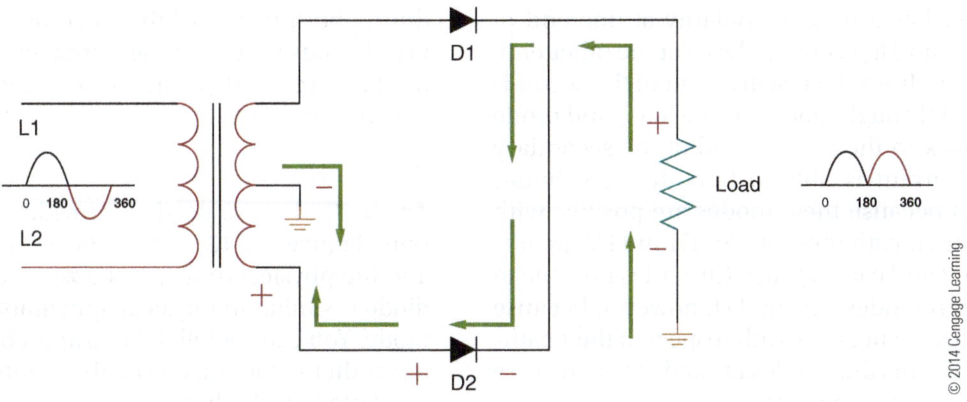

FIGURE 23-9 Current flow with diode D1 reverse biased and diode D2 forward biased.

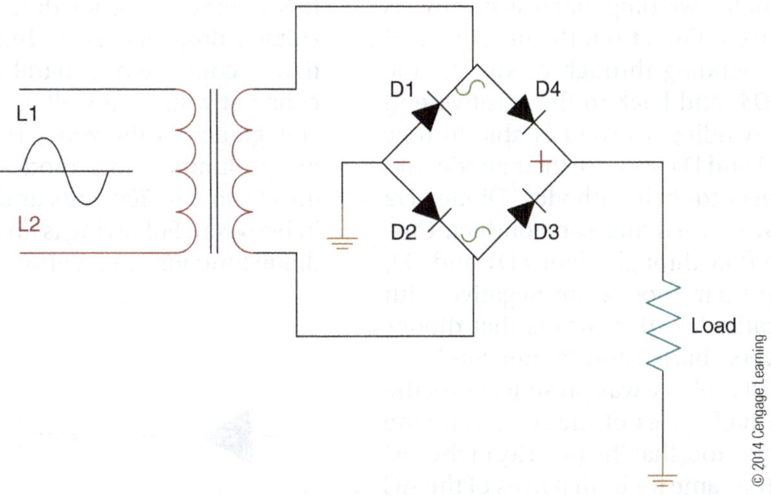

FIGURE 23-10 A full-wave bridge rectifier circuit. Note the absence of the center-tapped transformer.

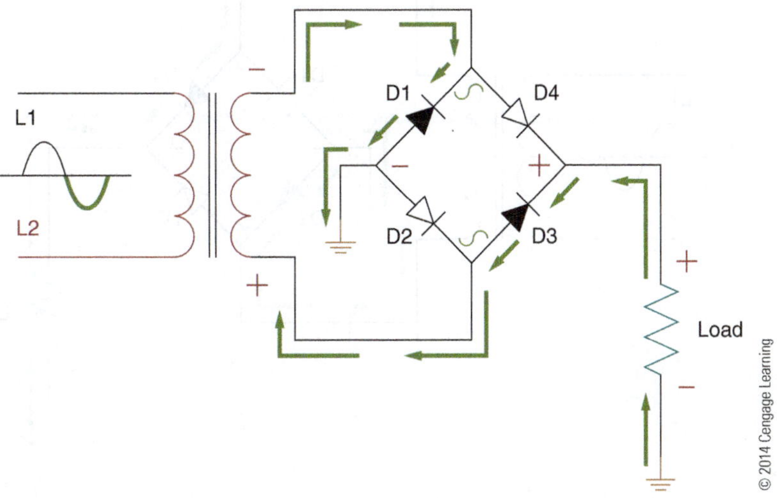

FIGURE 23-11 Current flow with diodes D1 and D3 forward biased and diodes D2 and D4 reverse biased.

transformer has a negative polarity at one end of the winding and a positive polarity at the other end. Current flows from the negative end of the secondary winding through diode D1, the load, and diode D3, and back to the positive end of the secondary winding. Current is able to flow through diodes D1 and D3 because their anodes are positive with respect to their cathodes. Diodes D1and D3 are forward biased and can conduct. Current is not able to flow through diodes D2 and D4, however, because their anodes are negative with respect to their cathodes. This means that diodes D2 and D4 are reverse biased and cannot conduct.

Refer to **Figure 23-12**. During the next half-cycle of the AC input sine wave, the secondary of the transformer polarity flips. This means that the end of the winding that was negative is now positive, and the end of the winding that was positive is now negative. Current flows from the negative end of the secondary winding through diode D2, the load, and diode D4, and back to the positive end of the secondary winding. Current is able to flow through diodes D2 and D4 because their anodes are positive with respect to their cathodes. Diodes D2 and D4 are forward biased and can conduct. Current is not able to flow through diodes D1 and D3, however, because their anodes are negative with respect to their cathodes. This means that diodes D1 and D3 are reverse biased and cannot conduct.

Notice that a DC voltage was present across the load during both half-cycles of the AC input sine wave voltage. Notice, too, that the polarity of the DC load voltage was the same for both halves of the AC input voltage. This circuit is called a full-wave rectifier because there was a DC output voltage present during both halves of the AC input voltage. However, because of the configuration of the four diodes in this circuit, this circuit is called a **full-wave bridge rectifier**.

Zener Diodes

Zener diodes are used to provide voltage regulation. **Figure 23-13** shows the schematic symbol and the physical drawing of a zener diode. A zener diode is similar in physical appearance to a regular diode. You cannot tell the two apart by simply looking at them. You must note the identifying number imprinted on the body of the diode and then contact the manufacturer to determine whether it is a regular diode or a zener diode.

When used in a circuit, the zener diode operates in the reverse breakdown region. When operated in this manner, a zener diode maintains a constant voltage drop across it. This means that any load that is connected in parallel with the zener has a constant voltage as well. Even if the load varies, the voltage across the zener remains constant, within design limits. Zener diodes are available with ratings from 2 to 200 volts and practically any voltage in between. Following is an example of how a zener diode functions as a **voltage regulator**.

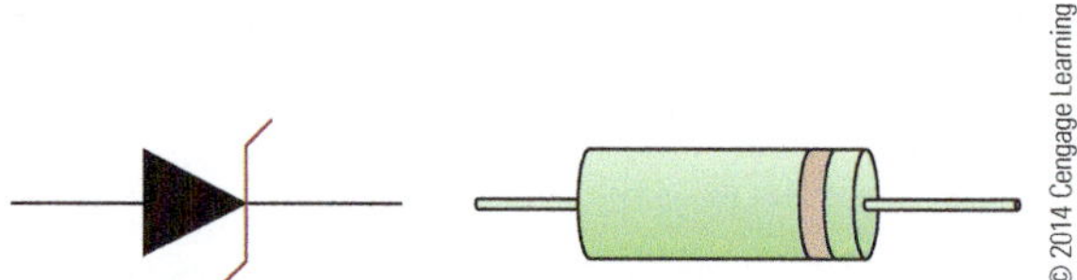

FIGURE 23-13 A zener diode: schematic and physical appearance. Note the band, which indicates the cathode.

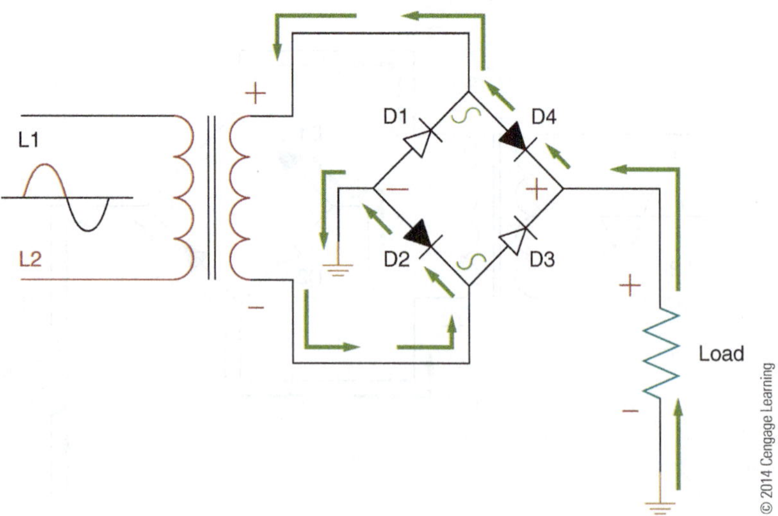

FIGURE 23-12 Current flow with diodes D1 and D3 reverse biased and diodes D2 and D4 forward biased.

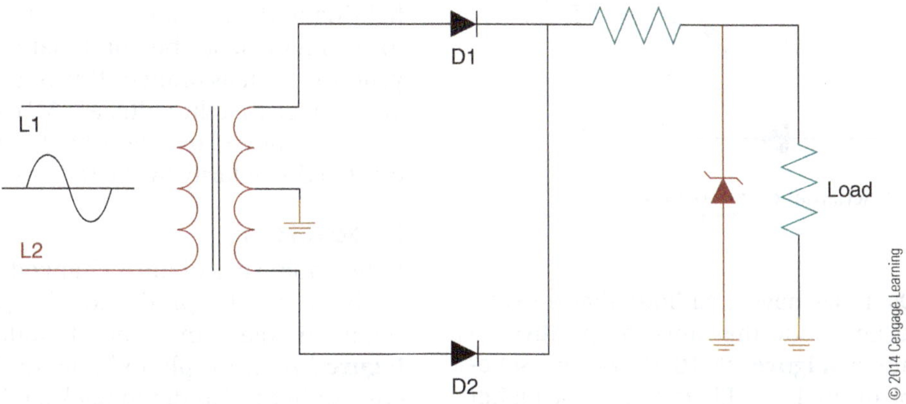

FIGURE 23-14 A zener diode is installed so that it is reverse biased. The load is placed in parallel with the zener diode.

Refer to **Figure 23-14**. Notice the manner in which the zener diode is connected in this circuit. The anode is connected to the negative portion of the circuit, whereas the cathode is connected to the positive portion. This is how a zener diode is installed so that it operates in the reverse breakdown region. When the circuit is energized, the zener diode does not conduct until the voltage drop across the zener is slightly more than its voltage rating. For example, a 12-volt zener diode does not conduct if the voltage across the zener is 11.5 volts. At this point, the load sees 11.5 volts as well. However, if the voltage drop across the zener increases to 12.1 volts, the zener diode will conduct. When the zener conducts, the voltage drop across the zener is a constant 12 volts, and the load sees 12 volts as well. Should the input voltage to the zener diode increase to 15 volts, the voltage drop across the zener will remain at 12 volts. This means that the load still sees 12 volts as well. This is how a zener diode is used to provide voltage regulation. Notice, however, that should the voltage fall below the voltage rating of the zener, the load voltage will fall as well. In this respect, the zener diode only provides regulation for excess voltage, not for undervoltage conditions.

Light-Emitting Diodes (LEDs)

Light-emitting diodes (LEDs) are diodes that have been doped in such a fashion that, when forward biased, they emit light. **Figure 23-15** shows several

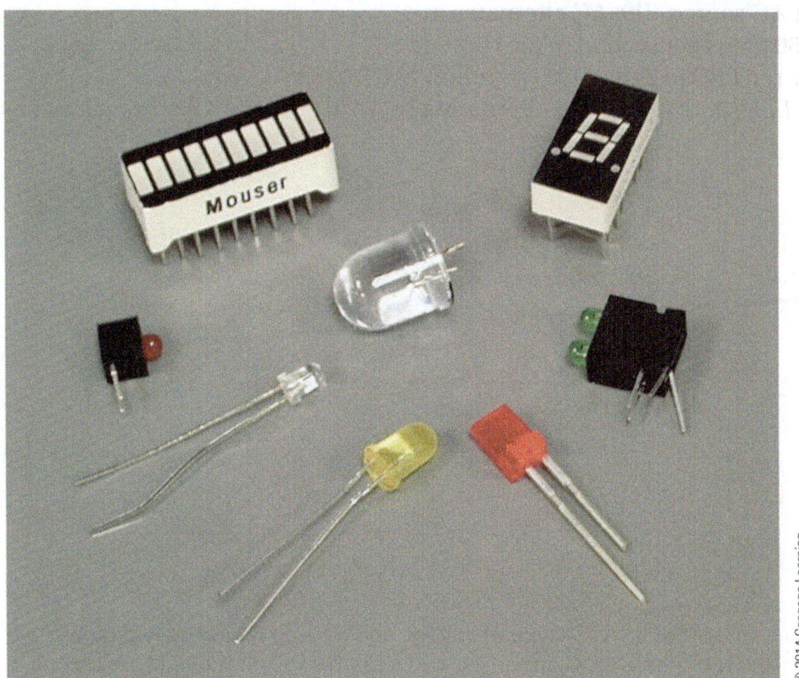

FIGURE 23-15 Various light-emitting diode (LED) packages.

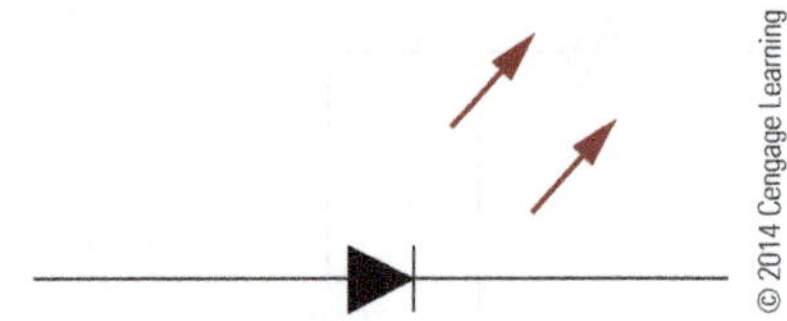

FIGURE 23-16 A schematic symbol of an LED.

styles of LEDs. LEDs have a cathode that must be negative, and an anode that must be positive, in order to conduct. **Figure 23-16** shows the schematic symbol of an LED. LEDs require a higher voltage and more current to conduct, however. Typically, LEDs require approximately 1.4 volts to forward bias the junction. When LEDs conduct, there is typically a current of approximately 20 milliamperes flowing through the junction.

LEDs are available in a wide variety of colors, such as red, yellow, green, and blue. Some LEDs have a clear lens, whereas others have a colored lens. In addition, **infrared LEDs (IRLEDs)** are available and are used in remote controls for TVs, VCRs, and other applications.

An interesting adaptation of the LED is the voltage indicator. Manufacturers package red and yellow LEDs in inverse parallel in the same package. See **Figure 23-17**. When the LED is connected to a DC source, either the red or the yellow LED light. Suppose that the red LED lit. If the polarity of the DC is reversed, the other LED will light. In this example, the yellow LED is now lit. This indicates the polarity of the DC. Now suppose that the LED is connected to an AC source. The AC alternates in polarity (typically 60 times per second in the United States), causing the red LED to light, then the yellow LED, then the red LED, and so on. The alternating lighting of the red and yellow LEDs occurs too fast to distinguish, so the combination of the red and yellow is seen as orange. This is a voltage indicator. You can tell that the voltage is AC by the orange light or DC by the red or yellow light; and if it is DC, you can tell the polarity by the red or yellow light.

Photodiode

If the doping of a diode is changed, yet another type of diode can be produced: the **photodiode**. The schematic diagram of the photodiode is shown in **Figure 23-18**. The photodiode conducts in the presence of light. The diode package has a small window that allows light to enter. Refer to **Figure 23-19**. The light enters the window and strikes the P-N junction, causing the photodiode to conduct in the reverse direction. If more light enters the window, the photodiode conducts more current. If less light enters the window, the photodiode conducts less current. Photodiodes are used to sense the presence or absence of light.

Testing Diodes

It is possible to perform a simple test on a diode to determine whether it is functioning properly. The test may be performed either in or out of circuit. However, when testing in circuit, be certain that power is removed and that the capacitors are fully

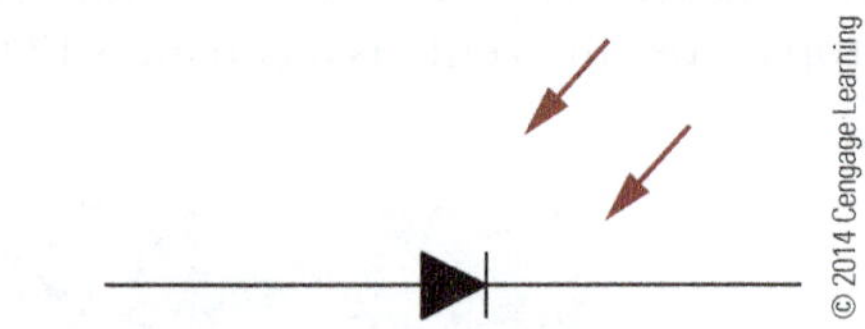

FIGURE 23-18 A schematic symbol of a photodiode.

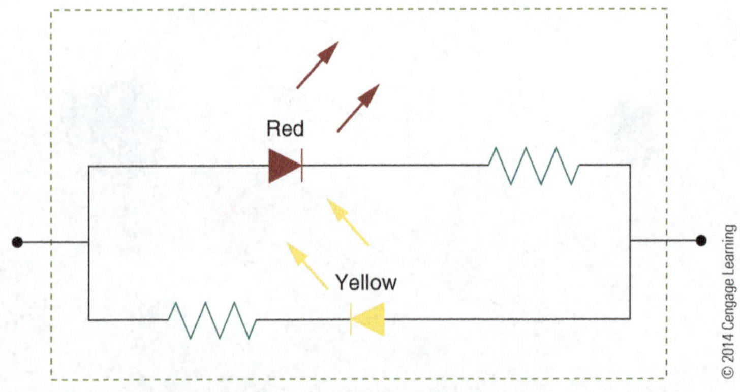

FIGURE 23-17 A type of voltage and polarity indicator. Red glows with positive DC on the left and negative DC on the right. Yellow glows with positive DC on the right and negative DC on the left. Orange glows with AC applied.

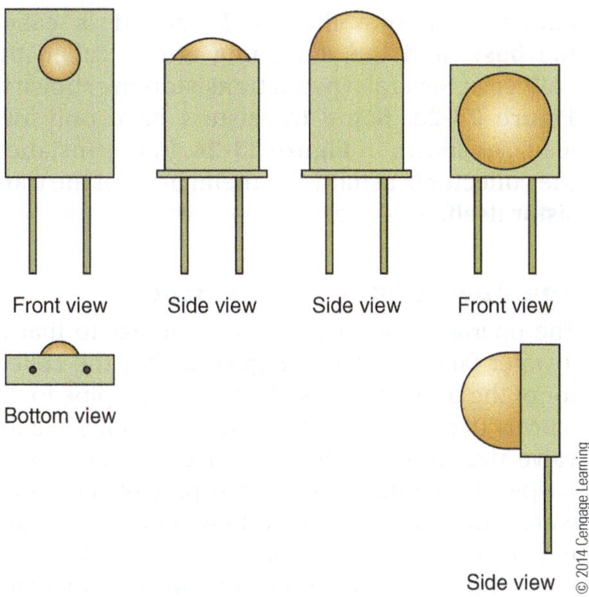

FIGURE 23-19 Various photodiode packages.

discharged. Also, if any of the readings are questionable, the diode should be removed and retested out of circuit.

This test uses nothing more than a simple multimeter. Refer to **Figure 23-20**. Most DMMs have a diode check position, as indicated by the diode symbol. Set the meter to this position. Place the negative lead of the DMM on the cathode of the diode to be tested. Connect the positive lead from the DMM to the anode of the diode under test. Notice the reading on the DMM. If the diode is connected properly (negative to the cathode and positive to the

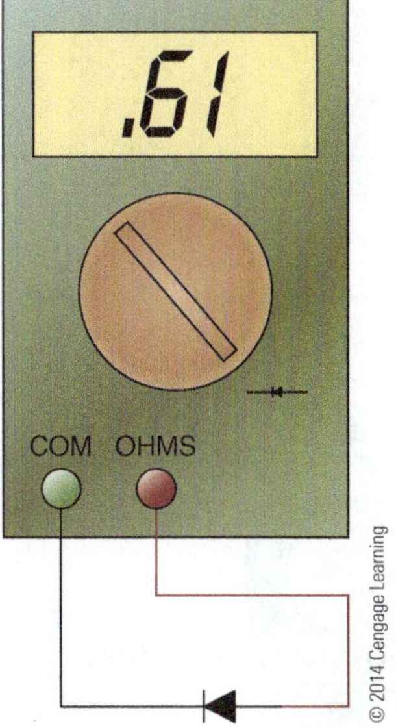

FIGURE 23-20 Testing a diode in the forward-bias direction with an ohmmeter on the diode check function. Note the low reading.

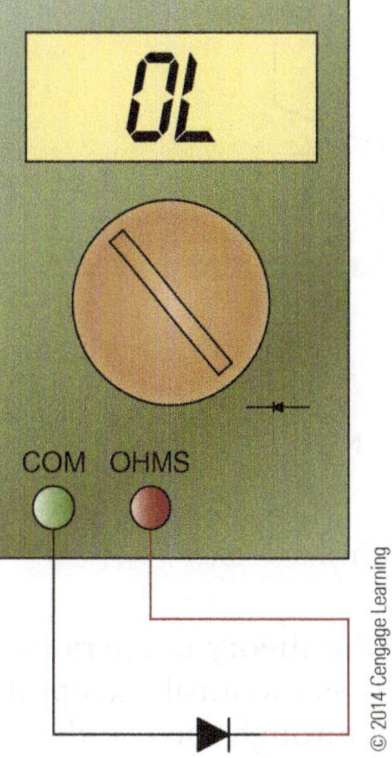

FIGURE 23-21 Testing a diode in the reverse-bias direction with an ohmmeter on the diode check function. Note the high reading.

anode), and if the diode is good, the DMM displays a reading of a few tenths of a volt. If the DMM does not display this reading, the diode may be defective or connected backwards (reverse biased), or the DMM may be defective. You can verify whether a the DMM is working properly by checking a known good diode. Switch the connections to the diode under test, as shown in **Figure 23-21**. This checks the diode for conduction in the reverse-bias direction. If the DMM now reads 0 volt, the diode is reverse biased and is probably good. If the DMM does not produce a reading with the diode connected either way, the diode is probably open and should be replaced. If the DMM produces the same reading with the diode connected either way, the diode is probably shorted and should be replaced. Even if the diode checked as good, it is still a possibility that under circuit conditions the diode may fail. This is a remote possibility, but a possibility nonetheless.

23-2 TRANSISTORS

Bipolar Junction Transistor

The **bipolar junction transistor (BJT)** is a three-terminal device and is available in two versions. The schematic symbols for both versions are shown in **Figure 23-22**. Notice that the direction of the arrowhead on one of the terminals is reversed. The symbol with the arrowhead pointing *in* is known as a **PNP (positive-negative-positive) transistor**. The symbol with the arrowhead pointing *out* is known as an **NPN (negative-positive-negative)**

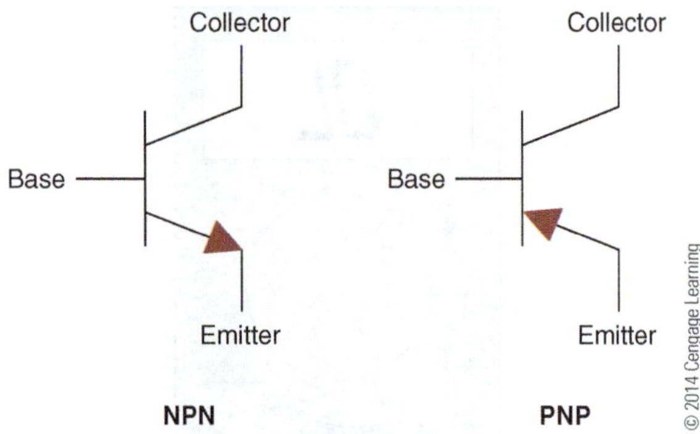

FIGURE 23-22 Schematic symbols of transistors: NPN (left); PNP (right).

transistor. The theory of operation of both types of transistors is identical, except the direction of current flow through the two devices is reversed. The three terminals are known by the following names: the one with the *arrowhead* is called the **emitter**; the one that is *T* shaped is called the **base**; and the remaining one is called the **collector**. Several types of transistors are shown in **Figure 23-23**. Some transistors have only two leads, as shown in **Figure 23-24**. In this instance, the collector is usually the metal body of the transistor itself.

The Operation of a Transistor

The operation of a transistor, is similar to that of a water faucet. Refer to **Figure 23-25**. The collector of the transistor acts like the supply pipe to the faucet. The base of the transistor functions like the valve that controls the flow of water (electrons). Lastly, the emitter acts as the part of the faucet where the water (electrons) flows out. In an actual transistor, the current flowing through the base and the emitter controls the current flowing through the collector and the emitter.

FIGURE 23-23 Various transistor packages.

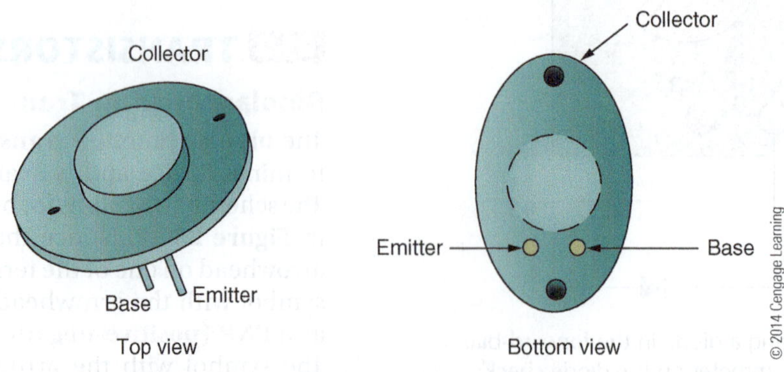

FIGURE 23-24 Some transistors use the metal case as the third lead.

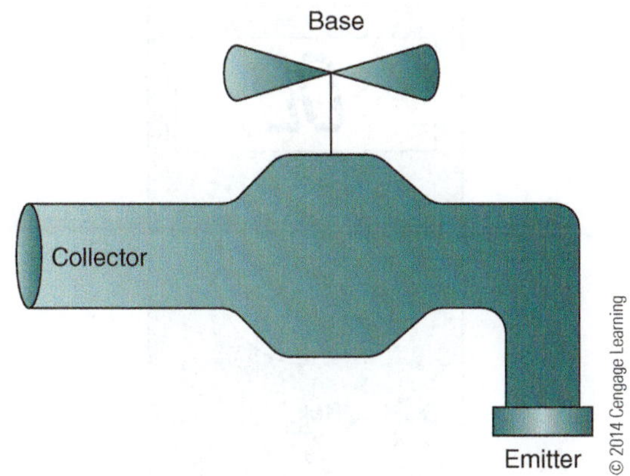

FIGURE 23-25 The current from the collector to the emitter is controlled by the base.

Transistors can be used as switches and amplifiers. When used as a switch, transistors have the capability of turning on and off several thousand times per second. This switching occurs without moving parts, making the transistor very efficient and reliable. When used as an amplifier, transistors are capable of reproducing and amplifying an input signal. It is not unusual for transistor amplifiers to have a gain of several hundred. In addition, transistors may be connected together in such a fashion as to cause an even greater gain to occur. When transistors are connected in this fashion, a **Darlington** amplifier is formed. **Figure 23-26** shows the schematic symbol of a Darlington amplifier. The gain of a Darlington transistor may exceed several thousand.

Testing Transistors

The transistor may be tested in or out of circuit. When testing in circuit, remember to de-energize the circuit and verify that the capacitors are fully discharged. It is important to note that if a transistor is *questionable* when tested in circuit, you must retest it out of circuit to be certain it is defective.

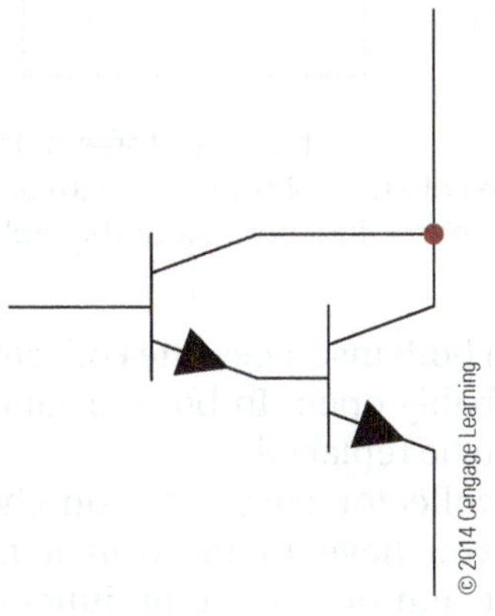

FIGURE 23-26 A schematic symbol of a Darlington transistor.

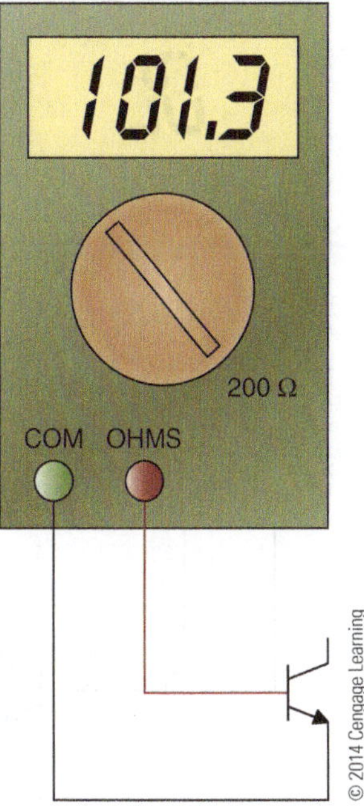

FIGURE 23-27 Testing the emitter–base junction (forward bias) of an NPN transistor with an ohmmeter on the diode check function. Note the low reading.

Refer to **Figure 23-27**. Begin by setting the DMM to a low-resistance range, such as the 200-ohm or 2-kilohm position. Do not use the low-ohm or the diode check position because these will not supply sufficient current to accurately check the transistor under test.

Notice that an NPN transistor is connected to a DMM. The negative lead of the DMM is connected to the emitter of the transistor. The positive lead of the DMM is connected to the base of the transistor. If the transistor is good, the DMM will read a low resistance (typically less than 1 kilohm). Now reverse the DMM connections, as shown in **Figure 23-28**. Put the positive lead of the DMM on the emitter and the nega tive lead from the DMM on the base. A good transistor typically measures in excess of 100 kilohms. If you measured a low value in both in stances, the emitter-base junction is probably shorted. If the measurement was high in both instances, the emitter-base junction is probably open. In both instances, the transistor must be replaced.

If the emitter–base junction check was satisfactory, you must now check the collector–base junction. Refer to **Figure 23-29**. Connect the negative lead of the DMM to the collector of the transistor. Connect the positive lead of the DMM to the base of the transistor. If the collector–base junction is good, you should measure less than 1 kilohm. Now reverse the connections to the transistor, as shown in **Figure 23-30**. Connect the positive lead from the

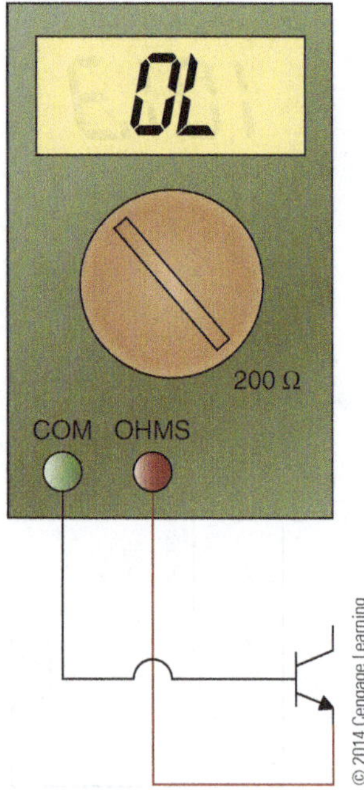

FIGURE 23-28 Testing the emitter–base junction (reverse bias) of an NPN transistor with an ohmmeter on the diode check function. Note the high reading.

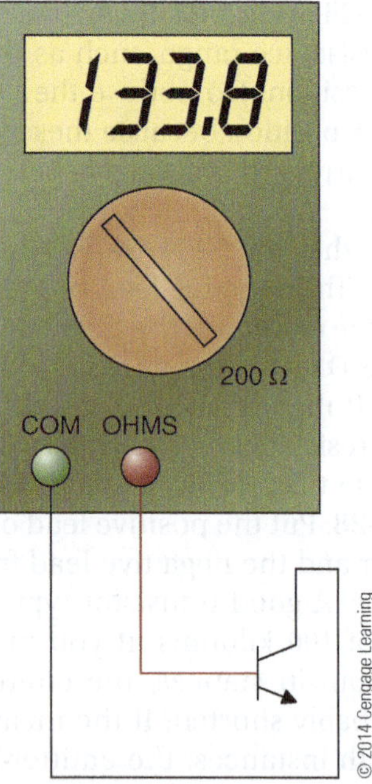

FIGURE 23-29 Testing the collector–base junction (forward bias) of an NPN transistor with an ohmmeter on the diode check function. Note the low reading.

DMM to the collector of the transistor. Connect the negative lead of the DMM to the base of the transistor. If the collector-base junction is good, you should measure greater than 100 kilohms. If you measured a low value in both instances, the collector-base junction is probably shorted. If the measurement

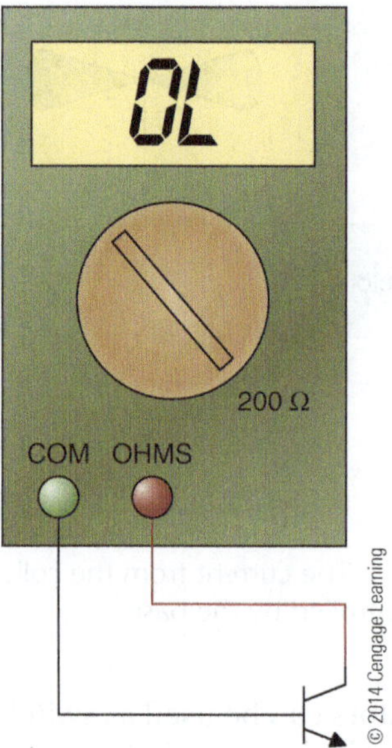

FIGURE 23-30 Testing the collector–base junction (reverse bias) of an NPN transistor with an ohmmeter on the diode check function. Note the high reading.

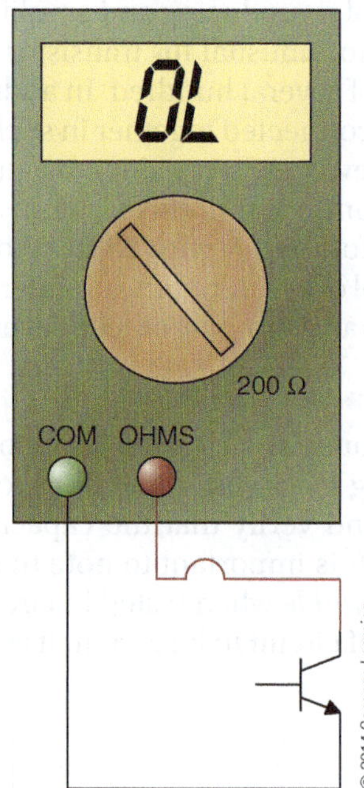

FIGURE 23-31 Testing the collector–emitter junction (forward bias) of an NPN transistor with an ohmmeter on the diode check function. Note the high reading.

was high in both instances, the collector-base junction is probably open. In both instances, the transistor has to be replaced.

If the collector-base junction check was satisfactory, you have to perform a final check—that of the emitter-collector junction. Refer to **Figure 23-31**. Begin by connecting the negative lead of the DMM to the emitter terminal. Next connect

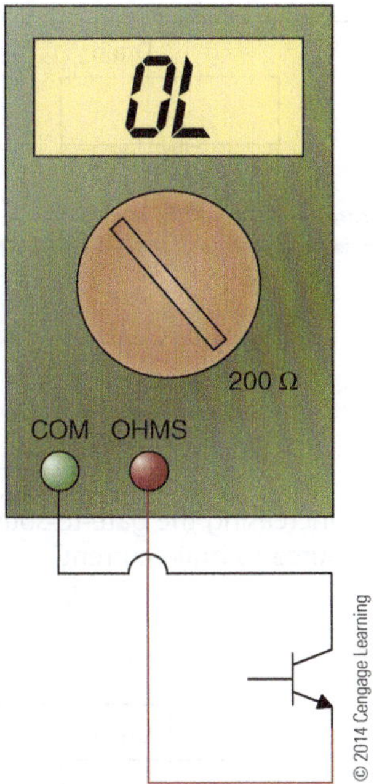

FIGURE 23-32 Testing the collector–emitter junction (reverse bias) of an NPN transistor with an ohmmeter on the diode check function. Note the high reading.

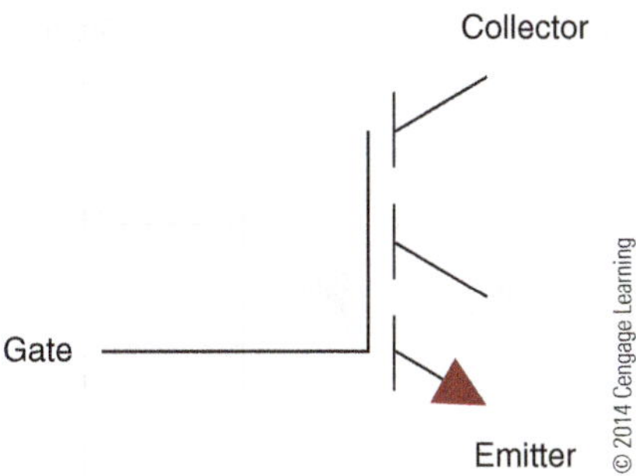

FIGURE 23-33 One form of a schematic symbol used to represent an IGBT.

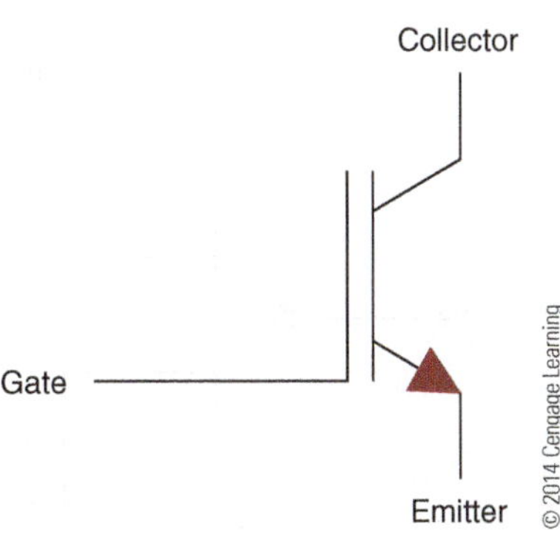

FIGURE 23-34 The more common schematic symbol used to represent an IGBT.

the positive lead of the DMM to the collector. If the transistor is good, the DMM will indicate more than 100 kilohms. Now reverse the connections to the transistor, as shown in **Figure 23-32**. Connect the positive lead of the DMM to the emitter and the negative lead of the DMM to the collector. Again the DMM should indicate more than 100 kilohms. If the DMM indicated a low resistance in both tests, the emitter to collector junction is probably shorted and the transistor has to be replaced.

If the transistor passed all of the above tests, it is more than likely good. However, it is possible that under circuit voltages, the transistor may break down and fail. It is also common for transistors to fail after they have heated up. Therefore, the DMM test is not 100% reliable, but it can be an effective and quick troubleshooting tool.

Insulated Gate Bipolar Transistors

Another three-terminal device is called the **insulated gate bipolar transistor (IGBT)**. The two schematic symbols in use for the IGBT are shown in **Figure 23-33** and **Figure 23-34**. In Figure 23-33, the terminal with the arrowhead is the emitter, the *L*-shaped terminal is the gate, and the remaining terminal is the collector. In Figure 23-34, the emitter is the terminal with the arrowhead symbol, the *L*-shaped lead is the **gate**, and the remaining lead is the collector.

IGBTs are finding more and more usage in industry due to their high-speed switching ability. In addition, they have very low internal resistance, which decreases any losses they inject into the circuit.

Because of the way an IGBT is constructed, a reliable check cannot be performed with an ohmmeter. A better understanding of electronic circuits is needed if one is to determine if an IGBT is defective.

Field-Effect Transistors

There are two types of field-effect transistors: the **junction field-effect transistor (JFET)**, and the **metal-oxide semiconductor field-effect transistor (MOSFET)**.

Junction Field-Effect Transistors

JFETs are divided into two types: the **n-channel JFET** and the **p-channel JFET**. The three leads of a JFET are labeled *source*, *gate*, and *drain*. Notice the schematic symbols of an n-channel JFET (**Figure 23-35**) and a p-channel JFET (**Figure 23-36**). The difference between the two is the direction of the arrowhead on the gate lead. The arrowhead points *inward* on the n-channel JFET and *outward*

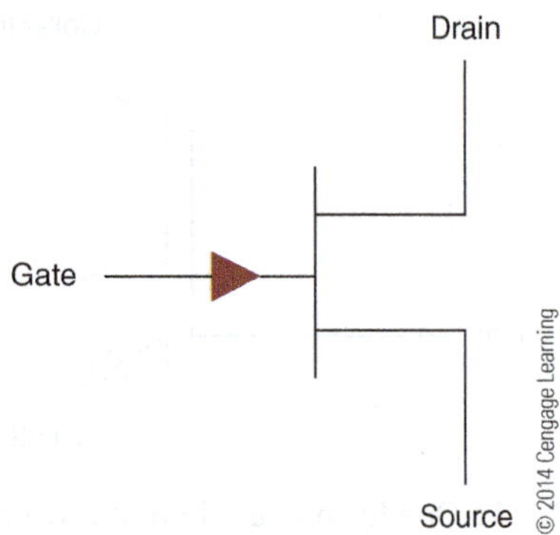

FIGURE 23-35 An n-channel JFET schematic symbol.

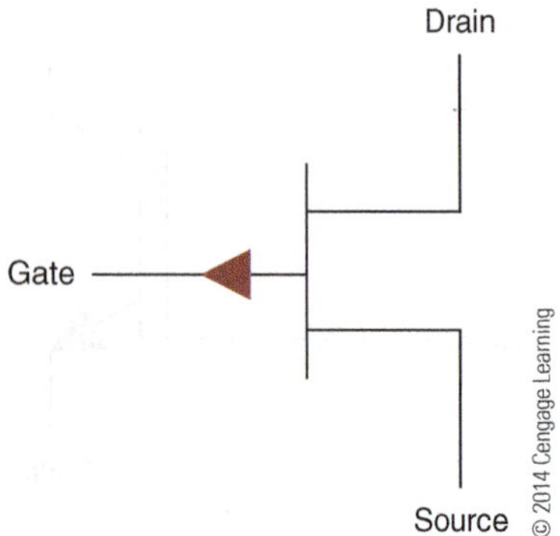

FIGURE 23-36 A p-channel JFET schematic symbol.

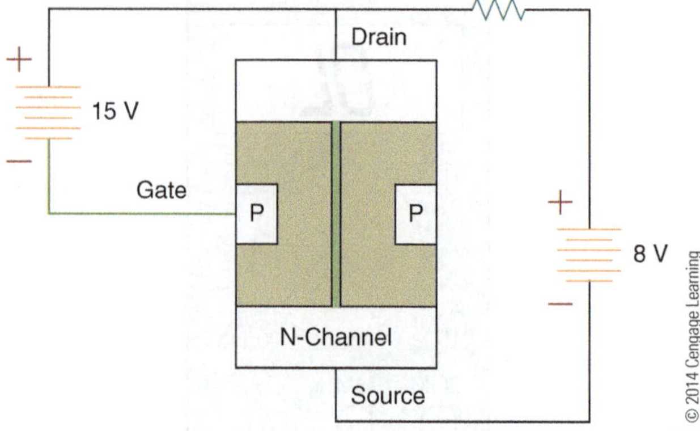

FIGURE 23-37 Increasing the gate-to-source voltage decreases the source-to-drain current.

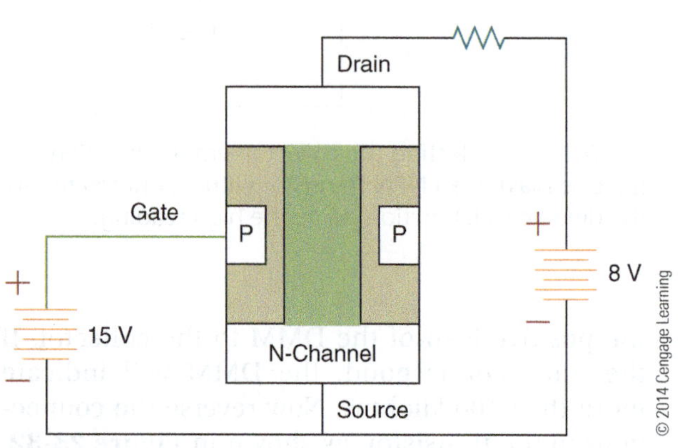

FIGURE 23-38 Decreasing the gate-to-source voltage increases the source-to-drain current.

on the p-channel JFET. The operation of the n- and p-channel JFETs is identical with the exception that the polarities of the voltages are reversed.

Unlike BJTs, *JFETs are voltage-controlled devices*. To control the conduction of current from the source to the drain, the gate voltage must be more negative than the source voltage. Increasing the amount of negative voltage (notice the size of the battery) applied to the gate will reduce the current flow from source to drain, as shown in **Figure 23-37**. Decreasing the amount of negative voltage (again notice the size of the battery) applied to the gate will cause the current flow from source to drain to increase, as shown in **Figure 23-38**. JFETs have a higher internal impedance than a BJT. Typically, the impedance of a JFET is in the neighborhood of 20,000 megohms.

JFETs can be tested with an ohmmeter with reasonable accuracy in or out of circuit. When testing in circuit, remember to de-energize the circuit and verify that the capacitors are fully discharged. It is important to note that if a JFET is *questionable* when tested in circuit, you will need to retest the JFET out-of circuit to be certain it is defective.

Refer to **Figure 23-39**. We begin by setting the DMM to the diode check position. Notice that an n-channel JFET is connected to a DMM. The negative lead of the DMM is connected to the source of the JFET. The positive lead of the DMM is connected to the gate of the JFET. If the JFET is good, the DMM will read a few tenths of a volt (typically 0.5 to 0.7 volt). Now reverse the DMM connections, as shown in **Figure 23-40**. Place the positive lead of the DMM on the source and the negative lead from the DMM on the gate. A good JFET will typically measure 0 volt. If you measured a low value in both instances, the source-gate junction is probably shorted. If the measurement was high in both instances, the source-gate junction is probably open. In both instances, the JFET will need to be replaced.

If the source-gate junction check was satisfactory, you must now check the drain-gate junction. Refer to **Figure 23-41**. Connect the negative lead of the DMM to the drain of the JFET. Connect the positive lead of the DMM to the gate of the JFET. If the drain-gate junction is good, you should measure a few tenths of a volt (typically 0.5 to 0.7 volt). Now reverse the connections to the JFET,

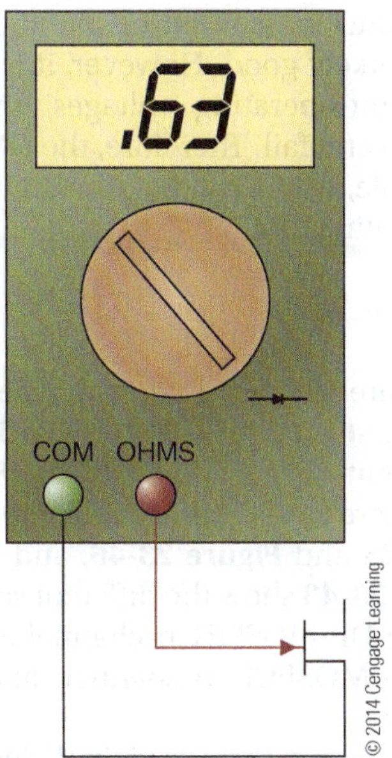

FIGURE 23-39 Testing the source-gate junction (forward bias) of an N-channel JFET with an ohmmeter on the diode check function. Note the low reading.

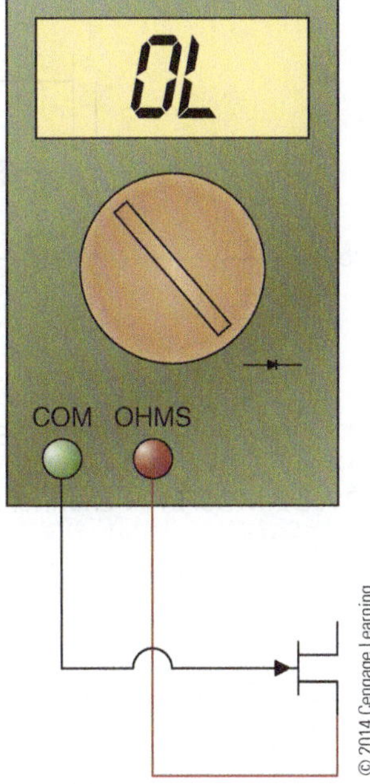

FIGURE 23-40 Testing the source-gate junction (reverse bias) of an N-channel JFET with an ohmmeter on the diode check function. Note the high reading.

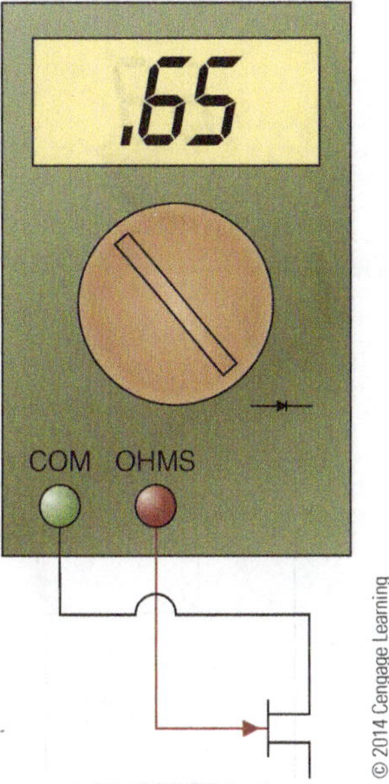

FIGURE 23-41 Testing the drain-gate junction (forward bias) of an N-channel JFET with an ohmmeter on the diode check function. Note the low reading.

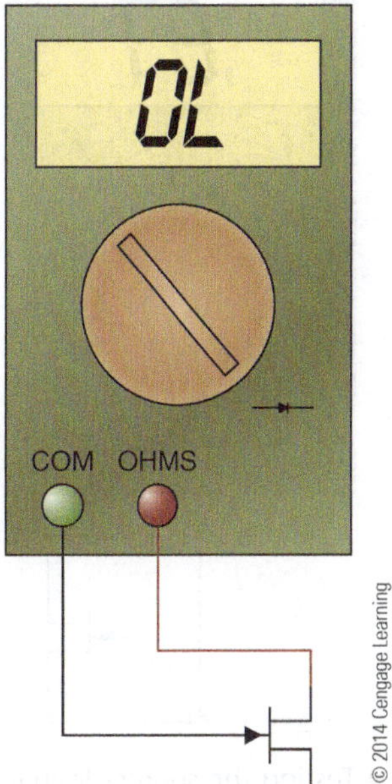

FIGURE 23-42 Testing the drain-gate junction (reverse bias) of an N-channel JFET with an ohmmeter on the diode check function. Note the high reading.

as shown in **Figure 23-42**. Connect the positive lead from the DMM to the drain of the JFET. Connect the negative lead of the DMM to the gate of the JFET. If the drain-gate junction is good, you should not measure any voltage. If you measured a low value in both instances, the drain-gate junction is probably shorted. If the measurement was high in both instances, the drain-gate junction is probably open. In both instances, the JFET will need to be replaced.

If the drain-gate junction check was satisfactory, you will need to perform a final check—that of the source-drain junction. Refer to **Figure 23-43**. Begin by connecting the negative lead of the DMM

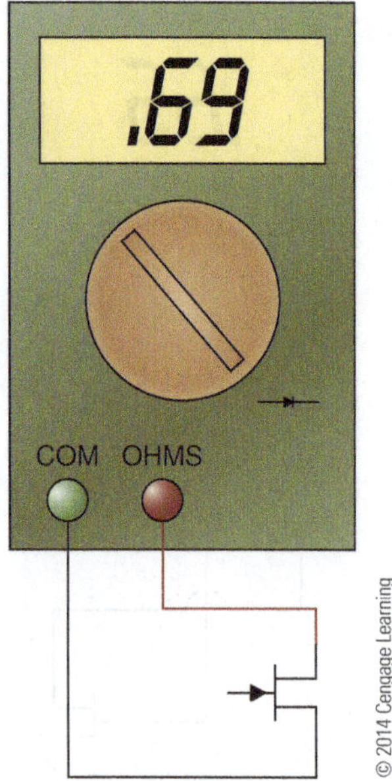

FIGURE 23-43 Testing the source-drain junction (forward bias) of an N-channel JFET with an ohmmeter on the diode check function. Note the low reading.

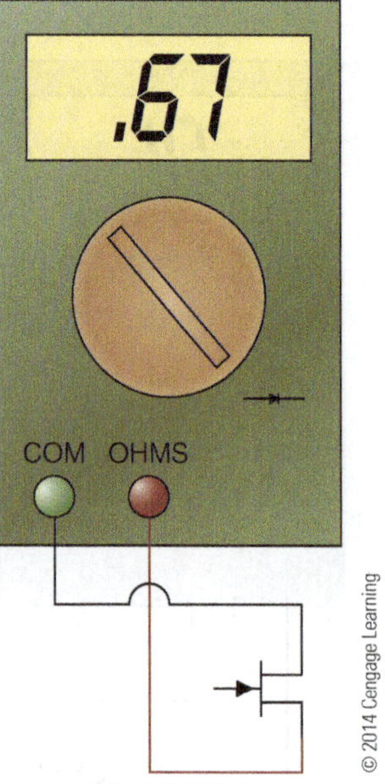

FIGURE 23-44 Testing the source-drain junction (reverse bias) of an N-channel JFET with an ohmmeter on the diode check function. Note the low reading.

to the source terminal. Next, connect the positive lead of the DMM to the drain. If the JFET is good, the DMM will indicate a few tenths of a volt. The actual amount of voltage will vary from JFET to JFET. Now reverse the connections to the JFET, as shown in **Figure 23-44**. Connect the positive lead of the DMM to the source and the negative lead of the DMM to the drain. Again, the DMM should indicate a few tenths of a volt.

If the JFET passed all of the above tests, it is more than likely good. However, it is possible that under circuit-operating voltages, the JFET may break down and fail. Therefore, the DMM test is not 100% reliable, but it can be an effective and quick troubleshooting tool.

Metal-Oxide Semiconductor Field-Effect Transistors

MOSFETs are divided into two types: **depletion-enhancement MOSFET (DE-MOSFET)** and **enhancement-only MOSFET (E-MOSFET)**. Within each type there are n-channel and p-channel devices. **Figure 23-45** and **Figure 23-46**, and **Figure 23-47**, and **Figure 23-48** show the different schematic symbols for the DE-MOSFET n-channel and p-channel and the E-MOSFET n-channel and p-channel, respectively.

MOSFETs are three-terminal devices. One of the terminals, called the *source*, is the one with the

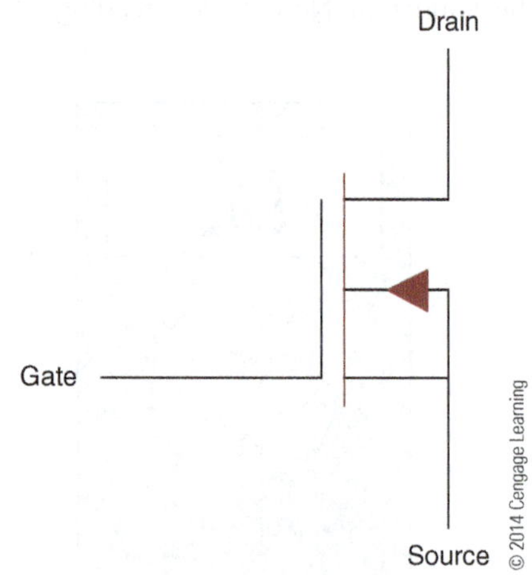

FIGURE 23-45 A DE-MOSFET N-channel schematic symbol.

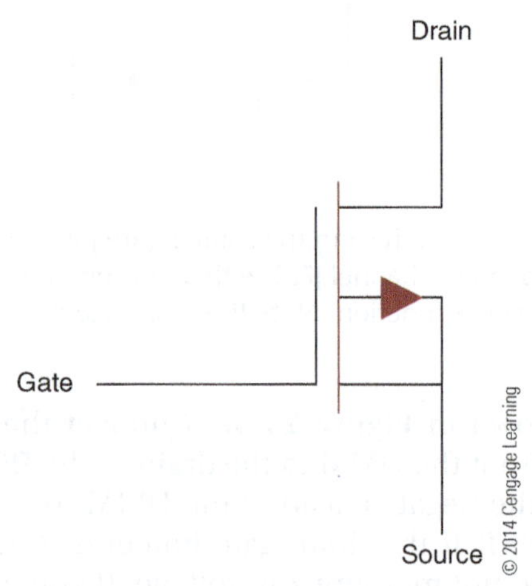

FIGURE 23-46 A DE-MOSFET P-channel schematic symbol.

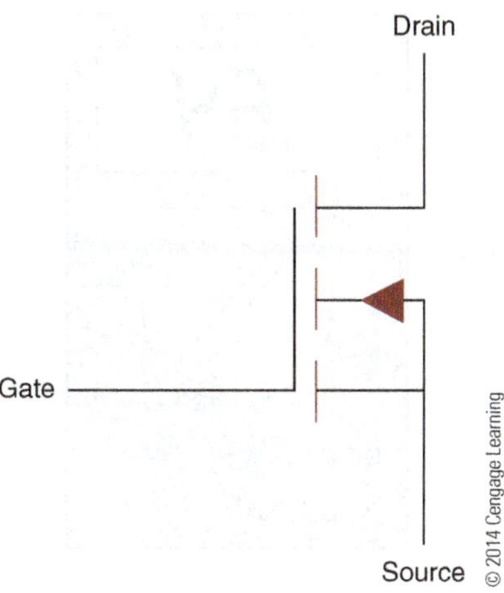

FIGURE 23-47 An E-MOSFET N-channel schematic symbol.

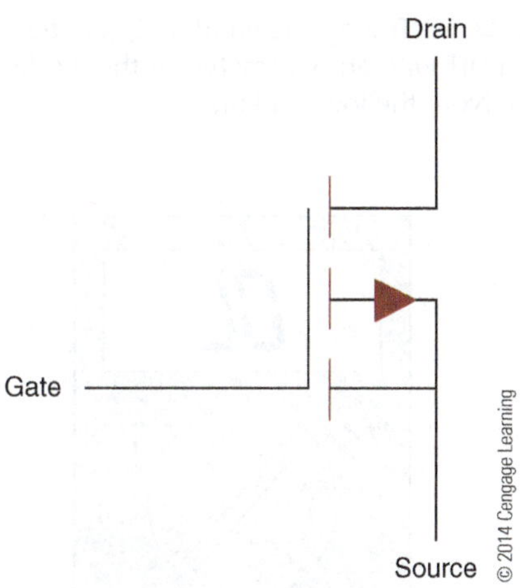

FIGURE 23-48 An E-MOSFET P-channel schematic symbol.

arrowhead. Another terminal is the *gate*, which is the one that is *L* shaped on the schematic. The remaining terminal is the *drain*.

MOSFETs have a much higher input resistance than the JFETs. Typically, the input resistance of a MOSFET is around 2 gigohms. As a result, MOSFETs are considered by some to be a perfect switch. This means that when the MOSFET is turned on, there is very little internal resistance. When the MOSFET is turned off, its input resistance is very high.

Due to the nature of a MOSFET, a reliable check cannot be performed with an ohmmeter. A better understanding of electronic circuits is needed if one is to try to determine if a MOSFET is defective. Also, most MOSFETs are sensitive to static electricity, there fore, special handling precautions are required so that these devices are not damaged by static discharge.

23-3 THYRISTORS

The family of devices known as *thyristors* includes unijunction transistors, silicon-controlled rectifiers, diacs, and triacs.

Unijunction Transistor

The **unijunction transistor (UJT)** has one emitter and two bases, as shown in **Figure 23-49**. UJTs are considered *digital* devices because they have only two states: off or on. To turn on a UJT, a voltage must be applied to the emitter. This voltage must be approximately 10 volts more positive than the voltage applied to base B_1. Base B_2 must be more positive than the emitter for the UJT to conduct. Current flows from base B_1 to base B_2 until the emitter voltage drops to 3 volts more positive than base B_1. At this point, the UJT turns off. The UJT is very useful in producing pulses.

UJTs can be tested with a digital multimeter with reasonable accuracy in or out of circuit. When testing in circuit, remember to de-energize the circuit and verify that all capacitors are fully discharged. It is important to note that if a UJT is *questionable* when tested in circuit, you need to retest the UJT out of circuit to be certain it is defective.

Refer to **Figure 23-50**. Begin by setting the DMM to the diode check position. The negative lead of the DMM is connected to base B_1 of the UJT. The positive lead of the DMM is connected to the emitter of the UJT. If the UJT is good, the DMM will read a few tenths of a volt (typically 0.5 to 0.7 volt). Now reverse the DMM connections, as shown in **Figure 23-51**. Place the positive lead of the DMM on base B_1 and the negative lead from the DMM on the emitter. A good UJT will typically measure 0 volt. If you measured a low value in both instances, the B_1-emitter junction is probably shorted. If the measurement was high in both instances, the B_1-emitter junction is

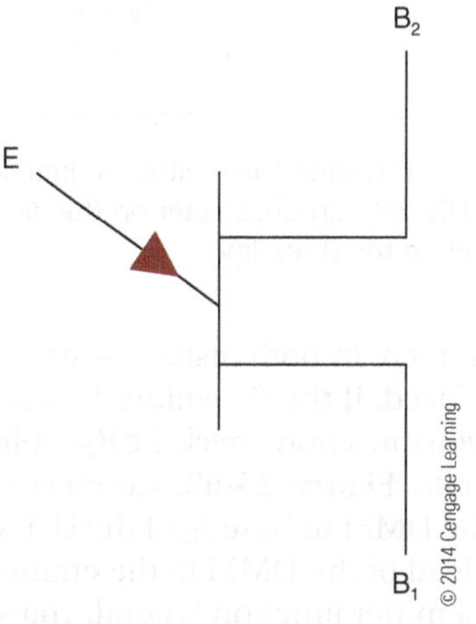

FIGURE 23-49 A schematic symbol of a unijunction transistor (UJT).

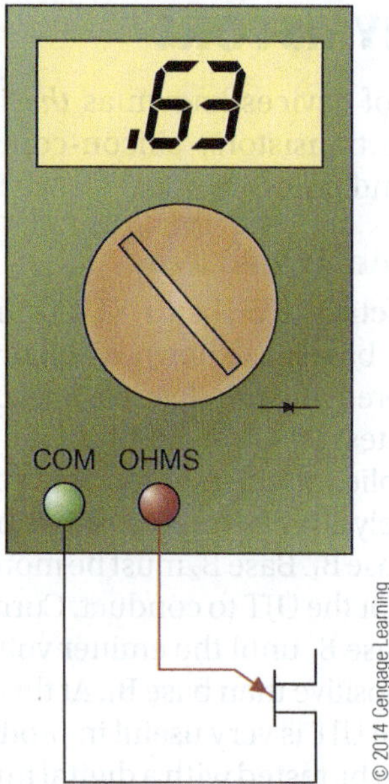

FIGURE 23-50 Testing the emitter–B_1 junction (forward bias) of a UJT with an ohmmeter on the diode check function. Note the low reading.

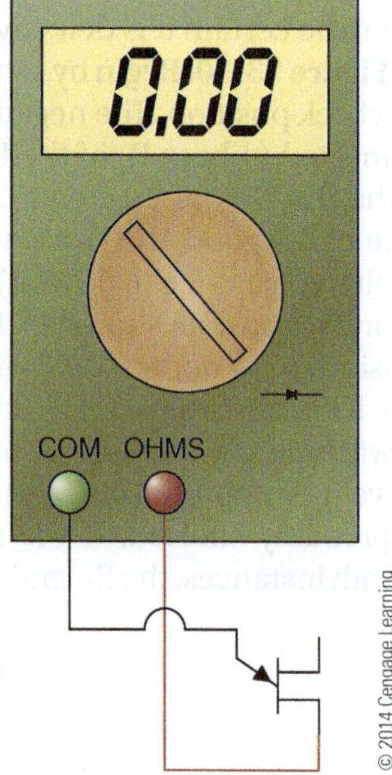

FIGURE 23-51 Testing the emitter–B_1 junction (reverse bias) of a UJT with an ohmmeter on the diode check function. Note the 0 reading.

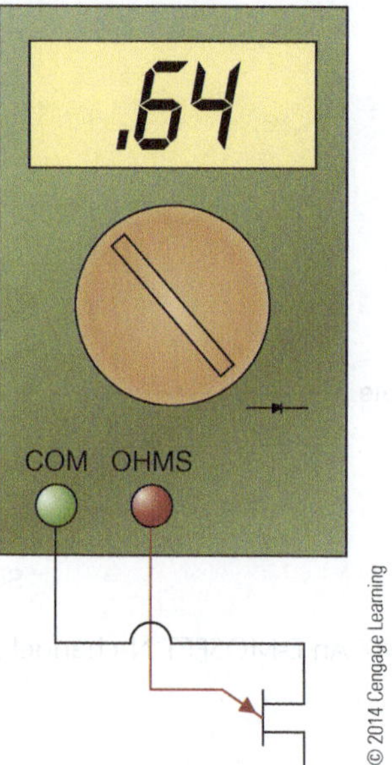

FIGURE 23-52 Testing the emitter–B_2 junction (forward bias) of a UJT with an ohmmeter on the diode check function. Note the low reading.

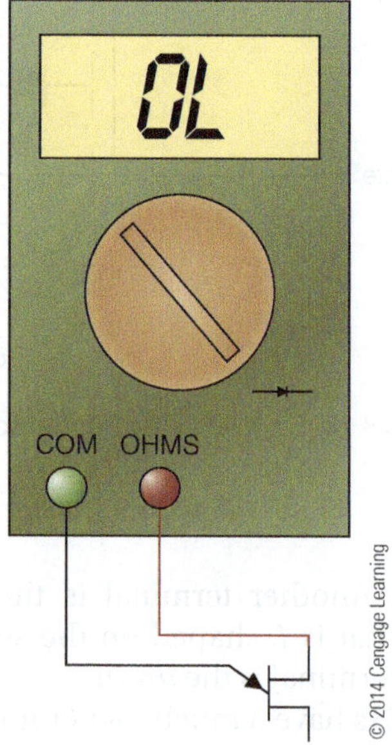

FIGURE 23-53 Testing the emitter–B_2 junction (reverse bias) of a UJT with an ohmmeter on the diode check function. Note the high reading.

probably open. In both instances, the UJT will need to be replaced. If the B_1–emitter junction was satisfactory, you must now check the B_2–emitter junction.

Refer to **Figure 23-52**. Connect the negative lead of the DMM to base B_2 of the UJT. Connect the positive lead of the DMM to the emitter of the UJT. If the B_2–emitter junction is good, you should measure a few tenths of a volt (typically 0.5 to 0.7 volt). Now reverse the connections to the UJT, as shown in **Figure 23-53**. Connect the positive lead from the DMM to base B_2 of the UJT. Connect the negative lead of the DMM to the emitter of the UJT. If the B_2–emitter junction is good, you should not measure any voltage. If you measured a low value in both instances, the B_2–emitter junction is probably shorted. If the measurement was high in both instances, the B_2–emitter junction is probably open. In both instances, the UJT will need to be replaced.

If the B_2-emitter junction check was satisfactory, you will need to perform a final check—that of the B_1–B_2 junction. Refer to **Figure 23-54**. Begin by connecting the negative lead of the DMM to base B_2 terminal. Next, connect the positive lead of the DMM to base B_1. If the UJT is good, the DMM will indicate a few tenths of a volt. The actual amount of voltage will vary from UJT to UJT. Now reverse the connections to the UJT, as shown in **Figure 23-55**. Connect the positive lead of the DMM to base B_2 and the negative lead of the DMM to base B_1. Again, the DMM should indicate a few tenths of a volt.

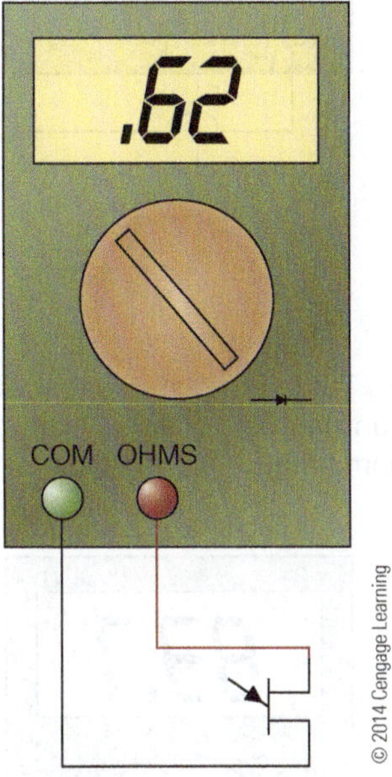

FIGURE 23-54 Testing the B_1–B_2 junction (forward bias) of a UJT with an ohmmeter on the diode check function. Note the low reading.

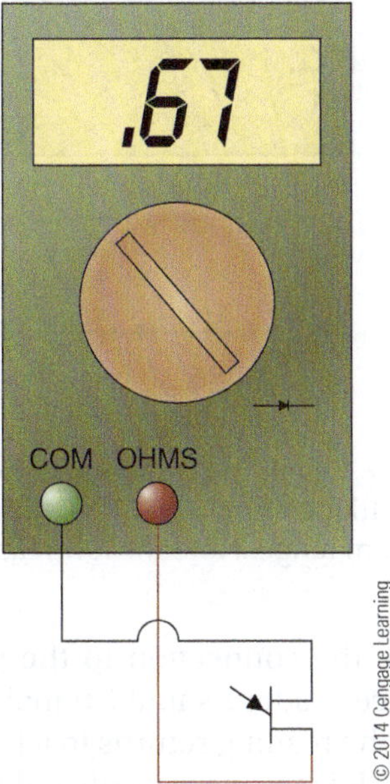

FIGURE 23-55 Testing the B_1–B_2 junction (reverse bias) of a UJT with an ohmmeter on the diode check function. Note the low reading.

If the UJT passed all of the tests, it is more than likely good. However, it is possible that under circuit-operating voltages, the UJT may break down and fail. Therefore, the DMM test is not 100% reliable, but it can be an effective and quick trouble shooting tool.

Silicon-Controlled Rectifier

Figure 23-56 shows the schematic symbol of a **silicon-controlled rectifier (SCR)**. You will notice that the SCR looks very similar to a diode except that there is an extra terminal. The terminals of the SCR are known as the anode which is the arrowhead symbol; the cathode, which is the *T*-shaped symbol; and the gate or **trigger**, which is the remaining terminal.

SCRs function similarly to a diode with the exception of the gate circuit. As shown earlier, a diode only conducts current when forward biased. The current flow is blocked when the diode is reverse biased. The same is true for the SCR with one exception. The SCR conducts current when forward biased (cathode negative with respect to the anode), but it does not conduct current in the forward direction until the proper gate voltage is applied. This means that it is possible to control *when* the SCR conducts.

In order for the SCR to conduct, you must apply a negative voltage to the cathode and a positive voltage to the anode. At this point, the SCR is not conducting. By applying a positive voltage to the gate, the SCR is triggered into conduction. If the gate voltage is removed, the SCR continues to conduct as long as current flows through the cathode-anode junction. The SCR turns off if the current drops below its *minimum holding current* rating. It does not turn on again until another positive gate voltage is applied. Different SCRs have different values of minimum holding current.

It is possible to perform a simple check of an SCR with a DMM either in or out of circuit. How ever, when testing in circuit, be certain that power is removed

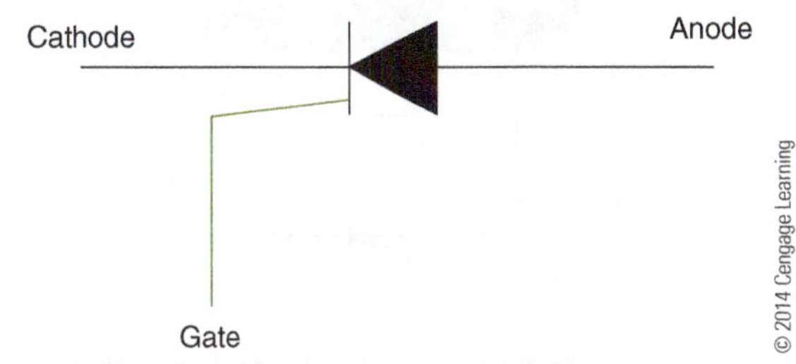

FIGURE 23-56 A schematic symbol of a silicon-controlled rectifier (SCR).

and the capacitors are fully discharged. Also, if any of the readings is questionable, the SCR should be removed from the circuit and retested out of circuit.

Refer to **Figure 23-57**. Begin by testing the *diode* portion of the SCR first. Set the DMM to either the 200-ohm or 2-kilohm position. Do not use the low-ohm or diode check position because these do not supply sufficient current to accurately check the SCR under test. Connect the negative lead from the DMM to the cathode of the SCR. Connect the positive lead from the DMM to the anode of the SCR. A good SCR will typically measure in excess of 100 kilohms. Now reverse your connections, as shown in **Figure 23-58**. Connect the negative lead from the DMM to the anode of the SCR, and connect the positive lead from the DMM to the cathode of the SCR. You should still measure in excess of 100 kilohms if the SCR is good.

Reverse your connections again so that the negative lead from the DMM is connected to the cathode of the SCR, and the positive lead from the DMM is connected to the anode of the SCR, as shown in Figure 23-57. Now take another test lead, and, while leaving the DMM connected, connect the positive lead of the DMM to the gate terminal of the SCR, as shown in **Figure 23-59**. The positive lead of the DMM is now connected to both the anode and the gate of the SCR. If the SCR is good, the resistance reading should drop to less than 1 kilohm. If there was no change in the resistance reading, the SCR is probably defective and has to be replaced. If you now remove the connection to the gate terminal, the resistance reading should remain at less than 1 kilohm. If the reading returns to a high-resistance value, the SCR is not necessarily defective because some DMMs do not supply sufficient holding current for the SCR to maintain conduction. This is especially true with larger, high-current SCRs.

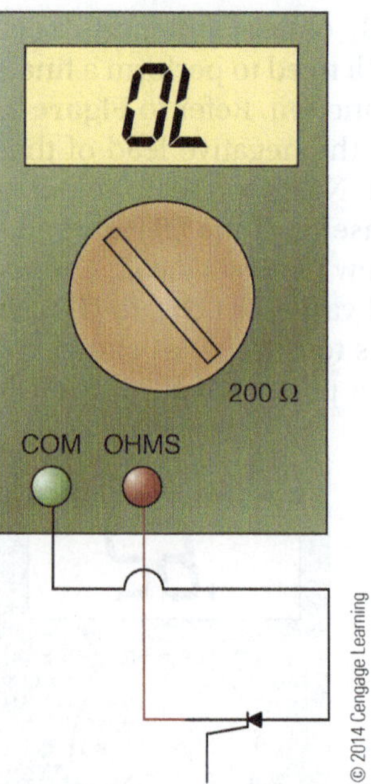

FIGURE 23-58 Testing the cathode–anode junction (reverse-bias untriggered) of an SCR with the ohm meter on the 200-ohm range. Note the high reading.

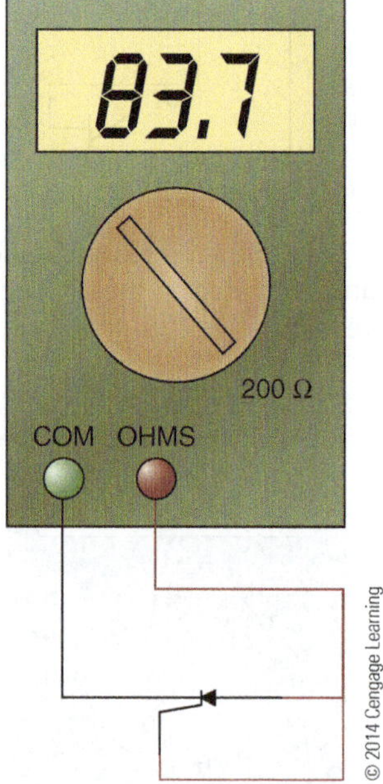

FIGURE 23-59 Testing the cathode–anode junction (forward-bias triggered) of an SCR with the ohmmeter on the 200-ohm range. Note the low reading.

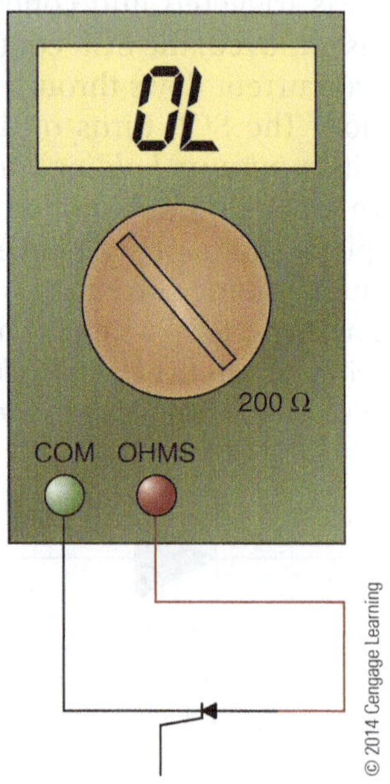

FIGURE 23-57 Testing the cathode-anode junction (forward-bias untriggered) of an SCR with the ohmmeter on the 200-ohm range. Note the high reading.

Breaking the connection to either the anode or the cathode turns off the SCR. The test can now be repeated.

DIAC

The **diac**, sometimes called a **bilateral trigger diode**, is a bidirectional diode. This means that it can operate in an AC circuit. The diac has two terminals called **main terminal 1 (MT 1)** and **main terminal 2 (MT 2)**, as shown by the schematic symbol in **Figure 23-60**. The diac functions like an AC version of a UJT; that is, a diac does not conduct until a threshold voltage is reached. The diac remains in conduction until the voltage falls below its minimum conduction voltage. At this point, the diac turns off. For example, a diac may have a turn-on voltage rating of 32 volts and a turn-off voltage rating of 25 volts. This means that when power is first applied to the circuit, the diac does not conduct. If the voltage rises to ±32 volts, the diac turns on and allows current flow. If the voltage drops to ±30 volts, the diac remains on. However, if the voltage drops below ±25 volts, the diac turns off and current flow stops. Diacs are available with different threshold voltages. You can think of the diac as a voltage-sensitive AC switch.

The only test that can be performed on a diac with an ohmmeter is to check for a shorted diac. Under normal conditions, a good diac indicates an open circuit when tested with an ohmmeter, as shown in **Figure 23-61** and **Figure 23-62**. A low-resistance reading in both directions indicates that the diac is shorted and has to be replaced.

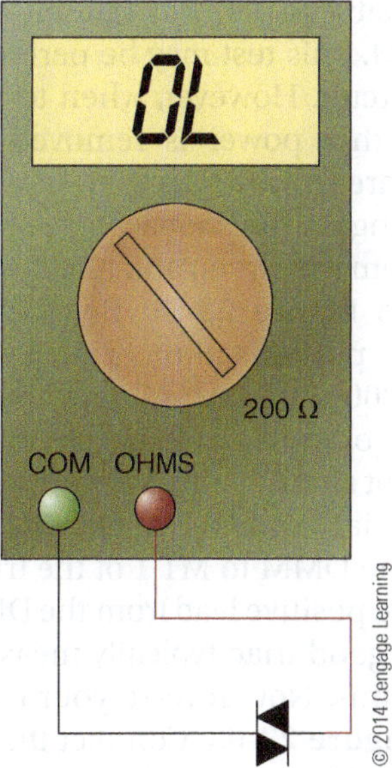

FIGURE 23-61 Testing a diac with the ohmmeter on the 200-ohm range. Note the high reading.

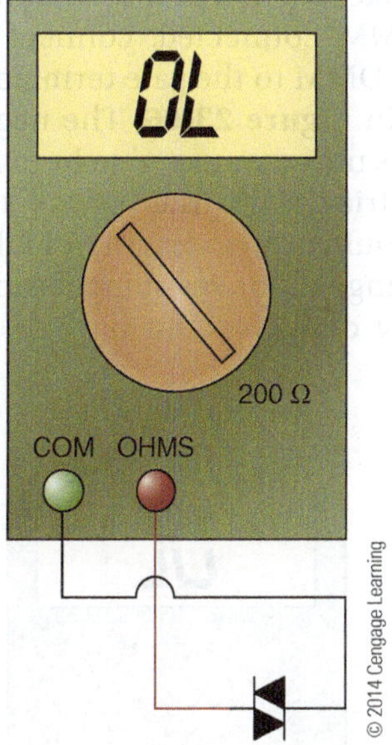

FIGURE 23-62 Testing a diac with the ohmmeter on the 200-ohm range. Note that the connections have been reversed, but the reading remains high.

TRIAC

A **triac** is a device that operates like an SCR for an AC circuit. The triac conducts during both halves of the AC waveform. This means that the output of a triac is AC, not DC as is the case with an SCR. The schematic symbol of a triac is shown in **Figure 23-63**. The triac is a three-terminal device. One terminal is called the *gate*, the terminal closest to the gate is called *main terminal 1* (*MT 1*), and the remaining terminal is called *main terminal 2* (*MT 2*). The gate voltage must be the same polarity as MT 2 to turn on the triac. The triac continues to conduct until the current flowing through it decreases below its minimum holding current level.

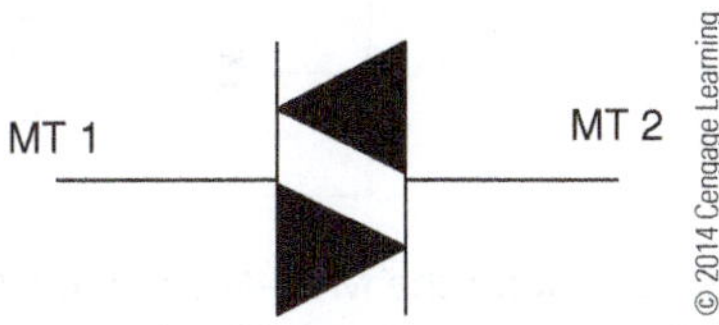

FIGURE 23-60 A schematic symbol of a diac.

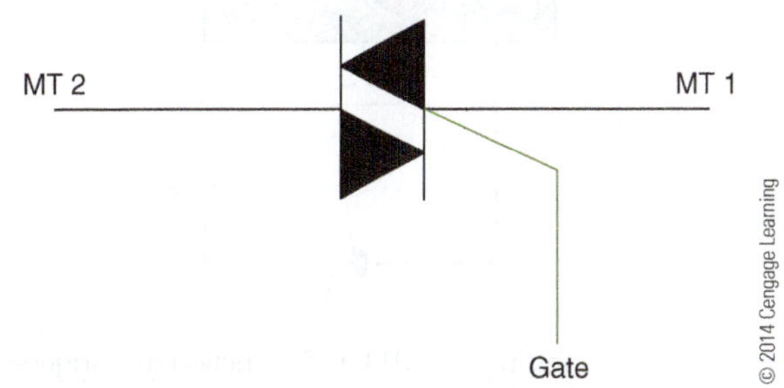

FIGURE 23-63 A schematic symbol of a triac.

It is possible to perform a simple check of a triac with a DMM. This test may be performed either in or out of circuit. However, when testing in circuit, be certain that power is removed and that the capacitors are fully discharged. In addition, if any of the readings is questionable, the triac should be removed from the circuit and retested out of circuit.

Refer to **Figure 23-64**. Begin by testing one of the diode portions of the triac. Set the DMM to either the 200-ohm or 2-kilohm position. Do not use the low-ohm or diode check position because these will not supply sufficient current to accurately check the triac under test. Connect the negative lead from the DMM to MT 1 of the triac under test. Con nect the positive lead from the DMM to MT 2 of the triac. A good triac typically measures in excess of 100 kilohms. Now reverse your connections, as shown in **Figure 23-65**. Connect the negative lead from the DMM to MT 2 of the triac. Connect the positive lead from the DMM to MT 1 of the triac. You should still measure in excess of 100 kilohms if the triac is good.

Now take another test lead, and while leaving the DMM connected, connect the negative lead of the DMM to the gate terminal of the triac, as shown in **Figure 23-66**. The negative lead of the DMM is now connected to both MT 2 and the gate of the triac. If the triac is good, the resistance reading should drop to less than 1 kilohm. If there was no change in the resistance reading, the triac is probably defective and have to be replaced. If you remove the connection to the gate terminal now, the resistance reading should remain at less than 1 kilohm. If the reading returns to a high-resistance value, the triac is not necessarily defective. Some DMMs do not supply sufficient holding current for the triac to maintain

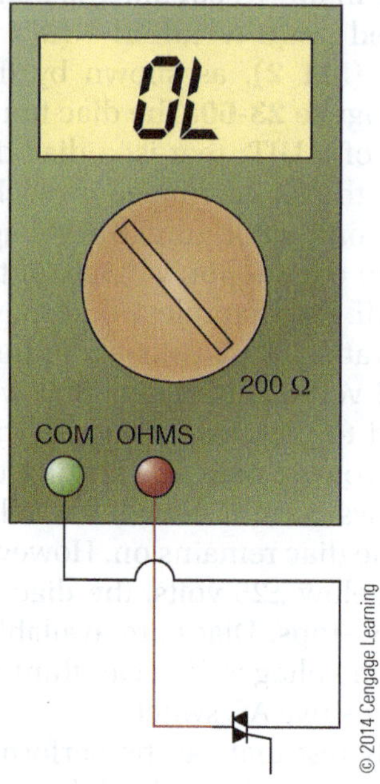

FIGURE 23-65 Testing the MT1–MT2 junction (untriggered) of a triac with the ohmmeter on the 200-ohm range. Note that the connections have been reversed, but the reading remains high.

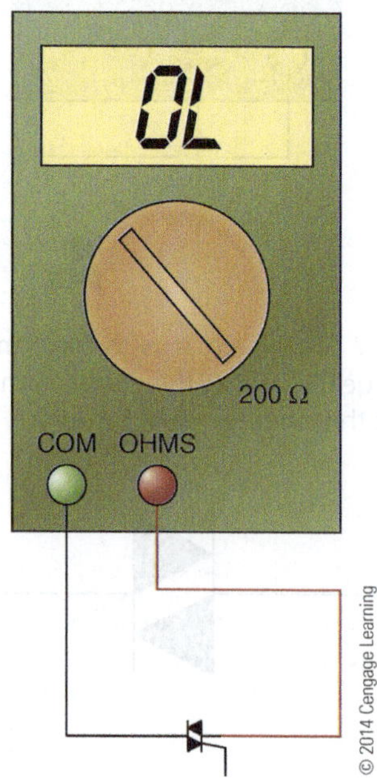

FIGURE 23-64 Testing the MT1–MT2 junction (untriggered) of a triac with the ohmmeter on the 200-ohm range. Note the high reading.

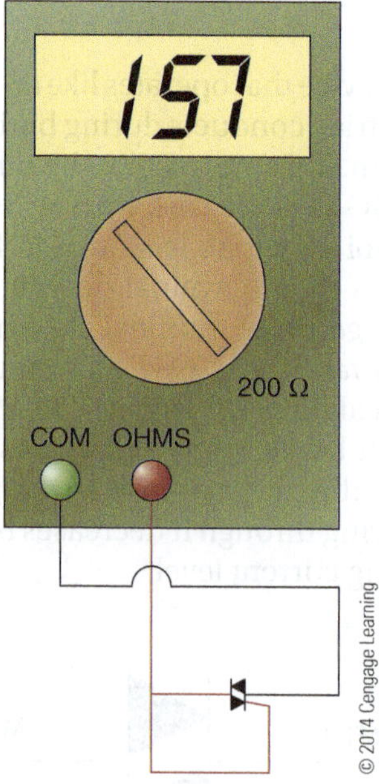

FIGURE 23-66 Testing the MT1–MT2 junction (triggered) of a triac with the ohmmeter on the 200-ohm range. Note the low reading.

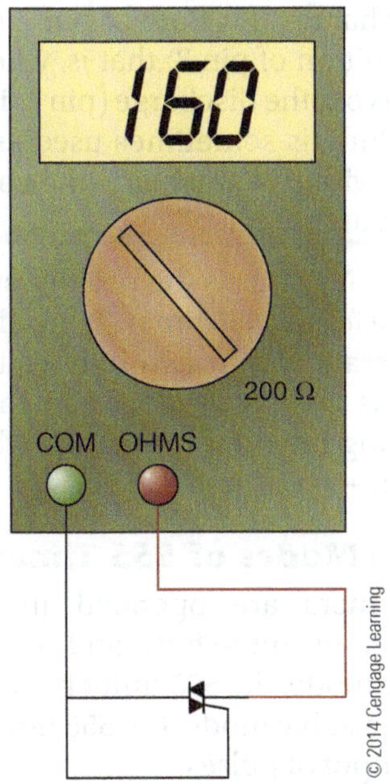

FIGURE 23-67 Testing the MT1–MT2 junction (triggered) of a triac with the ohmmeter on the 200-ohm range. Note that the connections have been reversed, but the reading remains low.

conduction. This is especially true with larger, high-current triacs. Breaking the connection to either MT 1 or MT 2 turns off the triac.

The test can now be repeated on the other half of the triac. Refer to **Figure 23-67**. Connect the negative lead from the DMM to MT 1, and connect the positive lead of the DMM to MT 2. Also, connect the positive lead of the DMM to the gate terminal of the triac. The positive lead of the DMM is now connected to both MT 2 and the gate of the triac. If the triac is good, the resistance reading should drop to less than 1 kilohm. If there was no change in the resistance reading, the triac is probably defective and have to be replaced. If you remove the connection to the gate terminal now, the resistance reading should remain at less than 1 kilohm. If the reading returns to a high-resistance value, the triac is not necessarily defective. Some DMMs do not supply sufficient holding current for the triac to maintain conduction. This is especially true with larger, high-current triacs. Breaking the connection to either MT 1 or MT 2 will turn off the triac.

23-4 555 TIMER

The **555 timer** is a member of a family of devices known as **integrated circuits (ICs)**. An IC is a device that contains anywhere from a few components to several thousand components in one package. These components are connected to form complete circuits. The circuits are miniaturized and contained within a device known as an IC.

The most common package for an IC is known as a **dual-in-line pin (DIP)** package. The 555 timer is typically found in an 8-pin DIP package, although it may also appear in a TO-5 package. In addition, there is a 556 timer that has two 555 timers in a 14-pin DIP package. There is also a 558 timer, which contains four 555 timers in a 14-pin DIP package. Because the 8-pin package is the most common, you will learn how the pins of the 8-pin DIP and TO-5 packages are numbered. **Figure 23-68** shows the pin numbers for an 8-pin DIP package, whereas **Figure 23-69** shows those for an 8-pin TO-5 package.

The 555 timers are used to perform many types of timing, pulsing, and delaying operations. It is possible to create different timing circuits to meet different needs by changing the connections and components associated with the 555 timer. After you have learned the function of each of the pins of a 555 timer, you will learn some of these timing circuits.

Pin Assignments

Each pin of the 555 timer has a function.

Pin 1—Ground. This pin is connected to the most negative potential of the external circuit. This is typically the circuit ground when the circuit uses a positive voltage supply.

Pin 2—Trigger. The trigger is used to fire or start the timing operation. The 555 timer is triggered when the voltage on pin 2 is lowered to less than one-third of the supply voltage to the 555 timer, or if the voltage on pin 2 is one-half of the voltage

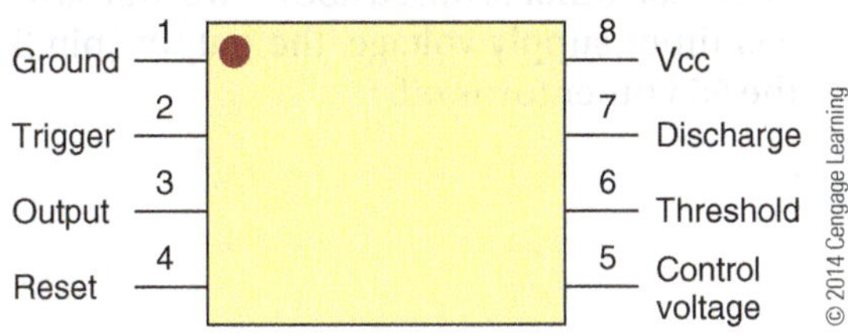

FIGURE 23-68 Pin-outs of a 555 timer, DIP package.

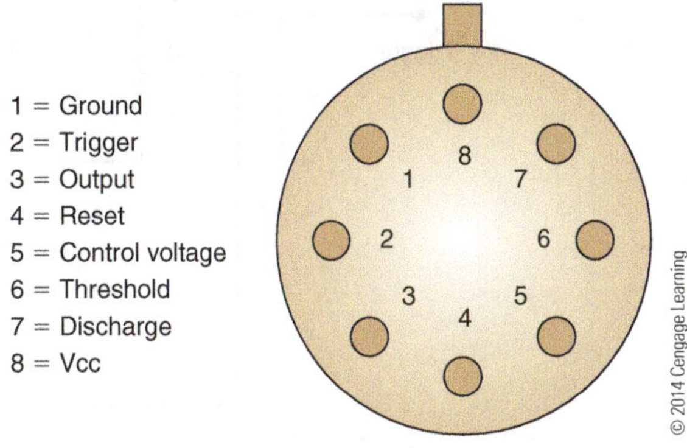

FIGURE 23-69 Pin-outs of a 555 timer, TO-5 package.

present on pin 5. Typically, pin 2 is connected to the circuit ground to trigger the 555 timer. It is important to realize that this connection must be momentary. If the connection to ground is maintained, the 555 timer will not operate.

Pin 3—Output. This is the output pin for the 555 timer. When the trigger (pin 2) voltage is lowered, the output is on. To turn the output off, either the voltage at pin 6 must be raised above two-thirds of the supply voltage to the 555 timer, or the voltage at pin 4 must be lowered to less than 0.7 volt.

Pin 4—Reset. Lowering the voltage at pin 4 to less than 0.7 volt resets or turns off the output (pin 3). This occurs regardless of the state or condition of the trigger (pin 2), the threshold (pin 6), or the discharge (pin 8). If the reset function is not used, pin 4 should be connected to the 555 timer supply voltage to avoid false resetting.

Pin 5—Control Voltage. Two-thirds of the 555 timer supply voltage appears at this pin. By connecting a variable resistor from pin 5 to the 555 timer supply voltage, the *on time* can be varied. This has no effect on the *off time*. By connecting a variable resistor from pin 5 to ground, the *off time* can be varied. This has no effect on the *on time*. If the control voltage function is not used, a 0.01-microfarad capacitor should be connected between pin 5 and ground. This filters any electrical noise from entering pin 5.

Pin 6—Threshold. When the voltage applied to pin 6 of the 555 timer is raised above two-thirds of the 555 timer supply voltage, the output (pin 3) of the 555 timer turns off.

Pin 7—Discharge. The state of pin 7 is opposite of the condition of pin 3; that is, when the output (pin 3) is on, the discharge (pin 7) is off, and vice versa. Pin 7 is sometimes used as an auxiliary output, with its state being the complement of the output (pin 3).

Pin 8—Vcc or V+. This is the positive supply voltage terminal for the 555 timer. Typically, a 555 timer can operate within a supply voltage range of +4.5 volts to +16 volts; however, the accuracy of the timing function is ensured when operated between +5 volts and +15 volts.

Operation Modes of 555 Timers

The 555 timers are operated in two modes: **monostable**, or **one-shot**, and **astable**. In the monostable mode, the 555 timer has a single, stable state. In the astable mode, the 555 timer generates a constant output of pulses.

Monostable Mode

Refer to **Figure 23-70**. The single stable state of a 555 timer is the off state. When triggered, the 555 timer switches to its on state and remains in it for a length of time determined by the values of a resistor-capacitor network. After the time has expired, the 555 timer returns to its off state. As a result, an output pulse is generated at pin 3. The on time of the output pulse is determined by the values of the resistor-capacitor network. The 555 timer must be retriggered manually to generate another output pulse.

Astable Mode

A 555 timer that is operating in the astable mode is also called a **multivibrator**. When used in this

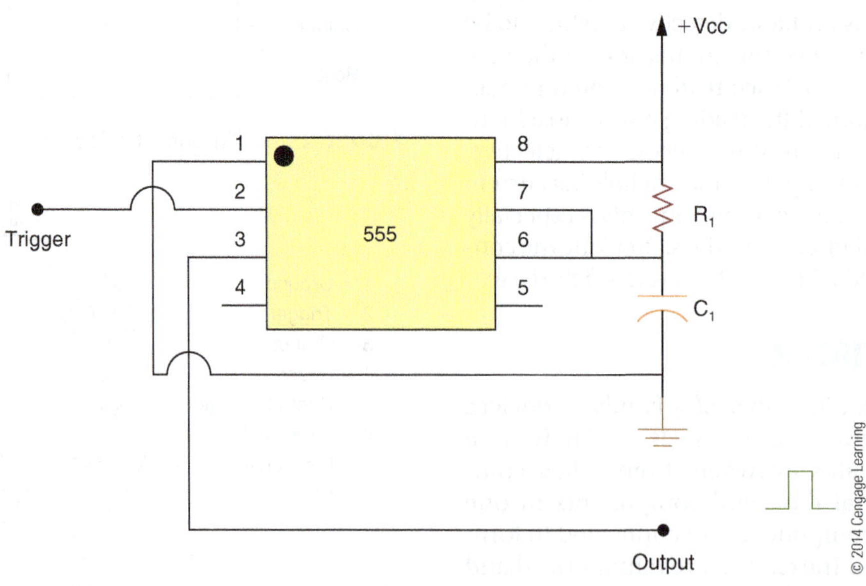

FIGURE 23-70 A 555 timer connected in the mono stable mode.

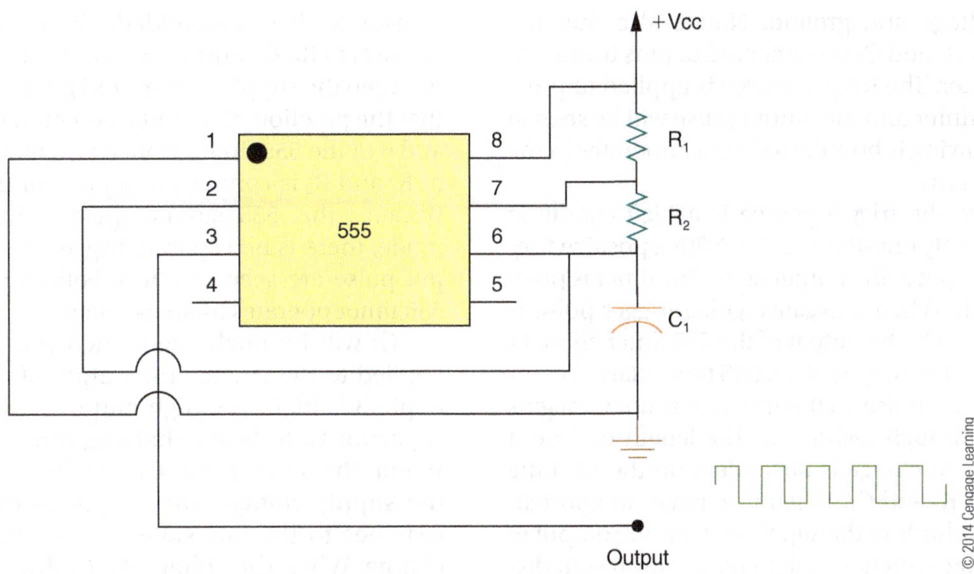

FIGURE 23-71 A 555 timer connected in the astable mode.

mode, the 555 timer produces a string of pulses at the output (pin 3). To produce the string of pulses, the 555 timer must be constantly retriggered. This is done automatically by connecting the trigger (pin 2) to the threshold (pin 6), as shown in **Figure 23-71**. If a variable resistor is connected between the supply voltage and the control voltage (pin 5) of the 555 timer, the on time can be made longer or shorter. The variable resistor does not affect the off time, or time between pulses. If a variable resistor is connected between the control voltage (pin 5) of the 555 timer and ground, the off time, or time between pulses, can be made longer or shorter. The variable resistor does not affect the on time. A resistor-capacitor network is used to control the frequency of the pulses.

555 Circuits

The 555 timer is a very versatile and useful device and has many circuits and applications. The following three topics show a few of these applications. The circuit diagram as well as an explanation of the circuit and example of use are presented.

One-Shot Configuration

Figure 23-72 shows a 555 timer configured as a one-shot. Notice the RC network consisting of R_1 and C_1. This network is connected between the

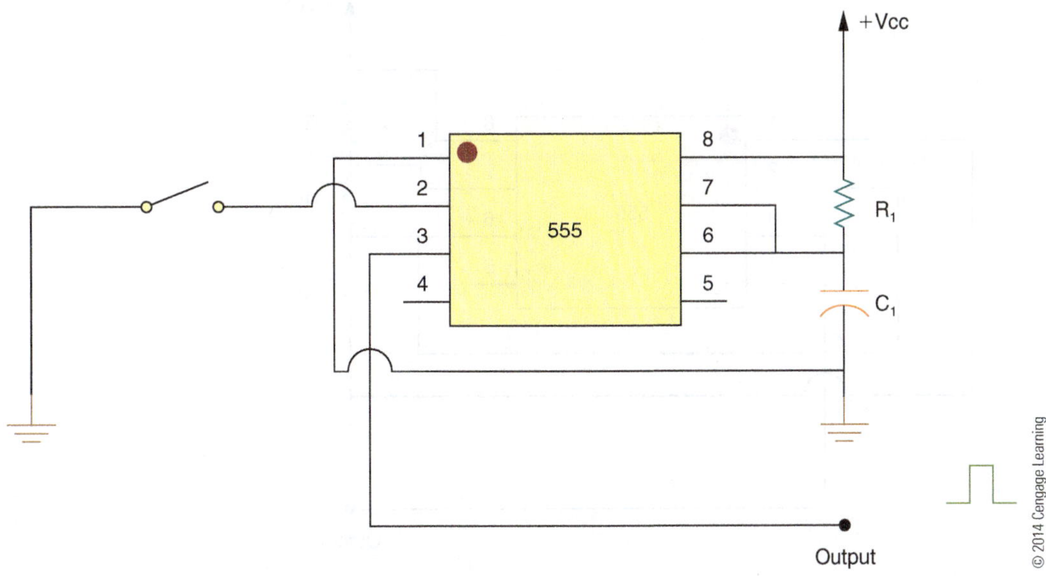

FIGURE 23-72 A 555 timer connected as a one-shot.

supply voltage and ground. Notice also that the junction of R_1 and C_1 is connected to pins 6 and 7 of the 555 timer. The trigger source is applied to pin 2 of the 555 timer and the output pulse will be seen at pin 3. Following is how the 555 timer operates in the one-shot mode.

Initially, the trigger source is a *high* condition (approximately one-third of V+). With a positive trigger signal at pin 2, the output of the 555 timer is practically 0 volt. When a negative going trigger pulse is applied to pin 2, the output of the 555 timer changes state; that is, the output of the 555 timer starts to produce a positive pulse. At this time, capacitor C_1 begins to charge through resistor R_1. The length of time it takes for C_1 to charge is dependent on the RC time constant of R_1 and C_1. When C_1 charges to approximately two-thirds of the supply voltage, the output of the 555 timer switches back to practically 0 volt, discharging capacitor C_1. The 555 timer is now reset and awaiting the next negative going trigger pulse to pin 2.

The length of time for the output pulse can be varied by either changing the value of C_1 or R_1. This affects the RC time constant. In this manner, a longer or shorter duration output pulse can be generated. It is possible to produce output pulses from as short as 10 microseconds to as long as practical, based on available values for R_1 and C_1.

A 555 timer used as a one-shot can be used as a triggering device. This is helpful when a well defined pulse is needed for triggering purposes.

Oscillator

Figure 23-73 shows a 555 timer configured as an oscillator. Notice in this circuit that pins 2 and 6 are connected together. Also notice that a second resistor, R_2, has been added. The RC network now consists of R_1, R_2, and C_1. This network is connected between the supply voltage and ground. Notice also that the junction of R_2 and C_1 is connected to pins 2 and 6 of the 555 timer. Notice also that the junction of R_1 and R_2 is connected to pin 7 of the 555 timer. Because this 555 circuit operates in the astable mode, there is no separate trigger source. The output pulse are seen at pin 3. Following is how the 555 timer operates as an oscillator.

C_1 will be uncharged when power is initially applied to the circuit. The output of the 555 timer at pin 3 is high. The high output condition causes capacitor C_1 to begin charging through R_1 and R_2. When the charge on C_1 reaches two-thirds of the supply voltage, the output of the 555 timer switches to the low state. C_1 now begins to discharge. When the voltage on C_1 drops to approximately one-third of the supply voltage, the output of the 555 timer switches to the high state. This causes C_1 to again begin charging, causing the entire cycle to repeat. This produces a continuous stream of rectangular pulses at the output of the 555 timer at pin 3.

The frequency of the output pulses is dependent on the values of R_1, R_2, and C_1. The time that the output of the 555 timer is *on* is dependent on the values of R_1 and R_2, whereas the time that the output of the 555 timer is *off* is dependent on the value of R_2 only.

A 555 timer that is used as an oscillator can produce various sounds when connected to an amplifier stage. This could be used to provide an audible indication of continuity or in some other application requiring an audible alert.

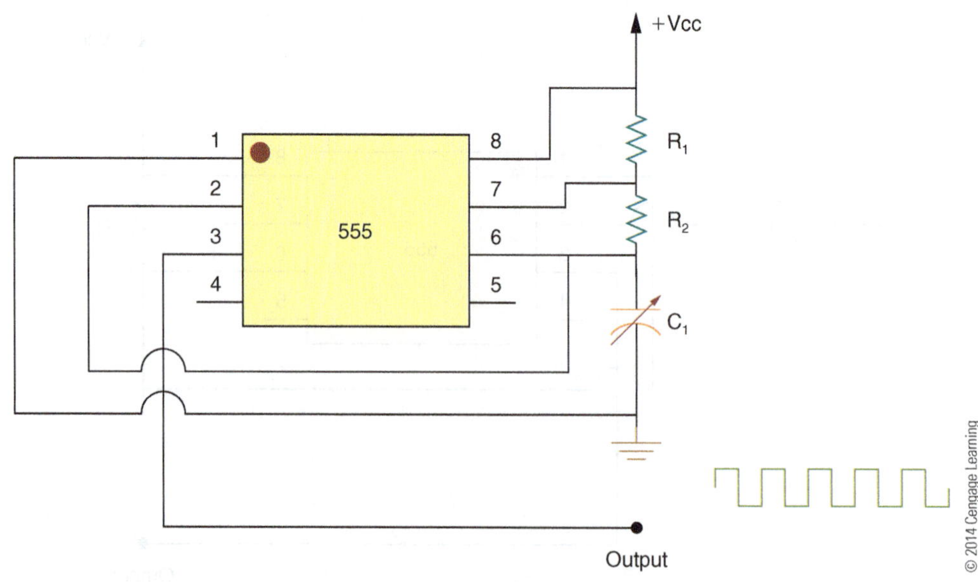

FIGURE 23-73 A 555 timer connected as an oscillator.

On-Delay Timer

Figure 23-74 shows a 555 timer connected to provide an on-delay timer function. A 555 timer used in this fashion can be used to energize a relay after a preset time delay. This is the basis of most electronic time-delay relays.

First, notice that a capacitor, C_2, has been added from pin 5 of the 555 timer to ground. This capacitor provides electrical noise immunity to the 555 timer. Notice also that the 555 timer is configured for astable operation. However, notice that pin 4 is connected to the collector of Q_2. This is used to reset the 555 timer. Also notice diode D_1, called a **freewheeling diode**. It is used to dissipate the voltage spike produced by the collapsing magnetic field in the coil of relay K_1 when the relay de-energizes. Assume that all capacitors are in a discharged state before power is first applied to the circuit.

When power is applied, capacitors C_1 and C_3 begin to charge. The charge time for capacitor C_3 is dependent on the RC time constant of C_3 and R_4. This time constant provides a short time delay before C_3 becomes charged. During this time delay, the output of the 555 timer is in the *on* state. Because the output of the 555 timer is *on*, transistor Q_1 is turned *on* as well. As a result of Q_1 being *on*, Q_2 is *off*. Because Q_2 is *off*, the relay, K_1 is *de-energized*.

Capacitor C_1 charge time is determined by the values of C_1, R_1, and R_2. When C_1 charges to approximately two-thirds of the supply voltage, the output of the 555 timer changes from *on* to *off*. When the output of the 555 timer changes to the *off* state, transistor Q_1 turns *off* as well. Because transistor Q_1 has turned *off*, transistor Q_2 turns *on*. This allows current to flow through the coil of relay K_1, *energizing*

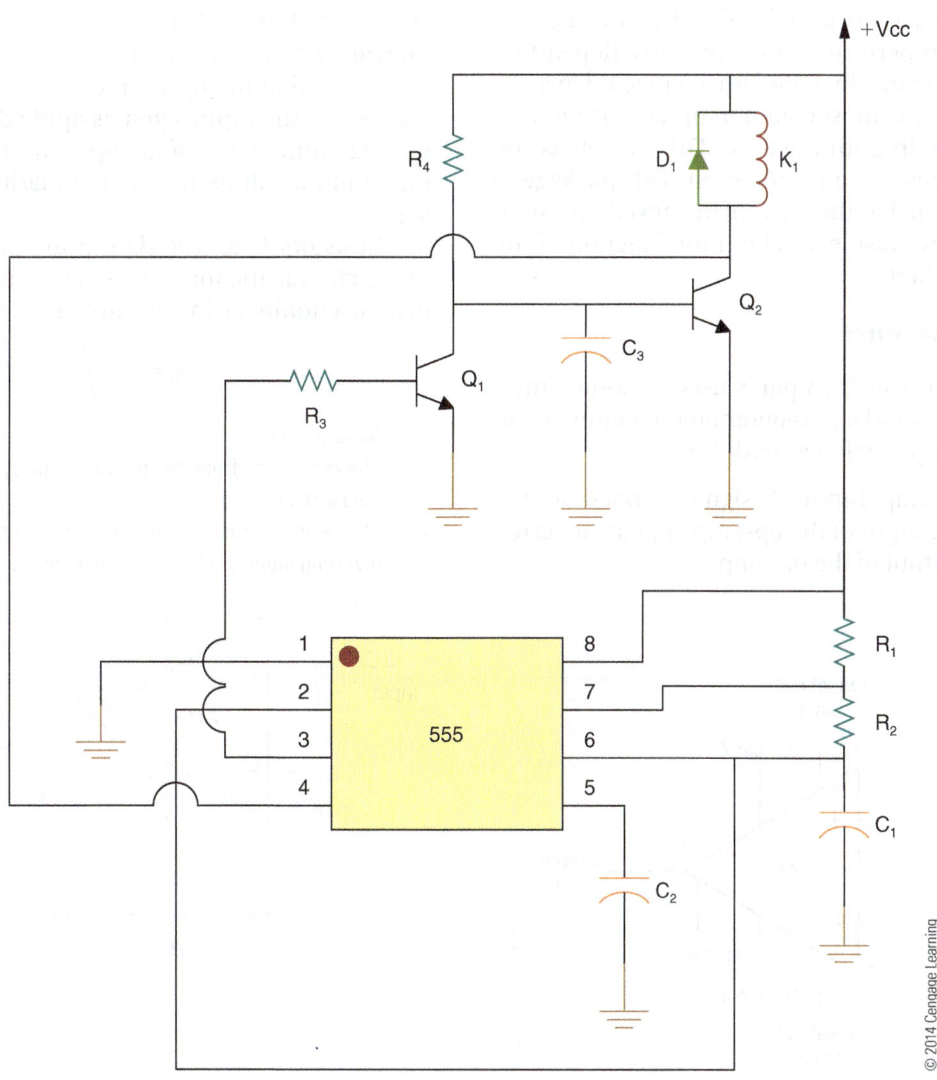

FIGURE 23-74 A 555 timer time-delay relay circuit.

the relay. Notice that relay K_1 energized *after* a time delay from when power was applied to the circuit.

By replacing resistor R_1 with a variable resistor, you can adjust the time for C_1 to charge. By varying R_1 you could vary the delay before relay K_1 energizes. Thus, you would have a *variable on-delay time-delay relay*.

As mentioned earlier, notice that the collector of Q_2 is connected to pin 4 of the 555 timer. This connection keeps the 555 timer from operating when the voltage at pin 4 is less than two-thirds of the supply voltage. This occurs when transistor Q_2 is turned on. The result of this is that power must be removed from the circuit to reset the timer and allow the delay function to be available again.

23-5 OPERATIONAL AMPLIFIERS

Another popular type of integrated circuit is the **operational amplifier**, or **op-amp**. The schematic symbol of an op-amp is shown in **Figure 23-75**. An op-amp is also a very useful and widely used device. Op-amps can perform many functions, depending on the manner in which they are connected in a circuit. Perhaps the most common op-amp is the 741. It is available in both the 8-pin DIP as well as the TO-5 packages. Because the 8-pin DIP package is more common, it is the focus of the next discussion. However, you must learn about the function of the various pins first.

Pin Assignments

Pin 1—Offset Null. This pin is used in conjunction with pin 5 and a potentiometer to eliminate an offset output voltage condition.

Pin 2—Inverting Input. A signal applied to the inverting input of the op-amp appears inverted at the output of the op-amp.

Pin 3—Noninverting Input. A signal applied to the noninverting input of the op-amp is not inverted at the output of the op-amp.

Pin 4—V−. This pin may be connected in two ways: one is to connect a negative supply voltage to it, and the other is to connect a circuit ground to it.

Pin 5—Offset Null. This pin is used in conjunction with pin 1 and a potentiometer to eliminate an offset output voltage condition.

Pin 6—Output. The output signal appears at this pin.

Pin 7—V+. This pin may be connected in two ways: one is to connect a positive supply voltage to it, and the other is to connect a circuit ground to it.

Pin 8—NC. There is no connection to pin 8. This is a dummy or dead pin and is not used.

Noninverting Amplifier

Figure 23-76 shows an op-amp circuit that is configured so that the op-amp acts as a **noninverting amplifier**. The input signal is applied to the noninverting input, pin 3, of the op-amp. The output signal at pin 6 will be the same polarity as the input signal.

Resistors R_1 and R_2 determine the voltage gain of the circuit. The formula to determine the voltage gain of a noninverting op-amp is

$$A_v = 1 + \frac{R_2}{R_1}$$

A_v = Voltage gain
R_1 = The resistance from the inverting input, pin 2, to ground, measured in ohms
R_2 = The feedback resistance from the output, pin 6, to the inverting input, pin 2, measured in ohms

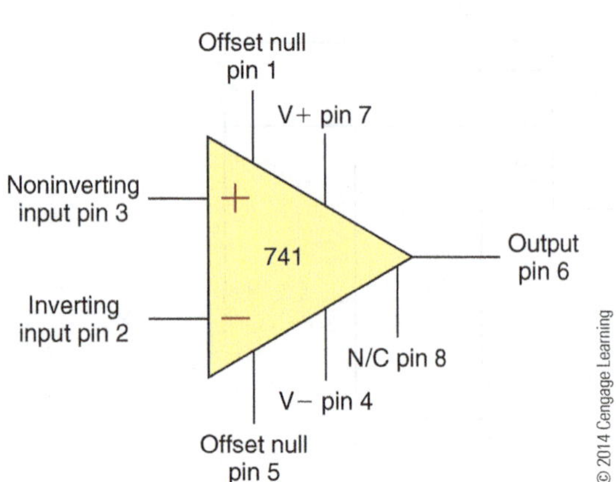

FIGURE 23-75 741 op-amp pin-outs.

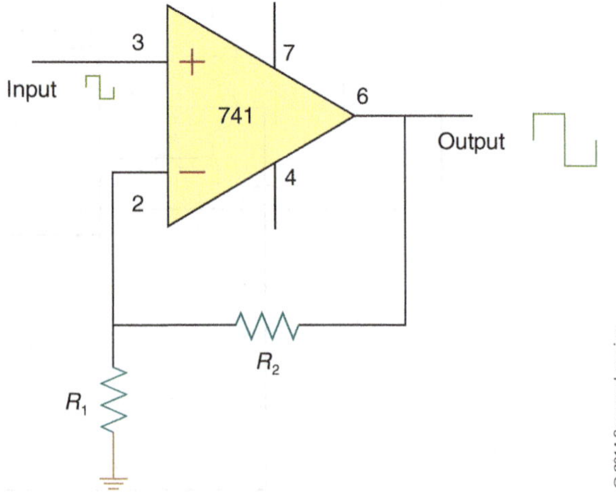

FIGURE 23-76 A noninverting amplifier.

For example, what is the voltage gain of a noninverting op-amp with R_1 = 33 kΩ, and R_2 = 1 MΩ? Use the preceding formula to determine the voltage gain of this circuit.

$$A_v = 1 + \frac{R_2}{R_1}$$

What values are known?

$$R_1 = 33 \text{ k I}$$

$$R_2 = 1 \text{ M I}$$

What value is not known?

$$A_v = ?$$

Now plug the known values into the formula and solve for A_v.

$$A_v = 1 + \frac{R_2}{R_1}$$

$$A_v = 1 + \frac{1 \text{ M I}}{33 \text{ k I}}$$

$$A_v = 31.3$$

Therefore, the noninverting op-amp circuit has a voltage gain of 31.3. This means that a 1-volt signal applied to the noninverting input, pin 3, of the op-amp produces a 31.3-volt signal at the output, pin 6.

Inverting Amplifier

Figure 23-77 shows an op-amp circuit that is configured so that the op-amp acts as an **inverting amplifier**. The input signal is applied to the inverting input, pin 2, of the op-amp. The output signal, at pin 6, will be of the opposite polarity of the input signal.

Resistors R_1 and R_2 determine the voltage gain of the circuit. The formula to determine the voltage gain of an inverting op-amp is

$$A_v = -\frac{R_2}{R_1}$$

A_v = Voltage gain

R_1 = The input resistance to the inverting input, pin 2, measured in ohms

R_2 = The feedback resistance from the output, pin 6, to the inverting input, pin 2, measured in ohms

For example, what is the voltage gain of an inverting op-amp with R_1 = 22 kΩ, and R_2 = 150 kΩ? Use the preceding formula to determine the voltage gain of this circuit.

$$A_v = -\frac{R_2}{R_1}$$

What values are known?

$$R_1 = 22 \text{ k I}$$

$$R_2 = 150 \text{ k I}$$

What value is not known?

$$A_v = ?$$

Now plug the known values into the formula and solve for A_v.

$$A_v = -\frac{R_2}{R_1}$$

$$A_v = -\frac{150 \text{ k I}}{22 \text{ k I}}$$

$$A_v = -6.82$$

Therefore, the inverting op-amp circuit has a voltage gain of –6.82. This means that a 1-volt signal applied to the inverting input, pin 2, of the op-amp produces a –6.82-volt signal at the output, pin 6. Notice that the output voltage is of an opposite polarity than that of the input voltage, hence the name, inverting op-amp.

Buffer Amplifier

Figure 23-78 shows an op-amp circuit that is configured so that the op-amp acts as a **buffer amplifier**. The input signal is applied to the non

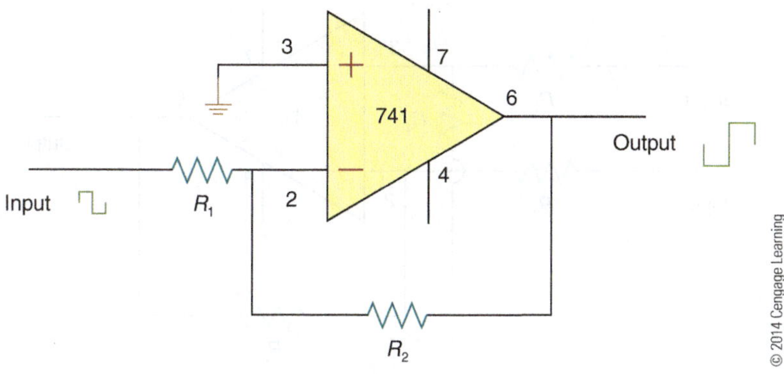

FIGURE 23-77 An inverting amplifier.

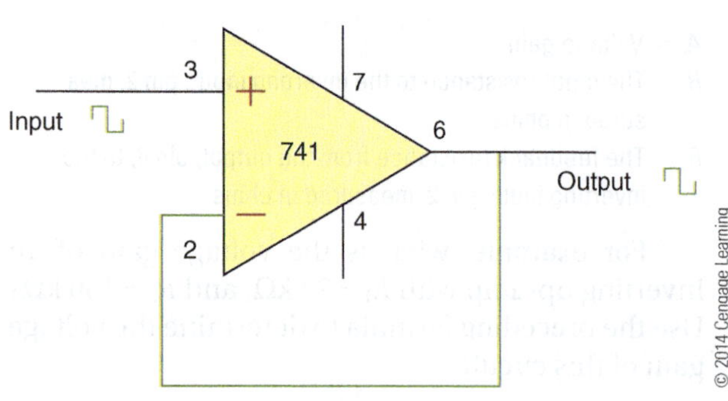

FIGURE 23-78 A buffer amplifier.

inverting input, pin 3, of the op-amp. The output signal, at pin 6, will be of the same polarity and magnitude as the input signal. You might ask, "Why use this circuit if the output is exactly the same as the input?" This circuit is used to provide impedance matching between the output of a previous circuit and the input to the following circuit. An op-amp buffer amplifier typically has a high impedance input and a low impedance output. The voltage gain of an op-amp buffer amplifier is 1.

Difference or Differential Amplifier

Figure 23-79 shows an op-amp circuit that is configured so that the op-amp acts as a **difference amplifier**. This means that the difference between the two input signals is amplified, resulting in the output signal being proportional to the difference between the input signals. One input signal is applied to the noninverting input, pin 3, of the op-amp, and another input signal is applied to the inverting input, pin 2. The difference between these two input signals is amplified and appears at the output, pin 6.

R_3 is the feedback resistor, whereas R_1 and R_2 are the input resistors. R_1 and R_2 will be of equal value. The formula to determine the voltage gain of a differential op-amp is:

$$A_v = \frac{R_f}{R_i} \times (E_1 - E_2)$$

A_v = Voltage gain

R_i = The resistance value of one of the input resistors, measured in ohms

R_f = The feedback resistance, measured in ohms

E_1 = The voltage applied to the inverting input, pin 2, of the op-amp

E_2 = The voltage applied to the noninverting input, pin 3, of the op-amp

For example, what is the voltage gain of a differential op-amp with $R_i = 5\ k\Omega$, $R_f = 80\ k\Omega$, $E_1 = 1.6\ V$, and $E_2 = 1.2\ V$? Use the preceding formula to determine the voltage gain of this circuit.

$$A_v = \frac{R_f}{R_i} \times (E_1 - E_2)$$

What values are known?

$$R_f = 80\text{ k I}$$

$$R_i = 5\text{ k I}$$

$$E_1 = 1.6\ V$$

$$E_2 = 1.2\ V$$

What value is not known?

$$A_v = ?$$

Now plug the known values into the formula and solve for A_v.

$$A_v = \frac{R_f}{R_i} \times (E_1 - E_2)$$

$$A_v = \frac{80\text{ k I}}{5\text{ k I}} \times (1.6\ V - 1.2\ V)$$

$$A_v = \frac{80\text{ k I}}{5\text{ k I}} \times (0.4\ V)$$

$$A_v = 6.4$$

Therefore, the differential op-amp circuit has a voltage gain of 6.4.

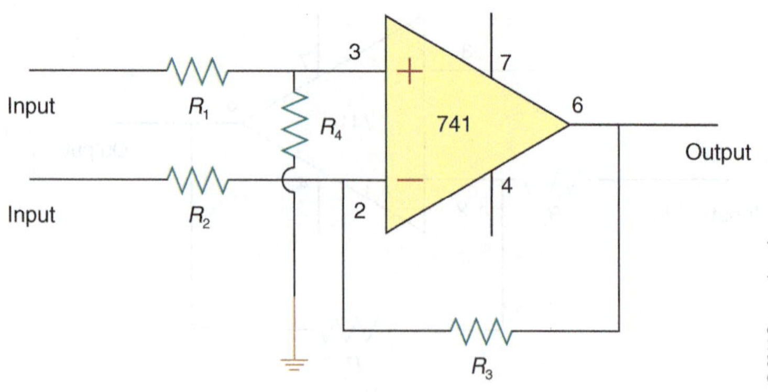

FIGURE 23-79 A difference amplifier.

Summing Amplifier

Figure 23-80 shows an op-amp circuit that is configured so that the op-amp acts as a **summing amplifier**. This means that the sum of several input signals is amplified, resulting in the output signal being the algebraic sum of all input signals. If the input signals are applied to the noninverting input, pin 3, of the op-amp, the output signal is of the same polarity. If the input signals are applied to the inverting input, pin 2, of the op-amp, the output signal is of the opposite polarity.

Comparator

Figure 23-81 shows an op-amp circuit that is configured as a **comparator**, or **level detector**. In this figure, the inverting input of the op-amp is connected to a fixed reference voltage, and the non inverting input is connected to a varying signal. If the varying signal on the noninverting input is less than the reference voltage applied to the inverting input of the op-amp, the output of the op-amp will be positive. When the varying input signal rises to a level above the reference voltage that is applied to the inverting input of the op-amp, the output of the op-amp is driven negative. In this fashion, a comparator can indicate when a varying input signal exceeds a reference voltage level.

Figure 23-82 shows a variation of the comparator shown in Figure 23-81. The circuit in Figure 23-82 shows a comparator with a variable resistor connected to the inverting input of the op-amp. This allows the reference voltage to be varied. The varying input signal is then compared to whatever voltage is set by the potentiometer.

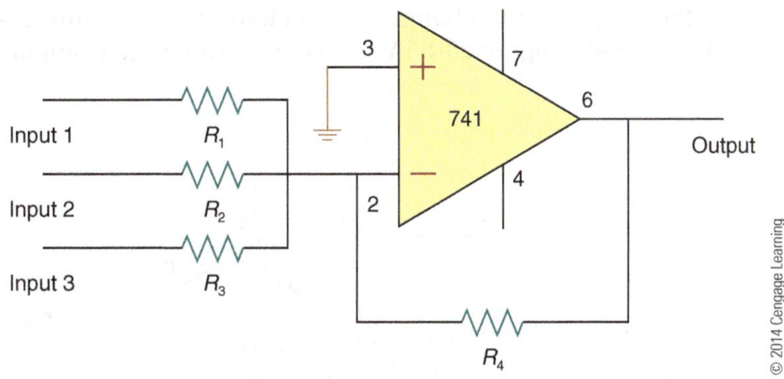

FIGURE 23-80 A summing amplifier.

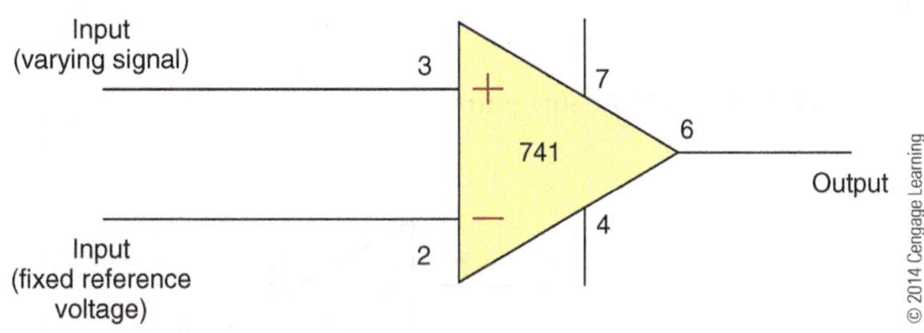

FIGURE 23-81 A comparator with fixed reference voltage.

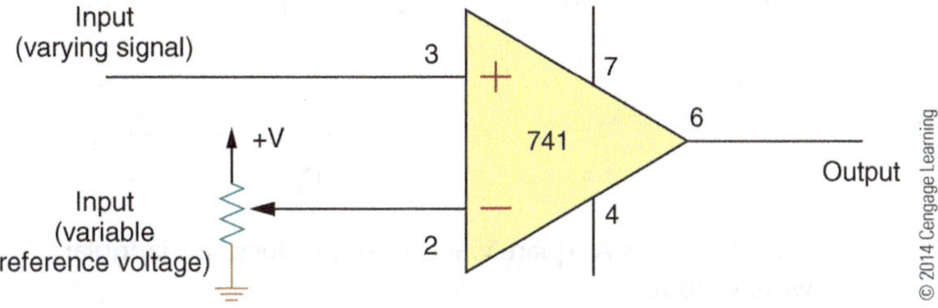

FIGURE 23-82 A comparator with variable reference voltage.

Integrator

An op-amp **integrator** is shown in **Figure 23-83**. Notice the resistor that is connected to the inverting input of the op-amp. Also, notice the capacitor that is connected between the inverting input and the output of the op-amp. An integrator produces an output that is proportional to the *integral* of the input. This means that the output voltage is proportional to the area of the input waveform.

Refer to **Figure 23-84**. If you were to apply a square wave signal to the inverting input of the integrator, the output would be a triangular wave. *An integrator is a square wave to triangular wave converter.* A triangular output wave is produced because of the charge-discharge time of the capacitor. Now consider the area under the positive portion of the square wave. As this area varies, the output will vary proportionally. For example, if this area becomes smaller (smaller integral), the output must change faster. Should the area become larger (larger integral), the output must change more slowly. Refer to **Figure 23-85**. Figure 23-85A shows the relationship between the square wave input signal and the triangular wave output signal. Figure 23-84B shows the relationship with a smaller integral, and Figure 23-85C shows the relationship with a larger integral.

Differentiator

Figure 23-86 shows a **differentiator** op-amp circuit. Compare the placement of the resistor and capacitor in this circuit with Figure 23-83. A differentiator produces an output that is proportional to the rate of change of the input signal. This means that an output is present only while the input signal is changing.

Refer to **Figure 23-87**. If you were to apply a triangular wave signal to the inverting input of the differentiator, the output would be a square wave. *A differentiator is a triangle wave to square wave converter.* A square wave output is produced because of the switching action of the op-amp. Refer to **Figure 23-88**. Figure 23-88A shows the relationship between the triangular wave input signal and

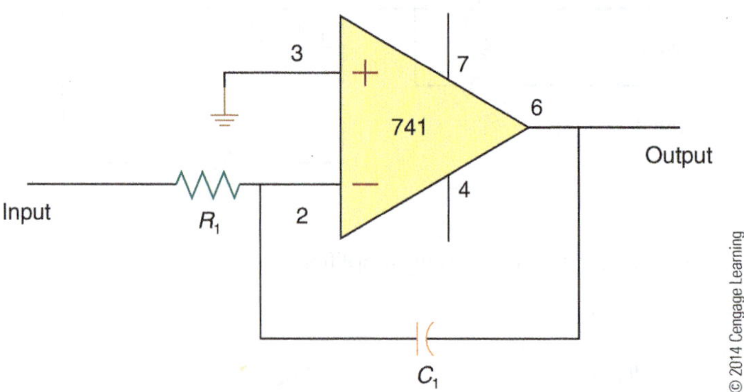

FIGURE 23-83 An integrator.

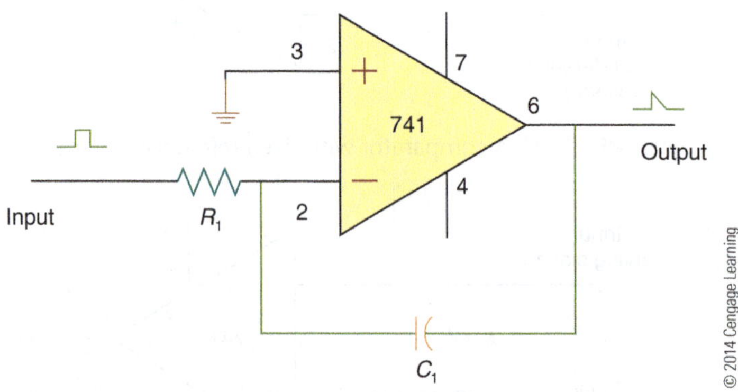

FIGURE 23-84 A square wave input produces a triangular wave output.

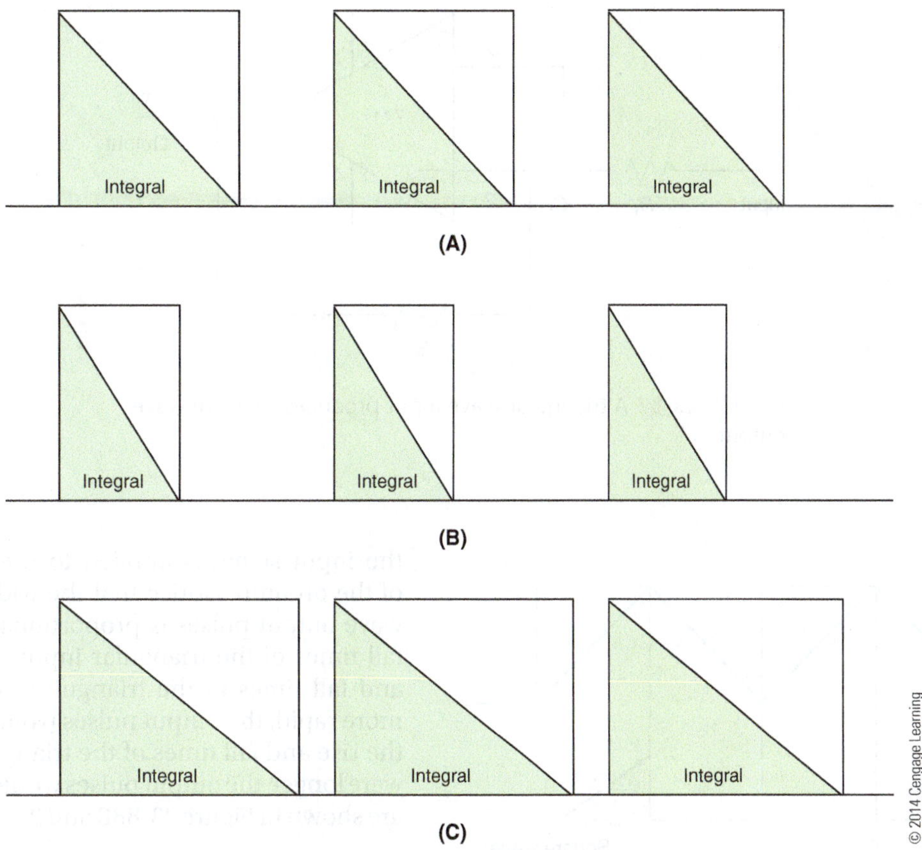

FIGURE 23-85 The integral varies as the width of the square wave varies: (A) the relationship between the square wave input signal and the triangular wave output signal; (B) the relationship with a smaller integral; (C) the relationship with a larger integral.

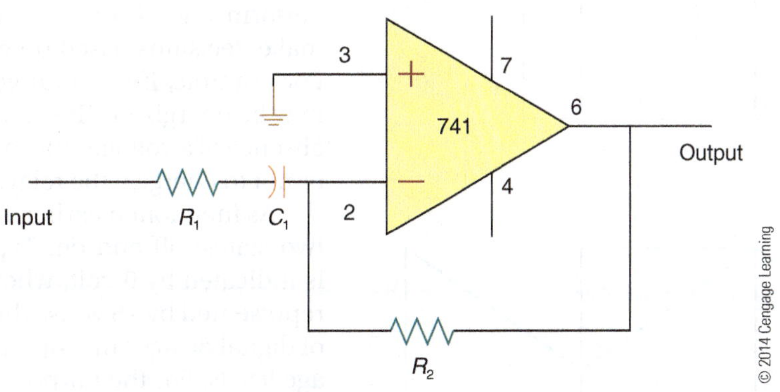

FIGURE 23-86 A differentiator.

the square wave output signal. Notice the rate of change in the triangular input signal between time T_0 and time T_1. This rate of change is constant and rising in the positive direction. During this time, the output of the differentiator is negative because the input signal is applied to the inverting input of the op-amp. When the triangular input signal begins to fall in the negative direction, the output of the op-amp switches from negative to positive. Because the rate of change between times T_1 and T_2 is constant and falling in the negative direction, the output of the differentiator is positive because

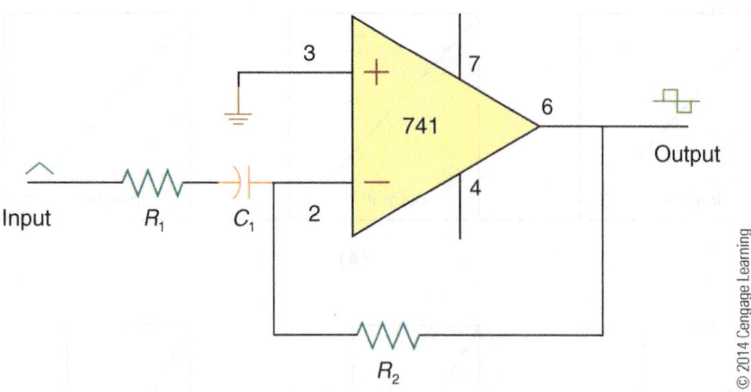

FIGURE 23-87 A triangular wave input produces a square wave output.

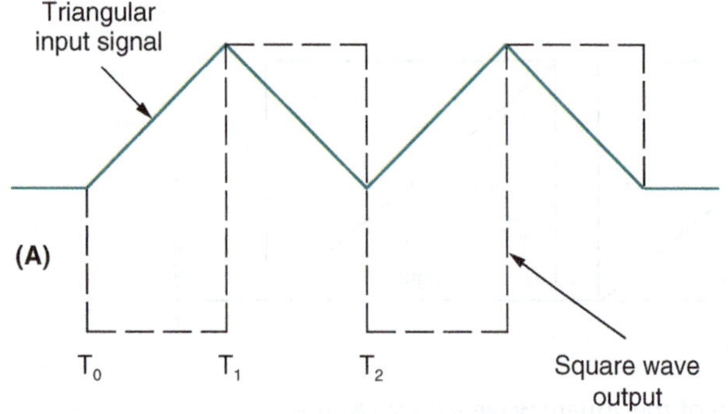

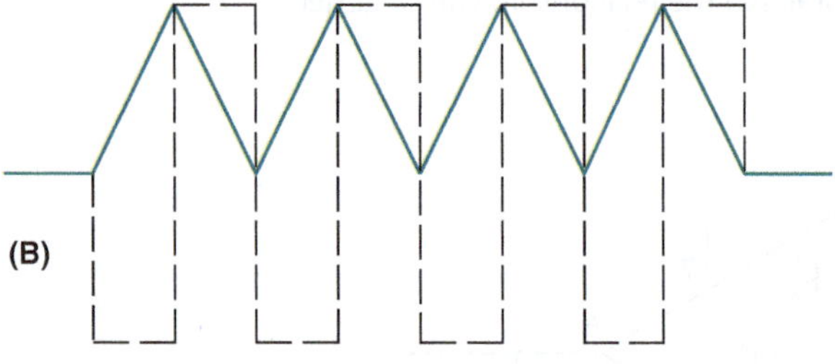

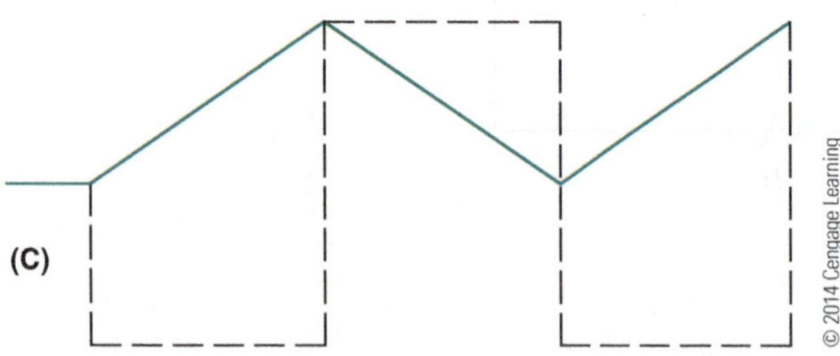

FIGURE 23-88 The width of the output square wave varies as the rate of change of the triangular input wave varies: (A) the relationship between the triangular wave input signal and the square wave output signal; (B) the rise and fall times of the triangular input signal are more rapid, thus, the narrower output pulses; (C) the rise and fall times of the triangular input signal are longer, thus, the wider output pulses.

the input signal is applied to the inverting input of the op-amp. Notice that the width of the square wave output pulses is proportional to the rise and fall times of the triangular input signal. If the rise and fall times of the triangular input signal were more rapid, the output pulses would be narrower. If the rise and fall times of the triangular input signal were longer, the output pulses would be wider. These are shown in Figure 23-88B and 23-88C respectively.

23-6 DIGITAL LOGIC

A **digital device** is a device that has only two conditions: off or on. A light switch on the wall is an example of a digital device. Increasingly, more electronic devices are using digital components to perform **logic functions**. This means circuits that make decisions based on certain input conditions. For example, if an input voltage is present, a relay may be energized. The condition is the presence or absence of a voltage. The logical decision is whether or not to energize the relay.

As mentioned earlier, digital devices operate in two states: off and on. Typically, an off condition is indicated by 0 volt, whereas the on condition is represented by +5 volts. There are different families of digital devices that operate at different logic voltage levels. For the purposes of this topic, a 0 (zero) and a 1 (one) will be used to represent off and on, respectively. This means that the logic is consistent regardless of the family of devices that is used.

The following discussion is about a family of ICs known as **digital logic gates**. These gates will make logical decisions based on certain input conditions. However, because these are digital devices, the inputs can only be one of two states: off (0) or on (1). Likewise, the output can only be one of two states: off (0) or on (1). You will learn how to use a *truth table* to indicate all possible input and output conditions.

The Inverter Gate

The simplest logic gate is the **inverter**, also called a **NOT gate**. An inverter gate inverts the input signal. Sometimes, the output of an inverter is referred to as the **complement** of the input. This means that if the input signal is 0, the output will be 1. Likewise, if the input signal is 1, the output will be 0. An easy way to remember the function of an inverter is to say that you will have an output when you do *not* have an input.

An inverter has only one input terminal and one output terminal. **Figure 23-89** shows the schematic symbol and the input and output terminals of the inverter. The schematic symbol does not show the supply voltage terminal or the supply ground terminal for the integrated circuit. These are found by identifying the device and obtaining the specifications and terminal connections or *pin-outs* for the particular IC that is being used.

Table 23-1 shows a truth table for an inverter. A truth table is a handy tool to use when evaluating a logic gate's response to different input conditions. It lists all possible input conditions and shows all possible output conditions for those inputs. The letter A represents the input terminal and the letter Y represents the output terminal. Because the inverter has only one input terminal, there can be only two possible input conditions: 0 and 1. Because the inverter has only one output terminal, there can be only two possible output conditions: 0 and 1. Notice that when the input is a 0, the output is a 1. Like wise, when the input is a 1, the output is a 0.

The operation of an inverter can be compared to that of a switch in a control circuit. Because the inverter provides an output when there is no input, it functions similarly to a normally closed switch. **Figure 23-90** shows a ladder diagram with a normally closed pushbutton switch and a pilot light. Notice that if the pushbutton switch is *not* pressed, the pilot light is on. If the pushbutton switch is pressed, the pilot light is off. **Table 23-2** shows a truth table for the operation of this circuit. Notice that the truth table for this circuit is identical to the truth table for an inverter. This means that an inverter performs the same logical operation as a normally closed switch.

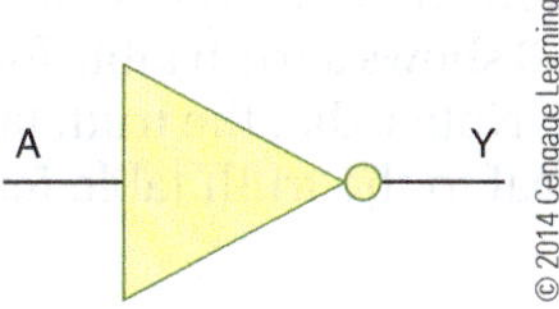

FIGURE 23-89 An inverter.

Table 23-1

INVERTER TRUTH TABLE

Input	Output
A	*Y*
0	1
1	0

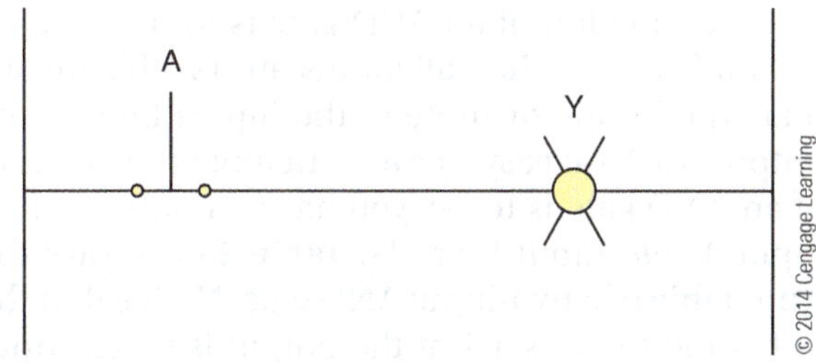

FIGURE 23-90 A ladder diagram circuit performing an inverter function.

Table 23-2

INVERTER LADDER DIAGRAM TRUTH TABLE

Input	Output
A	*Y*
0 (Not depressed)	1 (On)
1 (Depressed)	0 (Off)

The AND Gate

The **AND gate** has a minimum of two input terminals but always has one output terminal. AND gates are available with more than two inputs—some have three inputs, some four, and some more. Two-input and three-input AND gates, shown in **Figure 23-91**, are discussed next. Figure 23-91A shows the schematic symbol of a two-input AND gate, whereas Figure 23-91B shows the schematic symbol of a three-input AND gate. Once you have studied these, you should be able to determine the truth table for an AND gate with additional inputs.

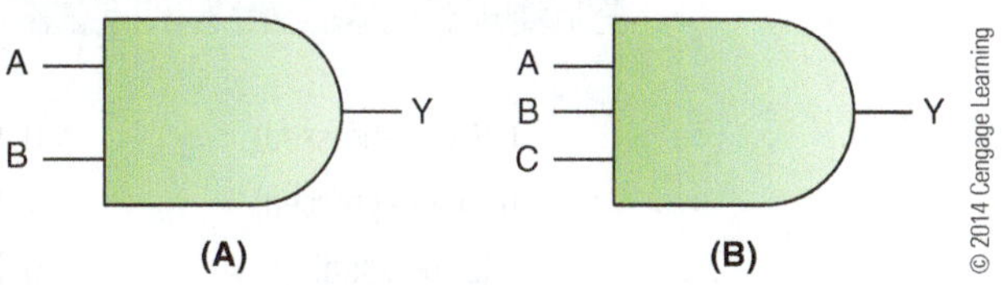

FIGURE 23-91 AND gates: (A) two-input AND gate; (B) three-input AND gate.

The function of an AND gate is to produce an output (1) only when all inputs are 1s. This means that should one or more of the inputs be a 0, the output is a 0. An easy way to remember the function of an AND gate is to say you have an output when input A *and* input B are 1s. **Table 23-3** shows the truth table of a two-input AND gate. Notice that the only time there is a 1 at the output is when input A *and* input B are 1s. **Table 23-4** shows the truth table for a three-input AND gate. Again, notice that the only time there is a 1 at the output occurs when inputs A, B, and C are 1s.

Table 23-3

TWO-INPUT AND GATE TRUTH TABLE

Inputs		Output
A	*B*	*Y*
0	0	0
0	1	0
1	0	0
1	1	1

Table 23-4

THREE-INPUT AND GATE TRUTH TABLE

Inputs			Output
A	*B*	*C*	*Y*
0	0	0	0
0	0	1	0
0	1	0	0
0	1	1	0
1	0	0	0
1	0	1	0
1	1	0	0
1	1	1	1

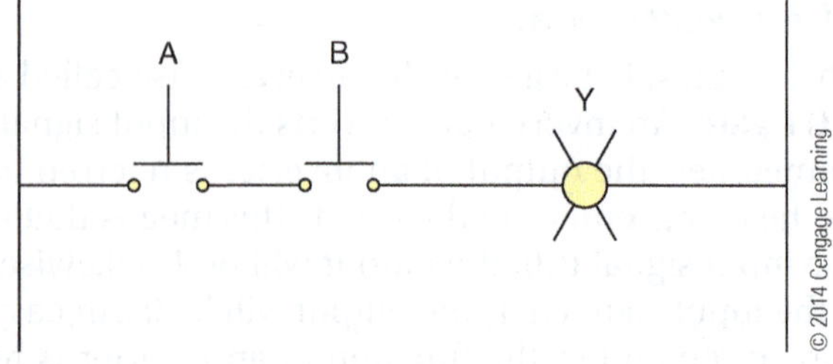

FIGURE 23-92 A ladder diagram circuit performing a two-input AND function.

The operation of a two-input AND gate can be compared to the operation of two switches in a control circuit. Because the AND gate provides an output only when both inputs are 1s, the two-input AND gate functions similarly to two normally open switches connected in series. **Figure 23-92** shows a ladder diagram with two normally open pushbutton switches connected in series and a pilot light. Notice that if none of the pushbuttons is pressed, the pilot light is off. If only switch A is pressed, the pilot light is off. Likewise, if only switch B is pressed, the pilot light is off. However, if both switch A *and* switch B are pressed, the pilot light is on. **Table 23-5** shows a truth table for the operation of this circuit. Notice that the truth table for this circuit is identical to the truth table for a two-input AND gate. This means that a two-input AND gate performs the same logical operation as two normally open switches wired in series.

Figure 23-93 shows a ladder diagram with three normally open pushbutton switches connected in series and a pilot light. Notice that if none of the pushbuttons are pressed, the pilot light is off. If only switch A is pressed, the pilot light is off. Likewise, if only switch B is pressed, the pilot light is off. If only switch C is pressed, the pilot light is off. However, if switches A, B, and C are pressed, the pilot light is on. **Table 23-6** shows a truth table for the operation of this circuit. Notice that the truth table for this circuit is identical to the truth table for a three-input

Table 23-5

TWO-SWITCH AND FUNCTION LADDER DIAGRAM TRUTH TABLE

Inputs		Output
A	*B*	*Y*
0 (Not depressed)	0 (Not depressed)	0 (Off)
0 (Not depressed)	1 (Depressed)	0 (Off)
1 (Depressed)	0 (Not depressed)	0 (Off)
1 (Depressed)	1 (Depressed)	1 (On)

Table 23-6

THREE-SWITCH AND FUNCTION LADDER DIAGRAM TRUTH TABLE

Inputs			Output
A	*B*	*C*	*Y*
0 (Not depressed)	0 (Not depressed)	0 (Not depressed)	0 (Off)
0 (Not depressed)	0 (Not depressed)	1 (Depressed)	0 (Off)
0 (Not depressed)	1 (Depressed)	0 (Not depressed)	0 (Off)
0 (Not depressed)	1 (Depressed)	1 (Depressed)	0 (Off)
1 (Depressed)	0 (Not depressed)	0 (Not depressed)	0 (Off)
1 (Depressed)	0 (Not depressed)	1 (Depressed)	0 (Off)
1 (Depressed)	1 (Depressed)	0 (Not depressed)	0 (Off)
1 (Depressed)	1 (Depressed)	1 (Depressed)	1 (On)

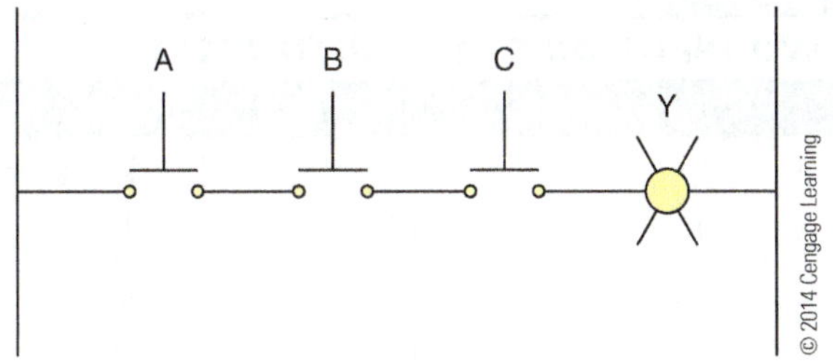

FIGURE 23-93 A ladder diagram circuit performing a three-input AND function.

AND gate. This means that a three-input AND gate performs the same logical operation as three normally open switches wired in series.

The NAND Gate

Logic gates can be combined to provide different logical operations. For example, if an inverter is added to the output of an AND gate, a different set of logical conditions will be created. This new circuit is called a **NAND gate**, as shown in **Figure 23-94**. The word *NAND* is a combination of *NOT* and *AND*. This means that this gate functions as an AND gate with an inverted output. Figure 23-94A shows a two-input AND gate with an inverter added to the output. Figure 23-94B shows the more common schematic for a two input NAND gate. Notice that the inverter is now represented by the small circle at the output of the AND gate. The small circle will be used in future schematics to represent the inverter function when added to another gate.

Table 23-7 shows the truth table for a two-input NAND gate. Notice that the only time there is not an output is when both inputs are 1s. Compare this truth table to that of a two-input AND gate in Table 23-3. The truth table for a two-input NAND gate is the exact opposite of the one for a two-input AND gate. This is because a NAND gate is an AND gate with an inverted output. Although the discussion has focused on a two input NAND gate, you should realize that NAND gates may have three, four, or more inputs, just like the AND gate.

Table 23-7

TWO-INPUT NAND GATE TRUTH TABLE

Inputs		Output
A	*B*	*Y*
0	0	1
0	1	1
1	0	1
1	1	0

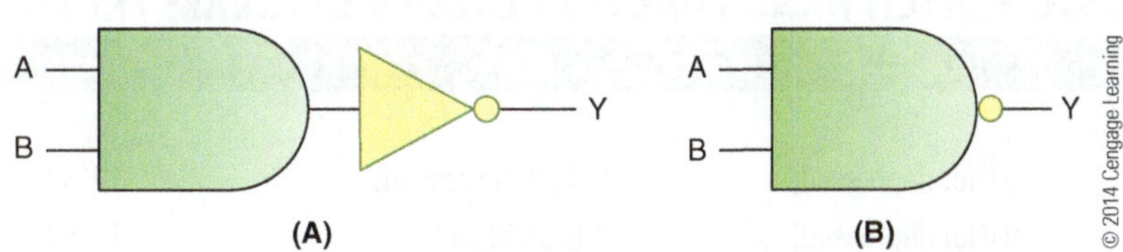

FIGURE 23-94 NAND gate: (A) AND gate with inverter; (B) AND gate with inverted output.

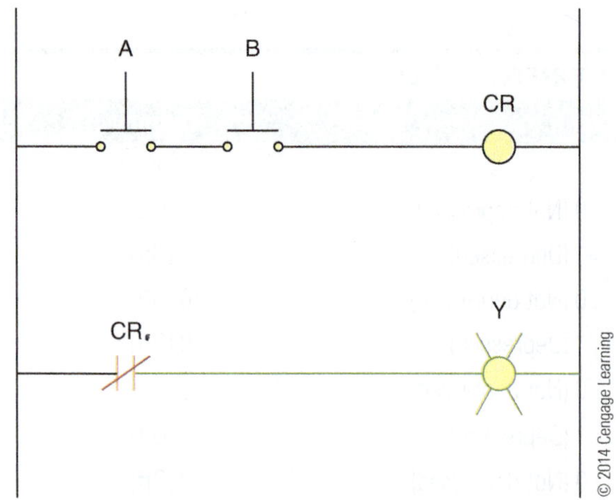

FIGURE 23-95 A ladder diagram circuit performing a two-input NAND function.

Figure 23-95 shows a ladder diagram with two normally open pushbutton switches, a relay, and a pilot light. The pushbuttons are both normally open switches and are wired in series with the relay coil. Notice that if none of the pushbuttons is pressed, the pilot light is on. If only switch A is pressed, the pilot light is on. Likewise, if only switch B is pressed, the pilot light is on. However, when both switch A and switch B are pressed, the pilot light is off. **Table 23-8** shows a truth table for the operation of this circuit. Notice that the truth table for this circuit is identical to the one for a two-input NAND gate.

The OR Gate

An **OR gate** has a minimum of two input terminals but will always have one output terminal. OR gates are available with more than two inputs—some have three inputs, some four, and some more. Two-input and three-input OR gates, shown in **Figure 23-96**, are discussed next. Figure 23-96A shows the schematic symbol of a two-input OR gate, whereas Figure 23-96B shows the schematic symbol of a three-input OR gate. Once you have studied these, you should be able to determine the truth table for an OR gate with additional inputs.

The function of an OR gate is to produce an output (1) when any input is a 1. This means that when

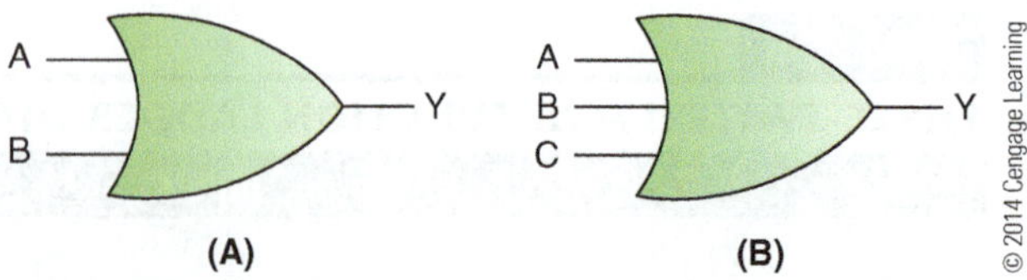

FIGURE 23-96 OR gates: (A) two-input OR gate; (B) three-input OR gate.

both inputs are 0, the output will be 0. An easy way to remember the function of an OR gate is to say you will have an output when input A *or* input B is a 1. **Table 23-9** shows the truth table of a two-input OR gate. Notice that the only time there is a 1 at the output is when input A or input B is a 1. **Table 23-10**

Table 23-9

TWO-INPUT OR GATE TRUTH TABLE

Inputs		Output
A	*B*	*Y*
0	0	0
0	1	1
1	0	1
1	1	1

Table 23-10

THREE-INPUT OR GATE TRUTH TABLE

Inputs			Output
A	*B*	*C*	*Y*
0	0	0	0
0	0	1	1
0	1	0	1
0	1	1	1
1	0	0	1
1	0	1	1
1	1	0	1
1	1	1	1

Table 23-8

TWO-SWITCH NAND FUNCTION LADDER DIAGRAM TRUTH TABLE

Inputs		Output
A	*B*	*Y*
0 (Not depressed)	0 (Not depressed)	1 (On)
0 (Not depressed)	1 (Depressed)	1 (On)
1 (Depressed)	0 (Not depressed)	1 (On)
1 (Depressed)	1 (Depressed)	0 (Off)

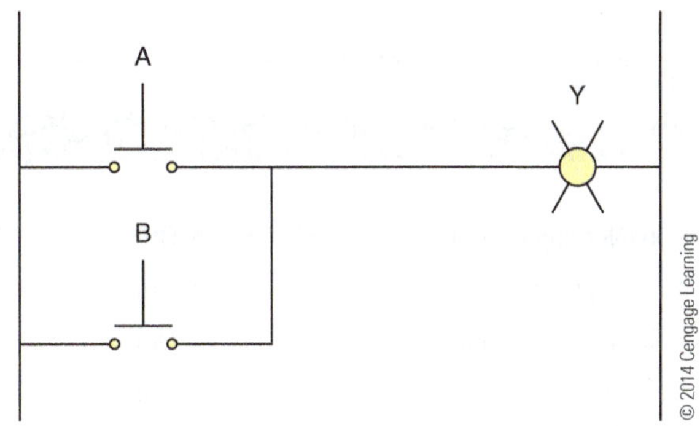

FIGURE 23-97 A ladder diagram circuit performing a two-input OR function.

shows the truth table for a three-input OR gate. Again, notice that the only time there is a 1 at the output occurs when input A, B, or C is a 1.

The operation of a two-input OR gate can be compared to the operation of two switches in a control circuit. Because the OR gate provides an output only when either input is a 1, the two-input OR gate functions similarly to two normally open switches connected in parallel. **Figure 23-97** shows a ladder diagram with two normally open pushbutton switches connected in parallel and a pilot light. Notice that if none of the pushbuttons is pressed, the pilot light is off. If only switch A is pressed, the pilot light is on. Likewise, if only switch B is pressed, the pilot light is on. If both switch A and switch B are pressed, the pilot light is on. **Table 23-11** shows a truth table for the operation of this circuit. Notice that the truth table for this circuit is identical to the one for a two-input OR gate. This means that a two-input OR gate performs the same logical operation as two normally open switches wired in parallel.

Figure 23-98 shows a ladder diagram with three normally open pushbutton switches connected in parallel and a pilot light. Notice that if none of the pushbuttons is pressed, the pilot light is off. If only switch A is pressed, the pilot light is on. If only switch B is pressed, the pilot light is on.

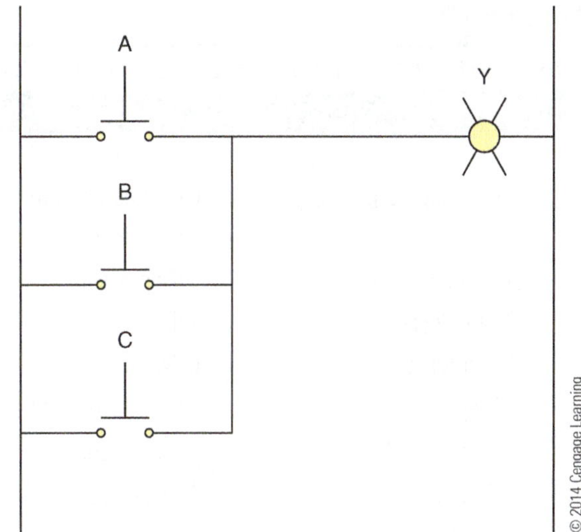

FIGURE 23-98 A ladder diagram circuit performing a three-input OR function.

If only switch C is pressed, the pilot light is on. Pressing switches A and B turns the pilot light on. Likewise, pressing switches B and C turns the pilot light on. Also, pressing switches A and C turns the pilot light on. Finally, if all three switches, A, B, and C, are pressed, the pilot light is on. **Table 23-12** shows a truth table for the operation of this circuit. Notice that the truth table for this circuit is identical to the one for a three-input OR gate. This means that a three-input OR gate performs the same logical operation as three normally open switches wired in parallel.

The NOR Gate

If an inverter is added to the output of an OR gate, a different set of lo\gical conditions is created. This new circuit is called a **NOR gate**. The word *NOR* is a combination of *NOT* and *OR*. This means that this gate functions as an OR gate with an inverted output. **Figure 23-99** shows a two-input OR gate with an inverter added to the output. Recall that the inverter is represented by the small circle at the output of the OR gate.

Table 23-11

TWO-SWITCH OR FUNCTION LADDER DIAGRAM TRUTH TABLE

Inputs		Output
A	*B*	*Y*
0 (Not depressed)	0 (Not depressed)	0 (Off)
0 (Not depressed)	1 (Depressed)	1 (On)
1 (Depressed)	0 (Not depressed)	1 (On)
1 (Depressed)	1 (Depressed)	1 (On)

Table 23-12

THREE-SWITCH OR FUNCTION LADDER DIAGRAM TRUTH TABLE

Inputs			Output
A	*B*	*C*	*Y*
0 (Not depressed)	0 (Not depressed)	0 (Not depressed)	0 (Off)
0 (Not depressed)	0 (Not depressed)	1 (Depressed)	1 (On)
0 (Not depressed)	1 (Depressed)	0 (Not depressed)	1 (On)
0 (Not depressed)	1 (Depressed)	1 (Depressed)	1 (On)
1 (Depressed)	0 (Not depressed)	0 (Not depressed)	1 (On)
1 (Depressed)	0 (Not depressed)	1 (Depressed)	1 (On)
1 (Depressed)	1 (Depressed)	0 (Not depressed)	1 (On)
1 (Depressed)	1 (Depressed)	1 (Depressed)	1 (On)

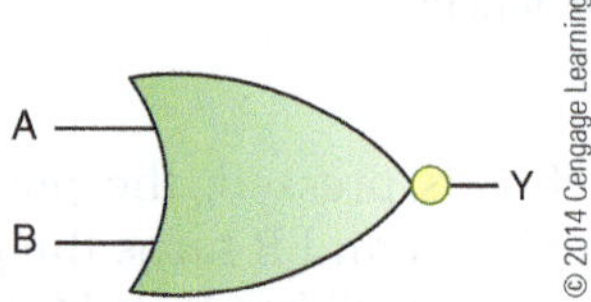

FIGURE 23-99 A two-input NOR gate.

Table 23-13

TWO-INPUT NOR GATE TRUTH TABLE

Inputs		Output
A	*B*	*Y*
0	0	1
0	1	0
1	0	0
1	1	0

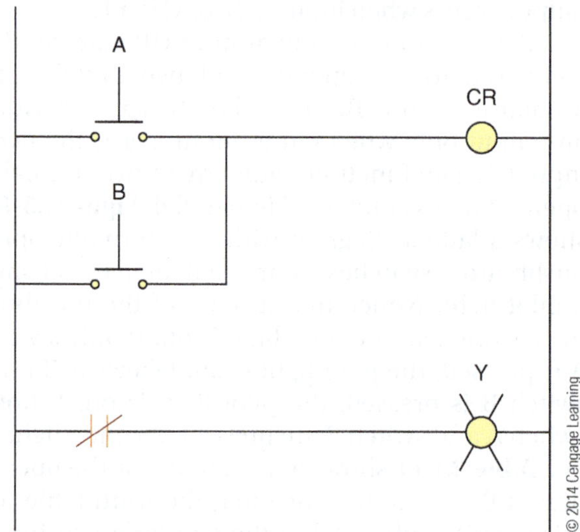

FIGURE 23-100 A ladder diagram circuit performing a two-input NOR gate function.

Table 23-13 shows the truth table for a two-input NOR gate. Notice that the only time there is an output is when both inputs are 0s. Compare this truth table to that of the two-input OR gate in Table 23-9. The truth table for a two-input NOR gate is the exact opposite of the one for a two-input OR gate. This is because a NOR gate is an OR gate with an inverted output. Although the discussion has focused on a two-input NOR gate, you should realize that NOR gates may have three, four, or more inputs, just like OR gates.

Figure 23-100 shows a ladder diagram with two normally open pushbutton switches, a relay, and a pilot light. The pushbuttons are both normally open switches and are wired in parallel with the relay coil. Notice that if none of the pushbuttons are pressed, the pilot light is on. If only switch A is pressed, the pilot light is off. Likewise, if only switch B is pressed, the pilot light is off. When both switch A and switch B are pressed, the pilot light is off. **Table 23-14** shows a truth table for the operation of this circuit. Notice that the truth table for this circuit is identical to the one for a two-input NOR gate.

The Exclusive OR Gate

Figure 23-101 shows the schematic symbol of the **exclusive OR gate**, also called an **XOR gate**. The XOR gate produces an output when either input is a 1, but not when both inputs are 0s or 1s. The truth table for an XOR gate is shown in **Table 23-15**.

The operation of an XOR gate can be compared to the operation of two switches in a control circuit. Because the XOR gate provides an output only when either input is a 1, the XOR gate functions

Table 23-14

TWO-SWITCH NOR FUNCTION LADDER DIAGRAM TRUTH TABLE

Inputs		Output
A	*B*	*Y*
0 (Not depressed)	0 (Not depressed)	1 (On)
0 (Not depressed)	1 (Depressed)	0 (Off)
1 (Depressed)	0 (Not depressed)	0 (Off)
1 (Depressed)	1 (Depressed)	0 (Off)

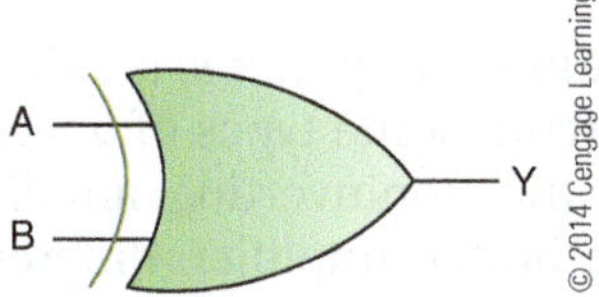

FIGURE 23-101 An exclusive OR (XOR) gate.

Table 23-15

TWO-INPUT XOR GATE TRUTH TABLE

Inputs		Output
A	*B*	*Y*
0	0	0
0	1	1
1	0	1
1	1	0

similarly to two switches, each with a normally open and a normally closed contact connected as shown in **Figure 23-102.** Notice that if none of the pushbuttons is pressed, the pilot light is off. If only switch A is pressed, the pilot light is on. Likewise, if only switch B is pressed, the pilot light is on. If both switch A and switch B are pressed, the pilot light is off. **Table 23-16** shows a truth table for the

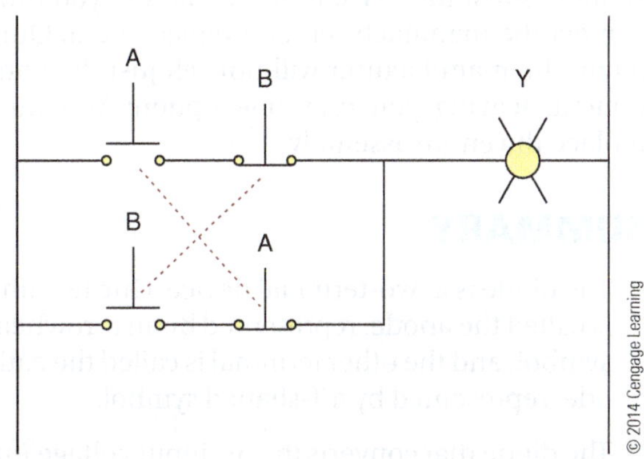

FIGURE 23-102 A ladder diagram circuit performing an XOR gate function.

operation of this circuit. Notice that the truth table for this circuit is identical to the one for an XOR gate, as shown in Table 23-15.

23-7 REPLACING SOLID-STATE DEVICES

When replacing a defective solid-state device, you should identify it by the number that is imprinted on its body. You can then contact the manufacturer for a replacement. There are also several companies that manufacture replacement or substitute devices

Table 23-16

TWO-SWITCH XOR FUNCTION LADDER DIAGRAM TRUTH TABLE

Inputs		Output
A	*B*	*Y*
0 (Not depressed)	0 (Not depressed)	0 (Off)
0 (Not depressed)	1 (Depressed)	1 (On)
1 (Depressed)	0 (Not depressed)	1 (On)
1 (Depressed)	1 (Depressed)	0 (Off)

that meet or exceed the original manufacturer's specifications. These companies typically publish a cross-reference guide. Simply look up the original manufacturer's part number in a listing of all replacements available. Once you have found your component, you can find the part number of the substitute part. Usually you also find the specifications for the substitute part. This helps ensure that the substitute will work as a replacement for the defective manufacturer's device.

Some devices are printed with the manufacturer's own unique number and cannot be cross-referenced. This makes it impossible to find a suitable substitute part. In this situation, you must contact the manufacturer for a replacement. Quite often, the manufacturer will not sell just the component, leaving you only one option: You must replace the entire assembly.

SUMMARY

- The diode is a two-terminal device. One terminal is called the anode, represented by an arrowhead symbol, and the other terminal is called the cathode, represented by a T-shaped symbol.
- The diode that converts the AC input voltage into a DC output voltage is called a rectifier.
- Zener diodes are used to regulate voltage.
- Light-emitting diodes (LEDs) are diodes that emit light when forward biased. LEDs have a cathode, which must be negative, and an anode, which must be positive, in order to conduct.
- A bipolar junction transistor (BJT) is a three-terminal device. The terminal with the arrowhead is called the emitter, the one with the T shape is called the base, and the remaining one is called the collector.
- An insulated gate bipolar transistor (IGBT) is another three-terminal device. It is being used more in industry due to its high-speed switching ability.
- There are two types of field-effect transistors: the junction field-effect transistor and the metal-oxide semiconductor field-effect transistor.
- A family of devices known as thyristors includes unijunction transistors, silicon-controlled rectifiers, diacs, and triacs.
- A 555 timer is a member of a family of devices known as integrated circuits (ICs). An IC is a device that contains anywhere from a few components to several thousand components in one package. These components are connected to form complete circuits, which, in turn, are miniaturized and contained within a device known as an IC.
- The 555 timer is typically found in an 8-pin DIP package, although it may appear in a TO-5 package.
- The 555 timer is is used to perform many types of timing, pulsing, and delaying operations.
- An operational amplifier is another type of integrated circuit. Some types of operational amplifiers are the noninverting amplifier, inverting amplifier, buffer amplifier, difference amplifier, summing amplifier, comparator, integrator, and differentiator.
- A digital device operates in two states: off, which is indicated by a 0 (zero), and on, which is indicated by a 1 (one).
- When replacing a solid-state device, look for the number that is imprinted on its body, and contact the manufacturer for a replacement. There are, however, several companies that manufacture replacement devices that meet or exceed the manufacturer's specifications. These companies publish cross-reference guides that include the manufacturer's part number and the available replacements.

REVIEW QUESTIONS

1. Describe how you would connect a diode to a DC source so that the diode is forward biased.
2. Explain the procedure you would follow to determine if an NPN transistor were defective.
3. List the three circuit configurations for an operational amplifier.
4. Which two-input digital logic gate only produces an output when both inputs are 1s?
5. Which two-input digital logic gate produces an output when either input is a 1, but not when both inputs are a 1 or a 0?

CHAPTER

24

Electronic Variable-Speed Drives

Motors have been used in industry for many years. Typically, these motors have been operated at full speed or slightly below. When necessary, the speed of the motor may be varied, usually by adding series resistance. Although this practice works well and is easy to accomplish, it is inefficient, wasting a large amount of energy in the form of heat given off by the resistors. Electronic variable-speed drives are now used to vary the speed of these motors with greater efficiency and more precise speed control than resistors offer. This chapter introduces you to DC electronic variable-speed drives.

OBJECTIVES

After studying this chapter, the student should be able to

- Discuss the general operating principle of a DC drive.
- Identify problems and troubleshoot a DC drive.
- Discuss the general operating principle of an AC drive.
- Identify problems and troubleshoot an AC drive.

24-1 DC DRIVES

An electronic variable-speed drive has two basic sections—the **control section** and the **power section**. The control section governs or controls the power section, while the power section supplies controlled power to the DC motor. (See **Figure 24-1**.)

The control section allows us to control not only the motor's speed but also its **torque**. (Recall that torque is the turning force that a motor produces.) Motor speed and torque control can be accomplished by one of two methods—we can vary either the voltage to the armature of the DC motor or the current to the field. When we vary the armature voltage, the motor produces full torque, but the speed is varied. However, if the field current is varied, both the motor speed and the torque vary. Because of the need to vary the armature voltage or the field current, a separately excited DC motor, which allows very precise control over speed and torque, is the most commonly used type of motor.

Attaining precise control over motor speed and torque requires a means of evaluating the motor's performance and automatically compensating for any variations from the desired levels. The control section of a DC drive uses three types of signals to evaluate motor performance—the **command**, **feedback**, and **error** signals.

FIGURE 24-1 Eurotherm model 590SP Digital DC drive.

A **command signal**, sometimes called the **set point** or **reference signal**, is programmed into the DC drive and sets the desired operating speed of a DC motor. While the motor is operating, a **feedback signal** from the motor indicates the motor's performance. The feedback signal can originate either from the **counter-electromotive force** produced by the motor or from a **tachometer-generator** or **encoder** mounted on the motor's shaft. Counter-electromotive force (CEMF) is the voltage produced in the motor's rotating armature, which cuts the magnetic lines of force in the field as it revolves. This armature voltage is called counter-EMF because it opposes the applied voltage from the electronic variable-speed drive. As a result of this opposing voltage, the amount of armature current is limited. A difference exists between the control signal and the feedback signal. This difference is the **error signal**. The electronic variable-speed drive's controller automatically adjusts the motor's performance until the error signal is reduced practically to zero. This function is an ongoing process.

A controller's **regulation** determines how well it responds to changes in motor performance and is usually expressed as a percentage. Different types of feedback signals result in variable degrees of regulation. For instance, controllers using counter-EMF feedback, also called **armature voltage feedback**, typically have a regulation of 5% to 8%. This means that the speed of a DC motor set to operate at 1800 RPM can vary from 1944 RPM to 1656 RPM (1800 RPM ± 8%). On the other hand, when shaft-coupled encoders are used to provide the feedback signal, regulation is much tighter. A typical shaft-coupled encoder produces a regulation of 0.01%. Thus, the speed variation of the same motor operating at 1800 RPM would drop to between 1799.92 RPM and 1800.18 RPM (1800 RPM ± 0.01%).

In addition to managing motor speed and torque, the control section of a DC drive determines the direction of motor rotation and controls motor braking as well.

Switching Amplifier Field Current Controller

Before you can understand how a switching amplifier field current controller works, you must understand what is meant by **open-loop control** and **closed-loop control**. In open-loop control, also called **manual control**, any variations in motor speed must be compensated by a manual adjustment. As you can imagine, paying someone to monitor and adjust motor speed constantly under varying load conditions would be very costly and inefficient. Therefore, open-loop control is limited to applications where the motor load is fairly constant.

When the motor load varies considerably or frequently, closed-loop control is used. Closed-loop control, also called **automatic control**, uses feedback information to monitor the performance of a motor. The information that is fed back automatically causes the control circuit to adjust the motor speed to varying load conditions.

A switching amplifier field current controller, as shown in **Figure 24-2**, receives and responds to a feedback signal from a DC motor that provides information about the speed of the motor. The controller first compares the feedback signal to a reference signal. Depending on the result of this comparison, the controller automatically increases or decreases the motor speed until it reaches the level indicated by the reference signal. To accomplish this adjustment, the controller switches the shunt field current on and off at varying rates.

To understand the following detailed explanation, refer to Figure 24-2. We begin with the feedback device.

In the upper right corner of Figure 24-2 is a schematic view of tachometer-generator (tach. gen.), a DC generator attached to the DC shunt motor shaft. As the DC shunt motor turns, the DC tachometer-generator turns and produces a positive DC voltage as a result. The faster the DC shunt motor turns, the more positive DC voltage the DC tachometer-generator produces. Thus, the output of the DC tachometer-generator is proportional to the speed of the DC shunt motor.

Next, we will consider the feedback section, shown in detail in **Figure 24-3**. The positive voltage produced by the DC tachometer-generator and fed through R1 into buffer amplifier U1 is the feedback signal. Resistor R1 is known as a scaling resistor. The value of R1 is adjusted to provide the proper amount of DC voltage from the tachometer-generator for a given RPM. The output of U1 is fed to the inverting input of U2, which inverts the polarity of the tachometer-generator voltage from positive to negative. The resulting negative voltage is then fed through R7 to the noninverting input of U3 in the preamplifier section. Note that one end of R8 is connected to the noninverting input of U3 and the other end of R8 is connected to variable resistor R6. Resistor R6 allows us to set the reference voltage level, which determines at what speed the DC shunt motor runs. This reference voltage may also be called the command signal or set point. The reference voltage (positive) and the tachometer-generator voltage (negative) are added together at the junction of R7 and R8, called the **summing point**. The difference of these two voltages, which appears at the noninverting input of U3, is the error signal.

To understand how this process works, assume that we wish the DC shunt motor to turn at 1800 RPM. We find that R6 must be set to 10 volts, the reference voltage required to attain a motor speed of 1800 RPM. Assume also that at a speed of 1800 RPM, the DC tachometer-generator produces a positive output of 5 volts. This value is inverted by U2. Thus, at the junction of R7 and R8, we have a positive reference voltage of 10 volts and a negative DC tachometer-generator feedback voltage of 5 volts. The sum of these two voltages is a positive error voltage of 5 volts $(10 + (-5) = 5)$.

If the speed of the DC shunt motor decreases below the reference speed, the tachometer-generator produces less positive DC voltage. Now, assume that the DC tachometer-generator produces a positive voltage of 3 volts DC, which results in a negative voltage of 3 volts at R7 from the output of U2. Because the reference voltage does not change, we still have a positive voltage of 10 volts at R8 from R6. The sum of these two voltages at the junction of R7 and R8 is a positive error voltage of 7 volts $(10 + (-3) = 7)$. Therefore, we can conclude that decreases in motor speed produce higher positive voltage at the noninverting input of U3. Likewise, increases in motor speed produce lower positive voltage at the noninverting input of U3.

Figure 24-4 (see in page 554) illustrates U3, a preamplifier stage that takes the voltage at its noninverting input and provides an amplified positive output voltage of the proper level for the comparator to work with. Before examining the comparator stage, we first look at the sawtooth generator stage.

Inverters U6 and U7, together with R20 and C2, form an oscillator circuit, as shown in **Figure 24-5** (see in page 555). Assume that the circuit's frequency of oscillation is 3 kHz. This oscillator circuit produces a rectangular pulse that is applied to U10, which is a pulse-shaping circuit. The output of U10, along with that of Q3, produces a 3 kHz sawtooth ramp voltage that is applied to the inverting input of U4 in the comparator stage.

In the comparator, shown in **Figure 24-6** (see in page 556), the noninverting input of U4 receives a positive voltage from the output of the preamplifier stage. Simultaneously, the inverting input of U4 receives a positive 3 kHz sawtooth ramp voltage from the sawtooth generator stage. This 3 kHz sawtooth voltage causes U4 to switch on and off. Whenever the noninverting input has a higher positive value than the inverting input from the sawtooth generator, U4 produces an output voltage. Conversely, when the inverting input from the sawtooth generator has a higher positive value than

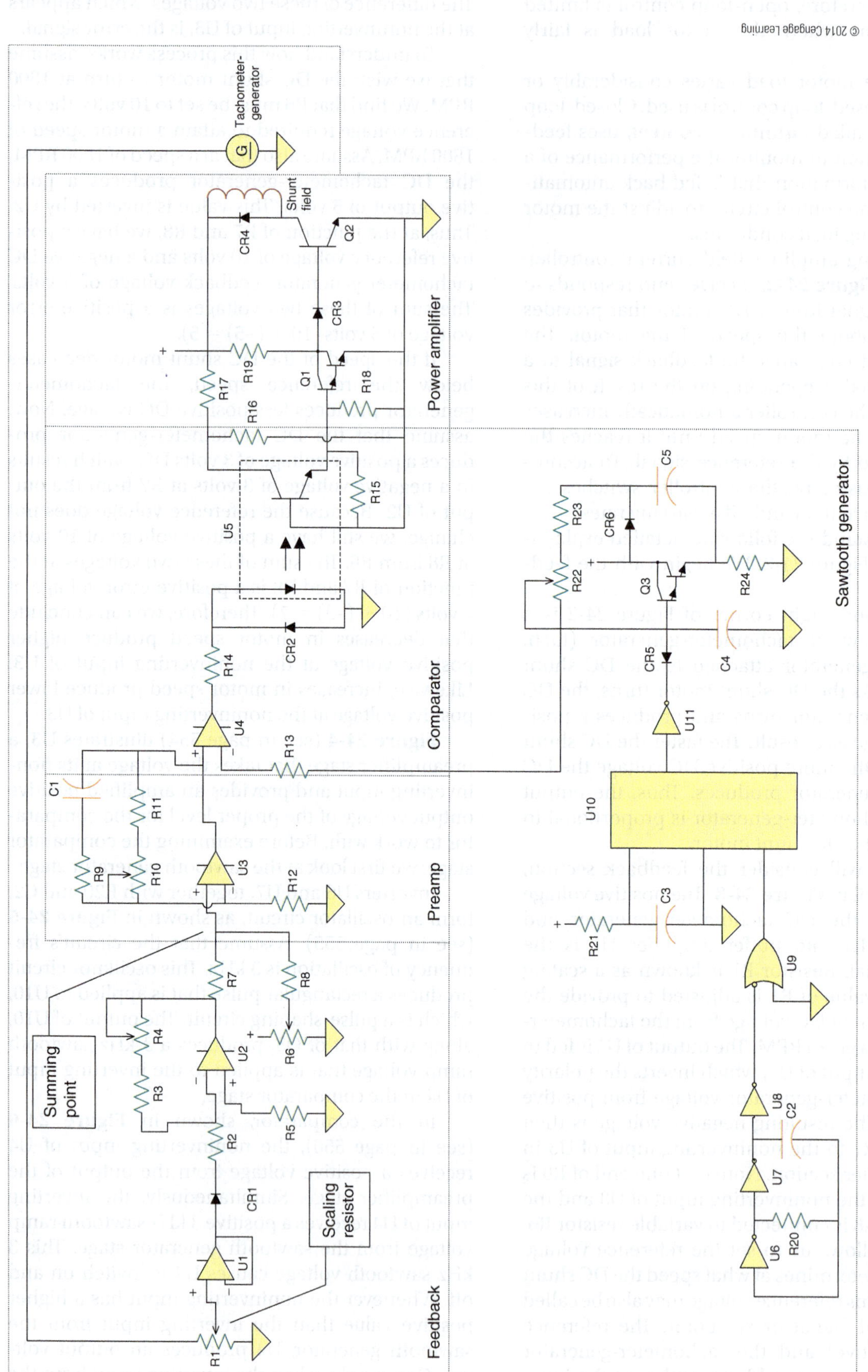

FIGURE 24-2 Schematic of a switching amplifier field current controller.

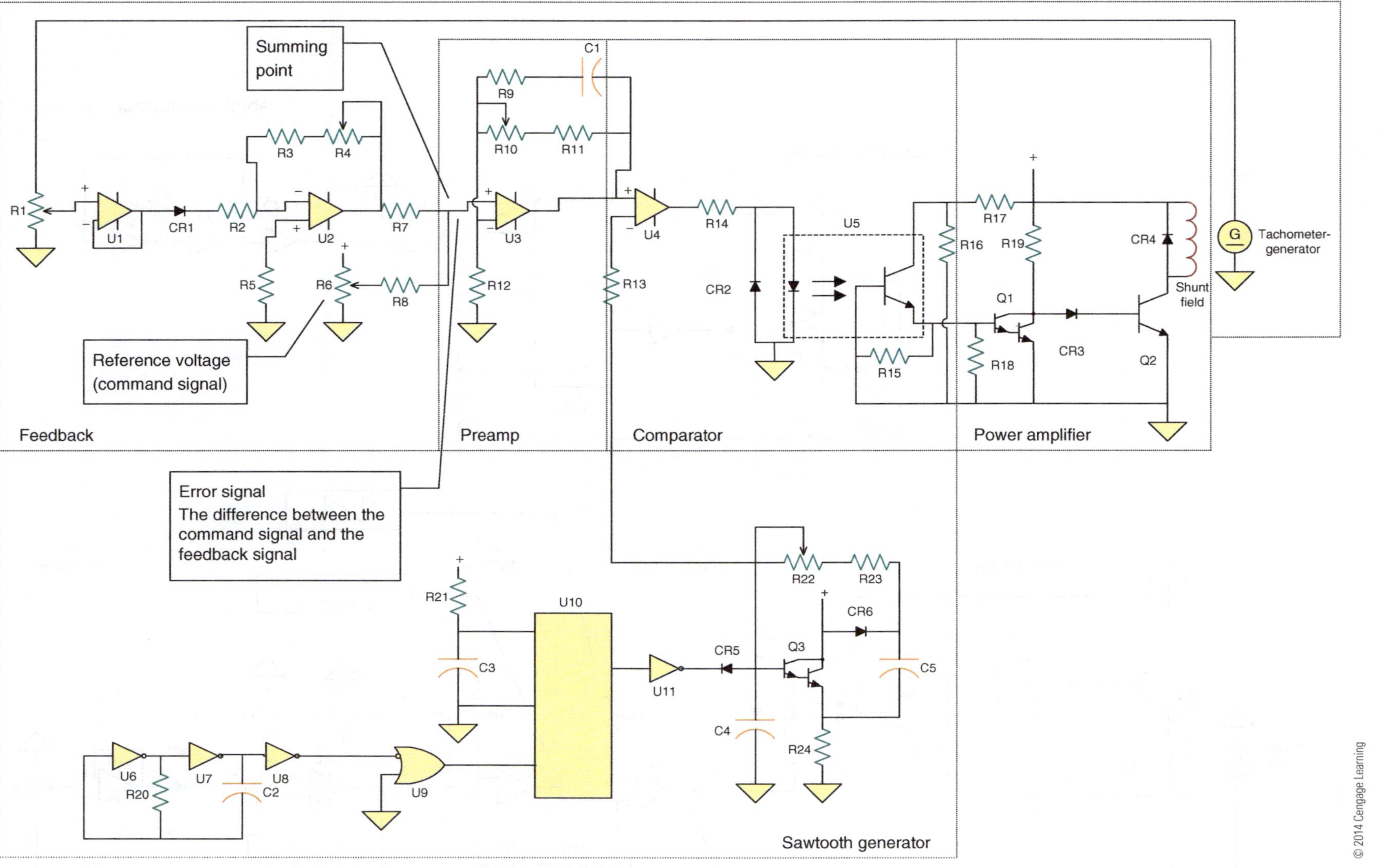

FIGURE 24-3 Feedback section.

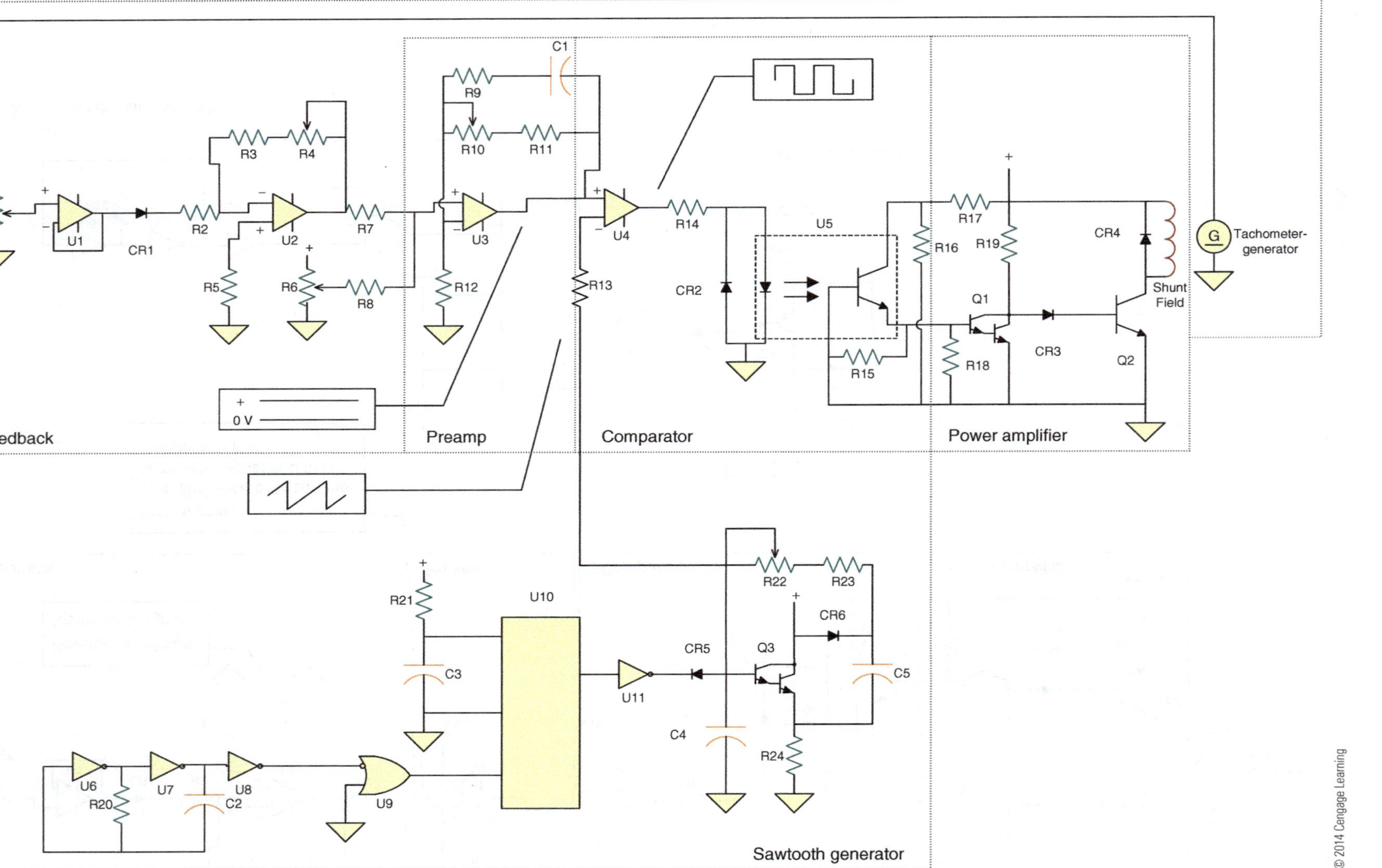

FIGURE 24-4 Preamplifier stage.

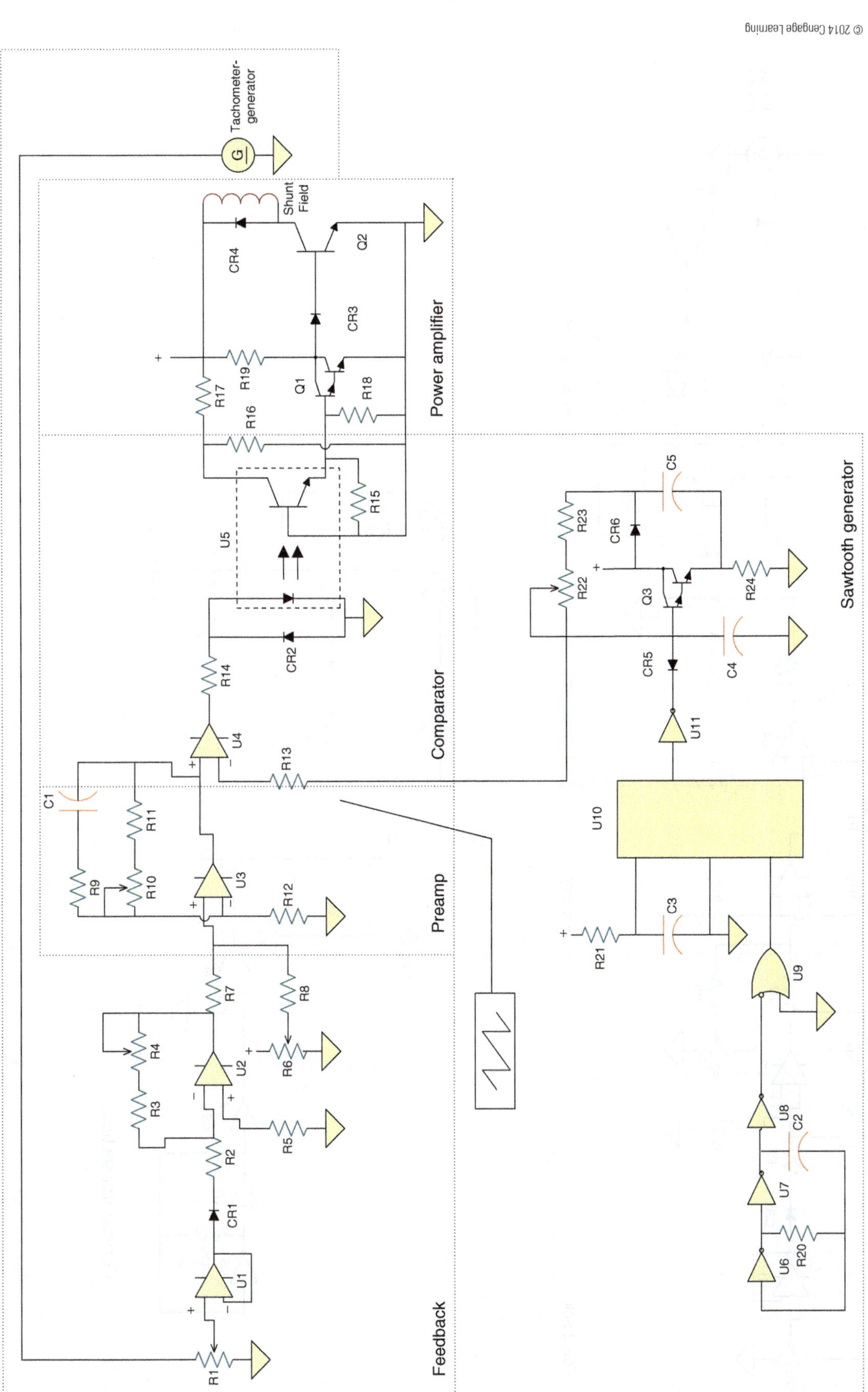

FIGURE 24-5 Sawtooth generator section.

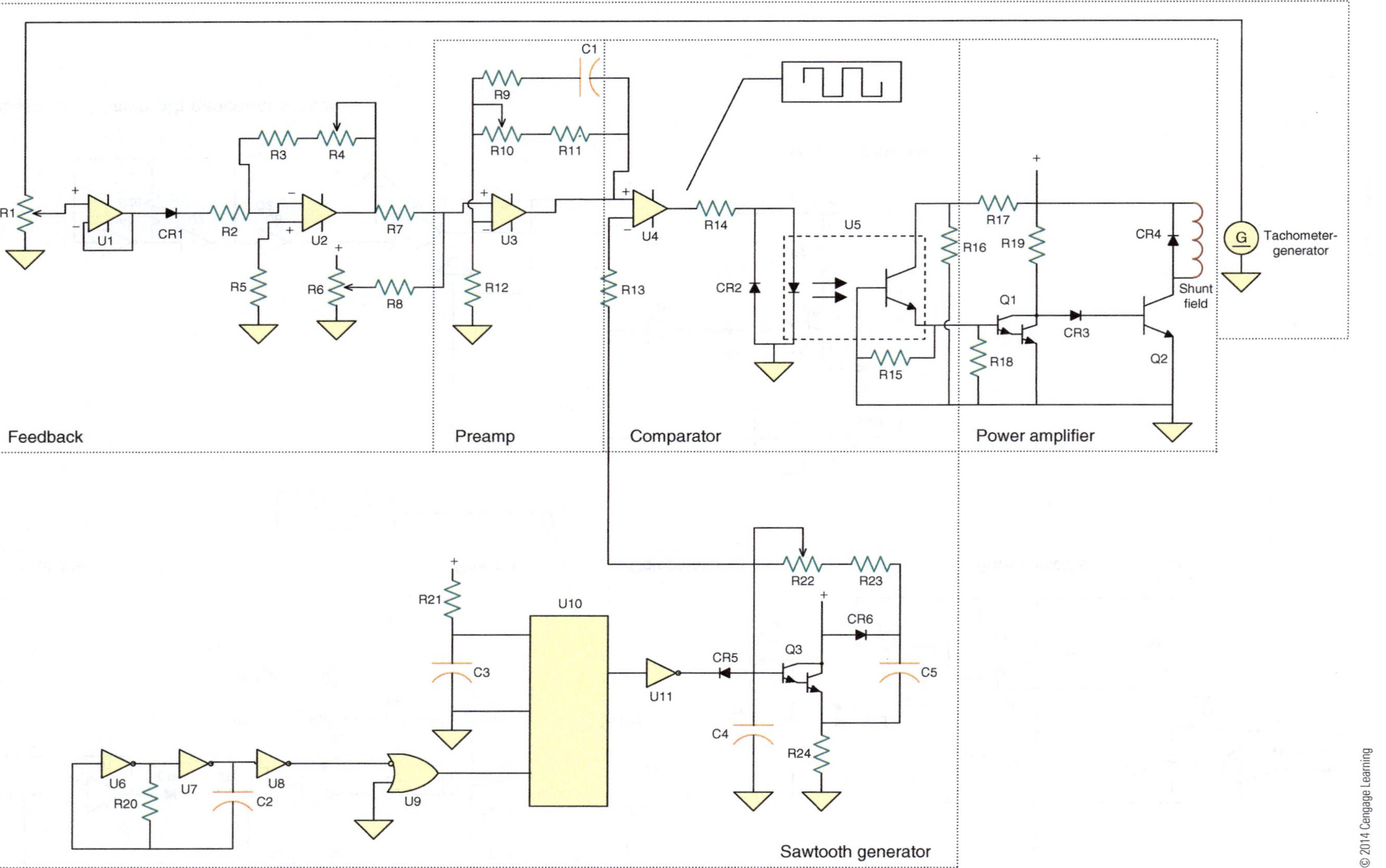

FIGURE 24-6 Comparator section.

the noninverting input, the output of U4 is turned off, and a rectangular pulse is produced at the output of U4 as a result, as shown in **Figure 24-7**. The width of U4's output pulse is controlled by the reference voltage from R6. If R6 is adjusted for a higher DC reference voltage, then U3 produces a higher positive voltage for the noninverting input of U4. In turn, the sawtooth ramp voltage must increase to a higher positive value at the inverting input of U4. Because the inverting input takes more time to attain a higher positive value than the noninverting input does, the output pulse of U4 becomes wider, as shown in **Figure 24-8**. Likewise, if R6 is adjusted for a lower DC reference voltage, U3 produces a lower positive DC voltage at the noninverting input of U4. Consequently, the sawtooth ramp voltage does not need to reach as high a positive value, and U4 turns off sooner. In this case, U4's output pulse is narrower, as shown in **Figure 24-9**. The effect is called pulse-width modulation, or PWM. The output pulse of U4 is applied to optocoupler (or optoisolator) U5. When the output of U4 is on, optocoupler U5 is on. When the output of U4 is off, optocoupler U5 is also off. Optocoupler U5 isolates the low-voltage logic circuits from the higher-voltage power circuits in the power amplifier stage and also prevents electrical noise that may be induced onto the power circuits from entering the control section of the drive.

Figure 24-10 (see in page 560) illustrates how the output of optocoupler U5 is fed to the base of Q1 in the power amplifier stage. The power amplifier stage consists primarily of Darlington transistor Q1 and power transistor Q2. Transistor Q1 receives

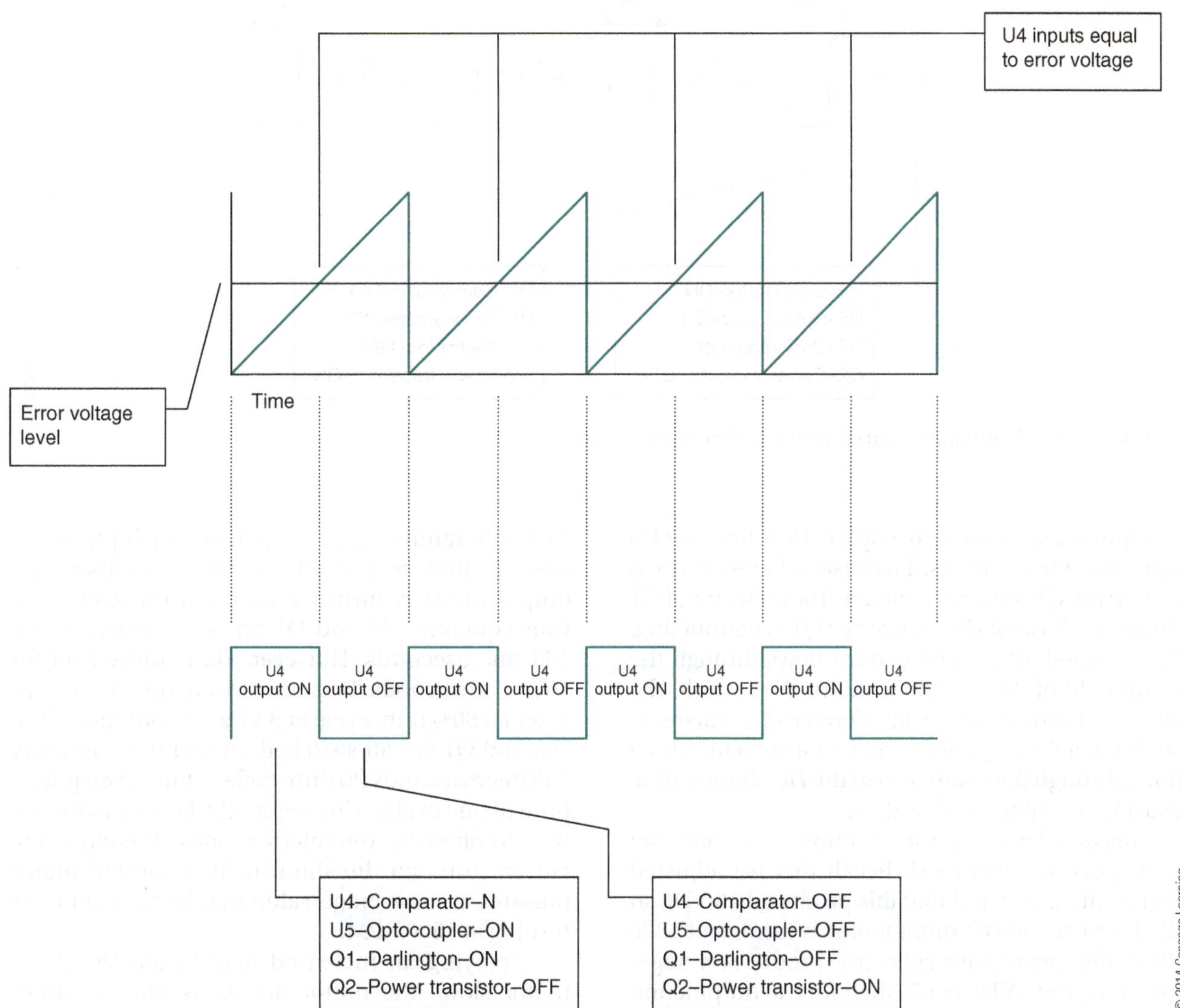

FIGURE 24-7 Output of the comparator section.

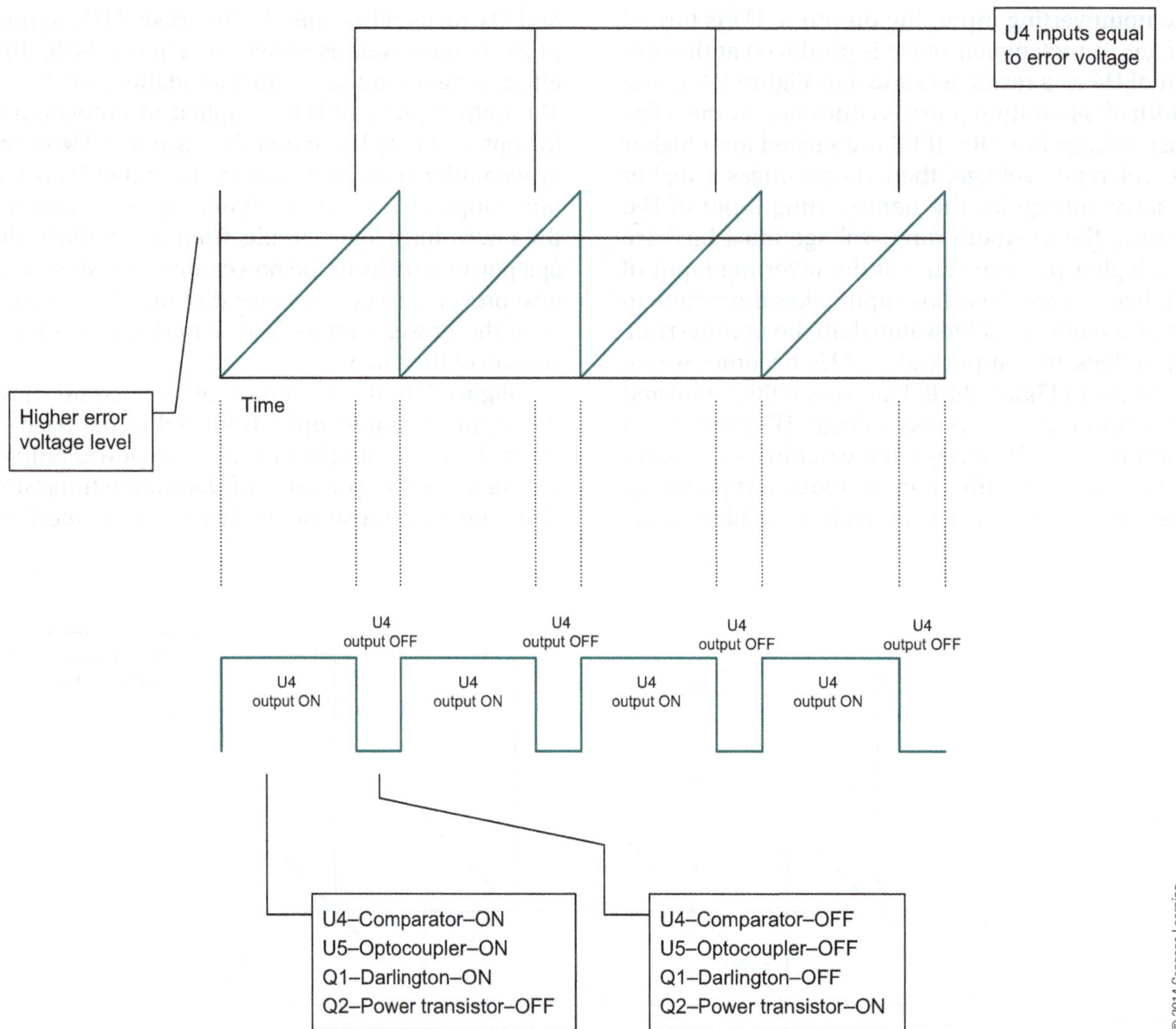

FIGURE 24-8 Comparator output with a higher error voltage.

its input signal from optocoupler U5. Whenever U5 turns on, Q1 conducts. Likewise, whenever U5 is turned off, Q1 does not conduct. The collector of Q1 drives the base of Q2. Whenever Q1 is conducting, Q2 is turned off, and no current flows through the shunt field of the DC shunt motor. As a result, the DC shunt motor speeds up. Conversely, whenever Q1 is turned off, Q2 does conduct a current, which flows through the shunt field of the DC shunt motor, causing the motor to slow down.

Reconsider our earlier example of a motor set to a speed of 1800 RPM. Recall that we adjusted R6 for this speed and that this produced 10 VDC at R8. Because the DC shunt motor is turning at 1800 RPM, the tachometer-generator produces a negative voltage of 5 VDC at R7. Therefore, at the junction of R7 and R8 we have a positive voltage of 5 VDC. This voltage is applied to the noninverting input of the comparator, U4, and then compared with the sawtooth ramp voltage at the inverting input of U4. Assume that as a result of this comparison the output of U4 is turned on for 166 microseconds. Consequently, U5 and Q1 are also turned on for 166 microseconds. However, Q2 is turned off for 166 microseconds. Because 166 microseconds represent a 50% duty cycle at 3 kHz, the outputs of U4, U5, and Q1 are all switched on and then off every 166 microseconds (% duty cycle = time of *on* pulse/time of one cycle). Consequently, Q2 is also turned on and off every 166 microseconds. Therefore, the current through the shunt field of the DC motor pulsates at an average value that keeps the motor turning at 1800 RPM.

Applying an increased load to the DC shunt motor slows the motor down. As the DC shunt motor turns more slowly, the tachometer-generator produces less output voltage. This reduction in voltage results in a lower negative voltage at R7.

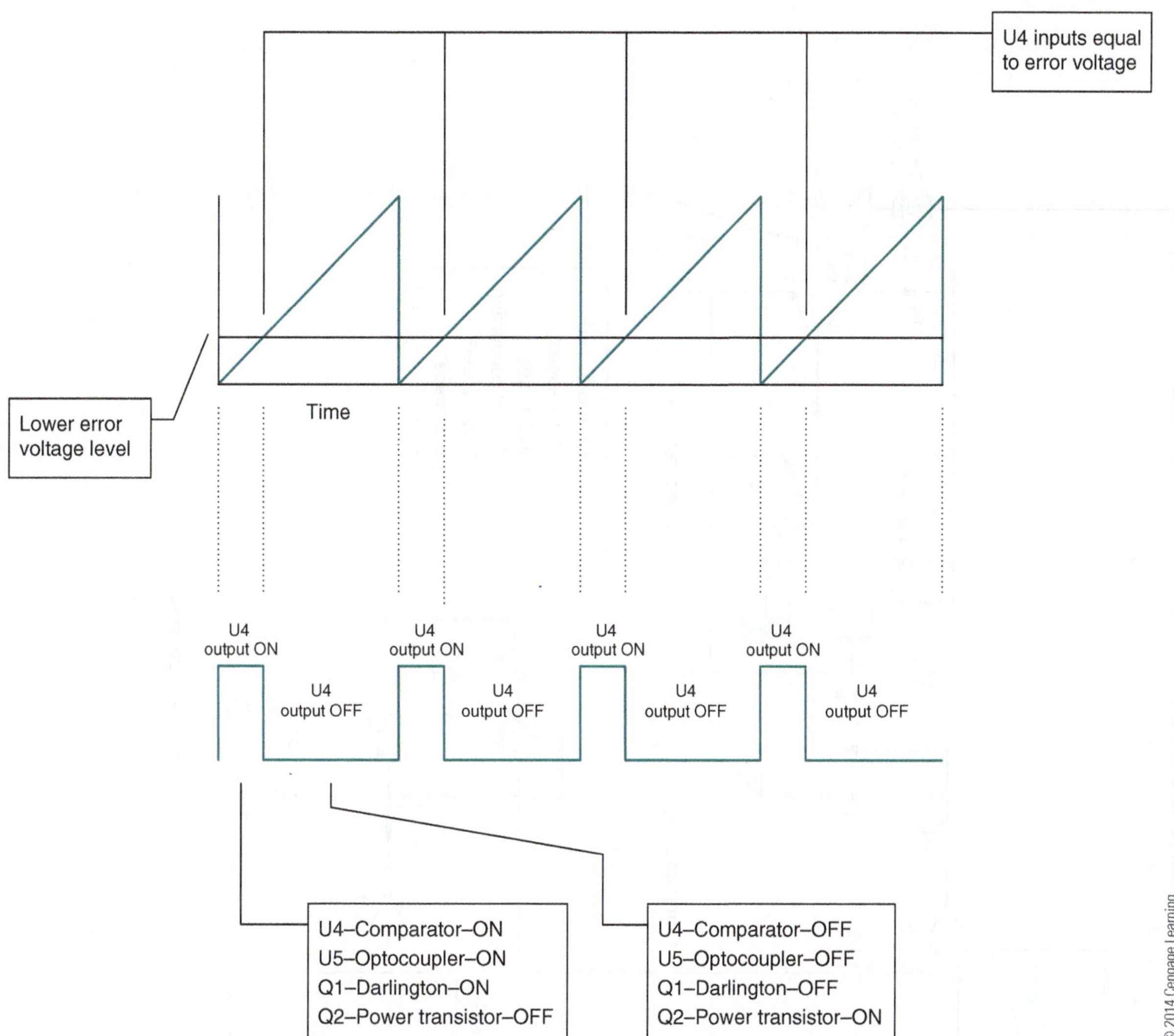

FIGURE 24-9 Comparator output with a lower error voltage.

Suppose that the tachometer-generator voltage drops from 5 volts to 3 volts. The resulting voltage at the junction of R7 and R8 will be positive 7 volts (10 + (–3) = 7). Consequently, a higher voltage (7 volts) is applied to the noninverting input of U4. When this voltage is compared to the sawtooth ramp voltage, the value of the sawtooth ramp voltage must increase to a higher value before U4 is turned off. Thus, the output of U4 is turned on for a longer period of time than it was previously. Assume that U4 is now turned on for 233 microseconds. This implies that the output of U4 is turned off for a shorter period of time (100 microseconds). The longer U4 is turned on and Q2 is turned off, the longer the time during which no current flows through the shunt field of the DC shunt motor. Therefore, the average value of the current flowing through the shunt field is lower. Reducing the shunt field current causes the DC shunt motor to speed up. The reverse is true if the load on the DC shunt motor is lessened. In this way, an increase or decrease in load on a DC shunt motor is automatically compensated. The process by which this compensation is accomplished is closed-loop, or automatic, control.

SCR Armature Voltage Controller

An SCR armature voltage controller is another method of DC motor speed control. The speed of a DC motor may be varied by controlling either the shunt field current or the armature voltage. Armature voltage control is the type most commonly used in DC drives. By controlling the firing of SCRs, we can control the voltage of the armature of a DC motor. A typical SCR armature voltage controller is shown in **Figure 24-11**.

As you look at Figure 24-11, try to recognize some of its similarities to Figure 24-2. Both units use the same type of circuit for closed-loop feedback

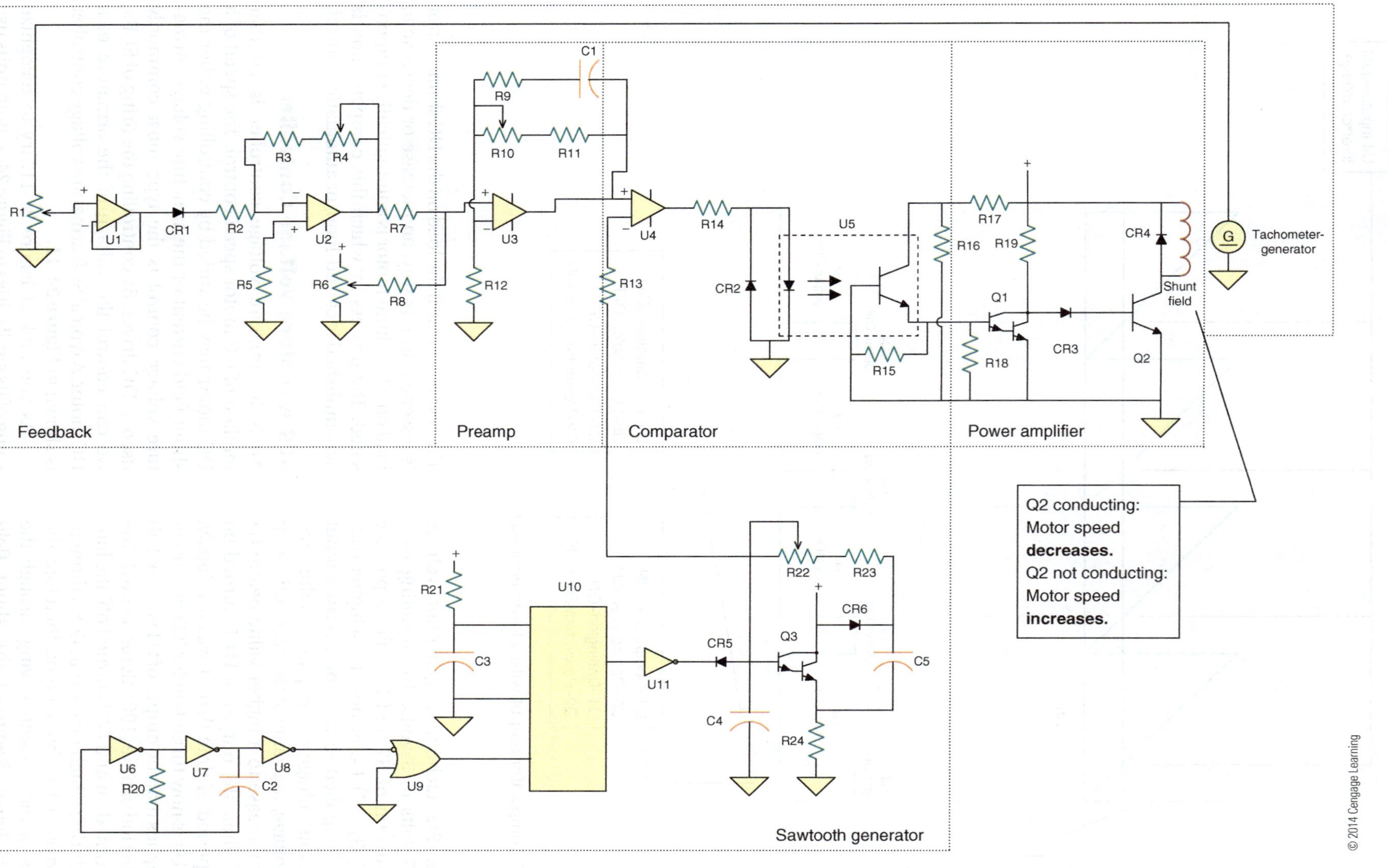

FIGURE 24-10 Power amplifier section.

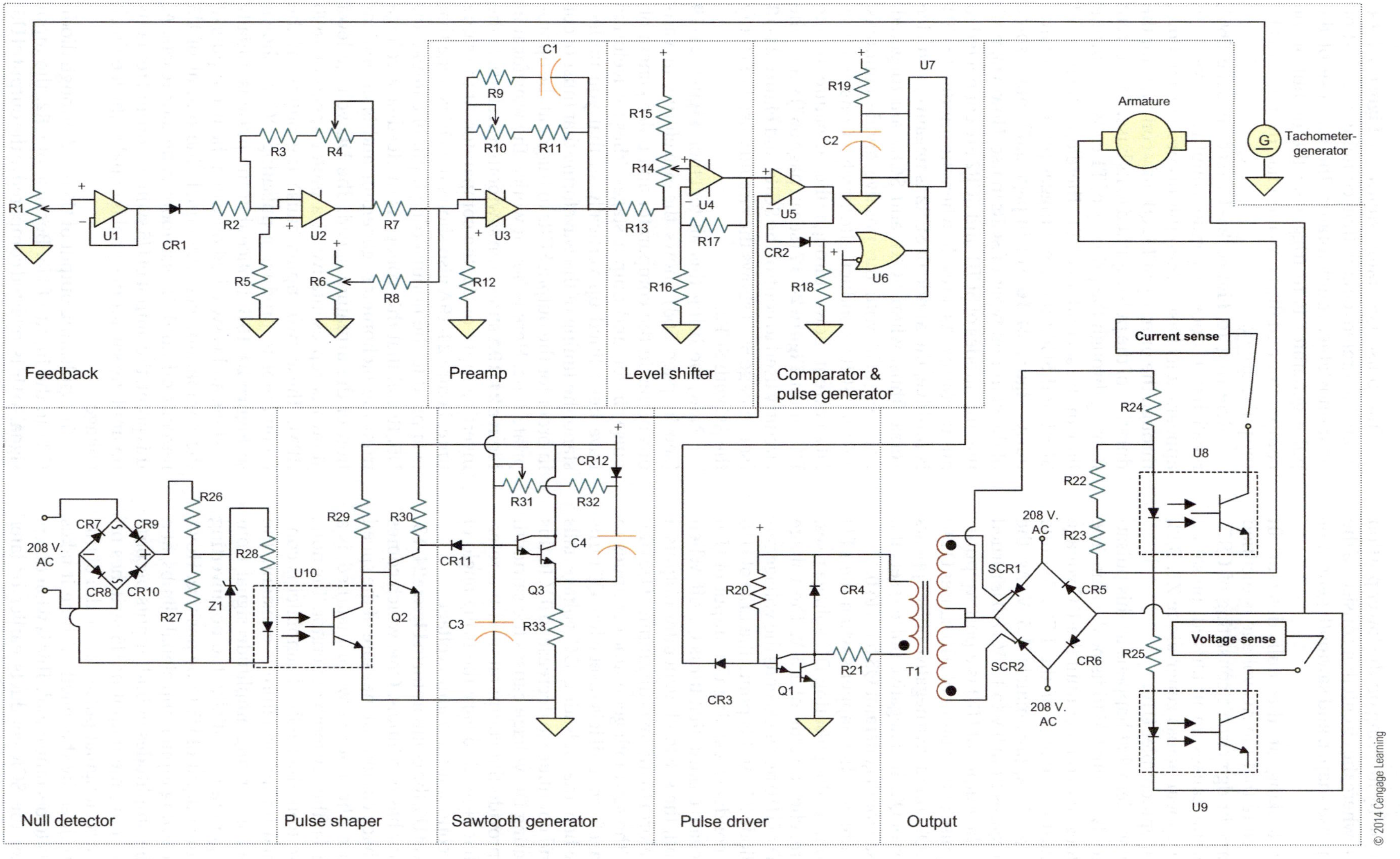

FIGURE 24-11 Schematic of an SCR armature voltage controller.

control. We begin by focusing on the lower portion of Figure 24-11, where the circuit that consists of the null detector, pulse shaper, and sawtooth generator are located.

Begin by looking at the null detector in **Figure 24-12**. Note that the AC voltage is rectified by the full-wave bridge circuit consisting of CR7–CR10. This circuit causes an unfiltered, pulsating DC voltage to appear across zener diode Z1, R28, and the LED of optocoupler U10.

First, we consider what happens as this pulsating DC voltage rises from 0 V to its peak value. The LED of U10 does not conduct until the DC voltage reaches the turn-on voltage of the LED. Assume that this voltage is approximately 1.5 V. As the DC voltage increases from 0 V to 1.5 V, U10 is turned off. Therefore, the output of U10 is a positive pulse. This positive pulse at the output of U10 causes Q2 to turn on. When Q2 conducts, a negative pulse appears at the collector of Q2. This negative pulse resets the output of the sawtooth generator Q3 back to 0.

Now, we look at what happens when the LED of U10 turns on. As the pulsating DC voltage rises, eventually it reaches a value of 1.5 V. The voltage across the LED of U10 never exceeds the voltage rating of zener diode Z1. At this point, the LED of U10 turns on and conducts, causing the output of U10 to supply a negative pulse that turns Q2 off. When Q2 is turned off, the sawtooth generator outputs a sawtooth waveform to the inverting input of U5 (the comparator). The sawtooth generator continues to output this waveform until it is reset when Q2 turns on—that is, when the pulsating DC voltage falls below 1.5 V and continues to decrease to 0 V. In this way, the pulsating DC voltage causes the sawtooth generator to provide the comparator with a reference signal that is used to vary the firing angle of the SCRs that control the armature voltage.

Next, look at U5, the comparator in **Figure 24-13**. The comparator has two inputs. One, which comes from the feedback circuit, is a measure of the speed of the armature. The other, as we just learned, is a ramp signal from the sawtooth generator. The noninverting input from the feedback amplifier is constant if the load on the armature is constant. The inverting input is a rising amplitude signal from the sawtooth generator. Initially, the noninverting input causes the output of U5 to be positive. However, as the inverting input ramp signal climbs, at a given point the amplitudes of both inputs become equal. At this point, the output of U5 switches off and becomes a rectangular pulse. The width of this pulse is determined by the length of time it takes for both inputs to become equal. This is what determines whether the SCR conducts earlier or later within the positive half of the cycle.

The pulse generator, shown in **Figure 24-14**, is an integrated circuit that contains a one-shot, or monostable, multivibrator. The purpose of the pulse generator is to trigger on the output of the comparator and thus provide a narrow pulse to Q1, the pulse driver.

Referring to **Figure 24-15** (see in page 566), recall that Q1 does not conduct until a positive pulse appears at its base. When this occurs, current flows through the primary of T1. Q1 only conducts for the duration of the pulse applied to its base. The current flowing through the primary of T1 causes a current flow in the secondary of T1. Therefore, the gates of SCR1 and SCR2 receive a trigger pulse simultaneously. **Figure 24-16** (see in page 567) shows some of the common types of SCRs in use. However, note that the anodes of SCR1 and SCR2 are connected to an AC source. As a result, when the anode of SCR1 is positive, the anode of SCR2 is negative, and vice versa. Thus, when SCR1 and SCR2 are triggered, only the SCR with the positive anode conducts, so the SCRs conduct alternately. Consequently, DC current flows through the armature of the DC motor. **Figure 24-17** (see in page 567) shows the circuitry used to control the SCRs, and **Figure 24-18** (see in page 567) shows the circuitry used to rectify the AC with SCRs.

Now, we bring this all together. **Figure 24-19** (see in page 568) shows the waveforms at various points in the comparator and pulse generator, pulse driver, and output stages. These waveforms have been lined up vertically so that you understand the timing of the waveforms that must occur in order for the output SCRs to trigger at the appropriate time. Here is how it works. The waveform in **Figure 24-19A** shows the sawtooth waveform at the inverting input of U5, the comparator. At the same time, Figure 24-19A also shows the DC voltage that is applied to the noninverting input of comparator U5. Recall that this voltage is the feedback voltage from the tachometer- generator and is an indication of the armature speed. The DC voltage level shown is representative of a motor speed of 1800 RPM. These two inputs cause the output of the comparator to produce a pulsating DC, as shown in **Figure 24-19B**. Notice that the pulse is initially positive and is switched off when the two inputs of the comparator become equal. The output of U5 remains off until the sawtooth generator resets. At this point, the output of U5 switches on and remains on until the sawtooth signal is equal to the feedback voltage.

The pulsating output of U5 is fed through diode CR2 to the input of U6. Because of CR2, the input signal to U6 is inverted compared to the output of U5. This is shown in **Figure 24-19C**. The positive pulses

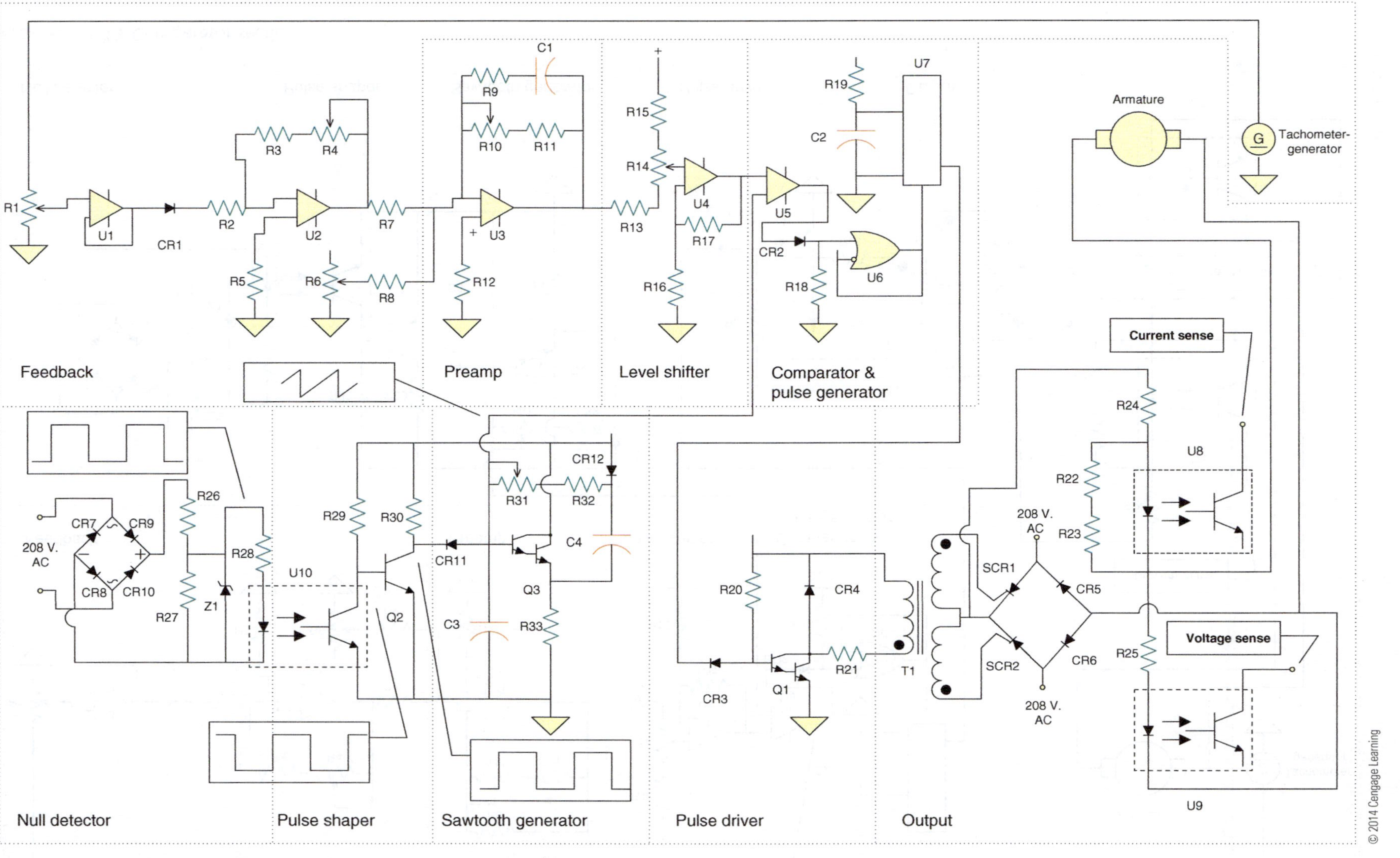

FIGURE 24-12 Null detector, pulse shaper, and sawtooth generator sections.

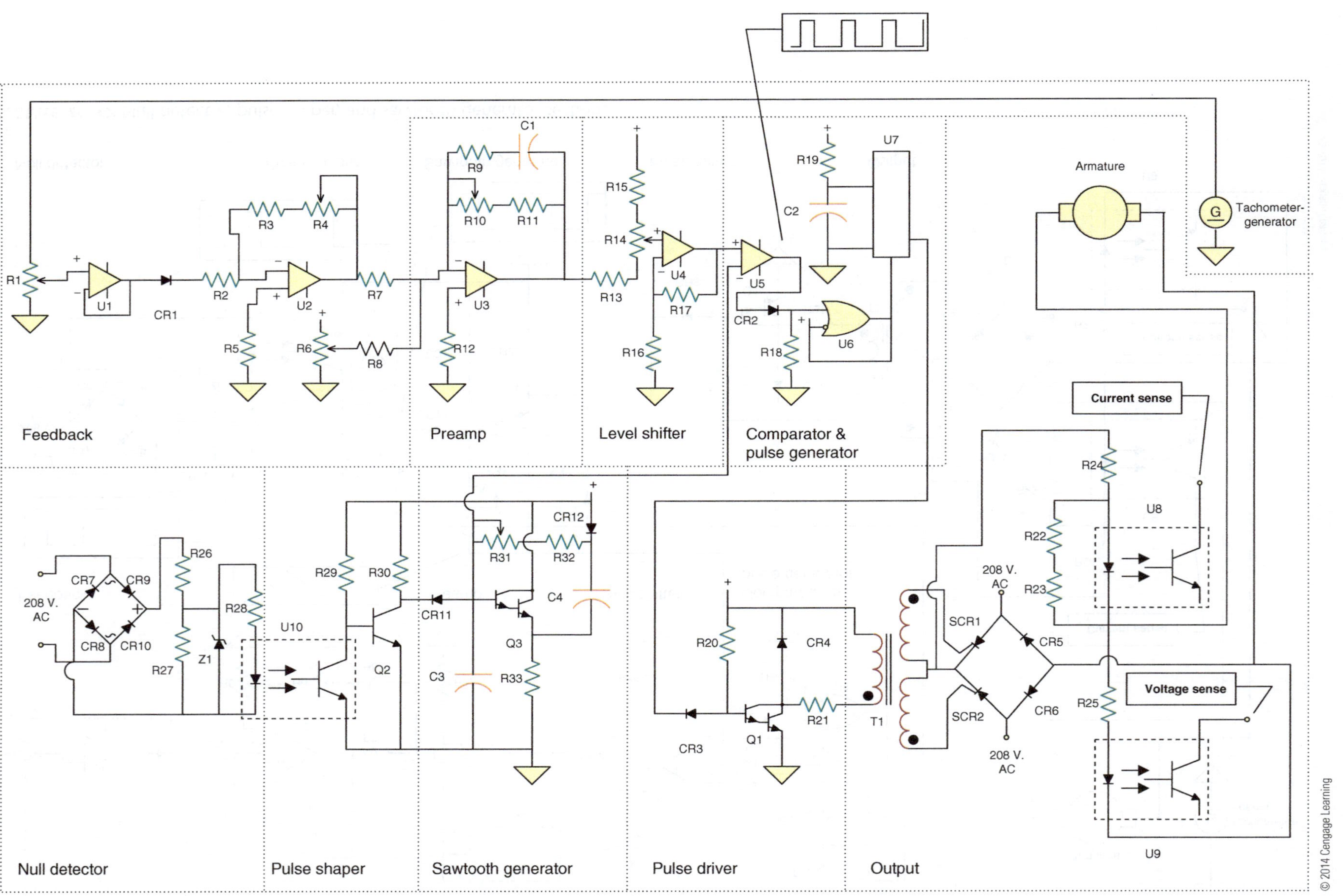

FIGURE 24-13 Comparator section.

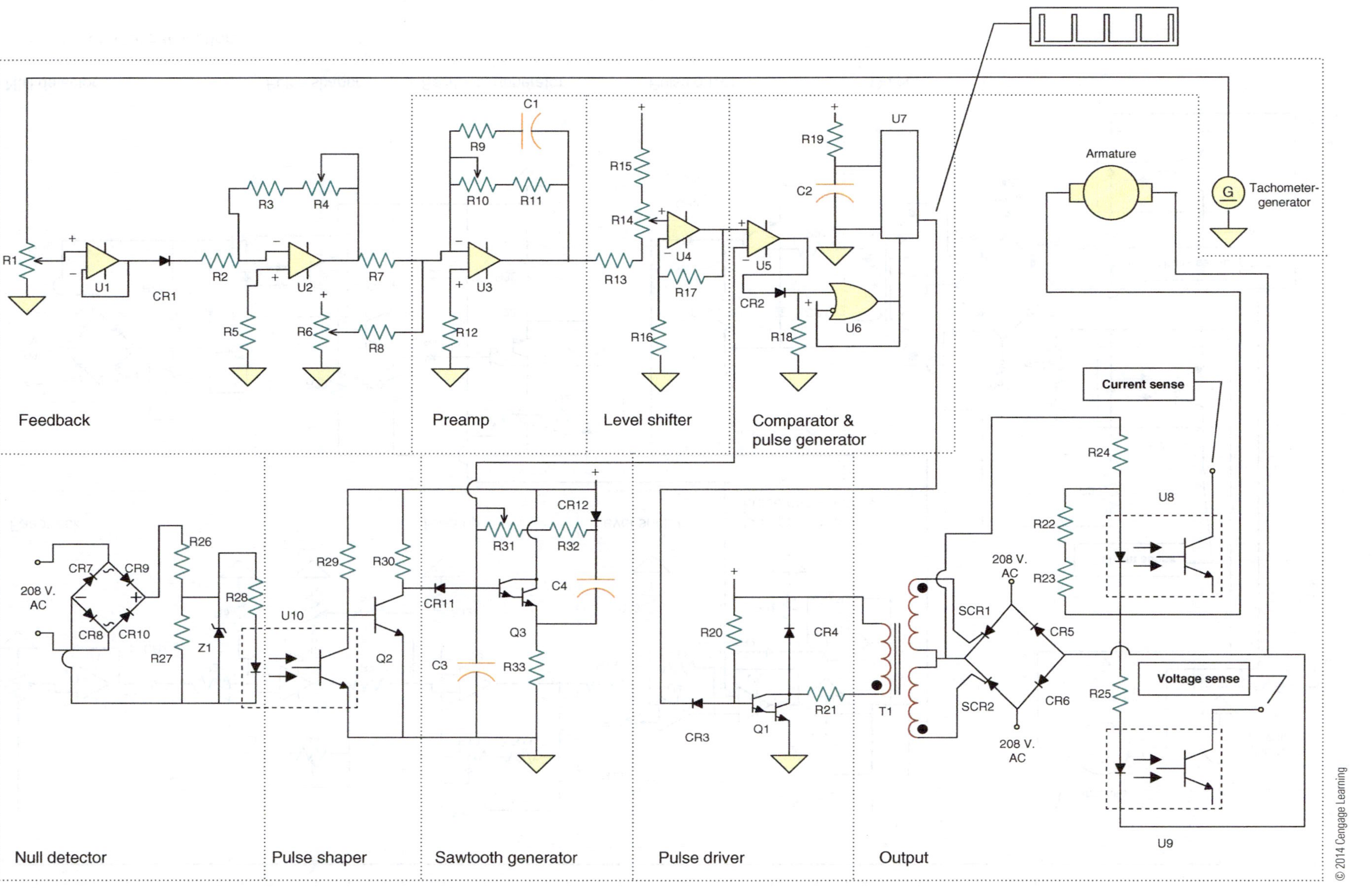

FIGURE 24-14 Pulse-generator section.

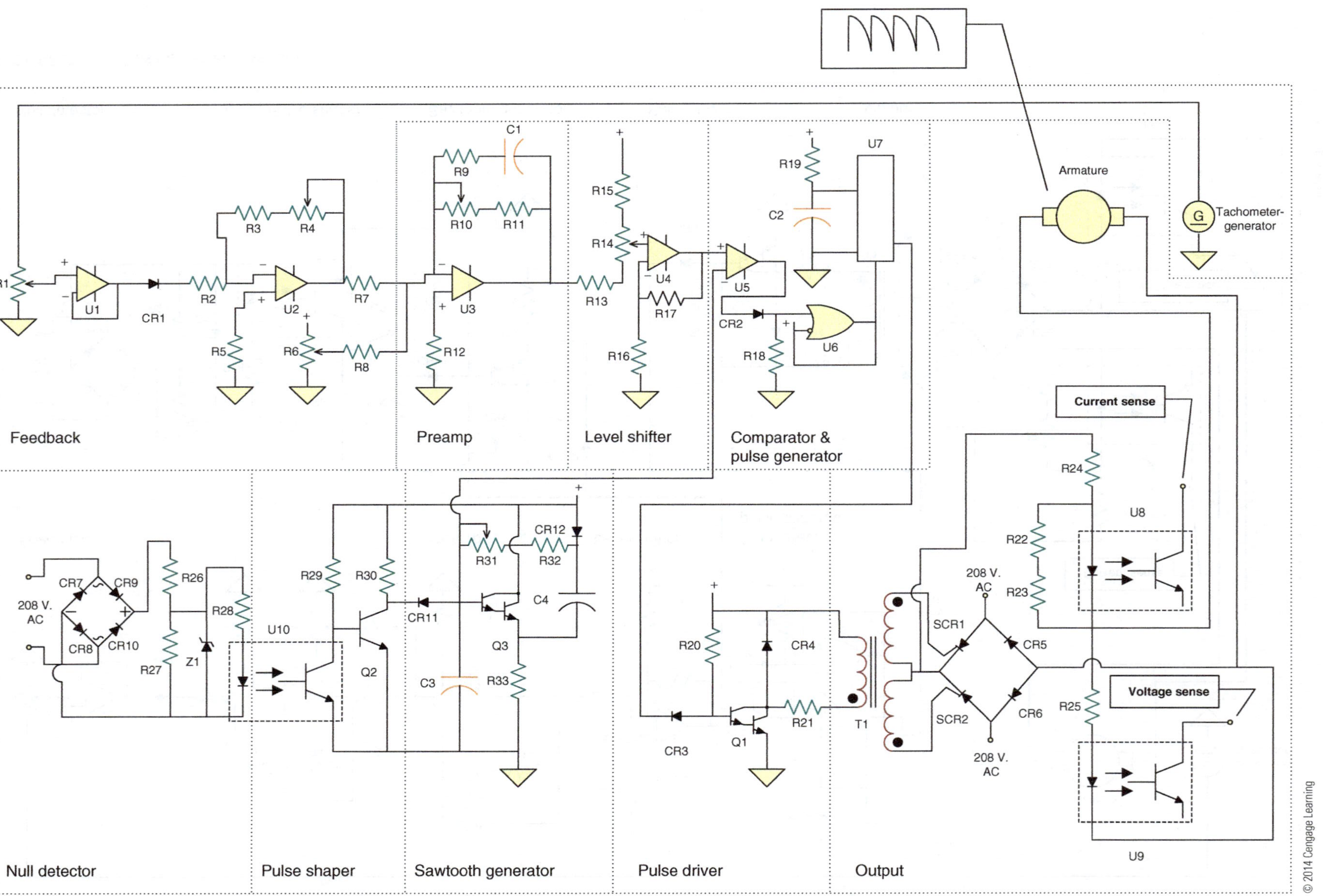

FIGURE 24-15 Output section.

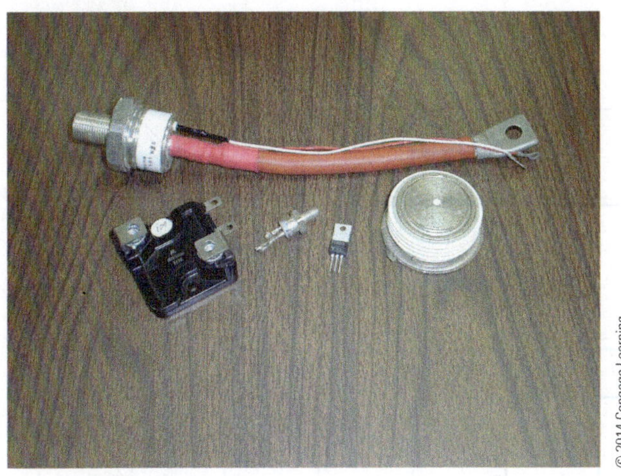

FIGURE 24-16 Various SCR packages.

FIGURE 24-18 Power rectifier section.

FIGURE 24-17 Control board assembly.

from the output of U6 trigger the one-shot, U7. The output of the one-shot, U7, appears in phase with the input pulses, as seen in **Figure 24-19D**.

The output of the one-shot, U7, is fed through diode CR3 to the base of the Darlington transistor, Q1. Notice in **Figure 24-19E** that the collector of Q1 is 180° out of phase with the output of the one-shot. When Q1 conducts, current flows through the pulse transformer, T1. When current flows through pulse transformer T1, a pulse appears at the gates of SCR1 and SCR2. This is shown in **Figure 24-19F**. Recall that only the SCR that is properly biased conducts. The SCR that is triggered and properly biased conducts until the AC voltage drops to the zero crossing point. The next pulse from T1 causes the other SCR to conduct. Therefore, the SCRs conduct alternately, as shown in **Figure 24-19G**.

Let us assume that an increase in load has caused the motor speed to decrease to 1000 RPM. Refer to **Figure 24-20**. The waveform in **Figure 24-20A** shows the sawtooth waveform at the inverting input of U5, the comparator. At the same time, Figure 24-20A shows the DC voltage that is applied to the noninverting input of comparator U5. The DC voltage level shown is representative of a motor speed of 1000 RPM. Notice that this level is lower than that shown in Figure 24-19A. This is a result of the slower motor speed and lower output voltage from the tachometer-generator. These two inputs cause the output of the comparator to produce a pulsating DC, as shown in **Figure 24-20B**. Notice that the pulse is initially positive and is switched off when the two inputs of the comparator become equal. The output of U5 remains off until the sawtooth generator resets. At this point, the output of U5 switches on and remains on until the sawtooth signal is equal to the feedback voltage. Compare the width of the positive pulses in Figure 24-20B with those in Figure 24-19B. Notice that the positive pulses are narrower in Figure 24-20B.

The pulsating output of U5 is fed through diode CR2 to the input of U6. Because of CR2, the input signal to U6 is inverted compared to the output of U5, as shown in **Figure 24-20C**. Again, compare the positive pulses in Figure 24-20C with those shown in Figure 24-19C. Notice that the positive pulses are wider in Figure 24-20C than those shown in Figure 24-19C. The positive pulses from the output

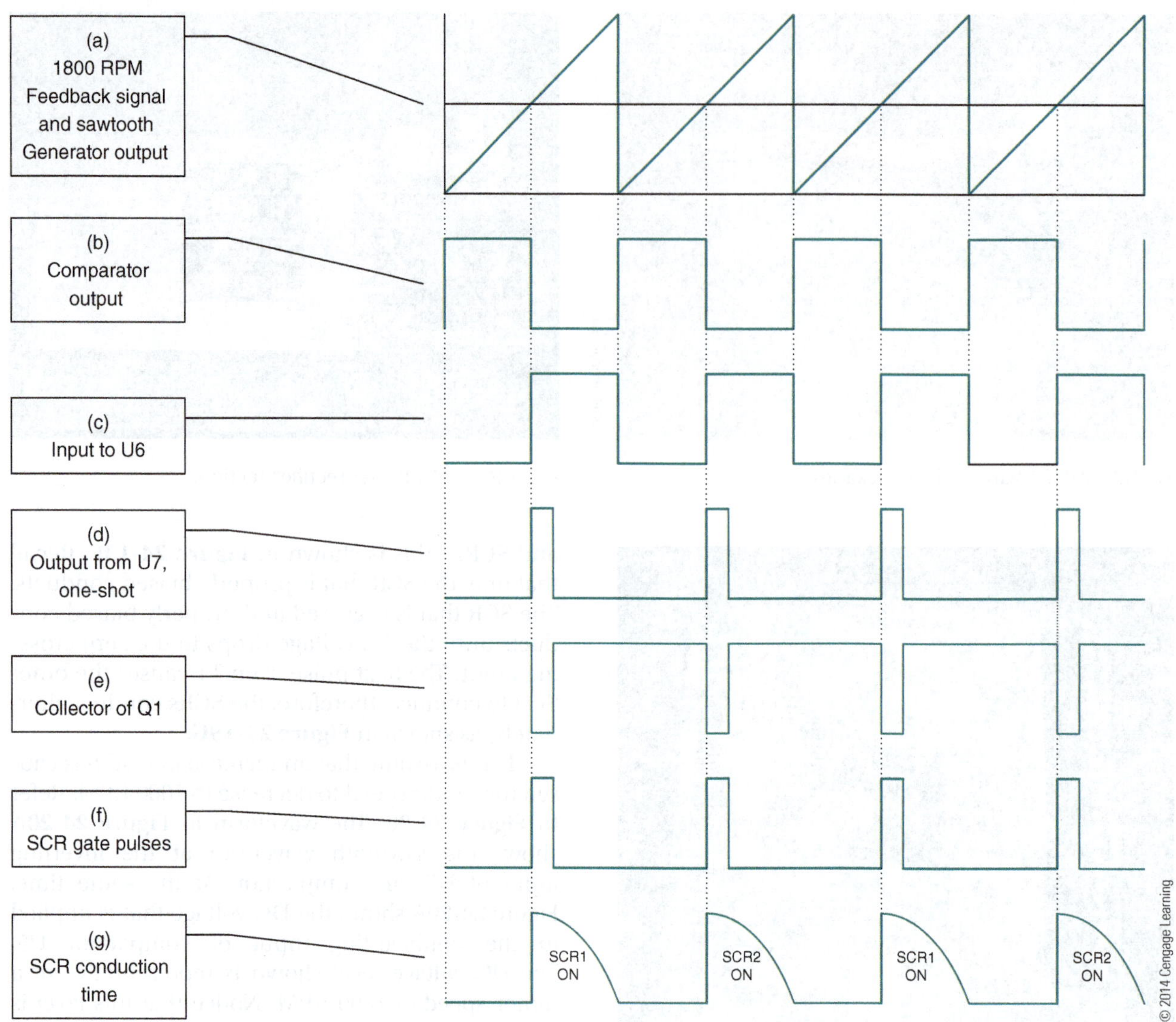

FIGURE 24-19 Waveform timing and SCR conduction time at 1800 rpm.

of U6 trigger the one-shot, U7. The output of the one-shot, U7, will appear in phase with the input pulses, as seen in **Figure 24-20D**. Notice that the output pulse of the one-shot now occurs earlier than it did in Figure 24-19D, which causes the SCRs to fire sooner and conduct for a longer period of time. The following example tests whether this is what occurs. The output of the one-shot, U7, is fed through diode CR3 to the base of the Darlington transistor, Q1. Notice in **Figure 24-20E** that the collector of Q1 is 180° out of phase with the output of the one-shot. When Q1 conducts, current flows through the pulse transformer, T1. When current flows through pulse transformer T1, a pulse appears at the gates of SCR1 and SCR2, as shown in **Figure 24-20F**. Recall that only the SCR that is properly biased conducts. The SCR that is triggered and properly biased conducts until the AC voltage drops to the zero crossing point. The next pulse from T1 causes the other SCR to conduct. Therefore, the SCRs conduct alternately, as shown in **Figure 24-20G**. Notice that the SCRs conduct for a longer amount of time as compared to the conduction of the SCRs in Figure 24-19G. This results in a higher average voltage applied to the armature of the DC motor and causes the speed of the DC motor to increase.

Let us assume that a decrease in load has caused the motor speed to increase to 2600 RPM. Refer to **Figure 24-21**. The waveform in **Figure 24-21A** shows the sawtooth waveform at the inverting input of U5, the comparator. At the same time, Figure 24-21A also shows the DC voltage that is applied to the noninverting input of comparator U5. The DC voltage level shown is representative of a motor speed of 2600 RPM. Notice that this level is higher than that shown in Figure 24-19A. This is a

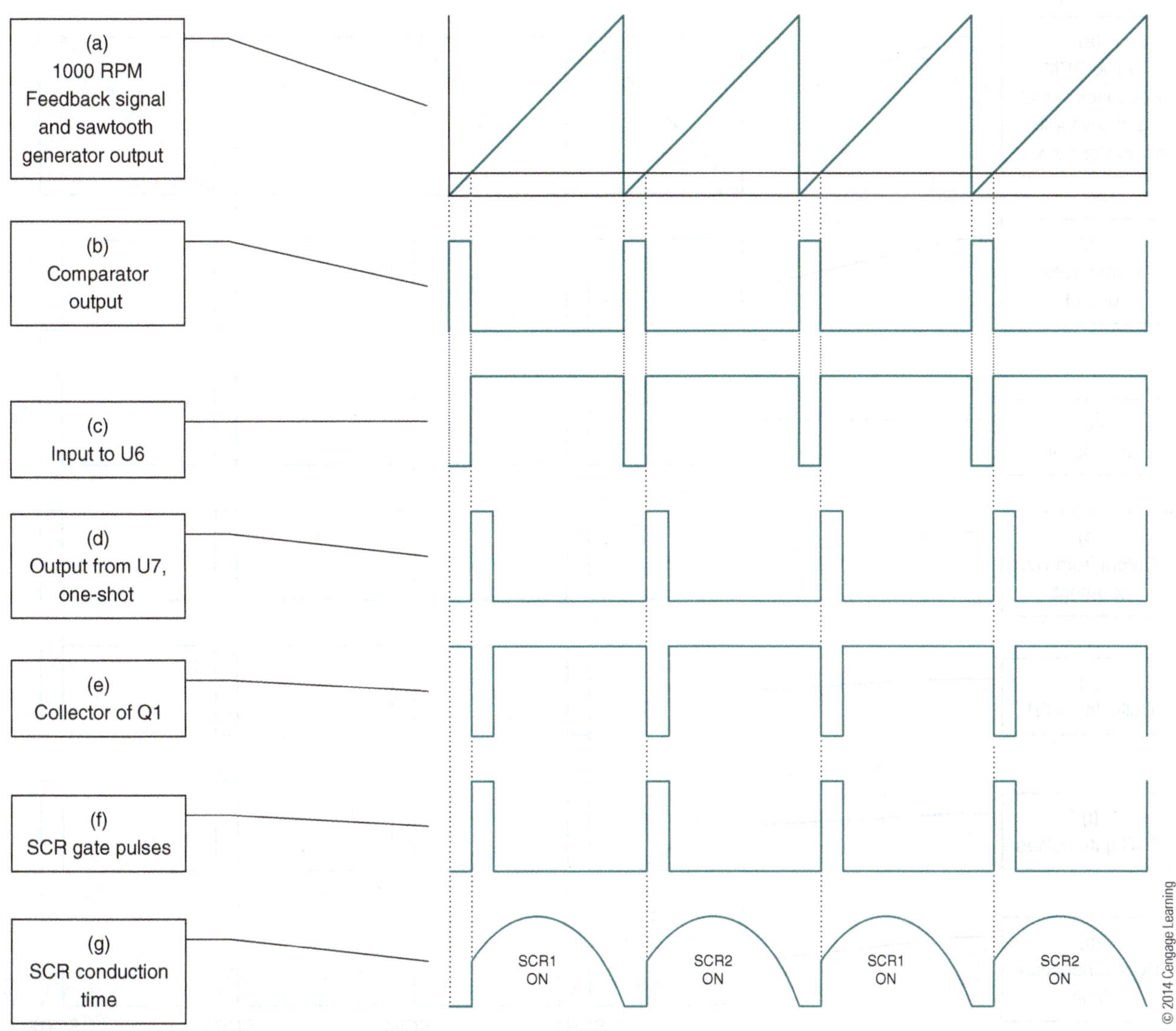

FIGURE 24-20 Waveform timing and SCR conduction time at 1000 rpm.

result of the faster motor speed and higher output voltage from the tachometer-generator. These two inputs cause the output of the comparator to produce a pulsating DC, as shown in **Figure 24-21B**. Notice that the pulse is initially positive and is switched off when the two inputs of the comparator become equal. The output of U5 remains off until the sawtooth generator resets. At this point, the output of U5 switches on and remains on until the sawtooth signal is equal to the feedback voltage. Compare the width of the positive pulses in Figure 24-21B with those in Figure 24-19B. Notice that the positive pulses are wider in Figure 24-21B.

The pulsating output of U5 is fed through diode CR2 to the input of U6. Because of CR2, the input signal to U6 is inverted compared to the output of U5. This is shown in **Figure 24-21C**. Again, compare the positive pulses in Figure 24-21C with those shown in Figure 24-19C. Notice that the positive pulses are narrower in Figure 24-21C. The positive pulses from the output of U6 trigger the one-shot, U7. The output of the one-shot, U7, appears in phase with the input pulses, as seen in **Figure 24-21D**. Notice that the output pulse of the one-shot now occurs later than it did in Figure 24-19D, which causes the SCRs to fire later and conduct for a shorter period of time. The following example tests whether this is what occurs. The output of the one-shot, U7, is fed through diode CR3 to the base of the Darlington transistor, Q1. Notice in **Figure 24-21E** that the collector of Q1 is 180° out of phase with the output of the one-shot. When Q1 conducts, current flows through the pulse transformer, T1. When current flows through pulse transformer T1, a pulse appears at the gates of SCR1 and SCR2, as shown in **Figure 24-21F**. The SCR that is triggered and properly biased conducts until the

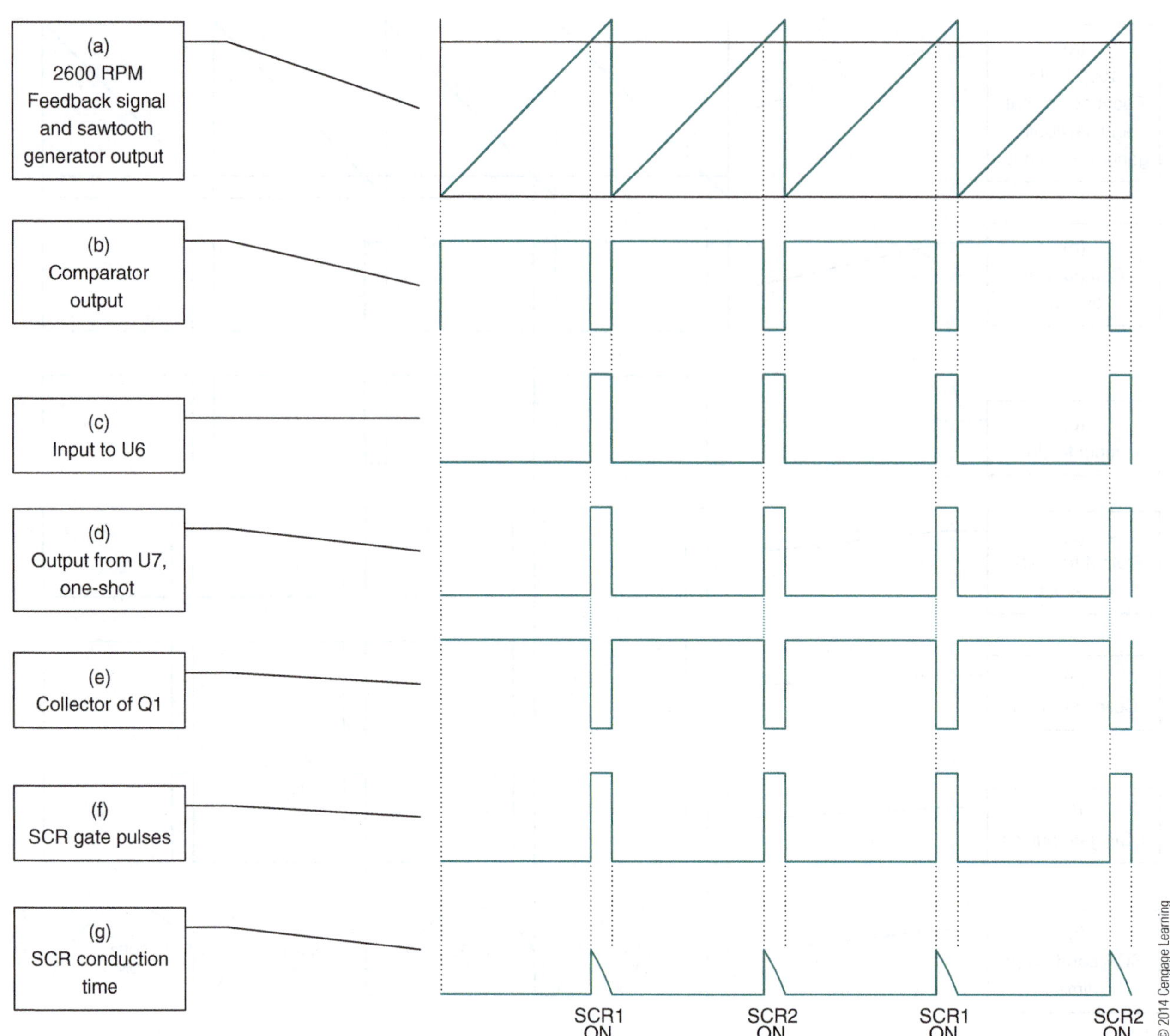

FIGURE 24-21 Waveform timing and SCR conduction time at 2600 rpm.

AC voltage drops to the zero crossing point. The next pulse from T1 will cause the other SCR to conduct. Therefore, the SCRs conduct alternately, as shown in **Figure 24-21G**. Notice that the SCRs conduct for a shorter amount of time as compared to the conduction of the SCRs in Figure 24-19G. This results in a lower average voltage applied to the armature of the DC motor and causes the speed of the DC motor to decrease.

Choppers

A **chopper** is a circuit that uses very fast electronic switches. These switches consist of transistors, SCRs, or metal-oxide semiconductor field-effect transistors (MOSFETs). MOSFETs switch the applied DC on and off more rapidly. By using them, we can create a DC output level that is either lower or higher than the applied DC. If the chopper's output is lower than the applied DC, the chopper is called a **buck chopper** or **step-down chopper**. If the chopper output is higher than the applied DC, the chopper is called a **boost chopper** or **step-up chopper**.

In **Figure 24-22**, notice that the gate of MOSFET Q1 is connected to a control signal source. The control signal emitted by this source is a **pulse-width modulated (PWM) signal**. We will begin by setting the duty cycle of the PWM signal to 50%. (Remember that the duty cycle is the ratio of the time of the *on* pulse to the time of one cycle. Therefore, a pulse set at a 50% duty cycle is on for the duration of one half cycle.)

When the control signal pulse is on, it is applied to the gate of the MOSFET, and Q1 conducts as a result. However, Q1 only conducts for half of one cycle. While Q1 conducts, the armature of the DC motor is connected to the DC supply. This connection causes

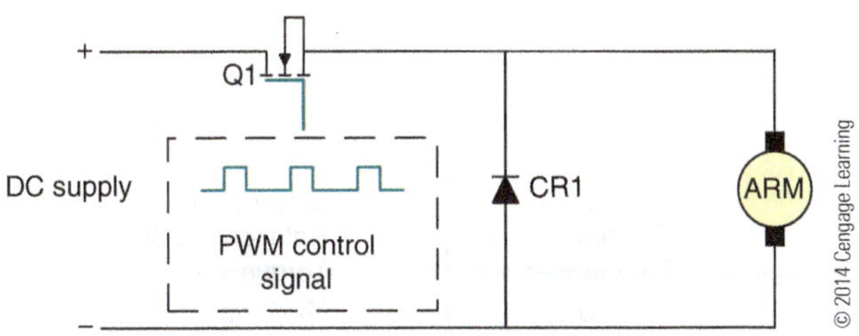

FIGURE 24-22 Buck or step-down chopper.

diode CR1, a freewheeling diode, to appear as an open circuit because it is reverse biased. Therefore, the current flowing through the armature of the DC motor increases. During the next half cycle, this pulse to the gate of Q1 is switched off, so Q1 is also turned off. As a result, the armature of the DC motor is no longer connected to the DC supply, and diode CR1 provides a discharge path for the collapsing magnetic field of the armature.

The output voltage of a buck chopper is proportional to the duty cycle of the control signal. If we increase the duty cycle (thus making the *on* pulse wider), the average value of the chopper's output voltage also increases. Likewise, if we decrease the duty cycle (making the *on* pulse narrower), the average value of the chopper's output voltage decreases as well. Varying the frequency of the control signal has no effect on the level of output voltage and current. However, a higher frequency control signal produces less ripple frequency in the output voltage and current, and thus results in smoother motor operation. Because of losses in the MOSFET, at no time does the output voltage of the buck chopper exceed the DC supply voltage. That is the reason why this particular chopper is known as a buck, or step-down, chopper.

The boost chopper shown in **Figure 24-23** again uses a PWM signal to turn the gate of MOSFET Q1 on and off. Let us begin with a 50% duty cycle and a charge on capacitor C1. When the signal pulse is on, it is applied to the gate of MOSFET Q1, and Q1 is also turned on. With Q1 turned on, a very low voltage develops across the source to drain leads of Q1, which in turn allows current to flow through inductor L1 and thus builds up a magnetic field.

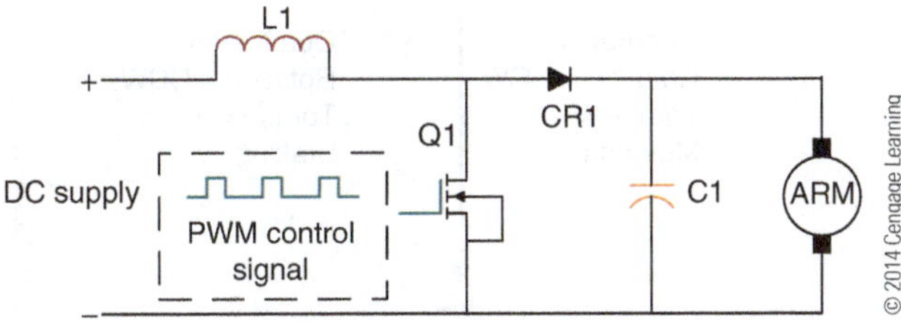

FIGURE 24-23 Boost or step-up chopper.

Because diode CR1 is now reverse biased, no current flows through it. Therefore, capacitor C1 discharges through the armature of the DC motor, causing the armature current, voltage, and motor speed to decrease as the charge on C1 decreases.

When the signal pulse to Q1 is switched off during the next half cycle, Q1 is also turned off, causing the source-to-drain voltage of Q1 to increase. As a result, CR1 now becomes forward biased, allowing the magnetic field of inductor L1 to collapse. The discharge current from L1, along with the supply current, recharges C1 and also increases the amount of current flowing through the armature of the DC motor. In turn, this increased armature current causes corresponding increases in armature voltage and motor speed.

To sum up, if Q1 is turned on for a longer period of time, motor speed decreases. If, on the other hand, Q1 is turned on for a shorter period of time, motor speed increases. Again, varying the frequency of the control signal has no effect on the level of output voltage and current. However, a higher frequency control signal produces less ripple frequency in the output voltage and current, resulting in smoother motor operation. Because the inductor voltage boosts, or supplements, the supply voltage, the output voltage can be higher than the input voltage. That is the reason why this circuit is known as a boost, or step-up, chopper.

We must realize that current only flows in one direction, from source to load, in both of these drives. Thus, both of these controls have some shortcomings. Most notable is the fact that speed control is difficult with high-inertia loads, which may cause **overhauling**.

Overhauling occurs when the load has enough momentum to cause the motor rotor to continue rotating even after the motor has been disconnected. A flywheel is an example of this type of load. The type of drive in which this phenomenon occurs is a one-quadrant drive.

The Four Quadrants of Motor Operation

First, look at the four quadrants shown in **Figure 24-24**. Note the polarities of the current and voltage in each quadrant. In quadrant 1, the current and the voltage are both positive. When a motor is operating in quadrant 1, we say that the motor is motoring, that is, driving a load and turning in the forward direction.

Look at quadrant 3 in **Figure 24-25**. Note that in this quadrant the current and the voltage are both negative. A motor operating in quadrant 3 is still motoring, but it is now operating in reverse.

Sometimes a motor must operate in both quadrants 1 and 3. This is necessary when a motor must

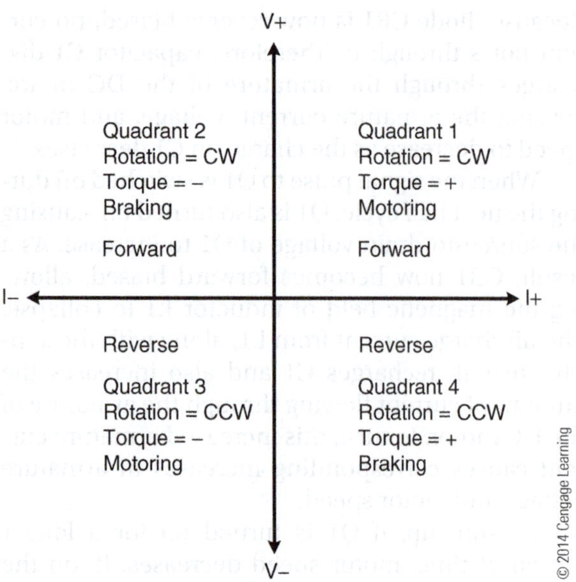

FIGURE 24-24 The four quadrants of motor operation.

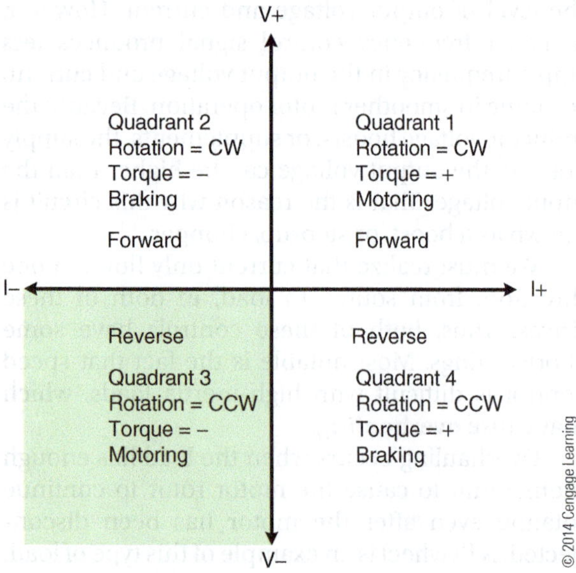

FIGURE 24-25 Quadrant 3: Motoring in reverse.

drive a load both forward and in reverse. Examples are the motor in an electric vehicle that runs forward and backward or the motor in a crane that must raise and lower a load.

In **Figure 24-26**, look first at quadrant 2. Note that although the voltage is still positive, the current is negative. We say that a motor operating in quadrant 2 is **regenerating**. During this time, the motor provides energy back to the power source while rotating forward. In other words, the motor acts like a generator. This regenerated energy provides braking for the motor, affording us better control over the motor's speed. However, operating in quadrant 2 does not allow for braking when the motor turns in reverse.

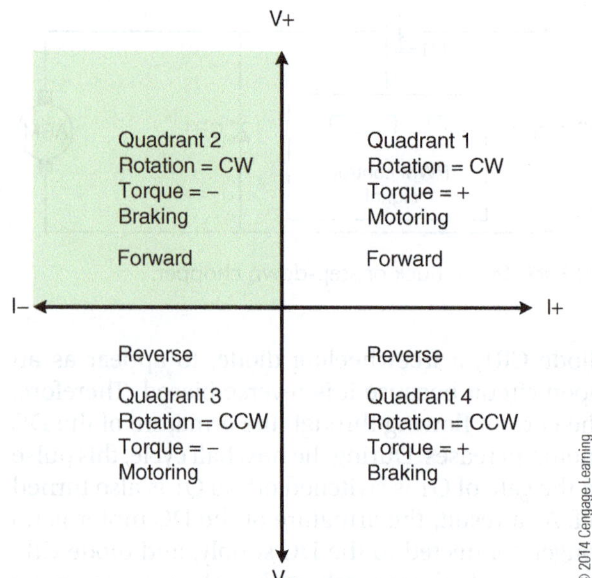

FIGURE 24-26 Quadrant 2: Braking in forward.

Now, look at quadrant 4 in **Figure 24-27**. Here, the voltage is negative, and the current is positive. Motor operation in quadrant 4 produces regenerative braking action when the motor turns in reverse. Think about the electric vehicle example. We want the vehicle to be able not only to travel forward and backward but also to stop in either direction of travel. Therefore, a DC motor for an electric vehicle must be able to operate in all four quadrants. Likewise, in the example of a crane, not only must we be able to lift and lower a load, we must also be able to control the speed at which it raises and lowers the load. Again, the DC motor must operate in all four quadrants.

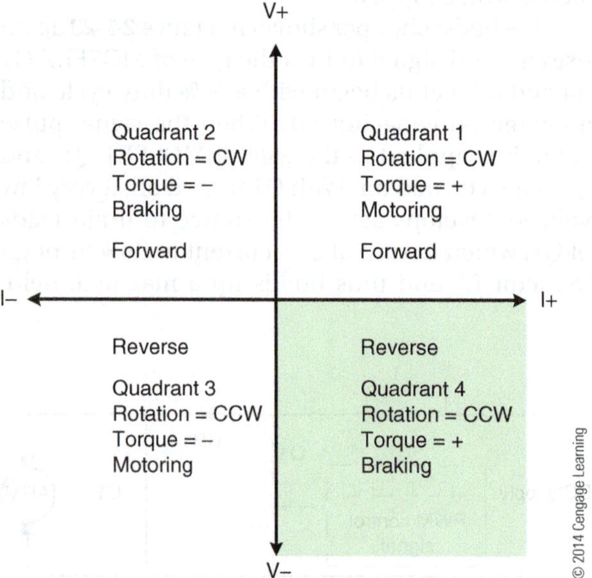

FIGURE 24-27 Quadrant 4: Braking in reverse.

As its name implies, a four-quadrant chopper operates in all four quadrants. That is, the motor operates in both forward and reverse, and it also has regenerative braking capabilities in both directions.

Regenerative braking occurs when the motor still rotates even after power has been removed. When power is removed from a motor, the motor continues to rotate until friction and windage slow it to a stop. While the motor is rotating, the magnetic fields in the armature and stator are collapsing. The motion of the armature through these collapsing magnetic fields generates a voltage and a current. The polarity of this generated voltage and current is the opposite of the polarity of the voltage and current that were originally applied to the DC motor. This opposite polarity tries to make the motor turn in the opposite direction, in effect braking the motor. Because the braking force is a result of the generated voltage and current, it is called regenerative braking.

In **Figure 24-28**, notice the four MOSFETs: Q1, Q2, Q3, and Q4. Also, notice that each MOSFET has an associated freewheeling diode: CR1, CR2, CR3, and CR4. The gates of the four MOSFETs are controlled by a switching control circuit that switches the MOSFETs on and off in pairs: Q1 with Q4 and Q2 with Q3. When one pair is turned on, the other is turned off.

We begin by assuming that Q1 and Q4 are turned on. Therefore, current flows through the armature of the DC motor and through Q1 and Q4. We also assume that at this time the motor is turning clockwise—that is, the motor is operating in quadrant 1 and is now motoring.

Look at **Figure 24-29**, and assume that we now turn off Q1 and Q4. The armature of the motor continues to rotate within the now collapsing

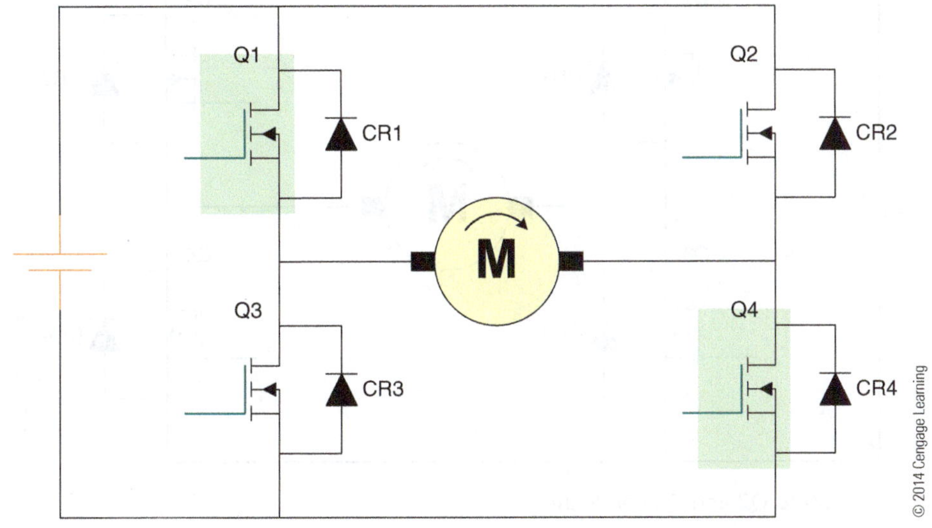

FIGURE 24-28 Four-quadrant chopper.

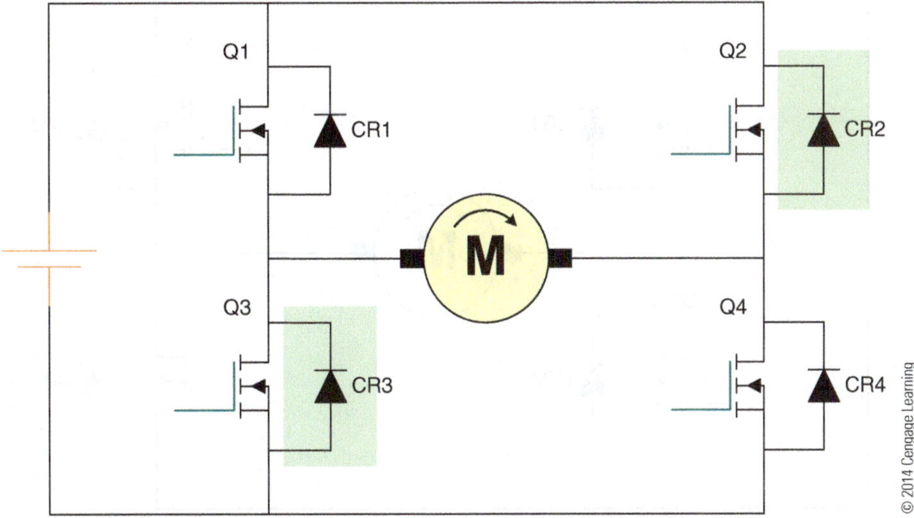

FIGURE 24-29 CR2 and CR3 conducting.

magnetic field. As this rotation continues, magnetic lines of force are cut by the armature. This produces a voltage that opposes the applied voltage and causes a current to flow from the armature, through CR3, to the supply, through CR2, and back to the armature. Because this voltage is opposite to the applied voltage, a counter-torque is produced, which provides a braking action to the motor. Notice, however, that the generated energy is now being applied to the power source. This is known as regenerative braking, and the motor is now operating in quadrant 2.

A motor operating in quadrant 3, as shown in **Figure 24-30**, must rotate in the opposite direction from a motor operating in quadrant 1. In our example, therefore, the motor now rotates counterclockwise. Turning on Q2 and Q3 in this case allows current flow in the reverse direction compared to quadrant 1, which causes the motor to turn counterclockwise when it operates in quadrant 3.

Let us now turn off Q2 and Q3, as shown in **Figure 24-31**. As before, the armature of the motor continues to rotate in the reverse direction within the now collapsing magnetic field. As this rotation continues, magnetic lines of force are cut by the armature, which produces a voltage that opposes the applied voltage. This voltage causes a current to flow from the armature, through CR4, to the supply, through CR1, and back to the armature. Because this voltage is opposite to the applied voltage,

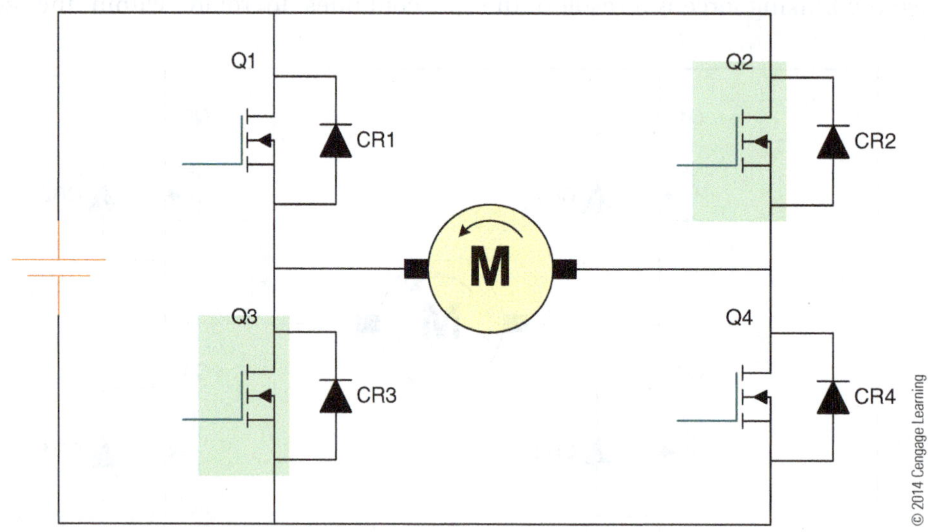

FIGURE 24-30 Q2 and Q3 conducting.

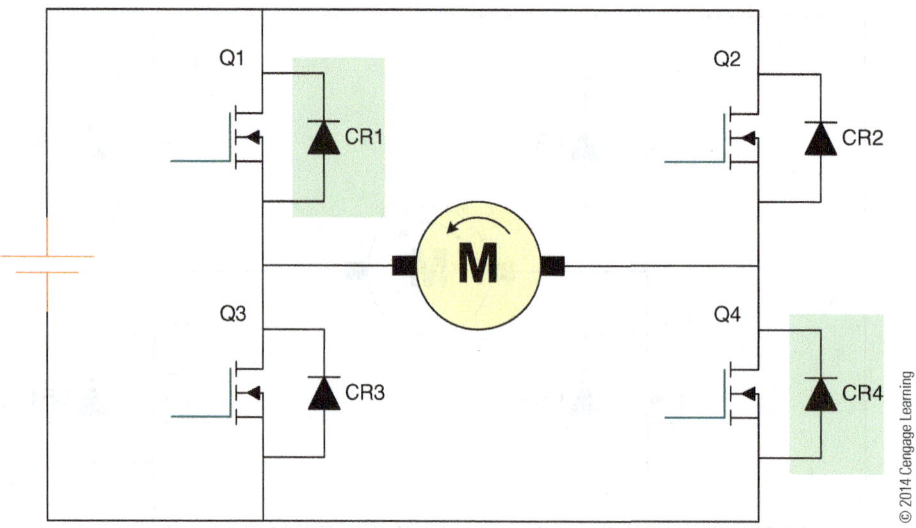

FIGURE 24-31 CR1 and CR4 conducting.

a counter-torque is produced. This counter-torque provides a braking action to the motor. Notice, however, that the generated energy is now being applied to the power source. This provides regenerative braking from the reverse direction, and the motor is now operating in quadrant 4.

We can control a motor's speed quite easily in all four quadrants. Remember that each MOSFET is controlled by a pulse-width modulation signal applied to its gate. If the width of the *on* pulse is wider, the MOSFET conducts for a longer period of time, and the motor will operate at a higher average speed. If we make the *on* pulse narrower, the MOSFET conducts for a shorter period of time, and the average speed of the motor will be lower. Varying the frequency of the control signal has no effect on the level of output voltage and current. However, a higher frequency control signal produces less ripple frequency in the output voltage and current, resulting in smoother motor operation. Therefore, a four-quadrant chopper controls not only the quadrant of operation but also the speed of the motor.

24-2 TROUBLESHOOTING DC DRIVES

All too often, a maintenance technician who is introduced to something new automatically assumes that any problems are located in the new item. As we shall see, this is not always true.

Assumption 1—If the motor does not run, the problem must be the drive.

First, note that in a DC drive system there are four main areas to check for possible problems:

1. The electrical supply to the motor and the drive
2. The motor and/or its load
3. The feedback device and/or the sensors that provide signals to the drive
4. The drive itself

Even though the problem may be in any one or several of these areas, the best place to begin troubleshooting is at the drive unit itself, because most drives have some type of display that aids in troubleshooting. This display may be simply an LED that illuminates to indicate a specific fault condition, or it could be an error or fault code that may be deciphered by looking in the operator's manual. For this reason, it is strongly suggested that a copy of the fault codes be made and fastened to the inside of the drive cabinet, where it will be readily accessible to the maintenance technician. The original should be placed in a safe location, such as the maintenance supervisor's office.

Assumption 2—After checking the drive and reading the code, it is time to make the repairs.

Wrong! Before going any further, you must consider safety factors. First, stop and think about what you are doing! Before working on any electrical circuit, disconnect all power. At times when this is not possible or permissible, work carefully and wear the appropriate safety equipment. Do not rely on safety interlocks, fuses, or circuit breakers to provide personal protection. Always use a voltmeter to verify that the equipment is de-energized, and tag and lock out the circuit!

Even when the power has been disconnected, you are still subject to shock and burn hazards. Most drives have high-power resistors inside, and these can and do get hot. Give them time to cool down before touching them. Most drives also have large electrolytic capacitors that can and do store an electrical charge. Usually, the capacitors have a bleeder circuit to dissipate this charge. However, be aware that this circuit may have failed. Always verify that electrolytic capacitors are fully discharged by carefully measuring any voltage present across the capacitor terminals. If voltage is present, use an approved shorting device to discharge the capacitor completely.

Having followed these steps, what should you do next? Use your senses! Most problems can be identified by using your senses.

Look! Are there any charred or blackened components? Have you noticed any arcing? Do fuses or circuit breakers appear to have been blown or tripped? Do you see any discoloration around wires, terminals, or components? A good visual inspection can save a lot of troubleshooting time.

Listen! Did you hear any funny or unusual noises? A frying or buzzing sound may indicate arcing. A hum may be normal or an indication of loose laminations in a transformer core. A rubbing or chafing sound may indicate that a cooling fan is not rotating freely.

Smell! Do you notice any unusual odors? Burnt components and wires give off a distinctive odor when overheated. Metal smells hot if there is too much friction.

Touch! (But very carefully!) If components feel cool, perhaps no current is flowing through the device. If components feel warm, chances are that everything is normal. If components feel hot, everything may be normal too, although it is more likely that the current is too high or the cooling device is not working properly. In any event, there may be a problem worth further investigation.

The point is that by being observant, you have a good chance of discovering the problem or problem area. We now discuss in more detail the four main areas mentioned earlier and some of the problems that we may encounter in each.

The Electrical Supply to the Motor and the Drive

Most maintenance technicians believe that the power distribution in an industrial environment is reliable, stable, and free of interference. Nothing could be further from reality. Power outages, voltage spikes and sags, and electrical noise are frequent occurrences. The effect of these phenomena is not as detrimental to motor performance as it can be to the operation of the drive itself. Most DC drives are designed to operate within a range of variation of supply voltages. Typically, the incoming power can vary as much as ±10%, with no noticeable change in drive performance. However, in the real world it is not unusual for power line fluctuations to exceed 10%. These fluctuations may occasionally cause a controller to trip, and if they occur repeatedly, a power line regulator may be required to hold the power at a constant level. A power line regulator is of little use, however, should the power supply to the controller fail. In this situation, an uninterruptible power supply (UPS) is needed. Several manufacturers produce complete power line conditioning units that combine a UPS with a power line regulator.

Quite often, controllers are connected to an inappropriate supply voltage. For example, it is not unusual to find a drive rated 208 V connected to a 240 V supply. Likewise, a drive rated 440 V may be connected to a 460 V or even a 480 V source. Usually, the source voltage should not exceed the voltage rating of the drive by more than 10%. For a drive rated 208 V, the maximum supply voltage should not exceed 229 V ($208 \times 10\% = 20.8 + 208 = 229$).

Obviously, a 208 V drive connected to a 240 V power supply is subjected to an overvoltage condition and should not be used under such circumstances. For a 440 V drive, the maximum supply voltage should not exceed 484 V ($440 \times 10\% = 44 + 440 = 484$). Furthermore, although this value is within acceptable limits, a potential problem still exists. Suppose that the power line voltage fluctuates by 10%. If the 440 V source suffers a 10% spike, the voltage increases to 484 V. This value falls within the permissible design parameters of the drive. But consider what can happen if we connect a 440 V drive to a 460 V or 480 V power line. If we experience that same 10% spike, the voltage will increase to 506 V in the 460 V line ($460 \times 10\% = 46 + 460 = 506$) and to 528 V in the 480 V line ($480 \times 10\% = 48 + 480 = 528$).

Exceeding the voltage rating of the drive to this extent will probably damage some internal drive components. Most susceptible to excessive voltage and spikes or transients are the SCRs, MOSFETs, and power transistors. Premature failure of capacitors can also occur. As you can see, it is very important to match the line voltage to the voltage rating of the drive.

An equally serious problem occurs when the phase voltages are unbalanced. Typically, during construction, care is taken to balance the electrical loads on the individual phases. As time goes by and new construction and remodeling occur, it is not unusual for the loading to become unbalanced. This imbalance causes intermittent tripping of the controller, which can result in premature failure of certain components.

To determine whether phase imbalance exists, you must do the following:

1. Measure and record the phase voltages (L1 to L2, L2 to L3, and L1 to L3).
2. Add the three voltage measurements obtained in step 1, and record the sum of the phase voltages.
3. Divide the sum obtained in step 2 by 3, and record the resulting average phase voltage.
4. Now, subtract the average phase voltage obtained in step 3 from each phase voltage measurement taken in step 1, and record the results. (Treat any negative answers as positive values.) These values are the individual phase imbalances.
5. Add the individual phase imbalances obtained in step 4, and record the resulting total phase imbalance.
6. Divide the total phase imbalance obtained in step 5 by 2, and record the adjusted total phase imbalance.
7. Next, divide the adjusted total phase imbalance from step 6 by the average phase voltage found in step 3, and record the resulting calculated phase imbalance.
8. Finally, multiply the calculated phase imbalance from step 7 by 100, and record this percentage of total phase imbalance.

Let us work through an example involving a 440 V, three-phase supply to a DC drive to see how this procedure works:

1. Assume that L1 to L2 = 437 V, L2 to L3 = 443 V, and L1 to L3 = 444 V.
2. The sum of these phase voltages equals 437 V + 443 V + 444 V, or 1324 V.

3. The average phase voltage equals 1324 V ÷ 3, or 441.3 V.
4. To find the individual phase imbalances, we subtract the average phase voltage from the individual phase voltages and treat any negative values as positive. Therefore, L1 to L2 = 437 V–441.3 V, or 4.3 V; L2 to L3 = 443 V–441.3 V, or 1.7 V; and L1 to L3 = 444 V–441.3 V, or 2.7 V.
5. Now, we find the total phase imbalance by adding together these individual phase imbalances: 4.3 V + 1.7 V + 2.7 V = 8.7 V.
6. To find the adjusted total phase imbalance, we divide the total phase imbalance by 2; therefore, 8.7 V ÷ 2 = 4.35 V.
7. Next, we divide the adjusted total phase imbalance by the average phase voltage to find the calculated phase imbalance—4.35 V ÷ 441.3 V = 0.0099.
8. Finally, we multiply the calculated phase imbalance by 100 to find the % total phase imbalance—0.0099 × 100 = 0.99%.

In this example, we are within tolerances, and the differences in the phase voltages should not cause any problems. In fact, as long as the % total phase imbalance does not exceed 2%, we should not experience any difficulties as a result of the differences in phase voltages.

What to Check When the Motor or the Load Is the Suspected Problem

Probably the most common cause of motor failure is heat. Excess heat can be simply a result of the motor's operating environment. Many motors are operated in areas of high ambient temperature. If steps are not taken to keep the motor within its operating temperature limits, the motor will ultimately fail.

Some motors have an internal fan to provide cooling. If such a motor is operated at reduced speed, the internal fan may not turn fast enough to cool the motor sufficiently. In these instances, an external fan may be needed to provide additional cooling to the motor. Typically, these fans are interlocked with the motor operation in such a way that the motor will not operate unless the fan operates as well. Therefore, it is possible for a fault in an external fan control to prevent a motor from operating.

The temperature sensor used in motors generally consists of a nonadjustable thermostatic switch that is normally closed and opens only when the temperature rises beyond an acceptable level. Therefore, it may be necessary to wait for an overheated motor to cool down sufficiently before the temperature sensor can be reset and the motor restarted.

Periodic inspection of the motor and any external cooling fans is strongly recommended. The fans should be checked for missing or bent vanes. All openings in the motor's and fan's housing intended to promote cooling should be kept free of obstructions. Any accumulation of dirt, grease, or oil there or elsewhere should be removed. Any filters used in the motor or fan must be cleaned or replaced on a routine schedule.

Heat may also cause other problems. When motor windings become overheated, the insulation on the wires may break down, causing a short circuit that may lead to an "open" condition. A common practice used to find shorts or opens in motor windings is to megger the windings with a megohmmeter. Extreme caution must be taken when using a Megger. When using a Megger on the motor leads, be certain that you have disconnected the leads from the drive. Failure to do so causes the Megger to apply a high voltage into the output section of the drive, and damage to the power semiconductors results. You may decide to megger the motor leads at the drive cabinet. Again, be certain that you have disconnected the leads from the drive unit, and megger the motor leads and motor winding. Never megger the output of the drive itself!

Because the motor drives a load of some kind, the load may also create problems in motor operation. The drive may trip out if the load causes the motor to draw an excessive amount of current for too long a time. Most drives display some type of fault indication when this phenomenon occurs. This problem may be a result of excessive motor operating speed. Quite often, a minor reduction in speed is all that is necessary to prevent the repetitive tripping of the drive. The same effect occurs if the motor is truly overloaded by too large a load. Obviously, in this case either the motor size must be increased or the size of the load decreased to prevent the drive from tripping.

Some loads have a high inertia. They require not only a lot of energy to move them, but once moving, a lot of energy to stop them. If the drive cannot provide sufficient braking action to match the inertia of the load, the drive may trip, or overhaul. A drive with greater braking capacity is needed to prevent the occurrence of overhauling.

Problems Associated with Feedback Devices and Sensors

Mechanical vibration may loosen the mounting or alter the alignment of feedback devices. Periodic inspections are necessary to verify that these devices are aligned and mounted properly.

It is also important to verify that the wiring to these devices is in good condition and that the terminations are clean and tight. Another consideration related to the wiring of feedback devices is electrical interference. Feedback devices produce low-voltage, low-current signals that are applied to the drive. If the signal wires from these devices are routed next to high-power cables, interference can occur. This interference may result in improper drive operation. To eliminate the possibility of interference, you must follow several steps. First, make certain that the signal wires from the feedback device are installed in a separate conduit. Do not install power wiring and signal wiring in the same conduit. The signal wires should consist of shielded cable, and the shield wire should be properly grounded at the drive cabinet only. Do not ground both ends of a shielded cable. When routing the shielded signal cable to its terminals in the drive cabinet, do not run or bundle the signal cable parallel to any power cables. The signal cable should be routed at a right angle to power cables. Also, do not route the signal cable near any high-power contactors or relays. When the coils of a contactor or relay are energized and de-energized, a spike is produced that can also create interference in the drive. To suppress this spike, it may be necessary to install a freewheeling diode across the DC coils or a snubber circuit across the AC coils.

Problems That Exist in the Drive Itself

First, look for any fault codes or fault indicators. Most drives provide some form of diagnostics, and this can be a great time-saver. The operator's manual interprets the fault codes and gives instructions for clearing the fault condition. However, there are other problems related to the drive that we should be aware of.

Heat can produce problems in the drive unit and elsewhere. The drive cabinet may have one or more cooling fans, with or without filters. These fans are often interlocked with the drive power in such a way that the fan must operate in order for the drive to operate. Make certain that the fans are operational and the filters are cleaned or replaced regularly. The drive's power semiconductors are typically mounted on heatsinks. A small thermostat may be mounted on a heatsink to detect excessive temperatures in the power semiconductors. If the heatsink becomes too hot, the thermostat opens and the drive trips. Usually, these thermostats reset themselves. You must wait for them to cool down and reset before the drive will operate. If overheating occurs repeatedly, a more serious problem exists that requires further investigation.

If the drive is newly installed, problems are often the result of improper adjustments. On the other hand, if the drive has been in operation for some time, it is unlikely that readjustments are needed. All too often, an untrained individual tries to readjust a setting to see whether that fixes it. Usually, such readjustments only make things worse! This is not to say that adjustments are never needed. For example, changes in the process being controlled or replacement of some component of the equipment will probably require changes in the drive settings. It is therefore very important to record the current drive settings and any changes made to them over the years. This record should be placed in a safe location, and a copy of it should be made and placed in the drive cabinet for easy access by maintenance personnel.

Consider the more common adjustments that may be performed on typical DC drives. Not all drives permit all of these adjustments. Some drives require small adjustments performed with a screwdriver, whereas others allow you to program adjustments using a keypad or jumpers:

- Acceleration/deceleration rates, or ramp—This adjustment controls how rapidly the motor speeds up or slows down, as shown in **Figure 24-32**.

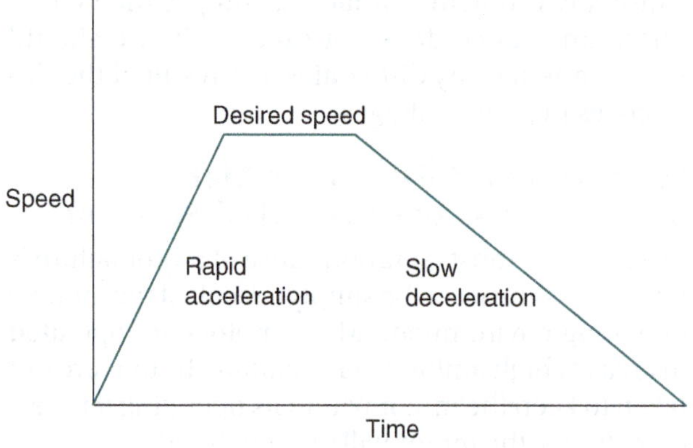

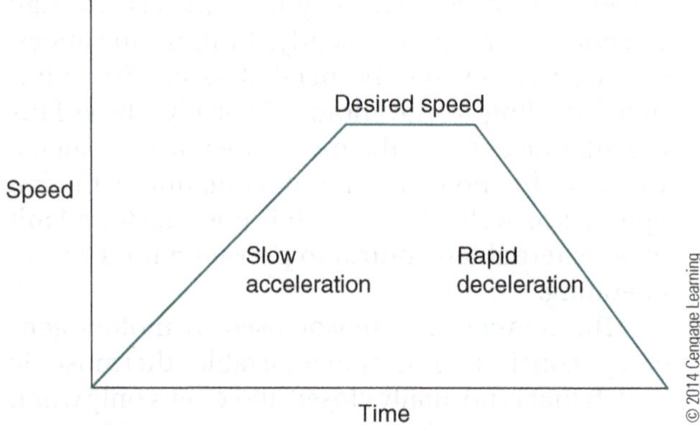

FIGURE 24-32 Variations in acceleration and deceleration rates.

If the motor must respond more quickly than its given load allows, the drive will trip.

- Field voltage—This adjustment sets the field voltage when the motor is not running. It not only saves energy but also lowers the winding temperature and increases motor life expectancy.
- IR compensation—Used to sense CEMF from the motor as an indication of motor speed, this adjustment matches the motor's characteristics to the drive. Therefore, readjustment should not be needed unless the motor or the drive is replaced. This adjustment is more commonly found on older drives and is rarely seen on drives that use feedback devices to sense motor speed.
- Jogging or inching speed—This is the speed of the motor expressed in small increments, usually 10% of the motor's full speed.
- Maximum current—This setting allows the motor to draw 150–300% more current than the motor's maximum rating for a short period of time. The higher the maximum current setting, the shorter the period of time that the motor can draw this current.
- Minimum current—This setting prevents the DC motor from overspeeding in the event that the shunt field circuit opens or the shunt field current becomes too small to produce a magnetic field strong enough to generate sufficient torque.
- Overspeeding—Typically, this setting trips the drive if the motor speed exceeds the desired speed by more than 10%.
- Watchdog circuit—Adjusted to detect certain levels of electrical interference or "noise," this setting also trips the drive in the event of a voltage sag, spike, or single phasing.

Remember, you should never adjust these settings unless you have been properly trained and know their effects.

What If the Drive Still Does Not Work?

If the drive still does not work, you might need a new one. Replacing the drive should clearly be the last choice. More often than not, even if the drive is defective, something external to the drive is the reason for the drive's failure. Replacing the drive without determining the cause of its failure may cause damage to the replacement drive. However, whenever a drive fails, regardless of the cause, some possibility exists that you can get it to work again.

Most drive failures occur in the power section, where you find the power SCRs, transistors, MOSFETs, and so on. In some drives, these devices are individual components. You can test these devices with reasonable accuracy using nothing more than an ohmmeter. In other drives, the SCRs, transistors, and MOSFETs are contained within a power module.

An SCR power module is shown in **Figure 24-33**. One of these modules is used for each phase of the three-phase AC. Notice that this module contains two SCRs. We will call the SCR to the left "A" and the SCR to the right "B." You can use an ohmmeter to test an SCR module with reasonable accuracy. To do so, you must understand what each terminal of the module represents. A schematic diagram of the SCR module is shown in **Figure 24-34**.

Notice in Figure 24-33 that there are a total of five terminals on the SCR module. Referring to Figure 24-34, you see the same five terminals. Two of the terminals (the small terminals in Figure 24-33) represent the gate connections for each SCR. The gate terminal toward the back of the module is connected to the gate of SCR B, and the gate terminal at the front of the module is connected to the gate of SCR A. The large terminal closest to the gate terminals is the negative terminal of the module, which is also the anode of SCR B. The large terminal farthest

FIGURE 24-33 SCR power module.

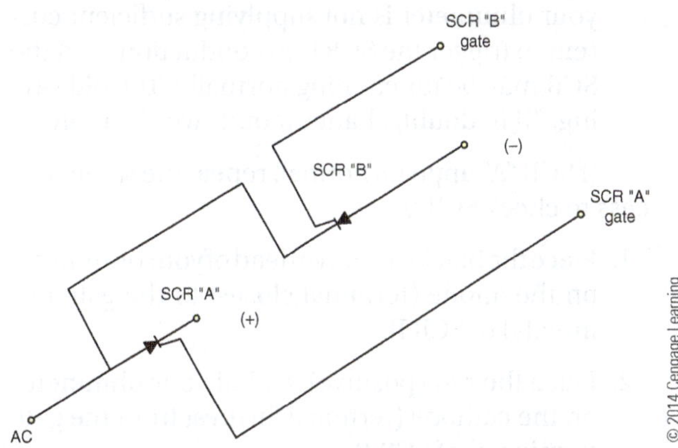

FIGURE 24-34 SCR power module schematic.

from the gate terminals is the AC terminal for the module, which is also the anode of SCR A. Notice that the anode of SCR "A" is internally connected to the cathode of SCR B. The large terminal in the center is the positive terminal of the module, which is also the cathode of SCR A.

To test the module with an ohmmeter, follow these steps:

1. Place the ohmmeter in the Ohms Ω position, if using a digital multimeter. (Use the 200 Ω position if using an analog multimeter.)
2. Place the black (negative) lead of your ohmmeter on the anode (terminal farthest from the gate terminals) of SCR A.
3. Place the red (positive) lead of your ohmmeter on the cathode (center terminal) of SCR A.
4. Your meter should indicate a very high or infinite resistance. A low-resistance reading indicates a faulty SCR. Replace the module.
5. Reverse your meter connections. Place the black (negative) lead of your ohmmeter on the cathode (center terminal) of SCR A.
6. Place the red (positive) lead of your ohmmeter on the anode (terminal farthest from the gate terminals) of SCR A.
7. Your meter should indicate a very high or infinite resistance. A low-resistance reading indicates a faulty SCR. Replace the module.
8. While leaving your ohmmeter connected as in steps 5 and 6, use a clip lead to connect the gate of SCR A (small terminal at the front of the module) to the red (positive) lead of your ohmmeter.
9. You should notice a drop in the resistance reading. This is a result of triggering SCR A into conduction. If your resistance reading does not drop, the SCR may be faulty, and you should replace the module. However, it is possible that your ohmmeter is not supplying sufficient current to trigger the SCR into conduction, and the SCR may be functioning normally. The old saying, "If in doubt, change it out!" would apply.

If SCR "A" appears normal, repeat the same process to check SCR B:

1. Place the black (negative) lead of your ohmmeter on the anode (terminal closest to the gate terminals) of SCR B.
2. Place the red (positive) lead of your ohmmeter on the cathode (terminal farthest from the gate terminals) of SCR B.
3. Your meter should indicate a very high or infinite resistance. A low-resistance reading indicates a faulty SCR. Replace the module.
4. Reverse your meter connections. Place the black (negative) lead of your ohmmeter on the cathode (terminal farthest from the gate terminals) of SCR B.
5. Place the red (positive) lead of your ohmmeter on the anode (terminal closest to the gate terminals) of SCR B.
6. Your meter should indicate a very high or infinite resistance. A low-resistance reading indicates a faulty SCR. Replace the module.
7. While leaving your ohmmeter connected as in steps 5 and 6, use a clip lead to connect the gate of SCR B (small terminal at the back of the module) to the red (positive) lead of your ohmmeter.
8. You should notice a drop in the resistance reading. This is a result of triggering SCR B into conduction. If your resistance reading does not drop, the SCR may be faulty and you should replace the module. However, it is possible that your ohmmeter is not supplying sufficient current to trigger the SCR into conduction, and the SCR may be functioning normally. Again, "If in doubt, change it out!"

While we are on the subject of power modules, there is another type of power module that you may find in the drive that you are working on. This module is technically not part of the power output section, although it does have a power function in the drive. The module is a three-phase bridge rectifier module. It is used to convert three-phase AC into rectified DC. A picture of this module appears in **Figure 24-35**, and a schematic of this module appears in **Figure 24-36**. Notice that this module

FIGURE 24-35 Three-phase bridge rectifier module.

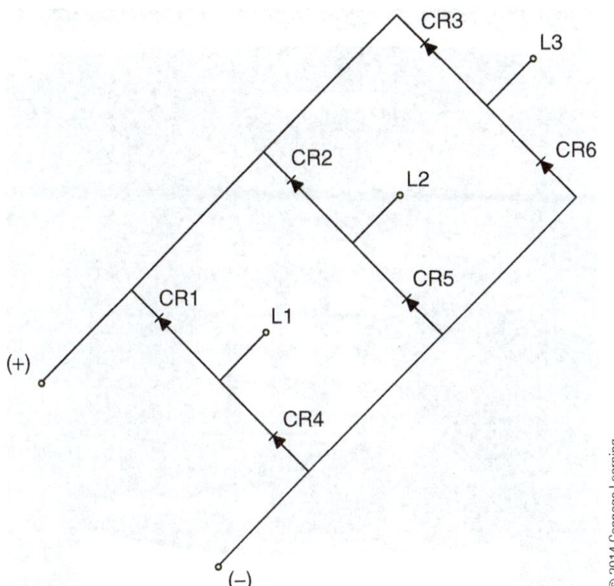

FIGURE 24-36 Three-phase bridge rectifier module schematic.

has five terminals. The two horizontal terminals at the left end of the module are the (+) and (–) DC connections. The three vertical terminals are the connections for the three-phase AC (L1, L2, and L3).

To test the module with an ohmmeter, follow these steps:

1. Place the ohmmeter in the Diode Test position, if using a digital multimeter. (Use the 200 Ω position if using an analog multimeter.)
2. Place the black (negative) lead of your ohmmeter on the (–) terminal of the module.
3. Place the red (positive) lead of your ohmmeter on AC terminal L1 (this is actually the cathode of diode CR4).
4. Your meter should indicate a very high or infinite resistance. A low-resistance reading indicates a faulty diode. Replace the module.
5. Move the red (positive) lead of your ohmmeter from AC terminal L1 to AC terminal L2 (this is actually the cathode of diode CR5).
6. Your meter should indicate a very high or infinite resistance. A low-resistance reading indicates a faulty diode. Replace the module.
7. Move the red (positive) lead of your ohmmeter from AC terminal L2 to AC terminal L3 (this is actually the cathode of diode CR6).
8. Your meter should indicate a very high or infinite resistance. A low-resistance reading indicates a faulty diode. Replace the module.
9. Reverse your meter connections. Place the red (positive) lead of your ohmmeter on the (–) terminal of the module.
10. Place the black (negative) lead of your ohmmeter on AC terminal L1 (this is actually the cathode of diode CR4).
11. Your meter should indicate a low resistance. A high- or infinite-resistance reading indicates a faulty diode. Replace the module.
12. Move the black (negative) lead of your ohmmeter from AC terminal L1 to AC terminal L2 (this is actually the cathode of diode CR5).
13. Your meter should indicate a low resistance. A high- or infinite-resistance reading indicates a faulty diode. Replace the module.
14. Move the black (negative) lead of your ohmmeter from AC terminal L2 to AC terminal L3 (this is actually the cathode of diode CR6).
15. Your meter should indicate a low resistance. A high- or infinite-resistance reading indicates a faulty diode. Replace the module.
16. Place the red (positive) lead of your ohmmeter on the (+) terminal of the module.
17. Place the black (negative) lead of your ohmmeter on AC terminal L1 (this is actually the anode of diode CR1).
18. Your meter should indicate a very high or infinite resistance. A low-resistance reading indicates a faulty diode. Replace the module.
19. Move the black (negative) lead of your ohmmeter from AC terminal L1 to AC terminal L2 (this is actually the anode of diode CR2).
20. Your meter should indicate a very high or infinite resistance. A low-resistance reading indicates a faulty diode. Replace the module.
21. Move the black (negative) lead of your ohmmeter from AC terminal L2 to AC terminal L3 (this is actually the anode of diode CR3).
22. Your meter should indicate a very high or infinite resistance. A low-resistance reading indicates a faulty diode. Replace the module.
23. Reverse your meter connections. Place the black (negative) lead of your ohmmeter on the (+) terminal of the module.
24. Place the red (positive) lead of your ohmmeter on AC terminal L1 (this is actually the anode of diode CR1).

25. Your meter should indicate a low resistance. A high- or infinite-resistance reading indicates a faulty diode. Replace the module.
26. Move the red (positive) lead of your ohmmeter from AC terminal L1 to AC terminal L2 (this is actually the anode of diode CR2).
27. Your meter should indicate a low resistance. A high- or infinite-resistance reading indicates a faulty diode. Replace the module.
28. Move the red (positive) lead of your ohmmeter from AC terminal L2 to AC terminal L3 (this is actually the anode of diode CR3).
29. Your meter should indicate a low resistance. A high- or infinite-resistance reading indicates a faulty diode. Replace the module.

Once you have completed testing the module, if you determine that the module is defective, you can usually obtain a substitute part from a local electronics parts supplier. If the part is not available, you will have to return the drive to the manufacturer for service or call a service technician for on-site repairs.

If the problem is not in the drive's power section, then it must be located in its control section. The electronics in the control section are more complex; therefore, troubleshooting is not recommended. In this event, you should definitely return the drive to the manufacturer for repair or call a service technician for on-site repairs.

24-3 AC (INVERTER) DRIVES

For many years, the mainstay motor of industry has been the three-phase, squirrel-cage induction motor. This motor has the advantages of low cost and low maintenance. Its biggest disadvantage has been its fixed operating speed. If you needed a three-phase induction motor with variable speed, you had to use a wound-rotor induction motor with a potentiometer. This configuration entailed added expense and increased maintenance.

Fortunately, we now have AC drives, more commonly called **inverters** (**Figure 24-37**). Inverters are now used not only to vary the speed of the squirrel-cage motor but also to vary its torque, start the motor slowly and smoothly, and increase the motor's efficiency.

An AC drive has three basic sections—the **converter**, the **DC filter**, and the inverter. **Figure 24-38** shows the inside view of an inverter drive. The converter rectifies the applied AC into DC. The DC filter (also called the **DC link** or **DC bus**) provides a smooth, rectified DC. The inverter switches the DC on and off so rapidly that the motor

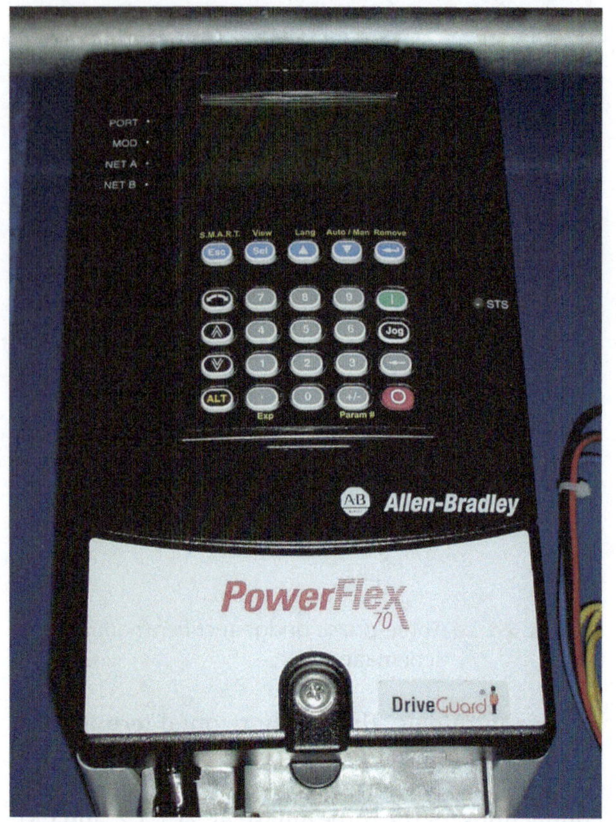

FIGURE 24-37 Allen-Bradley PowerFlex 70 AC drive (inverter).

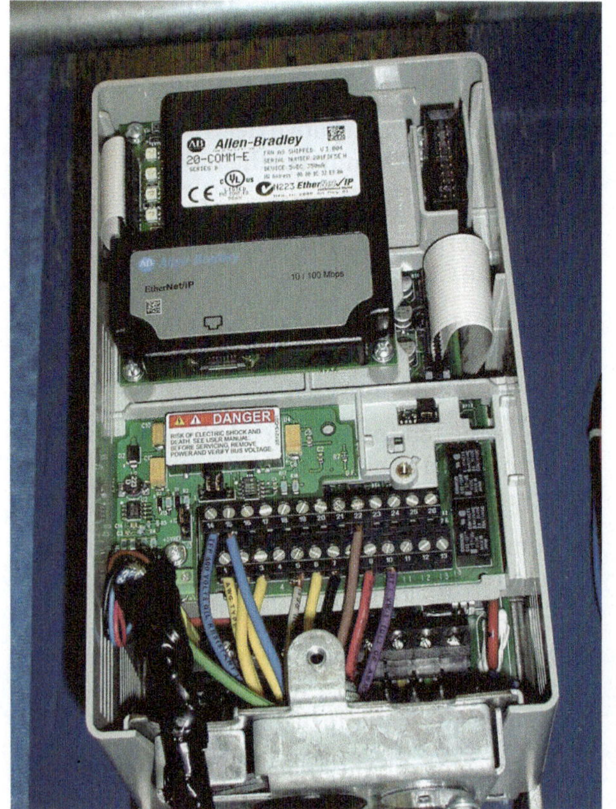

FIGURE 24-38 Allen-Bradley PowerFlex 70 AC drive (inverter) with cover removed.

receives a pulsating DC that appears similar to AC. Because this DC switching is controlled, we can vary the frequency of the artificial AC that is applied to the motor, something that we are normally unable to do.

The AC line frequency is set by the electric utility companies across the United States at a standard rate of 60 Hz. Because the AC line frequency could not normally be varied, the speed of a squirrel-cage motor was, for the most part, fixed. The number of the motor's poles could be increased or decreased, causing the motor to slow down or speed up, respectively. However, this could only be done as the motor was being built. We could also vary the stator voltage. However, doing so reduces the motor's ability to drive a load at low speeds, so this method has limited usefulness.

Today, we can use inverters to vary the frequency of the "AC" (the pulsating DC) applied to the squirrel-cage motor, and thus vary the motor's speed while maintaining constant torque. As we will see, in today's inverter drives, the stator voltage and the frequency are both variable.

The different methods used to create the pulsating DC constitute the basic differences among the various types of inverter drives. Another area of design variation is that of the ratio of **volts to hertz (V/Hz)**. The V/Hz ratio should be maintained at a constant value. This means that a motor turning at 1800 RPM, operating from 208 V at 60 Hz, would have to operate from 104 V at 30 Hz to attain a speed of 900 RPM, because these values do not change the V/Hz ratio from its original value.

There are two major types of inverter drives: the **voltage source inverter**, also known as the **voltage fed inverter (VSI)**, and the **current source inverter (CSI)**.

Voltage Source Inverter

The voltage source inverter may be further subdivided into the **variable voltage inverter** and the **pulse-width modulated inverter**.

Variable Voltage Inverter

A variable voltage inverter (VVI) is one of two categories of adjustable frequency, variable-speed AC drives. In a VVI, the DC voltage is controlled, and the DC current is free to respond to the motor needs. The VVI uses a converter, a DC link, and an inverter to vary the frequency of the applied AC voltage.

A converter is a circuit that changes the incoming AC power (fixed voltage, fixed frequency) into DC power. The converter circuit is simply a rectifier circuit that produces an unfiltered, pulsating DC. The converter can be either single-phase or three-phase, depending on the type of power that the AC induction motor you are using needs.

A typical DC link, also called a DC filter or DC bus, is simply a filter circuit composed of an inductor and a capacitor. The purpose of the DC link is to filter or smooth the AC ripple from the output of the converter stage. The filtered DC from the DC link is fed via the DC bus to the input of the inverter stage.

An inverter converts the applied DC voltage to a pulsating DC voltage. Because we can vary the magnitude and the frequency of this pulsating DC, we can therefore control the speed of an AC induction motor. This pulsating DC power acts as artificial AC power on the induction motor.

The VVI shown in the schematic diagram in **Figure 24-39** is divided into three basic parts: the converter, the DC link, and the inverter. Basically, three-phase AC is applied to the converter stage of the VVI. Naturally, this AC has both fixed amplitude and fixed frequency. The converter rectifies the AC into DC, which is then smoothed by the DC link. This

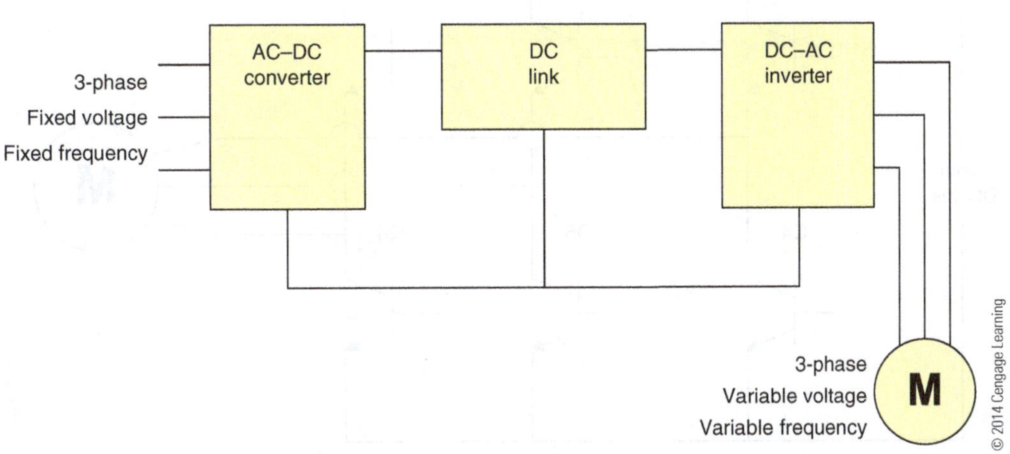

FIGURE 24-39 Variable Voltage Inverter (VVI) block diagram.

filtered DC is applied to the inverter, which chops the DC into pulsating AC. This AC is then applied to the motor. The chopping rate applied to the DC varies the frequency of the AC applied to the motor.

We can look at this in more detail in **Figures 24-40A** through **24-40D**, which are simplified schematics of the inverter stage of a six-step VVI. The inverter stage functions in the same way regardless of whether you use the phase control or the chopper control method, both of which we will study later. In the first one-sixth of the cycle, transistors Q1, Q5, and Q6 conduct. In the second one-sixth of the cycle, transistors Q1 and Q6 continue to conduct; however, transistor Q5 turns off, and transistor Q2 turns on. During the third one-sixth of the cycle, transistors Q2 and Q6 continue to conduct, but transistor Q1 turns off, and transistor Q4 conducts. During the fourth one-sixth of the cycle, transistors Q2 and Q4 remain on, while transistor Q6 turns off and Q3 turns on. After two more steps, the motor has completed one revolution. This six-step cycle is the reason for the inverter's name.

By turning the transistors on at a slower or faster rate, we can vary the frequency of the voltage applied to the motor. As the frequency varies, the inductive reactance of the motor windings also varies proportionally. An increase in reactance causes a decrease in current at higher speeds, and as a result the motor does not function properly. Therefore, the voltage must be increased proportionally to the frequency, in a ratio known as the volts/hertz ratio, or V/Hz. The V/Hz ratio must be kept constant, and this can be accomplished by one of two methods—phase control or chopper control.

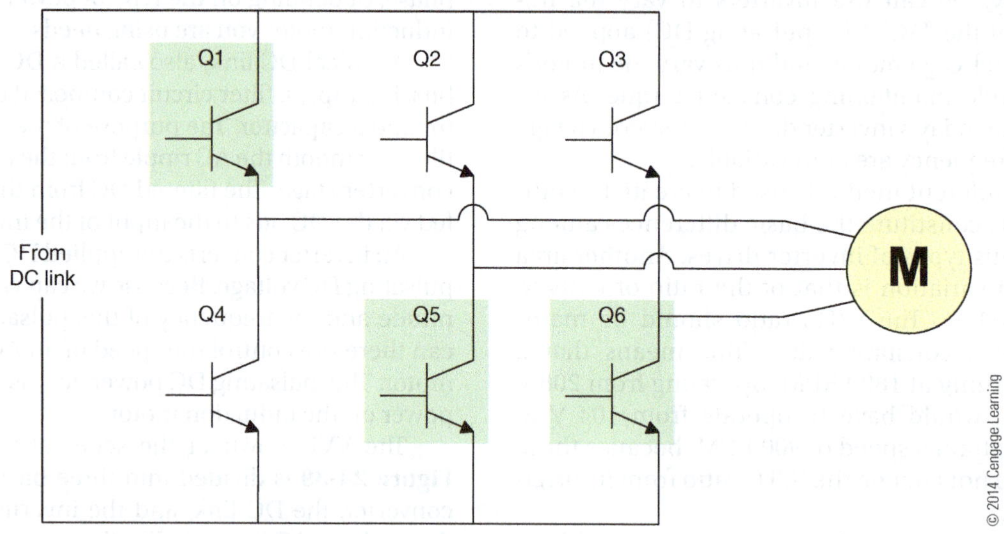

FIGURE 24-40A A six-step inverter with Q1, Q5, and Q6 conducting.

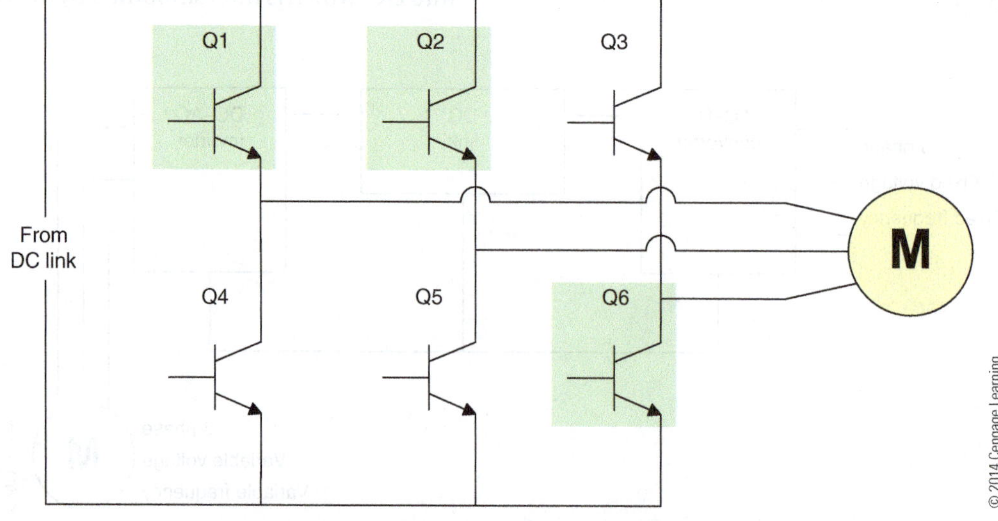

FIGURE 24-40B A six-step inverter with Q1, Q2, and Q6 conducting.

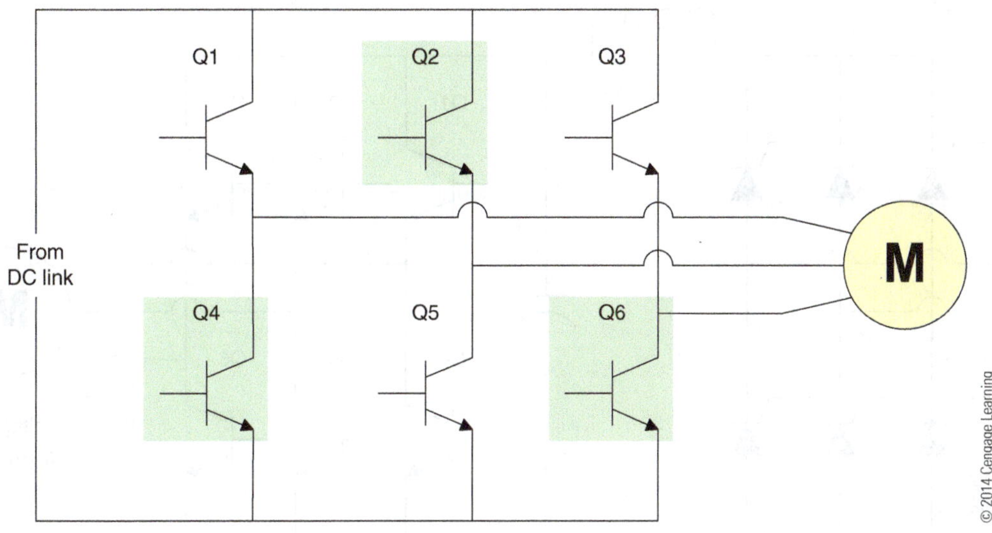

FIGURE 24-40C A six-step inverter with Q2, Q4, and Q6 conducting.

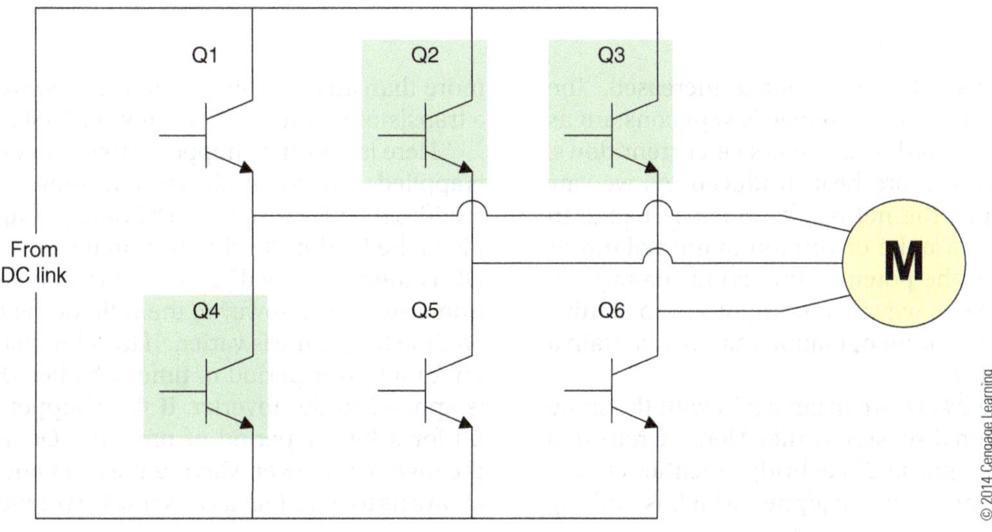

FIGURE 24-40D A six-step inverter with Q2, Q3, and Q4 conducting.

In **Figure 24-41**, the converter stage is composed of a three-phase SCR bridge circuit. Remember that we can control when and for how long an SCR conducts. If we turn the SCR on early in the cycle, we can provide a higher average voltage to the inverter because the SCR conducts for a longer portion of the cycle. If the SCR is turned on late in the cycle, the SCR conducts for a shorter period of time; therefore, the inverter will receive a lower average voltage. By varying the DC bus voltage level in this way and simultaneously varying the frequency of the output from the inverter section, we can maintain a constant V/Hz ratio.

The V/Hz ratio (also called the **V/f ratio**) is very important for a number of reasons. Remember that we need to be concerned with many characteristics when dealing with AC and inductors. We not only have to be aware of voltage, current, and resistance, but we must also recognize the importance of reactance, hysteresis, and eddy currents. We will see what could happen if we failed to maintain a constant V/Hz ratio.

Suppose that we could increase the voltage applied to the motor without adjusting the frequency of the applied voltage. What would happen? The increased voltage produces increased magnetic flux, which in turn saturates the iron components of the motor. This flux causes increased iron losses in the form of hysteresis and eddy currents. It also increases the stator current and could possibly damage the motor windings as a result.

If we examine what happened to the ratio of volts to hertz (V/Hz) in the preceding example, we see that the ratio increased because the frequency

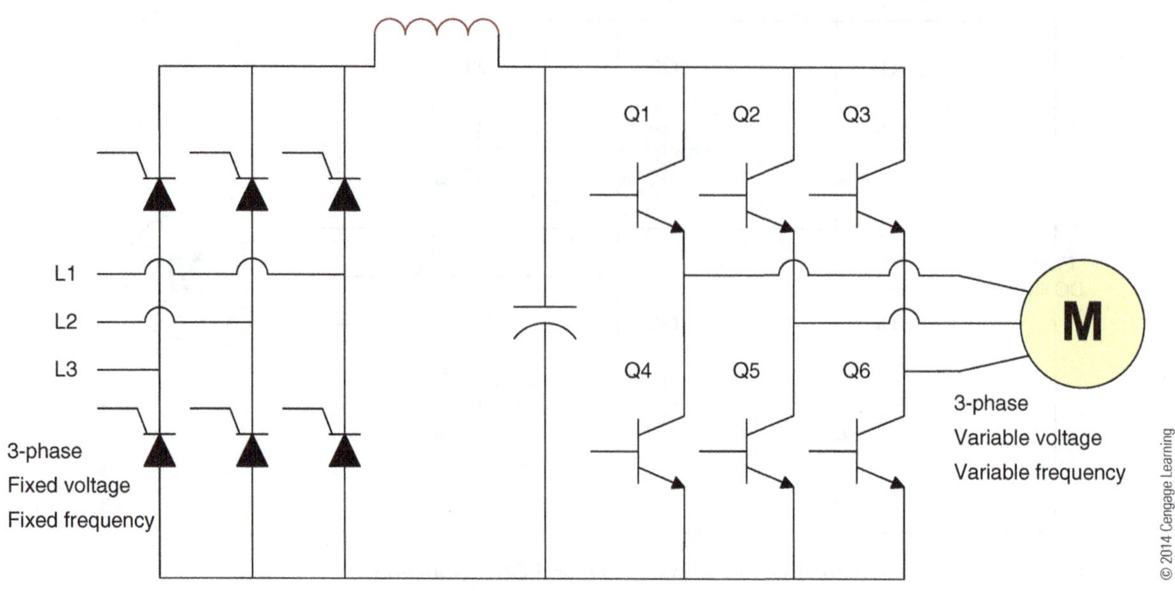

FIGURE 24-41 A variable voltage inverter with phase control.

remained constant as the voltage increased. The same effect occurs if the voltage is kept constant as the frequency is reduced. Excessive current flows, which results in more heat produced. As we saw in the example, the net results of such change to the V/Hz ratio are the disruption of normal motor operation and the potential for serious damage to the motor. That is why it is so important to continued and proper motor operation that we maintain a constant V/Hz ratio.

In **Figure 24-42**, we begin again with the same inverter section discussed earlier. Here we return to a simple three-phase diode bridge rectifier circuit. However, we now add a chopper, which is nothing more than an electronic switch. This switch may be a transistor or, more commonly, a MOSFET.

Here is how the chopper works. Three-phase AC is applied to the three-phase diode bridge. The bridge rectifies the AC into a fixed DC voltage. Capacitor C1 filters the DC for the chopper circuit. The smoothed DC is then "chopped" at a fixed frequency by electronic switches; however, the ratio of the chopper's *on* time to *off* time is varied. If the chopper is turned on for a longer period of time, a higher DC voltage is applied to the inverter. If the chopper is turned off for a longer period of time, the DC applied to the inverter is lower. Varying these *on* and *off* times allows us to maintain a constant V/Hz ratio.

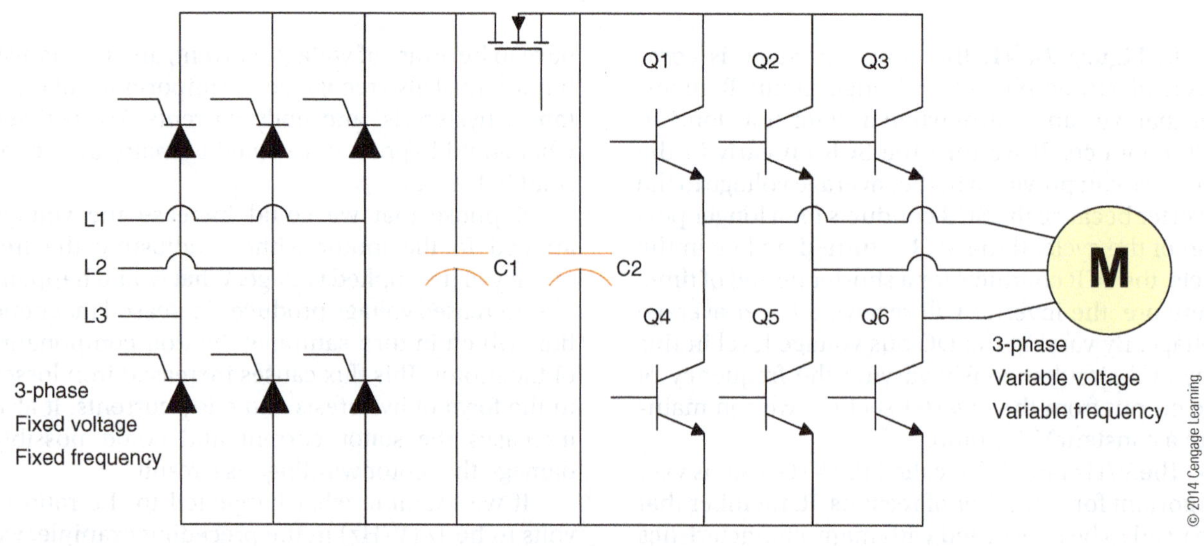

FIGURE 24-42 A variable voltage inverter with chopper control.

All types of VVIs provide open-circuit protection, can handle multiple motor applications and undersized motors, and do not require high-speed switching devices. These drives are quite light in weight, utilize relatively simple control circuits, and exhibit good efficiency at low speeds.

On the negative side, all VVIs lack short-circuit protection, and they cannot handle oversized motors. Another disadvantage that both types of VVIs share is that they do not operate smoothly at low frequencies (<6 Hz). Regenerative braking is not possible with a chopper-controlled VVI but is possible with a phase-controlled VVI. On the other hand, a chopper-controlled VVI can be operated from batteries, whereas a phase-controlled VVI cannot.

Pulse-Width Modulated/Variable Voltage Inverter

Another form of variable voltage inverter uses pulse-width modulation (PWM) to vary the voltage applied to an AC induction motor. Many similarities exist between this type of inverter and the DC drives discussed earlier.

Notice in **Figure 24-43** that the feedback device used is a tachometer-generator (you will find it located in the upper right corner of the schematic.) Recall that the tachometer-generator is a DC generator attached to an AC induction motor's shaft. As the AC induction motor turns, the DC tachometer-generator also turns, producing a positive DC voltage. The faster the AC induction motor turns, the more positive DC voltage the DC tachometer-generator produces. In other words, the output of the DC tachometer-generator is proportional to the speed of the AC induction motor.

Next, we consider the feedback section, shown in **Figure 24-44**. The positive voltage, or feedback signal, produced by the DC tachometer-generator is fed through R1 into buffer amplifier U1. The output of U1 is then fed to the inverting input of U2, which inverts the polarity of the tachometer-generator voltage from positive to negative. The resulting negative voltage is then fed through R7 to the noninverting input of U3 in the preamplifier section. Notice that one end of R8 is also connected to the noninverting input of U3 and that the other end of R8 is connected to variable resistor R6. Resistor R6 allows us to set the reference voltage level, sometimes called the command signal, which is the voltage necessary for the AC induction motor to run at a certain speed. The reference voltage (positive) and the tachometer-generator voltage (negative) are added together at the junction of R7 and R8, which is known as the summing point. The difference of these two voltages is the value of the voltage that appears at the noninverting input of U3. This value is known as the error signal. Now, let us leave this section for a moment and move on to the null detector and sawtooth generator circuits.

Notice that in the null detector shown in **Figure 24-45** (see in page 590) the applied AC voltage is rectified by the full-wave bridge circuit consisting of CR7–CR10. This process causes an unfiltered, pulsating DC voltage to appear across zener diode Z1 and across R28 and the LED of optocoupler U10.

What happens as the pulsating DC voltage rises from 0 V to its peak value? The LED of U10 does not conduct until the DC voltage reaches the turn-on voltage of the LED. Assume that this value is approximately 1.5 V. As the DC voltage increases from 0 V to 1.5 V, U10 is off. Therefore, the output of U10 is positive, causing Q2 to be on. When Q2 conducts, a negative pulse appears at the collector of Q2. This negative pulse is used to reset the output of the sawtooth generator back to 0.

What happens when the LED of U10 turns on? The pulsating DC voltage rises, eventually reaching 1.5 V. The voltage across the LED of U10 never exceeds the voltage rating of zener diode Z1. At this point, the LED of U10 turns on and conducts, causing the output of U10 to decrease to a level that turns Q2 off. With Q2 turned off, the sawtooth generator outputs a sawtooth waveform to the inverting input of U5 (the comparator). The sawtooth generator continues to output this waveform until it is reset when Q2 turns on—that is, when the pulsating DC voltage falls below 1.5 V and decreases to 0 V. Therefore, the pulsating DC voltage causes the sawtooth generator to provide a reference signal to the comparator. This reference signal is used to vary the firing angle of the SCRs that control the stator voltage.

Consider U5, the comparator, in **Figure 24-46** (see in page 591). The comparator has two inputs. One comes from the feedback circuit and is a measure of the speed of the armature. The other input, as we just learned, is a ramp signal from the sawtooth generator. The noninverting input from the feedback amplifier is constant if the load on the armature is constant. In contrast, the inverting input is a rising amplitude signal from the sawtooth generator. Initially, the noninverting input causes the output of U5 to be positive. However, as the inverting input ramp signal climbs, a point is reached where the amplitudes of both inputs are equal. At this point, the output of U5 switches off, becoming a rectangular pulse. The width of this pulse is determined by the length of time it takes for both inputs to become equal. This is what determines whether the SCRs conduct earlier or later within each half cycle.

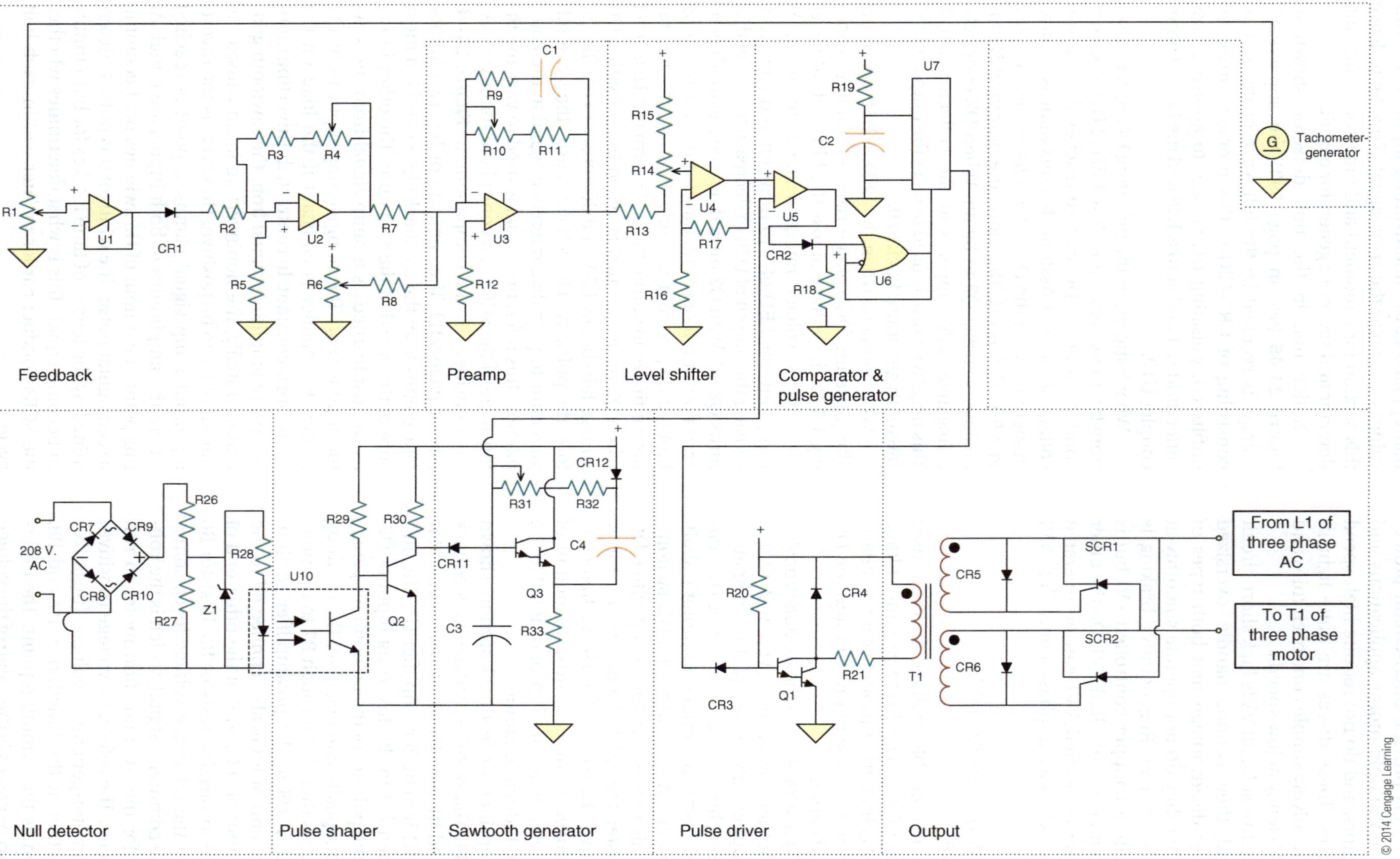

FIGURE 24-43 A variable voltage inverter with pulse-width modulation.

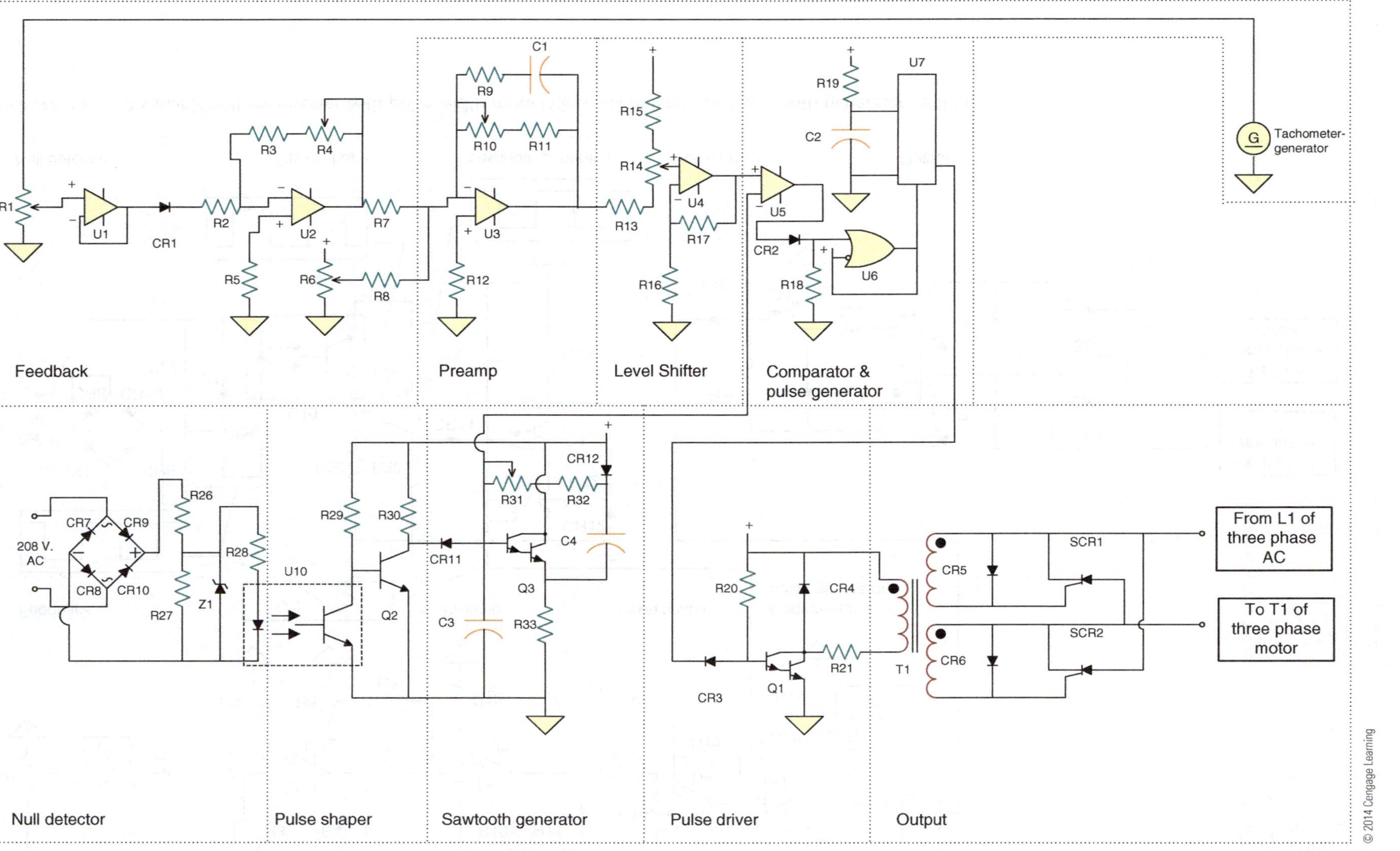

FIGURE 24-44 A variable voltage inverter with pulse-width modulation feedback section.

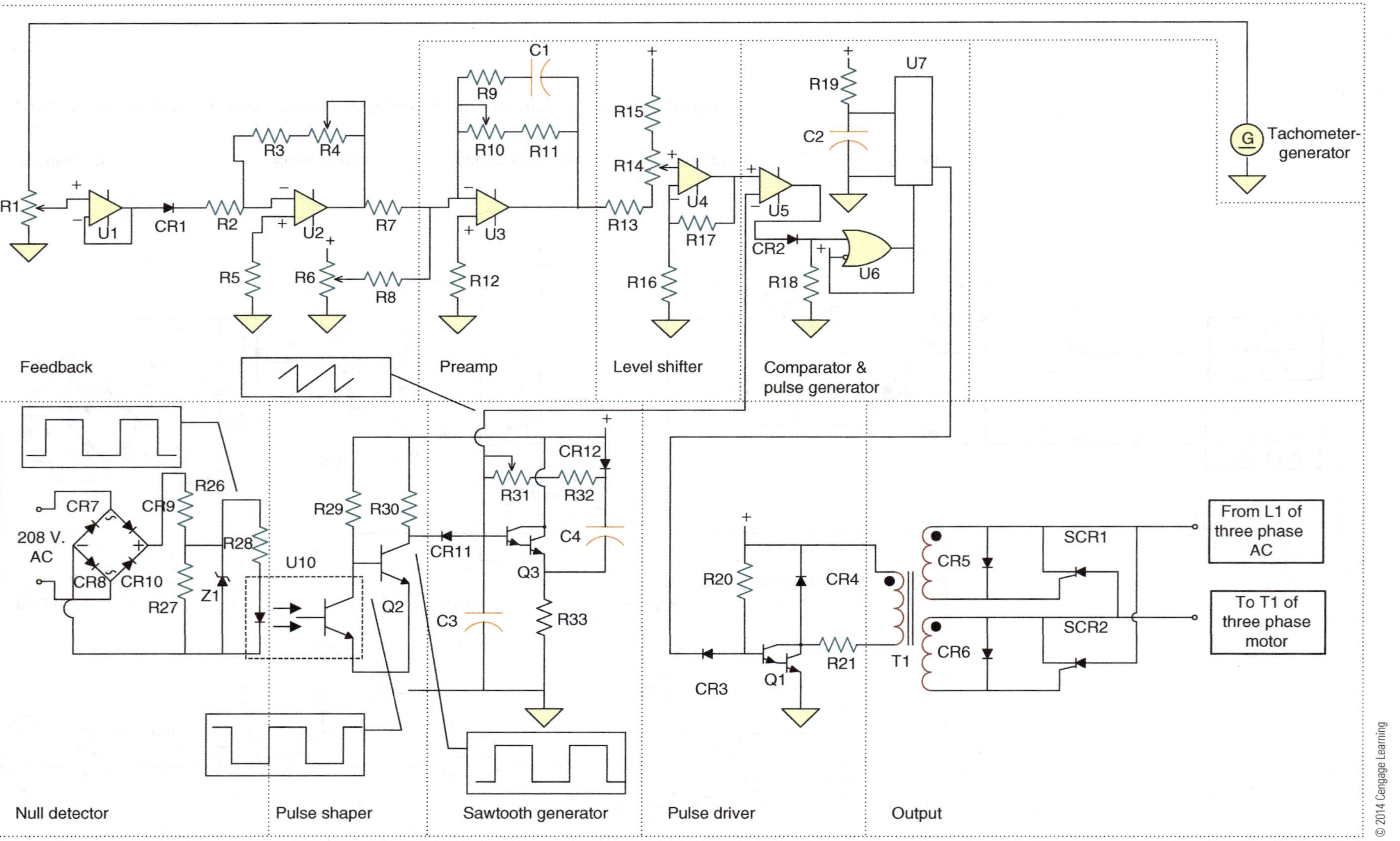

FIGURE 24-45 A variable voltage inverter with pulse-width modulation null detector and sawtooth generator section.

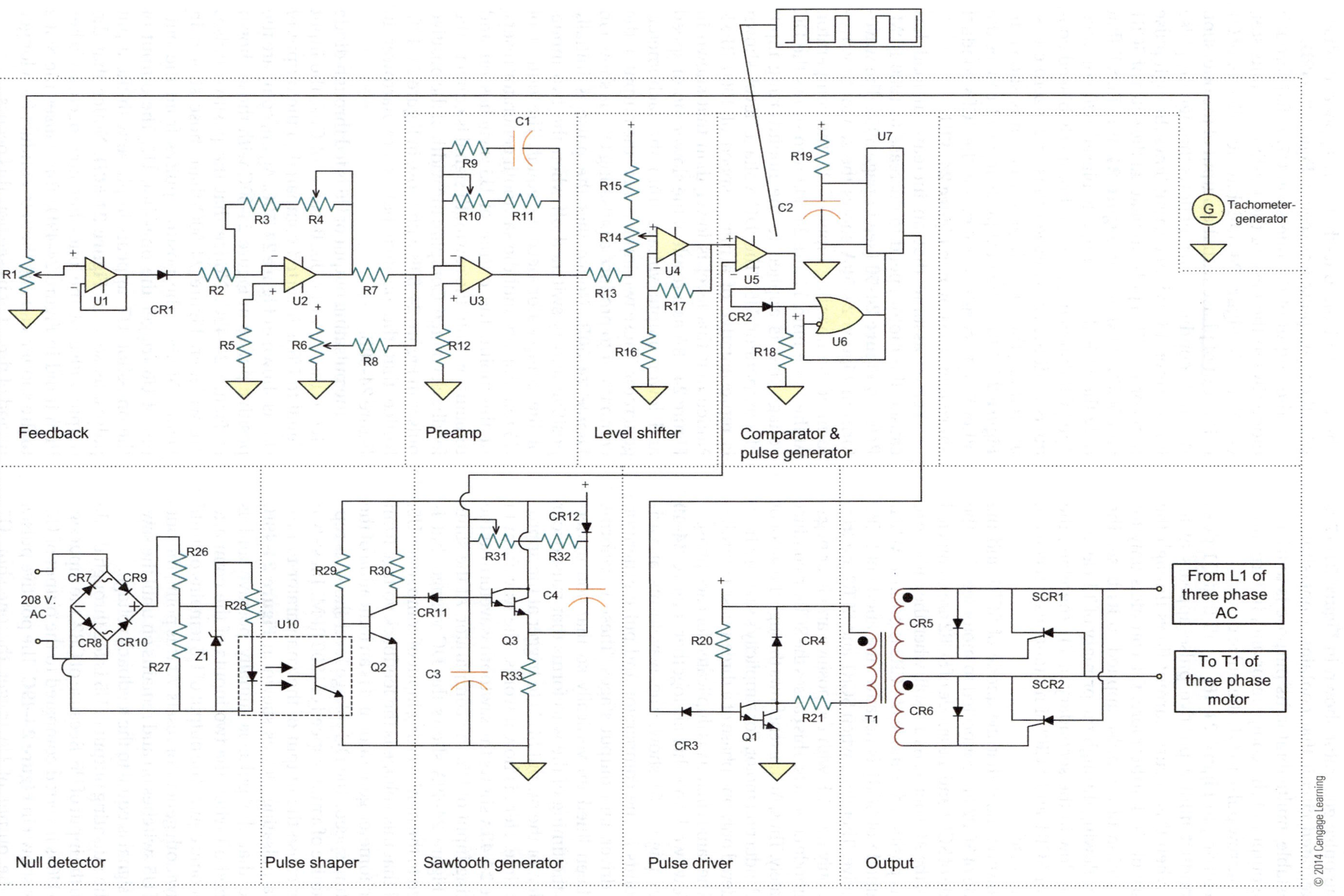

FIGURE 24-46 A variable voltage inverter with pulse-width modulation comparator section.

The pulse generator shown in **Figure 24-47** is an integrated circuit that contains a one-shot, or monostable, multivibrator. Its purpose is to trigger on the output of the comparator and, in turn, provide a narrow pulse to Q1, the pulse driver.

Referring to **Figure 24-48**, recall that Q1 does not conduct until a positive pulse appears at its base. When this occurs, current flows through the primary of T1. Furthermore, Q1 conducts only for the duration of the pulse applied to its base. The current flowing through the primary of T1 causes a current flow in the secondary of T1. Therefore, the gates of SCR1 and SCR2 simultaneously receive a trigger pulse.

However, note that the anode of SCR1 and the cathode of SCR2 are connected to motor lead T1. The cathode of SCR1 and the anode of SCR2 are connected to L1 of the AC source. As a result, when the cathode of SCR1 is positive, the anode of SCR2 is positive. When the cathode of SCR1 is negative, the anode of SCR2 is negative. Therefore, when SCR1 and SCR2 are triggered, only the SCR with the positive anode or negative cathode conducts. This causes the SCRs to conduct alternately. Thus, AC current flows through the stator of the AC induction motor. For simplicity's sake, we have considered only one phase of a three-phase controller here. The same circuitry is duplicated for each phase.

Now, we bring this all together. **Figure 24-49** (see in page 595) shows the waveforms at various points in the comparator and pulse generator, pulse driver, and output stages. These waveforms have been lined up vertically so that you understand the timing of the waveforms that must occur in order for the output SCRs to trigger at the appropriate time. Here is how it works. The waveform in **Figure 24-49A** shows the sawtooth waveform at the inverting input of U5, the comparator. At the same time, Figure 24-49A shows the DC voltage that is applied to the noninverting input of comparator U5. Recall that this voltage is the feedback voltage from the tachometer-generator and is an indication of the armature speed. The DC voltage level shown is representative of a motor speed of 1800 RPM. These two inputs cause the output of the comparator to produce a pulsating DC, as shown in **Figure 24-49B**. Notice that the pulse is initially positive and is switched off when the two inputs of the comparator become equal. The output of U5 remains off until the sawtooth generator resets. At this point, the output of U5 switches on and remains on until the sawtooth signal is equal to the feedback voltage.

The pulsating output of U5 is fed through diode CR2 to the input of U6. Because of CR2, the input signal to U6 is inverted compared to the output of U5. This is shown in **Figure 24-49C**. The positive pulses from the output of U6 trigger the one-shot, U7. The output of the one-shot, U7, appear in phase with the input pulses, as seen in **Figure 24-49D**.

The output of the one-shot, U7, is fed through diode CR3 to the base of the Darlington transistor, Q1. Notice in **Figure 24-49E** that the collector of Q1 is 180° out of phase with the output of the one-shot. When Q1 conducts, current flows through the pulse transformer, T1. When current flows through pulse transformer T1, a pulse appears at the gates of SCR1 and SCR2, as shown in **Figure 24-49F**. Recall that only the SCR that is properly biased will conduct. The SCR that is triggered and properly biased conducts until the AC voltage drops to the zero crossing point. (The AC voltage waveform is shown in **Figure 24-49G**.) The next pulse from T1 causes the other SCR to conduct. Therefore, the SCRs conduct alternately, as shown in **Figure 24-49H**.

Let us assume that an increase in load has caused the motor speed to decrease to 1000 RPM. Refer to **Figure 24-50** (see in page 596). The waveform in **Figure 24-50A** shows the sawtooth waveform at the inverting input of U5, the comparator. At the same time, Figure 24-50A also shows the DC voltage that is applied to the noninverting input of comparator U5. The DC voltage level shown is representative of a motor speed of 1000 RPM. Notice that this level is lower than that shown in Figure 24-49A as a result of the slower motor speed and lower output voltage from the tachometer-generator. These two inputs cause the output of the comparator to produce a pulsating DC, as shown in **Figure 24-50B**. Notice that the pulse is initially positive and is switched off when the two inputs of the comparator become equal. The output of U5 remains off until the sawtooth generator resets. At this point, the output of U5 switches on and remains on until the sawtooth signal is equal to the feedback voltage. Compare the width of the positive pulses in Figure 24-50B with those in Figure 24-49B. Notice that the positive pulses are narrower in Figure 24-50B.

The pulsating output of U5 is fed through diode CR2 to the input of U6. Because of CR2, the input signal to U6 is inverted compared to the output of U5, as shown in **Figure 24-50C**. Again, compare the positive pulses in Figure 24-50C with those shown in Figure 24-49C. Notice that the positive pulses are wider in Figure 24-50C than those shown in Figure 24-49C. The positive pulses from the output of U6 trigger the one-shot, U7. The output of the one-shot, U7, appears in phase with the input pulses, as seen in **Figure 24-50D**. Notice that the output pulse of the one-shot now occurs earlier than it did in Figure 24-49D. This causes the SCRs to fire sooner, causing them to conduct for a longer period of time. Let us see how this occurs.

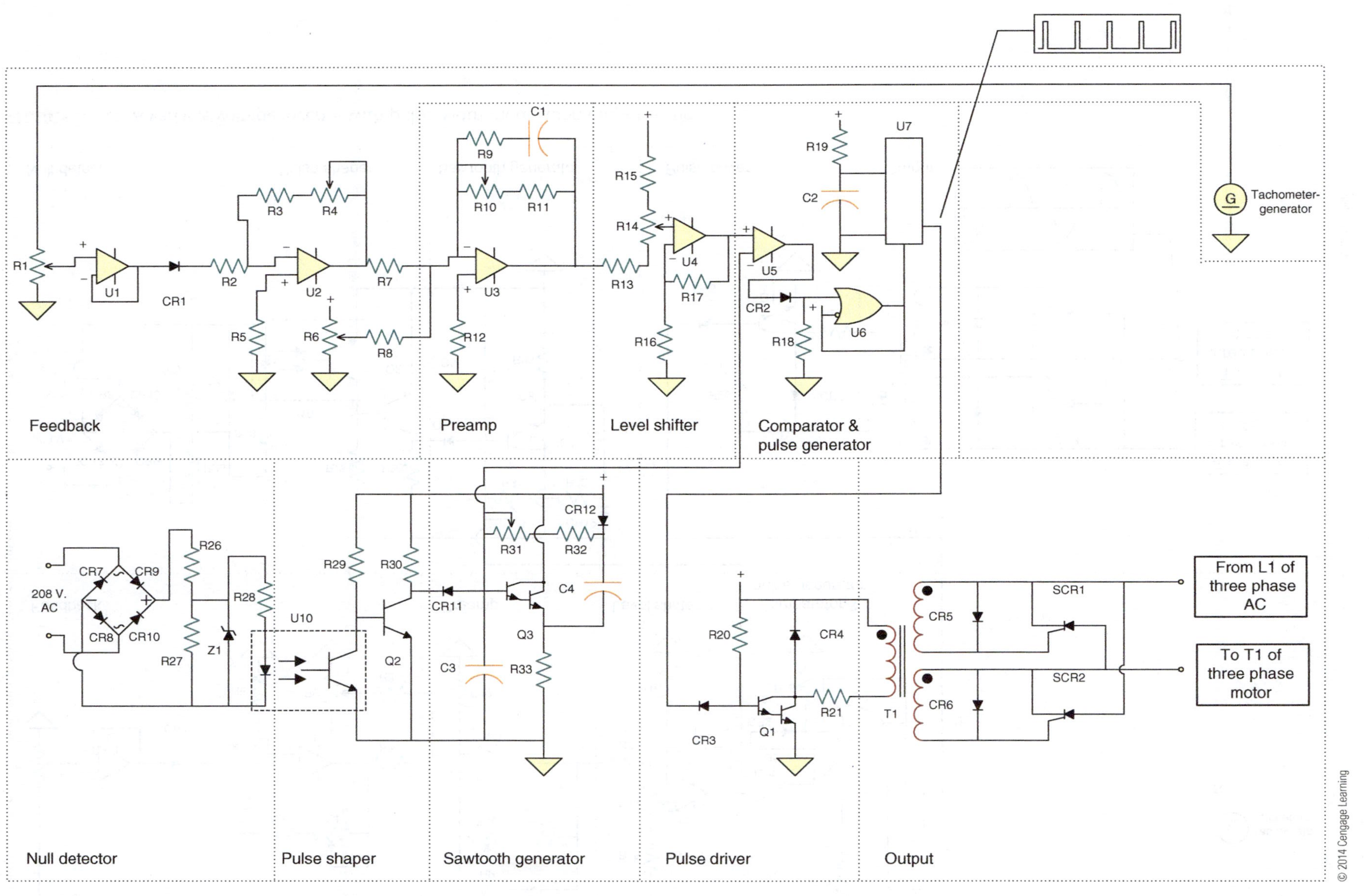

FIGURE 24-47 A variable voltage inverter with pulse-width modulation pulse generator section.

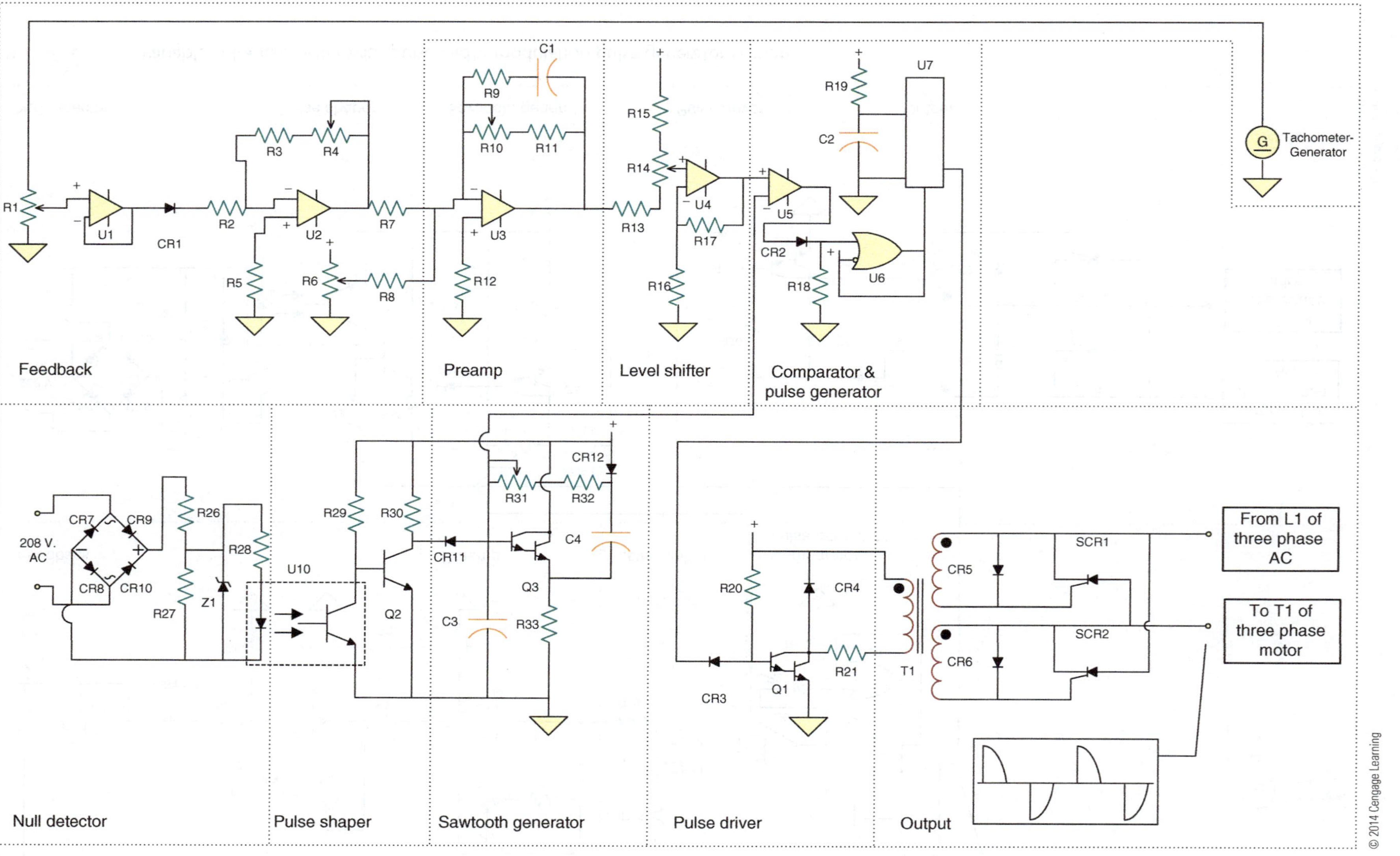

FIGURE 24-48 A variable voltage inverter with pulse-width modulation output section.

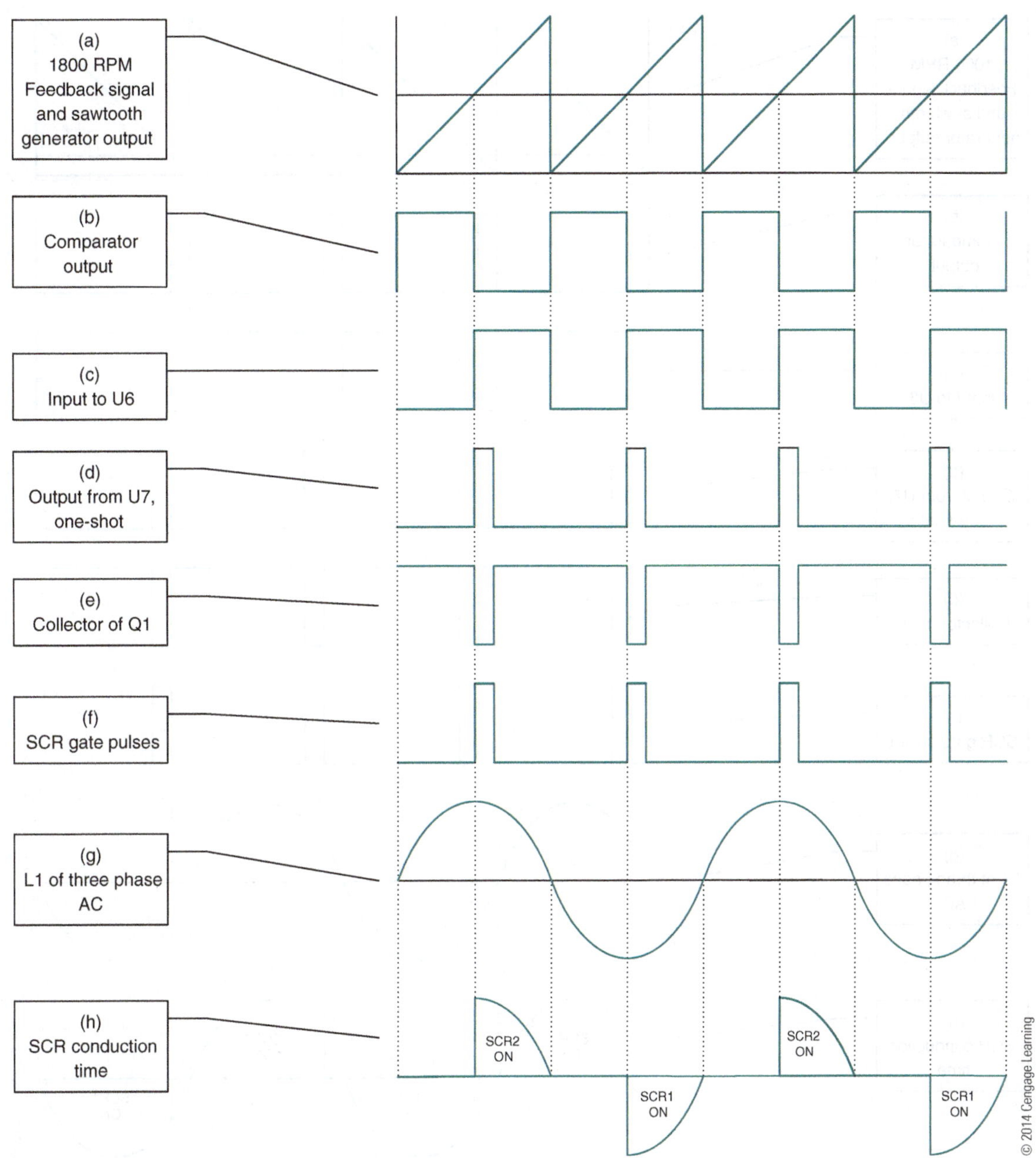

FIGURE 24-49 Waveform timing and SCR conduction time at 1800 rpm.

The output of the one-shot, U7, is fed through diode CR3 to the base of the Darlington transistor, Q1. Notice in **Figure 24-50E** that the collector of Q1 is 180° out of phase with the output of the one-shot. When Q1 conducts, current flows through the pulse transformer, T1. When current flows through pulse transformer T1, a pulse appears at the gates of SCR1 and SCR2, as shown in **Figure 24-50F**. Recall that only the SCR that is properly biased will conduct. The SCR that is triggered and properly biased conducts until the AC voltage drops to the zero crossing point. The next pulse from T1 causes the other SCR to conduct. Therefore, the SCRs conduct alternately, as shown in **Figure 24-50G**. Notice that the SCRs conduct for a longer amount of time as compared to the conduction of the SCRs in Figure 24-49G. This results in a higher average voltage applied to the AC motor. This will cause the speed of the AC motor to increase.

Let us assume that a decrease in load has caused the motor speed to increase to 2600 RPM. Refer to **Figure 24-51**. The waveform in **Figure 24-51A** shows the sawtooth waveform at the inverting input of U5, the comparator. At the same time, Figure 24-51A also shows the DC voltage that is applied to the noninverting input of comparator U5.

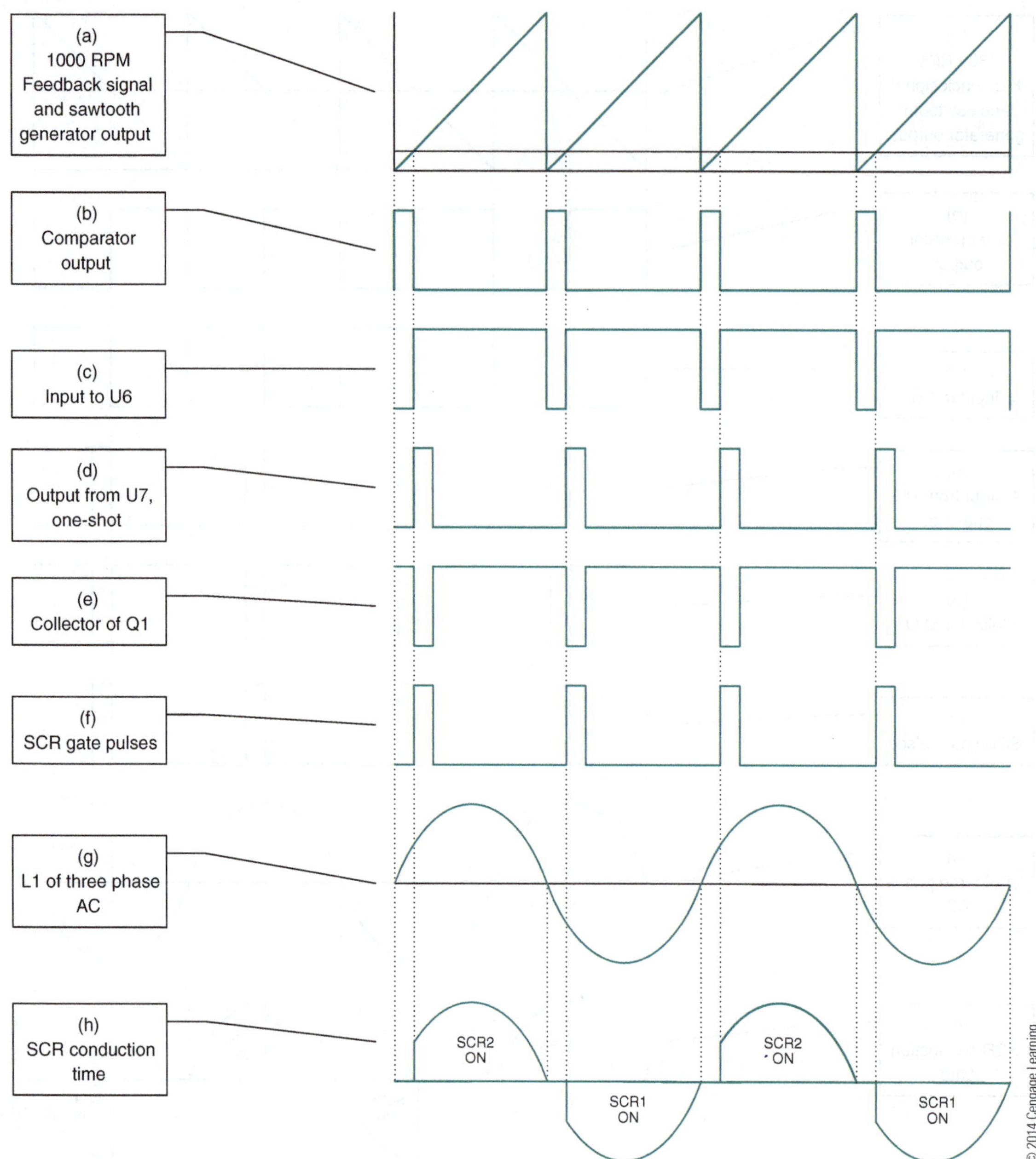

FIGURE 24-50 Waveform timing and SCR conduction time at 1000 rpm.

The DC voltage level shown is representative of a motor speed of 2600 RPM. Notice that this level is higher than that shown in Figure 24-49A. This is a result of the faster motor speed and higher output voltage from the tachometer-generator. These two inputs cause the output of the comparator to produce a pulsating DC, as shown in **Figure 24-51B**. Notice that the pulse is initially positive and is switched off when the two inputs of the comparator become equal. The output of U5 remains off until the sawtooth generator resets. At this point, the output of U5 switches on and remains on until the sawtooth signal is equal to the feedback voltage. Compare the width of the positive pulses in Figure 24-51B with those in Figure 24-49B. Notice that the positive pulses are wider in Figure 24-51B.

The pulsating output of U5 is fed through diode CR2 to the input of U6. Because of CR2, the input signal to U6 is inverted compared to the output of U5, as shown in **Figure 24-51C**. Again, compare the positive pulses in Figure 24-51C with those shown in Figure 24-49C. Notice that the positive pulses are narrower in Figure 24-51C than those shown in Figure 24-49C. The positive pulses from the

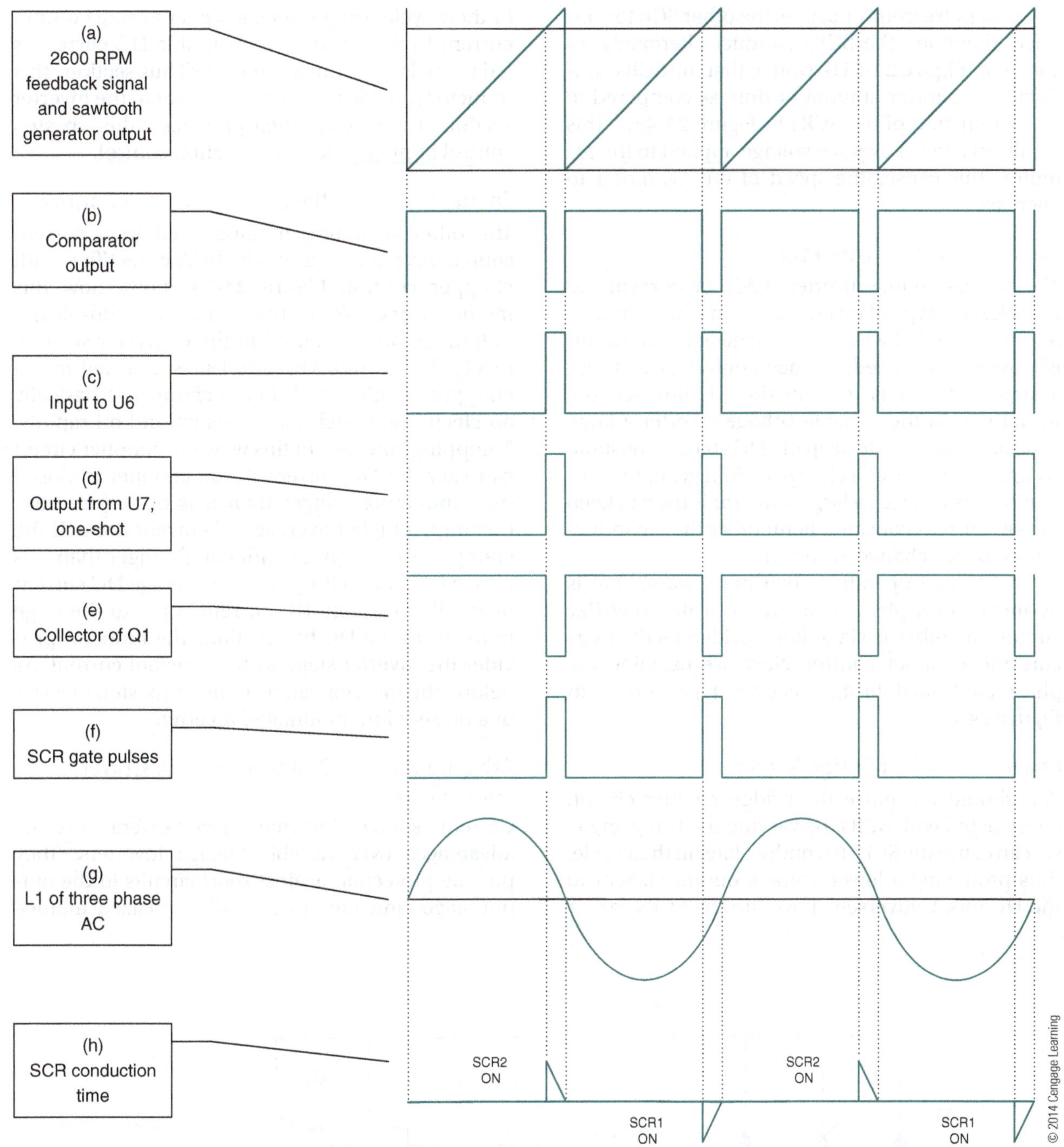

FIGURE 24-51 Waveform timing and SCR conduction time at 2600 rpm.

output of U6 trigger the one-shot, U7. The output of the one-shot, U7, appears in phase with the input pulses, as seen in **Figure 24-51D**. Notice that the output pulse of the one-shot now occurs later than it did in Figure 24-49D. This makes the SCRs fire later, causing them to conduct for a shorter period of time. Let us see how this occurs.

The output of the one-shot, U7, is fed through diode CR3 to the base of the Darlington transistor, Q1. Notice in **Figure 24-51E** that the collector of Q1 is 180° out of phase with the output of the one-shot. When Q1 conducts, current flows through the pulse transformer, T1. When current flows through pulse transformer T1, a pulse appears at the gates of SCR1 and SCR2, as shown in **Figure 24-51F**. Recall that only the SCR that is properly biased conducts. The SCR that is triggered and properly biased conducts until the AC voltage drops to the zero crossing point.

The next pulse from T1 causes the other SCR to conduct. Therefore, the SCRs conduct alternately, as shown in **Figure 24-51G**. Notice that the SCRs conduct for a shorter amount of time as compared to the conduction of the SCRs in Figure 24-49G. This results in a lower average voltage applied to the AC motor. This causes the speed of the AC motor to decrease.

Current Source Inverter

The current source inverter (CSI), or **current fed inverter**, is a type of inverter drive in which the current is controlled while the voltage is varied to satisfy the motor's needs. To accomplish these tasks, a large inductor is used in the DC link section. Recall that in the variable voltage inverter a large capacitor was used to keep the DC voltage constant. A capacitor opposes a change in voltage. In the current source inverter, a large inductor is used to keep the DC current constant. Remember that an inductor opposes a change in current.

A CSI can operate in two basic ways. One is by means of a **phase-controlled bridge rectifier** circuit; the other is via a diode bridge rectifier circuit and chopper control. Next, we examine the phase-controlled bridge rectifier type shown in **Figure 24-52**.

Phase-Controlled Bridge Rectifier

You should recognize the bridge rectifier circuit, constructed with SCRs. By varying the firing angle, we can cause the SCRs to conduct later in their cycle, thus producing a lower average output current to the DC bus. Conversely, if we fire the SCRs earlier in their cycle, we produce a higher average output current to the DC bus. The variable DC current is fed to the large inductor in the DC bus section. This inductor provides a constant current to the inverter section. The inverter then provides either six-step control or pulse-width modulation control.

Diode Bridge Rectifier with Chopper Control

The other operating method used in a current source inverter is a diode bridge rectifier with chopper control. **Figure 24-53** shows how this method works. Notice that a standard, run-of-the-mill diode bridge is used in the converter stage to rectify the AC into DC. The DC is then fed into a chopper circuit. Recall that a chopper is basically an electronic switch that turns on and off rapidly, "chopping" the DC. In this way, the chopper circuit can vary the DC current. If the chopper is closed (or conducting) longer than it is open (not conducting), a higher average DC current flows. If the chopper is open (not conducting) longer than it is closed (or conducting), a lower average DC current flows. This variable DC current is fed to the large inductor in the DC bus section. The inductor provides the inverter stage with a constant current. As before, the inverter can be either a six-step inverter or a pulse-width modulated inverter.

Advantages and Disadvantages of Current Source Inverters

Current source inverters have several distinct advantages over variable voltage inverters. They provide protection against short circuits in the output stage, and they can handle oversized motors.

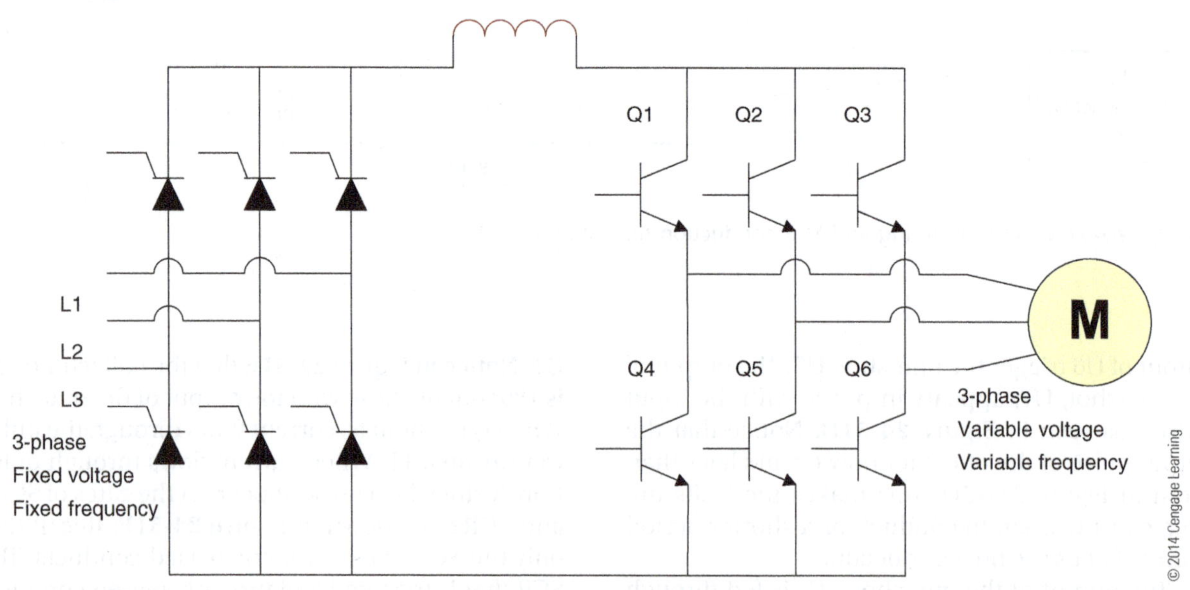

FIGURE 24-52 A current source inverter with a phase-controlled bridge rectifier.

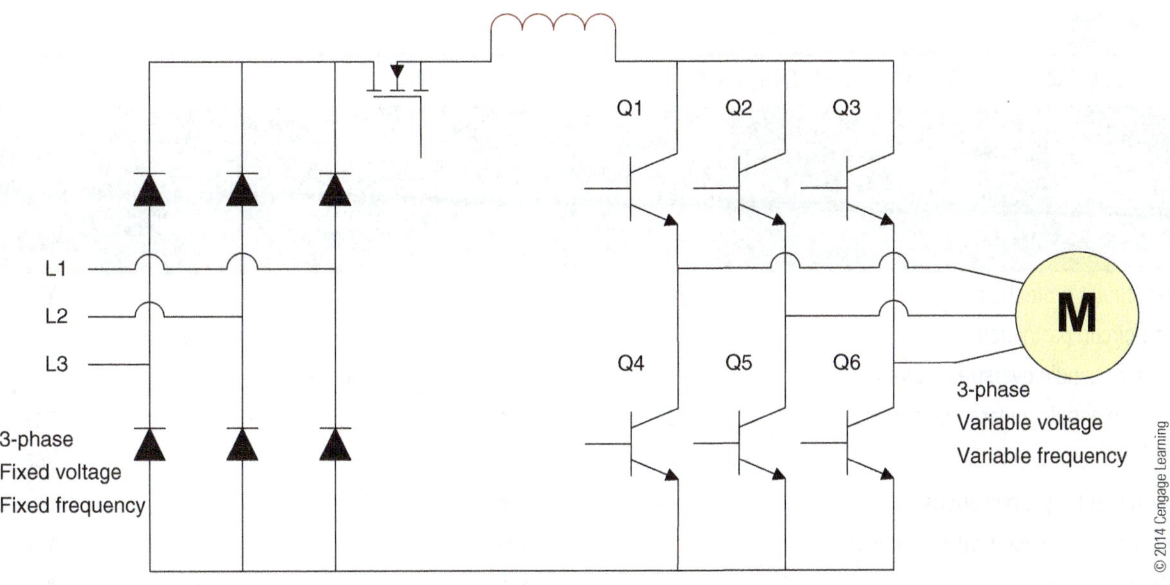

FIGURE 24-53 A current source inverter with a diode bridge rectifier and chopper control.

In addition, CSIs have relatively simple control circuits and good efficiency. As to their disadvantages, CSIs produce torque pulsations at low speed, cannot handle undersized motors, and are large and heavy. The phase-controlled bridge rectifier CSI is less noisy than its chopper-controlled counterpart, does not need high-speed switching devices, and cannot operate from batteries. The chopper-controlled CSI can operate from batteries and produces more noise as a result of its need for high-speed switching devices. Refer to the chart in **Table 24-1** for a quick comparison between the various types of variable voltage and current source inverters.

Flux Vector Drives

Because this text was written with the maintenance technician in mind, care has been taken to approach the theory behind the operation of electronic variable-speed drives in a simple, straightforward manner. However, one rather complicated type of drive, the flux vector drive, is gaining popularity. **Figure 24-54** shows a SECO model VR flux vector drive. **Figure 24-55** shows the control section of a SECO VR flux vector drive, and **Figure 24-56** shows the power section of a SECO VR flux vector drive. Due to the complexity of its nature, the theory behind this drive is somewhat beyond the scope of this text. However, for the individual exposed to flux vector drives, this section provides a basic understanding of their operation. We begin our discussion of flux vector drives with a slightly different look at general motor theory.

To explain the operation of a typical three-phase AC induction motor, we always speak in terms of the rotating magnetic field. In order to understand flux vector drives, we must delve a little deeper into the characteristics of the rotating magnetic field.

Figure 24-57 shows a typical three-phase sine wave. Notice also the vectors drawn to represent the balanced three-phase currents. (Recall that a vector represents both magnitude and direction.) Observe that these vectors are drawn 120° apart to represent the normal phase shift in a three-phase system. Note, too, that at 30° vectors, L1 and L2 are drawn with their arrowheads pointing outward, away from the center, or neutral point. Vector L3 is drawn with its arrowhead pointing inward, or toward the neutral point. The direction of these arrowheads corresponds to the polarity of the instantaneous current. If the current is in the positive portion of the sine wave, the arrowhead is drawn pointing outward, or away from the neutral. If the current is in the negative portion of the sine wave, the arrowhead will be drawn pointing inward, or toward the neutral point. Also, the length of the vectors varies to represent the magnitude of the current.

Now, look at **Figure 24-58** (see in page 602), which includes another set of drawings that show the addition of the current vectors and the resulting magnetic flux. (Recall that vectors are added by positioning them head to tail.) Notice that, as we advance from 30° to 90°, the current vectors appear to rotate in a counterclockwise direction, causing the flux to rotate in the same direction. As we continue to advance from 90° to 150°, the current vectors and the flux continue to rotate counterclockwise. If we continue to advance to 360°, the current vectors and the flux will complete

Table 24-1

INVERTER COMPARISON CHART

Features	Variable Voltage Inverter with Phase Control	Variable Voltage Inverter with Chopper Control	Current Source Inverter with Phase Control	Current Source Inverter with Chopper Control	Pulse-Width Modulation
Open-circuit protection	Yes	Yes			Yes
Short-circuit protection			Yes	Yes	
Ability to handle oversized motors			Yes	Yes	
Ability to handle undersized motors	Yes	Yes			Yes
Multiple motor applications	Yes	Yes			Yes
Low-speed torque pulsations	Yes	Yes	Yes	Yes	
Requires high-speed switching devices		Yes		Yes	Yes
Battery operation		Yes		Yes	Yes
Regenerative operation	Yes		Yes		
Low-speed efficiency	Good	Good	Good	Good	Medium
Complex control circuit	Medium	Medium	Medium	Medium	High
Size and weight	Medium	Medium	High	High	Low

FIGURE 24-54 A SECO model VR flux vector drive.

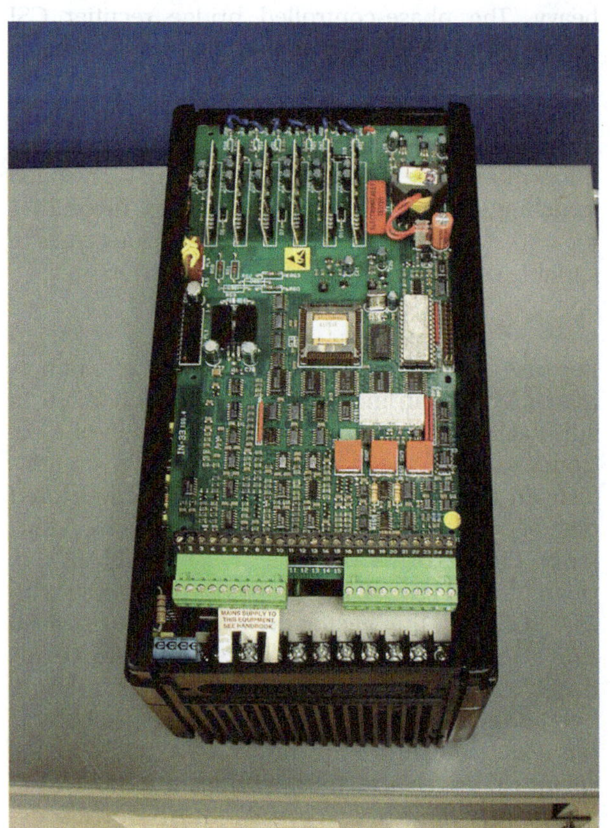

FIGURE 24-55 The control section of a SECO VR flux vector drive.

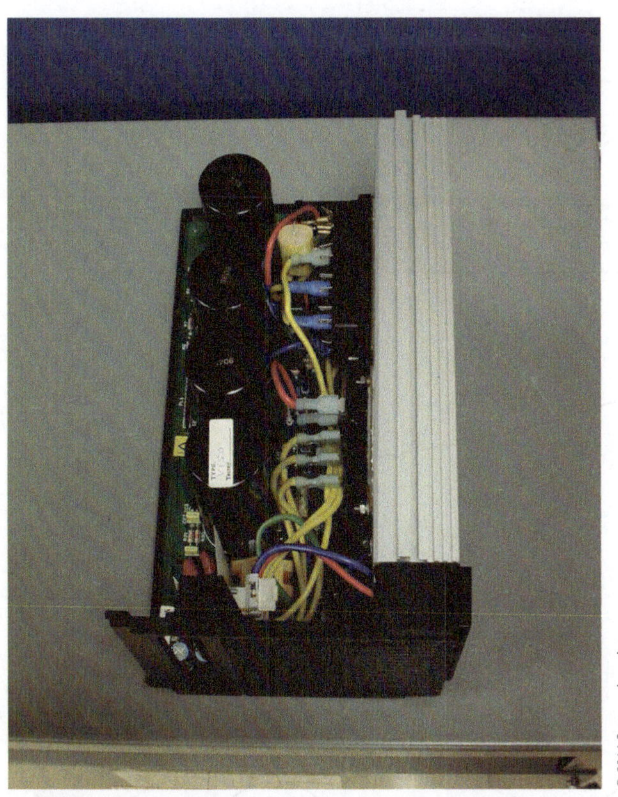

FIGURE 24-56 The power section of a SECO VR flux vector drive.

one full revolution. All of this occurs in a predictable manner because the three-phase supply is balanced and sinusoidal, and it has a constant amplitude and a constant frequency. What happens when we introduce some changes to this configuration?

Before we discuss changes to the frequency, amplitude, or phase rotation, be aware that the AC sine wave supplied to the motor is artificially produced by an inverter stage in the flux vector drive. This inverter stage allows us to exercise control over the synthetic AC and to make the changes described in the remainder of this chapter.

In **Figure 24-59**, we have caused the devices that produce the artificial AC to "jump ahead." In other words, the AC was advanced from 60° to 90° instantaneously. This change has no effect on the direction of rotation or on the strength of the magnetic flux. This does, however, cause the motor suddenly to advance 30° in rotation. Because we can control the firing sequence of the devices that produce the synthetic AC, we can also control the amount of advancement of the flux field. This jump is known as a step change.

Given this control over the devices that produce the synthetic AC, what happens when these same devices are held on for an extended period of

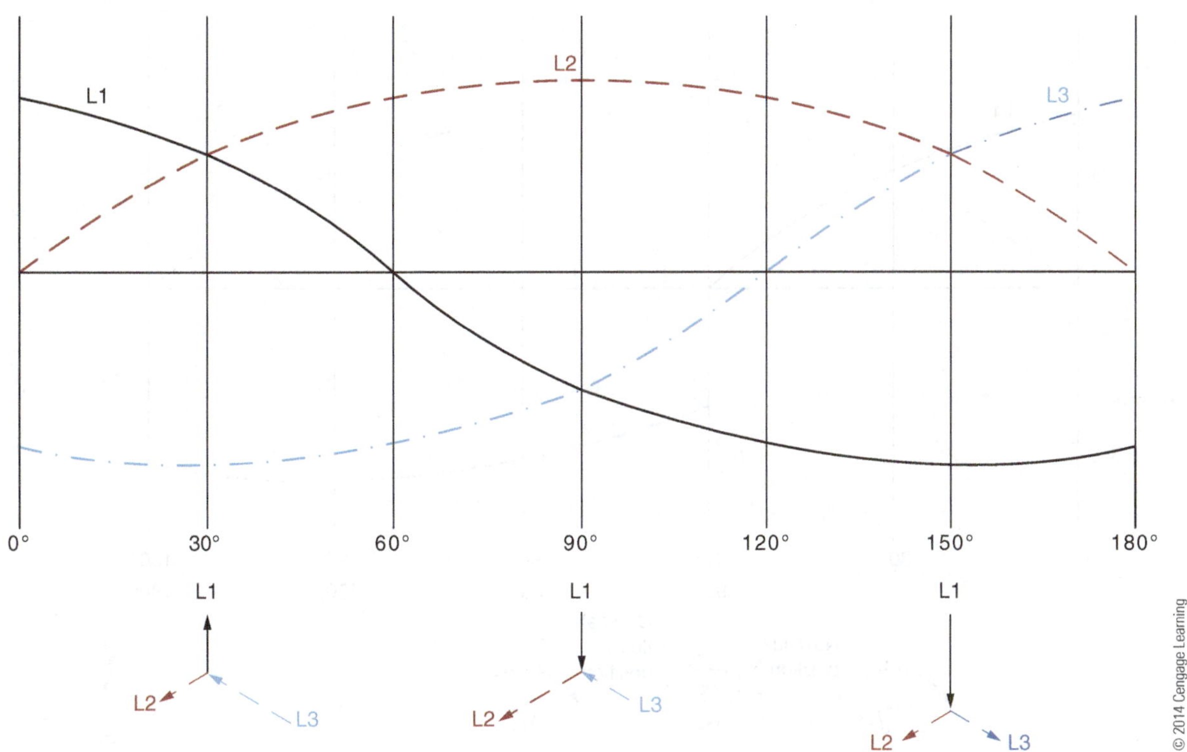

FIGURE 24-57 A typical three-phase AC with current vectors.

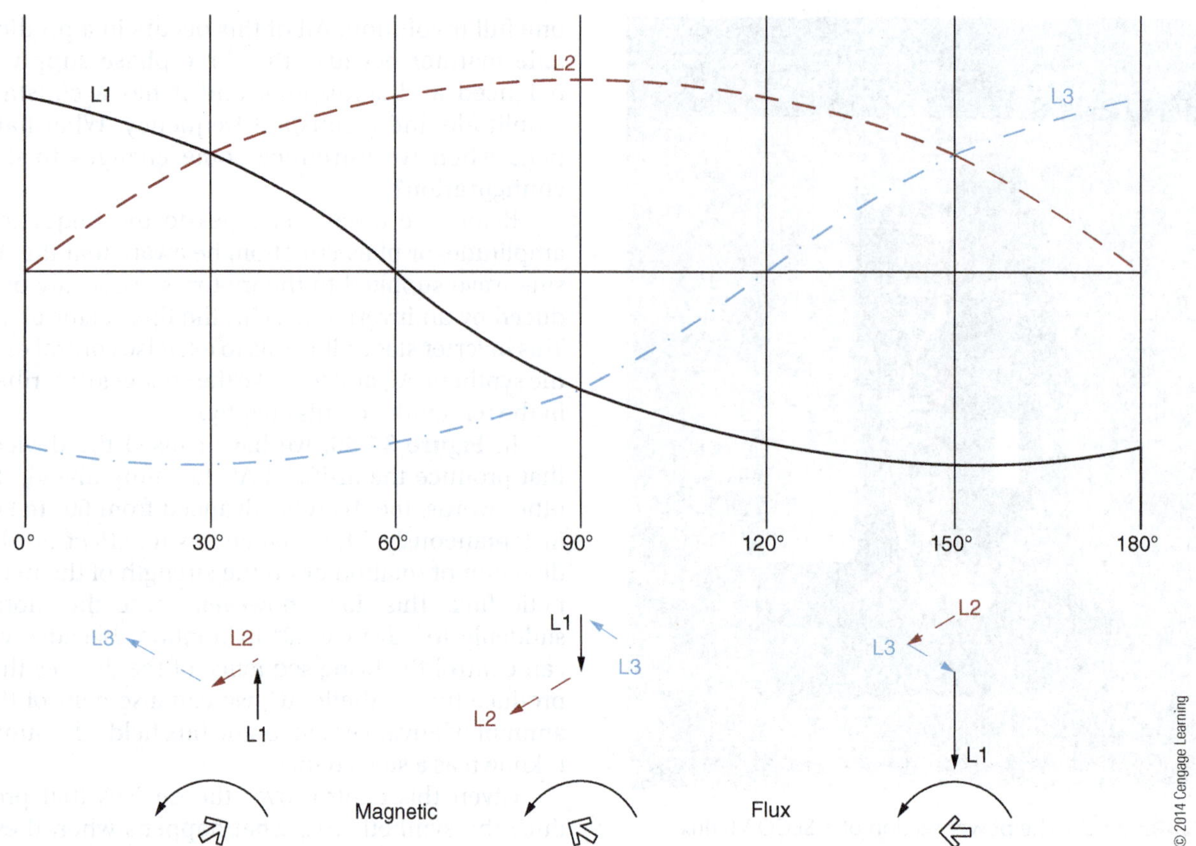

FIGURE 24-58 A typical three-phase AC showing addition of the current vectors and the resulting magnetic flux.

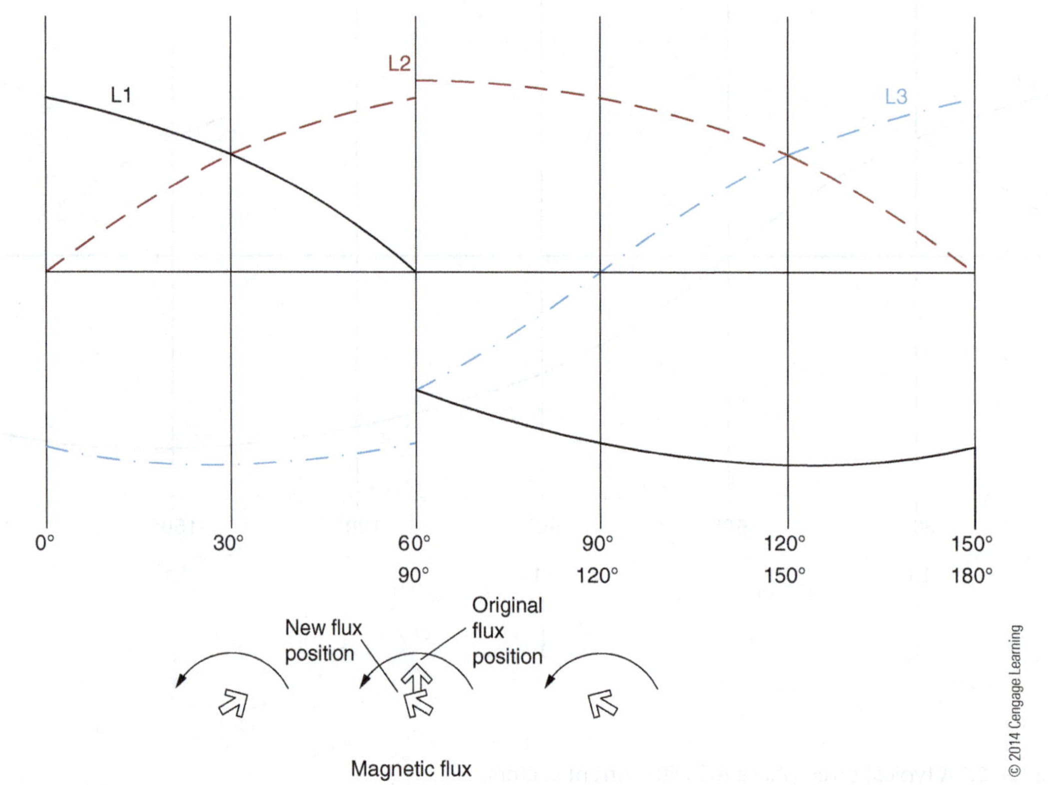

FIGURE 24-59 Change in magnetic flux position due to a step change.

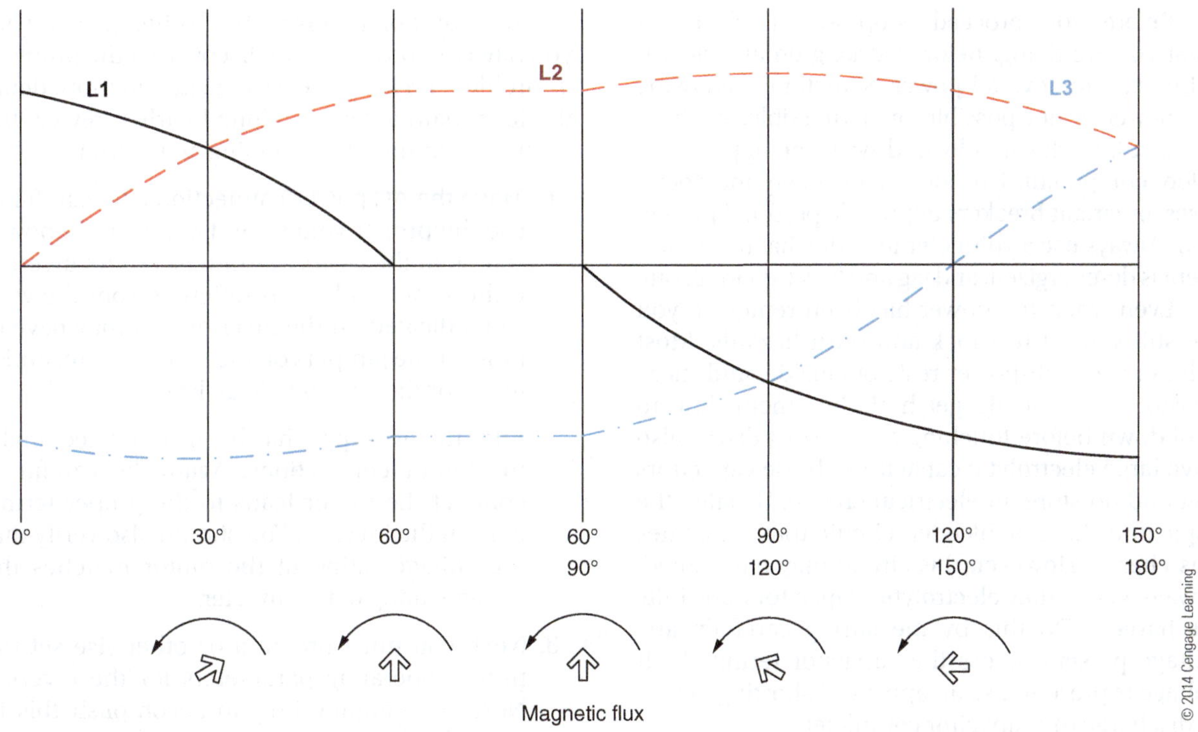

FIGURE 24-60 Magnetic flux held constant for 30°.

time? Referring to **Figure 24-60**, notice that at the 60° point the step change discussed previously does not occur, nor has the current to advance normally. In this instance, the current was held constant for 30°. As a result, the flux rotation was stopped or held stationary for the 30° period. You should also recognize that by taking advantage of our control over the AC, we can change the phase rotation and thus the direction of rotation of the motor as well.

Flux vector drives operate by monitoring rotor position and changing the rotor's position as compared to its orientation to the stator field. To know the rotor's position, encoders are typically used as feedback devices. In addition, current feedback from two of the three phases is used. This amount and variety of feedback is necessary to achieve the variety of control mechanisms that a flux vector drive offers. For instance, if we vary the amplitude of the current, we vary the magnitude of the current vectors. If we vary the phase rotation, we vary the direction of rotation. Causing a step change to occur results in an instantaneous advance in motor position. Finally, if we hold the phase current constant, we cause the motor to hold a constant position.

The flux vector drive must constantly monitor the motor's performance and allow for high-speed corrections to occur to maintain accuracy in motor positioning and performance. For this reason, flux vector drives are matched to the particular motor that they operate. Although this makes flux vector drives more expensive, the resulting improved performance in the areas of rapid response and precise control makes these drives well worth their price.

24-4 TROUBLESHOOTING INVERTER DRIVES

In an AC inverter drive, the four main areas of possible problems are the same as those in a DC drive:

1. The electrical supply to the motor and the drive
2. The motor and/or its load
3. The feedback device and/or sensors that provide signals to the drive
4. The drive itself

Even though the problem may be in any one or several of these areas, the best place to begin troubleshooting is the drive unit itself. The reason is that most drives have some type of display that aids in troubleshooting. This display may be simply an LED that illuminates to indicate a specific fault condition, or it may be an error or fault code that can be looked up in the operator's manual. For this reason, it is strongly suggested that a copy of the fault codes be made and fastened to the inside of the drive cabinet, where it will be readily available to the maintenance technician. The original should be placed in a safe location such as the maintenance supervisor's office.

Before you proceed, stop and think about what you are doing. Before working on any electrical circuit, remove all power. Sometimes removing the power is not possible or permissible. In these instances, work carefully and wear the appropriate safety equipment. Do not rely on safety interlocks, fuses, or circuit breakers to provide personal protection. Always use a voltmeter to verify that the equipment is de-energized, and tag and lock the circuit out!

Even when the power has been removed, you are still subject to shock and burn hazards. Most drives have high-power resistors inside, and these resistors can and do get hot! Give them time to cool down before touching them. Most drives also have large electrolytic capacitors. These capacitors can and do store an electrical charge. Usually, the capacitors have a bleeder circuit that dissipates this charge. However, this circuit may have failed. Always verify that electrolytic capacitors are fully discharged. Do this by measuring carefully any voltage present across the capacitor terminals. If voltage is present, use an approved shorting device to discharge the capacitor completely.

Now, slow down, take your time, and use your senses, even though production has stopped and you are under pressure to fix the equipment. Sometimes a little extra initial time to take stock of the situation can save considerable time later. So do the following:

Look! Do you see any charred or blackened components? Have you noticed any arcing? Do fuses or circuit breakers appear blown or tripped? Do you see any discoloration around wires, terminals, or components? A good visual inspection can save a lot of troubleshooting time.

Listen! Did you hear any funny or unusual noises? A frying or buzzing sound may indicate arcing. A hum may be normal or an indication of loose laminations in a transformer core. A rubbing or chafing sound may indicate that a cooling fan is not rotating freely.

Smell! Do you notice any unusual odors? Burnt components and wires give off a distinctive odor. Metal smells hot if subjected to too much friction.

Touch! (Be very careful with this!) If components feel cool, that may indicate that no current is flowing through the device. If components feel warm, chances are that everything is normal. If components feel hot, things may be normal, although more than likely too much current is flowing, or too little cooling is taking place. In any event, there may be a problem worth further investigation.

The point of all this is that by being observant you have a good chance of discovering the problem or problem area. Before we consider in more detail the four main areas mentioned earlier, review the proper techniques for installing an inverter.

1. Make the AC power connections. Be careful to use the proper voltage and connect the power source to the correct terminals on the inverter. If the supply voltage is different from the voltage indicated on the inverter, you may have to move some jumpers on the inverter to reconfigure it for the proper voltage levels.
2. After the AC supply has been connected, make the motor connections. Again, be careful to connect the motor leads to the proper terminals on the inverter. You should also verify that the voltage rating of the motor matches the voltage rating of the inverter.
3. Next, you must program or otherwise set the proper operating parameters for the inverter. Next, we examine how to accomplish this in more detail.
4. Depending on the model, you may need to set the inverter for remote control (start/stop/adjust speed from a remote location) or local control (start/stop/adjust speed from the inverter's control panel).
5. If you are using remote control, you must wire the control circuit. You also have to wire the speed/torque control potentiometer.

The preceding instructions are general guidelines for installing an inverter. Check the manual that comes with the particular inverter that you are using and follow the procedures presented there precisely.

Next, we examine some of the parameters that can be programmed into an inverter. Keep in mind that not all inverters share these parameters. Some manufacturers list the same parameters in different order, and certain manufacturers have other names for the settings discussed here. Also, keep in mind that some inverters have more parameters and others have fewer parameters than those mentioned here. Again, familiarize yourself with the particular inverter that you are using. In general, programmable inverter parameters include the following:

analog input select Sets either a 0–10 V or 4–20 mA reference that is proportional to the frequency input.

analog output select Provides a 4–20 mA output signal that is proportional to the motor speed.

basic set-up Can be set for constant torque or constant speed.

current The output current of the inverter.

current limit Sets the maximum current available to the motor. If the setting is at the maximum permissible value, the motor has maximum starting torque. This value can be on the order of 160% of nominal motor current. If the current limit is set too low, the inverter can trip out.

DC brake time If selected, this parameter will provide additional braking torque at low motor speeds.

digital input select If selected, this value bypasses the ramp time down setting, and the inverter decelerates the motor in the shortest possible amount of time as a result.

frequency The output frequency of the inverter.

jogging speed Sets the speed of the motor when jogging.

local/remote Can be set for local control from the inverter's control panel or remote control from a start/stop station located away from the inverter cabinet.

maximum speed Depending on the setting, it may be possible to attain a speed higher than the rated speed of the motor. This parameter must be set higher than the minimum speed setting. If it is set lower than the minimum speed, the motor will not run.

minimum speed Slowest speed setting at which the motor will run.

Motor magnetization Set to the no load current rating on the motor's nameplate.

Motor current Set to the full load current nominal rating on the motor's nameplate.

Motor nominal frequency The rated frequency of the motor from the motor nameplate. This value should be set as closely as possible to the specified value.

Motor nominal voltage The rated line voltage of the motor from the motor nameplate. This value should be set as closely as possible to the specified value.

Motor power This is the motor's power rating expressed either in horsepower (Hp) or in kilowatts (kW). Some drives will accept the value in either unit of measurement while others require converting the units from one measurement to the other.

ramp time down This is the deceleration time (the time required to get from maximum speed to minimum speed) expressed in seconds. If this time is set too short, the inverter can trip out.

ramp time up This is the acceleration time (the time required to get from minimum speed to maximum speed) expressed in seconds. If this time is too short, the inverter can trip out.

relay output select Provides a contact closure when the inverter is placed in the run mode.

slip compensation Typically, the factory setting for this value should be adequate. This setting is affected by the motor power, motor nominal voltage, and motor nominal frequency values.

start compensation Typically, the factory setting for this value should be adequate. This setting is affected by the motor power, motor nominal voltage, and motor nominal frequency values.

start voltage Typically, the factory setting for this value should be adequate. This setting is affected by the motor power, motor nominal voltage, and motor nominal frequency values.

start–stop mode This value is programmed for the various types of start–stop circuits used. For example, a two-wire start–stop, a three-wire start–stop, a three-wire start–stop with a jog, and so on.

thermal motor protect Depending on the setting chosen, this parameter will either flash the display when the motor's critical temperature is reached or trip the inverter.

torque Calculated motor torque. Dependent on the programmed settings of the motor nominal current and the motor magnetization.

trip reset mode If selected, this parameter will prevent the inverter from restarting automatically after a trip.

V/f ratio Typically, the factory setting for this value should be adequate. This setting is affected by the motor power, motor nominal voltage, and motor nominal frequency values.

voltage The output voltage of the inverter.

In order to begin troubleshooting, we need to understand the general, step-by-step sequence of events that most inverter drive systems follow on start-up and shutdown. We begin with the AC power applied to the inverter. We must adjust the reference or set point for the desired speed and torque characteristics of the motor. Next the start–stop circuit is placed in the start or run mode. Instantly, the main control components begin a diagnostic routine. We will examine some of the routine's fault codes shortly. If no faults are detected, the driver activates the power semiconductors. These produce the output frequency and V/f ratio programmed into the inverter to match the speed and torque settings. The programmed setting for the ramp time up controls how long it takes the motor to reach the desired speed. While the motor speed is ramping up, the motor current is monitored. Should the current exceed the programmed current limit setting, the inverter may (depending on the manufacturer) automatically adjust the ramp time up program or simply trip. If tripping occurs repeatedly, the ramp time up program may need to be modified to accommodate a longer acceleration time. The ramp time up setting causes the inverter output voltage and frequency to increase, accelerating the motor to the desired speed set point.

When it becomes necessary to stop the motor, most inverters offer several options. One option is basically to let the motor coast to a stop. This can take a very long time for high-inertia loads. The ramp down time setting allows the inverter to slow the motor gradually. This deceleration is accomplished by allowing the motor to feed its self-generated energy back into the inverter. The inverter uses large resistors to absorb this energy. This process, called dynamic braking, is usually insufficient to bring the motor to a controlled and rapid stop, because as the motor slows, less energy is self-generated. Therefore, it is common practice to use a mechanical brake in conjunction with dynamic braking to provide additional braking action at slow speeds. Another method of stopping the motor without using a mechanical brake is plugging, or DC injection. When the operator wishes to stop the motor, DC current is fed into the motor winding. This current replaces the rotating magnetic field with a fixed magnetic field. The rotor soon becomes locked with this fixed magnetic field, which, in effect, stops the motor rotation. This method should not be used repeatedly because heat can build up in the motor and damage the windings. Now that we have a basic understanding of the processes that an inverter follows on start-up and shutdown, let us investigate the types of problems that can be encountered in the four main areas of an inverter drive system.

The Electrical Supply to the Motor and the Drive

Most maintenance technicians believe that the power distribution in an industrial environment is reliable, stable, and free of interference. Nothing could be further from reality! Frequent outages, voltage spikes and sags, and electrical noise are normal operating occurrences. The effect of these is not as detrimental to motor performance as it can be to the operation of the drive itself. Most AC inverters are designed to operate despite variations in supply voltages. Typically the incoming power can vary as much as ±10% with no noticeable change in drive performance. However, in the real world, it is not unusual for power line fluctuations to exceed 10%. These fluctuations may occasionally cause a controller to trip. If tripping occurs repeatedly, a power line regulator may be required to hold the power at a constant level.

A power line regulator is of little use, however, should the power supply to the controller fail. In this situation, a UPS (uninterruptible power supply) is needed. Several manufacturers produce a complete power line conditioning unit. These units combine a UPS with a power line regulator.

Quite often, controllers are connected to an inappropriate supply voltage. For example, it is not unusual for a drive rated 208 V to be connected to a 240 V supply. Likewise, a 440 V rated drive may be connected to a 460 V or even a 480 V source. Usually, the source voltage should not exceed the voltage rating of the drive by more than 10%. For a drive rated at 208 V, the maximum supply voltage is 229 V ($208 \times 10\% = 20.8 + 208 = 229$). Obviously, the 208 V drive, when connected to the 240 V supply, is receiving excess voltage and should not be used. For our 440 V drive, the maximum supply voltage is 484 V ($440 \times 10\% + 440 = 484$). Although this value appears to fall within permissible limits, another potential problem exists. Suppose that the power line voltage fluctuates by 10%. If the 440 V source suffers a 10% spike, the voltage increases to 484 V. This value is within the design limits of the drive. But what happens when we connect the 440 V drive to a 460 V or a 480 V power line? If we experience that same 10% spike, the 460 V line increases to 506 V ($460 \times 10\% = 46 + 460 = 506$), and the 480 V line increases to 528 V ($480 \times 10\% = 48 + 480 = 528$). We have thus exceeded the voltage rating of the drive and probably damaged some internal components! Most susceptible to excess voltage and spikes or transients are the SCRs, MOSFETs, and power transistors. Premature failure of capacitors can also occur. As you can see, it is very important to match the line voltage to the voltage rating of the drive.

An equally serious problem occurs when the phase voltages are unbalanced. Typically, during construction care is taken to balance the electrical loads on the individual phases. As time goes by and new construction and remodeling occurs, it is not unusual for the loading to become imbalanced, causing intermittent tripping of the controller and perhaps premature failure of components. To determine whether an imbalanced phase condition exists, you must do the following:

1. Measure and record the phase voltages (L1 to L2, L2 to L3, and L1 to L3).
2. Add the three voltage measurements from step 1, and record the sum of all phase voltages.
3. Divide the sum from step 2 by 3, and record the average phase voltage.
4. Now, subtract the average phase voltage obtained in step 3 from each phase voltage measurement in step 1, and record the results. (Treat any negative answers as positive answers.) These values are the individual phase imbalances.
5. Add the individual phase imbalances from step 4, and record the total phase imbalance.
6. Divide the total phase imbalance from step 5 by 2, and record the adjusted total phase imbalance.
7. Now, divide the adjusted total phase imbalance from step 6 by the average phase voltage from step 3, and record the calculated phase imbalance.
8. Finally, multiply the calculated phase imbalance from step 7 by 100, and record the percent of total phase imbalance.

Consider an example involving a 440 V, three-phase supply to an AC inverter drive to see how this process works:

1. L1 to L2 = 432 V; L2 to L3 = 435 V; and L1 to L3 = 440 V.
2. The sum of all phase voltages equals 432 V + 435 V + 440 V, or 1307 V.
3. The average phase voltage is equal to 1307 V ÷ 3, or 435.7 V.
4. To find the individual phase imbalances, we subtract the average phase voltage from the individual phase voltages and treat any negative values as positive. So L1 to L2 = 432 V–435.7 V, or 3.7 V, L2 to L3 = 435 V–435.7 V, or 0.7 V; and L1 to L3 = 440 V–435.7 V, or 4.3 V.
5. Now, we find the total phase imbalances by adding the individual phase imbalances: 3.7 V + 0.7 V + 4.3 V = 8.7 V.
6. To find the adjusted total phase imbalance, we divide the total phase imbalance by 2: 8.7 V ÷ 2 = 4.35 V.
7. Next, we find the calculated phase imbalance by dividing the adjusted total phase imbalance by the average phase voltage: 4.35 V ÷ 435.7 V = 0.00998.
8. Finally, we multiply the calculated phase imbalance by 100 to find the% total phase imbalance: 0.00998 × 100 = 0.998%.

In this example, we are within tolerances, and the differences in the phase voltages should not cause any problems. In fact, as long as the% total phase imbalance does not exceed 2%, we should not experience any difficulties as a result of the differences in phase voltages.

The Motor or Its Load

Probably the most common cause of motor failure is heat. Heat can occur simply as a result of the operating environment of a motor. Many motors are operated in areas of high ambient temperature. If steps are not taken to keep the motor within its operating temperature limits, the motor will fail. Some motors have an internal fan that cools the motor. If the motor is operated at reduced speed, this internal fan may not turn fast enough to cool the motor sufficiently. In these instances, an additional external fan may be needed to provide additional cooling to the motor. Typically, such fans are interlocked with the motor operation in such a way that the motor will not operate unless the fan operates as well. Therefore, a fault in the external fan control may prevent the motor from operating.

The sensors used to sense motor temperatures consist simply of a nonadjustable thermostatic switch that is normally closed and opens when the temperature rises to a certain level. In this event, you must wait for the motor to cool down sufficiently before resetting the temperature sensor and restarting the motor.

Periodic inspection of the motor and any external cooling fans is strongly recommended. The fans should be checked for missing or bent vanes. All openings for cooling should be kept free of obstructions. Any accumulation of dirt, grease, or oil should be removed. If filters are used, these must be cleaned or replaced on a routine schedule.

Heat may also cause other problems. When motor windings become overheated, the insulation on the wires may break down. This breakdown may cause a short, which may lead to an open condition. A common practice used to find shorts or opens in motor windings is to megger the windings with a megohmmeter. Extreme caution must be taken when using a Megger on the motor leads. Be certain that you have disconnected the motor leads from the drive. Failure to do this causes the Megger to apply a high voltage to the output section of the drive, resulting in damage to the power semiconductors. You may also decide to megger the motor leads at the drive cabinet. Again, be certain that you have disconnected the leads from the drive unit, and megger only the motor leads and motor winding. Never megger the output of the drive itself!

Because the motor drives a load of some kind, it is also possible for the load to create problems. The drive may trip out if the load causes the motor to draw an excessive amount of current for too long a time. When this occurs, most drives display some type of fault indication. The problem may be a result of the motor operating at too high a speed. Quite often, a minor reduction in speed is all that is necessary to prevent repeated tripping of the drive. The same effect occurs if the motor is truly overloaded. Obviously, in this case, either the motor size needs to be increased or the size of the load decreased to prevent the drive from tripping.

Some loads have a high inertia. They require not only a large amount of energy to move, but once moving, a large amount of energy to stop. If the drive cannot provide sufficient braking action to match the inertia of the load, the drive may trip, or overhaul. A drive with greater braking capacity is needed in such cases to prevent tripping from recurring.

The Feedback Device and Sensors

Mechanical vibration may cause the mounting of feedback devices to loosen and their alignment to vary. Periodic inspections are necessary to verify that these devices are aligned and mounted properly.

It is also important to verify that the wiring to these devices is in good condition and that the terminations are clean and tight. Another consideration regarding the wiring of feedback devices is electrical interference. Feedback devices produce low-voltage/low-current signals that are applied to the drive. If the signal wires from these devices are routed next to high-power cables, interference can occur. This interference may result in improper drive operation. To eliminate the possibility that this will occur, several steps must be taken. The signal wires from the feedback device should be installed in their own conduits. Do not install power wiring and signal wiring in the same conduit. The signal wires should be shielded cable, with the shield wire grounded to a good ground at the drive cabinet only. Do not ground both ends of a shielded cable. When the shielded signal cable is routed to its terminals in the drive cabinet, the cable should not be run or bundled parallel to any power cables, but instead at right angles to such power cables. Furthermore, the signal cable should not be routed near any high-power contactors or relays. When the coils of a contactor or relay are energized and de-energized, a spike is produced. This spike can create interference with the drive. To suppress this spike, it may be necessary to install a freewheeling diode across any DC coils or a snubber circuit across any AC coils.

The Drive

First, look for fault codes or fault indicators. Most drives provide some form of diagnostics, and this can be a great time-saver. Next, we look at some of the fault codes, symptoms, probable causes, and fixes for some common problems:

- The inverter is inoperable, and no LED indicators are illuminated. Possibly no incoming power is present. You can verify this by measuring the voltage at the power supply input terminals in the inverter cabinet. The problem may be caused by a blown fuse, an open switch, an open circuit breaker, or an open disconnect. There are several things to check if the fuses are blown. One item to look for is a shorted metal-oxide varistor (MOV), a device that provides surge protection to the inverter. If a significant power surge occurred, the MOV may have shorted to protect the inverter. Because the MOV is located across the power supply lines (to provide protection), a shorted MOV can cause fuses to blow. Another reason why fuses blow is a shorted diode in the rectifier circuit. A shorted or leaky filter capacitor in the power supply may also cause fuses to blow.
- The inverter is powered up but does not work. There are indications of a fault condition. The "watchdog" circuit may have tripped. Remember that a watchdog circuit monitors the power lines for disturbances, and if the disturbance occurs for a long duration, the watchdog circuit trips the controller to protect it from damage. A heavy starting load may sag the power line voltage to such a point that the inverter receives insufficient voltage. This deficiency can trip the inverter. Likewise, if the load has high inertia, it is possible for the regenerative effect to provide excess voltage to the inverter. Another way that the inverter may receive excess voltage is use of the power factor correction capacitors while the load is removed.
- If the watchdog circuit has not tripped, other possible reasons exist for this fault condition. It is possible that there are interlocks on the cabinet or cooling fans, and one or more of these may be open. Likewise, a temperature sensor on a heatsink on the power semiconductors in the inverter or in the motor itself may have detected an excessive temperature condition and opened as a result.

- The inverter is energized, and a fault is indicated. The motor does not respond to any control signals. If the load is too high for the motor settings, the motor may fail to rotate. It may be necessary to increase the current limit or voltage boost settings to allow the motor to overcome the load. Another possibility is that the load is overhauling the motor. If this is the situation, it will be necessary to adjust the deceleration time to allow the motor to take longer to brake the load. It may also be necessary to add auxiliary braking in the form of a mechanical brake. If the motor leads have developed a short or the motor itself has a shorted winding or is overloaded or stalled, the current limit sensor may trip the inverter. Tripping may also occur as a result of a shorted power semiconductor.

As you can see, you need to be aware of many areas when dealing with an inoperable inverter. Fortunately the inverter itself can help a great deal by displaying fault codes. The operator's manual interprets the fault codes and gives instructions for clearing the fault condition. Next, we examine some fault codes and how they can help in your troubleshooting. Remember that not all manufacturers provide the same fault codes—some provide more, and others provide fewer.

As mentioned earlier, one fault that may be displayed indicates an overcurrent limit condition. This indicator should direct you to examine the motor for mechanical binding, jams, and so forth. To verify whether one of these conditions is the cause of the problem, disconnect the motor from the load and reset the inverter. If the fault clears, then you know that the load is the cause. If the fault reappears, you need to look further, perhaps at the motor itself.

Another fault code, overvoltage, may be the result of a high-inertia load that causes overhauling. This fault code may also be a result of setting the deceleration ramp down parameter for rapid deceleration. Lengthening the deceleration time may clear the fault. If lengthening the time is not possible, additional mechanical braking may be required to bring the load to a rapid stop.

The inverter overload fault code is an indication of electrical problems. Examples of these are shorted or grounded motor leads or windings and/or defective power semiconductors. If the motor is suspect, disconnect it from the inverter. If the fault clears when you reset the inverter, you can assume that the problem lies in the motor and/or its leads. To verify whether the problem is in the power semiconductors, disconnect the gate lead from one of the devices, and reset the inverter. If the fault clears, you have found the problem. If the fault is still present, reconnect the gate lead, move to the next device, and disconnect its gate lead. Reset the inverter. Again, if the fault clears, you have found the problem. If the fault is still present, repeat the preceding steps until you have tested all of the power semiconductors.

Another fault code indicates shorted control wiring. If this code is displayed, simply disconnect the control wiring and reset the inverter. The fault should clear. This result indicates problems in the control wiring. If the fault does not clear, try unplugging the control board and resetting the inverter. A cleared fault condition in this case indicates problems in the control board.

It is a good idea to maintain an inventory of spare PC boards for the various inverters at your plant. That way, if you determine that the problem is caused by a defective PC board, it should be a fairly simple matter to replace the board with a spare. This will minimize downtime and allow you to try to repair the PC board in the shop, under a lot less pressure. If the PC board cannot be repaired, it may be possible to return it to the manufacturer for repair or exchange.

Although fault codes are excellent troubleshooting tools and can save a great deal of time, you should be aware of other potential problems that may or may not show up as fault codes, depending on the manufacturer.

Heat can produce problems in the drive unit. The cabinet may have one or more cooling fans, with or without filters. These fans are often interlocked with the drive power in such a way that the fan must operate in order for the drive to operate. Make certain that the fans are operational and the filters are cleaned or replaced regularly. The power semiconductors are typically mounted on heatsinks. A heatsink may have a small thermostat mounted on it to detect an excess temperature condition in the power semiconductors. If the heatsink becomes too hot, the thermostat will open, and the drive will trip. Usually these thermostats are self-resetting. You must wait for them to cool down and reset themselves before the drive will operate. If tripping occurs repeatedly, a more serious problem exists that requires further investigation.

If the drive is newly installed, problems are often the result of improper adjustments to the drive. On the other hand, if the drive has been in operation for some time, it is unlikely that readjustments are needed. All too often, an untrained individual tries to adjust a setting to see whether this fixes it. Usually, such adjustments only make things worse! This is not to say that adjustments are never needed. For example, if the process being controlled is changed or some component of the equipment has been replaced, it probably will be necessary to change the drive settings. For this reason, it is very important to record the initial settings and any changes made over the years. This record should be placed in a safe location and a copy made and placed in the drive cabinet for easy access by maintenance personnel.

Replacing the drive should be the final choice. More often than not, if the drive is defective, something external to the drive is the reason for the drive's failure. Replacing the drive without determining what caused the failure may result in damage to the replacement drive. If the drive has failed, the possibility still exists that you can get it to work again.

Most drive failures occur in the power section, where you find the power SCRs, transistors, MOSFETs, and so on. In some drives, these devices are individual components. You can test these devices with reasonable accuracy using nothing more than an ohmmeter. In other drives, the SCRs, transistors, and MOSFETs are contained within a power module.

An SCR power module is shown in **Figure 24-61**. One of these modules would be used for each phase of the three-phase AC. Notice that this module contains two SCRs. We will call the SCR to the left "A" and the SCR to the right "B." You can use an ohmmeter to test an SCR module with reasonable accuracy. To do so, you must understand what each terminal of the module represents. A schematic diagram of the SCR module is shown in **Figure 24-62**.

Notice in Figure 24-61 that there are a total of five terminals on the SCR module. Referring to Figure 24-62, you see the same five terminals. Two of the terminals (the small terminals in Figure 24-61), represent the gate connections for each SCR.

© 2014 Cengage Learning

FIGURE 24-61 An SCR power module.

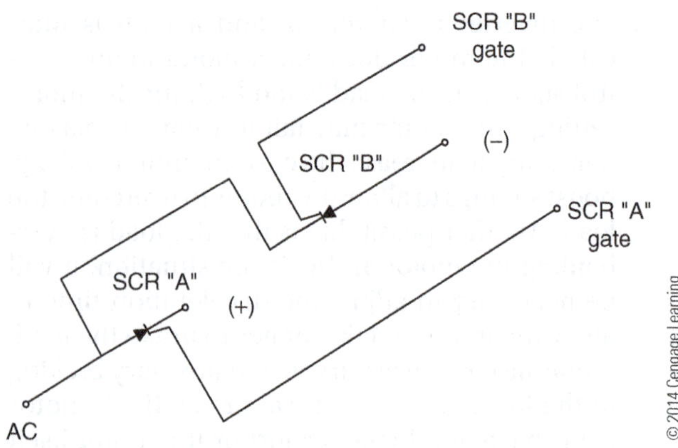

© 2014 Cengage Learning

FIGURE 24-62 An SCR power module schematic.

The gate terminal toward the back of the module is connected to the gate of SCR B, while the gate terminal at the front of the module is connected to the gate of SCR A. The large terminal closest to the gate terminals is the negative terminal of the module, which is also the anode of SCR B. The large terminal farthest from the gate terminals is the AC terminal for the module, which is also the anode of SCR A. Notice that the anode of SCR A is internally connected to the cathode of SCR B. The large terminal in the center is the positive terminal of the module, which is also the cathode of SCR A.

To test the module with an ohmmeter, follow these steps:

1. Place the ohmmeter in the Ohms position, if using a digital multimeter. (Use the 200 position if using an analog multimeter.)
2. Place the black (negative) lead of your ohmmeter on the anode (terminal farthest from the gate terminals) of SCR A.
3. Place the red (positive) lead of your ohmmeter on the cathode (center terminal) of SCR A.
4. Your meter should indicate a very high or infinite resistance. A low-resistance reading indicates a faulty SCR. Replace the module.
5. Reverse your meter connections. Place the black (negative) lead of your ohmmeter on the cathode (center terminal) of SCR A.
6. Place the red (positive) lead of your ohmmeter on the anode (terminal farthest from the gate terminals) of SCR A.
7. Your meter should indicate a very high or infinite resistance. A low-resistance reading indicates a faulty SCR. Replace the module.
8. While leaving your ohmmeter connected as in steps 5 and 6, use a clip lead to connect the gate

of SCR A (small terminal at the front of the module) to the red (positive) lead of your ohmmeter.

9. You should notice a drop in the resistance reading. This is a result of your triggering SCR A into conduction. If your resistance reading does not drop, the SCR may be faulty, and you should replace the module. However, it is possible that your ohmmeter is not supplying sufficient current to trigger the SCR into conduction, and the SCR may be functioning normally. The old saying, "If in doubt, change it out" would apply.

If SCR A appears normal, repeat the same process to check SCR B.

1. Place the black (negative) lead of your ohmmeter on the anode (terminal closest to the gate terminals) of SCR B.
2. Place the red (positive) lead of your ohmmeter on the cathode (terminal farthest from the gate terminals) of SCR B.
3. Your meter should indicate a very high or infinite resistance. A low-resistance reading indicates a faulty SCR. Replace the module.
4. Reverse your meter connections. Place the black (negative) lead of your ohmmeter on the cathode (terminal farthest from the gate terminals) of SCR B.
5. Place the red (positive) lead of your ohmmeter on the anode (terminal closest to the gate terminals) of SCR B.
6. Your meter should indicate a very high or infinite resistance. A low-resistance reading indicates a faulty SCR. Replace the module.
7. While leaving your ohmmeter connected as in steps 5 and 6, use a clip lead to connect the gate of SCR B (small terminal at the back of the module) to the red (positive) lead of your ohmmeter.
8. You should notice a drop in the resistance reading. This is a result of your triggering SCR B into conduction. If your resistance reading does not drop, the SCR may be faulty, and you should replace the module. However, it is possible that your ohmmeter is not supplying sufficient current to trigger the SCR into conduction, and the SCR may be functioning normally. Again, "If in doubt, change it out"

Figure 24-63 shows a transistor power module. Notice that this module has 17 terminals, 5 large and 12 small. Figure 24-64 shows the schematic diagram of the transistor power module. This module may look intimidating at first glance, but notice that the module consists simply of six transistors connected to perform an inverter function.

FIGURE 24-63 A transistor power module.

Looking at both Figure 24-63 and **Figure 24-64**, notice the row of six small terminals at the top of the module. Three of these terminals (BU, BV, and BW) are connected to the bases of transistors Q1, Q2, and Q3. The remaining three terminals (EU, EV, and EW) are connected to the emitters of the same transistors. Now, notice the row of six small terminals at the bottom of the module. Three of these terminals (BX, BY, and BZ) are connected to the bases of transistors Q4, Q5, and Q6. The remaining three terminals (EX, EY, and EZ) are connected to the emitters of the same transistors. You also see two large terminals at the left end of the module. The large terminal toward the top of the module is the positive (+) DC input to the module. This terminal is internally connected to the collectors of transistors Q1, Q2, and Q3. The large terminal toward the bottom of the module is the negative (–) DC input to the module. This terminal is internally connected to the emitters of transistors Q4, Q5, and Q6. Finally, you should see three large terminals in the center of the module (between the top and bottom rows of small terminals). The large terminal (W) at the right side of the module is connected to the emitter of Q3 and the collector of Q6. The next large terminal (V) to the left is connected to the emitter of Q2 and the collector of Q5. The last large terminal (U) to the left is connected to the emitter of Q1 and the collector of Q4.

You can perform a reasonably accurate test of this module with an ohmmeter. You should be aware, however, that the circuitry inside the module often is more complex than what is shown in Figure 24-64. The additional, but unknown, components may cause erroneous or unexpected readings on the ohmmeter. Therefore, you need to exercise

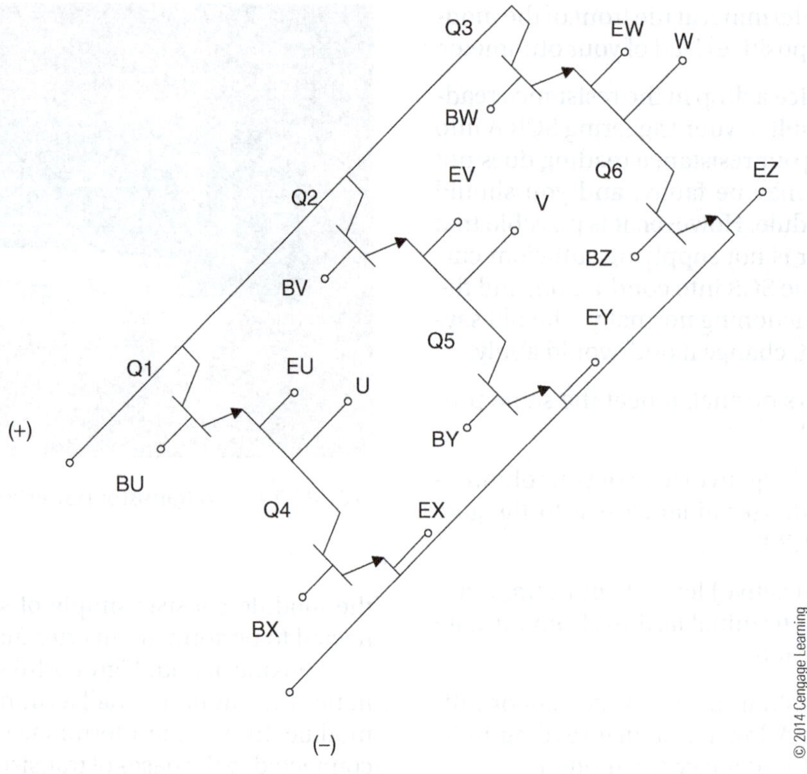

FIGURE 24-64 A transistor power module schematic.

some judgment in interpreting the ohmmeter readings to determine whether the module is faulty or not.

To test the module with an ohmmeter, proceed as follows:

1. Place the ohmmeter in the Ohms position, if using a digital multimeter. (Use the 200 position if using an analog multimeter.)
2. Place the black (negative) lead of your ohmmeter on the base (terminal BU) of transistor Q1.
3. Place the red (positive) lead of your ohmmeter on the emitter (terminal EU) of transistor Q1.
4. Your meter should indicate a very high or infinite resistance. A low-resistance reading indicates a faulty transistor. Replace the module.
5. Leave the black (negative) lead of your ohmmeter connected to the base (terminal BU) of transistor Q1.
6. Place the red (positive) lead of your ohmmeter on the (+) terminal of the module. (This connects your red lead to the collector of transistor Q1.)
7. Your meter should indicate a very high or infinite resistance. A low-resistance reading indicates a faulty transistor. Replace the module.
8. Reverse your meter connections. Place the black (negative) lead of your ohmmeter on the emitter (terminal EU) of transistor Q1.
9. Place the red (positive) lead of your ohmmeter on the base (terminal BU) of transistor Q1.
10. Your meter should indicate a low resistance. A high- or infinite-resistance reading indicates a faulty transistor. Replace the module.
11. Leave the red (positive) lead of your ohmmeter connected to the base (terminal BU) of transistor Q1.
12. Place the black (negative) lead of your ohmmeter on the (+) terminal of the module. (This connects your black lead to the collector of transistor Q1.)
13. Your meter should indicate a low resistance. A high- or infinite-resistance reading indicates a faulty transistor. Replace the module.
14. Place the red (positive) lead of your ohmmeter on the emitter (terminal EU) of transistor Q1.
15. Place the black (negative) lead of your ohmmeter on the (+) terminal of the module. (This connects your black lead to the collector of transistor Q1.)

16. Your meter should indicate a very high or infinite resistance. A low-resistance reading indicates a faulty transistor. Replace the module.
17. Reverse your meter connections. Place the red (positive) lead of your ohmmeter on the (+) terminal of the module. (This connects your red lead to the collector of transistor Q1.)
18. Place the black (negative) lead of your ohmmeter on the emitter (terminal EU) of transistor Q1.
19. Your meter should indicate a very high or infinite resistance. A low-resistance reading indicates a faulty transistor. Replace the module.
20. Repeat steps 1 through 19 for each of the remaining five transistors in the module. If any readings are questionable, remove any doubt by checking your measurements against a known good module, or simply replace the questionable module.

There is another type of power module that you may find in the drive upon which you are working. This module is technically not part of the power output section, although it does have a power function in the drive. The module is a three-phase bridge rectifier module. It is used to convert three-phase AC into rectified DC. A picture of this module appears in **Figure 24-65** and a schematic of this module appears in **Figure 24-66**.

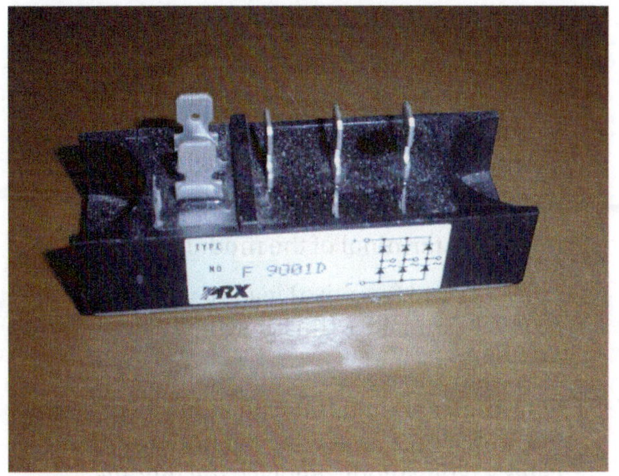

FIGURE 24-65 A three-phase bridge rectifier module.

Notice that this module has five terminals. The two horizontal terminals at the left end of the module are the (+) and (–) DC connections. The three vertical terminals are the connections for the three-phase AC (L1, L2, and L3).

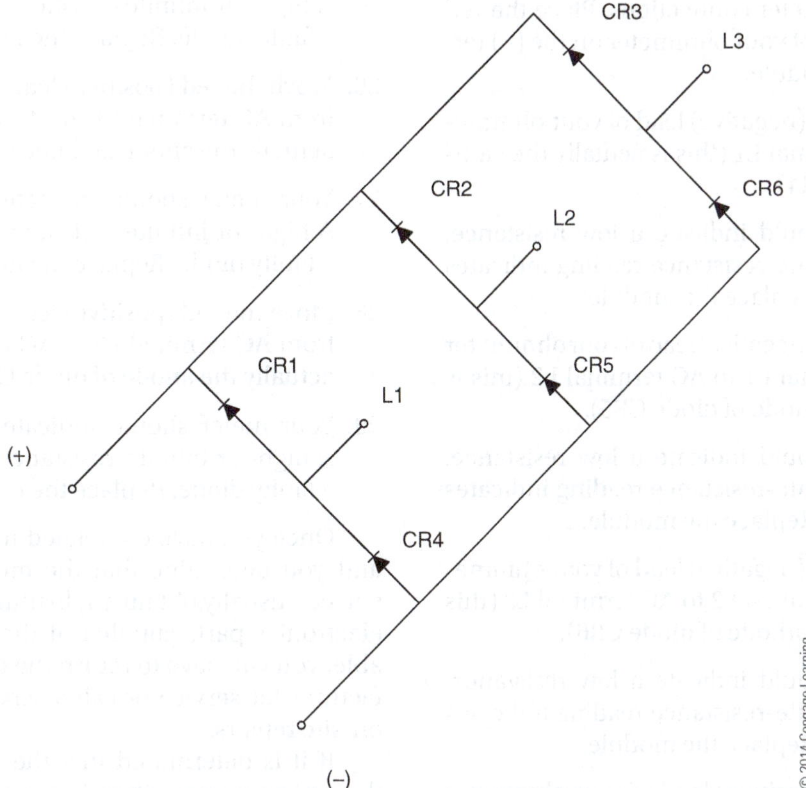

FIGURE 24-66 A three-phase bridge rectifier module schematic.

To test the module with an ohmmeter, follow these steps:

1. Place the ohmmeter in the Diode Test position, if using a digital multimeter. (Use the 200 Ω position if using an analog multimeter.)
2. Place the black (negative) lead of your ohmmeter on the (–) terminal of the module.
3. Place the red (positive) lead of your ohmmeter on AC terminal L1 (this is actually the cathode of diode CR4).
4. Your meter should indicate a very high or infinite resistance. A low-resistance reading indicates a faulty diode. Replace the module.
5. Move the red (positive) lead of your ohmmeter from AC terminal L1 to AC terminal L2 (this is actually the cathode of diode CR5).
6. Your meter should indicate a very high or infinite resistance. A low-resistance reading indicates a faulty diode. Replace the module.
7. Move the red (positive) lead of your ohmmeter from AC terminal L2 to AC terminal L3 (this is actually the cathode of diode CR6).
8. Your meter should indicate a very high or infinite resistance. A low-resistance reading indicates a faulty diode. Replace the module.
9. Reverse your meter connections. Place the red (positive) lead of your ohmmeter on the (–) terminal of the module.
10. Place the black (negative) lead of your ohmmeter on AC terminal L1 (this is actually the cathode of diode CR4).
11. Your meter should indicate a low resistance. A high- or infinite-resistance reading indicates a faulty diode. Replace the module.
12. Move the black (negative) lead of your ohmmeter from AC terminal L1 to AC terminal L2 (this is actually the cathode of diode CR5).
13. Your meter should indicate a low resistance. A high- or infinite-resistance reading indicates a faulty diode. Replace the module.
14. Move the black (negative) lead of your ohmmeter from AC terminal L2 to AC terminal L3 (this is actually the cathode of diode CR6).
15. Your meter should indicate a low resistance. A high- or infinite-resistance reading indicates a faulty diode. Replace the module.
16. Place the red (positive) lead of your ohmmeter on the (+) terminal of the module.
17. Place the black (negative) lead of your ohmmeter on AC terminal L1 (this is actually the anode of diode CR1).
18. Your meter should indicate a very high or infinite resistance. A low-resistance reading indicates a faulty diode. Replace the module.
19. Move the black (negative) lead of your ohmmeter from AC terminal L1 to AC terminal L2 (this is actually the anode of diode CR2).
20. Your meter should indicate a very high or infinite resistance. A low-resistance reading indicates a faulty diode. Replace the module.
21. Move the black (negative) lead of your ohmmeter from AC terminal L2 to AC terminal L3 (this is actually the anode of diode CR3).
22. Your meter should indicate a very high or infinite resistance. A low-resistance reading indicates a faulty diode. Replace the module.
23. Reverse your meter connections. Place the black (negative) lead of your ohmmeter on the (+) terminal of the module.
24. Place the red (positive) lead of your ohmmeter on AC terminal L1 (this is actually the anode of diode CR1).
25. Your meter should indicate a low resistance. A high- or infinite-resistance reading indicates a faulty diode. Replace the module.
26. Move the red (positive) lead of your ohmmeter from AC terminal L1 to AC terminal L2 (this is actually the anode of diode CR2).
27. Your meter should indicate a low resistance. A high- or infinite-resistance reading indicates a faulty diode. Replace the module.
28. Move the red (positive) lead of your ohmmeter from AC terminal L2 to AC terminal L3 (this is actually the anode of diode CR3).
29. Your meter should indicate a low resistance. A high- or infinite-resistance reading indicates a faulty diode. Replace the module.

Once you have completed testing the module, and you determine that the module is defective, you can usually obtain a substitute part from a local electronics parts supplier. If the part is not available, you will have to return the drive to the manufacturer for service or call a service technician for on-site repairs.

If it is determined that the problem is not in the power section, then it must be located in the control section of the drive. The electronics used in the control section are more complex; therefore, troubleshooting is not recommended. In this event, the drive must be returned to the manufacturer for repair or arrangements made for on-site repair by a factory-trained technician.

SUMMARY

- An electronic variable-speed drive has two basic sections: the control section and the power section.
- The control section governs or controls the power section, whereas the power section supplies controlled power to the motor.
- The control section not only controls the motor's speed but also its torque.
- The control section of a drive uses three types of signals: command, feedback, and error.
- A command signal is programmed into the drive and sets the desired operating speed of the motor. While the motor is operating, a feedback signal from the motor indicates its performance.
- The feedback signal can originate from the counter-electromotive force produced by the motor or from the tachometer-generator or encoder mounted on the motor's shaft.
- Counter-electromotive force (CEMF) is the voltage produced in the motor's rotating armature, which cuts the magnetic lines of force in the field as it revolves. The armature voltage is called CEMF because it opposes the applied voltage from the variable-speed drive.
- The difference that exists between the control and feedback signals is called the error signal.
- In open-loop control, any variations in motor speed must be compensated by manual adjustment. Open-loop control is limited to applications where the motor load is fairly constant.
- When the motor load varies, closed-loop control is used. It uses feedback information to monitor the performance of the motor.
- An SCR armature voltage controller is another method of DC motor speed control. The speed of the DC motor may be varied by controlling the shunt field current or armature voltage.
- A chopper is a circuit that uses very fast electronic switches. The switches consist of transistors, SCRs, or MOSFETs.
- In a drive system, the four areas to check for possible problems are the electrical supply to the motor and the drive, the motor or its load, the feedback device and sensors, and the drive itself.
- AC drives, or inverters, are used to vary the speed and torque of the squirrel-cage motor. They are also used to start the motor slowly and smoothly and increase its efficiency.
- The AC drive has three basic sections: the converter, the DC filter, and the inverter.
- The converter rectifies the applied AC into DC; the DC filter provides a smooth, rectified DC; and the inverter switches the DC on and off so rapidly that the motor receives a pulsating DC that appears similar to AC.

REVIEW QUESTIONS

1. Name the four parameters of a DC motor that a DC drive controls.
2. Describe each of the four quadrants of motor operation.
3. List the four main areas to check when a problem with a DC drive system occurs.
4. List the three basic sections of an AC inverter drive, and describe the purpose of each section.
5. Explain the term *phase imbalance.*

CHAPTER

25

Programmable Logic Controllers

The **programmable logic controller (PLC)** is an assembly of solid-state digital logic elements designed to make logical decisions and provide control. PLCs are used for the control and operation of manufacturing process equipment and machinery. The PLC is an industrially hardened computer designed to perform control functions in industrial environments. This means that unlike your desktop personal computer (PC), the PLC must be capable of operating in temperature extremes; with poor power conditions (spikes, sags, etc.); and in dusty, dirty, corrosive atmospheres, and withstand shock and vibration. In addition, PLCs are designed to be programmed by individuals who are familiar with motor control circuits. Therefore, most PLCs program in a language that resembles ladder diagrams, which makes learning PLCs very easy for most electricians. In addition to typical switching functions, PLCs can also perform counting, calculations, comparisons, processing of analog signals, and more.

OBJECTIVES

After studying this chapter, the student will be able to

- Identify the components of a PLC.
- Describe the function of a PLC.
- Correctly wire a PLC I/O module.
- Develop a simple PLC program.
- Develop a PLC I/O wiring diagram.
- Define various terms used in conjunction with PLCs.

25-1 PLC COMPONENTS

PLCs are available in three sizes—small, medium, and large. The size of a PLC is determined by the number of **I/O (input/output)** devices it can handle and the amount of program memory available. **Figure 25-1** shows a comparison between small, medium, and large PLCs. Most PLCs consist of an assortment of **modules** (input and output), a **CPU** (central processor unit), a power supply, and a rack in which to mount the aforementioned modules.

Figure 25-2 shows a fundamental block diagram of a PLC. Essentially, the PLC monitors input signals from various sensors and switches through an input module, as seen in **Figure 25-3**. The status of the input device is compared to the program stored within the PLC. **Figure 25-4** shows a complete PLC. Based on the program logic, the PLC activates or deactivates an output, located on an output module, as seen in **Figure 25-5**. The output is connected to some type of device, such as a relay or motor starter. The programming device is used to input the program and make modifications to an existing program. The programming device may be a handheld programmer, as seen in **Figure 25-6**, or it may be a PC (desktop or laptop). Typically, there is also some type of display, which may be as simple as an LCD display located on the programming

	Small	Medium	Large
Program memory size	Up to 18k	Up to 64k	Up to 100k
Digital I/O	Up to 84 (shared with analog I/O)	Up to 4096	Up to 50176
Analog I/O	Up to 84 (shared with digital I/O)	Up to 96	Up to 50176
Timers/counters	Limited by available memory	Limited by available memory	Limited by available memory
Scan time	As fast as 1 μSec	As fast as 0.225 mSec	As fast as 0.5 mSec
Expansion	Up to 4 expansion modules with up to 14 I/O points per module	Up to 3 expansion chassis with up to 30 I/O slots per chassis	Up to 125 expansion chassis with up to 16 I/O slots per chassis

I/O, input/output.

FIGURE 25-1 A comparison of small, medium, and large programmable logic controllers.

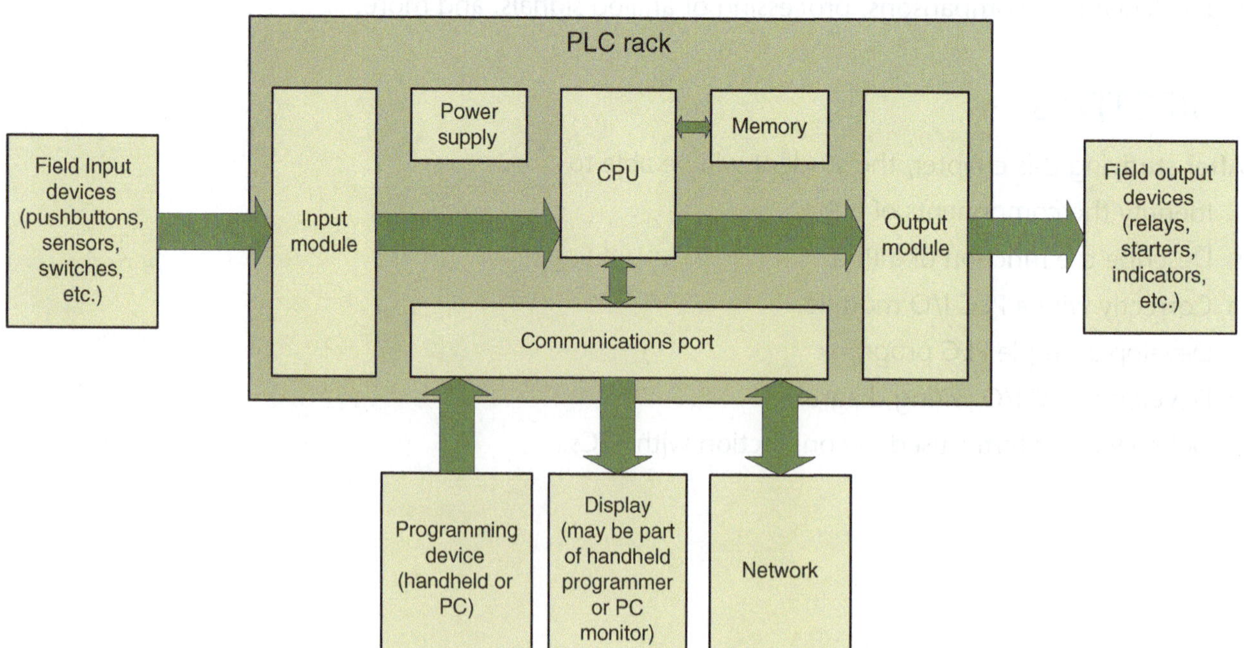

FIGURE 25-2 A fundamental block diagram of a PLC.

FIGURE 25-3 Digital input modules.

FIGURE 25-4 Three complete PLCs. Top – Siemens S7-200, middle – Allen-Bradley ControlLogix, bottom – GE 90-30.

device or may be the monitor of the PC that is used to program the PLC.

Input Module

The main purpose of the input module is to take the input signal from the field device (switch or sensor) and convert it to a signal that can be processed by

FIGURE 25-5 Output modules.

FIGURE 25-6 A handheld programmer (HHP).

the PLC central processing unit (CPU). Typically, this means converting the input signal to a 5 VDC level. In addition, the input module provides electrical isolation between the field device and the PLC.

There are many different types of input modules available. One type is the **digital input module** (Figure 25-3), which accepts input signals that are either off or on. Digital input modules are available for DC inputs, AC inputs, or a mix. Most digital inputs modules are available for 8, 16, or 32 inputs. In addition, digital input modules are available with a mix of inputs and outputs on the same module.

Another type of input module is the **analog input module**, as seen in **Figure 25-7**. This module accepts input signals that are either 0 to 10 VDC or 4 to 20 mA. These modules are very useful when used with instrumentation and control loops. Analog input modules are available for 4 inputs, 8 inputs,

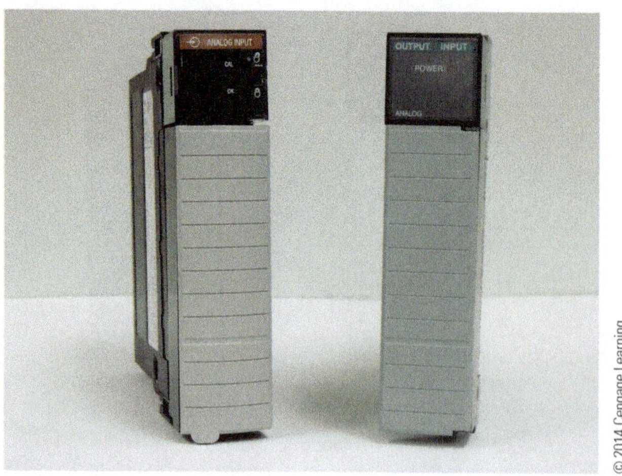

FIGURE 25-7 Analog input modules.

or 16 inputs. In addition, analog input modules are available with a mix of inputs and outputs on the same module.

There are many other types of input modules available. These include **high-speed counter input modules**, **thermocouple input modules**, **RTD input modules**, and **motion control modules**, to name a few.

Output Module

The main purpose of the output module is to take the signal from the PLC CPU and convert it to a signal for the field device (relay, motor starter, etc.). Typically, this means converting the 5 VDC level from the CPU to the required voltage for the output field device. In addition, the output module provides electrical isolation between the PLC and the field device.

There are many different types of output modules available. One type is the **digital output module** (**Figure 25-8**), which produces output signals that are either off or on. Digital output modules are available for DC outputs, AC outputs, or a mix. Most digital output modules are available for 8, 16, or 32 outputs. In addition, digital output modules are available with a mix of inputs and outputs on the same module.

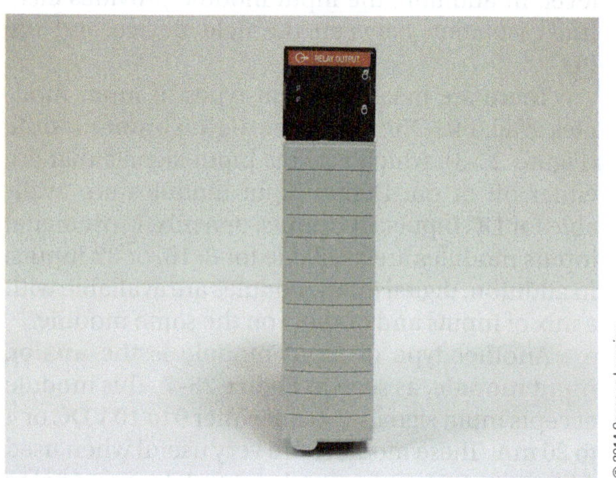

FIGURE 25-8 Digital output modules.

FIGURE 25-9 Analog output modules.

Another type of output module is the **analog output module**, as seen in **Figure 25-9.** This module produces output signals that are either 0 to 10 VDC or 4 to 20 mA. Like the analog input modules, these modules are very useful when used with instrumentation and control loops. Analog output modules are available for 4 outputs or a mix of inputs and outputs on the same module.

There are many other types of output modules available. These include *process* **control** and **positioning** to name a few.

25-2 I/O WIRING

One big advantage of using a PLC over point-to-point wiring of relay controls is that the wiring is greatly simplified and reduced. **Figure 25-10** shows a forward/reverse circuit, as it would typically be wired with relay-type controls. Compare this drawing to the one shown in **Figure 25-11**. Notice that there is less wiring in Figure 25-11. In addition, the wiring is easier to follow, which is the benefit of using a PLC. The PLC program performs the logical decisions. Therefore, the logic connections are made within the PLC program, *not* with the circuit wiring. In addition, modifications to the operation of the circuit do not require rewiring. Simply edit the PLC program for the desired operation.

Let us look at the wiring of the PLC I/O modules in more detail. Refer again to Figure 25-11. We begin with the input module, which is a 120-VAC

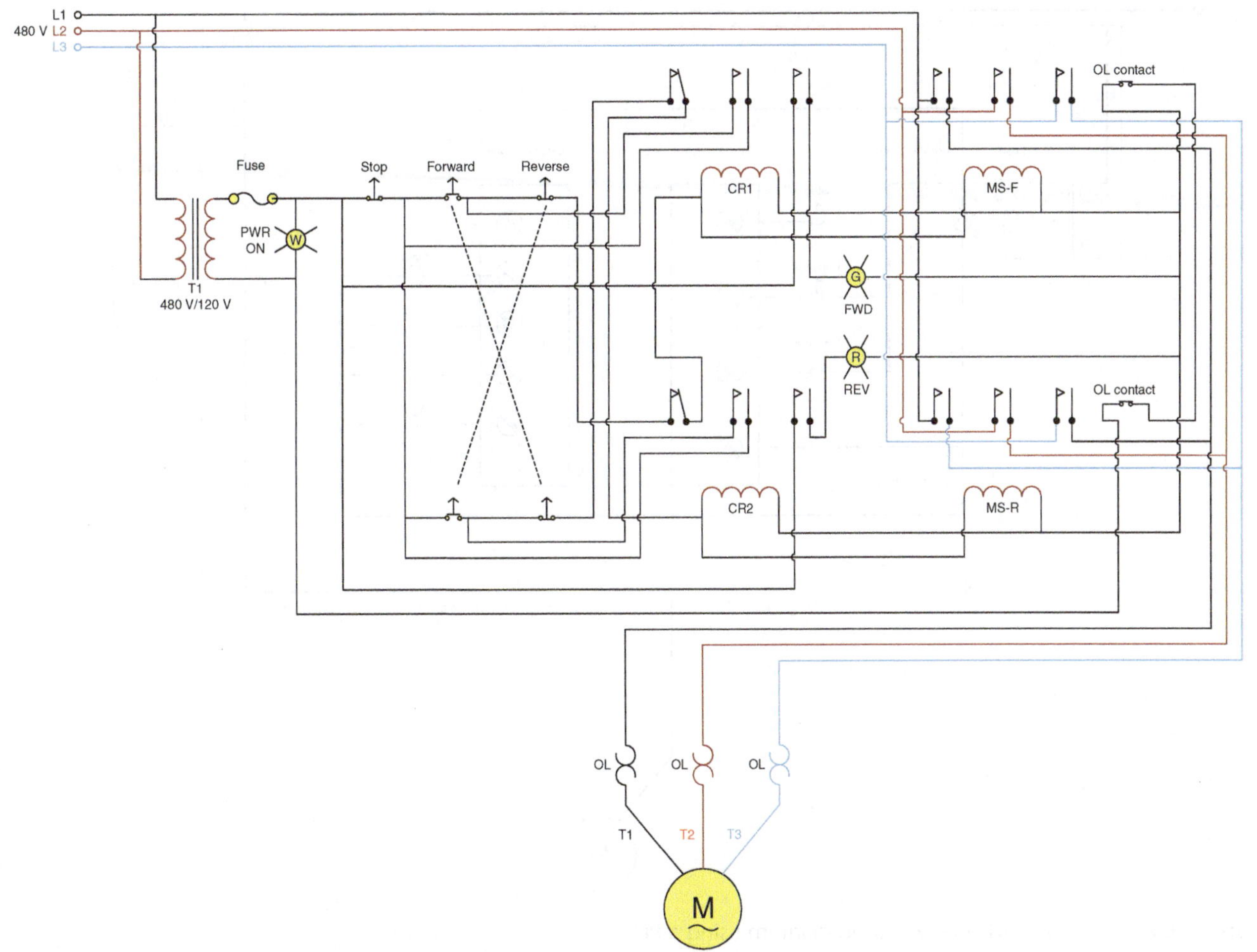

FIGURE 25-10 A forward/reverse wiring diagram using relay logic.

input module. This means that the input devices must supply 120 VAC to the input terminals of the input module.

Notice the stop, forward, and reverse pushbuttons. One side of each of these pushbuttons is connected to the 120-VAC line. Notice also that power is applied to the NC contacts of the stop pushbutton, while power is applied to the NO contacts of the forward and reverse pushbuttons. Each pushbutton is then connected to its own terminal on the 120-VAC input module.

Before we continue, we must draw attention to a few details. First, recall that the stop pushbutton is wired NC. This is the manner is which stop pushbuttons are generally wired in point-to-point wiring. Notice, also, that we are only using one set of contacts for the forward and reverse pushbuttons. This may seem odd at first, but remember that the PLC is capable of performing functions that were previously accomplished through the circuit wiring. Finally, notice the 120-VAC neutral wire that is connected to the input module. This is necessary for the module to operate properly.

This is all of the wiring needed for our input devices. We now turn our attention to the output module. This particular output module is a 120 VAC output module, which means that the module supplies 120 VAC to the output terminals of the output module.

Notice the forward and reverse motor starters. One side of each of these motor starters is connected to 120 VAC neutral. Each motor starter is then connected to its own terminal on the 120 VAC output module. The two pilot lights are connected in a similar manner. One side of each pilot light is connected to 120 VAC neutral, while the other side of each pilot light is connected to its own terminal on the 120 VAC output module. Notice, also, the 120 VAC line that is connected to the output module that is necessary for the module to operate properly.

We have now completed the wiring of our PLC forward/reverse control circuit. Notice how much

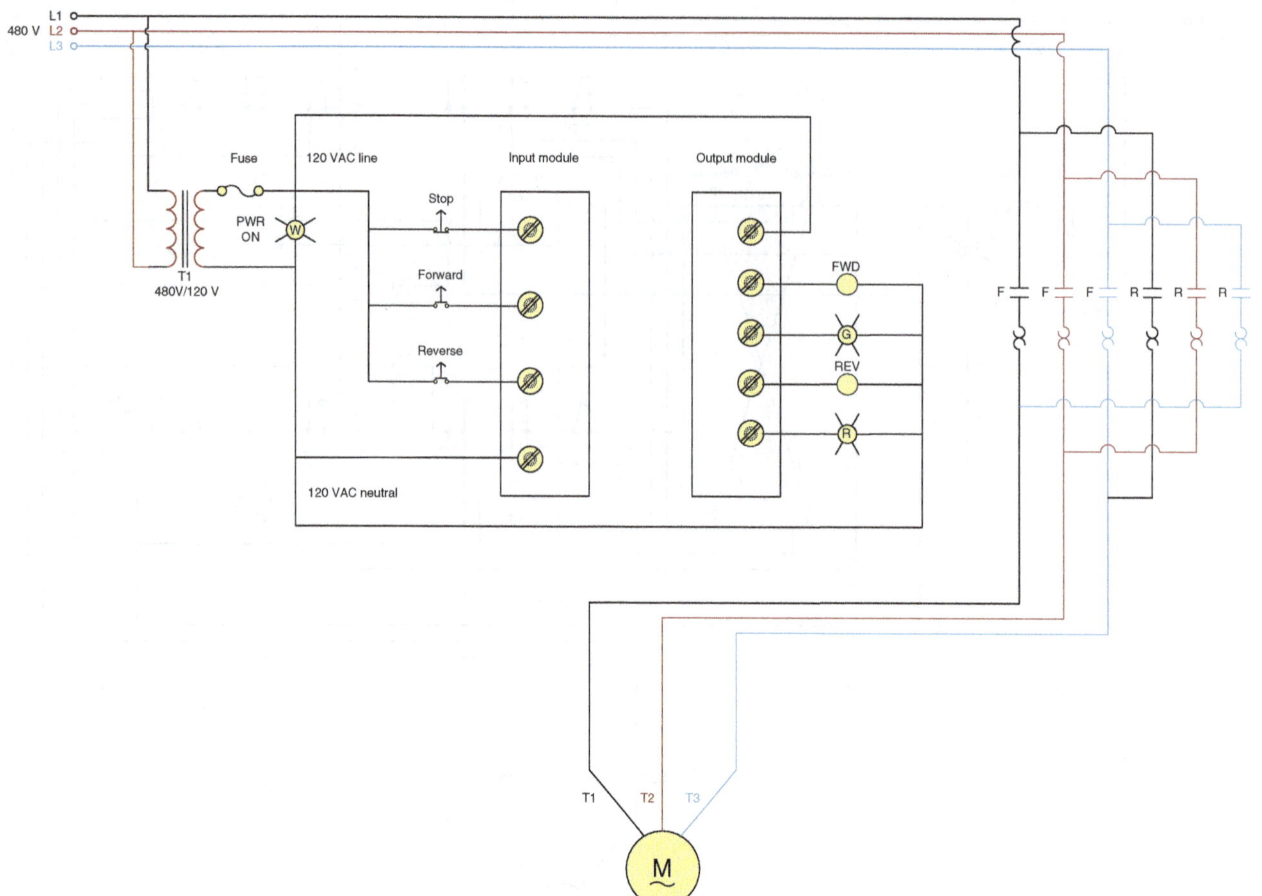

FIGURE 25-11 A forward/reverse wiring diagram using a PLC.

simpler the wiring is. **Figure 25-12** shows the panel for the forward/reverse control as it appeared when wired in a point-to-point fashion. **Figure 25-13** shows the same panel as wired for PLC control. Now, let us learn how to program this circuit. We will begin with a simple three-wire control circuit.

25-3 PROGRAMMING

Over the years, different PLC manufacturers have created different programming languages for their PLCs. This has created confusion and increased training costs. The **IEC (International Electrotechnical Commission)** has created a standard (IEC 1131-3) for five programming languages for PLCs. These five languages are known as:

1. Function Block Diagram (FBD)
2. Instruction List (IL)
3. Ladder Diagram (LD)
4. Sequential Function Chart (SFC)
5. Structured Text (ST)

- ***Function Block Diagram (FBD)***
This programming language is a **graphic language** that uses a library of functions (math, logic, etc.) in combination with custom functions (modem control, PID control, etc.) to create programs.

- ***Instruction List (IL)***
Instruction List is a **low-level program** best suited for small applications and fast execution.

- ***Ladder Diagram (LD)***
This is perhaps *the most popular programming language.* Ladder diagrams use graphics, discreet control, interlocking logic, and may incorporate function block instructions. This text focuses on the ladder diagram language.

- ***Sequential Function Chart (SFC)***
SFC is also a *graphical language.* While providing structure and coordination of sequential events, alternative and parallel sequences are supported as well.

- ***Structured Text (ST)***
Structured text programming language is a **BASIC style language.** ST is an excellent language for complex processes or calculations that are not graphic friendly.

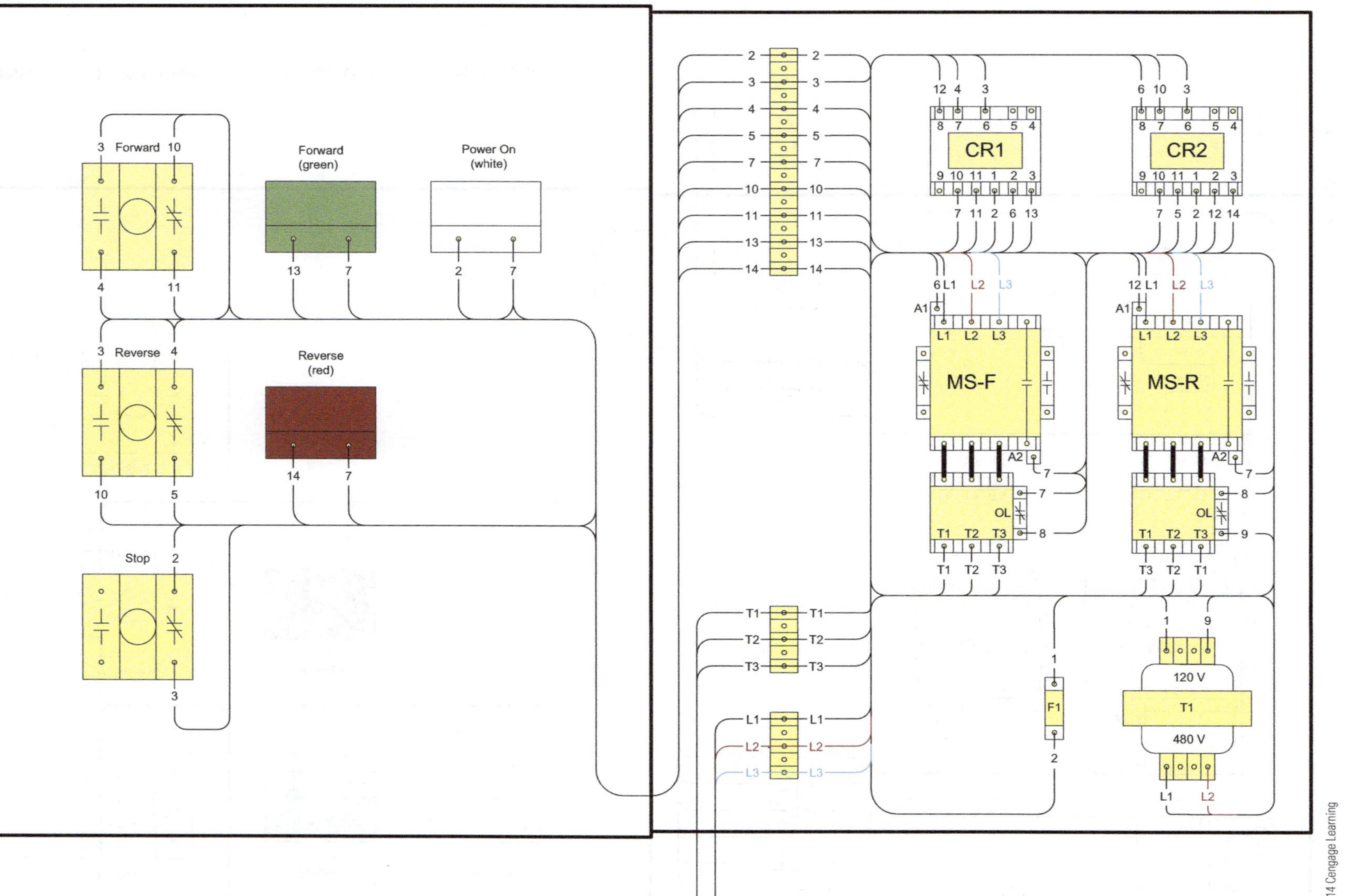

FIGURE 25-12 A forward/reverse panel wiring diagram using relay logic.

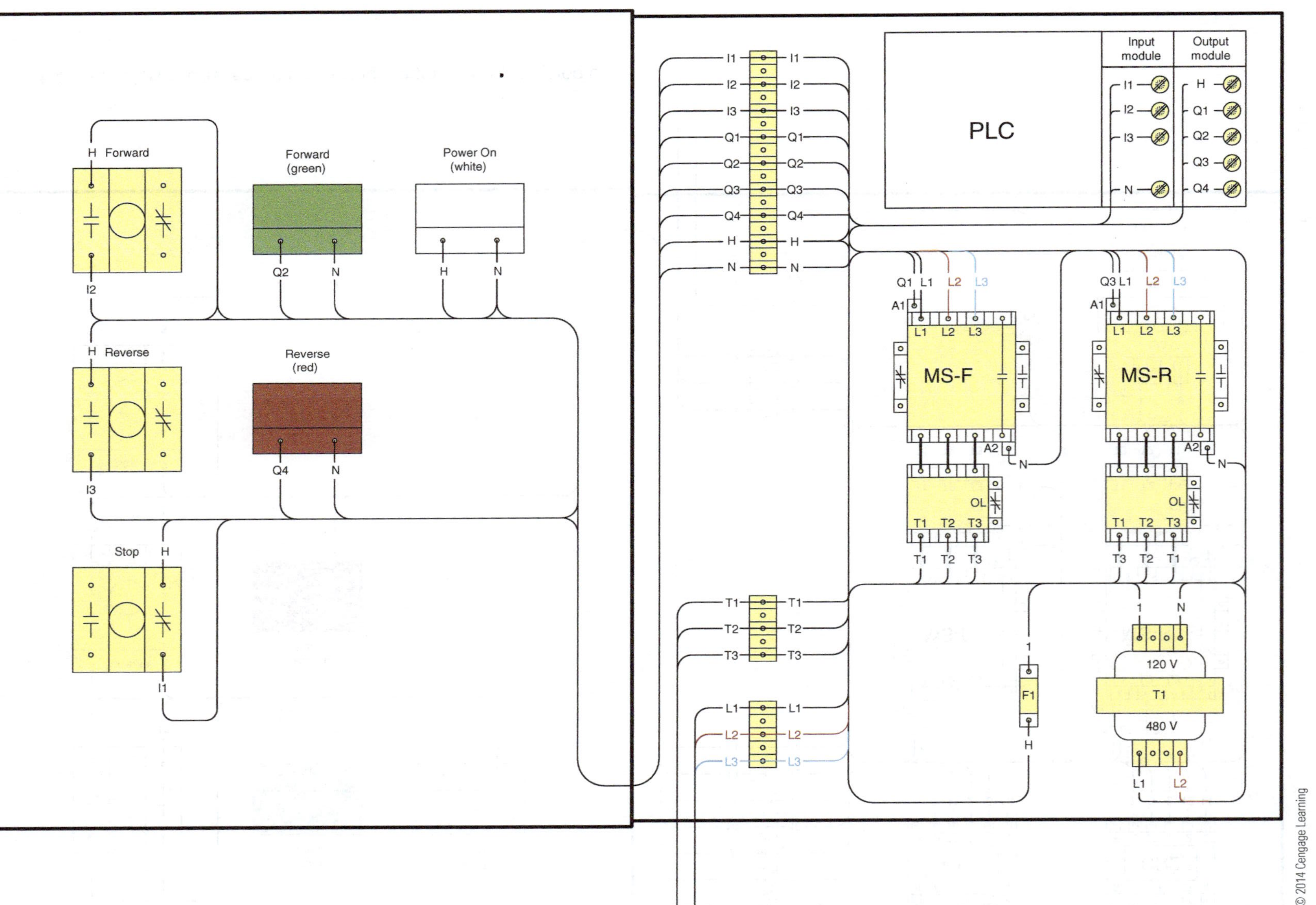

FIGURE 25-13 A forward/reverse panel wiring diagram using a PLC.

PLC Program

Before we begin to develop a PLC program, we need to realize that to a PLC, an input is an input, is an input and an output, is an output, is an output. This means that a PLC does not care what type of device provides the input signal to the input module. The input could come from a pushbutton, a limit switch, a contact from a relay, or whatever. A PLC treats them all the same. They are looked upon as either a normally open or a normally closed contact. The same holds true for an output. The output module may provide an output to a motor starter, control relay, pilot light, siren, or whatever. A PLC treats them all the same, and they are looked upon as a load. Therefore, a ladder diagram, written for relay logic, looks different when converted to a PLC ladder logic diagram.

Figure 25-14A shows a relay logic ladder diagram for a three-wire control. **Figure 25-14B** shows the PLC ladder logic program for the same circuit. Notice the different symbols used to represent the start, stop, and motor starter functions. Notice, also, the addressing of the start, stop, and motor starter functions. Each PLC manufacturer has their own methods of addressing functions. You need to become familiar with the method used by the PLC at your facility. For the purposes of this text, we take a somewhat generic approach to addressing. Input devices begin with the letter "I," and output devices will begin with the letter "Q."

Now, look at **Figure 25-15**. This is the same PLC program as shown in Figure 25-14B, except the input and output module wiring is shown as well. Next, we walk through the operation of this circuit.

When the PLC is running, it is constantly scanning the inputs of the input module looking for voltage. At this point, the PLC detects 120 VAC from the stop pushbutton at terminal I1 of the input module. This is a result of the stop pushbutton wired normally closed, causing 120 VAC line voltage to appear

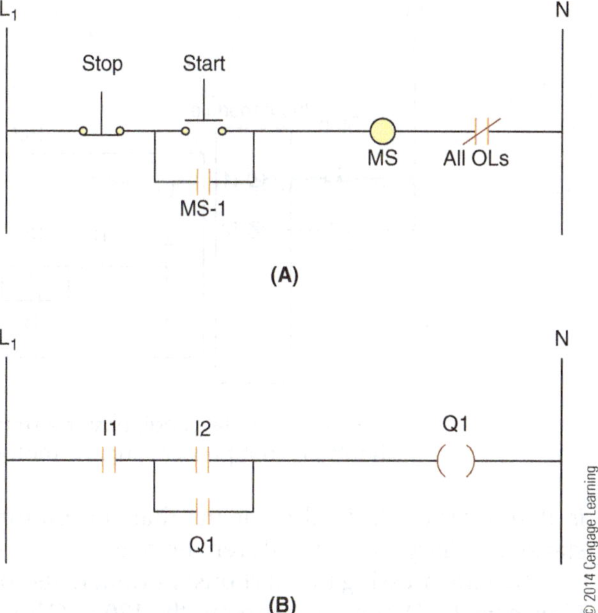

FIGURE 25-14 (A) Relay logic for a three-wire control; (B) PLC program for a three-wire control.

at this terminal. However, 120 VAC is not detected at terminal I2 of the input module because terminal I2 is connected to the start pushbutton, which is wired normally open. Therefore, the PLC program sees the stop pushbutton as passing power, but not the start pushbutton. On the PLC ladder logic program seen in **Figure 25-16**, the contact representing the stop pushbutton is highlighted to show power flow. Notice that the contact representing the start pushbutton is not highlighted because the start pushbutton is not passing power to the input module at this time. Notice also that the coil representing the motor starter is not highlighted. Because the start pushbutton has not been pressed, power cannot flow to the motor starter coil. The motor starter

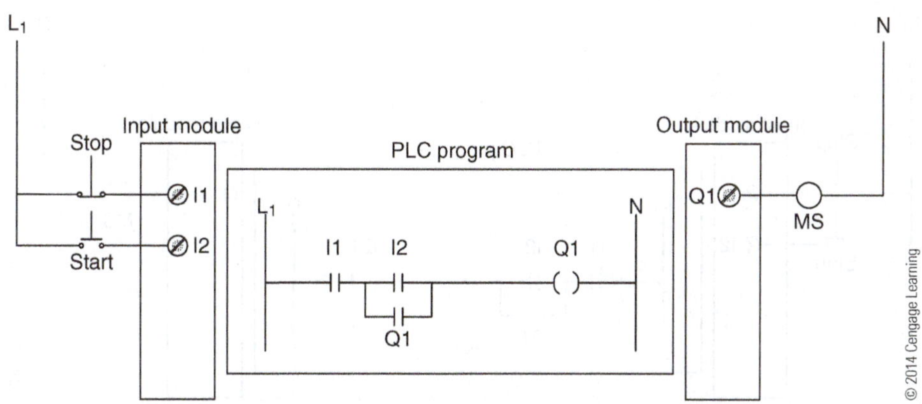

FIGURE 25-15 A PLC I/O wiring diagram and program for a three-wire control.

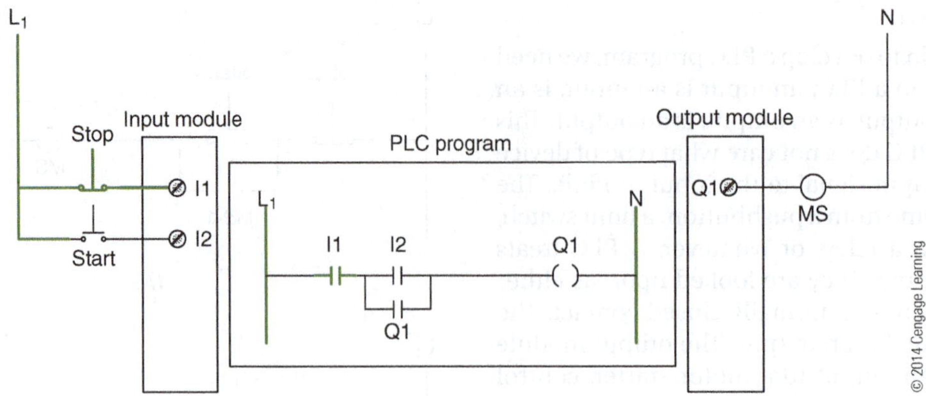

FIGURE 25-16 Initial conditions: the stop pushbutton is not pressed, the start pushbutton is not pressed, and the motor starter is de-energized.

is de-energized. This also means that the motor starter auxiliary contacts, Q1, remain open.

Imagine pressing the start pushbutton. Refer to **Figure 25-17**. When this happens, the 120 VAC line is connected to the second input terminal on the 120 VAC input module. The PLC detects the presence of voltage at terminal I2 and causes the associated rung element, I2, to close. This is indicated by the highlighting of the start pushbutton contact, I2, and causes the output element, Q1, to energize (highlighted). Now, two things occur simultaneously. The holding contact addressed Q1 closes (highlighted). This provides a latch or seal-in for the start pushbutton. In addition, terminal Q1 on the output module provides 120 VAC to the motor starter coil that is wired to terminal Q1, which results in the motor starter energizing.

Figure 25-18 shows the effects of releasing the start pushbutton. The PLC detects the absence of voltage at input module terminal I2. Therefore,

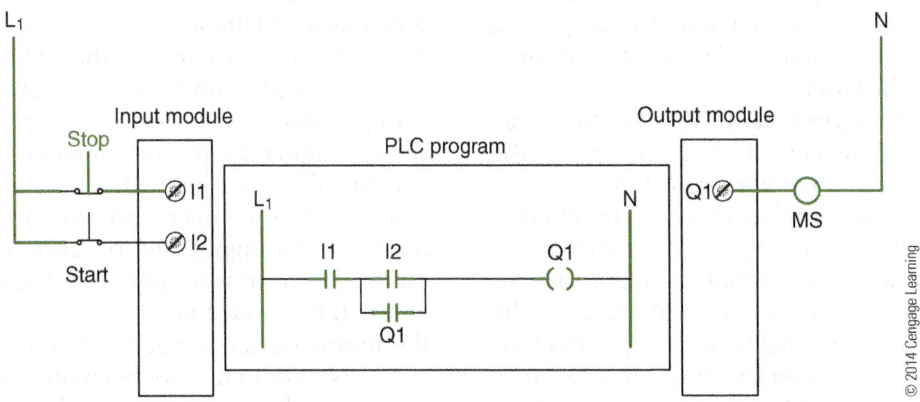

FIGURE 25-17 The start pushbutton is pressed: the motor starter is energized.

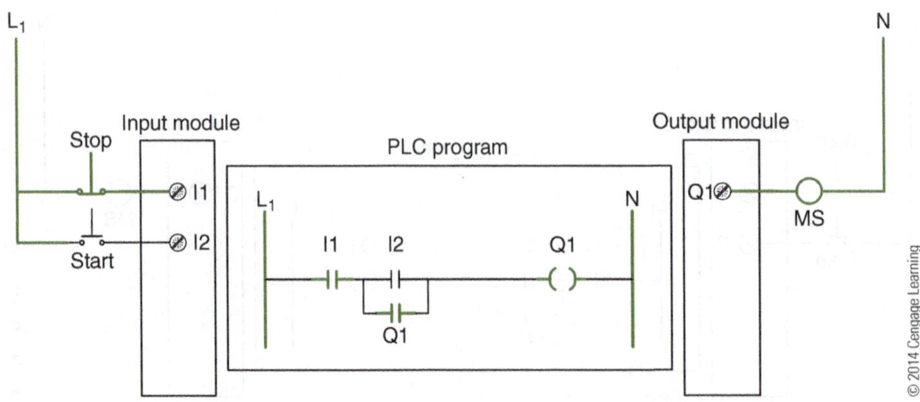

FIGURE 25-18 The start pushbutton is released: the motor starter remains energized.

the start pushbutton rung element, I2, is no longer highlighted. However, because this circuit contains a latch or seal-in circuit, all other elements in the PLC program remain highlighted. This means that there is power flow from the 120 VAC line, through the holding contacts, Q1, through the motor starter coil, Q1, to the 120 VAC neutral. The motor starter remains energized.

We will now see what happens when we press the stop pushbutton. Refer to **Figure 25-19**. When the stop pushbutton is pressed, the PLC detects the absence of voltage at input module terminal I1. Therefore, the stop pushbutton rung element, I1, is no longer highlighted. This means that there is no longer power flow from the 120 VAC line, through the stop pushbutton. As a result, the motor starter, Q1, de-energizes, causing the holding contacts, Q1, to open. In addition, terminal Q1 on the output module no longer provides 120 VAC to the motor starter coil that is wired to terminal Q1, which results in the motor starter de-energizing. This is indicated in the PLC program by the absence of the highlighting of the stop pushbutton, I1; the start pushbutton, I2; the motor starter coil, Q1; and the holding contacts, Q1. When the stop pushbutton is released, the circuit is returned to its original condition, as seen in **Figure 25-20**.

At this point, you may have two questions. First, the start pushbutton is wired as a *normally open* pushbutton to input terminal I2, and the program shows a *normally open* contact, I2. However, *the stop pushbutton is wired as a normally closed pushbutton,* wired to input terminal I1, and *the program shows a normally open contact,* I1. Is this correct? Your other question might be, what is the origin of the holding contacts, Q1? They do not appear as a set of contacts that have been wired into the input module. We will start with the first question.

Earlier, we stepped through the operation of the three-wire control circuit and found that it performs as expected. We now make a slight modification to the program so that the stop pushbutton rung element, I1, appears as a normally closed contact. This would appear to be the correct thing to do, because the start pushbutton is wired normally open and programmed normally open. It makes sense to wire the stop pushbutton normally closed and program

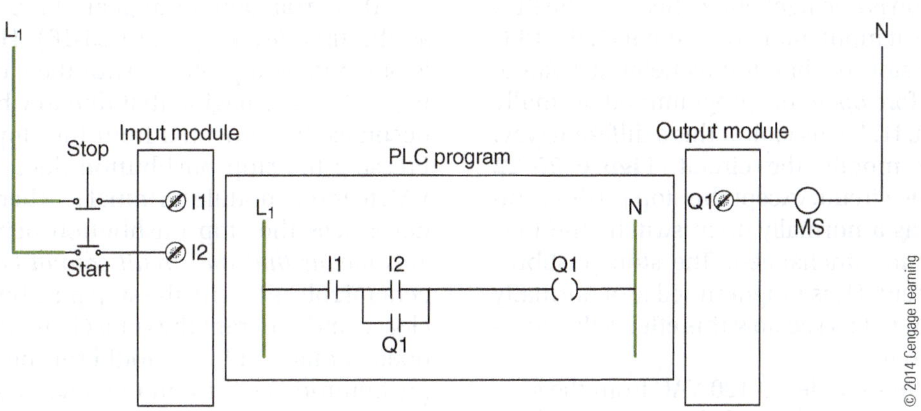

FIGURE 25-19 The stop pushbutton is pressed: the motor starter is de-energized.

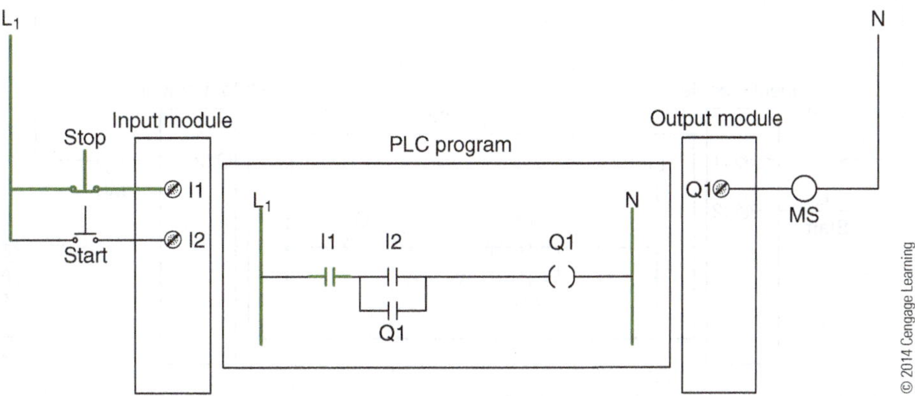

FIGURE 25-20 The stop pushbutton is released: the motor starter remains de-energized.

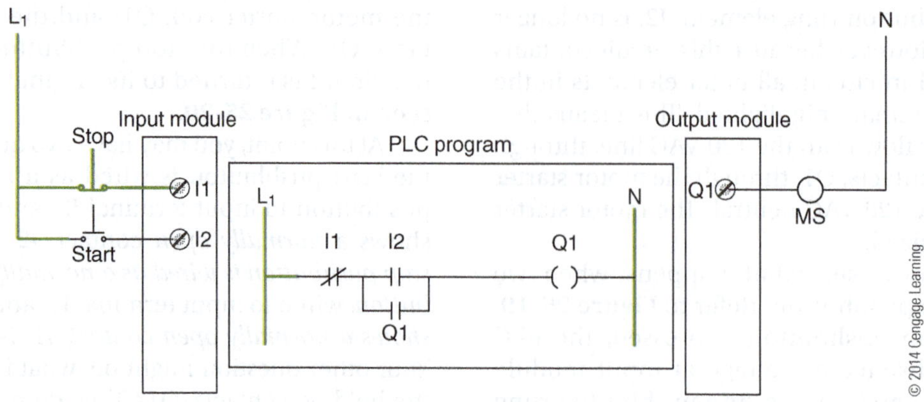

FIGURE 25-21 A stop pushbutton wired normally closed and programmed normally closed.

it normally closed as well. However, observe the results, as seen in **Figure 25-21**.

Recall that the PLC detects 120 VAC from the stop pushbutton at terminal I1 of the input module. This is a result of the stop pushbutton being wired normally closed and causing 120 VAC line voltage to appear at this terminal. Therefore, the PLC program sees the stop pushbutton as passing power. On the PLC ladder logic program seen in Figure 25-21, the stop pushbutton is *not* highlighted. Because the PLC senses voltage at input module terminal I1, the PLC interprets the stop pushbutton as being activated. The PLC therefore *opens* the programmed normally closed contact, I1. Let us look at this a different way.

We again modify the circuit. **Figure 25-22** shows the same circuit, except the stop pushbutton is now wired as a normally open switch. The PLC program remains unchanged. The stop pushbutton rung element, I1, is programmed as a normally closed contact. Let us see how this effects the operation of our circuit.

The PLC does *not* detect 120 VAC from the stop pushbutton at terminal I1 of the input module. This is a result of the stop pushbutton being wired normally open. Therefore, the PLC program sees the stop pushbutton as not passing power. On the PLC ladder logic program seen in Figure 25-22, the stop pushbutton *is* highlighted to show power flow due to the stop pushbutton rung element, I1, being programmed as a normally closed contact. Pressing the stop pushbutton causes the rung element, I1, to open, breaking the power flow.

This arrangement appears to function as well as the first circuit (Figure 25-16). However, there is one potential problem with the circuit shown in Figure 25-22. Imagine that due to vibration, a wire becomes disconnected from the stop pushbutton. Pressing the stop pushbutton does not apply 120 VAC to input module terminal I1. Therefore, the PLC never sees the stop pushbutton being depressed. *This means that the circuit cannot be stopped!* The general rule is to wire the stop pushbutton normally closed and program the rung element as a normally open contact. In fact, should this be a critical stop pushbutton, such as an emergency stop pushbutton, the general rule is to wire emergency stop

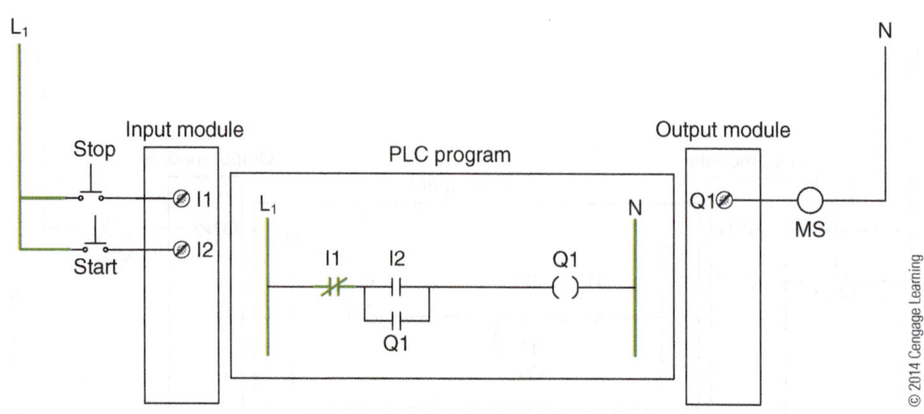

FIGURE 25-22 A stop pushbutton wired normally open and programmed normally closed.

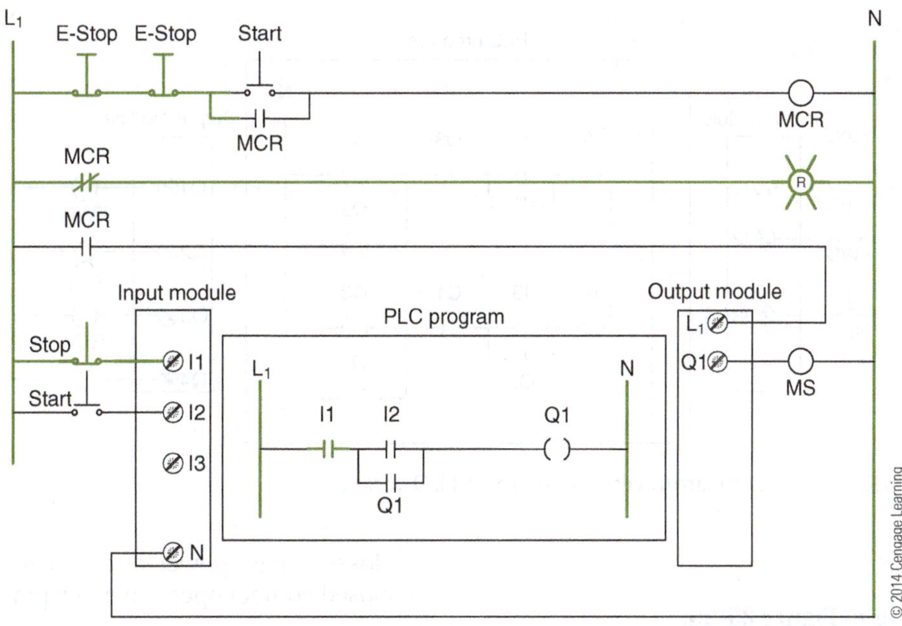

FIGURE 25-23 A typical PLC safety circuit.

pushbuttons directly to the circuit and not through the PLC logic. Should the PLC suffer a failure, you would still be able to stop the process or machine. **Figure 25-23** shows a typical PLC safety circuit. Notice that when either emergency stop pushbutton is pressed, power is removed from the output devices, but not the PLC or input devices. In this fashion, troubleshooting can be performed on the PLC while the machine or process is halted. Notice also that the relay that performs this function is called a master control relay (MCR). Do not confuse this relay with the PLC program instruction MCR (master control relay). The programmed MCR instruction should never be used in place of a hard-wired master control relay.

Now, we return to the second question: What is the origin of the holding contacts, Q1? They do not appear as a set of contacts that have been wired into the input module. In addition to performing logical decisions, PLCs have within them devices known as internal coils (relays), timers, counters, and more. When we programmed the output element, Q1, we really addressed an internal coil. This coil has practically an unlimited number of normally open and normally closed contacts available. By simply programming a normally open contact and addressing it the same as the coil, Q1, we have instructed the PLC to operate the contact whenever the coil is energized.

This should now draw attention to one of the greatest advantages of using PLCs over relay-type devices. You have available to you a large number of devices with practically unlimited contact arrangements. In addition, there are functions available to you that cannot be performed with relay-type devices.

Now, let us turn our attention back to the forward/reverse control that was seen previously. The circuit is shown again as **Figure 25-24**, only now we have added the PLC program as well. Here, we are only using one normally open set of contacts from the forward pushbutton. Notice that we are only using one set of normally open contacts from the reverse pushbutton. Looking at the PLC program, notice there are two contacts shown for the forward and reverse pushbuttons. These are internal contacts. When an input address is used, there are practically an unlimited number of normally open and normally closed contacts available to be assigned to that input address. This is what helps simplify the wiring in a circuit that uses a PLC.

Types of Instructions

PLCs are programmed by combining various instructions to form a logical control circuit. Many of these instructions are common to the different PLC manufacturers. There may be variations to the name, how they appear, or how they are programmed, but their basic function is essentially the same. Again, you need to familiarize yourself with the particular PLC at your facility. We now look at some of the common types of instructions used in PLC programming.

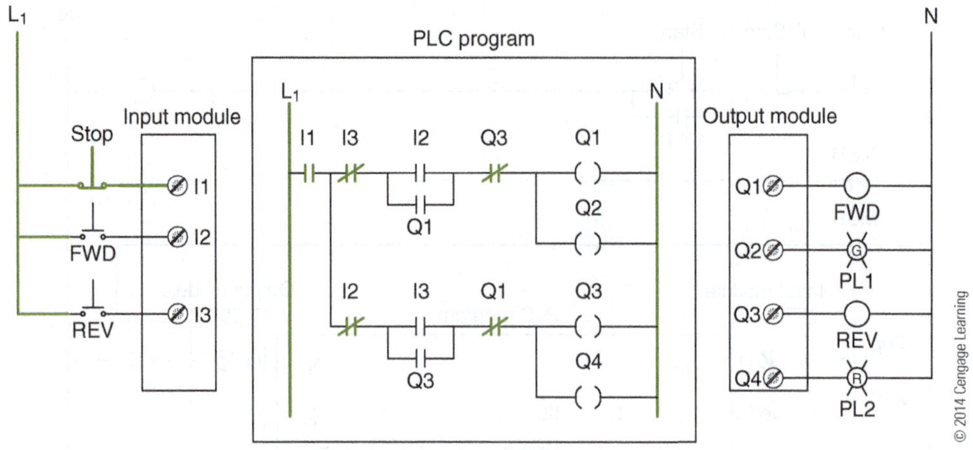

FIGURE 25-24 A forward/reverse program and I/O wiring.

Relay Functions

Contacts

Refer to **Figure 25-25**.

- Contacts are used to indicate the status of a referenced component. This may be a field-wired device (switch, sensor, etc.) or an internal element (coil). Contacts may be normally open or normally closed.
- A normally open contact acts as a switch to pass power when the status of its referenced component is on.
- A normally closed contact acts as a switch to pass power when the status of its referenced component is off.

Coils

Refer to **Figure 25-26**. There are two basic types of coils—*nonretentive* and *retentive*.

- **Nonretentive coils** revert to their de-energized state whenever the PLC is cycled from stop to run or whenever the power is cycled.
- **Retentive coils** retain their current state whenever the PLC is cycled from stop to run or whenever the power is cycled.
- A normal coil controls a referenced contact. When the coil receives power (becomes energized), a referenced normally open contact closes to pass power, and a referenced normally closed contact opens to block power.
- A negated coil operates inversely to a standard coil. A negated coil is energized when it does not receive power, and is de-energized when it receives power. When a negated coil is *not* receiving power (is *energized*), a referenced normally open contact will close to pass power, and a referenced normally closed contact will open to block power.
- A retentive coil operates in a similar fashion to a coil. However, its condition is not altered by a power failure.
- A negated retentive coil operates in a similar fashion to a negated coil. However, its condition is not altered by a power failure.
- A set coil is a nonretentive coil used to perform a latch function. When a set coil receives power (becomes energized), a referenced normally open contact closes to pass power, and a referenced normally closed contact opens to block power. These conditions remain, even if the set coil is de-energized. They do not remain if the PLC loses power. Set coils require the programming of a reset coil with the same address.

Type of contact	Display	Contact passes power:
Normally open	—┤├—	When referenced component is ON
Normally closed	—┤/├—	When referenced component is OFF

FIGURE 25-25 Contact instructions.

Type of Coil	Display	Power	Result
Normal	—( )—	ON	Set referenced component ON
Normal	—( )—	OFF	Set referenced component OFF
Negated	—(/)—	ON	Set referenced component OFF
Negated	—(/)—	OFF	Set referenced component ON
Retentive	—(M)—	ON	Set referenced component ON (retained)
Retentive	—(M)—	OFF	Set referenced component OFF (retained)
Negated retentive	—(/M)—	ON	Set referenced component OFF (retained)
Negated retentive	—(/M)—	OFF	Set referenced component ON (retained)
SET	—(S)—	ON	Set referenced component ON until reset by —(R)—
SET	—(S)—	OFF	Coil state unchanged
RESET	—(R)—	ON	Reset referenced component OFF until set by —(S)—
RESET	—(R)—	OFF	Coil state unchanged
Retentive SET	—(SM)—	ON	Set referenced component ON until reset by —(R)— (retained)
Retentive SET	—(SM)—	OFF	Coil state unchanged
Retentive RESET	—(RM)—	ON	Reset referenced component OFF until set by —(S)— (retained)
Retentive RESET	—(RM)—	OFF	Coil state unchanged

FIGURE 25-26 Coil instructions.

- A reset coil is used to negate a set coil's condition. The Reset coil must have the same address as the set coil to perform this function.
- A retentive set coil is used to perform a latch function. When a retentive set coil receives power (becomes energized), a referenced normally open contact closes to pass power, and a referenced normally closed contact opens to block power. These conditions remain, even if the set coil is de-energized or if the PLC loses power. Set coils require the programming of a reset coil with the same address.
- A retentive reset coil is used to negate a set coil's condition. The negated reset coil must have the same address as the set coil to perform this function.

Timers and Counters

Timers and counters Refer to **Figure 25-27**. There are two basic types of time delay functions—**on-delay** and **off-delay**.

- **On-delay timers** perform a timing function while receiving power. The state of the associated contact(s) does not change until the timer has timed out.
- **Off-delay timers** perform a timing function after becoming de-energized. The state of the associated contact(s) changes immediately upon the timer becoming energized. The contact(s) reverts to its original state after the timer has timed out after becoming de-energized.

Mnemonic	Function	Is Set When	Remains Set Until:
TMR or TON	On-delay timer	Rung conditions allow power flow, and accumulated value is equal to or greater than the preset value	Rung conditions break power flow
OFDT or TOF	Off-delay timer	Rung conditions allow power flow	Rung conditions break power flow, and accumulated value is equal to or greater than the preset value
ONDTR or RTO	On-delay timer (retentive)	Rung conditions allow power flow, and accumulated value is equal to or greater than the preset value	Associated RESET element is energized
CTU or UPCTR	Up counter	Rung element transitions from open to closed, and accumulated value is equal to or greater than the preset value	Associated RESET element is energized
CTD or DNCTR	Down counter	Rung element transitions from open to closed, and accumulated value is less than or equal to zero	Associated RESET element is energized

FIGURE 25-27 Timers and Counters Instructions.

There are both nonretentive and retentive timers.

- Nonretentive timers reset their timing function upon loss of power.
- Retentive timers retain their timed value during a loss of power.
- Retentive timers require an associated reset control to clear the timed value.

Counters There are two basic types of counter functions: **up-counters** and **down-counters.**

- Up-counters perform a counting function when the associated input element transitions from an off to on state. Up-counters begin at some preset value and increment upward. Up-counters are retentive and require an associated reset element to clear the counted values.
- Down-counters perform a counting function when the associated input element transitions from an off to on state. Down-counters begin at some preset value and decrement downward. Down-counters are retentive and require an associated reset element to clear the counted values.

Math Functions

Standard math functions Refer to **Figure 25-28.**

- Addition: Adds two integers
- Subtraction: Subtracts one integer from another
- Multiplication: Multiplies two integers
- Division: Divides one integer by another

Additional math functions Refer to **Figure 25-28.**

- Square Root: Finds the square root of an integer
- Trigonometric Functions

  Sine

  Cosine

  Tangent

  Arcsine

  Arccosine

  Arctangent
- Logarithmic/Exponential Functions

  Logarithm

  Natural logarithm

  Exponent
- Radian Conversion

  Radians

  Degrees

Relational Functions

Relational Functions Refer to **Figure 25-29.**

- Equal: Test two integers for equality

Mnemonic	Function	Description
ADD	Addition	Adds one integer to another
SUB	Subtraction	Subtracts one integer from another
MUL	Multiplication	Multiplies two integers
DIV	Division	Divides one integer by another
SQR	Square root	Finds the square root of an integer
SIN	Sine	Finds the sine of an integer
COS	Cosine	Finds the cosine of an integer
TAN	Tangent	Finds the tangent of an integer
ASIN	Arc sine	Finds the arc sine of an integer
ACOS	Arc cosine	Finds the arc cosine of an integer
ATAN	Arc tangent	Finds the arc tangent of an integer
LOG	Logarithm	Finds the base 10 logarithm of an integer
LN	Natural logarithm	Finds the natural logarithm of an integer
DEG	Degrees	Converts an integer in degree units to radians
RAD	Radians	Converts an integer in radian units to degrees

FIGURE 25-28 Math Instructions.

- Not Equal: Test two integers for non-equality
- Greater Than: Test for one integer greater than another
- Greater Than or Equal To: Test for one integer greater than or equal to another
- Less Than: Test for one integer less than another
- Less Than or Equal To: Test for one integer less than or equal to another
- Range: Test for one integer to have a value within a minimum and maximum value

Program Control Instructions

Program Control Instructions Refer to **Figure 25-30**.

- Master control relay (MCR): Denotes an area within a program that is controlled by a master control relay. The master control relay is located at the beginning of the area, and an ENDMCR is located at the end of the area. When an MCR is activated, any program elements located within the MCR area are ignored by the PLC.
- Master control relay END (ENDMCR): Denotes an area within a program that is controlled by a master control relay. The master control relay is located at the beginning of the area, and an ENDMCR is located at the end of the area. When an MCR is activated, any program elements located within the MCR area are ignored and deactivated by the PLC.
- Jump: When active, causes the PLC to jump to a specific location within the program as specified by a LABEL.
- Label: Designator used with a JUMP command.

There are many other functions available depending upon the manufacturer, model, and size of the PLC. Only by becoming familiar with the PLC at your facility can you fully realize the power and features available.

Mnemonic	Function	Description
EQ	Equal	Test whether two integers are equal
NE	Not equal	Test whether two integers are not equal
GT	Greater than	Test whether one integer is greater than another
GE	Greater than or equal	Test whether one integer is greater than or equal to another
LT	Less than	Test whether one integer is less than another
LE	Less than or equal	Test whether one integer is less than or equal to another
RANGE or LIMIT	Range or limit test	Test whether an integer is within a specified range or limit

FIGURE 25-29 Relational Instructions.

Mnemonic	Function	Description
MCR	Master control relay	Creates zone within program for controlling all non-retentive outputs within the zone
ENDMCR	End master control relay	Defines the end of a Master Control Relay zone
JUMP	Jump to label	Causes the PLC to jump over designated program rungs to specified label
LABEL	Label	Target rung for a JUMP command
CALL or JSR	Call or jump to specified subroutine	Causes the PLC to leave the main program and enter a subroutine program
RET	Return from subroutine	Causes the PLC to leave a subroutine program and return to the main program

FIGURE 25-30 Program Control Instructions.

SUMMARY

- The programmable logic controller (PLC) is an assembly of solid-state digital logic elements designed to make logical decisions and provide control.
- PLCs are available in small, medium, and large sizes.
- The size of a PLC is determined by the number of input/output devices it can handle and the amount of program memory available.
- Most PLCs consist of an assortment of input/output modules, a central processing unit, a power supply, and a rack on which to mount the modules.
- An input module takes the input signal from the field device and converts it to a signal that can be processed by the PLC CPU.
- There are many types of input modules, such as digital, analog, high-speed counter, thermocouple, RTD, and motion control.
- An output module takes the signal from the PLC CPU and converts it to a signal for the field device.
- There are many types of output modules, such as digital, analog, process control, and positioning.
- The IEC has created a standard for five programming languages used for PLCs: Function

Block Diagram, Instruction List, Ladder Diagram, Sequential Function Chart, and Structured Text.

- PLCs are programmed by combining various instructions to form a logical control circuit. There are instructions for relay functions, timers and counters, math functions, relational functions, and program control.

REVIEW QUESTIONS

1. A PLC is designed
   a. the same as an office computer.
   b. to operate in an industrial environment.
   c. the same as a home computer.
   d. to withstand heat.
2. Most PLCs are designed to be programmed using
   a. BASIC programming language.
   b. FORTRAN programming language.
   c. ladder logic programming language.
   d. special computer programming language.
3. The basic components of a PLC are
   a. the power supply and CPU.
   b. the programming terminal or device.
   c. the input/output modules.
   d. all of the above.
4. The CPU performs
   a. only special functions.
   b. all logic functions.
   c. only input functions.
   d. only output functions.
5. The programming terminal or device is used to program the
   a. power supply.
   b. input unit.
   c. output unit.
   d. CPU.

1. What is a PLC?
2. List three advantages of a PLC over electromagnetic control systems.
3. List two differences between a home or business PC and a PLC.
4. Name five components that form a PLC.
5. What is the purpose of an input module?
6. What is the most common PLC programming language in use today?
7. Describe the recommended method to wire and program a stop pushbutton.
8. What is the recommended method of wiring and programming an emergency stop pushbutton?
9. List one advantage of using a computer to program and monitor a PLC program as opposed to a handheld programmer.

CHAPTER 26

Lighting

Next to motors, perhaps the most common device or element found in any type of facility is lighting. Industrial lighting can involve many types and designs. The maintenance technician must become familiar with the different types of lighting and the various connection methods. This chapter introduces you to the most common types of lighting used in industry and shows you some of the common wiring methods used.

OBJECTIVES

After studying this chapter, the student should be able to

- Identify the lamps most commonly used in industry.
- Identify the most common types of lamp luminaires.
- Correctly wire several types of ballasts.
- Discuss the proper relamping procedures.

26-1 TYPES OF LAMPS

There are three basic types of lamps most commonly used in industry: the **incandescent lamp**, the **fluorescent lamp**, and the **high-intensity discharge (HID) lamp**. Each of these lamps offers advantages and disadvantages.

Incandescent Lamps

Incandescent lamps produce light by allowing a current to flow through a **filament**. The filament is attached to the bulb's base and encapsulated within a gas-filled glass envelope, as shown in **Figure 26-1**. A good filament exhibits electrical continuity. This can be verified with an ohmmeter, as shown in **Figure 26-2**. The actual resistance reading of a good filament is not important. What is important is that the filament has continuity. An open filament will not light. This indicates a defective lamp, which must be replaced.

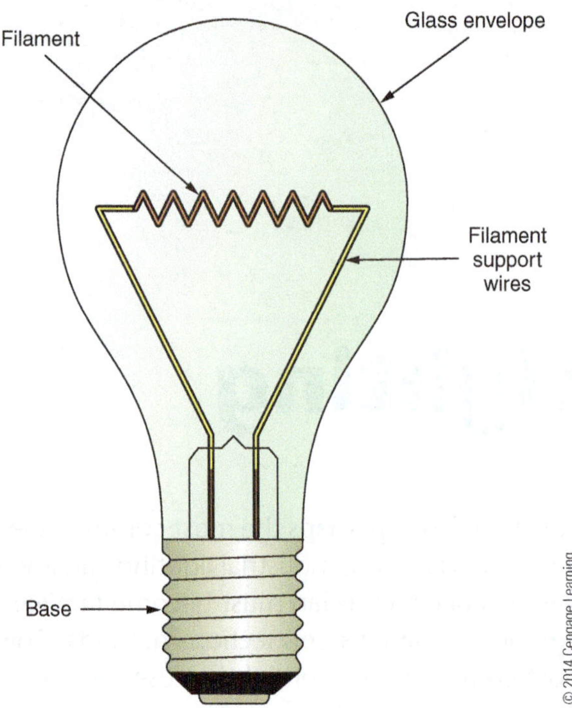

FIGURE 26-1 Construction of an incandescent lamp.

Fluorescent Lamps

Fluorescent lamps produce light by allowing an electrical current to flow through a gas. These lamps consist of a glass tube, which is coated on the inside with a **phosphor**. The glass tube is filled with a gas. There is no filament within a fluorescent lamp. However, there are small **heating elements** at each end of the glass tube. **Figure 26-3** shows the construction of a fluorescent lamp.

When power is applied to the fluorescent lamp, the heating elements heat up, lowering the resistance of the gas within the glass tube. At this time, high voltage is applied to the ends of the lamp, causing a current to flow through it. The gas within the glass tube acts as a conductor for the electrical current. As the electrical current flows through the gas, the phosphor coating on the inside surface of the glass tube begins to glow. At this time, the high voltage is removed so that the lamp continues to operate from normal voltage levels.

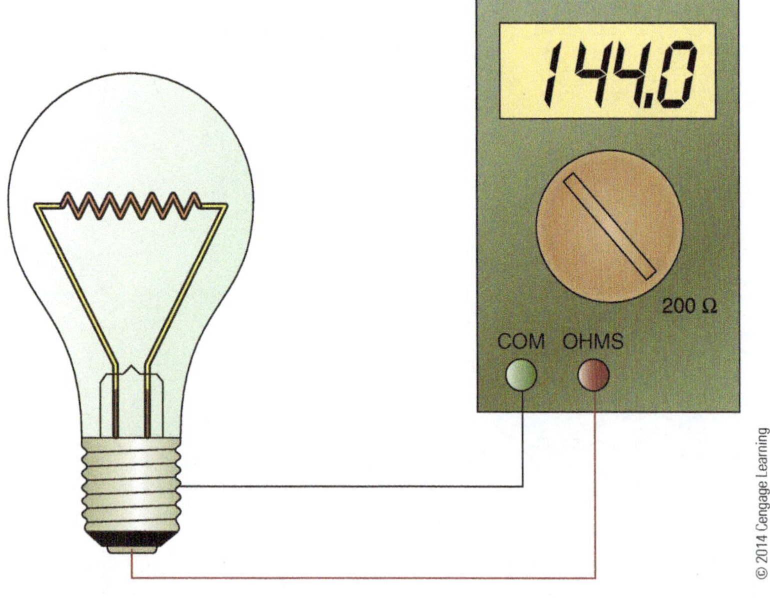

FIGURE 26-2 Testing an incandescent lamp with an ohmmeter.

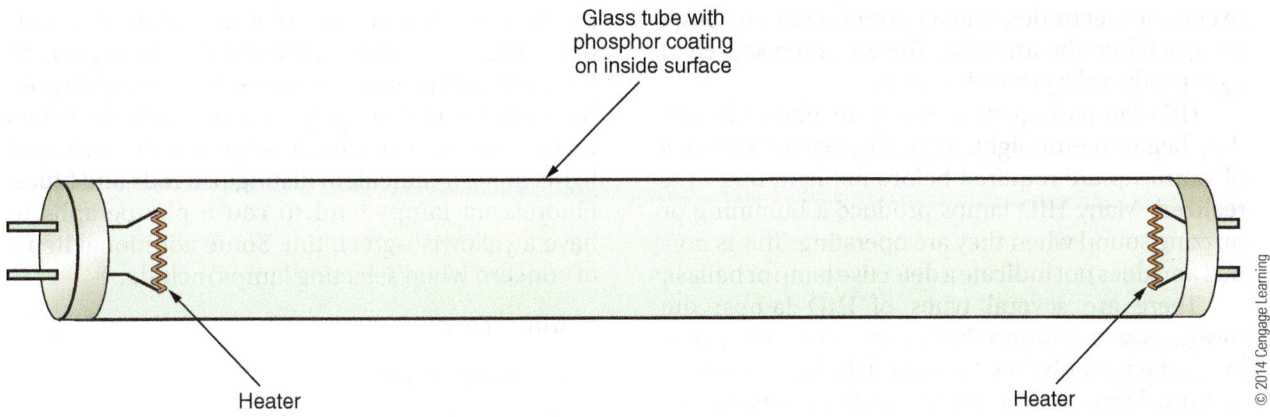

FIGURE 26-3 Construction of a fluorescent lamp.

The high voltage necessary for starting the fluorescent lamp is supplied by a **ballast**. A ballast may be a transformer or a solid-state circuit that is designed to provide the high voltage for starting while limiting the current. Ballasts are used by both fluorescent and high-intensity discharge lamps.

Over time, fluorescent lamps discolor at the ends. This is an indication that replacement is needed. Continued use eventually leads to total failure of the lamp and possibly the ballast. The possibility of saving a little money by continued use of a lamp with dark ends is negated by lower light output and more costly repairs when the ballast fails as well.

High-Intensity Discharge (HID) Lamps

HID lamps consist of a glass envelope, a base, and an **arc tube**, as shown in **Figure 26-4**. The arc tube is filled with a gas. Inside the arc tube is a set of **electrodes**. When the HID lamp is energized, a ballast provides high voltage to the electrodes within the arc tube. An electrical arc is created be

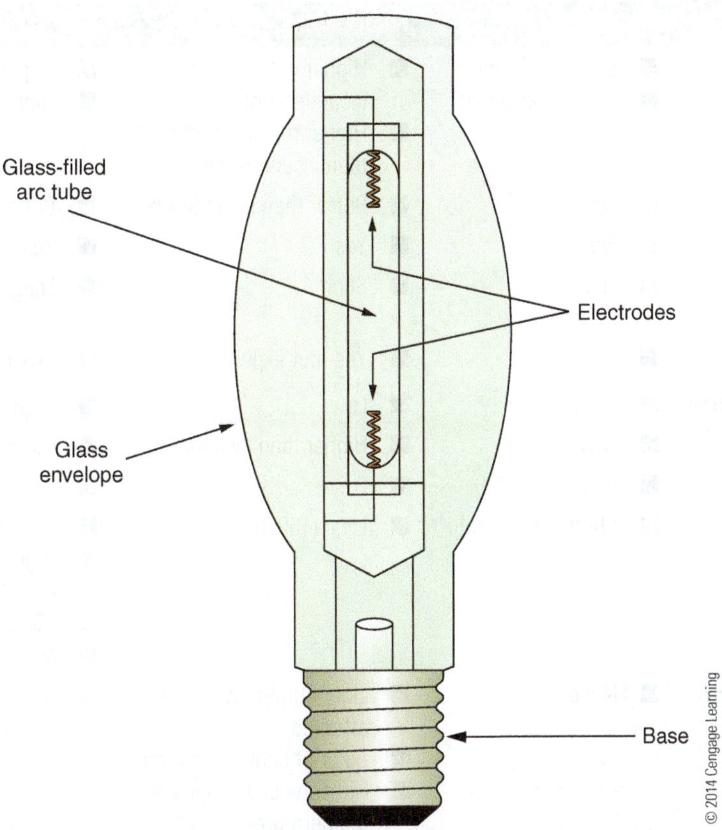

FIGURE 26-4 Construction of a high-intensity discharge (HID) lamp.

tween the electrodes, and a current passes through the gas filling the arc tube. The arc is the source of light produced by the HID lamp.

HID lamps require a warm-up period before they begin to emit light. Typically, several minutes of warm-up are required before full-light output is realized. Many HID lamps produce a humming or buzzing sound when they are operating. This is normal and does not indicate a defective lamp or ballast.

There are several types of HID lamps: the **low-pressure sodium lamp**, the **mercury-vapor lamp**, the **metal-halide lamp**, and the **high-pressure sodium lamp**. As the names imply, the basic differences among the types of HID lamps are the gas used to fill the arc tube and the pressure of the gas.

Comparison of the Types of Lamps

Table 26-1 shows a comparison between the incandescent, fluorescent, and HID types of lamps. Notice that one lamp is not best suited for all applications. Because of the different methods used to produce light, alterations in color rendition is a concern. For example, low-pressure sodium lights, which produce a yellow-orange color, are not used in locations where the appearance of colors is important. It is unlikely that low-pressure sodium lamps would be used in a photography studio because the colors would not appear true. Likewise, mercury-vapor lights make it difficult to distinguish reds and blues. Fluorescent lamps tend to cause photographs to have a yellowish-green tint. Some additional items of concern when selecting lamps include:

- Initial cost
- Ballast required
- Size and shape
- Warm-up required
- Efficiency
- Life expectancy
- Operating temperature
- Light output
- Ability to dim

Table 26-1

COMPARISON OF INCANDESCENT, FLUORESCENT, AND HIGH-INTENSITY DISCHARGE (HID) LAMPS

	Incandescent	Fluorescent	HID
Cost	■ Low initial cost ■ Low replacement cost	■ Higher initial cost than incandescent ■ Higher replacement cost than incandescent	■ Highest initial cost ■ Highest replacement cost
Efficiency	■ Low	■ Better than incandescent	■ Better than incandescent
Ballast required	■ No	■ Yes	■ Yes
Warm-up or re-start time required	■ No	■ Short	■ Long
Dimmable	■ Yes	■ Yes–but expensive	■ Expensive–or not possible
Operating temperature	■ High	■ Low	■ High
Light output	■ Low	■ Higher than incandescent	■ Highest
Operating voltage	■ Low	■ Low	■ High
Life expectancy	■ Short (900 hours)	■ Long (16000 hours)	■ Low pressure sodium (1900 hours) ■ High-pressure sodium (10000 hours) ■ Mercury-vapor (20000 hours) ■ Metal-Halide (11500 hours)
Ambient temperature effects	■ None	■ Light output levels are affected ■ Color of light is affected ■ Difficulty starting in cold temperatures	■ Difficulty starting in cold temperatures

- Color distortion
- Operating voltage
- Effects of ambient temperature

26-2 LUMINAIRES

The 2011 edition of the *National Electrical Code* defines a *luminaire* as:

A complete lighting unit consisting of a light source such as a lamp or lamps, together with the parts designed to position the light source and connect it to the power supply. It may also include parts to protect the light source or the ballast or to distribute the light. A lampholder itself is not a luminaire.

There are many shapes and styles of luminaires available today, as shown in **Figure 26-5**.

Incandescent luminaires

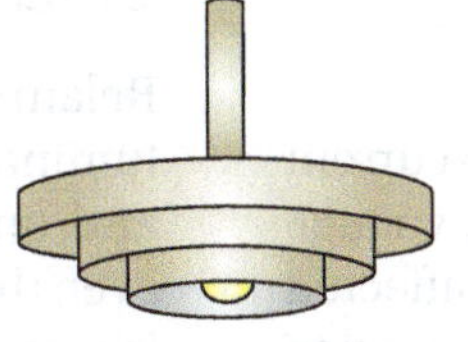

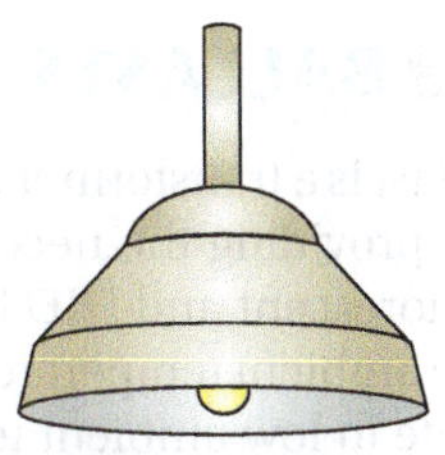

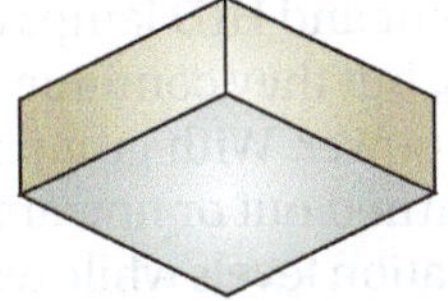

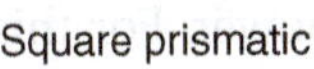

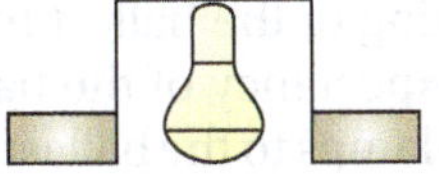

Hid luminaires

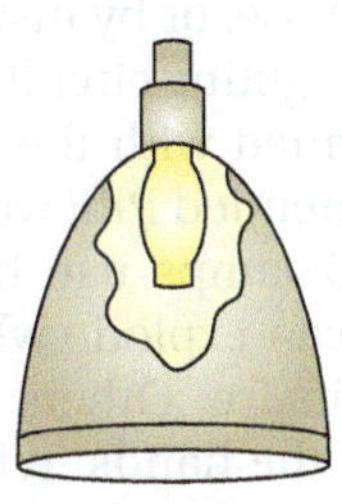

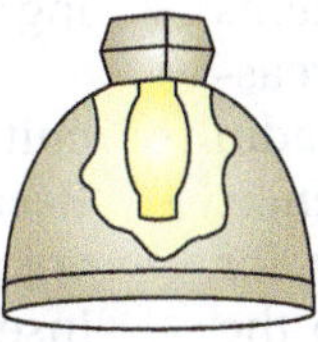

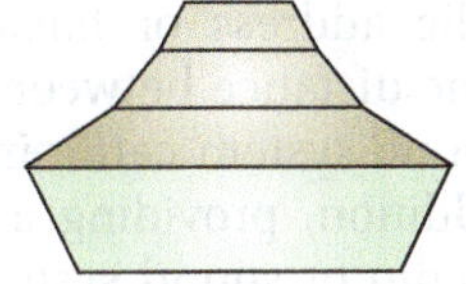

Fluorescent luminaires

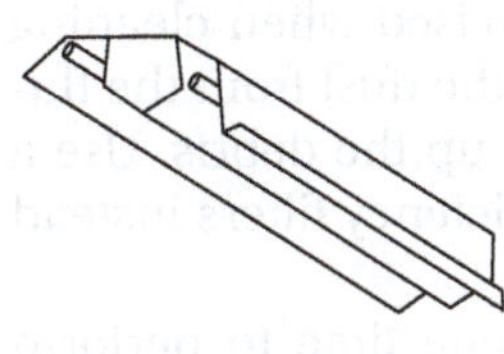

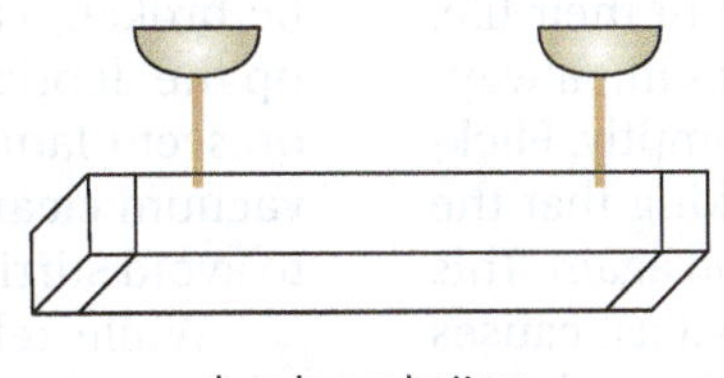

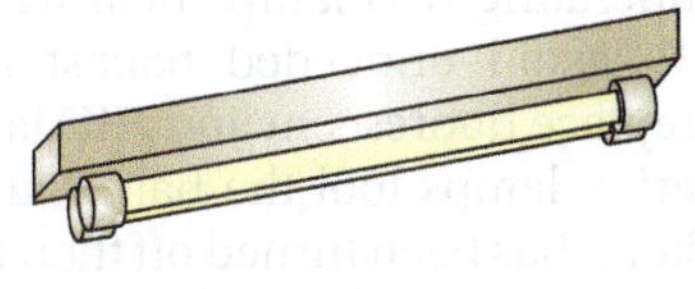

FIGURE 26-5 Various luminaires.

The different styles of luminaires use different types of lamps. Incandescent luminaires are designed to use incandescent lamps, fluorescent lamps must be installed in fluorescent lamp luminaires, and HID lamps must only be installed in luminaires designed for HID lamps. Just as the different types of lamps produce different types and qualities of light, the different luminaires are designed to be mechanically mounted by different methods. Some luminaires use a pendant style of mounting, some are surface mounted, some recessed mounted, and so forth. The variations in luminaires also allow for different illumination patterns.

26-3 BALLASTS

A ballast is a transformer that limits the lamp current while providing the necessary high-starting voltage for fluorescent and HID lamps. Ballasts are affected by the ambient temperature. If a ballast is required to operate in low ambient temperatures, lamp starting may be difficult if not impossible. In addition, even if the the lamps light, flickering may occur. Likewise, if a ballast is required to operate in high ambient temperatures, overheating of the ballast may occur. This shortens the life expectancy of the ballast. It is important to match the lamps to the ballast to ensure that high temperatures are not created.

Many ballasts produce a low-level hum or buzz. Although normal, it may be undesirable. There are special *quiet* ballasts, which are better suited for installation in office areas, or areas where the normal ballast noise may be objectionable. Ballasts are also known to produce radio interference, which can be a problem for AM radios and sound systems (public address or background music). Increasing the distance between the ballast and the radio or sound system can minimize radio interference. In addition, providing a good ground system for the radio or sound system, adding filtering to the power lines for the radio or sound system, or connecting the radio or sound system to a separate power circuit from the ballast can help minimize the interference.

Operating fluorescent lamps whose ends have darkened considerably or fail to light may damage ballasts. HID ballasts may also be damaged by operating HID lamps nearing the end of their life. To avoid unneeded ballast replacement, always replace fluorescent and HID lamps promptly. Flickering lamps fool the ballast into thinking that the lamp has been turned off then turned on again. This constant on-off-on-off cycling of the ballast causes it to overheat and eventually fail. Often when a ballast fails, a unique odor is produced. Once you have smelled an overheated ballast, you will always remember the distinctive odor. Follow ing the odor while looking for a luminaire that is no longer illuminated can greatly assist you in troubleshooting.

Different luminaires contain different numbers of lamps. In addition, different luminaires operate from different voltages. It is important to use the correct ballast for the number of lamps and operating voltage. **Figure 26-6** shows typical ballast wiring methods. These methods are based on the number of lamps and the operating voltage. Always follow the wiring diagram on the ballast that you are installing.

26-4 RELAMPING

Relamping is the process of replacing the lamps in a luminaire. Typically, lamps are replaced as needed, and only the burned-out ones are replaced. However, this may result in uneven lighting conditions. It is usually more cost-effective in the end to perform **group relamping**, which means that all lamps within an area are replaced when they have reached 60% to 80% of their rated life expectancy. Although this may seem wasteful, it is actually more efficient because as fluorescent and HID lamps age, their light output diminishes, but they continue to consume the same amount of power. With group relamping, all lamps, whether burned out or not, are replaced, increasing the illumination levels while using the same amount of electrical power. For this practice to pay off and make sense, documentation must be maintained to determine when the lamps have reached 60% to 80% of their average life. This can be accomplished by manually tracking the hours of use, or by installing an elapsed time meter on the lighting circuit.

Relamping should be performed with the circuit de-energized and all recommended PPE worn. Recall that fluorescent and HID lamps use high voltage to operate. Some lamps can explode when inserted into an energized luminaire. Others are damaged by the oils secreted by the hands. When handling the glass envelope with bare hands an oily residue is left on the glass. Then when the lamp is energized, the oil causes the envelope to fail and the lamp explodes. It is, therefore, good practice to wrap a clean cloth around the lamp or wear clean gloves when handling it. Should a fluorescent lamp be broken, care must be exercised when cleaning up the debris. Avoid inhaling the dust from the fluorescent lamp. Do not sweep up the debris. Use a vacuum cleaner with high-efficiency filters instead to avoid stirring up the dust.

While relamping, take some time to perform additional maintenance on the luminaires by removing dust, dirt, and grime. This can have a noticeable effect on the light output of the luminaire.

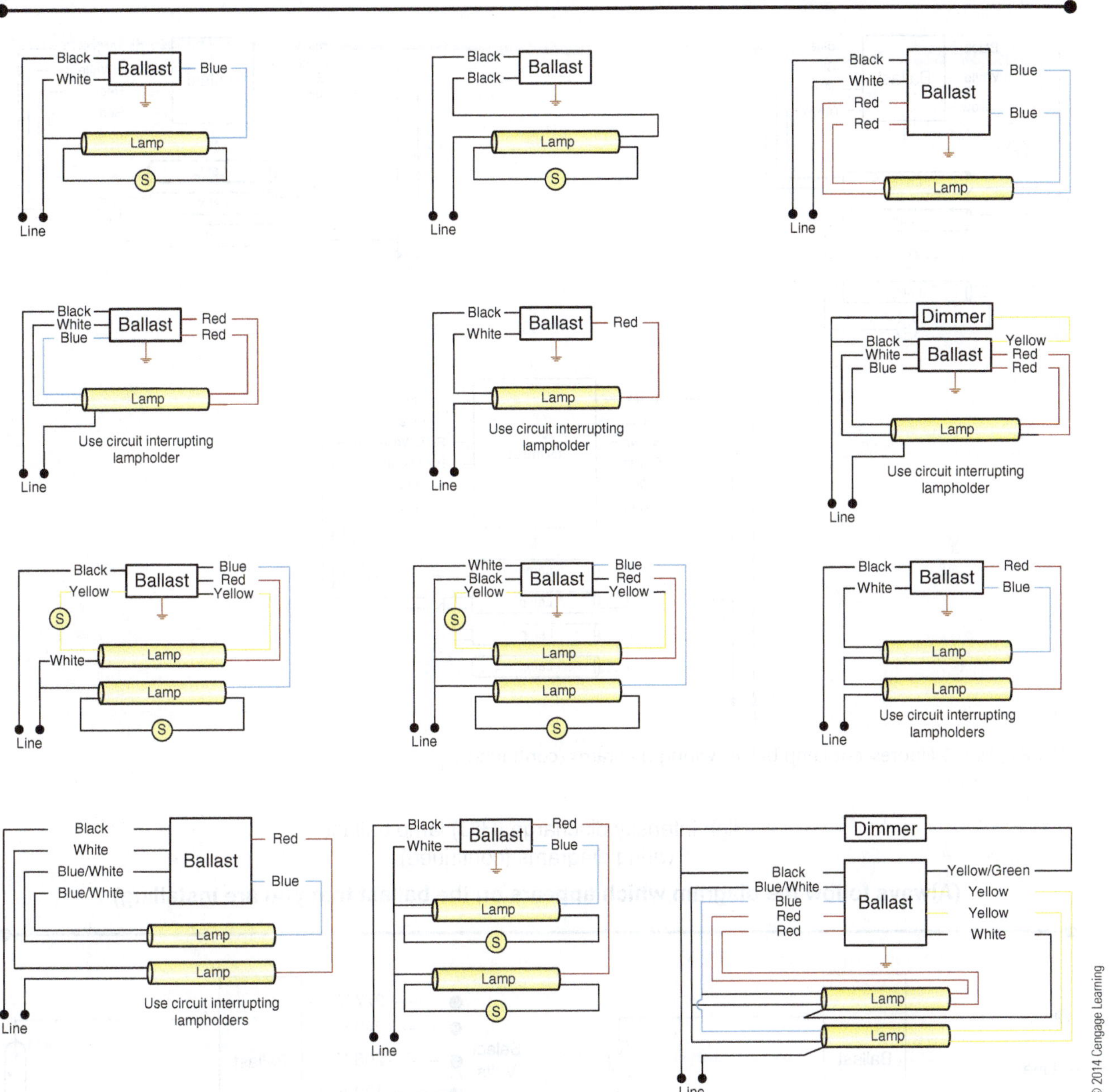

FIGURE 26-6A Fluorescent lamp ballast wiring diagrams.

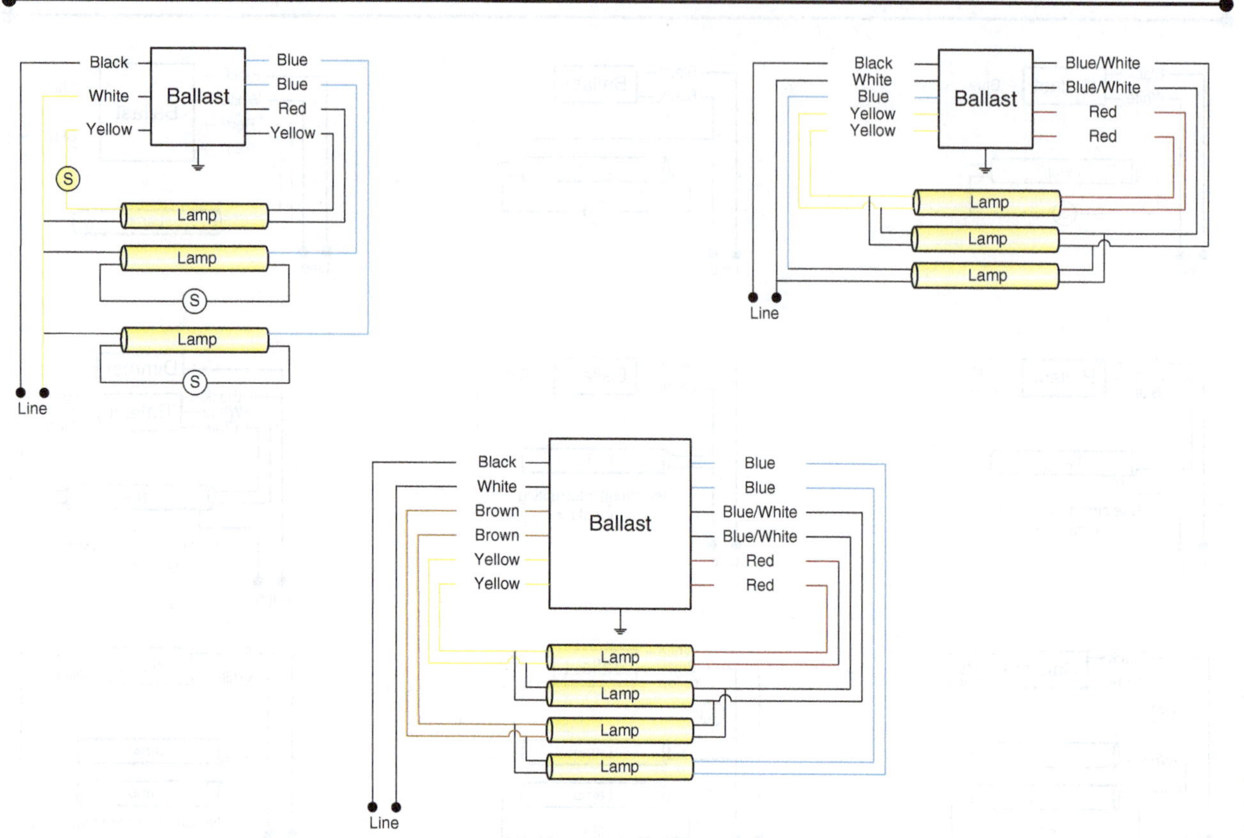

FIGURE 26-6B Fluorescent lamp ballast wiring diagrams (continued).

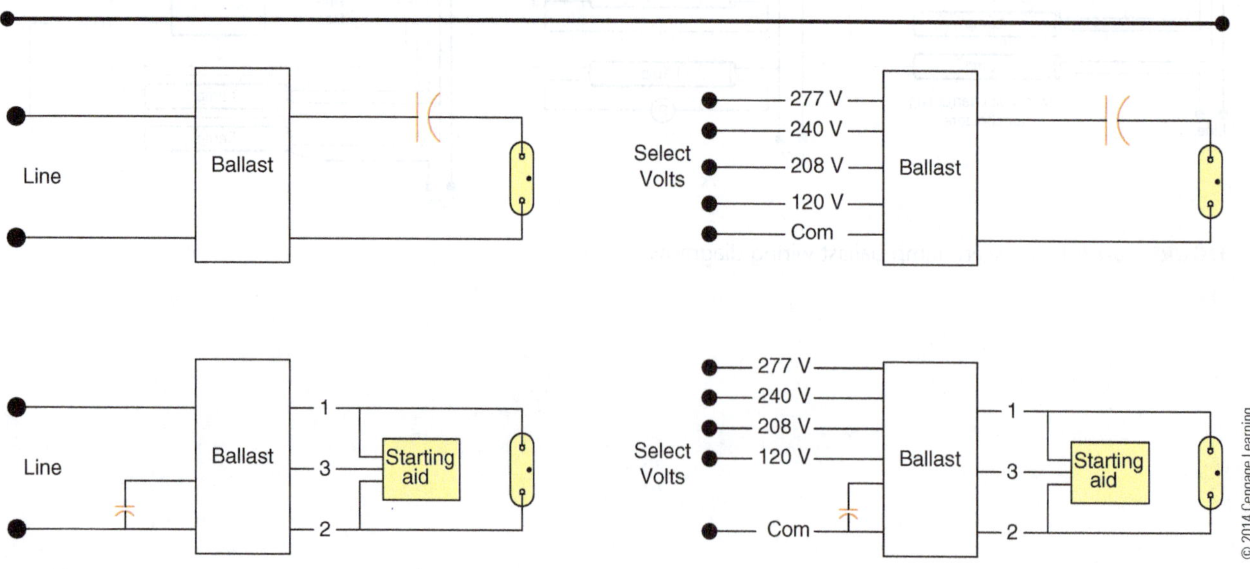

FIGURE 26-6C HID lamp ballast wiring diagrams.

SUMMARY

- The three basic types of lamps are the incandescent lamp, the fluorescent lamp, and the high-intensity discharge (HID) lamp.
- Incandescent lamps produce light when current flows through a filament. The filament is attached to the bulb's base and encapsulated within a gas-filled glass envelope.
- Fluorescent lamps consist of a glass tube, which is coated on the inside with a phosphor. The glass tube is filled with a gas. These lamps produce light by allowing an electrical current to flow through the gas. Fluorescent lamps do not have filaments; they have heating elements at each end of the glass tube.
- When power is applied to the fluorescent lamp, the heating elements heat up, lowering the resistance of the gas within the glass tube. At this time, high voltage is applied to the ends of the lamp, causing current to flow through it.
- HID lamps consist of a glass envelope, a base, and an arc tube. The arc tube is filled with a gas, and a set of electrodes is inside the arc tube.
- When an HID lamp is energized, a ballast provides high voltage to the electrodes within the arc tube. An electrical arc is created between the electrodes and a current passes through the gas, filling the arc tube. The arc is the source of light produced by the HID lamp.
- Luminaires are available in different styles and shapes. Incandescent luminaires are designed to use incandescent lamps, fluorescent lamps must use fluorescent luminaires, and HID lamps must use HID luminaires.
- A ballast is a transformer that limits the lamp current while providing the necessary high-starting voltage for fluorescent and HID lamps.
- Relamping means replacing the lamps in a luminaire. Group relamping means replacing all lamps within an area when they have reached 60% to 80% of their rated life expectancy.

REVIEW QUESTIONS

1. Name the three most common types of lamps used in industry.
2. List six items of concern when selecting the proper lamp for an application.
3. What is a luminaire?
4. What is the purpose of a ballast?
5. What is meant by the term *group relamping*?

SECTION 4

WELDING KNOWLEDGE

Gas Welding

Arc Welding

A TYPICAL DAY IN MAINTENANCE

Check It Out

You feel re-energized after a good lunch and a break. Upon returning to the press, you evaluate the crack in the pump mounting plate. You believe that the crack occurred because of the vibrations produced by the worn bearing.

You obtain a portable welder and all the necessary equipment and PPE. You also complete the required permits and documentation so that you can begin welding the plate. It is 1:30 P.M. when you finish.

Work It Out

1. How would you determine whether you should use gas or arc welding?
2. What type of PPE should you use when welding the pump mounting plate?
3. What special precautions or concerns should you have when performing the welding operation?

CHAPTER 27

Gas Welding

Gas welding equipment is typically used for bending, cutting, straightening, preheating, bronze welding, and welding of most low-grade steels. All of this is done by using a controlled mixture of oxygen and acetylene. This controlled mixture provides a workable flame that is easy to control while in use. The flame can be changed through the use of valves to accomplish many different tasks. There are many newer forms of welding, but some of them cannot do the same work as gas welding. Because of this, oxyacetylene welding has been around for a long time and probably will be for a long time to come. This chapter discusses the basics of oxyacetylene welding. As in many of the other chapters in this book, the scope cannot be too broad on this subject. There are many other things that the oxyacetylene torch can be used for in the realm of industry that are not mentioned in this chapter.

OBJECTIVES

After studying this chapter, the student should be able to

- Discuss the safety concerns of acetylene.
- Demonstrate the safe handling of cylinders.
- Visually recognize the difference between the oxygen system and the acetylene system.
- Properly perform the startup procedure.
- Properly perform the shutdown procedure.
- Describe five types of welds.
- Describe five types of joints.
- Discuss the four welding positions.

27-1 GAS WELDING SAFETY

Because oxygen and acetylene are dangerous gases, especially around a flame, it is important to go over some gas welding safety first.

Acetylene is a colorless gas that has a distinct and definitive odor. Some say that it has an odor similar to that of garlic. Acetylene is the combustible gas that is used in combination with oxygen to create the flame at the tip of the torch. When it is under low pressures, acetylene remains stable but becomes extremely unstable when put under pressures in excess of 15 psi. Because of this, all cylinders that are used to contain the acetylene have a **fusible plug** in the bottom that acts like a safety valve, which releases any excess pressure caused by too much heat or pressure within the cylinder.

The oxygen is also contained within a cylinder. The **cylinder valve**, which is at the top of the cylinder, controls the flow of acetylene or oxygen. It is generally acceptable to open the oxygen cylinder valve fully, whereas it is only acceptable to open the acetylene cylinder valve one turn. This is so that in the event of an accident, the acetylene cylinder valve can be shut off quickly. When the cylinders are not in use, **protector caps (Figure 27-1)** should be placed over the cylinder valves to protect them from being hit or damaged accidentally.

It is also required that all tanks be stored in the upright position and secured by a chain to prevent the cylinders from falling over. If the cylinders are being transported, it is imperative that the protector caps be on the cylinders.

FIGURE 27-1 Cylinders with protection caps.

Review the following common safety practices before working around oxygen and acetylene cylinders:

- Never lift a cylinder by the valve protector cap.
- Never slide or drag a cylinder because this may cause a spark and thus an explosion.
- Do not expose the cylinder to excessive heat.
- Turn off all the valves when moving the cylinders.
- Always crack a cylinder valve slowly when opening.
- Notify the supplier of any faulty cylinders or valves.
- Always secure the cylinders with a chain.

The following are safety concerns when using oxyacetylene equipment:

- Always make sure that the adjusting screw is backed out all the way, and close the regulator before opening the cylinder valve.
- Never stand over a cylinder valve when opening it.
- Never stand in front of a regulator when opening a cylinder valve.
- Do not use grease or oil to lubricate regulators because this may cause an explosion.
- Never exceed 15 psi on the acetylene side because this may also cause an explosion.

The following are safety concerns to keep in mind during the welding operation:

- Remove all flammable trash from the area where the welding will be performed.
- Always have a fire extinguisher available in the event that a fire starts accidentally.
- Always use a spark lighter to light a torch, not a cigarette lighter or a match.
- Always point the torch toward the floor when lighting it.
- Ensure proper ventilation when welding.

The most important thing to consider, however, is the personal safety of the welder.

- Always wear the appropriate PPE. This may include, but is not limited to, gloves, welding goggles, face shield, welder's cap, welder's apron, and steel-toed boots.
- Use a No. 5 shade lens when performing acetylene welding.

- Always ensure that your shirt is buttoned all the way up to the neck.
- Ensure that all pockets are closed and buttoned.
- Ensure that shirt sleeves are not cuffed so molten metal cannot get caught in them.
- Watch for hot metal. Use pliers to handle all hot metals, not your gloves, because this shortens the life of your gloves.
- Never use pressurized oxygen to blow off your clothes or workstation.

The **welding torch**, sometimes referred to as a blowpipe, is where the oxygen and the acetylene is mixed. **Figure 27-2** shows two types of gas torches. One is used for welding and the other is used for cutting.

The most common type of torch that is used is the equal pressure type. The oxygen and acetylene remain separated throughout the entire system until they are brought together in the torch handle. Oxygen and acetylene flow is controlled by an oxygen needle valve and an acetylene needle valve. These valves can be seen on both torches in **Figure 27-2**.

CAUTION

The gas system is color-coded for safety reasons. The components that are used on the oxygen side of the system are color coded green. The fittings that are used to connect the oxygen system together use right-hand threads. On the acetylene side of the system, the components that are used are color-coded red. The fittings that are used to connect the acetylene system together use left-hand threads. All of the fitting collars on the acetylene side also have a groove machined into them. It is important to know that the torch handle has an AC or F at the acetylene inlet port and an O or OX at the oxygen inlet port. This helps you recognize which hose connects to which gas, thus avoiding threading a left-handed threaded connector onto a right-handed thread inlet port, which damages the threads.

Regulators perform two functions:

1. They reduce the pressure of a gas that is coming from a cylinder to a much lower working pressure.
2. They allow the working pressure and flow to remain steady under changing cylinder pressure.

There are two gauges that are mounted on the regulator for both the oxygen and the acetylene. These gauges indicate the pressure within the cylinder and the working pressure. The regulator is connected to the output of the cylinder valve, which is at the top of the cylinder. These can be seen in **Figure 27-3**.

Closing this valve stops the gas from flowing out of the cylinder. The right-hand gauge on both regulators (oxygen and acetylene) shows the bottle pressure. The gauge on the left-hand side of the regulators

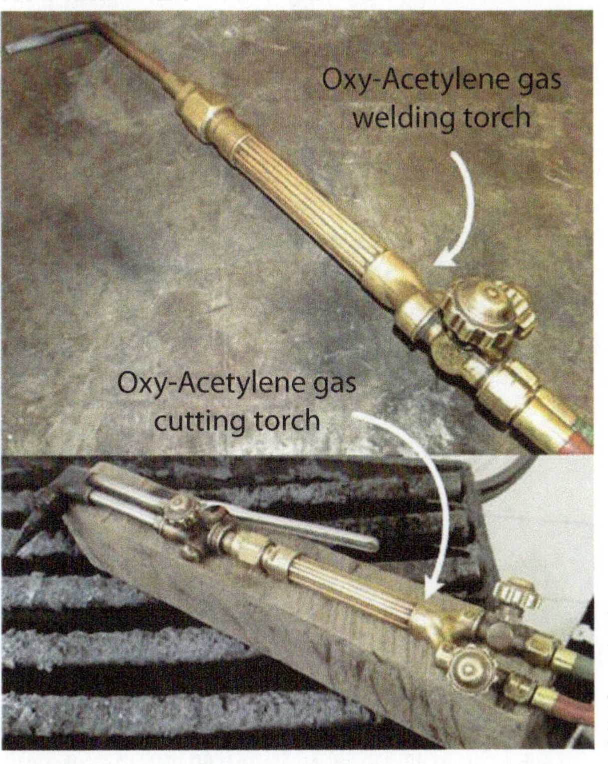

FIGURE 27-2 Gas torches.

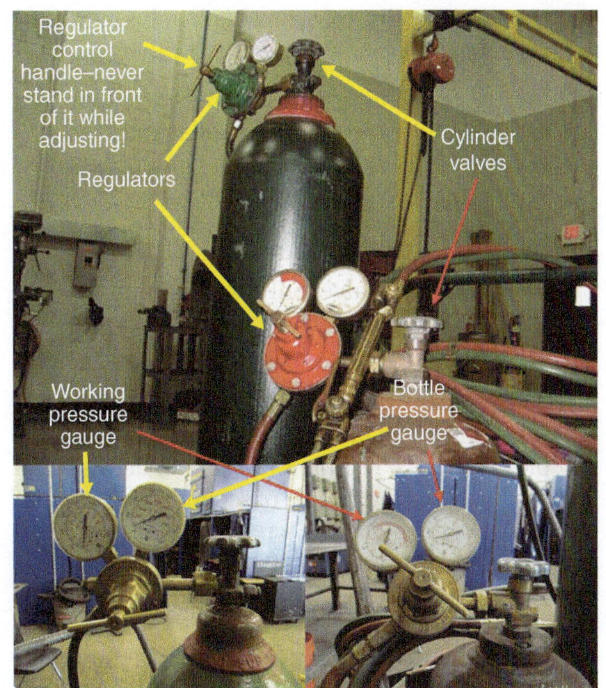

FIGURE 27-3 Oxygen and Acetylene gas regulators.

shows the working pressure. Notice in Figure 27-3 the color coding that is used throughout the system so that the regulators, bottles, hoses, and in some cases even the gauges can easily be identified as being in the oxygen system or the acetylene system.

27-2 SETUP PROCEDURES

It is imperative that the setup procedures be followed as a system is set up. Follow this list to ensure the safe operation of the system.

1. Inspect all of the equipment and threads before assembly.
2. Slowly crack the oxygen cylinder valve to blow out any debris that may be present in the valves. This should be a quick burst and should last for more than a second or two, then close the valve completely.
3. Repeat the same process for the acetylene cylinder.
4. Inspect both of the regulators to ensure that the adjustment valves are turned all the way out to close the valve. Once this has been verified, connect both of the regulators to their appropriate cylinders.
5. Connect the reverse flow check valves to the output of the regulators. Keep in mind that all of these components should be color-coded.
6. Connect the hoses to the check valves. Green connects to the oxygen check valve, and red connects to the acetylene check valve.
7. Before continuing, it is imperative that you never stand in front of a regulator as the cylinder valve is opened. If the regulator is faulty, it may explode, causing injury or even death. Stand to the side of the cylinder and slowly open the cylinder valves, remembering to open the acetylene valve only one complete turn. The oxygen cylinder valve can be completely opened at this time.
8. Slowly open the oxygen regulator to 3 or 4 psi. Allow the gas to blow through the hose for a couple of seconds. This blows any dirt or debris out of the hose. Once a couple of seconds have passed, back out the adjusting screw and close the regulator. Repeat the process for the acetylene side.
9. Connect the hoses to the torch handle.
10. Check the connections for leaks. It is important that all leaks be found before any welding starts. Close the needle valves that are on the torch handle, then open both of the regulators to about 5 psi. With the system pressurized, apply soapy water to all of the connecting points starting at the cylinder valve and working your way up to the needle valves on the torch handle. If a leak is present, bubbles form at the leak. If a leak is found, try to tighten the connection. If the leaking persists, replace the leaking component with a new one. Do not use a setup that leaks, because this could be extremely dangerous. If no leaks are found, proceed with the rest of the setup.
11. Select the size of the tip that is needed for the job at hand and locate the working pressure that is used for the chosen tip. Refer to the manufacturer's chart to get the proper operating pressures. An example of this chart is shown in **Table 27-1**.
12. After making sure that the needle valves are closed, insert the tip into the torch handle. Finger-tighten the connecting nut.
13. Adjust the oxygen regulator to the proper working pressure.
14. Adjust the acetylene regulator to the proper operating pressure.
15. Making sure that the oxygen needle valve is closed on the torch handle, open the acetylene needle valve about ¼ turn.
16. Point the tip of the torch toward the floor and ignite the acetylene with the spark lighter. You should notice a heavy yellow flame with a lot of thick, wispy smoke coming off the end of the flame.

Table 27-1

A MANUFACTURER'S SUGGESTED OPERATING PRESSURE CHART

Metal Thickness	Center Orifice Size: Drill Size	Center Orifice Size: Tip Cleaner	Oxygen Pressure in psi	Acetylene Pressure in psi
⅛	60	7	10	3
¼	60	7	15	3
⅜	55	11	20	3
½	55	11	25	4
¾	55	11	30	4
1	53	12	35	4
2	49	13	45	5
3	49	13	50	5
4	49	13	55	5
5	45	-	60	5

17. Continue to open the acetylene needle valve slowly until all of the smoke leaves the end of the flame.
18. Open the oxygen needle valve slowly until a well-defined white cone appears at the tip of the torch. A longer, translucent blue cone appears around the smaller white cone. This is referred to as a *neutral flame.*
19. Begin your work.

Depending on the welding job, different flames are used for different metals. The proper procedures to set up these flames are discussed later in this chapter.

27-3 SHUTDOWN PROCEDURES

You must follow a definite sequence when shutting down an oxyacetylene torch.

1. First, close the acetylene needle valve. This extinguishes the flame immediately.
2. Close the oxygen needle valve.
3. Close both of the cylinder valves.
4. Open the acetylene needle valve while watching the gauges on the acetylene regulator. The pressures should drop to zero. When this happens, close the acetylene needle valve and repeat this procedure for the oxygen side.
5. Release both of the adjusting screws on the regulator to close the regulators.
6. Neatly coil the hose out of the way and allow the torch tip time to cool. Store the cylinders in a safe place until they are needed again. Do not forget to secure the cylinders with a chain.

If it will be some time before the equipment is used again, disassemble it in the reverse order that it was assembled and store the parts in their appropriate areas.

27-4 TYPES OF FLAMES

There are three types of flames produced when using an oxyacetylene torch:

- Neutral flame
- Carburizing flame
- Oxidizing flame

Neutral Flame

To produce a **neutral flame**, adjust the acetylene regulator to the proper operating pressure. Making sure that the oxygen needle valve is closed on the torch handle, open the acetylene needle valve about ¼ turn. Point the tip of the torch toward the floor and ignite the acetylene with the spark lighter. A spark lighter can be seen in **Figure 27-4**. You should notice a heavy yellow flame with a lot of thick, wispy smoke coming off the end of the flame. Continue to open the acetylene needle valve slowly until all of the smoke leaves the end of the flame. Open the oxygen needle valve slowly until a well-defined white cone appears at the tip of the torch. A longer, translucent blue cone appears around the smaller white cone. This is referred to as a *neutral flame*. When the neutral flame has been achieved, a quieter purring noise should be heard rather than an irritating hissing noise. The temperature of the tip of the white cone of this type of flame is in excess of 6300°F. A neutral flame is used for welding all metals, including cast iron, malleable iron, wrought iron, and high-strength rods.

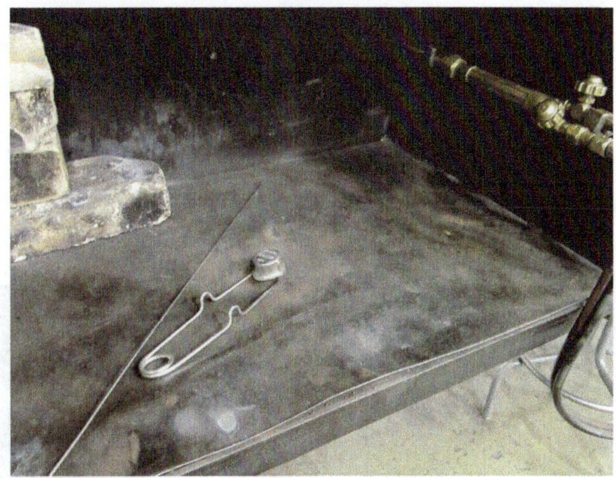

FIGURE 27-4 A spark lighter.

Carburizing Flame

The **carburizing flame**, or reducing flame, has three distinct flame zones, which are shown in **Figure 27-5**. Within the three different cones in the figure, you should recognize a small, bright, white cone at the tip, which is very similar to that of the neutral flame. It is referred to as the *inner cone.* The larger cone that surrounds it is the *intermediate cone.* Also notice the longer, translucent blue cone that surrounds the intermediate cone. If this flame is set up correctly, the intermediate cone is twice the length of the inner cone. This type of flame is set up with the same amount of acetylene as the neutral flame, but with less oxygen. The temperature of the tip of the inner cone on this type of flame is approximately 5700°F (3149°C). This type

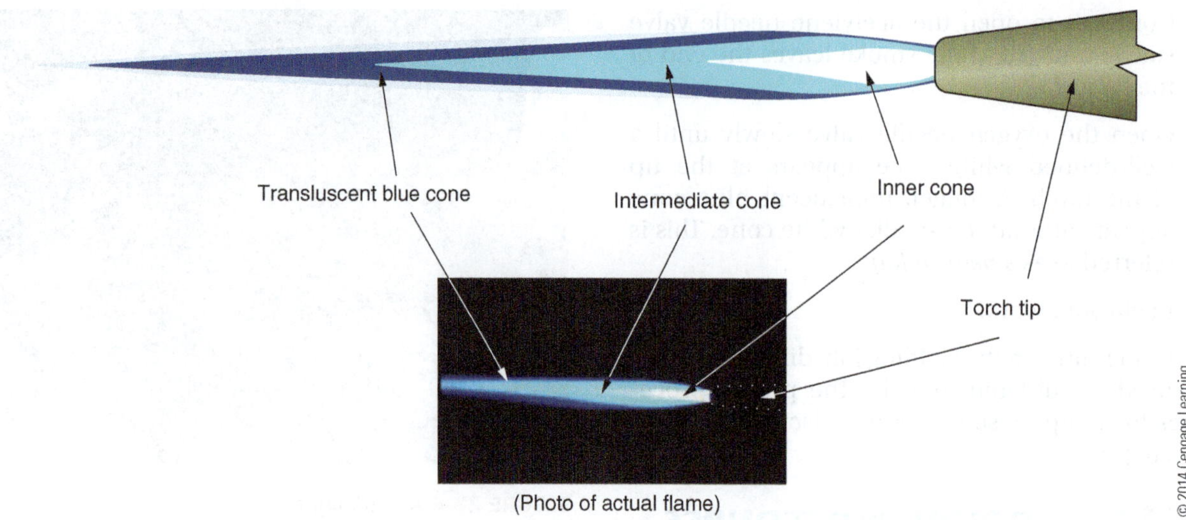

FIGURE 27-5 A carburizing flame.

of flame is used when brazing or soldering low-carbon steel, cast steel, nickel steel, chrome steel, copper, and brass.

Oxidizing Flame

The **oxidizing flame** is produced the same way as the neutral flame, with one exception. The oxidizing flame uses more oxygen. This type of flame is accompanied by a loud hissing noise. The cone of this flame is smaller than that of the neutral flame, and it has a sharper, more defined point at its tip. This is the least used flame of the three. It has a temperature of approximately 6300°F (3482°F) and is used for welding sheet brass, cast brass, or bronze.

27-5 PUDDLING AND RUNNING A BEAD

A **puddle** is a molten pool of metal that is formed when the tip of the flame causes the base metal to melt. To form a puddle without a filler rod, hold the tip of the neutral flame about 1/8 in. from (or off) the surface of the metal. The flame should be at a 45° angle, pointing in the direction in which you are going to weld. Start the puddle on the right side if you are right-handed, or the left side if you are left-handed, and work across the metal away from where you start the puddle. A good puddle is approximately 3/16 in. to 1/8 in. in diameter. As the puddle is formed, move across the base metal, making small circular movements, and never allowing the tip of the inner flame to touch the puddle. The speed of the flame is critical. If you move too fast, you lose the molten pool and cause the weld to be narrow. If you move the flame too slowly, the base metal may melt through and the weld will be too wide. Do not allow the flame to rise too high off the base metal because this causes the metal to oxidize. This makes for a weak weld, and it will appear to be scaly.

A **filler rod** may be used to build up a bead. A filler can be seen in Figure 27-4, lying behind the spark lighter. The filler rod should not be any thicker than the metal to be welded, and it should be made of the same material as the base metal. It is a good idea to bend the nonworking end of the filler rod to identify its safe end. When using a filler rod to build up a bead, a puddle should have already been formed before the filler rod is used. As you begin to move the puddle across the plate, momentarily insert the rod tip into the pool of molten metal, keeping the flame off the filler rod. As soon as the filler rod touches the molten pool, it begins to melt. The rod should be pulled back out to prevent an excess buildup of metal. Once the rod has been dipped into the pool and removed, make a small circle to move the puddle forward. Continue to insert the tip of the rod into the pool until the bead is finished. If the rod sticks while you are trying to weld, the puddle is not hot enough. Slow down the flame to allow the metal more time to melt.

27-6 TYPES OF WELDS

There are five common types of welds that are used when gas welding: the surfacing weld, the fillet weld, the groove weld, the plug weld, and the slot weld.

The **surfacing weld** is used to build up a surface by running one or more stringers or weave beads across an unbroken surface.

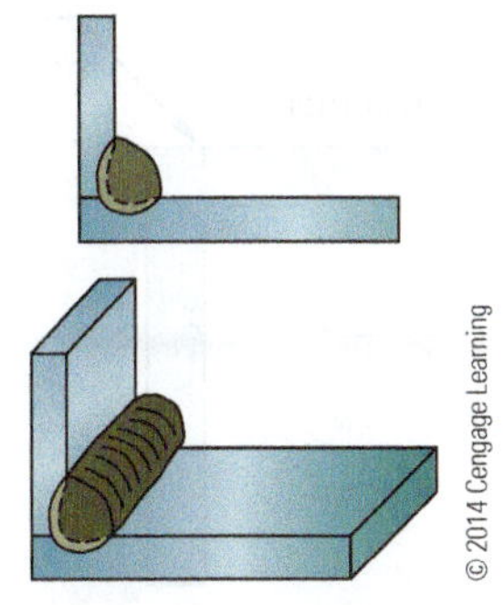

FIGURE 27-6 A fillet weld.

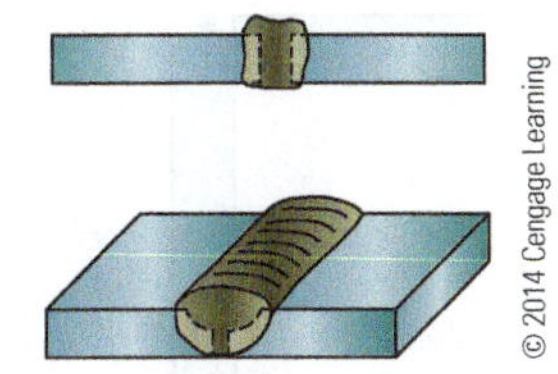

FIGURE 27-7 A groove weld.

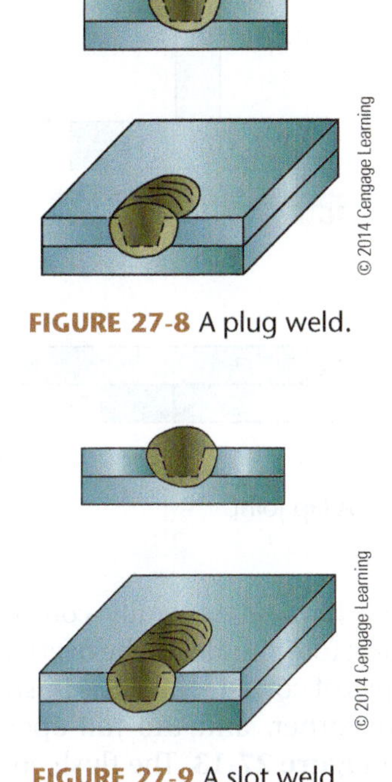

FIGURE 27-8 A plug weld.

FIGURE 27-9 A slot weld.

The **fillet weld** joins two surfaces at right angles to each other. This is illustrated in **Figure 27-6**.

The **groove weld** fills a groove between two pieces of metal that are to be joined. The ends of the pieces to be welded may be square, beveled, U-shaped, or J-shaped, as shown in **Figure 27-7**.

The **plug weld** is a little different because of the fact that it is accomplished by overlapping a piece of metal with a hole in it over another solid piece of metal, then welding the two pieces together through the hole. See **Figure 27-8**.

The **slot weld** is very similar to the plug weld, with one exception. It is accomplished by overlapping a piece of metal with a slot cut into it over another solid piece of metal, then welding the two pieces together through the slot. See **Figure 27-9**.

27-7 TYPES OF JOINTS

Five types of joints are used in welding today: the butt joint, tee joint, lap joint, corner joint, and edge joint.

The **butt joint** is when two pieces of metal are welded together at their edges as if they were lying on a table side-by-side. The edges must be close to parallel with a root opening between the plates. There are different methods to prepare the edges, such as the square, beveled, double-V, U- or double-U-type, or J-type butt joints. These are illustrated in **Figure 27-10**.

The **tee joint** is accomplished by placing the edge of one piece of metal on the surface of another, thus creating a right angle. See **Figure 27-11**. This type of joint requires a fillet weld on one or both sides. This type of weld is usually used for light to medium thick material.

The **lap joint** is the strongest joint. It is made by lapping one piece of metal over another and welding at the end of the top piece of metal. This is illustrated in **Figure 27-12**. The overlap of the metals should be at least four times the metal thickness. There are two basic types of lap joints: the single fillet and the double fillet. The single fillet is used for metals that are greater than ½ in. thick, whereas the double fillet is used for metals that are less than ½ in. thick.

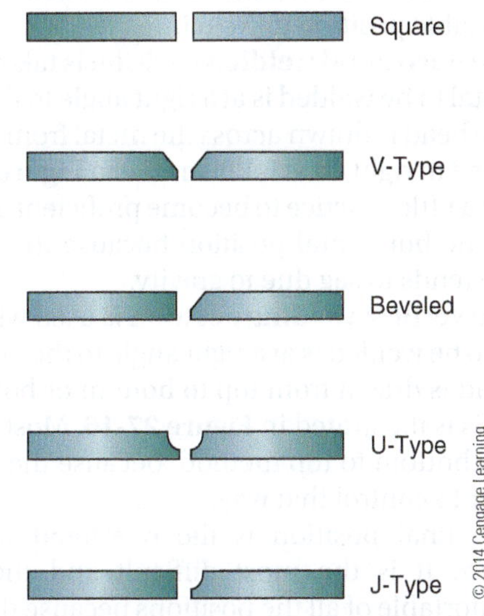

FIGURE 27-10 Different types of butt joints.

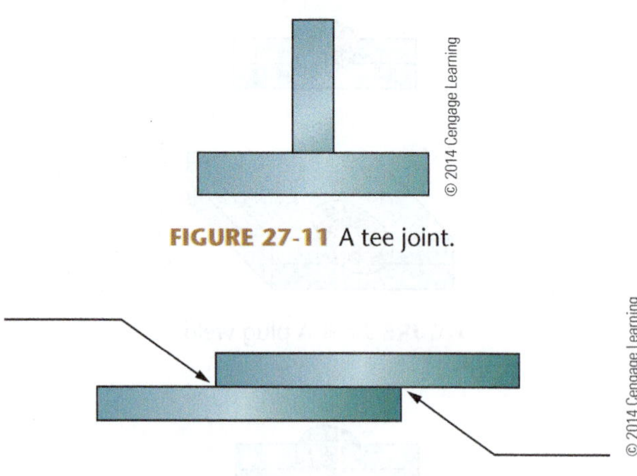

FIGURE 27-11 A tee joint.

FIGURE 27-12 A lap joint.

A **corner joint** is generally considered to be weak and therefore should not be overloaded. There are three types of corner joints: the flush corner, the half-opened corner, and the full-opened corner, as shown in **Figure 27-13**. The flush corner is considered to be the weakest of the three, whereas the full-opened joint is considered to be the strongest.

The **edge joint** is usually reserved for use on metals of ¼ in. thickness or less and should only be lightly loaded. It is also considered to be a very weak joint. An edge joint is illustrated in **Figure 27-14**.

27-8 WELDING POSITIONS

There are four different welding positions that a welder may use when welding: flat, horizontal, vertical, and overhead.

The **flat welding position** is probably the simplest of the four. When welding in this position, the material should be parallel to the floor as if it were lying on a table or workbench. This is often the most comfortable position to weld in.

The **horizontal welding position** is taken when the metal to be welded is at a right angle to the floor, and the bead is drawn across the metal from right to left or left to right. This is illustrated in **Figure 27-15**. It takes a little practice to become proficient at welding in the horizontal position because the molten puddle tends to sag due to gravity.

The **vertical welding position** is used when the metal to be welded is at a right angle to the floor and the bead is drawn from top to bottom or bottom to top. This is illustrated in **Figure 27-16**. Most people use the bottom to top method, because the puddle is easier to control that way.

The final position is the **overhead welding position**. It is the most difficult and the most uncomfortable of all the positions because the molten metal has a tendency to form drops and fall on the welder. Even many experienced welders find this position a challenge to weld in. It takes a lot of practice to become good at this position.

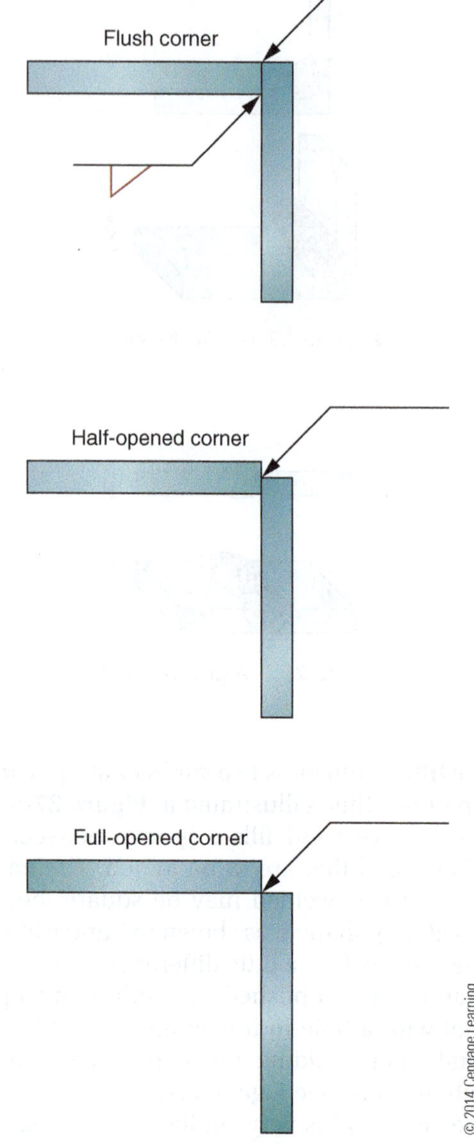

FIGURE 27-13 Three types of corner joints.

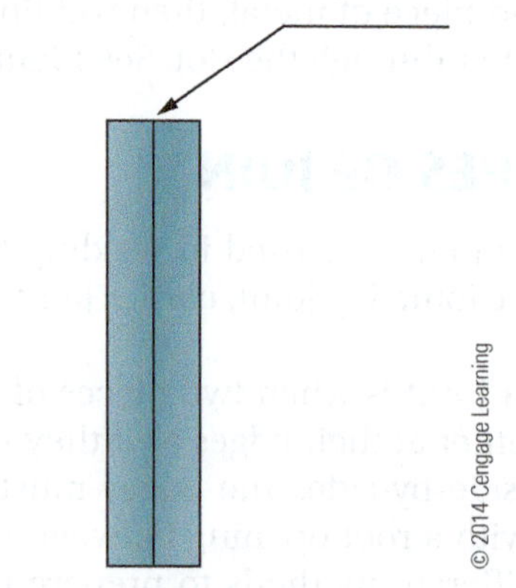

FIGURE 27-14 An edge joint.

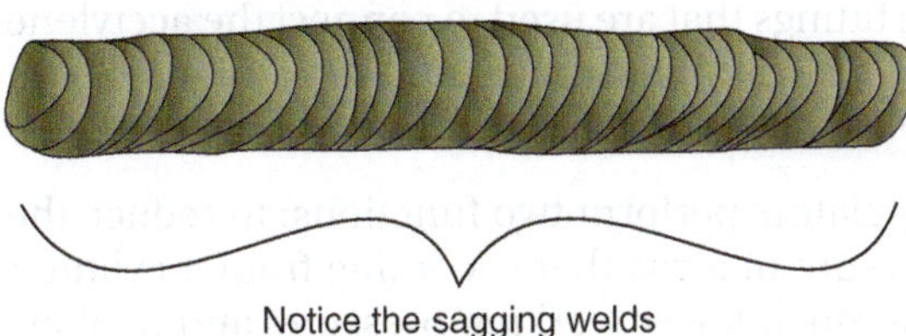
The bead shows signs that the puddle sagged while welding. This is due to gravity
Notice the sagging welds

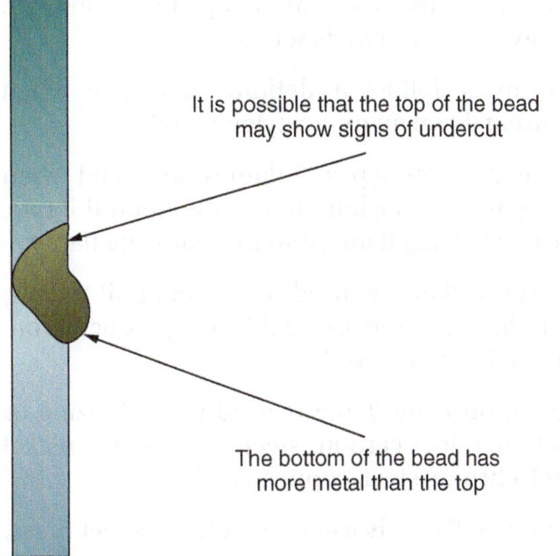
It is possible that the top of the bead may show signs of undercut
The bottom of the bead has more metal than the top
© 2014 Cengage Learning

FIGURE 27-15 The horizontal welding position (puddle sag).

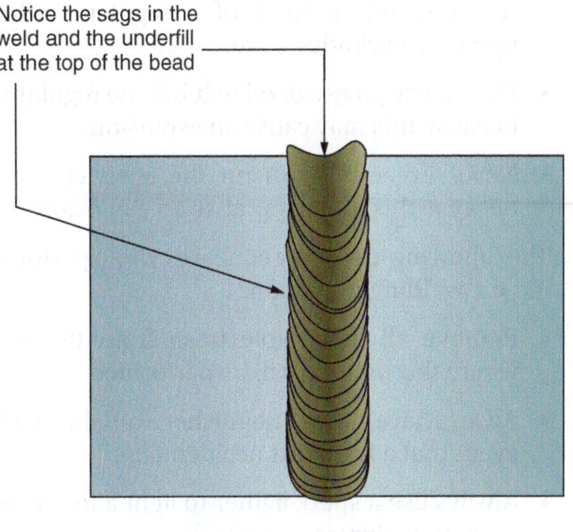
Notice the sags in the weld and the underfill at the top of the bead

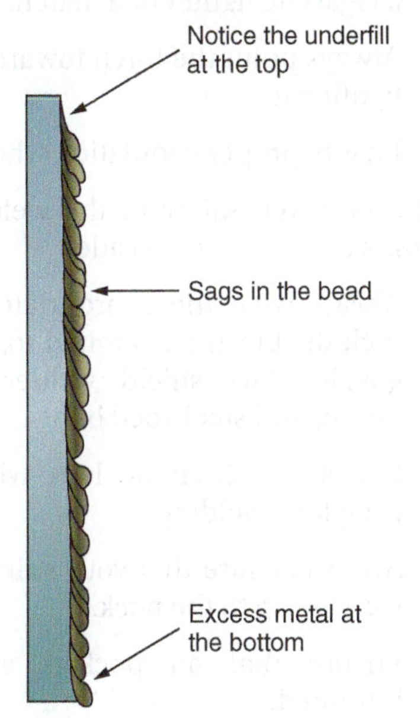
Notice the underfill at the top
Sags in the bead
Excess metal at the bottom
© 2014 Cengage Learning

FIGURE 27-16 Vertical position (welding from the bottom up).

SUMMARY

- Gas welding is used for bending, cutting, straightening, preheating, bronze welding, and welding of most low-grade steels.
- Acetylene is a colorless gas that has a distinct and definitive odor.
- It is generally acceptable to open the oxygen cylinder valve fully, whereas it is only acceptable to open the acetylene cylinder valve one turn. This is so that in the event of an accident, the acetylene cylinder valve can be shut off quickly.
- When the cylinders are not in use, protector caps should be placed over the valves.
- The following are some cylinder safety guidelines:
  - Never lift a cylinder by the valve protector cap.
  - Never slide or drag a cylinder because this may cause a spark and, thus, an explosion.
  - Do not expose the cylinder to excessive heat.
  - Turn off all the valves when moving the cylinders.
  - Always crack a cylinder valve slowly when opening.
  - Notify the supplier of any faulty cylinders or valves.
  - Always secure the cylinders with a chain.
- The following are safety concerns when using oxyacetylene equipment.
  - Always make sure that the adjusting screw is backed out all the way, closing the regulator before opening the cylinder valve.
  - Never stand over a cylinder valve when opening it.

- Never stand in front of a regulator when opening a cylinder valve.
- Do not use grease or oil to lubricate regulators because this may cause an explosion.
- Never exceed 15 psi on the acetylene side because this may also cause an explosion.

■ The following are safety concerns while performing the welding operation:

- Remove all flammable trash from the area where the welding will be performed.
- Always have a fire extinguisher available in the event that a fire starts accidentally.
- Always use a spark lighter to light a torch, not a cigarette lighter or a match.
- Always point the torch toward the floor when lighting it.
- Ensure proper ventilation when welding.

■ The personal safety of the welder is the most important thing to consider.

- Always wear the appropriate PPE. This may include, but is not limited to, gloves, welding goggles, face shield, welder's cap, welder's apron, and steel-toed boots.
- Use a No. 5 shade lens when performing acetylene welding.
- Always ensure that your shirt is buttoned all the way up to the neck.
- Ensure that all pockets are closed and buttoned.
- Ensure that shirt sleeves are not cuffed so that molten metal does not get caught in them.
- Watch for hot metal. Use pliers to handle all hot metals, not your gloves, because this shortens the life of your gloves.
- Never use pressurized oxygen to blow off your clothes or workstation.

■ The welding torch, sometimes referred to as a blowpipe, is where the oxygen and the acetylene is mixed. The most common type of torch that is used is the equal pressure type. The oxygen and acetylene remain separated throughout the entire system until they are brought together in the torch handle. Oxygen and acetylene flow is controlled by needle valves.

■ All of the components that are used on the oxygen side of the system are color-coded green, and the fittings that are used to connect the oxygen system together use right-hand threads. All of the components that are used on the acetylene side of the system are color-coded red, and the fittings that are used to connect the acetylene system together use left-hand threads and have a groove machined into them.

■ Regulators perform two functions: to reduce the pressure of a gas that is coming from a cylinder to a much lower working pressure, and to allow the working pressure and flow to remain steady under changing cylinder pressure.

■ It is imperative that the setup procedures are followed as a system is set up.

■ You must follow a definite sequence when shutting down an oxyacetylene torch.

■ There are three types of flames produced when using an oxyacetylene torch: the neutral flame, the carburizing flame, and the oxidizing flame.

■ A neutral flame is used for welding all metals, including cast iron, malleable iron, wrought iron, and high-strength rods.

■ The carburizing flame is used when brazing or soldering low-carbon steel, cast steel, nickel steel, chrome steel, copper, and brass.

■ Oxidizing flame is used for welding sheet brass, cast brass, or bronze.

■ There are five common types of welds that are used when gas welding: the surfacing weld, the fillet weld, the groove weld, the plug weld, and the slot weld.

■ There are five types of joints that are used in welding today: the butt joint, tee joint, lap joint, corner joint, and edge joint.

■ There are four welding positions that a welder may have to use when welding: flat, horizontal, vertical, and overhead.

REVIEW QUESTIONS

1. At what pressure does acetylene become unstable?
2. Referring to the cylinder valves, how should they be opened?
3. How can a person who is color-blind differentiate the oxygen fittings from the acetylene fittings?
4. Where are the gases mixed?
5. What two functions do the regulators perform?

CHAPTER 28

Arc Welding

Arc welding uses electric current to melt the metals together. It is less expensive to use when compared to gas welding. As the name implies, arc welding is performed by controlling an electric arc. The arc is produced when an electrode has a difference of potential with respect to the base metal and is held a small distance off the base metal. Current passes through the electrode and creates an arc over the gap of air, melting the metal in the process. Small particles of metal from the electrode also pass through the gap within the arc to bond with the base metal. Arc welding is widely used in industry. It is used for quick repair while creating a solid and secure joint. Welds that are made with the arc welder tend to be very strong because of the amount of penetration that is made by the arc.

OBJECTIVES

After studying this chapter, the student should be able to

- List several existing safety hazards caused by ultraviolet light.
- List the PPE that should be worn while arc welding.
- Explain the AWS numbering system used for electrode identification.
- Explain what the flux coating does for the weld.
- Demonstrate how to correctly set up a welding machine, and discuss the purpose of each adjustment.
- Demonstrate how to correctly strike an arc.
- Demonstrate how to correctly run a bead.
- Describe the types of welds that may be used while arc welding.
- Describe the types of joints that may be encountered while arc welding.
- Define *undercut* and *underfill.*

28-1 ARC WELDING SAFETY

Before discussing the operation of arc welding, some safety concerns must be mentioned. Any time an arc is produced, three types of light are produced: infrared light, white light, and ultraviolet light. The **white light**, which is a safe light, is what people are exposed to every day. This is the light that you can see. **Infrared light** waves are felt as heat. Because the infrared light can be felt, most people are not as susceptible to getting burned by it as they would be by ultraviolet light. **Ultraviolet light** is a type of light that can cause serious burns. It is dangerous because people cannot feel themselves getting burnt. The symptoms show up after exposure to ultraviolet light. Burning of the skin from arc welding is the same as burning of the skin from the sun. The ultraviolet light from the sun is what causes the skin to burn. It is important to completely protect yourself from these harmful rays of light. When welding, wear clothes that cover all of your skin. Another problem with ultraviolet light is that it is not good for the eyes. If a person were to look into an arc while welding, a flash burn or even blindness might occur. A **flash burn** is when the eye becomes burned due to the ultraviolet light entering the eye. For this reason, it is important to select the correct shade of lens when arc welding. When gas welding, a No. 5 lens is dark enough to protect the eyes. When arc welding, the light is so intense at the arc that a darker shade is required to protect the eyes. The lens should be dark enough to filter out the ultraviolet light, but not so dark that you should have to strain your eyes. New technologies have given birth to a new generation of welding hoods. When first introduced, they were extremely expensive but have become fairly economical in recent years. This new technology hood is known as a self-darkening hood. **Figure 28-1** shows an example of a self darkening hood.

Once the eyes and skin are protected, it is important to protect other people who may be in the welding area. They, too, can become exposed to the ultraviolet light and become burned. This can be accomplished by simply using a welding curtain that is placed around the welding area. If welding is performed out on the floor of the plant, a **portable welding curtain** should be erected around the job site to protect passersby.

Another safety precaution is to ventilate the welding area to provide fresh air and remove the fumes and smoke that are produced by arc welding. If there is no way of providing fresh air to the welding area, the welder should wear a breathing apparatus. A breathing apparatus provides a source of fresh air to welders as they are welding.

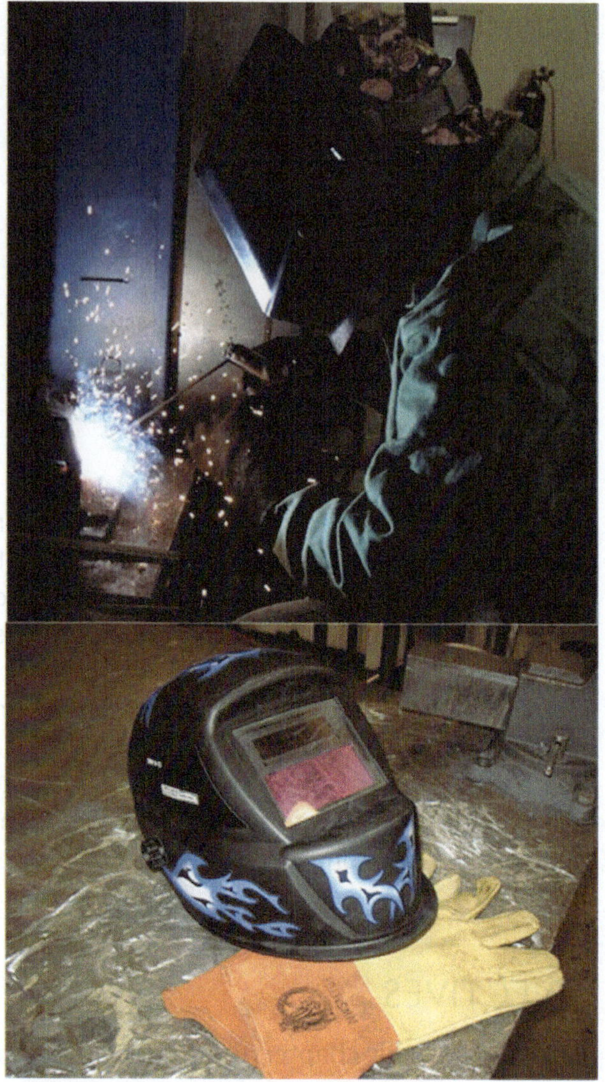

FIGURE 28-1 Auto-darkening (Self-darkening) hood.

As with the gas welding, remove all flammable trash from the welding area. Always have a fire extinguisher available in the event that a fire is started accidentally. Some facilities may be required to have a burn permit before any welding is to occur.

Always wear the appropriate PPE. This may include, but is not limited to, gloves, welding goggles, face shield, welder's cap, welder's apron, and steel-toed boots. Always ensure that your shirt is buttoned all the way up to your neck, that all pockets are closed and buttoned, and that shirt sleeves are not cuffed so that molten metal cannot get caught in them. Use pliers to handle all hot metals, not your gloves, because this shortens the life of your gloves. These are simple, but often forgotten, practices that protect you as you are welding.

The Role of Electricity in Arc Welding

In order to understand the arc, it is important to understand electricity. Voltage, current, resistance, and wattage are all components of electricity that play a role in arc welding. *Voltage* is the amount of electrical pressure in a circuit, *current* is the movement of the electrons through a conductor, *resistance* is the opposition to current flow, and **wattage** is an electrical power. **Power** is the amount of work that is being done in a given application. To find the total amount of power in an electrical circuit, multiply the total circuit voltage by the total circuit amperage. Now that the electrical quantities are known, analyze the arc that is used in arc welding.

The gap at the end of the electrode is referred to as the resistance. If the voltage is set high enough, there will be enough pressure to cause the electrons to jump the gap of air, thus producing a current flow from the electrode. As the current flows through the air, small particles of the electrode are carried through the arc and deposited in the molten pool of base metal. The molten pool, referred to as a *puddle*, is produced from the heat when current flows through the gap of air. Higher current means higher heat. This heat is high enough to cause the base metal to melt into a puddle at the point where the arc comes in contact with it.

To understand why the electrons jump the gap, you must understand polarity. Voltage is a difference in potential. This means that in order for voltage to be present, there must be a positive potential and another potential that is less positive. The less positive potential is often referred to as a negative. There are two types of current: alternating current (AC) and direct current (DC). AC changes its direction of current flow every 1/120 of a second. This means that the direction of current flow alternates continually. DC, however, is direct. It is either positive or negative and does not alternate.

To understand some of the settings on an arc-welding machine, you must know electron theory. Electron theory states that current flows from negative to positive. Because of this, the direction in which DC flows can be controlled by where the negative potential is placed. According to electron theory, if the negative is placed on the base metal, the electrode will be more positive and the current flows from the base metal to the electrode. This is known as direct current electrode positive (DCEP). DCEP is also referred to as reverse polarity. If the negative potential is placed on the electrode instead of the base metal, the electrode would be more negative and the current would flow from the electrode to the base metal. This is known as direct current electrode negative (DCEN). DCEN is sometimes referred to as straight polarity. The three voltage potentials that are most likely available on most welding units are AC, DCEP, and DCEN. The open-circuit voltage, which is the amount of voltage that is present on the end of the electrode when no arc is present, is usually 50 to 80 volts. The higher the voltage setting, the easier it is to strike an arc. When the arc is struck, however, the voltage level drops down to 17 to 40 volts. This is known as the operating voltage.

The amperage settings can be low, medium, or high. The low-current range is usually 0 to 150 amperes, the medium setting is usually 50 to 250 amperes, and the high setting is usually 175 to 425 amperes. One of these three ranges should be selected according to the task at hand. Once the desired current level has been selected, it is important to run a test bead or two, using the fine adjustment on the front of the welding machine to fine-tune the current to its best operating level.

28-2 ARC WELDING COMPONENTS

There are not as many components used in arc welding as there are in gas welding. The components that are used in arc welding are the welding machine, the cables, the work clamp (often referred to as the ground clamp), the electrode holder, and the electrode. The welding machine is the heart of the arc-welding system. It is responsible for delivering power to the electrode. The cables provide a path for the current to flow from the machine to the electrode holder and the work clamp. The electrode holder (**Figure 28-2**), as the name implies, is the device that holds the electrode. The **electrode** is a metal rod (core wire) that is coated with a flux material and used to produce the arc **Figure 28-3**. The electrode is the most complex part of the system because the type of electrode used may vary

FIGURE 28-2 The electrode holder and the grounding clamp.

FIGURE 28-3 The electrode in the electrode holder.

from job to job. The core wire of an electrode is the primary source of metal for the weld. The core wire used should correspond with the base metal being welded. The covering over the electrode has several functions. One is to supply the much-needed shielding gases that prevent the atmosphere from contaminating the weld metal as it transfers from the electrode to the puddle. These gases also protect the weld pool as it cools to a solid metal. There also are elements in the flux covering that pick up contaminants in the puddle and carry them to the surface of the weld pool to become trapped within the slag.

Most electrodes that are used in arc welding have a code printed somewhere on the flux with important information about the filler metal. An example would be an E6010 electrode. The code begins with a letter prefix. The letter E is the most commonly used prefix for arc welding because it indicates that the electrode was designed for electric arc welding. The next two numbers indicate the minimum tensile strength of the weld on completion. This number must be multiplied by 1000 and is referred to in pounds per square inch (psi). Using the E6010 rod as an example, the minimum tensile strength of the completed weld is 60,000 psi. The next number (1, 2, or 4) designates the welding position in which this rod should be used. For example, the E6010 rod can be used for all welding positions. Refer to **Figure 28-4**.

The last number indicates the type of current that the electrode should be used with. This is shown in **Table 28-1**. Notice in the table that the E6010 electrode is used for DCEP.

There may be times when a suffix is added to the code on rods other than the E series rods. This indicates the type of alloy or material that is used in the coating.

Selecting the proper rod for the proper job gets easier as you become more experienced. There are many variables to be considered when selecting the proper rod for a job. However, it is not within the scope of this textbook to list all of those possible variables. A few are listed, such as the type of metal to be welded, the type of current to be used, the thickness of the metal, the condition of the surface on which the weld will be placed, and the number of passes.

28-3 STRIKING THE ARC

Before striking an arc, you must know the signs of an inaccurate current setting. When the current setting is too low, the arc length is very short, causing the electrode to *stick* to the work. Turning the current up just a little using the fine-tune adjustment prevents the electrode from sticking. If the current is set too high, an excess amount of spatter occurs and the tip of the electrode covering will be discolored approximately 1/8 in. up from the tip. This is a sign of too much heat. Another way to determine whether the current is too high or too low is to watch the trailing edge of the molten pool. If the current is set too low, the shape of the molten pool will be narrow and oblong. If the current is set too high, the ripples will be pointed on the trailing end of the weld pool as it cools. It is a good idea to run a few test beads to ensure that the current setting is correct before beginning the critical weld. It is easier to correct a problem when it is performed on a piece of scrap metal instead of on the actual material that will be welded.

Before striking the first arc, the surface on which the weld will be placed should be cleaned. This should be taken care of with the flux coating on the rod; however, taking a little extra time to clean the surface before welding produces less contaminants in the weld pool, thus providing a cleaner weld. After the machine has been set up, the proper rod has been selected, and the proper PPE is worn, then the welding may begin. Now you are ready to strike an arc!

To begin, hold the tip of the electrode over the base metal, keeping the electrode steady and being careful not to touch the metal before you lower your helmet. If you accidentally touch the metal before lowering your helmet, your eyes may become flash burned. Once the electrode is over the starting point of the weld, quickly lower your helmet. Scratch the electrode across the surface of the base metal as if you were trying to strike a long match. Do not let the electrode travel too far from the original starting point while striking the arc. Once the arc has been started, raise the electrode to the desired

CAUTION

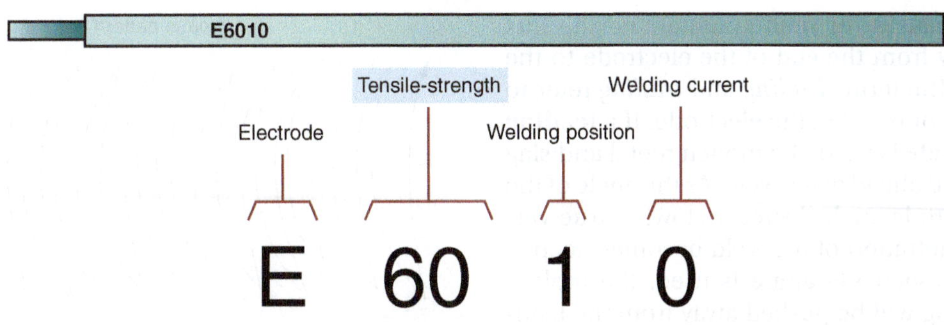

AWS classification	Welding position
E6010	F, V, OH, H
E6011	F, V, OH, H
E6012	F, V, OH, H
E6013	F, V, OH, H
E6020	H-Fillets
E6022	F
E6027	H-Fillets, F
E7014	F, V, OH, H
E7015	F, V, OH, H
E7016	F, V, OH, H
E6018	F, V, OH, H
E7024	H-Fillets, F
E7027	H-Fillets, F
E7028	H-Fillets, F
E7048	F, OH, H, V-Down

F — Flat
H — Horizontal
H-Fillets — Horizontal fillets
V-Down — Vertical down
V — Vertical
OH — Overhead

FIGURE 28-4 AWS rod identification code.

Table 28-1

WELDING CURRENTS

Welding Current Code	
EXXX0	DCRP only
EXXX1	AC and DCRP
EXXX2	AC and DCSP
EXXX3	AC and DC
EXXX4	AC and DC
EXXX5	DCRP only
EXXX6	AC and DCRP
EXXX8	AC and DCRP

arc length. Hold the arc in one place long enough for a molten pool of metal to form to the desired size. It is important to know that as the arc burns, the electrode is consumed. This means the overall length of the electrode gets shorter and shorter as it is used. If the electrode is not lowered as it burns off, the arc becomes too long. It is important to continually feed the electrode into the molten pool at a speed that keeps the tip of the electrode producing the desired length of arc as it is being consumed. Once the molten pool has formed, move the arc forward to start the bead.

28-4 RUNNING A BEAD

A **bead** is made by moving the puddle of molten metal in a forward direction with the electrode as the arc is present. It is important to hold the electrode in the proper position and angle to achieve the desired bead shape, contour, and penetration. The angle of the electrode affects the outcome of the bead. There are two angles used when arc welding: the leading angle and the trailing angle. This is important because if the electrode is angled as the arc is drawn, it causes the molten metal and the flux to blow away from the end of the electrode to the base metal. The terms *leading* and *trailing* refer to the direction of travel of the electrode. If a **leading electrode angle** is used, the molten metal and slag will be pushed ahead of the weld. As the angle of the leading electrode angle decreases toward true vertical, the penetration of the weld becomes deeper. If a **trailing electrode angle** is used, the molten metal and slag will be pushed away from the leading edge of the molten pool toward the back of the pool, where it solidifies. The effect of changing the trailing electrode angle is opposite to that of the leading angle. As the electrode is brought closer to true vertical, the penetration of the weld becomes shallower.

As you move the molten pool forward, try moving the electrode in a circular or weaving motion. This makes it easier to control the bead penetration, buildup, width, and overlap. The circular and the weave pattern are the most commonly used patterns when running a bead. To run a bead in a circular motion, allow the molten pool to be created first. Once a molten pool is created, then manipulate the electrode holder in a manner that will cause the tip of the electrode to move in small, controlled circles, stopping for a count of one every 360° of rotation. Sometimes it helps to use a counting rhythm to keep the bead uniform throughout the whole weld. Start the circular movement in the molten pool. Move the electrode toward the rear of the molten pool, then across the back to the other side of the molten pool. Return to the front edge of the molten pool. Hold the electrode at the front edge of the molten pool for about a count of one, allowing the base metal to melt before repeating the circular movement. Both movements are shown in **Figure 28-5**. If the correct movements were made with the electrode, the ripples on the bead will be even, consistent, and uniform in shape. When terminating the weld, let the electrode set in the molten pool for a count of one, then pull it straight up in a quick motion. This eliminates making a crater at the end of the bead because you have allowed time for the metal from the electrode to fill the molten pool before removal.

It is important to know that a **stringer bead** has little or no side-to-side movement of the electrode.

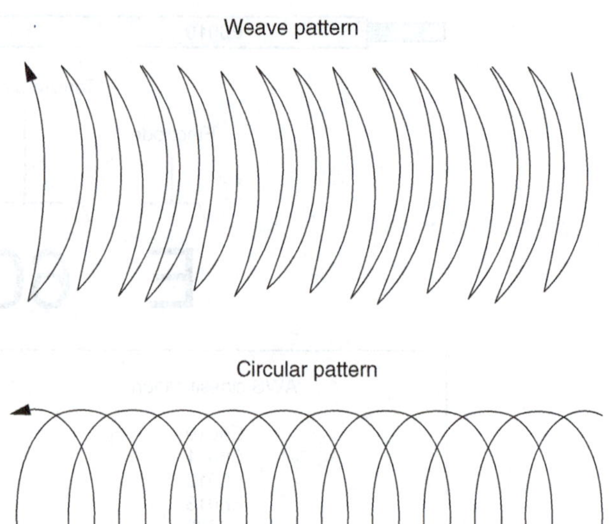

FIGURE 28-5 Circular movement of an electrode.

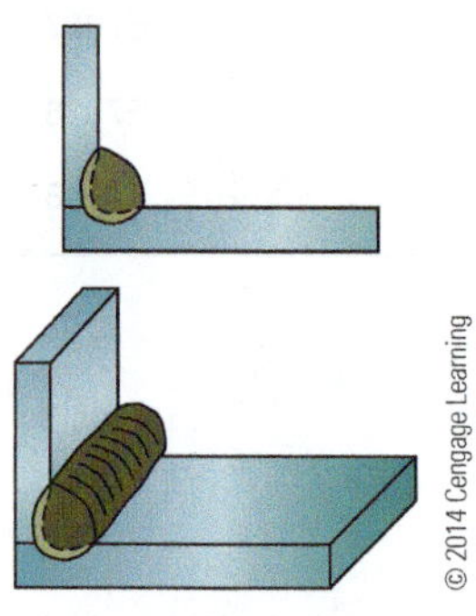

FIGURE 28-6 A fillet weld.

28-5 TYPES OF WELDS

This material was covered in Chapter 26; however, because the material is pertinent to arc welding as well, the information is repeated in this chapter. There are three common types of welds used when arc welding: the surfacing weld, the fillet weld, and the groove weld.

The *surfacing weld* is used to build up a surface by running one or more stringers or weave beads across an unbroken surface.

The *fillet weld* joins two surfaces at right angles to each other. This is illustrated in **Figure 28-6**.

The *groove weld* fills a groove between two pieces of metal that are to be joined. The ends of the pieces to be welded may be square, beveled, U-shaped, or J-shaped, as shown in **Figure 28-7**.

28-6 TYPES OF JOINTS

There are three types of joints used in welding today: the butt joint, the tee joint, and the lap joint.

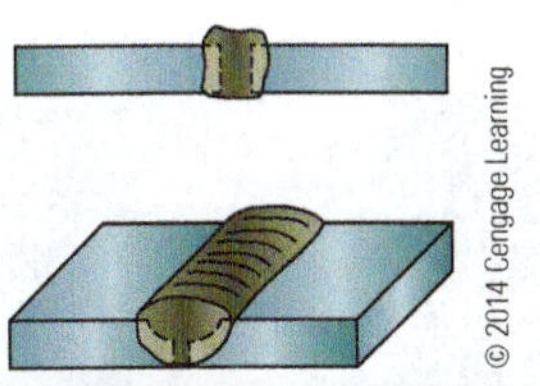

FIGURE 28-7 A groove weld.

The *butt joint* is when two pieces are welded together at their edges as if they were lying on a table side by side. The edges must be close to parallel, with a root opening between the plates. Different methods are used to prepare the edges. These may be the square butt joint, the beveled butt joint, the double-V butt joint, the U-type butt joint, or the double U-type, or J-type, butt joint, all of which are illustrated in **Figure 28-8**.

The *tee joint* is accomplished by placing the edge of one piece of metal on the surface of another, thus creating a right angle. See **Figure 28-9**. This type of joint requires a fillet weld on one or both sides. It is usually used for light to medium thick material.

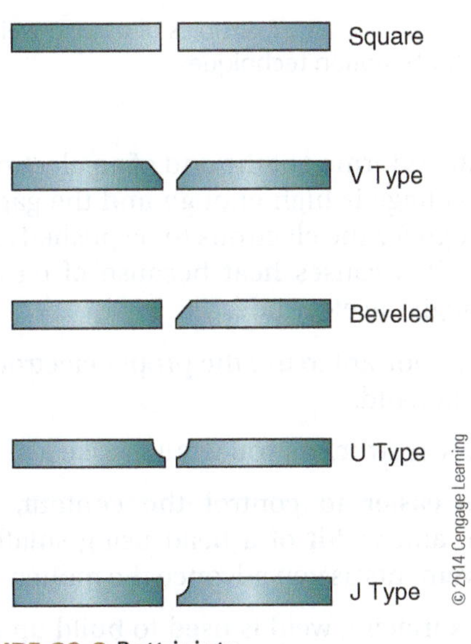

FIGURE 28-8 Butt joints.

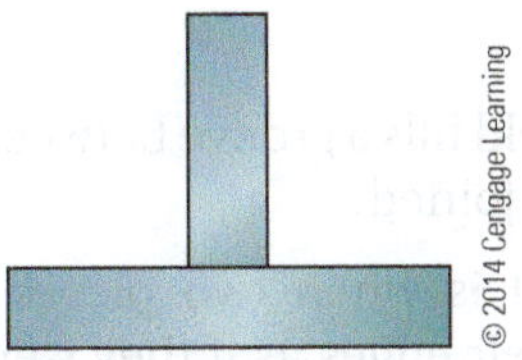

FIGURE 28-9 A tee joint.

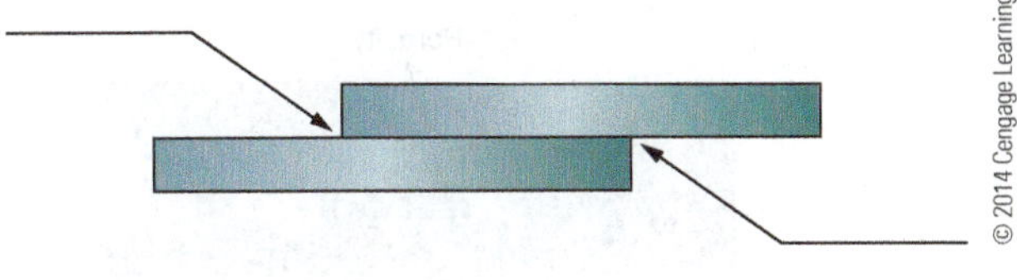

FIGURE 28-10 A lap joint.

The *lap joint* is the strongest joint. It is made by lapping one piece of metal over another and welding at the end of the top piece of metal. This is illustrated in **Figure 28-10**. The overlap of the metals should be at least four times the thickness of the metal. There are two basic types of lap joints: the single fillet and the double fillet. The single fillet is used for metals that are thicker than ½ in., whereas the double fillet is used for metals that are less than ½ in. thick.

28-7 COMMON PROBLEMS OF ARC WELDING

Two of the most common problems, other than spatter and inconsistency, are undercut and underfill.

Undercut is the result of having more metal removed from the base metal than what the electrode is replacing. This is commonly caused by excessive current, incorrect electrode angle, or excessive weave. This becomes less of a problem as you learn to manipulate the electrode.

Underfill occurs on a groove weld when the completed weld is below the original surface of the base metal. It occurs in a fillet weld when the weld deposit has an insufficient effective throat. This problem usually occurs because the travel rate is excessive. Slowing the travel rate or making several passes corrects this problem. **Figure 28-11** shows porosity, undercut, and underfill.

Porosity is another problem that occurs quite often as the result of welding on an unclean surface. If the surface that is to be welded is not properly cleaned, contaminants will be present in the molten puddle as the bead is being run. These contaminants become trapped in the metal as the weld cools. Many times this leaves pits (pores) and inconsistencies in the weld and can cause the weld to fail if severe. **Figure 28-12** shows examples of good welds using the circular and weave methods. The weave was done on a vertical uphill fillet weld, whereas the circular method of running a bead was used on the flat fillet weld.

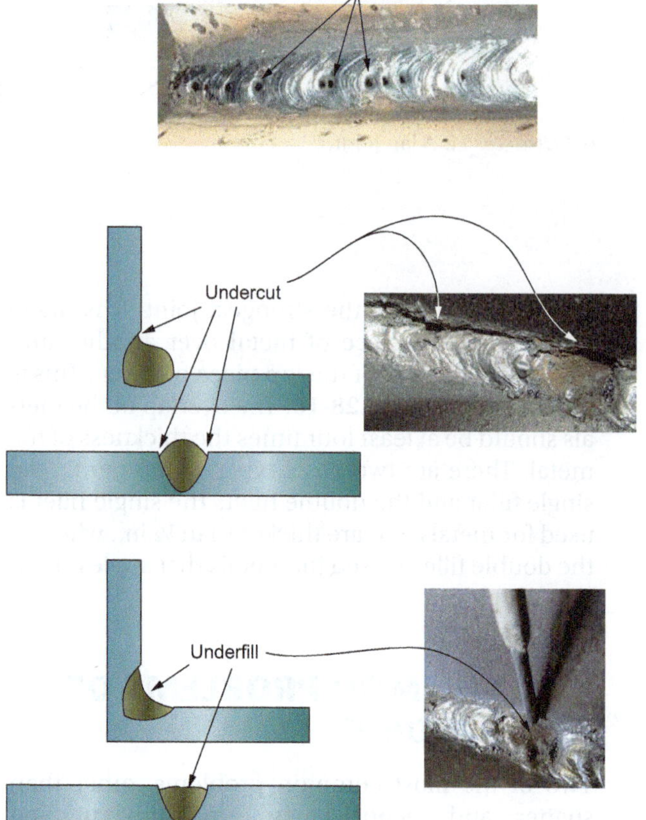

FIGURE 28-11 Common problems from improper cleaning or welding procedures.

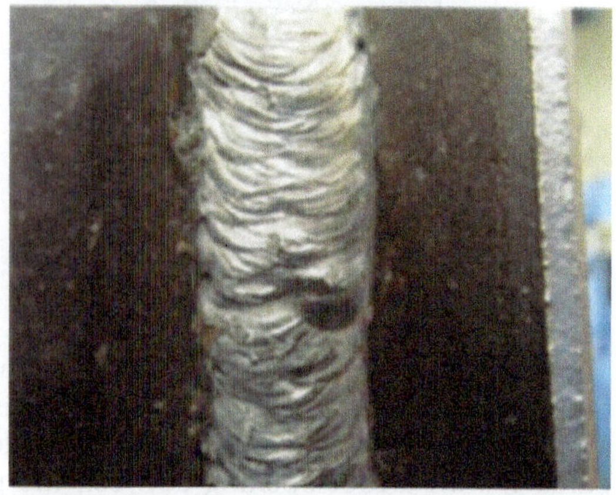

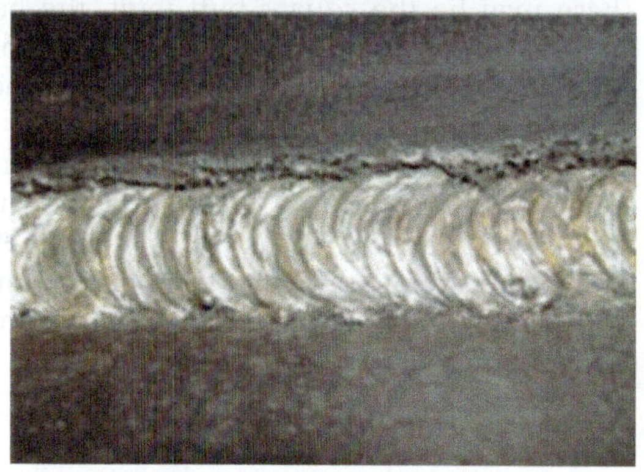

FIGURE 28-12 Examples of a good weld using both motion techniques.

SUMMARY

- There are three types of light: infrared, white, and ultraviolet.
- Ultraviolet light is the most dangerous of the three types of light.
- Flash burn or blindness may occur from looking at an arc with the naked eye.
- Voltage is the amount of pressure in a circuit.
- Current is the movement of electrons.
- Resistance is the opposition to current flow.
- Power is the total voltage times the total current produced by the welder.
- AC stands for alternating current, which means that current changes direction from positive to negative.
- DCEP stands for direct current electrode positive. It is also referred to as reverse polarity.
- DCEN stands for direct current electrode negative. It is also referred to as straight polarity.
- An arc is formed at the end of an electrode when the voltage is high enough and the gap is small enough for the electrons to be pushed across the gap. This causes heat because of current flow through the electrode.
- It is important to use the proper electrode for the task at hand.
- An arc is struck as you would strike a match.
- It is easier to control the contour, penetration, and width of a bead using small circular movements as you advance the molten pool.
- The surfacing weld is used to build up a surface by running one or more stringers or weave beads across an unbroken surface.
- The fillet weld joins two surfaces at right angles to each other.
- The groove weld fills a groove between two pieces of metal to be joined.
- The butt joint is when two pieces are welded together at their edges as if they were lying on a table side by side.

- The tee joint is accomplished by placing the edge of one piece of metal on the surface of another, thus creating a right angle.
- The lap joint is the strongest joint. It is made by lapping one piece of metal over another and welding at the end of the top piece of metal.
- Undercut is the result of having more metal removed from the base metal than what the electrode is replacing.
- Underfill occurs on a groove weld when the completed weld is below the original surface of the base metal.
- If the surface that is to be welded is not properly cleaned, contaminants will be present in the molten puddle as the bead is being run. This causes porosity.

REVIEW QUESTIONS

1. List the dangers caused by ultraviolet light.
2. What is the primary resistance of an arc welding system?
3. What type of current should an E7018 electrode be used on?
4. In what positions can an E6011 rod be used?
5. What does moving the electrode in small circular movements do for the bead?

SECTION 5

PREVENTIVE MAINTENANCE

Preventive Maintenance—Developing and Implementing

Mechanical PM

Electrical PM

A TYPICAL DAY IN MAINTENANCE

Check It Out

Because you have approximately 1½ hours before your shift ends, and because the press is already down for repairs, you decide to make the most of the remaining time by doing some PM work on the press.

You review the PM schedule for the press and then clean, lubricate, align, and tighten some of its parts according to PM procedures.

After performing a final thorough check, you remove your lockout/tagout and energize the press. You cycle the press through its operation and, satisfied, declare it back in service.

You clean up your area, gather all your tools and equipment, and return to the shop at 2:55 P.M. with a good day's work behind you.

Work It Out

1. Why not just call it a day and leave or head back to the shop?
2. Why is it important to perform a PM on a machine?
3. What is the benefit of operating the press before cleaning up for the day?

CHAPTER

29

Preventive Maintenance—Developing and Implementing

Preventive maintenance is a vital part of maintenance. If it is done correctly and often, the facility should experience very little downtime caused by machine or component failure. There are many facilities that do not believe in implementing a PM schedule because they are too concerned with losing money. These types of facilities run a machine to the point of destruction before they shut it down. When the machine breaks, they then fix everything that is broken at one time. This often leads to excessive downtime. If a regular PM schedule is maintained, the damage to the machine will never be so severe that it loses more than one shift's worth of downtime in comparison to many days or weeks. A facility that has an aggressive PM schedule will have less downtime over a longer period.

OBJECTIVES

After studying this chapter, the student should be able to:

- Understand the importance of maintaining a history log.
- Plan and coordinate an effective PM inspection.
- Understand the importance of retrieving proper permits and procedures for hazardous and confined spaces.
- Learn how to safely execute preventive maintenance.

29-1 HISTORY

The history of a machine is very important when referring to preventive maintenance. A maintenance log should be kept on every machine in the facility. A **maintenance log** contains entries of machine breakdowns over a period of time. Assume that the cycle for a scheduled PM is once a month every month for a given machine. A log should be kept of all of the breakdowns during the month between the scheduled PM inspections. This log should be reviewed before creating the coordination efforts of the PM. If it is noticed while reviewing the log that a machine has repeatedly broken down for the same reason, then the problem that is causing the repeated breakdown should be found and repaired during the PM. It is best to correct as many problems as possible during the regular shift maintenance as they arise; however, there are times when a problem is allowed to remain until it can be repaired during a scheduled PM. There are also times when a machine has to be shutdown because of the nature of the breakdown. If this is the case, it is a good time to get the entire PM crew on the machine because it is already down. This is sometimes referred to as an emergency PM. An emergency PM is a PM that is not scheduled. The PM coordinator should allow the PM scheduling to be flexible to make time for emergencies.

Refer to **Figure 29-1**. Upon studying the maintenance log in the figure, what would be the first thing that you, as the PM coordinator, would list as a task during the PM of this machine?

Upon studying the maintenance log, you should have noticed that there are seven entries on this one page of the maintenance log for #1 Finish Frame. Notice that there are three separate maintenance personnel who have worked on this machine on all three shifts. It should be recognized that four of the seven entries relate to the right turnstile on the machine. Notice that all four of these shutdowns occurred within a 7-day period. The nature of the breakdown and the work performed describe the symptoms of a bearing that is failing or one that has failed within the turnstile. Although each of the maintenance personnel got the machine running, none of them truly fixed the problem. They only fixed the symptom of the problem by applying more grease to the turnstile bearing, thus reducing the friction produced by the failing bearing. This type of breakdown should be recognized as one that would be critical enough to cause an emergency shutdown and PM inspection. While the machine is shutdown for the bearing replacement, the entire PM crew should inspect the rest of the machine for any other problems that may exist. The PM crew for this machine should have two mechanics to replace the bearing on the turnstile, the lubrication technician, two electricians, and one more mechanic to check the rest of the mechanical system on the machine. Two mechanics are needed to work on the bearing because the turnstile is heavy and needs two mechanics to lift the turnstile housing. The lubrication technician should be present during all PMs. As for the electricians, there should always be at least two working together for safety reasons. One mechanic is all that is needed to simply go over the rest of the machine. If the mechanic locates another big problem, the supervisor should decide whether or not another mechanic is required. This is just a suggested manning of a PM crew. It may not be possible for all facilities to release this many maintenance personnel to one machine. In this case, the PM coordinator should be present at all times during the shutdown in the event of an injury or an emergency.

29-2 PLANNING AND COORDINATING

It is important to plan the PM duties for a machine before it is ever shutdown. The first thing that a PM coordinator must do is to survey all of the potential hazards that may exist on the machine during the PM before the machine is shutdown. The PM coordinator should look for hazardous or dangerous chemicals that may be in the vicinity of the PM crew and identify all of the potential energy sources. The PM coordinator should also look for any flooring or catwalk grating that may need to be removed during the PM and recognize any pinch points that may cause injury. These are just a few of the things that PM coordinators should recognize as they survey the machines. Many more safety concerns exist and, therefore, need to be recognized upon the survey of the machine. This is why it is imperative for the PM coordinator to perform a thorough survey of the machine.

After recognizing all of the safety hazards, the PM coordinator should then survey the machine for faulty components and systems that must be changed during the shutdown. It is also a good idea to talk to the operators of each shift to find out if they have recognized any problems that may need the attention of the PM crew. It is always a good idea to include the operators in the planning of the PM of their machine. This accomplishes several things. It establishes a good working relationship between the maintenance personnel and the operators and removes any complaints or excuses that the operators may have that would prevent

Maintenance Log

Date	Time	Mach. Identification	Total Downtime	Work Performed/Maintenance Personnel
3/3/12	4:32 PM	#1 Finish Frame	1 Hour	Found that the right turnstile needed to be lubricated. Lubricated the turnstile and monitored the machine. Finish frame did not shut down. Released the machine to the operator. Jerry Jackson
Nature of Breakdown: Machine not running. Notice a lot of heat coming from the right turnstile on the exit end of the machine. After the machine has time to cool, the machine can be restarted. It runs awhile and then stops running again.				

Date	Time	Mach. Identification	Total Downtime	Work Performed/Maintenance Personnel
3/7/12	8:19 AM	#1 Finish Frame	45 Min.	Only one burner was running. Found faulty high temperature limit switch in the oven. Replaced the limit and fired the oven up to 280 degrees. All is well. Freddy Frazier
Nature of Breakdown: Oven temperature not getting hot enough.				

Date	Time	Mach. Identification	Total Downtime	Work Performed/Maintenance Personnel
3/8/12	2:20 AM	#1 Finish Frame	15 minutes	Noticed some metal filings on the floor below the turnstile. Checked chain on the turnstile. Lubricated the chain & bearing. Runs good now. Ricky Rodgers
Nature of Breakdown: Machine locked up. Noise coming from right turnstile just before it stopped running.				

Date	Time	Mach. Identification	Total Downtime	Work Performed/Maintenance Personnel
3/9/12	7:32 PM	#1 Finish Frame	34 minutes	Vane switch on main drum roll malfunctioned. Removed the bad vane switch and replaced with a new one. Monitored the counter for about 10 minutes. Released the machine to the operator. Jerry Jackson
Nature of Breakdown: Yardage counter not running.				

Date	Time	Mach. Identification	Total Downtime	Work Performed/Maintenance Personnel
3/10/12	10:45 AM	#1 Finish Frame	2 Hrs	The sprocket on the right turnstile was locked up. Noticed a lot of heat. Applied some grease to the bearing and checked the oiler on the chain. Started the machine up. Ran good. All is well. Freddy Frazier
Nature of Breakdown: Machine shut down by itself. Was running great when a loud screeching started. Seconds later, the machine shut down.				

Date	Time	Mach. Identification	Total Downtime	Work Performed/Maintenance Personnel
3/10/12	7:23 PM	#1 Finish Frame	20 minutes	Faulty valve. Replaced valve and filled dye tank. Monitored the system. Seems to be running OK. Jerry Jackson
Nature of Breakdown: Can't get any fluid down from the mezzanine to the dye tank on the entry end of the frame.				

Date	Time	Mach. Identification	Total Downtime	Work Performed/Maintenance Personnel
3/11/12	5:20 AM	#1 Finish Frame	1 hour 15 minutes	Noticed more metal filings on the floor below the turnstile. Checked chain on the turnstile again. Lubricated the bearing. Runs good now. Ricky Rodgers
Nature of Breakdown: Machine locked up. Right turnstile making a lot of noise.				

FIGURE 29-1 A maintenance log.

them from running their machine. It also shows the operators that their knowledge and opinions are appreciated and valued. This makes the facility more efficient and the working environment more pleasant for everybody.

On completion of the machine survey, a list needs to be made of the things that should be inspected and replaced during the PM shutdown. This should include changing all filters, visually inspecting all bearings, checking all belts and couplings, and so forth. It is important to have enough maintenance personnel working on the machine to be able to finish the PM in a safe and timely manner. During the planning stage, it should be determined how long each item on the list takes to complete. The PM coordinator also must take into account that upon inspection of the machine during the PM, other problems may be found that were not recognized during the preliminary survey of the machine. These other problems may cause the PM shutdown to be longer than anticipated. It is important to have enough maintenance personnel to complete all of the tasks listed for a PM shutdown. There should be no less than two mechanics, two electricians, and one lubrication technician on each PM crew.

The PM coordinator should also coordinate the jobs that need to be done on the machine in an organized manner. For example, while the mechanics are replacing a faulty bearing on the front of the machine, the electricians should be working on a faulty button on the opposite side of the machine. This prevents crowding, which can lead to accidents. The PM coordinator should coordinate the lubrication technician to arrive after the mechanics have finished inspecting and replacing all of the bad bearings on the machine. This prevents the lubrication technician from having to come to the machine twice. These are the types of coordinations that need to be considered before the first tool is placed on the machine. If a PM coordinator plans a PM effectively, the PM shutdown will be a smooth operation, with everyone completing their tasks in a safe and timely manner.

29-3 HAZARDOUS AND CONFINED LOCATIONS

Before any PM work begins, a survey of the machine or equipment should be implemented. If upon surveying a machine, it is noticed that a portion of it is in a hazardous location or within a confined space, the proper procedures must be followed when implementing PM on that machine or equipment.

A **confined space** is defined in OSHA Standard 29 1910.146 as a "space that is large enough and so configured that an employee can bodily enter and perform assigned work; has limited or restricted means for entry or exit (for example, tanks, vessels, silos, storage bins, hoppers, vaults, and pits are spaces that may have limited means of entry); and is not designed for continuous employee occupancy." OSHA Standard 29 CFR Part 1926.21 (b)(6)(i) & (ii) under Subpart-C also states:

> (6) (i) All employees required to enter into confined or enclosed spaces shall be instructed as to the nature of the hazards involved, the necessary precautions to be taken, and in the use of personal and emergency equipment required. The employer shall comply with any specific regulations that may apply to work in dangerous or potentially dangerous areas.
>
> (ii) For purposes of paragraph (b)(6)(i) of this section, "confined or enclosed space" means any space having a limited means of egress, which is subject to the accumulation of toxic or flammable contaminants or has an oxygen deficient atmosphere. Confined or enclosed spaces include, but are not limited to, storage tanks, process vessels, bins, boilers, ventilation or exhaust ducts, sewers, under ground utility vaults, tunnels, pipelines, and open top spaces more than 4 feet in depth such as pits, tubs, vaults, and vessels.

OSHA Standard 29 CFR 1910.146 states the procedures that *must* be followed before entering a confined space. It is very important that all safety procedures be followed during the PM of a machine. A confined space should be clearly identified. Every facility that has confined or enclosed spaces where maintenance personnel must work should have an aggressive policy regarding entering these spaces without a permit. The safety officer within a facility should be the only person who can issue the confined space entry permit and only after the appropriate forms are filled out and the enclosed space has been tested for the accumulation of toxic or flammable contaminants and oxygen levels. It is imperative that these request forms be filed in the safety office records. These forms may vary from facility to facility. They should all contain certain key pieces of information such as who can enter the confined space, the purpose for entering the confined space, how long the individual anticipates to be in the confined space, and the testing results that were taken before the individual enters the confined space. All confined spaces must be tested for the accumulation of toxic or flammable contaminants and oxygen levels before the maintenance personnel enters the confined space. All individuals should have a lifeline attached to them in the unfortunate event that they are rendered unconscious while in the confined space. It should also be the policy of the facility to ensure that another person remains present at the entry point of the confined space in the event that an individual must be pulled out because of losing consciousness while in the confined space. This can occur during PM duty if toxic or flammable contaminants accumulate or oxygen levels become depleted. It is also important to ventilate the confined space with a source of fresh air. This is important, especially if the PM includes welding. Welding would quickly fill the confined space with toxic fumes, which would render the maintenance personnel unconscious if proper ventilation were not supplied.

If a PM is to be performed in a hazardous location, it is imperative for the individual performing the PM to be knowledgeable of the hazards that are involved, the necessary precautions to be taken, and the proper use of personal and emergency equipment. The proper PPE should protect the maintenance personnel from any harm; however, they should not just rely on the PPE to protect them while working in a hazardous location. They should keep safety in the forefront of their minds and work safely. It is also a good idea to review the MSDS with the safety officer before entering the hazardous location. The MSDS provides the necessary steps that must be taken in the event that an individual becomes contaminated while performing the PM duty.

29-4 EXECUTING THE PM

A safety meeting should be held before every PM shutdown to advise the PM crew of the safety concerns of that particular shutdown and to assign the duties to each individual. An example of safety concerns would be to remind each person who will be working on the machine to have his or her lockout/tagout on the main energy sources to the machine. This includes electrical energy, pneumatic energy, hydraulic energy, and any gas supply that may be on the machine. The coordinator should also talk about the hazardous or dangerous chemicals that may be within the work area and inform the crew of any PPE that may be needed while working in these hazardous environments. Pinch points also need to be identified during this safety discussion. It is important to work safely during a PM. Always remember to wear the proper PPE, such as a hardhat, gloves, or steel-toed boots while working on a machine. While working on a PM crew, it is easy to hurry and cut corners due to time constraints; however, every PM crew member must remain vigilant to recognize any potential safety hazard while working. Safety must always be first.

Once all of the safety items are covered, the order of execution should be laid out for the maintenance personnel. There should be no surprises while working a scheduled PM. It is very important that all personnel be informed of their duties and responsibilities. Once the items of repair and inspection have been discussed, it is important to issue any permits that are required to the proper personnel. This includes any confined or enclosed space entry permits, welding permits, open flame permits, and so on.

A leadperson should be designated in the presence of those who will be working on the machine at this time. This is so that every crew member knows whom to turn to if a problem arises during the PM. Besides having the responsibility of monitoring the progress of the PM as it is being executed, the leadperson should also monitor the safety and actions of the other crew members. This includes discouraging such things as horseplay.

Before any work is started on the machine, all personnel who are working on the machine should place their lockouts/tagouts on all of the energy supplies that are present on the machine on which they will be working. When there are many people working on a given machine, some facilities place one lockout/tagout on each of the potential energy sources, and then place the keys to those lockout/tagouts in a lockbox. The lockbox is then closed and locked with a multi-hasp clamp on which everybody on the PM crew places their lockout. It should be noted that all individuals who work on a machine should use their lockout/tagouts to ensure their own safety even if they are not actual members of the PM crew. Once every person has placed his or her lockout on the lockbox, the lockbox should then be taken to the plant engineer's office or the safety coordinator's office. This ceremony should be performed before each PM work begins. The work may begin only after the lockbox has been secured in the coordinator's office or engineer's office and everybody's lockout/tagout key is in his or her own pockets.

The leadperson also needs to make sure that the area where the work will be performed is secured. This means that the area must be cordoned off so as not to allow any personnel other than the PM crew into the area where the PM is occurring. It is also important to rope off any gratings or floor panels that are removed during the PM to warn other PM crew members of the missing floor section. This prevents a fall that may cause injury or even death. If a crew member is working alone, it is always important for the leadperson to keep a constant eye on that individual. If possible, everyone should be working in pairs. It is also the responsibility of the leadperson to ensure that everyone is wearing the proper PPE at all times. This includes a hardhat, gloves, safety glasses, steel-toed boots, and so forth. Anyone not working safely should be notified of the possible safety hazards that exist. If the individual does not respond, then that individual should be reported to the PM coordinator, who, in turn, should report it to the safety officer. If any PM work requires a PM crew member to work off the ground, then the leadperson should be alert and ensure that the individual is wearing safety harnesses and that they are tied off. The leadperson has to be ever vigilant to ensure crew safety.

Upon starting the execution of the PM, it is a good idea to always start with the largest jobs that will take the longest. It is best to go ahead and get them out of the way instead of waiting until later and possibly running out of time, thus causing the PM shutdown to be extended. This should be avoided if possible. As each job that is listed on the assignment sheet is completed, it needs to be reported to the leadperson. The leadperson should at this time assign a new job to the individual to keep the work moving along. Once all of the work has been completed for each specialty (electrical, mechanical, etc.), the leadperson should check whether the remaining PM members could use extra help. If so, then it is important to brief the crew members on the status of the job and the goal of completion before starting.

The leadperson should run through the list of duties, ensuring that all of them have been completed. Once the PM has been totally completed, the lockbox should be retrieved from the plant engineer's office or the safety coordinator's office and brought to the machine. All personnel should remove their own lockout/tagout from the multi-hasp clamp on the lockout box. If a person is absent and cannot remove his or her lockout/tagout, the plant engineer must be the one who removes the lockout/tagout. This can, however, only be done in the presence of the leadperson and the plant safety officer. Only when every lockout/tagout is removed can the keys from within the lockout box be removed. This ensures that no energy sources are returned to their normal state until everyone has agreed that they are clear from all danger that might exist from reapplying these sources of energy.

The machine should be run for a short time once the energy sources are returned to their full potential to test all of the systems that were changed during the PM shutdown. It is a good idea to run the machine for no less than 15 minutes at its normal operating speed to ensure that it operates properly once it is released to the operator. After the PM is completed, all of the PM crew members should gather together to go over what was done to the machine. The goal is always to have completed all of the tasks and to do this without anyone being injured. If this is accomplished, then the PM was a successful one.

SUMMARY

- A maintenance log should be kept for every machine.
- The maintenance log can assist in planning the PM.
- The machine should be surveyed before the PM by the PM coordinator for potential dangers and hazardous conditions.
- The operators should be consulted upon this survey.
- A list should be generated of the things that should be inspected and replaced during the PM shutdown. This should include things such as changing all filters, visually inspecting all bearings, checking all belts and couplings, and so on.
- It is important to have enough maintenance personnel working on the machine to be able to finish the PM in a safe and timely manner.
- The PM coordinator should coordinate the jobs that need to be done on the machine in an organized manner so that the crew members work on opposite sides of the machine. This prevents crowding, which can lead to accidents.
- The PM coordinator should coordinate the lubrication technician to arrive after the mechanics have inspected and replaced all of the bad bearings on the machine.
- If a PM coordinator plans a PM effectively, the PM shutdown will run smoothly and all crew members will complete their tasks in a safe and timely manner.
- A confined space should be clearly identified. Every facility that has confined or enclosed spaces where maintenance personnel must work should have an aggressive policy regarding entering these spaces without a permit.
- The safety officer within a facility should be the only person who should issue the confined space entry permit and only after the appropriate forms

have been filled out and the enclosed space has been tested for the accumulation of toxic or flammable contaminants and oxygen levels.

- If a PM is to be performed in a hazardous location, it is imperative that the individual performing the PM be knowledgeable of the hazards that are involved, the necessary precautions to be taken, and the proper use of the required personal and emergency equipment.
- Every PM crew member must remain vigilant and recognize any potential safety hazard while working. Safety must always be first.
- A leadperson should be designated in the presence of those who will be working on the machine, to let every crew member know whom to turn to if a problem arises during the PM.
- Before any work is started on the machine, all personnel who are working on the machine should place their lockout/tagouts on all of the energy supplies that are present on the machine on which they will be working.
- All personnel should remove their own lockout/tagout.
- The machine should be run for a short time once the energy sources are returned to their full potential.
- The goal is to always complete the tasks and to do this without anyone being injured.

REVIEW QUESTIONS

1. List at least three things that the PM coordinator should do before the PM shutdown.
2. What OSHA standard discusses the responsibility of the employer to inform the employee of all hazardous conditions?
3. Define *confined space.*
4. What are some responsibilities of the leadperson?
5. What is the goal of a PM schedule?

CHAPTER 30

Mechanical PM

A machine is nothing more than smaller components used in tandem with each other to complete a task. If any one of the components fails, the process will fail altogether. It is therefore necessary to maintain all of the components individually to keep the machine functioning properly. These individual tasks, if performed correctly, ensure the proper operation of each component. Because of the difference in construction, each component may require different maintenance duties, although each is equally as important to the overall performance of the machine.

OBJECTIVES

After studying this chapter, the student should be able to

- Identify some visual symptoms that indicate imminent bearing failure.
- Identify some visual symptoms that indicate imminent gearbox failure.
- Diagnosis gear wear and eliminate as many of the conditions as possible that cause premature gear wear.
- Identify some common problems with belt drives and learn how to prevent these problems from occurring again.
- Identify some common problems with chains and sprockets, and learn how to prevent these problems from occurring again.
- Identify the signs of coupling wear.

30-1 BEARINGS

If properly maintained and not overloaded, the bearing should never fail until it simply succumbs to metal fatigue. Bearing failure is generally caused by a lack of maintenance or overloading. There are, however, some other extraordinary instances that may cause bearing failure. Once a bearing has failed, there are signals that indicate the cause of failure. Some of these signals can be noticed during inspection of the bearing before failure occurs. If a bearing were replaced without correcting the condition that caused it to fail, chances are that the next bearing would suffer the same fate. Changing the bearing is only fixing the symptom, not the problem.

Overheating is a result of high temperatures, which could be caused by different conditions. Another early indication of overheating is the presence of solid or caked lubricant. Rusting surfaces are an indication of moisture. Rusting is a form of oxidation that occurs when moisture is present and lubrication is lacking. A properly maintained bearing can operate in minimal moisture, however. If a rusted bearing is found on inspection and is not replaced, small particles of rust may flake off into the bearing, thus causing damage to the races and the rolling elements. This, in turn, eventually causes high temperatures, resulting in premature bearing failure. If on inspection you notice that an excess amount of dirt has accumulated on a bearing, remove the dirt to check whether the bearing is a sealed bearing. If it is not, replace it immediately with one that is.

Underlubrication is a common cause of bearing failure. Once a bearing has been installed, it is usually forgotten. It is for this reason that many bearings go unlubricated. Lack of lubrication causes friction and overheating within the bearing. This causes the bearing to fail prematurely. Overlubrcation can place excessive internal pressure on the bearing because the rolling elements have to move the excess amount of lubrication within the bearing as well as the load.

30-2 GEARBOXES

Gearboxes usually contain oil to keep the gears lubricated and cooled. This is the first thing that needs to be checked during a PM inspection of a gearbox. When the oil is allowed to run low, the gears within the gearbox begin to show signs of wear in the teeth of the gears. A symptom of this is the presence of small metal particles in the bottom of the oil reservoir. In gearboxes that have brass gearing, the oil begins to have a metallic look to it as a sign of gearbox wear. This is the reason why the oil should be checked and drained on a regular basis. It is also a good idea to use new oil when refilling the gearbox to ensure proper cooling of the gears. It is not uncommon for some mechanics to use the same oil that was taken out of the gearbox if no signs of wear are found. This, however, is not a good idea for the simple fact that old oil loses some of its cooling and lubricating characteristics during its first usage due to contaminants. This would be like putting the old oil back in your car engine during an oil change. Always replace old oil with new oil. Another thing to check for during inspection of the gearbox is the tooth wear on the gear. If excessive wear is noticed, replace the gear before putting the gearbox back into service. If a replacement gear cannot be found, replace the entire gearbox. It is a good idea to check all of the seals as well.

30-3 SEALS

On inspection of a seal, you may notice that it is leaking. It is important at this time to identify the type of seal that is leaking to determine the method that should be used to reduce or eliminate the leakage. It is, on some occasions, desirable to have a small amount of leakage through the packing material of a packing seal to keep the surface of the shaft or rod lubricated as it rubs or rotates against the packing material. Some wear usually occurs on the packing material and the shaft or rod during normal operation because there is friction present between the packing material and the shaft or rod. Because of the friction, both packing and shaft material may wear off. Over time the leakage becomes excessive if it is not frequently inspected and properly maintained. Making frequent adjustments to the follower ensures that excessive leakage does not occur. Remember to allow a small amount of leakage through the packing material to lubricate and to slow the wear of both the packing and shaft material down to a minimum.

It is important to check the lantern ring, which should be placed in the stuffing box in such a manner that it can be seen in the lubrication port. If it cannot be seen in the lubrication port, then the lantern ring has probably moved forward past the lubrication port. This does not allow the proper lubrication of the packing material. If this is noticed, then the packing has been compressed too much and needs to be replaced.

There should be no leakage when inspecting a mechanical seal. If leakage is found on inspection, the seal should be replaced at that time and not be allowed to remain in service. No leakage should be found upon checking an O-ring seal. If there is leakage, the O-ring has probably become fatigued or brittle. It may even have broken. The O-ring should be replaced at this time. The radial lip seal should also be inspected on a regular basis. If the lip of the seal has many nicks or small cuts, it too should be replaced. It is an extremely common occurrence with radial lip seals that have been used for exclusion. A seal that is properly maintained allows minimal leakage, thus saving the facility money in lubrication and oil.

30-4 BELTS

Flat Belt Maintenance

Proper care must be taken to ensure the quality and the life of a flat belt. For example, always check the belt to ensure that it is tight. This prevents slippage, allowing the belt to run at its optimum level. Another thing that could be done to ensure a longer life for a new belt is to run it in before putting the machine into service to stretch and flex the belt. As the belt stretches, it has to be tightened again. This may need to be done two or three times before turning a machine over to the operator. A belt that has been properly run in will not stretch very much once the product load is placed on the machine. It is always important to keep all belts clean and free of debris. Dirt and debris can cause severe damage to a belt, thus causing it to wear out prematurely. From time to time, some flat belts may need some belt dressing applied to them. Belts tend to lose their softness as they get older. This causes slippage. By dressing a belt, traction can be regained and the need to change the belt occurs less often. Once an excess amount of slippage begins to occur on a regular basis, it may be time to consider replacing the belt. Finally, one of the most important things is to ensure the proper operation of all belt guards. These are in place to protect you and your fellow workers from harm when the belt is in motion. A faulty guard could cause injury or death! A simple occasional check would prevent this possibility. It is worth it!

V-Belt Maintenance

It is not uncommon for a belt to become overloaded and cause the tension members to break. When this occurs, the belt loses its load capability and begins to stretch. Usually when one tension member breaks, the others will break shortly thereafter. As more tension members break, the belt begins to ride lower in the pulley groove. This is the improper way to transmit power through the V-belt. As mentioned earlier, a V-belt should transmit power through the wedging action of the tapered sides of the belt, not the bottom. Eventually, the belt simply begins to slip. This is the time to change the belt. There are some maintenance mechanics who just tighten the belt and spray it with belt dressing just to get it going again; however, this is not solving the problem. The problem reoccurs as the belt continues to stretch. If this continues, the belt eventually breaks. Another problem is that, over time, the pulley walls become worn. This is shown in **Figure 30-1**. In Figure 30-1A, the belt sets in the pulley in such a way that the top of the belt is flush with the top surface of the pulley wall. This is an ideal setup. As the pulley becomes worn, the belt begins to sit lower in the pulley, as shown in Figure 30-1B.

When the pulley has worn out enough to allow the belt to drop into it by 1⁄16 in., it is time to replace the pulley. If it is not replaced, wear will continue to the point where the belt is no longer transferring power through the walls, as shown in Figure 30-1C. At this point the wear is so bad that the belt is actually riding on the bottom of the groove, as shown in Figure 30-1D. This is not how the manufacturer intended for the belt or pulley to be used. When this occurs, the belt will begin to produce a lot of slippage, ineffectively transmitting power. Pulley and belt wear can be checked with a pulley gauge.

Positive-Drive Belt Maintenance

When a positive-drive belt becomes overloaded, the tension members will most likely begin to break. When this occurs, the belt loses its load capability and begins to stretch. Usually when one tension member breaks, the others break shortly thereafter. The belt loses tension as more tension members break. This causes the teeth to eventually jump out of their mating groove, often referred to as skipping. When skipping begins, the positive traction is compromised. Skipping also causes teeth to break off the belt. When these things occur, the belt can no longer keep time or transmit power efficiently. The belt should therefore be changed before the damage reaches this level. If the belts are inspected on a regular basis, problems will be noticed before too much damage occurs. It is important to keep an eye on pulley wear as well. The mating grooves in a

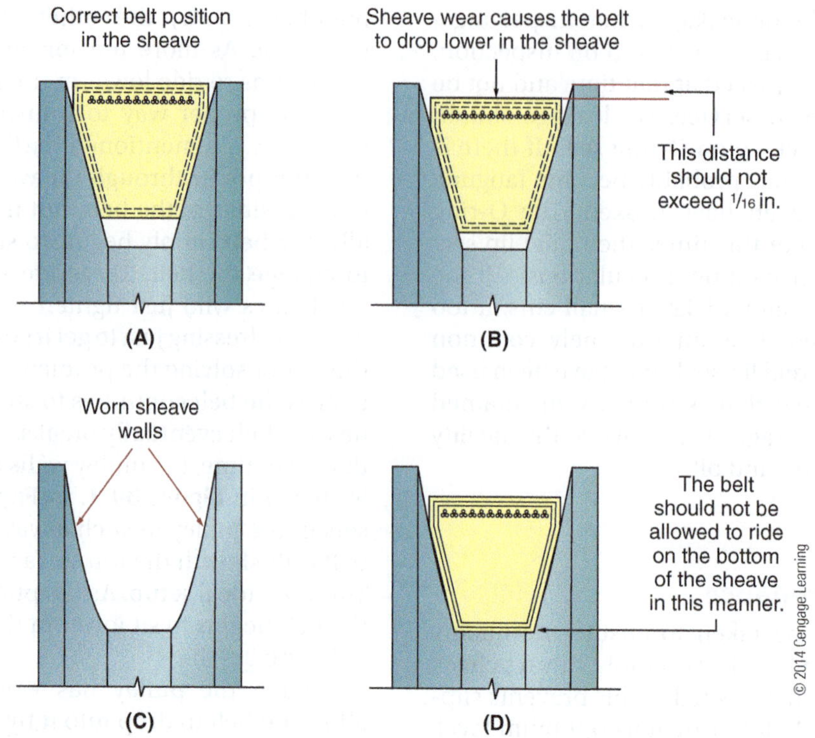
Correct belt position in the sheave
Sheave wear causes the belt to drop lower in the sheave
This distance should not exceed 1/16 in.
(A)
(B)
Worn sheave walls
The belt should not be allowed to ride on the bottom of the sheave in this manner.
(C)
(D)
© 2014 Cengage Learning

FIGURE 30-1 The V-belt is too low in the sheave.

positive-drive pulley may become worn and also cause slippage. This will eventually damage the belt as well. If pulley wear is recognized, the pulley should be changed as soon as possible to prevent damage to the belt.

30-5 CHAINS AND SPROCKETS

The most common problem with chain drives is that the chains are not checked and lubricated often enough. Because the rollers are riding on bushings, there is metal-to-metal contact. It is important to provide lubrication any time there is metal-to-metal contact. This can be done by removing the chain from the drive system and soaking it in oil for a few moments. Soaking allows time for the oil to seep into the roller and bushing area. If this is not done, the bushings and rollers start to wear, causing elongation. Elongation is the stretching of the chain. Once this occurs, the overall length of the chain increases and eventually causes the chain to jump teeth on the sprocket. As the teeth stretch, they ride higher on the sprocket teeth. If this is allowed to continue without repair, the sprocket eventually wears to the point that the chain constantly jumps. The sprocket, along with the chain, must be replaced at this time.

30-6 COUPLINGS

There are many types of couplings in use today. The two most common types are rigid and flexible coupling. Flexible couplings usually wear at the flexible component (called a filler) of the coupling. The filler is usually made of rubber or a synthetic material. The filler does wear out over time. Fillers are designed to absorb a lot of shock, thus preventing the equipment from receiving it. It is easy to recognize when a flexible coupling has had some wear because there is usually a small pile of a black powdery substance (for black rubber coupling fillers) on the floor just below the coupling assembly. This powder is actually small pieces of rubber that have worn off the coupling filler during operation. If this symptom is recognized during a PM inspection, turn off the equipment and check to see whether the coupling has worn out enough to cause any play to be present within the coupling. If play is present, there will be a small amount of independent movement between the two flanges that are in the filler. If this is the case, the coupling filler must be replaced.

Rigid couplings do not have flexible inserts; however, just as flexible couplings, they also show signs of wear. This is usually indicated by a small pile of metal filings on the floor just below the coupling assembly. These filings indicate that the metal surfaces have worn out during operation. Once again, if this symptom is recognized during inspection, turn off the equipment and check to see whether the coupling has worn out enough to cause any play to be present within the coupling. If play is present, there will be a small amount of independent movement between the two flanges of the rigid coupling. If this is the case, replace the coupling. Also perform a coupled-alignment test before replacing the coupling, because misalignment is the leading cause of coupling wear.

SUMMARY

- An early indication of overheating is the presence of solid or caked lubricant.
- Rusting surfaces are an indication of moisture. Rusting is a form of oxidation that occurs when moisture is present and lubrication is lacking.
- A sealed bearing should be used where an excess amount of dirt can accumulate on a bearing.
- Lack of lubrication cause frictions and overheating within the bearing. This causes the bearing to fail prematurely.
- Overlubrication can place excessive internal pressure on the bearing because the rolling elements have to move the excess amount of lubrication within the bearing as well as the load.
- Gearboxes must have oil in them to keep the gears lubricated and cooled. This is the first thing that needs to be checked during a PM inspection of a gearbox.
- Gear wear shows when the teeth of the gears begin to wear out. A symptom of gear wear is the presence of small metal particles in the bottom of the oil reservoir. In gearboxes that have brass gearing, the oil will begin to look metallic in color.
- Always replace old oil with new oil.
- Over time leakage becomes excessive if it is not frequently inspected and properly maintained.
- Remember to allow a small amount of leakage through the packing material to lubricate and to slow the wear of both the packing and shaft material down to a minimum.
- It is important to check the lantern ring.
- Proper care must be taken to ensure the quality and the life of a flat belt. For example, always check the belt to ensure that it is tight. This prevents slippage, allowing the belt to run at its optimum level.
- Dirt and debris can cause the belt to wear out prematurely.
- By dressing a belt, traction can be regained and the need to change the belt occurs less often. Once an excess amount of slippage begins to occur on a regular basis, it may be time to consider replacing the belt.
- It is not uncommon for a belt to become overloaded, causing the tension members to break. When this occurs, the belt loses its load capability and begins to stretch.
- Over time, pulley walls become worn and may need replacing.
- When a positive-drive belt becomes overloaded, its tension members will most likely begin to break also.
- As the positive-drive belt loses tension, the teeth begin to jump out of their mating groove, often referred to as skipping. When skipping begins, the positive traction is compromised, and the teeth break off the belt.
- The most common problem with chain drives is that the chains are not checked and lubricated often enough.
- As a chain stretches, its overall length increases. This eventually causes the chain to jump teeth on the sprocket because as the teeth stretch, they ride higher on the sprocket teeth. If this continues without repair, the sprocket eventually wears out to the point that the chain constantly jumps. The sprocket, along with the chain, has to be replaced at this time.
- Flexible couplings usually wear out at the flexible component (called a filler) of the coupling.
- A small pile of a black powdery substance (for black rubber coupling fillers) will be on the floor just below the coupling assembly when a flexible coupling begins to have some wear.
- Rigid couplings do not have flexible inserts; however, just as flexible couplings, they also will show signs of wear. This is usually indicated by a small pile of metal filings on the floor just below the coupling assembly.
- Perform a coupled-alignment test before replacing any coupling, because misalignment is the leading cause of coupling wear.

REVIEW QUESTIONS

1. What does it mean when a bearing begins to oxidize?
2. What is the first thing that needs to be checked during the PM of a gearbox?
3. What usually occurs after a tension member breaks in a belt?
4. What is the most common problem with chain drives?
5. What is an indication that a coupling is beginning to wear out?

CHAPTER

31

Electrical PM

In today's high-tech world, an effective electrical PM program is an absolute necessity. All too often, the mentality, "If it ain't broke, don't fix it!" prevails. Unfortunately, this thinking can result in increased downtime and repair costs. A PM program identifies potential problems and allows repairs to be scheduled while minimizing production downtime.

OBJECTIVES

After studying this chapter, the student should be able to

- Define the term *preventive maintenance.*
- List five types of records that should be maintained.
- List four rules of electrical maintenance.
- Identify who is responsible for electrical maintenance.

31-1 GENERAL REQUIREMENTS

The key to an effective maintenance program begins with the equipment and installation. Good quality equipment that is properly suited to the job must be selected. However, the equipment is only part of the equation. Once the proper equipment is selected, it must be installed properly. If the equipment is inappropriate for the application, there will be many failures and downtime. Likewise, if the equipment installation is done poorly, there will be problems with the wiring, mechanical installation, and more. In addition, there will be increased labor spent to correct mistakes created by poor workmanship. If these conditions are identified, they should be remedied as soon as possible.

After equipment selection and installation, the next most important factor in having an effective maintenance program is the maintenance personnel. Training of maintenance personnel is paramount, especially in today's age of computers and automation. Industry tends to follow two beliefs when identifying the roles of maintenance technicians. One belief is that maintenance technicians should be specialists. These individuals might have one or two areas of expertise as their responsibility. This means that they know these areas of responsibility inside and out. On the other hand, there is a tendency to have multicrafted individuals as maintenance technicians. Multicrafted technicians must be able to perform all aspects of maintenance within their areas of responsibility. Regardless of the type of maintenance technician employed, training is essential.

Once the equipment is selected and installed and maintenance personnel are trained, it is important to establish a PM program. This is a process that allows for the continual inspection of equipment, reporting and recording observations, creating a schedule for routine PM tasks, and scheduling the repair of defective equipment.

PM is a system of routine inspections of equipment. Information gathered from these inspections is recorded and maintained in a log for future reference. By acquiring and maintaining this information, a history of the equipment, its maintenance, and its failures is developed. A review of this information can identify a trend that may signal impending failure. By recognizing the danger signs, preparation for the failure can be made. This may mean that repairs can be made during an unrelated scheduled outage, thus preventing future failure at an inopportune time.

31-2 INSPECTION RECORDS

Maintaining inspection records is essential to a working PM program. The records provide data that may be lost as a result of an individual's memory lapse. Where multiple shifts of maintenance technicians are employed, the records provide a library of data that is available to everyone all the time. This increases the efficiency of the technicians because they can benefit from the findings and solutions of other technicians.

Today, many PM systems are maintained through computer software. There are several companies that market programs designed to make the recordkeeping, scheduling, and other aspects of maintenance easier and more automated. Whether your facility uses a paper and pencil system or software, there are five essential records that must be maintained:

1. **Equipment information**—This record contains information about the particular piece of equipment. Information such as make, model, serial number, manufacturer, contact information for the manufacturer, installation date, location, specifications (operating voltage, operating current, etc.), and so forth are found on this record.
2. **Inspection checklist**—This record lists all items for a particular piece of equipment that should be inspected on a scheduled basis. This record should list the date, time, item to be inspected, items needed to perform the inspection, results of the inspection, recommended action to be taken, and the name of the inspector.
3. **Repair information**—This record contains a running history of all repairs made to a particular piece of equipment. Information such as the date, time of day, the symptom, operator comments, what was done to repair the equipment, and the name(s) of the individual(s) who made the repairs should be listed on this record. Over time, this record will become extremely valuable because it will show symptoms and cures. This should help lower diagnostic and repair time and costs. In addition, a line of communication will be established between a technician currently working on a problem and someone who may have experienced a similar problem in the past.
4. **Preventive maintenance schedule**—This is a "to do" list for the maintenance technician to follow on a daily basis. This schedule tells maintenance technicians which inspections they need to perform on that particular day. The inspections could cover different areas of

one particular piece of equipment or consist of the same type of inspection to be performed on different machines.

5. **Parts inventory**—This record lists all spare parts and assemblies on hand. As parts or assemblies are used, the inventory is updated to reflect the change. Indicators are set to alert the responsible person that inventory of a particular item is low and replacements must be ordered. This record is invaluable when scheduling an outage or downtime. An accurate inventory can avert the disaster of having a scheduled outage only to find out that key parts are not on hand. By maintaining an accurate inventory and planning for a scheduled outage, replacement parts can be ordered and be on hand when the equipment is brought down for repair.

31-3 FOUR RULES OF ELECTRICAL MAINTENANCE

When considering the maintenance of electrical equipment, look into the wide variety of operating conditions in which electrical equipment is found. These conditions vary from normal room temperature and humidity to below freezing and above boiling temperatures. Electrical equipment operates in normal atmospheres, in explosive atmospheres, under water, in dust, and in other environments.

To minimize downtime and keep electrical equipment operating relatively trouble-free, you only need to remember four rules:

1. *Keep it clean.*
2. *Keep it dry.*
3. *Keep it tight.*
4. *Keep it friction-free (if appropriate).*

Keep It Clean

The most common cause of electrical failure is dirt. Dirt could be an accumulation of daily dust, it could be metallic filings from a nearby process, or it could be in the form of fibers, chemicals, and other materials. Often, dirt buildup provides an undesired path for current flow. This current leakage can cause equipment breakdown and failure. Also, dirt buildup can clog ventilation openings, causing heat buildup. Built-up heat can cause insulation failure and equipment breakdown as well. Whenever an inspection is performed, a thorough cleaning of the equipment is essential.

Keep It Dry

Keeping electrical equipment dry is essential for the proper operation of the equipment. Corrosion or rust can form from high humidity or moisture. Although this may cause mechanical binding, corrosion or rust of conductors may cause a high-resistance connection, resulting in heat buildup and failure. In addition, moisture may act as a current path, causing an inadvertent short circuit. Some liquids react with the material used as insulation on conductors, eventually breaking down and destroying the insulating qualities. Also, moisture tends to attract dirt. As previously mentioned, dirt buildup can also lead to failure.

Keep It Tight

Mechanical vibrations cause electrical connections to loosen over time. Loose electrical connections can create high-resistance connections that result in heat buildup and failure of the termination. Often the termination becomes intermittent. Loose connections can be very difficult and time-consuming to find. To the naked eye, the connection looks good—the screw is in place, and the wire is between the screw and the terminal. Unfortunately, the mechanical connection is loose. It is good practice while performing an inspection to spend a few extra minutes tightening all connections to save hours later trying to locate the loose ones.

Keep It Friction-Free

With the exception of braking functions, electrical equipment operates relatively friction-free. Failure of electrical equipment may occur if a mechanical bind or higher-than-normal friction exists. This may result in an overload condition, which may create nuisance tripping of overcurrent protective devices. It is good practice when performing an inspection to verify that all electrical equipment can operate with low friction.

Inspection Checklist

When all is said and done, a simple checklist that is suitable for use on any type of machine can be a good reminder to help ensure that the proper checks are made. Before beginning any work, the power should be removed from the machine to be inspected. The power source should be locked out and tagged out in accordance with your facility's policies and procedures. When inspecting a machine, the following items should be inspected and then cleaned, dried, and tightened, as appropriate:

- Remove dust, dirt, and debris from within enclosures, boxes, cabinets, and wherever else it is appropriate. This can be done by wiping with approved cleaning agents or solvents or by vacuuming.
- Ensure that all unused openings in enclosures, boxes, and cabinets, are properly closed. Repair or replace as necessary.
- Ensure that safe, proper, and required working clearances are maintained around all enclosures, cabinets, equipment, and machinery. Do whatever is necessary to restore required clearances.
- Inspect components, conductors, and terminations for damage, discoloration, and evidence of overheating. Repair or replace as necessary.
- Inspect for loose components and terminations. Tighten to specifications as necessary.
- Check all pushbuttons, switches, circuit breakers, contactors, motor starters, and so on. Verify proper working condition with no mechanical binding. Repair or replace as necessary.
- Verify that all overcurrent protective devices and thermal overload devices are properly sized and correct for the application. Repair or replace as necessary.
- Verify that all overcurrent protective devices, disconnects, and so forth, are correctly, legibly, and properly labeled. Repair or replace as necessary.
- Inspect the condition and proper support of wire raceways. Repair or replace as necessary.
- Inspect the flexible cords for condition and proper strain reliefs. Repair or replace as necessary.
- Verify that all documentation is up to date and accurately reflects any changes, modifications, and additions to the equipment or installation. Update or modify documentation as necessary.
- Remove lockout/tagout and restore power.
- Verify that all indicators (lights, alarms, etc.) are functioning properly. Repair or replace as necessary after again performing a power shutdown, lockout, and tagout.

31-4 MAINTENANCE RESPONSIBILITY

The responsibility for the continued operation of a facility rests with the maintenance technician. Quite often, maintenance technicians yield to pressure by production personnel. This means that equipment is allowed to run until a catastrophic failure occurs. Often this is very costly. The damage done is more serious, the repair time is longer, and management is more upset. It is in the best interest of the maintenance technician to educate all other plant personnel on the value of an effective PM program. Good maintenance practices benefit everyone.

SUMMARY

- Preventive maintenance is a system of routine inspections of equipment.
- Information gathered from these inspections is recorded and maintained in a log for future reference.
- Today, many PM systems are maintained through computer software.
- Five essential records must be maintained to have an effective PM system.
- There is a wide variety of operating conditions in which electrical equipment is found.
- These conditions vary from normal room temperature and humidity to below freezing and above boiling temperatures.
- Electrical equipment operates in normal atmospheres, in explosive atmospheres, under water, in dust, and in other environments.
- It is important to have a checklist when inspecting a machine to ensure that the proper checks are made.
- Before beginning any work, the power should be removed from the machine that will be inspected.
- The power source should be locked out and tagged out according to the facility's policies and procedures.
- The maintenance technician should educate all plant personnel on the values of an effective PM program.

REVIEW QUESTIONS

1. Explain what preventive maintenance means.
2. List five types of records that should be maintained.
3. List four rules to follow when dealing with electrical maintenance.
4. Who is responsible for electrical maintenance?

APPENDIX A

Table A-1

ALLOWABLE AMPACITIES OF INSULATED CONDUCTORS RATED UP TO AND INCLUDING 2000 VOLTS, 60°C THROUGH 90°C (140°F THROUGH 194°F) NOT MORE THAN THREE CURRENT-CARRYING CONDUCTORS IN RACEWAY, CABLE OR EARTH (DIRECTLY BURIED) BASED ON AMBIENT TEMPERATURE OF 30°C (86°F).

Table 310.15(B)(16) (formerly Table 310.16) Allowable Ampacities of Insulated Conductors Rated Up to and Including 2000 Volts, 60°C Through 90°C (140°F Through 194°F), Not More Than Three Current-Carrying Conductors in Raceway, Cable, or Earth (Directly Buried), Based on Ambient Temperature of 30°C (86°F)*

	Temperature Rating of Conductor [See Table 310.104(A).]						
	60°C (140°F)	75°C (167°F)	90°C (194°F)	60°C (140°F)	75°C (167°F)	90°C (194°F)	
Size AWG or kcmil	Types TW, UF	Types RHW, THHW, THW, THWN, XHHW, USE, ZW	Types TBS, SA, SIS, FEP, FEPB, MI, RHH, RHW-2, THHN, THHW, THW-2, THWN-2, USE-2, XHH, XHHW, XHHW-2, ZW-2	Types TW, UF	Types RHW, THHW, THW, THWN, XHHW, USE	Types TBS, SA, SIS, THHN, THHW, THW-2, THWN-2, RHH, RHW-2, USE-2, XHH, XHHW, XHHW-2, ZW-2	
	COPPER			ALUMINUM OR COPPER-CLAD ALUMINUM			Size AWG or kcmil
18	—	—	14	—	—	—	—
16	—	—	18	—	—	—	—
14**	15	20	25	—	—	—	—
12**	20	25	30	15	20	25	12**
10**	30	35	40	25	30	35	10**
8	40	50	55	35	40	45	8
6	55	65	75	40	50	55	6
4	70	85	95	55	65	75	4
3	85	100	115	65	75	85	3
2	95	115	130	75	90	100	2
1	110	130	145	85	100	115	1
1/0	125	150	170	100	120	135	1/0
2/0	145	175	195	115	135	150	2/0
3/0	165	200	225	130	155	175	3/0
4/0	195	230	260	150	180	205	4/0
250	215	255	290	170	205	230	250
300	240	285	320	195	230	260	300
350	260	310	350	210	250	280	350
400	280	335	380	225	270	305	400
500	320	380	430	260	310	350	500
600	350	420	475	285	340	385	600
700	385	460	520	315	375	425	700
750	400	475	535	320	385	435	750
800	410	490	555	330	395	445	800
900	435	520	585	355	425	480	900
1000	455	545	615	375	445	500	1000
1250	495	590	665	405	485	545	1250
1500	525	625	705	435	520	585	1500
1750	545	650	735	455	545	615	1750
2000	555	665	750	470	560	630	2000

*Refer to 310.15(B)(2) for the ampacity correction factors where the ambient temperature is other than 30°C (86°F).
**Refer to 240.4(D) for conductor overcurrent protection limitations.

Table A-2

ALLOWABLE AMPACITIES OF SINGLE-INSULATED CONDUCTORS RATED 0 THROUGH 2000 VOLTS IN FREE AIR, BASED ON AMBIENT TEMPERATURE OF 30°C (86°F).

Table 310.15(B)(17) (formerly Table 310.17) Allowable Ampacities of Single-Insulated Conductors Rated Up to and Including 2000 Volts in Free Air, Based on Ambient Temperature of 30°C (86°F)*

	Temperature Rating of Conductor [See Table 310.104(A).]						
	60°C (140°F)	75°C (167°F)	90°C (194°F)	60°C (140°F)	75°C (167°F)	90°C (194°F)	
	Types TW, UF	Types RHW, THHW, THW, THWN, XHHW, ZW	Types TBS, SA, SIS, FEP, FEPB, MI, RHH, RHW-2, THHN, THHW, THW-2, THWN-2, USE-2, XHH, XHHW, XHHW-2, ZW-2	Types TW, UF	Types RHW, THHW, THW, THWN, XHHW	Types TBS, SA, SIS, THHN, THHW, THW-2, THWN-2, RHH, RHW-2, USE-2, XHH, XHHW, XHHW-2, ZW-2	
Size AWG or kcmil	COPPER			ALUMINUM OR COPPER-CLAD ALUMINUM			Size AWG or kcmil
18	—	—	18	—	—	—	—
16	—	—	24	—	—	—	—
14**	25	30	35	—	—	—	—
12**	30	35	40	25	30	35	12**
10**	40	50	55	35	40	45	10**
8	60	70	80	45	55	60	8
6	80	95	105	60	75	85	6
4	105	125	140	80	100	115	4
3	120	145	165	95	115	130	3
2	140	170	190	110	135	150	2
1	165	195	220	130	155	175	1
1/0	195	230	260	150	180	205	1/0
2/0	225	265	300	175	210	235	2/0
3/0	260	310	350	200	240	270	3/0
4/0	300	360	405	235	280	315	4/0
250	340	405	455	265	315	355	250
300	375	445	500	290	350	395	300
350	420	505	570	330	395	445	350
400	455	545	615	355	425	480	400
500	515	620	700	405	485	545	500
600	575	690	780	455	545	615	600
700	630	755	850	500	595	670	700
750	655	785	885	515	620	700	750
800	680	815	920	535	645	725	800
900	730	870	980	580	700	790	900
1000	780	935	1055	625	750	845	1000
1250	890	1065	1200	710	855	965	1250
1500	980	1175	1325	795	950	1070	1500
1750	1070	1280	1445	875	1050	1185	1750
2000	1155	1385	1560	960	1150	1295	2000

*Refer to 310.15(B)(2) for the ampacity correction factors where the ambient temperature is other than 30°C (86°F).

**Refer to 240.4(D) for conductor overcurrent protection limitations.

Table A-3

ALLOWABLE AMPACITIES OF INSULATED CONDUCTORS RATED 0 THROUGH 2000 VOLTS, 150°C THROUGH 250°C (302°F THROUGH 482°F). NOT MORE THAN THREE CURRENT-CARRYING CONDUCTORS IN RACEWAY OR CABLE, BASED ON AMBIENT AIR TEMPERATURE OF 40°C (104°F).

Table 310.15(B)(18) (formerly Table 310.18) Allowable Ampacities of Insulated Conductors Rated Up to and Including 2000 Volts, 150°C Through 250°C (302°F Through 482°F). Not More Than Three Current-Carrying Conductors in Raceway or Cable, Based on Ambient Air Temperature of 40°C (104°F)*

	Temperature Rating of Conductor [See Table 310.104(A).]				
	150°C (302°F)	200°C (392°F)	250°C (482°F)	150°C (302°F)	
	Type Z	Types FEP, FEPB, PFA, SA	Types PFAH, TFE	Type Z	
Size AWG or kcmil	COPPER		NICKEL OR NICKEL-COATED COPPER	ALUMINUM OR COPPER-CLAD ALUMINUM	Size AWG or kcmil
14	34	36	39	—	14
12	43	45	54	30	12
10	55	60	73	44	10
8	76	83	93	57	8
6	96	110	117	75	6
4	120	125	148	94	4
3	143	152	166	109	3
2	160	171	191	124	2
1	186	197	215	145	1
1/0	215	229	244	169	1/0
2/0	251	260	273	198	2/0
3/0	288	297	308	227	3/0
4/0	332	346	361	260	4/0

*Refer to 310.15(B)(2) for the ampacity correction factors where the ambient temperature is other than 40°C (104°F).

Table A-4

ALLOWABLE AMPACITIES OF SINGLE-INSULATED CONDUCTORS, RATED 0 THROUGH 2000 VOLTS, 150°C THROUGH 250°C (302°F THROUGH 482°F), IN FREE AIR, BASED ON AMBIENT AIR TEMPERATURE OF 40°C (104°F).

Table 310.15(B)(19) (Formerly Table 310.19) Allowable Ampacities of Single-Insulated Conductors, Rated Up to and Including 2000 Volts, 150°C Through 250°C (302°F Through 482°F), in Free Air, Based on Ambient Air Temperature of 40°C (104°F)*

	Temperature Rating of Conductor [See Table 310.104(A).]				
	150°C (302°F)	200°C (392°F)	250°C (482°F)	150°C (302°F)	
	Type Z	Types FEP, FEPB, PFA, SA	Types PFAH, TFE	Type Z	
Size AWG or kcmil	COPPER		NICKEL, OR NICKEL-COATED COPPER	ALUMINUM OR COPPER-CLAD ALUMINUM	Size AWG or kcmil
14	46	54	59	—	14
12	60	68	78	47	12
10	80	90	107	63	10
8	106	124	142	83	8
6	155	165	205	112	6
4	190	220	278	148	4
3	214	252	327	170	3
2	255	293	381	198	2
1	293	344	440	228	1
1/0	339	399	532	263	1/0
2/0	390	467	591	305	2/0
3/0	451	546	708	351	3/0
4/0	529	629	830	411	4/0

*Refer to 310.15(B)(2) for the ampacity correction factors where the ambient temperature is other than 40°C (104°F).

APPENDIX 

Table B-1

ADJUSTMENT FACTORS FOR MORE THAN THREE CURRENT-CARRYING CONDUCTORS IN A RACEWAY OR CABLE.

Table 310.15(B)(3)(a) Adjustment Factors for More Than Three Current-Carrying Conductors in a Raceway or Cable

Number of Conductors[1]	Percent of Values in Table 310.15(B)(16) through Table 310.15(B)(19) as Adjusted for Ambient Temperature if Necessary
4–6	80
7–9	70
10–20	50
21–30	45
31–40	40
41 and above	35

[1]Number of conductors is the total number of conductors in the raceway or cable adjusted in accordance with 310.15(B)(5) and (6).

Table 310.15(B)(2)(a) Ambient Temperature Correction Factors Based on 30°C (86°F)

For ambient temperatures other than 30°C (86°F), multiply the allowable ampacities specified in the ampacity tables by the appropriate correction factor shown below.

Ambient Temperature (°C)	Temperature Rating of Conductor			Ambient Temperature (°F)
	60°C	75°C	90°C	
10 or less	1.29	1.20	1.15	50 or less
11–15	1.22	1.15	1.12	51–59
16–20	1.15	1.11	1.08	60–68
21–25	1.08	1.05	1.04	69–77
26–30	1.00	1.00	1.00	78–86
31–35	0.91	0.94	0.96	87–95
36–40	0.82	0.88	0.91	96–104
41–45	0.71	0.82	0.87	105–113
46–50	0.58	0.75	0.82	114–122
51–55	0.41	0.67	0.76	123–131
56–60	—	0.58	0.71	132–140
61–65	—	0.47	0.65	141–149
66–70	—	0.33	0.58	150–158
71–75	—	—	0.50	159–167
76–80	—	—	0.41	168–176
81–85	—	—	0.29	177–185

Table 310.15(B)(2)(b) Ambient Temperature Correction Factors Based on 40°C (104°F)

For ambient temperatures other than 40°C (104°F), multiply the allowable ampacities specified in the ampacity tables by the appropriate correction factor shown below.

Ambient Temperature (°C)	Temperature Rating of Conductor						Ambient Temperature (°F)
	60°C	75°C	90°C	150°C	200°C	250°C	
10 or less	1.58	1.36	1.26	1.13	1.09	1.07	50 or less
11–15	1.50	1.31	1.22	1.11	1.08	1.06	51–59
16–20	1.41	1.25	1.18	1.09	1.06	1.05	60–68
21–25	1.32	1.2	1.14	1.07	1.05	1.04	69–77
26–30	1.22	1.13	1.10	1.04	1.03	1.02	78–86
31–35	1.12	1.07	1.05	1.02	1.02	1.01	87–95
36–40	1.00	1.00	1.00	1.00	1.00	1.00	96–104
41–45	0.87	0.93	0.95	0.98	0.98	0.99	105–113
46–50	0.71	0.85	0.89	0.95	0.97	0.98	114–122
51–55	0.50	0.76	0.84	0.93	0.95	0.96	123–131
56–60	—	0.65	0.77	0.90	0.94	0.95	132–140
61–65	—	0.53	0.71	0.88	0.92	0.94	141–149
66–70	—	0.38	0.63	0.85	0.90	0.93	150–158
71–75	—	—	0.55	0.83	0.88	0.91	159–167
76–80	—	—	0.45	0.80	0.87	0.90	168–176
81–90	—	—	—	0.74	0.83	0.87	177–194
91–100	—	—	—	0.67	0.79	0.85	195–212
101–110	—	—	—	0.60	0.75	0.82	213–230
111–120	—	—	—	0.52	0.71	0.79	231–248
121–130	—	—	—	0.43	0.66	0.76	249–266
131–140	—	—	—	0.30	0.61	0.72	267–284
141–160	—	—	—	—	0.50	0.65	285–320
161–180	—	—	—	—	0.35	0.58	321–356
181–200	—	—	—	—	—	0.49	357–392
201–225	—	—	—	—	—	0.35	393–437

APPENDIX C

Table C-1

CONDUIT AND TUBING FILL TABLES FOR CONDUCTORS AND FIXTURE WIRES OF THE SAME SIZE.

Table C.1 Maximum Number of Conductors or Fixture Wires in Electrical Metallic Tubing (EMT) (*Based on Table 1, Chapter 9*)

CONDUCTORS											
	Conductor Size	Metric Designator (Trade Size)									
Type	(AWG kcmil)	16 (½)	21 (¾)	27 (1)	35 (1¼)	41 (1½)	53 (2)	63 (2½)	78 (3)	91 (3½)	103 (4)
RHH, RHW, RHW-2	14	4	7	11	20	27	46	80	120	157	201
	12	3	6	9	17	23	38	66	100	131	167
	10	2	5	8	13	18	30	53	81	105	135
	8	1	2	4	7	9	16	28	42	55	70
	6	1	1	3	5	8	13	22	34	44	56
	4	1	1	2	4	6	10	17	26	34	44
	3	1	1	1	4	5	9	15	23	30	38
	2	1	1	1	3	4	7	13	20	26	33
	1	0	1	1	1	3	5	9	13	17	22
	1/0	0	1	1	1	2	4	7	11	15	19
	2/0	0	1	1	1	2	4	6	10	13	17
	3/0	0	0	1	1	1	3	5	8	11	14
	4/0	0	0	1	1	1	3	5	7	9	12
	250	0	0	0	1	1	1	3	5	7	9
	300	0	0	0	1	1	1	3	5	6	8
	350	0	0	0	1	1	1	3	4	6	7
	400	0	0	0	1	1	1	2	4	5	7
	500	0	0	0	0	1	1	2	3	4	6
	600	0	0	0	0	1	1	1	3	4	5
	700	0	0	0	0	0	1	1	2	3	4
	750	0	0	0	0	0	1	1	2	3	4
	800	0	0	0	0	0	1	1	2	3	4
	900	0	0	0	0	0	1	1	1	3	3
	1000	0	0	0	0	0	1	1	1	2	3
	1250	0	0	0	0	0	0	1	1	1	2
	1500	0	0	0	0	0	0	1	1	1	1
	1750	0	0	0	0	0	0	1	1	1	1
	2000	0	0	0	0	0	0	1	1	1	1
TW	14	8	15	25	43	58	96	168	254	332	424
	12	6	11	19	33	45	74	129	195	255	326
	10	5	8	14	24	33	55	96	145	190	243
	8	2	5	8	13	18	30	53	81	105	135
RHH*, RHW*, RHW-2*, THHW, THW, THW-2	14	6	10	16	28	39	64	112	169	221	282
RHH*, RHW*, RHW-2*, THHW, THW	12	4	8	13	23	31	51	90	136	177	227
	10	3	6	10	18	24	40	70	106	138	177
RHH*, RHW*, RHW-2*, THHW, THW, THW-2	8	1	4	6	10	14	24	42	63	83	106

(*Continues*)

Table C.1 *Continued*

CONDUCTORS											
	Conductor Size	Metric Designator (Trade Size)									
Type	(AWG kcmil)	16 (½)	21 (¾)	27 (1)	35 (1¼)	41 (1½)	53 (2)	63 (2½)	78 (3)	91 (3½)	103 (4)
RHH*, RHW*, RHW-2*, TW, THW, THHW, THW-2	6	1	3	4	8	11	18	32	48	63	81
	4	1	1	3	6	8	13	24	36	47	60
	3	1	1	3	5	7	12	20	31	40	52
	2	1	1	2	4	6	10	17	26	34	44
	1	1	1	1	3	4	7	12	18	24	31
	1/0	0	1	1	2	3	6	10	16	20	26
	2/0	0	1	1	1	3	5	9	13	17	22
	3/0	0	1	1	1	2	4	7	11	15	19
	4/0	0	0	1	1	1	3	6	9	12	16
	250	0	0	1	1	1	3	5	7	10	13
	300	0	0	1	1	1	2	4	6	8	11
	350	0	0	0	1	1	1	4	6	7	10
	400	0	0	0	1	1	1	3	5	7	9
	500	0	0	0	1	1	1	3	4	6	7
	600	0	0	0	1	1	1	2	3	4	6
	700	0	0	0	0	1	1	1	3	4	5
	750	0	0	0	0	1	1	1	3	4	5
	800	0	0	0	0	1	1	1	3	3	5
	900	0	0	0	0	0	1	1	2	3	4
	1000	0	0	0	0	0	1	1	2	3	4
	1250	0	0	0	0	0	1	1	1	2	3
	1500	0	0	0	0	0	1	1	1	1	2
	1750	0	0	0	0	0	0	1	1	1	2
	2000	0	0	0	0	0	0	1	1	1	1
THHN, THWN, THWN-2	14	12	22	35	61	84	138	241	364	476	608
	12	9	16	26	45	61	101	176	266	347	443
	10	5	10	16	28	38	63	111	167	219	279
	8	3	6	9	16	22	36	64	96	126	161
	6	2	4	7	12	16	26	46	69	91	116
	4	1	2	4	7	10	16	28	43	56	71
	3	1	1	3	6	8	13	24	36	47	60
	2	1	1	3	5	7	11	20	30	40	51
	1	1	1	1	4	5	8	15	22	29	37
	1/0	1	1	1	3	4	7	12	19	25	32
	2/0	0	1	1	2	3	6	10	16	20	26
	3/0	0	1	1	1	3	5	8	13	17	22
	4/0	0	1	1	1	2	4	7	11	14	18
	250	0	0	1	1	1	3	6	9	11	15
	300	0	0	1	1	1	3	5	7	10	13
	350	0	0	1	1	1	2	4	6	9	11
	400	0	0	0	1	1	1	4	6	8	10
	500	0	0	0	1	1	1	3	5	6	8
	600	0	0	0	1	1	1	2	4	5	7
	700	0	0	0	1	1	1	2	3	4	6
	750	0	0	0	0	1	1	1	3	4	5
	800	0	0	0	0	1	1	1	3	4	5
	900	0	0	0	0	1	1	1	3	3	4
	1000	0	0	0	0	1	1	1	2	3	4
FEP, FEPB, PFA, PFAH, TFE	14	12	21	34	60	81	134	234	354	462	590
	12	9	15	25	43	59	98	171	258	337	430
	10	6	11	18	31	42	70	122	185	241	309
	8	3	6	10	18	24	40	70	106	138	177
	6	2	4	7	12	17	28	50	75	98	126
	4	1	3	5	9	12	20	35	53	69	88
	3	1	2	4	7	10	16	29	44	57	73
	2	1	1	3	6	8	13	24	36	47	60

(Continues)

Table C.1 *Continued*

CONDUCTORS											
	Conductor Size	Metric Designator (Trade Size)									
Type	(AWG kcmil)	16 (½)	21 (¾)	27 (1)	35 (1¼)	41 (1½)	53 (2)	63 (2½)	78 (3)	91 (3½)	103 (4)
PFA, PFAH, TFE	1	1	1	2	4	6	9	16	25	33	42
PFAH, TFE PFA, PFAH, TFE, Z	1/0	1	1	1	3	5	8	14	21	27	35
	2/0	0	1	1	3	4	6	11	17	22	29
	3/0	0	1	1	2	3	5	9	14	18	24
	4/0	0	1	1	1	2	4	8	11	15	19
Z	14	14	25	41	72	98	161	282	426	556	711
	12	10	18	29	51	69	114	200	302	394	504
	10	6	11	18	31	42	70	122	185	241	309
	8	4	7	11	20	27	44	77	117	153	195
	6	3	5	8	14	19	31	54	82	107	137
	4	1	3	5	9	13	21	37	56	74	94
	3	1	2	4	7	9	15	27	41	54	69
	2	1	1	3	6	8	13	22	34	45	57
	1	1	1	2	4	6	10	18	28	36	46
XHH, XHHW, XHHW-2, ZW	14	8	15	25	43	58	96	168	254	332	424
	12	6	11	19	33	45	74	129	195	255	326
	10	5	8	14	24	33	55	96	145	190	243
	8	2	5	8	13	18	30	53	81	105	135
	6	1	3	6	10	14	22	39	60	78	100
	4	1	2	4	7	10	16	28	43	56	72
	3	1	1	3	6	8	14	24	36	48	61
	2	1	1	3	5	7	11	20	31	40	51
XHH, XHHW, XHHW-2	1	1	1	1	4	5	8	15	23	30	38
	1/0	1	1	1	3	4	7	13	19	25	32
	2/0	0	1	1	2	3	6	10	16	21	27
	3/0	0	1	1	1	3	5	9	13	17	22
	4/0	0	1	1	1	2	4	7	11	14	18
	250	0	0	1	1	1	3	6	9	12	15
	300	0	0	1	1	1	3	5	8	10	13
	350	0	0	1	1	1	2	4	7	9	11
	400	0	0	0	1	1	1	4	6	8	10
	500	0	0	0	1	1	1	3	5	6	8
	600	0	0	0	1	1	1	2	4	5	6
	700	0	0	0	0	1	1	2	3	4	6
	750	0	0	0	0	1	1	1	3	4	5
	800	0	0	0	0	1	1	1	3	4	5
	900	0	0	0	0	1	1	1	3	3	4
	1000	0	0	0	0	0	1	1	2	3	4
	1250	0	0	0	0	0	1	1	1	2	3
	1500	0	0	0	0	0	1	1	1	1	3
	1750	0	0	0	0	0	0	1	1	1	2
	2000	0	0	0	0	0	0	1	1	1	1

(Continues)

Table C.1 *Continued*

FIXTURE WIRES							
	Conductor Size	Metric Designator (Trade Size)					
Type	(AWG/ kcmil)	16 (½)	21 (¾)	27 (1)	35 (1¼)	41 (1½)	53 (2)
FFH-2, RFH-2, RFHH-3	18	8	14	24	41	56	92
	16	7	12	20	34	47	78
SF-2, SFF-2	18	10	18	30	52	71	116
	16	8	15	25	43	58	96
	14	7	12	20	34	47	78
SF-1, SFF-1	18	18	33	53	92	125	206
RFH-1, RFHH-2, TF, TFF, XF, XFF	18	14	24	39	68	92	152
RFHH-2, TF, TFF, XF, XFF	16	11	19	31	55	74	123
XF, XFF	14	8	15	25	43	58	96
TFN, TFFN	18	22	38	63	108	148	244
	16	17	29	48	83	113	186
PF, PFF, PGF, PGFF, PAF, PTF, PTFF, PAFF	18	21	36	59	103	140	231
	16	16	28	46	79	108	179
	14	12	21	34	60	81	134
ZF, ZFF, ZHF, HF, HFF	18	27	47	77	133	181	298
	16	20	35	56	98	133	220
	14	14	25	41	72	98	161
KF-2, KFF-2	18	39	69	111	193	262	433
	16	27	48	78	136	185	305
	14	19	33	54	93	127	209
	12	13	23	37	64	87	144
	10	8	15	25	43	58	96
KF-1, KFF-1	18	46	82	133	230	313	516
	16	33	57	93	161	220	362
	14	22	38	63	108	148	244
	12	14	25	41	72	98	161
	10	9	16	27	47	64	105
XF, XFF	12	4	8	13	23	31	51
	10	3	6	10	18	24	40

Notes:

1. This table is for concentric stranded conductors only. For compact stranded conductors, Table C.1(A) should be used.

2. Two-hour fire-rated RHH cable has ceramifiable insulation which has much larger diameters than other RHH wires.

Consult manufacturer's conduit fill tables.

*Types RHH, RHW, and RHW-2 without outer covering.

Table C-2

CONDUIT AND TUBING FILL TABLES FOR CONDUCTORS AND FIXTURE WIRES OF THE SAME SIZE *(CONTINUED)*.

Table C.2 Maximum Number of Conductors or Fixture Wires in Electrical Nonmetallic Tubing (ENT) (*Based on Table 1, Chapter 9*)

CONDUCTORS							
	Conductor Size	Metric Designator (Trade Size)					
Type	(AWG/ kcmil)	16 (½)	21 (¾)	27 (1)	35 (1¼)	41 (1½)	53 (2)
RHH, RHW, RHW-2	14	3	6	10	19	26	43
	12	2	5	9	16	22	36
	10	1	4	7	13	17	29
	8	1	1	3	6	9	15
	6	1	1	3	5	7	12
	4	1	1	2	4	6	9
	3	1	1	1	3	5	8
	2	0	1	1	3	4	7
	1	0	1	1	1	3	5
	1/0	0	0	1	1	2	4
	2/0	0	0	1	1	1	3
	3/0	0	0	1	1	1	3
	4/0	0	0	1	1	1	2
	250	0	0	0	1	1	1
	300	0	0	0	1	1	1
	350	0	0	0	1	1	1
	400	0	0	0	1	1	1
	500	0	0	0	0	1	1
	600	0	0	0	0	1	1
	700	0	0	0	0	0	1
	750	0	0	0	0	0	1
	800	0	0	0	0	0	1
	900	0	0	0	0	0	1
	1000	0	0	0	0	0	1
	1250	0	0	0	0	0	0
	1500	0	0	0	0	0	0
	1750	0	0	0	0	0	0
	2000	0	0	0	0	0	0
TW	14	7	13	22	40	55	92
	12	5	10	17	31	42	71
	10	4	7	13	23	32	52
	8	1	4	7	13	17	29
RHH*, RHW*, RHW-2*, THHW, THW, THW-2	14	4	8	15	27	37	61
RHH*, RHW*, RHW-2*, THHW, THW	12	3	7	12	21	29	49
	10	3	5	9	17	23	38
RHH*, RHW*, RHW-2*, THHW, THW, THW-2	8	1	3	5	10	14	23

(Continues)

Table C.2 *Continued*

CONDUCTORS							
Type	Conductor Size (AWG/ kcmil)	Metric Designator (Trade Size)					
		16 (½)	21 (¾)	27 (1)	35 (1¼)	41 (1½)	53 (2)
RHH*, RHW*, RHW-2*, TW, THW, THHW, THW-2	6	1	2	4	7	10	17
	4	1	1	3	5	8	13
	3	1	1	2	5	7	11
	2	1	1	2	4	6	9
	1	0	1	1	3	4	6
	1/0	0	1	1	2	3	5
	2/0	0	1	1	1	3	5
	3/0	0	0	1	1	2	4
	4/0	0	0	1	1	1	3
	250	0	0	1	1	1	2
	300	0	0	0	1	1	2
	350	0	0	0	1	1	1
	400	0	0	0	1	1	1
	500	0	0	0	1	1	1
	600	0	0	0	0	1	1
	700	0	0	0	0	1	1
	750	0	0	0	0	1	1
	800	0	0	0	0	1	1
	900	0	0	0	0	0	1
	1000	0	0	0	0	0	1
	1250	0	0	0	0	0	1
	1500	0	0	0	0	0	0
	1750	0	0	0	0	0	0
	2000	0	0	0	0	0	0
THHN, THWN, THWN-2	14	10	18	32	58	80	132
	12	7	13	23	42	58	96
	10	4	8	15	26	36	60
	8	2	5	8	15	21	35
	6	1	3	6	11	15	25
	4	1	1	4	7	9	15
	3	1	1	3	5	8	13
	2	1	1	2	5	6	11
	1	1	1	1	3	5	8
	1/0	0	1	1	3	4	7
	2/0	0	1	1	2	3	5
	3/0	0	1	1	1	3	4
	4/0	0	0	1	1	2	4
	250	0	0	1	1	1	3
	300	0	0	1	1	1	2
	350	0	0	0	1	1	2
	400	0	0	0	1	1	1
	500	0	0	0	1	1	1
	600	0	0	0	1	1	1
	700	0	0	0	0	1	1
	750	0	0	0	0	1	1
	800	0	0	0	0	1	1
	900	0	0	0	0	1	1
	1000	0	0	0	0	0	1
FEP, FEPB, PFA, PFAH, TFE	14	10	18	31	56	77	128
	12	7	13	23	41	56	93
	10	5	9	16	29	40	67
	8	3	5	9	17	23	38
	6	1	4	6	12	16	27
	4	1	2	4	8	11	19
	3	1	1	4	7	9	16
	2	1	1	3	5	8	13
PFA, PFAH, TFE	1	1	1	1	4	5	9

(Continues)

Table C.2 *Continued*

CONDUCTORS							
	Conductor Size	Metric Designator (Trade Size)					
Type	(AWG/ kcmil)	16 (½)	21 (¾)	27 (1)	35 (1¼)	41 (1½)	53 (2)
PFA, PFAH, TFE, Z	1/0	0	1	1	3	4	7
	2/0	0	1	1	2	4	6
	3/0	0	1	1	1	3	5
	4/0	0	1	1	1	2	4
Z	14	12	22	38	68	93	154
	12	8	15	27	48	66	109
	10	5	9	16	29	40	67
	8	3	6	10	18	25	42
	6	1	4	7	13	18	30
	4	1	3	5	9	12	20
	3	1	1	3	6	9	15
	2	1	1	3	5	7	12
	1	1	1	2	4	6	10
XHH, XHHW, XHHW-2, ZW	14	7	13	22	40	55	92
	12	5	10	17	31	42	71
	10	4	7	13	23	32	52
	8	1	4	7	13	17	29
	6	1	3	5	9	13	21
	4	1	1	4	7	9	15
	3	1	1	3	6	8	13
	2	1	1	2	5	6	11
XHH, XHHW, XHHW-2	1	1	1	1	3	5	8
	1/0	0	1	1	3	4	7
	2/0	0	1	1	2	3	6
	3/0	0	1	1	1	3	5
	4/0	0	0	1	1	2	4
	250	0	0	1	1	1	3
	300	0	0	1	1	1	3
	350	0	0	1	1	1	2
	400	0	0	0	1	1	1
	500	0	0	0	1	1	1
	600	0	0	0	1	1	1
	700	0	0	0	0	1	1
	750	0	0	0	0	1	1
	800	0	0	0	0	1	1
	900	0	0	0	0	1	1
	1000	0	0	0	0	0	1
	1250	0	0	0	0	0	1
	1500	0	0	0	0	0	1
	1750	0	0	0	0	0	0
	2000	0	0	0	0	0	0

(Continues)

Table C.2 *Continued*

FIXTURE WIRES							
	Conductor Size	Metric Designator (Trade Size)					
Type	(AWG/ kcmil)	16 (½)	21 (¾)	27 (1)	35 (1¼)	41 (1½)	53 (2)
FFH-2, RFH-2,	18	6	12	21	39	53	88
	16	5	10	18	32	45	74
RFHH-3 SF-2, SFF-2	18	8	15	27	49	67	111
	16	7	13	22	40	55	92
	14	5	10	18	32	45	74
SF-1, SFF-1	18	15	28	48	86	119	197
RFH-1, RFHH-2, TF, TFF, XF, XFF	18	11	20	35	64	88	145
RFHH-2, TF, TFF, XF, XFF	16	9	16	29	51	71	117
XF, XFF	14	7	13	22	40	55	92
TFN, TFFN	18	18	33	57	102	141	233
	16	13	25	43	78	107	178
PF, PFF, PGF, PGFF, PAF, PTF, PTFF, PAFF	18	17	31	54	97	133	221
	16	13	24	42	75	103	171
	14	10	18	31	56	77	128
ZF, ZFF, ZHF, HF, HFF	18	22	40	70	125	172	285
	16	16	29	51	92	127	210
	14	12	22	38	68	93	154
KF-2, KFF-2	18	31	58	101	182	250	413
	16	22	41	71	128	176	291
	14	15	28	49	88	121	200
	12	10	19	33	60	83	138
	10	7	13	22	40	55	92
KF-1, KFF-1	18	38	69	121	217	298	493
	16	26	49	85	152	209	346
	14	18	33	57	102	141	233
	12	12	22	38	68	93	154
	10	7	14	24	44	61	101
XF, XFF	12	3	7	12	21	29	49
	10	3	5	9	17	23	38

Notes:

1. This table is for concentric stranded conductors only. For compact stranded conductors, Table C.2(A) should be used.

2. Two-hour fire-rated RHH cable has ceramifiable insulation which has much larger diameters than other RHH wires. Consult manufacturer's conduit fill tables.

*Types RHH, RHW, and RHW-2 without outer covering.

Table C-3

CONDUIT AND TUBING FILL TABLES FOR CONDUCTORS AND FIXTURE WIRES OF THE SAME SIZE *(CONTINUED).*

Table C.3 Maximum Number of Conductors or Fixture Wires in Flexible Metal Conduit (FMC) *(Based on Table 1, Chapter 9)*

CONDUCTORS											
	Conductor Size (AWG/ kcmil)	Metric Designator (Trade Size)									
Type		16 (½)	21 (¾)	27 (1)	35 (1¼)	41 (1½)	53 (2)	63 (2½)	78 (3)	91 (3½)	103 (4)
RHH, RHW, RHW-2	14	4	7	11	17	25	44	67	96	131	171
	12	3	6	9	14	21	37	55	80	109	142
	10	3	5	7	11	17	30	45	64	88	115
	8	1	2	4	6	9	15	23	34	46	60
	6	1	1	3	5	7	12	19	27	37	48
	4	1	1	2	4	5	10	14	21	29	37
	3	1	1	1	3	5	8	13	18	25	33
	2	1	1	1	3	4	7	11	16	22	28
	1	0	1	1	1	2	5	7	10	14	19
	1/0	0	1	1	1	2	4	6	9	12	16
	2/0	0	1	1	1	1	3	5	8	11	14
	3/0	0	0	1	1	1	3	5	7	9	12
	4/0	0	0	1	1	1	2	4	6	8	10
	250	0	0	0	1	1	1	3	4	6	8
	300	0	0	0	1	1	1	2	4	5	7
	350	0	0	0	1	1	1	2	3	5	6
	400	0	0	0	0	1	1	1	3	4	6
	500	0	0	0	0	1	1	1	3	4	5
	600	0	0	0	0	1	1	1	2	3	4
	700	0	0	0	0	0	1	1	1	3	3
	750	0	0	0	0	0	1	1	1	2	3
	800	0	0	0	0	0	1	1	1	2	3
	900	0	0	0	0	0	1	1	1	2	3
	1000	0	0	0	0	0	1	1	1	1	3
	1250	0	0	0	0	0	0	1	1	1	1
	1500	0	0	0	0	0	0	1	1	1	1
	1750	0	0	0	0	0	0	1	1	1	1
	2000	0	0	0	0	0	0	0	1	1	1
TW	14	9	15	23	36	53	94	141	203	277	361
	12	7	11	18	28	41	72	108	156	212	277
	10	5	8	13	21	30	54	81	116	158	207
	8	3	5	7	11	17	30	45	64	88	115
RHH*, RHW*, RHW-2*, THHW, THW, THW-2	14	6	10	15	24	35	62	94	135	184	240
RHH*, RHW*, RHW-2*, THHW, THW	12	5	8	12	19	28	50	75	108	148	193
	10	4	6	10	15	22	39	59	85	115	151
RHH*, RHW*, RHW-2*, THHW, THW, THW-2	8	1	4	6	9	13	23	35	51	69	90

(Continues)

Table C.3 *Continued*

		CONDUCTORS									
	Conductor Size	Metric Designator (Trade Size)									
Type	**(AWG/ kcmil)**	**16 (½)**	**21 (¾)**	**27 (1)**	**35 (1¼)**	**41 (1½)**	**53 (2)**	**63 (2½)**	**78 (3)**	**91 (3½)**	**103 (4)**
RHH*, RHW*, RHW-2*, TW, THW, THHW, THW-2	6	1	3	4	7	10	18	27	39	53	69
	4	1	1	3	5	7	13	20	29	39	51
	3	1	1	3	4	6	11	17	25	34	44
	2	1	1	2	4	5	10	14	21	29	37
	1	1	1	1	2	4	7	10	15	20	26
	1/0	0	1	1	1	3	6	9	12	17	22
	2/0	0	1	1	1	3	5	7	10	14	19
	3/0	0	1	1	1	2	4	6	9	12	16
	4/0	0	0	1	1	1	3	5	7	10	13
	250	0	0	1	1	1	3	4	6	8	11
	300	0	0	1	1	1	2	3	5	7	9
	350	0	0	0	1	1	1	3	4	6	8
	400	0	0	0	1	1	1	3	4	6	7
	500	0	0	0	1	1	1	2	3	5	6
	600	0	0	0	0	1	1	1	3	4	5
	700	0	0	0	0	1	1	1	2	3	4
	750	0	0	0	0	1	1	1	2	3	4
	800	0	0	0	0	1	1	1	1	3	4
	900	0	0	0	0	0	1	1	1	3	3
	1000	0	0	0	0	0	1	1	1	2	3
	1250	0	0	0	0	0	1	1	1	1	2
	1500	0	0	0	0	0	0	1	1	1	1
	1750	0	0	0	0	0	0	1	1	1	1
	2000	0	0	0	0	0	0	1	1	1	1
THHN, THWN, THWN-2	14	13	22	33	52	76	134	202	291	396	518
	12	9	16	24	38	56	98	147	212	289	378
	10	6	10	15	24	35	62	93	134	182	238
	8	3	6	9	14	20	35	53	77	105	137
	6	2	4	6	10	14	25	38	55	76	99
	4	1	2	4	6	9	16	24	34	46	61
	3	1	1	3	5	7	13	20	29	39	51
	2	1	1	3	4	6	11	17	24	33	43
	1	1	1	1	3	4	8	12	18	24	32
	1/0	1	1	1	2	4	7	10	15	20	27
	2/0	0	1	1	1	3	6	9	12	17	22
	3/0	0	1	1	1	2	5	7	10	14	18
	4/0	0	1	1	1	1	4	6	8	12	15
	250	0	0	1	1	1	3	5	7	9	12
	300	0	0	1	1	1	3	4	6	8	11
	350	0	0	1	1	1	2	3	5	7	9
	400	0	0	0	1	1	1	3	5	6	8
	500	0	0	0	1	1	1	2	4	5	7
	600	0	0	0	0	1	1	1	3	4	5
	700	0	0	0	0	1	1	1	3	4	5
	750	0	0	0	0	1	1	1	2	3	4
	800	0	0	0	0	1	1	1	2	3	4
	900	0	0	0	0	0	1	1	1	3	4
	1000	0	0	0	0	0	1	1	1	3	3
FEP, FEPB, PFA, PFAH, TFE	14	12	21	32	51	74	130	196	282	385	502
	12	9	15	24	37	54	95	143	206	281	367
	10	6	11	17	26	39	68	103	148	201	263
	8	4	6	10	15	22	39	59	85	115	151
	6	2	4	7	11	16	28	42	60	82	107
	4	1	3	5	7	11	19	29	42	57	75
	3	1	2	4	6	9	16	24	35	48	62
	2	1	1	3	5	7	13	20	29	39	51

(Continues)

Table C.3 *Continued*

CONDUCTORS											
	Conductor Size	Metric Designator (Trade Size)									
Type	(AWG/ kcmil)	16 (½)	21 (¾)	27 (1)	35 (1¼)	41 (1½)	53 (2)	63 (2½)	78 (3)	91 (3½)	103 (4)
PFA, PFAH, TFE	1	1	1	2	3	5	9	14	20	27	36
PFA, PFAH, TFE, Z	1/0	1	1	1	3	4	8	11	17	23	30
	2/0	1	1	1	2	3	6	9	14	19	24
	3/0	0	1	1	1	3	5	8	11	15	20
	4/0	0	1	1	1	2	4	6	9	13	16
Z	14	15	25	39	61	89	157	236	340	463	605
	12	11	18	28	43	63	111	168	241	329	429
	10	6	11	17	26	39	68	103	148	201	263
	8	4	7	11	17	24	43	65	93	127	166
	6	3	5	7	12	17	30	45	65	89	117
	4	1	3	5	8	12	21	31	45	61	80
	3	1	2	4	6	8	15	23	33	45	58
	2	1	1	3	5	7	12	19	27	37	49
	1	1	1	2	4	6	10	15	22	30	39
XHH, XHHW, XHHW-2, ZW	14	9	15	23	36	53	94	141	203	277	361
	12	7	11	18	28	41	72	108	156	212	277
	10	5	8	13	21	30	54	81	116	158	207
	8	3	5	7	11	17	30	45	64	88	115
	6	1	3	5	8	12	22	33	48	65	85
	4	1	2	4	6	9	16	24	34	47	61
	3	1	1	3	5	7	13	20	29	40	52
	2	1	1	3	4	6	11	17	24	33	44
XHH, XHHW, XHHW-2	1	1	1	1	3	5	8	13	18	25	32
	1/0	1	1	1	2	4	7	10	15	21	27
	2/0	0	1	1	2	3	6	9	13	17	23
	3/0	0	1	1	1	3	5	7	10	14	19
	4/0	0	1	1	1	2	4	6	9	12	15
	250	0	0	1	1	1	3	5	7	10	13
	300	0	0	1	1	1	3	4	6	8	11
	350	0	0	1	1	1	2	4	5	7	9
	400	0	0	0	1	1	1	3	5	6	8
	500	0	0	0	1	1	1	3	4	5	7
	600	0	0	0	0	1	1	1	3	4	5
	700	0	0	0	0	1	1	1	3	4	5
	750	0	0	0	0	1	1	1	2	3	4
	800	0	0	0	0	1	1	1	2	3	4
	900	0	0	0	0	0	1	1	1	3	4
	1000	0	0	0	0	0	1	1	1	3	3
	1250	0	0	0	0	0	1	1	1	1	3
	1500	0	0	0	0	0	1	1	1	1	2
	1750	0	0	0	0	0	0	1	1	1	1
	2000	0	0	0	0	0	0	1	1	1	1

(Continues)

Table C.3 *Continued*

FIXTURE WIRES							
	Conductor Size	Metric Designator (Trade Size)					
Type	(AWG/ kcmil)	16 (½)	21 (¾)	27 (1)	35 (1¼)	41 (1½)	53 (2)
FFH-2, RFH-2, RFHH-3	18	8	14	22	35	51	90
	16	7	12	19	29	43	76
SF-2, SFF-2	18	11	18	28	44	64	113
	16	9	15	23	36	53	94
	14	7	12	19	29	43	76
SF-1, SFF-1	18	19	32	50	78	114	201
RFH-1, RFHH-2, TF, TFF, XF, XFF	18	14	24	37	58	84	148
RFHH-2, TF, TFF, XF, XFF	16	11	19	30	47	68	120
XF, XFF	14	9	15	23	36	53	94
TFN, TFFN	18	23	38	59	93	135	237
	16	17	29	45	71	103	181
PF, PFF, PGF, PGFF, PAF, PTF, PTFF, PAFF	18	22	36	56	88	128	225
	16	17	28	43	68	99	174
	14	12	21	32	51	74	130
ZF, ZFF, ZHF, HF, HFF	18	28	47	72	113	165	290
	16	20	35	53	83	121	214
	14	15	25	39	61	89	157
KF-2, KFF-2	18	41	68	105	164	239	421
	16	28	48	74	116	168	297
	14	19	33	51	80	116	204
	12	13	23	35	55	80	140
	10	9	15	23	36	53	94
KF-1, KFF-1	18	48	82	125	196	285	503
	16	34	57	88	138	200	353
	14	23	38	59	93	135	237
	12	15	25	39	61	89	157
	10	10	16	25	40	58	103
XF, XFF	12	5	8	12	19	28	50
	10	4	6	10	15	22	39

Notes:

1. This table is for concentric stranded conductors only. For compact stranded conductors, Table C.3(A) should be used.

2. Two-hour fire-rated RHH cable has ceramifiable insulation which has much larger diameters than other RHH wires. Consult manufacturer's conduit fill tables.

*Types RHH, RHW, and RHW-2 without outer covering.

Table C-4

CONDUIT AND TUBING FILL TABLES FOR CONDUCTORS AND FIXTURE WIRES OF THE SAME SIZE *(CONTINUED)*.

Table C.4 Maximum Number of Conductors or Fixture Wires in Intermediate Metal Conduit (IMC) *(Based on Table 1, Chapter 9)*

CONDUCTORS											
	Conductor Size	Metric Designator (Trade Size)									
Type	(AWG/ kcmil)	16 (½)	21 (¾)	27 (1)	35 (1¼)	41 (1½)	53 (2)	63 (2½)	78 (3)	91 (3½)	103 (4)
RHH, RHW, RHW-2	14	4	8	13	22	30	49	70	108	144	186
	12	4	6	11	18	25	41	58	89	120	154
RHH, RHW, RHW-2	10	3	5	8	15	20	33	47	72	97	124
	8	1	3	4	8	10	17	24	38	50	65
	6	1	1	3	6	8	14	19	30	40	52
	4	1	1	3	5	6	11	15	23	31	41
	3	1	1	2	4	6	9	13	21	28	36
	2	1	1	1	3	5	8	11	18	24	31
	1	0	1	1	2	3	5	7	12	16	20
	1/0	0	1	1	1	3	4	6	10	14	18
	2/0	0	1	1	1	2	4	6	9	12	15
	3/0	0	0	1	1	1	3	5	7	10	13
	4/0	0	0	1	1	1	3	4	6	9	11
	250	0	0	1	1	1	1	3	5	6	8
	300	0	0	0	1	1	1	3	4	6	7
	350	0	0	0	1	1	1	2	4	5	7
	400	0	0	0	1	1	1	2	3	5	6
	500	0	0	0	1	1	1	1	3	4	5
	600	0	0	0	0	1	1	1	2	3	4
	700	0	0	0	0	1	1	1	2	3	4
	750	0	0	0	0	1	1	1	1	3	4
	800	0	0	0	0	0	1	1	1	3	3
	900	0	0	0	0	0	1	1	1	2	3
	1000	0	0	0	0	0	1	1	1	2	3
	1250	0	0	0	0	0	1	1	1	1	2
	1500	0	0	0	0	0	0	1	1	1	1
	1750	0	0	0	0	0	0	1	1	1	1
	2000	0	0	0	0	0	0	1	1	1	1
TW	14	10	17	27	47	64	104	147	228	304	392
	12	7	13	21	36	49	80	113	175	234	301
	10	5	9	15	27	36	59	84	130	174	224
	8	3	5	8	15	20	33	47	72	97	124
RHH*, RHW*, RHW-2*, THHW, THW, THW-2	14	6	11	18	31	42	69	98	151	202	261
RHH*, RHW*, RHW-2*, THHW, THW	12	5	9	14	25	34	56	79	122	163	209
	10	4	7	11	19	26	43	61	95	127	163
RHH*, RHW*, RHW-2*, THHW, THW, THW-2	8	2	4	7	12	16	26	37	57	76	98
RHH*, RHW*, RHW-2*, TW, THHW, THW, THW-2	6	1	3	5	9	12	20	28	43	58	75
	4	1	2	4	6	9	15	21	32	43	56

(Continues)

Table C.4 *Continued*

CONDUCTORS											
	Conductor Size	Metric Designator (Trade Size)									
Type	(AWG/ kcmil)	16 (½)	21 (¾)	27 (1)	35 (1¼)	41 (1½)	53 (2)	63 (2½)	78 (3)	91 (3½)	103 (4)
RHH*,	3	1	1	3	6	8	13	18	28	37	48
RHW*,	2	1	1	3	5	6	11	15	23	31	41
RHW-2*,	1	1	1	1	3	4	7	11	16	22	28
TW,	1/0	1	1	1	3	4	6	9	14	19	24
THW,	2/0	0	1	1	2	3	5	8	12	16	20
THHW,	3/0	0	1	1	1	3	4	6	10	13	17
THW-2	4/0	0	1	1	1	2	4	5	8	11	14
	250	0	0	1	1	1	3	4	7	9	12
	300	0	0	1	1	1	2	4	6	8	10
	350	0	0	1	1	1	2	3	5	7	9
	400	0	0	0	1	1	1	3	4	6	8
	500	0	0	0	1	1	1	2	4	5	7
	600	0	0	0	1	1	1	1	3	4	5
	700	0	0	0	0	1	1	1	3	4	5
	750	0	0	0	0	1	1	1	2	3	4
	800	0	0	0	0	1	1	1	2	3	4
	900	0	0	0	0	1	1	1	2	3	4
	1000	0	0	0	0	0	1	1	1	3	3
	1250	0	0	0	0	0	1	1	1	1	3
	1500	0	0	0	0	0	1	1	1	1	2
	1750	0	0	0	0	0	0	1	1	1	1
	2000	0	0	0	0	0	0	1	1	1	1
THHN,	14	14	24	39	68	91	149	211	326	436	562
THWN,	12	10	17	29	49	67	109	154	238	318	410
THWN-2	10	6	11	18	31	42	68	97	150	200	258
	8	3	6	10	18	24	39	56	86	115	149
	6	2	4	7	13	17	28	40	62	83	107
	4	1	3	4	8	10	17	25	38	51	66
	3	1	2	4	6	9	15	21	32	43	56
	2	1	1	3	5	7	12	17	27	36	47
	1	1	1	2	4	5	9	13	20	27	35
	1/0	1	1	1	3	4	8	11	17	23	29
	2/0	1	1	1	3	4	6	9	14	19	24
	3/0	0	1	1	2	3	5	7	12	16	20
	4/0	0	1	1	1	2	4	6	9	13	17
	250	0	0	1	1	1	3	5	8	10	13
	300	0	0	1	1	1	3	4	7	9	12
	350	0	0	1	1	1	2	4	6	8	10
	400	0	0	1	1	1	2	3	5	7	9
	500	0	0	0	1	1	1	3	4	6	7
	600	0	0	0	1	1	1	2	3	5	6
	700	0	0	0	1	1	1	1	3	4	5
	750	0	0	0	1	1	1	1	3	4	5
	800	0	0	0	0	1	1	1	3	4	5
	900	0	0	0	0	1	1	1	2	3	4
	1000	0	0	0	0	1	1	1	2	3	4
FEP,	14	13	23	38	66	89	145	205	317	423	545
FEPB,	12	10	17	28	48	65	106	150	231	309	398
PFA,	10	7	12	20	34	46	76	107	166	221	285
PFAH,	8	4	7	11	19	26	43	61	95	127	163
TFE	6	3	5	8	14	19	31	44	67	90	116
	4	1	3	5	10	13	21	30	47	63	81
	3	1	3	4	8	11	18	25	39	52	68
	2	1	2	4	6	9	15	21	32	43	56
PFA, PFAH, TFE	1	1	1	2	4	6	10	14	22	30	39

(Continues)

Table C.4 *Continued*

CONDUCTORS											
	Conductor Size	Metric Designator (Trade Size)									
Type	(AWG/ kcmil)	16 (½)	21 (¾)	27 (1)	35 (1¼)	41 (1½)	53 (2)	63 (2½)	78 (3)	91 (3½)	103 (4)
PFA, PFAH, TFE, Z	1/0	1	1	1	4	5	8	12	19	25	32
	2/0	1	1	1	3	4	7	10	15	21	27
	3/0	0	1	1	2	3	6	8	13	17	22
	4/0	0	1	1	1	3	5	7	10	14	18
Z	14	16	28	46	79	107	175	247	381	510	657
	12	11	20	32	56	76	124	175	271	362	466
	10	7	12	20	34	46	76	107	166	221	285
	8	4	7	12	21	29	48	68	105	140	180
	6	3	5	9	15	20	33	47	73	98	127
	4	1	3	6	10	14	23	33	50	67	87
	3	1	2	4	7	10	17	24	37	49	63
	2	1	1	3	6	8	14	20	30	41	53
	1	1	1	3	5	7	11	16	25	33	43
XHH, XHHW, XHHW-2, ZW	14	10	17	27	47	64	104	147	228	304	392
	12	7	13	21	36	49	80	113	175	234	301
	10	5	9	15	27	36	59	84	130	174	224
	8	3	5	8	15	20	33	47	72	97	124
	6	1	4	6	11	15	24	35	53	71	92
	4	1	3	4	8	11	18	25	39	52	67
	3	1	2	4	7	9	15	21	33	44	56
	2	1	1	3	5	7	12	18	27	37	47
XHH, XHHW, XHHW-2	1	1	1	2	4	5	9	13	20	27	35
	1/0	1	1	1	3	5	8	11	17	23	30
	2/0	1	1	1	3	4	6	9	14	19	25
	3/0	0	1	1	2	3	5	7	12	16	20
	4/0	0	1	1	1	2	4	6	10	13	17
	250	0	0	1	1	1	3	5	8	11	14
	300	0	0	1	1	1	3	4	7	9	12
	350	0	0	1	1	1	3	4	6	8	10
	400	0	0	1	1	1	2	3	5	7	9
	500	0	0	0	1	1	1	3	4	6	8
	600	0	0	0	1	1	1	2	3	5	6
	700	0	0	0	1	1	1	1	3	4	5
	750	0	0	0	1	1	1	1	3	4	5
	800	0	0	0	0	1	1	1	3	4	5
	900	0	0	0	0	1	1	1	2	3	4
	1000	0	0	0	0	1	1	1	2	3	4
	1250	0	0	0	0	0	1	1	1	2	3
	1500	0	0	0	0	0	1	1	1	1	2
	1750	0	0	0	0	0	1	1	1	1	2
	2000	0	0	0	0	0	0	1	1	1	1

(Continues)

Table C.4 *Continued*

FIXTURE WIRES							
	Conductor Size	Metric Designator (Trade Size)					
Type	(AWG/ kcmil)	16 (½)	21 (¾)	27 (1)	35 (1¼)	41 (1½)	53 (2)
FHH-2, RFH-2, RFHH-3	18	9	16	26	45	61	100
	16	8	13	22	38	51	84
SF-2, SFF-2	18	12	20	33	57	77	126
	16	10	17	27	47	64	104
	14	8	13	22	38	51	84
SF-1, SFF-1	18	21	36	59	101	137	223
RFH-1, RFHH-2, TF, TFF, XF, XFF	18	15	26	43	75	101	165
RFH-2, TF, TFF, XF, XFF	16	12	21	35	60	81	133
XF, XFF	14	10	17	27	47	64	104
TFN, TFFN	18	25	42	69	119	161	264
	16	19	32	53	91	123	201
PF, PFF, PGF, PGFF, PAF, PTF, PTFF, PAFF	18	23	40	66	113	153	250
	16	18	31	51	87	118	193
	14	13	23	38	66	89	145
ZF, ZFF, ZHF, HF, HFF	18	30	52	85	146	197	322
	16	22	38	63	108	145	238
	14	16	28	46	79	107	175
KF-2, KFF-2	18	44	75	123	212	287	468
	16	31	53	87	149	202	330
	14	21	36	60	103	139	227
	12	14	25	41	70	95	156
	10	10	17	27	47	64	104
KF-1, KFF-1	18	52	90	147	253	342	558
	16	37	63	103	178	240	392
	14	25	42	69	119	161	264
	12	16	28	46	79	107	175
	10	10	18	30	52	70	114
XF, XFF	12	5	9	14	25	34	56
	10	4	7	11	19	26	43

Notes:

1. This table is for concentric stranded conductors only. For compact stranded conductors, Table C.4(A) should be used.
2. Two-hour fire-rated RHH cable has ceramifiable insulation which has much larger diameters than other RHH wires. Consult manufacturer's conduit fill tables.

*Types RHH, RHW, and RHW-2 without outer covering.

Table C-5

CONDUIT AND TUBING FILL TABLES FOR CONDUCTORS AND FIXTURE WIRES OF THE SAME SIZE *(CONTINUED)*.

Table C.5 Maximum Number of Conductors or Fixture Wires in Liquidtight Flexible Nonmetallic Conduit (Type LFNC-B*) *(Based on Table 1, Chapter 9)*

CONDUCTORS								
	Conductor Size	Metric Designator (Trade Size)						
Type	(AWG/ kcmil)	12 (3/8)	16 (1/2)	21 (3/4)	27 (1)	35 (1¼)	41 (1½)	53 (2)
RHH, RHW, RHW-2	14	2	4	7	12	21	27	44
	12	1	3	6	10	17	22	36
	10	1	3	5	8	14	18	29
	8	1	1	2	4	7	9	15
	6	1	1	1	3	6	7	12
	4	0	1	1	2	4	6	9
	3	0	1	1	1	4	5	8
	2	0	1	1	1	3	4	7
	1	0	0	1	1	1	3	5
	1/0	0	0	1	1	1	2	4
	2/0	0	0	1	1	1	1	3
	3/0	0	0	0	1	1	1	3
	4/0	0	0	0	1	1	1	2
	250	0	0	0	0	1	1	1
	300	0	0	0	0	1	1	1
	350	0	0	0	0	1	1	1
	400	0	0	0	0	1	1	1
	500	0	0	0	0	1	1	1
	600	0	0	0	0	0	1	1
	700	0	0	0	0	0	0	1
	750	0	0	0	0	0	0	1
	800	0	0	0	0	0	0	1
	900	0	0	0	0	0	0	1
	1000	0	0	0	0	0	0	1
	1250	0	0	0	0	0	0	0
	1500	0	0	0	0	0	0	0
	1750	0	0	0	0	0	0	0
	2000	0	0	0	0	0	0	0
TW	14	5	9	15	25	44	57	93
	12	4	7	12	19	33	43	71
	10	3	5	9	14	25	32	53
	8	1	3	5	8	14	18	29
RHH†, RHW†, RHW-2†, THHW, THW, THW-2	14	3	6	10	16	29	38	62
RHH†, RHW†, RHW-2†, THHW, THW	12	3	5	8	13	23	30	50
	10	1	3	6	10	18	23	39
RHH†, RHW†, RHW-2†, THHW, THW, THW-2	8	1	1	4	6	11	14	23
RHH†, RHW†, RHW-2†, TW, THW, THHW, THW-2	6	1	1	3	5	8	11	18
	4	1	1	1	3	6	8	13
	3	1	1	1	3	5	7	11

(Continues)

Table C.5 *Continued*

CONDUCTORS								
	Conductor Size	Metric Designator (Trade Size)						
Type	(AWG/ kcmil)	12 (3/8)	16 (1/2)	21 (3/4)	27 (1)	35 (1¼)	41 (1½)	53 (2)
RHH†, RHW†, RHW-2†, TW, THW, THHW, THW-2	2	0	1	1	2	4	6	9
	1	0	1	1	1	3	4	7
	1/0	0	0	1	1	2	3	6
	2/0	0	0	1	1	2	3	5
	3/0	0	0	1	1	1	2	4
	4/0	0	0	0	1	1	1	3
	250	0	0	0	1	1	1	3
	300	0	0	0	1	1	1	2
	350	0	0	0	0	1	1	1
	400	0	0	0	0	1	1	1
	500	0	0	0	0	1	1	1
	600	0	0	0	0	1	1	1
	700	0	0	0	0	0	1	1
	750	0	0	0	0	0	1	1
	800	0	0	0	0	0	1	1
	900	0	0	0	0	0	0	1
	1000	0	0	0	0	0	0	1
	1250	0	0	0	0	0	0	1
	1500	0	0	0	0	0	0	0
	1750	0	0	0	0	0	0	0
	2000	0	0	0	0	0	0	0
THHN, THWN, THWN-2	14	8	13	22	36	63	81	133
	12	5	9	16	26	46	59	97
	10	3	6	10	16	29	37	61
	8	1	3	6	9	16	21	35
	6	1	2	4	7	12	15	25
	4	1	1	2	4	7	9	15
	3	1	1	1	3	6	8	13
	2	1	1	1	3	5	7	11
	1	0	1	1	1	4	5	8
	1/0	0	1	1	1	3	4	7
	2/0	0	0	1	1	2	3	6
	3/0	0	0	1	1	1	3	5
	4/0	0	0	1	1	1	2	4
	250	0	0	0	1	1	1	3
	300	0	0	0	1	1	1	3
	350	0	0	0	1	1	1	2
	400	0	0	0	0	1	1	1
	500	0	0	0	0	1	1	1
	600	0	0	0	0	1	1	1
	700	0	0	0	0	1	1	1
	750	0	0	0	0	0	1	1
	800	0	0	0	0	0	1	1
	900	0	0	0	0	0	1	1
	1000	0	0	0	0	0	0	1
FEP, FEPB, PFA, PFAH, TFE	14	7	12	21	35	61	79	129
	12	5	9	15	25	44	57	94
	10	4	6	11	18	32	41	68
	8	1	3	6	10	18	23	39
	6	1	2	4	7	13	17	27
	4	1	1	3	5	9	12	19
	3	1	1	2	4	7	10	16
	2	1	1	1	3	6	8	13
PFA, PFAH, TFE	1	0	1	1	2	4	5	9
PFA, PFAH	1/0	0	1	1	1	3	4	7

(Continues)

Table C.5 *Continued*

CONDUCTORS								
	Conductor Size	Metric Designator (Trade Size)						
Type	(AWG/ kcmil)	12 (3/8)	16 (1/2)	21 (3/4)	27 (1)	35 (1¼)	41 (1½)	53 (2)
TFE, Z	2/0	0	1	1	1	3	4	6
	3/0	0	0	1	1	2	3	5
	4/0	0	0	1	1	1	2	4
Z	14	9	15	26	42	73	95	156
	12	6	10	18	30	52	67	111
	10	4	6	11	18	32	41	68
	8	2	4	7	11	20	26	43
	6	1	3	5	8	14	18	30
	4	1	1	3	5	9	12	20
	3	1	1	2	4	7	9	15
	2	0	1	1	3	6	7	12
	1	0	1	1	2	5	6	10
XHH, XHHW, XHHW-2, ZW	14	5	9	15	25	44	57	93
	12	4	7	12	19	33	43	71
	10	3	5	9	14	25	32	53
	8	1	3	5	8	14	18	29
	6	1	1	3	6	10	13	22
	4	1	1	2	4	7	9	16
	3	1	1	1	3	6	8	13
	2	1	1	1	3	5	7	11
XHH, XHHW, XHHW-2	1	0	1	1	1	4	5	8
	1/0	0	1	1	1	3	4	7
	2/0	0	0	1	1	2	3	6
	3/0	0	0	1	1	1	3	5
	4/0	0	0	1	1	1	2	4
	250	0	0	0	1	1	1	3
	300	0	0	0	1	1	1	3
	350	0	0	0	1	1	1	2
	400	0	0	0	0	1	1	1
	500	0	0	0	0	1	1	1
	600	0	0	0	0	1	1	1
	700	0	0	0	0	1	1	1
	750	0	0	0	0	0	1	1
	800	0	0	0	0	0	1	1
	900	0	0	0	0	0	1	1
	1000	0	0	0	0	0	0	1
	1250	0	0	0	0	0	0	1
	1500	0	0	0	0	0	0	1
	1750	0	0	0	0	0	0	0
	2000	0	0	0	0	0	0	0

(Continues)

Table C.5 *Continued*

FIXTURE WIRES								
	Conductor Size	Metric Designator (Trade Size)						
Type	(AWG/ kcmil)	12 (⅜)	16 (½)	21 (¾)	27 (1)	35 (1¼)	41 (1½)	53 (2)
FFH-2, RFH-2	18	5	8	15	24	42	54	89
	16	4	7	12	20	35	46	75
SF-2, SFF-2	18	6	11	19	30	53	69	113
	16	5	9	15	25	44	57	93
	14	4	7	12	20	35	46	75
SF-1, SFF-1	18	11	19	33	53	94	122	199
RFH-1, RFHH-2, TF, TFF, XF, XFF	18	8	14	24	39	69	90	147
RFHH-2, TF, TFF, XF, XFF	16	7	11	20	32	56	72	119
XF, XFF	14	5	9	15	25	44	57	93
TFN, TFFN	18	14	23	39	63	111	144	236
	16	10	17	30	48	85	110	180
PF, PFF, PGF, PGFF, PAF, PTF, PTFF, PAFF	18	13	21	37	60	105	136	223
	16	10	16	29	46	81	105	173
	14	7	12	21	35	61	79	129
HF, HFF, ZF, ZFF, ZHF	18	17	28	48	77	136	176	288
	16	12	20	35	57	100	129	212
	14	9	15	26	42	73	95	156
KF-2, KFF-2	18	24	40	70	112	197	255	418
	16	17	28	49	79	139	180	295
	14	12	19	34	54	95	123	202
	12	8	13	23	37	65	85	139
	10	5	9	15	25	44	57	93
KF-1, KFF-1	18	29	48	83	134	235	304	499
	16	20	34	58	94	165	214	350
	14	14	23	39	63	111	144	236
	12	9	15	26	42	73	95	156
	10	6	10	17	27	48	62	102
XF, XFF	12	3	5	8	13	23	30	50
	10	1	3	6	10	18	23	39

Notes:

1. This table is for concentric stranded conductors only. For compact stranded conductors, Table C.5(A) should be used.

2. Two-hour fire-rated RHH cable has ceramifiable insulation which has much larger diameters than other RHH wires. Consult manufacturer's conduit fill tables.

*Corresponds to 356.2(2).

†Types RHH, RHW, and RHW-2 without outer covering.

Table C-6

CONDUIT AND TUBING FILL TABLES FOR CONDUCTORS AND FIXTURE WIRES OF THE SAME SIZE *(CONTINUED)*.

Table C.6 Maximum Number of Conductors or Fixture Wires in Liquidtight Flexible Nonmetallic Conduit (Type LFNC-A*) *(Based on Table 1, Chapter 9)*

CONDUCTORS								
	Conductor Size	Metric Designator (Trade Size)						
Type	**(AWG/ kcmil)**	**12 (3/8)**	**16 (1/2)**	**21 (3/4)**	**27 (1)**	**35 (1¼)**	**41 (1½)**	**53 (2)**
RHH, RHW, RHW-2	14	2	4	7	11	20	27	45
	12	1	3	6	9	17	23	38
	10	1	3	5	8	13	18	30
	8	1	1	2	4	7	9	16
	6	1	1	1	3	5	7	13
	4	0	1	1	2	4	6	10
	3	0	1	1	1	4	5	8
	2	0	1	1	1	3	4	7
	1	0	0	1	1	1	3	5
	1/0	0	0	1	1	1	2	4
	2/0	0	0	1	1	1	1	4
	3/0	0	0	0	1	1	1	3
	4/0	0	0	0	1	1	1	3
	250	0	0	0	0	1	1	1
	300	0	0	0	0	1	1	1
	350	0	0	0	0	1	1	1
	400	0	0	0	0	1	1	1
	500	0	0	0	0	0	1	1
	600	0	0	0	0	0	1	1
	700	0	0	0	0	0	0	1
	750	0	0	0	0	0	0	1
	800	0	0	0	0	0	0	1
	900	0	0	0	0	0	0	1
	1000	0	0	0	0	0	0	1
	1250	0	0	0	0	0	0	0
	1500	0	0	0	0	0	0	0
	1750	0	0	0	0	0	0	0
	2000	0	0	0	0	0	0	0
TW	14	5	9	15	24	43	58	96
	12	4	7	12	19	33	44	74
	10	3	5	9	14	24	33	55
	8	1	3	5	8	13	18	30
RHH†, RHW†, RHW-2†, THHW, THW, THW-2	14	3	6	10	16	28	38	64
RHH†, RHW†, RHW-2†, THHW, THW	12	3	4	8	13	23	31	51
	10	1	3	6	10	18	24	40
RHH†, RHW†, RHW-2†, THHW, THW, THW-2	8	1	1	4	6	10	14	24

(Continues)

Table C.6 *Continued*

CONDUCTORS								
	Conductor Size	Metric Designator (Trade Size)						
Type	(AWG/ kcmil)	12 (3/8)	16 (1/2)	21 (3/4)	27 (1)	35 (1¼)	41 (1½)	53 (2)
RHH†, RHW†, RHW-2†, TW, THW, THHW, THW-2	6	1	1	3	4	8	11	18
	4	1	1	1	3	6	8	13
	3	1	1	1	3	5	7	11
	2	0	1	1	2	4	6	10
	1	0	1	1	1	3	4	7
	1/0	0	0	1	1	2	3	6
	2/0	0	0	1	1	1	3	5
	3/0	0	0	1	1	1	2	4
	4/0	0	0	0	1	1	1	3
	250	0	0	0	1	1	1	3
	300	0	0	0	1	1	1	2
	350	0	0	0	0	1	1	1
	400	0	0	0	0	1	1	1
	500	0	0	0	0	1	1	1
	600	0	0	0	0	1	1	1
	700	0	0	0	0	0	1	1
	750	0	0	0	0	0	1	1
	800	0	0	0	0	0	1	1
	900	0	0	0	0	0	0	1
	1000	0	0	0	0	0	0	1
	1250	0	0	0	0	0	0	1
	1500	0	0	0	0	0	0	1
	1750	0	0	0	0	0	0	0
	2000	0	0	0	0	0	0	0
THHN, THWN, THWN-2	14	8	13	22	35	62	83	137
	12	5	9	16	25	45	60	100
	10	3	6	10	16	28	38	63
	8	1	3	6	9	16	22	36
	6	1	2	4	6	12	16	26
	4	1	1	2	4	7	9	16
	3	1	1	1	3	6	8	13
	2	1	1	1	3	5	7	11
	1	0	1	1	1	4	5	8
	1/0	0	1	1	1	3	4	7
	2/0	0	0	1	1	2	3	6
	3/0	0	0	1	1	1	3	5
	4/0	0	0	1	1	1	2	4
	250	0	0	0	1	1	1	3
	300	0	0	0	1	1	1	3
	350	0	0	0	1	1	1	2
	400	0	0	0	0	1	1	1
	500	0	0	0	0	1	1	1
	600	0	0	0	0	1	1	1
	700	0	0	0	0	1	1	1
	750	0	0	0	0	0	1	1
	800	0	0	0	0	0	1	1
	900	0	0	0	0	0	1	1
	1000	0	0	0	0	0	0	1
FEP, FEPB, PFA, PFAH, TFE	14	7	12	21	34	60	80	133
	12	5	9	15	25	44	59	97
	10	4	6	11	18	31	42	70
	8	1	3	6	10	18	24	40
	6	1	2	4	7	13	17	28
	4	1	1	3	5	9	12	20
	3	1	1	2	4	7	10	16
	2	1	1	1	3	6	8	13

(Continues)

Table C.6 *Continued*

CONDUCTORS								
	Conductor Size	Metric Designator (Trade Size)						
Type	(AWG/ kcmil)	12 (3/8)	16 (1/2)	21 (3/4)	27 (1)	35 (1¼)	41 (1½)	53 (2)
PFA, PFAH, TFE	1	0	1	1	2	4	5	9
PFA, PFAH, TFE, Z	1/0	0	1	1	1	3	5	8
	2/0	0	1	1	1	3	4	6
	3/0	0	0	1	1	2	3	5
	4/0	0	0	1	1	1	2	4
Z	14	9	15	25	41	72	97	161
	12	6	10	18	29	51	69	114
	10	4	6	11	18	31	42	70
	8	2	4	7	11	20	26	44
	6	1	3	5	8	14	18	31
	4	1	1	3	5	9	13	21
	3	1	1	2	4	7	9	15
	2	1	1	1	3	6	8	13
	1	1	1	1	2	4	6	10
XHH, XHHW, XHHW-2, ZW	14	5	9	15	24	43	58	96
	12	4	7	12	19	33	44	74
	10	3	5	9	14	24	33	55
	8	1	3	5	8	13	18	30
	6	1	1	3	5	10	13	22
	4	1	1	2	4	7	10	16
	3	1	1	1	3	6	8	14
	2	1	1	1	3	5	7	11
XHH, XHHW, XHHW-2	1	0	1	1	1	4	5	8
	1/0	0	1	1	1	3	4	7
	2/0	0	0	1	1	2	3	6
	3/0	0	0	1	1	1	3	5
	4/0	0	0	1	1	1	2	4
	250	0	0	0	1	1	1	3
	300	0	0	0	1	1	1	3
	350	0	0	0	1	1	1	2
	400	0	0	0	0	1	1	1
	500	0	0	0	0	1	1	1
	600	0	0	0	0	1	1	1
	700	0	0	0	0	1	1	1
	750	0	0	0	0	0	1	1
	800	0	0	0	0	0	1	1
	900	0	0	0	0	0	1	1
	1000	0	0	0	0	0	0	1
	1250	0	0	0	0	0	0	1
	1500	0	0	0	0	0	0	1
	1750	0	0	0	0	0	0	0
	2000	0	0	0	0	0	0	0

(Continues)

Table C.6 *Continued*

FIXTURE WIRES								
	Conductor Size	Metric Designator (Trade Size)						
Type	(AWG/ kcmil)	12 (3/8)	16 (1/2)	21 (3/4)	27 (1)	35 (1¼)	41 (1½)	53 (2)
FFH-2, RFH-2, RFHH-3	18	5	8	14	23	41	55	92
	16	4	7	12	20	35	47	77
SF-2, SFF-2	18	6	11	18	29	52	70	116
	16	5	9	15	24	43	58	96
	14	4	7	12	20	35	47	77
SF-1, SFF-1	18	12	19	33	52	92	124	205
RFH-1, RFHH-2, TF, TFF, XF, XFF	18	8	14	24	39	68	91	152
RFHH-2, TF, TFF, XF, XFF	16	7	11	19	31	55	74	122
XF, XFF	14	5	9	15	24	43	58	96
TFN, TFFN	18	14	22	39	62	109	146	243
	16	10	17	29	47	83	112	185
PF, PFF, PGF, PGFF, PAF, PTF, PTFF, PAFF	18	13	21	37	59	103	139	230
	16	10	16	28	45	80	107	178
	14	7	12	21	34	60	80	133
HF, HFF, ZF, ZFF, ZHF	18	17	27	47	76	133	179	297
	16	12	20	35	56	98	132	219
	14	9	15	25	41	72	97	161
KF-2, KFF-2	18	25	40	69	110	193	260	431
	16	17	28	48	77	136	183	303
	14	12	19	33	53	94	126	209
	12	8	13	23	36	64	86	143
	10	5	9	15	24	43	58	96
KF-1, KFF-1	18	29	48	82	131	231	310	514
	16	21	33	57	92	162	218	361
	14	14	22	39	62	109	146	243
	12	9	15	25	41	72	97	161
	10	6	10	17	27	47	63	105
XF, XFF	12	3	4	8	13	23	31	51
	10	1	3	6	10	18	24	40

Notes:

1. This table is for concentric stranded conductors only. For compact stranded conductors, Table C.6(A) should be used.
2. Two-hour fire-rated RHH cable has ceramifiable insulation which has much larger diameters than other RHH wires. Consult manufacturer's conduit fill tables.

*Corresponds to 356.2(1).

†Types RHH, RHW, and RHW-2 without outer covering.

Table C-7

CONDUIT AND TUBING FILL TABLES FOR CONDUCTORS AND FIXTURE WIRES OF THE SAME SIZE *(CONTINUED).*

Table C.7 Maximum Number of Conductors or Fixture Wires in Liquidtight Flexible Metal Conduit (LFMC) *(Based on Table 1, Chapter 9)*

CONDUCTORS											
	Conductor Size	Metric Designator (Trade Size)									
Type	(AWG/ kcmil)	16 (½)	21 (¾)	27 (1)	35 (1¼)	41 (1½)	53 (2)	63 (2½)	78 (3)	91 (3½)	103 (4)
RHH, RHW, RHW-2	14	4	7	12	21	27	44	66	102	133	173
	12	3	6	10	17	22	36	55	84	110	144
	10	3	5	8	14	18	29	44	68	89	116
	8	1	2	4	7	9	15	23	36	46	61
	6	1	1	3	6	7	12	18	28	37	48
	4	1	1	2	4	6	9	14	22	29	38
	3	1	1	1	4	5	8	13	19	25	33
	2	1	1	1	3	4	7	11	17	22	29
	1	0	1	1	1	3	5	7	11	14	19
	1/0	0	1	1	1	2	4	6	10	13	16
	2/0	0	1	1	1	1	3	5	8	11	14
	3/0	0	0	1	1	1	3	4	7	9	12
	4/0	0	0	1	1	1	2	4	6	8	10
	250	0	0	0	1	1	1	3	4	6	8
	300	0	0	0	1	1	1	2	4	5	7
	350	0	0	0	1	1	1	2	3	5	6
	400	0	0	0	1	1	1	1	3	4	6
	500	0	0	0	1	1	1	1	3	4	5
	600	0	0	0	0	1	1	1	2	3	4
	700	0	0	0	0	0	1	1	1	3	3
	750	0	0	0	0	0	1	1	1	2	3
	800	0	0	0	0	0	1	1	1	2	3
	900	0	0	0	0	0	1	1	1	2	3
	1000	0	0	0	0	0	1	1	1	1	3
	1250	0	0	0	0	0	0	1	1	1	1
	1500	0	0	0	0	0	0	1	1	1	1
	1750	0	0	0	0	0	0	1	1	1	1
	2000	0	0	0	0	0	0	0	1	1	1
TW	14	9	15	25	44	57	93	140	215	280	365
	12	7	12	19	33	43	71	108	165	215	280
	10	5	9	14	25	32	53	80	123	160	209
	8	3	5	8	14	18	29	44	68	89	116
RHH*, RHW*, RHW-2*, THHW, THW, THW-2	14	6	10	16	29	38	62	93	143	186	243
RHH*, RHW*, RHW-2*, THHW, THW	12	5	8	13	23	30	50	75	115	149	195
	10	3	6	10	18	23	39	58	89	117	152
RHH*, RHW*, RHW-2*, THHW, THW, THW-2	8	1	4	6	11	14	23	35	53	70	91

(Continues)

Table C.7 *Continued*

		CONDUCTORS									
	Conductor Size	Metric Designator (Trade Size)									
Type	(AWG/ kcmil)	16 (½)	21 (¾)	27 (1)	35 (1¼)	41 (1½)	53 (2)	63 (2½)	78 (3)	91 (3½)	103 (4)
RHH*, RHW*, RHW-2*, TW, THW, THHW, THW-2	6	1	3	5	8	11	18	27	41	53	70
	4	1	1	3	6	8	13	20	30	40	52
	3	1	1	3	5	7	11	17	26	34	44
	2	1	1	2	4	6	9	14	22	29	38
	1	1	1	1	3	4	7	10	15	20	26
	1/0	0	1	1	2	3	6	8	13	17	23
	2/0	0	1	1	2	3	5	7	11	15	19
	3/0	0	1	1	1	2	4	6	9	12	16
	4/0	0	0	1	1	1	3	5	8	10	13
	250	0	0	1	1	1	3	4	6	8	11
	300	0	0	1	1	1	2	3	5	7	9
	350	0	0	0	1	1	1	3	5	6	8
	400	0	0	0	1	1	1	3	4	6	7
	500	0	0	0	1	1	1	2	3	5	6
	600	0	0	0	1	1	1	1	3	4	5
	700	0	0	0	0	1	1	1	2	3	4
	750	0	0	0	0	1	1	1	2	3	4
	800	0	0	0	0	1	1	1	2	3	4
	900	0	0	0	0	0	1	1	1	3	3
	1000	0	0	0	0	0	1	1	1	2	3
	1250	0	0	0	0	0	1	1	1	1	2
	1500	0	0	0	0	0	0	1	1	1	2
	1750	0	0	0	0	0	0	1	1	1	1
	2000	0	0	0	0	0	0	1	1	1	1
THHN, THWN, THWN-2	14	13	22	36	63	81	133	201	308	401	523
	12	9	16	26	46	59	97	146	225	292	381
	10	6	10	16	29	37	61	92	141	184	240
	8	3	6	9	16	21	35	53	81	106	138
	6	2	4	7	12	15	25	38	59	76	100
	4	1	2	4	7	9	15	23	36	47	61
	3	1	1	3	6	8	13	20	30	40	52
	2	1	1	3	5	7	11	17	26	33	44
	1	1	1	1	4	5	8	12	19	25	32
	1/0	1	1	1	3	4	7	10	16	21	27
	2/0	0	1	1	2	3	6	8	13	17	23
	3/0	0	1	1	1	3	5	7	11	14	19
	4/0	0	1	1	1	2	4	6	9	12	15
	250	0	0	1	1	1	3	5	7	10	12
	300	0	0	1	1	1	3	4	6	8	11
	350	0	0	1	1	1	2	3	5	7	9
	400	0	0	0	1	1	1	3	5	6	8
	500	0	0	0	1	1	1	2	4	5	7
	600	0	0	0	1	1	1	1	3	4	6
	700	0	0	0	1	1	1	1	3	4	5
	750	0	0	0	0	1	1	1	3	3	5
	800	0	0	0	0	1	1	1	2	3	4
	900	0	0	0	0	1	1	1	2	3	4
	1000	0	0	0	0	0	1	1	1	3	3
FEP, FEPB, PFA, PFAH, TFE	14	12	21	35	61	79	129	195	299	389	507
	12	9	15	25	44	57	94	142	218	284	370
	10	6	11	18	32	41	68	102	156	203	266
	8	3	6	10	18	23	39	58	89	117	152
	6	2	4	7	13	17	27	41	64	83	108
	4	1	3	5	9	12	19	29	44	58	75
	3	1	2	4	7	10	16	24	37	48	63
	2	1	1	3	6	8	13	20	30	40	52

(Continues)

Table C.7 *Continued*

CONDUCTORS											
	Conductor Size	Metric Designator (Trade Size)									
Type	(AWG/ kcmil)	16 (½)	21 (¾)	27 (1)	35 (1¼)	41 (1½)	53 (2)	63 (2½)	78 (3)	91 (3½)	103 (4)
PFA, PFAH, TFE	1	1	1	2	4	5	9	14	21	28	36
PFA, PFAH, TFE, Z	1/0	1	1	1	3	4	7	11	18	23	30
	2/0	1	1	1	3	4	6	9	14	19	25
	3/0	0	1	1	2	3	5	8	12	16	20
	4/0	0	1	1	1	2	4	6	10	13	17
Z	14	20	26	42	73	95	156	235	360	469	611
	12	14	18	30	52	67	111	167	255	332	434
	10	8	11	18	32	41	68	102	156	203	266
	8	5	7	11	20	26	43	64	99	129	168
	6	4	5	8	14	18	30	45	69	90	118
	4	2	3	5	9	12	20	31	48	62	81
	3	2	2	4	7	9	15	23	35	45	59
	2	1	1	3	6	7	12	19	29	38	49
	1	1	1	2	5	6	10	15	23	30	40
XHH, XHHW, XHHW-2, ZW	14	9	15	25	44	57	93	140	215	280	365
	12	7	12	19	33	43	71	108	165	215	280
	10	5	9	14	25	32	53	80	123	160	209
	8	3	5	8	14	18	29	44	68	89	116
	6	1	3	6	10	13	22	33	50	66	86
	4	1	2	4	7	9	16	24	36	48	62
	3	1	1	3	6	8	13	20	31	40	52
	2	1	1	3	5	7	11	17	26	34	44
XHH, XHHW, XHHW-2	1	1	1	1	4	5	8	12	19	25	33
	1/0	1	1	1	3	4	7	10	16	21	28
	2/0	0	1	1	2	3	6	9	13	17	23
	3/0	0	1	1	1	3	5	7	11	14	19
	4/0	0	1	1	1	2	4	6	9	12	16
	250	0	0	1	1	1	3	5	7	10	13
	300	0	0	1	1	1	3	4	6	8	11
	350	0	0	1	1	1	2	3	5	7	10
	400	0	0	0	1	1	1	3	5	6	8
	500	0	0	0	1	1	1	2	4	5	7
	600	0	0	0	1	1	1	1	3	4	6
	700	0	0	0	1	1	1	1	3	4	5
	750	0	0	0	0	1	1	1	3	3	5
	800	0	0	0	0	1	1	1	2	3	4
	900	0	0	0	0	1	1	1	2	3	4
	1000	0	0	0	0	0	1	1	1	3	3
	1250	0	0	0	0	0	1	1	1	1	3
	1500	0	0	0	0	0	1	1	1	1	2
	1750	0	0	0	0	0	0	1	1	1	2
	2000	0	0	0	0	0	0	1	1	1	2

(Continues)

Table C.7 *Continued*

FIXTURE WIRES							
	Conductor Size	Metric Designator (Trade Size)					
Type	(AWG/ kcmil)	16 (½)	21 (¾)	27 (1)	35 (1¼)	41 (1½)	53 (2)
FFH-2, RFH-2, RFHH-3	18	8	15	24	42	54	89
	16	7	12	20	35	46	75
SF-2, SFF-2	18	11	19	30	53	69	113
	16	9	15	25	44	57	93
	14	7	12	20	35	46	75
SF-1, SFF-1	18	19	33	53	94	122	199
RFH-1, RFHH-2, TF, TFF, XF, XFF	18	14	24	39	69	90	147
RFHH-2, TF, TFF, XF, XFF	16	11	20	32	56	72	119
XF, XFF	14	9	15	25	44	57	93
TFN, TFFN	18	23	39	63	111	144	236
	16	17	30	48	85	110	180
PF, PFF, PGF, PGFF, PAF, PTF, PTFF, PAFF	18	21	37	60	105	136	223
	16	16	29	46	81	105	173
	14	12	21	35	61	79	129
HF, HFF, ZF, ZFF, ZHF	18	28	48	77	136	176	288
	16	20	35	57	100	129	212
	14	15	26	42	73	95	156
KF-2, KFF-2	18	40	70	112	197	255	418
	16	28	49	79	139	180	295
	14	19	34	54	95	123	202
	12	13	23	37	65	85	139
	10	9	15	25	44	57	93
KF-1, KFF-1	18	48	83	134	235	304	499
	16	34	58	94	165	214	350
	14	23	39	63	111	144	236
	12	15	26	42	73	95	156
	10	10	17	27	48	62	102
XF, XFF	12	5	8	13	23	30	50
	10	3	6	10	18	23	39

Notes:

1. This table is for concentric stranded conductors only. For compact stranded conductors, Table C.7(A) should be used.

2. Two-hour fire-rated RHH cable has ceramifiable insulation which has much larger diameters than other RHH wires. Consult manufacturer's conduit fill tables.

*Types RHH, RHW, and RHW-2 without outer covering.

Table C-8

CONDUIT AND TUBING FILL TABLES FOR CONDUCTORS AND FIXTURE WIRES OF THE SAME SIZE *(CONTINUED)*.

Table C.8 Maximum Number of Conductors or Fixture Wires in Rigid Metal Conduit (RMC) *(Based on Table 1, Chapter 9)*

CONDUCTORS													
	Conductor Size	Metric Designator (Trade Size)											
Type	(AWG/ kcmil)	16 (½)	21 (¾)	27 (1)	35 (1¼)	41 (1½)	53 (2)	63 (2½)	78 (3)	91 (3½)	103 (4)	129 (5)	155 (6)
RHH, RHW, RHW-2	14	4	7	12	21	28	46	66	102	136	176	276	398
	12	3	6	10	17	23	38	55	85	113	146	229	330
	10	3	5	8	14	19	31	44	68	91	118	185	267
	8	1	2	4	7	10	16	23	36	48	61	97	139
	6	1	1	3	6	8	13	18	29	38	49	77	112
	4	1	1	2	4	6	10	14	22	30	38	60	87
	3	1	1	2	4	5	9	12	19	26	34	53	76
	2	1	1	1	3	4	7	11	17	23	29	46	66
	1	0	1	1	1	3	5	7	11	15	19	30	44
	1/0	0	1	1	1	2	4	6	10	13	17	26	38
	2/0	0	1	1	1	2	4	5	8	11	14	23	33
	3/0	0	0	1	1	1	3	4	7	10	12	20	28
	4/0	0	0	1	1	1	3	4	6	8	11	17	24
	250	0	0	0	1	1	1	3	4	6	8	13	18
	300	0	0	0	1	1	1	2	4	5	7	11	16
	350	0	0	0	1	1	1	2	4	5	6	10	15
	400	0	0	0	1	1	1	1	3	4	6	9	13
	500	0	0	0	1	1	1	1	3	4	5	8	11
	600	0	0	0	0	1	1	1	2	3	4	6	9
	700	0	0	0	0	1	1	1	1	3	4	6	8
	750	0	0	0	0	0	1	1	1	3	3	5	8
	800	0	0	0	0	0	1	1	1	2	3	5	7
	900	0	0	0	0	0	1	1	1	2	3	5	7
	1000	0	0	0	0	0	1	1	1	1	3	4	6
	1250	0	0	0	0	0	0	1	1	1	1	3	5
	1500	0	0	0	0	0	0	1	1	1	1	3	4
	1750	0	0	0	0	0	0	1	1	1	1	2	4
	2000	0	0	0	0	0	0	0	1	1	1	2	3
TW	14	9	15	25	44	59	98	140	216	288	370	581	839
	12	7	12	19	33	45	75	107	165	221	284	446	644
	10	5	9	14	25	34	56	80	123	164	212	332	480
	8	3	5	8	14	19	31	44	68	91	118	185	267
RHH*, RHW*, RHW-2* THHW, THW, THW-2	14	6	10	17	29	39	65	93	143	191	246	387	558
RHH*, RHW*, RHW-2*, THHW, THW	12	5	8	13	23	32	52	75	115	154	198	311	448
	10	3	6	10	18	25	41	58	90	120	154	242	350
RHH*, RHW*, RHW-2*, THHW, THW, THW-2	8	1	4	6	11	15	24	35	54	72	92	145	209

(Continues)

Table C.8 *Continued*

CONDUCTORS													
	Conductor Size	Metric Designator (Trade Size)											
Type	(AWG/ kcmil)	16 (½)	21 (¾)	27 (1)	35 (1¼)	41 (1½)	53 (2)	63 (2½)	78 (3)	91 (3½)	103 (4)	129 (5)	155 (6)
RHH*, RHW*, RHW-2*, TW, THW, THHW, THW-2	6	1	3	5	8	11	18	27	41	55	71	111	160
	4	1	1	3	6	8	14	20	31	41	53	83	120
	3	1	1	3	5	7	12	17	26	35	45	71	103
	2	1	1	2	4	6	10	14	22	30	38	60	87
	1	1	1	1	3	4	7	10	15	21	27	42	61
	1/0	0	1	1	2	3	6	8	13	18	23	36	52
	2/0	0	1	1	2	3	5	7	11	15	19	31	44
	3/0	0	1	1	1	2	4	6	9	13	16	26	37
	4/0	0	0	1	1	1	3	5	8	10	14	21	31
	250	0	0	1	1	1	3	4	6	8	11	17	25
	300	0	0	1	1	1	2	3	5	7	9	15	22
	350	0	0	0	1	1	1	3	5	6	8	13	19
	400	0	0	0	1	1	1	3	4	6	7	12	17
	500	0	0	0	1	1	1	2	3	5	6	10	14
	600	0	0	0	1	1	1	1	3	4	5	8	12
	700	0	0	0	0	1	1	1	2	3	4	7	10
	750	0	0	0	0	1	1	1	2	3	4	7	10
	800	0	0	0	0	1	1	1	2	3	4	6	9
	900	0	0	0	0	1	1	1	1	3	4	6	8
	1000	0	0	0	0	0	1	1	1	2	3	5	8
	1250	0	0	0	0	0	1	1	1	1	2	4	6
	1500	0	0	0	0	0	1	1	1	1	2	3	5
	1750	0	0	0	0	0	0	1	1	1	1	3	4
	2000	0	0	0	0	0	0	1	1	1	1	3	4
THHN, THWN, THWN-2	14	13	22	36	63	85	140	200	309	412	531	833	1202
	12	9	16	26	46	62	102	146	225	301	387	608	877
	10	6	10	17	29	39	64	92	142	189	244	383	552
	8	3	6	9	16	22	37	53	82	109	140	221	318
	6	2	4	7	12	16	27	38	59	79	101	159	230
	4	1	2	4	7	10	16	23	36	48	62	98	141
	3	1	1	3	6	8	14	20	31	41	53	83	120
	2	1	1	3	5	7	11	17	26	34	44	70	100
	1	1	1	1	4	5	8	12	19	25	33	51	74
	1/0	1	1	1	3	4	7	10	16	21	27	43	63
	2/0	0	1	1	2	3	6	8	13	18	23	36	52
	3/0	0	1	1	1	3	5	7	11	15	19	30	43
	4/0	0	1	1	1	2	4	6	9	12	16	25	36
	250	0	0	1	1	1	3	5	7	10	13	20	29
	300	0	0	1	1	1	3	4	6	8	11	17	25
	350	0	0	1	1	1	2	3	5	7	10	15	22
	400	0	0	1	1	1	2	3	5	7	8	13	20
	500	0	0	0	1	1	1	2	4	5	7	11	16
	600	0	0	0	1	1	1	1	3	4	6	9	13
	700	0	0	0	1	1	1	1	3	4	5	8	11
	750	0	0	0	0	1	1	1	3	4	5	7	11
	800	0	0	0	0	1	1	1	2	3	4	7	10
	900	0	0	0	0	1	1	1	2	3	4	6	9
	1000	0	0	0	0	1	1	1	1	3	4	6	8
FEP, FEPB, PFA, PFAH, TFE	14	12	22	35	61	83	136	194	300	400	515	808	1166
	12	9	16	26	44	60	99	142	219	292	376	590	851
	10	6	11	18	32	43	71	102	157	209	269	423	610
	8	3	6	10	18	25	41	58	90	120	154	242	350
	6	2	4	7	13	17	29	41	64	85	110	172	249
	4	1	3	5	9	12	20	29	44	59	77	120	174
	3	1	2	4	7	10	17	24	37	50	64	100	145
	2	1	1	3	6	8	14	20	31	41	53	83	120

(Continues)

Table C.8 *Continued*

CONDUCTORS													
	Conductor Size	Metric Designator (Trade Size)											
Type	(AWG/ kcmil)	16 (½)	21 (¾)	27 (1)	35 (1¼)	41 (1½)	53 (2)	63 (2½)	78 (3)	91 (3½)	103 (4)	129 (5)	155 (6)
PFA, PFAH, TFE	1	1	1	2	4	6	9	14	21	28	37	57	83
PFA, PFAH, TFE, Z	1/0	1	1	1	3	5	8	11	18	24	30	48	69
	2/0	1	1	1	3	4	6	9	14	19	25	40	57
	3/0	0	1	1	2	3	5	8	12	16	21	33	47
	4/0	0	1	1	1	2	4	6	10	13	17	27	39
Z	14	15	26	42	73	100	164	234	361	482	621	974	1405
	12	10	18	30	52	71	116	166	256	342	440	691	997
	10	6	11	18	32	43	71	102	157	209	269	423	610
	8	4	7	11	20	27	45	64	99	132	170	267	386
	6	3	5	8	14	19	31	45	69	93	120	188	271
	4	1	3	5	9	13	22	31	48	64	82	129	186
	3	1	2	4	7	9	16	22	35	47	60	94	136
	2	1	1	3	6	8	13	19	29	39	50	78	113
	1	1	1	2	5	6	10	15	23	31	40	63	92
XHH, XHHW, XHHW-2 ZW	14	9	15	25	44	59	98	140	216	288	370	581	839
	12	7	12	19	33	45	75	107	165	221	284	446	644
	10	5	9	14	25	34	56	80	123	164	212	332	480
	8	3	5	8	14	19	31	44	68	91	118	185	267
	6	1	3	6	10	14	23	33	51	68	87	137	197
	4	1	2	4	7	10	16	24	37	49	63	99	143
	3	1	1	3	6	8	14	20	31	41	53	84	121
	2	1	1	3	5	7	12	17	26	35	45	70	101
XHH, XHHW, XHHW-2	1	1	1	1	4	5	9	12	19	26	33	52	76
	1/0	1	1	1	3	4	7	10	16	22	28	44	64
	2/0	0	1	1	2	3	6	9	13	18	23	37	53
	3/0	0	1	1	1	3	5	7	11	15	19	30	44
	4/0	0	1	1	1	2	4	6	9	12	16	25	36
	250	0	0	1	1	1	3	5	7	10	13	20	30
	300	0	0	1	1	1	3	4	6	9	11	18	25
	350	0	0	1	1	1	2	3	6	7	10	15	22
	400	0	0	1	1	1	2	3	5	7	9	14	20
	500	0	0	0	1	1	1	2	4	5	7	11	16
	600	0	0	0	1	1	1	1	3	4	6	9	13
	700	0	0	0	1	1	1	1	3	4	5	8	11
	750	0	0	0	0	1	1	1	3	4	5	7	11
	800	0	0	0	0	1	1	1	2	3	4	7	10
	900	0	0	0	0	1	1	1	2	3	4	6	9
	1000	0	0	0	0	1	1	1	1	3	4	6	8
	1250	0	0	0	0	0	1	1	1	2	3	4	6
	1500	0	0	0	0	0	1	1	1	1	2	4	5
	1750	0	0	0	0	0	0	1	1	1	1	3	5
	2000	0	0	0	0	0	0	1	1	1	1	3	4

(Continues)

Table C.8 *Continued*

FIXTURE WIRES							
	Conductor Size	Metric Designator (Trade Size)					
Type	(AWG/ kcmil)	16 (½)	21 (¾)	27 (1)	35 (1¼)	41 (1½)	53 (2)
FFH-2, RFH-2, RFHH-3	18	8	15	24	42	57	94
	16	7	12	20	35	48	79
SF-2, SFF-2	18	11	19	31	53	72	118
	16	9	15	25	44	59	98
	14	7	12	20	35	48	79
SF-1, SFF-1	18	19	33	54	94	127	209
RFH-1, RFHH-2, TF, TFF, XF, XFF	18	14	25	40	69	94	155
RFHH-2, TF, TFF, XF, XFF	16	11	20	32	56	76	125
XF, XFF	14	9	15	25	44	59	98
TFN, TFFN	18	23	40	64	111	150	248
	16	17	30	49	84	115	189
PF, PFF, PGF, PGFF, PAF, PTF, PTFF, PAFF	18	21	38	61	105	143	235
	16	16	29	47	81	110	181
	14	12	22	35	61	83	136
HF, HFF, ZF, ZFF, ZHF	18	28	48	79	135	184	303
	16	20	36	58	100	136	223
	14	15	26	42	73	100	164
KF-2, KFF-2	18	40	71	114	197	267	439
	16	28	50	80	138	188	310
	14	19	34	55	95	129	213
	12	13	23	38	65	89	146
	10	9	15	25	44	59	98
KF-1, KFF-1	18	48	84	136	235	318	524
	16	34	59	96	165	224	368
	14	23	40	64	111	150	248
	12	15	26	42	73	100	164
	10	10	17	28	48	65	107
XF, XFF	12	5	8	13	23	32	52
	10	3	6	10	18	25	41

Notes:

1. This table is for concentric stranded conductors only. For compact stranded conductors, Table C.8(A) should be used.

2. Two-hour fire-rated RHH cable has ceramifiable insulation which has much larger diameters than other RHH wires. Consult manufacturer's conduit fill tables.

*Types RHH, RHW, and RHW-2 without outer covering.

Table C-9

CONDUIT AND TUBING FILL TABLES FOR CONDUCTORS AND FIXTURE WIRES OF THE SAME SIZE *(CONTINUED).*

Table C.9 Maximum Number of Conductors or Fixture Wires in Rigid PVC Conduit, Schedule 80 (*Based on Table 1, Chapter 9*)

CONDUCTORS													
	Conductor Size	Metric Designator (Trade Size)											
Type	(AWG/ kcmil)	16 (½)	21 (¾)	27 (1)	35 (1¼)	41 (1½)	53 (2)	63 (2½)	78 (3)	91 (3½)	103 (4)	129 (5)	155 (6)
RHH, RHW, RHW-2	14	3	5	9	17	23	39	56	88	118	153	243	349
	12	2	4	7	14	19	32	46	73	98	127	202	290
	10	1	3	6	11	15	26	37	59	79	103	163	234
	8	1	1	3	6	8	13	19	31	41	54	85	122
	6	1	1	2	4	6	11	16	24	33	43	68	98
	4	1	1	1	3	5	8	12	19	26	33	53	77
	3	0	1	1	3	4	7	11	17	23	29	47	67
	2	0	1	1	3	4	6	9	14	20	25	41	58
	1	0	1	1	1	2	4	6	9	13	17	27	38
	1/0	0	0	1	1	1	3	5	8	11	15	23	33
	2/0	0	0	1	1	1	3	4	7	10	13	20	29
	3/0	0	0	1	1	1	3	4	6	8	11	17	25
	4/0	0	0	0	1	1	2	3	5	7	9	15	21
	250	0	0	0	1	1	1	2	4	5	7	11	16
	300	0	0	0	1	1	1	2	3	5	6	10	14
	350	0	0	0	1	1	1	1	3	4	5	9	13
	400	0	0	0	0	1	1	1	3	4	5	8	12
	500	0	0	0	0	1	1	1	2	3	4	7	10
	600	0	0	0	0	0	1	1	1	3	3	6	8
	700	0	0	0	0	0	1	1	1	2	3	5	7
	750	0	0	0	0	0	1	1	1	2	3	5	7
	800	0	0	0	0	0	1	1	1	2	3	4	7
	1000	0	0	0	0	0	1	1	1	1	2	4	5
	1250	0	0	0	0	0	0	1	1	1	1	3	4
	1500	0	0	0	0	0	0	1	1	1	1	2	4
	1750	0	0	0	0	0	0	0	1	1	1	2	3
	2000	0	0	0	0	0	0	0	1	1	1	1	3
TW	14	6	11	20	35	49	82	118	185	250	324	514	736
	12	5	9	15	27	38	63	91	142	192	248	394	565
	10	3	6	11	20	28	47	67	106	143	185	294	421
	8	1	3	6	11	15	26	37	59	79	103	163	234
RHH*, RHW*, RHW-2*, THHW, THW, THW-2	14	4	8	13	23	32	55	79	123	166	215	341	490
RHH*, RHW*, RHW-2*, THHW, THW	12	3	6	10	19	26	44	63	99	133	173	274	394
	10	2	5	8	15	20	34	49	77	104	135	214	307
RHH*, RHW*, RHW-2*, THHW, THW, THW-2	8	1	3	5	9	12	20	29	46	62	81	128	184

(Continues)

Table C.9 *Continued*

CONDUCTORS													
	Conductor Size	Metric Designator (Trade Size)											
Type	(AWG/ kcmil)	16 (½)	21 (¾)	27 (1)	35 (1¼)	41 (1½)	53 (2)	63 (2½)	78 (3)	91 (3½)	103 (4)	129 (5)	155 (6)
RHH*, RHW*, RHW-2*, TW, THW, THHW, THW-2	6	1	1	3	7	9	16	22	35	48	62	98	141
	4	1	1	3	5	7	12	17	26	35	46	73	105
	3	1	1	2	4	6	10	14	22	30	39	63	90
	2	1	1	1	3	5	8	12	19	26	33	53	77
	1	0	1	1	2	3	6	8	13	18	23	37	54
	1/0	0	1	1	1	3	5	7	11	15	20	32	46
	2/0	0	1	1	1	2	4	6	10	13	17	27	39
	3/0	0	0	1	1	1	3	5	8	11	14	23	33
	4/0	0	0	1	1	1	3	4	7	9	12	19	27
	250	0	0	0	1	1	2	3	5	7	9	15	22
	300	0	0	0	1	1	1	3	5	6	8	13	19
	350	0	0	0	1	1	1	2	4	6	7	12	17
	400	0	0	0	1	1	1	2	4	5	7	10	15
	500	0	0	0	1	1	1	1	3	4	5	9	13
	600	0	0	0	0	1	1	1	2	3	4	7	10
	700	0	0	0	0	1	1	1	2	3	4	6	9
	750	0	0	0	0	0	1	1	1	3	4	6	8
	800	0	0	0	0	0	1	1	1	3	3	6	8
	900	0	0	0	0	0	1	1	1	2	3	5	7
	1000	0	0	0	0	0	1	1	1	2	3	5	7
	1250	0	0	0	0	0	1	1	1	1	2	4	5
	1500	0	0	0	0	0	0	1	1	1	1	3	4
	1750	0	0	0	0	0	0	1	1	1	1	3	4
	2000	0	0	0	0	0	0	0	1	1	1	2	3
THHN, THWN, THWN-2	14	9	17	28	51	70	118	170	265	358	464	736	1055
	12	6	12	20	37	51	86	124	193	261	338	537	770
	10	4	7	13	23	32	54	78	122	164	213	338	485
	8	2	4	7	13	18	31	45	70	95	123	195	279
	6	1	3	5	9	13	22	32	51	68	89	141	202
	4	1	1	3	6	8	14	20	31	42	54	86	124
	3	1	1	3	5	7	12	17	26	35	46	73	105
	2	1	1	2	4	6	10	14	22	30	39	61	88
	1	0	1	1	3	4	7	10	16	22	29	45	65
	1/0	0	1	1	2	3	6	9	14	18	24	38	55
	2/0	0	1	1	1	3	5	7	11	15	20	32	46
	3/0	0	1	1	1	2	4	6	9	13	17	26	38
	4/0	0	0	1	1	1	3	5	8	10	14	22	31
	250	0	0	1	1	1	3	4	6	8	11	18	25
	300	0	0	0	1	1	2	3	5	7	9	15	22
	350	0	0	0	1	1	1	3	5	6	8	13	19
	400	0	0	0	1	1	1	3	4	6	7	12	17
	500	0	0	0	1	1	1	2	3	5	6	10	14
	600	0	0	0	0	1	1	1	3	4	5	8	12
	700	0	0	0	0	1	1	1	2	3	4	7	10
	750	0	0	0	0	1	1	1	2	3	4	7	9
	800	0	0	0	0	1	1	1	2	3	4	6	9
	900	0	0	0	0	0	1	1	1	3	3	6	8
	1000	0	0	0	0	0	1	1	1	2	3	5	7
FEP, FEPB, PFA, PFAH, TFE	14	8	16	27	49	68	115	164	257	347	450	714	1024
	12	6	12	20	36	50	84	120	188	253	328	521	747
	10	4	8	14	26	36	60	86	135	182	235	374	536
	8	2	5	8	15	20	34	49	77	104	135	214	307
	6	1	3	6	10	14	24	35	55	74	96	152	218
	4	1	2	4	7	10	17	24	38	52	67	106	153
	3	1	1	3	6	8	14	20	32	43	56	89	127
	2	1	1	3	5	7	12	17	26	35	46	73	105

(Continues)

Table C.9 ***Continued***

CONDUCTORS													
	Conductor Size	Metric Designator (Trade Size)											
Type	(AWG/ kcmil)	16 (½)	21 (¾)	27 (1)	35 (1¼)	41 (1½)	53 (2)	63 (2½)	78 (3)	91 (3½)	103 (4)	129 (5)	155 (6)
PFA, PFAH, TFE	1	1	1	1	3	5	8	11	18	25	32	51	73
PFA, PFAH, TFE, Z	1/0	0	1	1	3	4	7	10	15	20	27	42	61
	2/0	0	1	1	2	3	5	8	12	17	22	35	50
	3/0	0	1	1	1	2	4	6	10	14	18	29	41
	4/0	0	0	1	1	1	4	5	8	11	15	24	34
Z	14	10	19	33	59	82	138	198	310	418	542	860	1233
	12	7	14	23	42	58	98	141	220	297	385	610	875
	10	4	8	14	26	36	60	86	135	182	235	374	536
	8	3	5	9	16	22	38	54	85	115	149	236	339
	6	2	4	6	11	16	26	38	60	81	104	166	238
	4	1	2	4	8	11	18	26	41	55	72	114	164
	3	1	2	3	5	8	13	19	30	40	52	83	119
	2	1	1	2	5	6	11	16	25	33	43	69	99
	1	0	1	2	4	5	9	13	20	27	35	56	80
XHH, XHHW, XHHW-2, ZW	14	6	11	20	35	49	82	118	185	250	324	514	736
	12	5	9	15	27	38	63	91	142	192	248	394	565
	10	3	6	11	20	28	47	67	106	143	185	294	421
	8	1	3	6	11	15	26	37	59	79	103	163	234
	6	1	2	4	8	11	19	28	43	59	76	121	173
	4	1	1	3	6	8	14	20	31	42	55	87	125
	3	1	1	3	5	7	12	17	26	36	47	74	106
	2	1	1	2	4	6	10	14	22	30	39	62	89
XHH, XHHW, XHHW-2	1	0	1	1	3	4	7	10	16	22	29	46	66
	1/0	0	1	1	2	3	6	9	14	19	24	39	56
	2/0	0	1	1	1	3	5	7	11	16	20	32	46
	3/0	0	1	1	1	2	4	6	9	13	17	27	38
	4/0	0	0	1	1	1	3	5	8	11	14	22	32
	250	0	0	1	1	1	3	4	6	9	11	18	26
	300	0	0	1	1	1	2	3	5	7	10	15	22
	350	0	0	0	1	1	1	3	5	6	8	14	20
	400	0	0	0	1	1	1	3	4	6	7	12	17
	500	0	0	0	1	1	1	2	3	5	6	10	14
	600	0	0	0	0	1	1	1	3	4	5	8	11
	700	0	0	0	0	1	1	1	2	3	4	7	10
	750	0	0	0	0	1	1	1	2	3	4	6	9
	800	0	0	0	0	1	1	1	1	3	4	6	9
	900	0	0	0	0	0	1	1	—	3	3	5	8
	1000	0	0	0	0	0	1	1	1	2	3	5	7
	1250	0	0	0	0	0	1	1	1	1	2	4	6
	1500	0	0	0	0	0	0	1	1	1	1	3	5
	1750	0	0	0	0	0	0	1	1	1	1	3	4
	2000	0	0	0	0	0	0	1	1	1	1	2	4

(Continues)

Table C.9 *Continued*

FIXTURE WIRES							
	Conductor Size (AWG/ kcmil)	Metric Designator (Trade Size)					
Type		16 (½)	21 (¾)	27 (1)	35 (1¼)	41 (1½)	53 (2)
FFH-2, RFH-2, RFHH-3	18	6	11	19	34	47	79
	16	5	9	16	28	39	67
SF-2, SFF-2	18	7	14	24	43	59	100
	16	6	11	20	35	49	82
	14	5	9	16	28	39	67
SF-1, SFF-1	18	13	25	42	76	105	177
RFH-1, RFHH-2, TF, TFF, XF, XFF	18	10	18	31	56	77	130
RFHH-2, TF, TFF, XF, XFF	16	8	15	25	45	62	105
XF, XFF	14	6	11	20	35	49	82
TFN, TFFN	18	16	29	50	90	124	209
	16	12	22	38	68	95	159
PF, PFF, PGF, PGFF, PAF, PTF, PTFF, PAFF	18	15	28	47	85	118	198
	16	11	22	36	66	91	153
	14	8	16	27	49	68	115
HF, HFF, ZF, ZFF, ZHF	18	19	36	61	110	152	255
	16	14	27	45	81	112	188
	14	10	19	33	59	82	138
KF-2, KFF-2	18	28	53	88	159	220	371
	16	19	37	62	112	155	261
	14	13	25	43	77	107	179
	12	9	17	29	53	73	123
	10	6	11	20	35	49	82
KF-1, KFF-1	18	33	63	106	190	263	442
	16	23	44	74	133	185	310
	14	16	29	50	90	124	209
	12	10	19	33	59	82	138
	10	7	13	21	39	54	90
XF, XFF	12	3	6	10	19	26	44
	10	2	5	8	15	20	34

Notes:

1. This table is for concentric stranded conductors only. For compact stranded conductors, Table C.9(A) should be used.

2. Two-hour fire-rated RHH cable has ceramifiable insulation which has much larger diameters than other RHH wires. Consult manufacturer's conduit fill tables.

*Types RHH, RHW, and RHW-2 without outer covering.

Table C-10

CONDUIT AND TUBING FILL TABLES FOR CONDUCTORS AND FIXTURE WIRES OF THE SAME SIZE *(CONTINUED)*.

Table C.10 Maximum Number of Conductors or Fixture Wires in Rigid PVC Conduit, Schedule 40 and HDPE Conduit (*Based on Table 1, Chapter 9*)

CONDUCTORS													
	Conductor Size	Metric Designator (Trade Size)											
Type	(AWG/ kcmil)	16 (½)	21 (¾)	27 (1)	35 (1¼)	41 (1½)	53 (2)	63 (2½)	78 (3)	91 (3½)	103 (4)	129 (5)	155 (6)
RHH, RHW, RHW-2	14	4	7	11	20	27	45	64	99	133	171	269	390
	12	3	5	9	16	22	37	53	82	110	142	224	323
	10	2	4	7	13	18	30	43	66	89	115	181	261
	8	1	2	4	7	9	15	22	35	46	60	94	137
	6	1	1	3	5	7	12	18	28	37	48	76	109
	4	1	1	2	4	6	10	14	22	29	37	59	85
	3	1	1	1	4	5	8	12	19	25	33	52	75
	2	1	1	1	3	4	7	10	16	22	28	45	65
	1	0	1	1	1	3	5	7	11	14	19	29	43
	1/0	0	1	1	1	2	4	6	9	13	16	26	37
	2/0	0	0	1	1	1	3	5	8	11	14	22	32
	3/0	0	0	1	1	1	3	4	7	9	12	19	28
	4/0	0	0	1	1	1	2	4	6	8	10	16	24
	250	0	0	0	1	1	1	3	4	6	8	12	18
	300	0	0	0	1	1	1	2	4	5	7	11	16
	350	0	0	0	1	1	1	2	3	5	6	10	14
	400	0	0	0	1	1	1	1	3	4	6	9	13
	500	0	0	0	0	1	1	1	3	4	5	8	11
	600	0	0	0	0	1	1	1	2	3	4	6	9
	700	0	0	0	0	0	1	1	1	3	3	6	8
	750	0	0	0	0	0	1	1	1	2	3	5	8
	800	0	0	0	0	0	1	1	1	2	3	5	7
	900	0	0	0	0	0	1	1	1	2	3	5	7
	1000	0	0	0	0	0	1	1	1	1	3	4	6
	1250	0	0	0	0	0	0	1	1	1	1	3	5
	1500	0	0	0	0	0	0	1	1	1	1	3	4
	1750	0	0	0	0	0	0	1	1	1	1	2	3
	2000	0	0	0	0	0	0	0	1	1	1	2	3
TW	14	8	14	24	42	57	94	135	209	280	361	568	822
	12	6	11	18	32	44	72	103	160	215	277	436	631
	10	4	8	13	24	32	54	77	119	160	206	325	470
	8	2	4	7	13	18	30	43	66	89	115	181	261
RHH*, RHW*, RHW-2*, THHW, THW, THW-2	14	5	9	16	28	38	63	90	139	186	240	378	546
RHH*, RHW*, RHW-2*, THHW, THW	12	4	8	12	22	30	50	72	112	150	193	304	439
	10	3	6	10	17	24	39	56	87	117	150	237	343
RHH*, RHW*, RHW-2*, THHW, THW, THW-2	8	1	3	6	10	14	23	33	52	70	90	142	205

(Continues)

Table C.10 *Continued*

		CONDUCTORS											
	Conductor Size	Metric Designator (Trade Size)											
Type	(AWG/ kcmil)	16 (½)	21 (¾)	27 (1)	35 (1¼)	41 (1½)	53 (2)	63 (2½)	78 (3)	91 (3½)	103 (4)	129 (5)	155 (6)
RHH*, RHW*, RHW-2* TW, THW, THHW, THW-2	6	1	2	4	8	11	18	26	40	53	69	109	157
	4	1	1	3	6	8	13	19	30	40	51	81	117
	3	1	1	3	5	7	11	16	25	34	44	69	100
	2	1	1	2	4	6	10	14	22	29	37	59	85
	1	0	1	1	3	4	7	10	15	20	26	41	60
	1/0	0	1	1	2	3	6	8	13	17	22	35	51
	2/0	0	1	1	1	3	5	7	11	15	19	30	43
	3/0	0	1	1	1	2	4	6	9	12	16	25	36
	4/0	0	0	1	1	1	3	5	8	10	13	21	30
	250	0	0	1	1	1	3	4	6	8	11	17	25
	300	0	0	1	1	1	2	3	5	7	9	15	21
	350	0	0	0	1	1	1	3	5	6	8	13	19
	400	0	0	0	1	1	1	3	4	6	7	12	17
	500	0	0	0	1	1	1	2	3	5	6	10	14
	600	0	0	0	0	1	1	1	3	4	5	8	11
	700	0	0	0	0	1	1	1	2	3	4	7	10
	750	0	0	0	0	1	1	1	2	3	4	6	10
	800	0	0	0	0	1	1	1	2	3	4	6	9
	900	0	0	0	0	0	1	1	1	3	3	6	8
	1000	0	0	0	0	0	1	1	1	2	3	5	7
	1250	0	0	0	0	0	1	1	1	1	2	4	6
	1500	0	0	0	0	0	1	1	1	1	1	3	5
	1750	0	0	0	0	0	0	1	1	1	1	3	4
	2000	0	0	0	0	0	0	1	1	1	1	3	4
THHN, THWN, THWN-2	14	11	21	34	60	82	135	193	299	401	517	815	1178
	12	8	15	25	43	59	99	141	218	293	377	594	859
	10	5	9	15	27	37	62	89	137	184	238	374	541
	8	3	5	9	16	21	36	51	79	106	137	216	312
	6	1	4	6	11	15	26	37	57	77	99	156	225
	4	1	2	4	7	9	16	22	35	47	61	96	138
	3	1	1	3	6	8	13	19	30	40	51	81	117
	2	1	1	3	5	7	11	16	25	33	43	68	98
	1	1	1	1	3	5	8	12	18	25	32	50	73
	1/0	1	1	1	3	4	7	10	15	21	27	42	61
	2/0	0	1	1	2	3	6	8	13	17	22	35	51
	3/0	0	1	1	1	3	5	7	11	14	18	29	42
	4/0	0	1	1	1	2	4	6	9	12	15	24	35
	250	0	0	1	1	1	3	4	7	10	12	20	28
	300	0	0	1	1	1	3	4	6	8	11	17	24
	350	0	0	1	1	1	2	3	5	7	9	15	21
	400	0	0	0	1	1	1	3	5	6	8	13	19
	500	0	0	0	1	1	1	2	4	5	7	11	16
	600	0	0	0	1	1	1	1	3	4	5	9	13
	700	0	0	0	0	1	1	1	3	4	5	8	11
	750	0	0	0	0	1	1	1	2	3	4	7	11
	800	0	0	0	0	1	1	1	2	3	4	7	10
	900	0	0	0	0	1	1	1	2	3	4	6	9
	1000	0	0	0	0	0	1	1	1	3	3	6	8
FEP, FEPB, PFA, PFAH, TFE	14	11	20	33	58	79	131	188	290	389	502	790	1142
	12	8	15	24	42	58	96	137	212	284	366	577	834
	10	6	10	17	30	41	69	98	152	204	263	414	598
	8	3	6	10	17	24	39	56	87	117	150	237	343
	6	2	4	7	12	17	28	40	62	83	107	169	244
	4	1	3	5	8	12	19	28	43	58	75	118	170
	3	1	2	4	7	10	16	23	36	48	62	98	142
	2	1	1	3	6	8	13	19	30	40	51	81	117

(Continues)

Table C.10 *Continued*

CONDUCTORS													
	Conductor Size	Metric Designator (Trade Size)											
Type	(AWG/ kcmil)	16 (½)	21 (¾)	27 (1)	35 (1¼)	41 (1½)	53 (2)	63 (2½)	78 (3)	91 (3½)	103 (4)	129 (5)	155 (6)
PFA, PFAH, TFE	1	1	1	2	4	5	9	13	20	28	36	56	81
PFA, PFAH, TFE, Z	1/0	1	1	1	3	4	8	11	17	23	30	47	68
	2/0	0	1	1	3	4	6	9	14	19	24	39	56
	3/0	0	1	1	2	3	5	7	12	16	20	32	46
	4/0	0	1	1	1	2	4	6	9	13	16	26	38
Z	14	13	24	40	70	95	158	226	350	469	605	952	1376
	12	9	17	28	49	68	112	160	248	333	429	675	976
	10	6	10	17	30	41	69	98	152	204	263	414	598
	8	3	6	11	19	26	43	62	96	129	166	261	378
	6	2	4	7	13	18	30	43	67	90	116	184	265
	4	1	3	5	9	12	21	30	46	62	80	126	183
	3	1	2	4	6	9	15	22	34	45	58	92	133
	2	1	1	3	5	7	12	18	28	38	49	77	111
	1	1	1	2	4	6	10	14	23	30	39	62	90
XHH, XHHW, XHHW-2 ZW	14	8	14	24	42	57	94	135	209	280	361	568	822
	12	6	11	18	32	44	72	103	160	215	277	436	631
	10	4	8	13	24	32	54	77	119	160	206	325	470
	8	2	4	7	13	18	30	43	66	89	115	181	261
	6	1	3	5	10	13	22	32	49	66	85	134	193
	4	1	2	4	7	9	16	23	35	48	61	97	140
	3	1	1	3	6	8	13	19	30	40	52	82	118
	2	1	1	3	5	7	11	16	25	34	44	69	99
XHH, XHHW, XHHW-2	1	1	1	1	3	5	8	12	19	25	32	51	74
	1/0	1	1	1	3	4	7	10	16	21	27	43	62
	2/0	0	1	1	2	3	6	8	13	17	23	36	52
	3/0	0	1	1	1	3	5	7	11	14	19	30	43
	4/0	0	1	1	1	2	4	6	9	12	15	24	35
	250	0	0	1	1	1	3	5	7	10	13	20	29
	300	0	0	1	1	1	3	4	6	8	11	17	25
	350	0	0	1	1	1	2	3	5	7	9	15	22
	400	0	0	0	1	1	1	3	5	6	8	13	19
	500	0	0	0	1	1	1	2	4	5	7	11	16
	600	0	0	0	1	1	1	1	3	4	5	9	13
	700	0	0	0	0	1	1	1	3	4	5	8	11
	750	0	0	0	0	1	1	1	2	3	4	7	11
	800	0	0	0	0	1	1	1	2	3	4	7	10
	900	0	0	0	0	1	1	1	2	3	4	6	9
	1000	0	0	0	0	0	1	1	1	3	3	6	8
	1250	0	0	0	0	0	1	1	1	1	3	4	6
	1500	0	0	0	0	0	1	1	1	1	2	4	5
	1750	0	0	0	0	0	0	1	1	1	1	3	5
	2000	0	0	0	0	0	0	1	1	1	1	3	4

(Continues)

Table C.10 *Continued*

FIXTURE WIRES							
	Conductor Size	Metric Designator (Trade Size)					
Type	(AWG/ kcmil)	16 (½)	21 (¾)	27 (1)	35 (1¼)	41 (1½)	53 (2)
FFH-2, RFH-2, RFHH-3	18	8	14	23	40	54	90
	16	6	12	19	33	46	76
SF-2, SFF-2	18	10	17	29	50	69	114
	16	8	14	24	42	57	94
	14	6	12	19	33	46	76
SF-1, SFF-1	18	17	31	51	89	122	202
RFHH-2, TF, TFF, XF, XFF RFH-1,	18	13	23	38	66	90	149
RFHH-2, TF, TFF, XF, XFF	16	10	18	30	53	73	120
XF, XFF	14	8	14	24	42	57	94
TFN, TFFN	18	20	37	60	105	144	239
	16	16	28	46	80	110	183
PF, PFF, PGF, PGFF, PAF, PTF, PTFF, PAFF	18	19	35	57	100	137	227
	16	15	27	44	77	106	175
	14	11	20	33	58	79	131
HF, HFF, ZF, ZFF, ZHF	18	25	45	74	129	176	292
	16	18	33	54	95	130	216
	14	13	24	40	70	95	158
KF-2, KFF-2	18	36	65	107	187	256	424
	16	26	46	75	132	180	299
	14	17	31	52	90	124	205
	12	12	22	35	62	85	141
	10	8	14	24	42	57	94
KF-1, KFF-1	18	43	78	128	223	305	506
	16	30	55	90	157	214	355
	14	20	37	60	105	144	239
	12	13	24	40	70	95	158
	10	9	16	26	45	62	103
XF, XFF	12	4	8	12	22	30	50
	10	3	6	10	17	24	39

Notes:

1. This table is for concentric stranded conductors only. For compact stranded conductors, Table C.10(A) should be used.

2. Two-hour fire-rated RHH cable has ceramifiable insulation which has much larger diameters than other RHH wires. Consult manufacturer's conduit fill tables.

*Types RHH, RHW, and RHW-2 without outer covering.

Table C-11

CONDUIT AND TUBING FILL TABLES FOR CONDUCTORS AND FIXTURE WIRES OF THE SAME SIZE *(CONTINUED)*.

Table C.11 Maximum Number of Conductors or Fixture Wires in Type A, Rigid PVC Conduit
(Based on Table 1, Chapter 9)

CONDUCTORS											
	Conductor Size	Metric Designator (Trade Size)									
Type	(AWG/ kcmil)	16 (½)	21 (¾)	27 (1)	35 (1¼)	41 (1½)	53 (2)	63 (2½)	78 (3)	91 (3½)	103 (4)
RHH, RHW, RHW-2	14	5	9	15	24	31	49	74	112	146	187
	12	4	7	12	20	26	41	61	93	121	155
	10	3	6	10	16	21	33	50	75	98	125
	8	1	3	5	8	11	17	26	39	51	65
	6	1	2	4	6	9	14	21	31	41	52
	4	1	1	3	5	7	11	16	24	32	41
	3	1	1	3	4	6	9	14	21	28	36
	2	1	1	2	4	5	8	12	18	24	31
	1	0	1	1	2	3	5	8	12	16	20
	1/0	0	1	1	2	3	5	7	10	14	18
	2/0	0	1	1	1	2	4	6	9	12	15
	3/0	0	1	1	1	1	3	5	8	10	13
	4/0	0	0	1	1	1	3	4	7	9	11
	250	0	0	1	1	1	1	3	5	7	8
	300	0	0	1	1	1	1	3	4	6	7
	350	0	0	0	1	1	1	2	4	5	7
	400	0	0	0	1	1	1	2	4	5	6
	500	0	0	0	1	1	1	1	3	4	5
	600	0	0	0	0	1	1	1	2	3	4
	700	0	0	0	0	1	1	1	2	3	4
	750	0	0	0	0	1	1	1	1	3	4
	800	0	0	0	0	1	1	1	1	3	3
	900	0	0	0	0	0	1	1	1	2	3
	1000	0	0	0	0	0	1	1	1	2	3
	1250	0	0	0	0	0	1	1	1	1	2
	1500	0	0	0	0	0	0	1	1	1	1
	1750	0	0	0	0	0	0	1	1	1	1
	2000	0	0	0	0	0	0	1	1	1	1
TW	14	11	18	31	51	67	105	157	235	307	395
	12	8	14	24	39	51	80	120	181	236	303
	10	6	10	18	29	38	60	89	135	176	226
	8	3	6	10	16	21	33	50	75	98	125
RHH*, RHW*, RHW-2*, THHW, THW, THW-2	14	7	12	20	34	44	70	104	157	204	262
RHH*, RHW*, RHW-2*, THHW, THW	12	6	10	16	27	35	56	84	126	164	211
	10	4	8	13	21	28	44	65	98	128	165
RHH*, RHW*, RHW-2*, THHW, THW, THW-2	8	2	4	8	12	16	26	39	59	77	98
RHH*, RHW*, RHW-2*, TW, THHW, THW, THW-2	6	1	3	6	9	13	20	30	45	59	75

(Continues)

Table C.11 *Continued*

CONDUCTORS											
	Conductor Size	Metric Designator (Trade Size)									
Type	(AWG/ kcmil)	16 (½)	21 (¾)	27 (1)	35 (1¼)	41 (1½)	53 (2)	63 (2½)	78 (3)	91 (3½)	103 (4)
RHH*, RHW*, RHW-2*, TW, THW, THHW, THW-2	4	1	2	4	7	9	15	22	33	44	56
	3	1	1	4	6	8	13	19	29	37	48
	2	1	1	3	5	7	11	16	24	32	41
	1	1	1	1	3	5	7	11	17	22	29
	1/0	1	1	1	3	4	6	10	14	19	24
	2/0	0	1	1	2	3	5	8	12	16	21
	3/0	0	1	1	1	3	4	7	10	13	17
	4/0	0	1	1	1	2	4	6	9	11	14
	250	0	0	1	1	1	3	4	7	9	12
	300	0	0	1	1	1	2	4	6	8	10
	350	0	0	1	1	1	2	3	5	7	9
	400	0	0	1	1	1	1	3	5	6	8
	500	0	0	0	1	1	1	2	4	5	7
	600	0	0	0	1	1	1	1	3	4	5
	700	0	0	0	1	1	1	1	3	4	5
	750	0	0	0	1	1	1	1	3	3	4
	800	0	0	0	0	1	1	1	2	3	4
	900	0	0	0	0	1	1	1	2	3	4
	1000	0	0	0	0	1	1	1	1	3	3
	1250	0	0	0	0	0	1	1	1	1	3
	1500	0	0	0	0	0	1	1	1	1	2
	1750	0	0	0	0	0	0	1	1	1	1
	2000	0	0	0	0	0	0	1	1	1	1
THHN, THWN, THWN-2	14	16	27	44	73	96	150	225	338	441	566
	12	11	19	32	53	70	109	164	246	321	412
	10	7	12	20	33	44	69	103	155	202	260
	8	4	7	12	19	25	40	59	89	117	150
	6	3	5	8	14	18	28	43	64	84	108
	4	1	3	5	8	11	17	26	39	52	66
	3	1	2	4	7	9	15	22	33	44	56
	2	1	1	3	6	8	12	19	28	37	47
	1	1	1	2	4	6	9	14	21	27	35
	1/0	1	1	2	4	5	8	11	17	23	29
	2/0	1	1	1	3	4	6	10	14	19	24
	3/0	0	1	1	2	3	5	8	12	16	20
	4/0	0	1	1	1	3	4	6	10	13	17
	250	0	1	1	1	2	3	5	8	10	14
	300	0	0	1	1	1	3	4	7	9	12
	350	0	0	1	1	1	2	4	6	8	10
	400	0	0	1	1	1	2	3	5	7	9
	500	0	0	1	1	1	1	3	4	6	7
	600	0	0	0	1	1	1	2	3	5	6
	700	0	0	0	1	1	1	1	3	4	5
	750	0	0	0	1	1	1	1	3	4	5
	800	0	0	0	1	1	1	1	3	4	5
	900	0	0	0	0	1	1	1	2	3	4
	1000	0	0	0	0	1	1	1	2	3	4
FEP, FEPB, PFA, PFAH, TFE	14	15	26	43	70	93	146	218	327	427	549
	12	11	19	31	51	68	106	159	239	312	400
	10	8	13	22	37	48	76	114	171	224	287
	8	4	8	13	21	28	44	65	98	128	165
	6	3	5	9	15	20	31	46	70	91	117
	4	1	4	6	10	14	21	32	49	64	82
	3	1	3	5	8	11	18	27	40	53	68
	2	1	2	4	7	9	15	22	33	44	56
PFA, PFAH, TFE	1	1	1	3	5	6	10	15	23	30	39

(Continues)

Table C.11 *Continued*

CONDUCTORS											
	Conductor Size (AWG/ kcmil)	Metric Designator (Trade Size)									
Type		16 (½)	21 (¾)	27 (1)	35 (1¼)	41 (1½)	53 (2)	63 (2½)	78 (3)	91 (3½)	103 (4)
PFA, PFAH, TFE, Z	1/0	1	1	2	4	5	8	13	19	25	32
	2/0	1	1	1	3	4	7	10	16	21	27
	3/0	1	1	1	3	3	6	9	13	17	22
	4/0	0	1	1	2	3	5	7	11	14	18
Z	14	18	31	52	85	112	175	263	395	515	661
	12	13	22	37	60	79	124	186	280	365	469
	10	8	13	22	37	48	76	114	171	224	287
	8	5	8	14	23	30	48	72	108	141	181
	6	3	6	10	16	21	34	50	76	99	127
	4	2	4	7	11	15	23	35	52	68	88
	3	1	3	5	8	11	17	25	38	50	64
	2	1	2	4	7	9	14	21	32	41	53
	1	1	1	3	5	7	11	17	26	33	43
XHH, XHHW, XHHW-2, ZW	14	11	18	31	51	67	105	157	235	307	395
	12	8	14	24	39	51	80	120	181	236	303
	10	6	10	18	29	38	60	89	135	176	226
	8	3	6	10	16	21	33	50	75	98	125
	6	2	4	7	12	15	24	37	55	72	93
	4	1	3	5	8	11	18	26	40	52	67
	3	1	2	4	7	9	15	22	34	44	57
	2	1	1	3	6	8	12	19	28	37	48
XHH, XHHW, XHHW-2	1	1	1	3	4	6	9	14	21	28	35
	1/0	1	1	2	4	5	8	12	18	23	30
	2/0	1	1	1	3	4	6	10	15	19	25
	3/0	0	1	1	2	3	5	8	12	16	20
	4/0	0	1	1	1	3	4	7	10	13	17
	250	0	1	1	1	2	3	5	8	11	14
	300	0	0	1	1	1	3	5	7	9	12
	350	0	0	1	1	1	3	4	6	8	10
	400	0	0	1	1	1	2	3	5	7	9
	500	0	0	1	1	1	1	3	4	6	8
	600	0	0	0	1	1	1	2	3	5	6
	700	0	0	0	1	1	1	1	3	4	5
	750	0	0	0	1	1	1	1	3	4	5
	800	0	0	0	1	1	1	1	3	4	5
	900	0	0	0	0	1	1	1	2	3	4
	1000	0	0	0	0	1	1	1	2	3	4
	1250	0	0	0	0	0	1	1	1	2	3
	1500	0	0	0	0	0	1	1	1	1	2
	1750	0	0	0	0	0	1	1	1	1	2
	2000	0	0	0	0	0	0	1	1	1	1

(Continues)

Table C.11 *Continued*

FIXTURE WIRES							
	Conductor Size	Metric Designator (Trade Size)					
Type	(AWG/ kcmil)	16 (½)	21 (¾)	27 (1)	35 (1¼)	41 (1½)	53 (2)
FFH-2, RFH-2, RFHH-3	18	10	18	30	48	64	100
	16	9	15	25	41	54	85
SF-2, SFF-2	18	13	22	37	61	81	127
	16	11	18	31	51	67	105
	14	9	15	25	41	54	85
SF-1, SFF-1	18	23	40	66	108	143	224
RFH-1, RFHH-2, TF, TFF, XF, XFF	18	17	29	49	80	105	165
RFHH-2, TF, TFF, XF, XFF	16	14	24	39	65	85	134
XF, XFF	14	11	18	31	51	67	105
TFN, TFFN	18	28	47	79	128	169	265
	16	21	36	60	98	129	202
PF, PFF, PGF, PGFF, PAF, PTF, PTFF, PAFF	18	26	45	74	122	160	251
	16	20	34	58	94	124	194
	14	15	26	43	70	93	146
HF, HFF, ZF, ZFF, ZHF	18	34	58	96	157	206	324
	16	25	42	71	116	152	239
	14	18	31	52	85	112	175
KF-2, KFF-2	18	49	84	140	228	300	470
	16	35	59	98	160	211	331
	14	24	40	67	110	145	228
	12	16	28	46	76	100	157
	10	11	18	31	51	67	105
KF-1, KFF-1	18	59	100	167	272	357	561
	16	41	70	117	191	251	394
	14	28	47	79	128	169	265
	12	18	31	52	85	112	175
	10	12	20	34	55	73	115
XF, XFF	12	6	10	16	27	35	56
	10	4	8	13	21	28	44

Notes:

1. This table is for concentric stranded conductors only. For compact stranded conductors, Table C.11(A) should be used.

2. Two-hour fire-rated RHH cable has ceramifiable insulation which has much larger diameters than other RHH wires. Consult manufacturer's conduit fill tables.

*Types RHH, RHW, and RWH-2 without outer covering.

Table C-12

CONDUIT AND TUBING FILL TABLES FOR CONDUCTORS AND FIXTURE WIRES OF THE SAME SIZE *(CONTINUED).*

Table C.12 Maximum Number of Conductors in Type EB, PVC Conduit
(Based on Table 1, Chapter 9)

CONDUCTORS							
	Conductor Size	Metric Designator (Trade Size)					
Type	(AWG/ kcmil)	53 (2)	78 (3)	91 (3½)	103 (4)	129 (5)	155 (6)
RHH, RHW, RHW-2	14	53	119	155	197	303	430
	12	44	98	128	163	251	357
	10	35	79	104	132	203	288
	8	18	41	54	69	106	151
	6	15	33	43	55	85	121
	4	11	26	34	43	66	94
	3	10	23	30	38	58	83
	2	9	20	26	33	50	72
	1	6	13	17	21	33	47
	1/0	5	11	15	19	29	41
	2/0	4	10	13	16	25	36
	3/0	4	8	11	14	22	31
	4/0	3	7	9	12	18	26
	250	2	5	7	9	14	20
	300	1	5	6	8	12	17
	350	1	4	5	7	11	16
	400	1	4	5	6	10	14
	500	1	3	4	5	9	12
	600	1	3	3	4	7	10
	700	1	2	3	4	6	9
	750	1	2	3	4	6	9
	800	1	2	3	4	6	8
	900	1	1	2	3	5	7
	1000	1	1	2	3	5	7
	1250	1	1	1	2	3	5
	1500	0	1	1	1	3	4
	1750	0	1	1	1	3	4
	2000	0	1	1	1	2	3
TW	14	111	250	327	415	638	907
	12	85	192	251	319	490	696
	10	63	143	187	238	365	519
	8	35	79	104	132	203	288
RHH*, RHW*, RHW-2*, THHW, THW, THW-2	14	74	166	217	276	424	603
RHH*, RHW*, RHW-2*, THHW, THW	12	59	134	175	222	341	485
	10	46	104	136	173	266	378
RHH*, RHW*, RHW-2*, THHW, THW, THW-2	8	28	62	81	104	159	227

(Continues)

Table C.12 *Continued*

CONDUCTORS							
	Conductor Size	Metric Designator (Trade Size)					
Type	(AWG/ kcmil)	53 (2)	78 (3)	91 (3½)	103 (4)	129 (5)	155 (6)
RHH*, RHW*, RHW-2*, TW, THW, THHW, THW-2	6	21	48	62	79	122	173
	4	16	36	46	59	91	129
	3	13	30	40	51	78	111
	2	11	26	34	43	66	94
	1	8	18	24	30	46	66
	1/0	7	15	20	26	40	56
	2/0	6	13	17	22	34	48
	3/0	5	11	14	18	28	40
	4/0	4	9	12	15	24	34
	250	3	7	10	12	19	27
	300	3	6	8	11	17	24
	350	2	6	7	9	15	21
	400	2	5	7	8	13	19
	500	1	4	5	7	11	16
	600	1	3	4	6	9	13
	700	1	3	4	5	8	11
	750	1	3	4	5	7	11
	800	1	3	3	4	7	10
	900	1	2	3	4	6	9
	1000	1	2	3	4	6	8
	1250	1	1	2	3	4	6
	1500	1	1	1	2	4	6
	1750	1	1	1	2	3	5
	2000	0	1	1	1	3	4
THHN, THWN, THWN-2	14	159	359	468	595	915	1300
	12	116	262	342	434	667	948
	10	73	165	215	274	420	597
	8	42	95	124	158	242	344
	6	30	68	89	114	175	248
	4	19	42	55	70	107	153
	3	16	36	46	59	91	129
	2	13	30	39	50	76	109
	1	10	22	29	37	57	80
	1/0	8	18	24	31	48	68
	2/0	7	15	20	26	40	56
	3/0	5	13	17	21	33	47
	4/0	4	10	14	18	27	39
	250	4	8	11	14	22	31
	300	3	7	10	12	19	27
	350	3	6	8	11	17	24
	400	2	6	7	10	15	21
	500	1	5	6	8	12	18
	600	1	4	5	6	10	14
	700	1	3	4	6	9	12
	750	1	3	4	5	8	12
	800	1	3	4	5	8	11
	900	1	3	3	4	7	10
	1000	1	2	3	4	6	9
FEP, FEPB, PFA, PFAH, TFE	14	155	348	454	578	888	1261
	12	113	254	332	422	648	920
	10	81	182	238	302	465	660
	8	46	104	136	173	266	378
	6	33	74	97	123	189	269
	4	23	52	68	86	132	188
	3	19	43	56	72	110	157
	2	16	36	46	59	91	129
PFA, PFAH, TFE	1	11	25	32	41	63	90

(Continues)

Table C.12 *Continued*

CONDUCTORS							
Type	Conductor Size (AWG/ kcmil)	Metric Designator (Trade Size) 53 (2)	78 (3)	91 (3½)	103 (4)	129 (5)	155 (6)
PFA, PFAH, TFE, Z	1/0	9	20	27	34	53	75
	2/0	7	17	22	28	43	62
	3/0	6	14	18	23	36	51
	4/0	5	11	15	19	29	42
Z	14	186	419	547	696	1069	1519
	12	132	297	388	494	759	1078
	10	81	182	238	302	465	660
	8	51	115	150	191	294	417
	6	36	81	105	134	206	293
	4	24	55	72	92	142	201
	3	18	40	53	67	104	147
	2	15	34	44	56	86	122
	1	12	27	36	45	70	99
XHH, XHHW, XHHW-2, ZW	14	111	250	327	415	638	907
	12	85	192	251	319	490	696
	10	63	143	187	238	365	519
	8	35	79	104	132	203	288
	6	26	59	77	98	150	213
	4	19	42	56	71	109	155
	3	16	36	47	60	92	131
	2	13	30	39	50	77	110
XHH, XHHW, XHHW-2	1	10	22	29	37	58	82
	1/0	8	19	25	31	48	69
	2/0	7	16	20	26	40	57
	3/0	6	13	17	22	33	47
	4/0	5	11	14	18	27	39
	250	4	9	11	15	22	32
	300	3	7	10	12	19	28
	350	3	6	9	11	17	24
	400	2	6	8	10	15	22
	500	1	5	6	8	12	18
	600	1	4	5	6	10	14
	700	1	3	4	6	9	12
	750	1	3	4	5	8	12
	800	1	3	4	5	8	11
	900	1	3	3	4	7	10
	1000	1	2	3	4	6	9
	1250	1	1	2	3	5	7
	1500	1	1	1	3	4	6
	1750	1	1	1	2	4	5
	2000	0	1	1	1	3	5

Notes:

1. This table is for concentric stranded conductors only. For compact stranded conductors, Table C.12(A) should be used.

2. Two-hour fire-rated RHH cable has ceramifiable insulation which has much larger diameters than other RHH wires. Consult manufacturer's conduit fill tables.

*Types RHH, RHW, and RHW-2 without outer covering.

APPENDIX

Table D-1

DIMENSIONS OF INSULATED CONDUCTORS AND FIXTURE WIRES.

Table 5 Dimensions of Insulated Conductors and Fixture Wires

Type	Size (AWG or kcmil)	Approximate Diameter		Approximate Area	
		mm	in.	mm^2	in^2
Type: FFH-2, RFH-1, RFH-2, RHH*, RHW*, RHW-2*, RHH, RHW, RHW-2, SF-1, SF-2, SFF-1, SFF-2, TF, TFF, THHW, THW, THW-2, TW, XF, XFF					
RFH-2, FFH-2	18	3.454	0.136	9.355	0.0145
	16	3.759	0.148	11.10	0.0172
RHH, RHW, RHW-2	14	4.902	0.193	18.90	0.0293
	12	5.385	0.212	22.77	0.0353
	10	5.994	0.236	28.19	0.0437
	8	8.280	0.326	53.87	0.0835
	6	9.246	0.364	67.16	0.1041
	4	10.46	0.412	86.00	0.1333
	3	11.18	0.440	98.13	0.1521
	2	11.99	0.472	112.9	0.1750
	1	14.78	0.582	171.6	0.2660
	1/0	15.80	0.622	196.1	0.3039
	2/0	16.97	0.668	226.1	0.3505
	3/0	18.29	0.720	262.7	0.4072
	4/0	19.76	0.778	306.7	0.4754
	250	22.73	0.895	405.9	0.6291
	300	24.13	0.950	457.3	0.7088
	350	25.43	1.001	507.7	0.7870
	400	26.62	1.048	556.5	0.8626
	500	28.78	1.133	650.5	1.0082
	600	31.57	1.243	782.9	1.2135
	700	33.38	1.314	874.9	1.3561
	750	34.24	1.348	920.8	1.4272
	800	35.05	1.380	965.0	1.4957
	900	36.68	1.444	1057	1.6377
	1000	38.15	1.502	1143	1.7719
	1250	43.92	1.729	1515	2.3479
	1500	47.04	1.852	1738	2.6938
	1750	49.94	1.966	1959	3.0357
	2000	52.63	2.072	2175	3.3719
SF-2, SFF-2	18	3.073	0.121	7.419	0.0115
	16	3.378	0.133	8.968	0.0139
	14	3.759	0.148	11.10	0.0172
SF-1, SFF-1	18	2.311	0.091	4.194	0.0065
RFH-1, XF, XFF	18	2.692	0.106	5.161	0.0080
TF, TFF, XF, XFF	16	2.997	0.118	7.032	0.0109
TW, XF, XFF, THHW, THW, THW-2	14	3.378	0.133	8.968	0.0139

Table D-2

COMPACT ALUMINUM BUILDING WIRE NOMINAL DIMENSIONS AND AREAS.

TABLE 5A Compact Copper and Aluminum Building Wire Nominal Dimensions* and Areas

Size (AWG or kcmil)	Bare Conductor		Types RHH**, RHW**, or USE				Types THW and THHW				Type THHN				Type XHHW				Size (AWG or kcmil)
	Diameter		Approximate Diameter		Approximate Area		Approximate Diameter		Approximate Area		Approximate Diameter		Approximate Area		Approximate Diameter		Approximate Area		
	mm	in.	mm	in.	mm^2	in^2	mm	in.	mm^2	in^2	mm	in.	mm^2	in^2	mm	in.	mm^2	in^2	
8	3.404	0.134	6.604	0.260	34.25	0.0531	6.477	0.255	32.90	0.0510	—	—	—	—	5.690	0.224	25.42	0.0394	8
6	4.293	0.169	7.493	0.295	44.10	0.0683	7.366	0.290	42.58	0.0660	6.096	0.240	29.16	0.0452	6.604	0.260	34.19	0.0530	6
4	5.410	0.213	8.509	0.335	56.84	0.0881	8.509	0.335	56.84	0.0881	7.747	0.305	47.10	0.0730	7.747	0.305	47.10	0.0730	4
2	6.807	0.268	9.906	0.390	77.03	0.1194	9.906	0.390	77.03	0.1194	9.144	0.360	65.61	0.1017	9.144	0.360	65.61	0.1017	2
1	7.595	0.299	11.81	0.465	109.5	0.1698	11.81	0.465	109.5	0.1698	10.54	0.415	87.23	0.1352	10.54	0.415	87.23	0.1352	1
1/0	8.534	0.336	12.70	0.500	126.6	0.1963	12.70	0.500	126.6	0.1963	11.43	0.450	102.6	0.1590	11.43	0.450	102.6	0.1590	1/0
2/0	9.550	0.376	13.72	0.540	147.8	0.2290	13.84	0.545	150.5	0.2332	12.57	0.495	124.1	0.1924	12.45	0.490	121.6	0.1885	2/0
3/0	10.74	0.423	14.99	0.590	176.3	0.2733	14.99	0.590	176.3	0.2733	13.72	0.540	147.7	0.2290	13.72	0.540	147.7	0.2290	3/0
4/0	12.07	0.475	16.26	0.640	207.6	0.3217	16.38	0.645	210.8	0.3267	15.11	0.595	179.4	0.2780	14.99	0.590	176.3	0.2733	4/0
250	13.21	0.520	18.16	0.715	259.0	0.4015	18.42	0.725	266.3	0.4128	17.02	0.670	227.4	0.3525	16.76	0.660	220.7	0.3421	250
300	14.48	0.570	19.43	0.765	296.5	0.4596	19.69	0.775	304.3	0.4717	18.29	0.720	262.6	0.4071	18.16	0.715	259.0	0.4015	300
350	15.65	0.616	20.57	0.810	332.3	0.5153	20.83	0.820	340.7	0.5281	19.56	0.770	300.4	0.4656	19.30	0.760	292.6	0.4536	350
400	16.74	0.659	21.72	0.855	370.5	0.5741	21.97	0.865	379.1	0.5876	20.70	0.815	336.5	0.5216	20.32	0.800	324.3	0.5026	400
500	18.69	0.736	23.62	0.930	438.2	0.6793	23.88	0.940	447.7	0.6939	22.48	0.885	396.8	0.6151	22.35	0.880	392.4	0.6082	500
600	20.65	0.813	26.29	1.035	542.8	0.8413	26.67	1.050	558.6	0.8659	25.02	0.985	491.6	0.7620	24.89	0.980	486.6	0.7542	600
700	22.28	0.877	27.94	1.100	613.1	0.9503	28.19	1.110	624.3	0.9676	26.67	1.050	558.6	0.8659	26.67	1.050	558.6	0.8659	700
750	23.06	0.908	28.83	1.135	652.8	1.0118	29.21	1.150	670.1	1.0386	27.31	1.075	585.5	0.9076	27.69	1.090	602.0	0.9331	750
900	25.37	0.999	31.50	1.240	779.3	1.2076	31.09	1.224	759.1	1.1766	30.33	1.194	722.5	1.1196	29.69	1.169	692.3	1.0733	900
1000	26.92	1.060	32.64	1.285	836.6	1.2968	32.64	1.285	836.6	1.2968	31.88	1.255	798.1	1.2370	31.24	1.230	766.6	1.1882	1000

*Dimensions are from industry sources.

**Types RHH and RHW without outer coverings.

Most aluminum building wire in Types THW, THHW, THWN/THHN, and XHHW conductors is compact stranded. Table 5A provides appropriate dimensions for these types of wire.

APPENDIX E

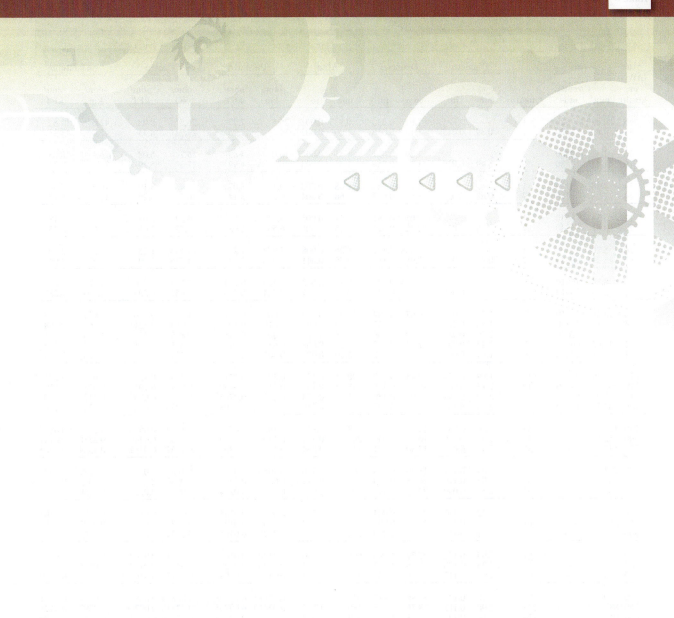

Table E-1

CONDUCTOR PROPERTIES.

Table 8 Conductor Properties

Size (AWG or kcmil)	Area		Conductors							Direct-Current Resistance at 75°C (167°F)					
			Stranding			Overall				Copper				Aluminum	
				Diameter		Diameter		Area		Uncoated		Coated			
	mm²	Circular mils	Quantity	mm	in.	mm	in.	mm²	in²	ohm/km	ohm/kFT	ohm/km	ohm/kFT	ohm/km	ohm/kFT
18	0.823	1620	1	—	—	1.02	0.040	0.823	0.001	25.5	7.77	26.5	8.08	42.0	12.8
18	0.823	1620	7	0.39	0.015	1.16	0.046	1.06	0.002	26.1	7.95	27.7	8.45	42.8	13.1
16	1.31	2580	1	—	—	1.29	0.051	1.31	0.002	16.0	4.89	16.7	5.08	26.4	8.05
16	1.31	2580	7	0.49	0.019	1.46	0.058	1.68	0.003	16.4	4.99	17.3	5.29	26.9	8.21
14	2.08	4110	1	—	—	1.63	0.064	2.08	0.003	10.1	3.07	10.4	3.19	16.6	5.06
14	2.08	4110	7	0.62	0.024	1.85	0.073	2.68	0.004	10.3	3.14	10.7	3.26	16.9	5.17
12	3.31	6530	1	—	—	2.05	0.081	3.31	0.005	6.34	1.93	6.57	2.01	10.45	3.18
12	3.31	6530	7	0.78	0.030	2.32	0.092	4.25	0.006	6.50	1.98	6.73	2.05	10.69	3.25
10	5.261	10380	1	—	—	2.588	0.102	5.26	0.008	3.984	1.21	4.148	1.26	6.561	2.00
10	5.261	10380	7	0.98	0.038	2.95	0.116	6.76	0.011	4.070	1.24	4.226	1.29	6.679	2.04
8	8.367	16510	1	—	—	3.264	0.128	8.37	0.013	2.506	0.764	2.579	0.786	4.125	1.26
8	8.367	16510	7	1.23	0.049	3.71	0.146	10.76	0.017	2.551	0.778	2.653	0.809	4.204	1.28
6	13.30	26240	7	1.56	0.061	4.67	0.184	17.09	0.027	1.608	0.491	1.671	0.510	2.652	0.808
4	21.15	41740	7	1.96	0.077	5.89	0.232	27.19	0.042	1.010	0.308	1.053	0.321	1.666	0.508
3	26.67	52620	7	2.20	0.087	6.60	0.260	34.28	0.053	0.802	0.245	0.833	0.254	1.320	0.403
2	33.62	66360	7	2.47	0.097	7.42	0.292	43.23	0.067	0.634	0.194	0.661	0.201	1.045	0.319
1	42.41	83690	19	1.69	0.066	8.43	0.332	55.80	0.087	0.505	0.154	0.524	0.160	0.829	0.253
1/0	53.49	105600	19	1.89	0.074	9.45	0.372	70.41	0.109	0.399	0.122	0.415	0.127	0.660	0.201
2/0	67.43	133100	19	2.13	0.084	10.62	0.418	88.74	0.137	0.3170	0.0967	0.329	0.101	0.523	0.159
3/0	85.01	167800	19	2.39	0.094	11.94	0.470	111.9	0.173	0.2512	0.0766	0.2610	0.0797	0.413	0.126
4/0	107.2	211600	19	2.68	0.106	13.41	0.528	141.1	0.219	0.1996	0.0608	0.2050	0.0626	0.328	0.100
250	127	—	37	2.09	0.082	14.61	0.575	168	0.260	0.1687	0.0515	0.1753	0.0535	0.2778	0.0847
300	152	—	37	2.29	0.090	16.00	0.630	201	0.312	0.1409	0.0429	0.1463	0.0446	0.2318	0.0707
350	177	—	37	2.47	0.097	17.30	0.681	235	0.364	0.1205	0.0367	0.1252	0.0382	0.1984	0.0605
400	203	—	37	2.64	0.104	18.49	0.728	268	0.416	0.1053	0.0321	0.1084	0.0331	0.1737	0.0529
500	253	—	37	2.95	0.116	20.65	0.813	336	0.519	0.0845	0.0258	0.0869	0.0265	0.1391	0.0424
600	304	—	61	2.52	0.099	22.68	0.893	404	0.626	0.0704	0.0214	0.0732	0.0223	0.1159	0.0353
700	355	—	61	2.72	0.107	24.49	0.964	471	0.730	0.0603	0.0184	0.0622	0.0189	0.0994	0.0303
750	380	—	61	2.82	0.111	25.35	0.998	505	0.782	0.0563	0.0171	0.0579	0.0176	0.0927	0.0282
800	405	—	61	2.91	0.114	26.16	1.030	538	0.834	0.0528	0.0161	0.0544	0.0166	0.0868	0.0265
900	456	—	61	3.09	0.122	27.79	1.094	606	0.940	0.0470	0.0143	0.0481	0.0147	0.0770	0.0235
1000	507	—	61	3.25	0.128	29.26	1.152	673	1.042	0.0423	0.0129	0.0434	0.0132	0.0695	0.0212
1250	633	—	91	2.98	0.117	32.74	1.289	842	1.305	0.0338	0.0103	0.0347	0.0106	0.0554	0.0169
1500	760	—	91	3.26	0.128	35.86	1.412	1011	1.566	0.02814	0.00858	0.02814	0.00883	0.0464	0.0141
1750	887	—	127	2.98	0.117	38.76	1.526	1180	1.829	0.02410	0.00735	0.02410	0.00756	0.0397	0.0121
2000	1013	—	127	3.19	0.126	41.45	1.632	1349	2.092	0.02109	0.00643	0.02109	0.00662	0.0348	0.0106

Notes:

1. These resistance values are valid **only** for the parameters as given. Using conductors having coated strands, different stranding type, and, especially, other temperatures changes the resistance.
2. Equation for temperature change: $R_2 = R_1\,[1 + \alpha\,(T_2 - 75)]$ where $\alpha_{cu} = 0.00323$, $\alpha_{AL} = 0.00330$ at 75°C.
3. Conductors with compact and compressed stranding have about 9 percent and 3 percent, respectively, smaller bare conductor diameters than those shown. See Table 5A for actual compact cable dimensions.
4. The IACS conductivities used: bare copper = 100%, aluminum = 61%.
5. Class B stranding is listed as well as solid for some sizes. Its overall diameter and area is that of its circumscribing circle.

GLOSSARY

1:1 transformer—a transformer that produces the same amount of voltage and current at its secondary as is applied to its primary; also called an isolation transformer.

10X probe—an oscilloscope probe that supplies an input signal that is attenuated by a factor of 10.

555 timer—an integrated circuit that can function as an astable or monostable multivibrator.

Absolute pressure—indicated by psia, it is a measure of the pressure above the absolute base of a perfect vacuum and must be measured with barometric-type instruments.

Accumulator—a device used to maintain pressure within a system when pressure or position is critical by filling with fluid only when the system pressure increases to a level that can overcome the pressure within the accumulator itself.

Acetylene—a colorless gas, that has a distinct and definite odor similar to that of garlic, used in combination with oxygen to create a flame at the tip of a torch.

Across-the-line starting—a starting method in which the motor is connected directly across the power lines, receiving full voltage and current.

Add-Alt-Chop control—oscilloscope vertical section control. The *ADD* position allows the signals from channel 1 and channel 2 to be algebraically added together, depending on the setting of the *CH 2 INVERT* switch. The *ALT* or *alternate* position allows the scope to show the channel 1 signal, then the channel 2 signal, then the channel 1 signal, etc. The *ALT* position allows both input signals to be displayed at the same time when viewing higher frequencies. The *CHOP* position is used when the scope is set to display lower frequencies.

Additives—substances that are used to improve the properties of lubrications.

Aluminum conductors—conductors drawn from an aluminum rod.

Ampere (amp)—the standard unit of measurement of electrical current. One ampere is equal to 6.25×10^{18} electrons (6,250,000,000,000,000,000) traveling past a given point in one second of time.

Analog input module—a module used in a PLC that converts an analog input signal from a field device into a digital signal.

Analog output module—A module used by a PLC which provides an analog (0–10 V or 4–20 mA) output signal.

Angle of loading—The inclination of a leg or branch of a sling measured from the horizontal or vertical plane.

AND gate—a digital logic device that produces an output only when all of its inputs are high.

Angle theta ($\angle\theta$)—used to represent the voltage-current phase shift angle.

Angular misalignment (pulley and sprocket)—a type of misalignment that occurs when two shafts are parallel to each other but are at a different angle with the horizontal plane.

Annunciator—a signaling apparatus. Annunciators may be audible or visual. Examples of an audible annunciator are a bell, klaxon, chime, horn, loudspeaker, siren, etc. Examples of a visible annunciator are an indicator light, strobe light, rotating beacon, etc.

Anode—in an electronic device, the name given to the terminal with the positive polarity.

Antifriction bearings—bearings that have some sort of rolling mechanism built within the bearing.

Apparent power—the combination of true power and reactive power.

Arc welding—using electric current flow to join metals together.

Area—The two-dimensional surface included within a set of lines, space, shape, or boundary.

Armature—the rotating portion of a motor or generator. Armatures are typically constructed with windings and a commutator.

Armature voltage feedback—another name for EMF feedback.

Astable—a mode of operation in which an oscillator circuit is capable of turning itself on and off.

Atom—the smallest part of a molecule. An atom is an element that, when combined with other elements, forms a molecule.

Automatic control—a type of machine control that does not require human intervention. An example is a thermostat that automatically turns a furnace on or off, depending on the temperature without the need for someone to operate the furnance.

Automatic three-point starting—a starting method in which two resistors are connected in series with each power line. (This could be the three power lines to the rotor of a wound rotor induction motor or the three power lines to a three phase stator.) Upon starting, all six resistors are connected to provide reduced voltage starting. This is the first point. After a delay, one set of resistors is shorted in each power line. This is the second point and causes the motor to receive a higher voltage. After a further delay, the remaining set of resistors is shorted. This is the third point. The motor is now receiving full voltage.

Autotransformer starter—a reduced voltage starting method in which an autotransformer is used to provide the reduction in voltage.

Autotransformers (also called autoformers)—a transformer constructed with only one winding.

Auxiliary contacts—additional contacts found on pushbuttons, rotary switches, motor starters, etc. These contacts are not the main (high-current) contacts but are used for additional, low-current switching applications.

Axial load—this occurs when the pressure from the load is parallel to the axis of the shaft.

Babbitt bearings—a bearing that is constructed from different types of Babbitt metals.

Background suppression sensor—a type of photoelectric switch in which an object can be sensed while the background is ignored or suppressed.

Backlash—the amount by which the width of the tooth space exceeds the thickness of the mating gear tooth.

Ball bearing—an antifriction bearing that contains spherical rolling elements.

Ballast—a transformer or solid-state circuit that provides the high voltage and current limiting for starting a fluorescent lamp.

Bar sag—exists when a bar or rod is supported at only one end. Gravity causes the unsupported end to droop lower than the supported end. This condition becomes greater when the length of the bar or rod is increased.

Base—In a transistor, the region between the collector and the emitter that controls the amount of collector-emitter current.

Basket hitch—A sling configuration whereby the sling is passed under the load and has both ends, end attachments, eyes, or handles, on the hook or a single master link.

Bead—made by moving a puddle of molten metal in a forward direction with the electrode as the arc is present.

BEAM FIND control—used to quickly locate the trace on an oscilloscope.

Bell—the end of a cast iron pipe in which the spigot will be inserted to join the pipes together.

Belt deflection—this can be found by pressing downward on the V-belt at the center of the span. The v-belt should not travel more than approximately 1/64 in. per inch of span between the two shaft centers.

Belt speed—this can be calculated by multiplying pi (π) by the pulley diameter in inches and the rpm of the pulley shaft divided by 12. The answer is expressed in ft/min.

Bevel gears—very similar to spur gears, the bevel gear has a conical shape instead of a cylindrical shape. These gears are used when the axes of the shafts on which they are mounted intersect. They are usually used in gearboxes where the input shaft and the output shaft are at a 90° angle to each other.

Bipolar junction transistor (BJT)—a transistor that consists of two p-n junctions; one is the emitter-base junction and the other is the collector-base junction.

Boost chopper (also called step-up chopper)—a chopper circuit in which the output voltage is higher than the input voltage.

Boundary lubrication—the condition of lubrication in which the friction between two surfaces, which are in motion, is determined by the properties of the surfaces and the properties of the lubrication that is used, excluding viscosity.

Box block—a block or box in which a stave bearing liner is placed.

Break-before-make contacts—a type of contact in which the normally closed circuit is opened before the normally opened circuit is closed.

Brush assembly—the mechanism on a motor or generator that is used to hold the brushes and supply the proper brush tension between the brush and slip rings or commutators.

Buck chopper (also called a step-down chopper)—a chopper circuit in which the output voltage is lower than the input voltage.

Buffer amplifier—an amplifier used to provide isolation from one circuit to another.

Butt joint—the type of joint that is formed when two pieces of metal are to be welded together at their edges as if they were lying on a table side by side.

Cages—A device that holds the rolling elements in place in a taper-roller bearing and evenly spaces them around the bearing. It is sometimes referred to as a separator.

Capacitance—a measure in farads of a capacitor's ability to store an electrical charge.

Capacitive proximity switch—a type of proximity switch in which an electrostatic field is used to sense the presence or absence of a metallic or nonmetallic object.

Capacitive reactance—a capacitor's opposition to alternating current flow.

Capacitor—two conductors, in close proximity, separated by some type of insulating material.

Capacitor-run motor—a single-phase motor that uses a capacitor continuously connected so as to reduce running noise and vibration.

Capacitor-start motor—a single-phase motor that uses a capacitor during startup to increase the starting torque. The start capacitor is switched out of the circuit at approximately 75% of normal running speed by a centrifugal switch.

Capacitor-start/capacitor-run motor—a single-phase motor that uses a capacitor during startup to increase the starting torque and a separate capacitor during normal running conditions to minimize noise and vibration. The start capacitor is switched out of the circuit by a centrifugal switch.

Carburizing flame—Sometimes referred to as a reducing flame, it has three distinct flame zones: a small, bright white cone at the tip, which is very similar to that of the neutral flame (the inner cone); a larger cone (the intermediate cone) that surrounds the inner cone; and a longer, translucent blue cone, which surrounds the intermediate cone. If this flame is set up correctly, the intermediate cone will be twice the length of the inner cone. This type of flame is set up with the same amount of acetylene as the neutral flame but with less oxygen. The temperature at the tip of the inner cone on this type of flame is approximately 5700°F. This type of flame is used when brazing or soldering low carbon steel, cast steel, nickel steel, chrome steel, copper, and brass.

Cathode—in an electronic device, the name given to the terminal with the negative polarity.

Cathode ray tube (CRT)—the section of an oscilloscope on which measurements are made. The CRT is part of the display section.

Cavitation—this occurs when air becomes entrained into the fluid as it is pumped and can be defined as a vacuum void within the fluid.

Center-tapped secondary—the secondary winding of a transformer with a third connection to the electrical center.

Center-to-center (C to C) measurement—a measurement that is taken from the center of one fitting, down the length of the pipe, to the center of the other fitting.

Central processing unit (CPU)—the part of a computer or PLC that performs all of the arithmetic and logic functions.

Centrifugal force—a force that tends to pull an object outward, away from the center of rotation, when the object is rotating rapidly around a center point.

Centrifugal pumps—typically used in lower pressure systems that move thinner fluids such as water. Centrifugal pumps rely on the centrifugal force that is developed within the fluid as it is being pumped through the propeller.

Centrifugal switch—a shaft-mounted switch which uses flyweights and the centrifugal forces generated by the rotating shaft to cause the switch contacts to open at a predetermined speed.

Centripetal force—An inward force that tends to draw an object toward the center of rotation.

CH 1 - BOTH - CH2 control—a control that allows the technician to determine which trace is displayed on the CRT of an oscilloscope.

CH 2 invert control—when this control is depressed, the signal displayed on channel 2 will be inverted.

Chain drives—a positive-drive system (meaning that they do not slip), which is a reliable form of power transmission.

Chain splitter—a tool that pushes against the pin of a pin link while holding the sidebar, enabling the sidebar to be removed, thus allowing the chain to be broken at that point.

Charge-coupled device (CCD)—A semiconductor device that receives light from an object as an input and then converts it into an electrical signal as an output.

Chemisorption—this occurs when weaker chemical bonds are formed between liquid or gas molecules and a solid surface providing a secondary boundary that further protects the metal in the event that the first boundary (film) is lost.

Choker hitch—A sling configuration where one end of the sling is passed under the load and through the end attachment, handle, or eye on the other end of the sling.

Chopper—a circuit used to break up or chop a steady DC into a series of pulses.

Circuit—an electrical path between two or more points. A circuit must contain a power source, conductors, and a load.

Circular pitch—the distance that is measured along the pitch circle from one tooth to the exact same point on the next tooth.

Clamp-on ammeter—allows the measurement of current without the need to de-energize or break apart the circuit.

Closed-loop control (also called automatic control)—a control system in which a signal indicating motor performance is fed back to the control section of the drive so that the drive can take corrective action automatically.

Collector—the portion of a transistor that must have the same polarity as the base in order for the transistor to conduct. On a schematic of a transistor, the collector is the lead identified by the lack of an arrow and is not T shaped.

Combination circuit—a circuit that has components connected in both series and parallel.

Command signal (also called set point or reference signal)—the voltage level set or programmed into a drive to ensure that the desired motor operating speed is maintained.

Common—an electrical connection shared by two or more conductors of the circuit.

Command signal—The signal that represents the desired value of the process being controlled. Also called the reference signal.

Commutator—a segmented ring of metal that surrounds the motor's shaft. Electrical connections are made from the armature windings to the commutator segments. Brushes ride on the commutator, thus providing an electrical connection between the brush and the rotating armature through the commutator.

Comparator (also called level detector)—a circuit that compares a reference signal to a detected signal and produces an output that is based on the results of the comparison.

Compensating the probe— when 10X probes must be matched to the oscilloscope input being used.

Complement—in a digital device, the reverse or opposite state. The complement of off is on, the complement of true is false, the complement of low is high, and the complement of 1 is 0 and vice versa.

Compound-wound DC generator—a DC generator in which both the series and shunt field windings are used.

Compound-wound DC motor—a DC motor in which both the series and shunt field windings are used.

Compression—this occurs when air is forced into a smaller space than it originally occupied.

Compression packing—the packing material used in packing seals.

Condensation—this occurs whenever air is heated through compression and then cools rapidly. Condensation introduces water and water vapor into a pneumatic system, which may be undesirable.

Cone—the inner ring of a taper-roller bearing, which is angled in reference to the bore axis line.

Confined space—defined by OSHA as a space that is large enough and so configured that an employee can bodily enter and perform assigned work, has limited or restricted means for entry or exit (for example, tanks, vessels, silos, storage bins, hoppers, vaults, and pits are spaces that may have limited means of entry), and is not designed for continuous employee occupancy.

Contactor—an electromagnetic switch or relay with heavy-duty contacts.

Control—a device used to switch the current flow on or off, such as a switch, a thermostat, or a relay.

Control section—the section of the drive responsible for evaluating motor performance against the desired performance and making corrections as necessary to achieve the desired results.

Converter—a device that changes AC power into DC power.

Copper-clad aluminum conductors—according to *NEC® Article 100*: "Conductors drawn from a copper-clad aluminum rod with the copper metallurgically bonded to an aluminum core. The copper forms a minimum of 10 percent of the cross-sectional area of a solid conductor or each strand of a stranded conductor."

Core—the material around which transformer windings are wound. Core material may be iron, plastic, air, and so on.

Corner joint—a joint that is placed together at the corners of both pieces of metal, thus forming a right angle.

Corrosion—deterioration of material caused by the presence of water and oxygen. This is sometimes referred to as oxidation.

Counter electromotive force (CEMF)—the induced voltage caused by the generator action of a revolving motor: Counter electromotive force opposes the voltage applied to the motor and thereby limits the armature current.

Crest—the highest point of the thread.

Cumulative wound—a method of wiring a compound-wound DC generator in which the series and shunt fields are connected so that their magnetic fields aid each other. Cumulative-wound DC generators have excellent voltage regulation.

Cup—the outer ring of a taper-roller bearing, which is angled in reference to the bore axis line.

Current—the movement of electrons.

Current source inverter (CSI) (also called current-fed inverter)—an inverter in which the DC bus current is controlled while the DC bus voltage varies to meet the demands of the motor.

Cut—the grooves cut into a file that produce a series of very sharp edges across its surface.

Cylinder valve—a valve located at the top of the cylinder that controls the flow of acetylene or oxygen leaving the cylinder.

Cylindrical roller bearing—a bearing that has rolling elements in the shape of a cylinder, which provide more surface contact, thus allowing the bearing load capacity to increase in comparison to the ball bearings.

Darlington—an arrangement of two transistors so that the collector of one transistor drives the base of the second transistor. This results in extremely high input impedance and current gain.

DC filter (also called DC link or DC bus)—provides smooth, rectified DC to the inverter stage.

DC generator—an electrical machine that converts mechanical energy into direct current.

Delivery—the rate at which the hydraulic fluid is supplied to a system from the pump in a specified amount of time. The delivery is expressed in gallons per minute (gpm).

Delivery rate—the rate at which a compressor transfers air from the inlet port to the outlet port. It is usually expressed in standard cubic feet per minute (scfm) or standard cubic meters per minute (scmm).

Delta connection—an electrical connection used in three-phase transformers. Delta-connected transformers have their phase voltage equal to their line voltage, but the line current will be 1.732 times greater than the phase current.

Delta-delta—the term used to describe a transformer whose primary and secondary windings are connected in a delta connection.

Delta-wye configuration (also called delta-star configuration)—the term used to describe a transformer whose primary is connected in a delta configuration, whereas its secondary is connected in a wye or star configuration.

Demulsifiers—additives that are used to remove water from within the lubrication.

Depletion-enhancement MOSFET (DE-MOSFET)—a mode of operation of a MOSFET in which a negative voltage is applied to the device.

Diac (also called bilateral trigger diode)—a solid state device that will conduct current in one of two directions when properly biased.

Dial caliper—a device with a dial and a set of calipers that is used to measure internal values, external values, and depth values.

Dial indicator—a device with a dial and a plunger that is used to indicate or measure very small amounts of variances that may be as small as a thousandth of an inch (0.001).

Diametral pitch—The distance along the pitch diameter in which gear teeth will be counted to give the tooth count-to-pitch diameter ratio.

Dielectric—insulating material found in a capacitor.

Difference amplifier—an amplifier with an output level that is proportional to the difference between the two input levels.

Difference of potential—a term used to describe voltage.

Differential measurements—the difference between two input signals on an oscilloscope.

Differential wound—a method of wiring a compound-wound DC generator in which the series and shunt fields are connected so that their magnetic fields oppose each other. Differential-wound DC generators have poorer voltage regulation but practically constant output current. Differential-compound-wound generators are most commonly used to power arc lights and in welding applications.

Differentiator—an electrical circuit whose output is the mathematical derivative of the input signal.

Diffuse—sensors that use the refl ectivity of the object being sensed to return the light from the transmitter to the receiver.

Digital device—an electronic device that has two operating modes: off and on.

Digital input module—a module used in a PLC that accepts a digital input signal from a field device.

Digital logic gates—electronic circuits that perform logical decisions based on certain input conditions.

Digital multimeter (DMM)—the most common multimeter in use today, it displays the measured value in a digital readout and is capable of measuring voltage, current, and resistance.

Digital output module—a module used in a PLC that provides a digital output signal to the field device.

Diode—an electronic device that allows current flow in one direction and blocks current flow in the other direction.

Direct coupled—the DC position of the input coupling switch on an oscilloscope represents, which allows the scope to display all components of the input signal.

Direct measurement (or 1X) probe—an oscilloscope probe that supplies an input signal of the same amplitude as the signal being measured.

Displacement—the amount of fluid that can be discharged from a pump in a single rotation or in one revolution of the pump.

Double seal—two inside seals that are in a back-to-back orientation.

Double throw—a switch that has two sets of contacts that can be operated. The switch can be positioned to operate one set of contacts (1 throw) or the other set of contacts (1 throw), hence double throw.

Down-counters—a PLC instruction that begins at a preset value and counts down incrementally whenever the input contact transitions from closed to open.

Drainpipe—the lowest pipe in a waste system, which will handle all of the wastewater and sewage in a structure.

Drip—a vertical extension at the lowest point of a piping system, using fittings, a nipple, and a cap, which is installed in a manner that will catch any moisture that may exist in the piping system due to condensation.

Drum switch—a manually operated switch that operates with a rotary motion and is used to reverse the direction of rotation of a motor.

Dual-in-line pin (DIP)—the style of packaging used in electronic devices where the connection terminals are arranged in two rows.

Dynamic braking—a braking method in which large resistors are used to absorb the energy produced by the generator action of a motor when the energizing force has been removed. The CEMF produced by the motor provides the counter torque, which, in turn, provides the braking action.

Eddy currents—undesirable currents produced in the iron core of a transformer as a result of the magnetic lines of force.

Edge joint—a joint that is usually reserved for use on metals of ¼ in. thickness or less, created by placing the two pieces side by side and welding along one or more edges.

Electrical energy—the product of power and time.

Electrical isolation—when two or more circuits are connected so as not to allow an electrical path to exist between circuits. A transformer with a primary and secondary winding has electrical isolation between the primary and secondary because an electrical path does not exist between the windings.

Electrical lockout—turning off the main disconnect that supplies power to the machine, and locking the handle of the disconnect in the off position so that it cannot be turned back on until all lockouts are removed.

Electrical pitting—pitting that occurs within a bearing when current flows through the bearing.

Electrodes—in a high-intensity discharge lamp, the elements between which the arc occurs. The electrodes are located within the arc tube.

Electromechanical switch—a type of switch that uses an electromagnetic field to cause the mechanical mechanism to operate.

Electromotive force—another term for voltage. Because voltage is the force that causes electrons to move, it is sometimes referred to as electromotive force.

Electronic variable-speed drives—method used to vary the speed or torque of DC and AC motors by the use of electronic devices.

Electrons—orbit the nucleus of an atom and have a negative charge.

Electrostatic field—the energy field produced by the oscillator of a capacitive proximity switch used to sense the presence or absence of a metallic or nonmetallic object.

Element—composed of atoms and forms the basis for all matter.

Elevated view—a view of an orthographic drawing from the side showing all of the vertical runs.

Emitter—the portion of a transistor that must have a different polarity from the base in order for the transistor to conduct. On a schematic of a transistor, the emitter is the lead identified by the arrow.

Encoder—an optical or magnetic device attached to the shaft of a motor to sense the motor shaft speed or position by generating voltage pulses.

End-to-center (E to C) measurement—a measurement that is taken from the end of a piece of pipe to the center of a fitting.

End-to-end (E to E) measurement—a measurement that is taken from one end to the other end of a single piece of pipe after it is cut and threaded. This measurement must be taken without any fittings on the pipe.

Engineering notation—a variation of scientific notation. Engineering notation works in a similar manner to scientific notation, except that engineering notation moves in steps of one thousand, not ten.

Enhancement-only MOSFET (E-MOSFET)—a mode of operation of a MOSFET in which a positive voltage is applied to the device.

Equipment information—one of five essential records that should be maintained for an effective maintenance program. This record contains information about the particular piece of equipment. Information such as make, model, serial number, manufacturer, contact information for the manufacturer, installation date, location, specifications (operating voltage, operating current, etc.), and so on, would be found on this record.

Error signal—the difference between the command or reference signal and the feedback signal.

Excitation current—the direct current that is applied to the field windings of a DC motor or generator or to the rotor of an alternator or synchronous motor. Excitation current is used to produce the magnetic fields needed for motor or generator operation.

Exclusive OR gate (also called XOR gate)—a digital logic device that produces an output only when its inputs are complements.

Face—the flat side of the thread that rises at an angle from the root to the crest.

Face alignment—when the readings are taken, using a dial indicator, from the face of the flange.

False Brinell—damage that occurs when continual impacting forces (such as vibrations) are passed from one ring to the other, through the rolling elements, when there is no rotation of the shaft.

Farads (F)—units of measurement that measure the amount of capacitance of a capacitor.

Feedback signal—a signal indicating motor performance.

Feeler gauge—thin strips of metal that have various precision thicknesses, which are used along with a straightedge for correcting any misalignment. Sometimes feeler gauges are referred to as thickness gauges.

Field windings—in a motor or generator, the windings used to produce the magnetic field required for operation.

Filament—in an incandescent lamp, the small wire-like component that produces the light when a current flows through it.

Filler rod—A thin rod made of the same material as the base metal to be welded which is used to build up a bead while welding.

Fillet weld—a type of weld that joins two surfaces that are at right angles to each other.

Film lubrication—this is accomplished when a lubricant creates a film that is thick enough to completely separate the two surfaces of metal.

Fittings—parts of a piping system used to provide an additional path from a main path, join pipes together, or change the direction in which the fluid is flowing.

Fixed-focus diffuse sensor—a sensor that uses a focusing lens to create a focal point. Light is transmitted from the transmitter, reflects off the object to be detected, and is sensed by the receiver. Should the object be too close or too far away, the reflected light will be out of focus and the object will not be sensed by the sensor.

Flashing the field—the procedure used to restore the residual magnetism in a self-excited DC motor or generator.

Flat belts—a belt that is flat in shape used to transmit power or convey material. Flat belts are usually made of materials that offer the most friction and flexibility, such as leather, canvas, or rubber.

Flat welding position—the simplest and most comfortable of the four welding positions; the material is parallel to the floor as if it were lying on a table or workbench.

Flats—the flat shoulders on a tap that are used to hold and drive the tap with a tap handle.

Flip-flop circuit—a circuit that uses a double-acting cylinder with a double-end rod to contact a limit switch on each side of the cylinder, causing the cylinder to oscillate back and forth automatically.

FLR—see *trio assembly.*

Fluid power—The act of using a fluid to transmit power from one location to another. There are two methods of transmitting power through a fluid: hydraulic and pneumatic.

Fluorescent lamp—a lamp that produces light by allowing an electrical current to flow through a gas.

Flutes—channels that are carved out along the axis of the tap, which remove the metal chips during the cutting of the threads, permitting the cutting fluid or coolant to reach the cutting edges.

Fluting—thin lines that are etched into the races when current flows through a bearing while it is rotating.

Flux—a paste that cleans the copper or brass pipe to be soldered and prevents it from oxidizing when the heat is applied.

FOCUS control—a control that is used to adjust the sharpness or crispness of the trace on the display.

Foot-pounds—the work performed when a force of one pound acts through a distance of one foot.

Forward biased—a condition in which the proper voltage levels and polarities are applied to a semiconductor, causing it to conduct.

Freewheeling diode—a diode installed in a circuit in such a fashion as to provide a path for current flow from a collapsing magnetic field caused by the removal of a DC current.

Full-wave bridge rectifier—a circuit that converts a sine wave AC into a pulsating DC by using four diodes and a transformer. An output is produced during the full 360° of the AC sine wave input.

Full-wave rectifier—a circuit that converts a sine wave AC into pulsating DC by using two diodes and a transformer with a center-tapped secondary. An output is produced during the full 360° of the sine wave input.

Fusible plug—located in the bottom of the cylinder, a plug that acts like a safety valve to release any excess pressure as a result of too much heat or too much pressure within the cylinder.

Galling—the bonding, shearing, and tearing away of material from two contacting or sliding metals.

Gate (also known as trigger)—(1) The lead of a field-effect transistor, SCR, or triac that allows control of the conduction of the device; (2) An electrical circuit that performs a logic function.

Gauge—used to show the pressure that is present in a fluid power system.

Gear—a toothed wheel or disc that is used to transmit mechanical power from one shaft to another through the meshing action of teeth. When used in drive systems, they provide positive-drive reaction and the ability to change direction, speed, or torque.

Gear ratio—the comparison between the pitch diameter of the drive gear versus the pitch diameter of the driven gear.

Gearbox—a housing that contains gears within it, usually having one input shaft and one output shaft, which is used to change the speed, direction, or torque of a mechanical system.

Grade markings—identification markings present on the head of the fastener that indicate a fastener's strength.

Graticule—the vertical and horizontal lines that form a grid pattern on the face of the CRT of an oscilloscope.

Groove weld—a type of weld that fills a groove between two pieces of metal that are to be joined. The ends of the pieces that are to be welded may be square, beveled, U-shaped, or J-shaped.

Group relamping—a maintenance practice that involves the replacing of all lamps within an area when the lamps have reached 60% to 80% of their rated life expectancy.

Half bearing—a plain bearing that is only used to handle radial loads.

Half-wave rectifier—a circuit that converts a sine wave AC into pulsating DC by using one diode. An output is produced during either the positive 180° or negative 180° of the sine wave input.

Hand-off-automatic (HOA) control—a control that allows a circuit to function under the control of either an automatic device or manual control. A selector switch with three positions—hand, off, and automatic—is generally used. With the selector switch in the off position, the circuit is de-energized. With the selector switch in the hand position, the circuit is energized and operating. With the selector switch in the automatic position, the circuit is energized but not operating until an automatic device activates the circuit.

Heating elements—elements that are found in fluorescent lamps and are used to lower the resistance of the gas within the lamp when power is applied.

Heavy duty—when referring to contacts, a rating that indicates the contacts are capable of switching higher amperage loads. Heavy-duty contacts are rated for heavier loads than standard duty or pilot duty.

Helical gears—a type of gear that is similar to the spur gear, except that its teeth are not parallel to the axis shaft but are slanted, causing the teeth to slide across the teeth of the mating gear. This gear produces a heavy side thrust because of the slanted teeth.

Henrys—unit of measurement used to measure inductance.

Herringbone gears—a gear that is constructed with two sets of teeth that are side by side, in a "V" configuration, thus eliminating the end thrust problem that existed with the helical gear.

High-intensity discharge (HID) lamp—lamps that consist of a glass envelope, a base, and an arc tube. The arc tube is filled with a gas. Inside the arc tube is a set of electrodes. When the HID lamp is energized, a ballast provides high voltage to the electrodes within the arc tube. An electrical arc is created between the electrodes, and a current passes through the gas filling the arc tube. The arc is the source of light produced by the HID lamp.

High-pressure sodium lamp—a type of HID lamp that uses sodium gas under high pressure.

High-speed counter input modules—a PLC module that accepts an input signal from a high-speed counter.

Horizontal alignment—a type of misalignment that exists in coupled shaft alignment and refers to any alignment, angular or offset, that can be found when looking downward over the top of a machine.

Horizontal Magnification—A control that expands or magnifies the displayed waveform in the horizontal direction.

Horizontal position control—a control that allows the technician to move the trace displayed on the CRT of an oscilloscope to the left or right.

Horizontal welding position—a welding position in which the metal to be welded is at a right angle to the floor and the bead is drawn across the metal from right to left or left to right.

Horsepower (hp)—a unit of power in the English system of units. One horsepower (hp) is equal to 33,000 foot-pounds (ft-lb.) per minute or 550 foot-pounds (ft-lb.) per second.

Hydraulic—a form of fluid power that uses liquid to transmit power from one location to another.

Hydrocarbon—any compound containing mostly hydrogen and carbon. A pure hydrocarbon contains only hydrogen and carbon.

IEC (International Electrotechnical commission)—is a not-for-profit, non-governmental organization, founded in 1906. The IEC is the leading global organization that publishes consensus-based International Standards and manages conformity assessment systems for electric and electronic products, systems and services, collectively known as electrotechnology.

Impedance—the combination of resistive and reactive opposition.

Incandescent lamp—a lamp that produces light by allowing a current to flow through a filament.

Inductance—in an inductor, the measure of the amount of induced voltage caused by the changing current.

Inductive proximity switches—a type of proximity switch that uses an electromagnetic field to sense the presence or absence of a ferrous object.

Inductive reactance—an inductor's opposition to alternating current flow.

Inductive shielded—an inductive proximity switch that is either cylindrical or shaped like a limit switch. It is designed to be flush mounted. It can detect ferrous and nonferrous metal. Its maximum sensing distance is 0.4 in.

Inductive unshielded—an inductive proximity switch that may be cylindrical or shaped like a limit switch, a small block, or a flat rectangle. It must have clearance around the target end; therefore, it cannot be flush mounted. As a result, the body of the switch will protrude and is subject to physical damage. The maximum sensing distance is 0.7 in. The unshielded version can detect ferrous and nonferrous metal.

Infrared LEDs (IRLEDs)—a light-emitting diode that emits light in the infrared region of the electromagnetic spectrum.

Infrared light—invisible light waves, that are felt as heat and which is found opposite ultra-violet light in the infrared frequency of the light spectrum.

Initial tension—the force that is developed within the fastener when it is tightened.

Inner race—a shallow groove that is cut into the inner ring in which the rolling elements will ride.

Inner rings—made of hardened steel, the part of the bearing that will mount to a shaft.

Input coupling control—a switch that has three positions: AC, GND, and DC. The DC position represents direct coupled, which allows the scope to display all components of the input signal. The AC position inserts a capacitor into the input circuit, blocking any DC component from the signal being measured. The scope will display only the AC component of the signal. The GND position disconnects the input signal from the scope and connects the scope input to the chassis ground of the scope. This prevents unwanted signals from being displayed while the trace is adjusted for a reference on the CRT. There are two input coupling controls on a dual-trace oscilloscope, one for each channel.

Input jacks—points on an oscilloscope where the scope probes are connected.

Input/output (I/O)—input (I) is the term used to identify devices (switches, sensors, modules, etc.) that provide a signal to a PLC. Output (O) is the term used to identify the device (relay, indicator light, module, etc.) that receives a signal from the PLC.

Input sensitivity control—a control that is typically labeled CH 1 VOLTS/DIV or CH 2 VOLTS/DIV that functions similarly to a range switch. It is used to adjust the displayed waveform for maximum vertical size while fitting within the graticule of the CRT.

Inside seal—a seal in which the rotating mating ring is internally mounted in the stuffing box.

Inspection checklist—one of five essential records that should be maintained for an effective maintenance program. This record should list all items for a particular piece of equipment that should be inspected on a scheduled basis. This record should list the date, time, and item to be inspected; the items needed to perform the inspection; the results of the inspection; the recommended action to be taken, and the name of the inspector.

Instruction List (IL)—a low-level PLC program best suited for small applications and fast execution.

Insulated gate bipolar transistor (IGBT)—a type of bipolar transistor with very high speed switching characteristics.

Insulators—elements that have seven or eight valence electrons and are considered poor conductors of electricity. This is because the valence electrons are very tightly held and are not so readily freed. Examples of insulators are rubber, plastic, glass, and fiberglass.

INTENSITY—An oscilloscope control which allows the brightness of the electron beam to be varied.

Integrated circuit (ICs)—an electronic circuit contained on a single semiconductor substrate.

Integrator—an electrical circuit whose output is the mathematical integral of the input signal.

Inverter—another name for an AC drive.

Inverter (also called NOT gate)—a digital logic device that produces an output that is the complement of the input.

Inverting amplifier—an electronic circuit whose output signal is 180° out of phase with its input signal.

Isolated inputs—on an oscilloscope, where the ground for channel 1 is electrically isolated from the ground for channel 2.

Isometric drawing—a three-dimensional type of drawing that has both the horizontal and the vertical runs of the piping system included on one drawing.

Jogging—a method of motor control in which slight movement of the motor is accomplished; also called bumping or inching.

Journal—part of a shaft that is set on a plain bearing and becomes a component of the bearing.

Junction field effect transistor (JFET)—a semiconductor device in which a channel is created for current flow. Voltage applied to the gate lead controls the current flow through the channel from the source to gate leads.

Kilo—used in engineering notation as an abbreviation for ×1000. The letter *k* is used to represent the word kilo.

Ladder chains—a chain that consists of a series of links made from precision bent wire with a loop on each end, which connects the links together.

Ladder diagram—shows the actual electrical connections, but not the actual physical location of components; a variation of the schematic diagram.

Laminated iron core—a construction method used on transformer iron cores. Thin sheets of iron are glued together (called laminating) to form the iron core. Laminating the iron core reduces the eddy currents.

Lands—the cutting components of the tap and the die.

Lantern ring—A device that provides spacing between the third and fourth packing rings to allow the packing material to be lubricated. This spacing can be filled with lubricant through a lubrication port, which should be in line with the lantern ring.

Lap joint—a joint that is made by lapping one piece of metal over another and welding at the end of the top piece of metal.

Laser shaft alignment—a method of alignment that uses a laser transmitter and receiver and is the most accurate and easiest method to accomplish coupled shaft alignment.

Latching—a name given to a circuit when wired in such a fashion as to provide a seal-in, memory, or holding circuit when a momentary device such as a pushbutton is released.

Law of charges—a law that states that opposite charges attract and like charges repel.

Leading electrode angle—this causes the molten metal and slag to be pushed ahead of the weld.

Light-emitting diodes (LEDs)—a solid-state device that functions similarly to a diode in that it only conducts current when forward biased. An LED is different from a diode in that when it conducts, light is emitted. The light may be visible or infrared.

Limit switches—a family of automatic switching devices that are used to sense the presence or absence of an object or the level of a liquid and more.

Line value—the value of voltage or current measured from line to line (L_1 to L_2, L_2 to L_3, and L_1 to L_3).

Line voltage—the value of voltage measured from line to line.

Linear actuators—a device, which when activated through the use of fluid power has a linear movement. An example would be a cylinder.

Load—any device that draws current from the power source, such as a resistor, a lightbulb, or a motor.

Logic functions—functions that refer to the principles, applications, and relationships of electrical signals in switching networks.

Long-range diffuse sensor—a sensor capable of detecting objects over a longer distance than the standard diffuse sensor. The long-range diffuse sensor works best when there is no background present, or the background is a considerable distance away from the object being detected.

Loop system—a method of wiring the distribution system in which the distribution lines form a complete loop, starting and ending at the service. With a loop system, the loads are supplied from both ends, and sections can be isolated in the event of trouble.

Low-pressure sodium lamp—a type of HID lamp that uses sodium gas under low pressure.

Lubricant—a substance that reduces friction by providing a smooth surface of film over parts that move against each other.

Magnetic field—the magnetic lines of force created when a current flows through a conductor.

Maiming—when some part of the body becomes mutilated, severed, disfigured, or seriously wounded.

Main terminal 1 (MT 1)—one of three leads found on a triac.

Main terminal 2 (MT 2)—one of three leads found on a triac. MT 2 must be connected to the same polarity as the gate terminal in order for the triac to conduct.

Maintenance log—a log that contains entries of machine breakdowns and repairs over a given period of time.

Major diameter—with regard to the external thread, this term refers to the outside diameter of the screw or bolt at the crests of the thread. When this term is applied to an internal thread, it refers to the maximum diameter that has been cut into the center of the nut, specifically, the root of the thread.

Make-before-break contacts—a type of contact in which the normally opened circuit is closed before the normally closed circuit is opened.

Manual—a type of machine control that requires human intervention; for example a light switch used to turn a lamp on or off.

Mating rings—machined faces that are used in the sealing action of the mechanical seal are primarily used for rotary sealing. Each mating ring, one rotating and one stationary, has a face that is machined to extremely high tolerances. The rotating mating ring is made to be static with respect to the shaft, whereas the stationary mating ring is made to be static to the housing.

Matter—any material or object that occupies space and has mass. Some examples of matter are plastic, metal, water, gasoline, air, or hydrogen. Matter can be found as a solid (the plastic and the metal), a liquid (the water and the gasoline), or a gas (the air and the hydrogen).

Mechanically held—a switch that uses a mechanical linkage to maintain the switch in the last operated position.

Mega—used in engineering notation as an abbreviation for ×1000000. The letter *M* is used to represent the word mega.

Megohmmeter—more commonly referred to as a *Megger®*, it allows the measurement of very high resistance values that is useful in determining the quality of insulation of wires, cables, transformers, motors, and generators. *Megger®* is a registered trademark of AVO International Limited.

Mercury-vapor lamp—a type of HID lamp that uses mercury-vapor gas.

Metal fatigue—this occurs over time when metal is continually placed under a load. This is caused by the constant flexing and stressing of the metal.

Metal-halide lamp—a type of HID lamp.

Metal-oxide semiconductor field effect transistor (MOSFET)—a field effect transistor (FET) with an insulated gate.

Micro—used in engineering notation as an abbreviation for ×0.000001. The Greek letter μ is used to represent the word micro.

Microfarads (μF)—units of measurement that capacitors typically have their values stated in because the farad is very large.

Milli—used in engineering notation as an abbreviation for ×0.001. The letter *m* is used to represent the word milli.

Minor diameter—with regard to the external thread, this term refers to the smallest diameter of the screw or bolt at the root of the thread. When this term is applied to an internal thread, it refers to the diameter of the hole in the nut at the crests of the thread.

Miter gears—similar to bevel gears, except that all miter gear sets have a 1:1 ratio and the axis of the shafts on which they are mounted is always at 90°.

Modules—plug-in units that provide flexibility and customization of a PLC. Examples of modules are a digital input module, digital output module, analog input module, etc.

Molecules—the smallest particle within matter that retains the same chemical properties of the matter.

Momentary—a type of switch in which the contacts will remain in their operated state only as long as the switch is operated.

Monostable (also called one-shot)—a device that has two states (off or on). Only one state presents a stable condition and the device can remain in that state for a long time. The other state is unstable and the device will remain in that state for a very short time.

Motion control modules—a PLC module specially designed for motion control applications.

Multiple taps—a transformer which has two or more additional connections or taps to its winding(s).

Multiple windings—a transformer that has two or more separate primary and/or secondary windings.

Multivibrator—an electrical circuit that functions as an oscillator by generating its own signal.

Mutual induction—the process by which a magnetic field created by current flowing through an inductor cuts across another inductor, thereby inducing a current in the second inductor.

NAND gate—an AND gate with an inverted output. An output is produced only when all of its inputs are low.

N-channel JFET—a junction field effect transistor in which the channel is made from n-type material.

Needle bearing—a type of antifriction bearing in which the rolling elements are long and thin.

Negative power—produced when current and voltage have opposite polarities caused by reactive components in the circuit. Reactive power is measured in units called VARs. VAR stands for volt-amps-reactive.

Neutral flame—a flame produced at the tip of the torch, which has a well-defined white cone at the tip of the torch and a longer, translucent blue cone surrounding the smaller white cone.

Neutrons—a component of the nucleus of an atom. A neutron has a neutral charge.

Nominal belt size—a standard belt size that is common to industry.

Noninverting amplifier—an amplifier whose output signal is in phase with its input signal.

Nonparallel misalignment—when two shafts are not parallel to each other.

NOR gate—a digital logic device that produces an output only when any of its inputs is high.

Normally closed (NC)—a set of contacts in the closed or conducting state without the application of power or being operated in any way.

Normally open (NO)—a set of contacts in the open or nonconducting state without the application of power or being operated in any way.

NPN (negative-positive-negative) transistor—a bipolar junction transistor constructed with n-type material for the emitter and collector regions and p-type material for the base region. The schematic symbol for an NPN transistor will show the emitter arrow pointing outward from the base.

Nucleus—located at the center of an atom. The nucleus contains protons and neutrons. Electrons orbit the nucleus.

Off-delay—a time-delay relay that changes state when power is removed and a preset time has elapsed.

Off-delay timers—a time-delay relay whose contacts change state immediately on power application, and, on removal of power, change state to the original conditions after a time delay has elapsed.

Offset misalignment (pulley and sprocket)—a type of misalignment that occurs when the two shafts are parallel to each other but the faces of the pulleys or sprockets are not on the same axis.

Ohms—the opposition to electrical current is called resistance. Resistance is measured in units called ohms. The Greek letter for omega, Ω, is used to represent ohms. In formulae, the letter *R* is used to represent resistance.

Ohm's law triangle—a memory device to help in the recollection of the Ohm's law formulae. A triangle is drawn and divided so that the letter *E* is located in the top compartment, the letter *I* is located in the lower left compartment, and the letter *R* is located in the lower right compartment. To use the Ohm's law triangle, cover up the compartment of the variable that you are trying to find. The remaining compartments show you the formula that you should use.

On-delay—a time-delay relay that changes state when power is applied and a preset time has elapsed.

On-delay timers—a time-delay relay whose contacts change state after a delay on power application, and, on removal of power, change state immediately to the original conditions.

Open-loop control (also called manual control)—a control system in which motor performance is not sensed electronically and, therefore, can vary under changing load conditions. Motor performance is monitored by personnel and corrective action is taken manually.

Operational amplifier (op-amp)—a high-gain, wide-band amplifier that is also capable of performing addition, subtraction, multiplication, integration, and differentiation.

OR gate—a digital logic gate which produces a logical 1 output when it receives a logical 1 at any or all of its inputs.

Orthographic drawing—a drawing that is made using two separate views. One view is from the top of the piping system (looking down) showing all of the horizontal runs that are made. The other view is from the side to show all of the vertical runs.

Oscillator—see *flip-flop circuit*.

Oscilloscope (or o-scope or scope)—measures voltage and allows the user to *see* a representation of the voltage as well.

Outer race—a shallow groove that is cut into the outer ring in which the rolling elements will ride.

Outer rings—made of hardened steel, the outermost part of the bearing.

Outside seal—a mechanical seal in which the rotating mating surface is mounted externally from the stuffing box.

Overhauling—a condition that can occur when a motor is driving a load with high inertia.

Overhead welding position—the most difficult and the most uncomfortable of all the positions because the metal to be welded is above the head.

Oxidizing flame—the flame that uses more oxygen. This type of flame is accompanied by a loud hissing noise. The cone of this flame is smaller than that of the neutral flame and has a sharper, more defined point at its tip. This is the least used flame of the three. It has a temperature of approximately 6300°F and is used for welding sheet brass, cast brass, or bronze.

Parallel circuit—a circuit that has two or more paths for current flow.

Parts inventory—one of five essential records that should be maintained for an effective maintenance program. This record lists all spare parts and assemblies on hand. As parts or assemblies are used, the inventory is updated to reflect the change. Indicators are set to alert the responsible person that inventory of a particular item is low and replacements must be ordered. This record is invaluable when scheduling an outage or downtime.

P-channel JFET—a junction field effect transistor in which the channel is made from p-type material.

Peak inverse voltage (PIV)—the reverse biased voltage rating of a diode.

Perfect vacuum—when atmospheric pressure causes mercury to rise in a glass tube to 30 in. Because the maximum atmospheric pressure at sea level is 14.7 psi, the maximum pressure that is available in a vacuum system is 30 in. of mercury; therefore, a perfect vacuum exists at 30 in. of mercury.

Permanent-magnet motor—a motor that uses permanent magnets to produce the magnetic field.

Personal protective equipment (PPE)—equipment and clothing that are meant, by design, to protect personnel from a hazardous condition or environment.

Phase value—the value of voltage or current measured at the phase winding of a transformer.

Phase voltage—the voltage measured across the phase winding of a transformer.

Phase-controlled bridge rectifier—a bridge rectifier circuit constructed with electronic devices that can be gated or triggered in synchronization with the proper bias conditions of the device.

Phasing dots—a method used to indicate the polarity of one transformer winding with respect to another winding on the same transformer. The dot indicates that the lead so marked will be in phase with the marked lead on the other winding.

Phosphor—the material that coats the inside of a fluorescent lamp tube.

Photodiode—a diode whose electrical conduction characteristics are affected by exposure to light.

Photoelectric switches—noncontact type of sensors that use some form of light to sense the presence or absence of an object. Photoelectric switches consist of three basic components: the transmitter, the receiver, and the switching device.

Photovoltaic—devices (also called solar cells) that produce electricity when struck by light.

Picofarads (pF)—units of measurement that capacitors typically have their values stated in because the farad is very large.

Piezo effect—present in certain types of crystals. Electricity can be produced when a pressure is exerted on these crystals. This pressure can be in the form of striking, twisting, or bending.

Pilot duty—when referring to contacts, a rating that indicates the contacts are capable of switching only low amperage loads. Pilot-duty-rated contacts are rated for lighter loads than standard duty or heavy duty.

Pin links—the component of a roller chain that slides into the bushing that is in the roller link.

Pinch points—points on a machine where two or more separate components meet or come together.

Pinion gear—the smaller gear in a gear set.

Piping—a system that uses pipes to contain the movement of a fluid or gas to direct it to a desired location.

PIRE wheel—a memory device to help in the recollection of the twelve formulae that relate to Ohm's law and the Power law. A circle is divided into four quadrants. Each quadrant contains the three formulae that relate to power (P), current (I), resistance (R), and voltage (E).

Pitch—the distance from any point on a thread to the next corresponding point of the next thread measured parallel along the axis.

Pitch circle—the point of a pulley or gear in which the power is transmitted; sometimes referred to as pitch diameter.

Pitch diameter—the diameter of an imaginary cylinder, the surface of which would pass through the threads at such points as to make equal the width of the threads and the width of the spaces cut by the surface of the cylinder.

Pitch line—the point where the two imaginary circles, which represent the pitch circle of the gears, meet.

Plain bearing—a sleeve or bushing that may support radial and axial loads.

Plan view—a view of an orthographic drawing from the top of the piping system (looking down) showing all of the horizontal runs that are made.

Plates—the conductive surfaces of a capacitor.

Plug weld—a weld that can be accomplished by overlapping a piece of metal with a hole in it over another solid piece of metal, then welding the two pieces together through the hole.

Plugging—a method of braking in which a motor turning in the forward direction is brought to a rapid stop by switching the running motor connections so as to make it turn in the reverse direction.

Pneumatic—a form of fluid power that uses compressed gas (compressed air) or a vacuum to transmit power from one location to another.

PNP (positive-negative-positive) transistor—a bipolar junction transistor constructed with p-type material for the emitter and collector regions and n-type material for the base region. The schematic symbol for a PNP transistor will show the emitter arrow pointing inward toward the base.

Polarized retroreflective—a type of photoelectric switch in which a polarized lens is placed over the receiver. The lens allows only light beams that are oriented properly to pass through to the receiver. The benefit of polarized retroreflective sensing is that shiny objects can be reliably sensed. The polarized retroreflector is also called an antiglare sensor.

Portable welding curtain—a curtain that surrounds the welder and the work as arc welding is occurring and protects any passersby who may otherwise be exposed to the ultraviolet light.

Porosity—Pits or pours that form within a weld bead which is cause from welding on a surface which has not been cleaned properly prior to welding the joint.

Positive displacement pump—a pump that allows very little leakage through its internal components because they fit so closely together; therefore, it can produce very high pressures in the fluid that is being pumped.

Positive-drive belt—a belt that provides positive traction while transmitting power; also referred to as a *timing belt* or a *gear belt.*

Potential difference—the voltage between two points in an electrical circuit.

Potentiometer (pot)—a variable resistor that is connected in a circuit in a fashion that allows for the control of voltage.

Power—the equivalent of how much work is being produced in a certain amount of time.

Power factor—the ratio of true power to apparent power. Power factor is a measure of efficiency and is usually expressed as a percent.

Power factor correction—the process of adding the required amount of capacitance to offset an inductive circuit, or adding the required amount of inductance to offset a capacitive circuit. The result is that circuit efficiency will be improved.

Power law—a law that shows the relationship among voltage, current, and power in an electrical circuit.

Power law triangle—a memory device to help in the recollection of the Power law formulae. A triangle is drawn and divided so that the letter *P* is located in the top compartment, the letter *I* is located in the lower left compartment, and the letter *E* is located in the lower right compartment. To use the Power law triangle, cover up the compartment of the variable that you are trying to find. The remaining compartments show you the formula that you should use.

Power section—the section of the drive that provides controlled power to the motor.

Power source—A battery, a power supply, or a generator.

P-P auto mode—a control that allows the scope to be triggered when a waveform is detected. When a signal is not present, a trigger signal is internally sent to the scope, causing a bright baseline to appear on the CRT. The scope is typically set for *P-P AUTO* mode.

Preload—the clamping force that is asserted on the parts when they are fastened together.

Pressing pinch point—the point where two separate parts of a machine that are pressed together meet.

Pressure—the application of force that is exerted within a fluid power system due to restrictions against the flow of the fluid. The total pressure within a fluid power system will be just the amount that is necessary to overcome the restrictions that are present in the system.

Pressure gauge readings—an indication of the pressure that is present within a fluid power system.

Preventive maintenance schedule—one of five essential records that should be maintained for an effective maintenance program. This is a "to-do" list for the maintenance technician to follow on a daily basis. This schedule tells the maintenance technician which inspections he or she is to perform on a particular day. The inspections could cover several different areas of one particular piece of equipment, or the inspections may be the same type of inspection to be performed on several different machines.

Primary distribution system—the system that consists of the wiring and associated transformers, substations, and generating stations that produce and transmit or distribute electrical energy from the generating station to the consumer (residential, commercial, or industrial).

Primary resistor starter—a reduced-voltage starting method in which resistors are connected in series with the motor on startup. This produces a voltage drop, leaving less voltage available for the motor. After a predetermined time, the resistors are shorted and the motor receives full voltage.

Primary winding—the winding of a transformer that is connected to the energy source.

Priming—this is accomplished by filling the pump and the supply lines with the fluid that is to be pumped before the pump is started.

Probe adjust—a test point on an oscilloscope that provides a test signal for compensating the probe.

Proof load—The load applied in performance of a proof test.

Proof test—A non-destructive tension test performed by the sling manufacturer or equivalent entity to verify the construction and workmanship of a sling.

Protector caps—caps that are placed over the cylinder valves to protect them.

Protons—a component of the nucleus of an atom. A proton has a positive charge.

Proximity switches—a noncontact type of switches. The typical use for a proximity switch is to sense the presence or absence of an object without actually contacting the object. Proximity switches combine high-speed switching with small physical size.

Puddle—a molten pool of metal that is formed when the tip of the flame causes the base metal to melt.

Pulleys—sometimes referred to as a sheave, the pulley transmits power from a belt to a shaft.

Pulse-width modulated (PWM) signal—a technique used by electronic variable-speed drives to vary the frequency of the voltage applied to a motor by varying the pulse width of the applied voltage.

Pulse-width modulated inverter—an inverter that uses pulse-width modulation to vary the voltage applied to an AC induction motor.

Pump—a mechanical device that changes a mechanical power into fluid power.

Pushbutton—a manual control switch that must be depressed in order to operate the switch. Pushbuttons may be either momentary or maintained and are available with a wide variety of contact types and switching arrangements.

Race—the shallow groove in which the rolling elements will ride.

Radial lip seals—often referred to as an oil seal, it is most commonly made of a molded synthetic rubber and used to prevent the escape of fluids or prevent the entry of contaminants. These seals are most commonly used in applications of a rotary, reciprocating, or oscillating nature.

Radial load—this occurs when the pressure from the load is perpendicular to the axis of the shaft.

Radial system—a method of wiring the distribution system in which the distribution lines appear as the hub in the spokes of a wheel.

Radial-axial loads—this occurs when pressures from the load are both perpendicular and parallel to the axis of the shaft at the same time.

Radial thrust—an antifriction bearing that is very similar to the single-row ball bearing, except that the shoulders on the inner and outer rings are raised slightly on one side. These bearings are sometimes referred to as angular-contact bearings and are always used in pairs.

Rated capacity (CAP)—The maximum working load of a rope (natural, synthetic, or wire), sling, or chain, that is permitted by the provision of OSHA regulations.

R-C circuits—circuits that contain both resistive and capacitive components.

Reactive power—the negative power produced by reactive components (inductors or capacitors).

Receiver—a part of a photoelectric switch that may also be called the photocell, photodiode, or photodetector. The function of the receiver is to detect the presence or absence of the light transmitted by the light source and then activate or deactivate the switching device.

Rectification—the process by which AC is converted into a pulsating DC.

Rectifier—a device or circuit used to convert AC into pulsating DC.

Rectifier diode—a diode used to convert AC into pulsating DC.

Regenerating—the process by which a motor functions as a generator, generating energy that can be returned to the power source.

Regulation—the ability of a system to provide near constant level of a quantity such as voltage, current, speed, or other.

Regulators—valves used to reduce the pressure of a gas that is coming from a cylinder to a much lower working pressure and to maintain the working pressure and flow under changing cylinder pressure.

Repair information—one of five essential records that should be maintained for an effective maintenance program. This record contains a running history of all repairs made to a particular piece of equipment. Information such as the date, the time of day, the symptom, operator comments, what was done to repair the equipment, and the name(s) of the individual(s) who made the repairs should be listed on this record. Over time, this record will become extremely valuable because it will show symptoms and cures. This should help lower diagnostic and repair time and costs. In addition, a line of communication will be established between a technician currently working on a problem and someone who may have experienced a similar problem in the past.

Reset coil—the coil that must be energized in order to reset or unlatch a latching relay.

Residual magnetism—the amount of magnetism remaining in an object after the magnetizing energy has been removed.

Resistance—The opposition to electrical current. Resistance is measured in units called ohms. The Greek letter for omega, Ω, is used to represent ohms. In formulae, the letter *R* represents resistance.

Resistors—devices used to limit the amount of electrical current in a circuit. Resistors can be made from a variety of materials to give them different heat handling capabilities or better precision.

Response time—in a photoelectric switch, the time difference that occurs between the time the light is detected (or not detected) and a change in the state of the switching device.

Retainer—a cage-like device that resides around the rolling elements, sometimes referred to as a separator, which holds the rolling elements in place and evenly spaces them around the bearing.

Retroreflective—a type of photoelectric switch also called reflective sensing or reflex sensing. A retroreflective sensor houses the transmitter and receiver in a single unit. A retroreflective device transmits the light to a reflector and detects the reflected light at its receiver.

Reverse biased—a condition in which the voltage levels and polarities are applied to a semiconductor, causing it not to conduct.

Reversing motor starter—a type of motor starter that uses two coils. When the forward coil is energized the forward contacts are closed, causing the motor to rotate in the forward direction. When the reverse coil is energized, the reverse contacts are closed, electrically switching the motor connections to L1 and L3, causing the motor to rotate in the reverse direction.

Rheostat—a variable resistor that is connected in a circuit in a fashion that allows for the control of current.

Rim alignment—when the readings are taken, using a dial indicator, from the rim of the coupling flange.

Rim-and-face alignment—when the readings are taken, using two dial indicators, from the rim of the flange and the face of the flange at the same time.

R-L circuits—circuits that contain resistive and inductive components.

R-L-C circuits—circuits that contain resistive, inductive, and capacitive components.

Roller chain—the most common type of chain used in power transmission, this chain transmits power through a series of roller links and pin links.

Roller links—consists of two rollers that are mounted on bushings. This configuration allows the roller to roll, thus lowering the amount of friction produced on the sprocket.

Rolling elements—the rolling mechanisms within the bearing that reduce the amount of friction.

Root—the lowest point of the thread.

Rotary switches—manually operated switches that require a turning motion to operate.

Rotating pinch point—the point where two rollers, gears, or anything that is rotating meet.

Rotor—the rotating portion of a motor or generator. Rotors are typically constructed without windings or a commutator.

Rotor slip—the difference in speed between the rotor and the synchronous speed of the motor.

RTD input modules—a PLC input module used to monitor information from an RTD.

Safe working load (SWL)—See "Rated capacity (CAP)".

Scientific notation—a type of shorthand that is used to make large and small numbers more easily managed. Powers of 10 are used to abbreviate the number.

Sealed bearing—a bearing that is exactly like the open-faced ball bearing, except that two seals (one metal, the other plastic, vinyl, or rubber) are inset between the inner and outer rings.

Secondary winding—the winding of a transformer that is connected to the load.

Seebeck effect—the phenomenon of producing electricity from heat.

Selector switches—another name for a rotary switch.

Self-excited, shunt-wound DC generator—a DC generator that uses residual magnetism to generate DC.

Selvage edge—The finished edge of a synthetic webbing which is designed to prevent unraveling.

Semiconductors—elements that have three, four, or five valence electrons. These elements are neither good conductors nor good insulators.

Semisolid—a gel-like substance that has the characteristics of both a solid and a liquid.

Separately excited, shunt-wound DC generator—a DC generator that uses an external source of energy to create a magnetic field for the generation of DC.

Series circuit—a circuit that has only one path for current flow.

Series-wound DC generator—a DC generator that uses the series field connected in series with the armature. Series-wound DC generators exhibit poor voltage regulation.

Series-wound DC motor—a DC motor that uses the series field connected in series with the armature. Series-wound DC motors exhibit poor speed regulation. Series-wound DC motors should never be connected to their loads with belts or chains that may break. They should always be directly coupled to their loads.

Set (saw blades)—the orientation of the alternating teeth on the saw blade.

Set coil—the coil that must be energized to set or latch a latching relay.

Shaft—the portion of a rotating machine on which the armature windings, shorting bars, centrifugal switch, commutator, and rotor windings may be mounted.

Shoulder—the flat portion of the rings between the rolling element and the face of the bearing.

Shunt-wound DC generator—a DC generator that uses the shunt field connected in parallel with the armature. Shunt-wound DC generators exhibit very good voltage regulation.

Shunt-wound DC motor—a DC motor that uses the shunt field connected in parallel with the armature. Shunt-wound DC motors exhibit very good speed regulation.

Side bar—a steel plate that exists on both sides of the roller link and the pin link. The bushings on the roller link and the pins on the pin link are pressed into the sidebar.

Silent chain—consists of a series of links that are joined together with bushings and pins, as is the roller chain; however, this chain is constructed in a manner that allows the chain to ride on the surface of the sprocket. This chain is very strong and quiet while it is in operation.

Silicon-controlled rectifier (SCR)—a three-terminal device that permits current flow in one direction when the gate terminal is turned on.

Single-throw—a switch that has one set of contacts that can be operated. The switch can be positioned to operate the set of contacts (1 throw) or not, hence single throw.

Single-line drawing—generally used to convey an overview of information, but not a lot of detail. This type of drawing does not show the actual electrical connections or the actual physical location of the devices; however, it shows that some type of connection exists between components.

Single-phase AC—an AC circuit that consists of two current-carrying conductors.

Single-row—an antifriction bearing that has only one row of rolling elements.

Sketch—a drawing that is made with minimal detail, which is used to convey information from the person who drew the sketch to the person who will interpret the sketch.

Shear load—A force that is applied to a material along a plane that is parallel to the direction of force.

Shear strength—The amount of strength that a material has to resist component failure under a shear load.

Sling—An assembly that connects the load to the material handling equipment.

Slip rings—Circular metal bands located on the rotating portion of a machine. Brushes are used to ride on the slip rings, making an electrical connection between the rotating member and an external circuit.

Slot weld—very similar to the plug weld, with one exception. It is accomplished by overlapping a piece of metal with a slot cut in it over another solid piece of metal, then welding the two pieces together through the slot.

Soft foot—a condition that exists when one or more of the feet on the machine that is being aligned is bent or was poorly manufactured.

Solid bearing—a plain bearing with a cylindrical shape that is made of a soft metal such as bronze, aluminum, or soft steel.

Source—a type of switching device used with photoelectric switches. A source device requires current to flow from the positive (+) through the output, through the load, to the negative (-). A source device is never paired with another source device. It must be used with a sink device.

Spalling—the flaking away of metal pieces caused by metal fatigue.

Spigot—the end of a cast iron pipe that will be inserted into the bell to join the pipes together.

Split bearing—a solid bearing that has been cut in half along its axis, creating two separate pieces.

Spur gear—the teeth on this gear are parallel to the axis of the shaft that the gear is mounted on.

Standard diffuse sensor—a type of photoelectric sensor that senses the reflectivity of an object.

Standard duty—when referring to contacts, a rating that indicates the contacts are capable of switching normal amperage loads. Standard-duty-rated contacts fall between pilot duty (lightest rating) and heavy duty (highest rated).

Standard retroreflective—a photoelectric switch that uses an industrial reflector to return the light from the transmitter to the receiver. Standard retroreflective sensors do not perform well in an environment with high moisture content in the air.

Static head pressure—the amount of pressure that is developed for every inch of rise in elevation above the point of measurement.

Stator—the stationary winding of an AC motor.

Step-down transformer—a transformer with more turns of wire on the primary side than on the secondary side. A step-down transformer will produce a lower secondary voltage than what is applied to the primary; however, the secondary current will be higher than the primary current.

Step-up transformer—a transformer with more turns of wire on the secondary side than on the primary side. A step-up transformer will produce a higher secondary voltage than what is applied to the primary; however, the secondary current will be lower than the primary current.

Stringer bead—a bead that has little or no side-to-side movement of the electrode.

Stuffing box seal—made up of three parts, the packing chamber, the packing rings, and the stuffing gland, the sealing action is accomplished by compressing the packing material between two surfaces.

Summing amplifier—an electrical circuit whose output is the sum of the input signals.

Summing point—when used in a feedback circuit, the junction where the algebraic sum of two or more signals is obtained.

Surfacing weld—a type of weld that is used to build up a surface by running one or more stringers or weave beads across an unbroken surface.

SWEEP SPEED—on an oscilloscope, the control that is used to vary the speed at which the electron beam sweeps across the face of the CRT.

Switching device—the portion of a photoelectric switch that operates in response to the transmitter/receiver portion. The switching device is the component that performs the switching function to the load.

Synchronous motor—a three-phase motor that uses DC applied to the rotor windings to cause the rotor to lock in step with the rotating magnetic field of the stator.

Synchronous speed—the speed at which the rotating magnetic field of an AC induction motor revolves.

Synthetic—something that is produced artificially by chemical means.

Tachometer-generator—a device mounted to the shaft of a motor. The tachometer-generator produces a DC voltage proportional to the speed of the motor.

Tagout—the placement of a tag on an energy isolation device to provide a prominent warning to others of one's presence on the machine. (*The tagout does not lock out the energy isolation device; it just warns of your presence on the machine.*)

Tap—a hardened shank, which commonly has four flat sides at the top of the tap, lands, and flutes, that is used for cutting threads into a predrilled hole.

Taper gauge—a flat, tapered tool with graduations marked along the edge (in thousandths of an inch) used for measuring misalignments.

Tapped secondary—additional electrical connections made to the secondary winding of a transformer to produce various voltages from the same secondary winding.

Tee joint—a type of joint that exists by placing the edge of one piece of metal on the surface of another, thus creating a right angle.

Television—a trigger mode of an oscilloscope that allows the scope to trigger on TV fields or lines. This function is not applicable in the industrial maintenance field.

Tensile strength—The amount of stress that a material can withstand while being pulled in opposite directions without tearing apart.

Tension members—the component within a V-belt that prevents stretching and is the main load-carrying component of the V-belt.

Tertiary winding—the name given to an additional winding on a transformer that already has a primary and a secondary winding.

Test probes—The leads (hand held or clipped into place) that are used to connect the test instrument to the device being tested.

Thermocouple input modules—a PLC module specially designed to accept information from a thermocouple.

Thread depth—the vertical distance between the root and the crest.

Threading die—a hardened tool, which commonly has four flat sides on the outside, lands, and flutes, that is used for cutting threads onto a shank.

Three-phase AC—an AC circuit that consists of three current-carrying conductors with each phase 120° out of phase with the other.

Three-phase squirrel-cage induction motor—a three-phase motor that uses a rotor constructed of end rings and shorting bars. Because of the simple design, rugged construction, and low-maintenance costs, the squirrel-cage induction motor is perhaps the most common motor used in industry.

Three-phase transformers—transformers that are used on a three-phase power system. They may consist of a single transformer or be formed from three single-phase transformers.

Three-wire controls—a type of control that uses a relay to provide a latch, memory, seal-in, or holding circuit for a momentary pushbutton switch. A three-wire control will not automatically reenergize after a power failure.

Thru-beam—a photoelectric switch also known as transmitted beam or opposed pair sensing. The transmitter is housed in one unit and the receiver is housed in another unit. The transmitter and receiver are located some distance apart. The item to be sensed is allowed to pass between the transmitter and receiver, thus breaking the light beam.

Thrust bearing—a plain bearing shaped like a washer meant to be used on axial or thrust loads.

Time-delay relays—a type of relay that uses a timing device to retard the operation of the contacts when the relay is energized (on-delay) or de-energized (off-delay).

Title block—usually placed in the corner of a drawing, it is a box that contains critical information about a machine, the drawing itself, and the engineering firm.

Torque—turning or rotating force produced by a motor.

Total indicator readings (TIR)—found by subtracting the lowest dial indicator reading from the highest dial indicator reading.

TRACE ROTATION control—a control that is adjusted until the display trace is aligned horizontally with the horizontal grid lines.

Tracking—the act of aligning two pulleys so a flat belt stays aligned.

Trailing electrode angle—this causes the molten metal and slag to be pushed away from the leading edge of the molten pool toward the back of the pool, where it will solidify.

Transformer—an electrical device used to change the value of the applied AC voltage to a different value of AC voltage.

Transmitter—in a photoelectric switch, the portion of the device that provides the light for the receiver portion. The transmitter may also be called the source, light source, or emitter. Transmitters have used incandescent light or LEDs to provide the light source.

Triac—a three-terminal device that permits current flow in either direction when the gate terminal is turned on.

Trigger (also called gate)—the lead of an SCR or triac that allows control of the conduction of the device.

Trigger coupling control—an oscilloscope control that is used when the SOURCE switch is set to the *EXT* position. This switch has three positions: AC, DC, and DC/10. The AC position inserts a capacitor into the external trigger input circuit, blocking any DC component from the signal that is being triggered. The scope will trigger only on the AC component of the signal. The DC position represents direct coupled, which allows the scope to trigger on all components of the input signal. The DC÷10 position also uses direct coupling, but attenuates the signal by a factor of 10. This is useful if the external trigger signal is too large.

Trigger level control—on an oscilloscope, this controls where the trigger point occurs on the signal. For most measurements this control is set to the mid position.

Trigger operating modes control—this consists of several controls on an oscilloscope: NORMAL, P-P AUTO, and TELEVISION modes and TRIGGER SOURCE switches.

Trigger slope control—on an oscilloscope, this controls whether the scope triggers on the rising or falling edge of a signal. Usually, the TRIGGER SLOPE switch is set for a rising slope.

Trigger source—composed of two switches, SOURCE and INT, that work in conjunction with one another.

Trigger system—the portion of the scope that controls *when* the waveform is displayed on the CRT.

Trio assembly—three devices that are installed in a pneumatic system side by side to filter, lubricate, and regulate the pneumatic system. A trio assembly is sometimes referred to as an FLR unit.

True power (also called resistive power)—power used by the circuit resistive component to perform work.

Turns ratio—the ratio of the number of turns of wire that forms the primary winding of a transformer compared to the number of turns of wire that forms the secondary winding of the same transformer.

Two-wire control—a type of control that uses mechanically held switches for the starting and stopping of a motor. A two-wire control will automatically reenergize after a power failure.

Ultraviolet light—an invisible type of light that can cause serious burns and is found opposite infrared light on the light spectrum.

Undercut—the result of more metal being removed from the base metal than is being replaced by the electrode.

Underfill—this occurs on a groove weld when the completed weld is below the original surface of the base metal.

Unijunction transistor (UJT)—a three-terminal semiconductor device constructed with only one p-n junction.

Unloading—this occurs in a pneumatic system once the pressure limit has been met. This is accomplished through the use of an unloading valve, which will vent the air that is being compressed during each cycle of the compressor into the atmosphere instead of turning off the motor.

Up-counters—a PLC instruction that begins at a preset value and counts up incrementally whenever the input contact transitions from closed to open.

Vacuum—a negative pressure that is measured in inches of mercury ("Hg). See also *perfect vacuum*.

Valence electron—The outermost electrons in any atom.

Variable resistors—A resistor whose value can be varied either by a rotary or a linear motion.

VARIABLE SEC/DIV—an oscilloscope control that is used to vary the amount of time represented by one division of the graticule.

Variable trigger holdoff control—on an oscilloscope, this control adjusts the delay time before the waveform is drawn on the CRT.

Variable voltage inverter—a type of AC drive in which the DC voltage is controlled and the DC current is free to respond to the motor needs.

Variable volts/div control—a variable control that allows independent adjustment of the vertical height of the displayed signal for channel 1 or channel 2. A dual-trace scope has two of these controls, one for channel 1 and one for channel 2.

V-belts—a belt made of rubber and canvas, which is molded with a V-shape used to transmit power.

Vertical alignment—a type of misalignment that exists in coupled shaft alignment and refers to any alignment, angular or offset, that can be found when looking at the side of a machine.

Vertical hitch—A method of supporting a load by a single, vertical part or leg of the sling.

Vertical operating mode control—on an oscilloscope, this consists of several controls: CH 1 - BOTH - CH 2, CH 2 INVERT, and ADD - ALT - CHOP.

Vertical position control—on an oscilloscope, this control allows the technician to move the trace displayed on the CRT up or down.

Vertical welding position—a welding position in which the metal to be welded is at a right angle to the floor and the bead is drawn from top to bottom or bottom to top.

Viscosity—the internal friction of a lubricant caused by the molecular attraction.

Volts—the unit of measurement of voltage. It is represented by the letter *V*.

Voltage—The pressure in an electrical system. Voltage is measured in units called volts and is abbreviated with the letter *V*.

Voltage regulator—an electronic circuit or device used to maintain a constant voltage level from a voltage source.

Voltage source inverter (VSI) (also called voltage-fed inverter)—an inverter in which the DC bus voltage is controlled while the DC bus current varies to meet the needs of the motor.

Volt-amperes (VA)—a combination of the true (resistive) power and the reactive power (VARs).

Volt-amps-reactive (VARs)—the reactive power in a circuit.

Volts to hertz (V/Hz) (also called V/f ratio)—the ratio of output voltage to frequency.

Volume—the size of an object using all three dimensions—height, width, and depth.

Volumetric efficiency—this can be found by dividing the actual displacement by the theoretical displacement (manufacturer's rating). Because efficiency is expressed as a percentage, multiply the answer by 100 to get the percentage of efficiency.

Wattage—unit of electrical power.

Welding torch—the component of a gas welding system in which the oxygen and the acetylene are mixed; sometimes referred to as a blowpipe.

White light—a visible, safe light in which one can see.

Wide-angle diffuse sensor—a type of sensing used in photoelectric switches. The wide-angle diffuse sensor uses a special lens. This lens widens the field of view of the sensor. This allows the sensor to sense objects in a wider area. This can present problems caused by insects, dust, and other airborne objects, which may cause false detections to occur.

Windage—the opposition to the turning of a rotor or armature caused by the surrounding air.

Wobble plate—a plate in which the driveshaft is connected within a wobble-plate pump. Sometimes referred to as a rotating cam plate, the wobble plate, because it is shaped like a cam, will cause the pistons to reciprocate.

Work—work is done when a force, acting upon an object, causes that object to move.

Working load limit (WLL)—See "Rated capacity (CAP)".

Worm gears—a gear that is used where there is a need for high-ratio speed reduction.

Wound-rotor induction motor—a three-phase induction motor with a rotor consisting of windings and slip rings. Brushes are used to provide an electrical connection to the rotor windings through the slip rings. This construction allows the strength of the rotor magnetic field to be varied, thus causing motor speed to vary as well.

Wye connection (also called star connection)—an electrical connection used in three-phase transformers. Wye-connected transformers have their phase current equal to their line current, but the line voltage will be 1.732 times greater than the phase voltage.

Wye-delta configuration (also called star-delta configuration)—the term used to describe a transformer whose primary is connected in a wye configuration and whose secondary windings are connected in a delta configuration.

Wye-delta starting—a type of starting method in which a three-phase motor is started with its windings connected in a wye configuration and then switched to a delta configuration for normal running.

Wye-wye connection (also called a star-star connection)—the term used to describe a transformer whose primary and secondary windings are connected in a wye connection.

Yield strength—The point at which a material begins to deform once stress is applied.

Zener diodes—a diode that has a specific designed breakdown voltage level in the reverse direction.

INDEX

Entries followed by *f* reference a figure; entries followed by *t* reference a table

D

E

Q

R

S

U

V